W9-AYB-536

Biology

Concepts and Applications 8e

Biology

Concepts and Applications 8e

Cecie Starr

Christine A. Evers

Lisa Starr

BROOKS/COLE

CENGAGE Learning™

Australia • Brazil • Japan • Korea • Mexico • Singapore • Spain • United Kingdom • United States

BROOKS/COLE
CENGAGE Learning

Biology: Concepts and Applications, Eighth Edition
Cecie Starr, Christine A. Evers, Lisa Starr

Senior Acquisitions Editor, Life Sciences:
 Peggy Williams

Publisher, Life Sciences: Yolanda Cossio

Editor in Chief: Michelle Julet

Assistant Editor: Elizabeth Momb

Editorial Assistant: Alexis Glubka

Managing Media Editor: Shelley Ryan

Media Editor: Lauren Oliveira

Senior Marketing Manager: Tom Ziolkowski

Marketing Coordinator: Elizabeth Wong

Senior Marketing Communications Manager:
 Linda Yip

Content Project Manager: Hal Humphrey

Senior Art Director: John Walker

Senior Print Buyer: Karen Hunt

Permissions Account Manager, Text: Bob Kauser

Senior Permissions Account Manager, Images:
 Dean Dauphinais

Production Service:
 Grace Davidson & Associates, Inc.

Text Designer: John Walker

Photo Researcher: Paul Forkner,
 Myrna Engler Photo Research Inc.

Copy Editor: Anita Wagner

Illustrators:
 ScEYEnce Studios, Lisa Starr,
 Precision Graphics

Cover Designer: John Walker

Cover Image: Florida panther (*Puma concolor coryi*). A South Florida population of about one hundred individuals is all that remains of this subspecies that once ranged across the southeastern United States (James Balog/Getty Images).

Compositor: Lachina Publishing Services

© 2011, 2008 Brooks/Cole, Cengage Learning

ALL RIGHTS RESERVED. No part of this work covered by the copyright herein may be reproduced, transmitted, stored, or used in any form or by any means graphic, electronic, or mechanical, including but not limited to photocopying, recording, scanning, digitizing, taping, Web distribution, information networks, or information storage and retrieval systems, except as permitted under Section 107 or 108 of the 1976 United States Copyright Act, without the prior written permission of the publisher.

For product information and technology assistance, contact us at
Cengage Learning Customer & Sales Support, 1-800-354-9706.
For permission to use material from this text or product,
submit all requests online at **cengage.com/permissions**.
Further permissions questions can be emailed to
permissionrequest@cengage.com.

ExamView® and *ExamView Pro*® are registered trademarks of FSCreations, Inc. Windows is a registered trademark of the Microsoft Corporation used herein under license. Macintosh and Power Macintosh are registered trademarks of Apple Computer, Inc. Used herein under license.

Library of Congress Control Number: 2010926063

Student Edition:

ISBN-13: 978-1-4390-4673-9

ISBN-10: 1-4390-4673-5

Loose-leaf Edition:

ISBN-13: 978-0-538-49389-5

ISBN-10: 0-538-49389-5

Brooks/Cole
20 Davis Drive
Belmont, CA 94002-3098
USA

Cengage Learning is a leading provider of customized learning solutions with office locations around the globe, including Singapore, the United Kingdom, Australia, Mexico, Brazil, and Japan. Locate your local office at **www.cengage.com/global**.

Cengage Learning products are represented in Canada by Nelson Education, Ltd.

To learn more about Brooks/Cole, visit **www.cengage.com/brookscole**

Purchase any of our products at your local college store or at our preferred online store **www.cengagebrain.com**

Printed in the United States of America
1 2 3 4 5 6 7 14 13 12 11 10

CONTENTS IN BRIEF

DETAILED CONTENTS

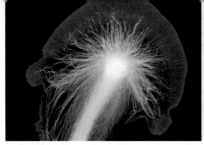
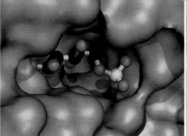
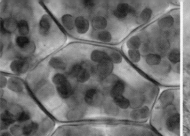

UNIT II GENETICS

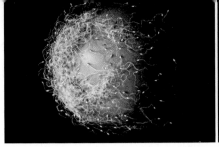

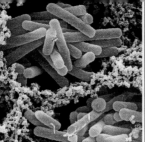
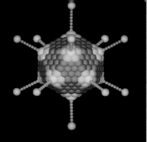
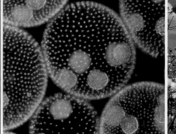

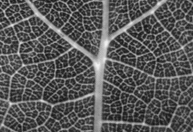
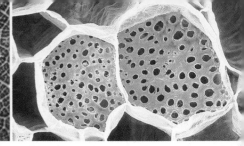

UNIT V HOW PLANTS WORK

25 Plant Tissues

26 Plant Nutrition and Transport

27 Plant Reproduction and Development

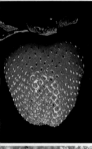

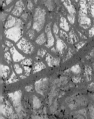

UNIT VI HOW ANIMALS WORK

28 Animal Tissues and Organ Systems

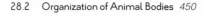

29 Neural Control

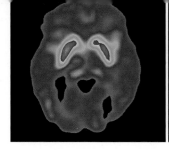
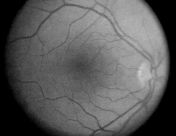

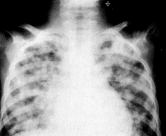

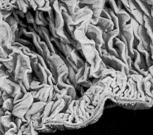

42 Ecosystems

43 The Biosphere

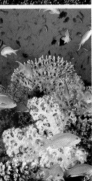

44 Human Effects on the Biosphere

PREFACE

Aptostichus stephencolberti

Since the last edition of this book was published, researchers have discovered ice on Mars and have documented rapid glacial melting here on Earth. They have identified thousands of new species, including a spider (*left*) now named after a television show host. They have also uncovered a wealth of information about extinct species, including some of our own close relatives. Biologists have created a vaccine that protects against cervical cancer, and can now make adult cells behave like embryonic stem cells.

In short, what we know about biological systems changes very rapidly. Given the pace of the changes, we would do well to help our students become lifelong learners. Most of the students who use this book will not be biologists, and many will never take another science course. Yet, as voting members of our society, all will face decisions that require a basic understanding of the principles of biology and the process of science in general. This book provides a foundation for such decisions. It teaches students how to draw connections between abstract ideas and everyday experiences, between recent discoveries and long-standing biological theories.

New To This Edition

Graphically Enhanced Key Concepts
The opening pages of each chapter introduces **Key Concepts**, concise statements that set the stage for the more detailed discussion that follows. In this edition, each key concept is accompanied by a photo or graphic that catches the student's eye. The student will encounter these images again within the chapter and in the **Illustrated Chapter Summary** as part of a visual message that threads through each chapter.

Improved Integration Among Chapters We retained the **Links to Earlier Concepts** on the opening spread of each chapter. This brief paragraph reminds students of relevant information that has been covered in previous chapters. We have also improved the usefulness of the in-chapter Links references by including a brief description of the relevant material. For example, the heading of the section that covers membrane potential in neurons notes that this material links to "Potential energy 5.2, Transport proteins 5.7."

Section-Based Glossary In addition to a full glossary of terms at the end of the book, each section now has a **Section-Based Glossary**. Important new terms that are boldfaced and defined in the text of a section are redefined in the Section-Based Glossary. A student can simply glance at this glossary for a quick check of a term's definition. As an additional aid, the Chapter Summary includes all glossary terms used in context. Here again, the terms appear in boldface.

Figure-Based Self-Assessment Questions Many figure captions now include a **Figure-It-Out Question** that allows

the student to check whether he or she understands the figure's content. For example, Figure 12.5, which shows the stages of meiosis, is accompanied by a question that asks the student to identify the stage of meiosis in which the chromosome number becomes reduced. Answers to Figure-It-Out questions are provided on the same page. Additional self-assessment material is provided in the Self-Quiz and Critical Thinking Questions at the end of the chapter.

Emphasis on Analyzing Scientific Data A new chapter end feature, the **Data Analysis Activity**, sharpens the student's analytical skills while reinforcing the process of science. Each activity asks the student to interpret data presented in graphic or tabular form. The data is related to the chapter material, and has usually been taken directly from a published scientific study or experiment. For example, the Data

Analysis Activity in Chapter 27, Plant Reproduction and Development, asks for analysis of data from a study that tested whether gerbils are the main pollinators of certain desert plants (*left*).

Chapter-Specific Changes This new edition contains 140 new photographs and almost 400 new or updated illustrations. In addition, the text of every chapter has been updated and revised for clarity. A page-by-page guide to new content and figures is available upon request, but we summarize the highlights here. Note that we have added a new final chapter, *Human Effects on the Biosphere*. We have also deleted a chapter (*Plants and Animals—Common Challenges*) that appeared in the previous edition. Material that had been in the *Common Challenges* chapter is now integrated into other chapters.

• *Chapter 1, Invitation to Biology* Revised and expanded coverage of critical thinking and the process of science; levels of life's organization now illustrated with the same organism.
• *Chapter 2, Life's Chemical Basis* New opening essay discusses mercury toxicity and prevalence; revisions emphasize electron behavior in atoms as it relates to ions and bonding.
• *Chapter 3, Molecules of Life* New opening essay discusses health impacts of *trans* fats; importance of protein structure now exemplified by prions.
• *Chapter 4, Cell Structure* New opening essay about *E. coli* O157:H7 contamination of food; expanded discussion and new photos of archaeans.
• *Chapter 5, Ground Rules of Metabolism* Example in opening essay is now ADH, to tie in with revisited section about ethanol metabolism, defects in the pathway that affect drinking behavior, and metabolic effects of alcohol abuse.
• *Chapter 6, Where It Starts—Photosynthesis* New opening essay about biofuels ties in to discussion regarding photosynthesis, carbon dioxide, and global climate change.
• *Chapter 7, How Cells Release Chemical Energy* Introductory section now discusses the relationship between the evolution of oxygenic photosynthesis and that of aerobic respiration.

• *Chapter 8, DNA Structure and Function*   Opening essay now discusses the cloning of 9/11 rescue dog Trakr; discussion of animal cloning techniques and applications expanded.

• *Chapter 9, From DNA to Protein*   Expanded introductory material now includes an overview of genetic information and gene expression; all new illustrations of translation.

• *Chapter 10, Controls Over Genes*   Theme of evolutionary connections strengthened throughout; for example, in the comparison of effects of mutations in the same homeotic gene in humans and flies.

• *Chapter 11, How Cells Reproduce*   Mitosis panel now compares micrographs of plant and animal cells; new section on cell cycle controls and neoplasms.

• *Chapter 12, Meiosis and Sexual Reproduction*   Opening essay updated; new photo shows pollen grains germinating on stigma; meiosis and segregation graphics updated.

• *Chapter 13, Observing Patterns in Inherited Traits*   Updated essay with current model of CF pathogenesis and greater emphasis on genetics behind the ubiquity of the allele.

• *Chapter 14, Human Inheritance*   New opening essay explores the genetics of skin color variation and discusses evolutionary advantages that reinforce small allelic differences.

• *Chapter 15, Biotechnology*   New opening essay discusses personal genetic testing; updated, expanded genomics section.

• *Chapter 16, Evidence of Evolution*   New graphics include a sheet from Darwin's evolution journal and "missing links" in cetacean evolutionary history; updated geologic time scale now graphically correlated with Grand Canyon stratigraphy.

• *Chapter 17, Processes of Evolution*   New illustrated examples include allopatric speciation in snapping shrimp; sympatric speciation in Lake Victoria cichlids; adaptive radiation of honeycreepers; coevolution of the large blue butterfly and *Maculinea arion* ant. Stasis and exaptation added.

• *Chapter 18, Life's Origin and Early Evolution*   New opening section now focuses on astrobiology; added information about biomarkers as historical evidence.

• *Chapter 19, Viruses, Bacteria, and Archaeans*   New opening section discusses evolution of HIV; new art of HIV replication cycle; new information about influenzas H5N1 and H1N1.

• *Chapter 20, The Protists*   Opening section now covers harmful algal blooms; new graphic showing relationships of protists to other groups; new art of *Plasmodium* life cycle.

• *Chapter 21, Plant Evolution*   Opening section now focuses on Nobel Prize winner Wangari Maathai's efforts on behalf of tropical forests; new section focusing on the ecological and economic importance of angiosperms.

• *Chapter 22, Fungi*   Opening section describes the threat wheat stem rust poses to world food supplies; more extensive coverage of fungi as plant and human pathogens.

• *Chapter 23, Animals I: Major Invertebrate Groups*   Updated evolutionary tree showing relationships among animal groups; major reorganization of coverage of insects and improved discussion of their ecological, economic, and health impacts.

• *Chapter 24, Animals II: The Chordates*   New evolutionary tree diagrams for chordates and for primates.

• *Chapter 25, Plant Tissues*   New opening essay about carbon sequestration in plant tissues; new section introduces stem specializations illustrated with photos of common food plants.

• *Chapter 26, Plant Nutrition and Transport*   Added function(s) to nutrients table; new photo of erosion in Providence Canyon, Georgia.

• *Chapter 27, Plant Reproduction and Development*   New opening essay about colony collapse disorder illustrated with photos showing superior fruit from insect-pollinated flowers.

• *Chapter 28, Animal Tissues and Organ Systems*   Opening essay about stem cells updated; improved coverage of embryonic tissues, development of body cavities.

• *Chapter 29, Neural Control*   Information about results of new studies of Ecstasy; material related to vertebrate nervous systems heavily reorganized to improve flow.

• *Chapter 30, Sensory Perception*   New graphics depicting effects of visual disorders; improved discussion of visceral sensations.

• *Chapter 31, Endocrine Control*   Added a subsection of neuroendocrine interactions; more coverage of adrenal gland disorders and how complications of diabetes arise.

• *Chapter 32, Structural Support and Movement*   Opening section now discusses effects of mutations that affect mysostatin; new graphic illustrating knee anatomy and opposing muscles of the arm; new material on classifying muscle fibers as red versus white and fast versus slow.

• *Chapter 33, Circulation*   Opening section now focuses on a young athlete saved by CPR and use of an automated external defibrillator; historical material about the first EKG deleted.

• *Chapter 34, Immunity*   New intro essay about HPV and cervical cancer, including Gardasil vaccine. New section details how antigens and immunity factor in transfusion reactions.

• *Chapter 35, Respiration*   New figure emphasizing the two sites of gas exchange; improved figure illustrating countercurrent flow in fishes; added information about pneumonia and asthma.

• *Chapter 36, Digestion and Human Nutrition*   New information about a common allele associated with obesity; discussion of stomach and intestines reorganized to improve flow; added information about celiac disease, heath effects of different lipids, and how to interpret nutrition labels.

• *Chapter 37, The Internal Environment*   Added discussion of invertebrate solute-regulating systems; new figure depicting urine formation.

• *Chapter 38, Reproduction and Development*   Heavy revision to shorten chapter; opening section now discusses the history of IVF and the "octomom" story.

• *Chapter 39, Animal Behavior*   New material on behavioral genetics and expanded coverage of mechanisms of learning.

• *Chapter 40, Population Ecology*   Opening essay now focuses on soaring numbers of Canada goose; human life table and age-structure diagrams updated.

• *Chapter 41, Community Ecology*   New description and photo of interspecific competition; more focused, concise coverage of exotic species; new subsection about herbivory.

• *Chapter 42, Ecosystems*   Opening section focuses on phosphate pollution of waterways; new diagrams of water, carbon, nitrogen, and phosphorus cycles.

• *Chapter 43, The Biosphere*   Discussion of deserts expanded; soil profiles integrated into biome descriptions.

• *Chapter 44, Human Effects on the Biosphere*   Includes material about declining biodiversity, desertification, deforestation, water shortages and pollution, biological accumulation and magnification, effects of trash on marine ecosystems, depletion of the ozone layer, ground level ozone pollution, effects of climate change, conservation biology and sustainable uses of resources.

Student and Instructor Resources

Test Bank   Nearly 4,000 test items, ranked according to difficulty and consisting of multiple-choice (organized by section heading), selecting the exception, matching, labeling, and short answer exercises. Includes selected images from the text. Also included in Microsoft® Word format on the PowerLecture DVD.

ExamView®   Create, deliver, and customize tests (both print and online) in minutes with this easy-to-use assessment and tutorial system. Each chapter's end-of-chapter material is also included.

Instructor's Resource Manual   Includes chapter outlines, objectives, key terms, lecture outlines, suggestions for presenting the material, classroom and lab enrichment ideas, discussion topics, paper topics, possible answers to critical thinking questions, answers to data analysis activities, and more. Also included in Microsoft® Word format on the PowerLecture DVD.

Resource Integration Guide   A chapter-by-chapter guide to help you use the book's resources effectively. Each chapter includes applied readings, all chapter animations, specific BBC and ABC video segments, hyperlink examples, a listing of the chapter's How Would You Vote? question, and pop-up tutor videos.

Student Interactive Workbook   Labeling exercises, self-quizzes, review questions, and critical thinking exercises help students with retention and better test results.

PowerLecture   This convenient tool makes it easy for you to create customized lectures. Each chapter includes the following features, all organized by chapter: lecture slides, all chapter art and photos, bonus photos, animations, videos, Instructor's Manual, Test Bank, ExamView testing software, and JoinIn polling and quizzing slides. This single disc places all the media resources at your fingertips.

The Brooks/Cole Biology Video Library 2009 featuring BBC Motion Gallery   Looking for an engaging way to launch your lectures? The Brooks/Cole series features short high-interest segments: Pesticides: Will More Restrictions Help or Hinder?; A Reduction in Biodiversity; Are Biofuels as Green as They Claim?; Bone Marrow as a New Source for the Creation of Sperm; Repairing Damaged Hearts with Patients' Own Stem Cells; Genetically Modified Virus Used to Fight Cancer; Seed Banks Helping to Save Our Fragile Ecosystem; The Vanishing Honeybee's Impact on Our Food Supply.

CengageNOW   Save time, learn more, and succeed in the course with CengageNOW, an online set of resources (including Personalized Study Plans) that give you the choices and tools you need to study smarter and get the grade. You will have access to hundreds of animations that clarify the illustrations in the text, videos, and quizzing to test your knowledge. You can also access live online tutoring from an experienced biology instructor. New to this edition are pop-up tutors, which help explain key topics with short video explanations. Get started today!

Webtutors for WebCT and BlackBoard   Jump-start your course with customizable, rich, text-specific content. Whether you want to Web-enable your class or put an entire course online, WebTutor delivers. WebTutor offers a wide array of resources including media assets, quizzing, web links, exercises, flashcards, and more. Visit webtutor.cengage.com to learn more. New to this edition are pop-up tutors, which help explain key topics with short video explanations.

Biology CourseMate   Cengage Learning's Biology Course-Mate brings course concepts to life with interactive learning, study, and exam preparation tools that support the printed textbook, or the included eBook. With CourseMate, professors can use the included Engagement Tracker to assess student preparation and engagement. Use the tracking tools to see progress for the class as a whole or for individual students.

Premium eBook   This complete online version of the text is integrated with multimedia resources and special study features, providing the motivation that so many students need to study and the interactivity they need to learn. New to this edition are pop-up tutors, which help explain key topics with short video explanations.

Acknowledgments

Thanks to our academic advisors for their ongoing impact on the book's content. We are especially grateful to Jean deSaix, David Rintoul, and Michael Plotkin for their ongoing advice and constructive criticism. This edition also reflects influential contributions of the instructors, listed on the following page, who helped shape our thinking. *Key Concepts, Data Analysis Activities,* custom videos—such features are direct responses to their suggestions.

Cengage Learning continues to prove why it is one of the world's foremost publishers. Michelle Julet and Yolanda Cossio, thank you again for allowing us to maintain our ideals and to express our creativity. Peggy Williams, we are, as always, grateful for your continuing guidance and encouragement; thank you for giving us the freedom to improve this edition while updating it. Producing this book would not have been possible without the organizational wizardry and unfailing patience of Grace Davidson. The talented John Walker deserves primary credit for this book's visual appeal; his inspiring design, together with Paul Forkner's dedicated photo research, were critical to our ongoing efforts to create the perfect marriage of text and illustration. Copyeditor Anita Wagner and proofreader Kathleen Dragolich helped us keep our text clear, concise, and correct; tireless editorial assistant Alexis Glubka organized meetings, reviews, and paperwork. Elizabeth Momb managed production of the book's many print supplements and Lauren Oliveira created a world-class technology package for both students and instructors.

LISA STARR, CHRIS EVERS, AND CECIE STARR   *2010*

Contributors to This Edition: Influential Class Tests and Reviews

Brenda Alston-Mills
North Carolina State University

Norris Armstrong
University of Georgia

Dave Bachoon
Georgia College & State University

Neil R. Baker
The Ohio State University

Andrew Baldwin
Mesa Community College

David Bass
University of Central Oklahoma

Lisa Lynn Boggs
Southwestern Oklahoma State University

Gail Breen
University of Texas at Dallas

Marguerite "Peggy" Brickman
University of Georgia

David William Bryan
Cincinnati State College

Uriel Buitrago-Suarez
Harper College

Sharon King Bullock
Virginia Commonwealth University

John Capehart
University of Houston - Downtown

Daniel Ceccoli
American InterContinental University

Tom Clark
Indiana University South Bend

Heather Collins
Greenville Technical College

Deborah Dardis
Southeastern Louisiana University

Cynthia Lynn Dassler
The Ohio State University

Carole Davis
Kellogg Community College

Lewis E. Deaton
University of Louisiana - Lafayette

Jean Swaim DeSaix
University of North Carolina - Chapel Hill

(Joan) Lee Edwards
Greenville Technical College

Hamid M. Elhag
Clayton State University

Patrick Enderle
East Carolina University

Daniel J. Fairbanks
Brigham Young University

Amy Fenster
Virginia Western Community College

Kathy E. Ferrell
Greenville Technical College

Rosa Gambier
Suffok Community College - Ammerman

Tim D. Gaskin
Cuyahoga Community College - Metropolitan

Stephen J. Gould
Johns Hopkins University

Marcella Hackney
Baton Rouge Community College

Gale R. Haigh
McNeese State University

John Hamilton
Gainesville State

Richard Hanke
Rose State Community College

Chris Haynes
Shelton St. Community College

Kendra M. Hill
South Dakota State University

Juliana Guillory Hinton
McNeese State University

W. Wyatt Hoback
University of Nebraska, Kearney

Kelly Hogan
University of North Carolina

Norma Hollebeke
Sinclair Community College

Robert Hunter
Trident Technical College

John Ireland
Jackson Community College

Thomas M. Justice
McLennan College

Timothy Owen Koneval
Laredo Community College

Sherry Krayesky
University of Louisiana - Lafayette

Dubear Kroening
University of Wisconsin - Fox Valley

Jerome Krueger
South Dakota State University

Jim Krupa
University of Kentucky

Mary Lynn LaMantia
Golden West College

Kevin T. Lampe
Bucks County Community College

Susanne W. Lindgren
Sacramento State University

Madeline Love
New River Community College

Dr. Kevin C. McGarry
Kaiser College - Melbourne

Jeanne Mitchell
Truman State University

Alice J. Monroe
St. Petersburg College - Clearwater

Brenda Moore
Truman State University

Erin L. G. Morrey
Georgia Perimeter College

Rajkumar "Raj" Nathaniel
Nicholls State University

Francine Natalie Norflus
Clayton State University

Alexander E. Olvido
Virginia State University

John C. Osterman
University of Nebraska, Lincoln

Bob Patterson
North Carolina State University

Shelley Penrod
North Harris College

Mary A. (Molly) Perry
Kaiser College - Corporate

John S. Peters
College of Charleston

Michael Plotkin
Mt. San Jacinto College

Ron Porter
Penn State University

Karen Raines
Colorado State University

Larry A. Reichard
Metropolitan Community College - Maplewood

Jill D. Reid
Virginia Commonwealth University

Robert Reinswold
University of Northern Colorado

Ashley E. Rhodes
Kansas State University

David Rintoul
Kansas State University

Darryl Ritter
Northwest Florida State College

Amy Wolf Rollins
Clayton State University

Sydha Salihu
West Virginia University

Jon W. Sandridge
University of Nebraska

Robin Searles-Adenegan
Morgan State University

Julie Shepker
Kaiser College - Melbourne

Rainy Shorey
Illinois Central College

Eric Sikorski
University of South Florida

Robert (Bob) Speed
Wallace Junior College

Tony Stancampiano
Oklahoma City Community College

Jon R. Stoltzfus
Michigan State University

Peter Svensson
West Valley College

Jeffrey L. Travis
University at Albany

Nels H. Troelstrup, Jr.
South Dakota State University

Allen Adair Tubbs
Troy University

Will Unsell
University of Central Oklahoma

Rani Vajravelu
University of Central Florida

Jack Waber
West Chester University of Pennsylvania

Kathy Webb
Bucks County Community College

Amy Stinnett White
Virginia Western Community College

Virginia White
Riverside Community College

Robert S. Whyte
California University of Pennsylvania

Kathleen Lucy Wilsenn
University of Northern Colorado

Penni Jo Wilso
Cleveland State Community College

Michael L. Womack
Macon State College

Maury Wrightson
Germanna Community College

Mark L. Wygoda
McNeese State University

Lan Xu
South Dakota State University

Poksyn ("Grace") Yoon
Johnson and Wales University

Muriel Zimmermann
Chaffey College

Biology

Concepts and Applications 8e

‹ Links to Earlier Concepts

The organization of topics in this book parallels life's levels of organization. In both cases, each level builds on the last. At the beginning of each chapter, we will remind you of concepts in previous chapters that will help you understand the material presented in the current chapter. Within chapters, cross-references will link you to the relevant sections in previous chapters.

Key Concepts

The Science of Nature

We understand life by studying it at different levels of organization, which extend from atoms and molecules to the biosphere. The quality we call "life" emerges at the level of cells.

Life's Unity

All organisms consist of one or more cells that take in energy and raw materials to stay alive; all sense and respond to stimuli; and all function and reproduce with the help of DNA.

1 Invitation to Biology

1.1 The Secret Life of Earth

In this era of Google Earth and global positioning systems, could there possibly be any unexplored places left on Earth? Well, yes, actually. In 2005, for instance, helicopters dropped a team of scientists into the middle of a vast and otherwise inaccessible Indonesian cloud forest (Figure 1.1). Within minutes, the explorers realized that their landing site, a dripping, moss-covered swamp, was home to plants and animals that had been previously unknown to science. Over the next month, they discovered dozens of new species there, including a plant with flowers the size of dinner plates and a frog the size of a pea. They also came across hundreds of species that are on the brink of extinction in other parts of the world, some that supposedly were extinct, and one that had not been seen for so many years that scientists had forgotten about it.

The animals in the forest had never learned to be afraid of humans, so they could be approached and even picked up. A few new species were discovered as they casually wandered through the campsite. Team member Bruce Beehler remarked, "Everywhere we looked, we saw amazing things we had never seen before. I was shouting. This trip was a once-in-a-lifetime series of shouting experiences."

New species are discovered all the time, often in places much more mundane than Indonesian cloud forests. How do we know what species a particular organism belongs to? What is a species, anyway, and why should discovering a new one matter to anyone other than a scientist? You will find the answers to such questions in this book. They are part of the scientific study of life, **biology**, which is one of many ways we humans try to make sense of the world around us.

Trying to understand the immense scope of life on Earth gives us some perspective on where we fit into it. For example, we routinely discover hundreds of species every year, but about

20 species become extinct *every minute* in rain forests alone—and those are only the ones we know about. The current rate of extinctions is about 1,000 times faster than normal. Human activities are responsible for the acceleration. At this rate, we

Figure 1.1 A peek inside the cloud forest of New Guinea's Foja Mountains, (*opposite*). Explorers recently discovered dozens of very rare species—and some new ones—in this forest. *Above*, a jungle hawk-owl (*Ninox theomacha*). This species is a not-so-rare resident of Indonesia, including the Foja Mountains.

will never know about most of the species that are alive on Earth today. Does that matter? Biologists think so. Whether or not we are aware of it, we are intimately connected with the world around us. Our activities are profoundly changing the entire fabric of life on Earth. The changes are, in turn, affecting us in ways we are only beginning to understand.

Ironically, the more we learn about the natural world, the more we realize we have yet to learn. But don't take our word for it. Find out what biologists know, and what they do not, and you will have a solid foundation upon which to base your own opinions about our place in this world. By reading this book, you are choosing to learn about the human connection—your connection—with all life on Earth.

biology The scientific study of life.

Life's Diversity
Observable characteristics vary tremendously among organisms. Various classification systems help us keep track of the differences.

The Nature of Science
Science helps us be objective about our observations by addressing only the observable. It involves making, testing, and evaluating hypotheses.

Experiments and Research
Researchers carefully design and carry out experiments in order to unravel cause-and-effect relationships in complex natural systems.

> Biologists study life by thinking about it at different levels of organization.

1 atom
Atoms are fundamental units of all substances, living or not. This image shows a model of a single atom.

2 molecule
Atoms joined in chemical bonds. This is a model of a water molecule. The molecules of life are much larger and more complex than water.

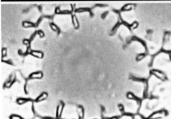

3 cell
The cell is the smallest unit of life. Some, like this plant cell, live and reproduce as part of a multicelled organism; others do so on their own.

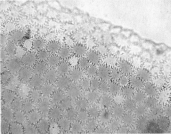

4 tissue
Organized array of cells and substances that interact in a collective task. This is epidermal tissue on the outer surface of a flower petal.

5 organ
Structural unit of interacting tissues. Flowers are the reproductive organs of many plants.

6 organ system
A set of interacting organs. The shoot system of this poppy plant includes its aboveground parts: leaves, flowers, and stems.

Figure 1.3 Animated Levels of life's organization.

Life Is More Than the Sum of Its Parts

What, exactly, is the property we call "life"? We may never actually come up with a good definition. Living things are too diverse, and they consist of the same basic components as nonliving things. When we try to define life, we end up only identifying the properties that differentiate living from nonliving things.

Complex properties, including life, often emerge from the interactions of much simpler parts. For an example, take a look at the drawings in Figure 1.2. The property of "roundness" emerges when the parts are organized one way, but not the other ways. Characteristics of a system that do not appear in any of the system's components are called **emergent properties**. The idea that structures with emergent properties can be assembled from the same basic building blocks is a recurring theme in our world, and also in biology.

A Pattern in Life's Organization

Biologists study all aspects of life, past and present. Through their work, we are beginning to understand a great pattern in life's organization. That organization occurs in successive levels, with new emergent properties appearing at each level (Figure 1.3).

Life's organization starts when atoms interact. **Atoms** are fundamental building blocks of all substances, living and nonliving **1**. There are no atoms unique to life, but there are unique molecules. **Molecules** are atoms joined

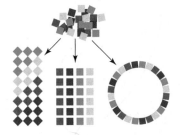

Figure 1.2 Animated
Example of how different objects can be assembled from the same parts. Roundness is an emergent property of the *rightmost* object.

atom Fundamental building block of all matter.
biosphere All regions of Earth where organisms live.
cell Smallest unit of life.
community All populations of all species in a given area.
ecosystem A community interacting with its environment.
emergent property A characteristic of a system that does not appear in any of the system's component parts.
molecule An association of two or more atoms.
organ In multicelled organisms, a grouping of tissues engaged in a collective task.
organism Individual that consists of one or more cells.
organ system In multicelled organisms, set of organs engaged in a collective task that keeps the body functioning properly.
population Group of individuals of the same species that live in a given area.
tissue In multicelled organisms, specialized cells organized in a pattern that allows them to perform a collective function.

in chemical bonds **2**. In today's world, only living things make the "molecules of life," which are complex carbohydrates and lipids, proteins, DNA, and RNA.

The emergent property of "life" appears at the next level, when many molecules of life become organized as a cell **3**. A **cell** is the smallest unit of life that can survive and reproduce on its own, given information in DNA, energy, and raw materials. Some cells live and reproduce independently. Others do so as part of a multicelled organism. An **organism** is an individual that consists of one or more cells. A poppy plant is an example of a multicelled organism **7**.

In most multicelled organisms, cells make up tissues **4**. Cells of a **tissue** are specialized and organized in a particular pattern. The arrangement allows the cells to collectively perform a special function such as movement (muscle tissue), fat storage (adipose tissue), and so on.

An **organ** is an organized array of tissues that collectively carry out a particular task or set of tasks **5**. For example, a flower is an organ of reproduction in plants; a heart, an organ that pumps blood in animals. An **organ system** is a set of organs and tissues that interact to keep the individual's body working properly **6**. Examples of organ systems include the aboveground parts of a plant (the shoot system), and the heart and blood vessels of an animal (the circulatory system).

A **population** is a group of individuals of the same type, or species, living in a given area **8**. An example would be all of the California poppies in California's Antelope Valley Poppy Reserve. At the next level, a **community** consists of all populations of all species in a given area. The Antelope Valley Reserve community includes California poppies and many other organisms such as microorganisms, animals, and other plants **9**. Communities may be large or small, depending on the area defined.

The next level of organization is the **ecosystem**, or a community interacting with its physical and chemical environment **10**. The most inclusive level, the **biosphere**, encompasses all regions of Earth's crust, waters, and atmosphere in which organisms live **11**.

Take-Home Message How does "life" differ from "nonlife"?

> All things, living or not, consist of the same building blocks—atoms. Atoms join as molecules.

> The unique properties of life emerge as certain kinds of molecules become organized into cells.

> Higher levels of life's organization include multicelled organisms, populations, communities, ecosystems, and the biosphere.

> Emergent properties occur at each successive level of life's organization.

7 multicelled organism
Individual that consists of different types of cells. The cells of this California poppy plant are part of its two organ systems: aboveground shoots and belowground roots.

8 population
Group of single-celled or multicelled individuals of a species in a given area. This population of California poppy plants is in California's Antelope Valley Poppy Reserve.

9 community
All populations of all species in a specified area. These flowering plants are part of the Antelope Valley Poppy Reserve community.

10 ecosystem
A community interacting with its physical environment through the transfer of energy and materials. Sunlight and water sustain the natural community in the Antelope Valley.

11 biosphere
The sum of all ecosystems: every region of Earth's waters, crust, and atmosphere in which organisms live. The biosphere is a finite system, so no ecosystem in it can be truly isolated from any other.

> Continual inputs of energy and the cycling of materials maintain life's complex organization.
> Organisms sense and respond to change.
> All organisms use information in the DNA they inherited from parents to function and to reproduce.

Even though we cannot define "life," we can intuitively understand what it means because all living things share some key features. All require ongoing inputs of energy and raw materials; all sense and respond to change; and all have DNA that guides their functioning.

Organisms Require Energy and Nutrients

Not all living things eat, but all require energy and nutrients on an ongoing basis. Both are essential to maintain life's organization and functioning. **Energy** is the capacity to do work. A **nutrient** is a substance that an organism needs for growth and survival but cannot make for itself.

Organisms spend a lot of time acquiring energy and nutrients. However, what type of energy and nutrients are acquired varies considerably depending on the type of organism. The differences allow us to classify all living things into two categories: producers and consumers. **Producers** make their own food using energy and simple raw materials they get directly from their environment. Plants are producers that use the energy of sunlight to make sugars from water and carbon dioxide (a gas in air), a process called **photosynthesis**. By contrast, **consumers** cannot make their own food. They get energy and nutrients by feeding on other organisms. Animals are consumers. So are decomposers, which feed on the wastes or remains of other organisms. The leftovers of consumers' meals end up in the environment, where they serve as nutrients for producers. Said another way, nutrients cycle between producers and consumers.

Energy, however, is not cycled. It flows through the world of life in one direction: from the environment, through organisms, and back to the environment. The flow of energy maintains the organization of individual organisms, and it is the basis of how organisms interact with one another and their environment. It is also a one-

sunlight energy

A Producers harvest energy from the environment. Some of that energy flows from producers to consumers.

Producers
plants and other self-feeding organisms

B Nutrients that become incorporated into the cells of producers and consumers are eventually released by decomposition. Some cycle back to producers.

Consumers
animals, most fungi, many protists, bacteria

C All of the energy that enters the world of life eventually flows out of it, mainly as heat released back to the environment.

Figure 1.4 Animated The one-way flow of energy and the cycling of materials in the world of life. The photo shows a producer acquiring energy and nutrients from the environment, and consumers acquiring energy and nutrients by eating the producer.

---

consumer Organism that gets energy and nutrients by feeding on tissues, wastes, or remains of other organisms.
development Multistep process by which the first cell of a new individual becomes a multicelled adult.
DNA Deoxyribonucleic acid; carries hereditary information that guides growth and development.
energy The capacity to do work.
growth In multicelled species, an increase in the number, size, and volume of cells.
homeostasis Set of processes by which an organism keeps its internal conditions within tolerable ranges.
inheritance Transmission of DNA from parents to offspring.
nutrient Substance that an organism needs for growth and survival, but cannot make for itself.
photosynthesis Process by which producers use light energy to make sugars from carbon dioxide and water.
producer Organism that makes its own food using energy and simple raw materials from the environment.
reproduction Processes by which parents produce offspring.

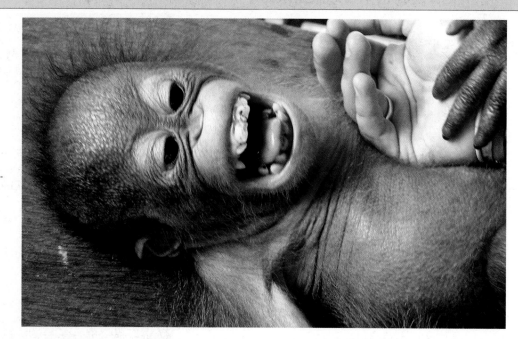

Figure 1.5 Organisms sense and respond to stimulation. This baby orangutan is laughing in response to being tickled. Apes and humans make different sounds when being tickled, but the airflow patterns are so similar that we can say apes really do laugh.

way flow, because with each transfer, some energy escapes as heat. Cells do not use heat to do work. Thus, all of the energy that enters the world of life eventually leaves it, permanently (Figure 1.4).

Organisms Sense and Respond to Change

An organism cannot survive for very long in a changing environment unless it adapts to the changes. Thus, every living thing has the ability to sense and respond to conditions both inside and outside of itself (Figure 1.5). For example, after you eat, the sugars from your meal enter your bloodstream. The added sugars set in motion a series of events that causes cells throughout the body to take up sugar faster, so the sugar level in your blood quickly falls. This response keeps your blood sugar level within a certain range, which in turn helps keep your cells alive and your body functioning.

The fluid in your blood is part of your body's internal environment, which consists of all body fluids outside of cells. Unless that internal environment is kept within certain ranges of composition, temperature, and other conditions, your body cells will die. By sensing and adjusting to change, you and all other organisms keep conditions in the internal environment within a range that favors cell survival. **Homeostasis** is the name for this process, and it is a defining feature of life. *Balance Internal*

Organisms Use DNA

With little variation, the same types of molecules perform the same basic functions in every organism. For example, information encoded in an organism's **DNA** (deoxyribonucleic acid) guides the ongoing metabolic activities that

sustain the individual through its lifetime. Such activities include **growth**: increases in cell number, size, and volume; **development**: the process by which the first cell of a new individual becomes a multicelled adult; and **reproduction**: processes by which parents produce offspring.

Individuals of every natural population are alike in certain aspects of their body form and behavior, an outcome of shared information encoded in DNA. Orangutans look like orangutans and not like caterpillars because they inherited orangutan DNA, which differs from caterpillar DNA in the information it carries. **Inheritance** refers to the transmission of DNA from parents to offspring. All organisms receive their DNA from parents.

Thus, DNA is the basis of similarities in form and function among organisms. However, the details of DNA molecules differ, and herein lies the source of life's diversity. Small variations in the details of DNA's structure give rise to differences among individuals, and among types of organisms. As you will see in later chapters, these differences are the raw material of evolution.

Take-Home Message How are all living things alike?

❯ A one-way flow of energy and a cycling of nutrients sustain life's organization.

❯ Organisms sense and respond to conditions inside and outside themselves. They make adjustments that keep conditions in their internal environment within a range that favors cell survival, a process called homeostasis.

❯ Organisms grow, develop, and reproduce based on information encoded in their DNA, which they inherit from their parents. DNA is the basis of similarities and differences in form and function.

❭ There is great variation in the details of appearance and other observable characteristics of living things.

Living things differ tremendously in their observable characteristics. Various classification schemes help us organize what we understand about this variation, which we call Earth's **biodiversity**.

For example, organisms can be classified into broad groups depending on whether they have a **nucleus**, a sac with two membranes that encloses and protects a cell's DNA. **Bacteria** (singular, bacterium) and **archaeans** are organisms whose DNA is not contained within a nucleus. All bacteria and archaeans are single-celled, which means each organism consists of one cell (Figure 1.6A,B). As a group, they are also the most diverse organisms: Different kinds are producers or consumers in nearly all regions of the biosphere. Some inhabit such extreme environments as frozen desert rocks, boiling sulfurous lakes, and nuclear reactor waste. The first cells on Earth may have faced similarly hostile challenges to survival.

Traditionally, organisms without a nucleus have been called prokaryotes, but this designation is an informal one. Despite their similar appearance, bacteria and archaeans are less related to one another than we had once thought. Archaeans are actually more closely related to **eukaryotes**, organisms whose DNA is contained within a nucleus. Some eukaryotes live as individual cells; others are multicelled (Figure 1.6C). Eukaryotic cells are typically larger and more complex than bacteria or archaeans.

Structurally, **protists** are the simplest eukaryotes, but as a group they vary dramatically. Protists range from single-celled amoebas to giant, multicelled seaweeds.

Most **fungi** (singular, fungus), such as the types that form mushrooms, are multicelled eukaryotes. Many are decomposers. All secrete enzymes that digest food outside the body, then their cells absorb the released nutrients.

Plants are multicelled eukaryotes that live on land or in freshwater environments. Most are photosynthetic producers. Besides feeding themselves, plants and other photosynthesizers also serve as food for most of the other organisms in the biosphere.

Animals are multicelled consumers that ingest tissues or juices of other organisms. Herbivores graze, carnivores eat meat, scavengers eat remains of other organisms, parasites withdraw nutrients from the tissues of a host, and so on. Animals grow and develop through a series of stages that lead to the adult form, and all kinds actively move about during at least part of their lives.

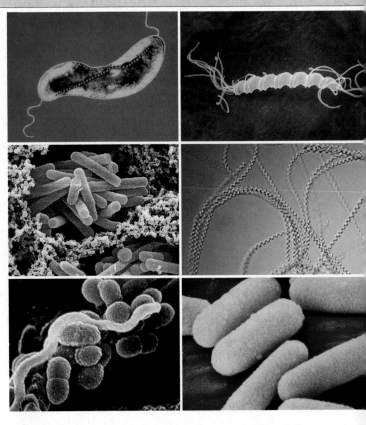

A Bacteria are the most numerous organisms on the planet. All are single-celled, but different types vary in shape and size. *Clockwise from upper left*, a bacterium with a row of iron crystals that acts like a tiny compass; *Helicobacter*, a common resident of cat and dog stomachs; spiral cyanobacteria; *E. coli*, a beneficial resident of human intestines; types found in dental plaque; *Lactobacillus* cells in yogurt.

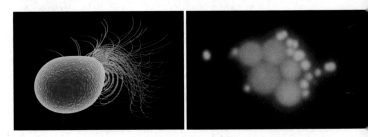

B Archaeans resemble bacteria, but are more closely related to eukaryotes. *Left*, an archaean from volcanic ocean sediments. *Right*, two types of archaeans from a hydrothermal vent on the sea floor.

Figure 1.6 Animated Representatives of life's diversity.

animal Multicelled consumer that develops through a series of stages and moves about during part or all of its life cycle.
archaean Member of a group of single-celled organisms that differ from bacteria.
bacterium Member of a large group of single-celled organisms.
biodiversity Variation among living organisms.
eukaryote Organism whose cells characteristically have a nucleus.
fungus Type of eukaryotic consumer that obtains nutrients by digestion and absorption outside the body.
nucleus Double-membraned sac that encloses a cell's DNA.
plant A multicelled, typically photosynthetic producer.
protist Member of a diverse group of simple eukaryotes.

Take-Home Message **How do organisms differ from one another?**
❭ Different types of organisms vary tremendously in observable characteristics. For example, some organisms have a nucleus; others do not.

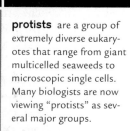

protists are a group of extremely diverse eukaryotes that range from giant multicelled seaweeds to microscopic single cells. Many biologists are now viewing "protists" as several major groups.

plants are multicelled eukaryotes, most of which are photosynthetic. Nearly all have roots, stems, and leaves. Plants are the primary producers in land ecosystems.

fungi are eukaryotes. Most are multicelled. Different kinds are parasites, pathogens, or decomposers. Without decomposers such as fungi, communities would be buried in their own wastes.

animals are multicelled eukaryotes that ingest tissues or juices of other organisms. All actively move about during at least part of their life.

C Eukaryotes are single-celled or multicelled organisms whose DNA is contained within a nucleus.

> Each type of organism, or species, is given a unique name.
> We define and group species based on shared traits.

Each time we discover a new **species**, or kind of organism, we name it. **Taxonomy**, a system of naming and classifying species, began thousands of years ago. However, doing it in a consistent way did not become a priority until the eighteenth century. At that time, European explorers and naturalists were beginning to discover the scope of life's diversity. They started having more and more trouble communicating with one another because species often had multiple names.

For example, one type of plant native to Europe, Africa, and Asia was alternately known as the dog rose, briar rose, witch's briar, herb patience, sweet briar, wild briar, dog briar, dog berry, briar hip, eglantine gall, hep tree, hip fruit, hip rose, hip tree, hop fruit, and hogseed—and those are only the English names! Species often had multiple scientific names too, Latin names that were descriptive but often cumbersome. For example, the scientific name of the dog rose was *Rosa sylvestris inodora seu canina* (odorless woodland dog rose), and also *Rosa sylvestris alba cum rubore, folio glabro* (pinkish white woodland rose with smooth leaves).

An eighteenth-century naturalist, Carolus Linnaeus, came up with a much simpler naming system that we still use. By the Linnaean system, every species is given a unique two-part scientific name. The first part is the name of the **genus** (plural, genera), a group of species that share a unique set of features. The second part is the **specific epithet**. Together, the genus name plus the specific epithet designate one species. Thus, the dog rose now has one official name: *Rosa canina*.

Genus and species names are always italicized. For example, *Panthera* is a genus of big cats. Lions belong to the species *Panthera leo*. Tigers belong to a different species in the same genus (*Panthera tigris*), and so do leopards (*P. pardus*). Note how the genus name may be abbreviated after it has been spelled out once.

Linnaeus ranked species into ever more inclusive categories. Each Linnaean category, or **taxon** (plural, taxa), is a group of organisms. The categories above species—genus, family, order, class, phylum, kingdom, and domain—are the higher taxa (Figure 1.7). Each higher taxon consists of a group of the next lower taxon. Using this system, we can sort all life into a few categories (Figure 1.8 and Table 1.1).

A Rose by Any Other Name . . .

The individuals of a species share a unique set of features, or traits. For example, giraffes normally have very long necks, brown spots on white coats, and so on. These are examples of morphological traits (*morpho*– means form). Individuals of a species also share physiological traits, such as metabolic activities, and they respond the same way to certain stimuli, as when hungry giraffes feed on tree leaves. These are behavioral traits.

DOMAIN	Eukarya	Eukarya	Eukarya	Eukarya	Eukarya
KINGDOM	Plantae	Plantae	Plantae	Plantae	Plantae
PHYLUM	Magnoliophyta	Magnoliophyta	Magnoliophyta	Magnoliophyta	Magnoliophyta
CLASS	Magnoliopsida	Magnoliopsida	Magnoliopsida	Magnoliopsida	Magnoliopsida
ORDER	Apiales	Rosales	Rosales	Rosales	Rosales
FAMILY	Apiaceae	Cannabaceae	Rosaceae	Rosaceae	Rosaceae
GENUS	*Daucus*	*Cannabis*	*Malus*	*Rosa*	*Rosa*
SPECIES	*carota*	*sativa*	*domesticus*	*acicularis*	*canina*
COMMON NAME	carrot	marijuana	apple	arctic rose	dog rose

Figure 1.7 Linnaean classification of five species that are related at different levels. Each species has been assigned to ever more inclusive groups, or taxa: in this case, from genus to domain.

>> **Figure It Out** Which of the plants shown here are in the same order?

Answer: Marijuana, apple, arctic rose, and dog rose are all in the order Rosales.

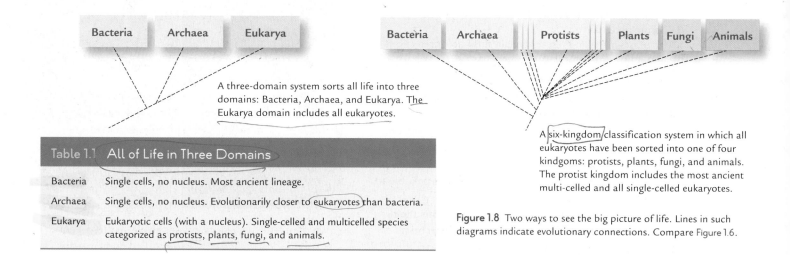

A three-domain system sorts all life into three domains: Bacteria, Archaea, and Eukarya. The Eukarya domain includes all eukaryotes.

Table 1.1	All of Life in Three Domains
Bacteria	Single cells, no nucleus. Most ancient lineage.
Archaea	Single cells, no nucleus. Evolutionarily closer to eukaryotes than bacteria.
Eukarya	Eukaryotic cells (with a nucleus). Single-celled and multicelled species categorized as protists, plants, fungi, and animals.

A six-kingdom classification system in which all eukaryotes have been sorted into one of four kingdoms: protists, plants, fungi, and animals. The protist kingdom includes the most ancient multi-celled and all single-celled eukaryotes.

Figure 1.8 Two ways to see the big picture of life. Lines in such diagrams indicate evolutionary connections. Compare Figure 1.6.

A species is assigned to higher taxa based on some subset of traits it shares with other species. That assignment may change as we discover more about the species and the traits involved. For example, Linnaeus grouped plants by the number and arrangement of reproductive parts, a scheme that resulted in odd pairings such as castor-oil plants with pine trees. Having more information today, we place these plants in separate phyla.

Traits vary a bit within a species, such as eye color does in people. However, there are often tremendous differences between species. Think of petunias and whales, beetles and emus, and so on. Such species look very different, so it is easy to tell them apart. Species that share a more recent ancestor may be much more difficult to distinguish (Figure 1.9).

How do we know if similar-looking organisms belong to different species or not? The short answer is that we rely on whatever information we have. For example, early naturalists studied anatomy and distribution—essentially the only techniques available at the time. Thus, species were named and classified according to what they looked like and where they lived. Today's biologists have at their disposal an array of techniques and machinery much more sophisticated than those of the eighteenth century. They are able to study traits that the early naturalists did not even know about—biochemistry, for example.

Evolutionary biologist Ernst Mayr defined a species as one or more groups of individuals that potentially can interbreed, produce fertile offspring, and do not interbreed with other groups. This "biological species concept"

is useful but it is not universally applicable. For example, not all populations of a species actually continue to interbreed. In many cases, we may never know whether populations separated by a great distance could interbreed successfully even if they did get together. Also, populations often continue to interbreed even as they become different, so the exact moment at which they become separate species is often impossible to pinpoint. We return to speciation and how it occurs in Chapter 17, but for now it is useful to remember that a "species" is a convenient but artificial construct of the human mind.

Figure 1.9 Four butterflies, two species: Which are which? Two forms of the species *Heliconius melpomene* are on the top row; two of *H. erato* are on the bottom row. These two species never cross-breed. Their alternate but similar patterns of coloration evolved as a shared warning signal to local birds that these butterflies taste terrible.

genus A group of species that share a unique set of traits; also the first part of a species name.
species A type of organism.
specific epithet Second part of a species name.
taxon Linnaean category; a grouping of organisms.
taxonomy The science of naming and classifying species.

Take-Home Message **How do we keep track of all the species we know about?**

❯ Each species has a unique, two-part scientific name.

❯ Various classification systems group species on the basis of shared traits.

> Critical thinking is judging the quality of information before accepting it.
> Scientists make and test potentially falsifiable predictions about how the natural world works.
> Science addresses only what is observable.

Thinking About Thinking

Most of us assume that we do our own thinking, but do we, really? You might be surprised to find out just how often we let others think for us. For instance, a school's job, which is to impart as much information as possible to students, meshes perfectly with a student's job, which is to acquire as much knowledge as possible. In this rapid-fire exchange of information, it is all too easy to forget about the quality of what is being exchanged. Any time you accept information without question, you allow someone else to think for you.

Critical thinking is the deliberate process of judging the quality of information before accepting it. "Critical" comes from the Greek *kriticos* (discerning judgment). When you use critical thinking, you move beyond the content of new information to consider supporting evidence, bias, and alternative interpretations. How does the busy student manage this? Critical thinking does not require extra time, just a bit of extra awareness. There are many ways to do it. For example, you might ask yourself some of the following questions while you are learning something new:

What message am I being asked to accept?
Is the message based on opinion or evidence?
Is there a different way to interpret the evidence?
What biases might the presenter have?
How do my own biases affect what I'm learning?

Such questions are simply a way of being conscious about learning. They will help you to decide whether to allow new information to guide your beliefs and actions.

How Science Works

Critical thinking is a big part of **science**, the systematic study of the observable world and how it works (Figure 1.10). A scientific line of inquiry usually begins with curiosity about something observable, such as, say, a decrease in the number of birds in a particular area. Typically, a scientist will read about what others have discovered before making a **hypothesis**, a testable explanation for a natural phenomenon. An example of a hypothesis would be, "The number of birds is decreasing because the number of cats is increasing." Making a hypothesis this way is an example of **inductive reasoning**, which means arriving at a conclusion based on one's observations. Inductive reasoning is the way we come up with new ideas about groups of objects or events.

A **prediction**, or statement of some condition that should exist if the hypothesis is correct, comes next. Making predictions is called the if–then process, in which the "if" part is the hypothesis, and the "then" part is the prediction. Using a hypothesis to make a prediction is a form of **deductive reasoning**, or logical process of using a general premise to draw a conclusion about a specific case.

Table 1.2 Example of the Scientific Method

1. Form a hypothesis	Observe some aspect of nature	Hangover symptoms vary in severity.
2. Test the hypothesis	Think of an explanation for the observation (a hypothesis)	Eating artichokes reduces the severity of hangover symptoms.
	Make a prediction based on the hypothesis	If eating artichokes reduces the severity of hangover symptoms, then taking artichoke extract will reduce the severity of a hangover after drinking alcohol.
	Test the prediction (experiments or surveys)	During a party at which alcohol is served, administer artichoke extract to half of the people who are drinking. The next day, ask everyone who drank alcohol at the party to rate the severity of their hangover symptoms.
3. Evaluate the hypothesis	Analyze the results of the tests (data) and make conclusions	See if there is a correlation between taking the artichoke extract and reduced hangover symptoms. Submit your results and conclusions to a peer-reviewed journal for publication.

Figure 1.10 Scientists doing research. *From left to right*, surveying wildlife in New Guinea; sequencing the human genome; looking for fungi in atmospheric dust collected in Cape Verde; improving the efficiency of biofuel production from agricultural wastes; studying the benefits of weedy buffer zones on farms.

Next, a scientist will devise ways to test a prediction. Tests may be performed on a **model**, or analogous system, if working with an object or event directly is not possible. For example, animal diseases are often used as models to investigate similar human diseases. Careful observations are one way to test predictions that flow from a hypothesis. So are **experiments**: tests designed to support or falsify a prediction. A typical experiment explores a cause and effect relationship.

Researchers investigate causal relationships by changing or observing **variables**, which are characteristics or events that can differ among individuals or over time. An **independent variable** is defined or controlled by the person doing the experiment. A **dependent variable** is an observed result that is supposed to be influenced by the independent variable. For example, an independent

variable in an investigation of hangover preventions may be the administration of artichoke extract before alcohol consumption. The dependent variable in this experiment would be the severity of the forthcoming hangover.

Biological systems are complex, with many interacting variables. It can be difficult to study one variable separately from the rest. Thus, biology researchers often test two groups of individuals simultaneously. An **experimental group** is a set of individuals that have a certain characteristic or receive a certain treatment. This group is tested side by side with a **control group**, which is identical to the experimental group except for one independent variable—the characteristic or the treatment being tested. Any differences in experimental results between the two groups should be an effect of changing the variable.

Test results—**data**—that are consistent with the prediction are evidence in support of the hypothesis. Data that are inconsistent with the prediction are evidence that the hypothesis is flawed and should be revised.

A necessary part of science is reporting one's results and conclusions in a standard way, such as in a peer-reviewed journal article. The communication gives other scientists an opportunity to check and confirm the work.

Forming, testing, and evaluating hypotheses are collectively called the **scientific method** (Table 1.2).

---

control group In an experiment, group of individuals who are not exposed to the independent variable that is being tested.
critical thinking Judging information before accepting it.
data Experimental results.
deductive reasoning Using a general idea to make a conclusion about a specific case.
dependent variable In an experiment, variable that is presumably affected by the independent variable being tested.
experiment A test designed to support or falsify a prediction.
experimental group In an experiment, group of individuals who are exposed to an independent variable.
hypothesis Testable explanation of a natural phenomenon.
independent variable Variable that is controlled by an experimenter in order to explore its relationship to a dependent variable.
inductive reasoning Drawing a conclusion based on obervation.
model Analogous system used for testing hypotheses.
prediction Statement, based on a hypothesis, about a condition that should exist if the hypothesis is correct.
science Systematic study of the observable world.
scientific method Making, testing, and evaluating hypotheses.
variable In an experiment, a characteristic or event that differs among individuals or over time.

Take-Home Message What is science?

> Science is concerned only with the observable—those objects or events for which objective evidence can be gathered.

> The scientific method consists of making, testing, and evaluating hypotheses. It is a way of critical thinking, or systematically judging the quality of information before allowing it to guide one's beliefs and actions.

> Experiments measure how changing an independent variable affects a dependent variable.

❯ Researchers unravel cause-and-effect relationships in complex natural processes by changing one variable at a time.

There are different ways to do research, particularly in biology. Some biologists do surveys; they observe without making hypotheses. Some make hypotheses and leave the experimentation to others. However, despite a broad range of subject matter, scientific experiments are typically designed in a consistent way. Experimenters try to change one independent variable at a time, and see what happens to a dependent variable.

To give you a sense of how biology experiments work, we summarize two published studies here.

Potato Chips and Stomachaches

In 1996 the FDA approved Olestra®, a fat replacement manufactured from sugar and vegetable oil, as a food additive. Potato chips were the first Olestra-containing food product on the market in the United States.

Controversy about the food additive soon raged. Many people complained of intestinal problems after eating the chips, and thought that the Olestra was at fault. Two years later, researchers at Johns Hopkins University School of Medicine designed an experiment to test whether Olestra causes cramps.

The researchers predicted that if Olestra indeed causes cramps, then people who eat Olestra will be more likely to get cramps than people who do not. To test the prediction, they used a Chicago theater as a "laboratory." They asked 1,100 people between the ages of thirteen and thirty-eight to watch a movie and eat their fill of potato chips. Each person got an unmarked bag that contained 13 ounces of chips.

In this experiment, the individuals who got Olestra-containing potato chips constituted the experimental group, and individuals who got regular chips were the control group. The independent variable was the presence or absence of Olestra in the chips.

A few days after the experiment was finished, the researchers contacted all of the people and collected any reports of post-movie gastrointestinal problems. Of 563 people making up the experimental group, 89 (15.8 percent) complained about cramps. However, so did 93 of the 529 people (17.6 percent) making up the control group—who had munched on the regular chips.

People were about as likely to get cramps whether or not they ate chips made with Olestra. These results did not support the prediction, so the researchers concluded that eating Olestra does not cause cramps (Figure 1.11).

Butterflies and Birds

Consider the peacock butterfly, a winged insect that was named for the large, colorful spots on its wings. In 2005, researchers published a report on their tests to identify factors that help peacock butterflies defend themselves against insect-eating birds. The researchers made two observations. First, when a peacock butterfly rests, it folds its wings, so only the dark underside shows (Figure 1.12A). Second, when a butterfly sees a predator approaching, it repeatedly flicks its wings open and closed, while also moving the hindwings in a way that produces a hissing sound and a series of clicks.

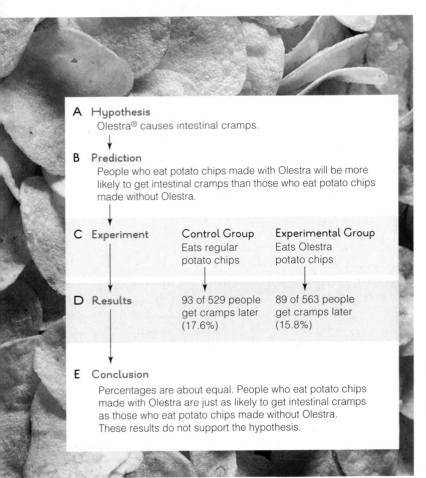

A Hypothesis
Olestra® causes intestinal cramps.

B Prediction
People who eat potato chips made with Olestra will be more likely to get intestinal cramps than those who eat potato chips made without Olestra.

C Experiment	Control Group	Experimental Group
	Eats regular potato chips	Eats Olestra potato chips
D Results	93 of 529 people get cramps later (17.6%)	89 of 563 people get cramps later (15.8%)

E Conclusion
Percentages are about equal. People who eat potato chips made with Olestra are just as likely to get intestinal cramps as those who eat potato chips made without Olestra. These results do not support the hypothesis.

Figure 1.11 The steps in a scientific experiment to determine if Olestra causes cramps. A report of this study was published in the *Journal of the American Medical Association* in January 1998.

❯❯ **Figure It Out** What was the dependent variable in this experiment?

Answer: Whether or not a person got cramps

A

B

C

Figure 1.12 Peacock butterfly defenses against predatory birds.
A With wings folded, a resting peacock butterfly looks a bit like a
dead leaf. B When a bird approaches, the butterfly repeatedly flicks
its wings open and closed, a behavior that exposes brilliant spots and
produces hissing and clicking sounds.

Researchers tested whether the butterfly's behavior deters blue tits C.
They painted over the spots of some butterflies, cut the sound-making
part of the wings on other butterflies, and did both to a third group;
then the biologists exposed each butterfly to a hungry bird.

The results, listed in Table 1.3, support the hypotheses that peacock
butterfly spots and sounds can deter predatory birds.

›› Figure It Out What percentage of butterflies with no spots and no
sound survived the test?

Answer: 20 percent.

Table 1.3 Results of Peacock Butterfly Experiment*

Wing Spots	Wing Sound	Total Number of Butterflies	Number Eaten	Number Survived
Spots	Sound	9	0	9 (100%)
No spots	Sound	10	5	5 (50%)
Spots	No sound	8	0	8 (100%)
No spots	No sound	10	8	2 (20%)

Proceedings of the Royal Society of London, Series B (2005) 272: 1203–1207.

The researchers were curious about why the peacock
butterfly flicks its wings. After they reviewed earlier studies, they came up with two hypotheses that might explain
the wing-flicking behavior:

1. Although wing-flicking probably attracts predatory
birds, it also exposes brilliant spots that resemble owl
eyes (Figure 1.12B). Anything that looks like owl eyes is
known to startle small, butterfly-eating birds, so exposing
the wing spots might scare off predators.
2. The hissing and clicking sounds produced when the
peacock butterfly moves its hindwings may be an additional defense that deters predatory birds.

The researchers used their hypotheses to make the following predictions:

1. If peacock butterflies startle predatory birds by exposing their brilliant wing spots, then individuals with wing
spots will be less likely to get eaten by predatory birds
than those without wing spots.
2. If peacock butterfly sounds deter predatory birds, then
sound-producing individuals will be less likely to get
eaten by predatory birds than silent individuals.

The next step was the experiment. The researchers
used a marker to paint the wing spots of some butterflies
black, and scissors to cut off the sound-making part of the
hindwings of others. A third group had their wing spots
painted and their hindwings cut. The researchers then
put each butterfly into a large cage with a hungry blue tit
(Figure 1.12C) and watched the pair for thirty minutes.

Table 1.3 lists the results of the experiment. All of the
butterflies with unmodified wing spots survived, regardless of whether they made sounds. By contrast, only half
of the butterflies that had spots painted out but could
make sounds survived. Most of the butterflies with neither spots nor sound structures were eaten quickly. The
test results confirmed both predictions, so they support
the hypotheses. Birds are deterred by peacock butterfly
sounds, and even more so by wing spots.

Take-Home Message Why do biologists perform experiments?

› Natural processes are often influenced by many interacting variables.

› Experiments help researchers unravel causes of complex natural processes
by focusing on the effects of changing a single variable.

> Science is, ideally, a self-correcting process because scientists check one another's work.

❶ Natalie, blindfolded, randomly plucks a jelly bean from a jar. There are 120 green and 280 black jelly beans in that jar, so 30 percent of the jelly beans in the jar are green, and 70 percent are black.

❷ The jar is hidden from Natalie's view before she removes her blindfold. She sees only one green jelly bean in her hand and assumes that the jar must hold only green jelly beans.

❸ Still blindfolded, Natalie randomly picks out 50 jelly beans from the jar. She ends up picking out 10 green and 40 black ones.

❹ The larger sample leads Natalie to assume that one-fifth of the jar's jelly beans are green (20 percent) and four-fifths are black (80 percent). The sample more closely approximates the jar's actual green-to-black ratio of 30 percent to 70 percent. The more times Natalie repeats the sampling, the greater the chance she will come close to knowing the actual ratio.

Figure 1.13 Animated Demonstration of sampling error.

The Trouble With Trends

Researchers can rarely observe all individuals of a group. For example, the explorers you read about in Section 1.1 did not—and could not—survey every uninhabited part of New Guinea. The cloud forest itself cloaks more than 2 million acres of the Foja Mountains, so surveying all of it would take unrealistic amounts of time and effort. Besides, tromping about even in a small area can damage delicate forest ecosystems.

Given such limitations, researchers often look at subsets of an area, a population, an event, or some other aspect of nature. They test or survey the subset, then use the results to make generalizations. However, generalizing from a subset is risky because a subset may not be representative of the whole.

For example, the golden-mantled tree kangaroo pictured on the *right* was first discovered in 1993 on a single forested mountaintop in New Guinea. For more than a decade, the species was never seen outside of that habitat, which is getting smaller every year because of human activities. Thus, the golden-mantled tree kangaroo was considered to be one of the most endangered animals on the planet. Then, in 2005, the New Guinea explorers discovered that this kangaroo species is fairly common in the Foja Mountain cloud forest. As a result, biologists now believe its future is secure, at least for the moment.

Problems With Probability

Making generalizations from testing or surveying a subset is risky because of sampling error. **Sampling error** is a difference between results obtained from a subset, and results from the whole (Figure 1.13).

Sampling error may be unavoidable, as illustrated by the example of the golden-mantled tree kangaroo. However, knowing how it can occur helps researchers design their experiments to minimize it. For example, sampling error can be a substantial problem with a small subset, so experimenters try to start with a relatively large sample, and they repeat their experiments.

To understand why such practices reduce the risk of sampling error, think about what happens when you flip a coin. There are two possible outcomes: The coin lands heads up, or it lands tails up. Thus, with each flip, the chance that the coin will land heads up is one in two (1/2), which is a proportion of 50 percent. However, when you flip a coin repeatedly, it often lands heads up, or tails up, several times in a row. With just 3 flips, the proportion of times that heads actually land up may not even be close

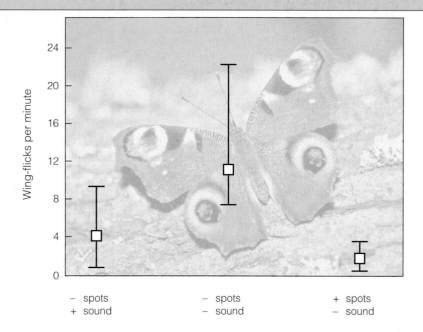

Figure 1.14 Example of error bars in a graph. This particular graph was adapted from the peacock butterfly research described in Section 1.7.

The researchers recorded the number of times each butterfly flicked its wings in response to an attack by a bird.

The squares represent average frequency of wing flicking for each sample set of butterflies. The error bars that extend above and below the squares indicate the range of values—the sampling error.

>> Figure It Out What was the fastest rate at which a butterfly with no spots or sound flicked its wings? *Answer: 22 times per minute.*

(graph y-axis: Wing-flicks per minute, values 0, 4, 8, 12, 16, 20, 24)

| – spots | – spots | + spots |
| + sound | – sound | – sound |

to 50 percent. With 1,000 flips, the overall proportion of times that the coin lands heads up is likely to be close to 50 percent.

In cases like flipping a coin, it is possible to calculate **probability**: the measure, expressed as a percentage, of the chance that a particular outcome will occur. That chance depends on the total number of possible outcomes. For instance, if 10 million people enter a drawing, each has the same probability of winning: 1 in 10 million, or (an extremely improbable) 0.00001 percent.

Analysis of experimental data often includes calculations of probability. If a result is very unlikely to have occurred by chance alone, it is said to be **statistically significant**. In this context, the word "significant" does not refer to the result's importance. It means that the result has been subjected to a rigorous statistical analysis that shows it has a very low probability (usually 5 percent or less) of being skewed by sampling error.

Variation in data is often shown as error bars on a graph (Figure 1.14). Depending on the graph, error bars may indicate variation around an average for one sample set, or the difference between two sample sets.

Bothering With Bias

Particularly when studying humans, experimenting with a single variable apart from all others is not often pos-

sible. For example, remember that the people who participated in the Olestra experiment were chosen randomly. That means the study was not controlled for gender, age, weight, medications taken, and so on. Such variables may well have influenced the results.

Human beings are by nature subjective, and scientists are no exception. Experimenters risk interpreting their results in terms of what they want to find out. That is why they often design experiments to yield quantitative results, which are counts or some other data that can be measured or gathered objectively. Such results minimize the potential for bias, and also give other scientists an opportunity to repeat the experiments and check the conclusions drawn from them.

This last point gets us back to the role of critical thinking in science. Scientists expect one another to recognize and put aside bias in order to test their hypotheses in ways that may prove them wrong. If a scientist does not, then others will, because exposing errors is just as useful as applauding insights. The scientific community consists of critically thinking people trying to poke holes in one another's ideas. Their collective efforts make science a self-correcting endeavor.

probability The chance that a particular outcome of an event will occur; depends on the total number of outcomes possible.
sampling error Difference between results derived from testing an entire group of events or individuals, and results derived from testing a susbet of the group.
statistically significant Refers to a result that is statistically unlikely to have occurred by chance.

Take-Home Message **How does science address the potential pitfalls of doing research?**

> Researchers minimize sampling error by using large sample sizes, or by repeating their experiments.

> Probability calculations can show whether a result is likely to have occurred by chance alone.

> Science is a self-correcting process because it is carried out by an aggregate community of people systematically checking one another's ideas.

> Scientific theories are our best descriptions of reality.
> Science helps us to be objective about our observations, in part because it is limited to the observable.

About the Word "Theory"

Suppose a hypothesis stands even after years of tests. It is consistent with all data ever gathered, and it has helped us make successful predictions about other phenomena. When a hypothesis meets these criteria, it is considered to be a **scientific theory** (Table 1.4).

To give an example, all observations to date have been consistent with the hypothesis that matter consists of atoms. Scientists no longer spend time testing this hypothesis for the compelling reason that, since we started looking 200 years ago, no one has discovered matter that doesn't consist of atoms. Thus, scientists use the hypothesis, now called atomic theory, to make other hypotheses about matter and the way it behaves.

Scientific theories are our best descriptions of reality. However, they can never be proven absolutely, because to do so would necessitate testing under every possible circumstance. For example, in order to prove atomic theory, the atomic composition of all matter in the universe would have to be checked—an impossible task even if someone wanted to try.

Like all hypotheses, a scientific theory can be disproven by a single observation or result that is inconsistent with it. For example, if someone discovers a form of matter that does not consist of atoms, atomic theory would have to be revised. The potentially falsifiable nature of scientific theories means that science has a built-in system of checks and balances. A theory is revised until no one can prove it to be incorrect. For example, the theory of evolution, which states that change occurs in a line of descent over time, still holds after a century of observations and testing. As with all other scientific theories, no one can be absolutely sure that it will hold under all possible conditions, but it has a very high probability of not being wrong. Few other theories have withstood as much scrutiny.

You may hear people apply the word "theory" to a speculative idea, as in the phrase "It's just a theory." This everyday usage of the word differs from the way it is used in science. Speculation is an opinion, belief, or personal conviction that is not necessarily supported by evidence. A scientific theory is different. By definition, it is supported by a large body of evidence, and it is consistent with all known facts.

A scientific theory also differs from a **law of nature**, which describes a phenomenon that has been observed to occur in every circumstance without fail, but for which we currently do not have a complete scientific explanation. The laws of thermodynamics, which describe energy, are examples. We know how energy behaves, but not exactly why it behaves the way it does.

The Limits of Science

Science helps us be objective about our observations in part because of its limitations. For example, science does not address many questions, such as "Why do I exist?" Answers to such questions can only come from within as an integration of the personal experiences and mental

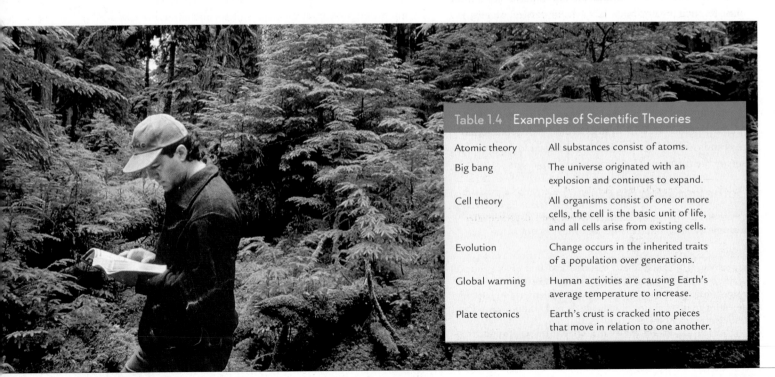

Table 1.4	Examples of Scientific Theories
Atomic theory	All substances consist of atoms.
Big bang	The universe originated with an explosion and continues to expand.
Cell theory	All organisms consist of one or more cells, the cell is the basic unit of life, and all cells arise from existing cells.
Evolution	Change occurs in the inherited traits of a population over generations.
Global warming	Human activities are causing Earth's average temperature to increase.
Plate tectonics	Earth's crust is cracked into pieces that move in relation to one another.

connections that shape our consciousness. This is not to say subjective answers have no value, because no human society can function for long unless its individuals share standards for making judgments, even if they are subjective. Moral, aesthetic, and philosophical standards vary from one society to the next, but all help people decide what is important and good. All give meaning to our lives.

Neither does science address the supernatural, or anything that is "beyond nature." Science neither assumes or denies that supernatural phenomena occur, but scientists often cause controversy when they discover a natural explanation for something that was thought to have none. Such controversy often arises when a society's moral standards are interwoven with its understanding of nature.

For example, Nicolaus Copernicus concluded in 1540 that Earth orbits the sun. Today that idea is generally accepted, but during Copernicus's time the prevailing belief system had Earth as the immovable center of the universe. In 1610, astronomer Galileo Galilei published evidence for the Copernican model of the solar system, an act that resulted in his imprisonment. He was publicly forced to recant his work, spent the rest of his life under house arrest, and was never allowed to publish again.

As Galileo's story illustrates, exploring a traditional view of the natural world from a scientific perspective might be misinterpreted as a violation of morality, even though the two are not the same. As a group, scientists are no less moral than anyone else. However, they follow a particular set of rules that do not necessarily apply to others: Their work concerns only the natural world, and their ideas must be testable in ways others can repeat.

Science helps us communicate our experiences without bias. As such, it may be as close as we can get to a universal language. We are fairly sure, for example, that the laws of gravity apply everywhere in the universe. Intelligent beings on a distant planet would likely understand the concept of gravity. We might well use gravity or another scientific concept to communicate with them, or anyone, anywhere. The point of science, however, is not to communicate with aliens. It is to find common ground here on Earth.

---

law of nature Generalization that describes a consistent natural phenomenon for which there is incomplete scientific explanation.
scientific theory Hypothesis that has not been disproven after many years of rigorous testing.

Take-Home Message Why does science work?

> Science has built-in checks and balances that help us to be objective about our observations.

> Because a scientific theory is revised until no one can prove it wrong, it is our best way of describing reality.

■ The Secret Life of Earth (revisited)

> We have discovered only a small fraction of the species that share Earth with us.

Of an estimated 100 billion species that have ever lived, at least 100 million are still with us today. That number is only an estimate because we are still discovering them—by the thousands every year. For example, a return expedition to New Guinea's Foja Mountains turned up a mouse-sized opossum and a cat-sized rat. Other surveys revealed lemurs (Figure 1.15) and sucker-footed bats in Madagascar; birds in the

Figure 1.15 It is traditional for the discoverer of a new species to have the honor of naming it. *Top*, this tiny mouse lemur, discovered in Madagascar in 2005, was named *Microcebus lehilahytsara* in honor of primatologist Steve Goodman (*lehilahytsara* is a combination of the Malagasy words for "good" and "man"). *Bottom*, Dr. Jason Bond holding a spider he discovered in California in 2008. Bond named the spider *Aptostichus stephencolberti*, after TV personality Stephen Colbert.

Philippines; monkeys in Tanzania, Brazil, and India; cave-dwelling spiders and insects in two of California's national parks; carnivorous sponges near Antarctica; whales, sharks, giant jellylike animals, fishes, and other aquatic wildlife; and scores of plants and single-celled organisms. Most were discovered by biologists who were simply trying to find out what lives where.

Biologists make discoveries every day, though we may never hear of them. Each new species they discover is another reminder that we do not yet know all of the organisms on our own planet. We don't even know how many to look for. The vast information about the 1.8 million species we do know about changes so quickly that collating it has been impossible—until recently. A new web site, titled the Encyclopedia of Life, is intended to be an online reference source and database of species information that is maintained by collaborative effort. See its progress at www.eol.org.

How Would You Vote? There is a possibility that substantial populations of some species currently listed as endangered may exist in unexplored areas. Should we wait to protect endangered species until all of Earth has been surveyed? See Cengage-Now for details, then vote online (cengagenow.com).

Summary

Section 1.1 Biology is the systematic study of life. We have encountered only a fraction of the organisms that live on Earth, in part because we have explored only a fraction of its inhabited regions.

Section 1.2 Biologists think about life at different levels of organization. **Emergent properties** appear at successively higher levels. Life emerges at the cellular level. All matter consists of **atoms**, which combine as **molecules**. **Organisms** are individuals that consist of one or more **cells**. Cells of larger multicelled organisms are organized as **tissues**, **organs**, and **organ systems**. A **population** is a group of individuals of a species in a given area; a **community** is all populations of all species in a given area. An **ecosystem** is a community interacting with its environment. The **biosphere** includes all regions of Earth that hold life.

Section 1.3 Life has underlying unity in that all living things have similar characteristics. (1) All organisms require **energy** and **nutrients** to sustain themselves: **producers** harvest energy from the environment to make their own food by processes such as **photosynthesis**; **consumers** eat other organisms, or their wastes and remains. (2) Organisms keep the conditions in their internal environment within ranges that their cells tolerate—a process called **homeostasis**. (3) **DNA** contains information that guides all of an organism's metabolic activities, including **growth**, **development**, and **reproduction**. The passage of DNA from parents to offspring is **inheritance**.

Section 1.4 The different types of organisms that currently exist on Earth differ greatly in details of body form and function. **Biodiversity** is the sum of differences among living things. **Bacteria** and **archaeans** are all single-celled, and their DNA is not contained within a **nucleus**. **Eukaryotes** (**protists**, **plants**, **fungi**, and **animals**) can be single-celled or multicelled. Their DNA is contained within a nucleus.

Section 1.5 Each type of organism has a two-part name. The first part is the **genus** name. When combined with the **specific epithet**, it designates a particular **species**. Linnean **taxonomy** ranks all species into successive **taxa** on the basis of shared traits.

Section 1.6 Critical thinking, the self-directed act of judging the quality of information as one learns, is an important part of **science**. Generally, a researcher observes something in nature, uses **inductive reasoning** to form a **hypothesis** (testable explanation) for it, then uses **deductive reasoning** to make a **prediction**

about what might occur if the hypothesis is not wrong. Predictions are tested with observations, **experiments**, or both. Experiments typically are performed on an **experimental group** as compared with a **control group**, and sometimes on **models**. Conclusions are drawn from experimental results, or **data**. A hypothesis that is not consistent with data is modified. Making, testing, and evaluating hypotheses is the **scientific method**.

Biological systems are usually influenced by many interacting **variables**. An **independent variable** influences a **dependent variable**.

Section 1.7 Scientific approaches differ, but experiments are typically designed in a consistent way. A researcher changes an independent variable, then observes the effects of the change on a dependent variable. This practice allows the researcher to unravel a cause-and-effect relationship in a complex natural system.

Section 1.8 Small sample size increases the potential for **sampling error** in experimental results. In such cases, a subset may be tested that is not representative of the whole. Researchers design experiments carefully to minimize sampling error and bias, and they use **probability** rules to check the **statistical significance** of their results. Science is ideally a self-correcting process because scientists check and test one another's ideas.

Section 1.9 Science helps us be objective about our observations because it is only concerned with testable ideas about observable aspects of nature. Opinion and belief have value in human culture, but they are not addressed by science. A **scientific theory** is a long-standing hypothesis that is useful for making predictions about other phenomena. It is our best way of describing reality. A **law of nature** describes something that occurs without fail, but for which we do not have a complete scientific explanation.

Self-Quiz   Answers in Appendix III

1. _atoms_ are fundamental building blocks of all matter.

2. The smallest unit of life is the _cell_.

3. _animals_ move around for at least part of their life.

4. Organisms require _energy_ and _nutrients_ to maintain themselves, grow, and reproduce.

5. _homeostasis_ is a process that maintains conditions in the internal environment within ranges that cells can tolerate.

6. DNA _D_.
   a. guides growth and development
   b. is the basis of traits
   c. is transmitted from parents to offspring
   d. all of the above

7. A process by which an organism produces offspring is called _reproduction_

Peacock Butterfly Predator Defenses
The photographs on the *right* represent the actual experimental and control groups used in the peacock butterfly experiment discussed in Section 1.7.

See if you can identify the experimental groups, and match them up with the relevant control group(s). *Hint:* Identify which variable is being tested in each group (each variable has a control).

A Wing spots painted out

B Wing spots visible; wings silenced

C Wing spots painted out; wings silenced

D Wings painted but spots visible

E Wings cut but not silenced

F Wings painted, spots visible; wings cut, not silenced

8. ____D____ is the transmission of DNA to offspring.
a. Reproduction   c. Homeostasis
b. Development   d. Inheritance

9. An animal is a(n) a, d, e (choose all that apply).
a. organism   e. consumer
b. domain   f. producer
c. species   g. hypothesis
d. eukaryote   h. trait

10. Plants are a, d, f (choose all that apply).
a. organisms   e. consumers
b. a domain   f. producers
c. a species   g. hypotheses
d. eukaryotes   h. traits

11. Science only addresses that which is ____B____ .
a. alive   c. variable
b. observable   d. indisputable

12. A control group is _____ .
a. a set of individuals that have a certain characteristic or receive a certain treatment
b. the standard against which an experimental group is compared
c. the experiment that gives conclusive results

13. Match the terms with the most suitable description.
C emergent property   a. statement of what a hypothesis leads you to expect
B species   b. type of organism
D scientific theory   c. occurs at a higher organizational level
E hypothesis   d. time-tested hypothesis
A prediction   e. testable explanation
F probability   f. measure of chance

Additional questions are available on **CENGAGENOW**.

Critical Thinking

1. A person is declared to be dead upon the irreversible cessation of spontaneous body functions: brain activity, or blood circulation and respiration. However, only about 1% of a person's cells have to die in order for all of these things to happen. How can someone be dead when 99% of his or her cells are still alive?

2. Why would you think twice about ordering from a cafe menu that lists the genus name but not the specific epithet of its offerings? *Hint:* Look up *Homarus americanus*, *Ursus americanus*, *Ceanothus americanus*, *Bufo americanus*, *Lepus americanus*, and *Nicrophorus americanus*.

3. Once there was a highly intelligent turkey that had nothing to do but reflect on the world's regularities. Morning always started out with the sky turning light, followed by the master's footsteps, which were always followed by the appearance of food. Other things varied, but food always followed footsteps. The sequence of events was so predictable that it eventually became the basis of the turkey's theory about the goodness of the world. One morning, after more than 100 confirmations of the goodness theory, the turkey listened for the master's footsteps, heard them, and had its head chopped off.

Any scientific theory is modified or discarded upon discovery of contradictory evidence. The absence of absolute certainty has led some people to conclude that "facts are irrelevant—facts change." If that is so, should we stop doing scientific research? Why or why not?

4. In 2005, researcher Woo-suk Hwang reported that he had made immortal stem cells from human patients. His research was hailed as a breakthrough for people affected by degenerative diseases, because stem cells may be used to repair a person's own damaged tissues. Hwang published his results in a peer-reviewed journal. In 2006, the journal retracted his paper after other scientists discovered that Hwang's group had faked their data. Does the incident show that results of scientific studies cannot be trusted? Or does it confirm the usefulness of a scientific approach, because other scientists discovered and exposed the fraud?

Animations and Interactions on **CENGAGENOW**:
❯ Life's building blocks; Life's levels of organization; Energy flow and materials cycling; Life's diversity; Three domains of life; Sampling error.

See an annotated scientific paper in Appendix II.

‹ Links to Earlier Concepts

Take a moment to review Section 1.2 because life's organization starts with atoms. Life's organization requires continuous inputs of energy (1.3). Organisms store that energy in chemical bonds between atoms. You will see examples of how the body's built-in mechanisms maintain homeostasis (1.3), and how scientists make major discoveries (1.6).

Key Concepts

Atoms and Elements
Atoms, the building blocks of all matter, differ in their numbers of protons, neutrons, and electrons.

Why Electrons Matter
How an atom interacts with other atoms depends on the number and arrangement of its electrons.

2 Life's Chemical Basis

2.1 Mercury Rising

Actor Jeremy Piven, best known for his Emmy-winning role on the television series *Entourage*, began starring in a Broadway play in 2008. He quit suddenly after two shows, citing medical problems. Piven said he was suffering from mercury poisoning caused by eating too much sushi. The play's producers and his co-actors were skeptical. The playwright ridiculed Piven, saying he was leaving to pursue a career as a thermometer. But mercury poisoning is no laughing matter.

Mercury is a naturally occurring metal. Most of it is safely locked away in rocky minerals, but volcanic activity and other geologic processes release it into the atmosphere. So do human activities, especially burning coal (Figure 2.1). Once airborne, mercury can drift long distances before settling to Earth's surface. There, microbes combine it with carbon to form a substance called methylmercury.

Unlike mercury alone, methylmercury easily crosses skin and mucous membranes. In water, it ends up in the tissues of aquatic organisms. All fish and shellfish contain methylmercury. Humans contain it too, mainly as a result of eating seafood.

When mercury enters the body, it damages the nervous system, brain, kidneys, and other organs. A dose as low as 3 micrograms per kilogram of body weight (about 200 micrograms for an average-sized adult) can cause tremors, itching or burning sensations, and loss of coordination. Exposure to larger amounts can result in thought and memory impairment, coma, and death.

The developing brain is particularly sensitive to mercury because the metal interferes with nerve formation. Thus, mercury is acutely toxic in infants, and it causes long-term neurological effects in children. Methylmercury in a pregnant woman's blood passes to her unborn child, along with a legacy of permanent developmental problems.

The U.S. Food and Drug Administration requires that foods contain less than 1 part per million of mercury, and for the most part they do. However, it takes months or even years for mer-

Figure 2.1 Atmospheric fallout from coal-fired power plant emissions is now the biggest cause of mercury pollution. *Opposite*, mercury can accumulate to toxic levels in the tissues of tuna and other large predatory fish.

cury to be cleared from the body, so it can build up to high levels if even small amounts are ingested on a regular basis. That is why large predatory fish have the most mercury in their tissues. It is also why the U.S. Environmental Protection Agency recommends that adults ingest less than 0.1 microgram of mercury per kilogram of body weight per day. For an average-sized person, that limit works out to be about 7 micrograms per day, which is not a big amount if you eat seafood. A two-ounce piece of sushi tuna typically contains about 40 micrograms of mercury, and the occasional piece has many times that amount. It doesn't matter if the fish is raw, grilled, or canned, because mercury is unaffected by cooking. Eat a medium-sized tuna steak, and you could be getting more than 700 micrograms of mercury along with it.

With this chapter, we turn to the first of life's levels of organization: atoms. Interactions between atoms make the molecules that sustain life, and also some that destroy it.

Atoms Bond
Atoms of many elements interact by acquiring, sharing, and giving up electrons. Interacting atoms may form ionic, covalent, or hydrogen bonds.

Water of Life
Water stabilizes temperature. It also has cohesion, and it can act as a solvent for many other substances. These properties make life possible.

The Power of Hydrogen
Most of the chemistry of life occurs in a narrow range of pH, so the fluids inside organisms are buffered to stay within that range.

❯ The behavior of elements, which make up all living things, starts with the structure of individual atoms.

❯ The number of protons in the atomic nucleus defines the element, and the number of neutrons defines the isotope.

❮ Link to Atoms 1.2

The idea that different structures can be assembled from the same basic building blocks is a recurring theme in our world. The theme is apparent in all levels of nature's organization, including atomic structure. Life's unique characteristics start with the properties of different **atoms**, tiny particles that are building blocks of all substances. Even though they are about 20 million times smaller than a grain of sand, atoms consist of even smaller subatomic particles: positively charged **protons** (p^+), uncharged **neutrons**, and negatively charged **electrons** (e^-). **Charge** is an electrical property. Opposite charges attract, and like charges repel. Protons and neutrons cluster in an atom's central core, or **nucleus**, and electrons move around the nucleus (Figure 2.2).

Atoms differ in their number of subatomic particles. The number of protons in an atom's nucleus is called the **atomic number**, and it determines the type of atom, or element. **Elements** are pure substances, each consisting only of atoms that have the same number of protons in their nucleus. For example, the atomic number of carbon is 6, so all atoms with six protons in their nucleus are carbon atoms, no matter how many electrons or neutrons they have. A chunk of carbon consists only of carbon atoms, and all of those atoms have six protons.

The same elements that make up a living body also occur in nonliving things, but their proportions differ. For example, a human body contains a much larger pro-

Figure 2.2 Atoms consist of electrons moving around a core, or nucleus, of protons and neutrons.

Models such as this diagram cannot show what atoms really look like. Electrons zoom around in fuzzy, three-dimensional spaces about 10,000 times bigger than the nucleus.

● proton
● neutron
○ electron

portion of carbon atoms than do rocks or seawater. Why? Unlike rocks or seawater, a body consists of a very high proportion of the molecules of life, which in turn consist of a high proportion of carbon atoms.

Knowing about the numbers of electrons, protons, and neutrons helps us predict how elements will behave. For example, elements in each vertical column of the periodic table behave in similar ways. The **periodic table** shows all of the known elements arranged in order of atomic number (Figure 2.3). Each of the 117 elements is listed in the table by a symbol that is usually an abbreviation of the element's Latin or Greek name. For instance, Pb (lead) is short for *plumbum;* the word "plumbing" is related— ancient Romans made their water pipes with lead. Carbon's symbol, C, is from *carbo*, the Latin word for coal (which is mostly carbon).

The first ninety-four elements occur in nature. We know about the rest because we have synthesized them a few atoms at a time. An atomic nucleus cannot be altered by heat or any other ordinary means, so synthesizing an element requires a particle accelerator and other equipment found only in nuclear physics laboratories.

atom Particle that is a fundamental building block of all matter.
atomic number Number of protons in the atomic nucleus; determines the element.
charge Electrical property. Opposite charges attract, and like charges repel.
electron Negatively charged subatomic particle that occupies orbitals around an atomic nucleus.
element A pure substance that consists only of atoms with the same number of protons.
isotopes Forms of an element that differ in the number of neutrons their atoms carry.
mass number Total number of protons and neutrons in the nucleus of an element's atoms.
neutron Uncharged subatomic particle in the atomic nucleus.
nucleus Core of an atom; occupied by protons and neutrons.
periodic table Tabular arrangement of the known elements by atomic number.
proton Positively charged subatomic particle that occurs in the nucleus of all atoms.
radioactive decay Process by which atoms of a radioisotope emit energy and/or subatomic particles when their nucleus spontaneously disintegrates.
radioisotope Isotope with an unstable nucleus.
tracer A molecule labelled with a detectable substance.

Figure 2.3 Animated The periodic table. The table was created in 1869 by chemist Dmitry Mendeleyev (*left*), who arranged the known elements by chemical properties. The arrangement turned out to be by atomic number. Until he came up with the table, Mendeleyev had been known mainly for his extravagant hair (he cut it only once per year).

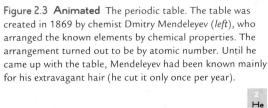

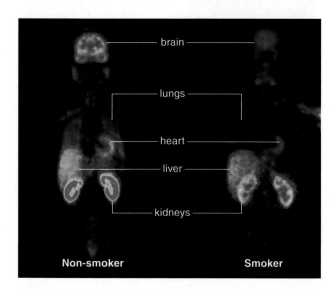

Figure 2.4 Animated PET scans. The result of a PET scan is a digital image of a process in the body's interior. These PET scans show the activity of a molecule called MAO-B in the body of a non-smoker (*left*) and a smoker (*right*). The activity is color coded from *red* (highest activity) to *purple* (lowest). Low MAO-B activity is associated with violence, impulsiveness, and other behavioral problems.

Isotopes and Radioisotopes

Atoms of an element that differ in their number of neutrons are called **isotopes**. All elements have isotopes. We define isotopes by their **mass number**, which is the total number of protons and neutrons in their nucleus. Mass number is written as a superscript to the left of an element's symbol. For example, the most common isotope of carbon has six protons and six neutrons, so it is designated ^{12}C, or carbon 12. The other naturally occurring isotopes of carbon are ^{13}C (six protons, seven neutrons), and ^{14}C (six protons, eight neutrons).

In 1896, physicist Henri Becquerel made a discovery after leaving some crystals of a uranium salt in a desk drawer, on top of a metal screen. Under the screen was an exposed film wrapped tightly in black paper. When Becquerel developed the film, he was surprised to see a negative image of the screen. He realized that "invisible radiations" coming from the uranium salts had passed through the paper and exposed the film around the screen. Uranium, like many other elements, has radioactive isotopes, or **radioisotopes**. The atoms of radioisotopes spontaneously emit subatomic particles or energy (radiation) when their nucleus breaks down. This process, **radioactive decay**, can transform one element into another. For example, carbon 14 is a radioisotope that decays when one of its neutrons splits into a proton and an electron. The nucleus emits the electron, so an atom with eight neutrons and six protons (^{14}C) becomes an atom with seven neutrons and seven protons, which is nitrogen (^{14}N).

Each radioisotope decays at a constant rate into certain products. This process occurs independently of external factors such as temperature, pressure, or whether the atoms are part of molecules, so it is very predictable. For example, we can reliably predict that about half of the atoms of ^{14}C in any sample will be ^{14}N atoms after 5,730 years. Thus, we can use the isotope content of a rock or fossil to estimate its age (we will return to this topic in Section 16.5).

Researchers and clinicians also use radioisotopes to track biological processes inside living organisms. Remember, isotopes are atoms of the same element. All isotopes of an element generally have the same chemical properties regardless of the number of neutrons in their atoms. This consistent chemical behavior means that organisms use atoms of one isotope (such as ^{14}C) the same way that they use atoms of another (such as ^{12}C). Thus, radioisotopes can be used in tracers. A **tracer** is any molecule with a detectable substance attached.

A typical radioactive tracer is a molecule in which radioisotopes have been swapped for one or more atoms. Researchers deliver radioactive tracers into a biological system such as a cell or a multicelled body. Instruments that can detect radioactivity let researchers follow the tracer as it moves through the system. For example, Melvin Calvin and his colleagues used a radioactive tracer to identify specific reaction steps of photosynthesis. The researchers made carbon dioxide with ^{14}C, then let green algae (simple aquatic organisms) take up the radioactive gas. Using instruments that detected the radioactive decay of ^{14}C, they tracked carbon through steps by which the algae—and all plants—make sugars.

Radioisotopes have medical applications as well. For example, PET (short for positron-emission tomography) helps us "see" a functional process inside the body. By this procedure, a radioactive sugar or other tracer is injected into a patient, who is then moved into a PET scanner. Inside the patient's body, cells with differing rates of activity take up the tracer at different rates. The scanner detects radioactive decay wherever the tracer is, then translates that data into an image (Figure 2.4).

Take-Home Message **What are the basic building blocks of all matter?**

> All matter consists of atoms, tiny particles that in turn consist of electrons moving around a nucleus of protons and neutrons.

> An element is a pure substance that consists only of atoms with the same number of protons. All elements occur as isotopes, which are forms of an element that have different numbers of neutrons.

> The unstable nuclei of radioisotopes disintegrate spontaneously (decay) at a predictable rate to form predictable products.

> Atoms acquire, share, and donate electrons.
> Whether an atom will interact with other atoms depends on how many electrons it has.
< Link to Atomic interactions 1.2

Energy Levels

Electrons are really, really small. How small are they? If they were as big as apples, you would be about 3.5 times taller than our solar system is wide. Simple physics explains the motion of, say, an apple falling from a tree, but electrons are so tiny that such everyday physics cannot explain their behavior. However, that behavior underlies atomic interactions.

A typical atom has about the same number of electrons and protons. In larger atoms, a lot of electrons may be zipping around one nucleus. Those electrons move at nearly the speed of light (300,000 kilometers per second, or 670 million miles per hour), but they never collide. Why not? Electrons avoid one another because they travel in different orbitals, which are defined volumes of space around the nucleus.

Imagine that an atom is a multilevel apartment building with a nucleus in the basement. Each "floor" of the building corresponds to a certain energy level, and each has a certain number of "rooms" (orbitals) available for rent to one or two electrons at a time. Electrons populate rooms from the ground floor up; in other words, they fill

orbitals from lower to higher energy levels. The farther an electron is from the nucleus in the basement, the greater its energy. An electron can move to a room on a higher floor if an energy input gives it a boost, but it immediately emits its extra energy and moves back down.

Shell models help us visualize how electrons populate atoms. In this model, nested "shells" correspond to successively higher energy levels. Thus, a shell includes all of the rooms on one floor of our atomic apartment building.

We draw a shell model of an atom by filling its shells with electrons (represented as balls or dots), from the innermost shell out, until there are as many electrons as the atom has protons. For example, there is only one room on the first floor, one orbital at the lowest energy level. It fills up first. In hydrogen, the simplest atom, a single electron occupies that room (Figure 2.5A). Helium, with two protons, has two electrons that fill its first shell. In larger atoms, more electrons rent the second-floor rooms (Figure 2.5B). When the second floor fills, more electrons rent third-floor rooms (Figure 2.3C), and so on.

A The first shell corresponds to the first energy level, and it can hold up to 2 electrons. Hydrogen has one proton, so it has 1 electron and 1 vacancy. A helium atom has 2 protons, 2 electrons, and no vacancies. The number of protons in each model is shown.

first shell

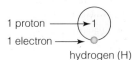

1 proton
1 electron

hydrogen (H)

helium (He)

B The second shell corresponds to the second energy level, and it can hold up to 8 electrons. Carbon has 6 protons, so its first shell is full. Its second shell has 4 electrons, and four vacancies. Oxygen has 8 protons and two vacancies. Neon has 10 protons and no vacancies.

second shell

carbon (C)

oxygen (O)

neon (Ne)

C The third shell, which corresponds to the third energy level, can hold up to 8 electrons. A sodium atom has 11 protons, so its first two shells are full; the third shell has one electron. Thus, sodium has seven vacancies. Chlorine has 17 protons and one vacancy. Argon has 18 protons and no vacancies.

third shell

sodium (Na)

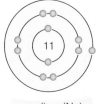

chlorine (Cl)

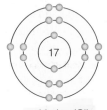

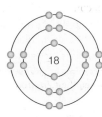

argon (Ar)

Figure 2.5 Animated Shell models. Each circle (shell) represents all orbitals at one energy level. A model is filled with electrons from the innermost shell out, until there are as many electrons as protons. Atoms with vacancies (room for additional electrons) in their outermost shell tend to interact with other atoms.

If an atom's outermost shell is full of electrons, we say that it has no vacancies. Helium is an example. Atoms of such elements are chemically inactive, which means they are most stable as single atoms. By contrast, if an atom's outermost shell has room for another electron, it has a vacancy. Hydrogen has one vacancy. Atoms with vacancies tend to interact with other atoms: They give up, acquire, or share electrons until they have no vacancies in their outermost shell. Any atom is in its most stable state when it has no vacancies.

Swapping Electrons

The negative charge of an electron is the same magnitude as the positive charge of a proton, so the two charges cancel one another. Thus, an atom with as many electrons as protons carries no overall (net) charge. An atom with different numbers of electrons and protons carries a charge. When an atom is charged, we call it an **ion**.

An atom acquires a positive charge by losing an electron. Conversely, an atom acquires a negative charge by pulling an electron away from another atom (Figure 2.6). **Electronegativity** is a measure of an atom's ability to pull electrons away from other atoms. Electronegativity is not the same as charge. Rather, an atom's electronegativity depends on its size and how many vacancies it has. For example, when a chlorine atom is uncharged, it has 17 protons and 17 electrons. Seven electrons are in its outer (third) shell, which can hold eight, so this atom has one vacancy. An uncharged chlorine atom is highly electronegative—it can easily pull an electron away from another atom to fill its third shell. When that happens, the atom becomes a chloride ion (Cl^-) with 17 protons, 18 electrons, and a net negative charge (Figure 2.6A).

As another example, an uncharged sodium atom has 11 protons and 11 electrons. This atom has one electron in its outer (third) shell, which can hold eight. It has seven vacancies. An uncharged sodium atom is weakly electronegative, so it cannot pull seven electrons from other atoms to fill its third shell. Instead, it tends to lose the single electron in its third shell. When that happens, two full shells—and no vacancies—remain. The atom has now become a sodium ion (Na^+), with 11 protons, 10 electrons, and a net positive charge (Figure 2.6B).

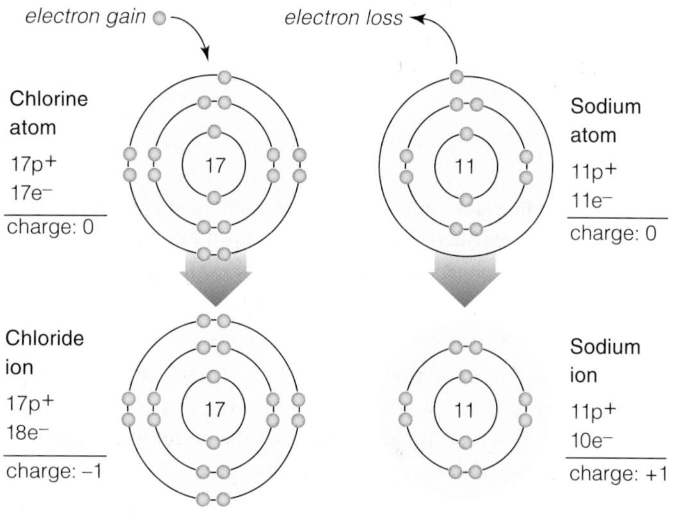

A A chlorine atom (Cl) becomes a negatively charged chloride ion (Cl^-) when it gains an electron and fills the vacancy in its third, outermost shell.

B A sodium atom (Na) becomes a positively charged sodium ion (Na^+) when it loses the electron in its third shell. The atom's full second shell is now its outermost, so it has no vacancies.

Figure 2.6 Animated Ion formation.

Sharing Electrons

An atom can get rid of vacancies by participating in a **chemical bond**, which is an attractive force that arises between two atoms when their electrons interact. A **molecule** forms when two or more atoms of the same or different elements join in chemical bonds. The next section explains the main types of bonds in biological molecules.

Compounds are molecules that consist of two or more different elements. The proportions of elements in a molecular substance do not vary. For example, all water molecules have one oxygen atom bonded to two hydrogen atoms. By contrast, a **mixture** is an intermingling of two or more substances. The proportions of substances in a mixture can vary because chemical bonds do not form. For example, sugar may dissolve in water in variable amounts because no chemical bonds form between the two substances. A liquid mixture is called a solution.

---

chemical bond An attractive force that arises between two atoms when their electrons interact.
compound Type of molecule that has atoms of more than one element.
electronegativity Measure of the ability of an atom to pull electrons away from other atoms.
ion Charged atom.
mixture An intermingling of two or more types of molecules.
molecule Group of two or more atoms joined by chemical bonds.
shell model Model of electron distribution in an atom.

Take-Home Message **Why do atoms interact?**

❯ An atom's electrons are the basis of its chemical behavior.

❯ Shells represent all electron orbitals at one energy level in an atom. When the outermost shell is not full of electrons, the atom has a vacancy.

❯ Atoms tend to get rid of vacancies by gaining or losing electrons (thereby becoming ions), or by sharing electrons with other atoms.

❯ Atoms with vacancies can form chemical bonds. Chemical bonds connect atoms into molecules.

❯ The characteristics of a bond arise from the properties
 of atoms that take part in it.
❮ Link to Molecules 1.2

Arranged in different ways, the same atomic building
blocks make different molecules. For example, carbon
atoms bonded one way form layered sheets of a soft,
slippery substance known as graphite. The same carbon
atoms bonded a different way form the rigid crystal lattice of the hardest substance, which is diamond. Carbon
bonded to oxygen and hydrogen atoms make sugar.

Although bonding applies to a range of interactions
among atoms, we can categorize most bonds into distinct
types based on their different properties. Which type
forms—an ionic, covalent, or hydrogen bond—depends
on the atoms that take part in it.

Ionic Bonds

Remember from the last section that a strongly electronegative atom tends to gain electrons until its outermost
shell is full. At that point it will be a negatively charged
ion. A weakly electronegative atom tends to lose electrons
until its outermost shell is full. At that point it will be a
positively charged ion.

An **ionic bond** is a strong mutual attraction of oppositely charged ions. Such bonds do not usually form
by the direct transfer of an electron from one atom to
another; rather, atoms that have already become ions stay
close together because of their opposite charges. Common
table salt offers an example. Each crystal of this substance
consists of a lattice of sodium and chloride ions interacting in ionic bonds (Figure 2.7).

A Each crystal of table salt is a cubic lattice of many sodium
and chloride ions locked in ionic bonds.

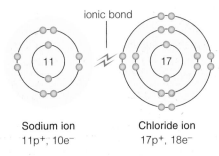

Sodium ion	Chloride ion
$11p^+$, $10e^-$	$17p^+$, $18e^-$

B The mutual attraction of opposite charges holds the two
kinds of ions together in the lattice.

Figure 2.7 Animated Ionic bonds.

Covalent Bonds

In a **covalent bond**, two atoms share a pair of electrons.
Such bonds typically form between atoms with similar
electronegativity and unpaired electrons. By sharing their
electrons, each atom's vacancy becomes partially filled
(Figure 2.8). Covalent bonds can be stronger than ionic
bonds, but they are not always so.

Structural formulas show how covalent bonds connect
atoms. A line between two atoms represents a single covalent bond, in which two atoms share one pair of electrons.
A simple example is molecular hydrogen (H_2), with one
covalent bond between hydrogen atoms (H—H).

Many atoms participate in more than one covalent
bond at the same time. The oxygen atom in a water molecule (H—O—H) is one example (Table 2.1). Two, three, or
even four covalent bonds may form between two atoms
when they share multiple pairs of electrons. For example,
two atoms sharing two pairs of electrons are connected by
two covalent bonds. Such double bonds are represented
by a double line between the atoms. A double bond links
the two oxygen atoms in molecular oxygen (O=O). Three
lines indicate a triple bond, in which two atoms share
three pairs of electrons. A triple covalent bond links the
two nitrogen atoms in molecular nitrogen (N≡N).

Some covalent bonds are nonpolar, meaning that the
atoms participating in the bond are sharing electrons
equally. There is no difference in charge between the two
ends of such bonds. The molecular hydrogen (H_2), oxygen

Table 2.1	Different Ways To Represent the Same Molecule	
Common name	Water	Familiar term.
Chemical name	Dihydrogen monoxide	Systematically describes elemental composition.
Chemical formula	H_2O	Indicates unvarying proportions of elements. Subscripts show number of atoms of an element per molecule. The absence of a subscript means one atom.
Structural formula	H—O—H $H^{\diagdown O \diagup}H$	Represents each covalent bond as a single line between atoms. Bond angles also may be represented.
Structural model		Shows the positions and relative sizes of atoms.
Shell model		Shows how pairs of electrons are shared in covalent bonds.

Molecular hydrogen (H — H)
Two hydrogen atoms, each with one proton, share two electrons in a single nonpolar covalent bond.

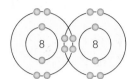

Molecular oxygen (O = O)
Two oxygen atoms, each with eight protons, share four electrons in a double covalent bond.

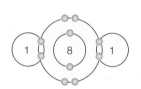

Water molecule (H — O — H)
Two hydrogen atoms share electrons with an oxygen atom in two polar covalent bonds. The oxygen exerts a greater pull on the shared electrons, so it has a slight negative charge. Each hydrogen has a slight positive charge.

Figure 2.8 Animated Covalent bonds, in which atoms with unpaired electrons in their outermost shell become more stable by sharing electrons. Two electrons are shared in each covalent bond. When sharing is equal, the bond is nonpolar. When one atom exerts a greater pull on the electrons, the bond is polar.

hydrogen bond

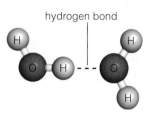

water molecule   water molecule

A A hydrogen (H) bond is an attraction between an electronegative atom and a hydrogen atom taking part in a separate polar covalent bond.

B Hydrogen bonds are individually weak, but many of them form. Collectively, they are strong enough to stabilize the structures of large biological molecules such as DNA, shown here.

Figure 2.9 Animated Hydrogen bonds. Hydrogen bonds form at a hydrogen atom taking part in a polar covalent bond. The hydrogen atom's slight positive charge weakly attracts an electronegative atom. As shown here, hydrogen (H) bonds can form between molecules or between different parts of the same molecule.

(O_2), and nitrogen (N_2) mentioned earlier are examples. These molecules are some of the gases that make up air.

Atoms participating in polar covalent bonds do not share electrons equally. Such bonds can form between atoms with a difference in electronegativity. The atom that is more electronegative pulls the electrons a little more toward its "end" of the bond, so that atom bears a slightly negative charge. The atom at the other end of the bond bears a slightly positive charge. The greater the difference in electronegativity between the atoms, the more polar is the covalent bond that forms between them.

For example, a water molecule has two polar covalent bonds. The oxygen atom carries a slight negative charge, and each of the hydrogen atoms carries a slight positive charge. Any such separation of charge into distinct positive and negative regions is called **polarity**. As you will see in the next section, the polarity of the water molecule is very important for the world of life.

Hydrogen Bonds

Hydrogen bonds form between polar regions of two molecules, or between regions of the same molecule.

covalent bond Chemical bond in which two atoms share a pair of electrons.
hydrogen bond Attraction that forms between a covalently bonded hydrogen atom and another atom taking part in a separate covalent bond.
ionic bond Type of chemical bond in which a strong mutual attraction forms between ions of opposite charge.
polarity Any separation of charge into distinct positive and negative regions.

A **hydrogen bond** is a weak attraction between a covalently bonded hydrogen atom and another atom taking part in a separate polar covalent bond. In a hydrogen bond, the atom interacting with the hydrogen is typically an oxygen, nitrogen, or other highly electronegative atom.

Like ionic bonds, hydrogen bonds form by the mutual attraction of opposite charges: The hydrogen atom carries a slight positive charge and the other atom carries a slight negative charge. However, unlike ionic bonds, hydrogen bonds do not make molecules out of atoms, so they are not chemical bonds.

Hydrogen bonds are individually weak. They form and break much more easily than covalent or ionic bonds. Even so, many of them form between molecules, or between different parts of a large one. Collectively, they are strong enough to stabilize the characteristic structures of large biological molecules such as DNA (Figure 2.9).

Take-Home Message **How do atoms interact?**

❯ A chemical bond forms when the electrons of two atoms interact. Depending on the atoms, the bond may be ionic or covalent.

❯ An ionic bond is a strong mutual attraction between two ions of opposite charge.

❯ Atoms share a pair of electrons in a covalent bond. When the atoms share electrons equally, the bond is nonpolar. When they share electrons unequally, the bond is polar.

❯ A hydrogen bond is a weak attraction between a highly electronegative atom and a hydrogen atom taking part in a separate polar covalent bond.

❯ Hydrogen bonds are individually weak, but collectively strong.

❭ Water is essential to life because of its unique properties.
❭ The unique properties of water are a result of the extensive hydrogen bonding among water molecules.

Life evolved in water. All living organisms are mostly water, many of them still live in it, and all of the chemical reactions of life are carried out in water. What is so special about water?

Each Water Molecule Is Polar

Water's special properties as a liquid begin with the two polar covalent bonds in each water molecule. Overall, the molecule has no charge, but the oxygen pulls the shared electrons a bit more than the hydrogen atoms do. Thus, each of the atoms in a water molecule carries a slight charge. The oxygen atom is slightly negative, and the hydrogen atoms are slightly positive:

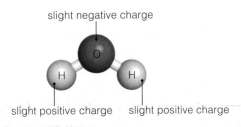

slight negative charge

slight positive charge   slight positive charge

The separation of charge means that the water molecule itself is polar. The polarity is very attractive to other water molecules, and hydrogen bonds form between them in tremendous numbers (Figure 2.10). Extensive hydrogen bonding between water molecules imparts unique properties to liquid water that make life possible.

Figure 2.10 Animated Water. Many hydrogen bonds (dashed lines) that quickly form and break keep water molecules clustered together tightly. The extensive hydrogen bonding in liquid water gives it unique properties.

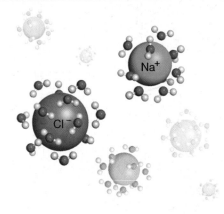

Figure 2.11 Animated Water molecules that surround an ionic solid pull its atoms apart, thereby dissolving them.

Water Is an Excellent Solvent

Water is a **solvent**—a substance, usually a liquid, that can dissolve other substances. When a substance dissolves, its individual molecules or ions become **solutes** as they disperse. Salts, sugars, and other compounds that dissolve easily in water are polar, so many hydrogen bonds form between them and water molecules.

A **salt** is a compound that dissolves easily in water and releases ions other than H^+ and OH^- when it does. Sodium chloride (NaCl) is an example of a salt. Water easily dissolves salts and other **hydrophilic** (water-loving) substances. Hydrogen bonds form between water and the polar molecules of such substances. These bonds dissolve solutes by pulling their molecules away from one another and keeping them apart (Figure 2.11).

You can see how water interacts with **hydrophobic** (water-dreading) substances if you shake a bottle filled with water and salad oil, then set it on a table and watch what happens. Salad oil consists of nonpolar molecules, and hydrogen bonds do not form between nonpolar molecules and water. Shaking breaks some of the hydrogen bonds that keep water molecules together, so the water breaks into small droplets that mix with the oil. However, the water quickly begins to cluster into larger and larger drops as new hydrogen bonds form among its molecules. The bonding excludes molecules of oil and pushes them together into drops that rise to the surface of the water. The same interactions occur at the thin, oily membrane that separates the watery fluid inside of cells from the

cohesion Tendency of molecules to resist separating from one another.
evaporation Transition of a liquid to a gas.
hydrophilic Describes a substance that dissolves easily in water.
hydrophobic Describes a substance that resists dissolving in water.
salt Compound that releases ions other than H^+ and OH^- when it dissolves in water.
solute A dissolved substance.
solvent Liquid that can dissolve other substances.
temperature Measure of molecular motion.

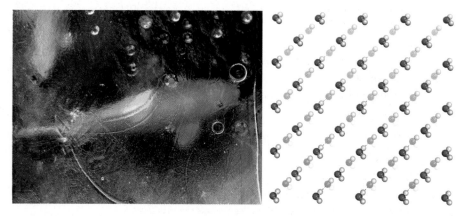

Figure 2.13 Ice floats on water. *Left,* a covering of ice can insulate water underneath it, thus keeping aquatic organisms from freezing during harsh winters. *Right,* ice forms below about 0°C (32°F), as hydrogen bonds lock water molecules in a rigid, three-dimensional lattice. It floats because the molecules pack less densely than in water.

Figure 2.12 Visible effect of cohesion: a wasp drinking, not sinking. Cohesion imparts surface tension to liquid water, which means that the surface of liquid water behaves a bit like a sheet of elastic.

watery fluid outside of them. The organization of membranes—and of life—starts with such interactions (you will read more about membranes in Chapter 4).

Cohesion

Another life-sustaining property of water is **cohesion**, which means that water molecules resist separating from one another. Hydrogen bonds collectively exert a continuous pull on individual water molecules. You can see the effect of cohesion as surface tension (Figure 2.12).

Cohesion is an important component of many processes that sustain multicelled bodies. As one example, water molecules constantly escape from the surface of liquid water as vapor, a process called **evaporation**. Evaporation is resisted by the hydrogen bonding that keeps water molecules together. In other words, overcoming water's cohesion takes energy. Thus, evaporation sucks energy in the form of heat from liquid water, lowering the water's surface temperature. Evaporative water loss can help you and some other mammals cool off when you sweat in hot, dry weather. Sweat, which is about 99 percent water, cools the skin as it evaporates.

Cohesion works inside organisms, too. For instance, plants continually absorb water from soil as they grow. Water molecules evaporate from leaves, and replacements are pulled upward from roots. Cohesion makes it possible for columns of liquid water to rise from roots to leaves inside narrow pipelines of vascular tissue. In some trees, these pipelines extend straight up for hundreds of feet. Section 26.4 returns to this topic.

Water Stabilizes Temperature

Temperature is a way to measure the energy of molecular motion. All molecules jiggle nonstop, and they jiggle faster as they absorb heat. However, extensive hydrogen bonding restricts the movement of water molecules—it keeps them from jiggling as much as they would otherwise. Thus, it takes more heat to raise the temperature of water compared with other liquids. Temperature stability is an important component of homeostasis, because most of the molecules of life function properly only within a certain range of temperature.

Below 0°C (32°F), water molecules do not jiggle enough to break hydrogen bonds, and they become locked in the rigid, lattice-like bonding pattern of ice. Individual water molecules pack less densely in ice than they do in water, so ice floats on water. Sheets of ice that form near the surface of ponds, lakes, and streams can insulate the water under them from subfreezing air temperatures. Such "ice blankets" protect aquatic organisms during extremely cold winters (Figure 2.13).

Take-Home Message **Why is water essential to life?**

> Extensive hydrogen bonding among water molecules imparts unique properties to water that make life possible.

> Hydrogen bonds form between water and polar molecules. This bonding dissolves hydrophilic substances easily. Hydrogen bonds do not form between water and nonpolar molecules of hydrophobic substances.

> Individual water molecules tend to stay together (cohesion).

> The temperature of water is more stable than that of other liquids.

> Ice is less dense than liquid water, so it floats. Ice insulates water beneath it.

❯ pH is a measure of the concentration of hydrogen ions.
❯ Most biological processes occur within a narrow range of pH, typically around pH 7.
❮ Link to Homeostasis 1.3

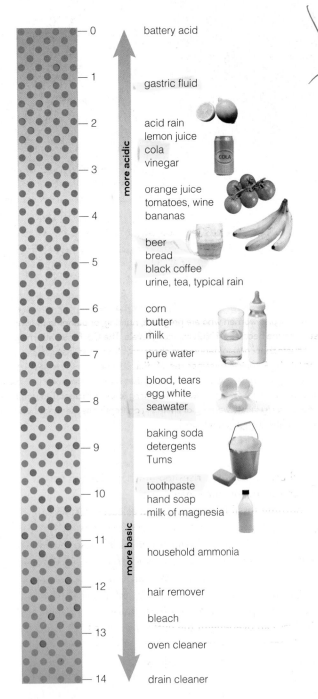

Figure 2.14 **Animated** A pH scale. Here, *red* dots signify hydrogen ions (H^+) and *blue* dots signify hydroxyl ions (OH^-). Also shown are approximate pH values for some common solutions.

This pH scale ranges from 0 (most acidic) to 14 (most basic). A change of one unit on the scale corresponds to a tenfold change in the amount of H^+ ions.

❯❯ **Figure It Out** What is the approximate pH of cola?

Answer: 2.5

In liquid water, water molecules spontaneously separate into hydrogen ions (H^+) and hydroxide ions (OH^-). These ions can combine again to form water:

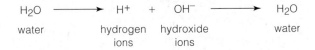

$$H_2O \longrightarrow H^+ + OH^- \longrightarrow H_2O$$
water   hydrogen   hydroxide   water
                 ions   ions

Concentration refers to the amount of a particular solute that is dissolved in a given volume of fluid. Hydrogen ion concentration is a special case. We measure the number of hydrogen ions in a solution using a value called **pH**. When the number of H^+ ions is the same as the number of OH^- ions, the pH of the solution is 7, or neutral. The pH of pure water (but not rainwater or seawater) is like this. The higher the number of hydrogen ions, the lower the pH. A one-unit decrease in pH corresponds to a tenfold increase in the number of H^+ ions, and a one-unit increase corresponds to a tenfold decrease in the number of H^+ ions (Figure 2.14).

One way to get a sense of pH is to taste dissolved baking soda (pH 9), distilled water (pH 7), and lemon juice (pH 2). Nearly all of life's chemistry occurs near pH 7. Most of your body's internal environment (tissue fluids and blood) stays between pH 7.3 and 7.5.

Substances called **acids** give up hydrogen ions when they dissolve in water, so they lower the pH of fluids and make them acidic (below pH 7). **Bases** accept hydrogen ions, so they can raise the pH of fluids and make them basic, or alkaline (above pH 7).

Acids and bases can be weak or strong. Weak acids are stingy H^+ donors. Strong acids give up more H^+ ions. Hydrochloric acid (HCl) is a strong acid that, when added to water, very easily separates into H^+ and Cl^-:

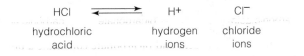

$$HCl \rightleftharpoons H^+ + Cl^-$$
hydrochloric   hydrogen   chloride
   acid   ions   ions

Inside your stomach, the H^+ from HCl makes gastric fluid acidic (pH 1–2). The acidity activates enzymes that digest proteins in your food.

Most biological molecules can function properly only within a narrow range of pH. Even a slight deviation from that range can halt cellular processes. Under normal circumstances, the fluids inside cells and bodies stay within

acid Substance that releases hydrogen ions in water.
base Substance that accepts hydrogen ions in water.
buffer Set of chemicals that can keep the pH of a solution stable by alternately donating and accepting ions that contribute to pH.
concentration The number of molecules or ions per unit volume of a solution.
pH Measure of the number of hydrogen ions in a fluid.

Figure 2.15 Corrosive effect of acid rain. Airborne pollutants dissolve in water vapor and form compounds that change the pH of rain.

a consistent range of pH because they are buffered. A **buffer** is a set of chemicals, often a weak acid or base and a salt, that can keep the pH of a solution stable. The two chemicals can alternately donate and accept ions that contribute to pH changes.

For example, when a base is added to an unbuffered fluid, the number of OH^- ions increases, so the pH rises. However, if the fluid is buffered, the addition of base causes the buffer to release H^+ ions. These combine with OH^- ions to form water, which has no effect on pH. So a buffered fluid's pH stays the same when base is added.

Carbon dioxide gas becomes a weak acid when it dissolves in the fluid portion of human blood:

$$H_2O + CO_2 \longrightarrow H_2CO_3$$

carbon dioxide   carbonic acid

Carbonic acid can separate into hydrogen ions and bicarbonate ions, which can in turn recombine to form carbonic acid:

$$H_2CO_3 \longrightarrow H^+ + HCO_3^- \longrightarrow H_2CO_3$$

carbonic acid   bicarbonate   carbonic acid

Together, carbonic acid and bicarbonate constitute a buffer. Any excess OH^- in blood combines with the H^+ to form water, which does not contribute to pH. Any excess H^+ in blood combines with the bicarbonate. Thus bonded, the hydrogen does not affect pH. The exchange of ions between carbonic acid and bicarbonate keeps the blood pH between 7.3 and 7.5, but only up to a point. A buffer can neutralize only so many ions. Even slightly more than that limit and the pH of the fluid changes dramatically.

Buffer failure can be catastrophic in a biological system. For example, too much carbonic acid forms in blood when breathing is impaired suddenly. The resulting decline in blood pH may cause an individual to enter a coma, which is a dangerous level of unconsciousness. Hyperventilation (sustained rapid breathing) causes the body to lose too much CO_2. The loss can result in a rise in

Mercury Rising (revisited)

Today, all ecosystems on Earth have detectable effects of air pollution. However, such effects are understudied, so they are probably underestimated. We do know that the amount of mercury in Earth's waters is rising (*inset*). The concentration of mercury in the Pacific Ocean is predicted to double within forty years. We also know that this rise is occurring as a consequence of human activities that release mercury into the atmosphere. In 2005 alone, our activities released more than 2,000 tons of mercury worldwide.

All human bodies, too, now have detectable amounts of mercury. Some of it comes from dental fillings, particularly when bleached. Imported skin-bleaching cosmetics and broken fluorescent lamps also contribute. However, most comes from eating contaminated fish and shellfish. Tuna harvested from the Pacific Ocean accounts for almost half of the mercury in human bodies.

How Would You Vote? A U.S. Food and Drug Administration advisory warns children, and women who are pregnant, nursing, or planning to conceive, to avoid eating tuna because of high mercury levels. The California attorney general requires products that contain dangerous chemicals from non-natural sources to carry warning labels. A typical 6-ounce can of albacore tuna contains about 60 micrograms of mercury. Canned tuna companies say that most of that mercury comes from natural sources, and consider the federal advisory to be sufficient. Do you think cans of tuna should carry labels warning of mercury content? See CengageNow for details, then vote online (cengagenow.com).

blood pH. If blood pH rises too much, prolonged muscle spasm (tetany) or coma may occur.

Burning fossil fuels such as coal releases sulfur and nitrogen compounds that affect the pH of rain and other forms of precipitation. Water is not buffered, so the addition of acids or bases has a dramatic effect. In places with a lot of fossil fuel emissions, the rain and fog can be more acidic than vinegar (Figure 2.15). This acid rain drastically changes the pH of water in soil, lakes, and streams. Such changes can overwhelm the buffering capacity of fluids inside organisms, with lethal effects. We will return to the topic of acid rain in Section 42.9.

Take-Home Message Why are hydrogen ions important in biological systems?

> pH reflects the number of hydrogen ions in a fluid. Most biological systems function properly only within a narrow range of pH.

> Acids release hydrogen ions in water; bases accept them. Salts release ions other than H^+ and OH^-.

> Buffers help keep pH stable. Inside organisms, they are part of homeostasis.

Summary

Section 2.1 At life's first level of organization, atoms interact with other atoms to form molecules. The properties of molecules depend on, but differ from, those of their atomic components.

Section 2.2 Atoms consist of **electrons**, which carry a negative **charge**, moving about a **nucleus** of positively charged **protons** and (except for hydrogen) uncharged **neutrons** (Table 2.2). The **periodic table** lists elements in order of their **atomic number. Isotopes** are atoms of an element that differ in the number of neutrons, so they also differ in **mass number.** Researchers can make **tracers** with **radioisotopes,** which spontaneously emit particles and energy by the process of **radioactive decay.**

Table 2.2 Players in the Chemistry of Life

Atoms	Particles that are basic building blocks of all matter; the smallest unit that retains an element's properties.
Proton (p$^+$)	Positively charged particle of an atom's nucleus.
Electron (e$^-$)	Negatively charged particle that can occupy a volume of space (orbital) around an atom's nucleus.
Neutron	Uncharged particle of an atom's nucleus.
Element	Pure substance that consists entirely of atoms with the same, characteristic number of protons.
Isotopes	Atoms of an element that differ in the number of neutrons.
Radioisotope	Unstable isotope that emits particles and energy when its nucleus disintegrates.
Tracer	Molecule that has a detectable substance (such as a radioisotope) attached. Used to track the movement or destination of the molecule in a biological system.
Ion	Atom that carries a charge after it has gained or lost one or more electrons. A single proton without an electron is a hydrogen ion (H$^+$).
Molecule	Two or more atoms joined in a chemical bond.
Compound	Molecule of two or more different elements in unvarying proportions (for example, water).
Mixture	Intermingling of two or more elements or compounds in proportions that can vary.
Solute	Molecule or ion dissolved in some solvent.
Hydrophilic	Refers to a substance that dissolves easily in water. Such substances consist of polar molecules.
Hydrophobic	Refers to a substance that resists dissolving in water. Such substances consist of nonpolar molecules.
Acid	Compound that releases H$^+$ when dissolved in water.
Base	Compound that accepts H$^+$ when dissolved in water.
Salt	Compound that releases ions other than H$^+$ or OH$^-$ when dissolved in water.
Solvent	Substance that can dissolve other substances.

Section 2.3 Up to two electrons occupy each orbital (volume of space) around a nucleus. Which orbital an electron occupies depends on its energy. A **shell model** represents successive energy levels as concentric circles. Atoms fill vacancies by gaining or losing electrons, or by sharing electrons with other atoms. **Electronegativity** is a measure of how strongly an atom attracts electrons from other atoms. **Ions** are charged atoms. A **chemical bond** is an attractive force that unites two atoms as a **molecule.** A **compound** is a molecule that consists of two or more elements. A **mixture** is an intermingling of substances.

Section 2.4 Atoms form different types of bonds depending on their electronegativity. An **ionic bond** is a strong association between oppositely charged ions; it arises from the mutual attraction of opposite charges. Atoms share a pair of electrons in a **covalent bond,** which is nonpolar if the sharing is equal, and polar if it is not. A molecule that has a separation of charge is said to show **polarity. Hydrogen bonds** collectively stabilize the structures of large molecules.

Section 2.5 Polar covalent bonds join two hydrogen atoms to one oxygen atom in each water molecule. The polarity invites extensive hydrogen bonding between water molecules, and this bonding is the basis of unique properties that sustain life: a capacity to act as a **solvent** for **salts** and other polar **solutes;** resistance to **temperature** changes; and **cohesion. Hydrophilic** substances dissolve easily in water; **hydrophobic** substances do not. **Evaporation** is the transition of a liquid to a gas.

Section 2.6 A solute's **concentration** refers to the amount of solute in a given volume of fluid. **pH** reflects the number of hydrogen ions (H$^+$) in a fluid. At neutral pH (7), the amounts of H$^+$ and OH$^-$ ions are the same. **Acids** release hydrogen ions in water; **bases** accept them. A **buffer** keeps a solution within a consistent range of pH. Most cell and body fluids are buffered because most molecules of life work only within a narrow range of pH.

Self-Quiz   Answers in Appendix III

Hydrogen is not in neutrons

1. Is this statement true or ~~false?~~ All atoms consist of electrons, protons, and neutrons.

2. In the periodic table, symbols for the elements are arranged according to _____ .
   - a. size
   - b. charge
   - c. mass number
   - d. atomic number

3. A(n) _____ is a molecule into which a radioisotope has been incorporated.
   - a. compound
   - b. tracer
   - c. salt
   - d. acid

Data Analysis Activities

Mercury Emissions By weight, coal does not contain much mercury, but we burn a lot of it. In addition to coal-fired power plants, several other industries contribute substantially to atmospheric mercury pollution. Figure 2.16 shows mercury emissions from different regions of the world in the year 2006.

1. About how many tons of mercury were released worldwide in 2006?
2. Which industry tops the list of mercury-pollution offenders? Which industry is next on the list?
3. Which region emits the most mercury from producing cement?
4. About how many tons of mercury were released from gold production in South America?

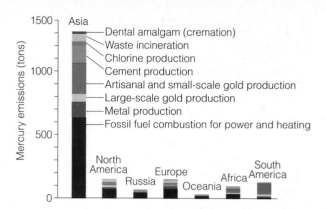

Figure 2.16 Global mercury emissions, 2006. *Source: Global Atmospheric Mercury Assessment: Sources, Emissions and Transport. United Nations Environmental Programme, Chemicals Branch. 2008*

4. An ion is an atom that has _____ .
   a. the same number of electrons and protons
   b. different numbers of electrons and protons

5. The measure of an atom's ability to pull electrons away from another atom is called _____ .
   a. electronegativity   c. charge
   b. polarity   d. concentration

6. The mutual attraction of opposite charges holds atoms together as molecules in a(n) _____ bond.
   a. ionic   c. polar covalent
   b. hydrogen   d. nonpolar covalent

7. Atoms share electrons unequally in a(n) _____ bond. *polar covalent*

8. A(n) _____ substance repels water.
   a. acidic   c. hydrophobic
   b. basic   d. polar

9. A salt releases ions other than _____ in water.

10. Hydrogen ions (H⁺) are _____ .
    a. indicated by a pH scale   c. in blood
    b. unbound protons   d. all of the above

11. When dissolved in water, a(n) _____ donates H⁺; a(n) _____ accepts H⁺.
    a. acid; base   c. buffer; solute
    b. base; acid   d. base; buffer

12. A(n) _____ is a chemical partnership between a weak acid or base and its salt.
    a. covalent bond   c. buffer
    b. hydrogen bond   d. pH

13. A(n) _____ is dissolved in a solvent.
    a. molecule   c. salt
    b. solute   d. acid

Animations and Interactions on **CENGAGENOW**:
❯ Isotopes; PET scans; Shell models; Types of bonds; Ions and ionic bonds; Hydrogen bonds; Water; Dissolution; pH.

14. Match the terms with their most suitable description.
    ___ hydrophilic   a. protons > electrons
    ___ atomic number   b. number of protons in nucleus
    ___ charged atom   c. polar; easily dissolves in water
    ___ mass number   d. ion
    ___ temperature   e. protons < electrons
    ___ uncharged   f. protons = electrons
    ___ negative charge   g. measure of molecular motion
    ___ positive charge   h. number of protons and neutrons in atomic nucleus

Additional questions are available on **CENGAGENOW**.

Critical Thinking

1. Alchemists were medieval scholars and philosophers who were the forerunners of modern-day chemists. Many spent their lives trying to transform lead (atomic number 82) into gold (atomic number 79). Explain why they never did succeed in that endeavor.

2. Draw a shell model of an uncharged nitrogen atom (nitrogen has 7 protons).

3. Polonium is a rare element with 33 radioisotopes. The most common one, ^{210}Po, has 82 protons and 128 neutrons. When ^{210}Po decays, it emits an alpha particle, which is a helium nucleus (2 protons and 2 neutrons). ^{210}Po decay is tricky to detect because alpha particles do not carry very much energy compared to other forms of radiation. They can be stopped by, for example, a sheet of paper or a few inches of air. That is one reason that authorities failed to discover toxic amounts of ^{210}Po in the body of former KGB agent Alexander Litvinenko until after he died suddenly and mysteriously in 2006. What element does an atom of ^{210}Po change into after it emits an alpha particle?

4. Some undiluted acids are not as corrosive as when they are diluted with water. That is why lab workers are told to wipe off splashes with a towel before washing. Explain.

‹ Links to Earlier Concepts

Having learned about atomic interactions (Section 2.3), you are now in a position to understand the structure of the molecules of life. Keep the big picture in mind by reviewing Section 1.2. You will be building on your knowledge of the nature of covalent bonding (2.4), acids and bases (2.6), and the effects of hydrogen bonds (2.4) as you learn about how the molecules of life are put together.

Key Concepts

Structure Dictates Function
We define cells partly by their capacity to build complex carbohydrates and lipids, proteins, and nucleic acids. All of these organic compounds have functional groups attached to a backbone of carbon atoms.

Carbohydrates
Carbohydrates are the most abundant biological molecules. They function as energy reservoirs and structural materials. Different types of carbohydrates are built from the same sugars, bonded in different patterns.

3 Molecules of Life

3.1 Fear of Frying

The human body requires only about a tablespoon of fat each day to stay healthy, but most people in developed countries eat far more than that. The average American eats about 70 pounds of fat per year, which may be part of the reason why the average American is overweight. Being overweight increases one's risk for many health conditions. However, the total quantity of fat may be less important than the kinds of fats we eat. Fats are more than just inert molecules that accumulate in strategic areas of our bodies. They are major constituents of cell membranes, and as such they have powerful effects on cell function.

The typical fat molecule has three fatty acids, each with a long chain of carbon atoms. Different fats consist of different fatty acids. Fats with a certain arrangement of hydrogen atoms around that carbon chain are called *trans* fats (Figure 3.1). Small amounts of *trans* fats occur naturally in red meat and dairy products, but most of the *trans* fats that humans eat come from partially hydrogenated vegetable oil, an artificial food product.

Hydrogenation, a manufacturing process that adds hydrogen atoms to a substance, changes liquid vegetable oils into solid fats. Procter & Gamble Co. developed partially hydrogenated soybean oil in 1908 as a substitute for the more expensive solid animal fats they had been using to make candles. However, the demand for candles began to wane as more households in the United States became wired for electricity, and P & G began to look for another way to sell its proprietary fat. Partially hydrogenated vegetable oil looks a lot like lard, so in 1911 the company began aggressively marketing it as a revolutionary new food: a solid cooking fat with a long shelf life, mild flavor, and lower cost than lard or butter.

By the mid-1950s, hydrogenated vegetable oil had become a major part of the American diet. It was (and still is) found in many manufactured and fast foods: french fries, butter substitutes, cookies, crackers, cakes and pancakes, peanut butter, pies, doughnuts, muffins, chips, granola bars, breakfast bars, chocolate, microwave popcorn, pizzas, burritos, chicken nuggets, fish sticks, and so on.

For decades, hydrogenated vegetable oil was considered more healthy than animal fats because it was made from plants, but we now know otherwise. The *trans* fats in hydrogenated vegetable oils raise the level of cholesterol in our blood more than any other fat, and they directly alter the function of our arteries and veins. The effects of such changes are quite serious. Eating as little as 2 grams a day of hydrogenated vegetable oils increases a person's risk of atherosclerosis (hardening of the arteries), heart attack, and diabetes. A small serving of french fries made with hydrogenated vegetable oil contains about 5 grams of *trans* fat.

All organisms consist of the same kinds of molecules, but small differences in the way those molecules are put together can have big effects in a living organism. With this concept, we introduce you to the chemistry of life. This is your chemistry. It makes you far more than the sum of your body's molecules.

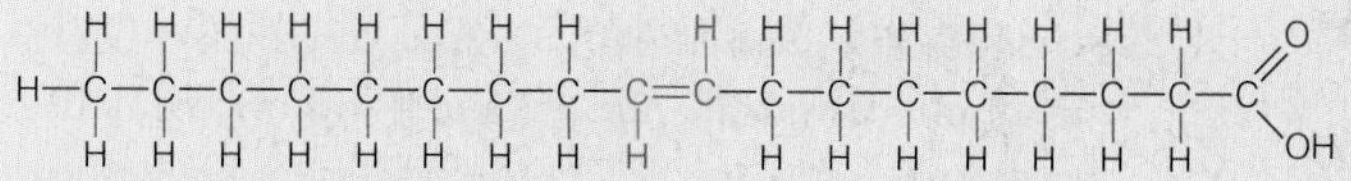

oleic acid (a *trans* fatty acid)

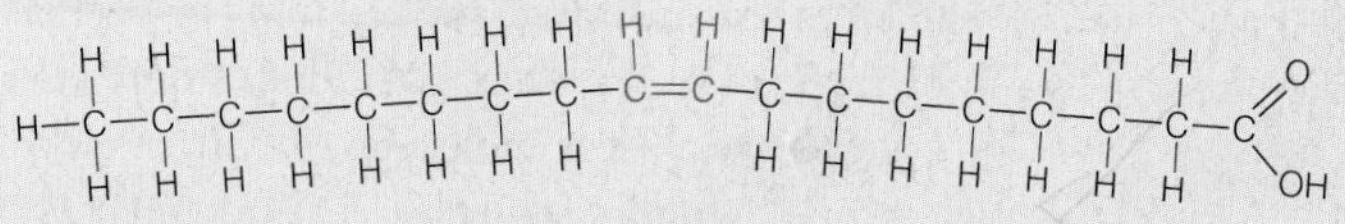

elaidic acid (a *cis* fatty acid)

Figure 3.1 *Trans* fats, an unhealthy food. The arrangement of hydrogen atoms around a double bond (in *red*) is what makes a fat *trans* (*top*) or *cis* (*bottom*). This seemingly small difference in structure makes a big difference in our bodies.

Lipids
Lipids function as energy reservoirs and as waterproofing or lubricating substances. Some are remodeled into other compounds such as vitamins. Lipids are the main structural component of all cell membranes.

Proteins
Structurally and functionally, proteins are the most diverse molecules of life. They include enzymes and structural materials. A protein's function arises from and depends on its structure.

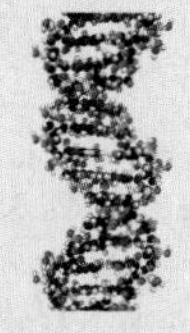

Nucleic Acids
Nucleotides are the building blocks of nucleic acids; some have additional roles in metabolism. DNA and RNA are part of a cell's system of storing and retrieving heritable information.

3.2 The Molecules of Life—From Structure to Function

> All of the molecules of life are built with carbon atoms.
> The function of organic molecules in biological systems begins with their structure.
< Links to Elements 2.2, Ions 2.3, Covalent bonds 2.4, Polarity 2.4, Acids and bases 2.6

Carbon accounts for a high proportion of the elements in living things, mainly because the molecules of life—complex carbohydrates and lipids, proteins, and nucleic acids—are organic. **Organic** compounds consist primarily of carbon and hydrogen atoms. The term is a holdover from a time when such molecules were thought to be made only by living things, as opposed to "inorganic" molecules that formed by nonliving processes. The term persists, even though we now know that organic compounds were present on Earth long before organisms were.

Carbon's importance to life starts with its versatile bonding behavior: Each carbon atom can form covalent bonds with up to four other atoms. Most organic compounds have a backbone, or chain, of carbon atoms, that may include rings. Using different models to represent organic compounds allows us to visualize different aspects of their structure.

Structural formulas show how all of the atoms connect (Figure 3.2A). Some atoms or bonds may be implied but not shown (Figure 3.2B). For further simplification, carbon ring structures are often represented as polygons that imply atoms at their corners (Figure 3.2C).

Molecular models show the positions of atoms in three dimensions. Atoms in such models are typically represented by colored balls, the size of which reflects relative atomic size. Ball-and-stick models show covalent bonds as a stick (Figure 3.2D). Space-filling models show overall shape (Figure 3.2E). To reduce visual complexity, other types of models omit individual atoms. Such models can reveal large-scale features, such as folds or pockets, that can be difficult to see when individual atoms are shown. For example, very large molecules are often shown as ribbons, which highlight structural features such as coils (see Figure 3.14 for an example).

Group	Character	Location	Structure	
hydroxyl	polar	amino acids; sugars, alcohols	—OH	
methyl	nonpolar	fatty acids, some amino acids	—C—H (with H above, below)	
carbonyl	polar, reactive	sugars, amino acids, nucleotides	—C—H with =O (aldehyde)	—C— with =O (ketone)
carboxyl	acidic	fatty acids, amino acids, carbohydrates	—C—OH with =O	—C—O⁻ with =O (ionized)
amine	basic	amino acids, some nucleotide bases	—N—H with H below	—NH⁺ with H above, H below (ionized)
phosphate	high energy, polar	nucleotides (e.g., ATP); DNA, RNA; phospholipids, many proteins	O⁻—P—O⁻ with =O below, O⁻ above	
sulfhydryl	forms disulfide bridges	cysteine (an amino acid)	—SH	—S—S— (disulfide bridge)

Figure 3.3 Animated Common functional groups. Such groups impart specific chemical characteristics to organic compounds.

Functional Groups

An organic molecule that consists only of hydrogen and carbon atoms is called a **hydrocarbon**. Methane, the simplest hydrocarbon, is one carbon atom bonded to four hydrogen atoms. Most of the molecules of life have at least one **functional group**, which is a cluster of atoms covalently bonded to a carbon atom of an organic molecule. Functional groups impart specific chemical proper-

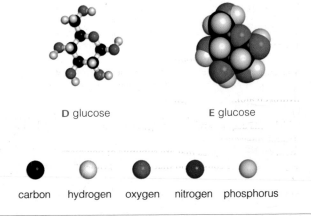

A glucose   B glucose   C glucose   D glucose   E glucose

carbon   hydrogen   oxygen   nitrogen   phosphorus

Figure 3.2 Modeling an organic molecule. **A** A structural formula shows atoms and bonds. **B,C** Structural formulas are often abbreviated to omit labels for some atoms such as the carbons at the corners of ring structures. **D** A ball-and-stick model shows the arrangement of atoms in three dimensions. **E** A space-filling model shows a molecule's overall shape. *Right*, typical color code for elements in molecular models.

ties such as polarity or acidity (Figure 3.3). The chemical behavior of the molecules of life arises largely from the number, kind, and arrangement of their functional groups. For example, hydroxyl groups (—OH) impart polarity to alcohols. Thus, alcohols (at least the small ones) dissolve quickly in water. Larger alcohols do not dissolve as easily because their long, hydrophobic hydrocarbon chains repel water.

What Cells Do to Organic Compounds

All biological systems are based on the same organic molecules, a similarity that is one of many legacies of life's common origin. However, the details of those molecules differ among organisms. Simple organic building blocks bonded in different numbers and arrangements form different versions of the molecules of life, just as atoms bonded in different numbers and arrangements form different molecules. Cells maintain reserves of small organic molecules that they can assemble into complex carbohydrates, lipids, proteins, and nucleic acids. When used as subunits of larger molecules, the small organic molecules (simple sugars, fatty acids, amino acids, and nucleotides) are called **monomers**. Molecules that consist of multiple monomers are called **polymers**.

Cells build polymers from monomers, and break down polymers to release monomers. **Metabolism** refers to activities by which cells acquire and use energy as they make and break apart organic compounds. These activities help cells stay alive, grow, and reproduce. Metabolism also requires **enzymes**, which are organic molecules that speed up reactions without being changed by them.

Table 3.1 lists some common metabolic reactions. For now, start thinking about two types of reactions. Large organic molecules are often built from smaller ones by **condensation**, a process in which an enzyme covalently bonds two molecules together. Water usually forms as a product of condensation when a hydroxyl group (—OH)

---

condensation Process by which enzymes build large molecules from smaller subunits; water also forms.
enzyme Compound (usually a protein) that speeds a reaction without being changed by it.
functional group A group of atoms bonded to a carbon of an organic compound; imparts a specific chemical property.
hydrocarbon Compound or region of one that consists only of carbon and hydrogen atoms.
hydrolysis Process by which an enzyme breaks a molecule into smaller subunits by attaching a hydroxyl group to one part and a hydrogen atom to the other.
metabolism All the enzyme-mediated chemical reactions by which cells acquire and use energy as they build and break down organic molecules.
monomers Molecules that are subunits of polymers.
organic Type of compound that consists primarily of carbon and hydrogen atoms.
polymer Molecule that consists of multiple monomers.

Table 3.1 What Cells Do to Organic Compounds

Type of Reaction	What Happens
Condensation	Two molecules covalently bond and become a larger molecule.
Hydrolysis	A molecule splits into two smaller molecules.
Functional group transfer	A functional group is transferred from one molecule to another.
Electron transfer	One molecule accepts electrons from another.
Rearrangement	Juggling of covalent bonds converts one organic compound into another.

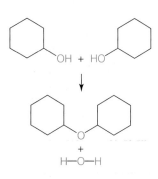

A Condensation. Cells build a large molecule from smaller ones by this reaction. An enzyme removes a hydroxyl group from one molecule and a hydrogen atom from another. A covalent bond forms between the two molecules, and water also forms.

B Hydrolysis. Cells split a large molecule into smaller ones by this water-requiring reaction. An enzyme attaches a hydroxyl group and a hydrogen atom (both from water) at the cleavage site.

Figure 3.4 Animated Two common metabolic processes by which cells build and break down organic molecules.

from one of the molecules combines with a hydrogen atom from the other molecule (Figure 3.4A). **Hydrolysis**, which is the reverse of condensation, breaks apart large organic molecules into smaller ones (Figure 3.4B). Hydrolysis enzymes break a bond by attaching a hydroxyl group to one atom and a hydrogen atom to the other. The —OH and —H come from a water molecule, so this reaction requires water.

Take-Home Message How do organic molecules function in living systems?

> The molecules of life are organic, which means they consist mainly of carbon and hydrogen atoms.

> Functional groups impart certain chemical characteristics to organic molecules. Such groups contribute to the particular function of a biological molecule.

> Cells assemble large polymers from smaller monomers. They also break apart polymers into component monomers.

3.3 Carbohydrates

> Carbohydrates are the most plentiful biological molecules.
> Cells use some carbohydrates as structural materials; they use others for fuel, or to store or transport energy.
< Link to Hydrogen bonds 2.4

Carbohydrates are organic compounds that consist of carbon, hydrogen, and oxygen in a 1:2:1 ratio. Cells use different kinds as structural materials, for fuel, and for storing and transporting energy. The three main types of carbohydrates in living systems are monosaccharides, oligosaccharides, and polysaccharides.

Simple Sugars

"Saccharide" is from *sacchar*, a Greek word that means sugar. Monosaccharides (one sugar unit) are the simplest type of carbohydrate, but they have extremely important roles as components of larger molecules. Common monosaccharides have a backbone of five or six carbon atoms, one carbonyl group, and two or more hydroxyl groups. Enzymes can easily break the bonds of monosaccharides to release energy (we will return to carbohydrate metabolism in Chapter 7). The solubility of these molecules also means that they move easily throughout the water-based internal environments of all organisms.

Monosaccharides that are components of the nucleic acids DNA and RNA have five carbon atoms. Glucose (at *left*) has six. Glucose can be used as a fuel to drive cellular processes, or as a structural material to build larger molecules. It can also be used as a precursor, or parent molecule, that is remodeled into other molecules. For example, cells of plants and many animals make vitamin C from glucose (human cells cannot, so we need to get our vitamin C from food).

glucose

Short-Chain Carbohydrates

An oligosaccharide is a short chain of covalently bonded monosaccharides (*oligo–* means a few). Disaccharides consist of two sugar monomers. The lactose in milk is a disaccharide, with one glucose and one galactose unit. Sucrose, the most plentiful sugar in nature, has a glucose and a fructose unit (Figure 3.5). Sucrose extracted from sugarcane or sugar beets is our table sugar. Oligosaccharides with three or more sugar units are often attached to lipids or proteins that have important functions in immunity.

Complex Carbohydrates

The "complex" carbohydrates, or polysaccharides, are straight or branched chains of many sugar monomers—often hundreds or thousands of them. There may be one type or many types of monomers in a polysaccharide. The most common polysaccharides are cellulose, glycogen, and starch. All consist of glucose monomers, but they differ dramatically in their chemical properties. Why? The answer begins with differences in patterns of covalent bonding that link their glucose monomers.

Cellulose, the major structural material of plants, is the most abundant biological molecule in the biosphere. It consists of long, straight chains of glucose monomers. Hydrogen bonds lock the chains into tight, sturdy bundles (Figure 3.6A). In plants, these tough cellulose fibers act like reinforcing rods that help stems resist wind and other forms of mechanical stress.

Cellulose does not dissolve in water, and it is not easily broken down. Some bacteria and fungi make enzymes that break it apart into its component sugars, but humans and other mammals do not. When we talk about dietary fiber, or "roughage," we are usually referring to the cellulose and other indigestible polysaccharides in our vegetable foods. Bacteria that live in the gut of termites and grazers such as cattle and sheep help these animals digest the cellulose in plants.

In starch, a different covalent bonding pattern between glucose monomers makes a chain that coils up into a spiral (Figure 3.6B). Starch is not as stable as cellulose, and it does not dissolve easily in water. Both properties make the molecule ideal for storing chemical energy in the

CH_2OH OH HO HO CH_2OH → CH_2OH OH HO HO $CH_2OH + H_2O$

HO OH HO O HO O

OH CH_2OH OH CH_2OH

glucose + fructose → sucrose + water

Figure 3.5 Animated The synthesis of a sucrose molecule is an example of a condensation reaction. You are already familiar with sucrose—it is common table sugar.

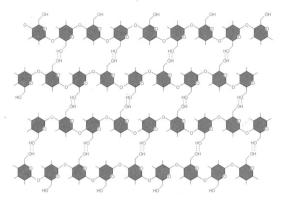

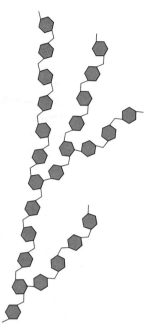

A Cellulose, a structural component of plants. Chains of glucose units stretch side by side and hydrogen bond at many —OH groups. The hydrogen bonds stabilize the chains in tight bundles that form long fibers. Very few types of organisms can digest this tough, insoluble material.

B In amylose, one type of starch, a series of glucose units form a chain that coils. Starch is the main energy reserve in plants, which store it in their roots, stems, leaves, fruits, and seeds (such as coconuts).

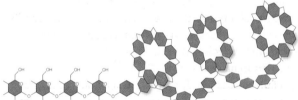

C Glycogen. In humans and other animals, this poly-saccharide functions as an energy reservoir. It is stored in muscles and in the liver.

Figure 3.6 Structure of A cellulose, B starch, and C glycogen, and their typical locations in a few organisms. All three carbohydrates consist only of glucose units, but the different bonding patterns that link the subunits result in substances with very different properties.

watery, enzyme-filled interior of plant cells. Most plants make much more glucose than they can use. The excess is stored as starch inside cells that make up roots, stems, and leaves. However, because it is insoluble, starch cannot be transported out of the cells and distributed to other parts of the plant. When sugars are in short supply, hydrolysis enzymes break the bonds between starch's monomers to release glucose subunits. Humans also have enzymes that hydrolyze starch, so this carbohydrate is an important component of our food.

The covalent bonding pattern in glycogen forms highly branched chains of glucose monomers (Figure 3.6C). In animals, glycogen is the sugar-storage equivalent of starch in plants. Muscle and liver cells store it in reserve to meet a sudden need for glucose. When the sugar level in blood falls, liver cells break down stored glycogen, and the released glucose subunits enter the blood.

---

carbohydrate Molecule that consists primarily of carbon, hydrogen, and oxygen atoms in a 1:2:1 ratio.

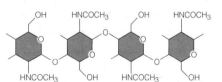

Figure 3.7 Chitin. This poly-saccharide strengthens the hard parts of many small animals, such as crabs.

Chitin is a polysaccharide similar to cellulose. Its monomers are glucose with a nitrogen-containing car-bonyl group (Figure 3.7). Long, unbranching chains of these monomers are linked by hydrogen bonds. As a structural material, chitin is durable, translucent, and flex-ible. It strengthens hard parts of many animals, including the outer cuticle of crabs, beetles, and ticks, and it rein-forces the cell wall of many fungi.

Take-Home Message What are carbohydrates?

❯ Subunits of simple carbohydrates (sugars), arranged in different ways, form various types of complex carbohydrates.

❯ Cells use carbohydrates for energy or as structural materials.

> Lipids are hydrophobic organic compounds.
> Common lipids include triglycerides, phospholipids, waxes, and steroids.

Lipids are fatty, oily, or waxy organic compounds. They vary in structure, but all are hydrophobic. Many lipids incorporate **fatty acids**, which are small organic molecules that consist of a hydrocarbon "tail" topped with a carboxyl group "head" (Figure 3.8). The tail of a fatty acid is hydrophobic (or fatty, hence the name), but the carboxyl group (the "acid" part of the name) makes the head hydrophilic. You are already familiar with the properties of fatty acids because these molecules are the main component of soap. The hydrophobic tails attract oily dirt, and the hydrophilic heads dissolve the dirt in water.

Fatty acids can be saturated or unsaturated. Saturated types have only single bonds in their tails. In other words, their carbon chains are fully saturated with hydrogen atoms (Figure 3.8A). Saturated fatty acid tails are flexible and they wiggle freely. The tails of unsaturated fatty acids have one or more double bonds that limit their flexibility (Figure 3.8B,C). These bonds are termed *cis* or *trans*, depending on the way the hydrogens are arranged around them. You can see in Figure 3.1 how a *cis* bond kinks the tail, and a *trans* bond keeps it straight.

Fats

The carboxyl group of a fatty acid easily forms bonds with other molecules. **Fats** are lipids with one, two, or three fatty acids bonded to a small alcohol called glycerol. A fatty acid attaches to a glycerol via its carboxyl group

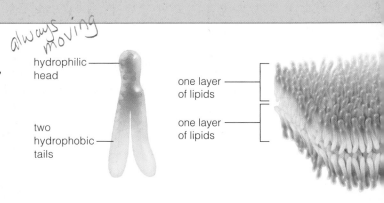

always moving

Figure 3.10 Phospholipids as components of cell membranes. *Left*, the head of a phospholipid is hydrophilic, and the tails are hydrophobic. *Right*, a double layer of phospholipids—the lipid bilayer—is the structural foundation of all cell membranes.

head. When it does, the fatty acid loses its hydrophilic character. When three fatty acids attach to a glycerol, the resulting molecule, which is called a **triglyceride**, is entirely hydrophobic (Figure 3.9A).

Because they are hydrophobic, triglycerides do not dissolve easily in water. Most "neutral" fats, such as butter and vegetable oils, are like this. Triglycerides are the most abundant and richest energy source in vertebrate bodies. They are concentrated in adipose tissue that insulates and cushions body parts.

Animal fats are saturated, which means they consist mainly of triglycerides with three saturated fatty acid tails. Saturated fats tend to remain solid at room temperature because their floppy saturated tails can pack

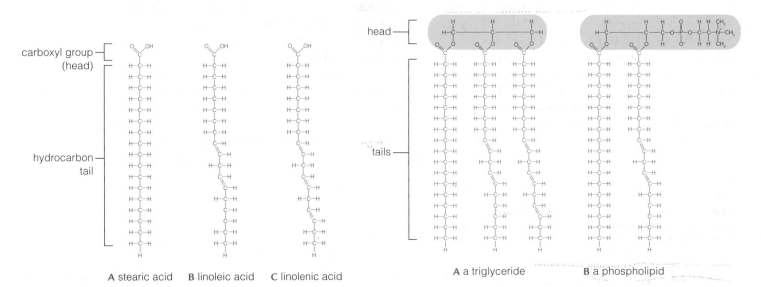

Figure 3.8 Fatty acids. A The tail of stearic acid is fully saturated with hydrogen atoms. **B** Linoleic acid, with two double bonds, is unsaturated. The first double bond occurs at the sixth carbon from the end, so linoleic acid is called an omega-6 fatty acid. Omega-6 and **C** omega-3 fatty acids are "essential fatty acids." Your body does not make them, so they must come from food.

A stearic acid   B linoleic acid   C linolenic acid

Figure 3.9 Animated Lipids with fatty acid tails. **A** Fatty acid tails of a triglyceride are attached to a glycerol head. **B** Fatty acid tails of a phospholipid are attached to a phosphate-containing head.

>> **Figure It Out** Is the triglyceride saturated or unsaturated?

Answer: Unsaturated

A a triglyceride   B a phospholipid

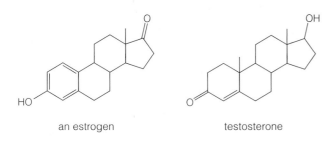

female
wood duck

male
wood duck

Figure 3.11 Estrogen and testosterone, steroid hormones that cause different traits to arise in males and females of many species such as wood ducks (*Aix sponsa*), pictured at *right*.

tightly together. Most vegetable oils are unsaturated, which means these fats consist mainly of triglycerides with one or more unsaturated fatty acid tails. Kinky tails do not pack tightly, so unsaturated fats are typically liquid at room temperature. The partially hydrogenated vegetable oils that you learned about in Section 3.1 are an exception. They are solid at room temperature. The special *trans* double bond that keeps fatty acid tails straight allows them to pack tightly, just like saturated fats do.

Phospholipids

A **phospholipid** has two fatty acid tails and a head that contains a phosphate group (Figure 3.9B). The tails are hydrophobic, but the highly polar phosphate group makes the head very hydrophilic. The opposing properties of a phospholipid molecule give rise to cell membrane structure. Phospholipids are the most abundant lipids in cell membranes, which have two layers of lipids (Figure 3.10). The heads of one layer are dissolved in the cell's watery interior, and the heads of the other layer are dissolved in the cell's fluid surroundings. In such lipid bilayers, all of the hydrophobic tails are sandwiched between the hydrophilic heads. You will read more about the structure of cell membranes in Chapter 4.

---

fat Lipid that consists of a glycerol molecule with one, two, or three fatty acid tails.
fatty acid Organic compound that consists of a chain of carbon atoms with an acidic carboxyl group at one end. Carbon chain of saturated types has single bonds only; that of unsaturated types has one or more double bonds.
lipid Fatty, oily, or waxy organic compound.
phospholipid A lipid with a phosphate group in its hydrophilic head, and two nonpolar fatty acid tails; main constituent of eukaryotic cell membranes.
steroid Type of lipid with four carbon rings and no fatty acid tails.
triglyceride A fat with three fatty acid tails.
wax Water-repellent mixture of lipids with long fatty acid tails bonded to long-chain alcohols or carbon rings.

Waxes

A **wax** is a complex, varying mixture of lipids with long fatty acid tails bonded to long-chain alcohols or carbon rings. The molecules pack tightly, so the resulting substance is firm and water-repellent. A layer of secreted waxes that covers the exposed surfaces of plants helps restrict water loss and keep out parasites and other pests. Other types of waxes protect, lubricate, and soften skin

and hair. Waxes, together with fats and fatty acids, make feathers waterproof. Bees store honey and raise new generations of bees inside honeycomb made from wax that they secrete.

Steroids

Steroids are lipids with a rigid backbone of four carbon rings and no fatty acid tails. All eukaryotic cell membranes contain them. Cholesterol, the most common steroid in animal tissue, is also a starting material that cells remodel into many molecules, such as bile salts (which help digest fats) and vitamin D (required to keep teeth and bones strong). Steroid hormones are also derived from cholesterol. Estrogens and testosterone, hormones that govern reproduction and secondary sexual traits, are steroid hormones (Figure 3.11).

Take-Home Message **What are lipids?**

> Lipids are fatty, waxy, or oily organic compounds. Common types include fats, phospholipids, waxes, and steroids.

> Triglycerides are lipids that serve as energy reservoirs in vertebrate animals.

> Phospholipids are the main lipid component of cell membranes.

> Waxes are lipid components of water-repelling and lubricating secretions.

> Steroids are lipids that occur in cell membranes. Some are remodeled into other molecules.

> Proteins are the most diverse biological molecule. All cellular processes involve them.
> Cells build thousands of different types of proteins by stringing together amino acids in different orders.
‹ Link to Covalent bonding 2.4

Of all biological molecules, proteins are the most diverse in both structure and function. Structural proteins support cell parts and, as part of tissues, multicelled bodies. Spiderwebs, feathers, hooves, and hair, as well as bones and other body parts, consist mainly of structural proteins. A tremendous number of different proteins, including some structural types, actively participate in all processes that sustain life. Most enzymes that drive metabolic reactions are proteins. Proteins move substances, help cells communicate, and defend the body.

Amino Acids

Amazingly, cells can make all of the thousands of different kinds of proteins they need from only twenty kinds of monomers called amino acids. **Proteins** are polymers of amino acids. An **amino acid** is a small organic compound with an amine group, a carboxyl group (the acid), and one or more atoms called an "R group." In most amino acids, all three groups are attached to the same carbon atom (Figure 3.12). The structures of the twenty amino acids used in eukaryotic proteins are shown in Appendix V.

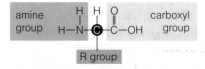

Figure 3.12 Generalized structure of amino acids. Appendix V has models of all twenty of the amino acids used in eukaryotic proteins.

Building Proteins

Protein synthesis involves covalently bonding amino acids into a chain. For each type of protein, instructions coded in DNA specify the order in which any of the twenty kinds of amino acids will occur at every place in the chain.

During protein synthesis, the amine group of one amino acid becomes bonded to the carboxyl group of the next. The bond that forms between the two amino acids is called a **peptide bond** (Figure 3.13). Enzymes repeat this bonding process hundreds or thousands of times, so a long chain of amino acids (a **polypeptide**) forms (Figure 3.14). The linear sequence of amino acids in the polypeptide is called the protein's primary structure ❶. You will learn more about protein synthesis in Chapter 9.

Protein Structure

You and all other organisms depend on working proteins: enzymes that speed metabolic processes, receptors that receive signals, hemoglobin that carries oxygen in your red blood cells, and so on. One of the fundamental ideas in biology is that structure dictates function. This idea is particularly appropriate as applied to proteins, because the shape of a protein defines its biological activity.

There are several levels of protein structure beyond amino acid sequence. Even before a polypeptide is finished being synthesized, it begins to twist and fold as hydrogen bonds form among the amino acids of the chain. This hydrogen bonding may cause parts of the polypeptide to form flat sheets or coils (helices), patterns that constitute a protein's secondary structure ❷. The primary structure of each type of protein is unique, but most proteins have sheets and coils.

Much as an overly twisted rubber band coils back upon itself, hydrogen bonding between different parts of a protein make it fold up even more into compact domains. A domain is a part of a protein that is organized

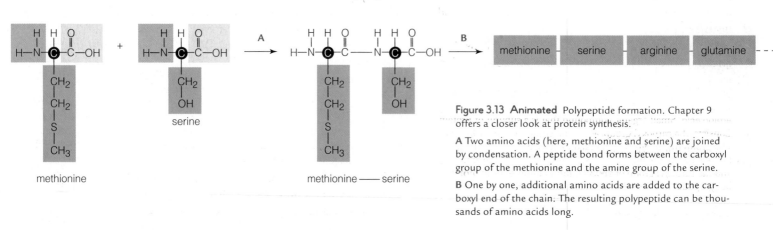

Figure 3.13 Animated Polypeptide formation. Chapter 9 offers a closer look at protein synthesis.

A Two amino acids (here, methionine and serine) are joined by condensation. A peptide bond forms between the carboxyl group of the methionine and the amine group of the serine.

B One by one, additional amino acids are added to the carboxyl end of the chain. The resulting polypeptide can be thousands of amino acids long.

| lysine | glycine | glycine | arginine |

1 A protein's primary structure consists of a linear sequence of amino acids (a polypeptide chain). Each type of protein has a unique primary structure.

Figure 3.14 Animated Protein structure.

2 Secondary structure arises as a polypeptide chain twists into a coil (helix) or sheet held in place by hydrogen bonds between different parts of the molecule. The same patterns of secondary structure occur in many different proteins.

3 Tertiary structure occurs when a chain's coils and sheets fold up into a functional domain such as a barrel or pocket. In this example, the coils of a globin chain form a pocket.

4 Some proteins have quaternary structure, in which two or more polypeptide chains associate as one molecule. Hemoglobin, shown here, consists of four globin chains (*green* and *blue*). Each globin pocket now holds a heme group (*red*).

as a structurally stable unit. Such units are part of a protein's overall three-dimensional shape, or tertiary structure **3**.

Tertiary structure is what makes a protein a working molecule. For example, the sheets of some proteins curl up into a barrel shape. A barrel domain often functions as a tunnel for small molecules, allowing them to pass, for example, through a cell membrane. Globular domains of enzymes form chemically active pockets that can make or break bonds of other molecules.

Many proteins also have quaternary structure, which means they consist of two or more polypeptide chains that are in close association or covalently bonded together. Most enzymes and many other proteins consist of two or more polypeptide chains that collectively form a roughly spherical shape **4**.

Some proteins aggregate by many thousands into much larger structures, with their polypeptide chains organized into strands or sheets. The keratin in your hair is an example **5**. Some fibrous proteins contribute to the structure and organization of cells and tissues. Others, such as the actin and myosin filaments in muscle cells, are part of the mechanisms that help cells, cell parts, and bodies move.

Enzymes often attach sugars or lipids to proteins. A glycoprotein forms when oligosaccharides are attached to

5 Many proteins aggregate by the thousands into much larger structures, such as the keratin filaments that make up hair.

a polypeptide. The molecules that allow a tissue or a body to recognize its own cells are glycoproteins, as are other molecules that help cells interact in immunity.

Some lipoproteins form when enzymes covalently bond lipids to a protein. Other lipoproteins are aggregate structures that consist of variable amounts and types of proteins and lipids. These aggregate molecules carry fats and cholesterol through the bloodstream. Low-density lipoprotein, or LDL, transports cholesterol out of the liver and into cells. High-density lipoprotein, or HDL, ferries the cholesterol that is released from dead cells back to the liver.

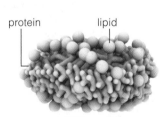

protein   lipid

an HDL particle

amino acid Small organic compound that is a subunit of proteins. Consists of a carboxyl group, an amine group, and a characteristic side group (R), all typically bonded to the same carbon atom.
peptide bond A bond between the amine group of one amino acid and the carboxyl group of another. Joins amino acids in proteins.
polypeptide Chain of amino acids linked by peptide bonds.
protein Organic compound that consists of one or more chains of amino acids (polypeptides).

Take-Home Message **What are proteins?**

❯ Proteins are chains of amino acids. The order of amino acids in a polypeptide chain dictates the type of protein.

❯ Polypeptide chains twist and fold into coils, sheets, and loops, which fold and pack further into functional domains.

❯ A protein's shape is the source of its function.

❭ Changes in a protein's shape may have drastic consequences to health.

Protein shape depends on many hydrogen bonds and other interactions that heat, some salts, shifts in pH, or detergents can disrupt. At such times, proteins **denature**, which means they unwind and otherwise lose their shape. Once a protein's shape unravels, so does its function.

You can see denaturation in action when you cook an egg. A protein called albumin is a major component of egg white. Cooking does not disrupt the covalent bonds of albumin's primary structure, but it does destroy the weaker hydrogen bonds that maintain the protein's shape. When a translucent egg white turns opaque, the albumin has been denatured. For very few proteins, denaturation is reversible if normal conditions return, but albumin is not one of them. There is no way to uncook an egg.

Prion diseases, including mad cow disease (bovine spongiform encephalitis, or BSE) in cattle, Creutzfeldt–Jakob disease in humans, and scrapie in sheep, are the dire aftermath of a protein that changes shape. These

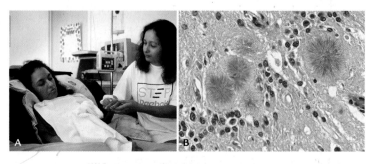

Figure 3.15 Variant Creutzfeldt–Jakob disease (vCJD).

A Charlene Singh, here being cared for by her mother, was one of three people who developed symptoms of vCJD disease while living in the United States. Like the others, Singh most likely contracted the disease elsewhere; she spent her childhood in Britain. She was diagnosed in 2001, and she died in 2004.

B Slice of brain tissue from a person with vCJD. Characteristic holes and prion protein fibers radiating from several deposits are visible.

infectious diseases may be inherited, but more often they arise spontaneously. All are characterized by relentless deterioration of mental and physical abilities that eventually causes the individual to die (Figure 3.15A).

All prion diseases begin with a protein that occurs normally in mammals. One such protein, PrP^C, is found in cell membranes throughout the body. This copper-binding protein is especially abundant in brain cells, but we still know very little about what it does. Very rarely, a PrP^C protein spontaneously misfolds so that it loses some of its

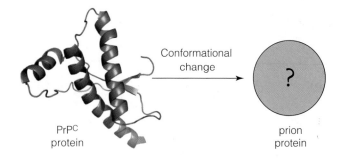

PrP^C protein   Conformational change   ?   prion protein

Figure 3.16 The PrP^C protein becomes a prion when it misfolds into an as yet unknown conformation. Prions cause other PrP^C proteins to misfold, and the misfolded proteins aggregate into long fibers.

coils. In itself, a single misfolded protein molecule would not pose much of a threat. However, when this particular protein misfolds, it becomes a **prion**, or infectious protein (Figure 3.16). The altered shape of a misfolded PrP^C protein somehow causes normally folded PrP^C proteins to misfold too. Because each protein that misfolds becomes infectious, the number of prions increases exponentially.

The shape of misfolded PrP^C proteins allows them to align tightly into long fibers. Fibers composed of aggregated prion proteins begin to accumulate in the brain as large, water-repellent patches (Figure 3.15B). The patches grow as more prions form, and they begin to disrupt brain cell function, causing symptoms such as confusion, memory loss, and lack of coordination. Tiny holes form in the brain as its cells die. Eventually, the brain becomes so riddled with holes that it looks like a sponge.

In the mid-1980s, an epidemic of mad cow disease in Britain was followed by an outbreak of a new variant of Creutzfeldt–Jakob disease (vCJD) in humans. Researchers isolated a prion similar to the one in scrapie-infected sheep from cows with BSE, and also from humans affected by the new type of Creutzfeldt–Jakob disease. How did the prion get from sheep to cattle to people? Prions are not denatured by cooking or typical treatments that inactivate other types of infectious agents. The cattle became infected by the prion after eating feed prepared from the remains of scrapie-infected sheep, and people became infected by eating beef from the infected cattle.

Two hundred people have died from vCJD since 1990. The use of animal parts in livestock feed is now banned, and the number of cases of BSE and vCJD has since declined. Cattle with BSE still turn up, but so rarely that they pose little threat to human populations.

denature To unravel the shape of a protein or other large biological molecule.
prion Infectious protein.

Take-Home Message Why is protein structure important?

❭ A protein's function depends on its structure, so changes in a protein's structure may also alter its function.

3.7 Nucleic Acids

> Nucleotides are subunits of DNA and RNA. Some have roles in metabolism.

< Links to Inheritance 1.3, Diversity 1.4, Hydrogen bonds 2.4

Nucleotides are small organic molecules that function as energy carriers, enzyme helpers, chemical messengers, and subunits of DNA and RNA. Each consists of a sugar with a five-carbon ring, bonded to a nitrogen-containing base and one or more phosphate groups. The nucleotide **ATP** (adenosine triphosphate) has a row of three phosphate groups attached to its ribose sugar (Figure 3.17A). When the outer phosphate group of an ATP is transferred to another molecule, energy is transferred along with it. You will read about such phosphate-group transfers and their important metabolic role in Chapter 5.

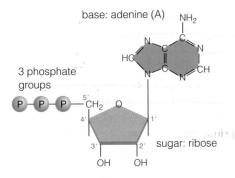

base: adenine (A)

3 phosphate groups

sugar: ribose

A ATP, a nucleotide monomer of RNA, and also an essential participant in many metabolic processes.

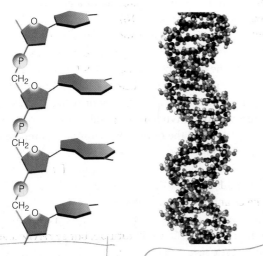

B A chain of nucleotides is a nucleic acid. The sugar of one nucleotide is covalently bonded to the phosphate group of the next, forming a sugar–phosphate backbone.

C DNA consists of two chains of nucleotides, twisted into a double helix. Hydrogen bonding maintains the three-dimensional structure of this nucleic acid.

Figure 3.17 Animated Nucleic acid structure.

Fear of Frying (revisited)

Trans fatty acids are relatively rare in unprocessed foods, so it makes sense from an evolutionary standpoint that our bodies may not have enzymes to deal with them efficiently. The enzymes that hydrolyze *cis* fatty acids have difficulty breaking down *trans* fatty acids, a problem that may be a factor in the ill effects of *trans* fats on our health.

How Would You Vote? All prepackaged foods in the United States are now required to list *trans* fat content, but may be marked "zero grams of *trans* fats" even when a single serving contains up to half a gram. Should hydrogenated oils be banned from all food? See CengageNow for details, then vote online (cengagenow.com).

Nucleic acids are polymers, chains of nucleotides in which the sugar of one nucleotide is joined to the phosphate group of the next (Figure 3.17B). An example is **RNA**, or ribonucleic acid, named after the ribose sugar of its component nucleotides. RNA consists of four kinds of nucleotide monomers, one of which is ATP. RNA molecules carry out protein synthesis, which we discuss in detail in Chapter 9.

DNA, or deoxyribonucleic acid, is a nucleic acid named after the deoxyribose sugar of its component nucleotides. A DNA molecule consists of two chains of nucleotides twisted into a double helix (Figure 3.17C). Hydrogen bonds between the nucleotides hold the two chains together.

Each cell starts out life with DNA inherited from a parent cell. That DNA contains all of the information necessary to build a new cell and, in the case of multicelled organisms, an entire individual. The cell uses the order of nucleotide bases in DNA—the DNA sequence—to guide production of RNA and proteins. Parts of the sequence are identical or nearly so in all organisms, but most is unique to a species or an individual (Chapter 8 returns to the topic of DNA structure and function).

ATP Adenosine triphosphate. Nucleotide that consists of an adenine base, a five-carbon ribose sugar, and three phosphate groups.
DNA Deoxyribonucleic acid. Nucleic acid that carries hereditary information about traits; consists of two nucleotide chains twisted in a double helix.
nucleic acid Single- or double-stranded chain of nucleotides joined by sugar–phosphate bonds; for example, DNA, RNA.
nucleotide Monomer of nucleic acids; has five-carbon sugar, nitrogen-containing base, and phosphate groups.
RNA Ribonucleic acid. Some types have roles in protein synthesis.

Take-Home Message **What are nucleic acids?**

> Nucleotides are monomers of the nucleic acids DNA and RNA. Many kinds, such as ATP, have other functions in metabolism.

> DNA's nucleotide sequence encodes heritable information.

> RNA molecules have roles in the processes by which a cell uses the information in its DNA.

Summary

Section 3.1 All organisms consist of the same kinds of molecules. Seemingly small differences in the way those molecules are put together can have big effects inside a living organism.

Section 3.2 Under present-day conditions in nature, only living things make the molecules of life—complex carbohydrates and lipids, proteins, and nucleic acids. All of these molecules are **organic**, which means they consist primarily of carbon and hydrogen atoms. **Hydrocarbons** have only carbon and hydrogen atoms.

Carbon chains or rings form the backbone of the molecules of life. **Functional groups** attached to the backbone influence the function of these compounds.

Metabolism includes all processes by which cells acquire and use energy as they make and break the bonds of organic compounds. By metabolic reactions such as **condensation**, **enzymes** build **polymers** from **monomers** of simple sugars, fatty acids, amino acids, and nucleotides. Reactions such as **hydrolysis** release the monomers by breaking apart the polymers.

Section 3.3 Enzymes assemble complex **carbohydrates** such as cellulose, glycogen, and starch from simple carbohydrate (sugar) subunits. Cells use carbohydrates for energy, and as structural materials.

Section 3.4 Lipids are fatty, oily, or waxy compounds. All are nonpolar. **Fats** and some other lipids have **fatty acid** tails; **triglycerides** have three.

Cells use lipids as major sources of energy and as structural materials. **Phospholipids** are the main structural component of cell membranes. **Waxes** are lipids that are part of water-repellent and lubricating secretions. **Steroids** occur in cell membranes, and some are remodeled into other molecules.

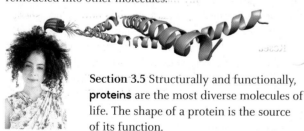

Section 3.5 Structurally and functionally, **proteins** are the most diverse molecules of life. The shape of a protein is the source of its function.

Protein structure begins as a linear sequence of **amino acids** linked by **peptide bonds** into a **polypeptide** (primary structure). Polypeptides twist into loops, sheets, and coils (secondary structure) that can pack further into functional domains (tertiary structure).

Many proteins, including most enzymes, consist of two or more polypeptides (quaternary structure). Fibrous proteins aggregate into much larger structures.

Section 3.6 A protein's structure dictates its function, so changes in a protein's structure may also alter its function. Shifts in pH or temperature, and exposure to detergent or some salts may disrupt hydrogen bonds and other molecular interactions that are responsible for the protein's shape. If that happens, the protein unravels, or **denatures**, and so loses its function. **Prion** diseases are a consequence of misfolded proteins.

Section 3.7 Nucleotides are small organic molecules consisting of a sugar, a phosphate group, and a nitrogen-containing base. Nucleotides are monomers of **DNA** and **RNA**, which are **nucleic acids**. Some nucleotides have additional functions. For example, **ATP** energizes many kinds of molecules by phosphate-group transfers. DNA encodes heritable information that guides the synthesis of RNA and proteins. RNAs interact with DNA and with one another to carry out protein synthesis.

Self-Quiz   Answers in Appendix III

1. Organic molecules consist mainly of ___C___ atoms.
   a. carbon
   b. carbon and oxygen
   c. carbon and hydrogen
   d. carbon and nitrogen

2. Each carbon atom can share pairs of electrons with as many as ___4___ other atom(s).

3. ___A___ groups impart polarity to alcohols.
   a. Hydroxyl ($-OH^-$)
   b. Phosphate ($-PO_4$)
   c. Methyl ($-CH_3$)
   d. Sulfhydryl ($-SH$)

4. ___E___ is a simple sugar (a monosaccharide).
   a. Glucose
   b. Sucrose
   c. Ribose
   d. Starch
   e. both a and c
   f. a, b, and c

5. Unlike saturated fats, the fatty acid tails of unsaturated fats incorporate one or more ___c___ .
   a. phosphate groups
   b. glycerols
   c. double bonds
   d. single bonds

6. Is this statement true or false? Unlike saturated fats, all unsaturated fats are beneficial to health because their fatty acid tails kink and do not pack together.

7. Steroids are among the lipids with no ___b___ .
   a. double bonds
   b. fatty acid tails
   c. hydrogens
   d. carbons

8. Name three kinds of carbohydrates that can be built using only glucose monomers. _Cellulose_ _Starch, glycogen_

9. Which of the following is a class of molecules that encompasses all of the other molecules listed?
   a. triglycerides
   b. fatty acids
   c. waxes
   d. steroids
   e. lipids
   f. phospholipids

10. ___c___ are to proteins as ___d___ are to nucleic acids.
    a. Sugars; lipids
    b. Sugars; proteins
    c. Amino acids; hydrogen bonds
    d. Amino acids; nucleotides

Data Analysis Activities

Effects of Dietary Fats on Lipoprotein Levels Cholesterol that is made by the liver or that enters the body from food does not dissolve in blood, so it is carried through the bloodstream by lipoproteins. Low-density lipoprotein (LDL) carries cholesterol to body tissues such as artery walls, where it can form deposits associated with cardiovascular disease. Thus, LDL is often called "bad" cholesterol. High-density lipoprotein (HDL) carries cholesterol away from tissues to the liver for disposal, so HDL is often called "good" cholesterol.

In 1990, Ronald Mensink and Martijn Katan published a study that tested the effects of different dietary fats on blood lipoprotein levels. Their results are shown in Figure 3.18.

1. In which group was the level of LDL ("bad" cholesterol) highest?
2. In which group was the level of HDL ("good" cholesterol) lowest?
3. An elevated risk of heart disease has been correlated with increasing LDL-to-HDL ratios. Which group had the highest LDL-to-HDL ratio?
4. Rank the three diets from best to worst according to their potential effect on heart disease.

	Main Dietary Fats			
	cis fatty acids	*trans* fatty acids	saturated fats	optimal level
LDL	103	117	121	<100
HDL	55	48	55	>40
ratio	1.87	2.44	2.2	<2

Figure 3.18 Effect of diet on lipoprotein levels. Researchers placed 59 men and women on a diet in which 10 percent of their daily energy intake consisted of *cis* fatty acids, *trans* fatty acids, or saturated fats.

Blood LDL and HDL levels were measured after three weeks on the diet; averaged results are shown in mg/dL (milligrams per deciliter of blood). All subjects were tested on each of the diets. The ratio of LDL to HDL is also shown.

---

11. A denatured protein has lost its _D_ .
    a. hydrogen bonds   c. function
    b. shape   d. all of the above

12. _D_ consists of nucleotides.
    a. Sugars   b. DNA   c. RNA   d. b and c

13. Which of the following is not found in DNA?
    a. amino acids   c. nucleotides
    b. sugars   d. phosphate groups

14. In the following list, identify the carbohydrate, the fatty acid, the amino acid, and the polypeptide:
    a. NH$_2$—CHR—COOH   c. (methionine)$_{20}$
    b. C$_6$H$_{12}$O$_6$ — carbohydrate   d. CH$_3$(CH$_2$)$_{16}$COOH — fatty acid

15. Match the molecules with the best description.
    C wax   a. protein primary structure
    E starch   b. an energy carrier
    F triglyceride   c. water-repellent secretions
    D DNA   d. carries heritable information
    B polypeptide   e. sugar storage in plants
    A ATP   f. richest energy source

16. Match each polymer with the most appropriate set of component monomers.
    G protein   a. glycerol, fatty acids, phosphate
    B phospholipid   b. amino acids, sugars
    D glycoprotein   c. glycerol, fatty acids
    F fat   d. nucleotides
    A nucleic acid   e. sugars
    C carbohydrate   f. sugar, phosphate, base
    E nucleotide   g. amino acids
    I lipoprotein   h. glucose, fructose
    H sucrose   i. lipids, amino acids

Additional questions are available on **CENGAGENOW**.

Critical Thinking

1. In 1976, a team of chemists in the United Kingdom was developing new insecticides by modifying sugars with chlorine (Cl$_2$), phosgene (Cl$_2$CO), and other toxic gases. One young member of the team misunderstood his verbal instructions to "test" a new molecule. He thought he had been told to "taste" it. Luckily, the molecule was not toxic, but it was very sweet. It became the food additive sucralose.

Sucralose has three chlorine atoms substituted for three hydroxyl groups of sucrose. It binds so strongly to the sweet-taste receptors on the tongue that the human brain perceives it as 600 times sweeter than sucrose (table sugar). Sucralose was originally marketed as an artificial sweetener called Splenda®, but it is now available under several other brand names.

Researchers proved that the body does not recognize sucralose as a carbohydrate by feeding sucralose labeled with ^{14}C to volunteers. Analysis of the radioactive molecules in the volunteers' urine and feces showed that 92.8 percent of the sucralose passed through the body without being altered. Many people are worried that the chlorine atoms impart toxicity to sucralose. How would you respond to that concern?

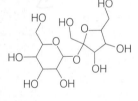

sucrose

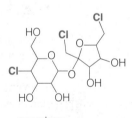

sucralose

Animations and Interactions on **CENGAGENOW**:
› Functional groups; Condensation and hydrolysis; Sucrose synthesis; Fatty acids; Triglyceride formation; Amino acids; Peptide bond formation.

‹ Links to Earlier Concepts

Reflect on the Section 1.2 overview of life's levels of organization. In this chapter, you will see how the properties of lipids (3.4) give rise to cell membranes; consider the location of DNA (3.7) and the sites where carbohydrates are built and broken apart (3.2, 3.3); and expand your understanding of the vital roles of proteins in cell function (3.5, 3.6). You will also see an application of tracers (2.2), and revisit scientific philosophy (1.6, 1.9).

Key Concepts

What Is a Cell?
A cell is the smallest unit of life. Each has a plasma membrane that separates its interior from the exterior environment. A cell's interior contains cytoplasm and DNA.

Microscopes
Most cells are too small to see with the naked eye. We use different types of microscopes to reveal different details of their structure.

4 Cell Structure

Food for Thought

We find bacteria at the bottom of the ocean, high up in the atmosphere, miles underground—essentially anywhere we look. Mammalian intestines typically harbor fantastic numbers of them, but bacteria are not just stowaways there. Intestinal bacteria make vitamins that mammals cannot, and they crowd out more dangerous germs. Cell for cell, bacteria that live in and on a human body outnumber the person's own cells by about ten to one.

Escherichia coli is one of the most common intestinal bacteria of warm-blooded animals. Only a few of the hundreds of types, or strains, of *E. coli*, are harmful. One that is called O157:H7 (Figure 4.1) makes a potent toxin that can severely damage the lining of the human intestine. After ingesting as few as ten O157:H7 cells, a person may become ill with severe cramps and bloody diarrhea that lasts up to ten days. In some people, complications of O157:H7 infection result in kidney failure, blindness, paralysis, and death. About 73,000 people in the United States become infected with *E. coli* O157:H7 each year, and more than 60 of them die.

E. coli O157:H7 can live in the intestines of other animals—mainly cattle, deer, goats, and sheep—apparently without sickening them. Humans are exposed to the bacteria when they come into contact with feces of animals that harbor it, for example, by eating contaminated ground beef. During slaughter, meat can come into contact with feces. Bacteria in the feces stick to the meat, then get thoroughly mixed into it during the grinding process. Unless contaminated meat is cooked to at least 71°C (160°F), live bacteria will enter the digestive tract of whoever eats it.

People also become infected by eating fresh fruits and vegetables that have come into contact with animal feces. Washing contaminated produce with water does not remove *E. coli* O157:H7, because the bacteria are sticky. In 2006, more than 200 people became ill and 3 died after eating fresh spinach grown in a field close to a cattle pasture. Water contaminated with manure may have been used to irrigate the field.

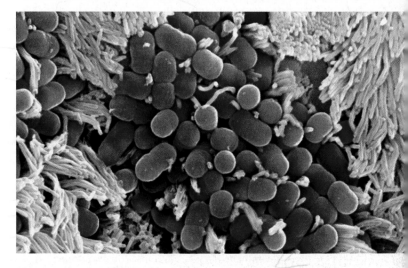

Figure 4.1 *E. coli* O157:H7 bacteria (*red*) clustering on intestinal cells of a small child. This type of bacteria can cause a serious intestinal illness in people who eat foods contaminated with it.

The economic impact of such outbreaks, which occur with some regularity, extends beyond the casualties. Growers lost $50–100 million recalling fresh spinach after the 2006 outbreak. In 2007, about 5.7 million pounds of ground beef were recalled after 14 people were sickened. More than 2.8 million pounds of meat products were recalled between November, 2009 and January, 2010. Food growers and processors are now using procedures that they hope will reduce *E. coli* O157:H7 outbreaks. Some meats and produce are now tested for pathogens before sale, and improved documentation should allow a source of contamination to be pinpointed more quickly.

What makes bacteria sticky? Why do people but not cows get sick with *E. coli* O157:H7? You will begin to find answers to these and many more questions that affect your health in this chapter, as you learn about cells and how they work.

Cell Membranes All cell membranes consist mainly of a lipid bilayer and different types of proteins. The proteins carry out various tasks, including control over which substances cross the membrane.

Bacteria and Archaeans Archaeans and bacteria have few internal membrane-enclosed compartments. In general, they are the smallest and structurally the simplest cells, but they are also the most numerous.

Eukaryotic Cells Cells of protists, plants, fungi, and animals are eukaryotic; they have a nucleus and other membrane-enclosed compartments. Cells differ in internal parts and surface specializations.

> All cells have a plasma membrane and cytoplasm, and all start out life with DNA.
< Links to Lipid structure 3.4, DNA 3.7

Traits Common to All Cells

A **cell** is the smallest unit that shows the properties of life. Cells vary dramatically in shape and in function (Figure 4.2). Despite their differences, however, all cells share certain organizational and functional features. Every cell has a **plasma membrane**, an outer membrane that separates the cell's contents from its environment. A plasma membrane is selectively permeable, which means it allows only certain materials to cross. Thus, it controls exchanges between the cell and its environment. All cell membranes, including the plasma membrane, consist mainly of lipids.

The plasma membrane encloses a fluid or jellylike mixture of water, sugars, ions, and proteins called **cytoplasm**. Some or all of a cell's metabolism occurs in the cytoplasm, and the cell's internal components, including organelles, are suspended in it. **Organelles** are structures that carry out special metabolic functions inside a cell.

All cells start out life with DNA, although a few types of cells lose it as they mature. We categorize cells based on whether their DNA is housed in a nucleus or not (Figure 4.3). Only eukaryotic cells have a **nucleus** (plural, nuclei), an organelle with a double membrane that contains the cell's DNA. In most bacteria and archaeans, the DNA is suspended directly in the cytoplasm.

Constraints on Cell Size

Almost all cells are too small to see with the naked eye. Why? The answer begins with the processes that keep a cell alive. A living cell must exchange substances with

Figure 4.2 Examples of cells. Each one of the cells pictured here is an individual organism; all are protists.

its environment at a rate that keeps pace with its metabolism. These exchanges occur across the plasma membrane, which can handle only so many exchanges at a time. Thus, cell size is limited by a physical relationship called the **surface-to-volume ratio**. By this ratio, an object's volume increases with the cube of its diameter, but its surface area increases only with the square.

Apply the surface-to-volume ratio to a round cell. As Figure 4.4 shows, when a cell expands in diameter, its volume increases faster than its surface area does. Imagine that a round cell expands until it is four times its original diameter. The volume of the cell has increased 64 times (4^3), but its surface area has increased only 16 times (4^2). Each unit of plasma membrane must now handle exchanges with four times as much cytoplasm ($64 = 16 \times 4$). If the cell gets too big, the inward flow of nutrients and the outward flow of wastes across that membrane will not be fast enough to keep the cell alive.

Surface-to-volume limits also affect the body plans of multicelled species. For example, small cells attach end

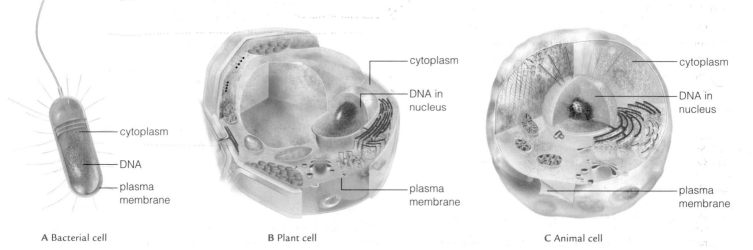

cytoplasm

DNA in nucleus

plasma membrane

cytoplasm

DNA

plasma membrane

cytoplasm

DNA in nucleus

plasma membrane

A Bacterial cell

B Plant cell

C Animal cell

Figure 4.3 Animated Overview of the general organization of bacteria and eukaryotic cells. Archaeans are similar to bacteria in overall structure. The three examples are not drawn to the same scale.

>> **Figure It Out:** Which of these cells is/are eukaryotic?

Answer: The plant and animal cells are eukaryotic.

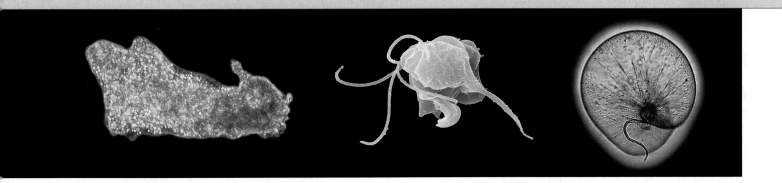

Diameter (cm)	2	3	6
Surface area (cm^2)	12.6	28.2	113
Volume (cm^3)	4.2	14.1	113
Surface-to-volume ratio	3:1	2:1	1:1

Figure 4.4 Animated Three examples of the surface-to-volume ratio. This physical relationship between increases in volume and surface area constrains cell size and shape.

to end to form strandlike algae, so that each can interact directly with its surroundings. Muscle cells in your thighs are as long as the muscle in which they occur, but each is thin, so it exchanges substances efficiently with fluids in the tissue surrounding it.

Cell Theory

Nearly all cells are so small that no one even knew they existed until after the first microscopes were invented. In 1665, Antoni van Leeuwenhoek, a Dutch draper, wrote about the tiny moving organisms he spied in rainwater, insects, fabric, sperm, feces, and other samples. In scrapings of tartar from his teeth, Leeuwenhoek saw "many very small animalcules, the motions of which were very pleasing to behold." He (incorrectly) assumed that movement defined life, and (correctly) concluded that the moving "beasties" he saw were alive. Perhaps Leeuwenhoek

cell Smallest unit that has the properties of life.
cell theory Theory that all organisms consist of one or more cells, which are the basic unit of life.
cytoplasm Semifluid substance enclosed by a cell's plasma membrane.
nucleus Organelle with two membranes that holds a eukaryotic cell's DNA.
organelle Structure that carries out a specialized metabolic function inside a cell.
plasma membrane A cell's outermost membrane.
surface-to-volume ratio A relationship in which the volume of an object increases with the cube of the diameter, but the surface area increases with the square.

was so pleased to behold his animalcules because he did not grasp the implications of what he was seeing: Our world, and our bodies, teem with microbial life. The term cell was coined when Robert Hooke, a contemporary of Leeuwenhoek, magnified a piece of thinly sliced cork. Hooke named the tiny compartments he observed "cellae," a Latin word for the small chambers that monks lived in.

In the 1820s, botanist Robert Brown was the first to identify a cell nucleus. Matthias Schleiden, another botanist, hypothesized that a plant cell is an independent living unit even when it is part of a plant. Schleiden compared notes with the zoologist Theodor Schwann, and both concluded that the tissues of animals as well as plants are composed of cells and their products. Together, the two scientists recognized that cells have a life of their own even when they are part of a multicelled body.

Later, physiologist Rudolf Virchow realized that all cells he studied descended from another living cell. These and many other observations yielded four generalizations that today constitute the **cell theory**:

1. Every living organism consists of one or more cells.
2. The cell is the structural and functional unit of all organisms. A cell is the smallest unit of life, individually alive even as part of a multicelled organism.
3. All living cells come from division of preexisting cells.
4. Cells contain hereditary material, which they pass to their offspring during division.

The cell theory, first articulated in 1839 by Schwann and Schleiden and later revised, remains an important foundation of modern biology.

Take-Home Message How are all cells alike?

> All cells start life with a plasma membrane, cytoplasm, and a region of DNA, which, in eukaryotic cells only, is enclosed by a nucleus.

> The surface-to-volume ratio limits cell size and influences cell shape.

> Observations of cells led to the cell theory: All organisms consist of one or more cells; the cell is the smallest unit of life; each new cell arises from another cell; and a cell passes hereditary material to its offspring.

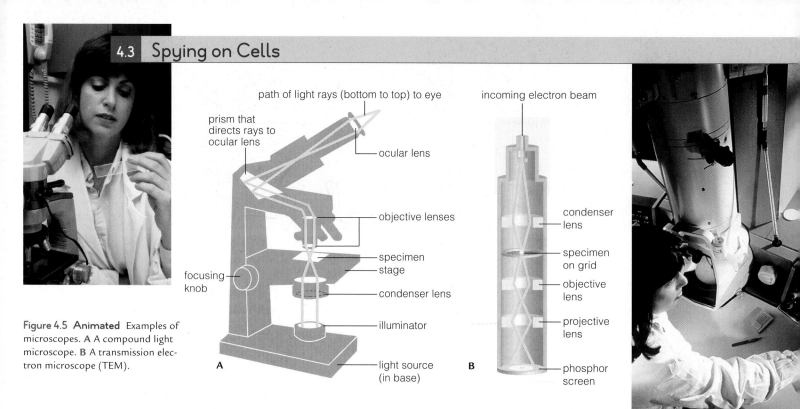

path of light rays (bottom to top) to eye

prism that directs rays to ocular lens

ocular lens

objective lenses

focusing knob

specimen stage

condenser lens

illuminator

light source (in base)

A

incoming electron beam

condenser lens

specimen on grid

objective lens

projective lens

phosphor screen

B

Figure 4.5 **Animated** Examples of microscopes. **A** A compound light microscope. **B** A transmission electron microscope (TEM).

❯ We use different types of microscopes to study different aspects of organisms, from the smallest to the largest.
❰ Link to Tracers 2.2

Microscopes allow us to study cells in detail. The ones that use visible light to illuminate objects are called light microscopes (Figure 4.5). All light travels in waves. This property makes light bend when it passes through curved glass lenses. Inside a light microscope, such lenses focus light that passes through a specimen, or bounces off of one, into a magnified image. Photographs of images enlarged with any microscope are called micrographs, and you will find many of them in this book.

Phase-contrast microscopes shine light through specimens. Most cells are nearly transparent, so their internal details may not be visible unless they are first stained, or exposed to dyes that only some cell parts soak up. Parts that absorb the most dye appear darkest. Staining results

in an increase in contrast (the difference between light and dark) that allows us to see a greater range of detail (Figure 4.6A). Surface details can be revealed by reflected light (Figure 4.6B).

With a fluorescence microscope, a cell or a molecule is the light source; it fluoresces, or emits energy in the form of light, when a laser beam is focused on it. Some molecules fluoresce naturally (Figure 4.6C). More typically, researchers attach a light-emitting tracer (Section 2.2) to the cell or molecule of interest.

Other types of microscopes can reveal finer details. For example, electron microscopes use electrons instead of visible light to illuminate samples. Because electrons travel in wavelengths that are much shorter than those

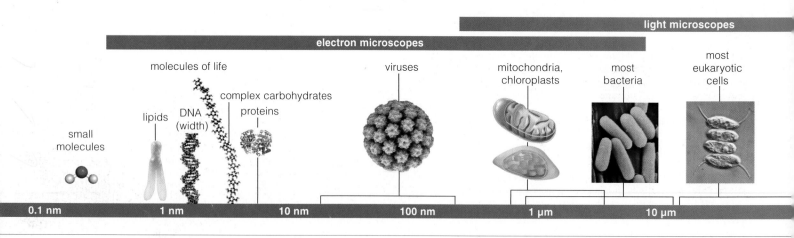

light microscopes

electron microscopes

molecules of life

complex carbohydrates

lipids

DNA (width)

proteins

small molecules

viruses

mitochondria, chloroplasts

most bacteria

most eukaryotic cells

| 0.1 nm | 1 nm | 10 nm | 100 nm | 1 μm | 10 μm |

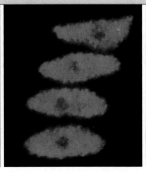
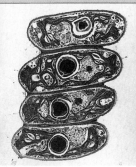

A Light micrograph. A phase-contrast microscope yields high-contrast images of transparent specimens. Dark areas have taken up dye.

B Light micrograph. A reflected light microscope captures light reflected from specimens.

C Fluorescence micrograph. This image shows fluorescent light emitted by chlorophyll molecules in the cells.

D A transmission electron micrograph reveals fantastically detailed images of internal structures.

E A scanning electron micrograph shows surface details. SEMs may be artificially colored to highlight specific details.

Figure 4.6 Different microscopes reveal different characteristics of the same organism, a green alga (*Scenedesmus*).

of visible light, electron microscopes can resolve details that are much smaller than you can see with light microscopes. Electron microscopes use magnetic fields to focus beams of electrons onto a sample. Transmission electron microscopes beam electrons through a thin specimen. The specimen's internal details appear on the resulting image as shadows (Figure 4.6D). Scanning electron microscopes direct a beam of electrons back and forth across a surface of a specimen, which has been coated with a thin layer of gold or another metal. The metal emits both electrons and x-rays, which are converted into an image of the sur-

face (Figure 4.6E). Both types of electron microscopes can resolve structures as small as 0.2 nanometer.

Figure 4.7 compares the resolving power of light and electron microscopes with that of the unaided human eye.

> ## Take-Home Message **How do we see cells?**
> ❯ Most cells are visible only with the help of microscopes.
> ❯ Different types of microscopes reveal different aspects of cell structure.

Figure 4.7 Relative sizes. *Below*, the diameter of most cells is in the range of 1 to 100 micrometers. *Right*, converting among units of length; also see Units of Measure, Appendix IX.

❯❯ **Figure It Out:** Which one is smallest: a protein, a lipid, or a water molecule?

Answer: A water molecule

1 centimeter (cm)	=	1/100 meter, or 0.4 inch
1 millimeter (mm)	=	1/1000 meter, or 0.04 inch
1 micrometer (μm)	=	1/1,000,000 meter, or 0.00004 inch
1 nanometer (nm)	=	1/1,000,000,000 meter, or 0.00000004 inch
1 meter	$= 10^2$ cm	$= 10^3$ mm $= 10^6$ μm $= 10^9$ nm

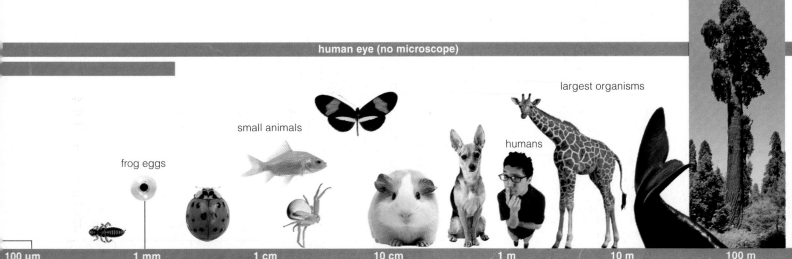

human eye (no microscope)

largest organisms

small animals

humans

frog eggs

| 100 μm | 1 mm | 1 cm | 10 cm | 1 m | 10 m | 100 m |

> A cell membrane is organized as a lipid bilayer with many proteins embedded in it and attached to its surfaces.
< Links to Emergent properties 1.2, Enzyme 3.2, Lipids 3.4, Proteins 3.5

Lipids—mainly phospholipids—make up the bulk of a cell membrane. A phospholipid consists of a phosphate-containing head and two fatty acid tails (Section 3.4). The polar head is hydrophilic, which means that it interacts with water molecules. The nonpolar tails are hydrophobic, so they do not interact with water molecules, but they do interact with the tails of other phospholipids. As a result of these influences, phospholipids swirled into water will spontaneously organize themselves into a **lipid bilayer**, two layers of lipids with nonpolar tails sandwiched between polar heads. Lipid bilayers are the basic structural and functional framework of all cell membranes (Figure 4.8A). In water, a lipid bilayer spontaneously shapes itself into a sheet or bubble. A cell is essentially a lipid bilayer bubble filled with fluid (Figure 4.9).

fluid

Figure 4.9 At its most basic, a cell is a lipid bilayer bubble filled with fluid.

A In a watery fluid, phospholipids spontaneously line up into two layers, tails to tails. This lipid bilayer spontaneously shapes itself into a sheet or a bubble. It is the basic structural and functional framework of all cell membranes. Many types of proteins intermingle among the lipids.

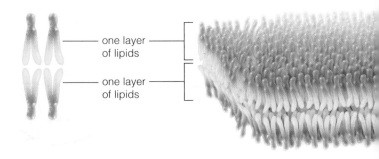

one layer of lipids

one layer of lipids

Figure 4.8 Animated Cell membrane structure. **A** Organization of phospholipids in cell membranes. **B–E** Examples of common membrane proteins.

Other molecules, including steroids and proteins, are embedded in or associated with the lipid bilayer of every cell membrane. Most of these molecules move around the membrane more or less freely. A cell membrane behaves like a two-dimensional liquid of mixed composition, so we describe it as a **fluid mosaic**. The "mosaic" part of the name comes from a cell membrane's mixed composition of lipids and proteins. The fluidity occurs because the phospholipids in a cell membrane are not bonded to one another. They stay organized as a bilayer as a result of collective hydrophobic and hydrophilic attractions, which, on an individual basis, are relatively weak. Thus, phospholipids in a bilayer can drift sideways and spin around their long axis, and their tails can wiggle.

Membrane Proteins

A cell membrane physically separates an external environment from an internal one, but that is not its only task. Many types of proteins are associated with a cell membrane, and each type adds a specific function to it (Table 4.1). Thus, different cell membranes can have different characteristics depending on which proteins are associated with them.

For example, a plasma membrane has certain proteins that no internal cell membrane has. Many plasma membrane proteins are enzymes. Others are **adhesion proteins**, which fasten cells together in animal tissues. **Recognition proteins** function as identity tags for a cell type, individual, or species (Figure 4.8B). Being able to

Table 4.1 Common Types of Membrane Proteins

Category	Function	Examples
Passive transporters	Allow ions or small molecules to cross a membrane to the side where they are less concentrated. Open or gated channels.	Porins; glucose transporter
Active transporters	Pump ions or molecules through membranes to the side where they are more concentrated. Require energy input, as from ATP.	Calcium pump; serotonin transporter
Receptors	Initiate change in a cell activity by responding to an outside signal (e.g., by binding a signaling molecule or absorbing light energy).	Insulin receptor; B cell receptor
Adhesion proteins	Help cells stick to one another and to protein matrixes that are part of tissues.	Integrins; cadherins
Recognition proteins	Identify cells as self (belonging to one's own body or tissue) or nonself (foreign to the body).	MHC molecules
Enzyme	Speeds a specific reaction. Membranes provide a relatively stable reaction site for enzymes that work in series with other molecules.	Cytochrome *c* oxidase of mitochondria

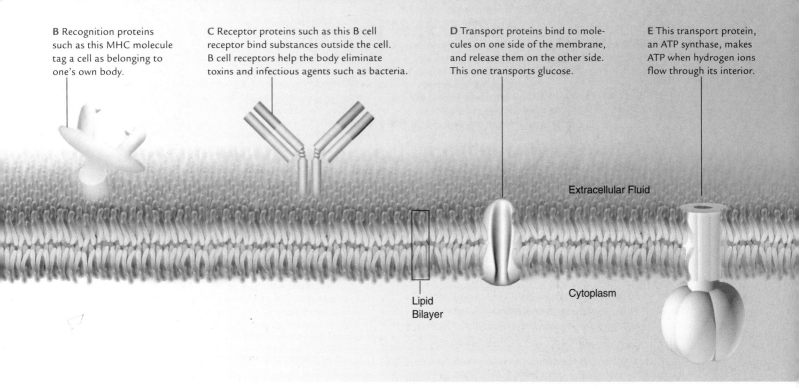

B Recognition proteins such as this MHC molecule tag a cell as belonging to one's own body.

C Receptor proteins such as this B cell receptor bind substances outside the cell. B cell receptors help the body eliminate toxins and infectious agents such as bacteria.

D Transport proteins bind to molecules on one side of the membrane, and release them on the other side. This one transports glucose.

E This transport protein, an ATP synthase, makes ATP when hydrogen ions flow through its interior.

Extracellular Fluid

Cytoplasm

Lipid Bilayer

recognize "self" means that foreign cells (harmful ones, in particular) can also be recognized. **Receptor proteins** bind to a particular substance outside of the cell, such as a hormone or toxin (Figure 4.8C). Binding triggers a change in the cell's activities that may involve metabolism, movement, division, or even cell death. Receptors for different types of substances occur on different cells, but all are critical for homeostasis.

Additional proteins occur on all cell membranes. **Transport proteins** move specific substances across a membrane, typically by forming a channel through it. These proteins are important because lipid bilayers are impermeable to most substances, including ions and polar molecules. Some transport proteins are open channels through which a substance moves on its own across a membrane (Figure 4.8D,E). Others use energy to actively pump a substance across. We return to the topic of membrane transport in the next chapter.

adhesion protein Membrane protein that helps cells stick together in tissues.
fluid mosaic Model of a cell membrane as a two-dimensional fluid of mixed composition.
lipid bilayer Structural foundation of cell membranes; double layer of lipids arranged tail-to-tail.
receptor protein Plasma membrane protein that binds to a particular substance outside of the cell.
recognition protein Plasma membrane protein that identifies a cell as belonging to self (one's own body).
transport protein Protein that passively or actively assists specific ions or molecules across a membrane.

Variations on the Model

Membranes differ in composition. For example, different kinds of cells may have different kinds of membrane phospholipids. Fatty acid tails of membrane phospholipids vary in length and saturation. Usually, at least one of the two tails is unsaturated (Section 3.4).

Some proteins stay put, such as those that cluster as rigid pores. Protein filaments inside the cell lock these and other proteins in place. For example, plasma membrane adhesion proteins connect the internal filaments of adjacent cells in animal tissues. This arrangement strengthens a tissue, and it can constrain certain membrane proteins to an upper or lower surface of the cells.

Archaeans do not build their phospholipids with fatty acids. Instead, they use molecules that have reactive side chains, so the tails of archaean phospholipids form covalent bonds with one another. As a result of this rigid crosslinking, archaean phospholipids do not drift, spin, or wiggle in a bilayer. Thus, the membranes of archaeans are far more rigid than those of bacteria or eukaryotes, a characteristic that may help these cells survive in extreme habitats that destroy other cells.

Take-Home Message **What is the function of a cell membrane?**

❯ A cell membrane selectively controls exchanges between the cell and its surroundings. It is a mosaic of different kinds of lipids and proteins.

❯ The foundation of cell membranes is the lipid bilayer—two layers of lipids (mainly phospholipids), with tails sandwiched between heads.

❯ All bacteria and archaeans are single-celled organisms with no nucleus.
❮ Links to Domains 1.4, Polysaccharides 3.3, ATP 3.7

All bacteria and archaeans are single-celled, and none have a nucleus. Outwardly, they appear so similar that archaeans were once thought to be an unusual group of bacteria. Both were classified as prokaryotes, a word that means "before the nucleus." By 1977, it had become clear that archaeans are more closely related to eukaryotes than to bacteria, so they were given their own separate domain. The term "prokaryote" is now being retired.

As a group, bacteria and archaeans are the smallest and most metabolically diverse forms of life that we know about. They inhabit nearly all of Earth's environments, including some very hostile places. The two kinds of cells differ in structure and metabolic details (Figure 4.10). Chapter 19 revisits them in more detail; here we present an overview of their structure (Figure 4.11).

Most bacteria and archaeans are not much bigger than a few micrometers. None has a complex internal framework, but protein filaments under the plasma membrane reinforce the cell's shape. Such filaments also act as scaffolding for internal structures. The cytoplasm of these cells contains many **ribosomes** (organelles upon which polypeptides are assembled), and in some species, additional organelles. **Plasmids** also occur in cytoplasm. These small circles of DNA carry a few genes (units of inheritance) that can provide advantages, such as resistance to antibiotics. The cell's single chromosome, a circular DNA molecule, is located in an irregularly shaped region of cytoplasm called the **nucleoid**. Some nucleoids are enclosed by a membrane, but most are not.

Many bacteria and archaeans have one or more flagella projecting from their surface. **Flagella** (singular, flagellum) are long, slender cellular structures used for motion. A bacterial flagellum rotates like a propeller that drives

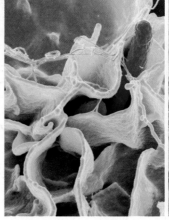
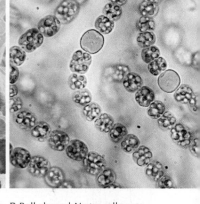

A Protein filaments, or pili, anchor bacterial cells to one another and to surfaces. Here, *Salmonella typhimurium* cells (*red*) use their pili to invade human cells.

B Ball-shaped *Nostoc* cells are a type of freshwater photosynthetic bacteria. The cells in each strand stick together in a sheath of their own jellylike secretions.

Figure 4.10 A sampling of bacteria (*this page*) and archaeans (*facing page*).

the cell through fluid habitats, such as an animal's body fluids. Some bacteria also have protein filaments called **pili** (singular, pilus) projecting from their surface. Pili help cells cling to or move across surfaces (Figure 4.10A). One kind, a "sex" pilus, attaches to another bacterium and then shortens. The attached cell is reeled in, and a plasmid is transferred from one cell to the other through the pilus.

A durable **cell wall** surrounds the plasma membrane of nearly all bacteria and archaeans. Dissolved substances easily cross this permeable layer on the way to and from the plasma membrane. The cell wall of most archaeans consists of proteins. The wall of most bacteria consists of a polymer of peptides and polysaccharides.

Sticky polysaccharides form a slime layer, or capsule, around the wall of many types of bacteria. The sticky capsule helps these cells adhere to many types of surfaces

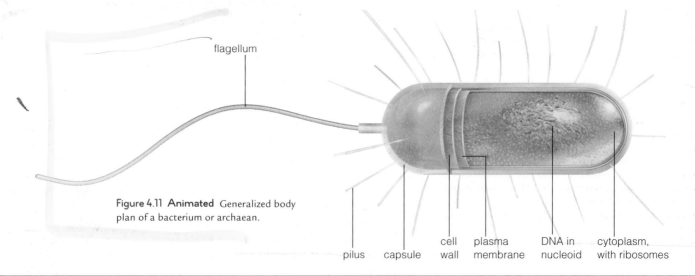

flagellum

Figure 4.11 Animated Generalized body plan of a bacterium or archaean.

pilus capsule cell wall plasma membrane DNA in nucleoid cytoplasm, with ribosomes

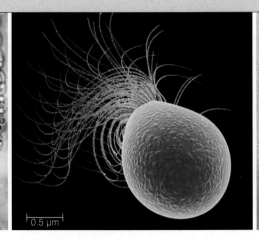

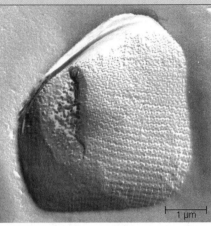

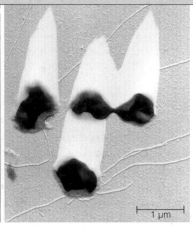

C The archaean *Pyrococcus furiosus* was discovered in ocean sediments near an active volcano. It lives best at 100°C (212°F), and it makes a rare kind of enzyme that contains tungsten atoms.

D *Ferroglobus placidus* prefers superheated water spewing from the ocean floor. The durable composition of archaean lipid bilayers (note the gridlike texture) keeps their membranes intact at extreme heat and pH.

E *Metallosphaera prunae*, an archaean discovered in a smoking pile of ore at a uranium mine, prefers high temperatures and low pH. (*White* shadows are an artifact of electron microscopy.)

(such as spinach leaves and meat), and it also offers some protection against predators and toxins.

The plasma membrane of all bacteria and archaeans selectively controls which substances move into and out of the cell, as it does for eukaryotic cells. The plasma membrane bristles with transporters and receptors, and it also incorporates proteins that carry out important metabolic processes. For example, part of the plasma membrane of cyanobacteria (Figure 4.10B) folds into the cytoplasm. Molecules that carry out photosynthesis are embedded in this membrane, just as they are in the inner membrane of chloroplasts, which are organelles specialized for photosynthesis in eukaryotic cells (we return to chloroplasts in Section 4.9).

Figure 4.12 A biofilm. A single species of bacteria, *Bacillus subtilis*, formed this biofilm. Note the distinct "neighborhoods."

0.5 cm

biofilm Community of microorganisms living within a shared mass of slime.
cell wall Semirigid but permeable structure that surrounds the plasma membrane of some cells.
flagellum Long, slender cellular structure used for motility.
nucleoid Region of cytoplasm where the DNA is concentrated inside a bacterium or archaean.
pilus A protein filament that projects from the surface of some bacterial cells.
plasmid Small circle of DNA in some bacteria and archaeans.
ribosome Organelle of protein synthesis.

Biofilms

Bacterial cells often live so close together that an entire community shares a layer of secreted polysaccharides and proteins. A communal living arrangement in which single-celled organisms live in a shared mass of slime is called a **biofilm**. In nature, a biofilm typically consists of multiple species, all entangled in their own mingled secretions. It may include bacteria, algae, fungi, protists, and archaeans. Participating in a biofilm allows the cells to linger in a favorable spot rather than be swept away by fluid currents, and to reap the benefits of living communally. For example, rigid or netlike secretions of some species serve as permanent scaffolding for others; species that break down toxic chemicals allow more sensitive ones to thrive in polluted habitats that they could not withstand on their own; and waste products of some serve as raw materials for others.

Like a bustling metropolitan city, a biofilm organizes itself into "neighborhoods," each with a distinct microenvironment that stems from its location within the biofilm and the particular species that inhabit it (Figure 4.12).

Take-Home Message **How are bacteria and archaeans alike?**

❯ Bacteria and archaeans do not have a nucleus. Most kinds have a cell wall around their plasma membrane. The wall is permeable, and it reinforces and imparts shape to the cell body.

❯ The structure of bacteria and archaeans is relatively simple, but as a group these organisms are the most diverse forms of life. They inhabit nearly all regions of the biosphere.

❯ Some metabolic processes occur at the plasma membrane of bacteria and archaeans. They are similar to complex processes that occur at certain internal membranes of eukaryotic cells.

> Eukaryotic cells carry out much of their metabolism inside organelles enclosed by membranes.

All protists, fungi, plants, and animals are eukaryotes. Some of these organisms are independent, free-living cells; others consist of many cells working together as a body. By definition, a eukaryotic cell starts out life with a nucleus (*eu–* means true; *karyon* means nut, or kernel).

Like many other organelles, a nucleus has a membrane. An organelle's outer membrane controls the types and amounts of substances that cross it. Such control maintains a special internal environment that allows the organelle to carry out its particular function. That function may be isolating toxic or sensitive substances from the rest of the cell, transporting substances through cytoplasm, maintaining fluid balance, or providing a favorable environment for a special process. For example, a mitochondrion makes ATP after concentrating hydrogen ions inside its membrane system.

Much as interactions among organ systems keep an animal body running, interactions among organelles keep a cell running. Substances shuttle from one kind of organelle to another, and to and from the plasma membrane. Some metabolic pathways take place in a series of different organelles.

Figure 4.13 shows two typical eukaryotic cells. In addition to a nucleus, these and most other eukaryotic cells contain an endomembrane system (ER, vesicles, and Golgi bodies), mitochondria, and a cytoskeleton, which is a dynamic "skeleton" of proteins (*cyto–* means cell). Certain cell types have additional organelles that allow them to do particular tasks (Table 4.2). We will discuss the main components of eukaryotic cells in the remaining sections of this chapter.

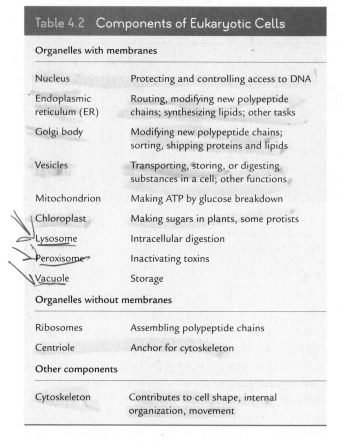

Table 4.2 Components of Eukaryotic Cells

Organelles with membranes	
Nucleus	Protecting and controlling access to DNA
Endoplasmic reticulum (ER)	Routing, modifying new polypeptide chains; synthesizing lipids; other tasks
Golgi body	Modifying new polypeptide chains; sorting, shipping proteins and lipids
Vesicles	Transporting, storing, or digesting substances in a cell; other functions
Mitochondrion	Making ATP by glucose breakdown
Chloroplast	Making sugars in plants, some protists
Lysosome	Intracellular digestion
Peroxisome	Inactivating toxins
Vacuole	Storage
Organelles without membranes	
Ribosomes	Assembling polypeptide chains
Centriole	Anchor for cytoskeleton
Other components	
Cytoskeleton	Contributes to cell shape, internal organization, movement

Take-Home Message **What do all eukaryotic cells have in common?**

> All eukaryotic cells start life with a nucleus and other membrane-enclosed organelles.

Figure 4.13 Transmission electron micrographs of eukaryotic cells.

A Human white blood cell.

B Photosynthetic cell from a blade of timothy grass.

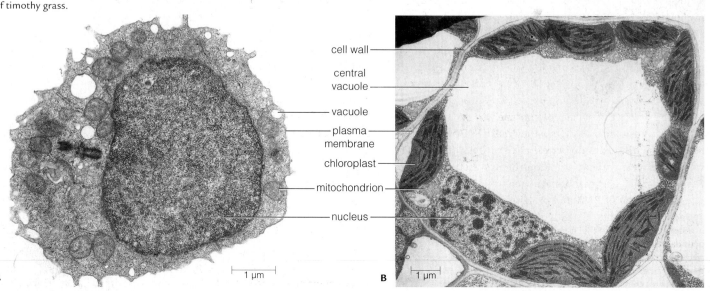

cell wall

central vacuole

vacuole

plasma membrane

chloroplast

mitochondrion

nucleus

1 µm

A

1 µm

B

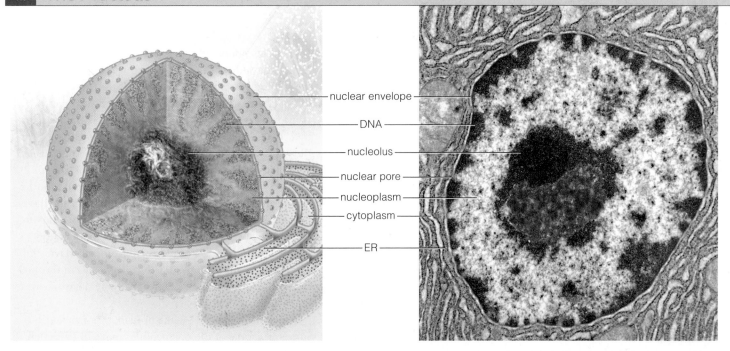

Figure 4.14 **Animated** The cell nucleus. TEM at *right*, nucleus of a mouse pancreas cell.

> A nucleus protects and controls access to a eukaryotic cell's DNA.

A eukaryotic cell's nucleus serves two important functions. First, it keeps the cell's genetic material—its DNA—safe and sound. Isolated in its own compartment, the cell's DNA stays separated from the bustling activity of the cytoplasm, and from metabolic processes that might damage it.

The second function of a nucleus is to control the passage of certain molecules between the nucleus and the cytoplasm. The nuclear membrane, which is called the **nuclear envelope**, carries out this function. A nuclear envelope consists of two lipid bilayers folded together as a single membrane. As Figure 4.14 shows, the outer bilayer of the membrane is continuous with the membrane of another organelle, the ER.

Receptors and transporters stud both sides of the nuclear envelope; other proteins cluster to form tiny pores that span it (Figure 4.15). These molecules and structures work as a system that selectively transports various molecules across the nuclear membrane.

Cells access their DNA when they make RNA and proteins, so the molecules involved in this process must pass into the nucleus and out of it. Control over their transport through the nuclear membrane is one way the cell regulates the amount of RNA and proteins it makes.

nuclear envelope A double membrane that constitutes the outer boundary of the nucleus.
nucleolus In a cell nucleus, a dense, irregularly shaped region where ribosomal subunits are assembled.
nucleoplasm Viscous fluid enclosed by the nuclear envelope.

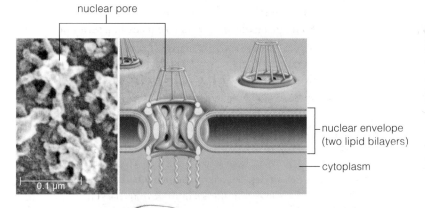

Figure 4.15 **Animated** Nuclear pores. Each is an organized cluster of membrane proteins that selectively allows substances to cross the nuclear membrane.

The nuclear envelope encloses **nucleoplasm**, a viscous fluid that is similar to cytoplasm. The nucleus also contains at least one **nucleolus** (plural, nucleoli), a dense, irregularly shaped region where subunits of ribosomes are assembled from proteins and RNA. The subunits pass through nuclear pores into the cytoplasm, where they join and become active in protein synthesis.

Take-Home Message **What is the function of a nucleus?**

> A nucleus protects and controls access to a eukaryotic cell's DNA.

> The nuclear envelope is a double lipid bilayer. Proteins embedded in it control the passage of molecules between the nucleus and cytoplasm.

> ❭ The endomembrane system is a set of organelles that makes, modifies, and transports proteins and lipids.
> ❬ Links to Lipids 3.4, Proteins 3.5

The **endomembrane system** is a series of interacting organelles between the nucleus and the plasma membrane. Its main function is to make lipids, enzymes, and proteins for secretion, or for insertion into cell membranes. It also destroys toxins, recycles wastes, and has other specialized functions. The system's components vary among different types of cells, but here we present the most common ones (Figure 4.16).

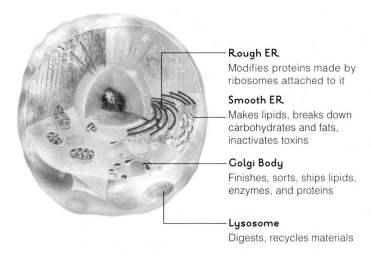

Rough ER
Modifies proteins made by ribosomes attached to it

Smooth ER
Makes lipids, breaks down carbohydrates and fats, inactivates toxins

Golgi Body
Finishes, sorts, ships lipids, enzymes, and proteins

Lysosome
Digests, recycles materials

Figure 4.16 Animated Endomembrane system, where many proteins are modified and lipids are built. These molecules are sorted and shipped to cellular destinations or to the plasma membrane for export.

Endoplasmic Reticulum

Part of the endomembrane system is an extension of the nuclear envelope called **endoplasmic reticulum**, or **ER**. ER forms a continuous compartment that folds into flattened sacs and tubes. The space inside the compartment is the site where many new polypeptide chains are modified. Two kinds of ER, rough and smooth, are named for their appearance in electron micrographs. Thousands of ribosomes are attached to the outer surface of rough ER. These ribosomes make polypeptides that thread into the interior of the ER as they are assembled ❶. Inside the ER, the polypeptides fold and take on their tertiary structure. Some of them become part of the ER membrane itself.

Cells that make, store, and secrete proteins have a lot of rough ER. For example, ER-rich gland cells in the pancreas make and secrete enzymes that help digest meals in the small intestine.

Smooth ER does not make protein, so it has no ribosomes ❷. Some of the polypeptides made in the rough ER end up as enzymes in the smooth ER. These enzymes assemble most of the lipids that form the cell's membranes. They also break down carbohydrates, fatty acids, and some drugs and poisons. In skeletal muscle cells, one type of smooth ER stores calcium ions and has a role in muscle contraction.

A Variety of Vesicles

Small, membrane-enclosed, saclike **vesicles** form in great numbers, in a variety of types, either on their own or by budding ❸. Many vesicles transport substances from one organelle to another, or to and from the plasma mem-

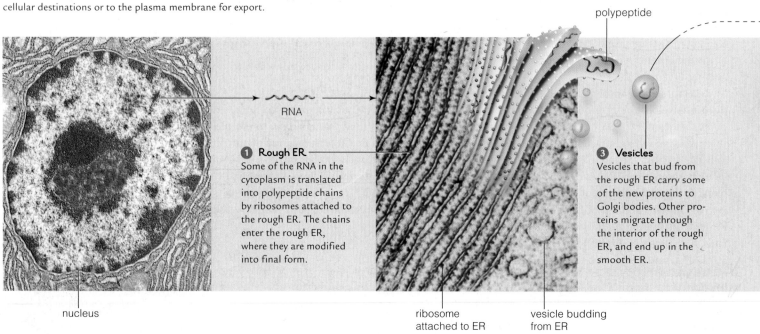

polypeptide

RNA

❶ **Rough ER**
Some of the RNA in the cytoplasm is translated into polypeptide chains by ribosomes attached to the rough ER. The chains enter the rough ER, where they are modified into final form.

❸ **Vesicles**
Vesicles that bud from the rough ER carry some of the new proteins to Golgi bodies. Other proteins migrate through the interior of the rough ER, and end up in the smooth ER.

nucleus

ribosome attached to ER

vesicle budding from ER

brane. *Endocytic* vesicles form as a patch of plasma membrane sinks into the cytoplasm. *Exocytic* vesicles bud from the ER or Golgi membranes and transport substances to the plasma membrane for export. Still other vesicles form on their own in the cytoplasm. Eukaryotic cells also contain **vacuoles**, which are vesicles that appear empty under a microscope. Amino acids, sugars, toxins, and ions accumulate in the fluid-filled interior of a plant cell's large **central vacuole** (Figure 4.13B). Fluid pressure in a central vacuole keeps plant cells plump, so stems, leaves, and other structures stay firm.

Other types of vesicles are a bit like trash cans that collect waste, debris, or toxins. These vesicles dispose of their contents by fusing with other vesicles called lysosomes. **Lysosomes** that bud from Golgi bodies take part in intracellular digestion. They contain powerful enzymes that

---

central vacuole Fluid-filled vesicle in many plant cells.
endomembrane system Series of interacting organelles (endoplasmic reticulum, Golgi bodies, vesicles) between nucleus and plasma membrane; produces lipids, proteins.
endoplasmic reticulum (ER) Organelle that is a continuous system of sacs and tubes; extension of the nuclear envelope. Rough ER is studded with ribosomes; smooth ER is not.
Golgi body Organelle that modifies polypeptides and lipids; also sorts and packages the finished products into vesicles.
lysosome Enzyme-filled vesicle that functions in intracellular digestion.
peroxisome Enzyme-filled vesicle that breaks down amino acids, fatty acids, and toxic substances.
vacuole A fluid-filled organelle that isolates or disposes of waste, debris, or toxic materials.
vesicle Small, membrane-enclosed, saclike organelle; different kinds store, transport, or degrade their contents.

can break down carbohydrates, proteins, nucleic acids, and lipids. Vesicles inside white blood cells or amoebas deliver ingested bacteria, cell parts, and other debris to lysosomes for destruction.

In plants and animals, vesicles called **peroxisomes** form and divide on their own, so they are not part of the endomembrane system. Peroxisomes contain enzymes that digest fatty acids and amino acids. They also break down hydrogen peroxide, a toxic by-product of fatty acid metabolism. Peroxisome enzymes convert hydrogen peroxide to water and oxygen, or use it in reactions that break down alcohol and other toxins.

Golgi Bodies

Some vesicles fuse with and empty their contents into a **Golgi body**. This organelle has a folded membrane that typically looks like a stack of pancakes ❹. Enzymes in a Golgi body put finishing touches on proteins and lipids that have been delivered from the ER. They attach phosphate groups or oligosaccharides, and cut certain polypeptides. The finished products are sorted and packaged into new vesicles that carry them to lysosomes or to the plasma membrane ❺.

Take-Home Message **What is the endomembrane system?**

❭ The endomembrane system includes rough and smooth endoplasmic reticulum, vesicles, and Golgi bodies.

❭ This series of organelles works together mainly to make and modify cell membrane proteins and lipids.

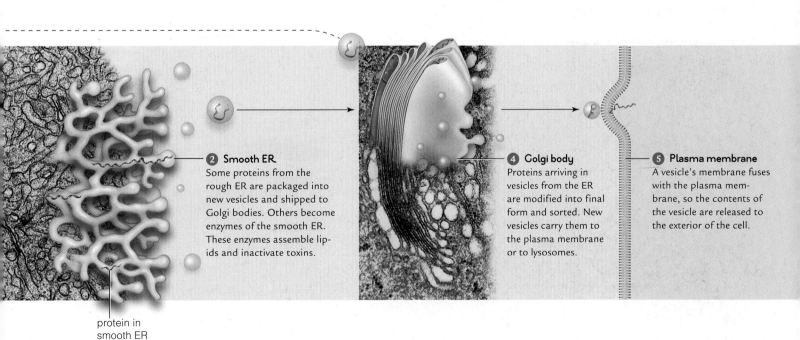

❷ **Smooth ER**
Some proteins from the rough ER are packaged into new vesicles and shipped to Golgi bodies. Others become enzymes of the smooth ER. These enzymes assemble lipids and inactivate toxins.

❹ **Golgi body**
Proteins arriving in vesicles from the ER are modified into final form and sorted. New vesicles carry them to the plasma membrane or to lysosomes.

❺ **Plasma membrane**
A vesicle's membrane fuses with the plasma membrane, so the contents of the vesicle are released to the exterior of the cell.

protein in smooth ER

> Eukaryotic cells make most of their ATP in mitochondria.
> Organelles called plastids function in storage and photo-synthesis in plants and some types of algae.
< Links to Metabolism 3.2, ATP 3.7

Mitochondria

The **mitochondrion** (plural, mitochondria) is a type of organelle that specializes in making ATP (Figure 4.17). Aerobic respiration, an oxygen-requiring series of reactions that proceeds inside mitochondria, can extract more energy from organic compounds than any other metabolic pathway. With each breath, you are taking in oxygen mainly for mitochondria in your trillions of aerobically respiring cells.

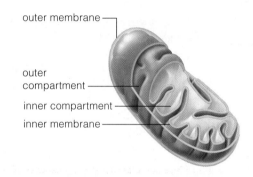

outer membrane

outer compartment

inner compartment

inner membrane

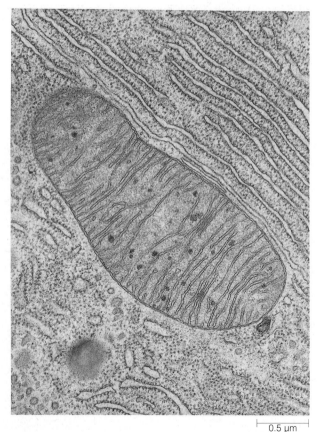

0.5 µm

Typical mitochondria are between 1 and 4 micrometers in length; a few are as long as 10 micrometers. Some are branched. Mitochondria can change shape, split in two, and fuse together. Each has two membranes, one highly folded inside the other. This arrangement creates two compartments. During aerobic respiration, hydrogen ions accumulate between the two membranes. The buildup causes the ions to flow across the inner mitochondrial membrane, through the interior of membrane transport proteins. That flow drives the formation of ATP.

Bacteria and archaeans have no mitochondria; they make ATP in their cell walls and cytoplasm. Nearly all eukaryotic cells have mitochondria, but the number varies by the type of cell and by the type of organism. For example, a single-celled yeast (a type of fungus) might have only one mitochondrion, but a human skeletal muscle cell may have a thousand or more. Cells that have a very high demand for energy tend to have many mitochondria.

Mitochondria resemble bacteria in size, form, and biochemistry. They have their own DNA, which is similar to bacterial DNA. They divide independently of the cell, and have their own ribosomes. Such clues led to a theory that mitochondria evolved from aerobic bacteria that took up permanent residence inside a host cell. By the theory of endosymbiosis, one cell was engulfed by another cell, or entered it as a parasite, but escaped digestion. The ingested cell kept its plasma membrane and reproduced inside its host. In time, the cell's descendants became permanent residents that offered their hosts the benefit of extra ATP. Structures and functions once required for independent life were no longer needed and were lost over time. Later descendants evolved into mitochondria. We will explore evidence for the theory of endosymbiosis in Section 18.6.

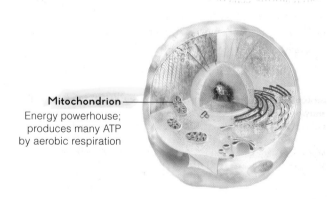

Mitochondrion
Energy powerhouse; produces many ATP by aerobic respiration

Figure 4.17 Animated The mitochondrion. This organelle specializes in producing large quantities of ATP.

>> **Figure It Out:** What organelle is visible in the upper right-hand corner of the TEM?

Answer: Rough ER

Chloroplasts and Other Plastids

Plastids are a category of membrane-enclosed organelles that function in photosynthesis or storage in plant and algal cells. Plastids called **chloroplasts** are organelles specialized for photosynthesis. In many ways, chloroplasts resemble photosynthetic bacteria, and like mitochondria they may have evolved by endosymbiosis.

Most chloroplasts are oval or disk-shaped (Figure 4.18). Each has two outer membranes enclosing a semifluid interior, the stroma, that contains enzymes and the chloroplast's own DNA. Inside the stroma, a third, highly folded membrane forms a single, continuous compartment. The folded membrane resembles stacks of flattened disks. The stacks are called grana (singular, granum). Photosynthesis takes place at this membrane, which is called the thylakoid membrane.

The thylakoid membrane incorporates many pigments. The most common ones are chlorophylls, which are green. The abundance of chlorophylls in plant cell chloroplasts is the reason most plants are green.

By the process of photosynthesis, chlorophylls and other molecules in the thylakoid membrane harness the energy in sunlight to drive the synthesis of ATP. The ATP is then used inside the stroma to build carbohydrates from carbon dioxide and water. (Chapter 6 describes the process of photosynthesis in more detail.)

Chromoplasts are plastids that make and store pigments other than chlorophylls. They typically have red, orange, and yellow pigments that color many flowers, leaves, fruits, and roots. For example, as a tomato ripens, its chlorophyll-containing (green) chloroplasts are converted to lycopene-containing (red) chromoplasts, so the color of the fruit changes.

Amyloplasts are unpigmented plastids. Typical amyloplasts store starch grains, and are notably abundant in starch-storing cells of stems, tubers (underground stems), and seeds. Starch-packed amyloplasts are dense and heavy compared to cytoplasm. In some plant cells, they function as gravity-sensing organelles.

---

chloroplast Organelle of photosynthesis in the cells of plants and many protists.
mitochondrion Double-membraned organelle that produces ATP by aerobic respiration in eukaryotes.
plastid An organelle that functions in photosynthesis or storage; for example, chloroplast, amyloplast.

Take-Home Message What are some organelles unique to specialized eukaryotic cells?

> Mitochondria are eukaryotic organelles that produce ATP from organic compounds in reactions that require oxygen.

> Chloroplasts are plastids that carry out photosynthesis.

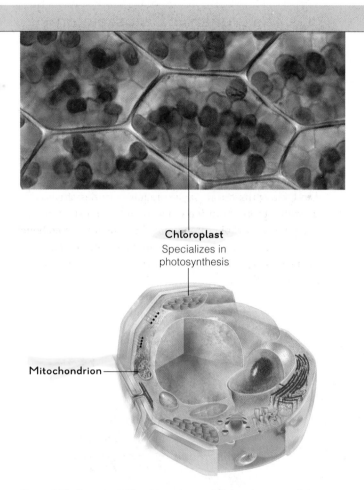

Chloroplast
Specializes in photosynthesis

Mitochondrion

Figure 4.18 Animated The chloroplast, a defining character of photosynthetic eukaryotic cells. *Bottom,* transmission electron micrograph of a chloroplast from a tobacco leaf (*Nicotiana tabacum*). The lighter patches are nucleoids where the chloroplast's own DNA is stored.

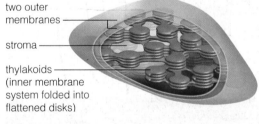

two outer membranes

stroma

thylakoids (inner membrane system folded into flattened disks)

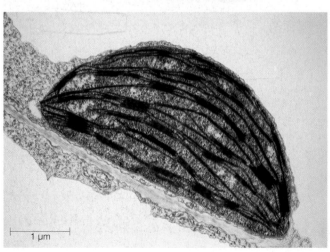

1 μm

> Eukaryotic cells have an extensive and dynamic internal framework called a cytoskeleton.
< Links to Protein structure and function 3.5, 3.6

Between the nucleus and plasma membrane of all eukaryotic cells is a system of interconnected protein filaments collectively called the **cytoskeleton**. Elements of the cytoskeleton reinforce, organize, and move cell structures, and often the whole cell (Figure 4.19). Some are permanent; others form only at certain times.

Microtubules are long, hollow cylinders that consist of subunits of the protein tubulin. They form a dynamic scaffolding for many cellular processes, rapidly assembling when they are needed and then disassembling when they are not. For example, before a eukaryotic cell divides, microtubules assemble, separate the cell's duplicated chromosomes, then disassemble. As another example, microtubules that form in the growing end of a young nerve cell support and guide its lengthening in a particular direction.

Microfilaments are fibers that consist primarily of subunits of the globular protein actin. They strengthen or change the shape of eukaryotic cells. Crosslinked, bundled, or gel-like arrays of them make up the **cell cortex**, which is a reinforcing mesh under the plasma membrane. Actin microfilaments that form at the edge of a cell drag or extend it in a certain direction. Myosin and actin microfilaments interact to bring about contraction of muscle cells.

Intermediate filaments that support cells and tissues are the most stable elements of the cytoskeleton. These filaments form a framework that lends structure and resilience to cells and tissues. Some kinds underlie and reinforce membranes. The nuclear envelope, for example, is supported by an inner layer of intermediate filaments called lamins. Other kinds connect to structures that lock cell membranes together in tissues.

Among the many accessory molecules associated with cytoskeletal elements are **motor proteins**, which move cell parts when energized by a phosphate-group transfer from ATP. A cell is like a bustling train station, with molecules and structures being transported continuously throughout its interior. Motor proteins are like freight trains, dragging their cellular cargo along tracks of dynamically assembled microtubules and microfilaments (Figure 4.20).

tubulin subunit   actin subunit   one polypeptide chain

8–12 nm

Intermediate filament

25 nm

Microtubule

6–7 nm

Microfilament

Figure 4.19 Animated Cytoskeletal elements. *Below*, a fluorescence micrograph shows microtubules (*yellow*) and actin microfilaments (*blue*) in the growing end of a nerve cell. These cytoskeletal elements support and guide the cell's lengthening.

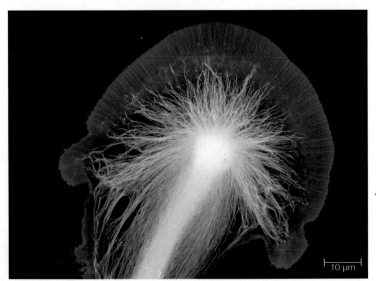

10 μm

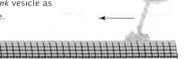

Figure 4.20 Animated Motor proteins. Here, kinesin (*tan*) drags a *pink* vesicle as it inches along a microtubule.

basal body Organelle that develops from a centriole.
cell cortex Reinforcing mesh of cytoskeletal elements under a plasma membrane.
centriole Barrel-shaped organelle from which microtubules grow.
cilium Short, movable structure that projects from the plasma membrane of some eukaryotic cells.
cytoskeleton Dynamic framework of protein filaments that support, organize, and move eukaryotic cells and their internal structures.
intermediate filament Stable cytoskeletal element that structurally supports cells and tissues.
microfilament Reinforcing cytoskeletal element; a fiber of actin subunits.
microtubule Cytoskeletal element involved in cellular movement; hollow filament of tubulin subunits.
motor protein Type of energy-using protein that interacts with cytoskeletal elements to move the cell's parts or the whole cell.

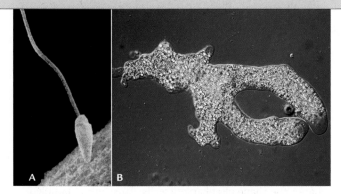

Figure 4.21 Examples of motile structures in cells.
A Flagellum of a human sperm, which is about to penetrate an egg.
B A predatory amoeba (*Chaos carolinense*) extending two pseudopods around its hapless meal: a single-celled green alga (*Pandorina*).

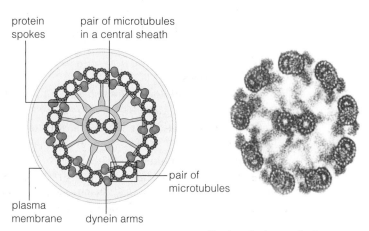

protein spokes — pair of microtubules in a central sheath — pair of microtubules — dynein arms — plasma membrane

A Sketch and micrograph of one eukaryotic flagellum, cross-section. Like a cilium, it contains a 9+2 array: a ring of nine pairs of microtubules plus one pair at its core. Stabilizing spokes and linking elements that connect to the microtubules keep them aligned in this radial pattern.

A motor protein called dynein interacts with organized arrays of microtubules to bring about movement of eukaryotic flagella and cilia. Eukaryotic flagella are structures that whip back and forth to propel cells such as sperm through fluid (Figure 4.21A). They have a different internal structure and type of motion than flagella of bacteria and archaeans. **Cilia** (singular, cilium) are short, hairlike structures that project from the surface of some cells. Cilia are usually more profuse than flagella. The coordinated waving of many cilia propels cells through fluid, and stirs fluid around stationary cells.

A special array of microtubules extends lengthwise through a flagellum or cilium. This 9+2 array consists of nine pairs of microtubules ringing another pair in the center (Figure 4.22). Protein spokes and links stabilize the array. The microtubules grow from a barrel-shaped organelle called the **centriole**, which remains below the finished array as a structure called a **basal body**.

Amoebas and other types of eukaryotic cells form pseudopods, or "false feet." As these temporary, irregular lobes bulge outward, they move the cell and engulf a target such as prey (Figure 4.21B). Elongating microfilaments force the lobe to advance in a steady direction. Motor proteins that are attached to the microfilaments drag the plasma membrane along with them.

B Projecting from each pair of microtubules in the outer ring are "arms" of dynein, a motor protein that has ATPase activity. Phosphate-group transfers from ATP cause the dynein arms to repeatedly bind the adjacent pair of microtubules, bend, and then disengage. The dynein arms "walk" along the microtubules. Their motion causes adjacent microtubule pairs to slide past one another

C Short, sliding strokes occur in a coordinated sequence around the ring, down the length of each microtubule pair. The flagellum bends as the array inside bends:

basal body, a microtubule organizing center that gives rise to the 9+2 array and then remains beneath it, inside the cytoplasm

Take-Home Message What is a cytoskeleton?

› A cytoskeleton of protein filaments is the basis of eukaryotic cell shape, internal structure, and movement.

› Microtubules organize the cell and help move its parts. Networks of microfilaments reinforce the cell surface. Intermediate filaments strengthen and maintain the shape of cells and tissues.

› When energized by ATP, motor proteins move along tracks of microtubules and microfilaments. As part of cilia, flagella, and pseudopods, they can move the whole cell.

Figure 4.22 Animated Mechanism of movement of eukaryotic flagella and cilia.

> ❯ Many cells secrete materials that form a covering or matrix outside their plasma membrane.
> ❮ Links to Tissue 1.2, Chitin 3.3

Matrixes Between and Around Cells

Most cells of multicelled organisms are surrounded and organized by a nonliving, complex mixture of fibrous proteins and polysaccharides called **extracellular matrix**, or **ECM**. Secreted by the cells it surrounds, ECM supports and anchors cells, separates tissues, and functions in cell signaling. Different types of cells secrete different kinds of ECM. For example, a waxy ECM secreted by plant cells forms a **cuticle**, or covering, that protects the plant's exposed surfaces and limits water loss (Figure 4.23). The cuticle of crabs, spiders, and other arthropods is mainly chitin, a polysaccharide. ECM in animals typically consists of various kinds of carbohydrates and proteins; it is the basis of tissue organization, and it provides structural support. Bone is mostly an extracellular matrix composed of collagen, a fibrous protein, hardened by mineral deposits.

The cell wall around the plasma membrane of plant cells is a type of ECM that is structurally different from the cell wall of bacteria and archaeans. Both types of wall protect, support, and impart shape to a cell. Both are also porous: Water and solutes easily cross it on the way to and from the plasma membrane. Cells could not live without exchanging these substances with their environment. A plant cell wall forms as a young cell secretes pectin and

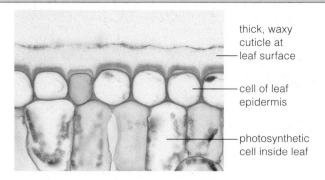

Figure 4.23 A plant ECM. Section through a plant leaf showing cuticle, a protective covering of deposits secreted by living cells.

thick, waxy cuticle at leaf surface

cell of leaf epidermis

photosynthetic cell inside leaf

other polysaccharides onto the outer surface of its plasma membrane. The sticky coating is shared between adjacent cells, and it cements them together. Each cell then forms a **primary wall** by secreting strands of cellulose into the coating. Some of the coating remains as the middle lamella, a sticky layer in between the primary walls of abutting plant cells (Figure 4.24A).

Being thin and pliable, a primary wall allows a growing plant cell to enlarge. Cells with only a thin primary wall can change shape as they develop. At maturity, cells in some plant tissues stop enlarging and begin to secrete material onto the primary wall's inner surface. These deposits form a firm **secondary wall** (Figure 4.24B). One of the materials deposited is **lignin**, an organic compound that makes up as much as 25 percent of the secondary

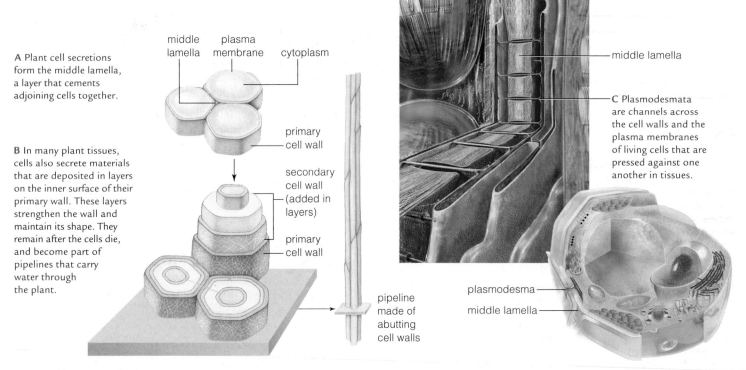

A Plant cell secretions form the middle lamella, a layer that cements adjoining cells together.

middle lamella   plasma membrane   cytoplasm

B In many plant tissues, cells also secrete materials that are deposited in layers on the inner surface of their primary wall. These layers strengthen the wall and maintain its shape. They remain after the cells die, and become part of pipelines that carry water through the plant.

primary cell wall

secondary cell wall (added in layers)

primary cell wall

pipeline made of abutting cell walls

middle lamella

C Plasmodesmata are channels across the cell walls and the plasma membranes of living cells that are pressed against one another in tissues.

plasmodesma

middle lamella

Figure 4.24 Animated Some characteristics of plant cell walls.

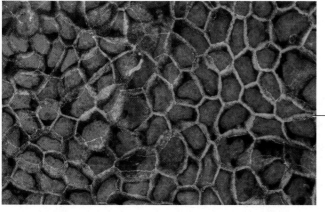

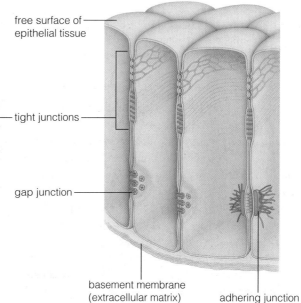

free surface of epithelial tissue

tight junctions

gap junction

basement membrane (extracellular matrix)

adhering junction

Figure 4.25 Animated Three types of cell junctions in animal tissues: tight junctions, gap junctions, and adhering junctions.

In the micrograph *above*, a profusion of tight junctions (*green*) seals abutting surfaces of kidney cell membranes and forms a waterproof tissue. The DNA in each cell nucleus appears *red*.

wall of cells in older stems and roots. Lignified plant parts are stronger, more waterproof, and less susceptible to plant-attacking organisms than younger tissues.

Cell Junctions

ECM does not prevent a cell from interacting with other cells or the surroundings. In multicelled species, such interaction occurs by way of **cell junctions**, which are structures that connect a cell to other cells and to the environment. Cells send and receive ions, molecules, or signals through some junctions. Other kinds help cells recognize and stick to each other and to extracellular matrix.

In most animal tissues, cells are connected to their neighbors and to ECM by cell junctions (Figure 4.25). In epithelial tissues that line body surfaces and internal cavities, rows of proteins that form **tight junctions** between plasma membranes prevent body fluids from seeping between adjacent cells. To cross these tissues, fluid must pass directly through the cells. Thus, transport proteins

embedded in the cell membranes help control which ions and molecules cross the tissue. For example, an abundance of tight junctions in the lining of the stomach normally keeps acidic fluid from leaking out. If a bacterial infection damages this lining, acid and enzymes can erode the underlying layers. The result is a painful peptic ulcer.

Adhering junctions composed of adhesion proteins snap cells to each other and anchor them to ECM. Skin and other tissues that are subject to abrasion or stretching have a lot of adhering junctions. These cell junctions also strengthen contractile tissues such as heart muscle. **Gap junctions** form channels that connect the cytoplasm of adjoining cells, thus permitting ions and small molecules to pass directly from the cytoplasm of one cell to another. By opening or closing, these channels allow entire regions of cells to respond to a single stimulus. Heart muscle and other tissues in which the cells perform a coordinated action have many of these communication channels.

In plants, open channels called **plasmodesmata** (singular, plasmodesma) extend across cell walls, connecting the cytoplasm of adjoining cells (Figure 4.24C). Substances such as water, nutrients, and signaling molecules can flow quickly from cell to cell through these cell junctions.

---

adhering junction Cell junction composed of adhesion proteins; anchors cells to each other and extracellular matrix.
cell junction Structure that connects a cell to another cell or to extracellular matrix.
cuticle Secreted covering at a body surface.
extracellular matrix (ECM) Complex mixture of cell secretions; supports cells and tissues; has roles in cell signaling.
gap junction Cell junction that forms a channel across the plasma membranes of adjoining animal cells.
lignin Material that stiffens cell walls of vascular plants.
plasmodesmata Cell junctions that connect the cytoplasm of adjacent plant cells.
primary wall The first cell wall of young plant cells.
secondary wall Lignin-reinforced wall that forms inside the primary wall of a plant cell.
tight junctions Arrays of fibrous proteins; join epithelial cells and collectively prevent fluids from leaking between them.

Take-Home Message **What structures form on the outside of eukaryotic cells?**

> Cells of many protists, nearly all fungi, and all plants have a porous wall around the plasma membrane. Animal cells do not have walls.

> Plant cell secretions form a waxy cuticle that helps protect the exposed surfaces of soft plant parts.

> Cell secretions form extracellular matrixes between cells in many tissues.

> Cells make structural and functional connections with one another and with extracellular matrix in tissues.

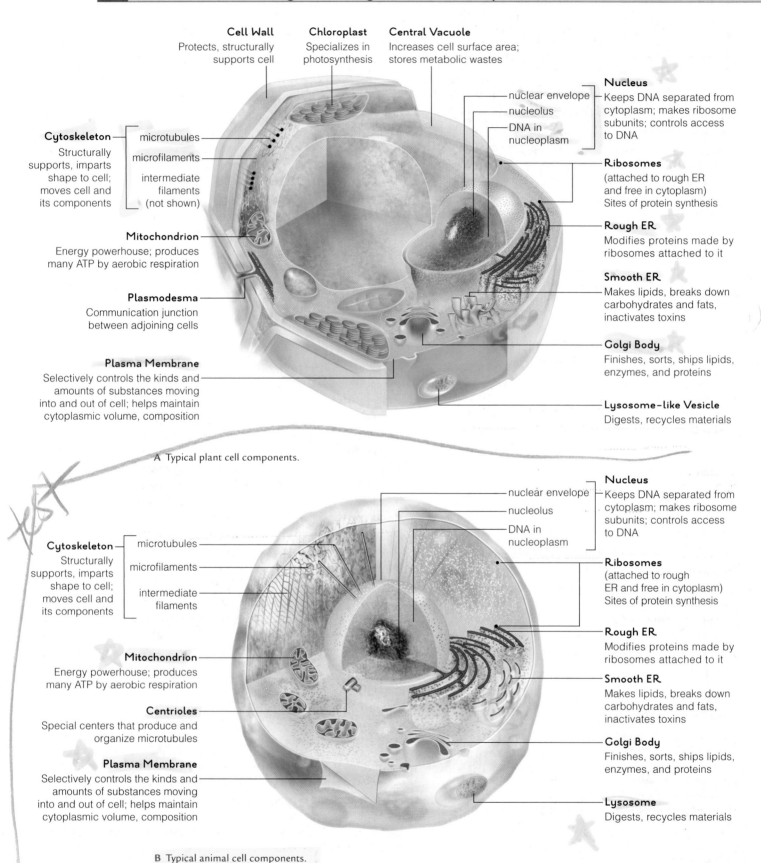

Cell Wall
Protects, structurally supports cell

Chloroplast
Specializes in photosynthesis

Central Vacuole
Increases cell surface area; stores metabolic wastes

nuclear envelope
nucleolus
DNA in nucleoplasm

Nucleus
Keeps DNA separated from cytoplasm; makes ribosome subunits; controls access to DNA

Cytoskeleton
microtubules
microfilaments
intermediate filaments (not shown)
Structurally supports, imparts shape to cell; moves cell and its components

Ribosomes
(attached to rough ER and free in cytoplasm) Sites of protein synthesis

Rough ER
Modifies proteins made by ribosomes attached to it

Mitochondrion
Energy powerhouse; produces many ATP by aerobic respiration

Smooth ER
Makes lipids, breaks down carbohydrates and fats, inactivates toxins

Plasmodesma
Communication junction between adjoining cells

Golgi Body
Finishes, sorts, ships lipids, enzymes, and proteins

Plasma Membrane
Selectively controls the kinds and amounts of substances moving into and out of cell; helps maintain cytoplasmic volume, composition

Lysosome-like Vesicle
Digests, recycles materials

A Typical plant cell components.

Nucleus
Keeps DNA separated from cytoplasm; makes ribosome subunits; controls access to DNA

nuclear envelope
nucleolus
DNA in nucleoplasm

Cytoskeleton
microtubules
microfilaments
intermediate filaments
Structurally supports, imparts shape to cell; moves cell and its components

Ribosomes
(attached to rough ER and free in cytoplasm) Sites of protein synthesis

Rough ER
Modifies proteins made by ribosomes attached to it

Mitochondrion
Energy powerhouse; produces many ATP by aerobic respiration

Smooth ER
Makes lipids, breaks down carbohydrates and fats, inactivates toxins

Centrioles
Special centers that produce and organize microtubules

Golgi Body
Finishes, sorts, ships lipids, enzymes, and proteins

Plasma Membrane
Selectively controls the kinds and amounts of substances moving into and out of cell; helps maintain cytoplasmic volume, composition

Lysosome
Digests, recycles materials

B Typical animal cell components.

Figure 4.26 Animated Organelles and structures typical of **A** plant cells and **B** animal cells.

4.13 The Nature of Life

> We define life by describing the set of properties that is unique to living things.

< Links to Life's levels of organization 1.2, Homeostasis 1.3, Philosophy of science 1.9

In this chapter, you learned about the structure of cells, which have at their minimum a plasma membrane, cytoplasm, and a region of DNA. Most cells have many other components. But what, exactly, makes a cell, or an organism that consists of them, alive? We say that life is a property that emerges from cellular components, but a collection of those components in the right amounts and proportions is not necessarily alive.

We know intuitively what "life" is, but defining it unambiguously is challenging, if not impossible. We can more easily describe what sets living things apart from nonliving things, but even that is tricky. For example, living organisms have a high proportion of the organic molecules of life, but so do the remains of dead organisms such as coal and other fossil fuels. Living things use energy to reproduce themselves, but computer viruses, which are arguably not alive, can do that too.

So how do biologists, who study life as a profession, describe it? The short answer is that their best description of life is very long. It consists of a list of properties associated with things we know to be alive. You already know about two of these properties of organisms:

1. They make and use the organic molecules of life.
2. They consist of one or more cells.

The remainder of this book details the other properties of living things:

3. They engage in self-sustaining biological processes such as metabolism and homeostasis.
4. They change over their lifetime, for example by growing, maturing, and aging.
5. They use DNA as their hereditary material.
6. They have the collective capacity to change over successive generations, for example by adapting to environmental pressures.

Collectively, these properties characterize living things as different from nonliving things.

Take-Home Message **What is life?**

> We describe the characteristic of "life" in terms of a set of properties. The set is unique to living things.

> In living things, the molecules of life are organized as one or more cells that engage in self-sustaining biological processes.

> Organisms use DNA as their hereditary material.

> Living things change over lifetimes, and over generations.

Food for Thought (revisited)

One food safety measure involves sterilization, which kills *E. coli* O157:H7 and other bacteria. Recalled, contaminated ground beef is typically cooked or otherwise sterilized, then processed into ready-to-eat products. Raw beef trimmings are effectively sterilized when sprayed with ammonia and ground to a paste. The resulting meat product is now routinely added to hamburger patties, fresh ground beef, hot dogs, lunch meats, sausages, frozen entrees, canned foods, and other items sold to quick service restaurants, hotel and restaurant chains, institutions, and school lunch programs. Meat, poultry, milk, and fruits sterilized by exposure to radiation are available in supermarkets. By law, irradiated foods must be marked with the symbol on the *left*, but foods sterilized with chemicals are not currently required to carry any disclosure. Some worry that sterilization may alter food or leave harmful chemicals in it. Whether any health risks are associated with consuming sterilized foods is unknown.

How Would You Vote? Some think the safest way to protect consumers from food poisoning is by sterilizing food with chemicals or radiation to kill any bacteria that may be in it. Others think we should make sure our food does not get contaminated in the first place. Would you choose sterilized food? See CengageNow for details, then vote online (cengagenow.com).

Summary

 Sections 4.1 and 4.2 The **cell** is the smallest unit of life. Although cells differ in size, shape, and function, each starts out life with a **plasma membrane**, **cytoplasm**, and a region of DNA—in eukaryotic cells, a **nucleus**. A lipid bilayer is the structural foundation of cell membranes, including **organelle** membranes.

The **surface-to-volume ratio** limits cell size. By the **cell theory**, all organisms consist of one or more cells; the cell is the smallest unit of life; each new cell arises from another, preexisting cell; and a cell passes hereditary material to its offspring.

 Section 4.3 Most cells are far too small to see with the naked eye. We use different types of microscopes and techniques to reveal cells and their internal and external details.

 Section 4.4 A cell membrane is a mosaic of lipids (mainly phospholipids) and proteins. It functions as a selectively permeable barrier that separates an internal environment from an external one. The lipids are organized as a **lipid bilayer**: a double layer of lipids in which the nonpolar tails of both layers are sandwiched between the polar heads.

The membranes of most cells can be described as a **fluid mosaic**. Proteins associated with a membrane carry out most membrane functions. All membranes have **transport proteins**. Plasma membranes also have **receptor proteins**, **adhesion proteins**, enzymes, and **recognition proteins**.

Summary

Section 4.5 Bacteria and archaeans have no nucleus, but they have **nucleoids** and **ribosomes**. Many have a **cell wall**, **flagella** or **pili**, and **plasmids**. Bacteria and other microbial organisms often share living arrangements in **biofilms**.

Sections 4.6, 4.7 Eukaryotic cells start out life with membrane-enclosed organelles, including a nucleus. A nucleus has a double-membraned **nuclear envelope** surrounding **nucleoplasm**. In the nucleus, ribosome subunits are assembled in dense regions called **nucleoli**.

Section 4.8 The **endomembrane system** includes rough and smooth **endoplasmic reticulum (ER)**, **vesicles**, and **Golgi bodies**. This system makes and modifies lipids and proteins; it also recycles and disposes of molecules and particles. **Peroxisomes**, **lysosomes**, and **vacuoles** (including **central vacuoles**) are vesicles.

Section 4.9 Mitochondria make ATP by breaking down organic compounds in the oxygen-requiring pathway of aerobic respiration. **Chloroplasts** are **plastids** that produce sugars by photosynthesis.

Section 4.10 A **cytoskeleton** includes **microtubules**, **microfilaments**, and **intermediate filaments**. A microfilament **cell cortex** reinforces plasma membranes. Interactions between **motor proteins** and microtubules move **cilia** and eukaryotic flagella. **Centrioles** give rise to microtubules, then become **basal bodies**.

Section 4.11 Cuticle is an example of **extracellular matrix (ECM)**. Pliable **primary walls** enclose **secondary walls** strengthened with **lignin**. **Cell junctions** connect cells to one another and to ECM. **Adhering junctions**, **gap junctions**, and **tight junctions** connect animal cells. **Plasmodesmata** connect plant cells.

Sections 4.12, 4.13 We describe the characteristic of life as a set of properties. The set is unique to living things.

Table 4.3 Summary of Typical Components of Cells

Cell Component	Main Functions	Bacteria, Archaeans	Eukaryotes			
			Protists	Fungi	Plants	Animals
Cell wall	Protection, structural support	✔*	✔*	✔	✔	None
Plasma membrane	Control of substances moving into and out of cell	✔	✔	✔	✔	✔
Nucleus	Physical separation of DNA from cytoplasm	None	✔	✔	✔	✔
DNA	Encoding of hereditary information	✔	✔	✔	✔	✔
RNA	Transcription, translation	✔	✔	✔	✔	✔
Nucleolus	Assembly of ribosome subunits	None	✔	✔	✔	✔
Ribosome	Protein synthesis	✔	✔	✔	✔	✔
Endoplasmic reticulum (ER)	Initial modification of polypeptide chains; lipid synthesis	None	✔	✔	✔	✔
Golgi body	Final modification of proteins, lipid assembly, and packaging of both for use inside cell or export	None	✔	✔	✔	✔
Lysosome	Intracellular digestion	None	✔	✔*	✔*	✔
Mitochondrion	ATP formation	**	✔	✔	✔	✔
Photosynthetic pigments	Light-to-energy conversion	✔*	✔*	None	✔	None
Chloroplast	Photosynthesis; some starch storage	None	✔*	None	✔	None
Central vacuole	Increasing cell surface area; storage	None	None	✔*	✔	None
Bacterial flagellum	Locomotion through fluid surroundings	✔*	None	None	None	None
Eukaryotic flagellum or cilium	Locomotion through or motion within fluid surroundings	None	✔*	✔*	✔*	✔
Cytoskeleton	Cell shape; internal organization; basis of cell movement and, in many cells, locomotion	Rudimentary***	✔*	✔*	✔*	✔

* Known to be present in cells of at least some groups.

** Many groups use oxygen-requiring (aerobic) pathways of ATP formation, but mitochondria are not involved.

*** Protein filaments form a simple scaffold that helps support the cell wall in at least some species.

Abnormal Motor Proteins Cause Kartagener Syndrome An abnormal form of a motor protein called dynein causes Kartagener syndrome, a genetic disorder characterized by chronic sinus and lung infections. Biofilms form in the thick mucus that collects in the airways, and the resulting bacterial activities and inflammation cause recurring respiratory infections that damage tissues.

Affected people are usually infertile. Men can produce sperm (Figure 4.27), and some have become fathers after a doctor injects their sperm cells directly into eggs. Review Figure 4.22, then explain how abnormal dynein could cause infertility in males.

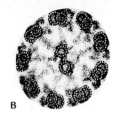
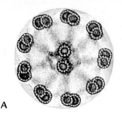

A **B**

Figure 4.27 Cross-section of the flagellum of a sperm cell from **A** a human male affected by Kartagener syndrome and **B** an unaffected male.

Self-Quiz   Answers in Appendix III

1. The **Cell** is the smallest unit of life.

2. Every cell is descended from another cell. This idea is part of _____ .
   a. evolution
   b. the theory of heredity
   c. the cell theory
   d. cell biology

3. True or false? Some protists have no nucleus.

4. Cell membranes consist mainly of a _____ .
   a. carbohydrate bilayer and proteins
   b. protein bilayer and phospholipids
   c. lipid bilayer and proteins

5. Unlike eukaryotic cells, bacterial cells _____ .
   a. have no plasma membrane   c. have no nucleus
   b. have RNA but not DNA   d. a and c

6. In a lipid bilayer, _____ of all the lipid molecules are sandwiched between all the _____ .
   a. hydrophilic tails; hydrophobic heads
   b. hydrophilic heads; hydrophilic tails
   c. hydrophobic tails; hydrophilic heads
   d. hydrophobic heads; hydrophilic tails

7. Enzymes contained in _____ break down worn-out organelles, bacteria, and other particles.

8. Put the following structures in order according to the pathway of a secreted protein:
   a. plasma membrane   c. endoplasmic reticulum
   b. Golgi bodies   d. post-Golgi vesicles

9. The main function of the endomembrane system is building and modifying _____ and _____ .

10. Is this statement true or false? The plasma membrane is the outermost component of all cells. Explain.

11. Most membrane functions are performed by _____ .
   a. proteins   c. nucleic acids
   b. phospholipids   d. hormones

12. No animal cell has a _____ .
   a. plasma membrane   c. lysosome
   b. flagellum   d. cell wall

13. _____ connect the cytoplasm of plant cells.
   a. Plasmodesmata   c. Tight junctions
   b. Adhering junctions   d. a and b

14. Match each cell component with its function.
   ___ mitochondrion   a. protein synthesis
   ___ chloroplast   b. associates with
   ___ ribosome   ribosomes
   ___ smooth ER   c. ATP production
   ___ Golgi body   d. sorts and ships
   ___ rough ER   e. assembles lipids
                            f. photosynthesis

Additional questions are available on CENGAGENOW.

Critical Thinking

1. In a classic episode of *Star Trek*, a gigantic amoeba engulfs an entire starship. Spock blows the cell to bits before it reproduces. Think of at least one problem a biologist would have with this particular scenario.

2. A student is examining different samples with a transmission electron microscope. She discovers a single-celled organism swimming in a freshwater pond (*below*). Which of this organism's structures can you identify? What type of cell do you think this is? Be as specific as you can. Look ahead to Section 20.3 to check your answers.

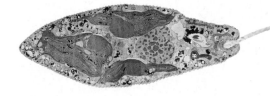

Animations and Interactions on CENGAGENOW:
❯ Overview of cells; Surface-to-volume ratio; Microscopes; Lipid bilayer; Membrane proteins; Bacterial structure; Eukaryotic organelles; The nuclear envelope; Endomembrane system; Mitochondria; Chloroplasts; Cytoskeletal elements; Motor proteins; Eukaryotic flagella and cilia; Plant cell walls; Animal cell junctions.

Key Concepts

Energy Flow
Organisms maintain their organization only by continually harvesting energy from their environment. ATP couples reactions that release usable energy with reactions that require it.

How Enzymes Work
Enzymes tremendously increase the rate of metabolic reactions. Cofactors assist enzymes, and environmental factors such as temperature, salt, and pH can influence enzyme function.

5 Ground Rules of Metabolism

5.1 A Toast to Alcohol Dehydrogenase

The next time someone asks you to have a drink, stop for a moment and think about the cells in your body that detoxify alcohol. It makes no difference whether a person drinks a bottle of beer, a glass of wine, or 1–1/2 ounces of vodka. Each holds the same amount of alcohol or, more precisely, ethanol. Ethanol molecules move quickly from the stomach and small intestine into the bloodstream. Almost all of the ethanol ends up in the liver, which is a large organ in the abdomen. The liver has impressive numbers of enzymes. One of them, alcohol dehydrogenase, helps rid the body of ethanol and other toxins (Figure 5.1).

Ethanol harms the liver. For one thing, breaking it down produces molecules that directly damage liver cells, so the more a person drinks, the fewer liver cells are left to do the breaking down. Ethanol also interferes with normal processes of metabolism. For example, oxygen that would normally take part in breaking down fatty acids is diverted to breaking down ethanol, so fats tend to accumulate as large globules in the tissues of heavy drinkers. A common outcome of such interference is alcoholic hepatitis, a disease characterized by inflammation and destruction of liver tissue. Long-term, heavy drinking can also lead to cirrhosis, or scarring of the liver. (The term cirrhosis is from the Greek *kirros*, meaning orange-colored, after the abnormal skin color of people with the disease.)

Eventually, the liver of a heavy drinker may just quit working, with dire health consequences. The liver is the largest gland in the human body, and it has many important functions. In addition to breaking down fats and toxins, it helps regulate the body's blood sugar level, and it makes proteins that are essential for blood clotting, immune function, and maintaining the solute balance of body fluids.

Heavy drinking can be dangerous in the short term too. Binge drinking, a self-destructive behavior that involves consuming large amounts of alcohol in a brief period of time, is currently the most serious drug problem on college campuses in the United States. Tens of thousands of undergraduate students have been polled about their drinking habits in recent surveys. More than half of them reported that they regularly drink five or more alcoholic beverages within a two-hour period.

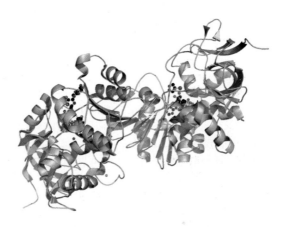

Figure 5.1 Alcohol metabolism. Alcohol dehydrogenase helps the body break down toxic alcohols such as ethanol. This enzyme makes it possible for humans to drink beer, wine, and other alcoholic beverages.

Binge drinking does far more than damage one's liver. Aside from the related 500,000 injuries from accidents, the 600,000 assaults by intoxicated students, 100,000 cases of date rape, and 400,000 incidences of unprotected sex among students, binge drinking is responsible for killing or causing the death of more than 1,700 college students every year. Ethanol is toxic: If you put more of it into your body than your enzymes can deal with, then you will die. With this sobering example, we invite you to learn about how and why your cells break down organic compounds, including toxic molecules such as ethanol.

The Nature of Metabolism
Metabolic pathways are energy-driven sequences of enzyme-mediated reactions. They build, convert, and dispose of materials in cells. Controls that govern steps in metabolic pathways can quickly shift cell activities.

Movement of Fluids
Gradients drive the directional movements of substances across membranes. Water tends to diffuse across selectively permeable membranes, including cell membranes, to regions where solute concentration is higher.

Membrane Trafficking
Transport proteins that work with or against gradients adjust or maintain solute concentrations. Large packets of substances move across the plasma membrane by processes of endocytosis and exocytosis.

> Energy input into living cells drives assembly of the molecules of life.

❮ Links to Life's organization 1.2, Energy 1.3, Laws of nature 1.9, Chemical bonding 2.4, ATP 3.7

Energy Disperses

We define **energy** as the capacity to do work, but this definition is not very satisfying. Even the brilliant physicists who study it cannot say what energy is, exactly. However, even without a perfect definition, we have an intuitive understanding of energy just by thinking about familiar forms of it, such as light, heat, electricity, and motion. We also understand intuitively that one form of energy can be converted to another. Think about how a light-bulb changes electricity into light, or how an automobile changes gasoline into the energy of motion, which is also called **kinetic energy** (Figure 5.2).

The formal study of heat and other forms of energy is called thermodynamics (*therm* is a Greek word for heat; *dynam* means energy). By making careful measurements, thermodynamics researchers discovered that the total amount of energy before and after every conversion is always the same. In other words, energy cannot be created or destroyed—a phenomenon that we call the **first law of thermodynamics**. Remember, a law of nature describes something that occurs without fail, but our explanation of why it occurs is incomplete (Section 1.9).

Energy also tends to spread out, or disperse, until no part of a system holds more than another part. In a kitchen, for example, heat always flows from a hot pan to cool air until the temperature of both is the same. We

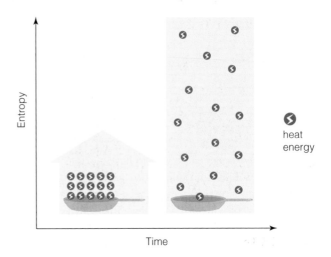

Figure 5.3 Entropy. Entropy tends to increase, but the total amount of energy in any system always stays the same.

never see cool air raising the temperature of a hot pan. **Entropy** is a measure of how much the energy of a particular system has become dispersed. Let's use the hot pan in a cool kitchen as an example of a system. As heat flows from the pan into the air, the entropy of the system increases (Figure 5.3). Entropy continues to increase until the heat is evenly distributed throughout the kitchen, and there is no longer a net (or overall) flow of heat from one area to another. Our system has now reached its maximum entropy with respect to heat. The tendency of entropy to increase is the **second law of thermodynamics.** This is just a fancy way of saying that energy tends to spread out spontaneously.

Biologists use the concept of entropy as it applies to chemical bonding, because energy flow in living things occurs mainly by the making and breaking of chemical bonds. How is entropy related to chemical bonding? Think about it just in terms of motion. Two unbound atoms can vibrate, spin, and rotate in every direction, so they are at high entropy with respect to motion. A covalent bond between the atoms restricts their movement, so they are able to move in fewer ways than they did before bonding. Thus, the entropy of two atoms decreases when a bond forms between them. Such entropy changes are part of the reason why some reactions occur spontaneously and others require an energy input, as you will see in the next section.

Energy's One-Way Flow

Work occurs as a result of an energy transfer. For example, making ATP is work, so it requires energy. A plant cell can do this work by transferring energy from the environment (light) to molecules that use the energy to

Figure 5.2 Demonstration of a familiar type of energy: motion, or kinetic energy.

Figure 5.4 It takes more than 10,000 pounds of soybeans and corn to raise a 1,000-pound steer. Where do the other 9,000 pounds go? About half of the steer's food is indigestible. The animal's body breaks down molecules in the remaining half to access energy stored in chemical bonds. Only about 15% of that energy goes toward building body mass. The rest is lost during energy conversions, as heat.

build ATP. This particular transfer involves an energy conversion (light energy to chemical energy). Most other types of cellular work occur by the transfer of chemical energy from one molecule to another. For example, cells make glucose by transferring chemical energy from ATP to other molecules.

As you learn about such processes, remember that every time energy is transferred, a bit of it disperses. The energy lost from the transfer is usually in the form of heat. As a simple example, a typical incandescent light-bulb converts about 5 percent of the energy of electricity into light. The remaining 95 percent of the energy ends up as heat that radiates from the bulb.

Dispersed heat is not very useful for doing work, and it is not easily converted to a more useful form of energy (such as electricity). Because some of the energy in every transfer disperses as heat, and heat is not useful for doing work, we can say that the total amount of energy available for doing work in the universe is always decreasing.

Is life an exception to this inevitable flow? An organized body is hardly dispersed. Energy becomes concentrated in each new organism as the molecules of life organize into cells. Even so, living things constantly use energy to grow, to move, to acquire nutrients, to reproduce, and so on. Inevitable losses occur during the energy transfers that maintain life (Figure 5.4). Unless those losses are replenished with energy from another source, the complex organization of life will end.

Most of the energy that fuels life on Earth comes from the sun. In our world, energy flows from the sun, through

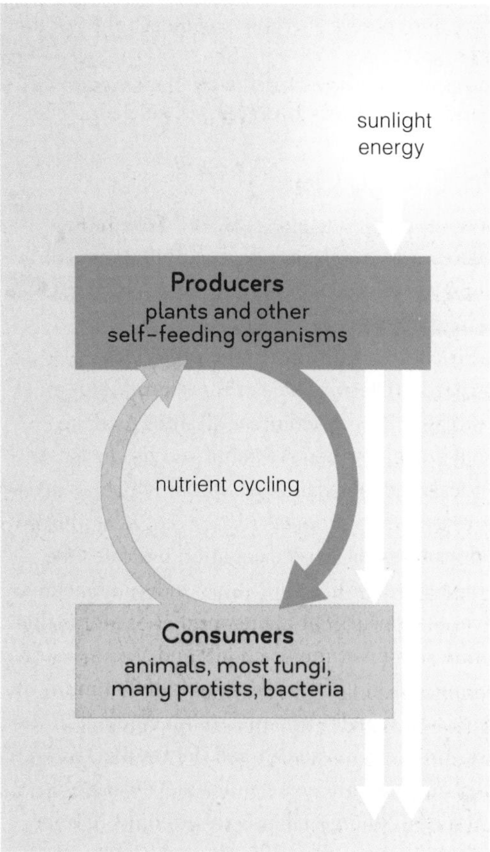

sunlight energy

Producers
plants and other self–feeding organisms

nutrient cycling

Consumers
animals, most fungi, many protists, bacteria

Energy In
Sunlight energy reaches environments on Earth. Producers in those environments capture some of the energy and convert it to other forms that can drive cellular work.

Energy Out
With each conversion, some energy escapes into the environment, mainly as heat. Living things do not use heat to drive cellular work, so the energy flows only in one direction.

Figure 5.5 **Animated** Energy flows from the environment into living organisms, and then back to the environment. The flow drives a cycling of materials among producers and consumers.

producers, then consumers (Figure 5.5). During this journey, the energy is transferred many times. With each transfer, some energy escapes as heat until, eventually, all of it is permanently dispersed. However, the second law of thermodynamics does not say how quickly the dispersal has to happen. Energy's spontaneous dispersal is resisted by chemical bonds. The energy in chemical bonds is a type of **potential energy**, because it can be stored. Think of all the bonds in the countless molecules that make up your skin, heart, liver, fluids, and other body parts. Those bonds hold the molecules, and you, together—at least for the time being.

---

energy The capacity to do work.
entropy Measure of how much the energy of a system is dispersed.
first law of thermodynamics Energy cannot be created or destroyed.
kinetic energy The energy of motion.
potential energy Stored energy.
second law of thermodynamics Energy tends to disperse spontaneously.

Take-Home Message What is energy?

> Energy is the capacity to do work. It can be transferred between systems or converted from one form to another, but it cannot be created or destroyed.

> Energy disperses spontaneously.

> Some energy is lost during every transfer or conversion.

> Organisms can maintain their complex organization only as long as they replenish themselves with energy they harvest from someplace else.

› All cells store and retrieve energy in chemical bonds of the molecules of life.
› ATP functions as an energy carrier in cellular reactions.
‹ Links to Bonding 2.4, Carbohydrates 3.3, Nucleotides 3.7

Energy In, Energy Out

Cells store energy in chemical bonds, and access the energy stored in chemical bonds by breaking them. Both processes change molecules. Any process by which such chemical change occurs is called a **reaction**.

During a reaction, one or more **reactants** (molecules that enter a reaction) become one or more **products** (molecules that remain at the reaction's end). Intermediate molecules may form between reactants and products. We show a chemical reaction as an equation in which an arrow points from reactants to products (Figure 5.6). A number before a chemical formula in the equation indicates the number of molecules. Atoms shuffle around in a reaction, but they never disappear: The same number of atoms that enter a reaction remain at the reaction's end.

Every chemical bond holds energy, and the amount of energy depends on which elements are taking part in the bond. For example, the covalent bond between an oxygen and hydrogen atom in any water molecule always holds the same amount of energy. That is the amount of energy required to break the bond, and it is also the amount of energy released when the bond forms. Bond energy and entropy both contribute to a molecule's free energy, which is the amount of energy that is available (free) to do work.

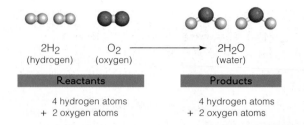

$$2H_2 \text{ (hydrogen)} \quad O_2 \text{ (oxygen)} \longrightarrow 2H_2O \text{ (water)}$$

Reactants	Products
4 hydrogen atoms + 2 oxygen atoms	4 hydrogen atoms + 2 oxygen atoms

Figure 5.6 Animated Chemical bookkeeping. In equations that represent chemical reactions, reactants are written to the left of an arrow that points to the products. A number before a formula indicates the number of molecules. Atoms shuffle around in a reaction, but they never disappear: The same number of atoms that enter a reaction remain at the reaction's end.

In most reactions, the free energy of reactants differs from the free energy of products (Figure 5.7). Reactions in which reactants have less free energy than products will not proceed without a net energy input. Such reactions are **endergonic**, which means "energy in" ❶. In other reactions, reactants have greater free energy than products. Such reactions are **exergonic**, which means "energy out," because they end with a net release of free energy ❷. Cells access the free energy of molecules by running exergonic reactions. An example is the overall process of aerobic respiration, which converts glucose and oxygen to carbon dioxide and water for a net energy output.

Why Earth Does Not Go Up in Flames

The molecules of life release energy when they combine with oxygen. For example, think of how a spark ignites tinder-dry wood in a campfire. Wood is mostly cellulose, which is a carbohydrate that consists of long chains of repeating glucose units. The spark initiates a reaction that converts cellulose in the wood and oxygen in air to water and carbon dioxide. The reaction is exergonic, and it releases enough energy to initiate the same reaction with other cellulose and oxygen molecules. That is why a campfire keeps burning once it has been lit.

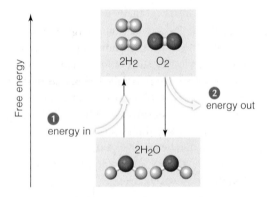

Figure 5.7 Energy inputs and outputs in chemical reactions.

❶ Endergonic reactions convert molecules with lower energy to molecules with higher energy, so they require a net energy input in order to proceed.

❷ Exergonic reactions convert molecules with higher energy to molecules with lower energy, so they end with a net energy output.

›› **Figure It Out** Which law of thermodynamics explains energy inputs and outputs in chemical reactions?

Answer: The first law

activation energy Minimum amount of energy required to start a reaction.
ATP Adenosine triposphate; an energy carrier that couples endergonic with exergonic reactions in cells.
ATP/ADP cycle Process by which cells regenerate ATP. ADP forms when ATP loses a phosphate group, then ATP forms again as ADP gains a phosphate group.
endergonic Type of reaction that requires a net input of free energy to proceed.
exergonic Type of reaction that ends with a net release of free energy.
phosphorylation Addition of a phosphate group to a molecule; occurs by the transfer of a phosphate group from a donor molecule such as ATP.
product A molecule that remains at the end of a reaction.
reactant Molecule that enters a reaction.
reaction Process of chemical change.

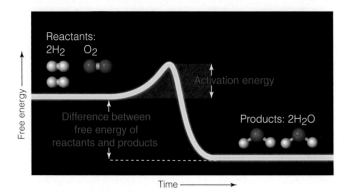

Figure 5.8 **Animated** Activation energy. Most reactions will not begin without an input of activation energy, which is shown here as a bump in an energy hill. In this example, the reactants have more energy than the products. Activation energy keeps this and other exergonic reactions from starting spontaneously.

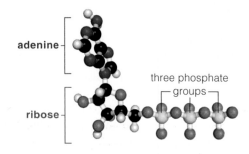

A Structure of ATP.

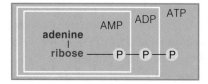

B After ATP loses one phosphate group, the nucleotide is ADP (adenosine diphosphate); after losing two phosphate groups, it is AMP (adenosine monophosphate).

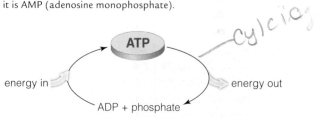

cylcic

C ATP forms by endergonic reactions. ADP forms again when ATP energy is transferred to another molecule along with a phosphate group. Energy from such transfers drives cellular work.

Figure 5.9 ATP, the energy currency of cells.

Earth is rich in oxygen—and in potential exergonic reactions. Why doesn't it burst into flames? Luckily, energy is required to break the chemical bonds of reactants, even in an exergonic reaction. **Activation energy** is the minimum amount of energy that will get a chemical reaction started (Figure 5.8). Activation energy is a bit like a hill that reactants must climb before they can coast down the other side to products.

Both endergonic and exergonic reactions have activation energy, but the amount varies with the reaction. Consider guncotton (nitrocellulose), a highly explosive derivative of cellulose. Christian Schönbein accidentally discovered a way to manufacture it when he used his wife's cotton apron to wipe up a nitric acid spill on his kitchen table, then hung it up to dry next to the oven. The apron exploded. Being a chemist in the 1800s, Schönbein immediately thought of marketing guncotton as a firearm explosive, but it proved to be too unstable to manufacture. So little activation energy is needed to make guncotton react with oxygen that it tends to explode unexpectedly. The substitute? Gunpowder, which has a higher activation energy for a reaction with oxygen.

ATP—The Cell's Energy Currency

ATP Cells pair reactions that require energy with reactions that release energy. ATP and other nucleotides are often part of that process. **ATP**, or adenosine triphosphate, functions as an energy carrier by accepting energy released by exergonic reactions, and delivering energy to endergonic reactions. ATP is the main currency in a cell's energy economy, so we use a cartoon coin to symbolize it.

ATP is a nucleotide with three phosphate groups (Figure 5.9A). The bonds that link those phosphate groups hold a lot of energy. When a phosphate group is trans-

ferred from ATP to another molecule, energy is transferred along with the phosphate. That energy contributes to the "energy in" part of an endergonic reaction. The transfer of a phosphate group is called **phosphorylation**.

Cells constantly use up ATP to drive endergonic reactions, so they constantly replenish it. When ATP loses a phosphate, ADP (adenosine diphosphate) forms (Figure 5.9B). ATP forms again when ADP binds phosphate in an endergonic reaction. The cycle of using and replenishing ATP is called the **ATP/ADP cycle** (Figure 5.9C).

> ### Take-Home Message How do cells use energy?
>
> ❯ Cells store and retrieve energy by making and breaking chemical bonds.
>
> ❯ Activation energy is the minimum amount of energy required to start a chemical reaction.
>
> ❯ Endergonic reactions cannot run without a net input of energy. Exergonic reactions end with a net release of energy.
>
> ❯ Energy carriers such as ATP couple exergonic reactions with endergonic ones.

❯ Enzymes make specific reactions occur much faster than they would on their own.

❮ Links to Temperature 2.5, pH 2.6, Protein structure 3.5, Denaturation 3.6

Centuries might pass before sugar would break down to carbon dioxide and water on its own, yet that same conversion takes just seconds inside your cells. Enzymes make the difference. In a process called **catalysis**, an enzyme makes a reaction run much faster than it would on its own. The enzyme is unchanged by participating in the reaction, so it can work again and again.

Most enzymes are proteins, but some are RNAs. Each kind recognizes specific reactants, or **substrates**, and alters them in a specific way. For instance, the enzyme hexokinase adds a phosphate group to the hydroxyl group on the sixth carbon of glucose. Such specificity occurs because an enzyme's polypeptide chains fold up into one or more **active sites**, which are pockets where substrates bind and where reactions proceed (Figure 5.10). An active

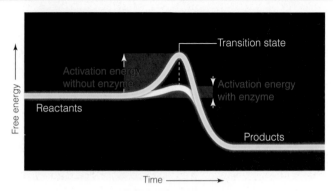

Figure 5.11 Animated An enzyme enhances the rate of a reaction by lowering its activation energy.

❯❯ **Figure It Out** Is this reaction endergonic or exergonic?

Answer: Exergonic

site is complementary in shape, size, polarity, and charge to the enzyme's substrate. This fit is the reason why each enzyme acts in a specific way on specific substrates.

When we talk about activation energy, we are really talking about the energy required to break the bonds of the reactants. Depending on the reaction, that energy may force substrates close together, redistribute their charge, or cause some other change. The change brings on the transition state, when substrate bonds reach their breaking point and the reaction will run spontaneously to product. Enzymes can help bring on the transition state by lowering activation energy (Figure 5.11). They do this by the following four mechanisms, which work alone or in combination.

Helping Substrates Get Together Binding at an active site brings two or more substrates close together. The closer the substrates are to one another, the more likely they are to react.

Orienting Substrates in Positions That Favor Reaction On their own, substrates collide from random directions. By contrast, binding at an active site positions substrates so they align appropriately for a reaction.

Inducing a Fit Between Enzyme and Substrate By the **induced–fit model**, an enzyme's active site is not quite complementary to its substrate. Interacting with a sub-

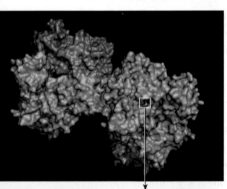

active site

enzyme

A Like other enzymes, hexokinase's active sites bind and alter specific substrates. A model of the whole enzyme is shown to the *left*.

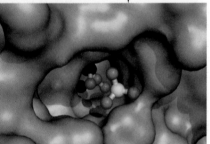

 reactant(s)

B A close-up shows glucose and phosphate meeting inside the enzyme's active site. The microenvironment of the site favors a reaction between the two substrate molecules.

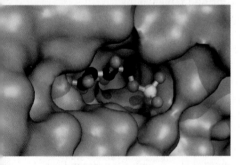

 product(s)

C Here, the glucose has bonded with the phosphate. The product of this reaction, glucose-6-phosphate, is shown leaving the active site.

Figure 5.10 An example of an active site. This one is in a hexokinase, an enzyme that phosphorylates glucose and other six-carbon sugars.

---

active site Pocket in an enzyme where substrates bind and a reaction occurs.

antioxidant Substance that prevents molecules from reacting with oxygen.

catalysis The acceleration of a reaction rate by a molecule that is unchanged by participating in the reaction.

coenzyme An organic molecule that is a cofactor.

cofactor A metal ion or a coenzyme that associates with an enzyme and is necessary for its function.

induced–fit model Substrate binding to an active site improves the fit between the two.

substrate A molecule that is specifically acted upon by an enzyme.

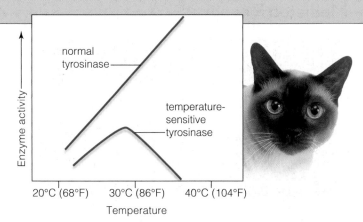

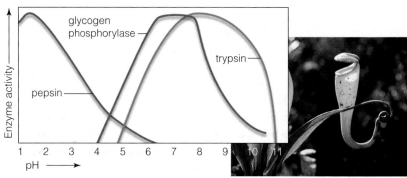

Figure 5.12 **Animated** Enzymes and temperature. Tyrosinase is involved in the production of melanin, a black pigment in skin cells. The form of this enzyme in Siamese cats is inactive above about 30°C (86°F), so the warmer parts of the cat's body end up with less melanin, and lighter fur.

Figure 5.13 Enzymes and pH. *Left*, how pH affects three enzymes. *Right*, carnivorous plants of the genus *Nepenthes* grow in nitrogen-poor habitats. They secrete acids and protein-digesting enzymes into a fluid-filled cup that consists of a modified leaf. The enzymes release nitrogen from insects that are attracted to odors from the fluid and then drown in it. One of these enzymes functions best at pH 2.6.

strate molecule causes the enzyme to change shape so that the fit between them improves. The improved fit may result in a stronger bond between enzyme and substrate, or it may better bring on the substrate's transition state.

Shutting Out Water Molecules Metabolism occurs in water-based fluids, but water molecules can interfere with certain reactions. The active sites of some enzymes repel water, and keep it away from the reactions.

Effects of Temperature, pH, and Salinity

Adding energy in the form of heat boosts free energy, which is one reason why molecular motion increases with temperature. The greater the free energy of reactants, the closer a reaction is to its activation energy. Thus, the rate of an enzymatic reaction typically increases with temperature. However, that rule holds only up to a point. An enzyme denatures above a characteristic temperature. Then, the reaction rate falls sharply as the shape of the enzyme changes and it stops functioning (Figure 5.12).

The pH tolerance of enzymes varies. Most enzymes in the human body have an optimal pH between 6 and 8. For instance, the hexokinase molecule shown in Figure 5.10A is most active in areas of the small intestine where the pH is around 8. Some enzymes work outside the typical range of pH. The enzyme pepsin functions only in stomach fluid, where it breaks down proteins in food. The fluid is very acidic, with a pH of about 2 (Figure 5.13).

An enzyme's activity is also influenced by the amount of salt in the surrounding fluid. Too much or too little salt can disrupt the hydrogen bonding that holds an enzyme in its three-dimensional shape.

Help From Cofactors

Cofactors are atoms or molecules (other than proteins) that associate with enzymes and are necessary for their

function. Some are metal ions. Organic molecules that are cofactors are called **coenzymes**. Vitamin C is a coenzyme, and many B vitamins are remodeled into coenzymes.

Coenzymes in some reactions are tightly bound to an enzyme. In others, they participate as separate molecules. Unlike enzymes, many coenzymes are modified by taking part in a reaction. They typically become regenerated in other reactions. For example, the coenzyme NAD^+ (nicotinamide adenine dinucleotide) becomes NADH by accepting electrons and hydrogen atoms in a reaction. NAD^+ is regenerated when the NADH gives up the electrons and hydrogen atoms in a different reaction.

We can use an enzyme called catalase as an example of how cofactors work. Like hemoglobin, catalase has four hemes. The iron atom at the center of each heme (*right*) is a cofactor. Iron is a metal, and metal atoms affect nearby electrons. Catalase works by holding a substrate molecule close to one of its iron atoms. The iron pulls on the substrate's electrons, which brings on the transition state.

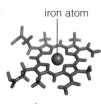

iron atom

a heme

Catalase is an **antioxidant**, which means its actions stop other molecules from reacting with oxygen. Oxygen reactions often produce free radicals: atoms or molecules with unpaired electrons. Free radicals form during normal metabolic reactions, but they are dangerous because they attack the structure of biological molecules. Antioxidants such as catalase are critical to health because they reduce the amount of free radicals that form in the body.

Take-Home Message How do enzymes work?

> Enzymes greatly enhance the rate of specific reactions. Binding at an enzyme's active site causes a substrate to reach its transition state. In this state, the substrate's bonds are at the breaking point, so the reaction can run spontaneously.

> Each enzyme works best at certain temperatures, pH, and salt concentration.

> Cofactors associate with enzymes and assist their function.

> ATP, enzymes, and other molecules interact in organized pathways of metabolism.
< Links to Electrons 2.3, Metabolism 3.2, Amino acids 3.5, Membrane proteins 4.4

Types of Metabolic Pathways

Metabolism, remember, refers to the activities by which cells acquire and use energy as they build and break down organic molecules. Building, rearranging, or breaking down an organic substance often occurs stepwise, in a series of reactions called a **metabolic pathway**.

Some metabolic pathways are linear, meaning that the reactions run straight from reactant to product:

reactant $\xrightarrow{\text{enzyme 1}}$ **intermediate** $\xrightarrow{\text{enzyme 2}}$ **intermediate** $\xrightarrow{\text{enzyme 3}}$ **product**

Other metabolic pathways are cyclic. In a cyclic pathway, the last step regenerates a reactant for the first step:

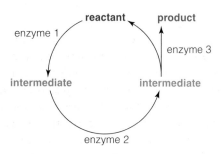

Controls Over Metabolism

Cells conserve energy and resources by making only what they need—no more, no less—at any given moment. How does a cell adjust the types and amounts of molecules it produces? Several mechanisms help a cell maintain, raise,

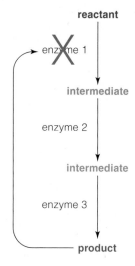

Figure 5.14 Animated Feedback inhibition. In this example, three kinds of enzymes act in sequence to convert a substrate to a product, which inhibits the activity of the first enzyme.

>> **Figure It Out** Is this an example of a cyclic or a linear metabolic pathway?

Answer: This is a linear pathway.

or lower its production of thousands of different substances. For example, reactions do not only run from reactants to products. Many also run in reverse at the same time, with some of the products being converted back to reactants. The rates of the forward and reverse reactions often depend on the concentrations of reactants and products: A high concentration of reactants pushes the reaction in the forward direction, and a high concentration of products pushes it in the reverse direction.

Other mechanisms more actively regulate enzymes. Certain molecules in a cell govern how fast enzyme molecules are made, or influence the activity of enzymes that have already been built. For example, the end product of a series of enzymatic reactions may inhibit the activity of one of the enzymes in the series, an effect called **feedback inhibition** (Figure 5.14).

Some regulatory molecules or ions activate or inhibit an enzyme by binding directly to its active site. Others bind to **allosteric** sites, which are regions of an enzyme (other than the active site) where regulatory molecules

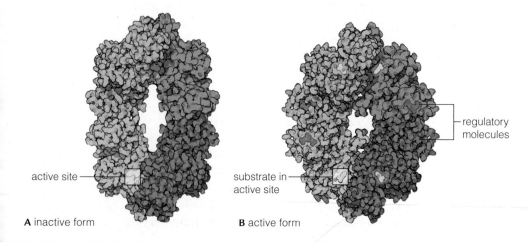

A inactive form　　**B** active form

Figure 5.15 Animated Allosteric effects.

A Pyruvate kinase is an enzyme that consists of four identical polypeptide chains. Each chain has an active site and a binding site for a regulatory molecule.

B When regulatory molecules (*red*) bind to the allosteric sites, the overall shape of the enzyme changes. The change makes the enzyme functional. Here, substrate molecules (*yellow*) are bound to the active sites. Small green balls are metal cofactors.

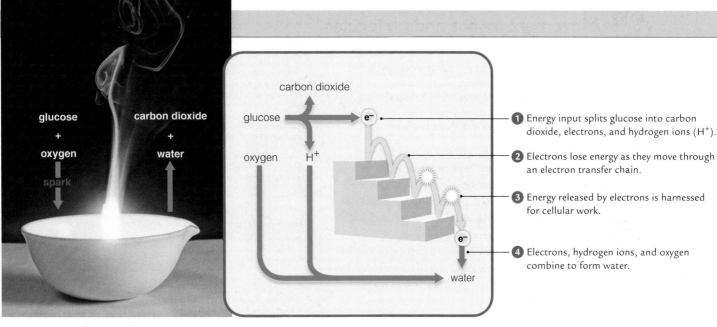

A Glucose and oxygen react (burn) when exposed to a spark. Energy is released all at once as light and heat when CO_2 and water form.

B The same overall reaction occurs in small steps with an electron transfer chain. Energy is released in amounts that cells can harness for cellular work.

① Energy input splits glucose into carbon dioxide, electrons, and hydrogen ions (H^+).

② Electrons lose energy as they move through an electron transfer chain.

③ Energy released by electrons is harnessed for cellular work.

④ Electrons, hydrogen ions, and oxygen combine to form water.

Figure 5.16 Animated Comparing uncontrolled and controlled energy release.

bind. *Allo–* means other, and *steric* means structure. Binding of an allosteric regulator alters the shape of the enzyme in a way that enhances or inhibits its function (Figure 5.15).

Redox Reactions

The bonds of organic molecules hold a lot of energy that can be released by a reaction with oxygen. One type of reaction with oxygen—burning—releases the energy of organic molecules all at once, explosively (Figure 5.16A).

Cells use oxygen to break the bonds of organic molecules. However, they have no way to harvest an explosive burst of energy. Instead, they break the molecules apart in several steps that release the energy in small, manageable increments. Most of these steps are oxidation–reduction reactions, or redox reactions for short. In a **redox reaction**, one molecule accepts electrons (it becomes reduced) from another molecule (which becomes oxidized). Redox

reactions are also called electron transfers. To remember what reduced means, think of how the negative charge of an electron "reduces" the charge of a recipient molecule.

In the next two chapters, you will learn about the importance of redox reactions in electron transfer chains. An **electron transfer chain** is an organized series of reaction steps in which membrane-bound arrays of enzymes and other molecules give up and accept electrons in turn. Electrons are at a higher energy level when they enter a chain than when they leave. Electron transfer chains can harvest the energy given off by an electron as it drops to a lower energy level (Figure 5.16B).

Many coenzymes deliver electrons to electron transfer chains in photosynthesis and aerobic respiration. Energy released at certain steps in those chains helps drive the synthesis of ATP. These pathways will occupy our attention in chapters to come.

---

allosteric Describes a region of an enzyme other than the active site that can bind regulatory molecules.
electron transfer chain Array of enzymes and other molecules that accept and give up electrons in sequence, thus releasing the energy of the electrons in usable increments.
feedback inhibition Mechanism by which a change that results from some activity decreases or stops the activity.
metabolic pathway Series of enzyme-mediated reactions by which cells build, remodel, or break down an organic molecule.
redox reaction Oxidation–reduction reaction in which one molecule accepts electrons (it becomes reduced) from another molecule (which becomes oxidized). Also called electron transfer.

Take-Home Message What is a metabolic pathway?

❭ A metabolic pathway is a sequence of enzyme-mediated reactions that builds, rearranges, or breaks down an organic molecule.

❭ Control mechanisms enhance or inhibit the activity of many enzymes. The adjustments help cells produce only what they require in any given interval.

❭ Many metabolic pathways involve oxidation–reduction (redox) reactions, which are also called electron transfers.

❭ Redox reactions occur in electron transfer chains, which are important processes of energy exchange in photosynthesis and aerobic respiration.

> Ions and molecules tend to move from one region to another in response to concentration gradients.
> Water diffuses across cell membranes by osmosis.
‹ Links to Homeostasis 1.3, Ions and molecules 2.3, Solutes 2.5, Cell membrane structure 4.2, Plant cell walls 4.11

Understanding how metabolism works in cells begins with the behavior of solutions. How much of a solute is dissolved in a given amount of fluid is the solute's **concentration**. A difference in solute concentration between adjacent regions of solution is a **concentration gradient**.

Solute molecules or ions tend to move "down" their concentration gradient, from a region of higher concentration to one of lower concentration. Why? Molecules and ions are always in motion; they collide at random and bounce off one another millions of times each second. The more crowded they are, the more often they collide. Rebounds from the collisions propel molecules away from one another. Thus, during any given interval, more molecules get bumped out of a region of higher concentration than get bumped into it.

Diffusion (*left*) is the net movement of molecules or ions in response to a concentration gradient. It is an essential way in which substances move into, through, and out of cells. The rate of diffusion depends on five factors:

1. *Size*. It takes more energy to move a large molecule than it does to move a small one. Thus, smaller molecules diffuse more quickly than larger ones.
2. *Temperature*. Molecules move faster at higher temperature, so they collide more often. Thus, the higher the temperature, the faster the diffusion.
3. *Steepness of the concentration gradient*. The rate of diffusion is higher with steeper gradients, because molecules collide more often in a region of greater concentration.
4. *Charge* can affect the rate and direction of diffusion between two regions. Each ion or charged molecule in a fluid contributes to the fluid's overall electric charge. A difference in charge between two regions can affect the rate and direction of diffusion between them. For exam-

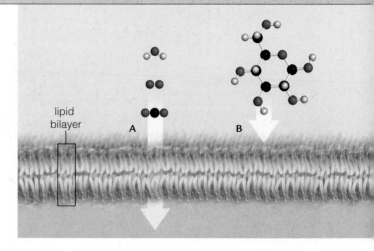

Figure 5.18 Animated Selective permeability of lipid bilayers. **A** Hydrophobic molecules, gases, and water molecules can cross a lipid bilayer on their own. **B** Ions in particular and most polar molecules such as glucose cannot; they cross cell membranes only with the help of transport proteins in the bilayer.

ple, positively charged substances, such as sodium ions, diffuse toward a region with an overall negative charge.
5. *Pressure*. Diffusion may be affected by a difference in pressure between two adjoining regions. Pressure squeezes molecules together, and molecules that are more crowded collide and rebound more frequently. Thus, diffusion occurs faster at higher pressures.

Diffusion Across Membranes

Tonicity refers to the total concentration of solutes in fluids separated by a selectively permeable membrane (such membranes allow some substances, but not others, to cross). When the overall solute concentrations of the two fluids differ, the fluid with the lower overall concentration of solutes is said to be **hypotonic** (*hypo–*, under). The other one, with the higher overall solute concentration, is **hypertonic** (*hyper–*, over). Fluids that are **isotonic** have the same overall solute concentration.

A typical selectively permeable membrane allows water to cross it. When this type of membrane separates two fluids that are not isotonic, water will diffuse across the membrane, and move from the hypotonic fluid into the hypertonic one (Figure 5.17). The diffusion will continue until the two fluids are isotonic, or until some

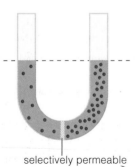 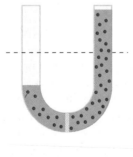

selectively permeable
membrane

Figure 5.17 Animated Osmosis. Water moves across a selectively permeable membrane that separates two fluids with differing concentration. The fluid volume changes in the two compartments as water follows its gradient and diffuses across the membrane.

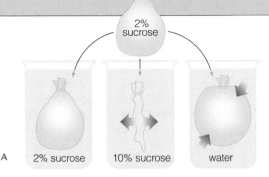

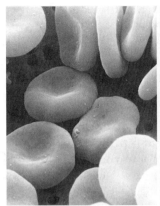

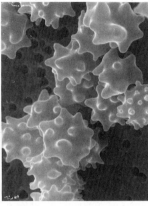

A

2% sucrose   10% sucrose   water

Figure 5.19 Animated Tonicity.

A Experiment that shows what happens when a selectively permeable membrane bag is immersed in solutions of different tonicity.

B–D The micrographs show human red blood cells immersed in fluids of different tonicity.

❯❯ **Figure It Out** Which of the three solutions in A is hypertonic with respect to the fluid in the bag?

Answer: The 10% sucrose solution

B Red blood cells immersed in an isotonic solution do not change in volume. The fluid portion of blood is typically isotonic with cytoplasm.

C Red blood cells immersed in a hypertonic solution shrivel up because more water diffuses out of the cells than into them.

D Red blood cells immersed in a hypotonic solution swell up because more water diffuses into the cells than out of them.

pressure against the hypertonic fluid counters it. The diffusion of water across a membrane is so important in biology that it is given a special name: **osmosis**.

A lipid bilayer is one type of selectively permeable membrane that ions and most polar molecules cannot cross (Figure 5.18). If a cell's cytoplasm is hypertonic with respect to the fluid outside of its plasma membrane, water diffuses into it. If the cytoplasm is hypotonic with respect to the fluid on the outside, water diffuses out. In either case, the solute concentration of the cytoplasm may change. If it changes enough, the cell's enzymes will stop working, with lethal results. Most free-living cells, and some that are part of multicelled organisms, have built-in mechanisms that compensate for differences in tonicity between cytoplasm and external fluid. In cells with no such mechanism, the volume—and solute concentration—of cytoplasm will change as water diffuses into or out of the cell (Figure 5.19).

concentration Number of molecules or ions per unit volume.
concentration gradient Difference in concentration between adjoining regions of fluid.
diffusion Net movement of molecules or ions from a region where they are more concentrated to a region where they are less so.
hypertonic Describes a fluid that has a high overall solute concentration relative to another fluid.
hypotonic Describes a fluid that has a low overall solute concentration relative to another fluid.
isotonic Describes two fluids with identical solute concentrations.
osmosis The diffusion of water across a selectively permeable membrane in response to a differing overall solute concentration.
osmotic pressure Amount of turgor that prevents osmosis into cytoplasm or other hypertonic fluid.
turgor Pressure that a fluid exerts against a wall, membrane, or other structure that contains it.

Turgor

Cell walls of plants and many protists, fungi, and bacteria can resist an increase in the volume of cytoplasm even in hypotonic environments. In the case of plant cells, cytoplasm usually contains more solutes than soil water does. Thus, water usually diffuses from soil into a plant—but only up to a point. Rigid walls keep plant cells from expanding very much. Thus, osmosis causes pressure to build up inside of these cells. Pressure that a fluid exerts against a structure that contains it is called **turgor**. When enough pressure builds up inside a plant cell, water stops diffusing into its cytoplasm. The amount of turgor that stops osmosis is called **osmotic pressure**.

Osmotic pressure keeps walled cells plump, just as high air pressure inside a tire keeps it inflated. A young land plant can resist gravity to stay erect because its cells are plump with cytoplasm. When soil dries out, it loses water but not solutes, so the concentration of solutes increases in whatever water remains in it. If soil water becomes hypertonic with respect to cytoplasm, water diffuses out of the plant's cells, so their cytoplasm shrinks. As turgor inside the cells decreases, the plant wilts.

Take-Home Message What influences the movement of ions and molecules?

❯ Molecules or ions tend to diffuse into an adjoining region of fluid in which they are not as concentrated.

❯ The steepness of a concentration gradient as well as temperature, molecular size, charge, and pressure affect the rate of diffusion.

❯ Osmosis is a net diffusion of water between two fluids that differ in water concentration and are separated by a selectively permeable membrane.

❯ Fluid pressure that a solution exerts against a membrane or wall influences the osmotic movement of water.

> Many types of molecules and ions diffuse across a lipid bilayer only with the help of transport proteins.
< Link to Transport proteins 4.4

You learned in the previous section that gases, water, and small nonpolar molecules can diffuse directly across a lipid bilayer. Most other molecules, and ions in particular, cross only with the help of membrane transport proteins. Each type of transport protein can move a specific ion or molecule. Glucose transporters only transport glucose; calcium pumps only pump calcium; and so on. The specificity of transport proteins means that the amounts and types of substances that cross a membrane depend on which transport proteins are embedded in it. It also allows cells to control the volume and composition of

their fluid interior by moving particular solutes one way or the other across their membranes.

For example, a glucose transporter in a plasma membrane can bind to a molecule of glucose, but not to a molecule of phosphorylated glucose. Enzymes in cytoplasm phosphorylate glucose as soon as it enters the cell. Phosphorylation prevents the molecule from moving back through the glucose transporter and leaving the cell.

Passive Transport

In **passive transport**, the movement of a solute (and the direction of the movement) through a transport protein is driven entirely by the solute's concentration gradient. For this reason, passive transport is also called facilitated diffusion. The solute simply binds to the passive transport protein, and the protein releases it to the other side of the membrane (Figure 5.20).

A glucose transporter is an example of a passive transport protein ❶. This protein changes shape when it binds to a molecule of glucose ❷. The shape change moves the solute to the opposite side of the membrane, where it detaches. Then, the transporter reverts to its original shape ❸. Some passive transporters do not change shape; they form permanently open channels through a membrane. Others are gated, which means they open and close in response to a stimulus such as a shift in electric charge or binding to a signaling molecule.

Active Transport

Solute concentrations may shift in extracellular fluid or in cytoplasm. Maintaining a particular solute's concentration at a certain level often means transporting the solute against its gradient, to the side of a membrane where it is more concentrated. Pumping a solute against its gradient

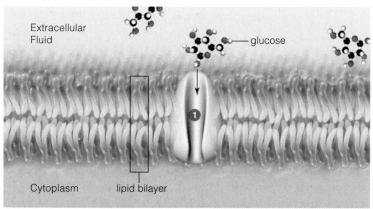

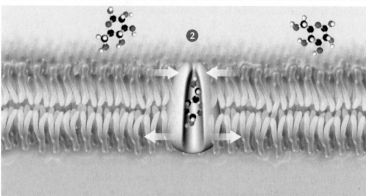

Extracellular Fluid

glucose

Cytoplasm

lipid bilayer

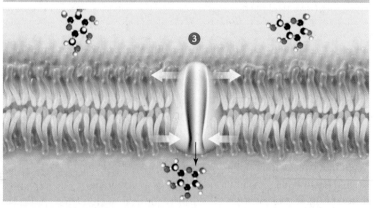

Figure 5.20 Animated Passive transport. This model shows a glucose transporter.

❶ A glucose molecule (here, in extracellular fluid) binds to a glucose transporter in the plasma membrane.

❷ Binding causes the transport protein to change shape.

❸ The glucose molecule detaches from the transport protein on the other side of the membrane (here, in cytoplasm), and the protein resumes its original shape.

>> **Figure It Out** In this example, which fluid is hypotonic: extracellular fluid or the cytoplasm?

Answer: Cytoplasm

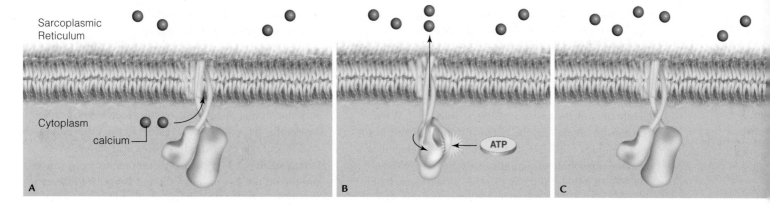

Sarcoplasmic Reticulum

Cytoplasm

calcium

A

ATP

B

C

Figure 5.21 **Animated** Active transport. This model shows a calcium pump embedded in a muscle cell membrane. **A** Two calcium ions bind to the transport protein. **B** Energy in the form of a phosphate group is transferred from ATP to the protein. The transfer causes the protein to change shape so that it ejects the calcium ions to the opposite side of the membrane. **C** After it loses the calcium ions, the transport protein resumes its original shape.

takes energy. In **active transport**, a transport protein uses energy to pump a solute against its gradient across a cell membrane. After a solute binds to an active transporter, an energy input (often in the form of a phosphate-group transfer from ATP) changes the shape of the protein. The change causes the transporter to release the solute to the other side of the membrane.

A calcium pump is an example of an active transporter. This protein moves calcium ions across cell membranes (Figure 5.21). Calcium ions act as potent messengers inside cells, and many enzymes have allosteric sites that bind these ions. Thus, their presence in cytoplasm is tightly regulated. Calcium pumps in the plasma membrane of all eukaryotic cells keep the concentration of calcium in cytoplasm 10,000 times higher than it is in extracellular fluid.

Cotransporters are active transport proteins that move two substances at the same time, in the same or opposite directions across a membrane. Nearly all of the cells in your body have cotransporters called sodium–potassium pumps (Figure 5.22). Sodium ions (Na+) in the cytoplasm diffuse into the pump's open channel and bind to its interior. A phosphate-group transfer from ATP causes the pump to change shape. Its channel opens to extracellular fluid, where it releases the Na+. Then, potassium ions (K+) from extracellular fluid diffuse into the channel and bind to its interior. The transporter releases the phosphate group and reverts to its original shape. The channel opens to the cytoplasm, where it releases the K+.

Bear in mind, the membranes of all cells, not just those of animals, have active transporters. For example, active transporters in plant leaf cells pump sugars into tubes that distribute them throughout the plant body.

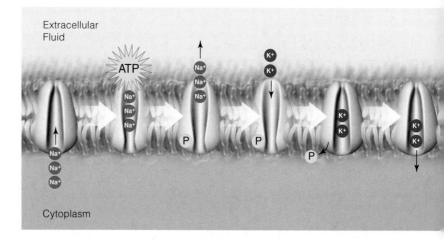

Extracellular Fluid

ATP

Na+
Na+
Na+
Na+

Na+
Na+
Na+

K+
K+

P

P

K+
K+

P

K+
K+

Na+
Na+
Na+

Cytoplasm

Figure 5.22 Cotransport. This model shows how a sodium–potassium pump transports sodium ions (Na+, *red*) from cytoplasm to extracellular fluid, and potassium ions (K+, *purple*) in the other direction across the plasma membrane. A phosphate-group transfer from ATP provides energy for the transport.

active transport Energy-requiring mechanism by which a transport protein pumps a solute across a cell membrane against its concentration gradient.
passive transport Mechanism by which a concentration gradient drives the movement of a solute across a cell membrane through a transport protein. Requires no energy input.

Take-Home Message How do molecules or ions that cannot diffuse through a lipid bilayer cross a cell membrane?

❯ Transport proteins help specific molecules or ions to cross cell membranes.

❯ In passive transport, a solute binds to a protein that releases it on the opposite side of the membrane. The movement is driven by a concentration gradient.

❯ In active transport, a protein pumps a solute across a membrane, against its concentration gradient. The transporter requires an energy input, as from ATP.

❯ By processes of exocytosis and endocytosis, cells take in and expel particles that are too big for transport proteins, as well as substances in bulk.

❮ Links to Lipoproteins 3.5, Vesicles 4.8, Motor proteins 4.10

Think back on the structure of a lipid bilayer. When a bilayer is disrupted, such as when part of the plasma membrane pinches off as a vesicle, it seals itself. Why? The disruption exposes the nonpolar fatty acid tails of the phospholipids to their watery surroundings. Remember, in water, phospholipids spontaneously rearrange themselves so that their tails stay together. When a patch of membrane buds, its phospholipid tails are repelled by water on both sides. The water "pushes" the phospholipid tails together, which helps round off the bud as a vesicle, and also seals the rupture in the membrane.

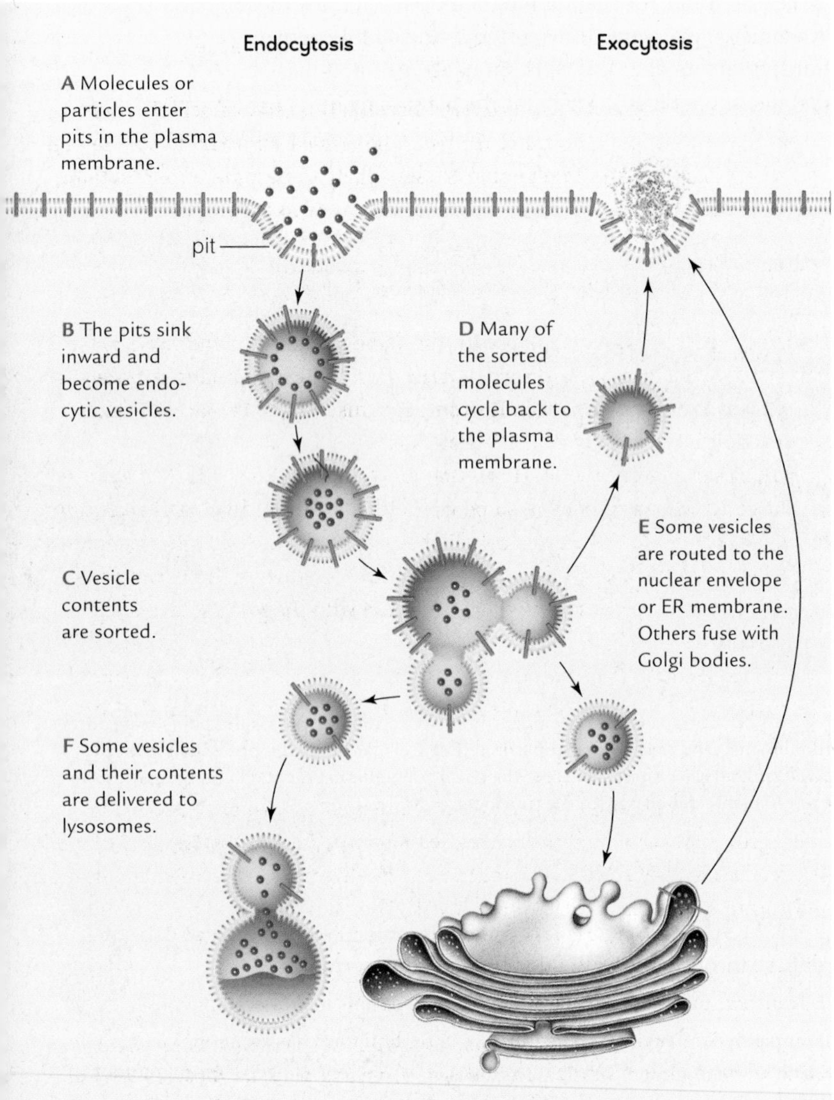

As part of vesicles, patches of membrane constantly move to and from the cell surface (Figure 5.23). The formation and movement of vesicles, which is called membrane trafficking, involves motor proteins and requires ATP.

By **exocytosis**, a vesicle moves to the cell surface, and the protein-studded lipid bilayer of its membrane fuses with the plasma membrane. As the exocytic vesicle loses its identity, its contents are released to the surroundings.

Endocytosis takes up substances near the cell's surface. A small patch of plasma membrane balloons inward, and then it pinches off after sinking farther into the cytoplasm (Figure 5.24). The membrane patch becomes the outer boundary of an endocytic vesicle, which delivers its contents to an organelle or stores them in cytoplasm.

As long as a cell is alive, exocytosis and endocytosis are continually replacing and withdrawing patches of its plasma membrane.

Phagocytosis ("cell eating") is an endocytic pathway in which phagocytic cells such as amoebas engulf microorganisms, cellular debris, or other particles. In animals, macrophages and other white blood cells engulf and digest pathogenic viruses and bacteria, cancerous body cells, and other threats. During phagocytosis, microfilaments form a mesh under the plasma membrane. When they contract, they force some cytoplasm and plasma membrane above it to bulge outward as a lobe called a pseudopod (Figure 5.25). Pseudopods engulf a target and merge as a vesicle that sinks into the cytoplasm and fuses with a lysosome. Enzymes in the lysosome break down the vesicle's contents. The resulting molecular bits may be recycled by the cell, or expelled by exocytosis.

Endocytosis   **Exocytosis**

A Molecules or particles enter pits in the plasma membrane.

pit

B The pits sink inward and become endocytic vesicles.

D Many of the sorted molecules cycle back to the plasma membrane.

C Vesicle contents are sorted.

E Some vesicles are routed to the nuclear envelope or ER membrane. Others fuse with Golgi bodies.

F Some vesicles and their contents are delivered to lysosomes.

Figure 5.23 Animated Endocytosis and exocytosis.

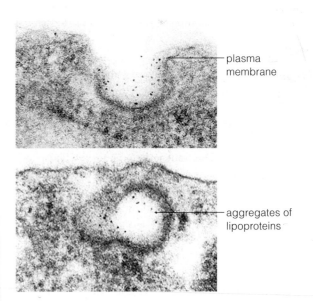

plasma membrane

aggregates of lipoproteins

Figure 5.24 Endocytosis of lipoproteins.

A Pseudopods surround a pathogen (*brown*).

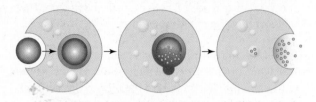

B Endocytic vesicle forms.

C Lysosome fuses with vesicle; enzymes digest pathogen.

D Cell uses the digested material or expels it.

Figure 5.25 Animated Phagocytosis. A phagocytic cell's pseudopods (extending lobes of cytoplasm) surround a pathogen. The plasma membrane above the bulging lobes fuses and forms an endocytic vesicle. Once inside the cytoplasm, the vesicle fuses with a lysosome, which digests its contents.

The composition of a plasma membrane begins in the ER. There, membrane proteins and lipids are made and modified, and both become part of vesicles that transport them to Golgi bodies for final modification. The finished proteins and lipids are repackaged as new vesicles that travel to the plasma membrane and fuse with it. The lipids and proteins of the vesicle membrane become part of the plasma membrane. This is the process by which new plasma membrane forms.

In a cell that is no longer growing, the total area of the plasma membrane remains more or less constant. Membrane is lost as a result of endocytosis, but it is replaced by membrane arriving as exocytic vesicles.

---

endocytosis Process by which a cell takes in a small amount of extracellular fluid by the ballooning inward of its plasma membrane.
exocytosis Process by which a cell expels a vesicle's contents to extracellular fluid.
phagocytosis "Cell eating"; an endocytic pathway by which a cell engulfs particles such as microbes or cellular debris.

Take-Home Message **How do cells take in large particles and bulk substances?**

❯ Exocytosis and endocytosis move materials in bulk across plasma membranes.

❯ In exocytosis, a cytoplasmic vesicle fuses with the plasma membrane and releases its contents to the outside of the cell.

❯ In endocytosis, a patch of plasma membrane sinks inward and forms a vesicle in the cytoplasm.

❯ Phagocytosis is an endocytic pathway by which cells engulf particles such as microorganisms.

■ A Toast to Alcohol Dehydrogenase (revisited)

In the human body, alcohol dehydrogenase (ADH) converts ethanol to acetaldehyde, an organic molecule even more toxic than ethanol and the most likely source of various hangover symptoms. A different enzyme, ALDH, very quickly converts acetaldehyde to nontoxic acetate. Thus, the overall pathway of ethanol metabolism in humans is:

$$\text{ethanol} \xrightarrow[\text{NAD}^+ \quad \text{NADH}]{\text{ADH}} \text{acetaldehyde} \xrightarrow[\text{NAD}^+ \quad \text{NADH}]{\text{ALDH}} \text{acetate}$$

In the average adult human body, this metabolic pathway can detoxify between 7 and 14 grams of ethanol per hour. The average alcoholic beverage contains between 10 and 20 grams of ethanol, which is why having more than one drink in any two-hour interval may result in a hangover.

Alcohol dehydrogenase detoxifies the tiny quantities of alcohols that form in some metabolic pathways. In animals, the enzyme also detoxifies alcohols made by gut-inhabiting bacteria, and those in foods such as ripe fruit.

Defects in ADH or ALDH affect alcohol metabolism. For example, if ADH is overactive, acetaldehyde accumulates faster than ALDH can detoxify it:

$$\text{ethanol} \xrightarrow{\text{ADH}} \begin{matrix}\text{acetaldehyde}\\\text{acetaldehyde}\\\text{acetaldehyde}\end{matrix} \xrightarrow{\text{ALDH}} \text{acetate}$$

People with an overactive form of ADH become flushed and feel ill after drinking even a small amount of alcohol. The unpleasant experience may be part of the reason that these people are less likely to become alcoholic than others. Underactive ALDH also causes acetaldehyde to accumulate:

$$\text{ethanol} \xrightarrow{\text{ADH}} \begin{matrix}\text{acetaldehyde}\\\text{acetaldehyde}\\\text{acetaldehyde}\end{matrix} \xrightarrow{\;\;X\;\;} \text{acetate}$$

Underactive ALDH is associated with the same effect—and the same protection from alcoholism—as overactive ADH. Both types of variant enzymes are common in people of Asian descent. For this reason, the alcohol flushing reaction is informally called "Asian flush."

Having an underactive ADH enzyme has the opposite effect. It results in slowed alcohol metabolism, so people with an underactive ADH may not feel the ill effects of drinking alcoholic beverages as much as other people do. When these people drink alcohol, they have a tendency to become alcoholics. One-quarter of undergraduate students who binge also have signs of alcoholism.

Alcohol abuse is the leading cause of cirrhosis of the liver in the United States. The liver becomes so scarred, hardened, and filled with fat that it loses its function. It stops making the protein albumin, so the solute balance of body fluids is disrupted, and the legs and abdomen swell with watery fluid. It cannot remove drugs and other toxins from the blood, so they accumulate in the brain—which impairs mental functioning and alters personality. Restricted blood flow through the liver causes veins to enlarge and rupture, so internal bleeding is a risk. The damage to the body results in a heightened risk of diabetes and liver cancer. Once cirrhosis has been diagnosed, a person has about a 50 percent chance of death within 10 years.

How Would You Vote? Some people have damaged their liver because they drank too much alcohol; others have had liver-damaging infections. Because there are not enough liver donors, should life-style be a factor in deciding who gets a liver transplant? See CengageNow for details, then vote online (cengagenow.com).

Summary

Section 5.1 Metabolic processes build and break down organic molecules such as ethanol and other toxins. Currently the most serious drug problem on college campuses is binge drinking.

Section 5.2 **Kinetic energy**, **potential energy**, and other forms of **energy** cannot be created or destroyed (**first law of thermodynamics**). Energy can be converted from one form to another and transferred between objects or systems. Energy tends to disperse spontaneously (**second law of thermodynamics**). Some energy disperses with every transfer, usually as heat. **Entropy** is a measure of how much the energy of a system is dispersed.

Living things maintain their organization only as long as they harvest energy from someplace else. Energy flows in one direction through the biosphere, starting mainly from the sun, then into and out of ecosystems. Producers and then consumers use energy to assemble, rearrange, and break down organic molecules that cycle among organisms throughout ecosystems.

Section 5.3 Cells store and retrieve free energy by making and breaking chemical bonds in metabolic **reactions**, in which **reactants** are converted to **products**. **Activation energy** is the minimum energy required to start a reaction. Phosphate-group transfers (**phosphorylations**) to and from **ATP** couple **exergonic** reactions with **endergonic** ones. Cells regenerate ATP by the **ATP/ADP cycle**.

Section 5.4 Enzymes enhance the rate of reactions without being changed by them, a process called **catalysis**. Enzymes lower activation energy by boosting local concentrations of **substrates**, orienting them in positions that favor reaction, inducing the fit between a substrate and the enzyme's **active site** (**induced-fit model**), or excluding water.

Each type of enzyme works best within a characteristic range of temperature, salt concentration, and pH. Most enzymes require **cofactors**, which are metal ions or organic **coenzymes**. Cofactors in some **antioxidants** help them stop reactions with oxygen that produce free radicals.

Section 5.5 Cells build, convert, and dispose of most substances in enzyme-mediated reaction sequences called **metabolic pathways**. Controls over enzymes allow cells to conserve energy and resources by producing only what they require. **Allosteric** sites are points of control by which a cell adjusts the types and amounts of substances it makes. **Feedback inhibition** is an example of enzyme control. **Redox** (oxidation–reduction) **reactions** in **electron transfer chains** allow cells to harvest energy in manageable increments.

Section 5.6 A **concentration gradient** is a difference in the **concentration** of a substance between adjoining regions of fluid. Molecules or ions tend to follow their own gradient and move toward the region where they are less concentrated, a behavior called **diffusion**. The steepness of the gradient, temperature, solute size, charge, and pressure influence the diffusion rate.

Osmosis is the diffusion of water across a selectively permeable membrane, from the region with a lower solute concentration (**hypotonic**) toward the region with a higher solute concentration (**hypertonic**). There is no net movement of water between **isotonic** solutions. **Osmotic pressure** is the amount of **turgor** (fluid pressure against a cell membrane or wall) that stops osmosis.

Section 5.7 Gases, water, and small nonpolar molecules can diffuse across a lipid bilayer. Most other molecules, and ions, cross only with the help of transport proteins, which allow a cell or membrane-enclosed organelle to control which substances enter and exit. The types of transport proteins in a membrane determine which substances cross it. **Active transport** proteins such as calcium pumps use energy, usually from ATP, to pump a solute against its concentration gradient. **Passive transport** proteins work without an energy input; solute movement is driven by the concentration gradient.

Section 5.8 Bulk substances and large particles move across plasma membranes by endocytosis and exocytosis. With **exocytosis**, a cytoplasmic vesicle fuses with the plasma membrane, and its contents are released to the outside of the cell. The vesicle's membrane lipids and proteins become part of the plasma membrane. With **phagocytosis** and other processes of **endocytosis**, a patch of plasma membrane balloons into the cell, and forms a vesicle that sinks into the cytoplasm.

Self-Quiz Answers in Appendix III

1. _____ is life's primary source of energy.
   a. Food   b. Water   (c. Sunlight)   d. ATP

2. If we liken a chemical reaction to an energy hill, then a(n) _____ reaction is an uphill run.
   (a. endergonic)   c. catalytic
   b. exergonic   d. both a and c

3. Which of the following statements is not correct?
   a. Energy cannot be created or destroyed.
   (b.) Energy cannot change from one form to another.
   c. Energy tends to disperse spontaneously.

4. Enzymes _____ .
   a. are proteins, except for a few RNAs
   b. lower the activation energy of a reaction
   c. are changed by the reactions they catalyze
   (d.) a and b

Effects of Artichoke Extract on Hangovers Ethanol is a toxin, so it makes sense that drinking it can cause various symptoms of poisoning: headache, stomachache, nausea, fatigue, impaired memory, dizziness, tremors, and diarrhea, among other ailments. All are symptoms of hangover, the common word for what happens as the body is recovering from a bout of heavy drinking.

The most effective treatment for a hangover is to avoid drinking in the first place. Folk remedies (such as aspirin, coffee, bananas, more alcohol, honey, barley grass, pizza, milkshakes, glutamine, raw eggs, charcoal tablets, asparagus, or cabbage) abound, but few have been studied scientifically. In 2003, Max Pittler and his colleagues tested one of them. The researchers gave 15 participants an unmarked pill containing either artichoke extract or a placebo (an inactive substance) just before or after drinking enough alcohol to cause a hangover. The results are shown in Figure 5.26.

1. How many participants experienced a hangover that was worse with the placebo than with the artichoke extract?
2. How many participants had a worse hangover with the artichoke extract?
3. Express the numbers you counted in questions 1 and 2 as a percentage of the total number of participants. How much difference is there between the percentages?
4. Do these results support the hypothesis that artichoke extract is an effective hangover treatment? Why or why not?

Participant (Age, Gender)	Severity of Hangover	
	Artichoke Extract	Placebo
1 (34, F)	1.9	3.8
2 (48, F)	5.0	0.6
3 (25, F)	7.7	3.2
4 (57, F)	2.4	4.4
5 (34, F)	5.4	1.6
6 (30, F)	1.5	3.9
7 (33, F)	1.4	0.1
8 (37, F)	0.7	3.6
9 (62, M)	4.5	0.9
10 (36, M)	3.7	5.9
11 (54, M)	1.6	0.2
12 (37, M)	2.6	5.6
13 (53, M)	4.1	6.3
14 (48, F)	0.5	0.4
15 (32, F)	1.3	2.5

Figure 5.26 Results of a study that tested artichoke extract as a hangover preventive. All participants were tested once with the placebo and once with the extract, with a week interval between. Each rated the severity of 20 hangover symptoms on a scale of 0 (not experienced) to 10 ("as bad as can be imagined"). The 20 ratings were averaged as an overall rating, which is listed here.

5. _____ are always changed by participating in a reaction. (Choose all that are correct.)
   a. Enzymes   c. Reactants
   b. Cofactors   d. Coenzymes

6. Name one environmental factor that typically influences enzyme function. *temperture, ph, salt concentration*

7. A metabolic pathway _____ .
   a. may build or break down molecules
   b. generates heat
   c. can include an electron transfer chain
   d. a and c
   e. all of the above

8. Diffusion is the movement of ions or molecules from a region where they are *more* (more/less) concentrated to another where they are *less* (more/less) concentrated.

9. Name one molecule that can readily diffuse across a lipid bilayer. *Water*

10. Transporters that require an energy boost help sodium ions across a cell membrane. This is a case of _____ .
    a. passive transport   c. facilitated diffusion
    b. active transport   d. a and c

11. Immerse a living human cell in a hypotonic solution, and water will tend to _____ .
    a. diffuse into the cell   c. show no net movement
    b. diffuse out of the cell   d. move in by endocytosis

12. Vesicles form during _____ .
    a. endocytosis ✓   d. halitosis
    b. exocytosis   e. a through c
    c. phagocytosis ✓   f. all of the above

13. Match each term with its most suitable description.
    C reactant   a. assists enzymes
    D enzyme   b. forms at reaction's end
    E first law of   c. enters a reaction
       thermodynamics   d. unchanged by participating in a reaction
    B product   e. energy cannot be created or destroyed
    A cofactor   f. basis of diffusion
    F gradient   g. no energy boost required
    G passive transport h. one cell engulfs another
    I active transport   i. requires energy boost
    H phagocytosis

Additional questions are available on **CENGAGENOW**.

Critical Thinking

1. Often, beginning physics students are taught the basic concepts of thermodynamics with two phrases: First, you can't win. Second, you can't break even. Explain.

2. Water molecules tend to diffuse in response to their own concentration gradient. How can water be more or less concentrated?

Animations and Interactions on **CENGAGENOW**:
> Energy flow and materials cycling; Chemical bookkeeping; Activation energy; How enzymes work; Enzymes and temperature; Feedback inhibition; Allosteric effects; Controlling energy release; Diffusion; Osmosis; Selective permeability; Tonicity; Passive and active transport; Endocytosis and exocytosis; Phagocytosis.

< Links to Earlier Concepts

A review of how energy flows through the biosphere (Sections 5.2, 5.3) will be useful before starting this chapter, which revisits electron energy levels (2.3), chemical bonds (2.4), carbohydrates (3.3), membrane proteins (4.4), and antioxidants (5.4). Chloroplasts (4.9) and surface specializations of plant cells (4.11) support photosynthesis. You will see an example of how cells harvest energy with electron transfer chains (5.5).

Key Concepts

The Rainbow Catchers

The flow of energy through the biosphere starts when chlorophylls and other photosynthetic pigments absorb the energy of visible light. In plants, some bacteria, and many protists, that energy ultimately drives the synthesis of glucose and other carbohydrates.

What Is Photosynthesis?

Photosynthesis has two stages in the chloroplasts of plants and many types of protists. In the first stage, sunlight energy is converted to chemical energy. Molecules that form in the first stage of photosynthesis power the formation of sugars in the second stage.

6 Where It Starts—Photosynthesis

Today, the expression "food is fuel" is not just about eating. With fossil fuel prices soaring, there is an increasing demand for biofuels, which are oils, gases, or alcohols made from organic matter that is not fossilized. Much of the material currently used for biofuel production consists of food crops—mainly corn, soybeans, and sugarcane. Growing these crops in large quantities is typically expensive and damaging to the environment, and using them to make biofuel competes with our food supply. The diversion of food crops to biofuel production contributes to increases in food prices worldwide.

How did we end up competing with our vehicles for food? We both run on the same fuel: energy that plants have stored in chemical bonds. Fossil fuels such as petroleum, coal, and natural gas formed from the remains of ancient swamp forests that decayed and compacted over millions of years. They consist mainly of molecules originally assembled by ancient plants. Biofuels—and foods—consist mainly of molecules originally assembled by modern plants.

Autotrophs harvest energy directly from the environment, and obtain carbon from inorganic molecules (*auto*– means self; –*troph* refers to nourishment). Plants and most other autotrophs make their own food by **photosynthesis**, a process in which they use the energy of sunlight to assemble carbohydrates from carbon dioxide and water. Directly or indirectly, photosynthesis also feeds most other life on Earth. Animals and other **heterotrophs** get energy and carbon by breaking down organic molecules assembled by other organisms (*hetero*– means other). We and almost all other organisms sustain ourselves by extracting energy from organic molecules originally assembled by photosynthesizers.

A lot of energy is locked up in the chemical bonds of molecules made by plants. That energy can fuel heterotrophs, as when an animal cell powers ATP synthesis by breaking the bonds of sugars. It can also fuel our cars, which run on energy released by burning biofuels or fossil fuels. Both processes are fundamentally the same: They release energy by breaking the bonds of organic molecules. Both use oxygen to break those bonds, and both produce carbon dioxide.

Corn and other food crops are rich in oils, starches, or sugars that can be easily converted to biofuels. The starch in corn kernels, for example, can be enzymatically broken down to glucose, which is converted to ethanol by heterotrophic bacteria or yeast. Making biofuels from other types of plant matter requires additional steps, because these materials contain a higher proportion of cellulose. Breaking down this tough, insoluble carbohydrate to its glucose monomers adds substantial cost to the biofuel product. Researchers are currently trying to find cost-effective ways to break down the abundant cellulose in fast-growing weeds such as switchgrass (Figure 6.1), and agricultural wastes such as wood chips, wheat straw, cotton stalks, and rice hulls.

Figure 6.1 Biofuels. *Opposite*, switchgrass (*Panicum virgatum*), a weed that grows wild in North American prairies. *Above*, researchers Ratna Sharma and Mari Chinn of North Carolina State University working to reduce the costs of producing biofuel from biomass such as switchgrass and agricultural wastes.

autotroph Organism that makes its own food using carbon from inorganic molecules such as CO_2, and energy from the environment.
heterotroph Organism that obtains energy and carbon from organic compounds assembled by other organisms.
photosynthesis Metabolic pathway by which most autotrophs capture light energy and use it to make sugars from CO_2 and water.

Making ATP and NADPH
ATP forms in the first stage of photosynthesis, which is light-dependent because the reactions run on the energy of light. The coenzyme NADPH forms in a noncyclic pathway that releases oxygen. ATP also forms in a cyclic pathway that does not release oxygen.

Making Sugars
The second stage is the "synthesis" part of photosynthesis. Sugars are assembled with carbon and oxygen atoms from CO_2. The reactions run on the chemical bond energy of ATP, and electrons donated by NADPH—molecules that formed in the first stage of photosynthesis.

Alternate Pathways
Details of light-independent reactions that vary among organisms are evolutionary adaptations to different environmental conditions.

> Photosynthetic organisms use pigments to capture the energy of sunlight.
< Links to Electrons 2.3, Chemical bonds 2.4, Carbohydrates 3.3, Plastids 4.9, Energy 5.2, Antioxidants 5.4

Properties of Light

Energy flow through nearly all ecosystems on Earth begins when photosynthesizers intercept energy from the sun. Harnessing the energy of sunlight for work is complicated business, or we would have been able to do it in an economically sustainable way by now. Plants do it by converting light energy to chemical energy, which they and most other organisms use to drive cellular work. The first step involves capturing light. In order to understand how that happens, you have to understand a little about the nature of light.

Most of the energy that reaches Earth's surface is in the form of visible light, so it should not be surprising that visible light is the energy that drives photosynthesis. Visible light is a very small part of a large spectrum of electromagnetic energy radiating from the sun. Like all other forms of electromagnetic energy, light travels in waves, moving through space a bit like waves moving across an ocean. The distance between the crests of two successive waves is the light's **wavelength**, which we measure in nanometers (nm). About 25 million nanometers are equal to one inch.

Visible light occurs between wavelengths of 380 and 750 nanometers. We see light of particular wavelengths in this range as different colors, and all wavelengths combined

as white. White light separates into its component colors when it passes through a prism. The prism bends the longer wavelengths more than it bends the shorter ones, so a rainbow of colors forms (Figure 6.2A).

Light travels in waves, but it is also organized in packets of energy called photons. A photon's energy and its wavelength are related, so all photons traveling at the same wavelength carry the same amount of energy. Photons that carry the least amount of energy travel in longer wavelengths; those that carry the most energy travel in shorter wavelengths (Figure 6.2B). Photons of wavelengths shorter than about 380 nanometers carry enough energy to alter or break the chemical bonds of DNA and other biological molecules. That is why UV (ultraviolet) light, x-rays, and gamma rays are a threat to life.

Pigments: The Rainbow Catchers

Photosynthesizers use pigments to capture light. A **pigment** is an organic molecule that selectively absorbs light of specific wavelengths. Wavelengths of light that are not absorbed are reflected, and that reflected light gives each pigment its characteristic color. For example, a pigment that absorbs violet, blue, and green light reflects the rest of the visible light spectrum—yellow, orange, and red light. This pigment appears orange to us.

Chlorophyll *a* is by far the most common photosynthetic pigment in plants, and also in photosynthetic protists and bacteria. Chlorophyll *a* absorbs violet and red light, so it appears green to us. Accessory pigments,

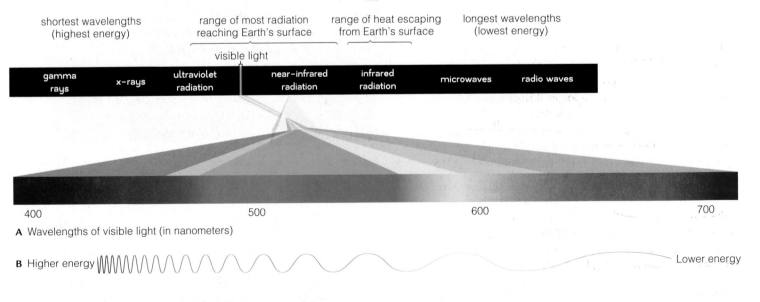

shortest wavelengths (highest energy) range of most radiation reaching Earth's surface range of heat escaping from Earth's surface longest wavelengths (lowest energy)

visible light

| gamma rays | x-rays | ultraviolet radiation | | near-infrared radiation | infrared radiation | microwaves | radio waves |

A Wavelengths of visible light (in nanometers)

400   500   600   700

B Higher energy Lower energy

Figure 6.2 Animated Properties of light. **A** Electromagnetic spectrum of radiant energy, which undulates across space as waves that we measure in nanometers. Visible light makes up a very small part of the spectrum. Raindrops or a prism can separate its different wavelengths, which we see as different colors. **B** Light is organized as packets of energy called photons. The shorter a photon's wavelength, the greater its energy.

Pigment	Color	Occurrence			
		Plants	Protists	Bacteria	Archaeans
Chlorophyll *a*	green	●	●	●	
Other chlorophylls	green	●	●	●	
Phycobilins					
phycocyanobilin	blue		●	●	
phycoerythrobilin	red		●	●	
phycoviolobilin	violet		●	●	
Carotenoids					
beta-carotene	orange	●	●	●	
lycopene	red	●	●	●	
lutein	yellow	●	●	●	
zeaxanthin	yellow	●	●	●	
fucoxanthin	brown	●	●	●	
Anthocyanins	red to blue	●	●	●	
Retinal	violet				●

A

chlorophyll *a*   beta-carotene

B

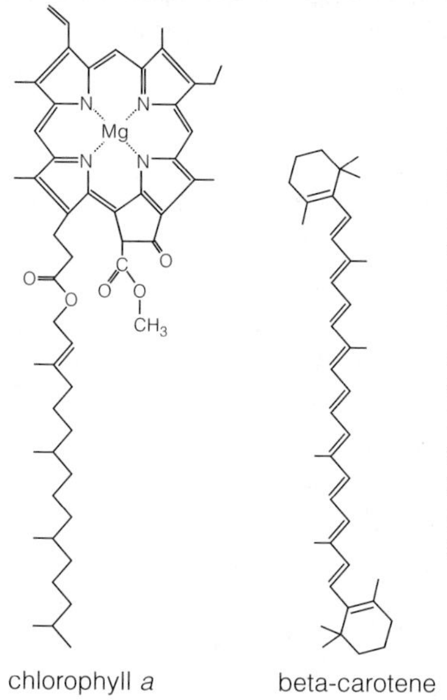

Figure 6.3 Examples of photosynthetic pigments.

A Collectively, photosynthetic pigments are capable of absorbing almost all visible light wavelengths.

B Structures of two photosynthetic pigments. The light-trapping ring structure of chlorophyll is almost identical to a heme group. Heme groups are part of hemoglobin, which is a red pigment.

including other chlorophylls, work together with chlorophyll *a* to harvest a wide range of light wavelengths for photosynthesis. The colors of a few of the 600 or so known accessory pigments are shown in Figure 6.3A.

Accessory pigments are typically multipurpose molecules. Their antioxidant properties protect plants and other organisms from the damaging effects of UV light in the sun's rays; their appealing colors attract animals to ripening fruit or pollinators to flowers. You may already be familiar with some of these molecules because they color familiar roots, fruits, and flowers. For example, carrots are orange because they have a lot of beta-carotene (which is also spelled β-carotene). The yellow color of corn comes from zeaxanthin. The color of a tomato changes from green to red as it ripens because its chlorophyll-containing chloroplasts develop into lycopene-containing chromoplasts (Section 4.9). Roses are red and violets are blue because of their anthocyanin content.

Most photosynthetic organisms use a combination of pigments for photosynthesis. In plants, chlorophylls are usually so abundant that they mask the colors of the other pigments, so leaves typically appear green. The green leaves of many plants change color during autumn because they stop making pigments in preparation for a period of dormancy. Chlorophyll breaks down faster than the other pigments, so the leaves turn red, orange, yellow, or violet as their chlorophyll content declines and their accessory pigments become visible.

The light-trapping part of a pigment is an array of atoms in which single bonds alternate with double bonds (Figure 6.3B). Electrons in such arrays easily absorb photons, so pigment molecules function a bit like antennas that are specialized for receiving light energy of only certain wavelengths.

Absorbing a photon excites electrons. Remember, an energy input can boost an electron to a higher energy level (Section 2.3). The excited electron returns quickly to a lower energy level by emitting the extra energy. As you will see, photosynthetic cells can capture energy emitted from an electron returning to a lower energy level. Arrays of chlorophylls and other photosynthetic pigments in these cells hold on to the energy by passing it back and forth between them. When the energy reaches a special pair of chlorophylls, the reactions of photosynthesis begin.

---

chlorophyll *a* Main photosynthetic pigment in plants.
pigment An organic molecule that can absorb light of certain wavelengths.
wavelength Distance between the crests of two successive waves of light.

Take-Home Message **How do photosynthesizers absorb light?**

❯ Energy radiating from the sun travels through space in waves and is organized as packets called photons.

❯ The spectrum of radiant energy from the sun includes visible light. Humans perceive different wavelengths of visible light as different colors. The shorter the wavelength, the greater the energy.

❯ Pigments absorb light at specific wavelengths. Photosynthetic species use pigments such as chlorophyll *a* to harvest the energy of light for photosynthesis.

❯ Photosynthetic pigments work together to harvest light of different wavelengths.

In 1882, botanist Theodor Engelmann designed an experiment to test his hypothesis that the color of light affects photosynthesis. It had long been known that photosynthesis releases oxygen, so Engelmann used the amount of oxygen released by photosynthetic cells as a measure of how much photosynthesis was occurring in them. He used a prism to divide a ray of light into its component colors, then directed the resulting spectrum across a single strand of *Chladophora*, a photosynthetic alga (Figure 6.4A), suspended in a drop of water.

Oxygen-sensing equipment had not yet been invented, so Engelmann used oxygen-requiring bacteria to show him where the oxygen concentration in the water was highest. The bacteria moved through the water and gathered mainly where violet or red light fell across the strand of algae (Figure 6.4B). Engelmann concluded that the algal cells illuminated by light of these colors were releasing the most oxygen—a sign that violet and red light are the best for driving photosynthesis.

Engelmann's experiment allowed him to correctly identify the colors of light (red and violet) that are most efficient at driving photosynthesis in *Chladophora*. His results constituted an absorption spectrum, which is a graph that shows how efficiently the different wavelengths of light are absorbed by a substance. Peaks in the graph indicate wavelengths of light that the substance absorbs best (Figure 6.4C). Engelmann's results represent the combined spectra of all the photosynthetic pigments in *Chladophora*.

Most photosynthetic organisms use a combination of pigments to drive photosynthesis, and the combination differs by species. Why? Different proportions of wavelengths in sunlight reach different parts of Earth. The particular set of pigments in each species is an adaptation that allows an organism to absorb the particular wavelengths of light available in its habitat.

For example, water absorbs light between wavelengths of 500 and 600 nm less efficiently than other wavelengths. Algae that live deep underwater have pigments that absorb light in the range of 500–600 nm, which is the range that water does not absorb very well. Phycobilins are the most common pigments in deep-water algae.

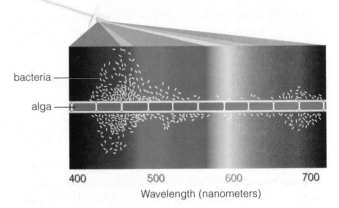

A Light micrograph of photosynthetic cells in a strand of *Chladophora*. Engelmann used this green alga to show that certain colors of light are best for photosynthesis.

bacteria

alga

400   500   600   700
Wavelength (nanometers)

B Engelmann directed light through a prism so that bands of colors crossed a water droplet on a microscope slide. The water held a strand of *Chladophora* and oxygen-requiring bacteria. The bacteria clustered around the algal cells that were releasing the most oxygen—the ones that were most actively engaged in photosynthesis. Those cells were under red and violet light.

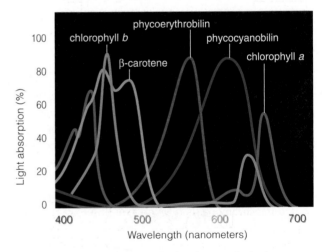

chlorophyll *b*   phycoerythrobilin   phycocyanobilin   chlorophyll *a*

β-carotene

400   500   600   700
Wavelength (nanometers)

C Absorption spectra of a few photosynthetic pigments. Line color is the characteristic color of each pigment.

Figure 6.4 Animated Discovery that photosynthesis is driven by particular wavelengths of light. Theodor Engelmann used the green alga *Chladophora* **A** in an early photosynthesis experiment **B**. His results constituted one of the first absorption spectra.

C Absorption spectra of chlorophylls *a* and *b*, β-carotene, and two phycobilins reveal the efficiency with which these pigments absorb different wavelengths of visible light.

❯❯ **Figure It Out** Of the five pigments represented in C, which three are the main photosynthetic pigments in *Chladophora*?

Answer: Chlorophyll *a*, chlorophyll *b*, and β-carotene

Take-Home Message **Why do cells use more than one photosynthetic pigment?**

❯ A combination of pigments allows a photosynthetic organism to most efficiently capture the particular range of light wavelengths that reaches the habitat in which it evolved.

❭ Photosynthesis occurs in two stages in the chloroplasts of plants and other photosynthetic eukaryotes.

The **chloroplast** is an organelle that specializes in photosynthesis in plants and many protists (Figure 6.5). Plant chloroplasts have two outer membranes, and are filled with a semifluid matrix called **stroma**. Stroma contains the chloroplast's own DNA, some ribosomes, and an inner, much-folded **thylakoid membrane**. The folds of a thylakoid membrane typically form stacks of disks (thylakoids) that are connected by channels. The space inside all of the disks and channels is one continuous compartment.

Photosynthesis is often summarized by this equation:

$$6CO_2 \;+\; 6H_2O \;\xrightarrow{\text{light energy}}\; C_6H_{12}O_6 \;+\; 6O_2$$

carbon dioxide   water   glucose   oxygen

However, photosynthesis is not one reaction. It is a series of many reactions that occur in two stages. The first stage occurs at the thylakoid membrane. It is driven by light, so the collective reactions of this stage are called the **light-dependent reactions**. Two different sets of light-dependent reactions constitute a noncyclic and a cyclic pathway. Both pathways convert light energy to chemical bond energy of ATP. The noncyclic pathway, which is the main one in chloroplasts, yields NADPH and O_2 in addition to ATP:

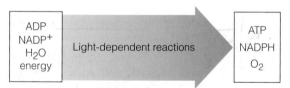

ADP / $NADP^+$ / H_2O / energy   Light-dependent reactions   ATP / NADPH / O_2

The reactions of the second stage of photosynthesis, which run in the stroma, make glucose and other carbohydrates from carbon dioxide and water. Light energy does not power them, so they are collectively called the **light-independent reactions**. They run on energy delivered by ATP and coenzymes produced in the first stage:

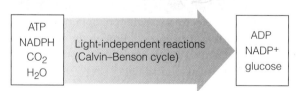

ATP / NADPH / CO_2 / H_2O Light-independent reactions (Calvin–Benson cycle)   ADP / NADP+ / glucose

chloroplast Organelle specialized for photosynthesis in plants and some protists.
light-dependent reactions First stage of photosynthesis; convert light energy to chemical energy of ATP and NADPH.
light-independent reactions Second stage of photosynthesis; use ATP and NADPH to assemble sugars from water and CO_2.
stroma Semifluid matrix between the thylakoid membrane and the two outer membranes of a chloroplast.
thylakoid membrane A chloroplast's highly folded inner membrane system; forms a continuous compartment in the stroma.

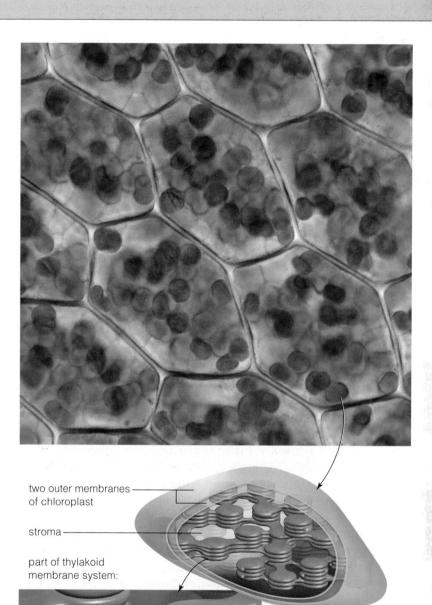

two outer membranes of chloroplast

stroma

part of thylakoid membrane system:

thylakoid compartment, cutaway view

Figure 6.5 Animated The chloroplast: site of photosynthesis in the cells of typical leafy plants. The micrograph shows chloroplast-stuffed cells of a moss, *Plagiomnium affine*.

Take-Home Message **What is photosynthesis and where in a eukaryotic cell does it take place?**

❭ In the first stage of photosynthesis, light energy drives the formation of ATP and NADPH, and oxygen is released. In eukaryotic cells, these light-dependent reactions occur at the thylakoid membrane of chloroplasts.

❭ The second stage of photosynthesis, the light-independent reactions, occur in the stroma. ATP and NADPH drive the synthesis of carbohydrates from water and carbon dioxide.

> The reactions of the first stage of photosynthesis convert the energy of light to the energy of chemical bonds.

‹ Links to Electrons and energy levels 2.3, Membrane proteins 4.4, Chloroplasts 4.9, Energy 5.2, Electron transfer chains 5.5, Gradients 5.6

Capturing Light for Photosynthesis

When a pigment in a thylakoid membrane absorbs a photon, the photon's energy boosts one of the pigment's electrons to a higher energy level. The electron quickly emits the extra energy and drops back to its unexcited state. That energy would be lost to the environment if nothing else were to happen, but in a thylakoid membrane the energy of excited electrons is not lost. Light-harvesting complexes keep it in play. Millions of these circular arrays of chlorophylls, accessory pigments, and proteins are embedded in each thylakoid membrane (Figure 6.6). Pigments that are part of a light-harvesting complex can hold on to energy by passing it back and forth, a bit like volleyball players pass a ball among team members. The energy gets volleyed from cluster to cluster until a photosystem absorbs it. **Photosystems** are groups of hundreds of chlorophylls, accessory pigments, and other molecules that work as a unit to begin the reactions of photosynthesis.

The Noncyclic Pathway

Thylakoid membranes contain two kinds of photosystems, type I and type II, which were named in the order of their discovery. They work together in a set of reactions called the noncyclic pathway of photosynthesis. These reactions begin when energy being passed among light-harvesting complexes reaches a photosystem II (Figure 6.7). At the center of each photosystem is a special pair of chlorophyll

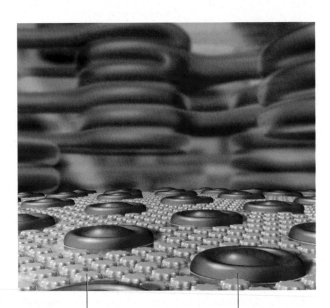

light-harvesting complex   photosystem

a molecules. When a photosystem absorbs energy, electrons are ejected from its special pair ❶. These electrons immediately enter an electron transfer chain in the thylakoid membrane.

Replacing Lost Electrons

A photosystem can donate only a few electrons to electron transfer chains before it must be restocked with more. Where do replacements come from? Photosystem II gets more electrons by pulling them off of water molecules in the thylakoid compartment. This reaction causes the water molecules to dissociate into hydrogen ions and oxygen ❷. The released oxygen diffuses out of the cell as O_2 gas. This and any other process by which a molecule is broken apart by light energy is called **photolysis**.

Harvesting Electron Energy

The actual conversion of light energy to chemical energy occurs when a photosystem donates electrons to an electron transfer chain ❸. Light does not take part in chemical reactions, but electrons do. In a series of redox reactions, electrons pass from one molecule of the chain to the next. With each reaction, the electrons release a bit of their extra energy.

The molecules of the electron transfer chain use the released energy to move hydrogen ions (H^+) across the membrane, from the stroma to the thylakoid compartment ❹. Thus, the flow of electrons through electron transfer chains sets up and maintains a hydrogen ion gradient across the thylakoid membrane. This gradient motivates hydrogen ions in the thylakoid compartment to move back into the stroma. However, ions cannot diffuse through lipid bilayers (Section 4.4). H^+ leaves the thylakoid compartment only by flowing through membrane transport proteins called ATP synthases.

Hydrogen ion flow through an ATP synthase causes this protein to attach a phosphate group to ADP ❼, so ATP forms in the stroma. The process by which the flow of electrons through electron transfer chains drives ATP formation is called **electron transfer phosphorylation**.

After the electrons have moved through the first electron transfer chain, they are accepted by a photosystem I. When this photosystem absorbs light energy, electrons

Figure 6.6 Artist's view of some of the components of the thylakoid membrane as seen from the stroma. Molecules of electron transfer chains and ATP synthases are also present, but not shown for clarity.

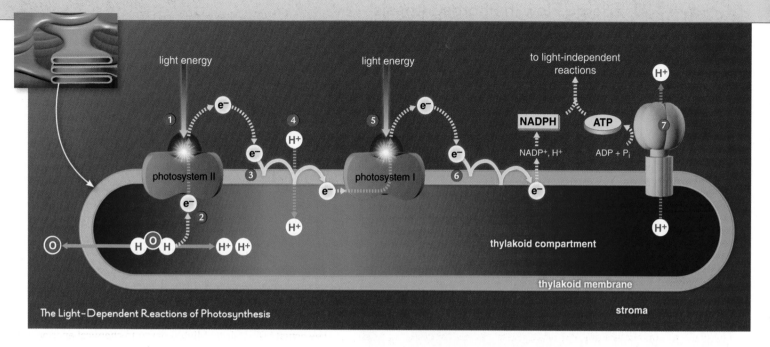

light energy

light energy

to light-independent
reactions

NADPH   **ATP**

NADP⁺, H⁺   ADP + P_i

photosystem II

photosystem I

thylakoid compartment

thylakoid membrane

stroma

The Light-Dependent Reactions of Photosynthesis

are ejected from its special pair of chlorophylls ❺. These electrons enter a second, different electron transfer chain. At the end of this chain, the coenzyme NADP⁺ accepts the electrons along with H⁺, so NADPH forms ❻:

$$NADP^+ + 2e^- + H^+ \longrightarrow NADPH$$

ATP and NADPH continue to form as long as electrons continue to flow through transfer chains in the thylakoid membrane, and electrons flow through the chains as long as water, NADP⁺, and light are plentiful. The flow of electrons slows—and so does ATP and NADPH production—at night, or when water or NADP⁺ is scarce.

The Cyclic Pathway

At high oxygen levels, or when NADPH accumulates in the stroma, the noncyclic pathway backs up and stalls. Even when the noncyclic pathway is not running, a cell can continue producing ATP by photosynthesis with the cyclic pathway. This pathway involves photosystem I and an electron transfer chain that cycles electrons back to it. The chain that acts in the cyclic pathway uses electron energy to move hydrogen ions into the thylakoid compartment. The resulting hydrogen ion gradient drives ATP formation, just as it does in the noncyclic pathway. However, NADPH does not form, because electrons at the end of this

Figure 6.7 **Animated** Light-dependent reactions of photosynthesis. This example shows the noncyclic reactions in a thylakoid membrane.

❶ Light energy ejects electrons from a photosystem II.

❷ The photosystem pulls replacement electrons from water molecules, which break apart into oxygen and hydrogen ions. The oxygen leaves the cell as O₂.

❸ The electrons enter an electron transfer chain in the thylakoid membrane.

❹ Energy lost by the electrons as they move through the transfer chain causes hydrogen ions to be pumped from the stroma into the thylakoid compartment. A hydrogen ion gradient forms across the thylakoid membrane.

❺ Light energy ejects electrons from a photosystem I. Replacement electrons come from an electron transfer chain.

❻ The electrons move through a second electron transfer chain, then combine with NADP⁺ and H⁺, so NADPH forms.

❼ Hydrogen ions in the thylakoid compartment are propelled through the interior of ATP synthases by their gradient across the thylakoid membrane. Hydrogen ion flow causes ATP synthases to attach phosphate to ADP, so ATP forms in the stroma.

chain are accepted by a photosystem I, not NADP⁺. Oxygen (O₂) does not form either, because photosystem I does not rely on photolysis to resupply itself with electrons.

Take-Home Message What happens during the light-dependent reactions of photosynthesis?

❯ In the light-dependent reactions of photosynthesis, chlorophylls and other pigments in the thylakoid membrane transfer the energy of light to photosystems.

❯ Absorbing energy causes electrons to leave photosystems and enter electron transfer chains in the membrane. The flow of electrons through the transfer chains sets up hydrogen ion gradients that drive ATP formation.

❯ In the noncyclic pathway, oxygen is released and electrons end up in NADPH.

❯ A cyclic pathway involving only photosystem I allows the cell to continue making ATP even when the noncyclic pathway is not running. NADPH does not form, and oxygen is not released.

---

electron transfer phosphorylation Process in which electron flow through electron transfer chains sets up a hydrogen ion gradient that drives ATP formation.
photolysis Process by which light energy breaks down a molecule.
photosystem Cluster of pigments and proteins that converts light energy to chemical energy in photosynthesis.

> Energy flow in the light-dependent reactions is an example of how organisms harvest energy from the environment.
< Links to Energy in metabolism 5.3, Redox reactions 5.5

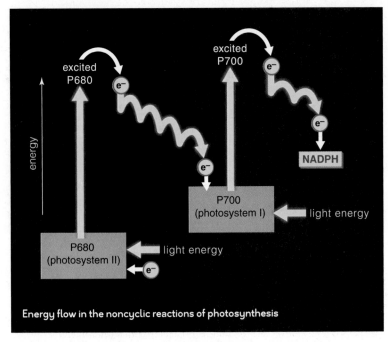

Energy flow in the noncyclic reactions of photosynthesis

A The noncyclic pathway is a one-way flow of electrons from water, to photosystem II, to photosystem I, to NADPH. As long as electrons continue to flow through the two electron transfer chains, H⁺ continues to be carried across the thylakoid membrane, and ATP and NADPH keep forming. Light provides the energy boosts that keep the pathway going.

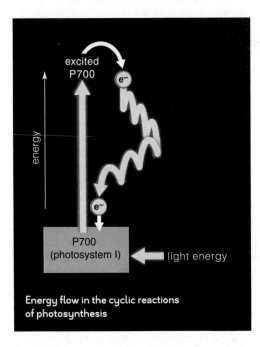

Energy flow in the cyclic reactions of photosynthesis

B In the cyclic pathway, electrons ejected from photosystem I are returned to it. As long as electrons continue to pass through its electron transfer chain, H⁺ continues to be carried across the thylakoid membrane, and ATP continues to form. Light provides the energy boost that keeps the cycle going.

Figure 6.8 Energy flow in the light-dependent reactions of photosynthesis. An energy input of an appropriate wavelength ejects electrons from a photosystem's special pair of chlorophylls. The special pair in photosystem I absorbs photons of a 680-nanometer wavelength, so it is called P680. The special pair in photosystem II absorbs photons of a 700-nanometer wavelength, so it is called P700.

One of the recurring themes in biology is that organisms use energy harvested from the environment to drive cellular processes. Energy flow in the light-dependent reactions of photosynthesis is a classic example of how that happens. Figure 6.8 compares energy flow in the two pathways of light-dependent reactions.

The simpler cyclic pathway evolved first, and still operates in nearly all photosynthesizers. Later, the photosynthetic machinery in some organisms became modified so that photosystem II became part of it. That modification was the beginning of a combined sequence of reactions that removes electrons from water molecules, with the release of hydrogen ions and oxygen.

In the noncyclic reactions, electrons that leave photosystem II do not return to it. Instead, they end up in NADPH, a powerful reducing agent (electron donor). In the cyclic reactions, electrons lost from photosystem I are cycled back to it. No NADPH forms, and no oxygen is released.

In both the cyclic and noncyclic reactions, molecules in electron transfer chains use electron energy to shuttle H⁺ across the thylakoid membrane. Hydrogen ions accumulate in the thylakoid compartment, forming a gradient that powers ATP synthesis.

The plasma membrane of different species of modern photosynthetic bacteria incorporates either type I or type II photosystems. Cyanobacteria, plants, and all photosynthetic protists use both types. Which of the two pathways predominates at any given time depends on the organism's immediate metabolic demands for ATP and NADPH.

Having the alternate pathways is efficient, because cells can direct energy to producing NADPH and ATP or to producing ATP alone. NADPH accumulates when it is not being used, such as when sugar production declines during cold snaps. The excess NADPH backs up the noncyclic pathway, so the cyclic pathway predominates. The cell still makes ATP, but not NADPH. When sugar production is in high gear, NADPH is being used quickly. It does not accumulate, and the noncyclic pathway is the predominant one.

Sunlight and water are essentially unlimited, free resources here on Earth. If we could use solar energy to split water molecules even a fraction as efficiently as photosystem II does, we would have a cheap source of hydrogen that we could use as clean-burning, renewable fuel.

Take-Home Message **How does energy flow during the reactions of photosynthesis?**

> Light provides energy inputs that keep electrons flowing through electron transfer chains.

> Energy lost by electrons as they flow through the chains sets up a hydrogen ion gradient that drives the synthesis of ATP alone, or ATP and NADPH.

❭ The chloroplast is a sugar factory operated by enzymes of the Calvin–Benson cycle. The cyclic, light-independent reactions are the "synthesis" part of photosynthesis.

❬ Links to Carbohydrates 3.3, ATP energy transfers and phosphorylation 5.3

The enzyme-mediated reactions of the **Calvin–Benson cycle** build sugars in the stroma of chloroplasts (Figure 6.9). These reactions are light-independent because light energy does not power them. Instead, they run on ATP and NADPH that formed in the light-dependent reactions.

Light-independent reactions use carbon atoms from CO_2 to make sugars. Extracting carbon atoms from an inorganic source and incorporating them into an organic molecule is a process called **carbon fixation**. In most plants, photosynthetic protists, and some bacteria, the enzyme **rubisco** fixes carbon by attaching CO_2 to five-carbon RuBP (ribulose bisphosphate) ❶.

The six-carbon intermediate that forms by this reaction is unstable, so it splits right away into two three-carbon molecules of PGA (phosphoglycerate). Each of the PGAs receives a phosphate group from ATP, and hydrogen and electrons from NADPH. Thus, ATP energy and the reducing power of NADPH convert each molecule of PGA into a molecule of PGAL (phosphoglyceraldehyde), a phosphorylated sugar ❷.

glucose

In later reactions, two or more of the three-carbon PGAL molecules can be combined and rearranged to form larger carbohydrates. Glucose, remember, has six carbon atoms. To make one glucose molecule, six CO_2 must be attached to six RuBP molecules, so twelve PGAL form. Two PGAL combine to form one glucose molecule ❸. The ten remaining PGAL regenerate the starting compound of the cycle, RuBP ❹.

Plants can use the glucose they make in the light-independent reactions as building blocks for other organic molecules, or they can break it down to access the energy held in its bonds. However, most of the glucose is converted at once to sucrose or starch by other pathways that conclude the light-independent reactions. Excess glucose is stored in the form of starch grains inside the stroma of chloroplasts. When sugars are needed in other parts of the plant, the starch is broken down to sugar monomers and exported from the cell.

Calvin–Benson cycle Light-independent reactions of photosynthesis; cyclic carbon-fixing pathway that forms sugars from CO_2.
carbon fixation Process by which carbon from an inorganic source such as carbon dioxide gets incorporated into an organic molecule.
rubisco Ribulose bisphosphate carboxylase. Carbon-fixing enzyme of the Calvin–Benson cycle.

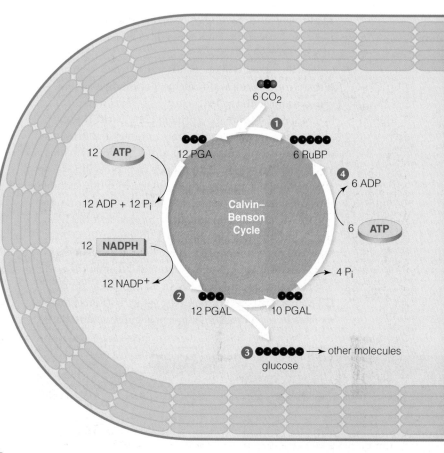

Figure 6.9 Animated Light-independent reactions of photosynthesis. The sketch shows a cross-section of a chloroplast with the light-independent reactions cycling in the stroma.

The steps shown are a summary of six cycles of the Calvin–Benson reactions. Black balls signify carbon atoms. Appendix VI details the reaction steps.

❶ Six CO_2 diffuse into a photosynthetic cell, and then into a chloroplast. Rubisco attaches each to a RuBP molecule. The resulting intermediates split, so twelve molecules of PGA form.

❷ Each PGA molecule gets a phosphate group from ATP, plus hydrogen and electrons from NADPH. Twelve PGAL form.

❸ Two PGAL combine to form one glucose molecule.

❹ The remaining ten PGAL receive phosphate groups from ATP. The transfer primes them for endergonic reactions that regenerate the 6 RuBP.

Take-Home Message **What happens during the light-independent reactions of photosynthesis?**

❭ The light-independent reactions of photosynthesis run on the bond energy of ATP and the energy of electrons donated by NADPH. Both molecules formed in the light-dependent reactions.

❭ Collectively called the Calvin–Benson cycle, these carbon-fixing reaction use hydrogen (from NADPH), and carbon and oxygen (from CO_2) to build sugars.

> Environments differ, and so do details of photosynthesis.
< Links to Surface specializations 4.11, Controls over metabolic reactions 5.5

Several adaptations allow plants to live where water is scarce or sporadically available. For example, a thin, waterproof coating called a cuticle prevents water loss by evaporation from aboveground plant parts. However, a cuticle also prevents gases from entering and exiting a plant by diffusing through cells at the surfaces of leaves and stems. Gases play a critical role in photosynthesis, so photosynthetic parts are often studded with tiny, closable gaps called **stomata** (singular, stoma). When stomata are open, carbon dioxide for the light-independent reactions can diffuse from air into the plant's photosynthetic tissues, and oxygen produced by the light-dependent reactions can diffuse from photosynthetic cells into the air.

Plants that use only the Calvin–Benson cycle are called **C3 plants**, because 3-carbon PGA is the first stable intermediate to form in the light-independent reactions. C3

plants typically conserve water on dry days by closing their stomata. However, when stomata are closed, oxygen produced by the light-dependent reactions cannot escape from the plant. Oxygen that accumulates in photosynthetic tissues limits sugar production. Why? At high oxygen levels, rubisco attaches oxygen (instead of carbon) to RuBP. This pathway, which is called **photorespiration**, produces carbon dioxide, so the plant loses carbon instead of fixing it (Figure 6.10A,B). In addition, ATP and NADPH are used to convert the pathway's intermediates to a molecule that can enter the Calvin–Benson cycle, so extra energy is required to make sugars on dry days. C3 plants compensate for rubisco's inefficiency by making a lot of it: Rubisco is the most abundant protein on Earth.

Some plants have adaptations that help them minimize photorespiration. A 4-carbon molecule is the first stable intermediate that forms in the light-independent reactions of corn, bamboo, and other **C4 plants** (Figure 6.10C). Such plants fix carbon twice, in two kinds of cells. In the first kind of cell, carbon is fixed by an enzyme that does not use oxygen even at high oxygen levels. The resulting intermediate is transported to another kind of cell, where it is converted to carbon dioxide that enters the Calvin–Benson cycle. The extra reactions keep the carbon dioxide level high near rubisco, thus minimizing photorespiration.

In cactuses and other **CAM plants**, the extra reactions run at a different time rather than in different cells: The C4 reactions run during the day, and the Calvin–Benson cycle runs at night. CAM stands for crassulacean acid metabolism, after the Crassulaceae family of plants in which this pathway was first studied (Figure 6.10D).

C3 plant Type of plant that uses only the Calvin–Benson cycle to fix carbon.
C4 plant Type of plant that minimizes photorespiration by fixing carbon twice, in two cell types.
CAM plant Type of C4 plant that conserves water by fixing carbon twice, at different times of day.
photorespiration Reaction in which rubisco attaches oxygen instead of carbon dioxide to ribulose bisphosphate.
stomata Gaps that open on plant surfaces; allow water vapor and gases to diffuse across the epidermis.

Take-Home Message How do carbon–fixing reactions vary?

> When stomata are closed, oxygen builds up inside leaves of C3 plants. Rubisco then can attach oxygen (instead of carbon dioxide) to RuBP. This reaction, photorespiration, reduces the efficiency of sugar production, so it can limit the plant's growth.

> Plants adapted to dry conditions limit photorespiration by fixing carbon twice. C4 plants separate the two sets of reactions in space; CAM plants separate them in time.

Figure 6.10 Animated Carbon-fixing adaptations. Most plants, including basswood (*Tilia americana* **A**), are C3 plants. **B** Photorespiration in C3 plants makes sugar production inefficient on dry days. Additional reactions minimize photorespiration in C4 plants such as corn (*Zea mays* **C**), and CAM plants such as jade plants (*Crassula argentea* **D**).

Green Energy (revisited)

Your body is about 9.5 percent carbon by weight, which means that you contain an enormous number of carbon atoms. Where did they all come from? You eat other organisms to get the carbon atoms your body uses for energy and for raw materials. Those atoms may have passed through other heterotrophs before you ate them, but at some point they were part of photosynthetic organisms. Photosynthesizers strip carbon from carbon dioxide, then use the atoms to build organic compounds. Your carbon atoms—and those of most other organisms—were recently part of Earth's atmosphere, in molecules of CO_2.

Photosynthesis removes carbon dioxide from the atmosphere, and locks its carbon atoms inside organic compounds. When photosynthesizers and other aerobic organisms break down the organic compounds for energy, carbon atoms are released in the form of CO_2, which then reenters the atmosphere. Since photosynthesis evolved, these two processes have constituted a balanced cycle of the biosphere. You will learn more about the carbon cycle in Section 42.8. For now, know that the amount of carbon dioxide that photosynthesis removes from the atmosphere is roughly the same amount that organisms release back into it. At least it was, until humans came along.

As early as 8,000 years ago, humans began burning forests to clear land for agriculture. When trees and other plants burn, most of the carbon locked in their tissues is released into the atmosphere as carbon dioxide. Fires that occur naturally release carbon dioxide the same way.

Today, we are burning a lot more than our ancestors ever did. In addition to wood, we are burning fossil fuels—coal, petroleum, and natural gas—to satisfy our greater and greater demands for energy (Figure 6.11). Fossil fuels are the organic remains of ancient organisms. When we burn these fuels, the carbon that has been locked in them for hundreds of millions of years is released back into the atmosphere, mainly as carbon dioxide.

Our activities have put Earth's atmospheric cycle of carbon dioxide out of balance. We are adding far more CO_2 to the atmosphere than photosynthetic organisms are removing from it. Today, we release about 28 billion tons of carbon dioxide into the atmosphere each year, more than ten times the amount we released in the year 1900. Most of it comes from burning fossil fuels. How do we know? Researchers can determine how long ago the carbon atoms in a sample of CO_2 were part of a living organism by measuring the ratio of different carbon isotopes in it (you will read more about radioisotope dating techniques in Section 16.5). These results are correlated with fossil fuel extraction, refining, and trade statistics.

Figure 6.11 Visible evidence of fossil fuel emissions in the atmosphere: the sky over New York City on a sunny day.

Researchers find pockets of our ancient atmosphere in Antarctica. Snow and ice have been accumulating in layers there, year after year, for the last 15 million years. Air and dust trapped in each layer reveal the composition of the atmosphere that prevailed when the layer formed. Thus, we now know that the atmospheric CO_2 level had been relatively stable for about 10,000 years before the industrial revolution. Since 1850, the CO_2 level has been steadily rising. In 2008, it was higher than it had been in *24 million years.*

The increase in atmospheric carbon dioxide is having dramatic effects on climate. CO_2 contributes to global climate change. We are seeing a warming trend that mirrors the increase in CO_2 levels: Earth is now the warmest it has been for 12,000 years. The trend is affecting biological systems everywhere. Life cycles are changing—birds are laying eggs earlier; plants are flowering earlier than usual; mammals are hibernating for shorter periods. Migration patterns and habitats are also changing. These changes may be too fast for many species, and the rate of extinctions is rising.

Under normal circumstances, extra carbon dioxide stimulates photosynthesis, which means extra carbon dioxide uptake. However, changes in temperature and moisture patterns as a result of global warming are offsetting this benefit because they are proving harmful to plants and other photosynthesizers.

Making biofuel production economically feasible is a high priority for today's energy researchers. Biofuels are a renewable source of energy: We can always make more of them simply by growing more biomass. Also, unlike fossil fuels, using plant matter for fuel recycles carbon that is already in the atmosphere, because plants remove carbon dioxide from the atmosphere as they grow.

How Would You Vote? Ethanol and other fuels manufactured from crops currently cost more than gasoline, but they are renewable energy sources and have fewer emissions. Would you pay a premium to drive a vehicle that runs on biofuels? See CengageNow for details, then vote online (cengagenow.com).

Summary

Section 6.1 Autotrophs make their own food using energy they get directly from the environment, and carbon from inorganic sources such as CO_2. By metabolic pathways of **photosynthesis**, plants and other autotrophs capture the energy of light and use it to build sugars from water and carbon dioxide. **Heterotrophs** get energy and carbon from molecules that other organisms have already assembled.

Sections 6.2, 6.3 Visible light is a very small part of the spectrum of electromagnetic energy radiating from the sun. That energy travels in waves, and it is organized as photons. Visible light drives photosynthesis, which begins when photons are absorbed by photosynthetic pigment molecules. **Pigments** are molecules that absorb light of particular **wavelengths** only; photons not captured by a pigment are reflected as its characteristic color. The main photosynthetic pigment—**chlorophyll a**—absorbs violet and red light, so it appears green. Accessory pigments absorb additional wavelengths.

Section 6.4 In **chloroplasts**, the **light-dependent reactions** of photosynthesis occur at a much-folded **thylakoid membrane**. The membrane forms a continuous compartment in the chloroplast's interior (**stroma**) where the **light-independent reactions** occur.

The following diagram summarizes the overall process of photosynthesis (with the noncyclic reactions):

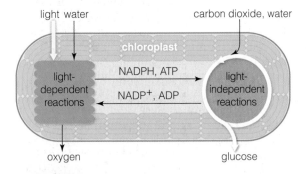

Sections 6.5, 6.6 In the light reactions of photosynthesis, light-harvesting complexes in the thylakoid membrane absorb photons and pass the energy to **photosystems**, which then release electrons.

In the noncyclic pathway, electrons released from photosystem II flow through an electron transfer chain, then to photosystem I. Photon energy causes photosystem I to release electrons, which end up in NADPH. Photosystem II replaces lost electrons by pulling them from water, which then dissociates into H^+ and O_2 (an example of **photolysis**).

In the cyclic pathway, the electrons released from photosystem I enter an electron transfer chain, then cycle back to photosystem I. NADPH does not form.

ATP forms by **electron transfer phosphorylation** in both pathways. Electrons flowing through electron transfer chains cause H^+ to accumulate in the thylakoid compartment. The H^+ follows its gradient back across the membrane through ATP synthases, driving ATP synthesis.

Section 6.7 Carbon fixation occurs in light-independent reactions. Inside the stroma, the enzyme **rubisco** attaches a carbon from CO_2 to RuBP to start the **Calvin–Benson cycle**. This cyclic pathway uses energy from ATP, carbon and oxygen from CO_2, and hydrogen and electrons from NADPH to make sugars.

Section 6.8 Environments differ, and so do details of the light-independent reactions. On dry days, plants conserve water by closing their **stomata**. However, when stomata are closed, O_2 from photosynthesis cannot escape the plant, and CO_2 for photosynthesis cannot enter it. In **C3 plants**, the resulting high O_2 level in the plant's tissues causes rubisco to attach O_2 instead of CO_2 to RuBP. This pathway, which is called **photorespiration**, reduces the efficiency of sugar production on dry days. In **C4 plants**, carbon fixation occurs twice. The first reactions release CO_2 near rubisco, and thus limit photorespiration when stomata are closed. **CAM plants** minimize photorespiration by opening their stomata and fixing carbon at night.

Self-Quiz   Answers in Appendix III

1. A cat eats a bird, which ate a caterpillar that chewed on a weed. Which organisms are autotrophs? Which ones are heterotrophs?

2. Photosynthetic autotrophs use _____ from the air as a carbon source and _____ as their energy source.

3. Chlorophyll *a* appears green because it absorbs mainly _____ light.
   a. violet and red   c. yellow
   b. green   d. blue

4. Light-dependent reactions in plants proceed in the _____ .
   a. thylakoid membrane   c. stroma
   b. plasma membrane   d. cytoplasm

5. When a photosystem absorbs light, _____ .
   a. sugar phosphates are produced
   b. electrons are transferred to ATP
   c. RuBP accepts electrons
   d. electrons are ejected from its special pair

6. In the light-dependent reactions, _____ .
   a. carbon dioxide is fixed   d. CO_2 accepts electrons
   b. ATP forms   e. b and c
   c. sugars form   f. a and c

7. What accumulates inside the thylakoid compartment during the light-dependent reactions?
   a. glucose   c. O_2
   b. hydrogen ions   d. CO_2

Energy Efficiency of Biofuel Production From Corn, Soy, and Prairie Grasses

Most corn is grown intensively in vast swaths, which means that farmers who grow it use fertilizers and pesticides, both of which are typically made from fossil fuels. Corn is an annual plant, and yearly harvests tend to cause runoff that depletes soil and pollutes rivers.

In 2006, David Tilman and his colleagues published the results of a 10-year study comparing the net energy output of various biofuels. The researchers grew a mixture of native perennial grasses without irrigation, fertilizer, pesticides, or herbicides, in sandy soil that was so depleted by intensive agriculture that it had been abandoned. They measured the usable energy in biofuels made from the grasses, from corn, and from soy. They also measured the energy it took to grow and produce each kind of biofuel. Some of their results are shown in Figure 6.12.

1. About how much energy did ethanol produced from one hectare of corn yield? How much energy did it take to grow the corn to make that ethanol?
2. Which of the biofuels tested had the highest ratio of energy output to energy input?
3. Which of the three crops would require the least amount of land to produce a given amount of biofuel energy?

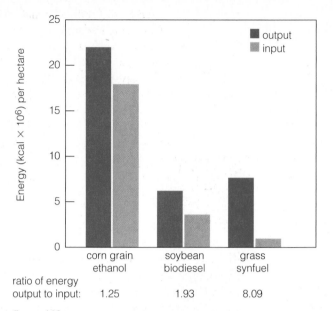

	corn grain ethanol	soybean biodiesel	grass synfuel
ratio of energy output to input:	1.25	1.93	8.09

Figure 6.12 Energy inputs and outputs of biofuels from corn and soy grown on fertile farmland, and grasses grown in infertile soil. One hectare is about 2.5 acres.

8. Light-independent reactions in plants proceed in the _____ .
   a. thylakoid membrane   c. stroma
   b. plasma membrane   d. cytoplasm

9. The Calvin–Benson cycle starts when _____ .
   a. light is available
   b. carbon dioxide is attached to RuBP
   c. electrons leave a photosystem II

10. Which of the following substances does *not* participate in the Calvin–Benson cycle?
    a. ATP   d. PGAL
    b. NADPH   e. O_2
    c. RuBP   f. CO_2

11. In the light-independent reactions, _____ .
    a. carbon dioxide is fixed   d. CO_2 accepts electrons
    b. ATP forms   e. b and c
    c. sugars form   f. a and c

12. Match each with its most suitable description.
    ___ PGAL formation   a. absorbs light
    ___ CO_2 fixation   b. converts light to
    ___ photolysis   chemical energy
    ___ ATP forms; NADPH   c. self-feeder
        does not   d. electrons cycle back
    ___ photorespiration   to photosystem I
    ___ photosynthesis   e. problem in C3 plants
    ___ pigment   f. ATP, NADPH
    ___ autotroph   required
                              g. water molecules split
                              h. rubisco function

Additional questions are available on **CENGAGENOW**.

Critical Thinking

1. About 200 years ago, Jan Baptista van Helmont wanted to know where growing plants get the materials necessary for increases in size. He planted a tree seedling weighing 5 pounds in a barrel filled with 200 pounds of soil and then watered the tree regularly. After five years, the tree weighed 169 pounds, 3 ounces, and the soil weighed 199 pounds, 14 ounces. Because the tree had gained so much weight and the soil had lost so little, he concluded that the tree had gained all of its additional weight by absorbing the water he had added to the barrel, but of course he was incorrect. What really happened?

2. While gazing into an aquarium, you observe bubbles coming from an aquatic plant (*left*). What are the bubbles and where do they come from?

3. A C3 plant absorbs a carbon radio-isotope (as part of $^{14}CO_2$). In which stable, organic compound does the labeled carbon appear first? Which compound forms first if a C4 plant absorbs the same radioisotope?

Animations and Interactions on **CENGAGENOW**:
❯ Wavelengths of light; Englemann's experiment; Sites of photosynthesis in a plant; Chloroplast structure; The noncyclic pathway; Calvin–Benson cycle; Photosynthesis in C3 and C4 plants compared.

‹ Links to Earlier Concepts

This chapter expands the picture of energy flow through the world of life (Section 5.2). It focuses on metabolic reactions and pathways (3.2, 5.3, 5.5) that make ATP by degrading carbohydrates (3.3). These reactions occur either in cytoplasm or in mitochondria (4.9). You will revisit lipids (3.4) and proteins (3.5), membrane transport (4.4, 5.7), free radicals (5.4), electron transfer chains (5.5, 6.5), and photosynthesis pathways (6.4, 6.7).

Key Concepts

Energy From Carbohydrates
Various pathways convert the chemical energy of glucose and other organic compounds to the chemical energy of ATP. Aerobic respiration yields the most ATP from each glucose molecule. In eukaryotes, this pathway ends in mitochondria.

Glycolysis
Glycolysis, the first stage of aerobic respiration and of anaerobic fermentation pathways, occurs in cytoplasm. Enzymes of glycolysis convert one molecule of glucose to two molecules of pyruvate for a net yield of two ATP.

7 How Cells Release Chemical Energy

7.1 When Mitochondria Spin Their Wheels

In the early 1960s, a Swedish physician, Rolf Luft, mulled over a patient's odd symptoms. The young woman felt weak and hot all the time. Even on the coldest winter days she could not stop sweating, and her skin was always flushed. She was thin, yet had a huge appetite. Luft inferred that his patient's symptoms pointed to a metabolic disorder: Her cells were very active, but much of their activity was being lost as metabolic heat. Luft checked the patient's rate of metabolism, the amount of energy her body was expending. Even while resting, her oxygen consumption was the highest that had ever been recorded.

Examination of a tissue sample revealed that the patient's skeletal muscles had plenty of mitochondria, the cell's ATP-producing powerhouses. But there were too many of them, and they were abnormally shaped. The mitochondria were making very little ATP despite working at top speed.

The disorder, now called Luft's syndrome, was the first to be linked to defective mitochondria. The cells of someone with Luft's syndrome are like cities that are burning tons of coal in many power plants but not getting much usable energy output. Skeletal and heart muscles, the brain, and other hardworking body parts with high energy demands are most affected.

More than forty disorders related to defective mitochondria are now known. One called Friedreich's ataxia causes loss of coordination (ataxia), weak muscles, and heart problems (Figure 7.1). Many of those affected die when they are young adults.

Like the chloroplasts described in the previous chapter, mitochondria have an internal folded membrane system that allows them to make ATP. By the process of aerobic respiration, electron transfer chains in the mitochondrial membrane set up hydrogen ion gradients that power ATP formation.

In Luft's syndrome, electron transfer chains in mitochondria work overtime, but too little ATP forms. In Friedreich's ataxia, a protein called frataxin does not work properly. This protein helps build some of the iron-containing enzymes of electron

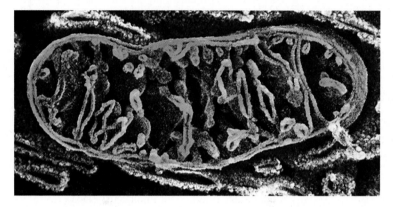

Figure 7.1 The mitochondrion, an ATP-producing powerhouse. *Above*, a mitochondrion's folded internal membrane is the source of its function.

Opposite, mitochondrial diseases such as Friedreich's ataxia cause serious health problems. Leah (*left*) started to lose her sense of balance and coordination at age five. Six years later she was in a wheelchair; now she is diabetic and partially deaf. Her brother Joshua (*right*) could not walk by the time he was eleven, and is now blind. Both have heart problems; both had spinal fusion surgery. Special equipment allows them to attend school and work part-time. Leah is a professional model.

transfer chains. When it malfunctions, iron atoms that were supposed to be incorporated into the enzymes accumulate inside mitochondria instead. Oxygen is present in mitochondria, and free radicals form when oxygen reacts with the iron atoms (Section 5.4). When too much iron accumulates in mitochondria, too many free radicals form. These free radicals destroy the molecules of life faster than they can be repaired or replaced. Eventually, the mitochondria stop working, and the cell dies.

You already have a sense of how cells harvest energy in electron transfer chains. Details of the reactions vary from one type of organism to the next, but all life relies on this ATP-forming machinery. When you consider mitochondria in this chapter, remember that without them, you would not make enough ATP to even read about how they do it.

Aerobic Respiration
The final stages of aerobic respiration break down pyruvate to CO_2. Many coenzymes that become reduced deliver electrons and hydrogen ions to electron transfer chains, where ATP forms by electron transfer phosphorylation.

Fermentation
Fermentation pathways start with glycolysis. Substances other than oxygen accept electrons at the end of the pathways. Compared with aerobic respiration, the net yield of ATP from fermentation is small.

Other Metabolic Pathways
Molecules other than carbohydrates are common energy sources in the animal body. Many different pathways can convert dietary lipids and proteins to molecules that may enter glycolysis or the Krebs cycle.

> Photoautotrophs use the ATP they produce by photosynthesis to make sugars.

> Most organisms, including photoautotrophs, make ATP by breaking down sugars and other organic compounds.

❮ Links to Energy flow 5.2, Free radicals 5.4, Metabolic pathways 5.5, Photosynthesis 6.4

Evolution of Earth's Atmosphere

The first cells on Earth did not use sunlight for energy. Like some modern archaeans, these ancient organisms extracted energy and carbon from simple molecules such as methane and hydrogen sulfide—gases that were plentiful in the nasty brew that constituted Earth's early atmosphere (Figure 7.2).

The first photosynthetic autotrophs, or **photoautotrophs**, evolved about 3.2 billion years ago, probably in shallow ocean waters. Sunlight offered these organisms an essentially unlimited supply of energy, and they were very successful. Oxygen gas (O_2) released from uncountable numbers of water molecules began seeping out of uncountable numbers of photosynthesizers, and it accumulated in the ocean and the atmosphere. From that time on, the world of life would never be the same.

Molecular oxygen had been a very small component of Earth's early atmosphere before photosynthesis evolved. The new abundance of atmospheric oxygen exerted tremendous selection pressure on all organisms. Why? Oxygen gas reacts with metals such as enzyme cofactors, and free radicals form during those reactions. Free radicals, remember, damage biological molecules, so they are dangerous to life (Section 5.4).

The ancient cells had no way to detoxify oxygen radicals, so most of them quickly died out. Only a few types persisted in deep water, muddy sediments, and other **anaerobic** (oxygen-free) habitats. Then, new metabolic pathways that detoxified oxygen radicals evolved in the survivors. Organisms with such pathways were the first

Figure 7.2 Then and now: An artist's conception of how Earth was permanently altered by the evolution of photosynthesis and aerobic respiration.

aerobic organisms—they could live in the presence of oxygen. One of the pathways, aerobic respiration, put the reactive properties of oxygen to good use. **Aerobic respiration** is one of several pathways by which organisms access the energy stored in carbohydrates. This equation summarizes aerobic respiration:

$$C_6H_{12}O_6 \ + \ O_2 \ \longrightarrow \ CO_2 \ + \ H_2O$$
$$\text{glucose} \qquad \text{oxygen} \qquad \text{carbon dioxide} \quad \text{water}$$

Note that aerobic respiration requires oxygen (a by-product of photosynthesis), and it produces carbon dioxide and water (the same raw materials from which photosynthesizers make sugars). With this connection, the cycling of carbon, hydrogen, and oxygen through the biosphere came full circle (Figure 7.3).

Carbohydrate Breakdown Pathways

Like those early organisms, photosynthetic autotrophs of today's world capture energy from the sun, and store it in the form of carbohydrates. They and most other organisms use energy stored in carbohydrates to run the diverse reactions that sustain life. However, carbohydrates rarely participate in such reactions, so how do cells harness their energy? In order to use the energy stored in carbohydrates, cells must first transfer it to energy-carriers such as ATP, which does participate in many of the energy-requiring reactions that a cell runs. The transfer occurs by breaking the bonds of the carbohydrates, which releases energy that drives ATP synthesis.

There are a few different pathways that break down carbohydrates, but aerobic respiration is the one that typical eukaryotic cells use at least most of the time. Aerobic respiration yields more ATP than other carbohydrate-breakdown pathways. You and other multicelled organisms could not live without its higher yield.

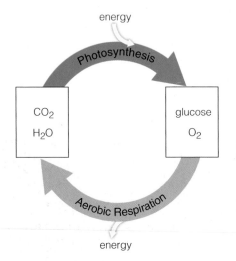

Figure 7.3 Animated
The connection between photosynthesis and aerobic respiration. Note the cycling of materials, and the one-way flow of energy (compare Figure 5.5).

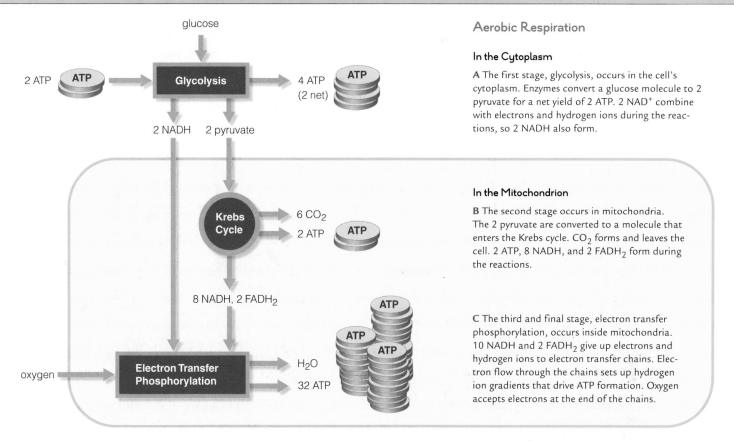

Figure 7.4 Animated Overview of aerobic respiration. The reactions start in the cytoplasm and end in mitochondria. **>> Figure It Out** What is aerobic respiration's typical net yield of ATP?

Answer: 38 − 2 = 36 ATP per glucose

Aerobic Respiration

In the Cytoplasm

A The first stage, glycolysis, occurs in the cell's cytoplasm. Enzymes convert a glucose molecule to 2 pyruvate for a net yield of 2 ATP. 2 NAD$^+$ combine with electrons and hydrogen ions during the reactions, so 2 NADH also form.

In the Mitochondrion

B The second stage occurs in mitochondria. The 2 pyruvate are converted to a molecule that enters the Krebs cycle. CO_2 forms and leaves the cell. 2 ATP, 8 NADH, and 2 FADH$_2$ form during the reactions.

C The third and final stage, electron transfer phosphorylation, occurs inside mitochondria. 10 NADH and 2 FADH$_2$ give up electrons and hydrogen ions to electron transfer chains. Electron flow through the chains sets up hydrogen ion gradients that drive ATP formation. Oxygen accepts electrons at the end of the chains.

Most types of eukaryotic cells either use aerobic respiration exclusively, or they use it most of the time. Many bacteria, archaeans, and protists use alternative pathways. Bacteria and single-celled protists that inhabit sea sediments, animal guts, improperly canned food, sewage treatment ponds, deep mud, and other anaerobic habitats use **fermentation**, a pathway by which cells harvest energy from carbohydrates anaerobically. Some of these organisms, including the bacteria that cause botulism, cannot tolerate aerobic conditions, and will die when they are exposed to oxygen.

Fermentation and aerobic respiration begin with the same reactions in the cytoplasm. These reactions, which are collectively called **glycolysis**, convert one six-carbon molecule of glucose into two molecules of **pyruvate**, an organic compound with a three-carbon backbone (Figure 7.4A). After glycolysis, the pathways of fermentation and aerobic respiration diverge. Aerobic respiration continues with two more stages that occur inside mitochondria (Figure 7.4B,C). It ends when oxygen accepts electrons at the end of electron transfer chains. Fermentation ends in the cytoplasm, where a molecule other than oxygen accepts electrons. Aerobic respiration is much more efficient than fermentation. You and other multicelled organisms could not live without its higher yield of ATP.

aerobic Involving or occurring in the presence of oxygen.
aerobic respiration Oxygen-requiring pathway that breaks down carbohydrates to produce ATP.
anaerobic Occurring in the absence of oxygen.
fermentation An anaerobic pathway by which cells harvest energy from carbohydrates to produce ATP.
glycolysis Set of reactions in which glucose or another sugar is broken down to two pyruvate for a net yield of two ATP.
photoautotroph Photosynthetic autotroph.
pyruvate Three-carbon end product of glycolysis.

Take-Home Message **How do cells access the chemical energy in carbohydrates?**

> Most cells convert the chemical energy of carbohydrates to the chemical energy of ATP by aerobic respiration or fermentation. Aerobic respiration and fermentation pathways start in cytoplasm, with glycolysis.

> Fermentation is anaerobic and ends in the cytoplasm.

> In eukaryotes, aerobic respiration requires oxygen and ends in mitochondria. This pathway evolved after a dramatic increase in oxygen content of the atmosphere brought about by the evolution of photosynthesis.

> The reactions of glycolysis convert one molecule of glucose to two molecules of pyruvate for a net yield of two ATP.
> An energy investment of ATP is required to start glycolysis.
< Links to Hydrolysis 3.2, Glucose 3.3, Endergonic reactions and phosphorylation 5.3, Redox reaction 5.5, Gradients 5.6, Glucose transporter 5.7, Calvin–Benson cycle 6.7

Glycolysis is a series of reactions that begins carbohydrate breakdown pathways. The reactions, which occur in the cytoplasm, convert one molecule of glucose to two molecules of pyruvate:

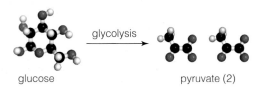

glucose   glycolysis →   pyruvate (2)

The word glycolysis (from the Greek words *glyk–*, sweet; and *–lysis,* loosening) refers to the release of chemical energy from sugars. Different sugars can enter glycolysis, but for clarity we focus here on glucose.

Glycolysis begins when a molecule of glucose enters a cell through a glucose transporter, a passive transport protein you encountered in Section 5.7. The cell invests two ATP in the endergonic reactions that begin the pathway (Figure 7.5). In the first reaction, a phosphate group is transferred from ATP to the glucose, thus forming glucose-6-phosphate ❶. A model of hexokinase, the enzyme that catalyzes this reaction, is pictured in Section 5.4.

Unlike glucose, glucose-6-phosphate does not pass through glucose transporters in the plasma membrane, so it is trapped inside the cell. Almost all of the glucose that enters a cell is immediately converted to glucose-6-phosphate. This phosphorylation keeps the glucose concentration in the cytoplasm lower than it is in the fluid outside of the cell. By maintaining this concentration gradient across the plasma membrane, the cell favors uptake of even more glucose.

Glycolysis continues as glucose-6-phosphate accepts a phosphate group from another ATP, then splits in two ❷, forming two PGAL (phosphoglyceraldehyde). This phosphorylated sugar also forms during the Calvin–Benson cycle (Section 6.7). A second phosphate group is attached to each PGAL, so two PGA (phosphoglycerate) form ❸. During the reaction, two electrons and a hydrogen ion are transferred from each PGAL to NAD+, so two NADH form.

Next, a phosphate group is transferred from each PGA to ADP, so two ATP form ❹. Two more ATP form when a phosphate group is transferred from another pair of inter-

mediates to two ADP ❺. This and any other reaction that transfers a phosphate group directly from a substrate to ADP is called a **substrate–level phosphorylation**.

Glycolysis ends with the formation of two three-carbon pyruvate molecules. These products may now enter the second-stage reactions of either aerobic respiration or fermentation. Remember, two ATP were invested to initiate the reactions of glycolysis. A total of four ATP form, so the net yield is two ATP per molecule of glucose that enters glycolysis ❻. Two NAD+ also pick up hydrogen ions and electrons, thereby becoming reduced to NADH:

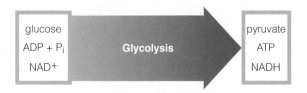

glucose
ADP + P_i
NAD+

Glycolysis

pyruvate
ATP
NADH

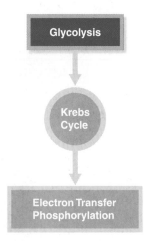

Glycolysis

Krebs Cycle

Electron Transfer Phosphorylation

Figure 7.5 Animated Glycolysis.

This first stage of carbohydrate breakdown starts and ends in the cytoplasm of all cells. *Opposite*, for clarity, we track only the six carbon atoms (*black* balls) that enter the reactions as part of glucose. Appendix VI has more details for interested students.

Cells invest two ATP to start glycolysis, so the net energy yield from one glucose molecule is two ATP. Two NADH also form, and two pyruvate molecules are the end products.

Take-Home Message What is glycolysis?

> Glycolysis is the first stage of carbohydrate breakdown in both aerobic respiration and fermentation.

> The reactions of glycolysis occur in the cytoplasm.

> Glycolysis converts one molecule of glucose to two molecules of pyruvate, with a net energy yield of two ATP. Two NADH also form.

---

substrate–level phosphorylation A reaction that transfers a phosphate group from a substrate directly to ADP, thus forming ATP.

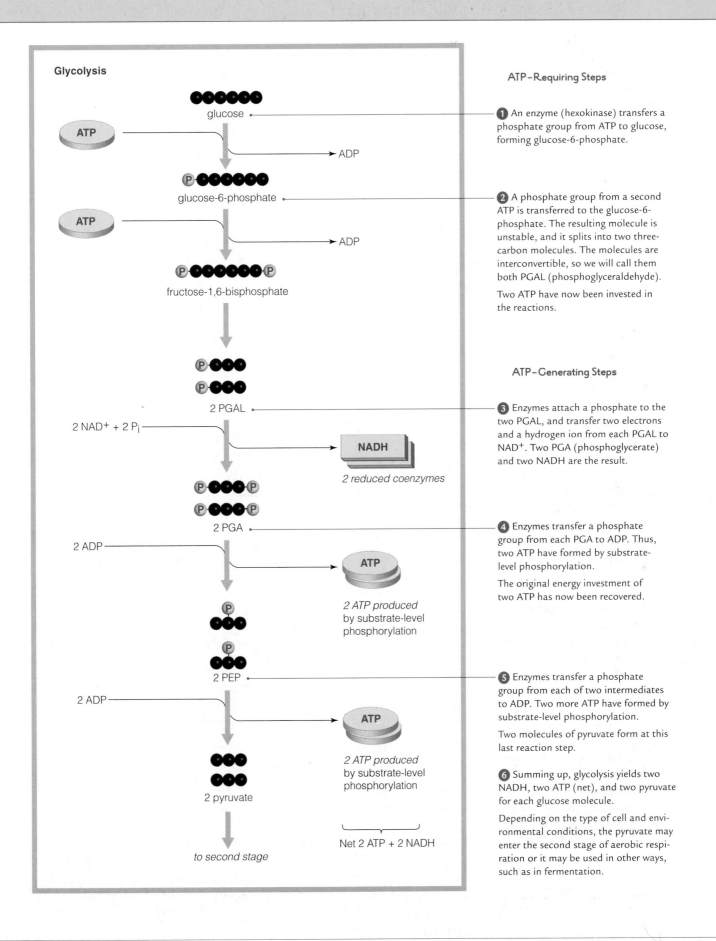

Glycolysis

glucose

ATP

ADP

glucose-6-phosphate

ATP

ADP

fructose-1,6-bisphosphate

2 PGAL

2 NAD⁺ + 2 Pᵢ

NADH

2 reduced coenzymes

2 PGA

2 ADP

ATP

2 ATP produced by substrate-level phosphorylation

2 PEP

2 ADP

ATP

2 ATP produced by substrate-level phosphorylation

2 pyruvate

to second stage

Net 2 ATP + 2 NADH

ATP–Requiring Steps

1 An enzyme (hexokinase) transfers a phosphate group from ATP to glucose, forming glucose-6-phosphate.

2 A phosphate group from a second ATP is transferred to the glucose-6-phosphate. The resulting molecule is unstable, and it splits into two three-carbon molecules. The molecules are interconvertible, so we will call them both PGAL (phosphoglyceraldehyde).

Two ATP have now been invested in the reactions.

ATP–Generating Steps

3 Enzymes attach a phosphate to the two PGAL, and transfer two electrons and a hydrogen ion from each PGAL to NAD⁺. Two PGA (phosphoglycerate) and two NADH are the result.

4 Enzymes transfer a phosphate group from each PGA to ADP. Thus, two ATP have formed by substrate-level phosphorylation.

The original energy investment of two ATP has now been recovered.

5 Enzymes transfer a phosphate group from each of two intermediates to ADP. Two more ATP have formed by substrate-level phosphorylation.

Two molecules of pyruvate form at this last reaction step.

6 Summing up, glycolysis yields two NADH, two ATP (net), and two pyruvate for each glucose molecule.

Depending on the type of cell and environmental conditions, the pyruvate may enter the second stage of aerobic respiration or it may be used in other ways, such as in fermentation.

> The second stage of aerobic respiration finishes the breakdown of glucose that began in glycolysis.
> Links to Mitochondria 4.9, Cyclic pathways 5.5

The second stage of aerobic respiration occurs inside mitochondria (Figure 7.6). It includes two sets of reactions, acetyl–CoA formation and the **Krebs cycle**, that break down pyruvate, the product of glycolysis. All of the carbon atoms that were once part of glucose end up in CO_2, which departs the cell. Only two ATP form. The big payoff is the formation of many reduced coenzymes that drive the third and final stage of aerobic respiration.

The second stage begins when the two pyruvate molecules formed by glycolysis enter a mitochondrion. Pyruvate is transported across the mitochondrion's inner membrane and into the inner compartment, which is called the matrix. Use Figure 7.7 to follow what happens next. In the first reaction, an enzyme splits each molecule of pyruvate into one molecule of CO_2 and a two-carbon acetyl group ❶. The CO_2 diffuses out of the cell, and the acetyl group combines with a molecule called coenzyme A (abbreviated CoA). The product of this reaction is acetyl–CoA. Electrons and hydrogen ions released by the reaction combine with NAD^+, so NADH also forms.

The Krebs Cycle

The Krebs cycle breaks down acetyl–CoA to CO_2. Remember from Section 5.5 that a cyclic pathway is not a physical object, such as a wheel. It is called a cycle because the last

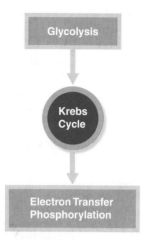

Figure 7.7 Animated Aerobic respiration's second stage: formation of acetyl–CoA and the Krebs cycle. The reactions occur in the mitochondrion's matrix.

Opposite, it takes two cycles of Krebs reactions to break down two pyruvate molecules. After two cycles, all six carbons that entered glycolysis in one glucose molecule have left the cell, in six CO_2.

Two ATP, eight NADH, and two $FADH_2$ form during the two cycles. See Appendix VI for details of the reactions.

reaction in the pathway regenerates the substrate of the first. In the Krebs cycle, a substrate of the first reaction—and a product of the last—is four-carbon oxaloacetate.

During each cycle of Krebs reactions, two carbon atoms of acetyl–CoA are transferred to four-carbon

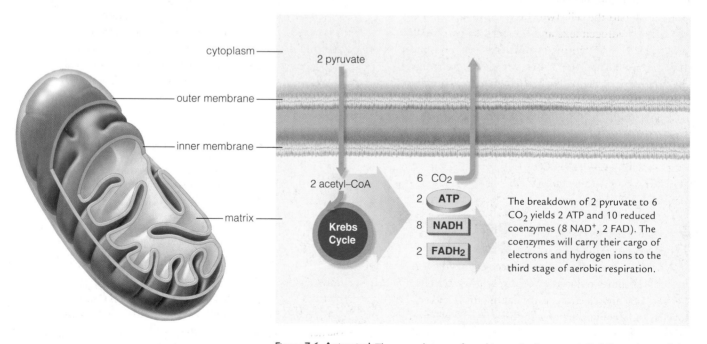

The breakdown of 2 pyruvate to 6 CO_2 yields 2 ATP and 10 reduced coenzymes (8 NAD^+, 2 FAD). The coenzymes will carry their cargo of electrons and hydrogen ions to the third stage of aerobic respiration.

a mitochondrion

Figure 7.6 Animated The second stage of aerobic respiration, acetyl–CoA formation and the Krebs cycle, occurs inside mitochondria. *Left*, an inner membrane divides a mitochondrion's interior into two fluid-filled compartments. *Right*, the second stage of aerobic respiration takes place in the mitochondrion's innermost compartment, or matrix.

Aerobic Res

7.5

❯ Many ATP are
aerobic res
❮ Links to
Ele

Acetyl–CoA Formation and the Krebs Cycle

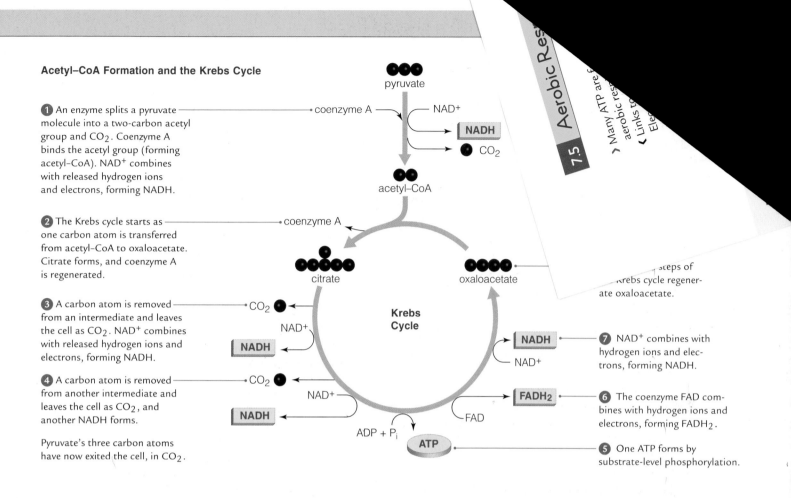

1 An enzyme splits a pyruvate molecule into a two-carbon acetyl group and CO_2. Coenzyme A binds the acetyl group (forming acetyl–CoA). NAD^+ combines with released hydrogen ions and electrons, forming NADH.

2 The Krebs cycle starts as one carbon atom is transferred from acetyl–CoA to oxaloacetate. Citrate forms, and coenzyme A is regenerated.

3 A carbon atom is removed from an intermediate and leaves the cell as CO_2. NAD^+ combines with released hydrogen ions and electrons, forming NADH.

4 A carbon atom is removed from another intermediate and leaves the cell as CO_2, and another NADH forms.

Pyruvate's three carbon atoms have now exited the cell, in CO_2.

7 NAD^+ combines with hydrogen ions and electrons, forming NADH.

6 The coenzyme FAD combines with hydrogen ions and electrons, forming $FADH_2$.

5 One ATP forms by substrate-level phosphorylation.

...steps of
Krebs cycle regener-
ate oxaloacetate.

oxaloacetate, forming citrate, the ionized form of citric acid **2**. The Krebs cycle is also called the citric acid cycle after this first intermediate. In later reactions, two CO_2 form and depart the cell. Two NAD^+ are reduced when they accept hydrogen ions and electrons, so two NADH form **3** and **4**. ATP forms by substrate-level phosphorylation **5**, and FAD **6** and another NAD^+ **7** are reduced. The final steps of the pathway regenerate oxaloacetate **8**.

Remember, glycolysis converted one glucose molecule to two pyruvate, and these were converted to two acetyl–CoA when they entered the matrix of a mitochondrion. There, the second-stage reactions convert the two molecules of acetyl–CoA to six CO_2. At this point in aerobic respiration, one glucose molecule has been broken down completely: Six carbon atoms have left the cell, in six CO_2. Two ATP formed, which adds to the small net yield of glycolysis. However, six NAD^+ were reduced to six NADH, and two FAD were reduced to two $FADH_2$.

What is so important about reduced coenzymes? A molecule becomes reduced when it receives electrons (Section 5.5), and electrons carry energy that can be used to drive endergonic reactions. In this case, the electrons picked up by coenzymes during the first two stages of

aerobic respiration carry energy that drives the reactions of the third stage.

In total, two ATP form and ten coenzymes (eight NAD^+ and two FAD) are reduced during acetyl–CoA formation and the Krebs cycle. Add in the two NAD^+ reduced in glycolysis, and the full breakdown of each glucose molecule has a big potential payoff. Twelve reduced coenzymes will deliver electrons (and the energy they carry) to the third stage of aerobic respiration:

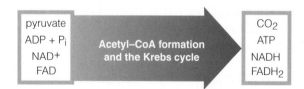

pyruvate		CO_2
$ADP + P_i$	**Acetyl–CoA formation**	ATP
NAD^+	**and the Krebs cycle**	NADH
FAD		$FADH_2$

Take-Home Message What happens during the second stage of aerobic respiration?

❯ The second stage of aerobic respiration, acetyl–CoA formation and the Krebs cycle, occurs in the inner compartment (matrix) of mitochondria.

❯ The pyruvate that formed in glycolysis is converted to acetyl–CoA and carbon dioxide. The acetyl–CoA enters the Krebs cycle, which breaks it down to CO_2.

❯ For each two pyruvate molecules broken down in the second-stage reactions, two ATP form, and ten coenzymes (eight NAD^+ and two FAD) are reduced.

Krebs cycle Cyclic pathway that, along with acetyl–CoA formation, breaks down pyruvate to carbon dioxide.

...ormed during the third and final stage of
...piration.

... Membrane proteins 4.4, Thermodynamics 5.2,
...tron transfer chains 5.5, Selective permeability of cell
...membranes 5.6, Electron transfer phosphorylation 6.5

Electron Transfer Phosphorylation

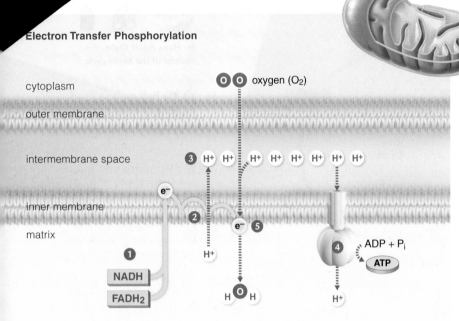

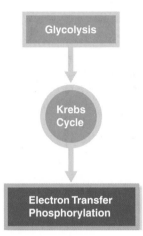

Figure 7.8 The third and final stage of aerobic respiration, electron transfer phosphorylation, occurs at the inner mitochondrial membrane.

❶ NADH and $FADH_2$ deliver electrons to electron transfer chains in the inner mitochondrial membrane.

❷ Electron flow through the chains causes hydrogen ions (H^+) to be pumped from the matrix to the intermembrane space.

❸ The activity of the electron transfer chains causes a hydrogen ion gradient to form across the inner mitochondrial membrane.

❹ Hydrogen ion flow back to the matrix through ATP synthases drives the formation of ATP from ADP and phosphate (P_i).

❺ Oxygen (O_2) accepts electrons and hydrogen ions at the end of mitochondrial electron transfer chains, so water forms.

» Figure It Out Which other metabolic pathway that you have learned about involves electron transfer phosphorylation?

Answer: The light-dependent reactions of photosynthesis

Electron Transfer Phosphorylation

The third stage of aerobic respiration, electron transfer phosphorylation, also occurs inside mitochondria. Remember that electron transfer phosphorylation is a process in which the flow of electrons through electron transfer chains ultimately results in the attachment of phosphate to ADP (Section 6.5).

The third-stage reactions take place at the inner mitochondrial membrane (Figure 7.8). They begin with the coenzymes NADH and $FADH_2$, which became reduced in the first two stages of aerobic respiration. These coenzymes donate their cargo of electrons and hydrogen ions to electron transfer chains embedded in the inner mitochondrial membrane ❶. As the electrons pass through the chains, they give up energy little by little (Section 5.5). Some molecules of the transfer chains harness that energy to actively transport hydrogen ions across the inner membrane, from the matrix to the intermembrane space ❷. The ions that accumulate in the intermembrane space set up a hydrogen ion gradient across the inner mitochondrial membrane ❸.

This gradient attracts hydrogen ions back toward the matrix. However, ions cannot diffuse through a lipid bilayer on their own (Section 5.6). H^+ can only cross the inner mitochondrial membrane by flowing through the interior of ATP synthases ❹. The flow causes these membrane transport proteins to attach phosphate groups to ADP, so ATP forms. The twelve coenzymes that were reduced in the first two stages can drive the synthesis of about thirty-two ATP in the third stage.

Oxygen accepts electrons at the end of the mitochondrial electron transfer chains ❺. Aerobic respiration, which literally means "taking a breath of air," refers to oxygen as the final electron acceptor in this pathway. When oxygen accepts electrons, it combines with H^+ to form water, which is one product of the third stage:

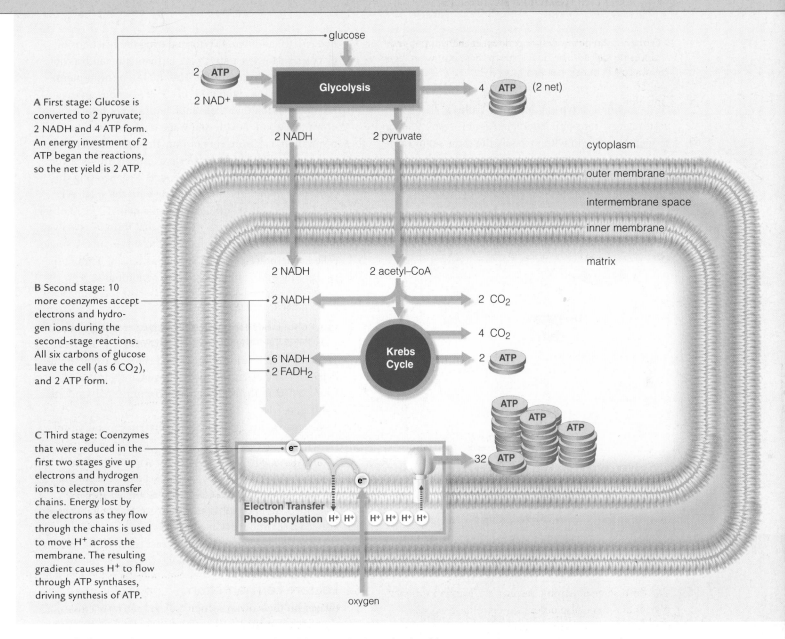

A First stage: Glucose is converted to 2 pyruvate; 2 NADH and 4 ATP form. An energy investment of 2 ATP began the reactions, so the net yield is 2 ATP.

B Second stage: 10 more coenzymes accept electrons and hydrogen ions during the second-stage reactions. All six carbons of glucose leave the cell (as 6 CO_2), and 2 ATP form.

C Third stage: Coenzymes that were reduced in the first two stages give up electrons and hydrogen ions to electron transfer chains. Energy lost by the electrons as they flow through the chains is used to move H^+ across the membrane. The resulting gradient causes H^+ to flow through ATP synthases, driving synthesis of ATP.

Figure 7.9 Animated Summary of the three stages of aerobic respiration in a mitochondrion.

Summing Up: The Energy Harvest

The breakdown of one glucose molecule in aerobic respiration typically yields thirty-six ATP (Figure 7.9). Four ATP form in the first and second stages, and thirty-two form in the third stage. However, the overall yield varies. Factors such as cell type affect it. For example, the typical yield of aerobic respiration in brain and skeletal muscle cells is thirty-eight ATP, not thirty-six.

Remember that some energy dissipates with every transfer (Section 5.2). Even though aerobic respiration is a very efficient way of retrieving energy from carbohydrates, about 60 percent of the energy harvested in this pathway disperses as metabolic heat.

Take-Home Message **What happens during the third stage of aerobic respiration?**

❯ In aerobic respiration's third stage, electron transfer phosphorylation, energy released by electrons flowing through electron transfer chains is ultimately captured in the attachment of phosphate to ADP. A typical net yield of aerobic respiration is thirty-six ATP per glucose.

❯ The reactions begin when coenzymes that were reduced in the first and second stages of reactions deliver electrons and hydrogen ions to electron transfer chains in the inner mitochondrial membrane.

❯ Energy released by electrons as they pass through electron transfer chains is used to pump H^+ from the mitochondrial matrix to the intermembrane space.

❯ The H^+ gradient that forms across the inner mitochondrial membrane drives the flow of hydrogen ions through ATP synthases, which results in ATP formation.

> Fermentation pathways break down carbohydrates without using oxygen. The final steps in these pathways regenerate NAD$^+$ but do not produce ATP.

Fermentation is a type of anaerobic pathway that harvests energy from carbohydrates. Aerobic respiration and fermentation begin with precisely the same set of reactions in the cytoplasm: glycolysis. Again, two pyruvate,

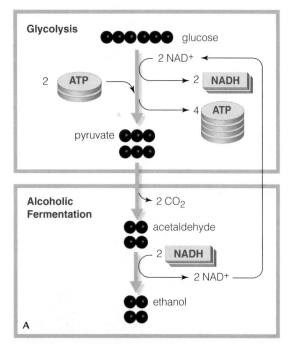

two NADH, and two ATP form during the reactions of glycolysis. However, after that, fermentation and aerobic respiration pathways differ. The final steps of fermentation occur in the cytoplasm. In these reactions, pyruvate is converted to other molecules, but it is not fully broken down to carbon dioxide and water as occurs in aerobic respiration. Electrons do not flow through transfer chains, so no more ATP forms. However, electrons are removed from NADH, so NAD$^+$ is regenerated. Regenerating this coenzyme allows glycolysis—along with the small ATP yield it offers—to continue. Thus, the net ATP yield of fermentation consists of the two ATP that form in glycolysis. That yield provides enough energy to sustain many single-celled anaerobic species. It also helps cells of aerobic species produce ATP under anaerobic conditions.

Alcoholic Fermentation

In **alcoholic fermentation**, the pyruvate from glycolysis is converted to ethyl alcohol, or ethanol. First, 3-carbon pyruvate is split into carbon dioxide and 2-carbon acetaldehyde. Then, electrons and hydrogen are transferred from NADH to the acetaldehyde, forming NAD$^+$ and ethanol (Figure 7.10).

Bakers use the alcoholic fermentation capabilities of one species of yeast, *Saccharomyces cerevisiae*, to make bread. These cells break down carbohydrates in bread dough, and release CO_2 by alcoholic fermentation. The dough expands (rises) as CO_2 forms bubbles in it.

Some wild and cultivated strains of *Saccharomyces* are also used to produce wine. Crushed grapes are left in vats along with large populations of yeast cells, which convert the sugars in the juice to ethanol.

Lactate Fermentation

In **lactate fermentation**, the electrons and hydrogen ions carried by NADH are transferred directly to pyruvate. This reaction converts pyruvate to 3-carbon lactate (the ionized form of lactic acid), and also converts NADH to NAD$^+$ (Figure 7.11A).

Some lactate fermenters spoil food, but we use others to preserve it. For instance, a bacterium called *Lactobacillus acidophilus* breaks down lactose in milk by fermentation. We use this bacteria to produce dairy products such

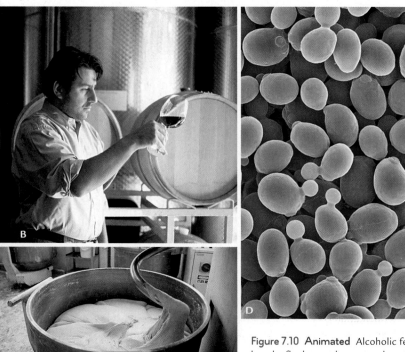

Figure 7.10 Animated Alcoholic fermentation. A Alcoholic fermentation begins with glycolysis, but the final steps do not produce ATP. They regenerate NAD$^+$. The net yield of these reactions is two ATP per molecule of glucose (from glycolysis). B A vintner examines a product of alcoholic fermentation. C A commercial vat of yeast dough rising with the help of D live *Saccharomyces* cells.

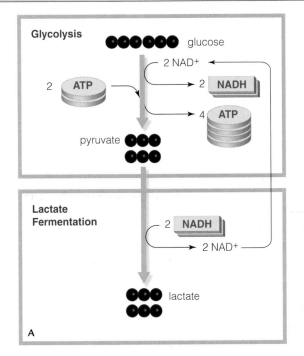

Figure 7.11 Animated Lactate fermentation. A Like alcoholic fermentation, the final steps of lactate fermentation regenerate NAD^+. The net yield is two ATP per molecule of glucose.

B Two types of muscle fibers, red and white, are visible in this cross-section through a human thigh muscle. The white fibers, which make ATP by lactate fermentation, support sprints and other intense bursts of activity C. The darker red fibers, which make ATP by aerobic respiration, sustain endurance activities D. Chickens spend most of their time walking, so their leg muscles consist mainly of red fibers.

as buttermilk, cheese, and yogurt. Yeast species ferment and preserve pickles, corned beef, sauerkraut, and kimchi.

Animal skeletal muscles, which move bones, consist of cells fused as long fibers. The fibers differ in how they make ATP. Red fibers have many mitochondria and produce ATP by aerobic respiration. These fibers sustain prolonged activity, such as marathon runs. They are red because they have an abundance of myoglobin, a protein that stores oxygen for aerobic respiration in these fibers (Figure 7.11B).

White muscle fibers contain few mitochondria and no myoglobin, so they do not carry out a lot of aerobic respiration. Instead, they make most of their ATP by lactate fermentation. This pathway makes ATP quickly but not for long, so it is useful for quick, strenuous activities such as sprinting or weight lifting. The low ATP yield does

not support prolonged activity. That is one reason why chickens cannot fly very far: Their flight muscles consist mostly of white fibers (thus, the "white" breast meat). Chickens fly only in short bursts. More often, a chicken walks or runs. Its leg muscles consist mostly of red muscle fibers, the "dark meat." Most human muscles consist of a mixture of white and red fibers, but the proportions vary among muscles and among individuals. Great sprinters tend to have more white fibers. Great marathon runners tend to have more red fibers. Section 32.6 offers a closer look at skeletal muscle fibers and how they work.

alcoholic fermentation Anaerobic carbohydrate breakdown pathway that produces ATP and ethanol. Begins with glycolysis; end reactions regenerate NAD^+ so glycolysis can continue.
lactate fermentation Anaerobic carbohydrate breakdown pathway that produces ATP and lactate.

Take-Home Message **What is fermentation?**

> ATP can form by carbohydrate breakdown in fermentation pathways, which are anaerobic.

> The end product of lactate fermentation is lactate. The end product of alcoholic fermentation is ethanol.

> Both pathways have a net yield of two ATP per glucose molecule. The ATP forms during glycolysis.

> Fermentation reactions regenerate the coenzyme NAD^+, without which glycolysis (and ATP production) would stop.

> Aerobic respiration can produce ATP from the breakdown of fats and proteins.
< Links to Metabolic reactions 3.2, Carbohydrates 3.3, Lipids 3.4, Proteins 3.5

As you eat, glucose and other breakdown products of digestion are absorbed across the gut lining, and blood transports these small organic molecules throughout the body. The concentration of glucose in the bloodstream rises, and in response the pancreas (an organ) increases its rate of insulin secretion. Insulin is a hormone (a type of signaling molecule) that causes cells to take up glucose, so an increase in insulin production causes cells to take up glucose faster.

In cytoplasm, glucose is immediately converted to glucose-6-phosphate, an intermediate of glycolysis. Unless ATP is being used quickly, its concentration rises in the cytoplasm. A high concentration of ATP causes glucose-6-phosphate to be diverted away from glycolysis and into a pathway that forms glycogen. Liver and muscle cells espe-cially favor the conversion of glucose to glycogen, and these cells maintain the body's largest stores of it.

What happens if you eat too many carbohydrates? When the blood level of glucose gets too high, acetyl–CoA is diverted away from the Krebs cycle and into a pathway that makes fatty acids. That is why excess dietary carbo-hydrate ends up as fat.

Between meals, the level of glucose declines in the blood. The pancreas responds by secreting glucagon, a hormone that causes liver cells to convert stored glycogen to glucose. The cells release glucose into the bloodstream, so the blood glucose level rises. Thus, hormones control whether the body's cells use glucose as an energy source immediately or save it for use at a later time.

Glycogen makes up about 1 percent of an average adult's total energy reserves, which is the energy equiva-lent of about two cups of cooked pasta. Unless you eat regularly, you will completely deplete your liver's glyco-gen stores in less than twelve hours.

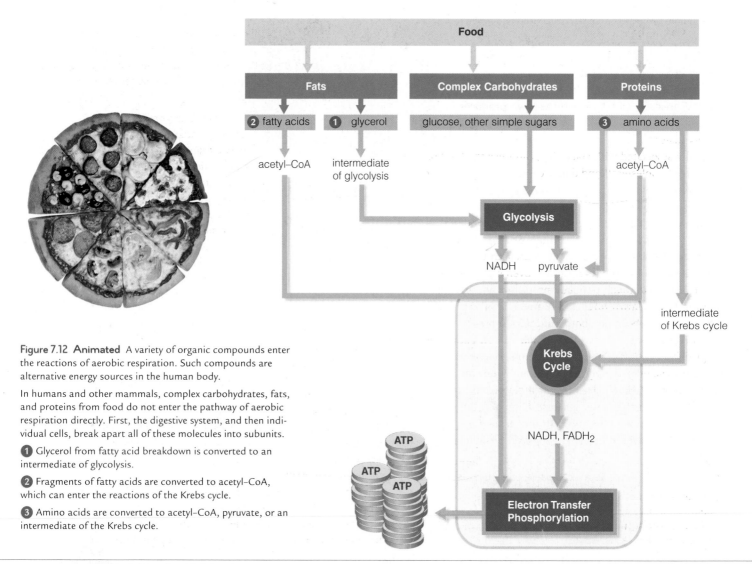

Figure 7.12 Animated A variety of organic compounds enter the reactions of aerobic respiration. Such compounds are alternative energy sources in the human body.

In humans and other mammals, complex carbohydrates, fats, and proteins from food do not enter the pathway of aerobic respiration directly. First, the digestive system, and then indi-vidual cells, break apart all of these molecules into subunits.

❶ Glycerol from fatty acid breakdown is converted to an intermediate of glycolysis.

❷ Fragments of fatty acids are converted to acetyl–CoA, which can enter the reactions of the Krebs cycle.

❸ Amino acids are converted to acetyl–CoA, pyruvate, or an intermediate of the Krebs cycle.

Energy From Fats

Of the total energy reserves in a typical adult who eats well, about 78 percent (about 10,000 kilocalories) is stored in body fat, and 21 percent in proteins. How does a body access its fat reservoir? A fat molecule has a glycerol head and one, two, or three fatty acid tails (Section 3.4). The body stores most fats as triglycerides, which have three fatty acid tails. Triglycerides accumulate in fat cells of adipose tissue. This tissue serves as an energy reservoir, and it also insulates and pads the buttocks and other strategic areas of the body.

When the blood glucose level falls, triglycerides are tapped to provide energy (Figure 7.12). Enzymes in fat cells break the bonds that connect glycerol with the fatty acids, and both are released into the bloodstream. Enzymes in liver cells convert the glycerol to another intermediate of glycolysis ❶. Nearly all cells of your body can take up the free fatty acids. Inside the cells, enzymes split the fatty acid backbones and convert the fragments to acetyl–CoA, which can enter the Krebs cycle ❷.

Compared to carbohydrate breakdown, fatty acid breakdown yields more ATP per carbon atom. Between meals or during steady, prolonged exercise, fatty acid breakdown supplies about half of the ATP that muscle, liver, and kidney cells require.

Diets that are extremely low in carbohydrates force the body to break down fats. The breakdown produces molecules called ketones, which most of your cells can use for energy instead of glucose. This metabolic state, which is normally associated with starvation, can raise the level of LDL ("bad" cholesterol) and damage the kidneys and liver.

Energy From Proteins

Some enzymes in your digestive system split dietary proteins into their amino acid subunits, which are then absorbed into the bloodstream. Cells use amino acids to build proteins or other molecules. Even so, when you eat more protein than your body needs, the amino acids become broken down further. The amino (NH_3^+) group is removed, and it becomes ammonia (NH_3), a waste product that the body eliminates in urine. Their carbon backbone is split, and acetyl–CoA, pyruvate, or an intermediate of the Krebs cycle forms, depending on the amino acid. Your cells can divert any of these organic molecules into the Krebs cycle ❸.

Take-Home Message **How does nutrition influence the metabolism of organic compounds?**

❯ In humans and other organisms, the entrance of organic compounds into an energy-releasing pathway depends on the kinds and proportions of carbohydrates, fats, and proteins in the diet.

When Mitochondria Spin Their Wheels (revisited)

At least 83 proteins are directly involved in the electron transfer chains of electron transfer phosphorylation in mitochondria. A defect in any one of them—or in any of the thousands of other proteins used by mitochondria, such as frataxin (*left*)—can wreak havoc in the body. About one in 5,000 people suffer from a known mitochondrial disorder. New research is showing that mitochondrial defects may be involved in many other illnesses such as diabetes, hypertension, Alzheimer's and Parkinson's disease, and even aging.

How Would You Vote? Developing new drugs is costly, which is why pharmaceutical companies tend to ignore Friedreich's ataxia and other disorders that affect relatively few people. Should governments fund private companies that research potential treatments? See CengageNow for details, then vote online (cengagenow.com).

Summary

Section 7.1 Mitochondria are the ATP-producing powerhouses of eukaryotic cells. Defects in mitochondrial function, such as occurs in Friedreich's ataxia, cause serious health problems.

Section 7.2 Most organisms convert chemical energy of carbohydrates to the chemical energy of ATP. **Anaerobic** and **aerobic** pathways of carbohydrate breakdown start in the cytoplasm with the same set of reactions, **glycolysis**, which converts glucose and other sugars to **pyruvate**. Anaerobic **fermentation** pathways end in cytoplasm and yield two ATP per molecule of glucose. **Aerobic respiration**, which uses oxygen and yields much more ATP than fermentation, evolved after oxygen released by early **photoautotrophs** changed Earth's atmosphere. In modern eukaryotes, aerobic respiration is completed inside mitochondria.

Section 7.3 Glycolysis, the first stage of aerobic respiration and fermentation, occurs in cytoplasm. In the reactions, enzymes use two ATP to convert one molecule of glucose or another six-carbon sugar to two molecules of pyruvate. Electrons and hydrogen ions are transferred to two NAD^+, which are thereby reduced to NADH. Four ATP also form by **substrate–level phosphorylation**, the direct transfer of a phosphate group from a reaction intermediate to ADP.

The net yield of glycolysis is two pyruvate, two ATP, and two NADH per glucose molecule. The pyruvate may continue in fermentation in the cytoplasm, or it may enter mitochondria and the next steps of aerobic respiration.

Summary

Section 7.4 A mitochondrion's inner membrane divides its interior into two fluid-filled spaces: the inner compartment, or matrix, and the intermembrane space. The second stage of aerobic respiration, acetyl–CoA formation and the Krebs cycle, takes place in the matrix. The first steps convert two pyruvate from glycolysis to two acetyl–CoA and two CO_2. The acetyl–CoA enters the Krebs cycle.

It takes two cycles of Krebs reactions to dismantle the two acetyl–CoA. At this stage, all of the carbon atoms in the glucose molecule that entered glycolysis have left the cell in CO_2. During these reactions, electrons and hydrogen ions are transferred to NAD^+ and FAD, which are thereby reduced to NADH and $FADH_2$. ATP forms by substrate-level phosphorylation.

In total, the breakdown of two pyruvate molecules in the second stage of aerobic respiration yields ten reduced coenzymes and two ATP.

Section 7.5 Aerobic respiration ends in mitochondria. In the third stage of reactions, electron transfer phosphorylation, coenzymes that were reduced in the first two stages deliver their cargo of electrons and hydrogen ions to electron transfer chains in the inner mitochondrial membrane. Electrons moving through the chains release energy bit by bit; molecules of the chain use that energy to move H^+ from the matrix to the intermembrane space.

Hydrogen ions that accumulate in the intermembrane space form a gradient across the inner membrane. The ions follow the gradient back to the matrix through ATP synthases. H^+ flow through these transport proteins drives ATP synthesis.

Oxygen combines with electrons and H^+ at the end of the transfer chains, thus forming water.

Overall, aerobic respiration typically yields thirty-six ATP for each glucose molecule.

Section 7.6 Anaerobic fermentation pathways begin with glycolysis and finish in the cytoplasm. A molecule other than oxygen accepts electrons at the end of these reactions. The end product of **alcoholic fermentation** is ethyl alcohol, or ethanol. The end product of **lactate fermentation** is lactate. The final steps of fermentation serve to regenerate NAD^+, which is required for glycolysis to continue, but they produce no ATP. Thus, the breakdown of one glucose molecule in either alcoholic or lactate fermentation yields only the two ATP that form in glycolysis reactions.

Skeletal muscle has two types of fibers: red and white fibers. ATP is produced primarily by aerobic respiration in red muscle fibers, so these fibers sustain activities that require endurance. Lactate fermentation in white fibers supports activities that occur in short, intense bursts.

Section 7.7 In humans and other mammals, the simple sugars from carbohydrate breakdown, glycerol and fatty acids from fat breakdown, and carbon backbones of amino acids from protein breakdown may enter aerobic respiration at various reaction steps.

Self-Quiz   Answers in Appendix III

1. Is the following statement true or false? Unlike animals, which make many ATP by aerobic respiration, plants make all of their ATP by photosynthesis.   *False*

2. Glycolysis starts and ends in the _____ .
   a. nucleus
   b. mitochondrion
   c. plasma membrane
   d. cytoplasm

3. Which of the following metabolic pathways require(s) molecular oxygen (O_2)?
   a. aerobic respiration
   b. lactate fermentation
   c. alcoholic fermentation
   d. all of the above

4. Which molecule does not form during glycolysis?
   a. NADH   b. pyruvate   c. $FADH_2$ d. ATP

5. In eukaryotes, aerobic respiration is completed in the _____ .
   a. nucleus
   b. mitochondrion
   c. plasma membrane
   d. cytoplasm

6. Which of the following reaction pathways is not part of the second stage of aerobic respiration?
   a. electron transfer phosphorylation
   b. acetyl–CoA formation
   c. Krebs cycle
   d. glycolysis
   e. a and d

7. After the Krebs reactions run through _____ cycle(s), one glucose molecule has been completely broken down to CO_2.
   a. one   b. two   c. three   d. six

8. In the third stage of aerobic respiration, _____ is the final acceptor of electrons.
   a. water   b. hydrogen   c. oxygen   d. NADH

9. In alcoholic fermentation, _____ is the final acceptor of electrons.
   a. oxygen
   b. pyruvate
   c. acetaldehyde
   d. sulfate

10. Fermentation pathways make no more ATP beyond the small yield from glycolysis. The remaining reactions serve to regenerate _____ .
    a. FAD
    b. NAD^+
    c. glucose
    d. oxygen

11. Most of the energy that is released by the full breakdown of glucose to CO_2 and water ends up in _____ .
    a. NADH
    b. ATP
    c. heat
    d. electrons

Mitochondrial Abnormalities in Tetralogy of Fallot Tetralogy of Fallot (TF) is a genetic disorder characterized by four major malformations of the heart. The circulation of blood is abnormal, so TF patients have too little oxygen in their blood. Inadequate oxygen levels result in damaged mitochondrial membranes, which in turn cause cells to self-destruct. In 2004, Sarah Kuruvilla and her colleages looked at abnormalities in the mitochondria of heart muscle in TF patients. Some of their results are shown in Figure 7.13.

1. Which abnormality was most strongly associated with TF?
2. Can you make any correlations between blood oxygen content and mitochondrial abnormalities in these patients?

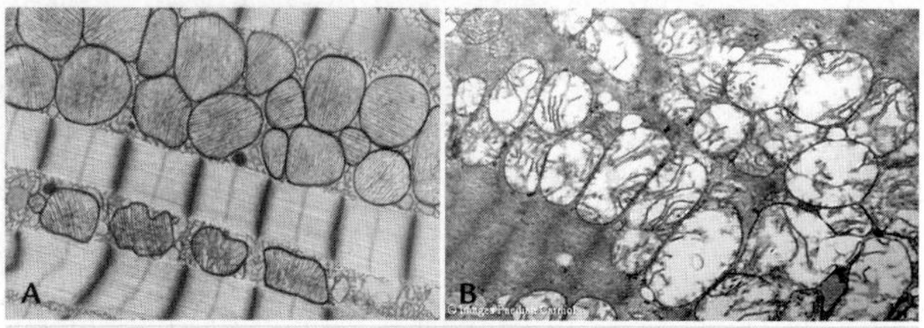

Figure 7.13 Mitochondrial changes in tetralogy of Fallot (TF).

A Normal heart muscle. Many mitochondria between the fibers provide muscle cells with ATP for contraction. **B** Heart muscle from a person with TF shows swollen, broken mitochondria.

C Mitochondrial abnormalities in TF patients. SPO_2 is oxygen saturation of the blood. A normal value of SPO_2 is 96%. Abnormalities are marked "+".

Patient (age)	SPO_2 (%)	Mitochondrial Abnormalities in TF			
		Number	Shape	Size	Broken
1 (5)	55	+	+	–	–
2 (3)	69	+	+	–	–
3 (22)	72	+	+	–	–
4 (2)	74	+	+	–	–
5 (3)	76	+	+	–	+
6 (2.5)	78	+	+	–	+
7 (1)	79	+	+	–	–
8 (12)	80	+	–	+	–
9 (4)	80	+	+	–	–
10 (8)	83	+	–	+	–
11 (20)	85	+	+	–	–
12 (2.5)	89	+	–	+	–

C

12. Your body cells can use _____ as an alternative energy source when glucose is in short supply.
   a. fatty acids   c. amino acids
   b. glycerol   (d. all of the above)

13. Which of the following is not produced by an animal muscle cell operating under anaerobic conditions? *has no oxygen*
   a. heat   d. ATP
   b. pyruvate   e. lactate
   c. NAD$^+$ (f. all are produced)

14. Match the event with its most suitable description.
   B glycolysis   a. ATP, NADH, FADH$_2$,
   C fermentation   and CO$_2$ form
   A Krebs cycle   b. glucose to two pyruvate
   d electron transfer   c. NAD$^+$ regenerated, little ATP
      phosphorylation   d. H$^+$ flows via ATP synthases

15. Match the term with the best description.
   B matrix   a. needed for glycolysis
   C pyruvate   b. inner space
   A NAD$^+$ c. makes many ATP
   D mitochondrion   d. end of glycolysis
   f intermembrane   e. reduced coenzyme
      space   f. hydrogen ions
   e NADH   accumulate here

Additional questions are available on **CENGAGENOW**.

Animations and Interactions on **CENGAGENOW**:
❯ Links with photosynthesis; Overview of aerobic respiration; Glycolysis; Second stage of aerobic respiration; Oxidative phosphorylation; Fermentation pathways; Alternative energy sources.

Critical Thinking

1. At high altitudes, the oxygen level in air is low. Climbers of very tall mountains risk altitude sickness, which is characterized by shortness of breath, weakness, dizziness, and confusion.

The early symptoms of cyanide poisoning are the same as those for altitude sickness. Cyanide binds tightly to cytochrome *c* oxidase, a protein complex that is the last component of mitochondrial electron transfer chains. Cytochrome *c* oxidase with bound cyanide can no longer transfer electrons. Explain why cyanide poisoning starts with the same symptoms as altitude sickness.

2. As you learned, membranes impermeable to hydrogen ions are required for electron transfer phosphorylation. Membranes in mitochondria serve this purpose in eukaryotes. Bacteria do not have this organelle, but they do make ATP by electron transfer phosphorylation. How do you think they do it, given that they have no mitochondria?

3. The bar-tailed godwit is a type of shorebird that makes an annual migration from Alaska to New Zealand and back. The birds make each 11,500-kilometer (7,145-mile) trip by flying over the Pacific Ocean in about nine days, depending on weather, wind speed, and direction of travel. One bird was observed to make the entire journey uninterrupted, a feat that is comparable to a human running a nonstop seven-day marathon at 70 kilometers per hour (43.5 miles per hour). Would you expect the flight (breast) muscles of bar-tailed godwits to be light or dark colored? Explain your answer.

‹ Links to Earlier Concepts

This chapter builds on your earlier introduction to nucleic acid structure (Section 3.7). You will revisit carbohydrates (3.3) as you explore the details of DNA's double helix structure, and phosphorylations (5.3) as you learn about DNA replication. You will also see another application of radioisotope tracers (2.2) in the research that led to the discovery that DNA, not protein (3.5), is the hereditary material of all organisms.

Key Concepts

Chromosomes
The DNA of a eukaryotic cell is divided among a character-istic number of chromosomes that differ in length and in shape. Sex chromosomes determine an individual's gender. Proteins associated with eukaryotic DNA help organize chromosomes so they can pack into a nucleus.

Discovery of DNA's Function
The work of many scientists over more than a century led to the discovery that DNA is the molecule that stores hereditary information.

8 DNA Structure and Function

8.1 A Hero Dog's Golden Clones

On September 11, 2001, an off-duty Canadian police officer, Constable James Symington, drove his search dog Trakr from Nova Scotia to Manhattan. Within hours of arriving, the dog led rescuers to the spot where the fifth and final survivor of the World Trade Center attacks was buried. She had been clinging to life, pinned under rubble from the building where she had worked. Symington and Trakr helped with the search and rescue efforts for three days nonstop, until Trakr collapsed from smoke and chemical inhalation, burns, and exhaustion.

Trakr survived the ordeal, but later lost the use of his limbs from a degenerative neurological disease probably linked to toxic smoke exposure at Ground Zero. The hero dog died in April 2009, but his DNA lives on in his genetic copies—his **clones**. Symington's essay about Trakr's superior nature and abilities as a search and rescue dog won the Golden Clone Giveaway, a contest to find the world's most clone-worthy dog. Trakr's DNA was shipped to Korea, where it was inserted into dog eggs, which were then implanted into surrogate mother dogs. Five puppies, all clones of Trakr, were delivered to Symington in July 2009 (Figure 8.1).

Many adult animals have been cloned besides Trakr, but cloning mammals is still unpredictable and far from routine. Typically, less than 2 percent of the implanted embryos result in a live birth. Of the clones that survive, many have serious health problems. Why the difficulty? Even though all cells of an individual inherit the same DNA, an adult cell uses only a fraction of it compared to an embryonic cell. To make a clone from an adult cell, researchers must reprogram its DNA to function like the DNA of an egg. Even though we are getting better at doing that, we still have a lot to learn.

So why do we keep trying? The potential benefits are enormous. Already, cells of cloned human embryos are helping researchers unravel the molecular mechanisms of human

clone Genetically identical copy of an organism.

genetic diseases. Such cells may one day be induced to form replacement tissues or organs for people with incurable diseases. Endangered animals might be saved from extinction; extinct animals may be brought back. Livestock and pets are already being cloned commercially.

Figure 8.1 Animal cloning. *Opposite*, James Symington and his dog Trakr at Ground Zero in 2001. *Above*, Symington with Trakr's clones in 2009.

Perfecting the methods for cloning animals brings us closer to the possibility of cloning humans, both technically and ethically. For example, if cloning a lost cat for a grieving pet owner is acceptable, why would it not be acceptable to clone a lost child for a grieving parent? Different people have very different answers to such questions, so controversy over cloning continues to rage even as the techniques improve. Understanding the basis of heredity—what DNA is and how it works—will help inform your own opinions about the issues surrounding cloning.

Structure of DNA
A DNA molecule consists of two long chains of nucleotides coiled into a double helix. Four kinds of nucleotides make up the chains: adenine, thymine, guanine, and cytosine. The order of these bases in DNA differs among individuals and among species.

DNA Replication
Before a cell divides, it copies its DNA so that each of its descendants gets a full complement of hereditary information. Newly forming DNA is monitored for errors, most of which are corrected quickly. Uncorrected errors may be perpetuated as mutations.

Cloning Animals
Several methods are now commonly used to produce clones of adult animals for research and agriculture. The techniques are far from perfect, and the practice continues to raise serious ethical questions.

❭ The DNA in a eukaryotic cell nucleus is organized as one or more chromosomes.

❬ Link to DNA 3.7

All organisms pass DNA to offspring when they reproduce. Inside a cell, each DNA molecule is organized as a structure called a **chromosome** (Figure 8.2). Eukaryotic cells typically have a number of chromosomes ❶. During most of a cell's life, each of its chromosomes consists of one DNA molecule. As it prepares to divide, the cell duplicates its chromosomes, so that both of its offspring get a full set. After the chromosomes are duplicated, each consists of two DNA molecules, which are then called **sister chromatids**. Sister chromatids attach to each other at a constricted region called a **centromere**:

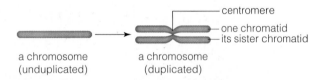

a chromosome (unduplicated) a chromosome (duplicated)

Stretched out end to end, the 46 chromosomes in a human cell would be about 2 meters (6.5 feet) long—a lot of DNA to fit into a nucleus that is typically less than 10 micrometers in diameter! Interactions between a DNA molecule and the proteins that associate with it structurally organize a chromosome and help it pack tightly.

At its most condensed, a duplicated chromosome consists of two long, tangled filaments (the sister chromatids) bunched into a characteristic X shape ❷. A closer look reveals that each filament is actually a hollow tube formed by coils, like a phone cord ❸. The coils them-

selves consist of a twisted fiber ❹ of DNA wrapped twice at regular intervals around "spools" of proteins called **histones** ❺. In micrographs, these DNA–histone spools look like beads on a string. Each "bead" is a **nucleosome**, the smallest unit of chromosomal organization in eukaryotes. As you will see Section 8.4, the DNA molecule consists of two strands twisted into a double helix ❻.

Chromosome Number

The genetic information of each eukaryotic species is distributed among some number of chromosomes, which differ in length and shape. The sum of all chromosomes in a cell of a given type is called the **chromosome number**. Each species has a characteristic chromosome number. For example, the chromosome number of oak trees is 12, so the nucleus of a cell from an oak tree contains 12 chromosomes. The chromosome number of king crab cells is 208, so they have 208 chromosomes. That of human body cells is 46, so human body cells have 46 chromosomes.

Actually, human body cells have two of each type of chromosome, which means that their chromosome number is **diploid** ($2n$). The 23 pairs of chromosomes are like two sets of books numbered 1 to 23. There are two versions of each book: a pair. Except for a pairing of sex chromosomes (XY) in males, the two members of each pair have the same length and shape, and they hold information about the same traits. Think of them as two sets of books on how to build a house. Your father gave you one set. Your mother had her own ideas about wiring, plumbing, and so on. She gave you an alternate set that says slightly different things about many of those tasks.

Figure 8.2 Animated Zooming in on chromosome structure. Tight packing allows a lot of DNA to fit into a very small nucleus.

❶ The DNA inside the nucleus of a eukaryotic cell is typically divided up into a number of chromosomes. *Inset*: a duplicated human chromosome.

❷ At its most condensed, a duplicated chromosome is packed tightly into an X shape.

❸ A chromosome unravels as a single fiber, a hollow cylinder formed by coiled coils.

❹ The coiled coils consist of a long molecule of DNA (*blue*) and the proteins that are associated with it (*purple*).

❺ At regular intervals, the DNA molecule is wrapped twice around a core of histone proteins. In this "beads-on-a-string" structure, the "string" is the DNA, and each "bead" is called a nucleosome.

❻ The DNA molecule itself has two strands that are twisted into a double helix.

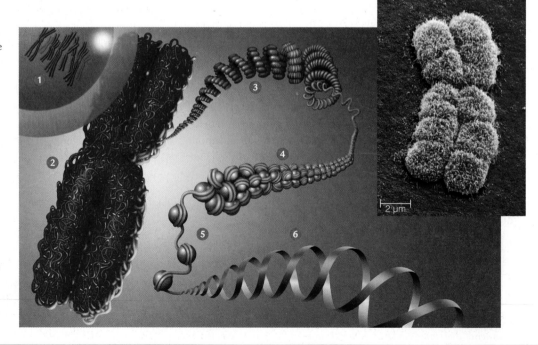

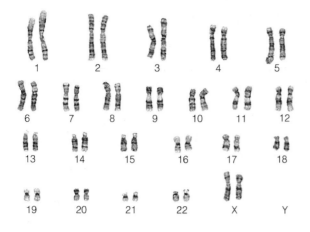

Figure 8.3 **Animated** A karyotype is an image of a single cell's diploid set of chromosomes. This human karyotype shows 22 pairs of autosomes and a pair of X chromosomes.

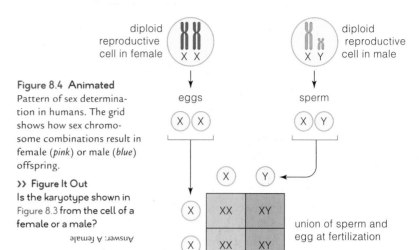

Figure 8.4 **Animated**
Pattern of sex determination in humans. The grid shows how sex chromosome combinations result in female (*pink*) or male (*blue*) offspring.

❯❯ **Figure It Out**
Is the karyotype shown in Figure 8.3 from the cell of a female or a male?

Answer: A female

union of sperm and egg at fertilization

Types of Chromosomes

Karyotyping reveals an individual's diploid complement of chromosomes. With this technique, cells taken from an individual are treated to make their chromosomes condense, and stained so the chromosomes become visible under a microscope. The microscope reveals the chromosomes in every cell. A micrograph of a single cell is digitally rearranged so the images of the chromosomes are lined up by centromere location, and arranged according to size, shape, and length (Figure 8.3). The finished array constitutes the individual's **karyotype**, which is compared with a normal standard. A karyotype shows how many chromosomes are in the individual's cells. Comparison with a standard can also reveal any extra or missing chromosomes, and some structural abnormalities.

All except one pair of a diploid cell's chromosomes are **autosomes**, which are the same in both females and males. The two autosomes of a pair have the same length, shape, and centromere location. Members of a pair of **sex chromosomes** differ between females and males. The differences determine an individual's sex.

The sex chromosomes of humans are called X and Y. Body cells of human females contain two X chromosomes (XX); those of human males contain one X and one Y chromosome (XY). XX females and XY males are the rule among fruit flies, mammals, and many other animals, but there are other patterns. In butterflies, moths, birds, and certain fishes, males have two identical sex chromosomes, and females do not. Environmental factors (not sex chromosomes) determine sex in some species of invertebrates, turtles, and frogs. As an example, the temperature of the sand in which sea turtle eggs are buried determines the sex of the hatchlings.

In humans, a new individual inherits a combination of sex chromosomes that dictates whether it will become a male or a female. All eggs made by a human female have one X chromosome. One-half of the sperm cells made by a male carry an X chromosome; the other half carry a Y chromosome. If an X-bearing sperm fertilizes an X-bearing egg, the resulting individual will develop into a female. If the sperm carries a Y chromosome, the individual will develop into a male (Figure 8.4).

---

autosome Any chromosome other than a sex chromosome.
centromere Constricted region in a eukaryotic chromosome where sister chromatids are attached.
chromosome A structure that consists of DNA and associated proteins; carries part or all of a cell's genetic information.
chromosome number The sum of all chromosomes in a cell of a given type.
diploid Having two of each type of chromosome characteristic of the species (2n).
histone Type of protein that structurally organizes eukaryotic chromosomes.
karyotype Image of an individual's complement of chromosomes arranged by size, length, shape, and centromere location.
nucleosome A length of DNA wound around a spool of histone proteins.
sex chromosome Member of a pair of chromosomes that differs between males and females.
sister chromatid One of two attached DNA molecules of a duplicated eukaryotic chromosome.

Take-Home Message **What are chromosomes?**

❯ A eukaryotic cell's DNA is divided among some characteristic number of chromosomes, which differ in length and shape.

❯ Members of a pair of sex chromosomes differ between males and females. Other chromosomes are autosomes—the same in males and females.

❯ Proteins that associate with DNA structurally organize chromosomes and allow them to pack tightly.

> Investigations that led to our understanding that DNA is the molecule of inheritance reveal how science advances.
< Links to Radioisotopes 2.2, Proteins 3.5, DNA 3.7

Early and Puzzling Clues

In 1865, a Swiss medical student, Johannes Miescher, became ill with typhus. The typhus left Miescher partially deaf, so becoming a doctor was no longer an option for him. He switched to organic chemistry instead. By 1869, he was collecting white blood cells from pus-filled bandages and sperm from fish so he could study the composition of the nucleus. Such cells do not contain much cytoplasm, which made isolating the substances in their nucleus easy. Miescher found that nuclei contain an acidic substance composed mostly of nitrogen and phosphorus. Later, that substance would be called deoxyribonucleic acid, or DNA.

Sixty years later, a British medical officer, Frederick Griffith, was trying to make a vaccine for pneumonia. He isolated two strains (types) of *Streptococcus pneumoniae*, a bacteria that causes pneumonia. He named one strain R, because it grows in Rough colonies. He named the other strain S, because it grows in Smooth colonies. Griffith used both strains in a series of experiments that did not lead to the development of a vaccine, but did reveal a clue about inheritance (Figure 8.5).

First, he injected mice with live R cells ❶. The mice did not develop pneumonia, so the R strain was harmless.

Second, he injected other mice with live S cells ❷. The mice died. Blood samples from them teemed with live S cells. The S strain was pathogenic; it caused pneumonia.

Third, he killed S cells by exposing them to high temperature. Mice injected with dead S cells did not die ❸.

Fourth, he mixed live R cells with heat-killed S cells. Mice injected with the mixture died ❹. Blood samples drawn from them teemed with live S cells.

What happened in the fourth experiment? If heat-killed S cells in the mix were not really dead, then mice injected with them in the third experiment would have died. If the harmless R cells had changed into killer cells, then mice injected with R cells in experiment 1 would have died.

The simplest explanation was that heat had killed the S cells, but had not destroyed their hereditary material, including whatever part specified "infect mice." Somehow, that material had been transferred from the dead S cells into the live R cells, which put it to use.

The transformation was permanent and heritable. Even after hundreds of generations, the descendants of transformed R cells were infectious. What had caused the transformation? Which substance encodes the information about traits that parents pass to offspring?

In 1940, Oswald Avery and Maclyn McCarty set out to identify that substance, which they termed the "transforming principle." They used a process of elimination that tested each type of molecular component of S cells. Avery and McCarty repeatedly froze and thawed S cells (ice crystals that form during freeze–thaw cycles disrupt membranes, thus releasing cell contents). The researchers then filtered any intact cells from the resulting slush. At the end of this process, the researchers had a fluid that contained the lipid, protein, and nucleic acid components of S cells.

The S cell extract could still transform R cells after it had been treated with lipid- and protein-destroying enzymes. Thus, the transforming principle could not be lipid or protein. Carbohydrates had been removed during the purification process, so Avery and McCarty realized that the substance they were seeking must be nucleic acid—RNA or DNA. The S cell extract could still transform R cells after treatment with RNA-degrading enzymes, but not after treatment with DNA-degrading enzymes. DNA had to be the transforming principle.

The result surprised Avery and McCarty, who, along with most other scientists, had assumed that proteins were the substance of heredity. After all, traits are diverse, and proteins were thought to be the most diverse biologi-

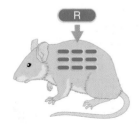

❶ Mice injected with live cells of harmless strain R do not die. Live R cells in their blood.

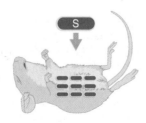

❷ Mice injected with live cells of killer strain S die. Live S cells in their blood.

❸ Mice injected with heat-killed S cells do not die. No live S cells in their blood.

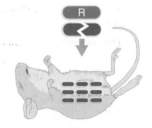

❹ Mice injected with live R cells plus heat-killed S cells die. Live S cells in their blood.

Figure 8.5 Animated Summary of results from Fred Griffith's experiments. The hereditary material of harmful *Streptococcus pneumoniae* cells transformed harmless cells into killers.

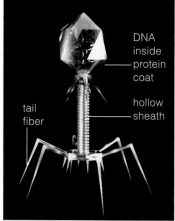

A *Top*, model of a bacteriophage. *Bottom*, micrograph of three viruses injecting DNA into an *E. coli* cell.

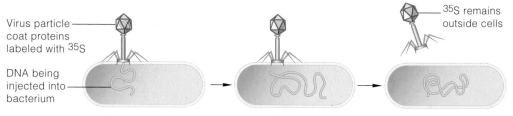

B In one experiment, bacteria were infected with virus particles that had been labeled with a radioisotope of sulfur (^{35}S). The sulfur had labeled only viral proteins. The viruses were dislodged from the bacteria by whirling the mixture in a kitchen blender. Most of the radioactive sulfur was detected in the viruses, not in the bacterial cells. The viruses had not injected protein into the bacteria.

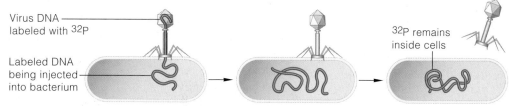

C In another experiment, bacteria were infected with virus particles that had been labeled with a radioisotope of phosphorus (^{32}P). The phosphorus had labeled only viral DNA. When the viruses were dislodged from the bacteria, the radioactive phosphorus was detected mainly inside the bacterial cells. The viruses had injected DNA into the cells—evidence that DNA is the genetic material of this virus.

Figure 8.6 Animated The Hershey–Chase experiments. Alfred Hershey and Martha Chase tested whether the genetic material injected by bacteriophage into bacteria is DNA, protein, or both. The experiments were based on the knowledge that proteins contain more sulfur (S) than phosphorus (P), and DNA contains more phosphorus than sulfur.

cal molecules. Other molecules just seemed too uniform. The two scientists were so skeptical that they published their results only after they had convinced themselves, by years of painstaking experimentation, that DNA was indeed hereditary material. They were also careful to point out that they had not proven DNA was the only hereditary material.

Confirmation of DNA's Function

By 1950, researchers had discovered **bacteriophage**, a type of virus that infects bacteria (Figure 8.6A). Like all viruses, these infectious particles carry hereditary information about how to make new viruses. After a virus infects a cell, the cell starts making new virus particles. Bacteriophages inject genetic material into bacteria, but was that material DNA, protein, or both?

Alfred Hershey and Martha Chase found the answer to that question by exploiting the long-known properties of protein (high sulfur content) and DNA (high phosphorus content). They cultured bacteria in growth medium containing an isotope of sulfur, ^{35}S. In this medium, the protein (but not the DNA) of bacteriophage that infected the bacteria became labeled with the ^{35}S tracer.

Hershey and Chase allowed the labeled viruses to infect a fresh culture of unlabeled bacteria. They knew from electron micrographs that phages attach to bacteria by their slender tails. They reasoned it would be easy to break this precarious attachment, so they poured the virus–bacteria mixture into a blender and turned it on. (At the time, kitchen appliances were commonly used as laboratory equipment.)

After blending, the researchers separated the bacteria from the virus-containing fluid, and measured the ^{35}S content of each separately. The fluid contained most of the ^{35}S. Thus, the viruses had not injected protein into the bacteria (Figure 8.6B).

Hershey and Chase repeated the experiment using an isotope of phosphorus, ^{32}P, which labeled the DNA (but not the proteins) of the bacteriophage. This time, infected bacteria contained most of the isotope. The viruses had injected DNA into the bacteria (Figure 8.6C).

Both of these experiments—and many others after them—supported the hypothesis that DNA, not protein, is the material of heredity common to all life on Earth.

bacteriophage Virus that infects bacteria.

Take-Home Message **What is the molecular basis of inheritance?**

❯ DNA is the material of heredity common to all life on Earth.

> Watson and Crick's discovery of DNA's structure was based on almost fifty years of research by other scientists.
‹ Links to Carbohydrate rings 3.3, Protein structure 3.5, Nucleic acids 3.7

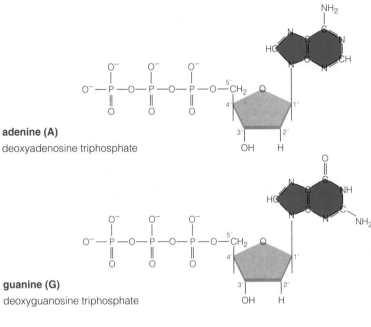

adenine (A)
deoxyadenosine triphosphate

guanine (G)
deoxyguanosine triphosphate

thymine (T)
deoxythymidine triphosphate

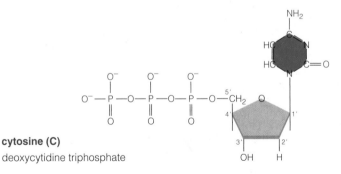

cytosine (C)
deoxycytidine triphosphate

Figure 8.7 Animated The four nucleotides in DNA. Each kind has three phosphate groups, a deoxyribose sugar (*orange*), and a nitrogen-containing base (*blue*) after which it is named. Biochemist Phoebus Levene identified the structure of these bases and how they are connected in DNA in the early 1900s. Levene worked with DNA for almost 40 years.

Numbering the carbons in the sugar rings (Section 3.3) allows us to keep track of the orientation of nucleotide chains, which is important in processes such as DNA replication. Compare Figure 8.8.

DNA's Building Blocks

Each strand of DNA is a polymer of nucleotides that have been linked into a chain. Even though a chain can be hundreds of millions of nucleotides long, it is composed of only four kinds of nucleotides. A DNA nucleotide has a five-carbon sugar, three phosphate groups, and one of four nitrogen-containing bases (Figure 8.7). Just how those four nucleotides—adenine (A), guanine (G), thymine (T), and cytosine (C)—are arranged in DNA was a puzzle that took over 50 years to solve. As molecules go, DNA is gigantic, and chromosomal DNA has a complex structural organization. Both factors made the molecule difficult to work with, given the laboratory methods of the time.

In 1950, Erwin Chargaff, one of many researchers who had been trying to solve the structure of DNA, made two discoveries. First, the amounts of thymine and adenine in all DNA are the same, as are the amounts of cytosine and guanine. We call this discovery Chargaff's first rule:

$$A = T \text{ and } G = C$$

Chargaff's second discovery, or rule, is that the proportion of adenine and guanine differs among the DNA of different species.

Meanwhile, American biologist James Watson and British biophysicist Francis Crick, both at Cambridge University, had been sharing their ideas about the structure of DNA. The helical pattern of secondary structure that occurs in many proteins (Section 3.5) had just been discovered, and Watson and Crick suspected that the DNA molecule was also a helix. They had spent many hours arguing about the size, shape, and bonding requirements of the four kinds of nucleotides that make up DNA. They had pestered chemists to help them identify bonds they might have overlooked. They had fiddled with cardboard cutouts, and made models from scraps of metal connected by suitably angled "bonds" of wire.

At King's College in London, biochemist Rosalind Franklin had also been working on the structure of DNA. Like Crick, Franklin specialized in x-ray crystallography, a technique in which x-rays are directed through a purified and crystallized substance. Atoms in the substance's molecules scatter the x-rays in a pattern that can be captured as an image. Researchers can use the pattern to calculate the size, shape, and spacing between any repeating elements of the molecules—all of which are details of molecular structure.

DNA is a large molecule, and difficult to crystallize. Also, as Franklin discovered, "wet" and "dry" DNA samples have different shapes. She made the first clear x-ray diffraction image of "wet" DNA, the form that occurs in cells. From the information in that image, she calculated that DNA is very long compared to its 2-nanometer diameter. She also identified a repeating pattern every 0.34 nanometer along its length, and another every 3.4 nanometers.

Franklin's image and data came to the attention of Watson and Crick, who now had all the information they needed to build a model of the DNA helix—one with two sugar–phosphate chains running in opposite directions, and paired bases inside (Figure 8.8). Bonds between the sugar of one nucleotide and the phosphate of the next form the backbone of each chain. Hydrogen bonds between the internally positioned bases hold the two strands together. Only two kinds of base pairings form: A to T, and G to C, which explains the first of Chargaff's rules. Most scientists had assumed (incorrectly) that the bases had to be on the outside of the helix, because they would be more accessible to DNA-copying enzymes that way. You will see in Section 8.6 how DNA replication enzymes access the bases on the inside of a double helix.

DNA's Base Pair Sequence

Just two kinds of base pairings give rise to the incredible diversity of traits we see among living things. How? Even though DNA is composed of only four bases, the order in which one base pair follows the next varies tremendously among species (which explains Chargaff's second rule). For example, a small piece of DNA from a tulip, a human, or any other organism might be:

one base pair

T	G	A	G	G	A	C	T	C	C	T	C
A	C	T	C	C	T	G	A	G	G	A	G

Notice how the two strands of DNA match up. They are complementary, which means each base on one strand is suitably paired with a partner base on the other. This bonding pattern (A to T, G to C) is the same in all molecules of DNA. However, which base pair follows the next differs among species, and among individuals of the same species. The information encoded by that sequence is the basis of visible traits that define species and distinguish individuals. Thus DNA, the molecule of inheritance in every cell, is the basis of life's unity. Variations in its base sequence are the foundation of life's diversity.

Take-Home Message What is the structure of DNA?

> A DNA molecule consists of two nucleotide chains (strands) running in opposite directions and coiled into a double helix. Internally positioned nucleotide bases hydrogen-bond between the two strands. A pairs with T, and G with C.

> The sequence of bases along a DNA strand is genetic information.

> DNA sequences vary among species and among individuals. This variation is the basis of life's diversity.

Figure 8.8 Animated Structure of DNA, as illustrated by a composite of different models.

Above, Watson and Crick with their model.

2-nanometer diameter

0.34 nanometer between each base pair

3.4-nanometer length of each full twist of the double helix

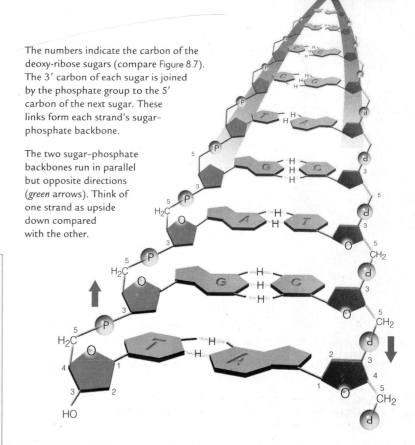

The numbers indicate the carbon of the deoxy-ribose sugars (compare Figure 8.7). The 3′ carbon of each sugar is joined by the phosphate group to the 5′ carbon of the next sugar. These links form each strand's sugar–phosphate backbone.

The two sugar–phosphate backbones run in parallel but opposite directions (*green* arrows). Think of one strand as upside down compared with the other.

> In science, as in other professions, public recognition for a discovery does not always include all contributors.

By the time she arrived at King's College, Rosalind Franklin was an expert x-ray crystallographer. She had solved the structure of coal, which is complex and unorganized (as are large biological molecules such as DNA). Like Pauling, she had built three-dimensional molecular models. Her assignment was to investigate DNA's structure.

Franklin had been told she would be the only one in the department working on the problem, so she did not know that Maurice Wilkins was already doing the same thing just down the hall. When Wilkins proposed a collaboration with her, Franklin thought that Wilkins was oddly overinterested in her work and declined bluntly.

Wilkins and Franklin had been given identical samples of DNA carefully prepared by Rudolf Signer. Franklin's meticulous work with hers yielded the first clear x-ray diffraction image of DNA as it occurs inside cells (Figure 8.9). She gave a presentation on this work in 1952. DNA, she said, had two chains twisted into a double helix, with a backbone of phosphate groups on the outside, and bases arranged in an unknown way on the inside. She had calculated DNA's diameter, the distance between its chains and between its bases, the angle of the helix, and the number of bases in each coil. Crick, with his crystallography background, would have recognized the significance of the work—if he had been there. Watson was in the audience but he was not a crystallographer, and he did not understand the implications of Franklin's x-ray diffraction image or her calculations.

Franklin started to write a research paper on her findings. Meanwhile, and perhaps without her knowledge, Watson reviewed Franklin's x-ray diffraction image with Wilkins, and Watson and Crick read a report detailing Franklin's unpublished data. Crick, who had more experience with molecular modeling than Franklin, immediately understood what the image and the data meant. Watson and Crick used that information to build their model of DNA.

On April 25, 1953, Franklin's paper appeared third in a series of articles about the structure of DNA in the journal *Nature*. It supported with solid experimental evidence Watson and Crick's theoretical model, which appeared in the first article of the series.

Rosalind Franklin died at age 37, of ovarian cancer probably caused by extensive exposure to x-rays. Because the Nobel Prize is not given posthumously, she did not share in the 1962 honor that went to Watson, Crick, and Wilkins for the discovery of the structure of DNA.

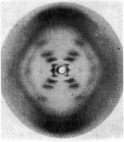

Figure 8.9 Rosalind Franklin and her x-ray diffraction image of DNA.

> A cell copies its DNA before it reproduces.
> DNA repair mechanisms correct most replication errors.
< Link to Phosphate-group transfers 5.3

During most of its life, a cell contains only one set of chromosomes. However, when the cell reproduces, it must contain two sets of chromosomes: one for each of its future offspring. By a process called **DNA replication**, a cell copies its DNA before it divides (Figure 8.10). Before DNA replication begins, each chromosome consists of one molecule of DNA—one double helix ❶. During replication, an enzyme called DNA helicase breaks the hydrogen bonds that hold the double helix together, so the two DNA strands unwind ❷. Another enzyme, **DNA polymerase**, assembles a complementary strand of DNA on each of the parent strands.

As each new DNA strand lengthens, it winds up with its "parent" strand into a double helix (Figure 8.11A). So, after replication, two double-stranded molecules of DNA have formed. One strand of each molecule is parental (old), and the other is new; hence the name of the process, semiconservative replication (Figure 8.11B).

As you will see, the order of nucleotide bases in a strand of DNA—the **DNA sequence**—is genetic information. Descendant cells must get an exact copy of that information, or inheritance will go awry. Because each new strand of DNA is complementary in sequence to a parent strand, both double-stranded molecules that result from DNA replication are duplicates of the parent.

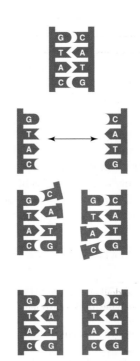

❶ The two strands of a DNA molecule are complementary: their nucleotides match up according to base-pairing rules (G to C, T to A).

❷ As replication starts, the two strands of DNA unwind at many sites along the length of the molecule.

❸ Each parent strand serves as a template for the assembly of a new DNA strand from nucleotides, according to base-pairing rules.

❹ DNA ligase seals any gaps that remain between bases of the "new" DNA, so a continuous strand forms. The base sequence of each half-old, half-new DNA molecule is identical to that of the parent.

Figure 8.10 Animated DNA replication. Both strands of the double helix serve as templates, so two double-stranded DNA molecules result.

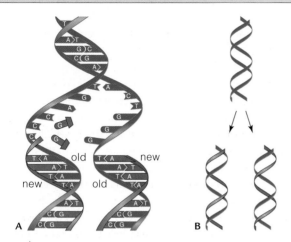

Figure 8.11 Semiconservative replication of DNA. **A** A parental DNA strand serves as a template for assembly of a new strand of DNA. The two parental DNA strands (*blue*) stay intact. A new strand (*magenta*) is assembled on each of the parental (old) strands. **B** One strand of each DNA molecule that forms is new.

Figure 8.12 DNA synthesis occurs in the 5′ to 3′ direction. Only one of two new DNA strands can be assembled in a single piece. The other forms in short segments, which are called Okazaki fragments after the two scientists who discovered them.

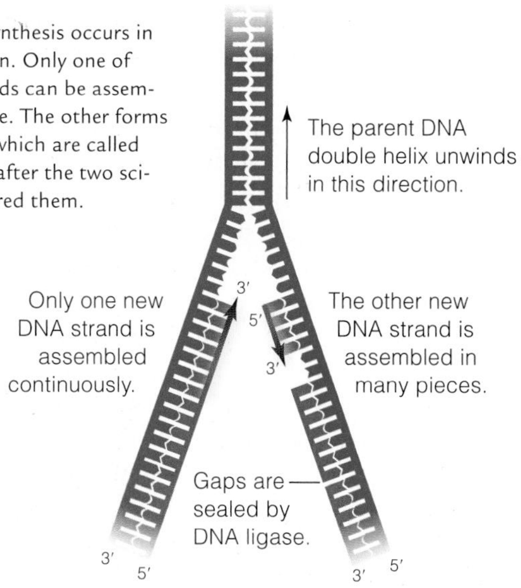

The parent DNA double helix unwinds in this direction.

Only one new DNA strand is assembled continuously.

The other new DNA strand is assembled in many pieces.

Gaps are sealed by DNA ligase.

The base sequence of each new strand of DNA is complementary to its parent because DNA polymerase follows base-pairing rules ❸. As the enzyme moves along a strand of DNA, it uses the sequence of bases as a template, or guide, to assemble a new strand of DNA from free nucleotides. The polymerase adds a T to the end of the new DNA strand when it reaches an A in the parent DNA sequence; it adds a G when it reaches a C; and so on. Phosphate-group transfers from the nucleotides provide energy for their own attachment to the end of a growing strand of DNA. The enzyme **DNA ligase** seals any gaps, so a new, continuous strand of DNA results ❹.

Numbering the carbons in nucleotides allows us to keep track of the DNA strands in a double helix, because each strand has an unbonded 5′ carbon at one end and an unbonded 3′ carbon at the other:

5′ IIIIIIIIIIIIIIIIIIIIIIIII 3′
3′ IIIIIIIIIIIIIIIIIIIIIIIII 5′

DNA polymerase can attach free nucleotides only to a 3′ carbon. Thus, it can replicate only one strand of a DNA molecule continuously (Figure 8.12). Synthesis of the other strand occurs in segments, in the direction opposite that of unwinding. DNA ligase joins those segments into a continuous strand of DNA.

DNA ligase Enzyme that seals gaps in double-stranded DNA.
DNA polymerase DNA replication enzyme. Uses a DNA template to assemble a complementary strand of DNA.
DNA repair mechanism Any of several processes by which enzymes repair damaged DNA.
DNA replication Process by which a cell duplicates its DNA before it divides.
DNA sequence Order of nucleotide bases in a strand of DNA.
mutation Permanent change in the sequence of DNA.

Proofreading

A DNA molecule is not always replicated with perfect fidelity. Sometimes the wrong base is added to a growing DNA strand; at other times, bases get lost, or extra ones are added. Either way, the new DNA strand will no longer match up perfectly with its parent.

Some of these errors occur after the DNA becomes damaged by exposure to radiation or toxic chemicals. DNA polymerases do not copy damaged DNA very well. In most cases, **DNA repair mechanisms** fix DNA by enzymatically excising and replacing any damaged or mismatched bases before replication begins.

Most DNA replication errors occur simply because DNA polymerases catalyze a tremendous number of reactions very quickly—up to 1,000 bases per second. Mistakes are inevitable, and some DNA polymerases make many of them. Luckily, most DNA polymerases proofread their own work. They correct any mismatches by reversing the synthesis reaction to remove a mismatched nucleotide, then resuming synthesis in the forward direction.

When proofreading and repair mechanisms fail, an error becomes a **mutation**, a permanent change in the DNA sequence. An individual or its offspring may not survive a mutation, because mutations can cause cancer in body cells. Mutations in cells that form eggs or sperm may lead to genetic disorders in offspring. However, not all mutations are dangerous. Some give rise to variations in traits that are the raw material of evolution.

Take-Home Message **How is DNA copied?**

❯ A cell replicates its DNA before it divides. Each strand of the double helix serves as a template for synthesis of a new, complementary strand of DNA.

❯ DNA repair mechanisms and proofreading maintain the integrity of a cell's genetic information. Unrepaired errors may become mutations.

❯ Reproductive cloning is a reproductive intervention that results in an exact genetic copy of an adult individual.

The word "cloning" means making an identical copy of something. In biology, cloning can refer to a laboratory method by which researchers copy DNA fragments (a technique we discuss in Chapter 15). It can also refer

A A cow egg is held in place by suction through a hollow glass tube called a micropipette. DNA is identified by a *purple* stain.

B Another micropipette punctures the egg and sucks out the DNA. All that remains inside the egg's plasma membrane is cytoplasm.

C A new micropipette prepares to enter the egg at the puncture site. The pipette contains a cell grown from the skin of a donor animal.

D The micropipette enters the egg and delivers the skin cell to a region between the cytoplasm and the plasma membrane.

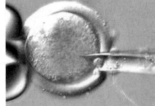

E After the pipette is withdrawn, the donor's skin cell is visible next to the cytoplasm of the egg. The transfer is now complete.

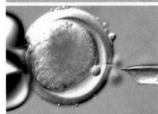

F The egg is exposed to an electric current. This treatment causes the foreign cell to fuse with and empty its nucleus into the cytoplasm of the egg. The egg begins to divide, and an embryo forms. After a few days, the embryo may be transplanted into a surrogate mother.

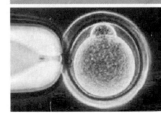

Figure 8.13 Animated Somatic cell nuclear transfer, using cattle cells. This series of micrographs was taken by scientists at Cyagra, a company that specializes in cloning livestock.

to interventions in reproduction that result in an exact genetic copy of an organism.

Genetically identical organisms occur all the time in nature. They arise mainly by the process of asexual reproduction, which we discuss in Chapter 12. Embryo splitting, another natural process, results in identical twins. The first few divisions of a fertilized egg form a ball of cells that sometimes splits spontaneously. If both halves continue to develop independently, identical twins result. Artificial embryo splitting has been routine in research and animal husbandry for decades. With this technique, a ball of cells is grown from a fertilized egg in a laboratory. The ball is teased apart into two halves, each of which goes on to develop as a separate embryo. The embryos are implanted in surrogate mothers, who give birth to identical twins. Artificial twinning and any other technology that yields genetically identical individuals is called **reproductive cloning**.

Twins get their DNA from two parents, which typically differ in their DNA sequence. Thus, although identical twins produced by embryo splitting are identical to one another, they are not identical to either parent. When animal breeders want an exact copy of a specific individual, they may turn to a cloning method that starts with a single cell taken from an adult organism. Such procedures present more of a technical challenge than embryo splitting. Unlike a fertilized egg, a body cell from an adult will not automatically start dividing. It must first be tricked into rewinding its developmental clock.

All cells descended from a fertilized egg inherit the same DNA. Thus, the DNA in each living cell of an individual is like a master blueprint that contains enough information to build an entirely new individual. As different cells in a developing embryo start using different subsets of their DNA, they differentiate, or become different in form and function. In animals, differentiation is usually a one-way path. Once a cell specializes, all of its descendant cells will be specialized the same way. By the time a liver cell, muscle cell, or other specialized cell forms, most of its DNA has been turned off, and is no longer used.

To clone an adult, scientists must first transform one of its differentiated cells into an undifferentiated cell by turning its unused DNA back on. In **somatic cell nuclear transfer** (SCNT), a researcher removes the nucleus from an unfertilized egg, then inserts into the egg a nucleus from an adult animal cell (Figure 8.13). A somatic cell is a body cell, as opposed to a reproductive cell (*soma* is a Greek word for body). If all goes well, the egg's cytoplasm reprograms the transplanted DNA to direct the development of an embryo, which is then implanted into a surrogate mother. The animal that is born to the surrogate is genetically identical with the donor of the nucleus.

Figure 8.14 Champion Holstein dairy cow Nelson's Estimate Liz (*right*) and her clone, Nelson's Estimate Liz II (*left*), who was produced by somatic cell nuclear transfer in 2003. Liz II had already begun to win championships by the time she was a yearling.

SCNT is now a common practice among people who breed prized livestock. Among other benefits, many more offspring can be produced in a given time frame by cloning than by traditional breeding methods. Cloned animals have the same championship features as their DNA donors (Figure 8.14). Offspring can also be produced after a donor animal is castrated or even dead.

The controversial issue with adult cloning is not necessarily about livestock. As the techniques become routine, cloning a human is no longer only within the realm of science fiction. Researchers are already using SCNT to produce human embryos for research, a practice called **therapeutic cloning**. The researchers harvest undifferentiated (stem) cells from the cloned human embryos. Cells from these embryos are being used to study, among many other things, how fatal diseases progress. For example, embryos created using cells from people with genetic heart defects will allow researchers to study how the defect causes developing heart cells to malfunction. Such research may ultimately lead to treatments for people who suffer from fatal diseases. (We return to the topic of stem cells and their potential medical benefits in Chapter 28.) Reproductive cloning of humans is not the intent of such research, but if it were, somatic cell nuclear transfer would indeed be the first step toward that end.

reproductive cloning Technology that produces genetically identical individuals.
somatic cell nuclear transfer (SCNT) Method of reproductive cloning in which genetic material is transferred from an adult somatic cell into an unfertilized, enucleated egg.
therapeutic cloning The use of SCNT to produce human embryos for research purposes.

A Hero Dog's Golden Clones (revisited)

Trakr's clones were produced using SCNT. The ability to clone dogs is a recent development, but the technique is not. Scottish geneticist Ian Wilmut made headlines in 1997 when he announced the first successful demonstration of SCNT. His team had removed the nucleus from an unfertilized sheep egg, then replaced it with the nucleus of a cell taken from the udder of a different sheep. The hybrid egg became an embryo, and then a lamb. The lamb, Dolly, was genetically identical to the sheep that had donated the udder cell.

At first, Dolly looked and acted like a normal sheep. But five years later, she was as fat and arthritic as a twelve-year-old sheep. The following year, Dolly contracted a lung disease that is typical of geriatric sheep, and was euthanized. Dolly's telomeres hinted that she had developed health problems because she was a clone. Telomeres are short, repeated DNA sequences at the ends of chromosomes. They become shorter and shorter as an animal ages. When Dolly was only two years old, her telomeres were as short as those of a six-year-old sheep—the exact age of the adult animal that had been her genetic donor.

Since then, SCNT has also been used to clone mice, rats, rabbits, pigs, cattle, goats, sheep, horses, mules, deer, cats, a camel, a ferret, a monkey, and a wolf. Dolly's telomere problem has not been seen in animals cloned since her, but other problems are common. Many clones are unusually overweight or have enlarged organs. Cloned mice develop lung and liver problems, and almost all die prematurely. Cloned pigs tend to limp and have heart problems. One never did develop a tail or, even worse, an anus.

How Would You Vote? Some view sickly or deformed clones as unfortunate but acceptable casualties of animal cloning research that also yields medical advances for human patients. Should animal cloning be banned? See CengageNow for details, then vote online (cengagenow.com).

Human eggs are difficult to come by. They also come with a hefty set of ethical dilemmas. Thus, researchers have started trying to make hybrid embryos using adult human cells and eggs from other animals, a technique called interspecies nuclear transfer, or iSCNT.

Take-Home Message What is cloning?

> Reproductive cloning technologies produce a clone: an exact genetic copy of an individual.

> The DNA inside a living cell contains all the information necessary to build a new individual.

> Somatic cell nuclear transfer (SCNT) is a reproductive cloning method in which nuclear DNA of an adult donor is transferred to an egg with no nucleus. The hybrid cell develops into an embryo that is genetically identical to the donor.

> Therapeutic cloning uses SCNT to produce human embryos for research.

Summary

Section 8.1 Making **clones**, or exact genetic copies, of adult animals is now a common practice. The techniques, while improving, are still far from perfect; many attempts are required to produce a clone, and clones that survive often have health problems. The practice continues to raise serious ethical questions.

Section 8.2 The DNA of eukaryotes is divided among a characteristic number of **chromosomes** that differ in length and shape. **Histone** proteins organize eukaryotic DNA into **nucleosomes**. When duplicated, a eukaryotic chromosome consists of two **sister chromatids** attached at a **centromere**. **Diploid** cells have two of each type of chromosome.

Chromosome number is the sum of all chromosomes in cells of a given type. A human body cell has twenty-three pairs of chromosomes. Members of a pair of **sex chromosomes** differ among males and females. All others are **autosomes**. Autosomes of a pair have the same length, shape, and centromere location, and they carry the same genes. A **karyotype** can reveal abnormalities in an individual's complement of chromosomes.

Section 8.3 Almost one hundred years of experiments with bacteria and **bacteriophage** offered solid evidence that deoxyribonucleic acid (DNA), not protein, is the hereditary material of life.

Sections 8.4, 8.5 A DNA molecule consists of two strands of DNA coiled into a helix. Nucleotide monomers are joined to form each strand. A DNA nucleotide has a five-carbon sugar (deoxyribose), three phosphate groups, and one of four nitrogen-containing bases after which the nucleotide is named: adenine, thymine, guanine, or cytosine.

Bases of the two DNA strands in a double helix pair in a consistent way: adenine with thymine (A–T), and guanine with cytosine (G–C). The order of bases along the strands varies among species and among individuals.

Section 8.6 The **DNA sequence** of an organism's chromosome(s) constitutes genetic information. A cell passes that information to offspring by copying its DNA before it divides, a process called **DNA replication**. A double-stranded molecule of DNA results in two double-stranded DNA molecules that are identical to the parent. One strand of each molecule is new, and the other is parental. During the replication process, the double helix unwinds. **DNA polymerase** uses each strand as a template to assemble new, complementary strands of DNA from free nucleotides. **DNA ligase** seals any gaps to form a continuous strand.

DNA **repair mechanisms** fix damaged DNA. Proofreading by DNA polymerases corrects most base-pairing errors. Uncorrected errors may become **mutations**.

Section 8.7 Various **reproductive cloning** technologies produce genetically identical individuals (clones). In **somatic cell nuclear transfer** (SCNT), one cell from an adult is fused with an enucleated egg. The hybrid cell is treated with electric shocks or another stimulus that provokes the cell to divide and begin developing into a new individual. SCNT with human cells, which is called **therapeutic cloning**, produces embryos that are used for stem cell research.

Self-Quiz   Answers in Appendix III

1. Chromosome number _____ .
   a. refers to a particular chromosome pair in a cell
   b. is an identifiable feature of a species
   c. is like a set of books
   d. all of the above

2. Sister chromatids connect at the _Centromere_

3. The basic unit that structurally organizes a eukaryotic chromosome is the _____ .
   a. higher-order coiling   c. base sequence
   b. double helix   d. nucleosome

4. Which is *not* a nucleotide base in DNA?
   a. adenine   c. uracil   e. cytosine
   b. guanine   d. thymine   f. All are in DNA.

5. What are the base-pairing rules for DNA?
   a. A–G, T–C   c. A–U, C–G
   b. A–C, T–G   d. A–T, G–C

6. One species' DNA differs from others in its _____ .
   a. sugars   c. base sequence
   b. phosphates   d. all of the above

7. When DNA replication begins, _____ .
   a. the two DNA strands unwind from each other
   b. the two DNA strands condense for base transfers
   c. two DNA molecules bond
   d. old strands move to find new strands

8. DNA replication requires _____ .
   a. template DNA   c. DNA polymerase
   b. free nucleotides   d. all of the above

9. Show the complementary strand of DNA that forms on this template DNA fragment during replication:

   5′—GGTTTCTTCAAGAGA—3′

10. _____ is an example of reproductive cloning.
    a. Somatic cell nuclear transfer (SCNT)
    b. Multiple offspring from the same pregnancy
    c. Artificial embryo splitting
    d. a and c
    e. all of the above

Data Analysis Activities

Hershey–Chase Experiments The graph in Figure 8.15 is reproduced from Hershey and Chase's original 1952 publication that showed DNA is the hereditary material of bacteriophage. The data are from the two experiments described in Section 8.3, in which bacteriophage DNA and protein were labeled with radioactive tracers and allowed to infect bacteria. The virus–bacteria mixtures were whirled in a blender to dislodge the viruses, and the tracers were tracked inside and outside of the bacteria.

1. Before blending, what percentage of ^{35}S was outside the bacteria? What percentage was inside? What percentage of ^{32}P was outside the bacteria? What percentage was inside?

2. After 4 minutes in the blender, what percentage of ^{35}S was outside the bacteria? What percentage was inside? What percentage of ^{32}P was outside the bacteria? What percentage was inside?

3. How did the researchers know that the radioisotopes in the fluid came from outside of the bacterial cells (extracellular) and not from bacteria that had been broken apart by whirling in the blender?

4. The extracellular concentration of which isotope, ^{35}S or ^{32}P, increased the most with blending? DNA contains much more phosphorus than do proteins; proteins contain much more sulfur than does DNA. Do these results imply that the viruses inject DNA or protein into bacteria? Why or why not?

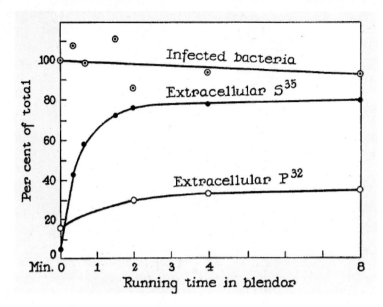

Figure 8.15 Detail of Alfred Hershey and Martha Chase's publication describing their experiments with bacteriophage. "Infected bacteria" refers to the percentage of bacteria that survived the blender.

Source: "Independent Functions of Viral Protein and Nucleic Acid in Growth of Bacteriophage." *Journal of General Physiology*, 36(1), Sept. 20, 1952.

11. The DNA of each species has unique _____ that set it apart from the DNA of all other species.
    a. nucleotides   c. sequences
    b. chromosomes   d. bases

12. _____ can be used to produce genetically identical organisms (clones).
    a. SCNT   c. Therapeutic cloning
    b. Embryo splitting   d. all of the above

13. A karyotype reveals the _____ of a single cell.
    a. base sequences   c. hereditary information
    b. chromosomes   d. clones

14. SCNT with human cells is called _____ .

15. Match the terms appropriately.
    ___ bacteriophage   a. nitrogen-containing base,
    ___ clone   sugar, phosphate group(s)
    ___ nucleotide   b. copy of an organism
    ___ diploid   c. does not determine sex
    ___ DNA ligase   d. only DNA and protein
    ___ DNA polymerase   e. fills in gaps, seals breaks
    ___ autosome   in a DNA strand
                         f. two chromosomes
                         of each type
                         g. adds nucleotides to a
                         growing DNA strand

Additional questions are available on **CENGAGENOW**.

Critical Thinking

1. Mutations are permanent changes in a cell's DNA base sequence. They typically have negative consequences, but they are also the original source of genetic variation and the raw material of evolution. How can mutations accumulate, given that cells have repair systems that fix changes or breaks in DNA strands?

2. Woolly mammoths have been extinct for about 10,000 years, but we occasionally find one that has been preserved in Siberian permafrost. Resurrecting these huge elephant-like mammals may be possible by cloning DNA isolated from such frozen remains. Researchers are now studying the DNA of a remarkably intact baby mammoth recently discovered frozen in a Siberian swamp. What are some of the pros and cons, both technical and ethical, of cloning an extinct animal?

Animations and Interactions on **CENGAGENOW**:
❯ Structural organization of eukaryotic chromosomes; Karyotyping; Sex determination in humans; Griffith's experiments; Hershey–Chase experiments; Subunits of DNA; Zooming in on DNA structure; Details of DNA replication; SCNT.

‹ Links to Earlier Concepts

In this chapter, you will see how cells use information in nucleic acids (Section 3.7) to build proteins (3.5, 3.6). You will use what you know about base pairing (8.4) to understand transcription, which is a bit like DNA replication (8.6). You will also revisit ribosomes (4.5) and other organelles (4.6–4.8), energy in metabolism (5.3), and enzymes (5.4). A review of electrons (2.3) and radiant energy (6.2) will be useful as you learn about mutations.

Key Concepts

DNA to RNA to Protein
The sequence of amino acids in a polypeptide chain corresponds to a sequence of nucleotide bases in DNA called a gene. The conversion of information in DNA to protein occurs in two steps: transcription and translation.

DNA to RNA: Transcription
During transcription, one strand of a DNA double helix serves as a template for assembling a single, complementary strand of RNA (a transcript). Each transcript is an RNA copy of a gene.

9 From DNA to Protein

9.1 Ricin and Your Ribosomes

Ricin is a naturally occurring protein that is highly toxic: A dose as small as a few grains of salt can kill an adult. Only botulinum and tetanus toxins are more deadly, and there is no antidote. Ricin effectively deters many animals from eating any part of the castor-oil plant (*Ricinus communis*), which grows wild in tropical regions worldwide and is widely cultivated for its seeds (Figure 9.1). Castor-oil seeds are the source of castor oil, an ingredient in plastics, cosmetics, paints, soaps, polishes, and many other items. After the oil is extracted from the seeds, the ricin is typically discarded along with the leftover seed pulp.

The lethal effects of ricin were known as long ago as 1888, but using ricin as a weapon is now banned by most countries under the Geneva Protocol. However, controlling its production is impossible, because it takes no special skills or equipment to manufacture the toxin from easily obtained raw materials. Thus, ricin appears periodically in the news.

For example, at the height of the Cold War, the Bulgarian writer Georgi Markov had defected to England and was working as a journalist for the BBC. As he made his way to a bus stop on a London street, an assassin used the tip of a modified umbrella to jam a small, ricin-laced ball into Markov's leg. Markov died in agony three days later.

More recently, police in 2003 acted on an intelligence tip and stormed a London apartment, where they found laboratory glassware and castor-oil beans. Traces of ricin were found in a United States Senate mailroom and State Department building, and also in an envelope addressed to the White House in 2004. In 2005, the FBI arrested a man who had castor-oil beans and an assault rifle stashed in his Florida home. Jars of banana baby food laced with ground castor-oil beans also made the news in 2005. In 2006, police found pipe bombs and a baby food jar full of ricin in a Tennessee man's shed. In 2008, castor beans, firearms, and several vials of ricin were found in a Las Vegas motel room after its occupant was hospitalized for ricin exposure.

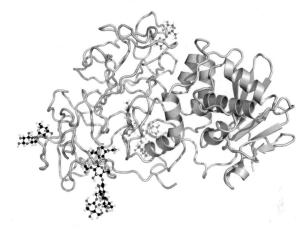

Figure 9.1 Ricin. Above, one of ricin's polypeptide chains (*green*) helps the molecule cross cell membranes. The other chain (*tan*) destroys a cell's capacity for protein synthesis. Sugars attached to the protein are shown. *Opposite*, seeds of the castor-oil plant, the source of ricin.

One of ricin's two polypeptide chains is an enzyme that inactivates ribosomes. This enzyme also occurs in wheat, barley, and other food plants, but it is not particularly toxic because it cannot cross cell membranes very well. However, ricin's second chain binds to plasma membranes, and the binding causes the cell to take up the ricin molecule by endocytosis. Once inside the cell, ricin's two polypeptide chains separate, and the enzyme begins to inactivate ribosomes. Ribosomes assemble amino acids into proteins. Proteins are critical to all life processes, so cells that cannot make them die very quickly. Someone who inhales ricin can die from low blood pressure and respiratory failure within a few days of exposure.

This chapter details how cells convert information encoded in their DNA to an RNA or a protein product. Even though it is extremely unlikely that your ribosomes will ever encounter ricin, protein synthesis is nevertheless worth appreciating for how it keeps you and all other organisms alive.

RNA
Messenger RNA (mRNA) carries DNA's protein-building instructions. Its nucleotide sequence is read three bases at a time. Sixty-four mRNA base triplets—codons—represent the genetic code. Two other types of RNA interact with mRNA during translation of that code.

RNA to Protein: Translation
Translation is an energy-intensive process by which a sequence of codons in mRNA is converted to a sequence of amino acids in a polypeptide chain. Transfer RNAs deliver amino acids to ribosomes, which catalyze the formation of peptide bonds between the amino acids.

Mutations
Small-scale, permanent changes in the nucleotide sequence of DNA may result from replication errors, the activity of transposable elements, or exposure to environmental hazards. Such mutations can change a gene's product.

> Transcription converts information in a gene to RNA; translation converts information in an mRNA to protein.
< Links to Proteins 3.5, Nucleic acids 3.7, Ribosomes 4.5, Enzymes 5.4, DNA sequence and replication 8.6

You learned in Chapter 8 that genetic information consists of the nucleotide base sequence of DNA. How does a cell convert that information into structural and functional components? Let's start with the nature of the information itself.

DNA is like a book, an encyclopedia that contains all of the instructions for building a new individual. You already know the alphabet used to write the book: the four letters A, T, G, and C, for the four nucleotide bases adenine, thymine, guanine, and cytosine. Each strand of DNA consists of a chain of those four kinds of nucleotides. The linear order, or sequence, of the four bases in the strand is the genetic information. That information occurs in subsets called **genes**.

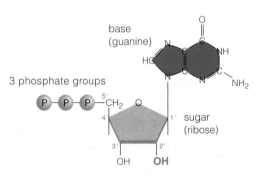

An RNA nucleotide: guanine (G),
or guanosine triphosphate (GTP)

A Guanine, one of the four nucleotides in RNA. The others (adenine, uracil, and cytosine) differ only in their component bases (*blue*). Three of the four bases in RNA nucleotides are identical to the bases in DNA nucleotides.

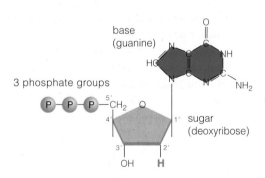

A DNA nucleotide: guanine (G),
or deoxyguanosine triphosphate (dGTP)

B Compare the DNA nucleotide guanine. The only difference between the RNA and DNA versions of guanine (or adenine, or cytosine) is the hydrogen atom or hydroxyl group at the 2' carbon of the sugar (shown in *green*).

Figure 9.2 Comparison between **A** an RNA nucleotide and **B** a DNA nucleotide.

Converting a Gene to an RNA

Converting the information encoded by a gene into a product starts with RNA synthesis, or **transcription**. By this process, enzymes use the nucleotide sequence of a gene as a template to synthesize a strand of RNA (ribonucleic acid):

$$\text{DNA} \xrightarrow{\textit{transcription}} \text{RNA}$$

Except for the double-stranded RNA that is the genetic material of some types of viruses, RNA usually occurs in single-stranded form. A strand of RNA is structurally similar to a single strand of DNA. For example, both are chains of four kinds of nucleotides. Like a DNA nucleotide, an RNA nucleotide has three phosphate groups, a sugar, and one of four bases. However, DNA and RNA nucleotides are slightly different (Figure 9.2). The two nucleic acids are named after their component sugars, ribose and deoxyribose, which differ in one functional group. Three of the bases (adenine, cytosine, and guanine) are the same in RNA and DNA nucleotides, but the fourth base in RNA is uracil, not thymine as it is in DNA.

Despite these small differences in structure, DNA and RNA have very different functions (Figure 9.3). DNA's only role is to store a cell's heritable information. By contrast, a cell transcribes several kinds of RNAs, and each kind has a different function. MicroRNAs are important in gene control, which is the subject of the next chapter. Three types of RNA have roles in protein synthesis. **Ribosomal RNA (rRNA)** is the main component of ribosomes, structures upon which polypeptide chains are built (Section 4.5). **Transfer RNA (tRNA)** delivers amino acids to ribosomes, one by one, in the order specified by a **messenger RNA (mRNA)**.

Converting mRNA to Protein

Messenger RNA is the only kind of RNA that carries a protein-building message. That message is encoded within the sequence of the mRNA by sets of three nucleotide bases, "genetic words" that follow one another along the length of the mRNA. Like the words of a sentence, a series

gene Part of a DNA base sequence; specifies an RNA or protein product.
gene expression Process by which the information in a gene becomes converted to an RNA or protein product.
messenger RNA (mRNA) A type of RNA that carries a protein-building message.
ribosomal RNA (rRNA) A type of RNA that becomes part of ribosomes.
transcription Process by which an RNA is assembled from nucleotides using the base sequence of a gene as a template.
transfer RNA (tRNA) A type of RNA that delivers amino acids to a ribosome during translation.
translation Process by which a polypeptide chain is assembled from amino acids in the order specified by an mRNA.

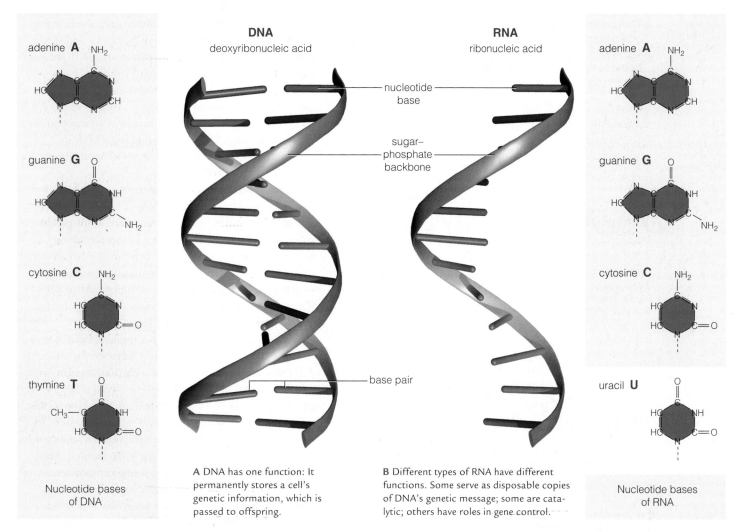

DNA
deoxyribonucleic acid

RNA
ribonucleic acid

adenine **A**

guanine **G**

cytosine **C**

thymine **T**

Nucleotide bases
of DNA

nucleotide
base

sugar–
phosphate
backbone

base pair

adenine **A**

guanine **G**

cytosine **C**

uracil **U**

Nucleotide bases
of RNA

A DNA has one function: It permanently stores a cell's genetic information, which is passed to offspring.

B Different types of RNA have different functions. Some serve as disposable copies of DNA's genetic message; some are catalytic; others have roles in gene control.

Figure 9.3 DNA and RNA compared.

of these genetic words can form a meaningful parcel of information—in this case, the sequence of amino acids of a protein.

By the process of **translation**, the protein-building information in an mRNA is decoded (translated) into a sequence of amino acids. The result is a polypeptide chain that twists and folds into a protein:

$$mRNA \xrightarrow{\textit{translation}} PROTEIN$$

Sections 9.4 and 9.5 describe how rRNA and tRNA interact to translate the sequence of base triplets in an mRNA into the sequence of amino acids in a protein.

The processes of transcription and translation are part of **gene expression**, a multistep process by which genetic information encoded by a gene is converted into a structural or functional part of a cell or a body:

$$DNA \xrightarrow{\textit{transcription}} mRNA \xrightarrow{\textit{translation}} PROTEIN$$

A cell's DNA sequence contains all of the information it needs to make the molecules of life. Each gene encodes an RNA, and different types of RNAs interact to assemble proteins from amino acids (Section 3.5). Some of those proteins are enzymes that assemble lipids and complex carbohydrates from simple building blocks (Section 3.2), replicate DNA (Section 8.6), and make RNA, as you will see in the next section.

Take-Home Message **What is the nature of genetic information carried by DNA?**

❯ The nucleotide sequence of a gene encodes instructions for building an RNA or protein product.

❯ A cell transcribes the nucleotide sequence of a gene into RNA.

❯ Although RNA is structurally similar to a single strand of DNA, the two types of molecules differ functionally.

❯ A messenger RNA (mRNA) carries a protein-building code in its nucleotide sequence. rRNAs and tRNAs interact to translate that sequence into a protein.

❯ RNA polymerase links RNA nucleotides into a chain, in the order dictated by the base sequence of a gene.

❯ A new RNA strand is complementary in sequence to the DNA strand from which it was transcribed.

❮ Links to Base pairing 8.4, DNA replication 8.6

Remember that DNA replication begins with one DNA double helix and ends with two DNA double helices (Section 8.6). The two double helices are identical to the

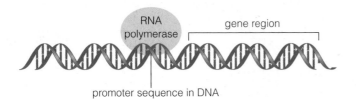

1 RNA polymerase binds to a promoter in the DNA. The binding positions the polymerase near a gene. In most cases, the base sequence of the gene occurs on only one of the two DNA strands. Only the DNA strand complementary to the gene sequence will be translated into RNA.

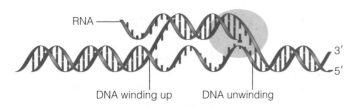

2 The polymerase begins to move along the DNA and unwind it. As it does, it links RNA nucleotides into a strand of RNA in the order specified by the base sequence of the DNA. The DNA winds up again after the polymerase passes. The structure of the "opened" DNA at the transcription site is called a transcription bubble, after its appearance.

parent molecule because base-pairing rules are followed during DNA replication. A nucleotide can be added to a growing strand of DNA only if it base-pairs with the corresponding nucleotide of the parent strand: G pairs with C, and A pairs with T (Section 8.4). Base-pairing rules also govern RNA synthesis during transcription. An RNA strand is structurally so similar to a DNA strand that the two can base-pair if their nucleotide sequences are complementary. In such hybrid molecules, G pairs with C; A pairs with U (uracil).

During transcription, a strand of DNA acts as a template upon which a strand of RNA—a transcript—is assembled from RNA nucleotides. A nucleotide can be added to a growing RNA only if it is complementary to the corresponding nucleotide of the parent DNA strand. Thus, each new RNA is complementary in sequence to the DNA strand that served as its template. As in DNA replication, each nucleotide provides the energy for its own attachment to the end of a growing strand.

Transcription is similar to DNA replication in that one strand of a nucleic acid serves as a template for synthesis of another. However, in contrast with DNA replication, only part of one DNA strand, not the whole molecule, is used as a template for transcription. The enzyme **RNA polymerase**, not DNA polymerase, adds nucleotides to the end of a growing transcript. Also, transcription results in a single strand of RNA, not two DNA double helices.

Transcription begins with a gene on a chromosome (Figure 9.4). The process gets under way when an RNA polymerase and several regulatory proteins attach to a specific binding site in the DNA called a **promoter** **1**. The binding positions the polymerase at a transcription

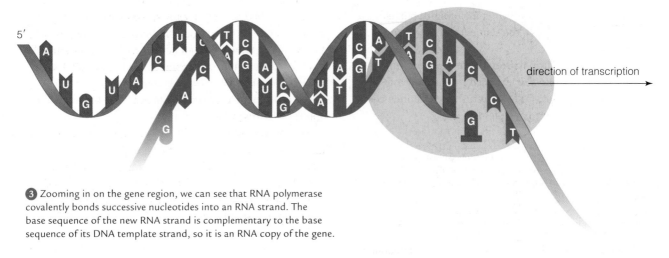

direction of transcription

3 Zooming in on the gene region, we can see that RNA polymerase covalently bonds successive nucleotides into an RNA strand. The base sequence of the new RNA strand is complementary to the base sequence of its DNA template strand, so it is an RNA copy of the gene.

Figure 9.4 Animated Transcription. By this process, a strand of RNA is assembled from nucleotides according to a template: a gene region in DNA.

❯❯ **Figure It Out** After the guanine, what is the next nucleotide that will be added to this growing strand of RNA?   Answer: Another guanine (G)

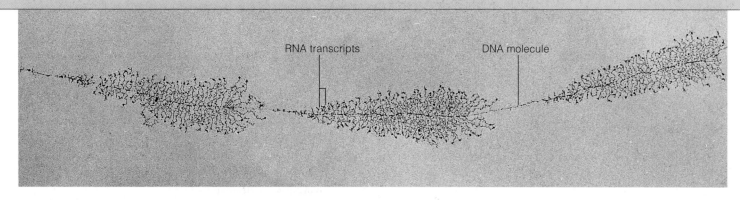

RNA transcripts   DNA molecule

Figure 9.5 Typically, many RNA polymerases simultaneously transcribe the same gene, producing a structure often called a "Christmas tree" after its shape. Here, three genes next to one another on the same chromosome are being transcribed.

>> **Figure It Out** Are the polymerases transcribing this DNA molecule moving from left to right or from right to left?

Answer: Left to right

start site close to a gene. The polymerase starts moving along the DNA, in the 3′ to 5′ direction over the gene ❷. As it moves, the polymerase unwinds the double helix just a bit so it can "read" the base sequence of the noncoding DNA strand. The polymerase joins free RNA nucleotides into a chain, in the order dictated by that DNA sequence. As in DNA replication, the synthesis is directional: An RNA polymerase adds nucleotides only to the 3′ end of the growing strand of RNA.

When the polymerase reaches the end of the gene, the DNA and the new RNA strand are released. RNA poly-

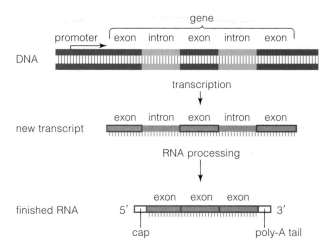

Figure 9.6 Animated Post-transcriptional modification of RNA. Introns are removed and exons spliced together. Messenger RNAs also get a poly-A tail and modified guanine "cap."

---

alternative splicing RNA processing event in which some exons are removed or joined in various combinations.
exon Nucleotide sequence that is not spliced out of RNA during processing.
intron Nucleotide sequence that intervenes between exons and is excised during RNA processing.
promoter In DNA, a sequence to which RNA polymerase binds.
RNA polymerase Enzyme that carries out transcription.

merase follows base-pairing rules, so the new RNA strand is complementary in base sequence to the DNA strand from which it was transcribed ❸. It is an RNA copy of a gene, the same way that a paper transcript of a conversation carries the same information in a different format. Typically, many polymerases transcribe a particular gene region at the same time, so many new RNA strands can be produced very quickly (Figure 9.5).

Post-Transcriptional Modifications

In eukaryotes, transcription takes place in the nucleus. Eukaryotes also modify their RNA inside the nucleus, then ship it to the cytoplasm. Just as a dressmaker may snip off loose threads or add bows to a dress before it leaves the shop, so do eukaryotic cells tailor their RNA before it leaves the nucleus.

For example, most eukaryotic genes contain **introns**, nucleotide sequences that are removed from a new RNA. Introns intervene between **exons**, sequences that stay in the RNA (Figure 9.6). Introns are transcribed along with exons, but they are removed before the RNA leaves the nucleus. Either all exons remain in the mature RNA, or some are removed and the remaining exons are spliced together. By such **alternative splicing**, one gene can encode different proteins.

New transcripts that will become mRNAs are further tailored after splicing. A modified guanine "cap" gets attached to the 5′ end of each. Later, the cap will help the mRNA bind to a ribosome. A tail of 50 to 300 adenines is also added to the 3′ end of a new mRNA; hence the name, poly-A tail.

Take-Home Message **How is RNA assembled?**

❯ In transcription, RNA polymerase uses the nucleotide sequence of a gene region in a chromosome as a template to assemble a strand of RNA.

❯ The new strand of RNA is a copy of the gene from which it was transcribed.

› Base triplets in an mRNA encode a protein-building message. Ribosomal RNA and transfer RNA translate that message into a polypeptide chain.

‹ Links to Polypeptides 3.5, Ribosomes 4.5, Catalysis 5.4

DNA stores heritable information about proteins, but making those proteins requires messenger RNA (mRNA), transfer RNA (tRNA), and ribosomal RNA (rRNA). The three types of RNA interact to translate DNA's information into a protein.

An mRNA is essentially a disposable copy of a gene. Its job is to carry DNA's protein-building information to the other two types of RNA for translation. That protein-building information consists of a linear sequence of genetic "words" spelled with an alphabet of the four bases A, C, G, and U. Each of the genetic "words" carried by an mRNA is three bases long, and each is a code—a **codon**—for a particular amino acid. There are four possible bases in each of the three positions of a codon, so there are a total of sixty-four (or 4^3) mRNA codons. Collectively, the sixty-four codons constitute the **genetic code** (Figure 9.7). Which of the four nucleotides is in the first, second, and third position of a base triplet determines which amino acid the codon specifies. For instance, the codon AUG codes for the amino acid methionine (met), and UGG codes for tryptophan (trp).

One codon follows the next along the length of an mRNA, so the order of codons in an mRNA determines the order of amino acids in the polypeptide that will be translated from it. Thus, the base sequence of a gene is transcribed into the base sequence of an mRNA, which is in turn translated into an amino acid sequence (Figure 9.8).

There are only twenty kinds of amino acids found in proteins. Sixty-four codons are more than are needed to specify twenty amino acids, so some amino acids are specified by more than one codon. For instance, GAA and GAG both code for glutamic acid. Other codons signal the beginning and end of a protein-coding sequence. For example, the first AUG in an mRNA is the signal to start translation in most species. AUG also happens to be the codon for methionine, so methionine is always the first amino acid in new polypeptides of such organisms. UAA, UAG, and UGA do not specify an amino acid. They are signals that stop translation, so they are called stop

first base	second base: U	second base: C	second base: A	second base: G	third base
U	UUU phe	UCU ser	UAU tyr	UGU cys	U
	UUC phe	UCC ser	UAC tyr	UGC cys	C
	UUA leu	UCA ser	UAA stop	UGA stop	A
	UUG leu	UCG ser	UAG stop	UGG trp	G
C	CUU leu	CCU pro	CAU his	CGU arg	U
	CUC leu	CCC pro	CAC his	CGC arg	C
	CUA leu	CCA pro	CAA gln	CGA arg	A
	CUG leu	CCG pro	CAG gln	CGG arg	G
A	AUU ile	ACU thr	AAU asn	AGU ser	U
	AUC ile	ACC thr	AAC asn	AGC ser	C
	AUA ile	ACA thr	AAA lys	AGA arg	A
	AUG met	ACG thr	AAG lys	AGG arg	G
G	GUU val	GCU ala	GAU asp	GGU gly	U
	GUC val	GCC ala	GAC asp	GGC gly	C
	GUA val	GCA ala	GAA glu	GGA gly	A
	GUG val	GCG ala	GAG glu	GGG gly	G

ala	alanine (A)	leu	leucine (L)
arg	arginine (R)	lys	lysine (K)
asn	asparagine (N)	met	methionine (M)
asp	aspartic acid (D)	phe	phenylalanine (F)
cys	cysteine (C)	pro	proline (P)
glu	glutamic acid (E)	ser	serine (S)
gln	glutamine (Q)	thr	threonine (T)
gly	glycine (G)	trp	tryptophan (W)
his	histidine (H)	tyr	tyrosine (Y)
ile	isoleucine (I)	val	valine (V)

Figure 9.7 Animated The genetic code. Each codon in mRNA is a set of three nucleotide bases. In the large chart, the *left* column lists a codon's first base, the *top* row lists the second, and the *right* column lists the third. Sixty-one of the triplets encode amino acids; the remaining three are signals that stop translation. The amino acid names that correspond to abbreviations in the chart are listed *above*.

›› Figure It Out Which codons specify the amino acid lysine (lys)?

Answer: AAA and AAG

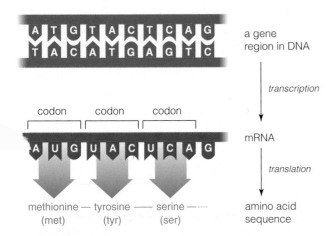

a gene region in DNA

transcription

codon   codon   codon

mRNA

A·U·G·U·A·C·U·C·A·G

translation

methionine — tyrosine —— serine —·····   amino acid
(met)   (tyr)   (ser)   sequence

Figure 9.8 Example of the correspondence between DNA, RNA, and proteins. A DNA strand is transcribed into mRNA, and the codons of the mRNA specify a chain of amino acids.

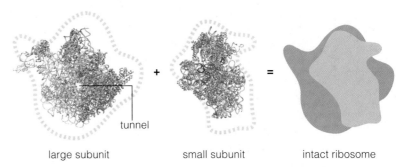

large subunit   +   small subunit   =   intact ribosome

tunnel

Figure 9.9 Animated Ribosome structure. An intact ribosome consists of a large and a small subunit. Structural protein components of the subunits are shown in *green*; catalytic rRNA components, in *tan*. Notice the tunnel through the interior of the large subunit. A polypeptide chain threads through this tunnel as it is being assembled by the ribosome. We show an mRNA (in *red*) attached to the small subunit.

codons. A stop codon marks the end of the protein-coding sequence in an mRNA.

The genetic code is highly conserved, which means that many organisms use the same code and probably always have. Bacteria, archaeans, and some protists have a few codons that differ from the typical code, as do mitochondria and chloroplasts. The variation was a clue that led to a theory of how organelles evolved (we will return to the topic of endosymbiosis in Section 18.6).

rRNA and tRNA—The Translators

A ribosome has one large and one small subunit. Each subunit consists mainly of rRNA, with some associated structural proteins. During translation, a large and a small ribosomal subunit converge as an intact ribosome on an mRNA (Figure 9.9).

Ribosomes and tRNAs interact to translate an mRNA into a polypeptide. Transfer RNAs deliver amino acids to ribosomes in the order specified by the mRNA. Each tRNA has two attachment sites. The first is an anticodon, a triplet of nucleotides that base-pairs with an mRNA codon (Figure 9.10). The other attachment site on a tRNA binds to an amino acid—the one specified by the codon. Transfer RNAs with different anticodons carry different amino acids. You will see in the next section how those tRNAs deliver amino acids, one after the next, to a ribosome during translation of an mRNA.

Ribosomal RNA is one of the few examples of RNA with enzymatic activity: The rRNA of a ribosome, not the protein, catalyzes the formation of a peptide bond

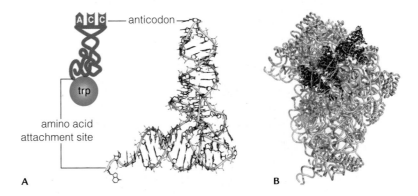

A·C·C —— anticodon

trp

amino acid attachment site

A   B

Figure 9.10 Animated tRNA structure. **A** Models of the tRNA that carries the amino acid tryptophan. Each tRNA's anticodon is complementary to an mRNA codon. Each also carries the amino acid specified by that codon. **B** During translation, tRNAs dock at an intact ribosome. Here, three tRNAs (*brown*) are docked at the small ribosomal subunit (the large subunit is not shown, for clarity). The anticodons of the tRNAs line up with complementary codons in an mRNA (shown in *red*).

between amino acids. As the amino acids are delivered, the ribosome joins them via peptide bonds into a new polypeptide (Section 3.5). Thus, the order of codons in an mRNA—DNA's protein-building message—becomes translated into a new protein.

---

codon In mRNA, a nucleotide base triplet that codes for an amino acid or stop signal during translation.
genetic code Complete set of sixty-four mRNA codons.

Take-Home Message What roles do mRNA, tRNA, and rRNA play during translation?

> mRNA carries protein-building information. The bases in mRNA are "read" in sets of three during protein synthesis. Most of these base triplets (codons) code for amino acids. The genetic code consists of all sixty-four codons.

> Ribosomes, which consist of two subunits of rRNA and proteins, assemble amino acids into polypeptide chains.

> A tRNA has an anticodon complementary to an mRNA codon, and it has a binding site for the amino acid specified by that codon. Transfer RNAs deliver amino acids to ribosomes.

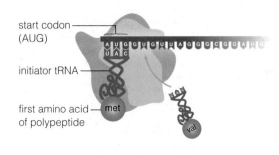

start codon (AUG)

initiator tRNA

first amino acid of polypeptide

met

val

1 Ribosome subunits and an initiator tRNA converge on an mRNA. A second tRNA binds to the second codon.

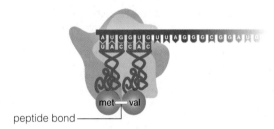

met — val

peptide bond

2 A peptide bond forms between the first two amino acids.

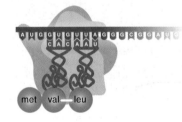

met   val

leu

3 The first tRNA is released and the ribosome moves to the next codon. A third tRNA binds to the third codon.

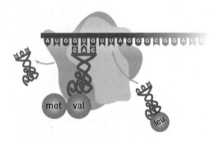

met   val — leu

4 A peptide bond forms between the second and third amino acids.

met   val   leu

gly

5 The second tRNA is released and the ribosome moves to the next codon. A fourth tRNA binds the fourth codon.

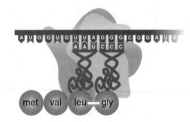

met   val   leu — gly

6 A peptide bond forms between the third and fourth amino acids.

The process repeats until the ribosome encounters a stop codon in the mRNA.

❭ Translation converts the information carried by an mRNA into a polypeptide.

❭ The order of the codons in an mRNA determines the order of amino acids in the translated polypeptide.

❬ Links to Peptide bonds 3.5, Endomembrane system 4.8, Nucleus 4.7, Energy in metabolism 5.3

Translation, the second part of protein synthesis, occurs in the cytoplasm, which has many free amino acids, tRNAs, and ribosomal subunits (Figures 9.11 and 9.12A). Translation proceeds in three stages: initiation, elongation, and termination. In eukaryotes, the initiation stage begins when an mRNA leaves the nucleus and a small ribosomal subunit binds to it. Next, the anticodon of a special tRNA called an initiator base-pairs with the first AUG codon of the mRNA. Then, a large ribosomal subunit joins the small subunit **1**.

In the elongation stage, the ribosome assembles a polypeptide chain as it moves along the mRNA. The initiator tRNA carries the amino acid methionine, so the first amino acid of the new polypeptide chain is methionine. Another tRNA brings the second amino acid to the complex as its anticodon base-pairs with the second codon in the mRNA. The ribosome catalyzes formation of a peptide bond between the first two amino acids **2**.

The first tRNA is released and the ribosome moves to the next codon. Another tRNA brings the third amino acid to the complex as its anticodon base-pairs with the third codon of the mRNA **3**. A peptide bond forms between the second and third amino acids **4**.

The second tRNA is released and the ribosome moves to the next codon. Another tRNA brings the fourth amino acid to the complex as its anticodon base-pairs with the fourth codon of the mRNA **5**. A peptide bond forms between the third and fourth amino acids **6**. The new polypeptide chain continues to elongate as the ribosome catalyzes peptide bonds between amino acids delivered by successive tRNAs.

Termination occurs when the ribosome reaches a stop codon in the mRNA. The mRNA and the polypeptide detach from the ribosome, and the ribosomal subunits separate from each other. Translation is now complete. The new polypeptide will either join the pool of proteins in the cytoplasm, or enter rough ER of the endomembrane system (Section 4.8).

In cells that are making a lot of protein, many ribosomes may simultaneously translate the same mRNA, in which case they are called polysomes (Figure 9.12B).

Figure 9.11 Animated Translation. Translation initiates when ribosomal subunits and an initiator tRNA converge on an mRNA. tRNAs deliver amino acids in the order dictated by successive codons in the mRNA. The ribosome links the amino acids together as it moves along the mRNA, so a polypeptide forms and elongates. Translation terminates when the ribosome reaches a stop codon.

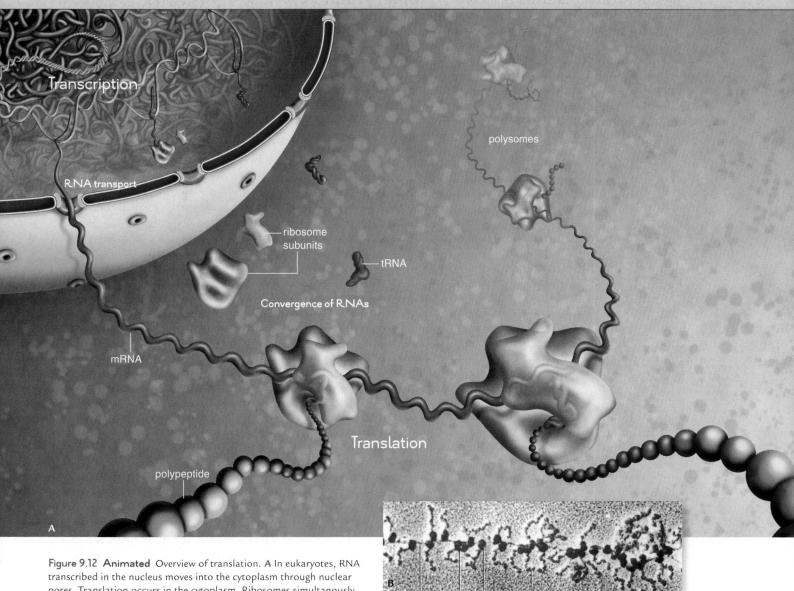

Transcription

RNA transport

ribosome subunits

tRNA

Convergence of RNAs

mRNA

polysomes

Translation

polypeptide

A

B

mRNA   polysomes   newly forming polypeptide

Figure 9.12 Animated Overview of translation. **A** In eukaryotes, RNA transcribed in the nucleus moves into the cytoplasm through nuclear pores. Translation occurs in the cytoplasm. Ribosomes simultaneously translating the same mRNA are called polysomes. **B** This micrograph shows polypeptides elongating from polysomes on an mRNA.

Transcription and translation both occur in the cytoplasm of bacteria, and these processes are closely linked in time and in space. Translation begins before transcription is done, so in these cells, a DNA transcription tree may be decorated with polysome "balls."

Given that many polypeptides are translated from one mRNA, why would a cell also make many copies of an mRNA? Compared with DNA, RNA is not very stable. An mRNA may last only a few minutes before it gets disassembled by enzymes in the cytoplasm. The fast turnover allows cells to adjust their protein synthesis quickly in response to changing needs.

Translation is energy intensive. That energy is provided mainly in the form of phosphate-group transfers from the RNA nucleotide GTP (shown in Figure 9.2A) to molecules involved in the process.

Take-Home Message How is mRNA translated into protein?

❭ Translation is an energy-requiring process that converts the protein-building information carried by an mRNA into a polypeptide.

❭ During initiation, an mRNA joins with an initiator tRNA and two ribosomal subunits.

❭ During elongation, amino acids are delivered to the complex by tRNAs in the order dictated by successive mRNA codons. As they arrive, the ribosome joins each to the end of the polypeptide chain.

❭ Termination occurs when the ribosome encounters a stop codon in the mRNA. The mRNA and the polypeptide are released, and the ribosome disassembles.

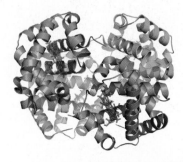

A Hemoglobin, an oxygen-binding protein in red blood cells. This protein consists of four polypeptides: two alpha globins (*blue*) and two beta globins (*green*). Each globin has a pocket that cradles a heme (*red*). Oxygen molecules bind to the iron atom at the center of each heme.

T	G	A	G	G	A	C	T	C	C	T	C	T	T	C
A	C	U	C	C	U	G	A	G	G	A	G	A	A	G

threonine —— proline —— glutamic acid — glutamic acid — lysine

B Part of the DNA (*blue*), mRNA (*brown*), and amino acid sequence (*green*) of human beta globin.

T	G	A	G	G	A	C	C	C	T	C	T	T	C	A
A	C	U	C	C	U	G	G	A	G	A	A	G	U

threonine —— proline —— glycine —— arginine —— serine

C A base-pair deletion causes the reading frame for the rest of the mRNA to shift, so a completely different protein product forms. The mutation shown results in a defective beta globin. The outcome is beta thalassemia, a genetic disorder in which a person has an abnormally low amount of hemoglobin.

T	G	A	G	G	A	C	A	C	C	T	C	T	T	C
A	C	U	C	C	U	G	U	G	G	A	G	A	A	G

threonine —— proline —— valine — glutamic acid — lysine

D A base-pair substitution replaces a thymine with an adenine. When the altered mRNA is translated, valine replaces glutamic acid as the sixth amino acid of the polypeptide. Hemoglobin with this form of beta globin is called HbS, or sickle hemoglobin.

E The substitution of valine for glutamic acid causes the HbS protein to aggregate into rod-like clumps at low oxygen levels. The clumps distort normally round red blood cells into sickle shapes (a sickle is a farm tool with a crescent-shaped blade).

Figure 9.13 Animated Examples of mutations in the human beta globin gene.

❯ If the nucleotide sequence of a gene changes, it may result in an altered gene product, with harmful effects.
❮ Links to Electron energy levels 2.3, Protein structure 3.6, Free radicals and hemes as cofactors 5.4, Radiant energy 6.2, DNA replication and mutations 8.6

Mutations are permanent changes in the sequence of a cell's DNA (Section 8.6). If a mutation changes the genetic instructions encoded in the DNA, an altered gene product may result. Remember, more than one codon can specify the same amino acid, so cells have some margin of safety. For example, a mutation that changes a UCU codon to UCC in an mRNA may not have further effects, because both codons specify serine. However, other mutations have drastic consequences.

The oxygen-binding properties of hemoglobin provide an example of how mutations can change the structure (and function) of a protein. As red blood cells circulate through the lungs, the hemoglobin proteins inside of them bind to oxygen molecules. The cells then travel to regions of the body where the oxygen level is low. There, the hemoglobin releases its oxygen cargo. When the red blood cells return to the lungs, the hemoglobin binds to more oxygen.

Hemoglobin's structure allows it to bind and release oxygen. The protein consists of four polypeptides called globins (Figure 9.13A). Each globin folds around a heme, a cofactor with an iron atom at its center (Section 5.4). Oxygen molecules bind to hemoglobin at those iron atoms.

In adult humans, two alpha globins and two beta globins make up each hemoglobin molecule. Defects in either polypeptide can cause a condition called anemia, in which a person's blood is deficient in red blood cells or in hemoglobin. Both outcomes limit the blood's ability to carry oxygen.

For example, the loss of a particular nucleotide—a **deletion**—from the DNA of the beta globin gene causes a type of anemia called beta thalassemia. Like many other deletions, this one causes the reading frame of mRNA codons to shift. The shift garbles the genetic message, just as incorrectly grouping a series of letters garbles the meaning of a sentence:

> The cat ate the rat.
> T hec ata tet her at.
> Th eca tat eth era t.

Expression of a beta globin gene with a beta thalassemia mutation results in a polypeptide that differs drastically from normal beta globin (Figure 9.13B,C). Hemoglobin molecules do not assemble correctly with the altered polypeptide, an outcome that is the source of the anemia.

Frameshifts may also be caused by **insertion** mutations, in which extra bases get inserted into a gene. Other types of mutations do not cause frameshifts. With

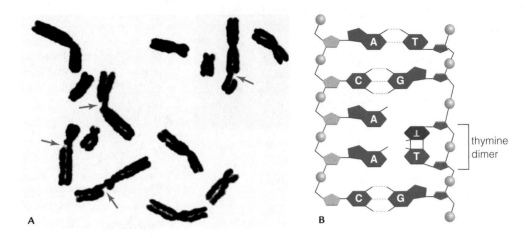

Figure 9.14 Two types of DNA damage that can lead to mutations. **A** Chromosomes from a human cell after exposure to gamma rays (a type of ionizing radiation). The broken pieces (*red arrows*) may get lost during DNA replication. **B** A thymine dimer.

A

B

thymine dimer

a **base–pair substitution**, a nucleotide and its partner are replaced by a different base pair. A substitution may result in an amino acid change in a gene's protein product, or a premature stop codon that shortens it. Sickle-cell anemia, a type of anemia that is most common in people of African ancestry, arises because of a base-pair substitution. The substitution causes valine to be the sixth amino acid of beta globin instead of glutamic acid (Figure 9.13D). Hemoglobin with this mutation in its beta globin is called sickle hemoglobin, or HbS.

Unlike glutamic acid, which carries a negative charge, valine carries no charge. As a result of that one substitution, a tiny patch of the beta globin polypeptide changes from hydrophilic to hydrophobic, which in turn causes the hemoglobin's behavior to change slightly. HbS molecules stick together and form large, rodlike clumps under certain conditions. Red blood cells that contain the clumps become distorted into a crescent (sickle) shape (Figure 9.13E). Sickled cells clog tiny blood vessels, thus disrupting blood circulation throughout the body. Over time, repeated episodes of sickling can damage organs and cause death.

What Causes Mutations?

Insertion mutations are often caused by activity of **transposable elements**, segments of DNA that can move spontaneously within or between chromosomes. Transposable elements can be hundreds or thousands of base pairs long, so when one interrupts a gene it becomes a major insertion that changes the gene's product. Transposable elements are common in the DNA of all species; about

45 percent of human DNA consists of transposable elements or their remnants.

Many mutations occur spontaneously during DNA replication (Section 8.6). DNA polymerases make mistakes at predictable rates, but most types fix errors as they occur. Errors that remain uncorrected are mutations.

Environmental agents also cause mutations. For example, x-rays and other forms of ionizing radiation can knock electrons out of atoms. Ionizing radiation breaks chromosomes into pieces that get lost during DNA replication (Figure 9.14A). It also damages DNA indirectly when it penetrates living tissue, because it leaves a trail of destructive free radicals in its wake (Section 5.4).

Nonionizing radiation boosts electrons to a higher energy level, but not enough to knock them out of an atom. DNA absorbs one kind, ultraviolet (UV) light. Exposure to UV light can cause two adjacent thymine bases to bond covalently to one another (Figure 9.14B). The resulting thymine dimer kinks the DNA. DNA polymerase may copy the kinked part incorrectly during replication, so a mutation becomes introduced into the DNA. Mutations that cause certain kinds of cancers begin with thymine dimers. Exposing unprotected skin to sunlight increases the risk of skin cancer because it causes thymine dimers to form in the DNA of skin cells.

Some natural or synthetic chemicals can also cause mutations. For instance, chemicals in cigarette smoke transfer small hydrocarbon groups to the bases in DNA. The altered bases mispair during replication, or stop replication entirely. Both events increase the chance of mutation.

base–pair substitution Type of mutation in which a single base-pair changes.
deletion Mutation in which one or more base pairs are lost.
insertion Mutation in which one or more base pairs become inserted into DNA.
transposable element Segment of DNA that can spontaneously move to a new location in a chromosome.

Take-Home Message **What is a mutation?**

❭ A base-pair substitution, insertion, or deletion are mutations that may alter a gene product.

❭ Most mutations arise during DNA replication as a result of unrepaired DNA polymerase errors. Some mutations occur as a result of transposable elements, or after exposure to harmful radiation or chemicals.

Ricin and Your Ribosomes (revisited)

The enzyme part of ricin inactivates ribosomes by removing a particular adenine base from one of the rRNAs in the heavy subunit. That adenine is part of an RNA binding site for proteins that help with elongation. After the base has been removed, a ribosome can no longer bind to those proteins, and elongation stops.

One molecule of ricin enzyme can inactivate more than 1,000 ribosomes per minute. If enough ribosomes are affected, protein synthesis grinds to a halt, and the cell quickly dies. A modified form of ricin is currently being tested as a treatment for some kinds of cancer. The ricin is attached to an antibody that can find cancer cells in a person's body. Researchers hope that the attached ricin will kill the cancer cells without harming normal ones.

How Would You Vote? Accidental exposure to ricin is unlikely, but terrorists may try to poison food or water supplies with it. Researchers have developed a vaccine against ricin. Do you want to be vaccinated? See CengageNow for details, then vote online (cengage.com).

genetic code. Each tRNA has an anticodon that can base-pair with a codon, and it binds to the amino acid specified by that codon. rRNA and proteins make up the two subunits of ribosomes.

Section 9.5 Genetic information carried by an mRNA directs the synthesis of a polypeptide during translation. First, an mRNA, an initiator tRNA, and two ribosomal subunits converge. The intact ribosome then joins successive amino acids, which are delivered by tRNAs in the order specified by the codons in the mRNA. Translation ends when the polymerase encounters a stop codon.

Section 9.6 Insertions, deletions, and **base–pair substitutions** are mutations that can arise by replication error, the activity of **transposable elements**, or exposure to environmental hazards such as radiation or chemicals. A mutation that changes a gene's product may have harmful effects. Sickle-cell anemia, which is caused by a base-pair substitution in the hemoglobin beta chain gene, is one example.

Summary

Section 9.1 The ability to make proteins is critical to all life processes. Ricin is toxic because it inactivates ribosomes.

Section 9.2 DNA's genetic information is encoded within its base sequence. **Genes** are subunits of that sequence. A cell uses the information in a gene to make an RNA or protein product. The process of **gene expression** involves **transcription** of a DNA sequence to an RNA, and **translation** of the information in an **mRNA**, or **messenger RNA**, to a protein product. Translation requires the participation of **tRNA (transfer RNA)** and **rRNA (ribosomal RNA).**

Section 9.3 During transcription, **RNA polymerase** binds to a **promoter** near a gene region of a chromosome. The polymerase assembles a strand of RNA by linking RNA nucleotides in the order dictated by the base sequence of the gene. Thus, the new RNA is complementary to the gene from which it was transcribed.

The RNA of eukaryotes is modified before it leaves the nucleus. **Introns** are removed. With **alternative splicing,** some **exons** may be removed also, and the remaining ones spliced in different combinations. A cap and a poly-A tail are also added to a new mRNA.

Section 9.4 mRNA carries DNA's protein-building information. The information consists of a series of **codons,** sets of three nucleotides. Sixty-four codons, most of which specify amino acids, constitute the

Self-Quiz   Answers in Appendix III

1. A chromosome contains many different gene regions that are transcribed into different _____ .
   a. proteins   c. RNAs
   b. polypeptides   d. a and b

2. A binding site for RNA polymerase is called a _____ .
   a. gene   c. codon
   b. promoter   d. protein

3. Energy that drives transcription is provided mainly by _____ .
   a. ATP   c. GTP
   b. RNA nucleotides   d. all are correct

4. An RNA molecule is typically _____ ; a DNA molecule is typically _____ .
   a. single-stranded; double-stranded
   b. double-stranded; single-stranded
   c. both are single-stranded
   d. both are double-stranded

5. RNAs form by _____ ; proteins form by _____ .
   a. replication; translation
   b. translation; transcription
   c. transcription; translation
   d. replication; transcription

6. Most codons specify a(n) _____ .
   a. protein   c. amino acid
   b. polypeptide   d. mRNA

7. Anticodons pair with _____ .
   a. mRNA codons   c. RNA anticodons
   b. DNA codons   d. amino acids

Data Analysis Activities

Herbicides and Chromosome Damage in Foresters Forestry workers in the U.S. routinely apply herbicides and pesticides as part of their work. A 2001 study by Vincent F. Garry looked at chromosome abnormalities inside white blood cells of forestry employees who worked with the herbicide 2,4-D (dichlorophenoxyacetic acid) more than five days in one year. Most of the workers in the study distributed the herbicide over large areas from a helicopter or airplane. The results, shown in Figure 9.15, were categorized by type of aberration.

1. Which group of workers, those who sprayed a low amount of 2,4-D or those who sprayed a midrange amount of the herbicide, had the lower frequency of chromosome breaks?
2. What group had the highest frequency of missing chromosome pieces? Breaks?
3. How many herbicide-wielding foresters were tested?

Type of Chromosome Aberration	Total Volume of Herbicide Applied			
	None	Low (1–100 gal)	Midrange (100–1,000 gal)	Heavy (>1,000 gal)
Rearrangements	0.65	1.20	1.00	2.22
Missing pieces	0.93	1.17	1.33	2.89
Breaks	1.29	1.03	1.33	3.00
Other	0.29	0.29	0.50	0.33
No. of subjects	14	7	6	9

Figure 9.15 Chromosome abnormalities and use of the herbicide 2,4-D. The table compares chromosome abnormalities in white blood cells of forestry workers who routinely apply herbicide as part of their job. The numbers indicate the average number of aberrations per 100 cells.

Results are categorized by the total volume of herbicide applied, and by type of chromosome damage.

8. What is the maximum length of a polypeptide encoded by an mRNA that is 45 nucleotides long?

9. _____ are removed from new mRNA transcripts.
   a. Introns
   b. Exons
   c. Telomeres
   d. Amino acids

10. Where does transcription take place in a typical eukaryotic cell?
   a. the nucleus
   b. ribosomes
   c. the cytoplasm
   d. b and c are correct

11. Where does translation take place in a typical eukaryotic cell?
   a. the nucleus
   b. ribosomes
   c. the cytoplasm
   d. b and c are correct

12. Each amino acid is specified by a set of _____ bases in an mRNA transcript.
   a. 3   b. 20   c. 64   d. 120

13. _____ different codons constitute the genetic code.
   a. 3   b. 20   c. 64   d. 120

14. _____ can cause mutations.
   a. Replication errors
   b. Transposons
   c. Ionizing radiation
   d. Nonionizing radiation
   e. b and c are correct
   f. all of the above

15. Match the terms with the best description.
   ___ genetic message
   ___ promoter
   ___ polysome
   ___ exon
   ___ genetic code
   ___ intron
   ___ transposable element

   a. protein-coding segment
   b. gets around
   c. read as base triplets
   d. removed before translation
   e. occurs only in groups
   f. complete set of 64 codons
   g. binding site for RNA polymerase

Additional questions are available on **CENGAGENOW**.

Critical Thinking

1. Antisense drugs help us fight some types of cancer and viral diseases. The drugs consist of short mRNA strands that are complementary in base sequence to mRNAs linked to the diseases. Speculate on how antisense drugs work.

2. An anticodon has the sequence GCG. What amino acid does this tRNA carry? What would be the effect of a mutation that changed the C of the anticodon to a G?

3. Each position of a codon can be occupied by one of four nucleotides. What is the minimum number of nucleotides per codon necessary to specify all 20 of the amino acids that are typical of eukaryotic proteins?

4. Using Figure 9.7, translate this nucleotide sequence into an amino acid sequence, starting at the first base:

   5′—GGUUUCUUGAAGAGA—3′

5. Translate the sequence of bases in the previous question, starting at the second base.

6. Cigarette smoke contains at least fifty-five different chemicals identified as carcinogenic (cancer-causing) by the International Agency for Research on Cancer (IARC). When these carcinogens enter the bloodstream, enzymes convert them to a series of chemical intermediates that are easier to excrete. Some of the intermediates bind irreversibly to DNA. Propose a hypothesis about why cigarette smoke causes cancer.

Animations and Interactions on **CENGAGENOW**:
❯ Transcription; RNA processing; The genetic code; Ribosome structure; tRNA structure; Translation; Differences between prokaryotic and eukaryotic protein synthesis; Substitutions; Frameshifts.

‹ Links to Earlier Concepts

A review of what you know about development (Section 1.3) and metabolic control (5.5) will be helpful as we revisit the concept of gene expression (9.2) in more depth. You will be applying what you know about the organization of chromosomal DNA (8.2) and mutations (9.6) as you learn about controls over transcription (9.3), translation (9.5), and other processes that affect gene expression. You will revisit carbohydrates (3.3) and fermentation (7.6) in context of gene control in bacteria.

Key Concepts

Gene Control in Eukaryotes
A variety of molecules and processes alter gene expression in response to changing conditions both inside and outside the cell. Selective gene expression also results in differentiation, by which cell lineages become specialized.

Mechanisms of Control
All cells in an embryo inherit the same genes, but they start using different subsets of those genes during development. The orderly, localized expression of master genes gives rise to the body plan of complex multicelled organisms.

10 Controls Over Genes

10.1 Between You and Eternity

You are in college, your whole life ahead of you. Your risk of developing cancer is as remote as old age, an abstract statistic that is easy to forget. "There is a moment when everything changes—when the width of two fingers can suddenly be the total distance between you and eternity." Robin Shoulla wrote those words after being diagnosed with breast cancer. She was seventeen. At an age when most young women are thinking about school, friends, parties, and potential careers, Robin was dealing with radical mastectomy: the removal of a breast, all lymph nodes under the arm, and skeletal muscles in the chest wall under the breast. She was pleading with her oncologist not to use her jugular vein for chemotherapy and wondering if she would survive to see the next year (Figure 10.1).

Robin's ordeal became part of a statistic, one of more than 200,000 new cases of breast cancer diagnosed in the United States each year. About 5,700 of those cases occur in women and men under thirty-four years of age.

Every second, millions of cells in your skin, bone marrow, gut lining, liver, and elsewhere are dividing and replacing their worn-out, dead, and dying predecessors. They do not divide at random. Many gene expression controls regulate cell growth and division. When those controls fail, cancer is the outcome. **Cancer** is a multistep process in which abnormally growing and dividing cells disrupt body tissues. Mechanisms that normally keep cells from getting overcrowded in tissues are lost, so cancer cell populations may reach extremely high densities. Unless chemotherapy, surgery, or another procedure eradicates them, cancer cells can put an individual on a painful road to death. Each year, cancers cause 15 to 20 percent of all human deaths in developed countries alone.

Cancer typically begins with a mutation in a gene whose product is part of a system of stringent controls over cell growth and division. Such controls govern when and how fast specific genes are transcribed and translated. The mutation may be

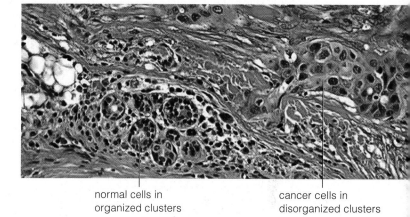

normal cells in organized clusters

cancer cells in disorganized clusters

Figure 10.1 A case of breast cancer. *Above*, this light micrograph shows irregular clusters of cancer cells in the milk ducts of human breast tissue. *Opposite*, Robin Shoulla. Diagnostic tests revealed abnormal cells such as these in her body.

inherited, or it may be a new one, as when DNA becomes damaged by environmental agents. If the mutation alters the gene's protein product so that it no longer works properly, one level of control over the cell's growth and division has been lost. You will be considering the impact of gene controls in chapters throughout the book, and in some chapters of your life.

Robin Shoulla survived. Although radical mastectomy is rarely performed today (a modified procedure is less disfiguring), it is the only option when cancer cells invade muscles under the breast. It was Robin's only option. Now, sixteen years later, she has what she calls a normal life: career, husband, children. Her goal as a cancer survivor: "To grow very old with gray hair and spreading hips, smiling."

cancer Disease that occurs when the uncontrolled growth of body cells physically and metabolically disrupts tissues.

Examples in Eukaryotes
One of the two X chromosomes is inactivated in every cell of female mammals. The Y chromosome carries a master gene that causes male traits to develop in the human fetus. Flower development is orchestrated by a set of homeotic genes.

Gene Control in Bacteria
Bacterial gene controls govern responses to short-term changes in nutrient availability and other aspects of the environment. The main gene controls bring about fast adjustments in the rate of transcription.

> Gene controls govern the kinds and amounts of substances that are present in a cell at any given interval.
< Links to Phosphorylation 5.3, Glycolysis 7.3, Histones 8.2, Gene expression 9.2, Promoters and transcription 9.3, Translation 9.5, Globin 9.6

All of the cells in your body are descended from the same fertilized egg, so they all contain the same DNA with the same genes. Some of the genes are transcribed by all cells; such genes affect structural features and metabolic pathways common to all cells.

In other ways, however, nearly all of your body cells are specialized. **Differentiation**, the process by which cells

Figure 10.3 Hypothetical part of a chromosome that contains a gene. Molecules that affect the rate of transcription of the gene bind at promoter (*yellow*) or enhancer (*green*) sequences.

become specialized, occurs as different cell lineages begin to express different subsets of their genes. Which genes a cell uses determines the molecules it will produce, which in turn determines what kind of cell it will be.

For example, most of your body cells express the genes that encode the enzymes of glycolysis (Section 7.3), but only immature red blood cells use the genes that code for globin (Section 9.6). Only your liver cells express genes for enzymes that neutralize certain toxins.

A cell rarely uses more than 10 percent of its genes at once. Which genes are expressed at any given time depends on many factors, such as conditions in the cytoplasm and extracellular fluid, and the type of cell. These factors affect controls governing all steps of gene expression, starting with transcription and ending with delivery of an RNA or protein product to its final destination (Figure 10.2). Such controls consist of processes that start, enhance, slow, or stop gene expression.

Control of Transcription Many controls affect whether and how fast certain genes are transcribed into RNA ❶. Those that prevent an RNA polymerase from attaching to a promoter near a gene also prevent transcription of the gene. Controls that help RNA polymerase bind to DNA also speed up transcription.

Some types of proteins affect the rate of transcription by binding to special nucleotide sequences in the DNA. For example, an **activator** speeds up transcription when it binds to a promoter. Activators also speed transcription by binding to DNA sequences called **enhancers**. An enhancer is not necessarily close to the gene it affects, and may even be on a different chromosome (Figure 10.3). As another example, a **repressor** slows or stops transcription when it binds to certain sites in DNA.

Regulatory proteins such as activators and repressors are called **transcription factors**. Whether and how fast a gene is transcribed depends on which transcription factors are bound to the DNA.

Interactions between DNA and the histone proteins it wraps around also affect transcription. RNA polymerase can only attach to DNA that is unwound from histones (Section 8.2). Attachment of methyl groups ($-CH_3$) causes DNA to wind tightly around histones; thus, molecules that methylate DNA prevent its transcription.

The number of copies of a gene also affects how fast its product is made. For example, in some cells, DNA is copied repeatedly with no cytoplasmic division between

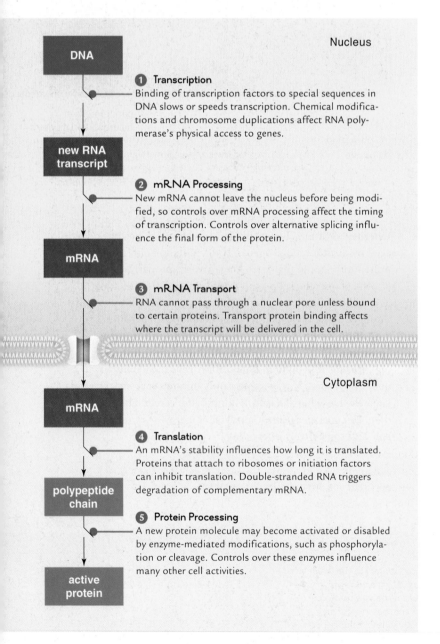

Nucleus

DNA

❶ **Transcription**
Binding of transcription factors to special sequences in DNA slows or speeds transcription. Chemical modifications and chromosome duplications affect RNA polymerase's physical access to genes.

new RNA transcript

❷ **mRNA Processing**
New mRNA cannot leave the nucleus before being modified, so controls over mRNA processing affect the timing of transcription. Controls over alternative splicing influence the final form of the protein.

mRNA

❸ **mRNA Transport**
RNA cannot pass through a nuclear pore unless bound to certain proteins. Transport protein binding affects where the transcript will be delivered in the cell.

Cytoplasm

mRNA

❹ **Translation**
An mRNA's stability influences how long it is translated. Proteins that attach to ribosomes or initiation factors can inhibit translation. Double-stranded RNA triggers degradation of complementary mRNA.

polypeptide chain

❺ **Protein Processing**
A new protein molecule may become activated or disabled by enzyme-mediated modifications, such as phosphorylation or cleavage. Controls over these enzymes influence many other cell activities.

active protein

Figure 10.2 Animated Points of control over eukaryotic gene expression.

promoter | exon1 | intron | exon2 | enhancer

→ transcription start site   ← transcription end

replications. The result is a cell full of polytene chromosomes, each of which consists of hundreds or thousands of side-by-side copies of the same DNA molecule. All of the DNA strands carry the same genes. Translation of one gene, which occurs simultaneously on all of the identical DNA strands, produces a lot of mRNA, which is translated quickly into a lot of protein. Polytene chromosomes are common in immature amphibian eggs, the storage tissues of some plants, and the saliva gland cells of some insect larvae (Figure 10.4).

mRNA Processing As you know, before eukaryotic mRNAs leave the nucleus, they are modified—spliced, capped, and finished with a poly-A tail (Section 9.3). Controls over these modifications can affect the form of a protein product and when it will appear in the cell ❷. For example, controls that determine which exons are spliced out of an mRNA affect which form of a protein will be translated from it.

mRNA Transport mRNA transport is another point of control ❸. For example, in eukaryotes, transcription occurs in the nucleus, and translation in the cytoplasm. A new RNA can pass through pores of the nuclear envelope only after it has been processed appropriately. Controls that delay the processing also delay an mRNA's appearance in the cytoplasm, and thereby delay its translation.

Controls also govern mRNA localization. A short base sequence near an mRNA's poly-A tail is like a zip code. Certain proteins that attach to the zip code drag the mRNA along cytoskeletal elements and deliver it to the organelle or area of the cytoplasm specified by the code. Other proteins that attach to the zip code region prevent the mRNA from being translated before it reaches its destination. mRNA localization allows cells to grow or move in specific directions. It is also crucial for proper embryonic development.

Translational Control Most controls over eukaryotic gene expression affect translation ❹. Many govern the production or function of the various molecules that carry out translation. Others affect mRNA stability: The longer an mRNA lasts, the more protein can be made from it. Enzymes begin to disassemble a new mRNA as soon as it arrives in the cytoplasm. The fast turnover allows cells to adjust their protein synthesis quickly in response to changing needs. How long an mRNA persists depends on its base sequence, the length of its poly-A tail, and which proteins are attached to it.

As a different example, microRNAs inhibit translation of other RNA. Part of a microRNA folds back on itself and forms a small double-stranded region. By a process called RNA interference, any double-stranded RNA (including a microRNA) is cut up into small bits that are taken up by special enzyme complexes. These complexes destroy every mRNA in a cell that can base-pair with the bits. So, expression of a microRNA complementary in sequence to a gene inhibits expression of that gene.

Post–Translational Modification Many newly synthesized polypeptide chains must be modified before they become functional ❺. For example, some enzymes become active only after they have been phosphorylated (Section 5.3). Such post-translational modifications inhibit, activate, or stabilize many molecules, including the enzymes that participate in transcription and translation.

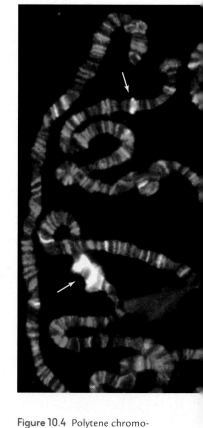

Figure 10.4 Polytene chromosomes in the salivary gland cells of fruit flies. These giant chromosomes form by repeated DNA replication without cell division.

Each of these chromosomes consists of hundreds or thousands of copies of the same DNA strand, aligned side by side. Transcription is visible as puffs (*white* arrows) where the DNA has loosened.

---

activator Regulatory protein that increases the rate of transcription when it binds to a promoter or enhancer.
differentiation The process by which cells become specialized.
enhancer Binding site in DNA for proteins that enhance the rate of transcription.
repressor Regulatory protein that blocks transcription.
transcription factor Regulatory protein that influences transcription; e.g., an activator or repressor.

Take-Home Message **What is gene control?**

❯ Most cells of multicelled organisms differentiate when they start expressing a unique subset of their genes. Which genes a cell expresses depends on the type of organism, its stage of development, and environmental conditions.

❯ Various control processes regulate all steps between gene and gene product.

> Research with fruit flies yielded the insight that body plans are a result of patterns of gene expression in embryos.
< Links to Development 1.3, Taxa 1.5, Genes 9.2

For about a hundred years, *Drosophila melanogaster* has been the subject of choice for many research experiments on eukaryotic gene expression. Why? It costs almost nothing to feed this fruit fly, which is only about 3 millimeters long and can live its entire life in a bottle. *D. melanogaster* also reproduces quickly and has a short life cycle. As well, experimenting on insects generally viewed as nuisance pests presents few ethical dilemmas.

Many important discoveries about how gene controls guide development have come from *Drosophila* research. The discoveries are clues to understanding similar processes in humans and other organisms, which have a shared evolutionary history.

Homeotic Genes

Homeotic genes control the formation of specific body parts (eyes, legs, segments, and so on) during the development of embryos. All homeotic genes encode transcription factors with a homeodomain, a region of about sixty amino acids that can bind to a promoter or some other DNA sequence in a chromosome.

Homeotic genes are a type of master gene. The products of **master genes** affect the expression of many other genes. Expression of a master gene causes other genes to be expressed, with the final outcome being the comple-

tion of an intricate task such as the formation of an eye during embryonic development.

Such processes begin long before body parts develop, as various master genes are expressed in local areas of the early embryo. The master gene products form in concentration gradients that span the entire embryo. Depending on where they are located within the gradients, embryonic cells begin to transcribe different homeotic genes. Products of the homeotic genes form in specific areas of the embryo. The different products cause cells to differentiate into tissues that form specific structures such as wings or a head.

The function of many homeotic genes has been discovered by manipulating their expression, one at a time. Researchers inactivate a homeotic gene by introducing a mutation or deleting it entirely, an experiment called a **knockout**. An organism that carries the knocked-out gene may differ from normal individuals, and the differences are clues to the function of the missing gene product.

Researchers often name homeotic genes based on what happens in their absence. For instance, fruit flies with a mutated *eyeless* gene develop with no eyes (Figure 10.5A,B). *Dunce* is required for learning and memory. *Wingless*, *wrinkled*, and *minibrain* are self-explanatory. *Tinman* is necessary for development of a heart. Flies with a mutated *groucho* gene have too many bristles above their eyes. One gene was named *toll*, after what its German discoverer exclaimed upon seeing the disastrous effects of the mutation (*toll* is a German slang word that means "cool!").

Figure 10.5 Eyes and *eyeless*. **A** A normal fruit fly has large, round eyes. **B** A fruit fly with a mutation in its *eyeless* gene develops with no eyes. **C** Eyes form wherever the *eyeless* gene is expressed in fly embryos—here, on the head and wing.

Humans, mice, squids, and other animals have a gene called *PAX6*. In humans, *PAX6* mutations result in missing irises, a condition called aniridia **D**. Compare a normal iris **E**. *PAX6* is so similar to *eyeless* that it triggers eye development when expressed in fly embryos.

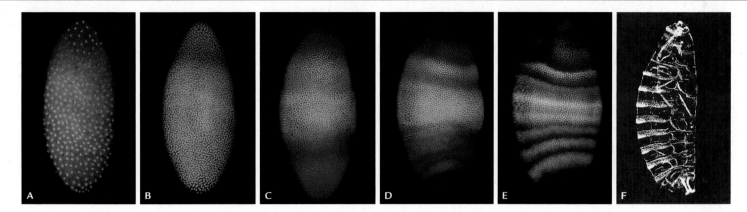

Figure 10.6 How gene expression control makes a fly, as illuminated by segmentation. The expression of different master genes is shown by different colors in fluorescence microscopy images of whole *Drosophila* embryos at successive stages of development. The bright dots are individual cell nuclei.

A, B The master gene *even-skipped* is expressed (in *red*) only where two maternal gene products (*blue* and *green*) overlap.

C–E The products of several master genes, including the two shown here in *green* and *blue*, confine the expression of *even-skipped* (*red*) to seven stripes.

F One day later, seven segments develop that correspond to the position of the stripes.

Homeotic genes control development by the same mechanisms in all eukaryotes, and many are interchangeable among different species. Thus, we can infer that they evolved in the most ancient eukaryotic cells. Homeodomains often differ among species only in conservative substitutions—one amino acid has replaced another with similar chemical properties.

Consider the *eyeless* gene. Eyes form in embryonic fruit flies wherever this gene is expressed, which is typically in tissues of the head. If the *eyeless* gene is expressed in another part of the developing embryo, eyes form there too (Figure 10.5C). Humans, squids, mice, fish, and many other animals have a gene called *PAX6*, which is very similar to the *eyeless* gene. In humans, mutations in *PAX6* cause eye disorders such as aniridia, in which a person's irises are underdeveloped or missing (Figure 10.5D,E). *PAX6* works across different species. For example, if the *PAX6* gene from a human or mouse is inserted into an *eyeless* mutant fly, it has the same effect as the *eyeless* gene: An eye forms wherever it is expressed. Such studies are evidence of a shared ancestor among these evolutionarily distant animals.

Filling in Details of Body Plans

As an embryo develops, its differentiating cells form tissues, organs, and body parts. Some cells that alternately migrate and stick to other cells develop into nerves, blood vessels, and other structures that weave through the tissues. Events like these fill in the body's details, and all are driven by cascades of master gene expression.

homeotic gene Type of master gene; its expression controls formation of specific body parts during development.
knockout An experiment in which a gene is deliberately inactivated in a living organism.
master gene Gene encoding a product that affects the expression of many other genes.
pattern formation Process by which a complex body forms from local processes during embryonic development.

Pattern formation is the process by which a complex body forms from local processes in an embryo. Pattern formation begins as maternal mRNAs are delivered to opposite ends of an unfertilized egg as it forms. The localized maternal mRNAs get translated right after the egg is fertilized, and their protein products diffuse away in gradients that span the entire embryo. Cells of the developing embryo translate different master genes, depending on where they fall within those gradients. The products of the master genes also form in overlapping gradients. Cells of the embryo translate still other master genes depending on where they fall within these gradients, and so on.

Such regional gene expression during development results in a three-dimensional map that consists of overlapping concentration gradients of master gene products. Which master genes are active at any given time changes, and so does the map. Some master gene products cause undifferentiated cells to differentiate, and specialized tissues are the outcome. The formation of body segments in a fruit fly embryo is an example of how pattern formation works (Figure 10.6).

Take-Home Message What controls gene expression?

❯ Research on fruit flies yielded many important discoveries about the mechanisms of gene control in eukaryotes.

❯ Development is orchestrated by cascades of master gene expression in embryos.

❯ The expression of homeotic genes during development governs the formation of specific body parts. Homeotic genes that function in similar ways across taxa are evidence of shared ancestry.

❯ Selective gene expression gives rise to many traits.
❮ Links to Chromosomes and sex determination 8.2, Mutations 9.6

Many of the traits that are characteristic of humans and other eukaryotic organisms arise as an outcome of gene expression controls, as the following examples illustrate.

X Chromosome Inactivation

In humans and other mammals, a female's cells each contain two X chromosomes, one inherited from her mother, the other one from her father. One X chromosome is always tightly condensed (Figure 10.7). We call the condensed X chromosomes "Barr bodies," after Murray Barr, who discovered them. RNA polymerase cannot access most of the genes on the condensed chromosome. This **X chromosome inactivation** ensures that only one of the two X chromosomes in a female's cells is active. According to a theory called **dosage compensation**, the inactivation equalizes expression of X chromosome genes between the sexes. The body cells of male mammals (XY) have one set of X chromosome genes. The body cells of female mammals (XX) have two sets, but only one is expressed. Normal development of female embryos depends on this control.

X chromosome inactivation occurs when an embryo is a ball of about 200 cells. In humans and many other mammals, it occurs independently in every cell of a female embryo. The maternal X chromosome may get inactivated in one cell, and the paternal or maternal X chromosome may get inactivated in a cell next to it. Once the selection is made in a cell, all of that cell's descen-

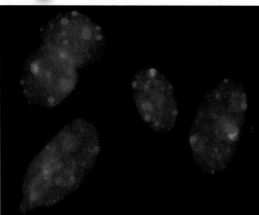
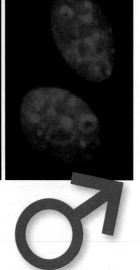

Figure 10.7 X chromosome inactivation. Barr bodies are visible as *red* spots in the nucleus of the four XX cells on the *left*. Compare the nucleus of two XY cells to the *right*.

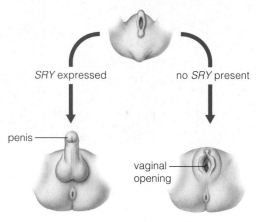

Structures that will give rise to external genitalia appear at seven weeks

SRY expressed   no SRY present

penis

vaginal opening

birth approaching

Figure 10.8 Development of reproductive organs in human embryos. An early human embryo appears neither male nor female. Gene expression determines what reproductive organs will form.

In an XY embryo, the SRY gene product triggers the formation of testes, male gonads that secrete testosterone. This hormone initiates development of other male traits. In an XX embryo, ovaries form in the absence of the Y chromosome and its SRY gene.

dants make the same selection as they continue dividing and forming tissues. As a result of the inactivation, an adult female mammal is a "mosaic" for the expression of genes on the X chromosome. She has patches of tissue in which genes of the maternal X chromosome are expressed, and patches in which genes of the paternal X chromosome are expressed.

How does just one of two X chromosomes get inactivated? An X chromosome gene called *XIST* does the trick. This gene is transcribed on only one of the two X chromosomes. The gene's product, a large RNA, sticks to the chromosome that expresses the gene. The RNA coats the chromosome and causes it to condense into a Barr body. Thus, transcription of the *XIST* gene keeps the chromosome from transcribing other genes. The other chromosome does not express *XIST*, so it does not get coated with RNA; its genes remain available for transcription. It is still unknown how the cell chooses which chromosome will express *XIST*.

Male Sex Determination in Humans

The human X chromosome carries 1,336 genes. Some of those genes are associated with sexual traits, such as the distribution of body fat and hair. However, most of the genes on the X chromosome govern nonsexual traits such as blood clotting and color perception. Such genes are expressed in both males and females. Males, remember, also inherit one X chromosome.

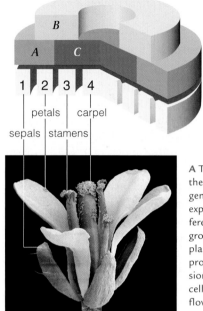

A The pattern in which the floral identity genes *A*, *B*, and *C* are expressed affects differentiation of cells growing in whorls in the plant's tips. Their gene products guide expression of other genes in cells of each whorl; a flower results.

B Mutations in *Arabidopsis* *ABC* genes result in malformed flowers.

Top left, right: *A* gene mutations lead to petal-less flowers with no structures in place of missing petals.

Bottom left: *B* gene mutations lead to flowers with sepals instead of petals.

Bottom right: *C* gene mutations lead to flowers with petals instead of sepals and carpels. Compare the normal flower in **A**.

Figure 10.9 Animated Control of flower formation, as revealed by mutations in *Arabidopsis thaliana*.

The human Y chromosome carries only 307 genes, but one of them is the *SRY* gene—the master gene for male sex determination in mammals. Its expression in XY embryos triggers the formation of testes, which are male gonads (Figure 10.8). Some of the cells in these primary male reproductive organs make testosterone, a sex hormone that controls the emergence of male secondary sexual traits such as facial hair, increased musculature, and a deep voice. We know that *SRY* is the master gene that controls emergence of male sexual traits because mutations in this gene cause XY individuals to develop external genitalia that appear female. An XX embryo has no Y chromosome, no *SRY* gene, and much less testosterone, so primary female reproductive organs (ovaries) form instead of testes. Ovaries make estrogens and other sex hormones that will govern the development of female secondary sexual traits, such as enlarged, functional breasts, and fat deposits around the hips and thighs.

Flower Formation

When it is time for a plant to flower, populations of cells that would otherwise give rise to leaves instead differentiate into floral parts—sepals, petals, stamens, and carpels. How does the switch happen? Studies of mutations in the common wall cress plant, *Arabidopsis thaliana*, elucidated how the specialized parts of a flower develop. Three sets of master genes called *A*, *B*, and *C* guide the process of flower formation. These genes are switched on by environmental cues such as seasonal changes in the length of night, as you will see in Section 27.9.

At the tip of a shoot, cells form whorls of tissue, one over the other like layers of an onion. Cells in each whorl give rise to different tissues depending on which of their *ABC* genes are activated (Figure 10.9A). In the outer whorl, only the *A* genes are switched on, and their products trigger events that cause sepals to form. Cells in the next whorl express both *A* and *B* genes; they give rise to petals. Cells farther in express *B* and *C* genes; they give rise to stamens, the structures that produce male reproductive cells. The cells of the innermost whorl express only the *C* genes; they give rise to carpels, the structures that produce female reproductive cells. The phenotypic effects of mutations in *ABC* genes support this model (Figure 10.9B).

dosage compensation Theory that X chromosome inactivation equalizes gene expression between males and females.
X chromosome inactivation Shutdown of one of the two X chromosomes in the cells of female mammals.

Take-Home Message What are some examples of gene control in eukaryotes?

❯ X chromosome inactivation balances expression of X chromosome genes between female (XX) and male (XY) mammals. The balance is vital for development of female embryos.

❯ *SRY* gene expression triggers the development of male traits in mammals.

❯ Gene control also guides flower formation. *ABC* master genes that are expressed differently in shoot tissues govern development of flower parts.

> Bacteria control gene expression mainly by adjusting the rate of transcription.

‹ Links to Carbohydrates 3.3, Controls over metabolism 5.5, Lactate fermentation 7.6

Bacteria and archaeans do not undergo development and become multicelled organisms, so these cells do not use master genes. However, they do use gene controls. By adjusting gene expression, they can respond to environmental conditions. For example, when a certain nutrient becomes available, a bacterial cell will begin transcribing genes whose products allow the cell to use that nutrient. When the nutrient is not available, transcription of those genes stops. Thus, the cell does not waste energy and resources producing gene products that are not needed at a particular moment.

Bacteria control their gene expression mainly by adjusting the rate of transcription. Genes that are used together often occur together on the chromosome, one after the other. All of them are transcribed together into a single RNA strand, so their transcription is controllable in a single step.

The Lactose Operon

Escherichia coli lives in the gut of mammals, where it dines on nutrients traveling past. Its carbohydrate of choice is glucose, but it can make use of other sugars, such as the lactose in milk. *E. coli* cells use a set of three enzymes in order to harvest the glucose subunit of lactose molecules. However, unless there is lactose in the gut, *E. coli* cells keep the three genes for those enzymes turned off. There is one promoter for all three genes. Flanking the promoter are two **operators**, regions of DNA that serve as binding sites for a repressor. (Repressors, remember, stop transcription.) A promoter and one or more operators that together control the transcription of multiple genes are collectively called an **operon**. Operons occur in bacteria, archaeans, and eukaryotes. The one that controls lactose metabolism in *E. coli* is called the lac operon (Figure 10.10 ❶).

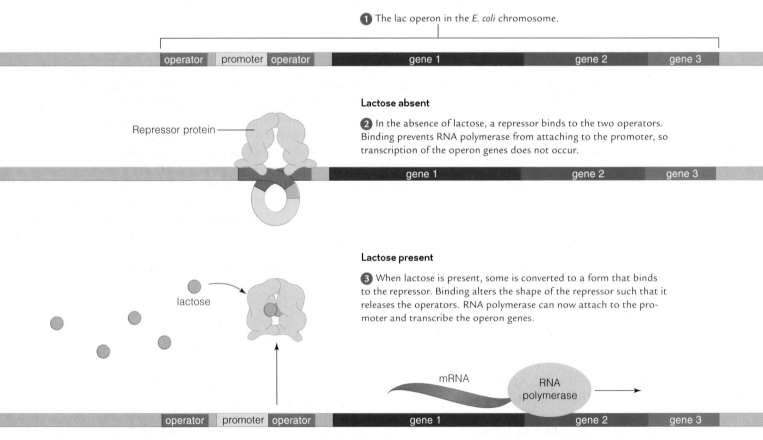

❶ The lac operon in the *E. coli* chromosome.

| operator | promoter | operator | gene 1 | gene 2 | gene 3 |

Lactose absent

❷ In the absence of lactose, a repressor binds to the two operators. Binding prevents RNA polymerase from attaching to the promoter, so transcription of the operon genes does not occur.

Repressor protein

gene 1 | gene 2 | gene 3

Lactose present

❸ When lactose is present, some is converted to a form that binds to the repressor. Binding alters the shape of the repressor such that it releases the operators. RNA polymerase can now attach to the promoter and transcribe the operon genes.

lactose

mRNA

RNA polymerase

| operator | promoter | operator | gene 1 | gene 2 | gene 3 |

Figure 10.10 Animated Example of gene control in bacteria: the lactose operon on a bacterial chromosome. The operon consists of a promoter flanked by two operators, and three genes for lactose-metabolizing enzymes.

›› **Figure It Out** What portion of the operon binds RNA polymerase when lactose is present?

Answer: The promoter.

When lactose is not present, the lac operon repressors bind to the *E. coli* DNA, and lactose-metabolizing genes stay switched off. One repressor molecule binds to both operators, so that the region of DNA with the promoter twists into a loop ❷. RNA polymerase cannot bind to the twisted promoter, so it cannot transcribe the operon genes.

When lactose is in the gut, some of it is converted to another sugar, allolactose. Allolactose binds to the repressor and changes its shape. The altered repressor can no longer bind to the operators. The looped DNA unwinds, the promoter is now accessible to RNA polymerase, and transcription begins ❸.

E. coli cells use extra enzymes to metabolize lactose compared with glucose, so it is more efficient for them to use glucose. Accordingly, when both sugars are present, the cells will use up all of the available glucose before switching to lactose metabolism.

Lactose Intolerance

Like infants of other mammals, human infants drink milk. Cells in the lining of the small intestine secrete lactase, an enzyme that cleaves the lactose in milk into its subunit monosaccharides. In most people, lactase production starts to decline at the age of five.

After that, it becomes more difficult to digest lactose in food—a condition called lactose intolerance. Lactose is not absorbed directly by the intestine. Thus, any that is not broken down in the small intestine ends up in the large intestine, which hosts *E. coli* and a variety of other bacteria. These resident organisms respond to the abundant sugar supply by switching on their lac operons. Carbon dioxide, methane, hydrogen, and other gaseous products of their various fermentation reactions accumulate quickly in the large intestine, distending its wall and causing pain. The other products of their metabolism (undigested carbohydrates) disrupt the solute–water balance inside the large intestine, and diarrhea results.

Not everybody is lactose intolerant. Many people carry a mutation in one of the genes responsible for

---

operator Part of an operon; a DNA binding site for a repressor.
operon Group of genes together with a promoter–operator DNA sequence that controls their transcription.

Between You and Eternity (revisited)

Mutations in some genes predispose individuals to develop certain kinds of cancer. Tumor suppressor genes are named because tumors are more likely to occur when these genes mutate. Two examples are *BRCA1* and *BRCA2:* A mutated version of one or both of these genes is often found in breast and ovarian cancer cells. Because mutations in genes such as *BRCA* can be inherited, cancer is not only a disease of the elderly, as Robin Shoulla's story illustrates. Robin is one of the unlucky people who carry mutations in both *BRCA1* and *BRCA2*.

If a *BRCA* gene mutates in one of three especially dangerous ways, a woman has an 80 percent chance of developing breast cancer before the age of seventy. *BRCA* genes are master genes whose protein products help maintain the structure and number of chromosomes in a dividing cell. The multiple functions of these proteins are still being unraveled. We do know they participate directly in DNA repair (Section 8.6), so any mutations that alter this function also alter the cell's capacity to repair damaged DNA. Other mutations are likely to accumulate, and that sets the stage for cancer.

The products of *BRCA* genes also bind to receptors for the hormones estrogen and progesterone, which are abundant on cells of breast and ovarian tissues. Binding suppresses transcription of growth factor genes in these cells. Among other things, growth factors stimulate cells to divide during normal, cyclic renewals of breast and ovarian tissues. When a mutation alters a *BRCA* gene so that its product cannot bind to hormone receptors, the cells overproduce growth factors. Cell division goes out of control, and tissue growth becomes disorganized. In other words, cancer develops.

Two groups of researchers, one at the Dana-Farber Cancer Institute at Harvard, the other at the University of Milan, recently found that the RNA product of the *XIST* gene localizes abnormally in breast cancer cells. In those cells, both X chromosomes are active. It makes sense that two active X chromosomes would have something to do with abnormal gene expression, but why the RNA product of an unmutated *XIST* gene does not localize properly in cancer cells remains a mystery.

Mutations in the *BRCA1* gene may be part of the answer. The Harvard researchers found that the protein product of the *BRCA1* gene physically associates with the RNA product of the *XIST* gene. They were able to restore proper *XIST* RNA localization—and proper X chromosome inactivation—by restoring the function of the *BRCA1* gene product in breast cancer cells.

How Would You Vote? Some women at high risk of developing breast cancer opt for preventive breast removal. Many of them never would have developed cancer. Should the surgery be restricted to cancer treatment? See CengageNow for details, then vote online (cengagenow.com).

---

the programmed lactase shutdown. These people make enough lactase to continue drinking milk without problems into adulthood.

Take-Home Message **Do bacteria control gene expression?**

❯ In bacteria, the main gene expression controls regulate transcription in response to shifts in nutrient availability and other outside conditions.

Summary

Section 10.1 Controls over gene expression are a critical part of the embryonic development and normal functioning of a multi-celled body. When gene controls fail, as occurs as a consequence of some mutations, **cancer** may be the outcome.

Section 10.2 Which genes a cell uses depends on the type of organism, the type of cell, factors inside and outside the cell, and, in complex multicelled species, the organism's stage of development.

Controls over gene expression are part of homeostasis in all organisms. They also drive development in multi-celled eukaryotes. All cells of an embryo share the same genes. As different cell lineages use different subsets of genes during development, they become specialized, a process called **differentiation**. Specialized cells form tissues and organs in the adult.

Different molecules and processes govern every step between transcription of a gene and delivery of the gene's product to its final destination. Most controls operate at transcription; **transcription factors** such as **activators** and **repressors** influence transcription by binding to promoters, **enhancers**, or other sequences in chromosomal DNA.

Section 10.3 **Knockouts** of **homeotic genes** in fruit flies (*Drosophila melanogaster*) revealed that local controls over gene expression govern the embryonic development of all complex, multicelled bodies, a process called **pattern formation**. Various **master genes** are expressed locally in different parts of an embryo as it develops. Their products diffuse through the embryo and affect expression of other master genes, which affect the expression of others, and so on. These cascades of master gene products form a dynamic spatial map of overlapping gradients that spans the entire embryo body. Cells differentiate according to their location on the map.

Section 10.4 In female mammals, most genes on one of the two X chromosomes are permanently inaccessible. **X chromosome inactivation** balances gene expression between the sexes. Such **dosage compensation** arises because the *XIST* gene gets transcribed on only one of the two X chromosomes. The gene's RNA product shuts down the chromosome that transcribes it.

Studies of mutations in *Arabidopsis thaliana* showed that three sets of master genes (*A*, *B*, and *C*) guide cell differentiation in the whorls of a floral shoot; sepals, petals, stamens, and carpels form.

Section 10.5 Bacterial cells do not have great structural complexity and do not undergo development. Most of their gene controls reversibly adjust transcription rates in response to environmental conditions, especially nutrient availability. **Operons** are examples of bacterial gene controls. The lactose operon governs expression of three genes, the three products of which allow the bacterial cell to metabolize lactose. Two **operators** that flank the promoter are binding sites for a repressor that blocks transcription.

Self-Quiz   Answers in Appendix III

1. The expression of a gene may depend on _____ .
   a. the type of organism   c. the type of cell
   b. environmental conditions   d. all of the above

2. Gene expression in multicelled eukaryotic cells changes in response to _____ .
   a. conditions outside the cell   c. operons
   b. master gene products   d. a and b

3. Binding of _____ to _____ in DNA can increase the rate of transcription of specific genes.
   a. activators; promoters   c. repressors; operators
   b. activators; enhancers   d. both a and b

4. Proteins that influence gene expression by binding to DNA are called _____ .

5. Polytene chromosomes form in some types of cells that _____ .
   a. have a lot of chromosomes   c. are differentiating
   b. are making a lot of protein   d. b and c are correct

6. Eukaryotic gene controls govern _____ .
   a. transcription   e. translation
   b. RNA processing   f. protein modification
   c. RNA transport   g. a through e
   d. mRNA degradation   h. all of the above

7. Controls over eukaryotic gene expression guide _____ .
   a. natural selection   c. development
   b. nutrient availability   d. all of the above

8. The expression of *ABC* genes _____ .
   a. occurs in layers
   b. controls flower formation
   c. causes mutations in flowers
   d. both a and b

9. Cell differentiation _____ .
   a. occurs in all complex multicelled organisms
   b. requires unique genes in different cells
   c. involves selective gene expression
   d. both a and c
   e. all of the above

10. During X chromosome inactivation _____ .
   a. female cells shut down   c. pigments form
   b. RNA coats chromosomes   d. both a and b

11. A cell with a Barr body is _____ .
   a. a bacterium   c. from a female mammal
   b. a sex cell   d. infected by Barr virus

Data Analysis Activities

BRCA Mutations in Women Diagnosed With Breast Cancer

Investigating a correlation between specific cancer-causing mutations and the risk of mortality in humans is challenging, in part because each cancer patient is given the best treatment available at the time. There are no "untreated control" cancer patients, and ideas about which treatments are best change quickly as new drugs become available and new discoveries are made.

Figure 10.11 shows results from a 2007 study by Pal Moller and his colleagues. The researchers looked for *BRCA* mutations in 442 women who had been diagnosed with breast cancer, and followed their treatments and progress over several years.

All of the women in the study had at least two affected close relatives, so their risk of developing breast cancer due to an inherited factor (such as a *BRCA* mutation) was estimated to be greater than that of the general population.

1. According to this study, what is a woman's risk of dying of cancer if two of her close relatives have breast cancer?
2. What is her risk of dying of cancer if she carries a mutated *BRCA1* gene?
3. According to these results, is a *BRCA1* or *BRCA2* mutation more dangerous in breast cancer cases?
4. What other data would you have to see in order to make a conclusion about the effectiveness of preventive mastectomy or oophorectomy?

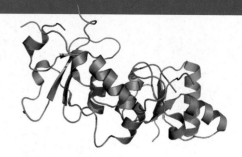

BRCA Mutations in Women Diagnosed With Breast Cancer				
	BRCA1	*BRCA2*	No *BRCA* Mutation	Total
Total number of patients	89	35	318	442
Avg. age at diagnosis	43.9	46.2	50.4	
Preventive mastectomy	6	3	14	23
Preventive oophorectomy	38	7	22	67
Number of deaths	16	1	21	38
Percent died	18.0	2.8	6.9	8.6

Figure 10.11 Results from a 2007 study investigating *BRCA* mutations in women diagnosed with breast cancer. All women in the study had a family history of the disease.

Some of the women underwent preventive mastectomy (removal of the noncancerous breast) during their course of treatment. Others had preventive oophorectomy (surgical removal of the ovaries) to prevent the possibility of getting ovarian cancer.

Top, model of the unmutated *BRCA1* protein.

---

12. Homeotic gene products _____ .
    a. flank a bacterial operon
    b. map out the overall body plan in embryos
    c. control the formation of specific body parts

13. A gene that is knocked out is _____ .
    a. deleted   c. expressed
    b. inactivated   d. either a or b

14. A promoter and a set of operators that control access to two or more genes is a(n) _____ .
    a. lactose molecule   c. dosage compensator
    b. operon   d. knockout

15. Match the terms with the most suitable description.
    ___ *ABC* genes   a. a big RNA is its product
    ___ *XIST* gene   b. binding site for repressor
    ___ operator   c. cells become specialized
    ___ Barr body   d. —CH_3 additions to DNA
    ___ differentiation   e. inactivated X chromosome
    ___ methylation   f. guide flower development

Additional questions are available on **CENGAGENOW.**

Animations and Interactions on **CENGAGENOW:**
❯ Points of control over gene expression; *ABC* model for flowering; X chromosome inactivation; Structure and function of the lac operon.

Critical Thinking

1. Why does a cell regulate its gene expression?

2. Do the same gene controls operate in bacterial cells and eukaryotic cells? Why or why not?

3. Unlike most rodents, guinea pigs are well developed at the time of birth. Within a few days, they can eat grass, vegetables, and other plant material.

Suppose a breeder decides to separate baby guinea pigs from their mothers three weeks after they were born. He wants to raise the males and the females in different cages. However, he has trouble identifying the sex of young guinea pigs. Suggest how a quick look through a microscope can help him identify the females.

4. Geraldo isolated an *E. coli* strain in which a mutation has hampered the capacity of the cAMP activator to bind the promoter of the lactose operon. How will this mutation affect transcription of the lactose operon when the *E. coli* cells are exposed to the following conditions?

a. Lactose and glucose are both available.

b. Lactose is available but glucose is not.

c. Both lactose and glucose are absent.

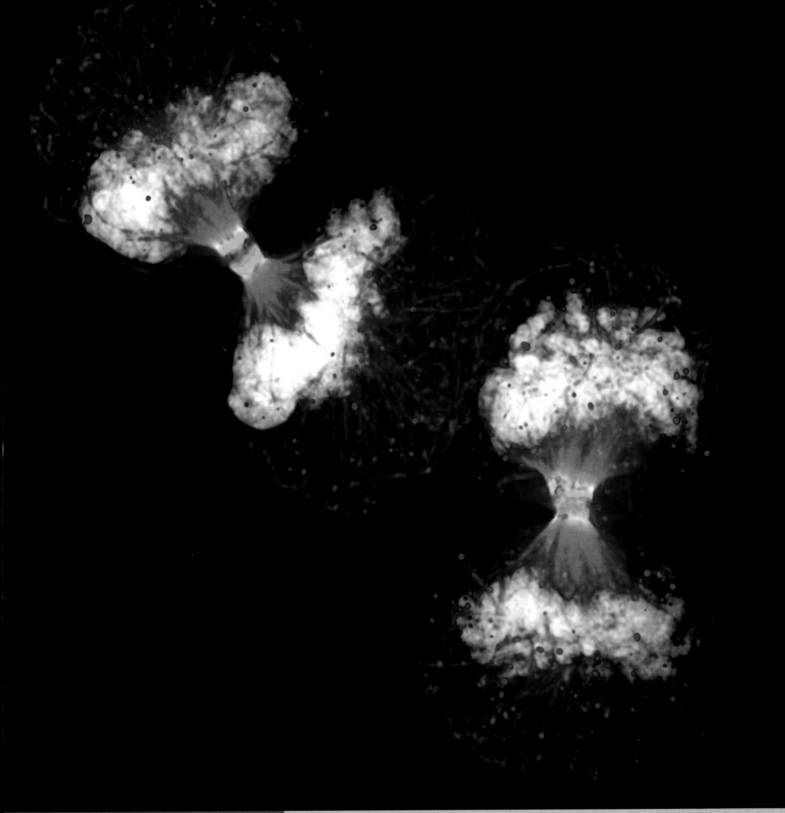

‹ Links to Earlier Concepts

Be sure you understand chromosome structure and chromosome number (Section 8.2) before reading this chapter. You will revisit eukaryotic cell structure (4.2, 4.10, 4.11), particularly the nucleus (4.7). What you know about free radicals (5.4) and mutations (9.6), receptors and recognition proteins (4.4), phosphorylation (5.3), fermentation (7.6), and gene control in eukaryotes (10.2) will help you understand cancer and how it develops.

Key Concepts

The Cell Cycle
A cell cycle starts when a new cell forms by division of a parent cell, and ends when the cell completes its own division. A typical cell cycle proceeds through intervals of interphase, mitosis, and cytoplasmic division.

Mitosis
Mitosis divides the nucleus and maintains the chromosome number. It has four sequential stages: prophase, metaphase, anaphase, and telophase. A spindle parcels the cell's duplicated chromosomes into two nuclei.

11 How Cells Reproduce

11.1 Henrietta's Immortal Cells

Each human starts out as a fertilized egg. By the time of birth, the human body consists of about a trillion cells, all descended from that single cell. Even in an adult, billions of cells divide every day as new cells replace worn-out ones. However, human cells tend to divide a few times and die within weeks when grown in the laboratory.

Researchers started trying to coax human cells to become immortal—to keep dividing outside of the body—in the mid-1800s. Why? Many human diseases occur only in human cells. Immortal cell lineages, or cell lines, would allow researchers to study human diseases (and potential cures for them) without experimenting on people.

At Johns Hopkins University, George and Margaret Gey were among the researchers trying to culture human cells. They had been working on the problem for almost thirty years when, in 1951, their assistant Mary Kubicek prepared a sample of human cancer cells. Mary named the cells HeLa, after the first and last names of the patient from whom the cells had been taken.

The HeLa cells began to divide, again and again. The cells were astonishingly vigorous, quickly coating the inside of their test tube and consuming the nutrient broth in which they were bathed. Four days later, there were so many cells that the researchers had to transfer them to more tubes. The cell populations increased at a phenomenal rate. The cells were dividing every twenty-four hours and coating the inside of the tubes within days.

Sadly, cancer cells in the patient were dividing just as fast. Just six months after she had been diagnosed with cervical cancer, malignant cells had invaded tissues throughout her body. Two months after that, Henrietta Lacks, a young African American woman from Baltimore, was dead.

Although Henrietta passed away, her cells lived on in the Geys' laboratory (Figure 11.1). The Geys were able to grow poliovirus in HeLa cells, a practice that enabled them to find out which strains of the virus cause polio. That work was a critical step in the development of polio vaccines, which have since saved millions of lives.

Figure 11.1 HeLa cells, a legacy of cancer victim Henrietta Lacks (*right*). *Opposite*, fluorescence micrograph of two HeLa cells in the process of dividing. *Blue* and *green* show two proteins that help microtubules (*red*) attach to chromosomes (*white*). Defects in these and other proteins that orchestrate cell division result in descendant cells with too many or too few chromosomes, an outcome that is a hallmark of cancer.

Henrietta Lacks's cells, frozen away in tiny tubes and packed in Styrofoam boxes, continue to be shipped among laboratories all over the world. Researchers use those cells to investigate cancer, viral growth, protein synthesis, and the effects of radiation. They helped several researchers win Nobel Prizes for research in medicine and chemistry. Some HeLa cells even traveled into space for experiments on the Discoverer XVII satellite.

Henrietta Lacks was just thirty-one, a wife and mother of five, when runaway cell divisions killed her. Her legacy continues to help people, through her cells that are still dividing, again and again, more than fifty years after she died. Understanding why cancer cells are immortal—and why we are not—begins with understanding the structures and mechanisms that cells use to divide.

Cytoplasmic Division
After nuclear division, the cytoplasm divides. Typically, one nucleus ends up in each of two new cells. The cytoplasm of an animal cell simply pinches in two. In plant cells, a cross-wall forms in the cytoplasm and divides it.

The Cell Cycle Gone Awry
Built-in mechanisms monitor and control the timing and rate of cell division. On rare occasions, the surveillance mechanisms fail, and cell division becomes uncontrollable. Tumor formation and cancer are outcomes.

> Cells reproduce by dividing.
> Division of a eukaryotic cell typically occurs in two steps: nuclear division followed by cytoplasmic division.
> The sequence of stages through which a cell passes during its lifetime is called the cell cycle.
‹ Links to Cell structure 4.2, Nucleus 4.7, Chromosomes 8.2, DNA replication 8.6, Gene expression controls 10.2

The Life of a Cell

Just as the series of events in an animal's life is called a life cycle, the events that occur from the time a cell forms until the time it divides is called a **cell cycle** (Figure 11.2). A typical cell spends most of its life in **interphase**. During this phase, the cell increases its mass, roughly doubles the number of its cytoplasmic components, and replicates its DNA in preparation for division. Interphase consists of three stages:

  1 G1, the first interval (or gap) of cell growth, before DNA replication
  2 S, the time of synthesis (DNA replication)
  3 G2, the second interval (or gap), when the cell prepares to divide

Gap intervals were named because outwardly they seem to be periods of inactivity. Actually, most cells going about their metabolic business are in G1. Cells preparing to divide enter S, when they copy their DNA. During G2, they make the proteins that will drive cell division. Once the S phase begins, DNA replication usually proceeds at a predictable rate and ends before the cell divides **4**.

The remainder of the cycle consists of the division process itself. When a cell divides, both of its two cellu-lar offspring end up with a blob of cytoplasm and some DNA. Each of the offspring of a eukaryotic cell inherits its DNA packaged inside a nucleus. Thus, a eukaryotic cell's nucleus has to divide **5** before its cytoplasm does **6**.

There are two processes by which cell nuclei divide. As you will discover in this chapter and the next, these two processes—mitosis and meiosis—have much in common, but their outcomes differ. **Mitosis** is a nuclear division mechanism that maintains the chromosome number. Remember from Section 8.2 that diploid cells have two sets of chromosomes. For example, human body cells have 46 chromosomes, two of each type. Except for a pairing of sex chromosomes (XY) in males, the chromosomes of each pair are homologous. **Homologous chromosomes** have the same length, shape, and genes (*hom*– means alike). Typically, each member of a pair was inherited from one of two parents.

With mitosis followed by cytoplasmic division, a diploid parent cell produces two diploid offspring. Both offspring have the same chromosome number as the parent. However, it is not just the number of chromosomes that matters. If only the total mattered, then one of the cell's offspring might get, say, two pairs of chromosome 22 and no pairs whatsoever of chromosome 9. A cell cannot function properly without a full complement of DNA, which means it needs to have one copy of each type of chromosome. Thus, each of a cell's descendants receives one copy of each chromosome.

When a cell is in G1, each of its chromosomes consists of one double-stranded DNA molecule (Figure 11.3A). The cell replicates its DNA in S, so by G2, each of its chromosomes consists of two double-stranded DNA molecules

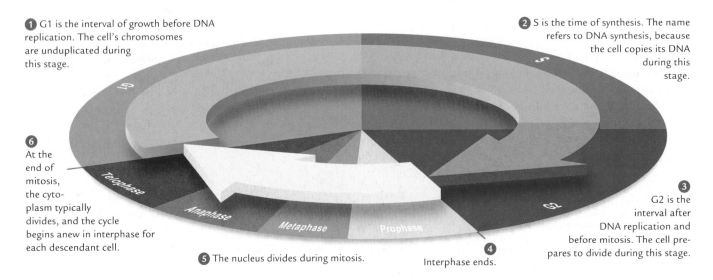

1 G1 is the interval of growth before DNA replication. The cell's chromosomes are unduplicated during this stage.

2 S is the time of synthesis. The name refers to DNA synthesis, because the cell copies its DNA during this stage.

6 At the end of mitosis, the cytoplasm typically divides, and the cycle begins anew in interphase for each descendant cell.

3 G2 is the interval after DNA replication and before mitosis. The cell prepares to divide during this stage.

5 The nucleus divides during mitosis.

4 Interphase ends.

Figure 11.2 Animated The eukaryotic cell cycle. The length of the intervals differs among cells. G1, S, and G2 are part of interphase.

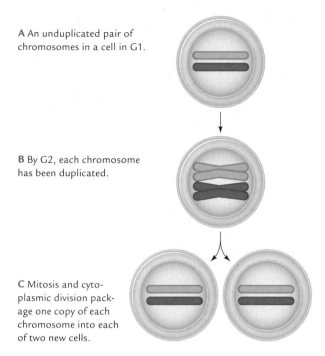

A An unduplicated pair of chromosomes in a cell in G1.

B By G2, each chromosome has been duplicated.

C Mitosis and cytoplasmic division package one copy of each chromosome into each of two new cells.

Figure 11.3 How mitosis maintains the chromosome number.

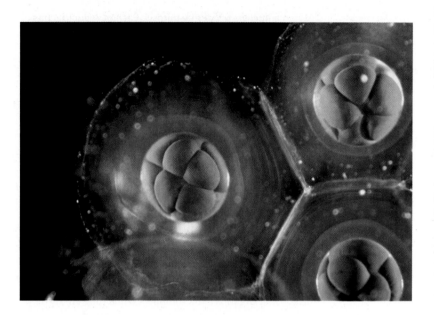

Figure 11.4 A multicelled eukaryote develops by repeated cell divisions. This photo shows early frog embryos, each a product of three mitotic divisions of one fertilized egg.

›› **Figure It Out** Each of these embryos consists of how many cells?　Answer: Eight

(Figure 11.3B). These molecules stay attached to one another at the centromere (as sister chromatids) until mitosis is almost over. The next section shows how mitosis parcels sister chromatids into separate nuclei. When the cytoplasm divides, these new nuclei are packaged into separate cells (Figure 11.3C). Each new cell has a full complement of unduplicated chromosomes, and each starts the cell cycle over again in G1 of interphase.

A Bigger Picture of Cell Division

Cell division is complicated business; it requires the coordinated participation of thousands of molecules. A host of gene expression controls (Section 10.2) orchestrates the process. As you will see in Section 11.5, many of these controls function as built-in brakes on the cell cycle. Apply the brakes that work in G1, and the cycle stalls in G1. Lift the brakes, and the cycle runs again. Sometimes the brakes

are not lifted. For example, the nerve cells of adult humans normally stay in G1 of interphase. Because the cell cycle of these cells cannot proceed, they do not divide. Thus, damaged nerve cells cannot be replaced. Other cells divide at rates that depend on cell type. The stem cells in your red bone marrow divide every 12 hours. Their descendants become red blood cells that replace 2 to 3 million worn-out ones in your blood each second. Cells in the tips of a bean plant root divide every 19 hours. The cells in a fruit fly embryo divide every 10 minutes.

Mitosis and cytoplasmic division are the basis of increases in body size during development (Figure 11.4), and ongoing replacements of damaged or dead cells. Individuals of many species of plants, animals, fungi, and protists reproduce by mitosis and cytoplasmic division, a process called **asexual reproduction**. Bacteria and archaeans also reproduce asexually, but they do it by binary fission, a separate mechanism that we will consider in Section 19.6.

asexual reproduction Reproductive mode by which offspring arise from a single parent only.
cell cycle A series of events from the time a cell forms until its cytoplasm divides.
homologous chromosomes Chromosomes with the same length, shape, and set of genes.
interphase In a eukaryotic cell cycle, the interval between mitotic divisions when a cell grows, roughly doubles the number of its cytoplasmic components, and replicates its DNA.
mitosis Nuclear division mechanism that maintains the chromosome number. Basis of body growth and tissue repair in multicelled eukaryotes; also asexual reproduction in some plants, animals, fungi, and protists.

Take-Home Message　What is cell division and why does it occur?

› The sequence of stages through which a cell passes during its lifetime (interphase, mitosis, and cytoplasmic division) is called the cell cycle.

› A eukaryotic cell reproduces by division: nucleus first, then cytoplasm. Each descendant cell receives a set of chromosomes and some cytoplasm.

› The nuclear division process of mitosis is the basis of body size increases, cell replacements, and tissue repair in multicelled eukaryotes. It is also the basis of asexual reproduction in single-celled and some multicelled eukaryotes.

› When a nucleus divides by mitosis, each new nucleus has the same chromosome number as the parent cell.

› The four main stages of mitosis are prophase, metaphase, anaphase, and telophase.

‹ Links to Cytoskeletal elements 4.10, Chromosome condensation 8.2, Transcriptional control 10.2

During interphase, a cell's chromosomes are loosened to allow transcription and DNA replication (Figure 11.5). In preparation for nuclear division, they begin to pack tightly ❶. Transcription and DNA replication stop as the chromosomes condense into their most compact "X" forms (Section 8.2). A cell reaches **prophase**, the first stage of mitosis, when its chromosomes have condensed so much that they are visible under a light microscope ❷. "Mitosis" is from the Greek word for thread, *mitos*, after the threadlike appearance of chromosomes during nuclear division.

Most animal cells have a centrosome, a region near the nucleus that organizes microtubules while they are forming. A centrosome usually includes two centrioles, barrel-shaped organelles that help microtubules assemble (Section 4.10). The centrosome gets duplicated just before prophase. Then, during prophase, one of the two centrosomes moves to the opposite side of the cell. Microtubules that begin to extend from both centrosomes form a **spindle**, a dynamic network of microtubules that moves chromosomes during nuclear division. Motor proteins traveling along the microtubules help the spindle extend in the appropriate directions. Plant cells have no centrosomes, but they do have spindles and structures that organize them.

In prophase, the spindle penetrates the nuclear region as the nuclear envelope breaks up. Some microtubules of the spindle stop lengthening when they reach the middle of the cell. Others lengthen until they reach a chromosome and attach to it at the centromere. By the end of prophase, one sister chromatid of each chromosome has become attached to microtubules extending from one spindle pole, and the other sister has become attached to microtubules extending from the other spindle pole ❸.

The opposing sets of microtubules then begin a tug-of-war by adding and losing tubulin subunits. As the microtubules extend and shrink, they push and pull the chromosomes. When all the microtubules are the same length, the chromosomes are aligned midway between

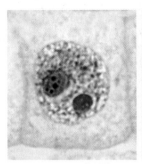
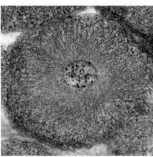

Onion root cell   Whitefish embryo cell

Figure 11.5 Animated Mitosis. Micrographs *here* and *opposite* show plant cells (onion root, *left*), and animal cells (whitefish embryo, *right*). *This page*, interphase cells are shown for comparison, but interphase is not part of mitosis.

Opposite page, the stages of mitosis. The drawings show a diploid (2*n*) animal cell. For clarity, only two pairs of chromosomes are illustrated, but nearly all eukaryotic cells have more than two. The two chromosomes of the pair inherited from one parent are *pink*; the two chromosomes from the other parent are *blue*.

the spindle poles ❹. The alignment marks **metaphase** (from *meta*, the ancient Greek word for between).

During **anaphase**, microtubules of the spindle separate the sister chromatids of each duplicated chromosome, and move them toward opposite spindle poles ❺. Each DNA molecule has now become a separate chromosome.

Telophase begins when the two clusters of chromosomes reach the spindle poles ❻. Each cluster has the same number and kinds of chromosomes as the parent cell nucleus had—two of each type of chromosome, if the parent cell was diploid. A new nucleus forms around each cluster as the chromosomes loosen up again. Once the two nuclei have formed, telophase is over, and so is mitosis.

anaphase Stage of mitosis during which sister chromatids separate and move to opposite spindle poles.

metaphase Stage of mitosis at which the cell's chromosomes are aligned midway between poles of the spindle.

prophase Stage of mitosis during which chromosomes condense and become attached to a newly forming spindle.

spindle Dynamically assembled and disassembled network of microtubules that moves chromosomes during nuclear division.

telophase Stage of mitosis during which chromosomes arrive at the spindle poles and decondense, and new nuclei form.

Take-Home Message **What is the sequence of events that take place during mitosis?**

› Each chromosome in a cell's nucleus was duplicated before mitosis begins, so each consists of two DNA molecules (sister chromatids).

› In prophase, the chromosomes condense and a spindle forms. The spindle microtubules attach to the chromosomes as the nuclear envelope breaks up.

› At metaphase, the (still duplicated) chromosomes are aligned midway between the spindle poles.

› In anaphase, microtubules separate the sister chromatids of each chromosome, and pull them toward opposite spindle poles. Each DNA molecule is now a separate chromosome.

› In telophase, two clusters of chromosomes reach the spindle poles. A new nuclear envelope forms around each cluster.

› Two new nuclei form at the end of mitosis. Each one has the same chromosome number as the parent cell's nucleus.

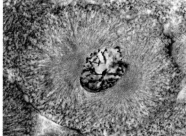

centrosome

① Early Prophase

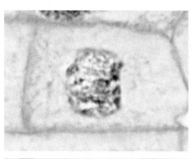  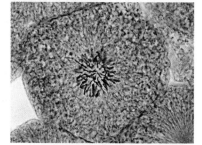

Mitosis begins. In the nucleus, the DNA begins to appear grainy as it starts to condense. The centrosome gets duplicated.

② Prophase

  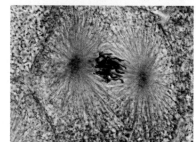

The duplicated chromosomes become visible as they condense. One of the two centrosomes moves to the opposite side of the nucleus. The nuclear envelope breaks up.

③ Transition to Metaphase

 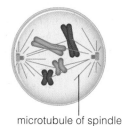 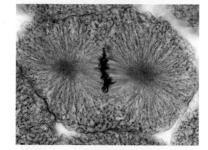

The nuclear envelope is gone, and the chromosomes are at their most condensed. Spindle microtubules assemble and bind to chromosomes at the centromere. Sister chromatids are attached to opposite spindle poles.

microtubule of spindle

④ Metaphase

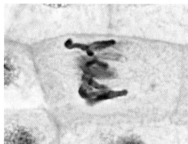  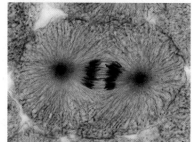

All of the chromosomes are aligned midway between the spindle poles.

⑤ Anaphase

Spindle microtubules separate the sister chromatids and move them toward opposite spindle poles. Each sister chromatid has now become an individual, unduplicated chromosome.

⑥ Telophase

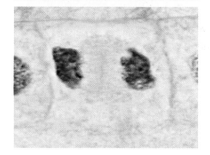  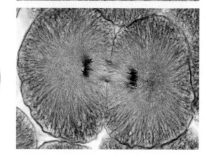

The chromosomes reach the spindle poles and decondense. A nuclear envelope forms around each cluster, and mitosis ends.

❯ In most eukaryotes, the cell cytoplasm divides between late anaphase and the end of telophase. The mechanism of division differs between plants and animals.
❮ Links to Cytoskeleton 4.10, Primary wall 4.11

A cell's cytoplasm usually divides after mitosis, so two cells form. The process of cytoplasmic division, which is called **cytokinesis**, differs among eukaryotes.

Typical animal cells pinch themselves in two after nuclear division (Figure 11.6). How? The spindle begins to disassemble during telophase ❶. The cell cortex, which is the mesh of cytoskeletal elements just under the plasma membrane, includes a band of actin and myosin filaments that wraps around the cell's midsection. The band is called a contractile ring because it contracts when its component proteins are energized by ATP. When the ring contracts, it shrinks, dragging the plasma membrane inward as it does ❷. The sinking plasma membrane becomes visible on the outside of the cell as an indentation between the former spindle poles ❸. The indentation is called a **cleavage furrow**. The cleavage furrow advances around the cell, deepening as it does, until the cytoplasm (and the cell) is pinched in two ❹. Each of the two cells formed by this division has its own nucleus and some of the parent cell's cytoplasm; each is enclosed by a plasma membrane.

Dividing plant cells face a particular challenge because they have stiff cell walls on the outside of their plasma membrane (Section 4.11). Accordingly, plant cells do not pinch themselves in two; they have a completely different mechanism of cytoplasmic division (Figure 11.7). By the end of anaphase in a plant cell, a set of short micro-

❺ The future plane of division was established before mitosis began. Vesicles cluster here when mitosis ends.

❻ As the vesicles fuse with each other, they form a cell plate along the plane of division.

❼ The cell plate expands outward along the plane of division. When it reaches the plasma membrane, it attaches to the membrane and partitions the cytoplasm.

❽ The cell plate matures as two new cell walls. These walls join with the parent cell wall, so each descendant cell becomes enclosed by its own cell wall.

Figure 11.7 Animated Cytoplasmic division of a plant cell.

tubules has formed on either side of the future plane of division. These microtubules now guide vesicles from Golgi bodies and the cell surface to the future plane of division ❺. There, the vesicles and their wall-building contents start to fuse into a disk-shaped **cell plate** ❻. The plate expands at its edges until it reaches the plasma membrane ❼. When the cell plate attaches to the membrane, it partitions the cytoplasm. In time, the cell plate will develop into a primary cell wall that merges with the parent cell's wall. Thus, by the end of division, each of the descendant cells will be enclosed by its own plasma membrane and its own cell wall ❽.

cell plate After nuclear division in a plant cell, a disk-shaped structure that forms a cross-wall between the two new nuclei.
cleavage furrow In a dividing animal cell, the indentation where cytoplasmic division will occur.
cytokinesis Cytoplasmic division.

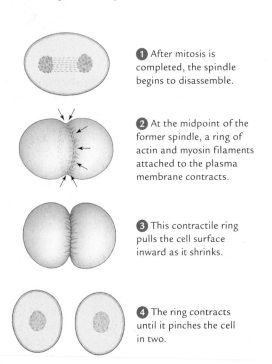

❶ After mitosis is completed, the spindle begins to disassemble.

❷ At the midpoint of the former spindle, a ring of actin and myosin filaments attached to the plasma membrane contracts.

❸ This contractile ring pulls the cell surface inward as it shrinks.

❹ The ring contracts until it pinches the cell in two.

Figure 11.6 Animated Cytoplasmic division of an animal cell.

Take-Home Message How do cells divide?

❯ After mitosis, the cytoplasm of the parent cell typically is partitioned into two descendant cells, each with its own nucleus.

❯ The process of cytoplasmic division differs between plants and animals.

❯ In animal cells, a contractile ring pinches the cytoplasm in two. In plant cells, a cell plate that forms midway between the spindle poles partitions the cytoplasm when it reaches and connects to the parent cell wall.

> On rare occasions, controls over cell division are lost.
< Links to Receptor proteins 4.4, Phosphorylation 5.3, Free radicals 5.4, UV light and mutations 9.6, Eukaryotic gene control 10.2

Every second, millions of cells in your skin, bone marrow, gut lining, liver, and elsewhere are dividing and replacing their worn-out, dead, and dying predecessors. They do not divide at random. Many gene controls govern DNA replication and cell division in eukaryotic cells.

What happens when something goes wrong? Suppose sister chromatids do not separate as they should during mitosis. As a result, one descendant cell ends up with too many chromosomes and the other with too few. Or suppose DNA gets damaged when a chromosome is being duplicated. A cell's DNA can also be damaged by free radicals (Section 5.4), or environmental assaults such as chemicals or ultraviolet radiation (Section 9.6). Such problems are frequent and inevitable, but a cell may not function properly unless they are countered quickly.

The cell cycle has built-in checkpoints that allow problems to be corrected before the cycle advances. Certain proteins, the products of "checkpoint" genes, can monitor whether a cell's DNA has been copied completely, whether it is damaged, and even whether enough nutrients to support division are available. Such proteins interact to delay or stop the cell cycle while simultaneously enhancing transcription of genes involved in chromosome repair (Figure 11.8). If the problem stays uncorrected, checkpoint gene products cause the cell to self-destruct. (You will read more about cell suicide, which is called apoptosis, in Section 28.9.)

Sometimes a checkpoint gene mutates so that its protein product no longer works properly. In other cases, the controls that regulate its production fail, and a cell makes too much or too little of its product. When enough checkpoint mechanisms fail, a cell loses control over its cell cycle. The cell may skip interphase, so division occurs over and over with no resting period. Signaling mechanisms that make an abnormal cell die may stop working. The problem is compounded because these checkpoint malfunctions are passed along to the cell's descendants, which form a **neoplasm**, an accumulation of cells that lost control over how they grow and divide.

Consider **growth factors**, which are molecules that stimulate cells to divide and differentiate. One kind, an epidermal growth factor (EGF), stimulates a cell to enter mitosis by binding to a receptor on the cell's plasma membrane. Binding to EGF changes the shape of the receptor so that it becomes enzymatic and phosophorylates itself. Phosphorylation activates the EGF receptor,

growth factor Molecule that stimulates mitosis.
neoplasm An accumulation of abnormally dividing cells.

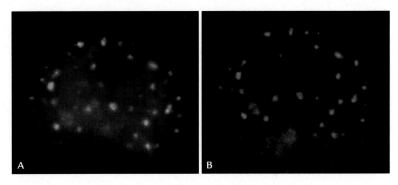

Figure 11.8 Checkpoint genes in action. Radiation damaged the DNA inside this nucleus. **A** *Green* dots pinpoint the location of the product of the *53BP1* gene, and **B** *red* dots pinpoint the location of the product of the *BRCA1* gene. Both proteins have clustered around the same chromosome breaks in the same nucleus; both function to recruit DNA repair enzymes. The integrated action of these and other checkpoint gene products blocks mitosis until the DNA breaks are fixed.

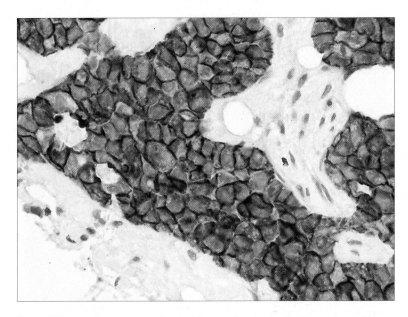

Figure 11.9 Neoplasms are associated with mutations in checkpoint genes. In this section of human breast tissue, phosphorylated EGF receptor is stained *brown*. Normal cells are the ones with lighter staining.

The heavily stained cells have formed a neoplasm; the abnormal overabundance of the phosphorylated EGF receptor means that mitosis is being continually stimulated in these cells. The EGF receptor is overproduced or overactive in most neoplasms.

and it starts a cascade of other intracellular events that ultimately moves the cell cycle out of interphase and into mitosis. The EGF receptor is the product of a checkpoint gene; cells of most neoplasms carry mutations resulting in its overactivity or overabundance (Figure 11.9).

Take-Home Message How does a cell "know" when to divide?

> Gene expression controls advance, delay, or block the cell cycle in response to internal and external conditions.

> The failure of cell cycle checkpoints results in uncontrolled cell divisions.

> Cancer develops as cells of a neoplasm become malignant.
< Links to Adhesion proteins 4.4, Fermentation 7.6

A neoplasm that forms a lump in the body is called a **tumor**, but the two terms are sometimes used interchangeably. Mutations that alter the products of checkpoint genes or the rate at which they are made are associated with an increased risk of tumor formation. Once such a tumor-causing mutation has occurred, the mutated gene is called an oncogene. An **oncogene** is any gene that transforms a normal cell into a tumor cell (from the Greek *onkos*, or bulging mass). Some mutations can be passed to offspring, which is one reason that some types of tumors tend to run in families.

Checkpoint genes encoding proteins that promote mitosis are called **proto-oncogenes** because mutations

can turn them into oncogenes. The gene that encodes the EGF receptor is an example of a proto-oncogene.

Checkpoint gene products that inhibit mitosis are called tumor suppressors because tumors form when they are missing. The products of the *BRCA1* and *BRCA2* genes (Chapter 10) are examples of tumor suppressors. These proteins regulate, among other things, the expression of DNA repair enzymes. Mutations in *BRCA* genes are often found in cells of neoplasms. As another example, viruses such as HPV (human papillomavirus) cause a cell to make proteins that interfere with its own tumor suppressors. Infection with HPV causes noncancerous skin growths called warts, and some kinds are associated with neoplasms that form on the cervix.

Benign neoplasms such as ordinary skin moles are not dangerous (Figure 11.10). They grow very slowly, and their cells retain the plasma membrane adhesion proteins that keep them properly anchored to the other cells in their home tissue **1**.

A malignant neoplasm is one that gets progressively worse, and is dangerous to health. The disease called **cancer** occurs when the abnormally dividing cells of a malignant neoplasm disrupt body tissues, both physically and metabolically. Malignant cells typically display the following three characteristics:

First, like cells of all neoplasms, malignant cells grow and divide abnormally. Controls that usually keep cells from getting overcrowded in tissues are lost, so their populations may reach extremely high densities with cell division occurring very rapidly. The number of small blood vessels, or capillaries, that transport blood to the growing cell mass also increases abnormally.

Second, the cytoplasm and plasma membrane of malignant cells are altered. The cytoskeleton may be shrunken, disorganized, or both. Malignant cells typically have an abnormal chromosome number, with some chromosomes present in multiple copies, and others missing or damaged. The balance of metabolism is often shifted, as in an amplified reliance on ATP formation by fermentation rather than by aerobic respiration.

Altered or missing proteins impair the function of the plasma membrane of malignant cells. For example, these cells do not stay anchored properly in tissues because their plasma membrane adhesion proteins are defective

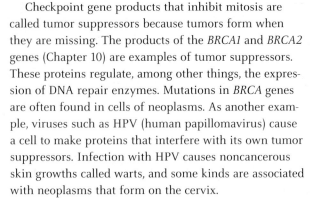

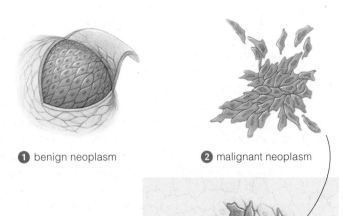

1 benign neoplasm   **2** malignant neoplasm

Figure 11.10 Animated Benign and malignant neoplasms.

1 Benign neoplasms grow slowly and stay in their home tissue.

2 Cells of a malignant neoplasm can break away from their home tissue.

3 The malignant cells become attached to the wall of a blood vessel or lymph vessel. They release digestive enzymes that create an opening in the wall, then enter the vessel.

4 The cells creep or tumble along inside blood vessels, then leave the bloodstream the same way they got in. They often start growing in other tissues, a process called metastasis.

---

cancer Disease that occurs when a neoplasm physically and metabolically disrupts body tissues.
metastasis The process in which cancer cells spread from one part of the body to another.
oncogene Gene that has the potential to transform a normal cell into a tumor cell.
proto-oncogene Gene that can become an oncogene.
tumor A neoplasm that forms a lump.

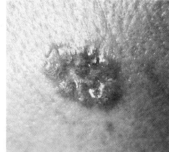

A Basal cell carcinoma is the most common type of skin cancer. This slow-growing, raised lump may be uncolored, reddish-brown, or black.

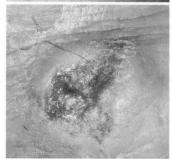

B The second most common form of skin cancer is a squamous cell carcinoma. This pink growth, firm to the touch, grows under the surface of skin.

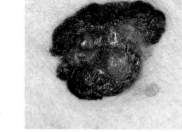

C Melanoma spreads fastest. Cells form dark, encrusted lumps that may itch or bleed easily.

Figure 11.11 Skin cancer is one type of cancer that can be detected early with periodic screening.

or missing ❷. Malignant cells can slip easily into and out of vessels of the circulatory and lymphatic systems ❸. By migrating through these vessels, the cells establish neoplasms elsewhere in the body ❹. The process in which malignant cells break loose from their home tissue and invade other parts of the body is called **metastasis**. Metastasis is the third hallmark of malignant cells.

Unless chemotherapy, surgery, or another procedure eliminates malignant cells from the body, they can put an individual on a painful road to death. Each year, cancer causes 15 to 20 percent of all human deaths in developed countries alone. The good news is that mutations in multiple checkpoint genes are required to transform a normal cell into a malignant one, and these mutations may take a lifetime to accumulate. Life-style choices such as not smoking and avoiding exposure of unprotected skin to sunlight can reduce one's risk of acquiring mutations in the first place. Some neoplasms can be detected with periodic screening procedures such as Pap tests or dermatology exams (Figure 11.11). Neoplasms that are detected early enough can often be removed before metastasis occurs.

Henrietta's Immortal Cells (revisited)

Cancer is a multistep process. Researchers already know about many of the mutations that contribute to the disease. They are working to identify drugs that target and destroy malignant cells or stop them from dividing. Such research may yield drugs that put the brakes on cancer.

HeLa cells have proven to be indispensable in cancer research. For example, they were used in early tests of taxol, a drug that keeps microtubules from disassembling and so interferes with mitosis. Frequent divisions of cancer cells make them more vulnerable to this poison than normal cells.

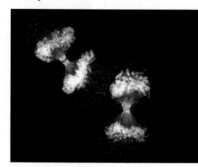

The photo on the *right* shows a more recent example of cancer research that relies on HeLa cells. In these telophase cells, the protein identified by the *blue* stain, INCENP, helps sister chromatids stay attached to one another at the centromere. In normal cells, INCENP associates with the enzyme identified by the *green* stain, Aurora B, only at specific times during mitosis. Aurora B helps attach spindle microtubules to centromeres, so defects in this enzyme or its expression result in unequal segregation of chromosomes into descendant cells. Researchers recently correlated overexpression of Aurora B in cancer cells with shortened patient survival rates. Thus, drugs that inhibit Aurora B function are now being tested as potential cancer therapies.

Despite the invaluable cellular legacy of Henrietta Lacks, her body rests in an unmarked grave in an unmarked cemetery. These days, physicians and researchers are required to obtain a signed consent form before they take tissue samples from a patient. No such requirement existed in the 1950s. It was common at that time for doctors to experiment on patients without their knowledge or consent. Thus, the young resident who was treating

Henrietta Lacks's cancerous cervix probably never even thought about asking permission before he took a sample of it. That sample was the one that the Geys used to establish the HeLa cell line. No one in Henrietta's family knew about the cells until 25 years after she died. HeLa cells are still being sold worldwide, but her family has not received any compensation to date.

How Would You Vote? You can legally donate—but not sell—your own organs and tissues. However, companies can profit from research on donated organs or tissues, and also from cell lines derived from these materials. Companies that do so are not obligated to share their profits with the donors. Should profits derived from donated tissues or cells be shared with the donors or their families? See CengageNow for details, then vote online (cengagenow.com).

Take-Home Message **What is cancer?**

> Cancer is a disease that occurs when the abnormally dividing cells of a neoplasm physically and metabolically disrupt body tissues.

> A malignant neoplasm results from mutations in multiple checkpoint genes.

> Although some mutations are inherited, life-style choices and early intervention can reduce one's risk of cancer.

Summary

Section 11.1 An immortal line of cancer cells is a legacy of cancer victim Henrietta Lacks. HeLa cells have been an invaluable research tool all over the world and even in space. Researchers trying to unravel the mechanisms of cancer continue to work with these cells.

Section 11.2 A **cell cycle** includes all the stages through which a eukaryotic cell passes during its lifetime: **interphase**, **mitosis**, and cytoplasmic division. Most of a cell's activities, including replication of the cell's **homologous chromosomes**, occur in interphase. A cell reproduces by dividing: nucleus first, then the cytoplasm. Each of the two descendant cells receives a complete set of chromosomes and a blob of cytoplasm.

Nuclear division mechanisms partition the duplicated chromosomes into new nuclei. Mitosis maintains the chromosome number. It is the basis of growth, cell replacements, and tissue repair in multicelled species, and **asexual reproduction** in many species.

Section 11.3 During mitosis, duplicated homologous chromosomes line up in the middle of the cell, then are pulled apart. Nuclear envelopes form around the two clusters of chromosomes, forming two new nuclei with the parental chromosome number. Mitosis proceeds in these four stages:

Prophase. Duplicated chromosomes start to condense. Microtubules assemble and form a **spindle**, and the nuclear envelope breaks up. Some microtubules that extend from one spindle pole attach to one chromatid of each chromosome; some that extend from the opposite spindle pole attach to its sister chromatid. Spindle microtubules drag each chromosome toward the center of the cell.

Metaphase. All chromosomes are aligned at the spindle's midpoint.

Anaphase. The sister chromatids of each chromosome detach from each other, and the spindle microtubules start moving them toward opposite spindle poles. Motor proteins drive the movements.

Telophase. A cluster of chromosomes that consists of a complete set of chromosomes reaches each spindle pole. A nuclear envelope forms around each cluster, forming two new nuclei. Both nuclei have the parental chromosome number. Mitosis is over when these nuclei form.

Section 11.4 In most cases, cells divide in two after their nucleus divides. Mechanisms of **cytokinesis** differ. In animal cells, a contractile ring of microfilaments that is part of the cell cortex pulls the plasma membrane inward, forming a **cleavage furrow**. Contraction continues until the cytoplasm is pinched in two. In plant cells, a band of microtubules and microfilaments forms around the nucleus before mitosis. This band marks the site where the **cell plate** forms. The cell plate expands until it fuses with the parent cell wall, thus becoming a cross-wall that partitions the cytoplasm.

Section 11.5 The products of checkpoint genes, including **growth factor** receptors, are part of a host of gene controls that govern the cell cycle. Such controls advance, pause, or stop the cycle in response to conditions inside or outside of the cell. Molecules that work together to monitor the integrity of the cell's DNA can pause the cycle until breaks or other problems are fixed. When checkpoint mechanisms fail, a cell loses control over its cell cycle, and the cell's descendants form a **neoplasm**.

Section 11.6 Mutations can turn **proto-oncogenes** into **oncogenes**. Such mutations typically disrupt checkpoint gene products or their expression, and can result in neoplasms. Neoplasms may form lumps called **tumors**. Mutations in multiple checkpoint genes can transform benign neoplasms into malignant ones. Cells of malignant neoplasms can break loose from their home tissues and colonize other parts of the body, a process called **metastasis**. **Cancer** occurs when malignant neoplasms physically and metabolically disrupt normal body tissues.

Self-Quiz   Answers in Appendix III

1. Mitosis and cytoplasmic division function in _____ .
    a. asexual reproduction of single-celled eukaryotes
    b. growth and tissue repair in multicelled species
    c. gamete formation in bacteria and archaeans
    d. both a and b

2. A duplicated chromosome has _____ chromatid(s).
    a. one   b. two   c. three   d. four

3. Except for a pairing of sex chromosomes, homologous chromosomes _____ .
    a. carry the same genes   c. are the same length
    b. are the same shape   d. all of the above

4. Most cells spend the majority of their lives in _____ .
    a. prophase   d. telophase
    b. metaphase   e. interphase
    c. anaphase   f. a and c

5. The chromosomes align at the midpoint of the spindle during _____ .
    a. prophase   d. telophase
    b. metaphase   e. interphase
    c. anaphase   f. cytokinesis

6. The spindle attaches to chromosomes at the _____ .
    a. centriole   c. centromere
    b. contractile ring   d. centrosome

7. Only _____ is not a stage of mitosis.
    a. prophase b. interphase   c. metaphase d. anaphase

8. In intervals of interphase, G stands for _____ .
    a. gap   b. growth   c. Gey   d. gene

Data Analysis Activities

HeLa Cells Are a Genetic Mess HeLa cells continue to be an extremely useful tool in cancer research. One early finding was that HeLa cells can vary in chromosome number. The panel of chromosomes in Figure 11.12, originally published in 1989 by Nicholas Popescu and Joseph DiPaolo, shows all of the chromosomes in a single metaphase HeLa cell.

1. What is the chromosome number of this HeLa cell?
2. How many extra chromosomes does this cell have, compared to a normal human body cell?
3. Can you tell that this cell came from a female? How?

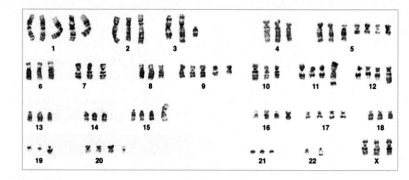

Figure 11.12 Chromosomes in a HeLa cell.

9. In the diagram of the nucleus *below*, fill in the blanks with the name of each interval.

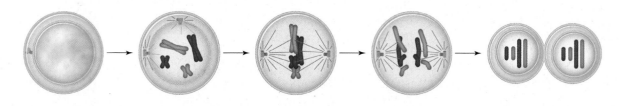

_____ _____ _____ _____ _____

10. Interphase is the part of the cell cycle when _____ .
   a. a cell ceases to function
   b. the spindle forms
   c. a cell grows and duplicates its DNA
   d. mitosis proceeds

11. After mitosis, the chromosome number of a descendant cell is _____ the parent cell's.
   a. the same as   c. rearranged compared to
   b. one-half of   d. doubled compared to

12. Name any checkpoint gene.

13. Which of the following encompasses the other two?
   a. cancer   b. neoplasm   c. tumor

14. Match each term with its best description.
   ___ cell plate   a. lump of cells
   ___ spindle   b. made of microfilaments
   ___ tumor   c. divides plant cells
   ___ cleavage furrow   d. organizes the spindle
   ___ contractile ring   e. metastatic cells
   ___ cancer   f. made of microtubules
   ___ centrosomes   g. indentation

15. Match each stage with the events listed.
   ___ metaphase   a. sister chromatids move apart
   ___ prophase   b. chromosomes start to condense
   ___ telophase   c. new nuclei form
   ___ anaphase   d. all duplicated chromosomes are
                       aligned at the spindle equator

Additional questions are available on **CENGAGENOW**.

Critical Thinking

1. When a cell reproduces by mitosis and cytoplasmic division, does its life end?

 2. The eukaryotic cell in the photo on the *left* is in the process of cytoplasmic division. Is this cell from a plant or an animal? How do you know?

3. Exposure to radioisotopes or other sources of radiation can damage DNA. Humans exposed to high levels of radiation face a condition called radiation poisoning. Why do you think that hair loss and damage to the lining of the gut are early symptoms of radiation poisoning? Speculate about why exposure to radiation is used as a therapy to treat some kinds of cancers.

4. Suppose you have a way to measure the amount of DNA in one cell during the cell cycle. You first measure the amount at the G1 phase. At what points in the rest of the cycle will you see a change in the amount of DNA per cell?

Animations and Interactions on **CENGAGENOW**:
❯ The cell cycle; Mitosis; Cytoplasmic division; Neoplasms; The cell cycle and cancer.

❮ Links to Earlier Concepts

Be sure you have a clear picture of the structural organization of chromosomes (Section 8.2) and how genes work (9.2) before you begin this chapter on the cellular basis of sexual reproduction. You will draw on your understanding of DNA replication (8.6), cytoplasmic division (11.4), and cell cycle controls (11.5) as we compare meiosis with mitosis (11.2). You also will be revisiting microtubules (4.10), genetically identical organisms (8.7), and the effects of mutation (9.6).

Key Concepts

Sexual Versus Asexual Reproduction
In asexual reproduction, one parent transmits its genes to offspring. In sexual reproduction, offspring inherit genes from two parents who usually differ in some number of alleles. Differences in alleles are the basis of differences in traits.

Stages of Meiosis
Meiosis is a nuclear division process that occurs only in cells set aside for sexual reproduction. Meiosis reduces the chromosome number by sorting a reproductive cell's chromosomes into four new nuclei.

12 Meiosis and Sexual Reproduction

12.1 Why Sex?

If the function of reproduction is the perpetuation of one's genes, then an asexual reproducer would seem to win the evolutionary race. In asexual reproduction, all of an individual's genes are passed to all of its offspring. **Sexual reproduction** mixes up the genes of two parents (Figure 12.1), so only about half of each parent's genetic information is passed to offspring.

So why sex? Variation in the forms and combinations of heritable traits is typical of sexually reproducing populations. Some traits allow their bearers to thrive in particular environments. All offspring of asexual reproducers are clones of their parent, so all are adapted the same way to an environment—and all are equally vulnerable to changes in it. By contrast, the offspring of sexual reproducers have unique combinations of many traits. As a group, their diversity offers them flexibility: a better chance of surviving environmental change than clones. Some of them may have a particular combination of traits that suits them perfectly to the changed environment.

Other organisms are part of the environment, and they, too, can change. Think of predator and prey, say, foxes and rabbits. If one rabbit is better than others at outrunning the foxes, it has a better chance of escaping, surviving, and passing to its offspring the genes that help it evade foxes. Thus, over many generations, rabbits may get faster. If one fox is better than others at outrunning faster rabbits, it has a better chance of eating, surviving, and passing to its offspring genes that help it catch faster rabbits. Thus, over many generations, the foxes may tend to get faster. As one species changes, so does the other—an idea called the Red Queen hypothesis, after Lewis Carroll's book *Through the Looking Glass*. In the book, the Queen of Hearts tells Alice, "It takes all the running you can do, to keep in the same place."

An adaptive trait tends to spread more quickly through a sexually reproducing population than through an asexually repro-

sexual reproduction Reproductive mode by which offspring arise from two parents and inherit genes from both.

ducing one. Why? In asexual reproduction, new combinations of traits can arise only by mutation. An adaptive trait is passed from one generation to the next along with the same set of other traits, adaptive or not. By contrast, sexual reproduction mixes up the genes of individuals that often have different forms of traits. It generates new combinations of traits in far fewer generations than does mutation alone.

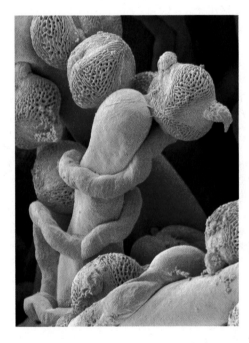

Figure 12.1
Moments in the stages of sexual reproduction of humans (*opposite*) and plants (*left*). Sexual reproduction mixes up the genetic material of two organisms.

In flowering plants, pollen grains (*orange*) germinate on flower carpels (*yellow*). Pollen tubes with male gametes inside grow from the grains down into tissues of the ovary, which house the flower's female gametes.

However, just because sexual reproducers are more genetically diverse does not mean that they win the evolutionary race. In terms of numbers of individuals and how long their lineages have endured, the most successful organisms on Earth—by a long shot—are bacteria, which reproduce by an entirely different mechanism.

Recombinations and Shufflings
During meiosis, homologous chromosomes come together and swap segments. Then they are randomly sorted into separate nuclei. Both processes lead to novel combinations of alleles among offspring.

Sexual Reproduction in the Context of Life Cycles
Gametes form by different mechanisms in males and females, but meiosis is part of both processes. In most plants, spore formation and other events intervene between meiosis and gamete formation.

Mitosis and Meiosis Compared
Similarities between mitosis and meiosis suggest that meiosis may have originated by evolutionary remodeling of mechanisms that already existed for mitosis and, before that, for repairing damaged DNA.

> Asexual reproduction produces clones.
> Sexual reproduction mixes up alleles from two parents.
> Meiosis, the basis of sexual reproduction, is a nuclear division mechanism that occurs in immature reproductive cells of eukaryotes.
< Links to Clones 8.1, Chromosomes in eukaryotes 8.2, DNA replication 8.6, Genetically identical organisms 8.7, Genes 9.2, Effects of mutations 9.6, Homologous chromosomes and asexual reproduction and mitosis 11.2

Introducing Alleles

An individual's genes collectively contain the information necessary to make a new individual (Section 8.7). All offspring of an asexual reproducer inherit the same number and kinds of genes; thus, mutations aside, all are clones of the parent (Section 8.1).

Inheritance gets more complicated with sexual reproduction because two parents contribute genes to offspring. The **somatic** (body) cells of multicelled organisms that reproduce sexually contain pairs of chromosomes. Typically, one chromosome of each pair is maternal and the other is paternal (Figure 12.2). Except for a pairing of nonidentical sex chromosomes, the two chromosomes of every pair carry the same set of genes.

If the DNA sequence of every gene pair were identical, then sexual reproduction would produce clones, just like asexual reproduction does. Just imagine: The entire human population might consist of clones, in which case everybody would look exactly alike.

But the two genes of a pair are often not identical. Why not? Mutations that inevitably accumulate in DNA change its sequence. Thus, the two genes of any pair might differ a bit. If the sequences differ enough, those genes will encode slightly different forms of the gene's product (Section 9.6). Different forms of the same gene are called **alleles**.

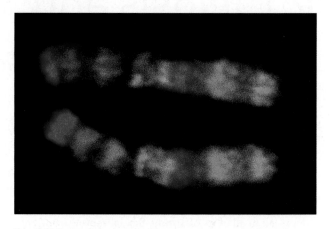

Figure 12.2 Homologous chromosomes. Typically, one chromosome of each pair is inherited from the mother; the other, from the father. Colored patches in this fluorescence micrograph indicate corresponding DNA sequences on the chromosomes.

These chromosomes carry the same series of genes, but the DNA sequence of any one of those genes might differ just a bit from that of its partner on the other chromosome. Different forms of a gene are called alleles.

Alleles influence thousands of traits. For example, the beta globin gene you encountered in Section 9.6 has more than 700 alleles: one that causes sickle-cell anemia, one that causes beta thalassemia, and so on. The beta globin gene is only one of about 30,000 human genes, and most genes have multiple alleles. Alleles are one reason that the individuals of a sexually reproducing species do not all look exactly the same. The offspring of sexual reproducers inherit new combinations of alleles, which is the basis of new combinations of traits.

What Meiosis Does

Sexual reproduction involves the fusion of reproductive cells from two parents. It requires **meiosis**, a nuclear divi-

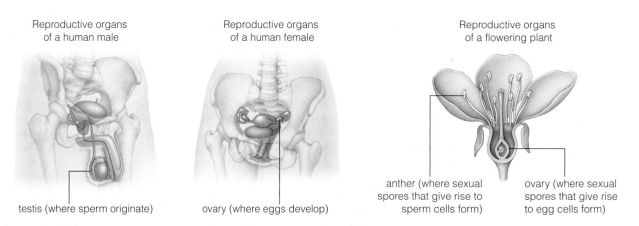

Reproductive organs of a human male

Reproductive organs of a human female

Reproductive organs of a flowering plant

testis (where sperm originate)

ovary (where eggs develop)

anther (where sexual spores that give rise to sperm cells form)

ovary (where sexual spores that give rise to egg cells form)

Figure 12.3 Animated Examples of reproductive organs. Meiosis of germ cells in reproductive organs gives rise to gametes: eggs and sperm in humans, egg cells and sperm cells in flowering plants.

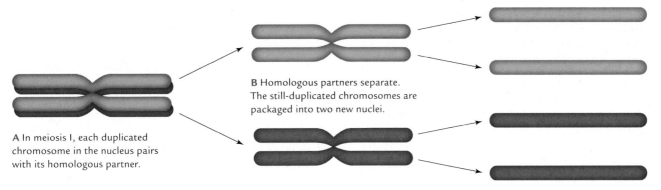

A In meiosis I, each duplicated chromosome in the nucleus pairs with its homologous partner.

B Homologous partners separate. The still-duplicated chromosomes are packaged into two new nuclei.

C Sister chromatids separate in meiosis II. The now unduplicated chromosomes are packaged into four new nuclei.

Figure 12.4 How meiosis halves the chromosome number.

sion mechanism that halves the chromosome number. The process of sexual reproduction begins with meiosis in **germ cells**, which are immature reproductive cells. Meiosis in germ cells produces mature reproductive cells called **gametes**. A sperm is an example of a male gamete. An egg is a female gamete. Gametes usually form inside special reproductive structures or organs (Figure 12.3).

Gametes have a single set of chromosomes, so they are **haploid** (n): Their chromosome number is half of the diploid ($2n$) number (Section 8.2). Human body cells are diploid, with 23 pairs of homologous chromosomes. Meiosis of a human germ cell ($2n$) normally produces gametes with 23 chromosomes: one of each pair (n). The diploid chromosome number is restored at **fertilization**, when two haploid gametes (one egg and one sperm, for example) fuse to form a **zygote**, the first cell of a new individual.

The first part of meiosis is similar to mitosis. A cell duplicates its DNA before either nuclear division process starts. As in mitosis, the microtubules of a spindle move the duplicated chromosomes to opposite spindle poles. However, meiosis sorts the chromosomes into new nuclei not once, but twice, so it results in the formation of four haploid nuclei. The two consecutive nuclear divisions are called meiosis I and meiosis II:

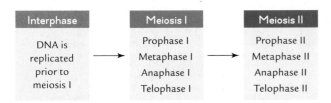

Interphase	Meiosis I	Meiosis II
DNA is replicated prior to meiosis I	Prophase I Metaphase I Anaphase I Telophase I	Prophase II Metaphase II Anaphase II Telophase II

In some cells, no resting period occurs between these two stages. In others, interphase with no DNA replication separates meiosis I and II.

During meiosis I, every duplicated chromosome aligns with its homologous partner (Figure 12.4A). Then the homologous chromosomes are pulled away from one another (Figure 12.4B). After homologous chromosomes separate, each ends up in one of two new nuclei. At this stage, the chromosomes are still duplicated (the sister chromatids are still attached to one another).

During meiosis II, the sister chromatids of each chromosome are pulled apart, so each becomes an individual, unduplicated chromosome (Figure 12.4C). The chromosomes are sorted into four new nuclei. With one unduplicated version of each chromosome, the new nuclei are all haploid (n).

Thus, meiosis partitions the chromosomes of one diploid nucleus ($2n$) into four haploid (n) nuclei. The next section zooms in on the details of this process.

alleles Forms of a gene that encode slightly different versions of the gene's product.
fertilization Fusion of two gametes to form a zygote.
gamete Mature, haploid reproductive cell; e.g., an egg or a sperm.
germ cell Diploid reproductive cell that gives rise to haploid gametes by meiosis.
haploid Having one of each type of chromosome characteristic of the species.
meiosis Nuclear division process that halves the chromosome number. Basis of sexual reproduction.
somatic Relating to the body.
zygote Cell formed by fusion of two gametes; the first cell of a new individual.

Take-Home Message **Why do populations that reproduce sexually tend to have the most variation in heritable traits?**

❯ Paired genes on homologous chromosomes may vary in sequence as alleles.

❯ Alleles are the basis of traits. Sexual reproduction mixes up alleles from two parents.

❯ The nuclear division process of meiosis is the basis of sexual reproduction in eukaryotes. It precedes the formation of gametes or spores.

❯ Meiosis halves the diploid ($2n$) chromosome number, to the haploid number (n). When two gametes fuse at fertilization, the chromosome number is restored. The resulting zygote has two sets of chromosomes, one from each parent.

❯ Meiosis halves the chromosome number.

❯ During meiosis, chromosomes of a diploid nucleus become distributed into four haploid nuclei.

❮ Links to Diploid chromosome number 8.2, DNA replication 8.6, Mitosis 11.2

DNA replication occurs prior to meiosis, so a cell's chromosomes are duplicated by the time meiosis I begins: Each chromosome consists of two sister chromatids. The nucleus is diploid (2*n*): It contains two sets of chromosomes, one from each parent. Let's now turn to the cellular events that occur during meiosis itself.

Figure 12.5 Animated Meiosis. Two pairs of chromosomes are illustrated in a diploid (2*n*) animal cell. Homologous chromosomes are indicated in *blue* and *pink*. Micrographs show meiosis in a lily plant cell (*Lilium regale*).

❯❯ **Figure It Out** During which phase of meiosis does the chromosome number become reduced?

Answer: Anaphase I

Meiosis I The first stage of meiosis I is prophase I (Figure 12.5). During this phase, the chromosomes condense, and homologous chromosomes align tightly and swap segments (more about segment-swapping in the next section). The centrosome gets duplicated along with its two centrioles. One centriole pair moves to the opposite side of the cell as the nuclear envelope breaks up. Spindle microtubules begin to extend from the centrosomes ❶.

By the end of prophase I, microtubules of the spindle connect the chromosomes to the spindle poles. Each chromosome is now attached to one spindle pole, and its homologous partner is attached to the other. The microtubules lengthen and shorten, pushing and pulling the chromosomes as they do. At metaphase I, all of the microtubules are the same length, and the chromosomes are aligned midway between the poles of the spindle ❷.

In anaphase I, the spindle microtubules separate the homologous chromosomes and pull them toward opposite

Meiosis I One diploid nucleus to two haploid nuclei

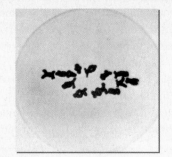
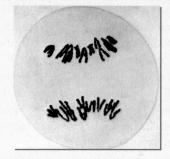
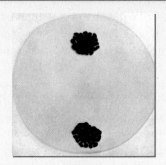

❶ Prophase I. Homologous chromosomes condense, pair up, and swap segments. Spindle microtubules attach to them as the nuclear envelope breaks up.

❷ Metaphase I. The homologous chromosome pairs are aligned midway between spindle poles.

❸ Anaphase I. The homologous chromosomes separate and begin heading toward the spindle poles.

❹ Telophase I. Two clusters of chromosomes reach the spindle poles. A new nuclear envelope forms around each cluster, so two haploid (*n*) nuclei form.

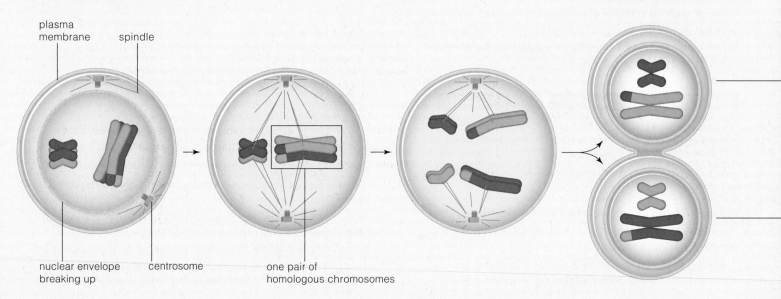

plasma membrane

spindle

nuclear envelope breaking up

centrosome

one pair of homologous chromosomes

spindle poles ❸. During telophase I, the chromosomes reach the spindle poles ❹. New nuclear envelopes form around the two clusters of chromosomes as the DNA loosens up. Each of the two haploid (*n*) nuclei that form contains one set of (duplicated) chromosomes. The cytoplasm may divide at this point to form two haploid cells. Interphase occurs in some cells at the end of meiosis I, but the DNA is not replicated before meiosis II begins.

Meiosis II During prophase II, the chromosomes condense as a new spindle forms. One centriole moves to the opposite side of each new nucleus, and the nuclear envelopes break up. By the end of prophase II, microtubules connect the chromosomes to the spindle poles. Each chromatid is now attached to one spindle pole, and its sister is attached to the other ❺. The microtubules lengthen and shorten, pushing and pulling the chromosomes as they do. At metaphase II, all of the microtubules are the same

length, and the chromosomes are aligned midway between the spindle poles ❻.

In anaphase II, the spindle microtubules pull the sister chromatids apart ❼. Each chromosome now consists of one molecule of DNA. During telophase II, the chromosomes (now unduplicated) reach the spindle poles ❽. New nuclear envelopes form around the four clusters of chromosomes as the DNA loosens up. Each of the four haploid (*n*) nuclei that form contains one set of unduplicated chromosomes. The cytoplasm may divide, so four haploid cells form.

> **Take-Home Message** **What happens to a cell during meiosis?**
>
> › During meiosis, the nucleus of a diploid (2*n*) cell divides twice. Four haploid (*n*) nuclei form, each with a full set of chromosomes—one of each type.

Meiosis II Two haploid nuclei to four haploid nuclei

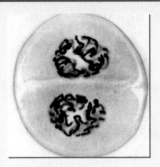  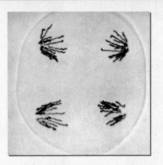 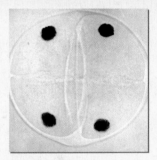

❺ Prophase II. The chromosomes condense. Spindle microtubules attach to each sister chromatid as the nuclear envelope breaks up.

❻ Metaphase II. The (still duplicated) chromosomes are aligned midway between poles of the spindle.

❼ Anaphase II. All sister chromatids separate. The now unduplicated chromosomes head to the spindle poles.

❽ Telophase II. A cluster of chromosomes reaches each spindle pole. A new nuclear envelope encloses each cluster, so four haploid (*n*) nuclei form.

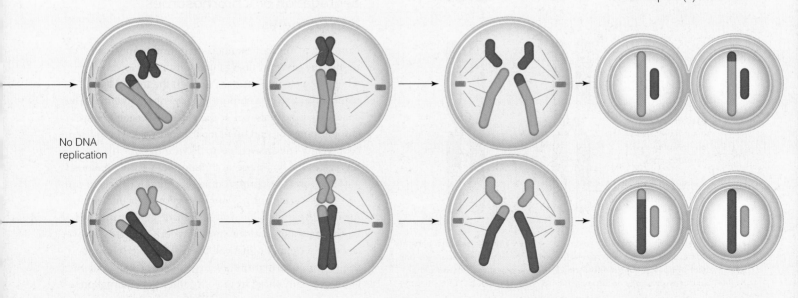

No DNA replication

> Crossovers and the random sorting of chromosomes into gametes result in new combinations of traits among offspring of sexual reproducers.

< Link to Chromosome structure 8.2

The previous section mentioned briefly that duplicated chromosomes swap segments with their homologous partners during prophase I. It also showed how each chromosome aligns with and then separates from its homologous partner during anaphase I. Both events introduce novel combinations of alleles into gametes. Along with fertilization, these events contribute to the variation in

A Here, we focus on only two of the many genes on a chromosome. In this example, one gene has alleles *A* and *a*; the other has alleles *B* and *b*.

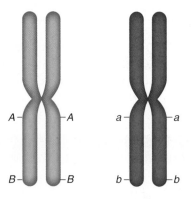

B Close contact between homologous chromosomes promotes crossing over between nonsister chromatids. Paternal and maternal chromatids exchange corresponding pieces.

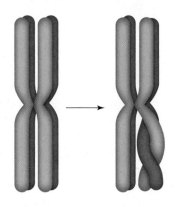

C Crossing over mixes up paternal and maternal alleles on homologous chromosomes.

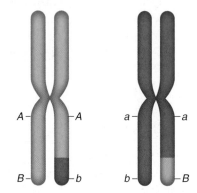

Figure 12.6 Animated Crossing over. *Blue* signifies a paternal chromosome, and *pink*, its maternal homologue. For clarity, we show only one pair of homologous chromosomes and one crossover, but more than one crossover may occur in each chromosome pair.

combinations of traits among the offspring of sexually reproducing species.

Crossing Over in Prophase I

Early in prophase I of meiosis, all chromosomes in a germ cell condense. When they do, each is drawn close to its homologue. The chromatids of one homologous chromosome become tightly aligned with the chromatids of the other along their length:

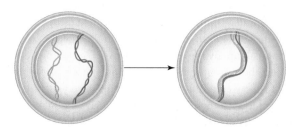

This tight, parallel orientation favors **crossing over**, a process in which a chromosome and its homologous partner exchange corresponding pieces of DNA (Figure 12.6). Homologous chromosomes may swap any segment or segments of DNA along their length, although crossovers tend to occur more frequently in certain regions.

Swapping segments of DNA shuffles alleles between homologous chromosomes. It breaks up the particular combinations of alleles that occurred on the parental chromosomes, and makes new ones on the chromosomes that end up in gametes. Thus, crossing over introduces novel combinations of traits among offspring. It is a normal and frequent process in meiosis, but the rate of crossing over varies among species and among chromosomes. In humans, between 46 and 95 crossovers occur per meiosis, so on average each chromosome crosses over at least once.

Segregation of Chromosomes Into Gametes

Normally, all of the new nuclei that form in meiosis I receive the same number of chromosomes. However, whether a new nucleus ends up with the maternal or paternal version of a chromosome is entirely random. The chance that the maternal or the paternal version of any chromosome will end up in a particular nucleus is 50 percent. Why? The answer has to do with the way the spindle segregates the homologous chromosomes during meiosis I.

The process of chromosome segregation begins in prophase I. Imagine one of your own germ cells undergoing meiosis. Crossovers have already made genetic mosaics of its chromosomes, but for simplicity let's put crossing over aside for a moment. Just call the twenty-three chromosomes you inherited from your mother the maternal ones, and the twenty-three from your father the paternal ones.

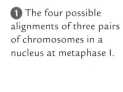 **1** The four possible alignments of three pairs of chromosomes in a nucleus at metaphase I.

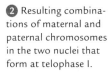 **2** Resulting combinations of maternal and paternal chromosomes in the two nuclei that form at telophase I.

3 Resulting combinations of maternal and paternal chromosomes in the four nuclei that form at telophase II. Eight different combinations are possible.

  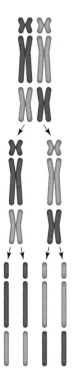  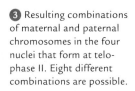

Figure 12.7 Animated Hypothetical segregation of three pairs of chromosomes in meiosis I. Maternal chromosomes are *pink*; paternal, *blue*. Which chromosome of each pair gets packaged into which of the two new nuclei that form at telophase I is random. For simplicity, no crossing over occurs in this example, so all sister chromatids are identical.

During prophase I, microtubules fasten your cell's chromosomes to the spindle poles. Chances are fairly slim that all of the maternal chromosomes get attached to one pole and all of the paternal chromosomes get attached to the other. Microtubules extending from a spindle pole bind to the centromere of the first chromosome they contact, regardless of whether it is maternal or paternal. Each homologous partner gets attached to the opposite spindle pole. Thus, there is no pattern to the attachment of the maternal or paternal chromosomes to a particular pole.

Now imagine that your germ cell has just three pairs of chromosomes (Figure 12.7). By metaphase I, those three pairs of maternal and paternal chromosomes are divvied up between the two spindle poles in one of four ways **1**. In anaphase I, homologous chromosomes separate and are pulled toward opposite spindle poles. In telophase I, a new nucleus forms around the chromosomes that cluster at each spindle pole. Each nucleus contains one of eight possible combinations of maternal and paternal chromosomes **2**.

In telophase II, each of the two nuclei divides and gives rise to two new haploid nuclei. The two new nuclei are identical because no crossing over occurred in our hypothetical example, so all of the sister chromatids were

___

crossing over Process in which homologous chromosomes exchange corresponding segments during prophase I of meiosis.

identical. Thus, at the end of meiosis in this cell, two (2) spindle poles have divvied up three (3) chromosome pairs. The resulting four nuclei have one of eight (2^3) possible combinations of maternal and paternal chromosomes **3**.

Cells that give rise to human gametes have twenty-three pairs of homologous chromosomes, not three. Each time a human germ cell undergoes meiosis, the four gametes that form end up with one of 8,388,608 (or 2^{23}) possible combinations of homologous chromosomes. That number does not even take into account crossing over, which mixes up the alleles on maternal and paternal chromosomes, or fusion with another gamete at fertilization. Are you getting an idea of why such fascinating combinations of traits show up among the generations of your own family tree?

Take-Home Message How does meiosis introduce variation in combinations of traits?

> Crossing over is recombination between nonsister chromatids of homologous chromosomes during prophase I. It makes new combinations of parental alleles.

> Homologous chromosomes can be attached to either spindle pole in prophase I, so each homologue can be packaged into either one of the two new nuclei. Thus, the random assortment of homologous chromosomes increases the number of potential combinations of maternal and paternal alleles in gametes.

> Details of gamete formation differ among plants and animals, but meiosis is always part of the process.
< Links to Flagella 4.10, Cytoplasmic division 11.4

Gametes are the specialized cells that are the basis of sexual reproduction. All are haploid, but they differ in other details. For example, human male gametes—sperm—have one flagellum (Section 4.10). Opossum sperm have two, and roundworm sperm have none. A flowering plant's male gamete consists simply of a nucleus. We leave details of reproduction for later chapters, but you will need to know a few concepts before you get there.

Gamete Formation in Plants Two kinds of multicelled bodies form during the life cycle of a plant: sporophytes and gametophytes. A **sporophyte** is typically diploid, and spores form by meiosis in its specialized parts. Spores consist of one or a few haploid cells. These cells undergo mitosis and give rise to **gametophytes**, multicelled haploid bodies inside which one or more gametes form. A sequoia tree is an example of a sporophyte (Figure 12.8A). Male and female gametophytes develop inside different types of cones that form on each tree. In flowering plants, gametophytes form in flowers, which are specialized reproductive shoots of the sporophyte body.

Gamete Formation in Animals In the life cycle of a typical animal, a zygote matures as a multicelled body that produces gametes (Figure 12.8B). Animal gametes arise by meiosis of diploid germ cells. In male animals (Figure 12.9), the germ cell develops into a primary spermatocyte ❶. Meiosis I in a primary spermatocyte results in two haploid secondary spermatocytes ❷, which undergo meiosis II and become four spermatids ❸. Each spermatid then matures as a **sperm** ❹.

In female animals (Figure 12.10), a germ cell becomes a primary oocyte, which is an immature egg ❺. This cell undergoes meiosis and division, as occurs with a primary spermatocyte. Two haploid cells form when the primary oocyte divides after meiosis I. However, the cytoplasm of a primary oocyte divides unequally, so the cells differ in size and function. One of the cells is called a first polar body. The other cell, the secondary oocyte, is much larger because it gets nearly all of the parent cell's cytoplasm ❻. This larger cell undergoes meiosis II and cytoplasmic division, which again is unequal ❼. One of the two cells that forms is a second polar body. The other cell gets most of the cytoplasm and matures into a female gamete, which is called an ovum (plural, ova), or **egg**.

Polar bodies are not nutrient-rich or plump with cytoplasm, and generally do not function as gametes. In time they degenerate. Their formation simply ensures that the egg will have a haploid chromosome number, and also will get enough metabolic machinery to support early divisions of the new individual.

Fertilization Meiosis of a diploid germ cell produces haploid gametes. When two gametes fuse at fertilization, the resulting zygote is diploid. Thus, meiosis halves the chromosome number, and fertilization restores it. If meiosis did not precede fertilization, the chromosome number would double with every generation. An individual's set

A Plant life cycle

zygote (2n) — mitosis — multicelled sporophyte (2n) — meiosis

diploid
haploid

gametes (n) — fertilization — spores (n)

multicelled gametophyte (n)

Figure 12.8 Comparing the life cycles of animals and plants.

A Generalized life cycle for most plants. A sequoia tree is a sporophyte.

B Generalized life cycle for animals. The zygote is the first cell to form when the nuclei of two gametes, such as a sperm and an egg, fuse at fertilization.

B Animal life cycle

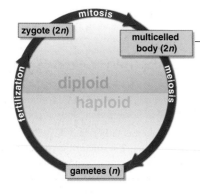

zygote (2n) — mitosis — multicelled body (2n) — meiosis

diploid
haploid

fertilization

gametes (n)

---

egg Mature female gamete, or ovum.
gametophyte A haploid, multicelled body in which gametes form during the life cycle of plants.
sperm Mature male gamete.
sporophyte Diploid, spore-producing stage of a plant life cycle.

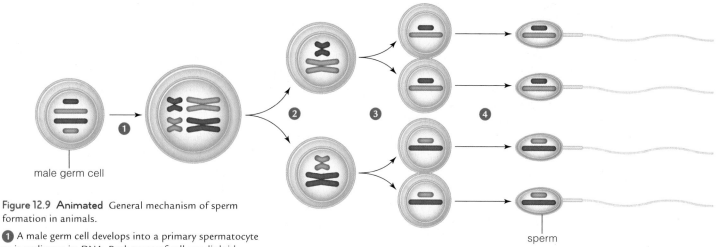

Figure 12.9 Animated General mechanism of sperm formation in animals.

1 A male germ cell develops into a primary spermatocyte as it replicates its DNA. Both types of cell are diploid.

2 Meiosis I in the primary spermatocyte results in two secondary spermatocytes, which are haploid.

3 Four haploid spermatids form when the secondary spermatocytes undergo meiosis II.

4 Spermatids mature as sperm (haploid male gametes).

male germ cell

sperm

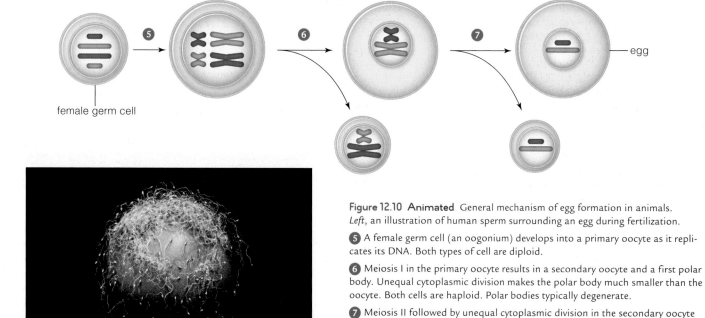

female germ cell

egg

Figure 12.10 Animated General mechanism of egg formation in animals. *Left*, an illustration of human sperm surrounding an egg during fertilization.

5 A female germ cell (an oogonium) develops into a primary oocyte as it replicates its DNA. Both types of cell are diploid.

6 Meiosis I in the primary oocyte results in a secondary oocyte and a first polar body. Unequal cytoplasmic division makes the polar body much smaller than the oocyte. Both cells are haploid. Polar bodies typically degenerate.

7 Meiosis II followed by unequal cytoplasmic division in the secondary oocyte results in a polar body and an ovum, or egg. Both cells are haploid.

of chromosomes is like a fine-tuned blueprint that must be followed exactly, page by page, in order to build a body that functions normally. As you will see in the next chapter, chromosome number changes can have drastic consequences, particularly in animals.

Take-Home Message **How does meiosis fit into the life cycle of plants and animals?**

❭ Meiosis and cytoplasmic division precede the development of haploid gametes in animals and spores in plants.

❭ The union of two haploid gametes at fertilization results in a diploid zygote.

> Though they have different results, mitosis and meiosis are fundamentally similar processes.
< Links to Mitosis 11.3, Cell cycle controls 11.5

This chapter opened with hypotheses about the evolutionary advantages of asexual and sexual reproduction. It seems like a giant evolutionary step from producing clones to producing genetically varied offspring, but was it really?

By mitosis and cytoplasmic division, one cell becomes two new cells. This process is the basis of growth and tissue repair in all multicelled species. Single-celled eukaryotes (and some multicelled ones) also reproduce asexually by way of mitosis and cytoplasmic division. Mitotic (asexual) reproduction results in clones, which are genetically identical copies of a parent.

Meiosis also begins with a diploid cell, one that is specialized for reproduction. Meiosis in this cell produces haploid gametes. Gametes of two parents fuse to form a zygote, which is a diploid cell of mixed parentage. Meiotic (sexual) reproduction results in offspring that are genetically different from the parent, and from one another.

Though their end results differ, there are striking parallels between the four stages of mitosis and meiosis II (Figure 12.11). As one example, a spindle forms and separates chromosomes during both processes. There are many more similarities at the molecular level.

Long ago, the molecular machinery of mitosis may have been remodeled into meiosis. Evidence for this hypothesis includes a host of molecules, including the products of the *BRCA* genes (Chapters 10 and 11), that are made by all modern eukaryotes. These molecules monitor and repair breaks in DNA, for example during DNA replication prior to mitosis. They actively maintain the integrity of a cell's chromosomes. It turns out that the very same set of molecules monitor and fix the breaks in homologous chromosomes during crossing over in prophase I of meiosis. Some of them function as checkpoint proteins in both mitosis and meiosis, so mutations that affect them or the rate at which they are made can affect the outcomes of both nuclear division processes.

In anaphase of mitosis, sister chromatids are pulled apart. What would happen if the connections between the sisters did not break? Each duplicated chromosome would be pulled to one or the other spindle pole—which is exactly what happens in anaphase I of meiosis.

Sexual reproduction probably originated by mutations that affected processes of mitosis. As you will see in later chapters, the remodeling of existing processes into new ones is a common evolutionary theme.

Take-Home Message Are the processes of mitosis and meiosis related?

> Meiosis may have evolved by the remodeling of existing mechanisms of mitosis.

Meiosis I One diploid nucleus to two haploid nuclei

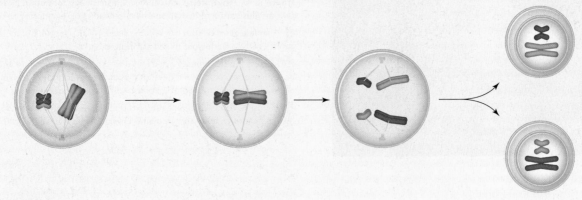

Prophase I
- Chromosomes condense.
- Homologous chromosomes pair.
- Crossovers occur (not shown).
- Spindle forms and attaches chromosomes to spindle poles.
- Nuclear envelope breaks up.

Metaphase I
- Chromosomes align midway between spindle poles.

Anaphase I
- Homologous chromosomes separate and move toward opposite spindle poles.

Telophase I
- Chromosome clusters arrive at spindle poles.
- New nuclear envelopes form.
- Chromosomes decondense.

Figure 12.11 Comparative summary of key features of mitosis and meiosis, starting with a diploid cell.

Only two paternal and two maternal chromosomes are shown. Both were duplicated in interphase, prior to nuclear division. A spindle of microtubules moves the chromosomes in mitosis as well as meiosis.

Mitosis maintains the parental chromosome number. Meiosis halves it, to the haploid number.

Mitotic cell division is the basis of asexual reproduction among eukaryotes. It also is the basis of growth and tissue repair of multicelled eukaryotic species.

Why Sex? (revisited)

There are a few all-female species of fishes, reptiles, and birds in nature, but not mammals. In 2004, researchers fused two mouse eggs in a test tube and made an embryo using no DNA from a male. The embryo developed into Kaguya, the world's first fatherless mammal (*right*). The mouse grew up healthy, engaged in sex with a male mouse, and gave birth to offspring. The researchers wanted to find out if sperm was required for normal development.

How Would You Vote? Researchers made a "fatherless" mouse from two mouse eggs. Should they be prevented from trying the process with human eggs? See CengageNow for details, then vote online (cengagenow.com).

Mitosis One diploid nucleus to two diploid nuclei

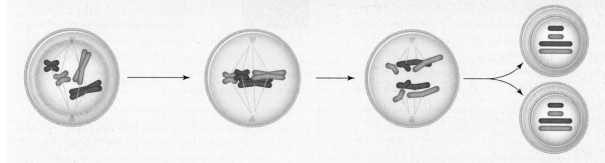

Prophase
- Chromosomes condense.
- Spindle forms and attaches chromosomes to spindle poles.
- Nuclear envelope breaks up.

Metaphase
- Chromosomes align midway between spindle poles.

Anaphase
- Sister chromatids separate and move toward opposite spindle poles.

Telophase
- Chromosome clusters arrive at spindle poles.
- New nuclear envelopes form.
- Chromosomes decondense.

Meiosis II Two haploid nuclei to four haploid nuclei

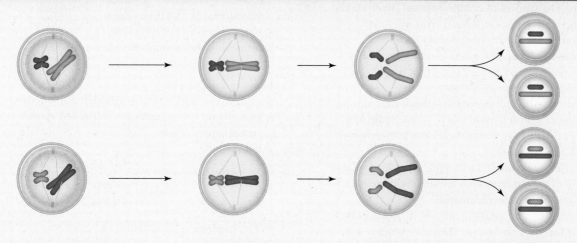

Prophase II
- Chromosomes condense.
- Spindle forms and attaches chromosomes to spindle poles.
- Nuclear envelope breaks up.

Metaphase II
- Chromosomes align midway between spindle poles.

Anaphase II
- Sister chromatids separate and move toward opposite spindle poles.

Telophase II
- Chromosome clusters arrive at spindle poles.
- New nuclear envelopes form.
- Chromosomes decondense.

Summary

Section 12.1 Sexual reproduction mixes up the genetic information of two parents. Under some conditions, the variation in traits among offspring of sexual reproducers provides them with an evolutionary advantage over genetically identical offspring.

Section 12.2 Offspring of asexual reproduction are genetically identical to their one parent: They are clones. The offspring of sexual reproduction differ from parents, and often from one another, in the details of shared traits. Offspring of most sexual reproducers inherit pairs of chromosomes, one of each pair from the mother and the other from the father. Paired genes on homologous chromosomes may vary in DNA sequence, in which case they are called **alleles**.

Meiosis, a nuclear division mechanism that occurs in eukaryotic **germ cells**, precedes the formation of **gametes**. Meiosis halves the parental chromosome number. It also shuffles parental alleles, so offspring inherit new combinations of alleles. The fusion of two **haploid** gametes during **fertilization** restores the parental chromosome number in the **zygote**, the first cell of the new individual.

Section 12.3 All chromosomes are duplicated during interphase, before meiosis. Two divisions, meiosis I and II, halve the parental chromosome number. In the first nuclear division, meiosis I, each duplicated chromosome lines up with its homologous partner; then the two are moved toward opposite spindle poles.

Prophase I. Chromosomes condense and align tightly with their homologues. Each pair of homologues typically undergoes crossing over. Microtubules form the spindle. One of two pairs of centrioles is moved to the other side of the nucleus. The nuclear envelope breaks up, so microtubules extending from each spindle pole can penetrate the nuclear region. The microtubules then attach to one or the other chromosome of each homologous pair.

Metaphase I. A tug-of-war between the microtubules from both poles has positioned all pairs of homologous chromosomes at the spindle equator.

Anaphase I. Microtubules separate each chromosome from its homologue and move one to each spindle pole. Anaphase I ends as a cluster of duplicated chromosomes nears each spindle pole.

Telophase I. Two nuclei form; typically the cytoplasm divides. All of the chromosomes are still duplicated; each still consists of two sister chromatids.

The second nuclear division, meiosis II, occurs in both nuclei that formed in meiosis I. The chromosomes condense in prophase II, and align in metaphase II. Sister chromatids of each chromosome are pulled apart from each other in anaphase II, so each becomes an individual chromosome. By the end of telophase II, four haploid nuclei have formed, each with one set of chromosomes.

Section 12.4 Events in prophase I and metaphase I produce nonparental combinations of alleles. The nonsister chromatids of homologous chromosomes undergo **crossing over** during prophase I: They exchange segments at the same place along their length, so each ends up with new combinations of alleles that were not present in either parental chromosome.

Crossing over during prophase I, and random segregation of maternal and paternal chromosomes into new nuclei, contribute to variation in traits among offspring. Microtubules can attach the maternal or the paternal chromosome of each pair to one or the other spindle pole. Either chromosome may end up in any new nucleus, and in any gamete. Such chromosome shufflings, along with crossovers during prophase I of meiosis, are the basis of variation in traits we see in sexually reproducing species.

Section 12.5 Multicelled diploid bodies are typical in life cycles of plants and animals. A diploid **sporophyte** is a multicelled plant body that makes haploid spores. Spores give rise to **gametophytes**, or multicelled plant bodies in which haploid gametes form. Germ cells in the reproductive organs of most animals give rise to **sperm** or **eggs**. Fusion of haploid gametes at fertilization results in a diploid zygote.

Section 12.6 Like mitosis, meiosis requires a spindle to move and sort duplicated chromosomes, but meiosis occurs only in cells that are set aside for sexual reproduction. Some mechanisms of meiosis resemble those of mitosis, and may have evolved from them.

Self-Quiz   Answers in Appendix III

1. The main evolutionary advantage of sexual over asexual reproduction is that it produces _____ .
   a. more offspring per individual
   b. more variation among offspring
   c. healthier offspring

2. Meiosis functions in _____ .
   a. asexual reproduction of single-celled eukaryotes
   b. growth and tissue repair in multicelled species
   c. sexual reproduction
   d. both a and b

3. Sexual reproduction in animals requires _____ .
   a. meiosis   c. spore formation
   b. fertilization   d. a and b

4. Meiosis _____ the parental chromosome number.
   a. doubles   c. maintains
   b. halves   d. mixes up

5. Crossing over mixes up _____ .
   a. chromosomes   c. zygotes
   b. alleles   d. gametes

BPA and Abnormal Meiosis In 1998, researchers at Case West-ern University were studying meiosis in mouse oocytes when they saw an unexpected and dramatic increase of abnormal meiosis events (Figure 12.12). Improper segregation of chro-mosomes during meiosis is one of the main causes of human genetic disorders, which we will discuss in Chapter 14.

The researchers discovered that the spike in meiotic abnor-malities began immediately after the mouse facility started washing the animals' plastic cages and water bottles in a new, alkaline detergent. The detergent had damaged the plastic, which began to leach bisphenol A (BPA). BPA is a synthetic chemical that mimics estrogen, the main female sex hormone in animals. BPA is used to manufacture polycarbonate plastic items (including baby bottles and water bottles) and epoxies (including the coating on the inside of metal cans of food).

1. What percentage of mouse oocytes displayed abnormalities of meiosis with no exposure to damaged caging?
2. Which group of mice showed the most meiotic abnormali-ties in their oocytes?
3. What is abnormal about metaphase I as it is occurring in the oocytes shown in Figure 12.12B, C, and D?

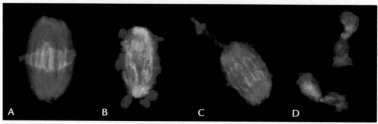

Caging materials	Total number of oocytes	Abnormalities
Control: New cages with glass bottles	271	5 (1.8%)
Damaged cages with glass bottles		
Mild damage	401	35 (8.7%)
Severe damage	149	30 (20.1%)
Damaged bottles	197	53 (26.9%)
Damaged cages with damaged bottles	58	24 (41.4%)

Figure 12.12 Meiotic abnormalities associated with exposure to dam-aged plastic caging. Fluorescent micrographs show nuclei of single mouse oocytes in metaphase I. **A** Normal metaphase; **B–D** examples of abnormal metaphase. Chromosomes are *red*; spindle fibers are *green*.

6. Crossing over happens during which phase of meiosis?

7. The stage of meiosis that makes descendant cells hap-loid is _____ .
   - a. prophase I
   - b. prophase II
   - c. anaphase I
   - d. anaphase II
   - e. metaphase I
   - f. metaphase II

8. Dogs have a diploid chromosome number of 78. How many chromosomes do their gametes have?
   - a. 39
   - b. 78
   - c. 156
   - d. 234

9. _____ contributes to variation in traits among the offspring of sexual reproducers.
   - a. Crossing over
   - b. Random attachment of chromosomes to spindle poles
   - c. Fertilization
   - d. both a and b
   - e. all are factors

10. The cell in the diagram to the *right* is in anaphase I, not anaphase II. I know this because _____ .

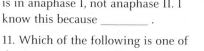

11. Which of the following is one of the very important differences between mitosis and meiosis?
   - a. Chromosomes align midway between spindle poles only in meiosis.
   - b. Homologous chromosomes pair up only in meiosis.
   - c. DNA is replicated only in mitosis.
   - d. Sister chromatids separate only in meiosis.
   - e. Interphase occurs only in mitosis.

12. Match each term with its description.
   - ___ interphase
   - ___ metaphase I
   - ___ allele
   - ___ sporophyte
   - ___ gamete
   - a. different molecular form of a gene
   - b. may be none between meiosis I and meiosis II
   - c. all chromosomes are aligned at spindle equator
   - d. haploid
   - e. does not occur in animals

Additional questions are available on **CENGAGENOW**.

Critical Thinking

1. Explain why you can predict that meiosis tends to give rise to greater genetic diversity among offspring in fewer generations than asexual reproduction does.

2. Make a simple sketch of meiosis in a cell with a diploid chromosome number of 4. Now try it when the chromo-some number is 3.

3. The diploid chromosome number for the body cells of a frog is 26. What would that number be after three gen-erations if meiosis did not occur before gamete formation?

Animations and Interactions on **CENGAGENOW**:
❯ Reproductive organs; Meiosis step-by-step; Crossing over; Random segregation of chromosomes in meiosis; Egg and sperm formation.

Lindsay, 22

Savannah, 19

Cody, 23

Brandon, 18

Ben, 23

Jeff, 21

‹ **Links to Earlier Concepts**

Before starting this chapter, you may want to review what you know about traits (Section 1.5), chromosomes (8.2), DNA (8.3), mutation (8.6), genes (9.2), and alleles (12.2). You will revisit probability and sampling error (1.8), laws of nature (1.9), protein structure (3.5), pigments (6.2), and clones (8.7). As you read, refer back to the the stages of meiosis (12.3). You will consider the roles that crossing over and the segregation of chromosomes (12.4) into gametes (12.5) play in inheritance.

Key Concepts

Where Modern Genetics Started
Gregor Mendel gathered evidence of the genetic basis of inheritance. His meticulous work gave him clues that heritable traits are specified in units. The units, which are distributed into gametes in predictable patterns, were later identified as genes.

Insights From Monohybrid Crosses
During meiosis, pairs of genes on homologous chromosomes separate and end up in different gametes. Inheritance patterns of alleles associated with different forms of a trait can be used as evidence of such gene segregation.

13 Observing Patterns in Inherited Traits

13.1 Menacing Mucus

In 1988, researchers discovered a gene that, when mutated, causes cystic fibrosis (CF). Cystic fibrosis is the most common fatal genetic disorder in the United States. The gene in question, *CFTR*, encodes a protein that moves chloride ions out of epithelial cells. Sheets of these cells line the passageways and ducts of the lungs, liver, pancreas, intestines, reproductive system, and skin. When the CFTR protein pumps chloride ions out of these cells, water follows the ions by osmosis. This two-step process maintains a thin film of water on the surface of the epithelial sheets. Mucus slides easily over the wet sheets of cells.

The most common mutation in CF is a deletion of three base pairs—one codon that specifies the 508th amino acid of the CFTR protein, a phenylalanine. This deletion, which is called Δ*F508*, disrupts membrane trafficking of CFTR so that newly assembled polypeptides are stranded in the endoplasmic reticulum. The protein itself can function properly, but it never reaches the cell surface to do its job.

One outcome is that the transport of chloride ions out of epithelial cells is disrupted. If not enough chloride ions leave the cells, not enough water leaves them either, so the surfaces of epithelial cell sheets are not as wet as they should be. Mucus that normally slips and slides through the body's tubes sticks to them instead. Thick globs of mucus accumulate and clog passageways and ducts throughout the body. Breathing becomes difficult as the mucus obstructs the smaller airways of the lungs.

The CFTR protein also functions as a receptor that alerts the body to the presence of bacteria. Bacteria bind to CFTR. The binding triggers endocytosis, which speeds the immune system's defensive responses. Without the CFTR protein on the surface of epithelial cell linings, disease-causing bacteria that enter the ducts and passageways of the body can persist there. Thus, chronic bacterial infections of the intestine and lungs are hallmarks of cystic fibrosis.

Daily routines of posture changes and thumps on the chest and back help clear the lungs of some of the thick mucus, and

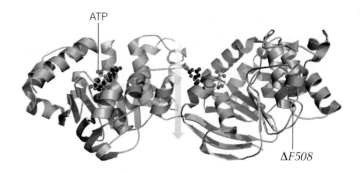

Figure 13.1 Cystic fibrosis. *Opposite*, a few of the many young victims of cystic fibrosis, which occurs most often in people of northern European ancestry. At least one young person dies every day in the United States from complications of this disease.

Above, model of the CFTR protein. The parts shown here are ATP-driven motors that widen or narrow a channel (*gray* arrow) across the plasma membrane. The tiny part of the protein that is deleted in most people with cystic fibrosis is shown on the ribbon in *green*.

antibiotics help control infections, but there is no cure. Even with a lung transplant, most cystic fibrosis patients live no longer than thirty years, at which time their tormented lungs usually fail (Figure 13.1).

More than 10 million people carry the Δ*F508* mutation in one of their two copies of the *CFTR* gene, but most of them do not realize it because they have no symptoms. Cystic fibrosis only occurs when a person inherits two mutated genes, one from each parent. This unlucky event occurs in about 1 of 3,300 births worldwide. Why is it that people with one copy of a mutated *CFTR* gene are healthy, but people with two copies are ill with cystic fibrosis? Why is the mutation that causes cystic fibrosis so common if its effects are so devastating? You will begin to find the answers to such questions in this chapter, in which we introduce principles of inheritance—how new individuals are put together in the image of their parents.

Insights From Dihybrid Crosses Pairs of genes on different chromosomes are typically distributed into gametes independently of how other gene pairs are distributed. Breeding experiments with alternative forms of two unrelated traits can be used as evidence of such independent assortment.

Variations on Mendel's Theme Not all traits appear in Mendelian inheritance patterns. An allele may be partly dominant over a nonidentical partner, or codominant with it. Multiple genes may influence a trait; some genes influence many traits. The environment also influences gene expression.

❭ Some traits are inherited in predictable patterns. Such patterns offer information about the alleles that influence those traits.

❬ Links to Traits 1.5, Chromosomes 8.2, Discovery of DNA's function 8.3, Mutations 8.6, Genes and gene expression 9.2, Mutations 9.6, Alleles 12.2, Gamete formation and fertilization 12.5

By the nineteenth century, most people had an idea that two parents contribute hereditary material to their offspring, but no one knew what that material was. Some thought that hereditary material must be some type of fluid, with fluids from both parents blending at fertilization like milk into coffee. However, the idea of "blending inheritance" failed to explain what people could see with their own eyes. Children sometimes had traits such as freckles that did not appear in either parent. A cross between a black horse and a white one did not produce only gray offspring.

The naturalist Charles Darwin did not accept the idea of blending inheritance, but he could not come up with an alternative even though inheritance was central to his theory of natural selection. Darwin saw that forms of traits often vary among individuals in a population. He realized that if some forms of traits help individuals survive and reproduce, then those forms would tend to appear more frequently in a population over generations (we return to Darwin and his theory of natural selection in Chapter 16). Despite these insights, however, neither Darwin nor anyone else at the time knew that hereditary information (DNA) is divided into discrete units (genes), an insight that is critical to understanding how heredity really works.

Even before Darwin presented his theory of natural selection, someone had been gathering evidence that would support it. Gregor Mendel, an Austrian monk (*left*), had been carefully breeding thousands of pea plants. By meticulously documenting how certain traits are passed from plant to plant, generation after generation, Mendel had been collecting evidence of how inheritance works.

Mendel's Experimental Approach

Mendel studied variation in the traits of the garden pea, *Pisum sativum*. This plant is naturally self-fertilizing, which means that its flowers produce male gametes (in pollen) and female gametes (in carpels) that form viable embryos when they meet up.

In order to study inheritance, Mendel had to breed particular individuals together, then observe and document the traits of their offspring. Control over the reproduction of an individual pea plant begins with preventing it from self-fertilizing. Mendel did this by removing a flower's pollen-bearing anthers, then brushing its carpel with pollen from another plant. When the plant set seed, Mendel collected the seeds, planted them, and recorded the traits of the new plants that grew. Figure 13.2 shows an example of this process.

Many of Mendel's experiments involved plants that "bred true" for a particular trait. Breeding true for a trait means that, mutations aside, all offspring have the same form of the trait as the parent(s), generation after generation. For example, all offspring of pea plants that breed true for white flowers also have white flowers.

Breeders cross-fertilize plants when they transfer pollen among individuals that have different traits. As you will see in the next section, Mendel discovered that the traits of the offspring of such cross-fertilized pea plants often appear in predictable patterns. Mendel's meticulous work tracking pea plant traits led him to conclude (correctly) that hereditary information is passed from one generation to the next in discrete units.

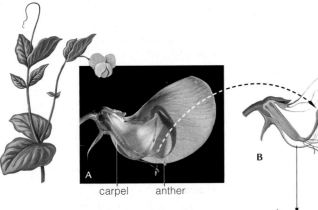

carpel   anther

A Garden pea flower, cut in half. Male gametes form in pollen grains produced by the anthers, and female gametes form in carpels. Experimenters can control the transfer of hereditary material from one flower to another by snipping off a flower's anthers (to prevent the flower from self-fertilizing), and then brushing pollen from another flower onto its carpel.

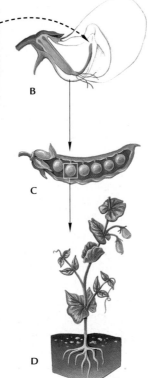

B In this example, pollen from a plant with purple flowers is brushed onto the carpel of a white-flowered plant.

C Later, seeds develop inside pods of the cross-fertilized plant. An embryo in each seed develops into a mature pea plant.

D Every plant that arises from this cross has purple flowers. Predictable patterns such as this offer evidence of how inheritance works.

Figure 13.2 Animated Breeding garden pea plants (*Pisum sativum*), which can self-fertilize or cross-fertilize.

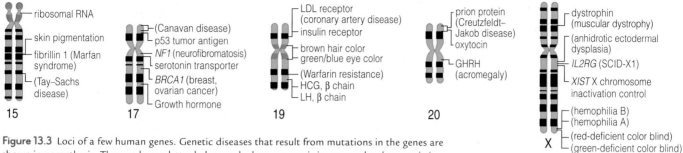

Figure 13.3 Loci of a few human genes. Genetic diseases that result from mutations in the genes are shown in parenthesis. The number or letter below each chromosome is its name; the characteristic banding patterns appear after staining. A map of all 23 human chromosomes is in Appendix VII.

Inheritance in Modern Terms

DNA was not proven to be hereditary material until the 1950s (Section 8.3), but Mendel discovered its units, which we now call genes, almost a century before then. Today, we know that individuals of a species share certain traits because their chromosomes carry the same genes. Offspring tend to look like their parents because they inherited their parents' genes.

The DNA sequence of each gene occurs at a specific location, or **locus** (plural, loci), on a particular chromosome (Figure 13.3). The somatic cells of humans and other animals are diploid, so they have pairs of genes, on pairs of homologous chromosomes (Figure 13.4). In most cases, both genes of a pair are expressed (Section 9.2).

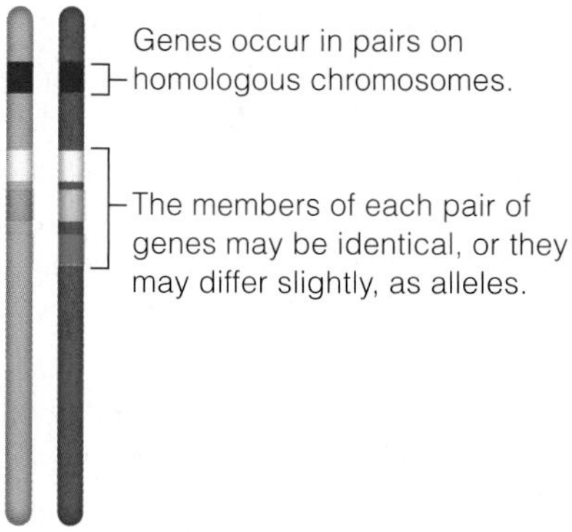

Genes occur in pairs on homologous chromosomes.

The members of each pair of genes may be identical, or they may differ slightly, as alleles.

Figure 13.4 Animated Genes on chromosomes. Any pair of genes on homologous chromosomes may vary as alleles. Different alleles may result in different versions of a trait.

dominant Refers to an allele that masks the effect of a recessive allele paired with it.
genotype The particular set of alleles carried by an individual.
heterozygous Having two different alleles of a gene.
homozygous Having identical alleles of a gene.
hybrid The offspring of a cross between two individuals that breed true for different forms of a trait; a heterozygous individual.
locus Location of a gene on a chromosome.
phenotype An individual's observable traits.
recessive Refers to an allele with an effect that is masked by a dominant allele on the homologous chromosome.

The two genes of a pair may be identical, or they may be slightly different. Alternative forms of a gene are called alleles (Section 12.2). Organisms that breed true for a specific trait probably have identical alleles governing that trait. An individual with identical alleles of a gene is said to be **homozygous** for the allele. The particular set of alleles that an individual carries is called **genotype**.

Alleles are the major source of variation in a trait. New alleles arise by mutation (Section 8.6). A mutation may cause a trait to change, as when a gene that causes flowers to be purple mutates so the resulting flowers are white. Flower color is an example of **phenotype**, which refers to an individual's observable traits. Any mutated gene is an allele, whether or not it affects phenotype.

The offspring of a cross, or mating, between individuals that breed true for different forms of a trait are **hybrids**. Hybrids carry different alleles of a gene, so they are said to be **heterozygous** for the alleles (*hetero–*, mixed). In many cases, the effect of one allele influences the effect of the other, and the outcome of this interaction is visible in the hybrid phenotype. An allele is **dominant** when its effect masks that of a **recessive** allele paired with it. Usually, italic capital letters such as *A* signify dominant alleles, and lowercase italic letters such as *a* signify recessive ones. Thus, a homozygous dominant individual carries a pair of dominant alleles (*AA*). A homozygous recessive individual carries a pair of recessive alleles (*aa*). A heterozygous, or hybrid, individual carries a pair of nonidentical alleles (*Aa*).

Take-Home Message How do alleles contribute to traits?

> Mendel discovered the role of genes in inheritance by carefully breeding pea plants and tracking observable traits of their offspring.

> Genotype refers to the particular set of alleles carried by an individual's somatic cells. Phenotype refers to the individual's set of observable traits.

> Genotype is the basis of phenotype.

> Homozygous individuals have identical alleles. Heterozygous individuals have nonidentical alleles.

> Dominant alleles mask the effects of recessive ones in heterozygous individuals.

> Pairs of genes on homologous chromosomes separate during meiosis, so they end up in different gametes.
< Links to Probability and sampling error 1.8, Laws of nature 1.9, Meiosis 12.3, Chromosome segregation 12.4

Mendel crossed plants that bred true for purple flowers with plants that bred true for white flowers. All of the offspring of these crosses had purple flowers, but Mendel did not know why this pattern occurred. We now understand that one gene governs purple flower color in pea plants. The allele that specifies purple (let's call it *P*) is dominant over the allele that specifies white (*p*). Thus, a pea plant with two *P* alleles (*PP*) has purple flowers, and one with two *p* alleles (*pp*) has white flowers.

When homologous chromosomes separate during meiosis, the gene pairs on those chromosomes separate too. Each gamete that forms carries only one of the two genes of a pair (Figure 13.5). Thus, plants homozygous for the dominant allele (*PP*) can only make gametes that carry the dominant allele *P* ❶. Plants homozygous for the recessive allele (*pp*) can only make gametes that carry the recessive allele *p* ❷. If these homozygous plants are crossed (*PP* × *pp*), only one outcome is possible: A gamete carrying a *P* allele meets up with a gamete carrying a *p* allele ❸. All of the offspring of this cross have one of each allele, so their genotype is *Pp*. A grid called a **Punnett square** makes it easier to predict the genetic outcomes of crosses ❹. Because all of the offspring of this cross carry the dominant allele *P*, all have purple flowers.

This pattern is so predictable that it can be used as evidence of a dominance relationship between alleles. Breeding experiments use such patterns to reveal genotype. In a **testcross**, an individual that has a dominant trait (but an unknown genotype) is crossed with an individual known to be homozygous recessive. The pattern of traits among the offspring of the cross can reveal whether the tested individual is heterozygous or homozygous.

For example, we may do a testcross between a purple-flowered pea plant (which could have a genotype of either *PP* or *Pp*) and a white-flowered pea plant (*pp*). If all of the offspring of this cross had purple flowers, we would know that the genotype of the purple-flowered parent was *PP*.

A **monohybrid cross** is a breeding experiment that checks for a dominance relationship between the alleles of a single gene. Individuals that are identically heterozygous for one gene—(*Pp*) for example—are bred together or self-fertilized. The frequency at which the two traits appear among the offspring of this cross may show that one of the alleles is dominant over the other.

To produce identically heterozygous individuals for a monohybrid cross, we would start with two individuals that breed true for two different forms of a trait (Figure 13.6A). In pea plants, purple or white flowers is one example of a trait with two distinct forms, but there are many others. Mendel investigated seven of them: stem length (tall and short), seed color (yellow and green), pod texture

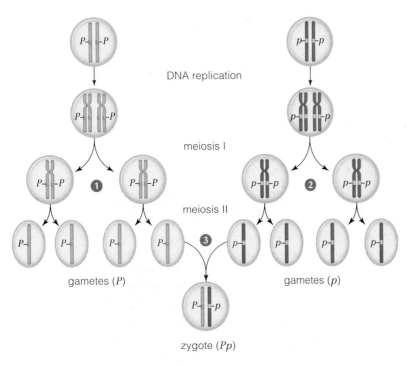

DNA replication

meiosis I

meiosis II

gametes (*P*)   gametes (*p*)

zygote (*Pp*)

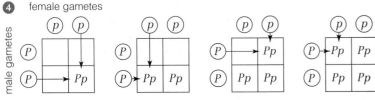

❹ female gametes

male gametes

Figure 13.5 Gene segregation. Homologous chromosomes separate during meiosis, so the pairs of genes they carry separate too. Each of the resulting gametes carries only one of the two members of each gene pair. For clarity, we show only one set of homologous chromosomes.

❶ All gametes made by a parent homozygous for a dominant allele carry that allele.

❷ All gametes made by a parent homozygous for a recessive allele carry that allele.

❸ If these two parents are crossed, the union of any of their gametes at fertilization produces a zygote with both alleles. All offspring of this cross will be heterozygous.

❹ This outcome is easy to see with a Punnett square. Parental gametes are listed in circles on the top and left sides of a grid. Each square is filled in with the combination of alleles that would result if the gametes in the corresponding row and column met up.

law of segregation The two members of each pair of genes on homologous chromosomes end up in different gametes during meiosis.

monohybrid cross Breeding experiment in which individuals identically heterozygous for one gene are crossed. The frequency of traits among the offspring offers information about the dominance relationship between the alleles.

Punnett square Diagram used to predict the genetic and phenotypic outcome of a cross.

testcross Method of determining genotype in which an individual of unknown genotype is crossed with one that is known to be homozygous recessive.

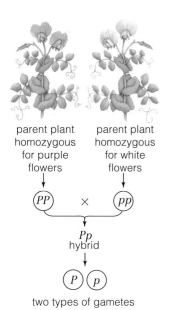

parent plant
homozygous
for purple
flowers

parent plant
homozygous
for white
flowers

(PP) × (pp)

Pp
hybrid

(P)(p)

two types of gametes

A All of the F₁ offspring of a cross between two plants that breed true for different forms of a trait are identically heterozygous. These offspring make two types of gametes: *P* and *p*.

Figure 13.6 Animated Example of a monohybrid cross. **>> Figure It Out** In this example, how many possible genotypes are there in the F₂ generation?

Answer: Three: PP, Pp, and pp

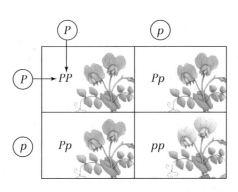

B A cross between the F₁ offspring is a monohybrid cross. The phenotype ratio in F₂ offspring in this example is 3:1 (3 purple to 1 white).

Table 13.1	Mendel's Seven Pea Plant Traits	
Trait	**Dominant Form**	**Recessive Form**
Seed Shape	Round	Wrinkled
Seed Color	Yellow	Green
Pod Texture	Smooth	Wrinkled
Pod Color	Green	Yellow
Flower Color	Purple	White
Flower Position	Along Stem	At Tip
Stem Length	Tall	Short

(smooth and wrinkled), and so on (Table 13.1). A cross between the two true-breeding individuals yields hybrid offspring: ones that are identically heterozygous for the alleles that govern the trait. When these F₁ (first generation) hybrids are crossed, the frequency at which the two traits appear in the F₂ (second generation) offspring offers information about dominance relationships. F is an abbreviation for filial, which means offspring.

A cross between two purple-flowered heterozygous individuals (*Pp*) offers an example. Each of these plants can make two types of gametes: ones that carry a *P* allele, and ones that carry a *p* allele. So, in a monohybrid cross between two *Pp* plants (*Pp* × *Pp*), the two types of gametes can meet up in four possible ways at fertilization:

Possible Event	Probable Outcome
Sperm *P* meets egg *P* ⟶	zygote genotype is *PP*
Sperm *P* meets egg *p* ⟶	zygote genotype is *Pp*
Sperm *p* meets egg *P* ⟶	zygote genotype is *Pp*
Sperm *p* meets egg *p* ⟶	zygote genotype is *pp*

Three out of four possible outcomes of this cross include at least one copy of the dominant allele *P*. Each time fertilization occurs, there are 3 chances in 4 that the resulting offspring will inherit a *P* allele, and have purple flowers. There is 1 chance in 4 that it will inherit two recessive *p* alleles, and have white flowers. Thus, the probability that a particular offspring of this cross will have purple or white flowers is 3 purple to 1 white, which we represent as a ratio of 3:1 (Figure 13.6B).

If the probability of one individual inheriting a particular genotype is difficult to imagine, think about probability in terms of the phenotypes of many offspring. In this example, there will be roughly three purple-flowered plants for every white-flowered one. The 3:1 pattern is an indication that purple and white flower color are specified by alleles with a clear dominant–recessive relationship: purple is dominant, and white is recessive.

The phenotype ratios in the F₂ offspring of Mendel's monohybrid crosses were all close to 3:1. These results became the basis of his **law of segregation**, which we state here in modern terms: Diploid cells carry pairs of genes, on pairs of homologous chromosomes. The two genes of each pair are separated from each other during meiosis, so they end up in different gametes.

Take-Home Message What is Mendel's law of segregation?

> Diploid cells carry pairs of genes, on pairs of homologous chromosomes. The two genes of each pair are separated from each other during meiosis, so they end up in different gametes.

> Mendel discovered patterns of inheritance in pea plants by tracking the results of many monohybrid crosses.

❯ Many gene pairs tend to sort into gametes independently of one another.

❮ Links to Meiosis 12.3, Crossing over and chromosome segregation 12.4

A monohybrid cross allows us to track alleles of one gene pair. What about alleles of two gene pairs? How two gene pairs get sorted into gametes depends partly on whether the two genes are on the same chromosome. When homologous chromosomes separate during meiosis, either one of the pair can end up in a particular nucleus. Thus, gene pairs on one chromosome get sorted into gametes independently of gene pairs on other chromosomes (Figure 13.7).

Punnett squares are particularly useful when predicting inheritance patterns of two or more genes simultaneously, such as with a dihybrid cross. A **dihybrid cross** tests for dominance relationships between alleles of two genes. In a typical dihybrid cross, individuals identically heterozygous for alleles of two genes (dihybrids) are crossed, and the traits of the offspring are observed.

To make a dihybrid cross, we would start with individuals that breed true for two different traits. Let's use genes for flower color (*P*, purple; *p*, white) and height (*T*, tall; *t*, short), and assume that these genes occur on separate chromosomes. Figure 13.8 shows a dihybrid cross starting with one parent plant that breeds true for purple flowers and tall stems (*PPTT*), and another that breeds true for white flowers and short stems (*pptt*). Each homo-

zygous plant makes only one type of gamete ❶. So, all of the offspring from a cross between these parent plants (*PPTT* × *pptt*), will be dihybrids (*PpTt*) and have purple flowers and tall stems ❷.

Four combinations of alleles are possible in the gametes of *PpTt* dihybrids ❸. If two *PpTt* plants are crossed (a dihybrid cross, *PpTt* × *PpTt*), the four types of gametes can combine in sixteen possible ways at fertilization ❹. Those sixteen genotypes would result in four different phenotypes. Nine would be tall with purple flowers, three would be short with purple flowers, three would be tall with white flowers, and one would be short with white flowers. Thus, the ratio of these phenotypes is 9:3:3:1.

Mendel discovered the 9:3:3:1 ratio of phenotypes among the offspring of his dihybrid crosses, although he had no idea what it meant. He could only say that "units" specifying one trait (such as flower color) are inherited independently of "units" specifying other traits (such as plant height). In time, Mendel's hypothesis became known as the **law of independent assortment**, which we state here in modern terms: Gene pairs are sorted into gametes independently of other gene pairs.

Mendel published his results in 1866, but apparently his work was read by few and understood by no one. In 1871 he became abbot of his monastery, and his pioneering experiments ended. He died in 1884, never to know that they would be the starting point for modern genetics.

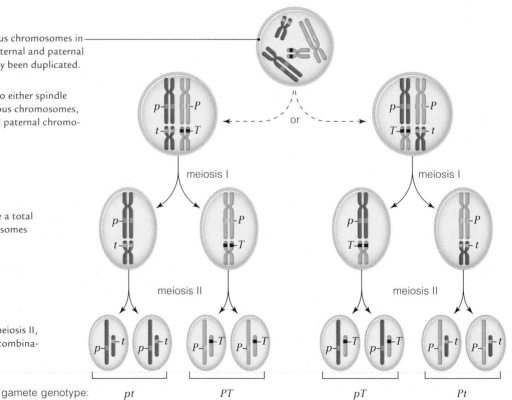

A This example shows just two pairs of homologous chromosomes in the nucleus of a diploid (2*n*) reproductive cell. Maternal and paternal chromosomes, shown in *pink* and *blue*, have already been duplicated.

B Either chromosome of a pair may get attached to either spindle pole during meiosis I. With two pairs of homologous chromosomes, there are two different ways that the maternal and paternal chromosomes can get attached to opposite spindle poles.

C Two nuclei form with each scenario, so there are a total of four possible combinations of parental chromosomes in the nuclei that form after meiosis I.

D Thus, when sister chromatids separate during meiosis II, the gametes that result have one of four possible combinations of maternal and paternal chromosomes.

meiosis I

meiosis II

gamete genotype: *pt* *PT* *pT* *Pt*

Figure 13.7 Independent assortment.

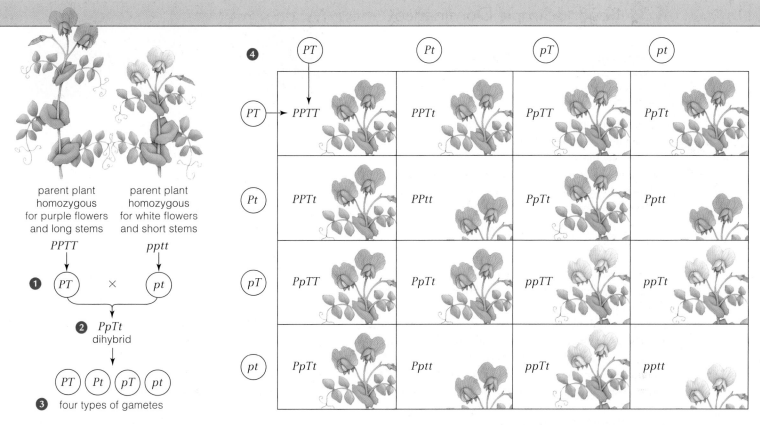

Figure 13.8 Animated A dihybrid cross between plants that differ in flower color and plant height. *P* and *p* stand for dominant and recessive alleles for flower color. *T* and *t* stand for dominant and recessive alleles for height.

❶ Meiosis in a homozygous individual results in one type of gamete.

❷ A cross between two homozygous individuals yields offspring with one possible genotype. All offspring that form in this example are dihybrids (heterozygous for two genes) with purple flowers and tall stems.

❸ Meiosis in dihybrid individuals results in four kinds of gametes.

❹ If two dihybrid individuals are crossed (a dihybrid cross), the four types of gametes can meet up in 16 possible ways. Out of 16 possible genotypes of the offspring, 9 will result in plants that are purple-flowered and tall; 3, purple-flowered and short; 3, white-flowered and tall; and 1, white-flowered and short. Thus, the ratio of phenotypes in a dihybrid cross is 9:3:3:1.

❯❯ **Figure It Out** What do the flowers inside the boxes represent?

Answer: Phenotypes of the F₂ offspring

The Contribution of Crossovers

It makes sense that gene pairs on different chromosomes would assort independently into gametes, but what about gene pairs on the same chromosome? Mendel studied seven genes in pea plants, which have seven chromosomes. Was he lucky enough to choose one gene on each of those chromosomes? As it turns out, some of the genes Mendel studied *are* on the same chromosome. The genes are far enough apart that crossing over occurs between them very frequently—so frequently that they tend to assort into gametes independently, just as if they were on different chromosomes.

By contrast, genes that are very close together on a chromosome do not assort independently, because crossing over does not happen very often between them. Such genes are said to be linked. Alleles of some linked genes stay together during meiosis more frequently than others, an effect due to the relative distance between the genes. Genes that are closer together get separated less frequently by crossovers. Thus, the closer together any two genes are on a chromosome, the more likely gametes

will be to receive parental combinations of alleles of those genes. Genes are said to be tightly linked if the distance between them is relatively small.

All of the genes on a chromosome are called a **linkage group**. Peas have 7 different chromosomes, so they have 7 linkage groups. Humans have 23 different chromosomes, so they have 23 linkage groups.

---

dihybrid cross Breeding experiment in which individuals identically heterozygous for two genes are crossed. The frequency of traits among the offspring offers information about the dominance relationships between the paired alleles.

law of independent assortment During meiosis, members of a pair of genes on homologous chromosomes get distributed into gametes independently of other gene pairs.

linkage group All genes on a chromosome.

Take-Home Message **What is Mendel's law of independent assortment?**

❯ Each member of a pair of genes on homologous chromosomes tends to be distributed into gametes independently of how other genes are distributed during meiosis.

> Mendel focused on traits that are based on clearly domi-
nant and recessive alleles. However, the expression patterns
of genes for some traits are not as straightforward.
< Links to Fibrous proteins 3.5, Pigments 6.2

The inheritance patterns in the last two sections offer
examples of simple dominance, in which a dominant
allele fully masks the expression of a recessive one. In
many other cases, both genes of a pair are expressed at
the same time. The "one gene equals one trait" equation
does not always apply, either. Several gene products may
influence the same trait, or the expression of a single gene
may influence multiple traits.

Codominance With **codominance**, two nonidentical
alleles of a gene are both fully expressed in heterozygotes,
so neither is dominant or recessive. Codominance may
occur in **multiple allele systems**, in which three or more
alleles of a gene persist among individuals of a population. The three alleles of the *ABO* gene are an example.
An enzyme encoded by the *ABO* gene modifies a carbohydrate on the surface of human red blood cell membranes.
The *A* and *B* alleles encode slightly different versions of

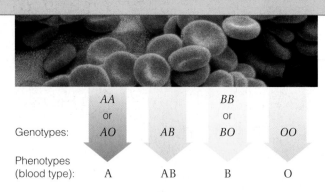

Genotypes:	*AA* or *AO*	*AB*	*BB* or *BO*	*OO*
Phenotypes (blood type):	A	AB	B	O

Figure 13.9 Animated Combinations of alleles that are the basis
of human blood type.

the enzyme, which in turn modify the carbohydrate differently. The *O* allele has a mutation that prevents its
enzyme product from becoming active at all.

The two alleles you carry for the *ABO* gene determine
the form of the carbohydrate on your blood cells, and that
carbohydrate is the basis of your blood type (Figure 13.9).
The *A* and the *B* allele are codominant when paired. If
your genotype is *AB*, then you have both versions of the
enzyme, and your blood is type AB. The *O* allele is recessive when paired with either the *A* or *B* allele. If your genotype is *AA* or *AO*, your blood is type A. If your genotype
is *BB* or *BO*, it is type B. If you are *OO*, it is type O.

Receiving incompatible blood cells in a transfusion
is very dangerous, because the immune system usually
attacks red blood cells bearing molecules that do not occur
in one's own body. The attack can cause the blood cells to
clump or burst, a transfusion reaction with potentially fatal
consequences. People with type O blood can donate blood
to anyone else, so they are called universal donors. However, they can receive transfusions of type O blood only.
People with type AB blood can receive a transfusion of any
blood type, so they are called universal recipients.

Incomplete Dominance With **incomplete dominance**,
one allele of a gene pair is not fully dominant over the
other, so the heterozygous phenotype is between the two
homozygous phenotypes. A gene that influences flower
color in snapdragon plants is an example. A cell has one
copy of this gene on each homologous chromosome,
and both copies are expressed. One allele (*R*) encodes an
enzyme that makes a red pigment. The enzyme encoded
by a mutated allele (*r*) cannot make any pigment. Plants
homozygous for the *R* allele (*RR*) make a lot of red pigment, so they have red flowers. Plants homozygous for
the *r* allele (*rr*) do not make any pigment at all, so their
flowers are white. Heterozygous plants (*Rr*) make only
enough red pigment to color their flowers pink (Figure
13.10A). A cross between two pink-flowered heterozygous
plants yields red-, pink-, and white-flowered offspring in a
1:2:1 ratio (Figure 13.10B).

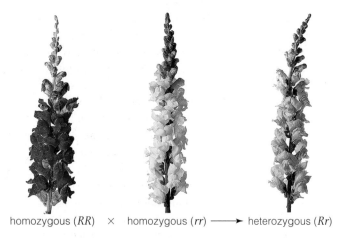

homozygous (*RR*) × homozygous (*rr*) ⟶ heterozygous (*Rr*)

A Cross a red-flowered with a white-flowered plant,
and all of the offspring will be pink heterozygotes.

B If two of the pink heterozygotes are crossed, the phenotypes of the resulting offspring
will occur in a 1:2:1 ratio.

	R	*r*
R	*RR*	*Rr*
r	*Rr*	*rr*

Figure 13.10 Animated Incomplete dominance in heterozygous (*pink*) snap-
dragons. An allele that affects red pigment is paired with a "white" allele.

>> **Figure It Out** Is the experiment in **B** a monohybrid or dihybrid cross?

Answer: A monohybrid cross

	Ⓔ⒝ EB	Ⓔ⒝ Eb	Ⓔ⒝ eB	Ⓔ⒝ eb
Ⓔ⒝ EB	EEBB	EEBb	EeBB	EeBb
Ⓔ⒝ Eb	EEBb	EEbb	EeBb	Eebb
Ⓔ⒝ eB	EeBB	EeBb	eeBB	eeBb
Ⓔ⒝ eb	EeBb	Eebb	eeBb	eebb

Figure 13.11 Animated Epistasis in dogs. Epistatic interactions among products of two gene pairs affect coat color in Laborador retrievers. *Left*, all dogs with an *E* and *B* allele have black fur. Those with an *E* and two recessive *b* alleles have brown fur. All dogs homozygous for the recessive *e* allele have yellow fur. *Right*: black, chocolate, and yellow Laborador retrievers.

Epistasis Some traits are affected by multiple gene products, an effect called polygenic inheritance or **epistasis**. Human skin color, which is a result of interactions among several gene products, is an example. Similar genes affect Labrador retriever coat color, which can be black, yellow, or brown (Figure 13.11). One gene is involved in the synthesis of the pigment melanin. A dominant allele of the gene (*B*) specifies black fur, and its recessive partner (*b*) specifies brown fur. A dominant allele of a different gene (*E*) causes melanin to be deposited in fur, and its recessive partner (*e*) reduces melanin deposition. Thus, a dog that carries an *E* and a *B* allele has black fur. One that carries an *E* allele and is homozygous for the *b* allele has brown fur. A dog homozygous for the *e* allele has yellow fur regardless of its *B* or *b* alleles.

Figure 13.12 Animated Marfan syndrome. Rising basketball star Haris Charalambous died suddenly in 2006 when his aorta burst during warm-up exercises. He was 21.

Charalambous was very tall and lanky, with long arms and legs—traits that are valued in professional athletes such as basketball players. These traits are also associated with Marfan syndrome.

About 1 in 5,000 people are affected by Marfan syndrome worldwide. Like many of them, Charalambous did not realize he had the syndrome.

codominant Refers to two alleles that are both fully expressed in heterozygous individuals.
epistasis Effect in which a trait is influenced by the products of multiple genes.
incomplete dominance Condition in which one allele is not fully dominant over another, so the heterozygous phenotype is between the two homozygous phenotypes.
multiple allele system Gene for which three or more alleles persist in a population.
pleiotropic Refers to a gene whose product influences multiple traits.

Pleiotropy A **pleiotropic** gene is one that influences multiple traits. Mutations in such genes are associated with complex genetic disorders such as sickle-cell anemia (Section 9.6) and cystic fibrosis. For example, thickened mucus in cystic fibrosis patients affects the entire body, not just the respiratory tract. The mucus clogs ducts that lead to the gut, which results in digestive problems. Male CF patients are typically infertile because their sperm flow is hampered by the thickened secretions.

Marfan syndrome is another example of a genetic disorder caused by mutation in a pleiotropic gene. In this case, the gene encodes fibrillin. Long fibers of this protein impart elasticity to tissues of the heart, skin, blood vessels, tendons, and other body parts. Mutations in the fibrillin gene result in tissues that form with defective fibrillin or none at all. The largest blood vessel leading from the heart, the aorta, is particularly affected. Muscle cells in the aorta's thick wall do not work very well, and the wall itself is not as elastic as it should be. The aorta expands under pressure, so the lack of elasticity eventually makes it thin and leaky. Calcium deposits accumulate inside. Inflamed, thinned, and weakened, the aorta can rupture abruptly during exercise.

Marfan syndrome is particularly difficult to diagnose. Affected people are often tall, thin, and loose-jointed, but there are plenty of tall, thin, loose-jointed people that do not have the syndrome. Symptoms may not be apparent, so many people die suddenly and early without ever knowing they had the disorder (Figure 13.12).

Take-Home Message Are all alleles clearly dominant or recessive?

> An allele may be fully dominant, incompletely dominant, or codominant with its partner on a homologous chromosome.

> In epistasis, two or more gene products influence a trait.

> The product of a pleiotropic gene influences two or more traits.

❯ Individuals of most species vary in some of their shared traits. Many traits show a continuous range of variation.
❮ Link to Genetically identical organisms 8.7

Most organic molecules are made in metabolic pathways that involve many enzymes. Genes encoding those enzymes can mutate in any number of ways, so their products may function within a spectrum of activity that ranges from excessive to not at all. Thus, the end product of a metabolic pathway can be produced within a range of concentration and activity. Environmental factors often add further variations on top of that. In the end, phenotype often results from complex interactions among gene products and the environment.

Continuous Variation The individuals of a species typically vary in many of their shared traits. Some of those traits appear in two or three distinct forms. Others occur in a range of small differences that is called **continuous variation**. The more genes and environmental factors that influence a trait, the more continuous is its variation.

How do we determine whether a trait varies continuously? First, we divide the total range of phenotypes into measurable categories, such as inches of height (Figure 13.13). The number of individuals that fall into each category gives the relative frequencies of phenotypes across

Figure 13.14 Example of environmental effects on animal phenotype. The color of the snowshoe hare's fur varies by season. In summer, the fur is brown (*left*); in winter, white (*right*). Both forms offer seasonally appropriate camouflage from predators.

our range of measurable values. Finally, we plot the data as a bar chart. A graph line around the top of the bars shows the distribution of values for the trait. If the line is a bell-shaped curve, or **bell curve**, the trait varies continuously.

Human eye color is another example of a trait that varies continuously. The colored part of the eye is the iris, a doughnut-shaped, pigmented structure. Iris color, like skin color, is the result of epistasis among gene products that make and distribute melanins. The more melanin deposited in the iris, the less light is reflected from it. Dark irises have dense melanin deposits that absorb almost all light, and reflect almost none. Melanin deposits are not as extensive in brown eyes, which reflect some light. Green and blue eyes have the least amount of melanin, so they reflect the greatest amount of light.

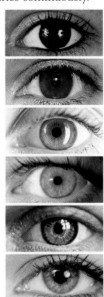

Environmental Effects on Phenotype Variations in traits are not always the result of differences in alleles. Environmental factors often affect gene expression, which in turn affects phenotype. For example, seasonal changes

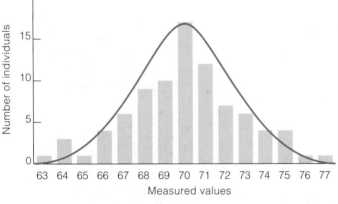

Figure 13.13 Animated Continuous variation in height among male biology students at the University of Florida. The students were divided into categories of one-inch increments in height and counted (*bottom*). A graph of the resulting data produces a bell-shaped curve (*top*), an indication that height varies continuously.

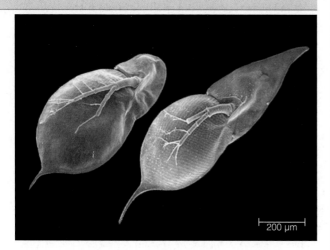

Figure 13.15 Environmental effects on the body form of *Daphnia*. In this animal, the form of the individual on the *left* develops in environments with few predators. The *Daphnia* on the *right* has a longer tail spine and a pointy head—traits that develop in response to chemicals emitted by predatory insects.

in temperature and the length of day affect the production of melanin and other pigments that color the skin and fur of many animals. These animals have different color phases in different seasons (Figure 13.14).

Factors such as the presence of predators can also influence phenotype. Consider *Daphnia pulex*, a microscopic freshwater relative of shrimp. Aquatic insects prey on these invertebrates. *Daphnia* living in ponds with few predators have rounded heads, but those in ponds with many predators have pointy heads (Figure 13.15). The predators emit chemicals that trigger the development of the pointier phenotype.

Yarrow plants offer another example of how environment influences phenotype. Yarrow is useful for genetics

A Plant grown at high elevation (3,060 meters above sea level)

B Plant grown at mid-elevation (1,400 meters above sea level)

C Plant grown at low elevation (30 meters above sea level)

Figure 13.16 Environmental effects on plant phenotype. Cuttings from the same yarrow plant (*Achillea millefolium*) grow to different heights at three different elevations.

bell curve Bell-shaped curve; typically results from graphing frequency versus distribution for a trait that varies continuously.
continuous variation In a population, a range of small differences in a shared trait.

Menacing Mucus (revisited)

The allele most commonly associated with cystic fibrosis, ΔF508, is eventually lethal in homozygous individuals, but not in those who are heterozygous. This allele is codominant with the normal one, so both copies of the gene are expressed in heterozygous individuals. Such individuals make enough of the normal CFTR protein to have normal chloride ion transport.

The ΔF508 allele is at least 50,000 years old and very common: up to 1 in 25 people carry it in some populations. Why has this allele persisted for so long and at such high frequency if it is dangerous? The ΔF508 allele may be the lesser of two evils because it offers heterozygous individuals a survival advantage against certain deadly infectious diseases. The unmutated CFTR protein triggers endocytosis when it binds to bacteria. This process is an essential part of the body's immune response to bacteria in the respiratory tract. However, the same function of CFTR allows bacteria to enter cells of the gastrointestinal tract, where they can be deadly. For example, endocytosis of *Salmonella typhi* (shown at *left*) into epithelial cells lining the gut results in a dangerous infection called typhoid fever. The ΔF508 mutation alters the CFTR protein so that bacteria can no longer be taken up by intestinal cells. People that carry it may have a decreased susceptibility to typhoid fever and other bacterial diseases that begin in the intestinal tract.

How Would You Vote? Tests for predisposition to genetic disorders are now available. Do you support legislation preventing discrimination based on the results of such tests? See CengageNow for details, then vote online (cengagenow.com).

experiments because it grows easily from cuttings. All cuttings of a plant have the same genotype, so experimenters know that genes are not the basis for any phenotypic differences among them. In one study, genetically identical yarrow plants had different phenotypes when grown at different altitudes (Figure 13.16).

The environment also affects human genes. One of our genes encodes a protein that transports serotonin across the membrane of brain cells. Serotonin lowers anxiety and depression during traumatic times. Some mutations in the serotonin transporter gene can reduce the ability to cope with stress. It is as if some of us are bicycling through life without an emotional helmet. Only when we crash does the mutation's phenotypic effect—depression—appear. Other human genes affect emotional state, but mutations in this one reduce our capacity to recover from emotional setbacks.

Take-Home Message **How does phenotype vary?**

> Some traits have a range of small differences, or continuous variation. The more genes and other factors that influence a trait, the more continuous the distribution of phenotype.

> Enzymes and other gene products control steps of most metabolic pathways. Environmental conditions can affect one or more steps, and thus contribute to variation in phenotypes.

Summary

Section 13.1 Cystic fibrosis, the most common fatal genetic disorder in the United States, is usually caused by a particular deletion in the *CFTR* gene. The allele that carries the mutation persists at high frequency despite its devastating effects. Only those homozygous for the allele have the disorder.

Section 13.2 Genetic information is passed to offspring in discrete units (genes). Each gene occurs at a particular **locus**, or location, on a chromosome. The somatic cells of diploid individuals have paired genes on their homologous chromosomes. Mutations give rise to alleles. Individuals with identical alleles are **homozygous** for the allele. Individuals that breed true for a trait are homozygous for alleles that affect the trait. **Heterozygous**, or **hybrid**, individuals carry two nonidentical alleles. A **dominant** allele masks the effect of a **recessive** allele on the homologous chromosome. **Genotype** (an individual's particular set of alleles) gives rise to **phenotype**, which refers to an individual's particular set of traits.

Section 13.3 Crossing individuals that breed true for two forms of a trait yields identically heterozygous F₁ offspring. A cross between such offspring is a **monohybrid cross**. The frequency at which the traits appear in the F₂ offspring can reveal dominance relationships among the alleles associated with those traits. Predictable patterns of inheritance can be used as evidence of genotype, as in **testcrosses**. **Punnett squares** are useful in determining the probability of the genotype and phenotype of the offspring of crosses. Mendel's monohybrid cross results led to his **law of segregation** (stated here in modern terms): During meiosis, the genes of each pair separate, so each gamete gets one or the other gene.

Section 13.4 Crossing individuals that breed true for two forms of two traits yields F₁ offspring identically heterozygous for alleles governing those traits. A cross between such offspring is a **dihybrid cross**. The frequency at which the traits appear in the F₂ offspring can reveal dominance relationships among the alleles associated with those traits. Mendel's dihybrid cross results led to his **law of independent assortment** (stated here in modern terms): During meiosis, gene pairs on homologous chromosomes tend to sort into gametes independently of other gene pairs. Crossovers can break up **linkage groups**.

Section 13.5 Not all alleles are clearly dominant or recessive. With **incomplete dominance**, an allele is not fully dominant over its partner on a homologous chromosome, so both are expressed. The combination of alleles gives rise to an intermediate phenotype.

Codominant alleles are both expressed at the same time in heterozygotes, as in **multiple allele systems** such as the one underlying ABO blood typing. In **epistasis**, interacting products of one or more genes often affect the same trait. A **pleiotropic** gene affects two or more traits.

Section 13.6 A trait that is influenced by the products of multiple genes often occurs in a range of small increments of phenotype called **continuous variation**. Continuous variation typically occurs as a **bell curve** in the range of values. An individual's phenotype may be influenced by environmental factors.

Self-Quiz   Answers in Appendix III

1. A heterozygous individual has a _____ for a trait being studied.
   - a. pair of identical alleles
   - b. pair of nonidentical alleles
   - c. haploid condition, in genetic terms

2. An organism's observable traits constitute its _____ .
   - a. phenotype   c. genotype
   - b. variation   d. pedigree

3. Filial means _____ .

4. The second-generation offspring of a cross between individuals who are homozygous for different alleles of a gene are called the _____ .
   - a. F₁ generation   c. hybrid generation
   - b. F₂ generation   d. none of the above

5. F₁ offspring of the cross *AA* × *aa* are _____ .
   - a. all *AA*   c. all *Aa*
   - b. all *aa*   d. 1/2 *AA* and 1/2 *aa*

6. Refer to question 4. Assuming complete dominance, the F₂ generation will show a phenotypic ratio of _____ .
   - a. 3:1   b. 9:1   c. 1:2:1   d. 9:3:3:1

7. A testcross is a way to determine _____ .
   - a. phenotype   b. genotype   c. both a and b

8. Assuming complete dominance, crosses between two dihybrid F₁ pea plants, which are offspring from a cross *AABB* × *aabb*, result in F₂ phenotype ratios of _____ .
   - a. 1:2:1   b. 3:1   c. 1:1:1:1   d. 9:3:3:1

9. The probability of a crossover occurring between two genes on the same chromosome _____ .
   - a. is unrelated to the distance between them
   - b. decreases with the distance between them
   - c. increases with the distance between them

10. A gene that affects three traits is _____ .
    - a. epistatic   c. pleiotropic
    - b. a multiple allele system   d. dominant

11. _____ alleles are both expressed.
    - a. Dominant   c. Pleiotropic
    - b. Codominant   d. Hybrid

12. A bell curve indicates _____ in a trait.

Data Analysis Activities

The Cystic Fibrosis Mutation and Typhoid Fever The ΔF508 mutation disables the receptor function of the CFTR protein, so it inhibits endocytosis of bacteria into epithelial cells. Endocytosis is an important part of the respiratory tract's immune defenses against common *Pseudomonas* bacteria, which is why *Pseudomonas* infections are a chronic problem in cystic fibrosis patients.

The ΔF508 mutation also inhibits endocytosis of *Salmonella typhi* into cells of the gastrointestinal tract, where internalization of this bacteria can cause typhoid fever. Typhoid fever is a common worldwide disease. Its symptoms include extreme fever and diarrhea, and the resulting dehydration causes delirium that may last several weeks. If untreated, it kills up to 30 percent of those infected. Around 600,000 people die annually from typhoid fever. Most of them are children.

In 1998, Gerald Pier and his colleagues compared the uptake of *S. typhi* by different types of epithelial cells: those homozygous for the normal allele, and those heterozygous for the ΔF508 mutation. (Cells homozygous for the mutation do not take up any *S. typhi* bacteria.) Some of their results are shown in Figure 13.17.

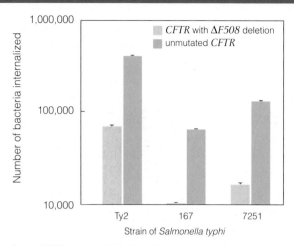

Figure 13.17 In epithelial cells, effect of the *CF* mutation on the uptake of three different strains of *Salmonella typhi* bacteria.

1. Regarding the Ty2 strain of *S. typhi*, about how many more bacteria were able to enter normal cells (those expressing unmutated *CFTR*) than cells expressing the gene with the ΔF508 deletion?
2. Which strain of bacteria entered normal epithelial cells most easily?
3. The ΔF508 deletion inhibited the entry of all three *S. typhi* strains into epithelial cells. Can you tell which strain was most inhibited?

---

13. Match the terms with the best description.

    ___dihybrid cross   a. *bb*
    ___monohybrid cross   b. *AABB × aabb*
    ___homozygous condition   c. *Aa*
    ___heterozygous condition   d. *Aa × Aa*

Additional questions are available on **CENGAGENOW**.

Genetics Problems   Answers in Appendix III

1. Mendel crossed a true-breeding pea plant with green pods and a true-breeding pea plant with yellow pods. All offspring had green pods. Which color is recessive?

2. Assuming that independent assortment occurs during meiosis, what type(s) of gametes will form in individuals with the following genotypes?
    a. *AABB*   b. *AaBB*   c. *Aabb*   d. *AaBb*

3. Refer to problem 2. Determine the frequencies of each genotype among offspring from the following matings:
    a. *AABB × aaBB*   c. *AaBb × aabb*
    b. *AaBB × AABb*   d. *AaBb × AaBb*

4. Suppose you identify a new gene in mice. One of its alleles specifies white fur, another specifies brown. You want to see if the two alleles interact in simple or incomplete dominance. What test would give you the answer?

5. Many genes are so vital for development that mutations in them are lethal. Even so, heterozygous individuals can perpetuate alleles that are lethal, such as the Manx (*M*)

allele in cats. Homozygous cats (*MM*) die before birth. In heterozygous cats (*Mm*), the spine develops abnormally, and these animals end up with no tail (*left*). What is the dominance relationship between the Manx allele and the normal allele? If two *Mm* cats mate, what is the probability that any one of their surviving kittens will be heterozygous (*Mm*)?

6. One gene encodes the second enzyme in a melanin-synthesizing pathway. An individual who is homozygous for a recessive allele of this gene cannot make or deposit melanin in body tissues. *Albinism*, the absence of melanin, is the result. Humans and many other organisms can have this phenotype. In the following situations, what are the probable genotypes of the father, the mother, and their children?

    a. Both parents have normal phenotypes; some but not all of their children have the albino phenotype.

    b. Both parents and all of their children have the albino phenotype.

    c. The mother is unaffected, the father has the albino phenotype, three children are unaffected and one child has the albino phenotype.

Additional genetics problems are available on **CENGAGENOW**.

Animations and Interactions on **CENGAGENOW**:
❯ Crossing garden pea plants; Genes on chromosomes; Monohybrid cross; Dihybrid cross; Codominance and ABO blood group; Incomplete dominance; Pleiotropy in Marfan syndrome; Epistasis in dogs; Continuous variation.

‹ Links to Earlier Concepts

Be sure you understand dominance relationships (Sections 13.2, 13.5, and 13.6), gene expression (9.2, 9.3), and mutations (9.6) before you start this chapter. You will use your knowledge of chromosomes (8.2), meiosis (12.3, 12.4), gametes (12.5), DNA (8.6) and sex determination (10.4). Sampling error (1.8), amino acids (3.5), lysosomes (4.8), the cell cortex (4.10), metabolic pathways (5.5), tyrosinase (5.4), pigments (6.2), and oncogenes (11.6) will turn up in context of genetic disorders.

Key Concepts

Tracking Traits in Humans

Inheritance patterns in humans are determined by following traits through generations of family trees. The types of traits followed in such studies include genetic abnormalities or syndromes associated with a genetic disorder.

Autosomal Inheritance

Many human traits can be traced to dominant or recessive alleles on autosomes. These alleles are inherited in characteristic patterns: dominant alleles tend to appear in every generation; recessive ones can skip generations.

14 Human Inheritance

The color of human skin begins with melanosomes. These skin cell organelles make two types of melanin pigments: one brownish-black; the other, reddish. Most people have about the same number of melanosomes in their skin cells. Variations in skin color occur because the kinds and amounts of melanins vary among people, as does the formation, transport, and distribution of the melanosomes.

Variations in skin color may have evolved as a balance between vitamin production and protection against harmful UV radiation. Dark skin would have been beneficial under the intense sunlight of the African savannas where humans first evolved. Melanin acts as a natural sunscreen because it prevents UV radiation in sunlight from breaking down folate, a vitamin essential for normal sperm formation and embryonic development. Children born to light-skinned women exposed to high levels of sunlight have a heightened risk of birth defects.

Early human groups that migrated to regions with colder climates were exposed to less sunlight. In these regions, lighter skin color would have been beneficial. Why? UV radiation stimulates skin cells to make a molecule the body converts to essential vitamin D. Where sunlight exposure is minimal, UV radiation damage is less of a risk than vitamin D deficiency, which has serious health consequences for developing fetuses and children. People with dark, UV-shielding skin have a high risk of this deficiency in regions where sunlight exposure is minimal.

Skin color, like most other human traits, has a genetic basis. More than 100 gene products are involved in the synthesis of melanin, and the formation and deposition of melanosomes. Mutations in at least some of these genes may have contributed to regional variations of human skin color. Consider a gene on chromosome 15, *SLC24A5*, that encodes a transport protein in melanosome membranes. Nearly all people of African, Native American, or east Asian descent carry the same allele of this gene. Between 6,000 and 10,000 years ago, a mutation gave rise to a different allele. The mutation, a single base–pair substitu-tion, changed the 111th amino acid of the transport protein from alanine to threonine. The change results in less melanin—and lighter skin color—than the original African allele does. Today, nearly all people of European descent carry this mutated allele.

A person of mixed ethnicity may make gametes that contain different combinations of alleles for dark and light skin. It is fairly rare that one of those gametes contains all of the alleles for dark skin, or all of the alleles for light skin, but it happens (Figure 14.1).

Skin color is only one of many human traits that vary as a result of single nucleotide mutations. The small scale of such changes offers a reminder that all of us share the genetic legacy of common ancestry.

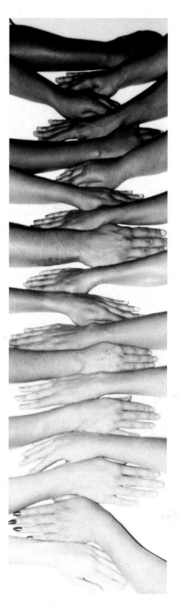

Figure 14.1 Variation in human skin color (*right*) begins with differences in alleles inherited from parents. *Opposite*, fraternal twin girls Kian and Remee, born in 2006. Both of the children's grand-mothers are of European descent, and have pale skin. Both of their grandfathers are of African descent, and have dark skin. The twins inherited different alleles of some of the genes that affect skin color from their mixed-race parents, who, given the appearance of their children, must be heterozygous for those alleles.

Sex–Linked Inheritance
The X chromosome holds about 10 percent of all human genes, so many traits are affected by alleles on this chromosome. Inheritance patterns of such X-linked alleles tend to differ between males and females.

Changes in Chromosome Structure and Number
A chromosome may undergo a large-scale, permanent change in its structure, or the number of autosomes or sex chromosomes may change. In humans, such changes usually result in a genetic disorder.

Genetic Testing
Genetic testing provides infor-mation about the risk of pass-ing a harmful allele to one's offspring. After conception, various methods of prenatal testing can reveal a genetic abnormality or dis-order in a fetus or embryo.

> Geneticists study inheritance patterns in humans by tracking genetic disorders and abnormalities through families.
> Charting genetic connections with pedigrees reveals inheritance patterns of certain traits.
< Links to Sampling error 1.8, Chromosomes 8.2, Dominance 13.2, Complex inheritance patterns 13.6

Some organisms, including pea plants and fruit flies, are ideal for genetic analysis. They have relatively few chromosomes, they reproduce quickly under controlled conditions, and breeding them poses few ethical problems. It does not take long to track a trait through many generations. Humans, however, are a different story. Unlike flies grown in laboratories, we humans live under variable

conditions, in different places, and we live as long as the geneticists who study us. Most of us select our own mates and reproduce if and when we want to. Our families tend to be on the small side, so sampling error (Section 1.8) is a major factor in human genetics studies.

Thus, inheritance patterns in humans are typically studied by tracking observable traits that crop up in families over many generations. Researchers graph such data as standardized charts of genetic connections called **pedigrees** (Figure 14.2). Pedigree analyses can reveal whether a trait is associated with a dominant or recessive allele, and whether the allele is on an autosome or a sex chromosome. Pedigree analysis also allows geneticists to determine the probability that a trait will recur in future generations of a family or a population.

Types of Genetic Variation

Some easily observed human traits follow Mendelian inheritance patterns. Like the flower color of pea plants, these traits are controlled by a single gene with two alleles, one dominant and the other recessive. For example, some people have earlobes that attach at their base, and others have earlobes that dangle free. The allele for unattached earlobes is dominant and the allele for attached earlobes is recessive. Similarly, the allele that specifies a cleft chin is dominant over the allele for a smooth chin, and the allele for dimples is dominant over that for no dimples. Someone who is homozygous for two recessive alleles of the *MC1R* gene makes the reddish kind of melanin but not the brownish-black kind, so this person has red hair.

Single genes on autosomes or sex chromosomes also govern more than 6,000 genetic abnormalities and disorders. Table 14.1 lists a few examples. A genetic abnormality

---

pedigree Chart showing the pattern of inheritance of a trait through generations in a family.

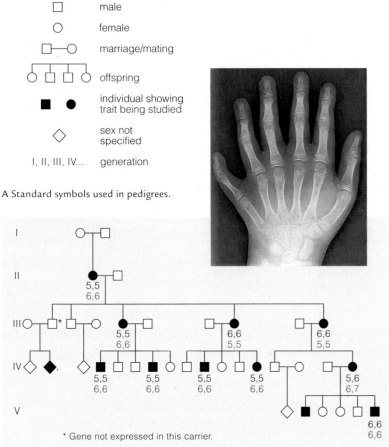

male □

female ○

marriage/mating □—○

offspring

individual showing trait being studied ■ ●

sex not specified ◇

I, II, III, IV... generation

A Standard symbols used in pedigrees.

B A pedigree for polydactyly, which is characterized by extra fingers, toes, or both. The *black* numbers signify the number of fingers on each hand; the *red* numbers signify the number of toes on each foot. Though it occurs on its own, polydactyly is also one of several symptoms of Ellis–van Creveld syndrome.

* Gene not expressed in this carrier.

C *Right*, pedigree for Huntington's disease, a progressive degeneration of the nervous system. Researcher Nancy Wexler and her team constructed this extended family tree for nearly 10,000 Venezuelans. Their analysis of unaffected and affected individuals revealed that a dominant allele on human chromosome 4 is the culprit. Wexler has a special interest in the disorder: It runs in her family.

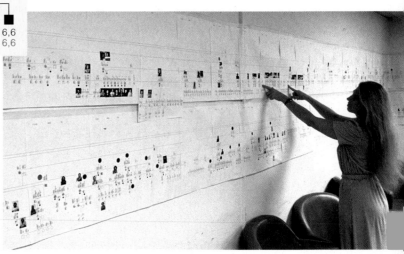

Figure 14.2 Animated Pedigrees.

is a rare or uncommon version of a trait, such as having six fingers on a hand or having a web between two toes. Genetic abnormalities are not inherently life-threatening, and how you view them is a matter of opinion. By contrast, a genetic disorder sooner or later causes medical problems that may be severe. A genetic disorder is often characterized by a specific set of symptoms (a syndrome). In general, much more research focuses on genetic disorders than on other human traits, because what we learn helps us develop treatments for affected people.

The next two sections of this chapter focus on inheritance patterns of human single-gene disorders, which affect about 1 in 200 people. Keep in mind that these inheritance patterns are the least common kind. Most human traits, including skin color, are polygenic (influenced by multiple genes) and often have environmental factors too. Many genetic disorders are like this, including diabetes, asthma, obesity, cancers, heart disease, and multiple sclerosis. The inheritance patterns of these disorders are extremely complex, and despite intense research our understanding of the alleles associated with them remains incomplete. For example, abnormalities on almost every chromosome have been found in people with autism, a developmental disorder, but a person who carries one or more of these abnormalities does not necessarily have autism. Mutations in specific regions of chromosomes 1, 2, 6, 7, 13, 15, and 22 increase an individual's chance of developing schizophrenia, a neurobiological disorder, but not everyone with those mutations develops schizophrenia. Appendix VII shows a map of human chromosomes with the locations of some alleles that are known to play a role in genetic disorders and other human traits.

Alleles that give rise to severe genetic disorders are generally rare in populations, because they compromise the health and reproductive ability of their bearers. Why do they persist? Mutations periodically reintroduce them. In some cases, a normal allele in heterozygotes masks the effects of a harmful one. In others, a codominant allele offers a survival advantage in a particularly hazardous environment. You will see examples of how this works in later chapters.

Take-Home Message How do we study inheritance patterns in humans?

> Inheritance patterns in humans are often studied by tracking traits through generations of families.

> A genetic abnormality is a rare version of an inherited trait. A genetic disorder is an inherited condition that causes medical problems.

> Some human genetic traits are governed by a single gene and inherited in a Mendelian fashion. Many others are influenced by multiple genes, as well as the environment.

Table 14.1 Patterns of Inheritance for Some Genetic Abnormalities and Disorders

Disorder or Abnormality	Main Symptoms
Autosomal dominant inheritance pattern	
Achondroplasia	One form of dwarfism
Aniridia	Defects of the eyes
Camptodactyly	Rigid, bent fingers
Familial hypercholesterolemia	High cholesterol level; clogged arteries
Huntington's disease	Degeneration of the nervous system
Marfan syndrome	Abnormal or missing connective tissue
Polydactyly	Extra fingers, toes, or both
Progeria	Drastic premature aging
Neurofibromatosis	Tumors of nervous system, skin
Autosomal recessive inheritance pattern	
Albinism	Absence of pigmentation
Hereditary methemoglobinemia	Blue skin coloration
Cystic fibrosis	Abnormal glandular secretions leading to tissue and organ damage
Ellis–van Creveld syndrome	Dwarfism, heart defects, polydactyly
Fanconi anemia	Physical abnormalities, bone marrow failure
Galactosemia	Brain, liver, eye damage
Hereditary hemochromatosis	Iron overload damages joints and organs
Phenylketonuria (PKU)	Mental impairment
Sickle-cell anemia	Adverse pleiotropic effects on entire body
Tay–Sachs disease	Deterioration of mental and physical abilities; early death
X-linked recessive inheritance pattern	
Androgen insensitivity syndrome	XY individual but having some female traits; sterility
Red–green color blindness	Inability to distinguish red from green
Hemophilia	Impaired blood clotting ability
Muscular dystrophies	Progressive loss of muscle function
X-linked anhidrotic dysplasia	Mosaic skin (patches with or without sweat glands); other effects
X-linked dominant inheritance pattern	
Fragile X syndrome	Intellectual, emotional disability
Changes in chromosome number	
Down syndrome	Mental impairment; heart defects
Turner syndrome (XO)	Sterility; abnormal ovaries and sexual traits
Klinefelter syndrome	Sterility; mild mental impairment
XXX syndrome	Minimal abnormalities
XYY condition	Mild mental impairment or no effect
Changes in chromosome structure	
Chronic myelogenous leukemia (CML)	Overproduction of white blood cells; organ malfunctions
Cri-du-chat syndrome	Mental impairment; abnormal larynx

> An allele is inherited in an autosomal dominant pattern if the trait it specifies appears in homozygous and heterozygous people.

> An allele is inherited in an autosomal recessive pattern if the trait it specifies appears only in homozygous people.

❬ Links to Lysosomes 4.8, Cytoskeletal elements 4.10, Tyrosinase 5.4, Autosomes 8.2, DNA replication and repair 8.6, Gene expression 9.2, RNA processing 9.3, Inheritance 13.2, Codominance and pleiotropy 13.5

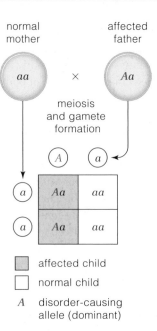

Figure 14.3 **Animated** Autosomal dominant inheritance, in which a dominant allele (*red*) is fully expressed in heterozygous people.

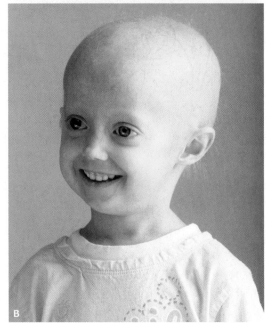

Figure 14.4 Examples of autosomal dominant disorders. **A** Achondroplasia affects Ivy Broadhead (*left*), as well as her brother, father, and grandfather. **B** Five-year-old Megan is already showing symptoms of Hutchinson–Gilford progeria.

The Autosomal Dominant Pattern

An allele on an autosome is inherited in a dominant pattern if it is expressed in both homozygotes and heterozygotes. A trait it specifies tends to appear in every generation. When one parent is heterozygous, and the other is homozygous for the recessive allele, each of their children has a 50 percent chance of inheriting the dominant allele and displaying the trait associated with it (Figure 14.3).

Achondroplasia is an example of an autosomal dominant disorder (a disorder caused by a dominant allele on an autosome). The allele responsible for achondroplasia interferes with formation of the embryonic skeleton. About 1 out of 10,000 people are heterozygous for this allele. As adults, they average about four feet, four inches tall, and have abnormally short arms and legs relative to other body parts (Figure 14.4A). Most homozygotes die before or shortly after birth.

Huntington's disease is also caused by an autosomal dominant allele. With this genetic disorder, involuntary muscle movements increase as the nervous system slowly deteriorates. Typically, symptoms do not start until after the age of thirty, and people affected by the syndrome die during their forties or fifties. The mutation that causes Huntington's alters a protein necessary for brain cell development. It is an expansion mutation, in which the same three nucleotides have been inserted into DNA many, many times. Hundreds of thousands of other expansion repeats occur harmlessly in and between other genes on human chromosomes. This one alters the function of a critical gene product.

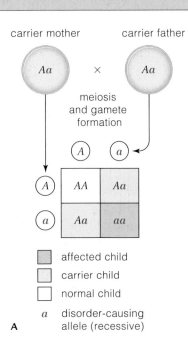

carrier mother   carrier father

$Aa \times Aa$

meiosis and gamete formation

A a

	A	a
A	AA	Aa
a	Aa	aa

■ affected child
□ carrier child
□ normal child

a disorder-causing allele (recessive)

A

B

C

Figure 14.5 Animated Autosomal recessive inheritance. Only homozygous people show the trait associated with a recessive allele on an autosome. **A** In this example, both parents are carriers of a recessive autosomal allele (*red*). Each of their children has a 25 percent chance of being homozygous for it. **B** The albino phenotype is associated with recessive alleles that cause a deficiency in a melanin-producing enzyme. **C** Conner Hopf was diagnosed with Tay–Sachs disease, an autosomal recessive disorder, at age 7-1/2 months. He died before his second birthday.

Hutchinson–Gilford progeria is an autosomal dominant disorder characterized by drastically accelerated aging. It usually results from a base-pair substitution in the gene for lamin A, a protein subunit of intermediate filaments that help organize chromosomes. The mutation adds a signal for an intron/exon splice. The resulting mis-spliced mRNA encodes a protein with a large deletion that cannot assemble into intermediate filaments, and instead accumulates on the nuclear membrane. The pleiotropic effects of the mutation include defects in transcription, mitosis, and division. Symptoms begin to appear before age two. Skin that should be plump and resilient starts to thin, muscles weaken, and bones that should lengthen and grow stronger soften. Premature baldness is inevitable. Most people with the disorder die in their early teens as a result of a stroke or heart attack brought on by hardened arteries, a condition typical of advanced age (Figure 14.4B).

Progeria does not run in families because affected people do not usually live long enough to reproduce. Other dominant alleles that cause severe problems can persist if their expression does not interfere with reproduction. The allele that causes achondroplasia is an example. With Huntington's disease and other late-onset disorders, people tend to reproduce before symptoms appear, so the allele may be passed unknowingly to children.

The Autosomal Recessive Pattern

An allele on an autosome is inherited in a recessive pattern if it is expressed only in homozygous people, so traits associated with the allele may skip generations. People heterozygous for the allele are carriers, which means that they have the allele but not the trait. Any child of two carriers has a 25 percent chance of inheriting the allele from

both parents (Figure 14.5A). Being homozygous for the allele, such children would have the trait. All children of homozygous parents are also homozygous.

Albinism, a lack of melanin, is inherited in an autosomal recessive pattern. The albino phenotype occurs in people homozygous for an allele that encodes a defective form of the enzyme tyrosinase. Melanocytes produce no melanin in the absence of this enzyme, so the hair, skin, and irises lack typical coloration (Figure 14.5B).

Tay–Sachs disease is an example of an autosomal recessive disorder. In the general population, about 1 in 300 people is a carrier for a Tay–Sachs allele, but the incidence is ten times higher in some groups, such as Jews of eastern European descent. Mutations associated with this disorder cause a deficiency or malfunction of a lysosomal enzyme that breaks down gangliosides, a lipid component of plasma membranes. These lipids can accumulate to toxic levels in nerve cells if they are not recycled properly by lysosomes. Affected infants typically seem normal for the first few months. Symptoms begin to appear as the gangliosides accumulate to higher and higher levels inside their nerve cells. Within three to six months the child becomes irritable, listless, and may have seizures. Blindness, deafness, and paralysis follow. Affected children usually die by the age of five (Figure 14.5C).

Take-Home Message How do we know when a trait is associated with an allele on an autosomal chromosome?

❯ With an autosomal dominant inheritance pattern, persons heterozygous for an allele have the associated trait. Thus, the trait appears in every generation.

❯ With an autosomal recessive inheritance pattern, only persons who are homozygous for an allele have the associated trait, which can skip generations.

> ❭ Traits associated with recessive alleles on the X chromosome appear more frequently in men than in women.
> ❭ A man cannot pass an X chromosome allele to a son.
> ❬ Links to Cell cortex 4.10, Pigments 6.2

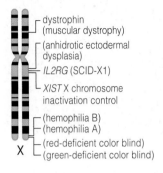

- dystrophin (muscular dystrophy)
- (anhidrotic ectodermal dysplasia)
- IL2RG (SCID-X1)
- XIST X chromosome inactivation control
- (hemophilia B)
- (hemophilia A)
- (red-deficient color blind)
- (green-deficient color blind)

X

The X chromosome (shown on the *left*) carries about 2,000 human genes, which is almost 10 percent of the total number. A recessive allele on this chromosome (an X-linked recessive allele) leaves certain clues when it causes a genetic disorder. First, more males than females are affected by the disorder. This is because heterozygous males are affected, but heterozygous females are not. Heterozygous females have a dominant, normal allele on one of their X chromosomes that masks the effects of the recessive allele on the other. Heterozygous males have only one X chromosome, so they are not similarly protected if they inherit an X-linked recessive allele (Figure 14.6).

Second, an affected father cannot pass his X-linked recessive allele to a son because all children who inherit their father's X chromosome are female. Thus, a heterozygous female must be the bridge between an affected male and his affected grandson.

X-linked dominant alleles that cause disorders are rarer than X-linked recessive ones, probably because they tend to be lethal in male embryos.

Red–Green Color Blindness

The pattern of X-linked recessive inheritance shows up among individuals who have some degree of color blindness (Figure 14.7). The term refers to a range of conditions in which an individual cannot distinguish among some or all colors in the spectrum of visible light. Color vision depends on the proper function of pigment-containing receptors in the eyes. Most of the genes involved in color vision are on the X chromosome, and mutations in those genes often result in altered or missing receptors. Normally, humans can sense the differences among 150 colors. A person who has red–green color blindness sees

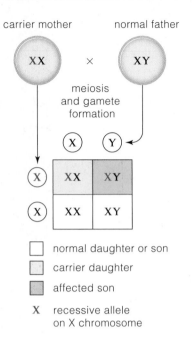

carrier mother   normal father

XX   ×   XY

meiosis and gamete formation

X   Y

X | XX | XY
X | XX | XY

☐ normal daughter or son
☐ carrier daughter
☐ affected son

X recessive allele on X chromosome

Figure 14.6 Animated X-linked recessive inheritance. In this case, the mother carries a recessive allele on one of her two X chromosomes (*red*).

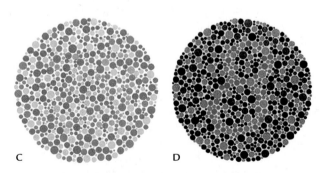

C   D

Figure 14.7 Color blindness. A View with red–green color blindness. The perception of blues and yellows is normal, but red and green appear similar. **B** Compare what a person with normal vision sees.

Above, two Ishihara plates, which are standardized tests for color blindness. **C** You may have one form of red–green color blindness if you see the number 7 instead of 29 in this circle. **D** You may have another form if you see a 3 instead of an 8.

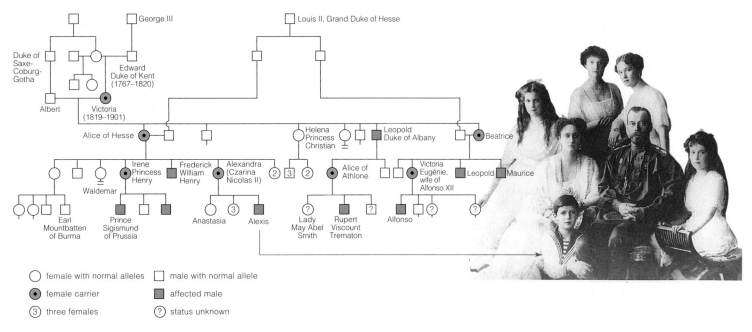

Figure 14.8 A classic case of X-linked recessive inheritance: a partial pedigree of the descendants of Queen Victoria of England. At one time, the recessive X-linked allele that resulted in hemophilia was present in eighteen of Victoria's sixty-nine descendants, who sometimes intermarried. Of the Russian royal family members shown, the mother (Alexandra Czarina Nicolas II) was a carrier.

❯❯ **Figure It Out** How many of Alexis's siblings were affected by hemophilia A? Answer: None

---

fewer than 25 colors because receptors that respond to red and green wavelengths are weakened or absent. Some people who have color blindness confuse red and green colors. Others see green as shades of gray, but perceive blues and yellows quite well.

Hemophilia A

Hemophilia A is an X-linked recessive disorder that interferes with blood clotting. Most of us have a blood clotting mechanism that quickly stops bleeding from minor injuries. That mechanism involves factor VIII, a protein product of a gene on the X chromosome. Bleeding can be prolonged in males who carry a mutation in this gene, or in females who are homozygous for one. Affected people tend to bruise very easily, but internal bleeding is their most serious problem. Repeated bleeding inside the joints disfigures them and causes chronic arthritis. Heterozygous females make enough factor VIII to have a clotting time that is close to normal.

In the nineteenth century, the incidence of hemophilia A was relatively high in royal families of Europe and Russia, probably because the common practice of inbreeding kept the allele in their family trees (Figure 14.8). Today, about 1 in 7,500 people is affected, but that number may be rising because the disorder is now a treatable one. More affected people are living long enough to transmit the mutated allele to children.

Duchenne Muscular Dystrophy

Duchenne muscular dystrophy (DMD) is one of several X-linked recessive disorders characterized by muscle degeneration. A gene on the X chromosome encodes dystrophin, a protein in muscle and nerve cells. Dystrophin is part of a complex of proteins that structurally and functionally links the cytoskeleton to extracellular matrix across the plasma membrane of these cells. By a process that is still unknown, abnormal or absent dystrophin causes muscle cells to die. The dead cells are eventually replaced by fat cells and connective tissue.

DMD affects about 1 in 3,500 people, almost all of them boys. Symptoms begin between age three and seven. Anti-inflammatory medication can sometimes slow the progression of this disorder, but there is no cure. When an affected boy is about twelve, he will begin to use a wheelchair and his heart muscle will start to fail. Even with the best care, he will probably die before he is thirty, from a heart disorder or respiratory failure (suffocation).

Take-Home Message How do we know when a trait is associated with an allele on an X chromosome?

❯ Men heterozygous for an X-linked recessive allele have the trait associated with the allele. Heterozygous women do not, because they have a normal allele on their second X chromosome. Thus, the trait appears more often in men.

❯ Men transmit an X-linked allele to their daughters, but not to their sons.

❯ Major changes in chromosome structure have been evolutionarily important. More frequently, such changes tend to result in genetic disorders.

❮ Links to Karyotyping 8.2, DNA sequence 8.6, Mutations and hemoglobin 9.6, *SRY* gene 10.4, Oncogenes 11.6, Meiosis 12.3, Crossing over 12.4

Large-scale changes in chromosome structure usually have drastic effects on health; about half of all miscarriages are due to chromosome abnormalities of the developing embryo. These changes are rare, but they do occur spontaneously in nature. They can also be induced by exposure to certain chemicals or radiation. Either way, the scale of such changes often allows them to be detected by karyotyping (Section 8.2).

Duplication Even normal chromosomes have DNA sequences that are repeated two or more times. These repetitions are called **duplications** (Figure 14.9A). Duplications happen during prophase I of meiosis, when crossing over occurs unequally between homologous chromosomes. When homologous chromosomes align side by

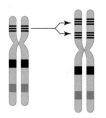
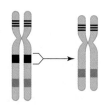

A With a duplication, a section of a chromosome gets repeated.

B With a deletion, a section of a chromosome gets lost.

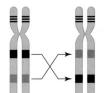

C With an inversion, a section of a chromosome gets flipped so it runs in the opposite orientation.

D With a translocation, a broken piece of a chromosome gets reattached in the wrong place. This example shows a reciprocal translocation, in which two chromosomes exchange chunks.

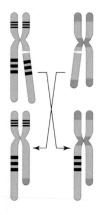

Figure 14.9 Large-scale changes in chromosome structure.

side, their DNA sequences may misalign at some point along their length. In this case, the crossover deletes a stretch of DNA from one chromosome and splices it into the homologous partner. The probability of misalignment is greater in regions where the same sequence of nucleotides is repeated many times. Some duplications, such as the expansion mutations that cause Huntington's, cause genetic abnormalities or disorders. Others have been evolutionarily important.

Deletion A **deletion** is the loss of some portion of a chromosome (Figure 14.9B). In mammals, deletions usually cause serious disorders and are often lethal. The loss of genes results in the disruption of growth, development, and metabolism. For instance, a small deletion in chromosome 5 causes mental impairment and an abnormally shaped larynx. Affected infants tend to make a sound like the meow of a cat, hence the name of the disorder, cri-du-chat, which is French for "cat's cry."

Inversion With an **inversion**, part of the sequence of DNA within the chromosome becomes oriented in the reverse direction, with no molecular loss (Figure 14.9C). An inversion may not affect a carrier's health if it does not disrupt a gene region, because the individual's cells have a full complement of genes. However, it may affect fertility, because inverted chromosomes tend to mispair during meiosis. Crossovers between mispaired chromosomes can produce large deletions or duplications that reduce the viability of forthcoming embryos. Some carriers do not know that they have an inversion until they are diagnosed with infertility and their karyotype is tested.

Translocation If a chromosome breaks, the broken part may get attached to a different chromosome, or to a different part of the same one. This structural change is called a **translocation**. Most translocations are reciprocal, or balanced, which means that two chromosomes exchange broken parts (Figure 14.9D). A reciprocal translocation between chromosomes 8 and 14 is the usual cause of Burkitt's lymphoma, an aggressive cancer of the immune system. This translocation moves a proto-oncogene to a region that is vigorously transcribed in immune cells, with disastrous results. Many other reciprocal translocations have no adverse effects on health, but, like inversions, they can affect fertility. Translocated chromosomes pair abnormally during meiosis. They segregate improperly about half of the time, so about half of the resulting gametes carry major duplications or deletions. If one of these gametes unites with a normal gamete at fertilization, the resulting embryo almost always dies. As with inversions, many people do not realize they carry a translocation until they have difficulty with fertility.

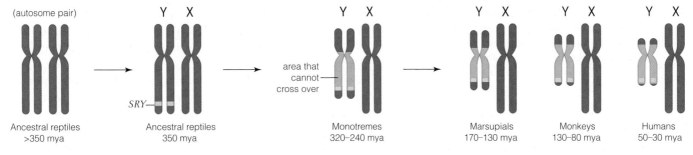

(autosome pair)   Y X   Y X   Y X   Y X   Y X

SRY—

| Ancestral reptiles >350 mya | Ancestral reptiles 350 mya | Monotremes 320–240 mya | Marsupials 170–130 mya | Monkeys 130–80 mya | Humans 50–30 mya |

area that cannot—cross over

A Before 350 mya, sex was determined by temperature, not by chromosome differences.

B The *SRY* gene begins to evolve 350 mya. The DNA sequences of the chromosomes diverge as other mutations accumulate.

C By 320–240 mya, the DNA sequences of the chromosomes are so different that the pair can no longer cross over in one region. The Y chromosome begins to get shorter.

D Three more times, the pair stops crossing over in yet another region. Each time, the DNA sequences of the chromosomes diverge, and the Y chromosome shortens. Today, the pair crosses over only at a small region near the ends.

Figure 14.10 Evolution of the Y chromosome. Today, the *SRY* gene determines male sex. Homologous regions of the chromosomes are shown in *pink*; mya, million years ago.

Chromosome Changes in Evolution

As you can see, large-scale alterations in chromosome structure may reduce an individual's fertility. Individuals who are heterozygous for such changes may not be able to produce offspring at all. However, individuals homozygous for an inversion sometimes become the founders of new species. It may seem as if this outcome would be exceedingly rare, but it is not. Speciation can and does occur by large-scale changes in chromosomes.

Karyotyping and DNA sequence comparisons show that the chromosomes of all species contain evidence of major structural alterations. For example, duplications have often allowed a copy of a gene to mutate while the original carried out its unaltered function. The multiple and strikingly similar globin chain genes of humans and other primates apparently evolved by this process. Four globin chains associate in each hemoglobin molecule (Section 9.6). Different alleles specify different versions of the chains. Which versions of the chains get assembled into a hemoglobin molecule determine the oxygen-binding characteristics of the resulting protein.

As another example, X and Y chromosomes were once homologous autosomes in reptilelike ancestors of mammals (Figure 14.10). Ambient temperature probably determined the gender of those organisms, as it still does in turtles and some other modern reptiles. Then, about 350 million years ago (mya), a gene on one of the two homologous chromosomes mutated. The change, which was the beginning of the male sex determination gene *SRY*,

interfered with crossing over during meiosis. A reduced frequency of crossing over allowed the chromosomes to diverge around the changed region. Mutations began to accumulate separately in the two chromosomes. Over evolutionary time, the chromosomes became so different that they no longer crossed over at all in the changed region, so they diverged even more. Today, the Y chromosome is much smaller than the X, and only retains about 5 percent homology with it. The Y crosses over mainly with itself—by translocating duplicated regions of its own DNA.

Some chromosome structure changes contributed to differences among closely related organisms, such as apes and humans. Human somatic cells have twenty-three pairs of chromosomes, but those of chimpanzees, gorillas, and orangutans have twenty-four. Thirteen human chromosomes are almost identical with chimpanzee chromosomes. Nine more are similar, except for some inversions. One human chromosome matches up with two in chimpanzees and the other great apes (Figure 14.11). During human evolution, two chromosomes evidently fused end to end and formed our chromosome 2. How do we know? The region where the fusion occurred contains the remnants of a telomere, which is a special DNA sequence that caps the ends of chromosomes.

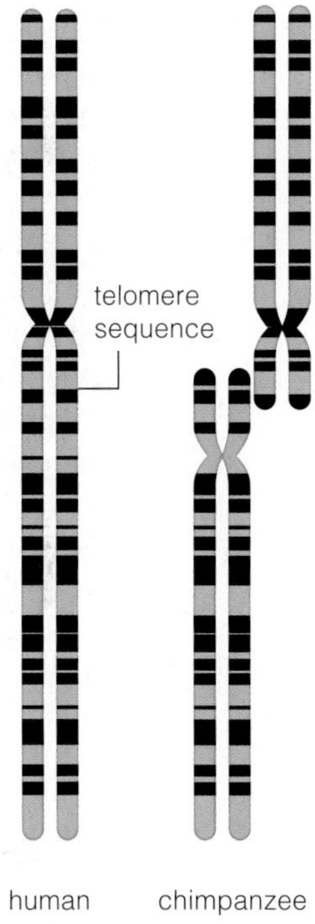

telomere sequence

human   chimpanzee

Figure 14.11
Human chromosome 2 compared with chimpanzee chromosomes 2A and 2B.

---

deletion Loss of part of a chromosome.
duplication Repeated section of a chromosome.
inversion Structural rearrangement of a chromosome in which part of it becomes oriented in the reverse direction.
translocation Structural change of a chromosome in which a broken piece gets reattached in the wrong location.

> **Take-Home Message** **Does chromosome structure change?**
>
> ❯ A segment of a chromosome may be duplicated, deleted, inverted, or translocated. Such a change is usually harmful or lethal, but may be conserved in the rare circumstance that it has a neutral or beneficial effect.

> Occasionally, abnormal events occur before or during meiosis, and new individuals end up with the wrong chromosome number. Consequences range from minor to lethal changes in form and function.

< Links to Sampling error and bias 1.8, Meiosis 12.3, Gamete formation 12.5

Less than 1 percent of children are born with a chromosome number that differs from their parents. In humans, such major changes to the genetic blueprint can have serious effects on an individual's health.

Chromosome number changes often arise through **nondisjunction**, in which a cell's chromosomes do not separate properly during nuclear division. Nondisjunction during meiosis (Figure 14.12) can affect the chromosome number at fertilization. For example, suppose that a normal gamete fuses with an $n+1$ gamete (one with an extra chromosome). The new individual will be trisomic ($2n+1$), having three of one type of chromosome and two of every other type. As another example, if an $n-1$ gamete fuses with a normal n gamete, the new individual will be $2n-1$, or monosomic.

Trisomy and monosomy are types of **aneuploidy**, a condition in which cells have too many or too few copies of a chromosome. Autosomal aneuploidy is usually fatal in humans. However, about 70 percent of flowering plant species, and some insects, fishes, and other animals, are **polyploid**, which means that their cells have three or more of each type of chromosome.

Autosomal Change and Down Syndrome A few trisomic humans are born alive, but only those that have trisomy 21 will survive infancy. A newborn with three chromosomes 21 will develop Down syndrome. This auto-somal disorder occurs once in 800 to 1,000 births, and it affects more than 350,000 people in the United States alone (Figure 14.13). Individuals with Down syndrome have upward-slanting eyes, a fold of skin that starts at the inner corner of each eye, a deep crease across the sole of each palm and foot, one (instead of two) horizontal furrows on their fifth fingers, slightly flattened facial features, and other symptoms.

Not all of the outward symptoms develop in every individual. That said, trisomic 21 individuals tend to have moderate to severe mental impairment and heart problems. Their skeleton grows and develops abnormally, so older children have short body parts, loose joints, and misaligned bones of the fingers, toes, and hips. The muscles and reflexes are weak, and motor skills such as speech develop slowly. With medical care, trisomy 21 individuals live about fifty-five years. Early training can help affected individuals learn to care for themselves and to take part in normal activities.

Change in the Sex Chromosome Number

Nondisjunction also causes alterations in the number of X and Y chromosomes, with a frequency of about 1 in 400 live births. Most often, such alterations lead to difficulties in learning and impaired motor skills such as a speech delay, but problems may be so subtle that the underlying cause is never diagnosed.

Female Sex Chromosome Abnormalities Individuals with Turner syndrome have an X chromosome and no corresponding X or Y chromosome (XO). The syndrome probably arises most frequently as an outcome of inheriting an unstable Y chromosome from the father: The

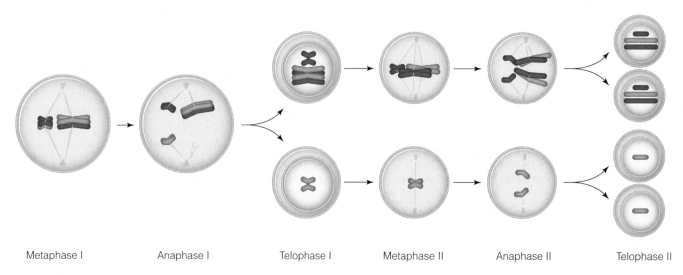

| Metaphase I | Anaphase I | Telophase I | Metaphase II | Anaphase II | Telophase II |

Figure 14.12 An example of nondisjunction during meiosis. Of the two pairs of homologous chromosomes shown here, one fails to separate during anaphase I. The chromosome number is altered in the resulting gametes.

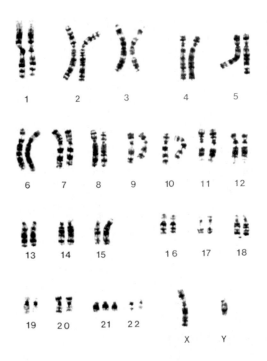

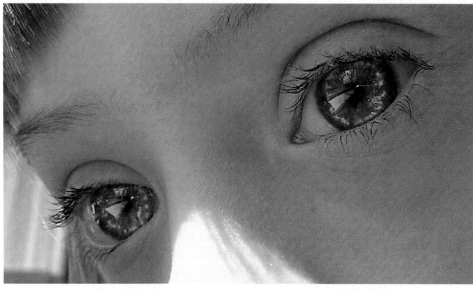

Figure 14.13 Down syndrome, genotype and phenotype.

>> **Figure It Out** Is the karyotype of an individual who is male or female?

Answer: Male (XY)

zygote was genetically male, but the Y chromosome broke up and got lost early in development.

The incidence of Turner syndrome does not rise with maternal age, and there are fewer people affected by it than other chromosome abnormalities: Only about 1 in 2,500 newborn girls has it. XO individuals grow up well proportioned but short (with an average height of four feet, eight inches). Their ovaries do not develop properly, so they do not make enough sex hormones to become sexually mature. The development of secondary sexual traits such as breasts is also inhibited.

A female may inherit multiple X chromosomes. XXX syndrome occurs in about 1 of 1,000 births. Only one X chromosome is typically active in female cells (Section 10.4), so having extra X chromosomes usually does not result in physical or medical problems.

Male Sex Chromosome Abnormalities About 1 out of every 500 males has an extra X chromosome (XXY). Most cases are an outcome of nondisjunction during meiosis. The resulting disorder, Klinefelter syndrome, develops at puberty. XXY males tend to be overweight, tall, and within a normal range of intelligence. They make more estrogen and less testosterone than normal males, and this hormone imbalance has feminizing effects. Affected men tend to have small testes and prostate glands, low sperm counts, sparse facial and body hair, high-pitched voices, and enlarged breasts. Testosterone injections during puberty can reverse these traits.

About 1 in 1,000 males is born with an extra Y chromosome (XYY). Adults tend to be taller than average and have mild mental impairment, but most are otherwise normal. XYY men were once thought to be predisposed to a life of crime. This misguided view was based on sampling error (too few cases in narrowly chosen groups such as prison inmates) and bias (the researchers who gathered the karyotypes also took the personal histories of the participants). That view was disproven in 1976, when a geneticist reported results from his study of 4,139 tall males, all twenty-six years old, who had registered for military service. Besides their data from physical examinations and intelligence tests, the records offered clues to social and economic status, education, and any criminal convictions. Only twelve of the males studied were XYY, which meant that the "control group" had more than 4,000 males. The only findings? Mentally impaired, tall males who engage in criminal deeds are more likely to get caught, irrespective of karyotype.

aneuploidy A chromosome abnormality in which an individual's cells carry too many or too few copies of a particular chromosome.
nondisjunction Failure of sister chromatids or homologous chromosomes to separate during nuclear division.
polyploid Having three or more of each type of chromosome characteristic of the species.

Take-Home Message **What are the effects of chromosome number changes?**

> Nondisjunction can change the number of autosomes or sex chromosomes in gametes. Such changes usually cause genetic disorders in offspring.

> Sex chromosome abnormalities are usually associated with some degree of learning difficulty and motor skill impairment.

> Our understanding of human inheritance can be used to provide prospective parents with information about the health of their future children.

‹ Links to Probability 1.8, Amino acids 3.5, Metabolic pathways 5.5

Studying human inheritance patterns has given us many insights into how genetic disorders arise and progress, and how to treat them. Surgery, prescription drugs, hormone replacement therapy, and dietary controls can minimize and in some cases eliminate the symptoms of a genetic disorder. Some disorders can be detected early enough to start countermeasures before symptoms develop. Most hospitals in the United States now screen newborns for mutations in the gene for phenylalanine hydroxylase, an enzyme that catalyzes the conversion of the amino acid phenylalanine to tyrosine. Without a functional form of this enzyme, the body becomes deficient in tyrosine, and phenylalanine accumulates to high levels. The imbalance inhibits protein synthesis in the brain, which in turn results in the severe neurological symptoms characteristic of *phenylketonuria*, or PKU. Restricting all intake of phenylalanine can slow the progression of PKU, so routine early screening has resulted in fewer individuals developing the disorder.

Prospective parents worried about the possibility that a future child of theirs might have a genetic disorder also benefit from human genetics studies. The probability that a child will inherit a genetic disorder can be estimated by checking parental karyotypes and pedigrees, and testing the parents for alleles known to be associated with genetic disorders. This type of genetic screening is typically done before pregnancy, to help the prospective parents make decisions about family planning. Most couples would choose to know if their future children face a high risk of inheriting a severe genetic disorder, but the information may come at a heavy price. Learning about a life-threatening allele in your DNA can be devastating.

Prenatal Diagnosis

Genetic screening can also be done post-conception, in which case it is called prenatal diagnosis. Prenatal means before birth. *Embryo* is a term that applies until eight weeks after fertilization, after which *fetus* is appropriate. Prenatal diagnosis checks for physical and genetic abnormalities in an embryo or fetus. It can often reveal the presence of a genetic disorder in an unborn child. More than 30 conditions, including aneuploidy, hemophilia, Tay–Sachs disease, sickle-cell anemia, muscular dystrophy, and cystic fibrosis, are detectable prenatally. If the disorder is treatable, early detection allows the newborn to receive prompt and appropriate treatment. A few defects are even surgically correctable before birth. Prenatal diagnosis also gives parents time to prepare for the birth of an

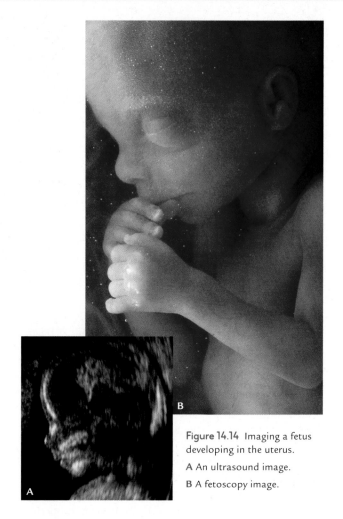

Figure 14.14 Imaging a fetus developing in the uterus.
A An ultrasound image.
B A fetoscopy image.

affected child, and an opportunity to decide whether to continue with the pregnancy or terminate it.

As an example of how prenatal diagnosis works, consider a thirty-five-year-old woman who becomes pregnant. Her doctor will probably use a noninvasive procedure, *obstetric sonography*, in which ultrasound waves directed across the woman's abdomen form images of the fetus's developing limbs and internal organs (Figure 14.14A). An ultrasound image is not very detailed, but there is no detectable risk to the pregnancy. The images may reveal physical defects associated with a genetic disorder, in which case a more invasive technique would be recommended for further diagnosis. *Fetoscopy* yields images of the fetus that are much higher in resolution than ultrasound images (Figure 14.14B). With this procedure, sound waves are pulsed from inside the mother's uterus. A sample of fetal blood is often drawn at the same time.

Human genetics studies show that our thirty-five-year-old woman has about a 1 in 80 chance that her baby will be born with a chromosomal abnormality—a risk more than six times higher than when she was twenty years old. Thus, even if no abnormalities were detected by

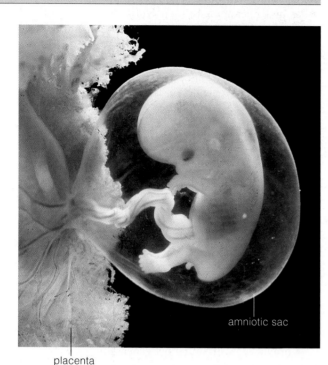

placenta

amniotic sac

Figure 14.15 An 8-week-old fetus. With amniocentesis, fetal cells shed into the fluid inside the amniotic sac are tested for genetic disorders. Chorionic villus sampling tests cells of the chorion, which is part of the placenta.

ultrasound, she probably will be offered a more thorough diagnostic procedure, *amniocentesis*, in which a small sample of fluid is drawn from the amniotic sac enclosing the fetus (Figure 14.15). The fluid contains cells shed by the fetus, and those cells can be tested for genetic disorders. *Chorionic villus sampling* (CVS) can be done earlier than amniocentesis. With this technique, a few cells from the chorion are removed and tested for genetic disorders. The chorion is a membrane that surrounds the amniotic sac and helps form the placenta, an organ that allows substances to be exchanged between mother and embryo.

An invasive procedure often carries a risk to the fetus. For example, if a punctured amniotic sac does not reseal itself quickly, too much fluid may leak out of it, resulting in miscarriage. The risks vary by the procedure. Amniocentesis has improved so much that, in the hands of a skilled physician, the procedure no longer increases the risk of miscarriage. CVS occasionally disrupts the placenta's development and thus causes underdeveloped or missing fingers and toes in 0.3 percent of newborns. Fetoscopy raises the miscarriage risk by 2 to 10 percent.

Preimplantation Diagnosis

Reproductive interventions such as preimplantation diagnosis offer an alternative to couples who discover they are at high risk of having a child with a genetic disorder.

Shades of Skin (revisited)

Chinese and Europeans do not share any skin pigmentation allele that does not also occur in other populations. However, most people of Chinese descent carry a particular allele of the *DCT* gene, the product of which helps convert tyrosine to melanin. Few people of European or African descent have this allele. Taken together, the distribution of the *SLC24A5* and *DCT* genes suggests that (1) an African population was ancestral to both the Chinese and Europeans, and (2) Chinese and European populations separated before their pigmentation genes mutated and their skin color changed.

How Would You Vote? Physical attributes such as skin color, which have a genetic basis, are often used to define race. Do twins such as Kian and Remee belong to different races? See CengageNow for details, then vote online (cengagenow.com).

Preimplantation diagnosis is a procedure that relies on *in vitro* fertilization, in which sperm and eggs taken from prospective parents are mixed in a test tube. If an egg becomes fertilized, the resulting zygote will begin to divide. In about forty-eight hours, it will have become an embryo that consists of a ball of eight cells (Figure 14.16). All of the cells in this ball have the same genes, but none has yet committed to being specialized one way or another. Doctors can remove one of these undifferentiated cells and analyze its genes. The withdrawn cell will not be missed. If the embryo has no detectable genetic defects, it is inserted into the woman's uterus to develop. Many of the resulting "test-tube babies" are born in good health.

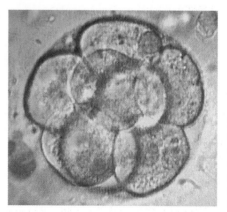

Figure 14.16 Clump of cells formed by three mitotic divisions after *in vitro* fertilization. All eight of the cells are identical and one can be removed for genetic analysis to determine whether the embryo carries any genetic defects.

Take-Home Message How do we use what we know about human inheritance?

> Genetic testing can provide prospective parents with information about the health of their future children.

Summary

 Section 14.1 Like most other human traits, skin color has a genetic basis. Minor differences in the alleles that govern melanin synthesis and the deposition of melanosomes affect skin color. The differences probably evolved as a balance between vitamin production and protection against harmful UV radiation.

 Section 14.2 Geneticists use **pedigrees** to track certain traits through generations of a family. Such studies can reveal inheritance patterns for alleles that can be predictably associated with specific phenotypes, including genetic abnormalities or disorders. A genetic abnormality is an uncommon version of a heritable trait that does not result in medical problems. A genetic disorder is a heritable condition that sooner or later results in mild or severe medical problems.

 Section 14.3 An allele on an autosome is inherited in an autosomal dominant pattern if the trait associated with the allele appears in heterozygous individuals. Such traits tend to appear in every generation of families that carry the allele. An allele on an autosome is inherited in an autosomal recessive pattern if the trait associated with the allele only appears in homozygous individuals. Such traits can skip generations.

 Section 14.4 An allele is inherited in an X-linked pattern when it occurs on the X chromosome. Most X-linked inheritance disorders are recessive, because X-linked dominant alleles tend to be lethal in male embryos. X-linked recessive disorders tend to appear in men more often than in women. This is because women have two X chromosomes, so they can be heterozygous for the recessive allele. Men can transmit an X-linked allele to their daughters, but not to their sons. Only a woman can pass an X-linked allele to a son.

 Section 14.5 Major changes in chromosome structure include **duplications**, **deletions**, **inversions**, and **translocations**. Most major alterations are harmful or lethal in humans. Even so, many major structural changes have accumulated in the chromosomes of all species over evolutionary time.

 Section 14.6 Changes in chromosome number are usually an outcome of **nondisjunction**, in which chromosomes fail to separate properly during meiosis. Such changes tend to cause genetic disorders among the resulting offspring. In **aneuploidy**, an individual's cells have too many or too few copies of a chromosome.

The most common aneuploidy, trisomy 21, causes Down syndrome. Most other cases of autosomal aneuploidy are lethal in embryos. **Polyploid** individuals have three or more of each type of chromosome. Polyploidy is lethal in humans, but many flowering plants, and some insects, fishes, and other animals, are polyploid.

A change in the number of sex chromosomes usually results in some degree of impairment in learning and motor skills. These problems can be so subtle that the underlying cause may not ever be diagnosed, as among XXY, XXX, and XYY children.

 Section 14.7 Prospective parents can estimate their risk of transmitting a harmful allele to offspring with genetic screening, in which their pedigrees and genotype are analyzed by a genetic counselor. Prenatal genetic testing of an embryo or fetus can reveal genetic abnormalities or disorders before birth.

Self-Quiz   Answers in Appendix III

1. Constructing a pedigree is useful when studying inheritance patterns in organisms that _____ .
   a. produce many offspring per generation
   b. produce few offspring per generation
   c. have a very large chromosome number
   d. reproduce sexually
   e. have a fast life cycle

2. Pedigree analysis is necessary when studying human inheritance patterns because _____ .
   a. humans have more than 20,000 genes
   b. of ethical problems with experimenting on humans
   c. inheritance in humans is more complicated than in other organisms
   d. genetic disorders occur in humans
   e. all of the above

3. A recognized set of symptoms that characterize a genetic disorder is a(n) _____ .
   a. syndrome   b. disease   c. abnormality

4. If one parent is heterozygous for a dominant allele on an autosome and the other parent does not carry the allele, any child of theirs has a _____ chance of being heterozygous.
   a. 25 percent   c. 75 percent
   b. 50 percent   d. no chance; it will die

5. Is this statement true or false? A son can inherit an X-linked recessive allele from his father.

6. A trait that is present in a male child but not in either of his parents is characteristic of _____ inheritance.
   a. autosomal dominant   d. It is not possible to
   b. autosomal recessive   answer this question
   c. X-linked recessive   without more information.

7. Color blindness is a case of _____ inheritance.
   a. autosomal dominant   c. X-linked dominant
   b. autosomal recessive   d. X-linked recessive

8. What do you think the pattern of inheritance of the human *SRY* gene is called?

Skin Color Survey of Native Peoples A 2000 study measured average skin color of people native to more than fifty regions, and correlated them to the amount of UV radiation received in those regions. Some of their results are shown in Figure 14.17.

1. Which country receives the most UV radiation? The least?
2. The people native to which country have the darkest skin? The lightest?
3. According to these data, how does the skin color of indigenous peoples correlate with the amount of UV radiation incident in their native regions?

Country	Skin Reflectance	UVMED
Australia	19.30	335.55
Kenya	32.40	354.21
India	44.60	219.65
Cambodia	54.00	310.28
Japan	55.42	130.87
Afghanistan	55.70	249.98
China	59.17	204.57
Ireland	65.00	52.92
Germany	66.90	69.29
Netherlands	67.37	62.58

Figure 14.17
Skin color of indigenous peoples and regional incident UV radiation. Skin reflectance measures how much light of 685 nanometers wavelength is reflected from skin; UVMED is the annual average UV radiation received at Earth's surface.

---

9. A female child inherits one X chromosome from her mother and one from her father. What sex chromosome does a male child inherit from each of his parents?

10. Nondisjunction may occur during _____ .
   a. mitosis
   b. meiosis
   c. fertilization
   d. both a and b

11. Nondisjunction can result in _____ .
   a. polyploidy
   b. aneuploidy
   c. crossing over
   d. pleiotropy

12. Nondisjunction can occur during _____ of meiosis.
   a. anaphase I
   b. telophase I
   c. anaphase II
   d. a or c

13. Is this statement true or false? Body cells may inherit three or more of each type of chromosome characteristic of the species, a condition called polyploidy.

14. Klinefelter syndrome (XXY) can be easily diagnosed by _____ .
   a. pedigree analysis
   b. aneuploidy
   c. karyotyping
   d. phenotypic treatment

15. Match the chromosome terms appropriately.
   ___ polyploidy
   ___ deletion
   ___ aneuploidy
   ___ translocation
   ___ syndrome
   ___ nondisjunction during meiosis

   a. symptoms of a genetic disorder
   b. segment of a chromosome moves to a nonhomologous chromosome
   c. extra sets of chromosomes
   d. results in gametes with the wrong chromosome number
   e. a chromosome segment lost
   f. one extra chromosome

Additional questions are available on **CENGAGENOW**.

Animations and Interactions on **CENGAGENOW**:
› A human pedigree; Autosomal dominant inheritance; Autosomal recessive inheritance; Duplications, deletions, inversions, and translocations; X-linked inheritance.

Genetics Problems   Answers in Appendix III

1. Does the phenotype indicated by the red circles and squares in this pedigree show an inheritance pattern that is autosomal dominant, autosomal recessive, or X-linked?

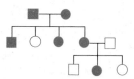

2. Human females are XX and males are XY.
   a. With respect to X-linked alleles, how many different types of gametes can a male produce?
   b. If a female is homozygous for an X-linked allele, how many types of gametes can she produce with respect to that allele?
   c. If a female is heterozygous for an X-linked allele, how many types of gametes can she produce with respect to that allele?

3. People homozygous for the *HbS* allele develop sickle-cell anemia (Section 9.6). Heterozygotes have fewer symptoms. A couple who are both heterozygous for the *HbS* allele plan to have children. For each of the pregnancies, state the probability that they will have a child who is:
   a. homozygous for the *HbS* allele
   b. homozygous for the normal allele
   c. heterozygous (having the normal and the *HbS* allele)

4. A few individuals with Down syndrome have forty-six chromosomes: two normal-appearing chromosomes 21, and a longer-than-normal chromosome 14. Speculate on how this chromosome abnormality arises.

5. An allele responsible for Marfan syndrome (Section 13.5) is inherited in an autosomal dominant pattern. What is the chance that any child will inherit it if one parent does not carry the allele and the other is heterozygous?

‹ Links to Earlier Concepts

This chapter builds on your understanding of DNA's structure (Sections 8.3, 8.4, and 13.2) and replication (8.6). Clones (8.1), gene expression (9.2, 9.3), and knockouts (10.3) are important in genetic engineering, particularly as they apply to research on human traits (13.6) and genetic disorders (14.2). You will revisit tracers (2.2), triglycerides (3.4), denaturation (3.6), β-carotene (6.2), the *lac* operon (10.5), and cancer (11.6).

Key Concepts

DNA Cloning
Researchers routinely make recombinant DNA by cutting and pasting together DNA from different species. Plasmids and other vectors can carry foreign DNA into host cells.

Finding Needles in Haystacks
DNA libraries, hybridization, and PCR are techniques that allow researchers to isolate and make many copies of a fragment of DNA they want to study.

15 Biotechnology

15.1 Personal DNA Testing

About 99 percent of your DNA is exactly the same as everyone else's. If you compared your DNA with your neighbor's, about 29.7 billion nucleotides of the two sequences would be identical. The remaining 30 million or so are sprinkled throughout your chromosomes, mainly as single nucleotide differences.

The sprinkling is not entirely random; some regions of DNA vary less than others. Such conserved regions are of particular interest to researchers because they are the ones most likely to have an essential function. When a conserved sequence does vary among people, the variation tends to be in particular nucleotides. A nucleotide difference carried by a measurable percentage of a population, usually above 1 percent, is called a single-nucleotide polymorphism, or SNP (pronounced "snip").

Alleles of most genes differ by single nucleotides, and differences in alleles are the basis of the variation in human traits that makes each individual unique (Section 12.2). Thus, SNPs account for many of the differences in the way humans look, and they also have a lot to do with differences in the way our bodies work—how we age, respond to drugs, weather assaults by pathogens and toxins, and so on.

Consider a gene, *APOE*, that specifies apolipoprotein E, a protein component of lipoprotein particles (Section 3.5). One allele of this gene, *ε4*, is carried by about 25 percent of people. Nucleotide 4,874 of this allele is a cytosine instead of the normal thymine, a SNP that results in a single amino acid change in the protein product of the gene. How this change affects the function of apolipoprotein E is unclear, but we do know that having the *ε4* allele increases one's risk of developing Alzheimer's disease later in life, particularly in people homozygous for it.

About 4.5 million SNPs in human DNA have been identified, and that number is growing every day. A few companies are now offering to determine some of the SNPs you carry (Figure 15.1). The companies extract your DNA from the cells in a few drops of spit, then analyze it for SNPs.

Personal genetic testing may soon revolutionize medicine by allowing physicians to customize treatments on the basis of an individual's genetic makeup. For example, an allele associated with a heightened risk of a particular medical condition could be identified long before symptoms actually appear. People with that allele could then be encouraged to make life-style changes

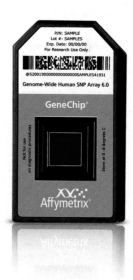

Figure 15.1 Personal genetic testing. *Right*, a SNP-chip. Personal DNA testing companies use chips like this one to analyze their customers' chromosomes for SNPs. This chip, shown actual size, reveals which versions of 906,600 SNPs occur in the individual's DNA.

known to delay the onset of the condition. For some conditions, treatment that begins early enough may prevent symptoms from developing at all. Physicians could design treatments to fit the way a condition is likely to progress in the individual, and also to prescribe only those drugs that will work in the person's body.

You are now at a time when geneticists hold molecular keys to the kingdom of inheritance. As you will see, what they are unlocking is already having an impact on all of us.

DNA Sequencing
Sequencing reveals the linear order of nucleotides in DNA. Comparing genomes offers insights into human genes and evolution. An individual can be identified by unique parts of their DNA.

Genetic Engineering
Genetic engineering, the directed modification of an organism's genes, is now a routine part of research and development. Genetically modified organisms are now quite common.

Gene Therapy
Genetic engineering continues to be tested in medical applications. It also continues to raise ethical questions.

> Researchers cut up DNA from different sources, then paste the resulting fragments together.
> Cloning vectors can carry foreign DNA into host cells.
< Links to Clones 8.1, Discovery of DNA structure 8.3, Base pairing and directionality of DNA strands 8.4, DNA ligase 8.6, mRNA 9.2, Introns 9.3, The *lac* operon 10.5

In the 1950s, excitement over the discovery of DNA's structure gave way to frustration: No one could determine the order of nucleotides in a molecule of DNA. Identifying a single base among thousands or millions of others turned out to be a huge technical challenge. A seemingly unrelated discovery offered a solution. Viruses called bacteriophages infect bacteria by injecting DNA into them (Section 8.3). Some bacteria are resistant to infection, and Werner Arber, Hamilton Smith, and their coworkers discovered why: Special enzymes inside these bacteria chop up any injected viral DNA before it has a chance to integrate into the bacterial chromosome. The enzymes restrict viral growth; hence their name, **restriction enzymes**. A restriction enzyme cuts DNA wherever a specific nucleotide sequence occurs (Figure 15.2). For example, the enzyme *Eco*RI (named after *E. coli,* the bacteria from which it was isolated) cuts DNA at the sequence GAATTC ❶. Other restriction enzymes cut different sequences.

The discovery of restriction enzymes allowed researchers to cut gigantic molecules of chromosomal DNA into manageable and predictable chunks. It also allowed them to combine DNA fragments from different organisms. How? Many restriction enzymes, including *Eco*RI, leave single-stranded tails on DNA fragments ❷. Researchers realized that complementary tails will base-pair ❸. Thus, the tails are called "sticky ends," because two fragments of DNA stick together when their matching tails base-pair. Regardless of the source of DNA, any two fragments will stick together, as long as their tails are complementary.

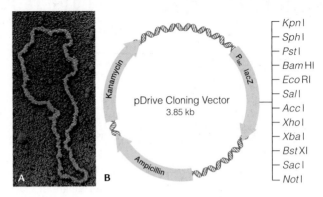

Figure 15.3 Plasmid cloning vectors. **A** Micrograph of a plasmid. **B** A commercial plasmid cloning vector. Restriction enzyme recognition sequences are indicated on the *right* by the name of the enzyme that cuts them. Researchers insert foreign DNA into the vector at these sites. Bacterial genes (*gold*) help researchers identify host cells that take up a vector with inserted DNA. This vector carries two antibiotic resistance genes and the *lac* operon (Section 10.5).

Base-paired sticky ends can be covalently bonded together with the enzyme DNA ligase ❹. Thus, using appropriate restriction enzymes and DNA ligase, researchers can cut and paste DNA from different organisms. The result, a hybrid molecule composed of DNA from two or more organisms, is called **recombinant DNA**.

Why make recombinant DNA? It is the first step in **DNA cloning**, a set of laboratory methods that uses living cells to mass-produce specific DNA fragments. For example, researchers often insert DNA fragments into **plasmids**, small circles of DNA independent of the chromosome. Before a bacterium divides, it copies any plasmids it carries along with its chromosome, so both descendant cells get one of each. If a plasmid carries a fragment of foreign DNA, that fragment gets copied and distributed to descendant cells along with the plasmid.

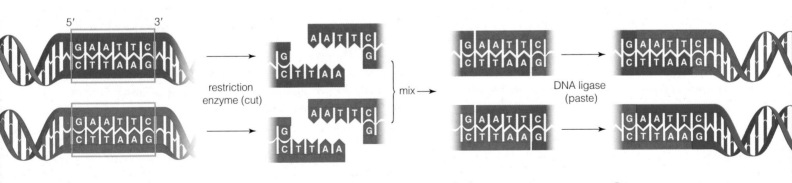

❶ A restriction enzyme recognizes a specific base sequence (*orange* boxes) in DNA from any source.

❷ The enzyme cuts DNA from two sources into fragments. This enzyme leaves sticky ends.

❸ When the DNA fragments from the two sources are mixed together, matching sticky ends base-pair with each other.

❹ DNA ligase joins the base-paired DNA fragments. Molecules of recombinant DNA are the result.

Figure 15.2 Animated Using restriction enzymes to make recombinant DNA.

>> **Figure It Out** Why did the enzyme cut both strands of DNA?

Answer: Because the recognition sequence occurs on both strands.

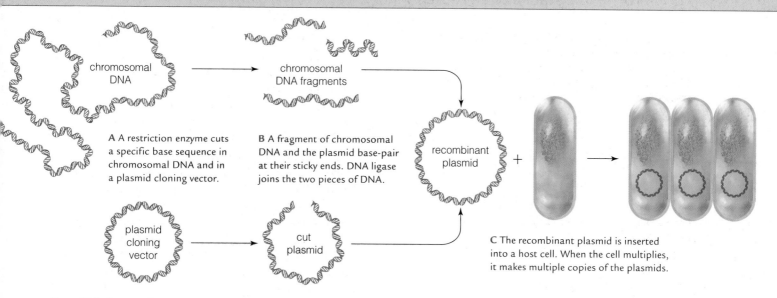

A A restriction enzyme cuts a specific base sequence in chromosomal DNA and in a plasmid cloning vector.

B A fragment of chromosomal DNA and the plasmid base-pair at their sticky ends. DNA ligase joins the two pieces of DNA.

C The recombinant plasmid is inserted into a host cell. When the cell multiplies, it makes multiple copies of the plasmids.

chromosomal DNA

chromosomal DNA fragments

recombinant plasmid

plasmid cloning vector

cut plasmid

Figure 15.4 Animated An example of cloning. Here, a fragment of chromosomal DNA is inserted into a plasmid.

Thus, plasmids can be used as **cloning vectors**, which are molecules that carry foreign DNA into host cells (Figure 15.3). A host cell into which a cloning vector has been inserted can be grown in the laboratory (cultured) to yield a huge population of genetically identical cells, or clones (Section 8.1). Each clone contains a copy of the vector and the fragment of foreign DNA it carries (Figure 15.4). Researchers can harvest the DNA fragment from the clones in large quantities.

cDNA Cloning

Cloning eukaryotic genes can be tricky, because eukaryotic DNA contains introns (Section 9.3). Unless you are a eukaryotic cell, it is not very easy to find the parts of the DNA that encode gene products. Thus, researchers who study eukaryotic genes and their expression work with mRNA, because the introns have already been snipped out of it. However, restriction enzymes and DNA ligase only work on double-stranded DNA, not single-stranded RNA. In order to study mRNA, researchers first make a DNA copy of it, then clone the DNA.

An mRNA can be transcribed into a molecule of double-stranded DNA by a process that is essentially the reverse of RNA transcription. In this process, researchers use **reverse transcriptase**, a replication enzyme made by certain types of viruses, to assemble a strand of complementary DNA, or **cDNA**, on an mRNA template:

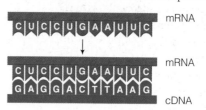

DNA polymerase added to the mixture strips the RNA from the hybrid molecule as it copies the cDNA into a second strand of DNA. The outcome is a double-stranded DNA copy of the original mRNA:

*Eco*RI recognition site

Like any other DNA, double-stranded cDNA may be cut with restriction enzymes and pasted into a cloning vector using DNA ligase.

cDNA DNA synthesized from an RNA template by the enzyme reverse transcriptase.
cloning vector A DNA molecule that can accept foreign DNA, be transferred to a host cell, and get replicated in it.
DNA cloning Set of procedures that uses living cells to make many identical copies of a DNA fragment.
plasmid Of many bacteria and archaeans, a small ring of nonchromosomal DNA replicated independently of the chromosome.
recombinant DNA A DNA molecule that contains genetic material from more than one organism.
restriction enzyme Type of enzyme that cuts specific nucleotide sequences in DNA.
reverse transcriptase A viral enzyme that uses mRNA as a template to make a strand of cDNA.

Take-Home Message **What is DNA cloning?**

❯ DNA cloning uses living cells to mass-produce particular DNA fragments. Restriction enzymes cut DNA into fragments, then DNA ligase seals the fragments into cloning vectors. Recombinant DNA molecules result.

❯ A cloning vector that holds foreign DNA can be introduced into a living cell. When the host cell divides, it gives rise to huge populations of genetically identical cells (clones), each of which contains a copy of the foreign DNA.

> DNA libraries and the polymerase chain reaction (PCR) help researchers isolate particular DNA fragments.
< Links to Tracers 2.2, Denaturation 3.6, Base pairing 8.4, DNA replication 8.6

A Individual bacterial cells from a DNA library are spread over the surface of a solid growth medium. The cells divide repeatedly and form colonies—clusters of millions of genetically identical descendant cells.

B A piece of special paper pressed onto the surface of the growth medium will bind some cells from each colony.

C The paper is soaked in a solution that ruptures the cells and releases their DNA. The DNA clings to the paper in spots mirroring the distribution of colonies.

D A probe is added to the liquid bathing the paper. The probe hybridizes (base-pairs) with the spots of DNA that contain complementary base sequences.

E The bound probe makes a spot. Here, one radioactive spot darkens x-ray film. The position of the spot is compared to the positions of the original bacterial colonies. Cells from the colony that made the spot are cultured, and the DNA they contain is harvested.

Figure 15.5 Animated Nucleic acid hybridization. In this example, a radioactive probe helps identify a bacterial colony that contains a targeted sequence of DNA.

Isolating Genes

The entire set of genetic material—the **genome**—of most organisms consists of thousands of genes. To study or manipulate a single gene, researchers must first separate the gene from all of the others. To do that, researchers often begin by cutting an organism's DNA into pieces, and then cloning all the pieces. The result is a genomic library, a set of clones that collectively contain all of the DNA in a genome. Researchers can also harvest mRNA, make cDNA copies of it, and then clone the cDNA to make a cDNA library. A cDNA library represents only those genes being expressed at the time the mRNA was harvested.

Genomic and cDNA libraries are **DNA libraries**—sets of cells that host various cloned DNA fragments. In such libraries, a cell that contains a particular DNA fragment of interest is mixed up with thousands or millions of others that do not. All the cells look the same, so researchers have to get tricky to find that one clone among all of the others—the needle in the haystack.

Using a probe is one trick. A **probe** is a fragment of DNA or RNA labeled with a tracer (Section 2.2). Researchers design probes to match a targeted DNA sequence. For example, they may synthesize an oligomer (a short chain of nucleotides) based on a known DNA sequence, then attach a radioactive phosphate group to it. The nucleotide sequence of a probe is complementary to that of the targeted gene, so the probe can base-pair with the gene. Base pairing between DNA (or DNA and RNA) from more than one source is called **nucleic acid hybridization**.

A probe mixed with DNA from a library base-pairs with (hybridizes to) the targeted gene (Figure 15.5). Researchers pinpoint a clone that hosts the gene by detecting the label on the probe. The clone is then cultured, and the DNA fragment of interest is extracted in bulk from the cultured cells.

PCR

The **polymerase chain reaction (PCR)** is a technique used to mass-produce copies of a particular section of DNA without having to clone it in living cells. The reaction can transform a needle in a haystack—that one-in-a-million fragment of DNA—into a huge stack of needles with a little hay in it (Figure 15.6).

DNA library Collection of cells that host different fragments of foreign DNA, often representing an organism's entire genome.
genome An organism's complete set of genetic material.
nucleic acid hybridization Base-pairing between DNA or RNA from different sources.
polymerase chain reaction (PCR) Method that rapidly generates many copies of a specific section of DNA.
primer Short, single strand of DNA designed to hybridize with a DNA fragment.
probe Short fragment of DNA labeled with a tracer; designed to hybridize with a nucleotide sequence of interest.

Figure 15.6 Animated Two rounds of PCR. Each cycle of this reaction can double the number of copies of a targeted section of DNA. Thirty cycles can make a billion copies.

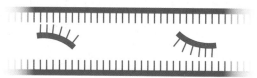

targeted section

1 DNA template (*blue*) is mixed with primers (*pink*), nucleotides, and heat-tolerant *Taq* DNA polymerase.

The starting material for PCR is any sample of DNA with at least one molecule of a target sequence. It might be DNA from a mixture of 10 million different clones, a sperm, a hair left at a crime scene, or a mummy. Essentially any sample that has DNA in it can be used for PCR.

The PCR reaction is based on DNA replication (Section 8.6). First, the starting material is mixed with DNA polymerase, nucleotides, and primers. **Primers** are short single strands of DNA that base-pair with a certain DNA sequence. In PCR, two primers are made. Each base-pairs with one end of the section of DNA to be amplified, or mass-produced **1**. Researchers expose the reaction mixture to repeated cycles of high and low temperature. High temperature disrupts the hydrogen bonds that hold the two strands of a DNA double helix together (Section 8.4). Thus, during a high-temperature cycle, every molecule of double-stranded DNA unwinds and becomes single-stranded **2**. During a low-temperature cycle, the single DNA strands hybridize with complementary partner strands, and double-stranded DNA forms again.

The DNA polymerases of most organisms denature at the high temperatures required to separate DNA strands. The kind that is used in PCR reactions, *Taq* polymerase, is from *Thermus aquaticus*. This bacterial species lives in hot springs and hydrothermal vents, so its DNA polymerase is necessarily heat-tolerant.

Taq polymerase recognizes hybridized primers as places to start DNA synthesis. During a low-temperature cycle, the enzyme starts replicating DNA where primers have hybridized with template **3**. Synthesis proceeds along the template strand until the temperature rises and the DNA separates into single strands **4**. The newly synthesized DNA is a copy of the targeted section.

When the mixture cools, the primers rehybridize, and DNA synthesis begins again. The number of copies of the targeted section of DNA can double with each cycle of heating and cooling **5**. Thirty PCR cycles may amplify that number a billionfold.

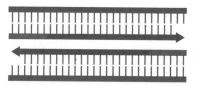

2 When the mixture is heated, the double-stranded DNA template separates into single strands. When it is cooled, some of the primers base-pair with the template DNA.

3 *Taq* polymerase begins DNA synthesis at the primers, so complementary strands of DNA form on the single-stranded templates.

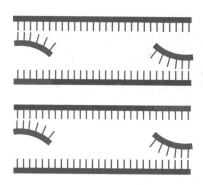

4 The mixture is heated again, and the double-stranded DNA separates into single strands. When it is cooled, some of the primers base-pair with the template DNA. The copied DNA also serves as a template.

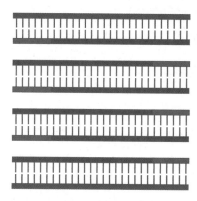

5 Each round of PCR reactions can double the number of copies of the targeted DNA section.

Take-Home Message How do researchers study one gene in the context of many others?

> Researchers isolate one gene from the many other genes in a genome by making DNA libraries or using PCR.

> Probes are used to identify one clone that hosts a DNA fragment of interest among many other clones in a DNA library.

> PCR, the polymerase chain reaction, quickly mass-produces copies of a particular section of DNA.

> DNA sequencing reveals the order of nucleotide bases in a section of DNA.

< Links to Tracers 2.2, Nucleotides 8.4, DNA replication 8.6

Researchers determine the order of the nucleotide bases in DNA with **DNA sequencing** (Figure 15.7). The most commonly used method is similar to DNA replication, in that DNA polymerase synthesizes a strand of DNA based on the nucleotide sequence of a template molecule. Researchers mix the DNA to be sequenced (the template) with nucleotides, DNA polymerase, and a primer that hybridizes to the DNA. Starting at the primer, the polymerase joins free nucleotides into a new strand of DNA, in the order dictated by the sequence of the template.

DNA polymerase joins a nucleotide to a DNA strand only at the hydroxyl group on the strand's 3' carbon (Section 8.6). The DNA sequencing reaction mixture includes four kinds of dideoxynucleotides, which have no hydroxyl group on their 3' carbon ❶. During the reaction, a polymerase randomly adds either a regular nucleotide or a dideoxynucleotide to the end of a growing DNA strand. If it adds a dideoxynucleotide, the 3' carbon of the strand will not have a hydroxyl group, so synthesis of the strand ends there ❷. After about 10 minutes, the reaction has produced millions of DNA fragments of all different lengths; most are incomplete copies of the starting DNA. All of the copies end with one of the four dideoxynucleotides ❸. For example, there will be many 10-base-pair-long copies of the template in the mixture. If the tenth base in the original DNA molecule was adenine, every one of those fragments will end with a dideoxyadenine.

The fragments are then separated by **electrophoresis**. With this technique, an electric field pulls the DNA fragments through a semisolid gel. DNA fragments of different sizes move through the gel at different rates. The shorter the fragment, the faster it moves, because shorter fragments slip through the tangled molecules of the gel faster than longer fragments do.

All fragments of the same length move through the gel at the same speed, so they gather into bands. All of the fragments in a given band have the same dideoxynucleotide at their ends. Each of the four types of dideoxynucleotides (A, C, G, or T) was labeled with a different colored pigment tracer, and those tracers now impart distinct colors to the bands ❹. Each color designates one of the four dideoxynucleotides, so the order of colored bands in the gel represents the DNA sequence ❺.

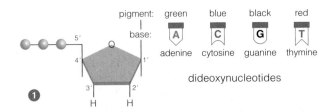

Figure 15.7 Animated DNA sequencing, in which DNA polymerase is used to incompletely replicate a section of DNA.

❶ Sequencing depends on dideoxynucleotides to terminate DNA replication. Each is labeled with a colored pigment. Compare Figure 8.7.

❷ DNA polymerase uses a section of DNA as a template to synthesize new strands of DNA. Synthesis of each new strand stops when a dideoxynucleotide is added.

❸ At the end of the reaction, there are many incomplete copies of the original DNA in the mixture.

❹ Electrophoresis separates the copied DNA fragments into bands according to their length. All of the DNA strands in each band end with the same dideoxynucleotide; thus, each band is the color of that dideoxynucleotide's tracer pigment.

❺ A computer detects and records the color of successive bands on the gel (see Figure 15.8 for an example). The order of colors of the bands represents the sequence of the template DNA.

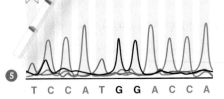

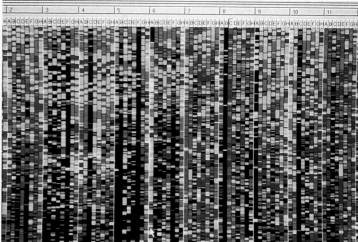

Figure 15.8 Human genome sequencing. *Left*, some of the supercomputers used to sequence the human genome at Venter's Celera Genomics in Maryland. Information in Celera's SNP database is the basis of many new genetic tests. *Right*, a human DNA sequence, raw data.

The Human Genome Project

The technique we just described was invented in 1975. Ten years later, DNA sequencing had become so routine that people were thinking about sequencing the entire human genome: a daunting proposition, given that it consists of about 3 billion bases. At the time, the task would have required at least 6 million sequencing reactions, all done by hand. Proponents insisted that sequencing the genome would have enormous payoffs for medicine and research. Opponents said it would divert funds from work that was more urgent—and had a better chance of succeeding. But sequencing techniques kept getting better, so every year more bases could be sequenced in less time. Automated (robotic) DNA sequencing and PCR had just been invented. Both of these techniques were still cumbersome and expensive, but many researchers sensed their potential. Waiting for faster technologies seemed the most efficient way to sequence 3 billion bases, but how fast did they need to be before the project could begin?

A few private companies decided not to wait, and started to sequence the human genome. One of them intended to patent the sequence after it was determined. This development provoked widespread outrage, but it also spurred commitments in the public sector. In 1988, the National Institutes of Health (NIH) effectively annexed the project by hiring James Watson (of DNA structure fame) to head the official Human Genome Project, and providing $200 million per year to fund it. A consortium formed between the NIH and international institutions that were sequencing different parts

of the genome. Watson set aside 3 percent of the funding for studies of ethical and social issues arising from the research. He later resigned over a patent disagreement, and geneticist Francis Collins took his place.

Amid the ongoing squabbles over patent issues, Celera Genomics formed in 1998. With biologist Craig Venter at its helm, the company intended to commercialize genetic information. Celera started to invent faster techniques for sequencing genomic DNA (Figure 15.8), because the first to have the complete sequence had a legal basis for patenting it. The competition motivated the public consortium to move its efforts into high gear.

Then, in 2000, U.S. President Bill Clinton and British Prime Minister Tony Blair jointly declared that the sequence of the human genome could not be patented. Celera kept on sequencing anyway. Celera and the public consortium separately published about 90 percent of the sequence in 2001. By 2003, fifty years after the discovery of the structure of DNA, the sequence of the human genome was officially completed. At this writing, about 99 percent of its coding regions have been identified. Researchers have not discovered what all of the genes encode, only where they are in the genome. What do we do with this vast amount of data? The next step is to find out what the sequence means.

DNA sequencing Method of determining the order of nucleotides in DNA.
electrophoresis Technique that separates DNA fragments by size.

Take-Home Message **How is the order of nucleotides in DNA determined?**

❯ With DNA sequencing, a strand of DNA is partially replicated. Electrophoresis is used to separate the resulting fragments of DNA, which are tagged with tracers, by length.

❯ Improved sequencing techniques and worldwide efforts allowed the human genome sequence to be determined.

> Comparing the sequence of the human genome with that of other species is helping us understand how the human body works.
> Unique sequences of genomic DNA can be used to distinguish an individual from all others.
< Links to Lipoproteins 3.5, DNA replication 8.6, Knockouts 10.3, Locus 13.2, Complex variation in traits 13.6

It took 15 years to sequence the human genome for the first time, but the techniques have improved so much that sequencing an entire genome now takes less than a month. Full genome sequencing is already available to the general public. However, even though we are able to determine the sequence of an individual's genome, it will be a long time before we understand all the information coded within that sequence.

The human genome contains a massive amount of seemingly cryptic data. Currently, the best way to decipher it is by comparing it to genomes of other organisms, the premise being that all organisms are descended from shared ancestors, so all genomes are related to some extent. We see evidence of such genetic relationships simply by comparing the raw sequence data, which, in some regions, is extremely similar across many species (Figure 15.9).

Comparing genomes is part of **genomics**, the study of genomes. Genomics is a broad field that encompasses whole-genome comparisons, structural analysis of gene products, and the study of small-scale variation. It is also providing powerful insights into evolution. For example, comparing primate genomes revealed how speciation can occur by structural changes in chromosomes (Section 14.5). Comparing genomes also revealed that changes in chromosome structure do not occur randomly. Rather, if a chromosome breaks, it tends to do it in a particular spot. Human, mouse, rat, cow, pig, dog, cat, and horse chromosomes have undergone several translocations at these breakage hot spots during evolution. In humans, chromosome abnormalities that contribute to the progression of cancer also occur at the very same hot spots.

Comparisons between coding regions of a genome are offering medical benefits. We have learned about the function of many human genes by studying their coun-

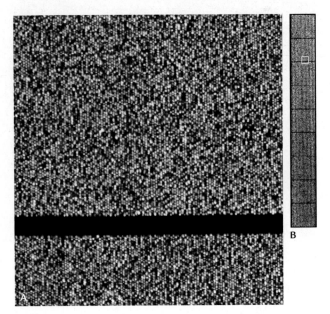

Figure 15.10 SNP–chip analysis. **A** Each spot is a region where the individual's genomic DNA has hybridized with one SNP. **B** The entire chip tests for 550,000 SNPs. The small *white* box indicates the magnified portion shown in **A**.

terpart genes in other species. For example, researchers comparing the human and mouse genomes discovered a human version of a mouse gene, *APOA5*, that encodes a lipoprotein (Section 3.5). Mice with an *APOA5* knockout have four times the normal level of triglycerides in their blood. The researchers then looked for—and found—a correlation between *APOA5* mutations and high triglyceride levels in humans. High triglycerides are a risk factor for coronary artery disease.

DNA Profiling

As you learned in Section 15.1, only about 1 percent of your DNA is unique. The shared part is what makes you human; the differences make you a unique member of the species. In fact, those differences are so unique that they can be used to identify you. Identifying an individual by his or her DNA is called **DNA profiling**.

```
758 GATAATCCTGTTTTGAACAAAAGGTCAAATTGCTGAATAGAAA-GTCTTGATTAACTAAAAGATGTACAAAGTGGAATTA 836  Human
752 GATAATCCTGTTTTGAACAAAAGGTCAAATTGCTGAATAGAAA-GTCTTGATTAACTAAAAGATGTACAAAGTGGAATTA 830  Mouse
751 GATAATCCTGTTTTGAACAAAAGGTCAAATTGCTGAATAGAAA-GTCTTGATTAACTAAAAGATGTACAAAGTGGAATTA 829  Rat
754 GATAATCCTGTTTTGAACAAAAGGTCAAATTGCTGAATAGAAA-GTCTTGATTAACTAAAAGATGTACAAAGTGGAATTA 832  Dog
782 GATAATCCTGTTTTGAACAAAAGGTCAAATTGCTGAATAGAAA-GTCTTGATTAACTAAAAGATGTACAAAGTGGAATTA 860  Chicken
758 GATAATCCTGTTTTGAACAAAAGGTCAAATTGCTGAATAGAAA-GTCTTGATTAAGTAAAAGATGTACAAAGTGGAATTA 836  Frog
823 GATAATCCTGTTTTGAACAAAAGGTCAGATTGCTGAATAGAAAAGGCTTGATTAAAGCAGAGATGTACAAAGTGGACGCA 902  Zebrafish
763 GATAATCCTGTTTTGAACAAAAGGTCAAATTGTTGAATAGAGACGCTTTGATAAAGCGGAGGAGGTACAAAGTGGGACC- 841  Fugu
```

Figure 15.9 Genomic DNA alignment. This is a region of the gene for a DNA polymerase. Differences are highlighted. The chance that any two of these sequences would randomly match is about 1 in 10^{46}.

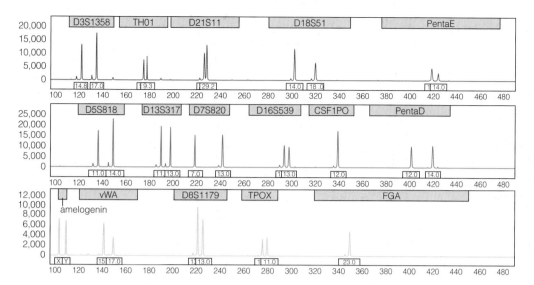

Figure 15.11 An individual's short tandem repeat profile. Each peak on the chart represents a test for one short tandem repeat. The size of a peak indicates the number of repeats at that locus.

For example, this individual has 14 repeats at the D3S1358 locus on chromosome 3, and 17 repeats on the other chromosome 3 (remember that human body cells are diploid, with two sets of chromosomes).

SNP analysis is one example of DNA profiling. SNP-chips are microscopic arrays (microarrays) of DNA samples that have been stamped in separate spots on small glass plates. Each sample is an oligomer with one SNP. In an SNP analysis, an individual's genomic DNA is washed over a SNP-chip. The DNA hybridizes only with oligomers that have a matching SNP sequence. Probes reveal where the genomic DNA has hybridized to an oligomer—and which SNPs are carried by the individual (Figure 15.10).

Another example of DNA profiling involves **short tandem repeats**, sections of DNA in which a series of 4 or 5 nucleotides is repeated several times in a row. Short tandem repeats tend to occur in predictable spots, but the number of repeats in each spot differs among individuals. For example, one person's DNA may have fifteen repeats of the bases TTTTC at a certain locus. Another person's DNA may have this sequence repeated only twice in the same locus. Such repeats slip spontaneously into DNA during replication, and their numbers grow or shrink over generations. Unless two people are identical twins, the chance that they have identical short tandem repeats in even three regions of DNA is 1 in 10^{18}, which is far more than the number of people on Earth. Thus, an individual's array of short tandem repeats is, for all practical purposes, unique.

Analyzing a person's short tandem repeats begins with PCR, which is used to copy ten to thirteen particular regions of chromosomal DNA known to have repeats. The sizes of the copied DNA fragments differ among most individuals, because the number of tandem repeats in those regions also differs. Thus, electrophoresis can be used to reveal an individual's unique array of short tandem repeats (Figure 15.11).

Analysis of short tandem repeats will soon be replaced by full genome sequencing, but for now it continues to be a common DNA profiling method. Geneticists compare short tandem repeats on Y chromosomes to determine relationships among male relatives, and to trace an individual's ethnic heritage. They also track mutations that accumulate in populations over time by comparing DNA profiles of living organisms with those of ancient ones. Such studies are allowing us to reconstruct population dispersals that happened long ago.

Short tandem repeat profiles are routinely used to resolve kinship disputes, and as evidence in criminal cases. Within the context of a criminal or forensic investigation, DNA profiling is called DNA fingerprinting. The Federal Bureau of Investigation maintains a database of DNA fingerprints. As of November 2009, the database contained the short tandem repeat profiles of 7.6 million offenders, and had been used in over 100,000 criminal investigations. DNA fingerprints have also been used to identify the remains of almost 300,000 people, including the individuals who died in the World Trade Center on September 11, 2001.

DNA profiling Identifying an individual by analyzing the unique parts of his or her DNA.
genomics The study of genomes.
short tandem repeats In chromosomal DNA, sequences of 4 or 5 bases repeated multiple times in a row.

Take-Home Message **What do we do with information about genomes?**

❯ Analysis of the human genome sequence is yielding new information about human genes and how they work.

❯ DNA profiling identifies individuals by the unique parts of their DNA.

15.6 Genetic Engineering

❯ Bacteria and yeast are the organisms most commonly subjected to genetic engineering.

❮ Link to Gene expression 9.2

Traditional cross-breeding methods can alter genomes, but only if individuals with the desired traits will interbreed. Genetic engineering takes gene-swapping to an entirely different level. **Genetic engineering** is a laboratory process by which an individual's genome is deliberately modified. A gene may be altered and reinserted into an individual of the same species, or a gene from one species may be transferred to another to produce an organism that is **transgenic**. Both methods result in a **genetically modified organism**, or **GMO**.

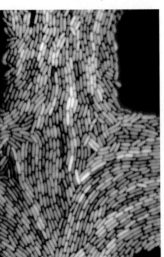

Genetically modified bacteria expressing a jellyfish gene emit green light.

The most common GMOs are bacteria and yeast. These cells have the metabolic machinery to make complex organic molecules, and they are easily modified. For example, the *E. coli* on the *left* have been modified to produce a fluorescent protein from jellyfish. The cells are genetically identical, so the visible variation in fluorescence among them reveals differences in gene expression. Such differences may help us discover why some bacteria of a population become dangerously resistant to antibiotics, and others do not.

Bacteria and yeast have been modified to produce medically important proteins. People with diabetes were among the first beneficiaries of such organisms. Insulin for their injections was once extracted from animals, but it provoked an allergic reaction in some people. Human insulin, which does not provoke allergic reactions, has been produced by transgenic *E. coli* since 1982. Slight modifications of the gene have also yielded fast-acting and slow-release forms of human insulin.

Engineered microorganisms also produce proteins used in food manufacturing. For example, cheese is traditionally made with an extract of calf stomachs, which contain the enzyme chymotrypsin. Most cheese manufacturers now use chymotrypsin made by genetically engineered bacteria. Other examples are GMO-produced enzymes that improve the taste and clarity of beer and fruit juice, slow bread staling, or modify fats.

15.7 Designer Plants

❯ Genetically engineered crop plants are widespread in the United States.

❮ Links to β-carotene 6.2, Promoters 9.3

Agrobacterium tumefaciens is a species of bacteria that infects many plants, including peas, beans, potatoes, and other important crops. It carries a plasmid with genes that cause tumors to form on infected plants; hence the name Ti plasmid (for Tumor-inducing). Researchers use the Ti plasmid as a vector to transfer foreign or modified genes into plants. They remove the tumor-inducing genes from the plasmid, then insert desired genes. Whole plants can be grown from plant cells that integrate the modified plasmid into their chromosomes (Figure 15.12).

Genetically modified *A. tumefaciens* bacteria are used to deliver genes into some food crop plants, including soybeans, squash, and potatoes. Researchers also transfer genes into plants by way of electric or chemical shocks, or by blasting them with DNA-coated pellets.

As crop production expands to keep pace with human population growth, it places unavoidable pressure on ecosystems everywhere. Irrigation leaves mineral and salt residues in soils. Tilled soil erodes, taking topsoil with it. Runoff clogs rivers, and fertilizer in it causes algae to grow so fast that fish suffocate. Pesticides can be harmful to humans, other animals, and beneficial insects.

Pressured to produce more food at lower cost and with less damage to the environment, many farmers have begun to rely on genetically modified crop plants. Some of these modified plants carry genes that impart resistance to devastating plant diseases. Others offer improved yields, such as a strain of transgenic wheat that has twice the yield of unmodified wheat. GMO crops such as Bt corn and soy help farmers use smaller amounts of toxic pesticides. Organic farmers often spray their crops with spores of Bt (*Bacillus thuringiensis*), a bacterial species that makes a protein toxic only to insect larvae. Researchers transferred the gene encoding the Bt protein into plants. The engineered plants produce the Bt protein, but otherwise they are identical to unmodified plants. Insect larvae die shortly after eating their first and only GMO meal. Farmers can use much less pesticide on crops that make their own (Figure 15.13).

Transgenic crop plants are also being developed for Africa and other drought-stricken, impoverished regions of the world. Genes that confer drought tolerance and

---

Take-Home Message What is genetic engineering?

❯ Genetic engineering is the deliberate alteration of an individual's genome, and it results in a genetically modified organism (GMO).

❯ A transgenic organism carries a gene from a different species. Transgenic bacteria and yeast are used in research, medicine, and industry.

genetic engineering Process by which deliberate changes are introduced into an individual's genome.

genetically modified organism (GMO) Organism whose genome has been modified by genetic engineering.

transgenic Refers to a genetically modified organism that carries a gene from a different species.

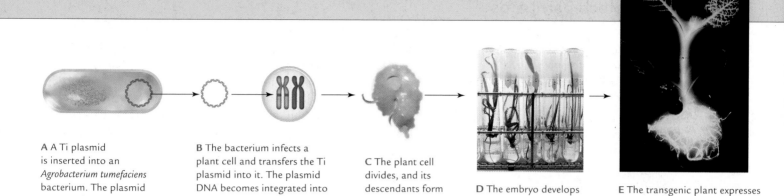

A A Ti plasmid is inserted into an *Agrobacterium tumefaciens* bacterium. The plasmid carries a foreign gene.

B The bacterium infects a plant cell and transfers the Ti plasmid into it. The plasmid DNA becomes integrated into one of the cell's chromosomes.

C The plant cell divides, and its descendants form an embryo.

D The embryo develops into a transgenic plant.

E The transgenic plant expresses the foreign gene. This tobacco plant is expressing a firefly gene.

Figure 15.12 Animated Using the Ti plasmid to make a transgenic plant.

insect resistance are being introduced into plants such as corn, beans, sugarcane, cassava, cowpeas, banana, and wheat. The resulting GMO crops may help people in those regions who rely on agriculture for food and income.

Genetic modifications can make food plants more nutritious. For example, rice plants have been engineered to make β-carotene in their seeds. β-carotene is an orange photosynthetic pigment (Section 6.2) that is remodeled by cells of the small intestine into vitamin A. Two genes in the β-carotene synthesis pathway were transferred into rice plants. One gene was from corn; the other, from bacteria. Both are under the control of a promoter that works in seeds. One cup of the engineered rice seeds—grains of Golden Rice—has enough β-carotene to satisfy a child's daily recommended amount of vitamin A.

The USDA Animal and Plant Health Inspection Service (APHIS) regulates the introduction of GMOs into the environment. At this writing, APHIS has deregulated seventy-four genetically modified crop plants, which means the plants are approved for unregulated use in the United States. The most widely planted GMO crops include corn, sorghum, cotton, soy, canola, and alfalfa engineered for resistance to glyphosate, an herbicide. Rather than tilling the soil to control weeds, farmers can spray their fields with glyphosate, which kills the weeds but not the engineered crops.

After long-term, widespread use of glyphosate, weeds resistant to the herbicide are becoming more common. The engineered gene is also appearing in wild plants and in nonengineered crops, which means that transgenes can (and do) escape into the environment. The genes are probably being transferred from transgenic plants to nontransgenic ones via pollen carried by wind or insects.

Many people are opposed to any GMO. Some worry that our ability to tinker with genetics has surpassed our ability to understand the impact of the tinkering. Controversy raised by such GMO use invites you to read the research and form your own opinions. The alternative is to be swayed by media hype (the term "Frankenfood,"

Figure 15.13 Genetically modified crops can help farmers use less pesticide. *Top*, the Bt gene conferred insect resistance to the genetically modified plants that produced this corn. *Bottom*, corn produced by unmodified plants is more vulnerable to insect pests.

for instance), or by reports from possibly biased sources (such as herbicide manufacturers).

> Take-Home Message **Are there genetically modified plants?**
> ❯ Plants with modified or foreign genes are now common farm crops.

> Genetically engineered animals are invaluable in medical research and in other applications.

< Links to Knockout experiments 10.3, Human genetic disorders Chapter 14

Traditional cross-breeding has produced animals so unusual that transgenic animals may seem a bit mundane by comparison (Figure 15.14A). Cross-breeding is also a form of genetic manipulation, but many transgenic animals would probably never have occurred without laboratory intervention (Figure 15.14B,C).

The first genetically modified animals were mice. Today, such mice are commonplace, and they are invaluable in research (Figure 15.15). For example, we have discovered the function of human genes (including the *APOA5* gene discussed in Section 15.5) by inactivating their counterparts in mice. Genetically modified mice are also used as models of human diseases. For example, researchers inactivated the molecules involved in the control of glucose metabolism, one by one, in mice. Studying the effects of the knockouts has resulted in much of our current understanding of how diabetes works in humans. Genetically modified animals also make proteins that have medical and industrial applications. Various transgenic goats produce proteins used to treat cystic fibrosis, heart attacks, blood clotting disorders, and even nerve gas

A

exposure. Milk from goats transgenic for lysozyme, an antibacterial protein in human milk, may protect infants and children in developing countries from acute diarrheal disease. Goats transgenic for a spider silk gene produce the silk protein in their milk; researchers can spin this protein into nanofibers that are useful in medical and electronics applications. Rabbits make human interleukin-2, a protein that triggers divisions of immune cells. Genetic engineering has also given us dairy goats with heart-healthy milk, pigs with heart-healthy fat and environmentally friendly low-phosphate feces, extra-large sheep, and cows that are resistant to mad cow disease. Many people think that genetically engineering livestock is unconscionable. Others see it as an extension of thousands of years of acceptable animal husbandry practices. The techniques have changed, but not the intent: We humans continue to have a vested interest in improving our livestock.

Knockouts and Organ Factories

Millions of people suffer with organs or tissues that are damaged beyond repair. In any given year, more than 80,000 of them are on waiting lists for an organ transplant in the United States alone. Human donors are in such short supply that illegal organ trafficking is now a common problem.

Pigs are a potential source of organs for transplantation, because pig and human organs are about the same in both size and function. However, the human immune system battles anything it recognizes as nonself. It rejects a pig organ at once, because it recognizes proteins and

Figure 15.14 Genetically modified animals. **A** Featherless chicken developed by traditional cross-breeding methods in Israel. Such chickens survive in hot deserts where cooling systems are not an option. **B** The pig on the *left* is transgenic for a yellow fluorescent protein; its nontransgenic littermate is on the *right*. **C** Mira the transgenic goat produces a human anticlotting factor in her milk.

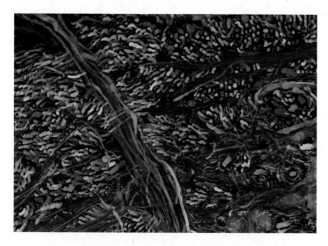

Figure 15.15 Example of how genetically engineered animals are useful in research. Mice transgenic for multiple pigments ("brainbow mice") are allowing researchers to map the complex neural circuitry of the brain. Individual nerve cells in the brainstem of a brainbow mouse are visible in this fluorescence micrograph.

carbohydrates on the plasma membrane of pig cells. Within a few hours, blood coagulates inside the organ's vessels and dooms the transplant. Drugs can suppress the immune response, but they also render organ recipients particularly vulnerable to infection. Researchers have produced genetically modified pigs that lack the offending molecules on their cells. The human immune system may not reject tissues or organs transplanted from these pigs.

Transferring an organ from one species into another is called **xenotransplantation**. Critics of xenotransplantation are concerned that, among other things, pig-to-human transplants would invite pig viruses to cross the species barrier and infect humans, perhaps catastrophically. Their concerns are not unfounded. Evidence suggests that some of the worst pandemics arose when animal viruses adapted to new hosts: humans.

Tinkering with the genes of animals raises a host of ethical dilemmas. For example, mice, monkeys, and other animals have been genetically modified to carry mutations associated with certain human diseases. These animals are allowing researchers to study—and test treatments for—conditions such as multiple sclerosis, cystic fibrosis, diabetes, cancer, and Huntington's disease without experimenting on humans. However, the engineered animals often suffer the same terrible symptoms of these conditions as humans do.

xenotransplantation Transplantation of an organ from one species into another.

15.9 Safety Issues

❭ The first transfer of foreign DNA into bacteria ignited an ongoing debate about potential dangers of transgenic organisms that enter the environment.

When James Watson and Francis Crick presented their model of the DNA double helix in 1953, they ignited a global blaze of optimism about genetic research. The very book of life seemed to be open for scrutiny. In reality, no one could read it. Scientific breakthroughs are not very often accompanied by the simultaneous discovery of the tools needed to study them. New techniques would have to be invented before that book would become readable.

Twenty years later, Paul Berg and his coworkers discovered how to make recombinant organisms by fusing DNA from two species of bacteria. By isolating DNA in manageable subsets, researchers now had the tools to be able to study its sequence in detail. They began to clone and analyze DNA from many different organisms. The technique of genetic engineering was born, and suddenly everyone was worried about it.

Researchers knew that DNA itself was not toxic, but they could not predict with certainty what would happen every time they fused genetic material from different organisms. Would they accidentally make a superpathogen? Could they make a new, dangerous form of life by fusing DNA of two normally harmless organisms? What if that new form escaped from the laboratory and transformed other organisms?

In a remarkably quick and responsible display of self-regulation, scientists reached a consensus on new safety guidelines for DNA research. Adopted at once by the NIH, these guidelines included precautions for laboratory procedures. They covered the design and use of host organisms that could survive only under the narrow range of conditions inside the laboratory. Researchers stopped using DNA from pathogenic or toxic organisms for recombinant DNA experiments until proper containment facilities were developed.

Now, all genetic engineering should be done under these laboratory guidelines, but the rules are not a guarantee of safety. We are still learning about escaped GMOs and their effects, and enforcement is a problem. For example, the expense of deregulating a GMO for release and importation is prohibitive for endeavors in the public sector. Thus, most commercial GMOs were produced by large, private companies—the same ones that typically wield tremendous political influence over the very government agencies charged with regulating them.

Take-Home Message Why do we genetically engineer animals?

❭ Animals that would be impossible to produce by traditional breeding methods are being created by genetic engineering.

Take-Home Message Is genetic engineering safe?

❭ Guidelines for DNA research have been in place for decades in the United States and other countries. Researchers are expected to comply, but the guidelines are not a guarantee of safety.

❯ We as a society continue to work our way through the ethical implications of applying new DNA technologies.

❯ The manipulation of individual genomes continues even as we are weighing the risks and benefits of this research.

❮ Links to Proto-oncogenes and cancer 11.6, Locus 13.2, Human genetic disorders 14.2

Getting Better We know of more than 15,000 serious genetic disorders. Collectively, they cause 20 to 30 percent of infant deaths each year, and account for half of all mentally impaired patients and a fourth of all hospital admissions. They also contribute to many age-related disorders, including cancer, Parkinson's disease, and diabetes. Drugs and other treatments can minimize the symptoms of some genetic disorders, but gene therapy is the only cure. **Gene therapy** is the transfer of recombinant DNA into an individual's body cells, with the intent to correct a genetic defect or treat a disease. The transfer, which occurs by way of lipid clusters or genetically engineered viruses, inserts an unmutated gene into an individual's chromosomes.

Human gene therapy is a compelling reason to embrace genetic engineering research. It is now being tested as a treatment for heart attack, sickle-cell anemia, cystic fibrosis, hemophilia A, Parkinson's disease, Alzheimer's disease, several types of cancer, and inherited diseases of the eye, the ear, and the immune system. The results are encouraging. For example, little Rhys Evans (Figure 15.16) was born with SCID-X1, a severe X-linked genetic disorder that stems from a mutated allele of the *IL2RG* gene. The gene encodes a receptor for an immune signaling molecule. Children affected by this disorder can survive only in germ-free isolation tents, because they cannot fight infections. In the late 1990s, researchers used a genetically engineered virus to insert unmutated copies of *IL2RG* into cells taken from the bone marrow of twenty boys with SCID-X1. Each child's modified cells were infused back into his bone marrow. Within months of their treatment eighteen of the boys left their isolation tents for good. Rhys was one of them. Gene therapy had permanently repaired their immune systems.

Getting Worse Manipulating a gene within the context of a living individual is unpredictable even when we know its sequence and locus. No one, for example, can predict where a virus-injected gene will become integrated into a chromosome. Its insertion might disrupt other genes. If it interrupts a gene that is part of the controls over cell division, then cancer might be the outcome. Five of the twenty boys treated with gene therapy for SCID-X1 have since developed a type of bone marrow cancer called leukemia, and one of them has died. The researchers had wrongly predicted that cancer related to the gene therapy would be rare. Research now implicates the very gene targeted for repair, especially when combined with the virus

Figure 15.16 Rhys Evans, who was born with SCID-X1. His immune system has been permanently repaired by gene therapy.

that delivered it. Apparently, integration of the modified viral DNA activated nearby proto-oncogenes (Section 11.6) in the children's chromosomes.

Other unanticipated problems sometimes occur with gene therapy. Jesse Gelsinger had a rare genetic deficiency of a liver enzyme that helps the body rid itself of ammonia, a toxic by-product of protein breakdown. Jesse's health was fairly stable while he was on a low-protein diet, but he had to take a lot of medication. In 1999, Jesse volunteered to be in a clinical trial testing a gene therapy for his condition. He had a severe allergic reaction to the viral vector, and four days after receiving the treatment, his organs shut down and he died. He was 18.

Getting Perfect The idea of selecting the most desirable human traits, **eugenics**, is an old one. It has been used as a justification for some of the most horrific episodes in human history, including the genocide of 6 million Jews during World War II. Thus, it continues to be a hotly debated social issue. For example, using gene therapy to cure human genetic disorders seems like a socially acceptable goal to most people. However, imagine taking this idea a bit further. Would it also be acceptable to engineer the genome of an individual who is within a normal range of phenotype in order to modify a particular trait? Researchers have already produced mice that have improved memory, enhanced learning ability, bigger muscles, and longer lives. Why not people?

eugenics Idea of deliberately improving the genetic qualities of the human race.
gene therapy The transfer of a normal or modified gene into an individual with the goal of treating a genetic defect or disorder.

Given the pace of genetics research, the eugenics debate is no longer about how we would engineer desirable traits, but how we would choose the traits that are desirable. Realistically, cures for many severe but rare genetic disorders will not be found, because the financial return will not even cover the cost of the research. Eugenics, however, might just turn a profit. How much would potential parents pay to be sure that their child will be tall or blue-eyed? Would it be okay to engineer "superhumans" with breathtaking strength or intelligence? How about a treatment that can help you lose that extra weight, and keep it off permanently? The gray area between interesting and abhorrent can be very different depending on who is asked. In a survey conducted in the United States, more than 40 percent of those interviewed said it would be fine to use gene therapy to make smarter and cuter babies. In one poll of British parents, 18 percent would be willing to use it to keep a child from being aggressive, and 10 percent would use it to keep a child from growing up to be homosexual.

Getting There Some people are adamant that we must never alter the DNA of anything. The concern is that gene therapy puts us on a slippery slope that may result in irreversible damage to ourselves and to the biosphere. We as a society may not have the wisdom to know how to stop once we set foot on that slope. One is reminded of our peculiar human tendency to leap before we look. And yet, something about the human experience allows us to dream of such things as wings of our own making, a capacity that carried us into space. In this brave new world, the questions before you are these: What do we stand to lose if serious risks are not taken? And, do we have the right to impose the consequences of taking such risks on those who would choose not to take them?

Take-Home Message Can people be genetically modified?

> Genes can be transferred into a person's cells to correct a genetic defect or treat a disease. However, the outcome of altering a person's genome remains unpredictable given our current understanding of how the genome works.

Personal DNA Testing (revisited)

The results of SNP analysis by a personal DNA testing company also include estimated risks of developing conditions associated with your particular set of SNPs. For example, the test will probably determine whether you are homozygous for one allele of the *MC1R* gene. If you are, the company's report will tell you that you have red hair. Very few SNPs have such a clear cause-and-effect relationship as the *MC1R* allele for red hair, however. Most human traits are polygenic, and many are also influenced by environmental factors such as life-style (Section 13.6). Thus, although a DNA test can reliably determine the SNPs in an individual's genome, it cannot reliably predict the effect of those SNPs on the individual.

For example, if you carry one ε4 allele of the *APOE* gene, a DNA testing company cannot tell you whether you will develop Alzheimer's disease later in life. Instead, the company will report your lifetime risk of developing the disease, which is about 29 percent, as compared with about 9 percent for someone who has no ε4 allele.

What does a 29 percent lifetime risk of developing Alzheimer's disease mean? The number is a probability statistic; it means that, on average, 29 of every 100 people who have the ε4 allele eventually get the disease. Having a high risk does not mean you are certain to end up with Alzheimer's, however. Not everyone who develops the disease has the ε4 allele, and not everyone with the ε4 allele develops Alzheimer's disease. Other as yet unknown alleles—some protective, some not—contribute to the disease.

How Would You Vote? The plunging cost of genetic testing has spurred an explosion of companies offering personal DNA sequencing and SNP profiling. The results of such testing may in some cases be of clinical use, for example in diagnosis of early-onset genetic disorders, or in predicting how an individual will respond to certain medications. However, we are still at an extremely early stage in our understanding of how genes contribute to most conditions, particularly age-related disorders such as Alzheimer's disease. Geneticists believe that it will be five to ten more years before we can use genotype to accurately predict an individual's risk of these conditions. Until then, should genetic testing companies be prohibited from informing clients of their estimated risk of developing such disorders based on SNPs? See CengageNow for details, then vote online (cengagenow.com).

Summary

Section 15.1 Personal DNA testing companies identify a person's unique array of single-nucleotide polymorphisms. Personal genetic testing may soon revolutionize the way medicine is practiced.

Section 15.2 The discovery of **restriction enzymes** allowed researchers to cut huge molecules of chromosomal DNA into manageable, predictable chunks. It also allowed them to combine DNA fragments from different organisms to make **recombinant DNA**. With **DNA cloning**, restriction enzymes cut DNA into pieces,

then DNA ligase splices the pieces into **plasmids** or other **cloning vectors**. The resulting hybrid molecules are inserted into host cells such as bacteria. When a host cell divides, it forms huge populations of genetically identical descendant cells. Each of these clones has a copy of the foreign DNA.

RNA cannot be cloned directly. **Reverse transcriptase**, a viral enzyme, is used to transcribe single-stranded RNA into **cDNA** for cloning.

 Section 15.3 A **DNA library** is a collection of cells that host different fragments of DNA, often representing an organism's entire **genome**. Researchers can use **probes** to identify cells that host a specific fragment of DNA. Base pairing between nucleic acids from different sources is called **nucleic acid hybridization**.

The **polymerase chain reaction (PCR)** uses **primers** and a heat-resistant DNA polymerase to rapidly increase the number of copies of a section of DNA.

 Section 15.4 DNA sequencing reveals the order of bases in a section of DNA. DNA polymerase is used to partially replicate a DNA template. The reaction produces a mixture of DNA fragments of all different lengths. **Electrophoresis** separates the fragments by length into bands. The entire genomes of several organisms have now been sequenced.

 Section 15.5 Genomics, or the study of genomes, is providing insights into the function of the human genome. Similarities between genomes of different organisms are evidence of evolutionary relationships, and can be used as a predictive tool in research.

DNA profiling identifies a person by the unique parts of his or her DNA. Examples include methods of determining an individual's array of SNPs or **short tandem repeats**. Within the context of a criminal investigation, a DNA profile is called a DNA fingerprint.

 Sections 15.6–15.9 Recombinant DNA technology is the basis of **genetic engineering**, the directed modification of an organism's genetic makeup with the intent to modify its phenotype. A gene is modified and reinserted into an individual of the same species, or a gene from one species is inserted into an individual of a different species to make a **transgenic** organism. The result of either process is a **genetically modified organism (GMO)**.

Transgenic bacteria and yeast produce medically valuable proteins. Transgenic crop plants are helping farmers produce food more efficiently. Genetically modified animals produce human proteins, and may one day provide a source of organs and tissues for **xenotransplantation** into humans.

Safety guidelines minimize potential risks to researchers in genetic engineering labs. Although these and other government regulations limit the release of genetically modified organisms into the environment, such laws are not guarantees against accidental releases or unforeseen environmental effects.

 Section 15.10 With **gene therapy**, a gene is transferred into body cells to correct a genetic defect or treat a disease. As with any new technology, the potential benefits of genetically modifying humans must be weighed against the potential risks. The practice raises ethical issues such as whether **eugenics** is desirable.

Self–Quiz Answers in Appendix III

1. _____ cut(s) DNA molecules at specific sites.
   a. DNA polymerase c. Restriction enzymes
   b. DNA probes d. Reverse transcriptase

2. A _____ is a small circle of bacterial DNA that contains a few genes and is separate from the chromosome.
   a. plasmid c. nucleus
   b. chromosome d. double helix

3. By reverse transcription, _____ is assembled on a(n) _____ template.
   a. mRNA; DNA c. DNA; ribosome
   b. cDNA; mRNA d. protein; mRNA

4. For each species, all _____ in the complete set of chromosomes is the _____ .
   a. genomes; phenotype c. mRNA; start of cDNA
   b. DNA; genome d. cDNA; start of mRNA

5. A set of cells that host various DNA fragments collectively representing an organism's entire set of genetic information is a _____ .
   a. genome c. genomic library
   b. clone d. GMO

6. _____ is a technique to determine the order of nucleotide bases in a fragment of DNA.

7. Fragments of DNA can be separated by electrophoresis according to _____ .
   a. sequence c. species
   b. length d. composition

8. PCR can be used _____ .
   a. to increase the number of specific DNA fragments
   b. in DNA fingerprinting
   c. to modify a human genome
   d. a and b are correct

9. An individual's set of unique _____ can be used as a DNA profile.
   a. DNA sequences c. SNPs
   b. short tandem repeats d. all of the above

10. Which of the following can be used to carry foreign DNA into host cells? Choose all correct answers.
   a. RNA e. lipid clusters
   b. viruses f. blasts of pellets
   c. PCR g. xenotransplantation
   d. plasmids h. sequencing

Enhanced Spatial Learning in Mice with Autism Mutation

Autism is a neurobiological disorder with a range of symptoms that include impaired social interactions, stereotyped patterns of behavior such as hand-flapping or rocking, and, occasionally, greatly enhanced intellectual abilities. Some autistic people have a mutation in neuroligin 3, a type of cell adhesion protein (Section 4.4) that connects brain cells to one another. One mutation changes amino acid 451 from arginine to cysteine.

Mouse and human neuroligin 3 are very similar. In 2007, Katsuhiko Tabuchi and his colleagues genetically modified mice to carry the same arginine-to-cysteine substitution in their neuroligin 3. The mutation caused an increase in transmission of some types of signals between brain cells. Mice with the mutation had impaired social behavior, and, unexpectedly, enhanced spatial learning ability (Figure 15.17).

1. In the first test, how many days did unmodified mice need to learn to find the location of a hidden platform within 10 seconds?

2. Did the modified or the unmodified mice learn the location of the platform faster in the first test?

3. Which mice learned faster the second time around?

4. Which mice showed the greatest improvement in memory between the first and the second test?

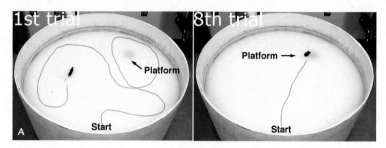

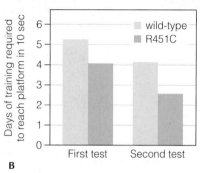

Figure 15.17 Enhanced spatial learning ability in mice with a mutation in neuroligin 3 (R451C), compared with unmodified (wild-type) mice.

A The mice were tested in a water maze, in which a platform is submerged a few millimeters below the surface of a deep pool of warm water. The platform is not visible to swimming mice. Mice do not particularly enjoy swimming, so they locate a hidden platform as fast as they can. When tested again, they can remember its location by checking visual cues around the edge of the pool.

B How quickly they remember the platform's location is a measure of spatial learning ability. The platform was moved and the experiment was repeated for the second test.

---

11. A transgenic organism _____ .
    a. carries a gene from another species
    b. has been genetically modified
    c. both a and b

12. _____ can be used to correct a genetic defect.
    a. Cloning vectors   d. Xenotransplantation
    b. Gene therapy   e. a and b
    c. Cloning   f. all of the above

13. Match the recombinant DNA method with the appropriate enzyme.
    ____ PCR   a. *Taq* polymerase
    ____ cutting DNA   b. DNA ligase
    ____ cDNA synthesis   c. reverse transcriptase
    ____ DNA sequencing   d. restriction enzyme
    ____ pasting DNA   e. DNA polymerase

14. Match the terms with the most suitable description.
    ____ DNA fingerprint   a. having a foreign gene
    ____ Ti plasmid   b. alleles have them
    ____ nucleic acid   c. a person's unique collection
         hybridization   of short tandem repeats
    ____ eugenics   d. base pairing of DNA or
    ____ SNP   DNA and RNA from
    ____ transgenic   different sources
    ____ GMO   e. selecting "desirable" traits
                           f. genetically modified
                           g. used in some gene transfers

Additional questions are available on **CENGAGENOW.**

Critical Thinking

1. Restriction enzymes in bacterial cytoplasm cut injected bacteriophage DNA wherever certain sequences occur. Why do you think the bacterial chromosome does not get chopped up too?

2. The *FOXP2* gene encodes a transcription factor associated with vocal learning in mice, bats, birds, and humans. The chimpanzee, gorilla, and rhesus FOXP2 proteins are identical; the human version differs in only 2 of 715 amino acids, a change thought to have contributed to the development of spoken language. In humans, loss-of-function mutations in *FOXP2* result in severe speech and language disorders. In mice, they hamper brain function and impair vocalizations. Mice genetically engineered to carry the human version of *FOXP2* show changes in their vocal patterns, and more growth and greater adaptability of neurons involved in memory and learning. Biologists do not anticipate that a similar experiment in chimpanzees would confer the ability to speak, because spoken language is a complex, epistatic trait (Section 13.5). What do you think might happen if their prediction is incorrect?

Animations and Interactions on **CENGAGENOW:**
❯ Recombinant DNA; Cloning; Using a radioactive probe; DNA sequencing; PCR; DNA fingerprinting; Genetically engineering plants.

‹ Links to Earlier Concepts

This chapter explores a clash between traditional beliefs and science. You may wish to review critical thinking (Section 1.6) before you begin. What you know about inheritance of alleles (13.2) will help you understand natural selection. Finding the age of ancient rocks and fossils depends on the properties of radioisotopes (2.2). You will see how master genes (10.3, 10.4) are evidence of shared ancestry, and revisit evolution by gene duplication (14.5).

Key Concepts

Emergence of Evolutionary Thought
Nineteenth-century naturalists started to think about the global distribution of species. They discovered similarities and differences among major groups, including those represented as fossils.

A Theory Takes Form
Evidence of evolution, or change in lines of descent, led Charles Darwin and Alfred Wallace to independently develop a theory of natural selection. The theory explains how traits that define each species change over time.

16 Evidence of Evolution

16.1 Reflections of a Distant Past

How do you think about time? Perhaps you can conceive of a few hundred years of human events, but how about a few million? Envisioning the distant past requires an intellectual leap from the familiar to the unknown.

One way to make that leap involves, surprisingly, asteroids. Asteroids are small planets hurtling through space. They range in size from 1 to 1,500 kilometers (roughly 0.5 to 1,000 miles) wide. Millions of them orbit the sun between Mars and Jupiter —cold, stony leftovers from the formation of our solar system. Asteroids are difficult to see even with the best telescopes, because they do not emit light. Many cross Earth's orbit, but most of those pass us by before we know about them. Some have not passed us at all.

The mile-wide Barringer Crater in Arizona is difficult to miss (Figure 16.1). A 300,000-ton asteroid made this impressive pockmark in the desert sandstone when it slammed into Earth 50,000 years ago. The impact was 150 times more powerful than the bomb that leveled Hiroshima.

No humans were in North America at the time of the impact. If there were no witnesses, how do we know what happened? We often reconstruct history by studying physical evidence of events that took place long ago. Geologists were able to infer the most probable cause of the Barringer Crater by analyzing tons of meteorites, melted sand, and other rocky clues at the site.

Similar evidence points to even larger asteroid impacts in the more distant past. For example, a **mass extinction**, or permanent loss of major groups of organisms, occurred 65.5 million years ago. The event is marked by an unusual, worldwide layer of rock called the K–T boundary layer. There are plenty of dinosaur fossils below this layer. Above it, in rock layers that were deposited more recently, there are no dinosaur fossils, anywhere. An impact crater off the coast of what is now the Yucatán Peninsula dates to about 65.5 million years ago. Coincidence? Many

---

mass extinction Simultaneous loss of many lineages from Earth.

Figure 16.1 From evidence to inference. What made the Barringer Crater (*opposite*)? Rocky evidence points to a 300,000-ton asteroid that collided with Earth 50,000 years ago. *Above*, bands that are part of a unique layer of rock that formed 65.5 million years ago, worldwide. The layer marks an abrupt transition in the fossil record that implies a mass extinction. The *red* pocketknife gives an idea of scale.

scientists say no. They have inferred from the evidence that the impact of an asteroid about 20 km (12 miles) wide caused a global catastrophe that wiped out the dinosaurs.

You are about to make an intellectual leap through time, to places that were not even known a few centuries ago. We invite you to launch yourself from this premise: Natural phenomena that occurred in the past can be explained by the very same physical, chemical, and biological processes that operate today. That premise is the foundation for scientific research into the history of life. The research represents a shift from experience to inference—from the known to what can only be surmised— and it gives us astonishing glimpses into the distant past.

Evidence From Fossils
The fossil record provides physical evidence of past changes in many lines of descent. We use the property of radioisotope decay to determine the age of rocks and fossils.

Evidence From Biogeography
Geologic events have influenced evolution. Correlating geologic and evolutionary events helps explain the distribution of species, past and present.

Evidence in Form and Function
Different lineages may have similar body parts that reflect descent from a shared ancestor. Lineages with common ancestry often develop in similar ways.

❭ Belief systems are influenced by the extent of our understanding of the natural world. Those that are inconsistent with systematic observations tend to change over time.

The seeds of biological inquiry were taking hold in the Western world more than 2,000 years ago. Aristotle, the Greek philosopher, was making connections between observations in an attempt to explain the order of the natural world. Like few others of his time, Aristotle viewed nature as a continuum of organization, from lifeless matter through complex plants and animals. Aristotle was one of the first **naturalists**, people who observe life from a scientific perspective.

By the fourteenth century, Aristotle's earlier ideas about nature had been transformed into a rigid view of life, in which a "great chain of being" extended from the lowest form (snakes), through humans, to spiritual beings. Each link in the chain was a species, and each was said to have been forged at the same time in a perfect state. The chain itself was complete and continuous. Because everything that needed to exist already did, there was no room for change. Once every species had been discovered, the meaning of life would be revealed.

European naturalists that embarked on globe-spanning survey expeditions brought back tens of thousands of plants and animals from Asia, Africa, North and South America, and the Pacific Islands. Each newly discovered species was carefully catalogued as another link in the chain of being.

By the late 1800s, naturalists such as Alfred Wallace were seeing patterns in where species live and how they might be related, and had started to think about the natural forces that shape life. These naturalists were pioneers in **biogeography**, the study of patterns in the geographic distribution of species. Some of the patterns raised questions that could not be answered within the framework of prevailing belief systems. For example, globe-trotting explorers had discovered plants and animals living in extremely isolated places. The isolated species looked suspiciously similar to species living across vast expanses of open ocean, or on the other side of impassable mountain ranges. Could different species be related? If so, how could the related species end up geographically isolated?

For example, the three birds in Figure 16.2 live on different continents, but they share a set of unusual features.

Figure 16.3 Similar-looking, unrelated species that are native to distant geographic realms: *above*, an American spiny cactus; and *left*, an African spiny spurge.

A Emu, native to Australia

B Rhea, native to South America

C Ostrich, native to Africa

Figure 16.2 Similar-looking, related species that are native to distant geographic realms. The three types of ratite birds are unlike most other birds in several traits, including their long, muscular legs and their inability to fly.

These flightless birds sprint about on long, muscular legs in flat, open grasslands about the same distance from the equator. All raise their long necks to watch for predators. Wallace thought that the shared set of unusual traits might mean that these three birds descended from a common ancestor (and he was right), but he had no idea how they could have ended up on different continents.

Naturalists of the time also had trouble classifying organisms that are very similar in some features, but different in others. For example, the plants in Figure 16.3 are native to different continents. Both live in hot deserts where water is seasonally scarce. Both have rows of sharp spines that deter herbivores, and both store water in their thick, fleshy stems. However, their reproductive parts are very different, so these plants cannot be as closely related as their outward appearance might suggest.

Observations like these are part of **comparative morphology**, the study of body plans and structures among groups of organisms. Organisms that are outwardly very similar may be quite different internally; think of fishes and porpoises. Others that differ greatly in outward appearance may be very similar in underlying structure. For example, a human arm, a porpoise flipper, an elephant leg, and a bat wing have comparable internal bones, as Section 16.8 will explain.

Comparative morphology in the nineteenth century revealed body parts that have no apparent function, an idea that added to the naturalists' confusion. According to prevailing beliefs, every species had been created in a perfect state. If that were so, then why were there useless parts such as leg bones in snakes (which do not walk), or the vestiges of a tail in humans (Figure 16.4)?

Fossils were puzzling too. A **fossil** is the remains or traces of an organism that lived in the ancient past—physical evidence of ancient life. Geologists mapping rock formations exposed by erosion or quarrying had discovered identical sequences of rock layers in different parts of the world. Deeper layers held fossils of simple marine life. Layers above them held similar but more intricate fossils. In higher layers, fossils that were similar but even more intricate looked like they belonged to modern species. The photos on the *right* show one such series, ten fossils of shelled protists, each from a successive layer of rock in a stack. What did these fossil sequences mean?

Fossils of many animals that had no living representatives were also being unearthed. If the animals had been perfect at the time of creation, then why had they become extinct?

Taken as a whole, the accumulating findings from biogeography, comparative morphology, and geology did not fit with prevailing beliefs of the nineteenth century. If species had not been created in a perfect state (and extinct species, fossil sequences and "useless" body parts implied that they had not), then perhaps species had indeed changed over time.

leg bones

coccyx

A Pythons and boa constrictors have tiny leg bones, but snakes do not walk.

B We humans use our legs, but not our coccyx (tail bones).

Figure 16.4 Animated Vestigial body parts.

---

biogeography Study of patterns in the geographic distribution of species and communities.
comparative morphology Study of body plans and structures among groups of organisms.
fossil Physical evidence of an organism that lived in the ancient past.
naturalist Person who observes life from a scientific perspective.

Take-Home Message **How did observations of the natural world change our thinking in the nineteenth century?**

❯ Increasingly extensive observations of nature in the nineteenth century did not fit with prevailing belief systems.

❯ The cumulative findings from biogeography, comparative morphology, and geology led to new ways of thinking about the natural world.

> In the 1800s, many scholars realized that life on Earth had changed over time, and began to think about what could have caused the changes.

⟨ Link to Critical thinking and how science works 1.6

Squeezing New Evidence Into Old Beliefs

In the nineteenth century, naturalists were faced with increasing evidence that life on Earth, and even Earth itself, had changed over time. Around 1800, Georges Cuvier, an expert in zoology and paleontology, was trying to make sense of the new information. He had observed abrupt changes in the fossil record, and knew that many fossil species seemed to have no living counterparts. Given this evidence, he proposed a startling idea: Many species that had once existed were now extinct. Cuvier also knew about evidence that Earth's surface had changed. For example, he had seen fossilized seashells on mountainsides far from modern seas. Like most others of his time, he assumed Earth's age to be in the thousands, not millions, of years. He reasoned that geologic forces unlike those known today would have been necessary to raise sea floors to mountaintops in this short time span. Catastrophic geological events would have caused

extinctions, after which surviving species repopulated the planet. Cuvier's idea came to be known as **catastrophism**. We now know it is incorrect; geologic processes have not changed over time.

Another scholar, Jean-Baptiste Lamarck, was thinking about processes that drive **evolution**, or change in a line of descent. A line of descent is also called a **lineage**. Lamarck thought that a species gradually improved over generations because of an inherent drive toward perfection, up the chain of being. The drive directed an unknown "fluida" into body parts needing change. By Lamarck's hypothesis, environmental pressures cause an internal need for change in an individual's body, and the resulting change is inherited by offspring. Try using Lamarck's hypothesis to explain why a giraffe's neck is very long. We might predict that some short-necked ancestor of the modern giraffe stretched its neck to browse on leaves beyond the reach of other animals. The stretches may have even made its neck a bit longer. By Lamarck's hypothesis, that animal's offspring would inherit a longer neck, and after many generations strained to reach ever loftier leaves, the modern giraffe would have been the result. Lamarck was correct in thinking that environmental factors affect a species' traits, but his understanding of how inheritance works was incomplete.

Darwin and the HMS *Beagle*

In 1831, the twenty-two-year-old Charles Darwin was wondering what to do with his life. Ever since he was eight, he had wanted to hunt, fish, collect shells, or watch insects and birds—anything but sit in school. After an attempt to study medicine in college, he earned a degree in theology from Cambridge. All through school, however, Darwin spent most of his time with faculty members and other students who embraced natural history. Botanist John Henslow arranged for Darwin to become a naturalist aboard the *Beagle*, a ship about to embark on a survey expedition to South America.

The *Beagle* set sail for South America in December, 1831 (Figure 16.5). The young man who had hated school and had no formal training in science quickly became an enthusiastic naturalist. During the *Beagle*'s five-year voyage, Darwin found many unusual fossils. He saw diverse species living in environments that ranged from the sandy shores of remote islands to the plains high in the Andes. Along the way, he read the first volume of a new and popular book, Charles Lyell's *Principles of Geology*.

Figure 16.5 Voyage of the HMS *Beagle*. With Darwin aboard as ship's naturalist, the vessel (*top*) originally set sail to map the coast of South America, but ended up circumnavigating the globe over a period of five years (*bottom*). Darwin's detailed observations of the geology, fossils, plants, and animals he encountered on this expedition changed the way he thought about evolution.

Figure 16.6 Charles Darwin, *left*. *Right*, this page from Darwin's 1836 notes on the "Transmutation of Species" reads, "Let a pair be introduced and increase slowly, from many enemies, so as often to intermarry who will dare say what result / According to this view animals on separate islands ought to become different if kept long enough apart with slightly differing circumstances. — Now Galapagos Tortoises, Mocking birds, Falkland Fox, Chiloe fox, — Inglish and Irish Hare."

Lyell was a proponent of what became known as the **theory of uniformity**, the idea that gradual, repetitive change had shaped Earth. For many years, geologists had been chipping away at the sandstones, limestones, and other types of rocks that form from accumulated sediments in lakebeds, river bottoms, and ocean floors. These rocks held evidence that gradual processes of geologic change operating in the present were the same ones that operated in the distant past.

The theory of uniformity held that strange catastrophes were not necessary to explain Earth's surface. Over great spans of time, gradual, everyday geologic processes such as erosion could have sculpted Earth's current landscape.

catastrophism Now-abandoned hypothesis that catastrophic geologic forces unlike those of the present day shaped Earth's surface.
evolution Change in a line of descent.
lineage Line of descent.
theory of uniformity Idea that gradual repetitive processes occurring over long time spans shaped Earth's surface.

The theory challenged the prevailing belief that Earth was 6,000 years old. According to traditional scholars, people had recorded everything that happened in those 6,000 years—and in all that time, no one had mentioned seeing a species evolve. However, by Lyell's calculations, it must have taken millions of years to sculpt Earth's surface. Darwin's exposure to Lyell's ideas gave him insights into the geologic history of the regions he would encounter on his journey. Was millions of years enough time for species to evolve? Darwin thought that it was (Figure 16.6).

Take-Home Message **How did new evidence change the way people in the nineteenth century thought about the history of life?**

❯ In the 1800s, fossils and other evidence led some naturalists to propose that Earth and the species on it had changed over time. The naturalists also began to reconsider the age of Earth.

❯ Darwin's detailed observations of nature during a five-year voyage around the world changed his ideas about how evolution occurs.

❭ Darwin's observations of species in different parts of the world helped him understand a driving force of evolution.
❬ Link to Alleles and traits 13.2

Old Bones and Armadillos

Darwin sent to England the thousands of specimens he had collected on his voyage. Among them were fossil glyptodons from Argentina. These armored mammals are extinct, but they have many traits in common with modern armadillos (Figure 16.7). For example, armadillos live only in places where glyptodons once lived. Like glyptodons, armadillos have helmets and protective shells that consist of unusual bony plates. Could the odd shared traits mean that glyptodons were ancient relatives of armadillos? If so, perhaps traits of their common ancestor had changed in the line of descent that led to armadillos. But why would such changes occur?

A Key Insight—Variation in Traits

Back in England, Darwin pondered his notes and fossils. He also read an essay by one of his contemporaries, economist Thomas Malthus. Malthus had correlated increases in human population size with famine, disease,

and war. He proposed that humans run out of food, living space, and other resources because they tend to reproduce beyond the capacity of their environment to sustain them. When that happens, the individuals of a population must either compete with one another for the limited resources, or develop technology to increase their productivity. Darwin realized that Malthus's ideas had wider application: All populations, not just human ones, must have the capacity to produce more individuals than their environment can support.

Darwin also knew that individuals of a species are not always identical; they have many traits in common, but they also vary in size, color, or other features. He saw such variation among many of the finch species that live on isolated islands of the Galápagos archipelago. This island chain is separated from South America by 900 kilometers (550 miles) of open ocean, so Darwin realized that most of the species living on the islands had been isolated there for a long time.

Darwin also knew about **artificial selection**, the process whereby humans choose traits that they favor in a domestic species. For example, he was familiar with dramatic variations in traits that breeders of dogs and horses had produced through selective breeding. He recognized that an environment could similarly select traits that make individuals of a population suited to it.

Finches living on individual islands of the Galápagos resembled species Darwin saw living in South America, but many of them had unique traits that suited them to their particular island habitat. It dawned on Darwin that having a particular version of a variable trait might give an individual an advantage over competing members of its species. The trait might enhance the individual's ability to survive and reproduce in its particular environment. Darwin realized that in any population, some individuals have traits that make them better suited to their environment than others. In other words, individuals of a natural population vary in fitness. We define **fitness** as the degree of adaptation to a specific environment, and measure it as relative genetic contribution to future generations. A trait that enhances an individual's fitness is called an evolutionary **adaptation**, or **adaptive trait**.

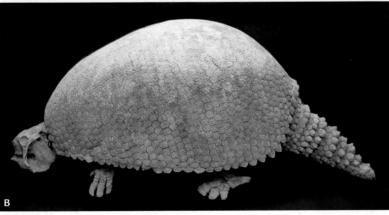

Figure 16.7 Ancient relatives. **A** A modern armadillo, about a foot long. **B** Fossil of a glyptodon, an automobile-sized mammal that lived between 2 million and 15,000 years ago.

Glyptodons and armadillos are widely separated in time, but they share a restricted distribution and unusual traits, including a shell and helmet of keratin-covered bony plates—a material similar to crocodile and lizard skin. (The fossil in **B** is missing its helmet.) Their unique shared traits were a clue that helped Darwin develop a theory of evolution by natural selection.

Figure 16.8 Alfred Wallace, codiscoverer of natural selection.

Table 16.1 Principles of Natural Selection

Observations About Populations

> Natural populations have an inherent reproductive capacity to increase in size over time.

> As a population expands, resources that are used by its individuals (such as food and living space) eventually become limited.

> When resources are limited, the individuals of a population compete for them.

Observations About Genetics

> Individuals of a species share certain traits.

> Individuals of a natural population vary in the details of those shared traits.

> Shared traits have a heritable basis, in genes. Alleles (slightly different forms of a gene) arise by mutation.

Inferences

> A certain form of a shared trait may make its bearer better able to survive.

> The individuals of a population that are better able to survive tend to leave more offspring.

> Thus, an allele associated with an adaptive trait tends to become more common in a population over time.

Over many generations, individuals with the most adaptive traits tend to survive longer and reproduce more than their less fit rivals. Darwin understood that this process, which he called **natural selection**, could be a process by which evolution occurs. If an individual has a form of a trait that makes it better suited to an environment, then it is better able to survive. If an individual is better able to survive, then it has a better chance of living long enough to produce offspring. If individuals that bear an adaptive, heritable trait produce more offspring than those that do not, then the frequency of that trait will tend to increase in the population over successive generations. Table 16.1 summarizes this reasoning in modern terms.

Great Minds Think Alike

Darwin wrote out his ideas about natural selection, but let ten years pass without publishing them. In the meantime, Alfred Wallace, who had been studying wildlife in the Amazon basin and the Malay Archipelago, wrote an essay and sent it to Darwin for advice. Wallace's essay outlined evolution by natural selection—the very same theory as Darwin's. Wallace had written earlier letters to Darwin and Lyell about patterns in the geographic distribution of species; he too had connected the dots. Wallace is now called the father of biogeography (Figure 16.8).

In 1858, just weeks after Darwin received Wallace's essay, the theory of evolution by natural selection was presented at a scientific meeting. Both Darwin and Wallace were credited as authors. Wallace was still in the field and knew nothing about the meeting, which Darwin did not attend. The next year, Darwin published *On the Origin of Species*, which laid out detailed evidence in support of the theory. Many scholars had already accepted the idea of descent with modification, or evolution. However, there was a fierce debate over the idea that evolution occurs by natural selection. Decades would pass before experimental evidence from the field of genetics led to its widespread acceptance in the scientific community.

As you will see in the remainder of this chapter, the theory of evolution by natural selection is supported by and helps explain the fossil record as well as similarities in the form, function, and biochemistry of living things.

adaptation (adaptive trait) A heritable trait that enhances an individual's fitness.
artificial selection Selective breeding of animals by humans.
fitness Degree of adaptation to an environment, as measured by an individual's relative genetic contribution to future generations.
natural selection A process in which environmental pressures result in the differential survival and reproduction of individuals of a population who vary in the details of shared, heritable traits.

Take-Home Message What is natural selection?

> Natural selection is a process in which individuals of a population who vary in the details of shared, heritable traits survive and reproduce with differing success as a result of environmental pressures.

> Traits favored by natural selection are said to be adaptive. An adaptive trait increases the chances that an individual bearing it will survive and reproduce.

> Fossils are remnants or traces of organisms that lived in the past. The fossil record holds clues to life's evolution.
< Link to Radioisotopes 2.2

Even before Darwin's time, fossils were recognized as stone-hard evidence of earlier forms of life. Most fos-

sils are mineralized bones, teeth, shells, seeds, spores, or other hard body parts. Trace fossils such as footprints and other impressions, nests, burrows, trails, eggshells, or feces (*left*) are evidence of an organism's activities.

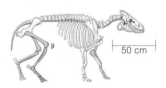

A 30-million-year-old *Elomeryx*. This small terrestrial mammal was a member of the same artiodactyl group that gave rise to hippopotamuses, pigs, deer, sheep, cows, and whales.

B *Rodhocetus*, an ancient whale, lived about 47 million years ago. Its distinctive ankle bones point to a close evolutionary connection to artiodactyls.

C *Dorudon atrox*, an ancient whale that lived about 37 million years ago. Its artiodactyl-like ankle bones were much too small to have supported the weight of its huge body on land, so this mammal had to be fully aquatic.

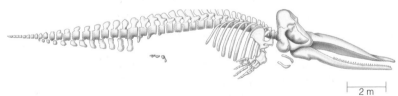

D Modern cetaceans such as the sperm whale have remnants of a pelvis and leg, but no ankle bones.

Figure 16.9 Links in the ancient lineage of whales. The ancestors of whales probably walked on land. The skull and lower jaw of cetaceans—which include whales, dolphins,

and porpoises—have distinctive features that are also characteristic of ancient carnivorous land animals. DNA sequence comparisons suggested that those animals were probably artiodactyls, hooved mammals with two or four toes on each foot. With their artiodactyl-like ankle bones, *Rodhocetus* and *Dorudon* were probably offshoots of the ancient artiodactyl-to-modern-whale lineage as it transitioned back to life in water. The photo compares the ankle bones of a *Rodhocetus* (*left*) with those of a modern artiodactyl, a pronghorn antelope (*right*).

The process of fossilization begins when an organism or its traces become covered by sediments or volcanic ash. Water seeps into the remains, and metal ions and other inorganic compounds dissolved in the water gradually replace the minerals in the bones and other hard tissues. Sediments that accumulate on top of the remains exert increasing pressure on them. After a very long time, the pressure and mineralization process transform the remains into rock.

Most fossils are found in layers of sedimentary rock such as mudstone, sandstone, and shale. These rocks form as rivers wash silt, sand, volcanic ash, and other materials from land to sea. Mineral particles in the materials settle on sea floors in horizontal layers that vary in thickness and composition. After hundreds of millions of years, the layers of sediments become compacted into layers of rock.

We study layers of sedimentary rock in order to understand the historical context of fossils we find in them. Usually, the deeper layers in a stack were the first to form, and those closest to the surface formed most recently. Thus, the deeper the layer of sedimentary rock, the older the fossils it contains. A layer's composition and thickness relative to other layers is also a clue about local and global events that were occurring as it formed. For instance, layers of sedimentary rock deposited during ice ages are thinner than other layers. Why? Tremendous volumes of water froze and became locked in glaciers during the ice ages. Rivers dried up, and sedimentation slowed. When the glaciers melted, sedimentation resumed and the layers became thicker.

The Fossil Record

We have fossils for more than 250,000 known species. Considering the current range of biodiversity, there must have been many millions more, but we will never know all of them. Why not?

The odds are against finding evidence of an extinct species, because fossils are relatively rare. Most of the time, an organism's remains are quickly obliterated by scavengers or decay. Organic materials decompose in the presence of oxygen, so remains can endure only if they are encased in an air-excluding material such as sap, tar, ice, or mud. Remains that do become fossilized are often deformed, crushed, or scattered by erosion and other geologic assaults. For us to find a fossil of an extinct species, at least one specimen had to be buried before it decomposed or something ate it. The burial site had to escape destructive geologic events, and it had to be a place accessible enough for us to find.

Despite these challenges, the fossil record is substantial enough to help us reconstruct patterns in the history of life. We have been able to find fossil evidence of the evolutionary history of many species (Figure 16.9).

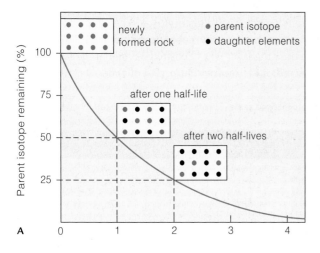

newly formed rock
- parent isotope
- daughter elements

after one half-life

after two half-lives

A

B Long ago, trace amounts of ^{14}C and a lot more ^{12}C were incorporated into the tissues of a nautilus. The carbon atoms were part of organic molecules in the nautilus's food. ^{12}C is stable and ^{14}C decays, but the proportion of the two isotopes in the nautilus's tissues remained the same. Why? As long as it was alive, the nautilus continued to gain both types of carbon atoms in the same proportions from its food.

C When the nautilus died, it stopped eating, so its body stopped gaining carbon. The ^{12}C atoms already in its tissues were stable, but the ^{14}C atoms (represented as *red* dots) were decaying into nitrogen atoms. Thus, over time, the amount of ^{14}C decreased relative to the amount of ^{12}C. After 5,370 years, half of the ^{14}C had decayed; after another 5,370 years, half of what was left had decayed, and so on.

D Fossil hunters discover the fossil and measure its ^{14}C and ^{12}C content—the number of atoms of each isotope. The ratio of those numbers can be used to calculate how many half-lives passed since the organism died. For example, if the ^{14}C to ^{12}C ratio is one-eighth of the ratio in living organisms, then three half-lives $(\frac{1}{2})^3$ must have passed since the nautilus died. Three half-lives of ^{14}C is 16,110 years.

Figure 16.10 Animated Radiometric dating. **A** Half-life, the time it takes for half of the atoms in a sample of radioisotope to decay.

B–D Using radiometric dating to find the age of a fossil. Carbon 14 (^{14}C) is a radioisotope of carbon that decays into nitrogen. It forms in the atmosphere and combines with oxygen to become CO_2, which enters food chains by way of photosynthesis.

❯❯ **Figure It Out** How much of any radioisotope remains after two of its half-lives have passed?

Answer: 25 percent.

Radiometric Dating

Remember from Section 2.2 that a radioisotope is a form of an element with an unstable nucleus. Atoms of a radioisotope become atoms of other elements as their nucleus disintegrates. The predictable products of this process, radioactive decay, are called daughter elements.

Radioactive decay is not influenced by temperature, pressure, chemical bonding state, or moisture; it is influenced only by time. Thus, like the ticking of a perfect clock, each type of radioisotope decays at a constant rate. The time it takes for half of a radioisotope's atoms to decay is a characteristic of the radioisotope called **half-life** (Figure 16.10A). For example, radioactive uranium 238 decays into thorium 234, which decays into something else, and so on until it becomes lead 206. The half-life of the decay of uranium 238 to lead 206 is 4.5 billion years.

The predictability of radioactive decay can be used to find the age of a volcanic rock—the date it cooled. Rock deep inside Earth is hot and molten, so atoms swirl and mix in it. Rock that reaches the surface cools and hardens. As the rock cools, minerals crystallize in it. Each kind of mineral has a characteristic structure and composition. For example, the mineral zircon consists primarily of

half-life Characteristic time it takes for half of a quantity of a radioisotope to decay.
radiometric dating Method of estimating the age of a rock or fossil by measuring the content and proportions of a radioisotope and its daughter elements.

ordered arrays of zircon silicate molecules ($ZrSiO_4$). Some of the molecules in a zircon crystal have uranium atoms substituted for zirconium atoms, but never lead atoms. Thus, new zircon crystals that form as molten rock cools contain no lead. However, uranium decays into lead at a predictable rate. Thus, over time, uranium atoms disappear from a zircon crystal, and lead atoms accumulate in it. The ratio of uranium atoms to lead atoms in a zircon crystal can be measured precisely. That ratio can be used to calculate how long ago the crystal formed (its age).

zircon

We have just described **radiometric dating**, a method that can reveal the age of a material by measuring its content of a radioisotope and daughter elements. The oldest known terrestrial rock, a tiny zircon crystal from Australia, is 4.404 billion years old.

Recent fossils that still contain carbon can be dated by measuring their carbon 14 content (Figure 16.10B–D). Most of the ^{14}C in a fossil will have decayed after about 60,000 years. The age of fossils older than that can be estimated only by dating volcanic rocks in lava flows above and below the fossil-containing rock.

Take-Home Message **What are fossils?**

❯ Fossils are evidence of organisms that lived in the remote past, a stone-hard historical record of life.

❯ Researchers use the predictability of radioisotope decay to estimate the age of rocks and fossils.

Putting Time Into Perspective

> Transitions in the fossil record are boundaries for great intervals of the geologic time scale.

Radiometric dating and fossils allow us to recognize similar sequences of sedimentary rock layers around the world. Transitions between layers mark boundaries between great intervals of time in the **geologic time scale**, which is a chronology of Earth's history (Figure 16.11). Each layer's composition offers clues about conditions on Earth during the time the layer was deposited. Fossils in the layers are a record of life during that period of time.

---

geologic time scale Chronology of Earth's history.

Eon	Era	Period	Epoch	mya	Major Geologic and Biological Events
Phanerozoic	Cenozoic	Quaternary	Recent	0.01	Modern humans evolve. Major extinction event is now under way.
			Pleistocene	1.8	
		Tertiary	Pliocene	5.3	Tropics, subtropics extend poleward. Climate cools; dry woodlands and grasslands emerge. Adaptive radiations of mammals, insects, birds.
			Miocene	23.0	
			Oligocene	33.9	
			Eocene	55.8	
			Paleocene		
	Mesozoic	Cretaceous		65.5 ◄	Major extinction event, perhaps precipitated by asteroid impact. Mass extinction of all dinosaurs and many marine organisms.
			Late	99.6	
			Early		Climate very warm. Dinosaurs continue to dominate. Important modern insect groups appear (bees, butterflies, termites, ants, and herbivorous insects including aphids and grasshoppers). Flowering plants originate and become dominant land plants.
		Jurassic		145.5	Age of dinosaurs. Lush vegetation; abundant gymnosperms and ferns. Birds appear. Pangea breaks up.
		Triassic		199.6 ◄	Major extinction event
					Recovery from the major extinction at end of Permian. Many new groups appear, including turtles, dinosaurs, pterosaurs, and mammals.
	Paleozoic	Permian		251 ◄	Major extinction event
					Supercontinent Pangea and world ocean form. Adaptive radiation of conifers. Cycads and ginkgos appear. Relatively dry climate leads to drought-adapted gymnosperms and insects such as beetles and flies.
		Carboniferous		299	High atmospheric oxygen level fosters giant arthropods. Spore-releasing plants dominate. Age of great lycophyte trees; vast coal forests form. Ears evolve in amphibians; penises evolve in early reptiles (vaginas evolve later, in mammals only).
		Devonian		359 ◄	Major extinction event
					Land tetrapods appear. Explosion of plant diversity leads to tree forms, forests, and many new plant groups including lycophytes, ferns with complex leaves, seed plants.
		Silurian		416	Radiations of marine invertebrates. First appearances of land fungi, vascular plants, bony fishes, and perhaps terrestrial animals (millipedes, spiders).
		Ordovician		443 ◄	Major extinction event
					Major period for first appearances. The first land plants, fishes, and reef-forming corals appear. Gondwana moves toward the South Pole and becomes frigid.
		Cambrian		488	Earth thaws. Explosion of animal diversity. Most major groups of animals appear (in the oceans). Trilobites and shelled organisms evolve.
Proterozoic				542	Oxygen accumulates in atmosphere. Origin of aerobic metabolism. Origin of eukaryotic cells, then protists, fungi, plants, animals. Evidence that Earth mostly freezes over in a series of global ice ages between 750 and 600 mya.
Archaean and earlier				2,500	3,800–2,500 mya. Origin of bacteria and archaeans.
					4,600–3,800 mya. Origin of Earth's crust, first atmosphere, first seas. Chemical, molecular evolution leads to origin of life (from protocells to anaerobic single cells).

Figure 16.11 Animated The geologic time scale correlated with sedimentary rock exposed by erosion in the Grand Canyon. **A** Transitions between layers of sedimentary rock mark great time spans in Earth's history (not to the same scale). mya: millions of years ago. Dates are from the International Commission on Stratigraphy, 2007.

B We can reconstruct some of the events in the history of life by studying rocky clues in the layers. Here, the *red* triangles mark times of great mass extinctions. "First appearance" refers to appearance in the fossil record, not necessarily the first appearance on Earth; we often discover fossils that are significantly older than previously discovered specimens.

Kaibab Limestone

Toroweap Formation

Permian

Coconino Sandstone

Hermit Shale

Esplanade Sandstone

Wescogame Formation

Carboniferous

Manakacha Formation

Watahomigi Formation

Redwall Limestone

Temple Butte Formation

Muav Limestone

Cambrian

Bright Angel Shale

Tapeats Sandstone

*Chuar Group**

*Nankoweap Formation**

*Unkar Group**

Proterozoic

Vishnu Basement Rocks

*Layers not visible in this
view of the Grand Canyon*

C Each rock layer has a composition and set of fossils that reflect events during its deposition. For example, Coconino Sandstone, which stretches from California to Montana, is mainly weathered sand. Ripple marks and reptile tracks are the only fossils in it. Many think it is the remains of a vast sand desert, like the Sahara is today.

Take-Home Message What is the geologic time scale?

> The geologic time scale is a chronology of Earth's history that correlates geologic and evolutionary events of the ancient past.

❯ Over billions of years, movements of Earth's outer layer have changed the land, atmosphere, and oceans, with profound effects on the evolution of life.

Wind, water, and other natural forces continuously sculpt the surface of Earth, but they are only part of a bigger picture of geologic change. Earth itself also changes dramatically. For instance, the Atlantic coasts of South America and Africa seem to "fit" like jigsaw puzzle pieces. By one theory, all continents that exist today were once part of a bigger supercontinent—**Pangea**—that had split into fragments and drifted apart. The idea explained why the same types of fossils occur in sedimentary rock on both sides of the Atlantic Ocean.

At first, most scientists did not accept this theory, which was called continental drift. Continents drifting about Earth seemed to be an outrageous idea, and no one knew what would drive such movement. However, evidence that supported the model kept turning up. For instance, molten rock deep inside Earth wells up and solidifies on the surface. Some iron-rich minerals become magnetic as they solidify, and their magnetic poles align with Earth's poles when they do. If continents never moved, then all of these ancient rocky magnets would be aligned north-to-south, like compass needles. Indeed, the magnetic poles of the rock formations are aligned—but not north-to-south. The poles of rock formations on different continents point in all different directions. Either Earth's magnetic poles veer dramatically from their north–south axis, or the continents wander.

Deep-sea explorers also discovered that ocean floors are not as static and featureless as had been assumed.

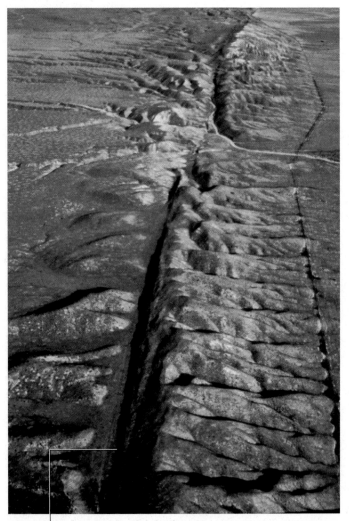

Figure 16.12 Animated Plate tectonics. Huge pieces of Earth's outer rock layer slowly drift apart and collide. As the plates move, they convey continents around the globe. The current configuration of the plates is shown in Appendix VIII.

❶ At oceanic ridges, huge plumes of molten rock welling up from Earth's interior drive the movement of tectonic plates. New crust spreads outward as it forms on the surface, forcing adjacent tectonic plates away from the ridge and into trenches elsewhere.

❷ At trenches, the advancing edge of one plate plows under an adjacent plate and buckles it.

❸ Faults are ruptures in Earth's crust where plates meet. The diagram shows a type of fault called a rift, in which plates move apart. The aerial photo on the *left* shows about 4.2 kilometers (2.6 miles) of the San Andreas Fault, which extends 1,300 km (800 miles) through California. This fault is a boundary between two tectonic plates slipping past one another.

❹ Plumes of molten rock rupture a tectonic plate at what are called "hot spots." The Hawaiian Islands have been forming from molten rock that continues to erupt from a hot spot under the Pacific (tectonic) Plate.

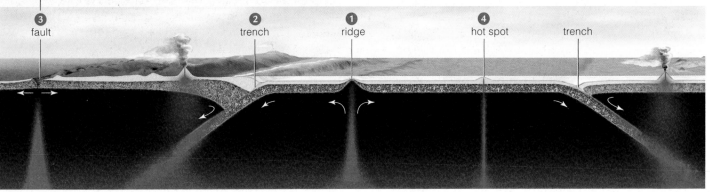

❸ fault   ❷ trench   ❶ ridge   ❹ hot spot   trench

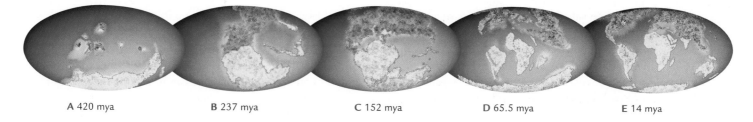

| A 420 mya | B 237 mya | C 152 mya | D 65.5 mya | E 14 mya |

Figure 16.13 A series of reconstructions of the drifting continents. **A** The supercontinent Gondwana (*yellow*) had begun to break up by the Silurian. **B** The supercontinent Pangea formed during the Triassic, then **C** began to break up in the Jurassic. **D** K–T boundary. **E** The continents reached their modern configuration in the Miocene.

Immense ridges stretch thousands of kilometers across the sea floor (Figure 16.12). Molten rock spewing from the ridges pushes old sea floor outward in both directions ❶, then cools and hardens into new sea floor. Elsewhere, older sea floor plunges into deep trenches ❷.

Such discoveries swayed the skeptics. Finally, there was a plausible mechanism for continental drift, which is now called **plate tectonics**. By this theory, Earth's relatively thin outer layer of rock is cracked into immense plates, a bit like a gigantic cracked eggshell. Molten rock emanating from an undersea ridge or continental rift at one edge of a plate pushes old rock at the opposite edge into a trench. The movement is like that of a colossal conveyer belt that transports continents on top of it to new locations. Each plate moves no more than 10 centimeters (4 inches) a year—about half as fast as your toenails grow—but that is enough to carry a continent all the way around the world after 40 million years or so.

Evidence of tectonic movement is all around us, in faults ❸ and other various geologic features of our landscapes. For example, volcanic island chains (archipelagos) form as a plate moves across an undersea hot spot. Hot spots are places where a narrow plume of molten rock wells up from deep inside Earth and ruptures a plate ❹.

Plate tectonics solved some long-standing puzzles. Consider an unusual geologic formation that occurs in a belt across Africa. The sequence of rock layers in this formation is so complex that it is quite unlikely to have formed more than once, but identical sequences also occur in huge belts that span India, South America, Africa, Madagascar, Australia, and Antarctica. The most likely explanation for the wide distribution is that the formations were deposited together on a single continent that later broke up. This explanation is supported by fossils in the rock layers: the remains of a type of fern (*Glossopteris*) whose seeds were too heavy to float or to be wind-blown over an ocean, and of an early reptile (*Lystrosaurus*) whose body was not built for swimming between continents.

Here is the puzzle: *Glossopteris* disappeared in the Permian–Triassic mass extinction event (251 million years ago), and *Lystrosaurus* disappeared 6 million years after that. Both organisms were extinct millions of years before Pangea formed. Could they have evolved together on a different supercontinent, one that predated Pangea? Evidence suggests that they did. The older supercontinent, which we now call **Gondwana**, included most of the land masses that currently exist in the Southern Hemisphere as well as India and Arabia (Figure 16.13). Many modern species, including the ratite birds pictured in Figure 16.2, live only in places that were once part of Gondwana.

After Gondwana formed, it drifted south, across the South Pole, then north until it merged with other continents to form Pangea. We now know that at least five times since Earth's outer layer of rock solidified 4.55 billion years ago, a single supercontinent with one ocean lapping at its coastline formed and then split up again. The resulting changes in Earth's surface, atmosphere, and waters have had a profound impact on the course of life's evolution. A continent's climate changes—often dramatically so—along with its position on Earth. Colliding continents physically separate organisms living in oceans, and bring together those that had been living apart on land. As continents break up, they separate organisms living on land, and bring together ones that had been living in separate oceans. Such changes are a major driving force of evolution, as you will see in Chapter 17. Lineages that cannot adapt to the changes die out, and new evolutionary opportunities open up for the survivors.

Gondwana Supercontinent that existed before Pangea, more than 500 million years ago.
Pangea Supercontinent that formed about 237 million years ago and broke up about 152 million years ago.
plate tectonics Theory that Earth's outer layer of rock is cracked into plates, the slow movement of which rafts continents to new locations over geologic time.

Take-Home Message **How has Earth changed over geologic time spans?**

❯ Over geologic time, movements of Earth's crust have caused dramatic changes in the continents, atmosphere, and oceans.

❯ The course of life's evolution has been influenced by these changes.

❯ Physical similarities may be evidence of shared ancestry.

How do we know about evolution that occurred in the ancient past? Like asteroid impacts, evolution leaves evidence. Fossils are one example, but organisms that are alive today provide others. To a biologist, remember, evolution means change in a line of descent. Clues about the history of a lineage may be encoded in body form, function, or biochemistry. For example, similarities in the structure of body parts often reflect shared ancestry. Comparative morphology can be used to unravel evolutionary relationships in such cases. Similar body parts that evolved in a common ancestor are called **homologous structures** (*hom–* means "the same"). Homologous structures may be used for different purposes in different groups, but the very same genes direct their development.

Morphological Divergence

A body part that appears very different in different lineages may be similar in some underlying aspect of form. For example, even though vertebrate forelimbs are not the same in size, shape, or function from one group to the next, they clearly are alike in the structure and positioning of bony elements. They also are alike in the patterns of nerves, blood vessels, and muscles that develop inside of them. Such similarities are evidence of shared ancestry.

As you will see in the next chapter, populations that are not interbreeding tend to diverge genetically, and in time they diverge morphologically. Change from the body form of a common ancestor is an evolutionary pattern called **morphological divergence**. We have evidence from fossilized limb bones that all modern land vertebrates are descended from a family of ancient "stem reptiles" that crouched low to the ground on four legs. Descendants of this ancestral family diversified into many new habitats on land, and gave rise to the groups we call reptiles, birds, and mammals. A few lineages that had become adapted to walking on land even returned to life in the seas.

Over millions of years, the stem reptile's five-toed limbs become adapted for very different purposes across many lineages (Figure 16.14). In extinct reptiles called pterosaurs, most birds, and bats, they were modified for flight. In penguins and porpoises, the limbs are now flippers useful for swimming. In humans, five-toed forelimbs became arms and hands, in which the thumb evolved in opposition to the fingers. An opposable thumb was the basis of more precise motions and a firmer grip. Among elephants, the limbs are now strong and pillarlike, capable of supporting a great deal of weight. Limbs degenerated to nubs in pythons and boa constrictors, and to nothing at all in other snakes.

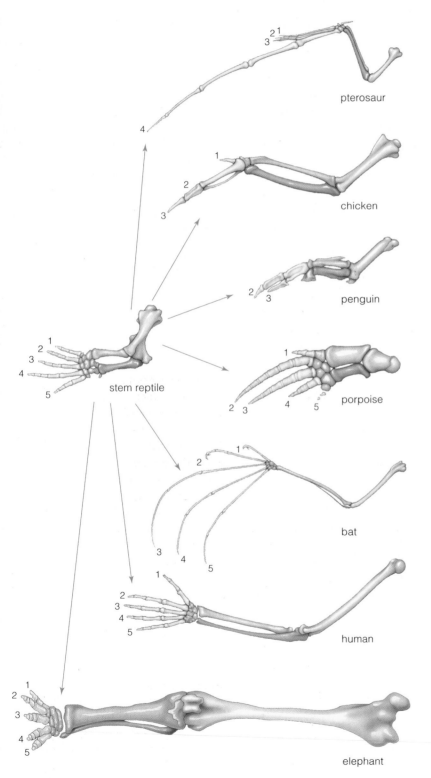

pterosaur

chicken

penguin

porpoise

stem reptile

bat

human

elephant

Figure 16.14 Morphological divergence among vertebrate forelimbs, starting with the bones of a stem reptile. The number and position of many skeletal elements were preserved when these diverse forms evolved; notice the bones of the forearms. Certain bones were lost over time in some of the lineages (compare the digits numbered 1 through 5). The drawings are not to the same scale.

A

B

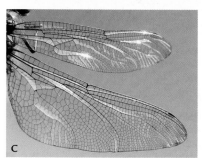

C

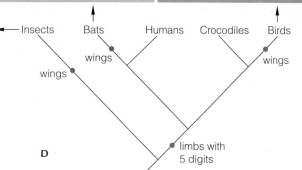

← Insects   Bats   Humans   Crocodiles   Birds

wings

wings

wings

D

limbs with
5 digits

Figure 16.15 Morphological convergence. The flight surfaces of a bat wing **A**, a bird wing **B**, and an insect wing **C** are analogous structures. **D** The independent evolution of wings in the three separate lineages that led to bats, birds, and insects. You will read more about diagrams that show evolutionary relationships in Section 17.14.

Morphological Convergence

Similar body parts are not always homologous. Similar structures may evolve independently in separate lineages as adaptations to the same environmental pressures. In this case they are called **analogous structures**. Analogous structures look alike in different lineages but did not evolve in a shared ancestor; they evolved independently after the lineages diverged. Evolution of similar body parts in different lineages is **morphological convergence**.

We can sometimes identify analogous structures by studying their underlying form. For example, bird, bat, and insect wings all perform the same function: flight. However, several clues tell us that the flight surfaces of these wings are not homologous. The wing surfaces are adapted to the same physical constraints that govern flight, but the adaptations are different. In the case of

birds and bats, the limbs themselves are homologous, but the adaptations that make those limbs useful for flight differ. The surface of a bat wing is a thin, membranous extension of the animal's skin. By contrast, the surface of a bird wing is a sweep of feathers, which are specialized structures derived from skin. Insect wings differ even more. An insect wing forms as a saclike extension of the body wall. Except at forked veins, the sac flattens and fuses into a thin membrane. The veins are reinforced with chitin, which structurally support the wing. The unique adaptations for flight are evidence that wing surfaces of birds, bats, and insects are analogous structures—they evolved after the ancestors of these modern groups diverged (Figure 16.15).

analogous structures Similar body structures that evolved separately in different lineages.
homologous structures Similar body parts that evolved in a common ancestor.
morphological convergence Evolutionary pattern in which similar body parts evolve separately in different lineages.
morphological divergence Evolutionary pattern in which a body part of an ancestor changes in its descendants.

Take-Home Message Are similar body parts indicative of an evolutionary relationship?

❯ In morphological divergence, a body part inherited from a common ancestor becomes modified differently in different lines of descent. Such parts are called homologous structures.

❯ In morphological convergence, body parts that appear alike evolved independently in different lineages, not in a common ancestor. Such parts are called analogous structures.

> Similar patterns of embryonic development may be evidence of evolutionary relationships.
< Links to Master genes in development 10.3, Floral identity gene mutations 10.4, Evolution by gene duplications 14.5

The development of an embryo into the body of a plant or animal is orchestrated by layer after layer of master gene expression. The failure of any single master gene to participate in this symphony of expression can result in a drastically altered body plan, typically with devastating consequences. Because a mutation in a master gene typically unravels development, these genes tend to be highly conserved, which means they have changed very little or not at all over evolutionary time. Thus, a master gene with a similar sequence and function across different lineages is strong evidence that those lineages are related.

Similar Genes in Plants

The master genes called homeotic genes guide formation of specific body parts during development. A mutation in one homeotic gene can disrupt details of the body's form. For example, any mutation that inactivates a floral identity gene, *Apetala1*, in wild cabbage plants (*Brassica oleracea*) results in mutated flowers. Such flowers form with male reproductive structures (stamens) where petals are supposed to be. At least in the laboratory, these abundantly stamened flowers are exceptionally fertile, but such alterations usually are selected against in nature. *Apetala1* mutations in common wall cress plants (*Arabidopsis thaliana*) also results in flowers that have no petals (Section 10.4). The *Apetala1* gene affects the formation of petals across many different lineages, so it is very likely that this gene evolved in a shared ancestor.

Developmental Comparisons in Animals

The embryos of many vertebrate species develop in similar ways. Their tissues form the same way, as embryonic cells divide, differentiate, and interact. For example, all vertebrates go through a stage in which they have four limb buds, a tail, and a series of somites—divisions of the body that give rise to a backbone (Figure 16.16). Given that

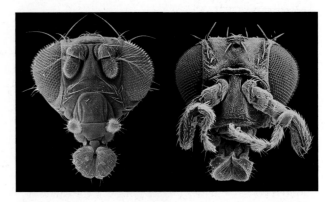

Figure 16.17 Expression of the *antennapedia* gene in the embryonic tissues of the insect thorax causes legs to form. Normally, the gene is never expressed in cells of any other tissue. A mutation that causes *antennapedia* to be expressed in the embryonic tissues of a *Drosophila*'s head (*left*) causes legs to form there too (*right*).

the same genes direct development in these lineages, how do the adult forms end up so different? Part of the answer is that there are differences in the onset, rate, or completion of early steps in development. These differences are brought about by variations in the expression patterns of master genes that govern development. The variation has apparently arisen mainly as a result of gene duplications followed by mutation, the same way that multiple globin genes evolved in primates (Section 14.5).

For example, master genes called *Hox* sculpt details of the body's form during embryonic development. The pattern of expression of these genes determines the identity of particular zones along the body axis. Insects and other arthropods have ten *Hox* genes. One of them, *antennapedia*, determines the identity of the thorax (the part with legs). Legs develop wherever *antennapedia* is expressed in an embryo (Figure 16.17). Vertebrates have four sets of the same ten *Hox* genes that occur in insects. A vertebrate version of *antennapedia*, the *Hoxc6* gene, determines the identity of the back (as opposed to the neck or tail). Expression of this gene causes ribs to develop on a vertebra (Figure 16.18). Vertebrae of the neck and tail normally develop with no *Hoxc6* expression, and no ribs.

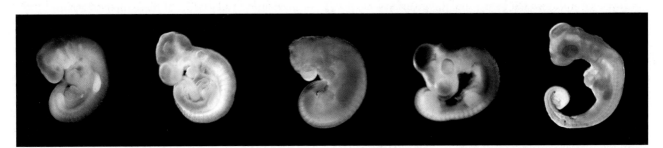

Figure 16.16 Visual comparison of vertebrate embryos. All vertebrates go through an embryonic stage in which they have four limb buds, a tail, and divisions called somites along their back. Embryos *left* to *right*: human, mouse, bat, chicken, alligator.

Figure 16.18 An example of comparative embryology. Expression of the *Hoxc6* gene is indicated by *purple* stain in two vertebrate embryos, chick (*left*) and garter snake (*right*). Expression of this gene causes a vertebra to develop ribs as part of the back. Chickens have 7 vertebrae in their back and 14 to 17 vertebrae in their neck; snakes have upwards of 450 back vertebrae and essentially no neck.

Hox genes also regulate limb formation. Body appendages as diverse as crab legs, beetle legs, sea star arms, butterfly wings, fish fins, and mouse feet start out as clusters of cells that bud from the surface of the embryo. The buds form wherever the homeotic gene *Dlx* is expressed. *Dlx* encodes a transcription factor that signals clusters of embryonic cells to "stick out from the body" and give rise to an appendage. *Hox* genes suppress *Dlx* expression in all parts of an embryo that will not have appendages.

Forever Young At an early stage, a chimpanzee skull and a human skull appear quite similar. As development continues, both skulls change shape as different parts grow at different rates (Figure 16.19). However, the human skull undergoes less pronounced differential growth than the chimpanzee skull does. As a result, a human adult has a rounder braincase, a flatter face, and a less protruding jaw compared with an adult chimpanzee.

In its proportions, a human adult skull is more like the skull of an infant chimpanzee than the skull of an adult chimpanzee. The similarity suggests that human evolution involved changes that slowed the rate of development, causing traits that were previously typical of juvenile stages to persist into adulthood.

Juvenile features also persist in other adult animals, notably salamanders called axolotls. The larvae of most species of salamander live in water and use external gills to breathe. Lungs that replace the gills as development continues allow the adult to breathe air and live on land. By contrast, axolotls never give up their aquatic life-style; their external gills and other larval traits persist into adulthood.

The closest relatives of axolotls are tiger salamanders. As you might expect, tiger salamander larvae resemble axolotls, although they are smaller.

Reflections of a Distant Past (revisited)

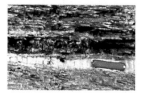

The K–T boundary layer (*left*) consists of an unusual clay that formed 65 million years ago, worldwide (the *red* pocketknife is shown for scale). The clay is rich in iridium, an element rare on Earth's surface but common in asteroids. After finding the iridium, researchers looked for evidence of an asteroid big enough to cover the entire Earth with its debris. They found a crater that is about 65 million years old, buried under sediments off the coast of Mexico's Yucatán Peninsula. It is so big—273.6 kilometers (170 miles) across and 1 kilometer (3,000 feet) deep—that no one had realized it was a crater. This crater is evidence of an asteroid impact 40 million times more powerful than the one that made the Barringer Crater, certainly big enough to have influenced life on Earth in a big way.

How Would You Vote? Many theories and hypotheses about events in the ancient past are necessarily based on traces left by those events, not on data collected by direct observations. Is indirect evidence ever enough to prove a theory about a past event? See CengageNow for details, then vote online (cengagenow.com).

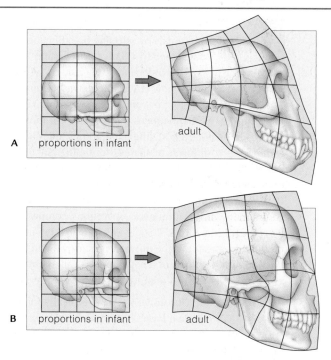

Figure 16.19 **Animated** Morphological differences between two primates. These skulls are depicted as paintings on a rubber sheet divided into a grid. Stretching the sheets deforms the grid. Differences in how they are stretched are analogous to different growth patterns. Shown here, proportional changes during skull development in **A** the chimpanzee and **B** the human. Chimpanzee skulls change more than human skulls, so the relative proportions in bones of adult and infant humans are more similar than those of adult and infant chimpanzees.

Take-Home Message Are similarities in development indicative of shared ancestry?

> Similarities in patterns of development are the result of master genes that have been conserved over evolutionary time.

> Some differences between closely related species arose as a result of changes in the rate of development.

Summary

Section 16.1 Events of the ancient past can be explained by the same physical, chemical, and biological processes that operate today. An asteroid impact may have caused a **mass extinction** 65.5 million years ago.

Section 16.2 Expeditions by nineteenth-century **naturalists** yielded increasingly detailed observations of nature. Geology, **biogeography**, and **comparative morphology** of organisms and their **fossils** led to new ways of thinking about the natural world.

Section 16.3 Prevailing belief systems may influence interpretation of the underlying cause of a natural event. The nineteenth-century naturalists proposed **catastrophism** and the **theory of uniformity** in their attempts to reconcile traditional beliefs with physical evidence of **evolution**, or change in a **lineage** over time.

Section 16.4 Humans select desirable traits in animals by selective breeding, or **artificial selection**. Charles Darwin and Alfred Wallace independently came up with a theory of how environments also select traits: A population tends to grow until it exhausts environmental resources. As that happens, competition for those resources intensifies among the population's individuals. Individuals with forms of shared, heritable traits that make them more competitive for the resources tend to produce more offspring. Thus, **adaptive traits** (**adaptations**) that impart greater **fitness** to an individual become more common in a population over generations, compared with less competitive forms. The process in which environmental pressures result in the differential survival and reproduction of individuals of a population is called **natural selection**. It is one of the processes that drives evolution.

Section 16.5 Fossils are typically found in stacked layers of sedimentary rock. Younger fossils usually occur in layers deposited more recently, on top of older fossils in older layers. Fossils are relatively scarce, so the fossil record will always be incomplete. The characteristic **half-life** of a radioisotope allows us to determine the age of rocks and fossils using **radiometric dating**.

Section 16.6 Transitions in the fossil record are the boundaries of great intervals of the **geologic time scale**, a chronology of Earth's history that correlates geologic and evolutionary events.

Section 16.7 By the theory of **plate tectonics**, the movements of Earth's tectonic plates carry land masses to new positions. Such movements had profound impacts on the directions of life's evolution. Several times in Earth's history, all land masses have converged as supercontinents. **Gondwana** and **Pangea** are examples.

Section 16.8 Comparative morphology can reveal evolutionary connections among lineages. **Homologous structures** are similar body parts that, by **morphological divergence**, became modified differently in different lineages. Such parts are evidence of a common ancestor. **Analogous structures** are body parts that look alike in different lineages but did not evolve in a common ancestor. By the process of **morphological convergence**, they evolved separately after the lineages diverged.

Section 16.9 Similarities among patterns of embryonic development reflect shared ancestry. Genes that affect development tend to be conserved. Mutations that alter the rate of development may allow juvenile traits to persist into adulthood.

Self-Quiz Answers in Appendix III

1. The number of species on an island depends on the size of the island and its distance from a mainland. This statement would most likely be made by _____ .
   a. an explorer   c. a geologist
   b. a biogeographer   d. a philosopher

2. Evolution _____ .
   a. is natural selection
   b. is heritable change in a line of descent
   c. can occur by natural selection
   d. b and c are correct

3. Which of the following is a fossil?
   a. An insect encased in 10-million-year-old tree sap
   b. A woolly mammoth frozen in Arctic permafrost for the last 50,000 years
   c. Mineral-hardened remains of a whale-like animal found in an Egyptian desert
   d. An impression of a plant leaf in a rock
   e. All of the above could be considered fossils

4. Did Pangea or Gondwana form first?

5. The bones of a bird's wing are similar to the bones in a bat's wing. This observation is an example of _____ .
   a. uniformity   c. comparative morphology
   b. evolution   d. a lineage

6. If the half-life of a radioisotope is 20,000 years, then a sample in which three-quarters of that radioisotope has decayed is _____ years old.
   a. 15,000   b. 26,667   c. 30,000   d. 40,000

7. Forces of geologic change include _____ (select all that are correct).
   a. erosion   e. tectonic plate movement
   b. fossilization   f. wind
   c. volcanic activity   g. asteroid impacts
   d. evolution   h. hot spots

8. The Cretaceous ended _____ million years ago.

Data Analysis Activities

Abundance of Iridium in the K–T Boundary Layer In the late 1970s, geologist Walter Alvarez was investigating the composition of the 1-centimeter-thick layer of clay that marks the Cretaceous–Tertiary (K–T) boundary all over the world. He asked his father, Nobel Prize–winning physicist Luis Alvarez, to help him analyze the elemental composition of the layer. The photo shows Luis and Walter Alvarez with a section of the K–T boundary layer.

The Alvarezes and their colleagues tested the layer in Italy and in Denmark. The researchers discovered that the K–T boundary layer had a much higher iridium content than the surrounding rock layers. Some of their results are shown in the table in Figure 16.20.

Iridium belongs to a group of elements (Appendix IV) that are much more abundant in asteroids and other solar system materials than they are in Earth's crust. The Alvarez group concluded that the K–T boundary layer must have originated with extraterrestrial material. They calculated that an asteroid 14 kilometers (8.7 miles) in diameter would contain enough iridium to account for the iridium in the K–T boundary layer.

Sample Depth	Average Abundance of Iridium (ppb)
+ 2.7 m	< 0.3
+ 1.2 m	< 0.3
+ 0.7 m	0.36
boundary layer	41.6
– 0.5 m	0.25
– 5.4 m	0.30

Figure 16.20 Abundance of iridium in and near the K–T boundary layer in Stevns Klint, Denmark. Many rock samples taken from above, below, and at the boundary layer were tested for iridium content. Depths are given as meters above or below the boundary layer.

The iridium content of an average Earth rock is 0.4 parts per billion (ppb) of iridium. An average meteorite contains about 550 parts per billion of iridium.

1. What was the iridium content of the K–T boundary layer?
2. How much higher was the iridium content of the boundary layer than the sample taken 0.7 meter above the layer?

---

9. Life originated in the _____ .

10. Through _____ , a body part of an ancestor is modified differently in different lines of descent.
    a. morphological divergence
    b. adaptive divergence
    c. morphological convergence
    d. homologous evolution

11. Homologous structures among major groups of organisms may differ in _____ .
    a. size　　　　　　c. function
    b. shape　　　　　 d. all of the above

12. Match the terms with the most suitable description.
    ___ fitness
    ___ fossils
    ___ homeotic genes
    ___ half-life
    ___ homologous structures
    ___ uniformity
    ___ analogous structures
    ___ natural selection

    a. evidence of life in distant past
    b. geologic change occurs continuously
    c. human arm and bird wing
    d. big role in development
    e. measured by reproductive success
    f. insect wing and bird wing
    g. survival of the fittest
    h. characteristic of radioisotope

Additional questions are available on CENGAGENOW.

Animations and Interactions on CENGAGENOW:
〉 Vestigial body parts; Geologic time scale; Half-life; Radiometric dating; Plate tectonics; Proportional changes in embryonic development.

Critical Thinking

1. Radiometric dating does not measure the age of an individual atom. It is a measure of the age of a quantity of atoms—a statistic. As with any statistical measure, its values may deviate around an average (see sampling error, Section 1.8). Imagine that one sample of rock is dated ten different ways. Nine of the tests yield an age close to 225,000 years. One test yields an age of 3.2 million years. Do the nine consistent results imply that the one that deviates is incorrect, or does the one odd result invalidate the nine that are consistent?

2. If you think of geologic time spans as minutes, life's history might be plotted on a clock such as the one shown on the *right*. According to this clock, the most recent epoch started in the last 0.1 second before noon. Where does that put you?

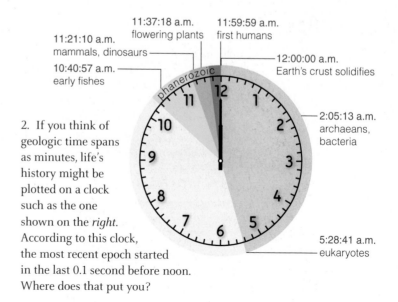

11:21:10 a.m.
mammals, dinosaurs

11:37:18 a.m.
flowering plants

11:59:59 a.m.
first humans

10:40:57 a.m.
early fishes

phanerozoic

12:00:00 a.m.
Earth's crust solidifies

2:05:13 a.m.
archaeans, bacteria

5:28:41 a.m.
eukaryotes

‹ Links to Earlier Concepts

This chapter builds on the theory of natural selection (Section 16.4). You may wish to review life's organization (1.2), the geologic time scale (16.6), and plate tectonics (16.7). You will revisit experiments (1.7), sampling error (1.8), alleles (12.2), cellular reproduction (11.3, 12.2, 12.3), the genetic basis of traits (10.3, 13.2, 13.5, 13.6), effects of genetic changes (9.6, 14.2, 14.6), bacteria (4.1, 4.5), melanin (13.5), and transgenic plants (15.7).

Key Concepts

Microevolution
Individuals of a population inherit different alleles, and so they differ in phenotype. Over generations, any allele may increase or decrease in frequency in a population. Such change is called microevolution.

Processes of Microevolution
Natural selection may maintain or shift the range of variation of a shared, heritable trait in a population. Gene flow counters the evolutionary effects of mutation, natural selection, and genetic drift.

17 Processes of Evolution

17.1 Rise of the Super Rats

Slipping in and out of the pages of human history are rats—*Rattus*—the most notorious of mammalian pests. Rats thrive in urban centers, where garbage is plentiful and natural predators are not (Figure 17.1). The average city in the United States sustains about one rat for every ten people. Part of their success stems from an ability to reproduce very quickly. Rat populations can expand within weeks to match the amount of garbage available for them to eat. Unfortunately for us, rats carry pathogens and parasites associated with infectious diseases such as bubonic plague and typhus. They chew their way through walls and wires, and eat or foul 20 to 30 percent of our total food production. Rats cost us about $19 billion per year.

For years, people have been fighting back with poisons, including arsenic and cyanide. Baits laced with warfarin, an organic compound that interferes with blood clotting, were popular in the 1950s. Rats that ate the poisoned baits died within days after bleeding internally or losing blood through cuts or scrapes. Warfarin was very effective, and compared to other rat poisons, it had much less impact on harmless species. It quickly became the rodenticide of choice.

In 1958, however, a Scottish researcher reported that warfarin was not working against some rats. Similar reports from other European countries followed. About twenty years later, about 10 percent of rats caught in urban areas of the United States were resistant to warfarin. What happened?

To find out, researchers compared rats that were resistant to warfarin with those who were not. The difference was traced to a gene on one of the rat chromosomes. Certain mutations in the gene were common among warfarin-resistant rat populations but rare among vulnerable ones. Warfarin inhibits the gene's product, an enzyme that recycles vitamin K after it has been used to activate blood clotting factors. The mutations made the enzyme less active, but also insensitive to warfarin.

"What happened" was evolution by natural selection. As warfarin exerted pressure on rat populations, the populations

Figure 17.1 Rats as pests. *Above*, rats infesting rice fields in the Philippine Islands ruin more than 20 percent of the crop. *Opposite*, rats thrive wherever people do. Dousing buildings and soil with poisons does not usually exterminate rat populations, which recover quickly. Rather, the practice selects for rats that are resistant to the poisons.

changed. The previously rare alleles became adaptive. Rats that had an unmutated gene died after eating warfarin. The lucky ones that had one of the warfarin-resistance alleles survived and passed it to their offspring. The rat populations recovered quickly, and a higher proportion of rats in the next generation carried the alleles. With each onslaught of warfarin, the frequency of the alleles in rat populations increased.

Selection pressures can and often do change. When warfarin resistance increased in rat populations, people stopped using warfarin. The frequency of warfarin-resistance alleles in rat populations declined, probably because rats with the alleles are not as healthy as normal rats. Now, savvy exterminators in urban areas know that the best way to control a rat infestation is to exert another kind of selection pressure: Remove their source of food, which is usually garbage. Then the rats will eat each other.

How Species Arise
Speciation varies in its details, but it always involves the end of gene flow between populations. Microevolutionary events that occur independently lead to genetic divergences, which are reinforced by reproductive isolation.

Macroevolution
Patterns of genetic change that involve more than one species are called macroevolution. Recurring patterns of macroevolution include the origin of major groups, one species giving rise to many, and mass extinction.

Cladistics
Evolutionary tree diagrams are based on the premise that all species interconnect through shared ancestors. Grouping species by shared ancestry better reflects evolutionary history than do traditional ranking systems.

> Evolution starts with mutations in individuals.
> Mutation is the source of new alleles.
> Sexual reproduction can quickly spread a mutation through a population.
< Links to Life's organization 1.2, Mutation 9.6, Alleles 12.2, Mendelian inheritance 13.2, Complex traits 13.5 and 13.6

Variation in Populations

A **population** is a group of interbreeding individuals of the same species in some specified area (Section 1.2). The individuals of a species—and a population—share certain features. For example, giraffes normally have very long necks, brown spots on white coats, and so on. These are examples of morphological traits (*morpho*– means form). Individuals of a species also share physiological traits, such as metabolic activities. They also respond the same way to certain stimuli, as when hungry giraffes feed on tree leaves. These are behavioral traits.

Individuals of a population have the same traits because they have the same genes. However, almost every shared trait varies a bit among individuals of a population (Figure 17.2). Alleles of the shared genes are the main source of this variation. Many traits have two or more distinct forms, or morphs. A trait with only two forms is dimorphic (*di*– means two). Purple and white flower color in the pea plants that Gregor Mendel studied is an example of a dimorphic trait (Section 13.3). Dimorphic flower color occurs in this case because the interaction of two alleles with a clear dominance relationship gives rise to the trait. Traits with more than two distinct forms are polymorphic (*poly*–, many). Human blood type, which is determined by the codominant *ABO* alleles, is an example (Section 13.5). Traits that vary continuously among the individuals of a population often arise by interactions among alleles of several genes, and may be influenced by environmental factors (Sections 13.5 and 13.6).

Table 17.1	Sources of Variation in Traits Among Individuals of a Species
Genetic Event	**Effect**
Mutation	Source of new alleles
Crossing over at meiosis I	Introduces new combinations of alleles into chromosomes
Independent assortment at meiosis I	Mixes maternal and paternal chromosomes
Fertilization	Combines alleles from two parents
Changes in chromosome number or structure	Transposition, duplication, or loss of chromosomes

In earlier chapters, you learned about the processes that introduce and maintain variation in traits among individuals of a species. Table 17.1 summarizes the key events involved. Mutation is the original source of new alleles. Other events shuffle alleles into different combinations, and what a shuffle that is! There are $10^{116,446,000}$ possible combinations of human alleles. Not even 10^{10} people are living today. Unless you have an identical twin, it is unlikely that another person with your precise genetic makeup has ever lived, or ever will.

An Evolutionary View of Mutations

Being the original source of new alleles, mutations are worth another look—this time within the context of their impact on populations. We cannot predict when or in which individual a particular gene will mutate. We can, however, predict the average mutation rate of a species, which is the probability that a mutation will occur in a given interval. In humans, that rate is about 2.2×10^{-9} mutations per base pair per year. That means almost 70 mutations accumulate in the human genome per decade.

Many mutations give rise to structural, functional, or behavioral alterations that reduce an individual's chances of surviving and reproducing. Even one biochemical change may be devastating. For instance, the skin, bones, tendons, lungs, blood vessels, and other vertebrate organs incorporate the protein collagen. If one of the genes for collagen mutates in a way that changes the protein's function, the entire body may be affected. A mutation such as this can change phenotype so drastically that it results in death, in which case it is a **lethal mutation**.

A **neutral mutation** changes the base sequence in DNA, but the alteration has no effect on survival or reproduction. It neither helps nor hurts the individual. For instance, if you carry a mutation that keeps your earlobes attached to your head instead of swinging freely, attached earlobes should not in itself stop you from surviving and

Figure 17.2 Sampling phenotypic variation in **A** (*opposite*) a type of snail found on islands in the Caribbean, and **B** humans. The variation in shared traits among individuals is mainly an outcome of variations in alleles that influence those traits.

reproducing as well as anybody else. So, natural selection does not affect the frequency of this particular mutation in a population.

Occasionally, a change in the environment favors a mutation that had previously been neutral or even somewhat harmful. The warfarin resistance gene in rats is an example. Even if a beneficial mutation bestows only a slight advantage, its frequency tends to increase in a population over time. This is because natural selection operates on traits with a genetic basis. With natural selection, remember, environmental pressures result in an increase in the frequency of a beneficial trait in a population over generations (Section 16.4).

Mutations have been altering genomes for billions of years, and they are still at it. Cumulatively, they have given rise to Earth's staggering biodiversity. Think about it: The reason you do not look like an avocado or an earthworm or even your next-door neighbor began with mutations that occurred in different lines of descent.

Allele Frequencies

Together, all the alleles of all the genes of a population comprise a pool of genetic resources called a **gene pool**. Members of a population breed with one another more often than they breed with members of other populations, so their gene pool is more or less isolated. **Allele frequency** refers to the abundance of a particular allele among the individuals of a population. Change in an allele's frequency in a population is the same thing as change in a line of descent—evolution. Evolution within a population or species is called **microevolution**.

A theoretical reference point, **genetic equilibrium**, occurs when the allele frequencies of a population do not change (in other words, the population is not evolving). Genetic equilibrium can only occur if every one of the following five conditions are met: (1) Mutations never occur; (2) the population is infinitely large; (3) the population is isolated from all other populations of the species; (4) mating is random; and (5) all individuals survive to produce the same number of offspring. As you can imagine, all five conditions are never met in nature, so natural populations are never in genetic equilibrium.

Microevolution is always occurring in natural populations because the processes that drive it are always operating. The remaining sections of this chapter explore microevolutionary processes—mutation, natural selection, genetic drift, and gene flow—and their effects. Remember, even though we can recognize patterns of evolution, none of them are purposeful. Evolution simply fills the nooks and crannies of opportunity.

---

allele frequency Abundance of a particular allele among members of a population.
gene pool All of the alleles of all of the genes in a population; a pool of genetic resources.
genetic equilibrium Theoretical state in which a population is not evolving.
lethal mutation Mutation that drastically alters phenotype; causes death.
microevolution Change in allele frequencies in a population or species.
neutral mutation A mutation that has no effect on survival or reproduction.
population A group of organisms of the same species who live in a specific location and breed with one another more often than they breed with members of other populations.

Take-Home Message What mechanisms drive evolution?

❭ We partly characterize a natural population by shared morphological, physiological, and behavioral traits.

❭ Different alleles are the basis of differences in the details of a population's shared traits.

❭ Alleles of all individuals in a population comprise the population's gene pool.

❭ Natural populations are always evolving, which means that allele frequencies in their gene pool are always changing over generations.

❭ Microevolution refers to evolutionary change in a population or species.

> ❯ Researchers know whether a population is evolving by tracking deviations from a baseline of genetic equilibrium.
> ❮ Link to Codominant alleles 13.5

The Hardy–Weinberg Formula Early in the twentieth century, Godfrey Hardy (a mathematician) and Wilhelm Weinberg (a physician) independently applied the rules of probability to sexually reproducing populations. They realized that gene pools can remain stable only when five conditions are being met:

1. Mutations do not occur.

2. The population is infinitely large.

3. The population is isolated from all other populations of the species (no gene flow).

4. Mating is random.

5. All individuals survive and produce the same number of offspring.

These conditions never occur all at once in nature. Thus, allele frequencies for any gene in the shared pool always change. However, we can think about a hypothetical situation in which the five conditions are being met and a population is not evolving.

Hardy and Weinberg developed a simple formula that can be used to track whether a population of any sexually reproducing species is in a state of genetic equilibrium. Consider a hypothetical gene that encodes a blue pigment in butterflies. Two alleles of this gene, *B* and *b*, are codominant. A butterfly homozygous for the *B* allele (*BB*) has dark-blue wings. A butterfly homozygous for the *b* allele (*bb*) has white wings. A heterozygous butterfly (*Bb*) has medium-blue wings (Figure 17.3).

At genetic equilibrium, the proportions of the wing-color genotypes are

$$p^2(BB) \ + \ 2pq(Bb) \ + \ q^2(bb) \ = \ 1.0$$

where *p* and *q* are the frequencies of alleles *B* and *b*. This equation became known as the Hardy–Weinberg equilibrium equation. It defines the frequency of a dominant allele (*B*) and a recessive allele (*b*) for a gene that controls a particular trait in a population.

The frequencies of *B* and *b* must add up to 1.0. To give a specific example, if *B* occupies 90 percent of the loci, then *b* must occupy the remaining 10 percent (0.9 + 0.1 = 1.0). No matter what the proportions,

$$p \ + \ q \ = \ 1.0$$

At meiosis, remember, paired alleles are assorted into different gametes. The proportion of gametes with the *B* allele is *p*, and the proportion with the *b* allele is *q*. The Punnett square *below* shows the genotypes possible in the next generation (*BB*, *Bb*, and *bb*). Note that the frequencies of the three genotypes add up to 1.0:

$$p^2 \ + \ 2pq \ + \ q^2 \ = \ 1.0$$

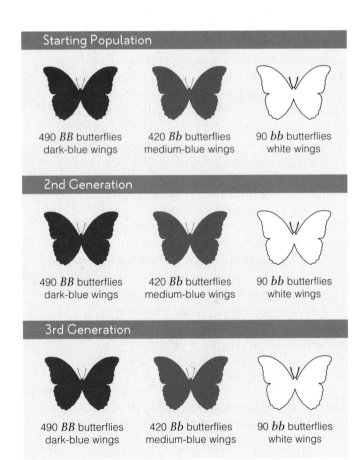

Starting Population

490 *BB* butterflies
dark-blue wings

420 *Bb* butterflies
medium-blue wings

90 *bb* butterflies
white wings

2nd Generation

490 *BB* butterflies
dark-blue wings

420 *Bb* butterflies
medium-blue wings

90 *bb* butterflies
white wings

3rd Generation

490 *BB* butterflies
dark-blue wings

420 *Bb* butterflies
medium-blue wings

90 *bb* butterflies
white wings

	p Ⓑ	*q* ⓑ
p Ⓑ	*BB* (p^2)	*Bb* (*pq*)
q ⓑ	*Bb* (*pq*)	*bb* (q^2)

Suppose that the population has 1,000 individuals and that each one produces two gametes:

490 *BB* individuals make 980 *B* gametes
420 *Bb* individuals make 420 *B* and 420 *b* gametes
90 *bb* individuals make 180 *b* gametes

Figure 17.3 Animated Finding out whether a population is evolving. The frequencies of wing-color alleles among all of the individuals in this hypothetical population of butterflies are not changing; thus, the population is not evolving.

The frequency of alleles *B* and *b* among 2,000 gametes is:

$$B = \frac{980 + 420}{2{,}000 \text{ alleles}} = \frac{1{,}400}{2{,}000} = 0.7 = p$$

$$b = \frac{180 + 420}{2{,}000 \text{ alleles}} = \frac{600}{2{,}000} = 0.3 = q$$

At fertilization, gametes combine at random and start a new generation. If the population size stays constant at 1,000, there will be 490 *BB*, 420 *Bb*, and 90 *bb* individuals. The frequencies of the alleles for dark-blue, medium-blue, and white wings are the same as they were in the original gametes. Thus, dark-blue, medium-blue, and white wings occur at the same frequencies in the new generation.

As long as the assumptions that Hardy and Weinberg identified continue to hold, the pattern persists. If traits show up in different proportions from one generation to the next, though, one or more of the five assumptions is not being met. The hunt can begin for the evolutionary forces driving the change.

Applying the Rule How does the Hardy–Weinberg formula work in the real world? Researchers can use it to estimate the frequency of carriers of alleles that cause genetic traits and disorders.

As an example, hereditary hemochromatosis (HH) is the most common genetic disorder among people of Irish ancestry. Affected individuals absorb too much iron from food. The symptoms of this autosomal recessive disorder include liver problems, fatigue, and arthritis. A study in Ireland found the frequency for one allele that causes HH to be 0.14. If $q = 0.14$, then p is 0.86. Based on this study, the carrier frequency ($2pq$) can be calculated to be about 0.24. Such information is useful to doctors and to public health professionals.

Another example: A mutation in the *BRCA2* gene has been linked to breast cancer in adults. A deviation from the birth frequencies predicted by the Hardy–Weinberg formula suggests that this mutation can also have effects even before birth. In one study, researchers looked at the mutation's frequency among newborn girls. They found fewer homozygotes than expected, based on the number of heterozygotes and the Hardy–Weinberg formula. Thus, it seems that in homozygous form the mutation impairs the survival of female embryos.

> Natural selection occurs in different patterns depending on the organisms involved and their environment.
< Link to Natural selection theory and fitness 16.4

The remainder of this chapter explores the mechanisms and effects of processes that drive evolution, including natural selection. **Natural selection** is a process in which environmental pressures result in the differential survival and reproduction of individuals of a population. It influences the frequency of alleles in a population by operating on phenotypes that have a genetic basis.

We observe different patterns of natural selection, depending on the selection pressures and the organisms involved. Sometimes, individuals with a trait at one extreme of a range of variation are selected against, and those at the other extreme are favored. We call this pattern directional selection. With stabilizing selection, midrange forms are favored, and the extremes are selected against. With disruptive selection, forms at the extremes of the range of variation are favored, and the intermediate forms are selected against. We will discuss these three modes of natural selection, which Figure 17.4 summarizes, in the following two sections.

Section 17.7 explores sexual selection, a mode of natural selection that operates on a population by influencing mating success. This section also discusses balanced polymorphism, a particular case of natural selection in which the fitness (Section 16.4) of heterozygous individuals is greater than that of homozygous individuals in a particular environment.

Natural selection and other processes of evolution can alter a population or species so much that it becomes a new species. We discuss mechanisms of speciation in later sections of this chapter.

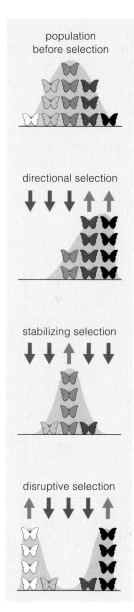

Figure 17.4
Overview of three modes of natural selection.

natural selection A process in which environmental pressures result in the differential survival and reproduction of individuals of a population who vary in the details of shared, heritable traits.

Take-Home Message How do we measure genetic change?

> Researchers measure genetic change by comparing it with a theoretical baseline of genetic equilibrium.

Take-Home Message Does evolution occur in recognizable patterns?

> Natural selection, the most influential process of evolution, occurs in patterns that depend on the organisms and their environment.

> › Directional selection favors a phenotype at one end of a range of variation.
> ‹ Links to Experimental design 1.7, Bacteria 4.1 and 4.5, Melanin deposition in fur 13.5, Continuous variation 13.6

Directional selection shifts an allele's frequency in a consistent direction, so forms at one end of a range of phenotypic variation become more common over time (Figure 17.5). The following examples show how field observations provide evidence of directional selection.

The Peppered Moth Peppered moths feed and mate at night, and rest motionless on trees during the day. Their behavior and coloration offer camouflage from day-flying,

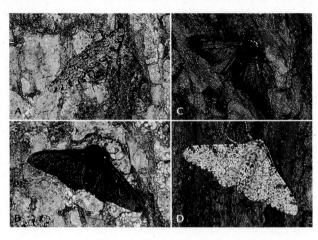

Figure 17.6 Animated Directional selection in the peppered moth. **A** Light peppered moths on a nonsooty tree trunk are hidden from predators. **B** Dark ones stand out. In places where soot darkens tree trunks, the dark color **C** is more adaptive than **D** the light color.

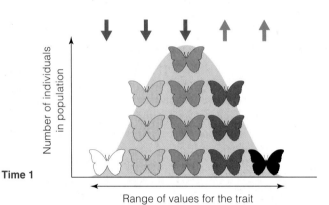

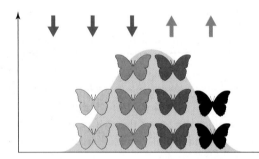

Figure 17.5 Animated Directional selection. The bell-shaped curves indicate continuous variation in a butterfly wing-color trait. *Red* arrows show which forms are being selected against; *green*, forms that are being favored.

moth-eating birds. Light-colored moths were the most common form in preindustrial England. A dominant allele that resulted in the dark color was rare. The air was clean, and light-gray lichens grew on the trunks and branches of most trees. Light moths were camouflaged when they rested on the lichens, but dark moths were not (Figure 17.6A,B). By the 1850s, the dark moths had become much more common. Why? The industrial revolution had begun, and smoke from coal-burning factories was beginning to change the environment. Air pollution was killing the lichens. Dark moths were better camouflaged on soot-darkened trees (Figure 17.6C,D). Researchers hypothesized that birds were selectively eliminating light moths from populations of peppered moths living in industrialized areas. In the 1950s, H. B. Kettlewell bred both moth forms in captivity and marked hundreds so that they could be easily identified. He released them near highly industrialized areas around Birmingham and near an unpolluted part of Dorset. His team recaptured more of the dark moths in the polluted area and more light ones near Dorset. They also observed predatory birds eating more light moths in Birmingham, and more dark moths in Dorset.

Pollution controls went into effect in 1952. As a result, tree trunks gradually became free of soot, and lichens made a comeback. Kettlewell observed that moth phenotypes shifted too: Wherever pollution decreased, the frequency of dark moths decreased as well. Many other researchers since Kettlewell have confirmed the rise and fall of the dark-colored form of the peppered moth.

Rock Pocket Mice Directional selection also affects the color of rock pocket mice in Arizona's Sonoran Desert. Rock pocket mice are small mammals that spend the day sleeping in underground burrows, emerging at night to forage for seeds.

Figure 17.7 Directional selection in populations of rock pocket mice. **A** Mice with light fur are more common in areas with light-colored granite. **B** Mice with dark fur are more common in areas with dark basalt. **C,D** Mice with coat colors that do not match their surroundings are more easily seen by predators, so they are preferentially eliminated from the populations.

The Sonoran Desert is dominated by outcroppings of light brown granite (Figure 17.7A). There are also patches of dark basalt rock, the remains of ancient lava flows (Figure 17.7B). Most of the mice in populations that inhabit the dark rock have dark gray coats. Most of the mice in populations that inhabit the light brown rock have light brown coats. The difference arises because mice that match the rock color in each habitat are camouflaged from their natural predators. Night-flying owls more easily see mice that do not match the rocks, and they preferentially eliminate easily seen mice from each population (Figure 17.7C,D). The preferential predation results in a directional shift in the frequency of alleles that affect coat color. Compared to granite-dwelling populations, populations living on the dark basalt have a much higher frequency of four alleles that cause increased melanin deposition in fur.

Antibiotic Resistance Human attempts to control the environment can result in directional selection, as is the case with the warfarin-resistant rats. The use of antibiotics is another example. Prior to the 1940s, scarlet fever, tuberculosis, and pneumonia caused one-fourth of the annual deaths in the United States alone. Since the 1940s, we have been relying on antibiotics such as penicillin to fight these and other dangerous bacterial diseases. We also use them in other, less dire circumstances. Antibiotics are used preventively, both in humans and in livestock. They are part of the daily rations of millions of cattle, pigs, chickens, fish, and other animals that are raised on factory farms.

Bacteria evolve at an accelerated rate compared with humans, in part because they reproduce very quickly. For example, the common intestinal bacteria *E. coli* can divide every 17 minutes. Each new generation is an opportunity for mutation, so the gene pool of a bacterial population varies greatly. Thus, in any population of bacteria, some cells are likely to carry alleles that allow them to survive an antibiotic treatment. When the survivors reproduce, the frequency of antibiotic-resistance alleles increases in the population. A typical two-week course of antibiotics can potentially exert selection pressure on over a thousand generations of bacteria, and antibiotic-resistant strains may be the outcome.

Antibiotic-resistant bacteria have plagued hospitals for many years, and now they are becoming similarly common in schools. Even as researchers scramble to find new antibiotics, this trend is bad news for the millions of people each year who contract cholera, tuberculosis, or another dangerous bacterial disease.

directional selection Mode of natural selection in which phenotypes at one end of a range of variation are favored.

Take-Home Message **What is the effect of directional selection?**

> Directional selection causes allele frequencies underlying a range of variation to shift in a consistent direction.

> Stabilizing selection is a form of natural selection that maintains an intermediate phenotype.
> Disruptive selection favors forms of a trait at both ends of a range of variation.

Natural selection can bring about a directional shift in a population's range of phenotypes. Depending on the environment and the organisms involved, the process may also favor a midrange form of a trait, or it may eliminate the midrange form and favor extremes.

Stabilizing Selection

With **stabilizing selection**, an intermediate form of a trait is favored, and extreme forms are not. This mode of natural selection is also called balancing selection because it tends to preserve the midrange phenotypes in a population (Figure 17.8). For example, the body weight of sociable weavers (*Philetairus socius*) is subject to stabilizing selection (Figure 17.9). Weaver birds build large communal nests in areas of the African savanna. Between 1993 and 2000, Rita Covas and her colleagues captured, tagged, weighed, and released birds living in communal nests before the breeding season began. The researchers then recaptured and weighed the surviving birds after the breeding season was over.

Covas's field studies indicated that body weight in sociable weavers is a trade-off between the risks of starvation and predation. Foraging is not easy in the sparse habitat of an African savanna, and leaner birds do not store enough fat to avoid starvation. A meager food supply selects against birds with low body weight. Fatter birds may be more attractive to predators, and not as agile when escaping. Predators select against birds of high body weight. Thus, birds of intermediate weight have the selective advantage, and make up the bulk of sociable weaver populations.

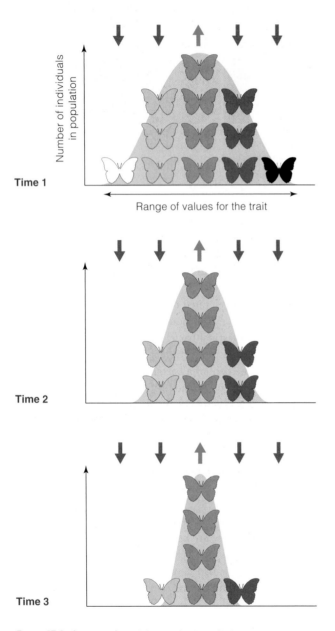

Time 1

Range of values for the trait

Time 2

Time 3

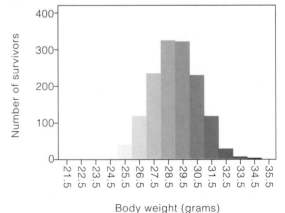

Figure 17.8 Animated Stabilizing selection eliminates extreme forms of a trait, and maintains the predominance of an intermediate phenotype in a population. *Red* arrows indicate which forms are being selected against; *green*, forms that are being favored. Compare the data set from a field experiment shown in Figure 17.9.

Figure 17.9 Stabilizing selection in sociable weavers. Graph shows the number of birds (out of 977) that survived a breeding season.

>> **Figure It Out** What is the optimal weight of a sociable weaver?

Answer: About 29 grams

Disruptive Selection

Conditions that favor forms of a trait at both ends of a range of variation drive **disruptive selection**. With this mode of natural selection, intermediate forms are selected against (Figure 17.10). Consider the black-bellied seedcracker (*Pyrenestes ostrinus*), a colorful finch species native to Cameroon, Africa. In these birds, there is a genetic basis for bill size. The bill of a typical black-bellied seedcracker, male or female, is either 12 millimeters wide, or wider than fifteen millimeters (Figure 17.11). Birds

that have a bill between 12 and 15 millimeters wide are uncommon. Seedcrackers with the large and small bill forms inhabit the same geographic range, and they breed randomly with respect to bill size. It is as if every human adult were four feet or six feet tall, with no one of intermediate height.

Environmental factors that affect seedcracker feeding performance maintain the dimorphism in bill size. The finches feed mainly on the seeds of two types of sedge, which is a grasslike plant. One sedge produces hard seeds; the other, soft seeds. Small-billed birds are better at opening the soft seeds, but large-billed birds are better at cracking the hard ones. All seeds are abundant during Cameroon's wet seasons, and all seedcrackers feed on both types of seeds. However, during the region's dry seasons, sedge seeds are scarce, and each bird focuses on eating the seeds that it opens most efficiently. Small-billed birds feed mainly on soft seeds, and large-billed birds feed mainly on hard seeds. Birds with intermediate-sized bills cannot open either type of seed as efficiently as the other birds, so they are less likely to survive the dry seasons.

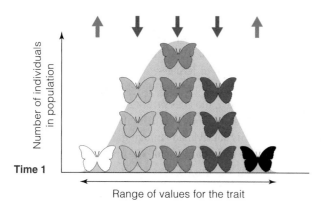

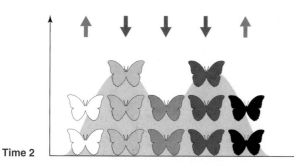

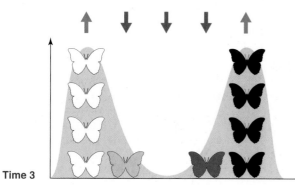

Figure 17.10 Animated Disruptive selection eliminates midrange forms of a trait, and maintains extreme forms. *Red* arrows indicate which forms are being selected against; *green*, forms that are being favored.

Figure 17.11 Animated Disruptive selection in African seedcracker populations. In these birds, a distinct dimorphism in bill size is a result of competition for scarce food during dry seasons. These conditions favor birds with bills that are **A** twelve millimeters wide or **B** fifteen to twenty millimeters wide. Birds with bills of intermediate size are selected against.

disruptive selection Mode of natural selection that favors forms of a trait at the extremes of a range of variation; intermediate forms are selected against.
stabilizing selection Mode of natural selection in which intermediate forms of a trait are favored over extremes.

Take-Home Message What types of natural selection favor intermediate or extreme forms of traits?

❭ With stabilizing selection, an intermediate phenotype is favored, and extreme forms are selected against.

❭ With disruptive selection, an intermediate form of a trait is selected against, and extreme phenotypes are favored.

> Individuals may be selective agents for their own species.
> Any mode of natural selection may maintain two or more alleles in a population.

⟨ Links to Sickle-cell anemia 9.6, Codominance 13.5

Selection pressures that operate on natural populations are often not as clear-cut as the examples in the previous sections might suggest. An allele may be adaptive in one circumstance but harmful in another, as the story about warfarin-resistance in rats illustrates. Even individuals of the same species can play a role.

Nonrandom Mating Not all natural selection occurs as a result of interactions between a species and its environment. Competition within a species also drives evolution. Consider how the individuals of many sexually reproducing species have a distinct male or female phenotype. Individuals of one sex (often males) tend to be more colorful, larger, or more aggressive than individuals of the other sex. These traits seem puzzling because they take energy and time away from an individual's survival activities. Some are probably maladaptive because they attract predators. Why do they persist?

The answer is **sexual selection**, in which the genetic winners outreproduce others of a population because they are better at securing mates. With this mode of natural selection, the most adaptive forms of a trait are those that help individuals defeat same-sex rivals for mates, or are the ones most attractive to the opposite sex.

For example, the females of some species cluster in defensible groups when they are sexually receptive.

Males of these species typically compete for access to the clusters of females. Competition for ready-made harems favors males that are combative (Figure 17.12A).

As another example, males or or females that are choosy about mates act as selective agents on their own species. The females of some species shop for a mate among males that display species-specific cues such as a specialized appearance or courtship behavior (Figure 17.12B). The cues often include flashy body parts or behaviors, traits that can be a physical hindrance and they may attract predators. However, a flashy male's survival despite his obvious handicap implies health and vigor, two traits that are likely to improve a female's chances of bearing healthy, vigorous offspring. The selected males pass alleles for their attractive traits to the next generation of males, and females pass alleles that influence mate preference to the next generation of females.

Sexual selection can give rise to highly exaggerated traits. For example, the eyes of the Malaysian stalk-eyed fly are on the tips of long, horizontal eyestalks that provide no obvious survival advantage to their bearers. Their adaptive value is sexual: Female flies prefer to mate with males sporting the longest eyestalks (Figure 17.12C).

---

balanced polymorphism Maintenance of two or more alleles for a trait at high frequency in a population as a result of natural selection against homozygotes.
sexual selection Mode of natural selection in which some individuals outreproduce others of a population because they are better at securing mates.

Figure 17.12
Sexual selection in action.

A Male elephant seals fight for sexual access to a cluster of females.

B A male bird of paradise engaged in a flashy courtship display has caught the eye (and, perhaps, the sexual interest) of a female. Female birds of paradise are choosy; a male mates with any female that accepts him.

C Stalk-eyed flies cluster on aerial roots to mate. Females prefer males with the longest eyestalks. This photo, taken in Malaysia, shows a male with very long eyestalks (*top*) that has captured the interest of the three females below him.

Balanced Polymorphism Any mode of natural selection may result in a **balanced polymorphism**. In this state, two or more alleles are maintained in a population at relatively high frequency by an environment that favors heterozygotes (individuals with nonidentical alleles).

Consider the gene that encodes the beta globin chain of hemoglobin, which is the oxygen-transporting protein in blood. *HbA* is the normal allele of this gene. The codominant *HbS* allele carries a mutation that causes sickle-cell anemia (Section 9.6).

Individuals homozygous for the *HbS* allele often die in their teens or early twenties. Despite being so harmful, the *HbS* allele persists at very high frequency among the human populations in tropical and subtropical regions of Asia and Africa. Why? Populations with the highest frequency of the *HbS* allele also have the highest incidence of malaria (Figure 17.13). Mosquitoes transmit the parasitic protist that causes malaria, *Plasmodium*, to human hosts. *Plasmodium* multiplies in the liver and then in red blood cells. The cells rupture and release new parasites during recurring bouts of severe illness.

It turns out that people who make both normal and sickle hemoglobin are more likely to survive malaria than people who make only normal hemoglobin. Several mechanisms are possible. For example, in *HbA/HbS* heterozygotes, *Plasmodium*-infected red blood cells sometimes take on a sickle shape. The abnormal shape brings the cells to the attention of the immune system, which destroys them—along with the parasites they harbor. By contrast, *Plasmodium*-infected red blood cells cells of *HbA/HbA* homozygotes do not sickle, so the parasite may remain hidden from the immune system.

In areas where malaria is common, the persistence of the *HbS* allele is a matter of relative evils. Malaria and sickle-cell anemia are both potentially deadly. *HbA/HbS* heterozygotes are more likely to survive malaria than *HbA/HbA* homozygotes. Heterozygotes are not completely healthy, but they do make enough normal hemoglobin to survive. With or without malaria, heterozygotes are more likely to live long enough to reproduce than *HbS/HbS* homozygotes. The result is that nearly one-third of the people living in the most malaria-ridden regions of the world are heterozygous for the *HbS* allele.

Take-Home Message How does natural selection maintain diversity?

› With sexual selection, a trait is adaptive if it gives an individual an advantage in securing mates. Sexual selection reinforces phenotypical differences between males and females, and sometimes gives rise to exaggerated traits.

› Environmental pressures that favor heterozygotes can lead to a balanced polymorphism.

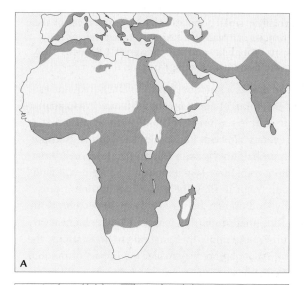

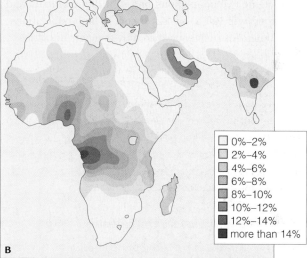

☐	0%–2%
☐	2%–4%
☐	4%–6%
☐	6%–8%
☐	8%–10%
☐	10%–12%
☐	12%–14%
☐	more than 14%

Figure 17.13 Malaria and sickle-cell anemia.

A Distribution of malaria cases (*orange*) reported in Africa, Asia, and the Middle East in the 1920s, before the start of programs to control mosquitoes, which transmit the parasitic protist that causes the disease.

B Distribution (by percentage) of people that carry the sickle-cell allele. Notice the correlation between the maps.

C Physician searching for mosquito larvae in Southeast Asia.

> Especially in small populations, random changes in allele frequencies can lead to a loss of genetic diversity.

< Links to Probability and sampling error 1.8, Locus 13.2, Ellis–van Creveld syndrome 14.2

Genetic drift is a random change in allele frequencies over time, brought about by chance alone. We explain genetic drift in terms of probability—the chance that some event will occur. Remember, sample size is important in probability (Section 1.8). For example, every time you flip a coin, there is a 50 percent chance it will land heads up. With 10 flips, the proportion of times heads actually land up may be very far from 50 percent. With 1,000 flips, that proportion is more likely to be near 50 percent. We can apply the same rule to populations: the larger the population, the smaller the impact of random changes in allele frequencies.

Imagine two populations, one with 10 individuals, and the other with 100. If allele X occurs in both populations at a 10 percent frequency, then only one person carries

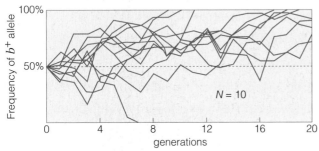

A The size of these populations of beetles was maintained at 10 breeding individuals. Allele b^+ was lost in one population (one graph line ends at 0).

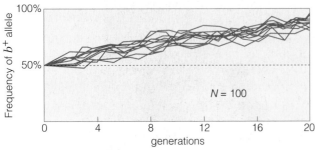

B The size of these populations was maintained at 100 individuals. Drift in these populations was less than the small populations in A.

Figure 17.14 Animated Genetic drift in flour beetles (*Tribolium castaneum*, shown *below left* on a flake of cereal). Randomly selected beetles heterozygous for alleles b^+ and b were maintained in populations of A 10 individuals or B 100 individuals for 20 generations.

 Graph lines in B are smoother than in A, indicating that drift was greatest in the sets of 10 beetles and least in the sets of 100. Notice that the average frequency of allele b^+ rose at the same rate in both groups, an indication that natural selection was at work too: Allele b^+ was weakly favored.

>> **Figure It Out** In how many populations did allele b^+ become fixed?   Answer: Six

the allele in the small population. If that person dies without reproducing, allele X will be lost from that population. However, ten people in the large population carry the allele. All ten would have to die without reproducing for the allele to be lost. Thus, the chance that the small population will lose allele X is greater than that for the large population. Steven Rich and his colleagues demonstrated this effect in populations of flour beetles (Figure 17.14).

Genetic drift can lead to the loss of genetic diversity in a population. This outcome is possible in all populations, but it is more likely to occur in small ones. When all individuals of a population are homozygous for an allele, we say that the allele is **fixed**. The frequency of an allele that is fixed will not change unless mutation or another process introduces a different allele into the population.

Bottlenecks Genetic drift can be dramatic when a few individuals rebuild a population or start a new one, such as occurs after a **bottleneck**. A bottleneck is a drastic reduction in population size brought about by severe selection pressure. For example, excessive hunting had reduced the population of northern elephant seals (shown in Figure 17.12A) to a mere twenty individuals by the late 1890s. The population has recovered to about 170,000 individuals since hunting restrictions were implemented, but every seal is homozygous at every gene locus analyzed to date. Genetic drift after the bottleneck has fixed all of the alleles in this population.

Bottlenecking can also occur when a small group of individuals founds a new population. If the group is not representative of the original population in terms of allele frequencies, then the new population will not be representative of it either. This outcome is called the **founder effect**. If a founding group is very small, the new population's genetic diversity may be quite reduced. Imagine that a seabird lands in the middle of a population of plants on a mainland. In this population's gene pool, half of the alleles governing flower color specify white flowers; the other half specify yellow flowers. A few seeds stick to the bird's feathers. The bird flies to a remote island and drops the seeds, which later sprout and form a new population on the island. If most of the seeds happened to be homozygous for the yellow flower allele, then the frequency of that allele in the new population will be much greater than 50 percent.

bottleneck Reduction in population size so severe that it reduces genetic diversity.
fixed Refers to an allele for which all members of a population are homozygous.
founder effect Change in allele frequencies that occurs when a small number of individuals establish a new population.
genetic drift Change in allele frequencies in a population due to chance alone.
inbreeding Mating among close relatives.

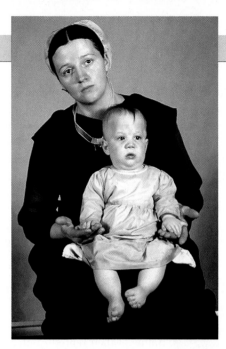

Figure 17.15 An Amish child with Ellis–van Creveld syndrome. The syndrome is characterized by dwarfism, polydactyly, and heart defects, among other symptoms. The recessive allele that causes it is common in the Old Order Amish of Lancaster County, an outcome of the founder effect and moderate inbreeding.

Genetic drift affects allele frequencies in inbred populations. **Inbreeding** is nonrandom breeding or mating between close relatives, which share more alleles than nonrelatives do. Inbreeding lowers a population's genetic diversity. Loss of diversity tends to be a bad thing, because it means more individuals in the population are homozygous for recessive alleles with harmful effects. This is why most societies discourage or forbid incest, or mating between parents and children or between siblings.

The Old Order Amish in Lancaster County, Pennsylvania, offer an example of the effects of inbreeding. Amish people marry only within their community. Intermarriage with other groups is not permitted, and no "outsiders" are allowed to join the community. As a result, Amish populations are moderately inbred, and many of their individuals are homozygous for harmful alleles. The Lancaster population has an unusually high frequency of a recessive allele that causes Ellis–van Creveld syndrome (Figure 17.15). This allele has been traced to a man and his wife, two of a group of 400 Amish who immigrated to the United States in the mid-1700s. As a result of the founder effect and inbreeding since then, about 1 of 8 people in the Lancaster population is now heterozygous for the allele, and 1 in 200 is homozygous for it.

17.9 Gene Flow

> Individuals, along with their alleles, move into and out of populations. This flow of alleles counters genetic change that tends to occur within a population.
< Link to Transgenic plants 15.7

Individuals tend to mate or breed most frequently with other members of their own population. However, most populations of a species are not completely isolated from one another, so there may be intermating among nearby populations. Also, individuals sometimes leave one population and join another. **Gene flow**, the movement of alleles among populations, occurs in both cases. Gene flow stabilizes allele frequencies, so it counters the effects of mutation, natural selection, and genetic drift.

Gene flow is typical among populations of animals, which tend to be more mobile, but it also occurs in plant populations. Consider the acorns that blue jays disperse when they gather nuts for the winter. Every fall, jays visit acorn-bearing oak trees repeatedly, then bury acorns in the soil of home territories that may be as much as a mile away (Figure 17.16). The jays transfer acorns—and the alleles inside them—among populations of oak trees that would otherwise be genetically isolated.

Figure 17.16 Blue jay, a mover of acorns that helps keep genes flowing between separate oak populations.

Gene flow in plants also occurs when pollen is transferred from one individual to another, often over great distances. Many opponents of genetic engineering cite gene flow from transgenic organisms into wild populations via pollen transfer. Herbicide-resistance genes and the Bt gene (Section 15.7) have been found in weeds and unmodified crop plants that are growing near fields of transgenic plants. The long-term effects of this gene flow are currently unknown.

gene flow The movement of alleles into and out of a population.

Take-Home Message How does the genetic diversity of a population become reduced?

> Genetic drift, or random change in allele frequencies, can reduce a population's genetic diversity. Its effect is greatest in small populations, such as one that endures a bottleneck.

Take-Home Message How does gene flow affect allele frequencies in a population?

> Gene flow is the physical movement of alleles into and out of a population. It tends to counter the evolutionary effects of mutation, natural selection, and genetic drift.

> Speciation differs in its details, but reproductive isolating mechanisms are always part of the process.

‹ Links to Zygote 12.2, Meiosis 12.3

Mutation, natural selection, and genetic drift operate on all natural populations, and they do so independently in populations that are not interbreeding. When gene flow does not keep populations alike, different genetic changes accumulate in each one. Sooner or later, the populations become so different that we call them different species. The evolutionary process by which new species arise is called **speciation.**

Evolution is a dynamic, extravagant, messy, and ongoing process that can be challenging for people who like their categories neat. Speciation offers a perfect example, because it rarely occurs at a precise moment in time. Individuals often continue to interbreed even as populations are diverging, and populations that have already diverged may come together and interbreed again.

Every time speciation happens, it happens in a unique way, because each species is a product of its own unique evolutionary history. However, we can identify some recurring patterns. For example, reproductive isolation is always part of speciation. **Reproductive isolation** refers to the end of gene flow between populations. It is part of the process by which sexually reproducing species attain and maintain their separate identities. By preventing successful interbreeding, reproductive isolation reinforces differences between diverging populations. We say the isolation is prezygotic if pollination or mating cannot occur, or if zygotes cannot form. If hybrids form but are unfit or infertile, the isolation is postzygotic (Figure 17.17).

Mechanisms of Reproductive Isolation

Temporal Isolation Some populations cannot interbreed because the timing of their reproduction differs. The periodical cicada (*inset*) offers an example. Cicadas feed on roots as they mature underground, then emerge to reproduce. Three species of cicada reproduce every 17 years. Each has a sibling species with nearly identical form and behavior, except that the siblings emerge on a 13-year cycle instead of a 17-year cycle. Sibling species have the potential to interbreed, but they can only get together once every 221 years!

Mechanical Isolation In some cases, the size or shape of an individual's reproductive parts prevent it from mating with members of another population. For example, black sage (*Salvia mellifera*) and white sage (*S. apiana*) grow in the same areas, but hybrids rarely form because the flowers of the two species have become specialized for different pollinators (Figure 17.18). Carpenter bees, hawkmoths, and other large insects pollinate white sage when they force open the petals to access nectar hidden inside the flowers. Honeybees seeking nectar are too small to touch the reproductive parts of a white sage flower, but they are just the right size to pollinate flowers of black sage. The weight of larger bees perching on the tiny flowers of black sage pulls the delicate petals closed. Large bees access the nectar of this species by piercing the petals, so they usually avoid touching the flower's reproductive parts.

Ecological Isolation Populations adapted to different microenvironments in the same region may be ecologically isolated. For example, two species of manzanita (a plant) native to the Sierra Nevada mountain range rarely hybridize. One species that is better adapted for conserving water inhabits dry, rocky hillsides high in the foothills. The other lives on lower slopes where water stress is not as intense. The physical separation makes cross-pollination unlikely.

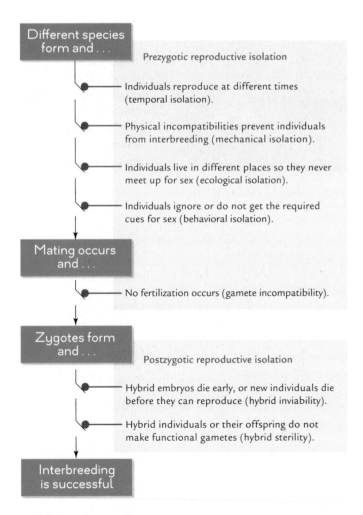

Figure 17.17 **Animated** How reproductive isolation prevents interbreeding.

anthers

stigma

A Black sage is pollinated mainly by honeybees and other small insects.

Figure 17.18
Mechanical isolation in sage.

B The flowers of black sage are too delicate to support larger insects. Big insects access the nectar of small sage flowers only by piercing from the outside, as this carpenter bee is doing. When they do so, they avoid touching the flower's reproductive parts.

C The reproductive parts (anthers and stigma) of white sage flowers are too far away from the petals to be brushed by honeybees, so honeybees cannot pollinate this species. White sage is pollinated mainly by larger bees and hawkmoths, which brush the flower's stigma and anthers as they pry apart the petals to access nectar.

Behavioral Isolation In animals, behavioral differences can stop gene flow between related species. For instance, males and females of some bird species engage in courtship displays before sex (Figure 17.19). A female recognizes the vocalizations and movements of a male of her species as an overture to sex, but females of different species usually do not.

Gamete Incompatibility Even if gametes of different species do meet up, they often have molecular incompatibilities that prevent them from fusing. Gamete incompatibility may be the primary speciation route of animals that fertilize their eggs by releasing free-swimming sperm in water.

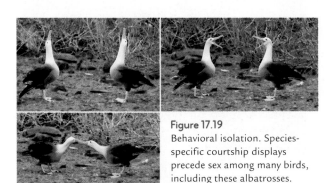

Figure 17.19
Behavioral isolation. Species-specific courtship displays precede sex among many birds, including these albatrosses.

Hybrid Inviability As populations begin to diverge, so do their genes. Even chromosomes of species that diverged recently may have major differences. Thus, a hybrid zygote may have extra or missing genes, or genes with incompatible products. If genetic incompatibilities disrupt development, a hybrid embryo may die. Hybrid offspring that survive may have reduced fitness. For example, ligers and tigons (offspring of lions and tigers) have more health problems and a shorter life expectancy than individuals of either parent species.

Hybrid Sterility Some interspecies crosses produce robust but sterile offspring. For example, mating a female horse (64 chromosomes) with a male donkey (62 chromosomes) produces a mule. The mule's 63 chromosomes cannot pair up evenly during meiosis, so this animal makes few viable gametes.

If hybrids are fertile, their offspring usually have lower and lower fitness with each successive generation. A mismatch between nuclear and mitochondrial DNA may be the cause (mitochondrial DNA is inherited from the mother only).

---

reproductive isolation Absence of gene flow between populations; always part of speciation.
speciation One of several processes by which new species arise.

Take-Home Message How do species attain and maintain separate identities?

❯ Speciation is an evolutionary process by which new species form. It varies in its details and duration.

❯ Reproductive isolation, which occurs by one of several mechanisms, is always a part of speciation.

> In allopatric speciation, a physical barrier arises and ends gene flow between populations.
< Links to Galápagos archipelago 16.4, Plate tectonics 16.7

Genetic changes that lead to a new species can begin with physical separation between populations, so **allopatric speciation** is one way that new species form (*allo*– means different; *patria*, fatherland). By this speciation mode, a physical barrier separates two population and ends gene flow between them. Then, reproductive isolating mechanisms arise, so even if the populations meet up again later, their individuals could not interbreed.

Populations of most species are separated by some distance, and gene flow between them is usually intermittent. Whether a geographic barrier can block that gene flow depends on whether and how an organism travels (such as by swimming, walking, or flying), and how it reproduces (for example, by internal fertilization or by pollen dispersal). A geographic barrier can arise in an instant, or over an eon. The Great Wall of China is an example of a barrier that arose abruptly. As it was being built, the wall cut off gene flow among nearby populations of insect-pollinated plants. DNA sequencing comparisons show that trees, shrubs, and herbs on either side of the wall are now diverging genetically.

Geographic isolation usually occurs much more slowly. For example, it took millions of years of tectonic plate movements (Section 16.7) to bring the two continents of

North and South America close enough to collide. The land bridge where the two continents now connect is called the Isthmus of Panama. When this isthmus formed about 4 million years ago, it cut off the flow of water—and gene flow among populations of aquatic organisms—as it separated one large ocean into what are now the Pacific and the Atlantic oceans (Figure 17.20).

Speciation in Archipelagos

The Florida Keys and some other island chains are so close to a mainland that gene flow is more or less unimpeded, so they foster little if any speciation. The Hawaiian Islands, the Galápagos Islands, and some other island chains are archipelagos. These remote, isolated islands were born of hot spots on the ocean floor. They are the tops of volcanoes, so we can assume that their fiery surfaces were initially barren and inhospitable to life.

Winds or currents sometimes carry a few individuals of a mainland species to an island in an archipelago. If the individuals reproduce, their descendants may establish a population on the island. The vast expanse of ocean that isolates the island from the mainland functions as a geographic barrier to gene flow. Thus, over generations, the island population diverges from the mainland species.

Individuals of the diverging population may in turn colonize other islands in the archipelago. Habitats and selection pressures that differ within and between the islands can foster even more divergences from the ances-

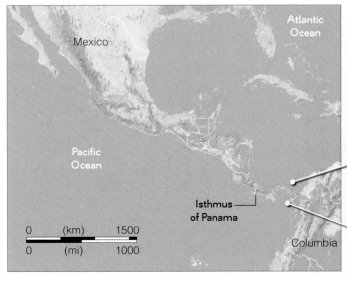

Figure 17.20 Allopatric speciation in snapping shrimp. The Isthmus of Panama (*above*) cut off gene flow among populations of these aquatic shrimp when it formed 4 million years ago. Today, individuals from opposite sides of the isthmus are so similar that they might interbreed, but they are behaviorally isolated: Instead of mating when they are brought together, they snap their claws at one another aggressively. The photos on the *right* show two of the many closely related species that live on opposite sides of the isthmus.

Alpheus nuttingi (Atlantic)

Alpheus millsae (Pacific)

Akepa
(*Loxops coccineus*)

Insects, spiders, nectar; high mountain rain forest

Akekee
(*Loxops caeruleirostris*)

Insects, spiders, nectar; high mountain rain forest

Nihoa finch
(*Telespiza ultima*)

Insects, buds, seeds, flowers, seabird eggs; rocky or shrubby slopes

Palila
(*Loxioides bailleui*)

Mamane seeds, buds, flowers, berries, insects; high mountain dry forests

Maui parrotbill
(*Pseudonestor xanthophrys*)

Insect larvae, pupae, caterpillars; mountain forests, dense underbrush

Apapane
(*Himatione sanguinea*)

Nectar, caterpillars and other insects, spiders; high mountain forests

Poouli
(*Melamprosops phaeosoma*)

Tree snails, insects in understory; last one died in 2004

Maui Alauahio
(*Paroreomyza montana*)

Bark or leaf insects, some nectar; high mountain rain forest

Kauai Amakihi
(*Hemignathus kauaiensis*)

Bark-picker; insects, spiders, nectar; high mountain rain forest

Akiapolaau
(*Hemignathus munroi*)

Probes, digs insects from big trees; high mountain rain forest

Akohekohe
(*Palmeria dolei*)

Mostly nectar from flowering trees, some insects, pollen; high mountain rain forest

Iiwi
(*Vestiaria coccinea*)

Mostly nectar (ohia flowers, lobelias, mints), some insects; high mountain rain forest

Figure 17.21 Animated Allopatric speciation on an archipelago.

Archipelagos such as the Hawaiian Islands (*right*) are separated from mainland continents by thousands of miles of open ocean—a geographic barrier that prevents gene flow between any island colonizers and mainland populations. Further divergences occur as colonizers spread to the other islands in the chain.

Above, a few of the known species of Hawaiian honeycreepers, with some of their dietary and habitat preferences. Specialized bills and behaviors adapt the honeycreepers to feed on certain insects, seeds, fruits, nectar, or other foods. DNA sequence comparisons suggest that the ancestor of all Hawaiian honeycreepers resembled the housefinch (*Carpodacus*) at *left*.

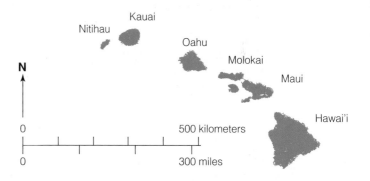

tral species. Even as the island populations become different species, some individuals may return to an island colonized by their ancestors.

The Hawaiian archipelago includes 19 islands and more than 100 atolls that stretch 1,500 miles in the Pacific Ocean. The habitats on the land masses of this archipelago range from lava beds, rain forests, and grasslands to dry woodlands and snow-capped peaks. The first birds to colonize it found a buffet of fruits, seeds, nectars, and tasty insects. The near absence of competitors and preda-

tors in an abundance of rich and vacant habitats spurred rapid speciation. Figure 17.21 hints at the variation among Hawaiian honeycreepers, descendants of one mainland finch species that arrived on the archipelago about 3.5 million years ago. Hawaiian honeycreepers and thousands of other species are unique to the Hawaiian archipelago.

allopatric speciation Speciation pattern in which a physical barrier that separates members of a population ends gene flow between them.

Take-Home Message What happens after a physical barrier arises and prevents populations from interbreeding?

> A physical barrier that intervenes between populations or subpopulations of a species prevents gene flow among them. As gene flow ends, genetic divergences give rise to new species. This process is called allopatric speciation.

❯ Populations sometimes speciate even without a physical barrier that bars gene flow between them.
❮ Links to Polyploidy 14.6, Fitness 16.4

Sympatric Speciation

In **sympatric speciation**, populations inhabiting the same geographic region speciate in the absence of a physical barrier between them (*sym–* means together).

Sympatric speciation can occur in an instant with a change in chromosome number. About 95 percent of ferns and 70 percent of flowering plant species are polyploid, as well as a few conifers, insects and other arthropods, mollusks, fishes, amphibians, and reptiles. Remember, chromosome number can change as a result of nondisjunction (Section 14.6). Common bread wheat originated after related species hybridized, and then the chromosome number multiplied in the offspring (Figure 17.22). In flowering plants, a chromosome number change can also originate in one parent. If a somatic cell fails to divide during mitosis, the resulting polyploid cells may proliferate to form shoots and flowers. If the flowers self-fertilize, a new species may result.

Sympatric speciation can also occur with no change in chromosome number. The mechanically isolated sage plants you learned about in Section 17.10 speciated with no physical barrier to gene flow. As another example, more than 500 species of cichlid, a freshwater fish, speciated in the shallow waters of Lake Victoria. This large freshwater lake sits isolated from river inflow on an elevated plain in Africa's Great Rift Valley. Since Lake Victoria formed about 400,000 years ago, it has dried up three times. DNA sequence comparisons indicate that almost all of the cichlid species in this lake arose since the last dry

spell, which was 12,400 years ago. How could hundreds of species arise so quickly? The answer begins with differences in the color of ambient light and water clarity in different parts of the lake. Water absorbs blue light, so the deeper it is, the less blue light penetrates it. The light in shallower, clear water is mainly blue; the light that penetrates deeper, muddier water is mainly red.

Lake Victoria cichlids vary in color and in patterning (Figure 17.23). Outside of captivity, female cichlids rarely mate with males of other species. Given a choice, they prefer brightly colored males of their own species. Their preference has a molecular basis in genes that encode light-sensitive pigments of the eye. The pigments made by species that live mainly in shallow, clear water are more sensitive to blue light. The males of these species are also the bluest. The pigments made by species that live mainly in deeper, murkier water are more sensitive to red light. Males of these species are redder. In other words, the colors that a female cichlid sees best are the same colors displayed by males of her species. Thus, mutations in genes that affect color perception are likely to affect the choice of mates and of habitats. Such mutations are probably the way sympatric speciation occurs in these fish.

Sympatric speciation has also occurred in greenish warblers of central Asia (*Phylloscopus trochiloides*). A chain of populations of this bird encircles the Tibetan plateau. Adjacent populations of greenish warblers interbreed easily, except for two populations in northern Siberia. Individuals of these two populations overlap in range, but they do not interbreed because they do not recognize one another's songs. They have become behaviorally isolated. Small genetic differences between the other populations have added up to major differences between

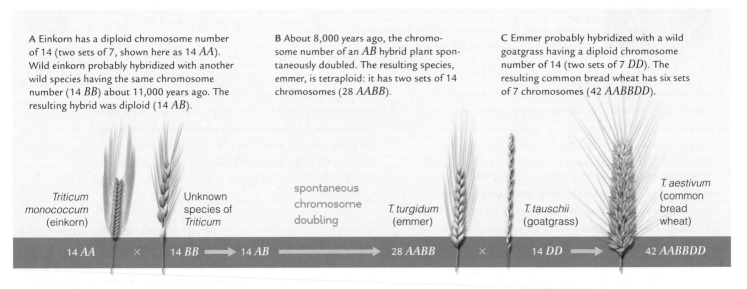

A Einkorn has a diploid chromosome number of 14 (two sets of 7, shown here as 14 *AA*). Wild einkorn probably hybridized with another wild species having the same chromosome number (14 *BB*) about 11,000 years ago. The resulting hybrid was diploid (14 *AB*).

B About 8,000 years ago, the chromosome number of an *AB* hybrid plant spontaneously doubled. The resulting species, emmer, is tetraploid: it has two sets of 14 chromosomes (28 *AABB*).

C Emmer probably hybridized with a wild goatgrass having a diploid chromosome number of 14 (two sets of 7 *DD*). The resulting common bread wheat has six sets of 7 chromosomes (42 *AABBDD*).

Triticum monococcum (einkorn)   Unknown species of *Triticum*   spontaneous chromosome doubling   *T. turgidum* (emmer)   *T. tauschii* (goatgrass)   *T. aestivum* (common bread wheat)

14 *AA*   ×   14 *BB* ⟶ 14 *AB* ⟶ 28 *AABB*   ×   14 *DD* ⟶ 42 *AABBDD*

Figure 17.22 Sympatric speciation in wheat.

Figure 17.23 Red fish, blue fish: Males of four closely related species of cichlid native to Lake Victoria, Africa. Hundreds of cichlids speciated in sympatry in this lake. Mutations in genes that affect females' perception of the color of ambient light in deeper or shallower regions of the lake also affect their choice of mates. Female cichlids prefer to mate with brightly colored males of their own species.

>> Figure It Out What form of natural selection has been driving sympatric speciation in Lake Victoria cichlids?

Answer: Sexual selection

the two populations at the ends of the chain. The chain of greenish warbler populations are collectively called a ring species. Such species present one of those paradoxes for people who like neat categories: Gene flow occurs continuously all around the chain, but the two populations at the ends of the chain are different species. Where should we draw the line that divides those two species?

Parapatric Speciation

Parapatric speciation may occur when one population extends across a broad region encompassing diverse habitats. The different habitats exert distinct selection pressures on parts of the population, and the result may be divergences that lead to speciation. Hybrids that form in a contact zone between habitats are less fit than individuals on either side of it.

Consider velvet walking worms, which resemble caterpillars but may be more related to spiders. These worms are predatory; they shoot streams of glue from their head at insects. Once entangled in the sticky glue, the insects are easy prey for the worms. Two rare species of velvet walking worm are native to the island of Tasmania. The giant velvet walking worm (*Tasmanipatus barretti*) and the blind velvet walking worm (*T. anophthalmus*) can interbreed, but they only do so in a tiny area where their habitats overlap. Hybrid offspring are sterile, which

parapatric speciation Speciation model in which different selection pressures lead to divergences within a single population.
sympatric speciation Pattern in which speciation occurs in the absence of a physical barrier.

Table 17.2 Comparison of Speciation Models

Speciation Model:	Allopatric	Sympatric	Parapatric
Original population	●	●	●
Initiating event:	barrier arises	genetic change	new niche entered
Reproductive isolation occurs			
New species arises:	in isolation	within population	in new niche

may be the main reason the two species are maintaining separate identities in the absence of a physical barrier between their populations. Table 17.2 compares parapatric speciation with other speciation models.

Take-Home Message Can speciation occur without a physical barrier to gene flow?

> By a sympatric speciation model, new species arise from a population even in the absence of a physical barrier.

> By a parapatric speciation model, populations maintaining contact along a common border evolve into distinct species.

> Macroevolution includes patterns of change such as one species giving rise to multiple species, the origin of major groups, and major extinction events.

‹ Links to Homeotic genes 10.3, Geologic time scale 16.6, Plate tectonics 16.7

Patterns of Macroevolution

Microevolution is change in allele frequencies within a single species or population. Macroevolution is our name for evolutionary patterns on a larger scale. Flowering plants evolved from seed plants, animals with four legs (tetrapods) evolved from fish, birds evolved from dinosaurs—all of these are examples of macroevolution that occurred over millions of years.

A central theme of macroevolution is that major evolutionary novelties have often stemmed from the adaptation of an existing structure for a completely different purpose. This theme is called preadaptation or **exaptation**. Some traits serve a very different purpose in modern species than they did when they first evolved. For example, the feathers that allow modern birds to fly are derived from feathers that first evolved in some dinosaurs. Those dinosaurs could not have used their feathers

Figure 17.24 An example of stasis. *Top*, 320-million-year-old coelacanth fossil found in Montana. *Bottom*, a live coelacanth (*Latimeria chalumnae*) caught off the waters of Sulawesi in 1998. The coelacanth lineage has changed very little over evolutionary time.

for flight, but they probably did use them for insulation. Thus, we say that flight feathers in birds are an exaptation of insulating feathers in dinosaurs.

Stasis With the simplest macroevolutionary pattern, **stasis**, lineages persist for millions of years with little or no change. Consider coelacanths, an order of ancient lobe-finned fish that had been assumed extinct for at least 70 million years until a fisherman caught one in 1938. The modern coelacanth species are very similar to fossil specimens hundreds of millions of years old (Figure 17.24).

Mass Extinctions By current estimates, more than 99 percent of all species that ever lived are now **extinct**, or irrevocably lost from Earth. In addition to continuing small-scale extinctions, the fossil record indicates that there have been more than twenty mass extinctions, which are simultaneous losses of many lineages. These include five catastrophic events in which the majority of species on Earth disappeared (Section 16.6).

Adaptive Radiation In an evolutionary pattern called **adaptive radiation**, a lineage rapidly diversifies into several new species. Adaptive radiation can occur after individuals colonize a new environment that has a variety of different habitats with few or no competitors. The adaptation of populations to different regions of the new environment produces new species. The Hawaiian honeycreepers that you read about in Section 17.11 arose this way (Figure 17.25).

Adaptive radiation may occur after a key innovation evolves. A **key innovation** is a new trait that allows its bearer to exploit a habitat more efficiently or in a novel way. The evolution of lungs offers an example, because

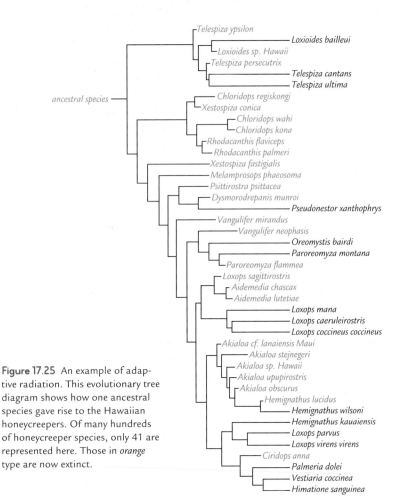

Figure 17.25 An example of adaptive radiation. This evolutionary tree diagram shows how one ancestral species gave rise to the Hawaiian honeycreepers. Of many hundreds of honeycreeper species, only 41 are represented here. Those in *orange* type are now extinct.

ancestral species

Telespiza ypsilon
Loxioides bailleui
Loxioides sp. Hawaii
Telespiza persecutrix
Telespiza cantans
Telespiza ultima
Chloridops regiskongi
Xestospiza conica
Chloridops wahi
Chloridops kona
Rhodacanthis flaviceps
Rhodacanthis palmeri
Xestospiza fastigialis
Melamprosops phaeosoma
Psittirostra psittacea
Dysmorodrepanis munroi
Pseudonestor xanthophrys
Vangulifer mirandus
Vangulifer neophasis
Oreomystis bairdi
Paroreomyza montana
Paroreomyza flammea
Loxops sagittirostris
Aidemedia chascax
Aidemedia lutetiae
Loxops mana
Loxops caeruleirostris
Loxops coccineus coccineus
Akialoa cf. lanaiensis Maui
Akialoa stejnegeri
Akialoa sp. Hawaii
Akialoa upupirostris
Akialoa obscurus
Hemignathus lucidus
Hemignathus wilsoni
Hemignathus kauaiensis
Loxops parvus
Loxops virens virens
Ciridops anna
Palmeria dolei
Vestiaria coccinea
Himatione sanguinea

Figure 17.26 An example of coevolved species.

A The large blue butterfly (*Maculinea arion*) parasitizes a species of red ant, *Myrmica sabuleti*.

B To an ant, a honey-exuding, hunched-up *Maculinea arion* caterpillar appears to be an ant larva. This deceived ant is preparing to carry the caterpillar back to its nest, where the caterpillar will eat ant larvae for the next 10 months until it pupates.

lungs were a key innovation that opened the way for an adaptive radiation of vertebrates on land.

Adaptive radiations also occur after geologic or climatic events eliminate some species from a habitat. The surviving species then can exploit resources from which they had been previously excluded. This is the way mammals were able to undergo an adaptive radiation after the dinosaurs disappeared.

Coevolution The process by which close ecological interactions between two species cause them to evolve jointly is called **coevolution**. One species acts as an agent of selection on the other; each adapts to changes in the other. Over evolutionary time, the two species may become so interdependent that they can no longer survive without one another.

Relationships between coevolved species can be quite intricate. For example, the large blue butterfly (*Maculinea arion*) parasitizes red ants (Figure 17.26A). After hatching, the larva (caterpillar) of a large blue butterfly feeds on wild thyme flowers and then drops to the ground. A red ant that finds the caterpillar strokes it, whereupon the caterpillar exudes honey. The ant eats the honey and continues to stroke the caterpillar, which secretes more honey. This interaction continues for hours, until the caterpillar suddenly hunches itself up into a shape that appears, to

an ant anyway, very much like an ant larva (Figure 17.26B). The beguiled ant then picks up the caterpillar and carries it back to the ant nest, where, in most cases, other ants kill it—except, however, if the ants are of one particular species, *Myrmica sabuleti*. Secretions of a *M. arion* caterpillar fool these ants into treating it just like a larva of their own. For the next 10 months, the caterpillar lives in the nest and grows to gigantic proportions by feeding on ant larvae. After it metamorphoses into a butterfly, it lays its eggs on wild thyme near another *M. sabuleti* nest, and the cycle starts anew. This coevolved relationship between ant and butterfly is extremely specific. Any increase in the ants' ability to identify a caterpillar in their nest selects for caterpillars that better deceive the ants, which in turn select for ants that can better identify the caterpillars. Each species exerts directional selection on the other.

Evolutionary Theory

Biologists do not doubt that macroevolution occurs, but many disagree about how it occurs. However we choose to categorize evolutionary processes, the very same genetic change may be at the root of all evolution—fast or slow, large-scale or small-scale. Dramatic jumps in morphology, if they are not artifacts of gaps in the fossil record, may be the result of mutations in homeotic or other regulatory genes. Macroevolution may include more processes than microevolution, or it may not. It may be an accumulation of many microevolutionary events, or it may be an entirely different process. Evolutionary biologists may disagree about these and other hypotheses, but all of them are trying to explain the same thing: how all species are related by descent from common ancestors.

---

adaptive radiation A burst of genetic divergences from a lineage gives rise to many new species.
coevolution The joint evolution of two closely interacting species; each species is a selective agent for traits of the other.
exaptation Adaptation of an existing structure for a completely different purpose; a major evolutionary novelty.
extinct Refers to a species that has been permanently lost.
key innovation An evolutionary adaptation that gives its bearer the opportunity to exploit a particular environment more efficiently or in a new way.
stasis Evolutionary pattern in which a lineage persists with little or no change over evolutionary time.

Take-Home Message **What is macroevolution?**

❯ Macroevolution comprises large-scale patterns of evolutionary change such as adaptive radiations, coevolution, and mass extinction.

❯ Cladistics allows us to organize our knowledge about how species are related.
❮ Link to Taxonomy 1.5

Ranking Versus Grouping

Linnaeus devised his system of taxonomy before anyone knew about evolution, so taxonomy rankings do not necessarily reflect evolutionary relationships. Our increasing understanding of evolution is prompting a major, ongoing overhaul of the way biologists view life's diversity. Instead of trying to divide that tremendous diversity into a series of ranks, most biologists are now focusing on evolutionary connections. Each species is viewed not as a member or representative of a rank in a hierarchy, but rather as part of a bigger picture of evolution.

Phylogeny is the evolutionary history of a species or a group of them, a kind of genealogy that follows a lineage's evolutionary relationships through time. The central question of phylogeny is, "Who is related to whom?" Methods of finding the answer to that question are an important part of phylogeny. One method, **cladistics**, groups species on the basis of their shared characters. A **character** is a quantifiable, heritable characteristic—any physical, behavioral, physiological, or molecular trait of a species (Table 17.3).

The result of a cladistic analysis is a **cladogram**, a diagram that shows a network of evolutionary relationships (Figure 17.27). Each line in a cladogram represents a lin-

eage, which may branch into two lineages at a node. The node represents a common ancestor of the two lineages. Every branch ends with a **clade** (from *klados*, a Greek word for twig or branch), a species or group of species that share a set of characters.

Ideally, each clade is a **monophyletic group** that comprises an ancestor and all of its descendants. However, as you learned in Section 17.10, evolution can be challenging for those who like neat categories. Each species has many characters, and researchers discover more species and more characters all the time. Because we do not yet know all species, or all characters, cladistic groupings are necessarily hypotheses. They may differ depending on which characters are used for the analysis, so clades often change when new discoveries are made.

Cladograms and other types of **evolutionary trees** summarize our best data-supported hypotheses about how a group of species evolved. We use these diagrams to visualize evolutionary trends and patterns. For instance, the two lineages that emerge from a node on a cladogram are called **sister groups**. Sister groups are, by default, the same age. We may not know what that age is, but we can compare sister groups on a cladogram and say something about their relative rates of evolution.

Like other hypotheses, evolutionary tree diagrams get revised. However, the diagrams are based on the solid premise that all species are interconnected by shared ancestry. Every living thing is related if you just go back far enough in time. An evolutionary biologist's job is to figure out where the connections are.

How We Use Evolutionary Biology

The first Polynesians arrived on the Hawaiian Islands sometime before 1000 A.D., and Europeans followed in 1778. Hawaii's rich ecosystem was hospitable to all newcomers, including the settlers' dogs, cats, pigs, cows, goats, deer, and sheep. Escaped livestock began to eat and trample rain forest plants that had provided Hawaiian honeycreepers with food and shelter. Entire forests were cleared to grow imported crops, and plants that escaped

Table 17.3	Examples of Characters			
	Multicellular	Backbone	Legs	Hair or Fur
Earthworm	✔	–	–	–
Tuna	✔	✔	–	–
Lizard	✔	✔	✔	–
Mouse	✔	✔	✔	✔
Human	✔	✔	✔	✔

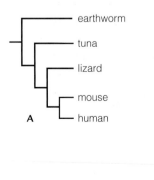

A

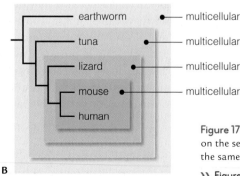

- multicellular
- multicellular with a backbone
- multicellular with a backbone and legs
- multicellular with a backbone, legs, and fur or hair

B

Figure 17.27 Animated Cladograms. **A** This example is based on the set of characters chosen in Table 17.3. **B** We can visualize the same cladogram as "sets within sets" of characters.

❯❯ **Figure It Out** In this cladogram, which is the sister group of the mouse?

Answer: Human

Figure 17.28 Three honeycreeper species: going, going, and gone.

A The palila has an adaptation that allows it to feed on mamane seeds, which are toxic to most other birds. The one remaining palila population is declining because mamane plants are being trampled by cows and gnawed to death by goats and sheep. Only about 2,640 palila remained in 2009.

B Avian malaria carried by mosquitoes to higher altitudes is decimating the last population of the akekee. Between 2000 and 2007, the number of akekee plummeted from 7,839 birds to 3,536.

C This male poouli—rare, old, and missing an eye—died in 2004 from avian malaria. There were two other poouli alive at the time, but neither has been seen since then.

cultivation began to crowd out native plants. Mosquitoes introduced in 1826 spread diseases from imported chickens to native bird species. Stowaway rats and snakes ate their way through populations of native birds and their eggs. Mongooses deliberately imported to eat the rats and snakes preferred to eat birds and eggs.

Ironically, the very isolation that spurred adaptive radiations made the honeycreepers vulnerable to extinction. The birds had no built-in defenses against predators or diseases of the mainland. Specializations such as extravagantly elongated beaks became hindrances when the birds' habitats suddenly changed or disappeared.

character Quantifiable, heritable characteristic or trait.
clade A species or group of species that share a set of characters.
cladistics Method of determining evolutionary relationships by grouping species into clades based on shared characters.
cladogram Evolutionary tree that shows a network of evolutionary relationships among clades.
evolutionary tree Type of diagram that summarizes evolutionary relationships among a group of species.
monophyletic group An ancestor and all of its descendants.
phylogeny Evolutionary history of a species or group of species.
sister groups The two lineages that emerge from a node on a cladogram.

Rise of the Super Rats (revisited)

The gene most commonly involved in warfarin resistance in rats is inherited in a simple dominance pattern. The gene's normal allele is recessive when paired with the mutated allele that confers resistance. In the presence of warfarin, the dominant allele is clearly adaptive. Rats with this allele require a lot of extra vitamin K, but being vitamin K–deficient is not so bad when compared with being dead from rat poison. However, in the absence of warfarin, individuals that have the warfarin resistance allele are at a serious disadvantage compared with those who do not. Rats with the allele cannot easily obtain enough vitamin K from their diet to sustain normal blood clotting and bone formation. Thus, the allele is adaptive when warfarin is present, and maladaptive when it is not, so periodic exposure to warfarin maintains a balanced polymorphism of the resistance gene in rat populations.

How Would You Vote? Antibiotic–resistant strains of bacteria are now widespread. One standard animal husbandry practice includes continually dosing healthy livestock with the same antibiotics prescribed for people. Should this practice stop? See CengageNow for details, then vote online (cengagenow.com).

Thus, at least 43 species of honeycreeper that had thrived on the islands before the arrival of humans were extinct by 1778. Today, 32 of the remaining 71 species are endangered, and 26 are extinct despite tremendous conservation efforts since the 1960s (Figure 17.28). Invasive, non-native species of plants and animals are now established, and the rise in global temperatures is allowing disease-bearing mosquitoes to invade high-altitude habitats that had previously been too cold for them.

The story of the Hawaiian honeycreepers is a dramatic illustration of how evolution works. It also shows how finding ancestral connections can help species that are still living. As more and more honeycreeper species become extinct, the group's reservoir of genetic diversity dwindles. The lowered diversity means the group as a whole is less resilient to change, and more likely to suffer catastrophic species losses. Deciphering their phylogeny can tell us which honeycreeper species are most different from the others—and those are the ones most valuable in terms of preserving genetic diversity. Such research allows us to concentrate our resources and conservation efforts on those species that hold the best hope for the survival of the entire group. For example, we now know the poouli (Figure 17.28C) to be the most distantly related member of the genus. Unfortunately, that knowledge came too late; the species is probably extinct.

Take-Home Message How do evolutionary biologists study life's diversity?

› Evolutionary biologists study phylogeny in order to understand how all species are connected by shared ancestry.

Summary

Section 17.1 Our efforts to control pests have resulted in directional selection for resistant populations. Populations tend to change along with the selection pressures that are acting on them.

Sections 17.2, 17.3 Individuals of a **population** share traits with a heritable basis. All of their alleles form a **gene pool**. New alleles that arise by mutation may be **neutral**, **lethal**, or advantageous. **Microevolution**, or changes in the **allele frequencies** of a population, occurs constantly by processes of mutation, natural selection, genetic drift, and gene flow. We use deviations from **genetic equilibrium** to study how a population is evolving.

Sections 17.4–17.6 **Natural selection** occurs in different patterns. **Directional selection** shifts the range of variation in traits in one direction. **Stabilizing selection** favors intermediate forms of a trait. **Disruptive selection** favors forms at the extremes of a range of variation.

Section 17.7 Traits that differ between males and females are an outcome of **sexual selection**, in which adaptive traits are ones that make their bearers better at securing mates. With **balanced polymorphism**, an environment that favors heterozygotes maintains two or more alleles at high frequency.

Sections 17.8, 17.9 Random change in allele frequencies, or **genetic drift**, can lead to the loss of genetic diversity by causing alleles to become **fixed**. Genetic drift is most pronounced in small or **inbreeding** populations. The **founder effect** may occur after an evolutionary **bottleneck**. **Gene flow** counters the effects of mutation, natural selection, and genetic drift.

Section 17.10 The details of **speciation** differ every time it occurs, but **reproductive isolation**, the absence of gene flow between populations, is always a part of the process. The exact moment at which two populations become separate species is often impossible to pinpoint.

Section 17.11 In **allopatric speciation**, a geographic barrier arises and interrupts gene flow between populations. After gene flow ends, genetic divergences then give rise to new species.

Section 17.12 Speciation can occur with no physical barrier to gene flow. In **sympatric speciation**, populations in physical contact speciate. Polyploid species of many plants (and a few animals) often originate in sympatry. With **parapatric speciation**, populations in contact along a common border speciate.

Section 17.13 Macroevolution refers to patterns of evolution above the species level: stasis, adaptive radiation, coevolution, and extinction. Major evolutionary novelty typically arises by **exaptation**, in which a lineage uses a structure for a different purpose than its ancestor did. With **stasis**, a lineage changes very little over evolutionary time. A **key innovation** can result in an **adaptive radiation**, or rapid diversification into new species. **Coevolution** occurs when two species act as agents of selection upon one another. A lineage that is permanently lost from Earth is **extinct**.

Section 17.14 **Cladistics** allows us to reconstruct evolutionary history (**phylogeny**). Species are grouped into **clades** based on shared **characters**. Ideally, a clade is a **monophyletic group**, but clades often change as a result of new information. A **cladogram** is a type of **evolutionary tree** in which each line represents one lineage. A lineage can branch into two **sister groups** at a node, which represents a shared ancestor.

Self-Quiz   Answers in Appendix III

1. Individuals don't evolve, _____ do.

2. Biologists define evolution as _____ .
   a. purposeful change in a lineage
   b. heritable change in a line of descent
   c. acquiring traits during the individual's lifetime

3. _____ is the original source of new alleles.
   a. Mutation   d. Gene flow
   b. Natural selection   e. All are original sources of
   c. Genetic drift   new alleles

4. Evolution can only occur in a population when _____ .
   a. mating is random
   b. there is selection pressure
   c. neither is necessary

5. A fire devastates all trees in a wide swath of forest. Populations of a species of tree-dwelling frog on either side of the burned area diverge to become separate species. This is an example of _____ .

6. Stabilizing selection tends to _____ (select all that apply).
   a. eliminate extreme forms of a trait
   b. favor extreme forms of a trait
   c. eliminate intermediate forms of a trait
   d. favor intermediate forms of a trait
   e. shift allele frequencies in one direction

7. Disruptive selection tends to _____ (select all that apply).
   a. eliminate extreme forms of a trait
   b. favor extreme forms of a trait
   c. eliminate intermediate forms of a trait
   d. favor intermediate forms of a trait
   e. shift allele frequencies in one direction

Data Analysis Activities

Resistance to Rodenticides in Wild Rat Populations Beginning in 1990, rat infestations in northwestern Germany started to intensify despite continuing use of rat poisons. In 2000, Michael H. Kohn and his colleagues analyzed the genetics of wild rat populations around Munich. For part of their research, they trapped wild rats in five towns, and tested those rats for resistance to warfarin and the more recently developed poison bromadiolone. The results are shown in Figure 17.29.

1. In which of the five towns were most of the rats susceptible to warfarin?
2. Which town had the highest percentage of poison-resistant wild rats?
3. What percentage of rats in Olfen were resistant to warfarin?
4. In which town do you think the application of bromadiolone was most intensive?

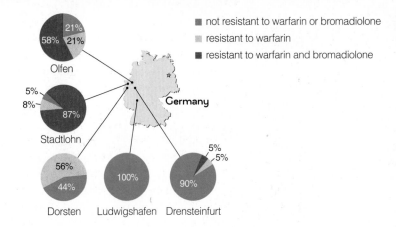

■ not resistant to warfarin or bromadiolone
■ resistant to warfarin
■ resistant to warfarin and bromadiolone

Figure 17.29 Resistance to rat poisons in wild populations of rats in Germany, 2000.

---

8. Directional selection tends to _____ (select all that apply).
   - a. eliminate extreme forms of a trait
   - b. favor an extreme form of a trait
   - c. eliminate intermediate forms of a trait
   - d. favor intermediate forms of a trait
   - e. shift allele frequencies in one direction

9. Sexual selection, such as occurs when males compete for access to fertile females, frequently influences aspects of body form and can lead to _____ .
   - a. male/female differences
   - b. male aggression
   - c. exaggerated traits
   - d. all of the above

10. The persistence of malaria and sickle-cell anemia in a population is a case of _____ .
    - a. bottlenecking
    - b. balanced polymorphism
    - c. natural selection
    - d. artificial selection
    - e. both b and c

11. _____ tends to keep populations of a species similar to one another.
    - a. Genetic drift
    - b. Gene flow
    - c. Mutation
    - d. Natural selection

12. In evolutionary trees, each node represents a(n) _____ .
    - a. single lineage
    - b. extinction
    - c. point of divergence
    - d. adaptive radiation

13. In cladograms, sister groups are _____ .
    - a. inbred
    - b. the same age
    - c. represented by nodes
    - d. in the same family

Animations and Interactions on **CENGAGENOW**:
❯ Hardy–Weinberg analysis; Directional selection; Directional selection in the peppered moth; Stabilizing selection; Disruptive selection; Disruptive selection in African finches; Genetic drift; Reproductive isolation; Allopatric speciation on an archipelago; Cladograms.

14. Match the evolution concepts.
    - ___ gene flow
    - ___ natural selection
    - ___ mutation
    - ___ genetic drift
    - ___ adaptive radiation
    - ___ coevolution
    - ___ phylogeny
    - ___ cladogram

    - a. can lead to interdependent species
    - b. changes in a population's allele frequencies due to chance alone
    - c. alleles enter or leave a population
    - d. evolutionary history
    - e. occurs in different patterns
    - f. burst of divergences from one lineage into many
    - g. source of new alleles
    - h. diagram of sets within sets

Additional questions are available on **CENGAGENOW**.

Critical Thinking

1. Rama the cama, a llama–camel hybrid, was born in 1997. The idea was to breed an animal that has the camel's strength and endurance and the llama's gentle disposition. However, instead of being large, strong, and sweet, Rama is smaller than expected and has a camel's short temper. The breeders plan to mate him with Kamilah, a female cama, but they wonder if offspring from such a match would be fertile. What does Rama's story tell you about the genetic changes required for reproductive isolation in nature? Explain why a biologist might not view Rama as evidence that llamas and camels are the same species.

2. Some theorists have hypothesized that many of our uniquely human traits arose by sexual selection. Over many thousands of years, women attracted to charming, witty men perhaps prompted the development of human intellect far beyond what was necessary for mere survival. Men attracted to women with juvenile features may have shifted the species as a whole to be less hairy and softer featured than any of our simian relatives. Can you think of a way to test these hypotheses?

❮ Links to Earlier Concepts

This chapter starts your survey of biodiversity, introduced in Section 1.4. Here, we will place the origin organic compounds (3.2), bacteria and archaea (4.5), and eukaryotes (4.6) on the time line of Earth history (16.6).

We return to connections between photosynthesis and aerobic respiration (Chapters 6 and 7) and consider how the nucleus, ER, mitochondria, and chloroplasts (4.7–4.9) might have originated.

Key Concepts

Setting the Stage for Life
Earth formed about 4 billion years ago from matter distributed in space by the big bang (the origin of the universe). The early Earth was an inhospitable place, where meteorite impacts and volcanic eruptions were common and the atmosphere held little or no oxygen.

Building Blocks of Life
All life is composed of the same organic subunits. Simulations of conditions on the early Earth show that these molecules could have formed by reactions in the atmosphere or sea. Organic subunits also form in space and could have been delivered to Earth by meteorites.

18 Life's Origin and Early Evolution

18.1 Looking for Life

The photo at the *left* shows the Eagle Nebula, an unimaginably huge cloud of gas and dust 7,000 light years away. We live in a vast universe that we have only begun to explore. So far, we know of only one planet that has life—Earth. In addition, biochemical, genetic, and metabolic similarities among Earth's species imply that all evolved from a common ancestor that lived billions of years ago. What properties of the ancient Earth allowed life to arise, survive, and diversify? Could similar processes occur on other planets? These are some of the questions posed by **astrobiology**, the study of life's origins and distribution in the universe.

Astrobiologists study Earth's extreme habitats to determine the range of conditions that living things can tolerate. One group found bacteria living about 30 centimeters (1 foot) below the soil surface of Chile's Atacama Desert, a place said to be the driest on Earth (Figure 18.1). Another group drilled 3 kilometers (almost 2 miles) beneath the soil surface in Virginia, where they found bacteria thriving at high pressure and temperature. They named their find *Bacillus infernus*, or "bacterium from hell."

Knowledge gained from studies of life on Earth inform the search for extraterrestrial life. On Earth, all metabolic reactions involve interactions among molecules in aqueous solution (dissolved in water). We assume that the same physical and chemical laws operate throughout the universe, so liquid water is considered an essential requirement for life. Thus, scientists were excited when a robotic lander discovered water frozen in the soil of Mars, our closest planetary neighbor. If there is life on Mars, it is likely to be underground. Mars has no ozone layer, so ultraviolet radiation would fry organisms at the planet's surface. However, Martian life may exist in deep rock layers just as it does on Earth.

Suppose scientists do find evidence of microbial life on Mars or another planet. Why would it matter? Such a discovery would

Figure 18.1 The Mars-like landscape of Chile's Atacama Desert. Scientist Jay Quade, visible in the distance at the *right*, was a member of a team that found bacteria living beneath this desert soil.

support the hypothesis that life on Earth arose as a consequence of physical and chemical processes that occur throughout the universe. The discovery of extraterrestrial microbes would also make the possibility of nonhuman intelligent life in the universe more likely. The more places life exists, the more likely it is that complex, intelligent life evolved on other planets in the same manner that it did on Earth.

---

astrobiology The scientific study of life's origin and distribution in the universe.

The First Cells Form
All cells have enzymes that carry out reactions, a plasma membrane, and a genome of DNA. Experiments provide insight into how cells arose through physical and chemical processes, such as the tendency of lipids to form membrane-like structures when mixed with water.

Life's Early Evolution
The first cells were probably anaerobic. An early divergence separated bacteria from archaeans and ancestors of eukaryotic cells. Evolution of oxygen-producing photosynthesis in bacteria altered Earth's atmosphere, creating conditions that favored aerobic organisms.

Eukaryotic Organelles
A nucleus, ER, and other membrane-enclosed organelles are defining features of eukaryotic cells. Some organelles may have evolved from infoldings of the plasma membrane. Mitochondria and chloroplasts probably descended from bacteria that lived inside other cells.

> Physical and geological forces produced Earth, its seas, and its atmosphere.
> The early atmosphere had little oxygen.
< Link to Elements 2.2

From the Big Bang to the Early Earth

No one was around to witness the birth of the universe, so our ideas about what happened will forever remain in the realm of conjecture. However, the ancient events that led to our universe and our planet left their signature in the energy and matter that exist today. Scientists who study stars and space continue to discover clues about how our universe originated.

The widely accepted **big bang theory** states that the universe began in a single instant, about 13 to 15 billion years ago. In that instant all existing matter and energy suddenly appeared and exploded outward from a single point. Simple elements such as hydrogen and helium formed within minutes. Then, over millions of years, gravity drew the gases together and they condensed to form giant stars (Figure 18.2).

Explosions of these early stars scattered the heavier elements from which today's galaxies formed. Our own galaxy, the Milky Way, probably began as a cloud of stellar debris trillions of kilometers wide. Some of that debris condensed and formed the galaxy's stars.

About 5 billion years ago, the star we call our sun was orbited by a cloud of dust and rocks (asteroids), but no planets. The asteroids collided and merged into bigger asteroids. The heavier these pre-planetary rocks became, the more gravitational pull they exerted, and the more material they gathered. By about 4.6 billion years ago, Earth and the other planets of our solar system had formed.

Figure 18.2 What the cloud of dust, gases, rocks, and ice around the early sun may have looked like.

Figure 18.3 Artist's depiction of early Earth, at a time when volcanic activity and meteor strikes were still common events.

Conditions on the Early Earth

Planet formation did not clear out all of the debris orbiting the sun, so the early Earth received a constant hail of meteorites. Earth's surface was molten, and more molten rock and gases spewed continually from volcanoes. Gases released by volcanoes and meteorite impacts were the main components of the early atmosphere.

What was Earth's early atmosphere like? Studies of volcanic eruptions, meteorites, ancient rocks, and other planets suggest that the air contained water vapor, carbon dioxide, and gaseous hydrogen and nitrogen. We know that there was little or no oxygen, because the oldest existing rocks show no evidence of iron oxidation (rusting). If oxygen had been present in Earth's early atmosphere, it would have caused rust to form. More important, it would have interfered with assembly of the organic compounds necessary for life. Oxygen would have reacted with and destroyed the compounds as fast as they formed.

At first, any water falling on Earth's molten surface evaporated immediately. As the surface cooled, rocks formed. Later, rains washed mineral salts out of these rocks and the salty runoff pooled in early seas (Figure 18.3). It was in these seas that life began.

---

big bang theory Model describing formation of the universe as a nearly instant distribution of matter through space.

Take-Home Message What were conditions like on the early Earth?

> Meteor impacts were common.
> The atmosphere had little or no oxygen.
> The seas contained mineral salts leached from the rocks.

The Source of Life's Building Blocks

> All living things are made from organic subunits: simple sugars, amino acids, fatty acids, and nucleotides. Where did the subunits that made up the first life come from? Here we look at three well-researched possibilities.
< Link to Molecules of life 3.2

Lightning–Fueled Atmospheric Reactions

In 1953, Stanley Miller and Harold Urey proposed that reactions in Earth's early atmosphere could have produced building blocks for the first life. At that time, many scientists thought Earth's early atmosphere consisted of methane, ammonia, and hydrogen gas. Miller and Urey placed water and these gases into a reaction chamber (Figure 18.4). As the mix circulated, sparks from electrodes simulated lightning. Within a week, a variety of amino acids and other small molecules formed.

The Miller–Urey experiment was initially hailed as a breakthrough that demonstrated the first step on the road to life. Then, the idea that Earth's early atmosphere consisted mainly of carbon dioxide and nitrogen dioxide gained favor. When Miller redid his experiment using these gases in his apparatus, he was unable to detect any amino acid formation.

Miller died in 2007, but experiments by other scientists have given new credence to his ideas. One experiment showed that amino acids do form in a simulated carbon dioxide and nitrogen atmosphere. Miller failed to detect amino acids when he used these gases because his experiment also formed compounds that break down amino acids. On the early Earth, rains would have washed amino acids formed by atmospheric reactions into the sea, where the breakdown reactions might not occur.

Reactions at Hydrothermal Vents

Reactions near deep-sea hydrothermal vents also produce organic building blocks. A **hydrothermal vent** is like an underwater geyser, a place where hot, mineral-rich water streams out through a rocky opening (Figure 18.5). The water is heated by geothermal energy. Günter Wächtershäuser and Claudia Huber simulated conditions near a hydrothermal vent by combining hot water with carbon monoxide (CO) and potassium cyanide (KCN) and metal ions like those in rocks near the vents. Their results showed amino acids formed within a week.

Delivery From Space

The presence of amino acids, sugars, and nucleotide bases in meteorites that fell to Earth suggests another possible origin for life's building blocks. These molecules may have formed in interstellar clouds of ice, dust, and gases and been delivered to Earth by meteorites. During Earth's early years, meteorites fell to Earth thousands of times more frequently than they do today.

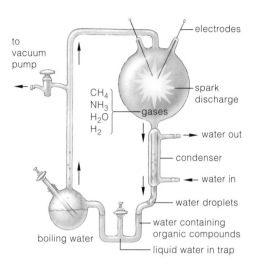

Figure 18.4 Animated Diagram of an apparatus designed by Stanley Miller and Harold Urey to test whether organic compounds could have formed by chemical interactions in Earth's early atmosphere. Water, hydrogen gas (H_2), methane (CH_4), and ammonia (NH_3) were kept circulating through the apparatus. Sparks from an electrode simulated lightning.

>> **Figure It Out** Which gas in this mixture provided the nitrogen for the amino group in the amino acids?   Answer: Ammonia

Figure 18.5 A hydrothermal vent on the sea floor. Mineral-rich water heated by geothermal energy streams out of the vent, into the cold ocean water. The drop in temperature causes dissolved minerals to come out of solution and form a chimney-like structure around the vent.

hydrothermal vent Rocky, underwater opening where mineral-rich water heated by geothermal energy streams out.

Take-Home Message Where did the simple organic building blocks of the first life come from?

> Simulation experiments support the hypothesis that simple organic compounds could have formed by chemical reactions in Earth's early atmosphere or in the sea near a hydrothermal vent.

> Observations and experiments also support the hypothesis that such compounds could have formed in space and been carried to Earth on meteorites.

> › Experiments demonstrate how traits and processes seen in all living cells could have begun with physical and chemical reactions among nonliving collections of molecules.
> ‹ Link to Cell structure 4.2

Steps on the Road to Life

In addition to sharing the same molecular components, all cells have a plasma membrane with a lipid bilayer. They have a genome of DNA that enzymes transcribe into RNA, and ribosomes that translate RNA into proteins. All cells replicate, and pass on copies of their genetic material to their descendants. The many similarities in structure, metabolism, and replication processes among all life are evidence of descent from a common cellular ancestor.

Time has erased all evidence of the earliest cells, but scientists can still investigate this first chapter in life's history. They use their knowledge of chemistry to design experiments that test whether a particular hypothesis about how life began is plausible. Such studies support the hypothesis that cells arose as a result of a stepwise process that began with inorganic materials (Figure 18.6). Each step on the road to life can be explained by familiar chemical and physical mechanisms that still occur today.

Origin of Metabolism

Modern cells take up organic subunits, concentrate them, and assemble them into organic polymers. Before there were cells, a nonbiological process that concentrated organic subunits in one place would have increased the chance that the subunits would combine.

By one hypothesis, this process occurred on clay-rich tidal flats. Clay particles have a slight negative charge, so positively charged molecules in seawater stick to them. At low tide, evaporation would have concentrated the subunits even more, and energy from the sun might have caused them to bond together as polymers. In simulations of tidal flat conditions, amino acids form short chains.

By another hypothesis, metabolic reactions began in the high-temperature, high-pressure environment near a hydrothermal vent. Rocks around the vents contain iron sulfide (pyrite) and are porous, with many tiny chambers

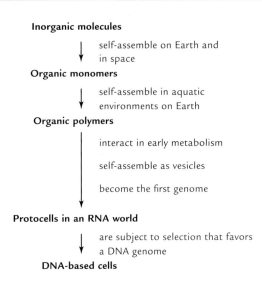

Inorganic molecules
↓ self-assemble on Earth and in space
Organic monomers
↓ self-assemble in aquatic environments on Earth
Organic polymers
→ interact in early metabolism
→ self-assemble as vesicles
→ become the first genome
↓
Protocells in an RNA world
→ are subject to selection that favors a DNA genome
↓
DNA-based cells

Figure 18.6 Proposed sequence for the evolution of cells. Scientists investigate this process by carrying out experiments and simulations that test hypotheses about feasibility of individual steps.

about the size of cells (Figure 18.7). Metabolism may have begun when iron sulfide in the rocks donated electrons to dissolved carbon monoxide, setting in motion reactions that formed larger organic compounds. Researchers who tested this hypothesis by simulating vent conditions found that organic compounds such as pyruvate, do form and accumulate in the chambers. In addition, all modern organisms require iron-sulfide cofactors to carry out some reactions. The universal requirement for these cofactors may be a legacy of life's rocky beginnings.

Origin of the Cell Membrane

Molecules formed by early synthetic reactions would have simply floated away from one another unless something enclosed them. In modern cells, a plasma membrane serves this function. If the first reactions took place in tiny rock chambers, the rock would have acted as a boundary. Over time, lipids produced by reactions inside a chamber could have accumulated and lined the chamber wall. Such lipid-enclosed collections of interacting molecules could have been the first **protocells**. A protocell is a membrane-enclosed collection of molecules that takes up material and replicates itself.

Experiments by Jack Szostak and others have shown that rock chambers are not necessary for protocell formation. Figure 18.8A illustrates one type of protocell that Szostak investigates. Figure 18.8B is a photo of a protocell that formed in his laboratory. A membrane of lipid bilayer encloses strands of RNA. The protocell "grows" by adding fatty acids to its membrane and nucleotides to its RNA. Mechanical force causes protocell division.

Figure 18.7 Cell-sized chambers in iron-sulfide-rich rocks formed by simulations of conditions near hydrothermal vents. Similar chambers could have served as protected environments in which the first metabolic reactions took place.

20 µm

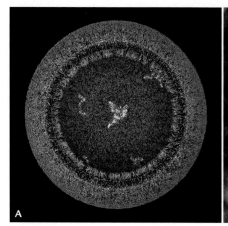
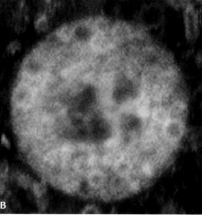

Figure 18.8 Protocells. **A** Illustration of a laboratory-produced protocell. It has a bilayer membrane of fatty acids and holds strands of RNA. Ribonucleotides that diffuse into the protocell become incorporated into complementary strands of RNA. The vesicle can enlarge by incorporating additional fatty acids. **B** Laboratory-formed protocell consisting of RNA-coated clay (*red*) surrounded by fatty acids and alcohols.

C Field-testing a hypothesis about protocell formation. David Deamer pours a mix of small organic molecules and phosphates into a hot acidic pool in Russia.

David Deamer studies protocell formation in both the laboratory and the field. In the lab, he has shown that the small organic molecules carried to Earth on meteorites can react with minerals and seawater to form vesicles with a bilayer membrane. However, Deamer has yet to locate a natural environment that facilitates the same process. In one experiment, he added a mix of organic subunits to the acidic waters of a clay-rich volcanic pool in Russia (Figure 18.8C). The organic subunits bound tightly to the clay, but no vesicle-like structures formed. Deamer concluded that hot acidic waters of volcanic springs do not provide the right conditions for protocell formation. He continues to carry out experiments to determine what naturally occurring conditions do favor this process.

Origin of the Genome

All modern cells have a genome of DNA. They pass copies of their DNA to descendant cells, which use instructions encoded in the DNA to build proteins. Some of these proteins are enzymes that synthesize new DNA, which is passed along to descendant cells, and so on. Thus, protein synthesis depends on DNA, which is built by proteins. How did this cycle begin?

In the 1960s, Francis Crick and Leslie Orgel addressed this dilemma by suggesting that RNA may have been the first molecule to encode genetic information. Since then, evidence for an early **RNA world**—a time when RNA both stored genetic information and functioned like an enzyme in protein synthesis—has accumulated. **Ribozymes**, or RNAs that function as enzymes, have been discovered in living cells. The rRNA in ribosomes speeds formation of peptide bonds during protein synthesis. Other ribozymes cut noncoding bits (introns) out of newly formed RNAs. Researchers have also produced self-replicating ribozymes that copy themselves by assembling free nucleotides.

If the earliest self-replicating genetic systems were RNA-based, then why do all organisms have a genome of DNA? The structure of DNA may hold the answer. Compared to a double-stranded DNA molecule, single-stranded RNA breaks apart more easily and mutates more often. Thus, a switch from RNA to DNA would make larger, more stable genomes possible.

---

protocell Membranous sac that contains interacting organic molecules; hypothesized to have formed prior to the earliest life forms.
ribozyme RNA that functions as an enzyme.
RNA world Hypothetical early interval when RNA served as the genetic information.

> ## Take-Home Message What have experiments revealed about the steps that led to the first cells?
>
> ❯ All living cells carry out metabolic reactions, are enclosed within a plasma membrane, and can replicate themselves.
>
> ❯ Concentration of molecules on clay particles or in tiny rock chambers near hydrothermal vents may have helped start metabolic reactions.
>
> ❯ Vesicle-like structures with outer membranes can form spontaneously.
>
> ❯ An RNA-based system of inheritance may have preceded DNA-based systems.

> Fossils and molecular comparisons among living species inform us about the history of life on Earth.

< Links to Bacteria and archaeans 4.5, Eukaryotic cells 4.6, Photosynthesis 6.4, Evolution of aerobic respiration 7.2

Origin of Bacteria and Archaea

How old is life on Earth? Different analyses provide slightly different answers. Given the genetic differences among living species and current mutation rates, scientists estimate that the common ancestor of all cells lived about 4.3 billion years ago. Some microscopic filaments from Australia that date back 3.5 billion years may be fossil cells (Figure 18.9A). Microfossils from another Australian location are widely accepted as evidence that cells were living around hydrothermal vents on the sea floor by 3.2 billion years ago.

The small size and simple structure of early fossil cells suggests that they were not eukaryotes. This finding is consistent with evidence from gene comparisons among living organisms that places bacteria and archaea near the base of the tree of life. Because Earth's early air and seas held little oxygen, the first cells were probably anaerobic.

Genetic analysis tells us that a divergence early in the history of life separated the domains Bacteria and Archaea. After the split, light-capturing pigments evolved in some members of both groups. The oxygen-releasing noncyclic pathway of photosynthesis evolved only in one bacterial lineage, the cyanobacteria (Figure 18.9B,C). Cyanobacteria are a relatively recent branch on the bacterial family tree, so noncyclic photosynthesis presumably arose through mutations that modified the cyclic pathway.

Cyanobacteria and other photosynthetic bacteria grew as dense mats in shallow sunlit water. The mats trapped minerals and sediments. Over many years, continual cell growth and deposition of minerals formed large dome-shaped, layered structures called **stromatolites** (Figure 18.9D,E). Such structures still form in some seas today.

Effects of Increasing Oxygen

By about 2.4 billion years ago, the oxygen produced by cyanobacteria had began to accumulate in Earth's waters and atmosphere. Here we pick up the story that we began in Section 7.2.

The rise in Earth's oxygen levels had three important consequences for life:

1. Oxygen interferes with the self-assembly of complex organic compounds, so life could no longer arise from nonliving materials.

2. The presence of oxygen put organisms that thrived in aerobic conditions at an advantage. Species that could not adapt to higher oxygen levels became extinct, or became restricted to the remaining low-oxygen environments, such as deep ocean sediments. Aerobic respiration evolved and became widespread. This pathway uses oxygen, and it is far more efficient at releasing energy than other reactions. Aerobic respiration would later allow the evolution of multicelled eukaryotes with high energy requirements.

Figure 18.9 Fossils of early life. **A** Strand of what may be some type of bacterial cells dates back 3.5 billion years. **B,C** Fossils of two types of cyanobacteria that lived approximately 850 million years ago in what is now Bitter Springs, Australia. **D** Artist's depiction of stromatolites in an ancient sea. **E** Cross-section through a fossilized stromatolite. Each layer formed when a mat of living cells trapped sediments. Descendant cells grew over the sediment layer, then trapped more sediment, forming the next layer.

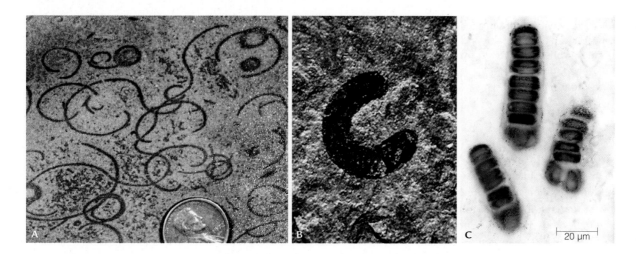

Figure 18.10 Fossil history of eukaryotes. **A** *Grypania spiralis*, dates to 2.1 billion years ago. It may be the oldest known eukaryote, but some scientists think the coils are colonial bacteria. **B** *Tawuia*, probably an early alga. **C** Fossils of a red alga, *Bangiomorpha pubescens*. This multicelled species lived 1.2 billion years ago. Some cells formed a holdfast that anchored the body. Other cells produced sexual spores.

3. As oxygen enriched the atmosphere, some oxygen molecules broke apart, then recombined as ozone (O_3). Formation of an ozone layer in the upper atmosphere reduced the amount of solar ultraviolet (UV) radiation that reached Earth's surface. UV radiation can damage DNA and other biological molecules. It does not penetrate deep into water, but without the protective effect of the ozone layer, life could not have moved onto land.

The Rise of Eukaryotes

The third domain of life arose when eukaryotic cells branched off from the archaean lineage. Trace amounts of lipids in 2.7-billion-year-old rocks give us hints about when this second great branching took place. The lipids are biomarkers for eukaryotes. A **biomarker** is a compound made only by a particular type of cell; it is like a molecular signature.

Fossilized coils so large that they can be seen with the naked eye may also be evidence of early eukaryotes (Figure 18.10A,B). The fossil in Figure 18.10C is certainly a eukaryote. It is a red alga that lived about 1.2 billion years ago. This alga also has the distinction of being the oldest species known to reproduce sexually. (Only eukaryotes reproduce sexually.) The alga grew as hairlike strands, with cells at one end forming a holdfast that held it in

place. Cells at the strand's other end were specialized to produce sexual spores by meiosis.

The evolution of sexual reproduction and multicellularity were milestones in the history of life. Sex gave some eukaryotic organisms a new way to exchange genes. Multicellularity coupled with cellular differentiation opened the way to evolution of larger bodies that have specialized parts adapted to specific functions.

Trace fossils and biomarkers indicate that sponge-like animals may have evolved by 870 million years ago. By 570 million years ago, animals with more complex bodies shared the oceans with bacteria, archaeans, fungi, and protists, including the lineage of green algae that would later give rise to land plants.

Animal diversity increased greatly during a great adaptive radiation in the Cambrian, 543 million years ago. When that period finally ended, all of the major animal lineages, including the vertebrates (animals with backbones), were represented in the seas.

---

biomarker Molecule produced only by a specific type of cell.
stromatolite Dome-shaped structures composed of layers of bacterial cells and sediments.

> ## Take-Home Message What do we know about events that occurred early in the history of life?
>
> ❯ The first cells evolved by 3.5 billion years ago. They did not have a nucleus and were probably anaerobic.
>
> ❯ An early diverge separated the bacteria and archaeans.
>
> ❯ After the noncyclic pathway of photosynthesis evolved, oxygen accumulated in the atmosphere and ended the further spontaneous chemical origin of life. The stage was set for the evolution of eukaryotic cells.

> Eukaryotic cells have a composite ancestry, with different components derived from archaea and bacteria.
> Scientists study modern cells to test hypotheses about how organelles evolved in the past.
< Links to Nucleus 4.7, Chloroplasts and mitochondria 4.9

Origin of the Nucleus

In all eukaryotes, the DNA resides in a nucleus. The outer layer of the nucleus, the nuclear envelope, consists of a double layer of membrane with protein-lined pores that control flow of material into and out of the nucleus. By contrast, the DNA of archaeans, the ancestors of eukaryotes, typically lies unenclosed in the cytoplasm.

The nucleus and endomembrane system probably evolved when the plasma membrane of an ancestral cell

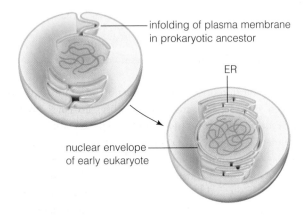

infolding of plasma membrane in prokaryotic ancestor

ER

nuclear envelope of early eukaryote

Figure 18.11 Animated One model for the origin of the nuclear envelope and the endoplasmic reticulum. These organelles may have formed when portions of the plasma membrane folded inward.

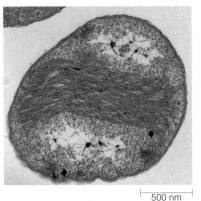

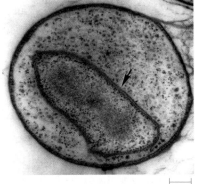

500 nm

0.2 μm

A Marine bacterium (*Nitrosococcus oceani*) with highly folded internal membranes visible across its midline.

B Freshwater bacterium (*Gemmata obscuriglobus*) with DNA enclosed by a two-layer membrane (indicated by the arrow).

Figure 18.12 Bacteria with internal membranes. Both photos are electron micrographs.

folded inward (Figure 18.11). In support of this hypothesis, a few modern bacteria do have some internal membrane-enclosed compartments. Membrane infoldings can be selectively advantageous because the folds increase the surface area available for membrane-associated reactions. For example, the marine bacterium *Nitrosococcus oceani* has a system of highly folded internal membranes (Figure 18.12A). Enzymes embedded in the membranes allow the cell to meet its energy needs by breaking down ammonia.

An infolded membrane that cordons off a cell's genetic material can help protect the genome from physical or biological threats. For example, *Gemmata obscuriglobus* is one of the few bacteria that has a membrane around its DNA (Figure 18.12B). Compared to typical bacteria, it can withstand much higher levels of mutation-causing radiation. Researchers attribute this cell's radiation resistance to the tight packing of its DNA within the membrane-enclosed compartment. In other bacteria, the DNA spreads out through a broader area of the cytoplasm. Enclosing the genetic material within a membrane could also help protect it from viruses that inject their genetic material into bacteria or from interference caused by bits of DNA absorbed from the environment.

Mitochondria and Chloroplasts

Mitochondria and chloroplasts are eukaryotic organelles that resemble bacteria in their size and structure. Like bacteria, these organelles have a genome arranged as a circle of DNA. The organelles also behave somewhat independently, duplicating their DNA and dividing at a different time than the cell that holds them. Taken together, these observations prompted the hypothesis that mitochondria and chloroplasts evolved as a result of **endosymbiosis**, a relationship in which one type of cell (the symbiont) lives and replicates inside another cell (the host). The host in an endosymbiotic relationship passes some symbionts along to its descendants when it divides.

Genetic similarities between mitochondria and modern aerobic bacteria called rickettsias (Figure 18.13A) suggests that the two groups share a common ancestor. Presumably, a rickettsia-like cell infected an early eukaryote. The host began to use ATP produced by its aerobic symbiont while the symbiont began to rely on the host for raw materials. Over time, genes that occurred in both the host and symbiont were free to mutate. If a gene lost its function in one partner, a gene from the other could take up the slack. Eventually, the host and symbiont both became incapable of living independently.

Similarly, chloroplasts are structurally and genetically similar to a group of modern oxygen-producing photosynthetic bacteria called cyanobacteria. These similarities cause biologists to infer that chloroplasts evolved from an ancient relative of these cells.

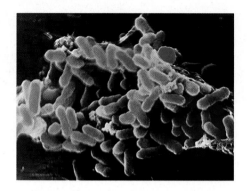

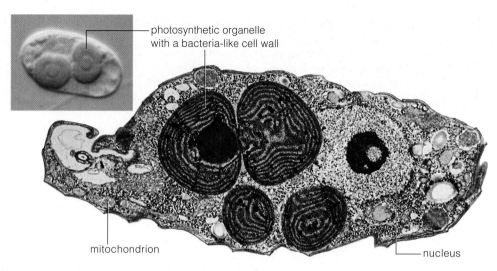

photosynthetic organelle
with a bacteria-like cell wall

mitochondrion

nucleus

A *Rickettsia prowazekii*, an aerobic bacterium that infects human cells and causes the disease typhus. Of all bacterial genomes sequenced so far, that of *R. prowazekii* is most similar to the mitochondrial genome. Like mitochondria, these bacteria take up pyruvate from the cytoplasm and break it down by aerobic respiration.

B *Cyanophora paradoxa*, one of the flagellated protists called glaucophytes. Its photosynthetic structures resemble cyanobacteria. They even have a wall similar in composition to the wall around a cyanobacterial cell.

Figure 18.13 Some modern cells that provide evidence in support of the endosymbiotic hypothesis for the origin of mitochondria and chloroplasts.

Additional Evidence of Endosymbiosis

A chance discovery made by microbiologist Kwang Jeon supports the hypothesis that bacteria can evolve into organelles. In 1966, Jeon was studying *Amoeba proteus*, a species of single-celled protist. By accident, one of his cultures became infected by a rod-shaped bacterium. Some infected amoebas died right away. Others kept growing, but only slowly. Intrigued, Jeon maintained those infected cultures to see what would happen. Five years later, the descendant amoebas were host to many bacterial cells, yet they seemed healthy. In fact, when these amoebas were treated with bacteria-killing drugs that usually do not harm amoebas, they died.

Experiments confirmed that the amoebas had come to rely on the bacteria. When Jeon swapped the nucleus from a bacteria-tolerant amoeba for the nucleus in a typical amoeba, the recipient cell died. Yet, when bacteria were included with the nucleus transplant, most cells survived. It seemed that the amoebas had come to require the bacteria for some life-sustaining function. Additional studies showed that the amoebas had lost the ability to make an essential enzyme. They now depended on their bacterial endosymbionts to make that enzyme for them.

We also have evidence to support the hypothesis that cyanobacteria can become organelles. The interior of the single-celled protists called glaucophytes is taken up largely by green photosynthetic organelles that resemble cyanobacteria (Figure 18.13B). The organelle even has a cell wall that contains peptidoglycan, a material made by some bacteria, but no eukaryotes. Many other aquatic protists have cyanobacteria living inside them. However, in most cases these bacteria are endosymbionts, not organelles; the bacteria can still live on their own if removed from their host. However, the photosynthetic organelles of glaucophytes, like chloroplasts, have evolved a dependence on their host. They cannot survive on their own.

However they arose, early eukaryotic cells had a nucleus, endomembrane system, mitochondria, and—in certain lineages—chloroplasts. These cells were the first protists. Over time, their many descendants came to include the modern protist lineages, as well as the plants, fungi, and animals. The next section provides a time frame for these pivotal evolutionary events.

endosymbiosis One species lives and reproduces inside another.

Take-Home Message How did eukaryotic organelles evolve?

❯ The nucleus and endomembrane system may have evolved from infoldings of the plasma membrane.

❯ Mitochondria and chloroplasts may have evolved when bacterial endosymbionts and their hosts became mutually dependent.

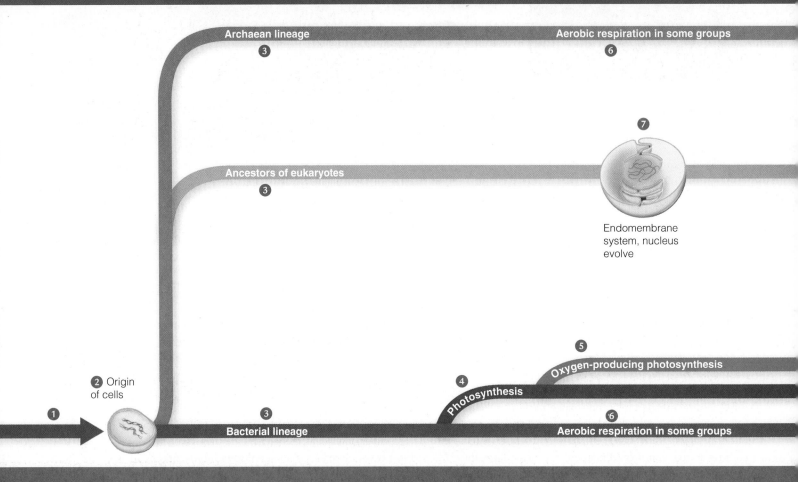

Hydrogen-rich, oxygen-poor atmosphere

Atmospheric oxygen level begins to increase

Archaean lineage ③

Aerobic respiration in some groups ⑥

⑦

Ancestors of eukaryotes ③

Endomembrane system, nucleus evolve

⑤

Oxygen-producing photosynthesis

④

Photosynthesis

② Origin of cells

① ③ Bacterial lineage

⑥ Aerobic respiration in some groups

3.8 billion years ago

3.2 billion years ago

2.7 billion years ago

Steps Preceding Cells

❶ Between 5 billion and 3.8 billion years ago, as an outcome of chemical and molecular evolution, complex carbohydrates, lipids, proteins, and nucleic acids formed from the simple organic compounds present on early Earth.

Origin of Cells

❷ The first living cells evolved by 3.8 billion years ago. They did not have a nucleus or other organelles. Atmospheric oxygen was low, so early cells probably made ATP by anaerobic pathways.

Three Domains of Life

❸ The first major divergence gave rise to bacteria and to the common ancestor of the archaeans and all eukaryotic cells.

Not long after, the ancestors of archaeans and eukaryotic cells diverged.

Photosynthesis, Aerobic Respiration Evolve

❹ A cyclic pathway of photosynthesis evolved in some bacterial groups.

❺ An oxygen-releasing noncyclic pathway evolved later in the cyanobacteria and, over time, changed the atmosphere.

❻ Aerobic respiration evolved independently in many bacterial groups.

Origin of Endomembrane System, Nucleus

❼ Cell sizes and the amount of genetic information continued to expand in ancestors of what would become the eukaryotic cells. The endomembrane system, including the nuclear envelope, arose through the modification of cell membranes between 3 and 2 billion years ago.

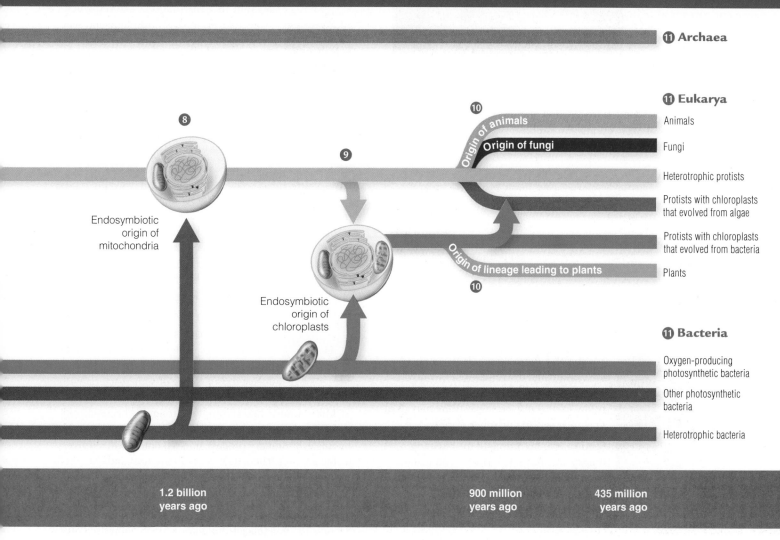

Atmospheric oxygen reaches current levels; ozone layer gradually forms

11 Archaea

11 Eukarya

8

Endosymbiotic origin of mitochondria

9

10

Origin of animals

Animals

Origin of fungi

Fungi

Heterotrophic protists

Protists with chloroplasts that evolved from algae

Endosymbiotic origin of chloroplasts

Origin of lineage leading to plants

Protists with chloroplasts that evolved from bacteria

Plants

10

11 Bacteria

Oxygen-producing photosynthetic bacteria

Other photosynthetic bacteria

Heterotrophic bacteria

| 1.2 billion years ago | 900 million years ago | 435 million years ago |

Endosymbiotic Origin of Mitochondria

8 An aerobic bacterium entered an anaerobic eukaryotic cell. Over many generations, the two species established a symbiotic relationship. Descendants of the bacterial cell became mitochondria.

Endosymbiotic Origin of Chloroplasts

9 A heterotrophic protist took in oxygen-producing bacteria (cyanobacteria). Descendants of the bacteria evolved into chloroplasts. Later, some photosynthetic protists would evolve into chloroplasts inside other protist hosts.

Plants, Fungi, and Animals Evolve

10 All major lineages—including fungi, animals, and the algae that would give rise to plants evolved in the seas.

Lineages That Have Endured to the Present

11 Today, organisms live in nearly all regions of Earth's waters, crust, and atmosphere. They are related by descent and share certain traits. However, each lineage encountered different selective pressures, and unique traits evolved in each one.

Figure 18.14 Animated Milestones in the history of life, based on the most widely accepted hypotheses. This figure also shows the evolutionary connections among all groups of organisms. The time line is not to scale.

» Figure It Out Which organelle evolved first, mitochondria or chloroplasts?   Answer: Mitochondria

Looking For Life (revisited)

When it comes to sustaining life, Earth is just the right size. If the planet were much smaller, it would not exert enough gravitational pull to keep atmospheric gases from drifting off into space. The photo at the *right* shows the relative sizes of Earth

and Mars. As you can see, Mars is only about half the size of Earth. As a result, it has a much thinner atmosphere. What atmosphere there is consists mainly of carbon dioxide, some nitrogen, and only traces of oxygen. Thus, if life exists on Mars it is almost certainly anaerobic.

How would you vote? Martian soil may contain microbes that could provide new information about the origin and evolution of life. Should we bring samples of Martian soil to Earth for analysis? See CengageNow for details, then vote online (cengagenow.com).

Summary

Section 18.1 Astrobiology is the study of life's origin and distribution in the universe. The presence of cells in deserts and deep below Earth's surface suggests life may exist in similar settings on other planets.

Section 18.2 According to the **big bang theory**, the universe formed in an instant 13 to 15 billion years ago. Earth and other planets formed more than 4 billion years ago. Early in Earth's history, there was little oxygen in the air, volcanic eruptions were common, and there was a constant hail of meteorites.

Section 18.3 Laboratory simulations offer indirect evidence that organic compounds self-assemble spontaneously under conditions like those in Earth's early atmosphere or in the hot, mineral-rich water around **hydrothermal vents**. Examination of meteorites shows that such compounds might have formed in deep space and reached Earth in meteorites.

Section 18.4 Proteins that speed metabolic reactions might have first formed when amino acids stuck to clay, then bonded under the heat of the sun. Or, reactants could have begun interacting in rocks near deep-sea hydrothermal vents. Membrane-like structures and vesicles form when proteins or lipids are mixed with

water. They serve as a model for **protocells**, which may have preceded cells. An **RNA world**, a time in which RNA was the genetic material, may have preceded DNA-based systems. RNA still is a part of ribosomes that carry out protein synthesis in all organisms. Discovery of **ribozymes**, RNAs that act as enzymes, lends support to the RNA world hypothesis. A later switch from RNA to DNA would have made the genome more stable.

Section 18.5 The first cells evolved when oxygen levels in the atmosphere and seas were low, so they probably were anaerobic. An early divergence separated bacteria from the common ancestor of archaeans and eukaryotes. An oxygen-releasing, non-cyclic pathway of photosynthesis evolved in one bacterial lineage (cyanobacteria). These bacteria grew in mats that collected sediment and, over countless generations, formed dome-shaped structures called **stromatolites**.

Over time, oxygen released by cyanobacteria changed Earth's atmosphere. The increased oxygen level prevented evolution of new life from nonliving molecules, created a protective ozone layer, and favored cells that carried out aerobic respiration. This ATP-forming metabolic pathway was a key innovation in the evolution of eukaryotic cells.

Protists were the first eukaryotes. Their **biomarkers** and fossils date back more than 2 billion years. Diversification of protists gave rise to plants, fungi, and animals.

Section 18.6 By one hypothesis, the internal membranes that are typical of eukaryotic cells may have evolved through infoldings of the plasma membrane of prokaryotic ancestors. Existence of some bacteria with internal membranes supports this hypothesis.

Mitochondria and chloroplasts resemble bacteria, and these organelles most likely evolved by **endosymbiosis**. By this evolutionary process, one cell enters and survives inside another. Then, over generations, host and guest cells come to depend on one another for essential metabolic processes. Some modern protists have bacterial symbionts inside them.

Section 18.7 Evidence from many sources allows scientists to reconstruct the order of events and make a hypothetical time line for the history of life.

Self-Quiz Answers in Appendix III

1. An abundance of _____ in the atmosphere would have prevented the spontaneous assembly of organic compounds on early Earth.
   a. hydrogen b. methane c. oxygen d. nitrogen

2. The prevalance of iron-sulfide cofactors in organisms supports the hypothesis that life arose _____ .
   a. in outer space   c. near deep-sea vents
   b. on tidal flats   d. in the upper atmosphere

Data Analysis Activities

A Changing Earth Modern conditions on Earth are unlike those when life first evolved. Figure 18.15 shows how the frequency of asteroid impacts and composition of the atmosphere have changed over time. Use this figure and information in the chapter to answer the following questions.

1. Which occurred first, a decline in asteroid impacts, or a rise in the atmospheric level of oxygen?
2. How do modern levels of carbon dioxide and oxygen compare to those at the time when the first cells arose?
3. Which is now more abundant, oxygen or carbon dioxide?

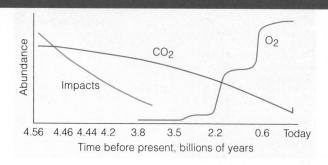

Figure 18.15 How asteroid impacts (*green*), atmospheric carbon dioxide concentration (*pink*), and oxygen concentration (*blue*) changed over geologic time.

---

3. Ribosomes can catalyze formation of peptide bonds. This supports the hypothesis that _____ .
   a. an RNA world preceded DNA-based genomes
   b. RNA can hold more information than DNA
   c. the first protists had RNA as their genetic material
   d. all of the above

4. By one hypothesis, clay _____ .
   a. facilitated assembly of early polypeptides
   b. was present at hydrothermal vents
   c. provided energy for early metabolism
   d. all of the above

5. The evolution of _____ resulted in an increase in the levels of atmospheric oxygen.
   a. sexual reproduction
   b. aerobic respiration
   c. the noncyclic pathway of photosynthesis
   d. the cyclic pathway of photosynthesis

6. Mitochondria most resemble _____ .
   a. archaeans   c. cyanobacteria
   b. aerobic bacteria   d. early eukaryotes

7. What was the energy source in the Miller–Urey simulation of conditions on the early Earth?

8. The first sexual reproducers were _____ .
   a. archaeans   c. cyanobacteria
   b. aerobic bacteria   d. eukaryotes

9. Oxygen released by _____ accumulated in the atmosphere and produced the ozone layer.
   a. archaeans   c. cyanobacteria
   b. aerobic bacteria   d. early eukaryotes

10. What is a ribozyme made of?

11. A rise in oxygen in Earth's air and seas put organisms that engaged in _____ at a selective advantage.
    a. aerobic respiration   c. photosynthesis
    b. fermentation   d. sexual reproduction

12. Which of the following was not present on Earth when mitochondria first evolved?
    a. archaeans   c. protists
    b. bacteria   d. animals

13. Chloroplasts most resemble _____ .
    a. archaeans   c. cyanobacteria
    b. aerobic bacteria   d. early eukaryotes

14. Which provides a more stable genome, DNA or RNA?

15. Arrange these events in order of occurence, with 1 being the earliest and 6 the most recent.
    ___1   a. emergence of the noncyclic
    ___2   pathway of photosynthesis
    ___3   b. origin of mitochondria
    ___4   c. origin of protocells
    ___5   d. emergence of the cyclic
    ___6   pathway of photosynthesis
                  e. origin of chloroplasts
                  f. the big bang

Additional questions are available on **CENGAGENOW.**

Critical Thinking

1. Researchers looking for fossils of the earliest life forms face many hurdles. For example, few sedimentary rocks date back more than 3 billion years. Review what you learned about plate tectonics (Section 16.7). Explain why so few remaining samples of these early rocks remain.

2. Craig Venter and Claire Fraser are working to create a "minimal organism." They are starting with *Mycoplasma genitalium*, a bacterium that has 517 genes. By disabling its genes one at a time, they discovered that 265–350 of them code for essential proteins. The scientists are synthesizing the essential genes and inserting them, one by one, into an engineered cell consisting only of a plasma membrane and cytoplasm. They want to see how few genes it takes to build a new life form. What properties would such a cell have to exhibit for you to conclude that it was alive?

Animations and Interactions on **CENGAGENOW**:
› Miller–Urey experiment; Milestones in history of life.

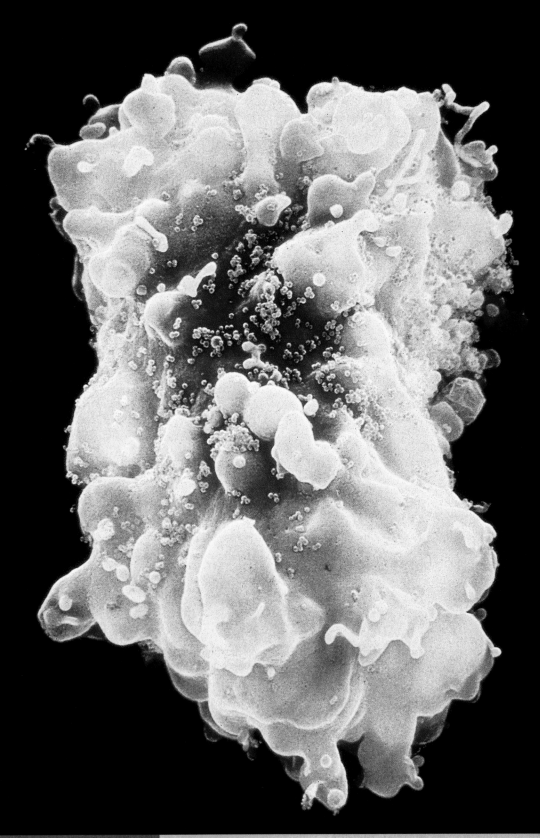

‹ Links to Earlier Concepts

This introduction to viruses and viroids touches on reverse transcription (Section 15.2), ribozymes (18.4), DNA repair (8.6), and cancer (10.1). The chapter also covers bacteria and archaeans (4.5). You will learn more about the three domain classification system (1.5), bacteria that gave rise to organelles (18.6), and antibiotic resistance (17.5). You will also draw on your knowledge of coevolution (17.13) and cladistics (17.14).

Key Concepts

Viruses and Viroids
Viruses are noncellular, with a protein coat and a genome of nucleic acid, but no metabolic machinery. Viruses must infect cells to replicate. Some infect humans and cause disease. Viroids are RNA bits that do not encode proteins. Even so, they can infect plant cells and replicate inside them.

Structure and Function of Bacteria
Bacteria are small cells with DNA and ribosomes, but no nucleus or typical eukaryotic organelles. They are also the most abundant and metabolically diverse organisms, with autotrophs (self-feeders) and heterotrophs (feeders on others) among them.

19 Viruses, Bacteria, and Archaeans

19.1 Evolution of a Disease

In this chapter, we explore the diversity of two of Earth's oldest lineages. Billions of years before there were plants or animals, Earth's seas were home to bacteria and archaeans. These small cells do not have a nucleus or other typical eukaryotic organelles. Viruses are simpler still, with no chromosomes, ribosomes, or metabolic machinery. By many definitions, viruses are not even alive. Despite their simplicity, viruses can evolve because they have genes that mutate.

For example, scientists have learned quite a bit about the origin and evolution of HIV (human immunodeficiency virus). This virus causes AIDS (acquired immunodeficiency syndrome). Researchers first isolated HIV in the early 1980s, and have since determined that there are two strains (subtypes), HIV-1 and the less prevalant HIV-2. By sequencing the HIV-1 genome and comparing it to the genomes of primate viruses, researchers found that the human virus evolved from simian immunodeficiency virus (SIV). SIV infects wild chimpanzees.

By one hypothesis, the first human infected was someone who butchered or ate meat from an SIV-infected chimp. To test the plausibility of this hypothesis, researchers looked for evidence of simian foamy virus (SFV), another primate virus, among people in an African village who commonly hunt and eat monkeys and apes. The researchers found that one percent of the villagers showed sign of prior infection by SFV—evidence that ape-to-human transmission of a virus is possible.

To find out when HIV jumped to humans, researchers have looked for the virus in old tissue samples stored from routine hospital tests. Samples from two people who lived in Africa's Democratic Republic of the Congo are the earliest evidence of HIV in humans. One is a sample of blood stored since 1959. The other, a woman's lymph node, was removed in 1960. Gene sequences of the two viral samples differ a bit, which implies that HIV had already been around and mutating by the time

these people became infected. Given what we know about the mutation rates for viruses, the common ancestor of the two genotypes (the earliest HIV-1) must have first infected humans in the early 1900s.

Comparing genes of HIV in stored and modern blood samples has allowed researchers to trace the movement of the virus out of Africa. This data shows that HIV-1 was carried from Africa to Haiti in about 1966. The virus diversified in Haiti. Then, in about 1969, one person infected by HIV with mutations that arose in Haiti brought the virus to the United States. Once there, it spread quietly for 12 years until AIDS was identified as a threat in 1981.

Today more than 20 million people have died from AIDS and about 30 million are infected with HIV. The virus infects and replicates inside white blood cells essential to immune responses (Figure 19.1). Eventually, the infected white blood cells die. Death of such cells destroys the body's ability to defend itself. As a result, many disease-causing organisms run rampant, causing symptoms of AIDS and health problems that can be fatal.

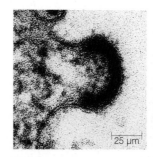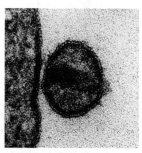

Figure 19.1 Micrographs of a new HIV particle budding from an infected white blood cell. The photo on the *opposite page* shows many viral particles (*blue* dots) on an infected cell.

Replication and Gene Exchange
Bacteria have a single chromosome and some also have one or more plasmids. They reproduce by fission, a type of asexual reproduction. Cells can exchange genes by swapping plasmids, and by other processes.

Bacterial Diversity
Bacteria are well studied and highly diverse. They put oxygen into the air, supply nutrients to plants, and break down wastes and remains. Some live in or on our bodies and have beneficial effects. Others are pathogens that cause human disease.

Archaean Diversity
Archaeans were discovered relatively recently. Many are adapted to life in very hot or very salty places. Others live in low-oxygen environments and make methane. Still others live beside bacteria in soils and seas. None cause human disease.

❯ Viruses hover near the border between living and nonliving things. They have genes, but none of the cellular machinery required to express those genes or replicate them.
❮ Link to Reverse transcription 15.2

Viral Traits and Diversity

A **virus** is a noncellular infectious agent composed of a protein coat wrapped around genetic material (RNA or DNA) and a few viral enzymes. A virus is far smaller than any cell and has no ribosomes or other metabolic machinery. To replicate, the virus must infect a cell of a specific organism, which we call its host.

Each type of virus has structural adaptations that allow it to infect and replicate in hosts of a particular type. **Bacteriophages** infect bacteria. One well-studied group has a complex coat (Figure 19.2A). DNA is encased in a protein "head." Attached to the head is a rodlike "tail" with fibers that attach the virus to its host.

The tobacco mosaic virus infects plants. It has a helical structure, with coat proteins arranged around a strand of RNA to form a rod (Figure 19.2B).

Many animal viruses have a 20-sided protein coat. In adenoviruses, this coat has a protein spike at each corner (Figure 19.2C). Adenoviruses are "naked," but most animal viruses are enveloped; a bit of membrane from a prior host encloses the virus. Herpesviruses are enveloped DNA viruses (Figure 19.2D). HIV is an enveloped RNA virus.

Viral Replication

Viral replication begins when a virus attaches to proteins in the host's plasma membrane. A virus cannot seek out a host, but rather relies on a chance encounter. The virus, or just its genetic material, enters the cell and hijacks the cell's metabolic machinery. Viral genes direct the replication of viral genetic material and the production of viral proteins. These components then self-assemble to form new viral particles. New virus buds from the infected host cell or is released when the host bursts.

Bacteriophage Replication Two replication pathways are common among bacteriophages (Figure 19.3). In the **lytic pathway**, viral genes enter a host and immediately direct it to make new viral particles. Soon the cell dies by lysis (breaks open), allowing new viral particles to escape.

In the **lysogenic pathway**, viral DNA becomes integrated into the host chromosome and a latent period precedes formation of new viruses. The viral DNA is copied along with host DNA, and is passed along to all descendants of the host cell. Like tiny time bombs, viral DNA inside these descendant cells awaits a signal to enter the lytic pathway.

HIV Replication HIV replicates inside a human white blood cell (Figure 19.4). Spikes of viral protein that extend beyond the envelope attach to proteins in the host cell's membrane ❶. The viral envelope fuses with this membrane, and viral enzymes and genetic material (RNA) enter the cell ❷. A viral enzyme (reverse transcriptase) converts viral RNA into double-stranded DNA that the cell's genetic machinery can read ❸. Viral DNA is moved to the nucleus, where a viral enzyme integrates it into a host chromosome ❹. Viral DNA is transcribed along with host genes ❺. Some of the resulting RNA is translated into viral proteins ❻, and some becomes the genetic material of new HIV particles ❼. The viral particles self-assemble at the plasma membrane ❽. As the virus buds

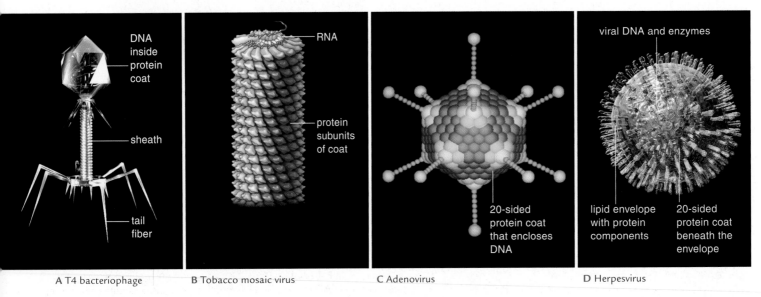

DNA inside protein coat

sheath

tail fiber

A T4 bacteriophage

RNA

protein subunits of coat

B Tobacco mosaic virus

20-sided protein coat that encloses DNA

C Adenovirus

viral DNA and enzymes

lipid envelope with protein components

20-sided protein coat beneath the envelope

D Herpesvirus

Figure 19.2 Models illustrating viral structure.   **❯❯ Figure It Out** Which virus acquired membrane from its host?   Answer: The herpesvirus

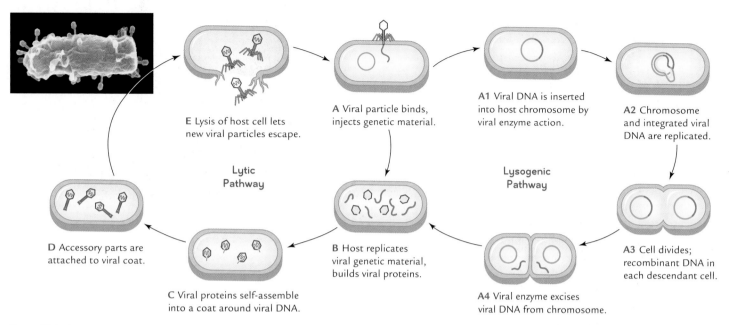

E Lysis of host cell lets new viral particles escape.

A Viral particle binds, injects genetic material.

A1 Viral DNA is inserted into host chromosome by viral enzyme action.

A2 Chromosome and integrated viral DNA are replicated.

Lytic Pathway

Lysogenic Pathway

D Accessory parts are attached to viral coat.

B Host replicates viral genetic material, builds viral proteins.

A3 Cell divides; recombinant DNA in each descendant cell.

C Viral proteins self-assemble into a coat around viral DNA.

A4 Viral enzyme excises viral DNA from chromosome.

Figure 19.3 Animated Bacteriophage replication pathways.

›› **Figure It Out** What does the blue circle in A represent?

Answer: The bacterial chromosome

Figure 19.4 Animated Replication cycle of HIV, an eveloped RNA virus.

❶ Virus binds to a host cell.

❷ Viral RNA and enzymes enter cell.

❸ Viral reverse transcriptase uses viral RNA to make double-stranded viral DNA.

❹ Viral DNA integrates into host genome.

❺ Transcription produces viral RNA.

❻ Some viral RNA is translated to produce viral proteins.

❼ Other viral RNA forms the new viral genome.

❽ Viral proteins and viral RNA self-assemble at the host membrane.

❾ New virus buds from the host cell, with an envelope of host plasma membrane.

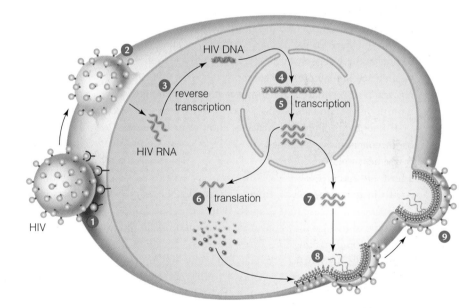

from the host cell, some of the host's plasma membrane becomes the viral envelope ❾. Each viral particle can now infect other white blood cells.

Drugs that fight HIV interfere with viral binding to the host, reverse transcription, integration of DNA, or processing of viral polypeptides to form viral proteins.

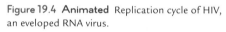

bacteriophage Virus that infects bacteria.
lysogenic pathway Bacteriophage replication path in which viral DNA becomes integrated into the host's chromosome and is passed to the host's descendants.
lytic pathway Bacteriophage replication pathway in which a virus immediately replicates in its host and kills it.
virus Noncellular, infectious particle of protein and nucleic acid; replicates only in a host cell.

Take-Home Message **What is a virus and how does it replicate?**

❯ A virus is a noncellular infectious particle that consists of nucleic acid enclosed in a protein coat and sometimes an outer envelope.

❯ A virus replicates by binding to a specific type of host cell, taking over the host's metabolic machinery, and using that machinery to produce viral components. These components self-assemble to form new viral particles.

> Viral particles far outnumber cells, and viruses have wide-ranging effects on all forms of life, including humans.
< Links to DNA repair 8.6, Cancer 10.1

Some viruses have a positive effect on human health, as when bacteriophages kill bacteria that could cause food poisoning. However, other viruses are **pathogens**, meaning they cause disease.

Common Viral Diseases

A variety of nonenveloped viruses, including adenoviruses, infect membranes of the upper respiratory system and cause common colds. Other nonenveloped viruses infect the lining of the intestine and cause viral gastroenteritis, commonly referred to as a stomach flu. Nonenveloped viruses also cause warts. Human papillomavirus (HPV), the cause of genital warts, is the most common sexually transmitted virus. Some strains of HPV can also cause cervical cancer.

Cold sores are caused by one herpesvirus (an enveloped virus). Another herpesvirus causes genital herpes. Still others cause infectious mononucleosis and chicken pox. Once a person has been infected by a herpesvirus, the virus persists in the body for life. It may enter a latent state, similar to that in the lysogenic cycle of a bacteriophage, only to resume activity later on. For example, the virus that causes chicken pox in children sometimes reemerge as shingles (a painful rash) at an older age.

Influenza (flu), mumps, measles, and German measles are also caused by enveloped viruses.

Emerging Viral Diseases

An **emerging disease** is a disease that suddenly expands its range, or a disease that is newly detected in humans. A new type of disease may appear when a mutation alters a viral genome, making the virus more easily spread or more deadly. RNA viruses have an especially high mutation rate because a host's DNA proofreading and repair mechanisms (Section 8.6) do not work on RNA. Viruses also evolve by picking up new genes from their host, or from another virus that infects a cell at the same time.

AIDS is one example of an emerging disease. Here we consider a few others.

West Nile Fever West Nile virus is an enveloped RNA virus that replicates in birds. Mosquitoes carry the virus from host to host, so we say they are the **vector** for this virus. Sometimes a bite from a virus-carrying mosquito causes a human infection. This is a dead end for the virus. Although it can replicate in human cells, not enough virus gets into the blood for the infection to be passed on.

Most people infected with West Nile virus are not sickened, but some develop West Nile fever, which causes

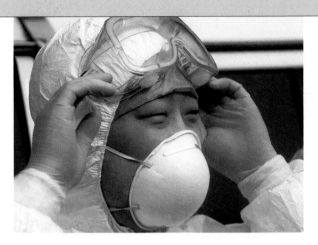

Figure 19.5 A health care worker putting on protective gear during the SARS epidemic.

flulike symptoms. In about 1 percent of West Nile fever cases, the virus attacks the nervous system, with results that can be fatal.

West Nile virus had long been present in Africa, the Middle East, and parts of Europe, but it was unknown in the Western Hemisphere until 1999, when it began killing people and birds in New York City. Over the next few years, infected migratory birds spread the virus across the country. Today, West Nile fever is an **endemic disease** throughout the continental United States, meaning the disease remains present, but at a low level.

SARS Sudden acute respiratory syndrome (SARS) first appeared in late 2002 in China (Figure 19.5). The disease became an **epidemic**, a disease that is widespread in one region. Then, with the help of air travelers, SARS became a **pandemic**, an outbreak of disease that encompasses many regions and poses a threat to human health. Over the course of 9 months, SARS sickened about 8,000 people in 37 countries and killed 774.

A previously unknown type of coronavirus (*right*) causes SARS. Coronaviruses are named for the "corona" or crown of protein spikes that extends through their envelope. Where did this new virus come from? Researchers found a SARS-like virus in Chinese horseshoe bats, but it does not infect humans. They hypothesize 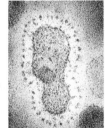 that bats with the SARS-like virus were captured and brought to wildlife markets, where they gave the virus to other animals. Inside those animals, the virus evolved the ability to infect humans.

Influenza H5N1 and H1N1 Influenzaviruses commonly cause flu outbreaks during the winter in temperate regions. Because of mutations, each year's flu virus is a bit different than that of previous years. However, a typical seasonal flu has a low mortality rate and causes deaths mainly among older people. Doctors recommend that people over 50 get a yearly flu shot that will protect against the anticipated version of the seasonal flu. Such shots do not protect people from dramatically different strains of influenzavirus, such as those described below.

Avian influenza H5N1, commonly called bird flu, occasionally infects people who have direct contact with birds. When the virus does infect people, the death rate is disturbingly high. From 2003 to 2009, the World Health Organization received reports of 417 human cases of influenza H5N1, mainly in Asia. Of these, 257 (about 60 percent) were fatal. Fortunately, person-to-person transmission of the H5N1 virus is exceedingly rare.

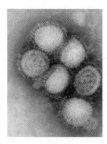

By contrast, influenza H1N1, shown in the micrograph at the *left*, is easily transmitted by a cough or a sneeze. It is commonly called swine flu. The H1N1 virus first appeared in Mexico in early 2009, where it caused severe respiratory symptoms. From Mexico, the virus spread to the United States, and then throughout the world. Fortunately, initial fears of a high death rate proved unfounded, antiviral drugs were released to treat the virus, and a vaccine was created.

Health officials continue to monitor H5N1 and H1N1 influenza. Either virus could mutate and their coexistence raises the possibility of a potentially disastrous gene exchange. The influenza H1N1 virus already has a composite genome, with material from a human flu virus, bird flu virus, and two different swine flu viruses. If it picked up genes from avian H5N1, the result could be an influenzavirus that is easily transmissible and deadly.

---

emerging disease A disease that was previously unknown or has recently begun spreading to a new region.
endemic disease Disease that persists at a low level in a region or population.
epidemic Disease outbreak limited to one region.
pandemic Outbreak of disease that affects many separate regions and poses a serious threat to human health.
pathogen Disease-causing agent.
vector Animal that carries a pathogen from one host to the next.

Take-Home Message How do viruses affect human health?

› Viruses cause many widespread, familiar diseases such as the common cold and cold sores.

› Other viruses cause emerging diseases such as AIDS and SARS that have only recently become threats.

› Tiny bits of RNA, just a few hundred nucleotides long, can use enzymes in a plant cell to replicate themselves.
‹ Links to Plasmodesmata 4.12, Ribozyme 18.4

In 1971, plant pathologist Theodor Diener announced the discovery of a new type of pathogen, a small RNA without a protein coat. He named it a **viroid**, because it seemed like a stripped-down version of a virus.

Diener had been investigating potato spindle tuber disease, an illness that stunts potato plants and causes them to produce only a few small, deformed potatoes. Diener expected to find a viral pathogen, but was forced to consider other options after discovering that the infectious agent passed through filters too fine to allow passage of even the smallest virus.

To identify components of the apparently minuscule pathogen, Diener treated extracts from infected plants with enzymes to see what would inactivate the pathogen. Extracts treated with enzymes that destroyed DNA, lipids, or protein still infected plants. Only RNA-digesting enzymes made the extracts harmless. Diener concluded that the pathogen must consist of RNA.

Plant pathologists have now described about thirty viroids that cause disease in commercially valuable plants, including citrus, apples, coconuts, avocados, and chrysanthemums. All known viroids are circular, single-stranded RNAs. Base pairing between different parts of a viroid usually causes it to fold up into a rodlike shape (*right*). The viroid is remarkably small, with fewer than 400 nucleotides. By comparison, even the smallest viral genome consists of thousands of nucleotides.

Unlike the genetic material of a virus, viroid RNA does not encode proteins. However, the viroid itself has enzymatic activity. In other words, it is a ribozyme (Section 18.4). Typically, viroid replication occurs when a host enzyme (RNA polymerase) moves along the circular RNA repeatedly. The result is a long RNA strand, with many copies of the viroid attached end to end. The strand then cuts itself up in appropriate places, forming new viroids. The viroids spread through the plant via plasmodesmata that connect cells, and phloem (food-carrying vessels).

---

viroid Small noncoding RNA that can infect plants.

Take-Home Message What are viroids and how do they differ from viruses?

› Viroids are small RNAs that infect plants.

› Unlike a virus, a viroid does not have a protein coat or protein-encoding genes.

> Bacteria are small, structurally simple, widely dispersed, and highly abundant cells.

‹ Links to Bacteria 4.5, Evolution of organelles 18.6

Cell Size, Structure, and Motility

The typical bacterial cell cannot be seen without a light microscope. It is far smaller than a eukaryotic cell, about the size of a mitochondrion. In fact, there is evidence that certain bacteria were the ancestors of mitochondria (Section 18.6).

Biologists describe bacteria by their shapes (Figure 19.6A). A spherical cell is called a coccus; a rod-shaped cell a bacillus; and a spiral cell a spirillum.

Nearly all bacteria have a semirigid, porous cell wall around the plasma membrane (Figure 19.6B). A secreted slime layer or a capsule may enclose the wall. Slime helps a cell stick to surfaces. A capsule is tougher and helps some bacterial pathogens evade the immune defenses of their vertebrate hosts.

Inside the cell, the **bacterial chromosome** is a circle of double-stranded DNA. It attaches to the plasma membrane and resides in a cytoplasmic region called the **nucleoid**. Ribosomes scattered through the cytoplasm make proteins. There is no endomembrane system like that of eukaryotes, although a few kinds of bacteria do have internal membranes of some sort (Section 18.6).

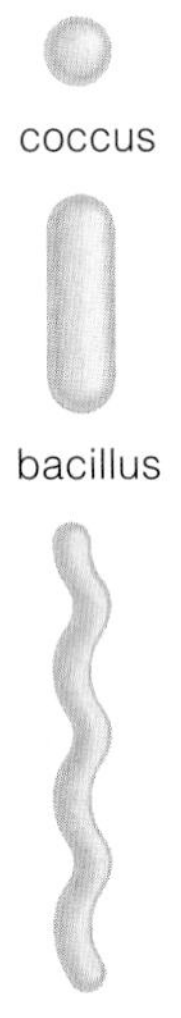

coccus

bacillus

spirillum

Many bacterial cells have one or more flagella. Unlike eukaryotic flagella, bacterial flagella do not contain microtubules and do not bend side to side. Instead, they rotate like a propeller.

Hairlike filaments called **pili** (singular, pilus) often extend from the cell surface. Some cells use pili to stick to surfaces. Others glide along by using their pili as grappling hooks. A pilus extends out to a surface, sticks to it, then shortens, drawing the cell forward. Another type of retractable pilus draws cells together for gene exchanges as described in the next section.

Abundance and Metabolic Diversity

In terms of sheer numbers the bacteria are unparalleled among cells. Biologists at the University of Georgia have estimated that 5 million trillion trillion bacterial cells live on Earth.

Metabolic diversity contributes to bacterial success. There are four known modes of nutrition and, as a group, bacteria use them all (Table 19.1).

Photoautotrophs are photosynthetic; they use light energy to build organic compounds from carbon dioxide and water. This group includes nearly all plants, and some protists, as well as many bacteria.

Chemoautotrophs get energy by removing electrons from inorganic molecules such as sulfides. They use this energy to build organic compounds from carbon dioxide and water. All are bacteria or archaeans.

Photoheterotrophs use light energy and get carbon by breaking down organic compounds in their environment. All are bacteria or archaeans.

Chemoheterotrophs get carbon and energy by breaking down organic compounds assembled by other organisms. Many bacteria are in this group, as are some archaeans, protists, and all animals and fungi. Some bacterial chemoheterotrophs feed on living organisms. Others are decomposers that break down organic wastes or remains.

---

bacterial chromosome Circle of double-stranded DNA that resides in the bacterial cytoplasm.
nucleoid Cytoplasmic region where prokaryotic chromosome lies.
pilus Hairlike extension from the cell wall of some bacteria.

pilus

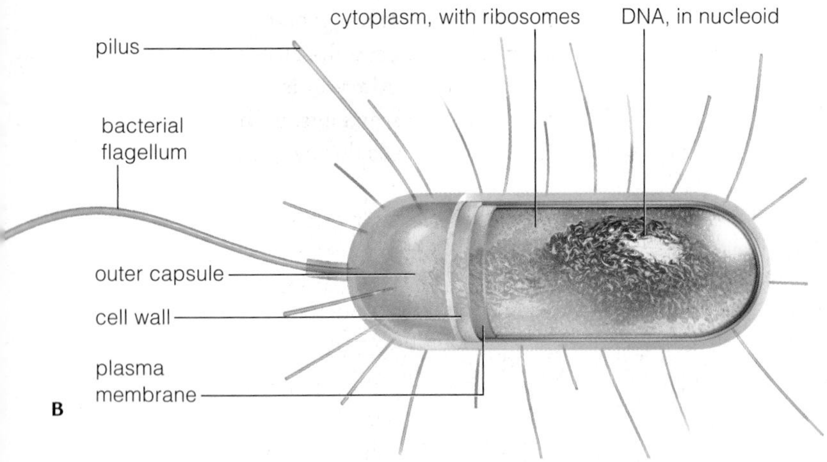

cytoplasm, with ribosomes

DNA, in nucleoid

bacterial flagellum

outer capsule

cell wall

plasma membrane

Figure 19.6 Animated Bacterial cell shapes **A** and body plan **B**.

Table 19.1 Nutritional Modes		
Mode of Nutrition	Energy Source	Carbon Source
Photoautotrophic	Sunlight	Carbon dioxide
Chemoautotrophic	Inorganic substances	Carbon dioxide
Photoheterotrophic	Sunlight	Organic compounds
Chemoheterotrophic	Organic compounds	Organic compounds

Take-Home Message What are are features of bacterial cells?

> Bacteria are small, typically walled cells with no nucleus. Their single chromosome lies in the cytoplasm and there is no endomembrane system.

> As a group, bacteria are the most numerous and the most metabolically diverse organisms.

> Bacteria can only reproduce asexually, but new gene combinations arise when they exchange genetic material.
< Link to Asexual reproduction 11.2

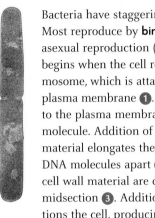

Bacteria have staggering reproductive potential. Most reproduce by **binary fission**, a type of asexual reproduction (Figure 19.7). The process begins when the cell replicates its single chromosome, which is attached to the inside of the plasma membrane ❶. The DNA replica attaches to the plasma membrane adjacent to the parent molecule. Addition of new membrane and wall material elongates the cell and moves the two DNA molecules apart ❷. Then, membrane and cell wall material are deposited across the cell's midsection ❸. Addition of this material partitions the cell, producing two identical cells ❹.

❶ A bacterium has one circular chromosome that attaches to the inside of the plasma membrane.

❷ The cell duplicates its chromosome, attaches the copy beside the original, and adds membrane and wall material between them.

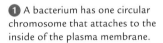

❸ When the cell has almost doubled in size, new membrane and wall are deposited across its midsection.

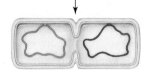

❹ Two genetically identical cells result.

Figure 19.7 Animated Binary fission. The micrograph *above* shows step 3 of this process in the bacteria *Bacillus cereus*.

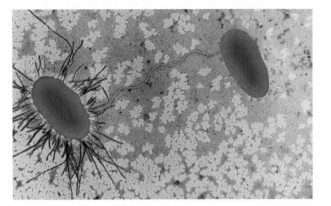

Figure 19.8 Prokaryotic conjugation. One cell extends a sex pilus out to another, draws it close, and gives it a copy of a plasmid.

>> **Figure It Out** Does conjugation increase the number of cells?

Answer: No. It is not a mode of reproduction.

Horizontal Gene Transfers

Besides inheriting DNA "vertically" from a parent cell, bacteria engage in **horizontal gene transfer**: the transfer of genetic material between existing individuals. In the process of **conjugation**, one cell gives a plasmid to the other. A **plasmid** is a small circular DNA molecule that is separate from the bacterial chromosome and has only a few genes. Cells get together for conjugation when one cell extends a sex pilus out to a prospective partner and reels it in (Figure 19.8). Once the cells are close together, the cell that made the sex pilus passes a copy of its plasmid to its partner. The cells then separate. Afterwards, each cell will pass the plasmid on to its descendants. Each cell can also donate the plasmid to other cells during another gene transfer.

Two other processes can also introduce new genes. First, a cell can take up DNA from its environment, a process called transformation. Second, viruses that infect bacteria sometimes move genes between their hosts.

The ability of bacteria to acquire new genes has important implications. Suppose a gene for antibiotic resistance arises by mutation in a bacterial cell. This gene can not only be passed on to that cell's descendants, but also be transferred to other existing cells. Gene transfers speed the rate at which a gene spreads through a population, thus accelerating the response to selective pressure.

binary fission Method of asexual reproduction that divides one bacterial or archaean cell into two identical descendant cells.
conjugation Mechanism of gene exchange in which one bacterial or archaean cell passes a plasmid to another.
horizontal gene transfer Transfer of genetic material.
plasmid Of many bacteria and archaeans, a small ring of nonchromosomal DNA replicated independently of the chromosome.

Take-Home Message How do bacteria reproduce and exchange genes among cells?

> Bacteria reproduce by binary fission, a type of asexual reproduction.

> Gene exchange occurs by conjugation. Bacteria also obtain new genes from viruses and directly from their environment.

> Bacteria serve as decomposers, cycle nutrients, and form partnerships with many other species.
< Links to PCR 15.3, Antibiotics 17.5, Chloroplast origin 18.6

Bacteria that cause human disease often get the spotlight, but most bacteria are either harmless or beneficial. As you will see, they live in many habitats and show an amazing degree of ecological diversity.

Heat-Loving Bacteria

If life emerged in thermal pools or near hydrothermal vents, the modern heat-loving bacteria may resemble those early cells. Biochemical comparisons put them near the base of the bacterial family tree. One species, *Thermus aquaticus*, was discovered in a volcanic spring in Yellowstone National Park. Biochemist Kary Mullis isolated a heat-stable DNA polymerase from *T. aquaticus* and put the enzyme to work in the first PCR reactions (Section 15.3). He won a Nobel Prize for inventing this process, which is now widely used in biotechnology.

Oxygen-Producing Cyanobacteria

Photosynthesis evolved in many bacterial lineages, but only the **cyanobacteria** release free oxygen by a noncyclic pathway, as plants do (Figure 19.9A). If, as evidence suggests, chloroplasts evolved from ancient cyanobacteria, we have cyanobacteria and their chloroplast descendants to thank for the oxygen in Earth's atmosphere (Section 18.6).

Some cyanobacteria partner with fungi and form lichens (Chapter 22), and some live on the surface of the soil, but most are aquatic. One of these, *Spirulina*,

is grown commercially and sold as a health food. Other aquatic cyanobacteria carry out the ecologically important task of **nitrogen fixation**: They incorporate gaseous nitrogen ($N\equiv N$) into ammonia (NH_3). Plants and algae need nitrogen but cannot use nitrogen gas because they cannot break its triple bond. They can, however, take up the ammonia produced and released by cyanobacteria.

Highly Diverse Proteobacteria

Proteobacteria, the most diverse bacterial lineage, also includes some nitrogen fixers. *Rhizobium* lives in roots of legumes, a group of plants that includes peas and beans. Nitrogen-fixing by *Rhizobium* benefits host plants and also enriches the soil.

Myxobacteria, another group of soil proteobacteria, show remarkable cooperative behavior. They glide about as a cohesive group, feeding on other bacteria. When food runs out, thousands of cells join together to form a multicelled fruiting body, a structure with spores (dormant cells) atop a stalk. Wind disperses spores to new habitats, where each germinates and releases a single cell.

The largest known bacterium is a marine proteobacterium, *Thiomargarita namibiensis* (Figure 19.9B). This chemoautotroph stores nitrogen and sulfur in a huge vacuole, making it big enough to be visible to the naked eye.

Escherichia coli is a chemoheterotroph that lives in the mammalian gut. It is part of the **normal flora**, a collection of microorganisms that typically live in and on a body. Most *E. coli* benefit their human host by producing vitamin K. However, the strain *E. coli* O157:H7 is among the top three causes of food poisoning. The other two, *Salmonella* and *Campylobacter*, are also proteobaceria.

Other proteobacterial pathogens that affect the gut include *Helicobactor pylori* (the main cause of stomach ulcers) and *Vibrio cholerae* (the agent of cholera).

Rickettsias are a proteobacterial subgroup of tiny cells that live as intracellular parasites and are transmitted by ticks or insects. Tick-borne rickettsias cause Rocky Mountain spotted fever. Rickettsias are also notable as the closest living relatives of the ancient cells that evolved into mitochondria.

The Thick-Walled Gram Positives

Gram-positive bacteria are a lineage characterized by thick cell walls that are tinted purple when prepared for microscopy by **Gram staining**. Thinner-walled bacteria such as cyanobacteria and proteobacteria are stained pink by this staining process, and are described as Gram-negative.

Most Gram-positive bacteria are chemoheterotrophs. For example, *Lactobacillus* is a lactate fermenter (Section 7.6). It is a common decomposer and sometimes spoils milk. We use it to produce yogurt, cheese, sauerkraut, and sour foods (Figure 19.9C). *L. acidophilus* is part of

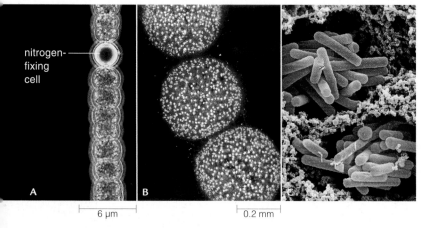

nitrogen-
fixing
cell

A   **B**   **C**

6 μm   0.2 mm

Figure 19.9 Ecologically important bacteria. **A** Chain of cyanobacteria, with many photosynthetic, oxygen-producing cells and one nitrogen-fixing cell.

B *Thiomargarita namibiensis*, a proteobacterium and the largest known bacterim. It has an enormous vacuole that holds sulfur and nitrate.

C Lactate-fermenting bacteria (*Lactobacillus*) in yogurt. Related cells are decomposers or part of the human normal flora.

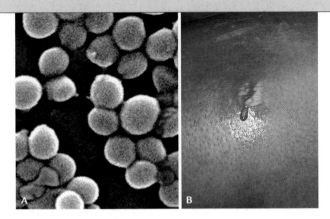

Figure 19.10 A Staphylococci, common skin bacteria.
B An abscess caused by an antibiotic-resistant staph infection.

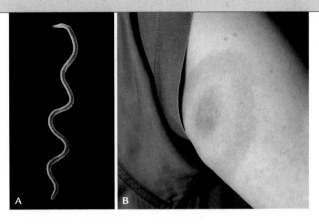

Figure 19.11 A Spirochete that causes Lyme disease. **B** Bull's-eye rash at the site of a tick bite is often the first sign of infection.

the normal flora on skin and in the gut and vagina. The lactate that the bacteria produce lowers the pH of the surroundings, and helps keep pathogenic bacteria in check. *L. acidophilus* spores are available in capsules to be taken to promote gut health.

Clostridium and *Bacillus* are Gram-positive soil bacteria that can form an **endospore**. Unlike a typical bacterial spore, an endospore can survive heating, freezing, radiation, and disinfectants.

Toxins made by some endospore-forming bacteria can be deadly. Inhale *Bacillus anthracis* endospores and you may get anthrax, a disorder in which the bacterial toxin interferes with breathing. *Clostridium tetani* endospores that germinate in wounds cause tetanus, in which toxins lock muscles in ongoing contraction. *C. botulinum*, an anaerobe that grows in improperly canned foods, makes a toxin that causes a paralyzing food poisoning known as botulism. The same toxin, prepared as Botox, can be injected to temporarily paralyze facial muscles that tug on the skin and cause wrinkling.

Worldwide, about one-third of the population is infected by *Mycobacterium tuberculosis*, the cause of tuberculosis. Droplets from coughs spread the disease, which kills about 1.6 million people each year. Bacteria-laden droplets also spread *Streptococcus*, the cause of strep throat. If these bacteria get into a wound they can become what the media calls "flesh-eating bacteria." The

result is fast-spreading infection that kills surface tissue and can be fatal. *Staphylococcus*, a common skin bacteria, can have the same effect if it enters a cut. More commonly, a staph infection will cause a boil or an abscess (Figure 19.10).

Most staph infections can be cured with methicillin, a type of antibiotic. However, directional selection has favored antibiotic resistance (Section 17.5). Antibiotic-resistant staph infections previously occurred mainly in hospitals and nursing homes. Now such infections are breaking out in schools and prisons. The bacteria are transmitted by contact with an infected person or something an infected person has touched, as by sharing towels and razors.

The sexually transmitted disease gonorrhea is caused by a diplococcus, a Gram-positive spherical bacterium that usually occurs as paired cells.

Spring–Shaped Spirochetes

Spirochetes look like a stretched-out spring (Figure 19.11). Some live in the cattle gut and help their host by breaking down cellulose. Others are aquatic decomposers and some fix nitrogen. A pathogenic spirochete causes the sexually-transmitted disease syphilis. Another spirochete, transmitted by ticks, causes Lyme disease.

Parasitic Chlamydias

Chlamydias are tiny cocci. Like the rickettsias, they can only live and replicate in eukaryotic host cells. Chlamydia infection is the most common sexually transmitted bacterial disease in the United States.

chlamydias Tiny round bacteria that are intracellular parasites of eukaryotic cells.
cyanobacteria Photosynthetic, oxygen-producing bacteria.
endospore Resistant resting stage of some soil bacteria.
Gram staining Process used to prepare bacterial cells for microscopy, and to distinguish groups based on cell wall structure.
nitrogen fixation Incorporation of nitrogen gas into ammonia.
normal flora Normally harmless or beneficial microorganisms that typically live in or on a body.
proteobacteria Largest bacterial lineage.
spirochetes Bacteria that resemble a stretched-out spring.

Take-Home Message How do bacteria affect other organisms?

> Most bacteria play beneficial roles in nutrient cycles by adding oxygen to the air, making nitrogen available to plants, or serving as decomposers.

> Some bacteria are human pathogens.

> Archaeans are more similar to eukaryotes than they are to bacteria. Many survive in extreme environments, but new species are turning up almost everywhere.
< Links to Three-domain system 1.5, Cladistics 17.14

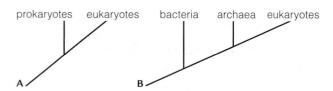

Figure 19.12 Comparison of **A** two-domain and **B** three-domain trees of life. The two-domain model was widely accepted until new evidence revealed previously unknown differences between bacteria and archaeans. The three-domain model is now in wide use.

Discovery of the Third Domain

Science is a self-correcting process in which new discoveries can overturn even long-held ways of thinking. For example, biologists historically divided all life into two groups, prokaryotes and eukaryotes. These groups were very different in size and structure, so scientists thought they represented distinct lineages (clades) that parted ways early in the history of life (Figure 19.12A).

Then, in the late 1970s Carl Woese began investigating evolutionary relationships among the prokaryotes. At that time, all were considered bacteria. By comparing ribosomal RNA gene sequences, Woese found that some methane-making cells were as similar to eukaryotes as they were to typical bacteria. Woese concluded that the methane makers were not bacteria, but rather a previously unrecognized branch on the tree of life. To accommodate this branch, he proposed a new classification system with three domains of life: bacteria, archaea, eukaryotes (Figure 19.12B).

Woese's ideas were greeted with skepticism. He had been trained as a physicist, and many biologists could not accept that he had found something they had missed. They were sure his methods were flawed. But as years went by, evidence in support of Woese's conclusions mounted. Archaeans and bacteria have different cell wall and membrane components. Archaeans and eukaryotes organize their DNA around histone proteins, which bacteria do not have. Sequencing the genome of *Methanococcus jannaschii* (*left*) provided the definitive evidence. Most of this archaean's genes have no counterpart in bacteria.

Today, the hypothesis that bacteria and archaeans constitute a single lineage has been discarded, and the three-domain classification system is in widespread use. Woese compares the discovery of archaeans to the discovery of a new continent, which he and others are now exploring.

0.5 µm

Archaean Diversity

On the basis of their physiology, many archaeans fall into one of three groups: methanogens, extreme halophiles, and extreme thermophiles.

Methanogens are organisms that produce methane (CH_4), commonly known as natural gas. They are adapted to anaerobic conditions, and exposure to oxygen inhibits their growth or kills them. Methane-making archaeans live near deep-sea hydrothermal vents, in soils, and in ocean sediments (Figure 19.13A). They also live in the gut of humans and grazers such as cows and sheep. Methane produced in the gut escapes as belches or flatulence.

By their metabolic activity, methanogens produce 2 billion tons of methane annually. Release of methane into the air has important environmental effects, because methane is a greenhouse gas (an atmospheric gas that traps heat near Earth, thus causing global warming).

Extreme halophiles are adapted to life in a highly salty environment. Salt-loving archaeans live in the Dead Sea, the Great Salt Lake, saltwater evaporation ponds, and other highly salty habitats (Figure 19.13B). Some have a purple pigment (bacteriorhodopsin) that allows them to use light energy to produce ATP.

Extreme thermophiles are adapted to life at a very high temperature. They live in hot springs (Figure 19.13C) and near deep-sea hydrothermal vents, where temperatures can exceed 110°C (230°F). Their existence is cited as evidence that life could have originated on the sea floor.

As biologists continue to explore archaean diversity, they are finding that archaeans are not restricted to extreme environments. Archaeans live alongside bacteria nearly everywhere and even exchange genes with them by conjugation. So far, scientists have not found any archaeans that pose a major threat to human health, although some that live in the mouth may encourage periodontal (gum) disease.

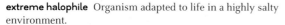

extreme halophile Organism adapted to life in a highly salty environment.
extreme thermophile Organism adapted to life in a very high-temperature environment.
methanogen Organism that produces methane gas as a metabolic by-product.

Take-Home Message What are archaeans?

> Archaeans belong to a lineage that is structurally and genetically distinct from bacteria.

> Many archaeans are adapted to extremely hot or salty conditions. Some produce methane as a metabolic by-product.

> Some archaeans live in human bodies, but none are known to be important as pathogens.

A Deep-sea sediments. Bubbles of methane rising from the floor of the Black Sea are evidence of methanogens in sediments below.

B Highly salty waters. Pigmented extreme halophiles color the brine in this California lake.

C Thermally heated waters. Pigmented archaeans color the rocks in waters of this hot spring in Nevada.

Figure 19.13 Examples of archaean habitats.

Evolution of a Disease (revisited)

❮ Links to Directional selection 17.5, Coevolution 17.13

As noted earlier, viruses have genes and can evolve. Just as bacteria evolve resistance to antibiotics, HIV has adapted to some antiviral drugs. For example, AZT, the first drug approved to fight AIDS, inhibits HIV replication by interfering with reverse transcription of viral RNA. When random mutations made some HIV particles resistant to AZT, directional selection favored those mutations. As a result of AZT treatment, evolution occurred: AZT-resistant strains of HIV became increasingly common.

Pathogens also coevolve with their hosts. For example, HIV seems to be adapting to human immune defenses. Our white blood cells have recognition proteins at their surface that allows them to detect HIV and fight it. There are different alleles for these recognition proteins and they vary among regions, with some alleles being common in Asia, others in Africa, and others in Europe. A recent study found a corresponding difference in the frequency of "escape mutations" in populations of HIV. Escape mutations help the virus evade detection by the recognition proteins of white blood cells. If escape mutations arose randomly and were not under selection, all types would be similarly prevalent in all HIV populations. Instead, escape mutations that evade common Asian recognition proteins prevail in Asia, while those that evade common African proteins predominate in Africa. Directional selection has apparently favored HIV mutations that evade the most common white blood cell defense in each region.

Selection also acts on hosts, favoring those that can fight off or evade a pathogen. For example, about 10 percent of people of European ancestry have a mutation that lessens the likelihood of infection by most HIV strains. The frequency of this mutation is highest among northern Europeans and declines with latitude. It is absent in American Indian, east Asian, and African populations. The protective mutation now enjoys a selective advantage as a result of the AIDS epidemic. However, the mutation did not arise as a result of AIDS. Mutation is a random process and studies of ancient remains tell us that this mutation has been in the northern European gene pool for thousands of years.

By one hypothesis, the mutation's current frequency and distribution reflect previous positive selection during epidemics of other diseases. In one test of this hypothesis, researchers compared the frequency of the mutation among inhabitants of small islands off the coast of Croatia. The researchers chose these islands because they knew from historical records that during the mid-1400s repeated outbreaks of an unknown disease occurred on some islands, but not others. Where epidemics did occur, the population declined by an average of 70 percent. Results supported the hypothesis. Inhabitants of islands affected by epidemics in the 1400s are significantly more likely to have the protective allele than inhabitants of islands that were unaffected.

How would you vote? Antiviral drugs help keep people with HIV healthy and lessen the likelihood of viral transmission. However, an estimated 25 percent of HIV–infected Americans do not know they are infected. Annual, voluntary HIV tests with drug treatment for those infected could help curtail the AIDS pandemic. Do you favor an expanded, voluntary testing program? See CengageNow for details, then vote online (cengagenow.com).

Summary

Section 19.1 Scientists use their knowledge of evolution to investigate how a new disease such as AIDS can arise and spread in the human population.

Section 19.2 A **virus** is a noncellular infectious agent that consists of a protein coat around a core of DNA or RNA. In some viruses, the coat is enveloped in a bit of plasma membrane derived from a previous host.

Because a virus lacks ribosomes and other metabolic machinery, it must replicate inside a host cell. Viruses attach to a host cell, then enter it or insert viral genetic material into it. Viral genes and enzymes direct the host to replicate viral genetic material and make viral proteins. New viral particles self-assemble and are released.

Bacteriophages, viruses that infect bacteria, have two types of replication pathways. In a **lytic pathway**, multiplication is rapid, and the new viral particles are released by lysis. In a **lysogenic pathway**, the virus enters a latent state that extends the cycle.

HIV is an enveloped RNA virus that replicates in human white blood cells. Viral RNA that enters the cell must be reverse transcribed to DNA to begin the process of replication. The virus acquires its envelope as it buds from the cell membrane.

Section 19.3 Some viruses are human **pathogens**, agents that cause disease. A **vector** is an animal that carries a pathogen between hosts. An **emerging disease** is new to humans or spreading to a new region. An **endemic disease** is one that is present but not spreading. An outbreak of disease in one region is an **epidemic**. If the outbreak affects many regions and threatens human health it is a **pandemic**. Changes to viral genomes as a result of mutation or gene exchanges can alter the properties of a viral disease.

Section 19.4 **Viroids** are bits of RNA that do not encode proteins. They infect plants and replicate inside them.

Section 19.5 Bacteria are small, structurally simple cells. They do not have a nucleus or cytoplasmic organelles typical of eukaryotes. The single **bacterial chromosome**, a circle of double-stranded DNA, resides in a cytoplasmic region called the **nucleoid**. Cell shapes vary. Typical surface structures include a cell wall, a protective capsule or slime layer, one or more flagella, and hairlike extensions called **pili**.

As a group, bacteria are metabolically diverse, including both autotrophs and heterotrophs.

Section 19.6 Bacteria reproduce by **binary fission**: replication of a single, circular chromosome and division of a parent cell into two genetically equivalent descendants. **Horizontal gene transfers** move genes between existing cells, as when **conjugation** moves a **plasmid** with a few genes from one cell into another.

Section 19.7 Bacteria are widespread, abundant, and diverse. Many have essential ecological roles. **Cyanobacteria** produce oxygen during photosynthesis. Some also carry out **nitrogen fixation**, producing ammonia that algae and plants need as a nutrient. **Proteobacteria**, the largest bacterial lineage, also includes nitrogen-fixers. In addition, it includes soil bacteria that show cooperative behavior, the closest relatives of mitochondria, cells that are part of our **normal flora**, and some pathogens. **Gram staining** is a method used to prepare bacteria for examination under a microscope. Gram-positive bacteria have thick walls. Some Gram-positive soil bacteria produce **endospores** that allow them to survive boiling and disinfectants. **Chlamydias** are tiny bacteria that live inside vertebrate cells. **Spirochetes** resemble a stretched-out spring and some are pathogens.

Section 19.8 Archaeans superficially resemble bacteria. However, comparisons of structure, function, and genetic sequences position archaeans in a separate domain, closer to eukaryotes than to bacteria. Ongoing research is showing that archaeans are more diverse and widely distributed than was previously thought. In their physiology, many archaeans are **methanogens** (methane makers), **extreme halophiles** (salt lovers), and **extreme thermophiles** (heat lovers). Archaeans coexist with bacteria in many habitats and can exchange genes with them.

Self-Quiz Answers in Appendix III

1. A(n) _____ may have a genome of RNA or DNA.
   a. bacterium   b. viroid   c. virus   d. archaean

2. Which is smallest?
   a. bacterium   b. viroid   c. virus   d. archaean

3. In _____ , viral DNA is integrated into a bacterial chromosome and passed to descendant cells.
   a. prokaryotic fission   c. the lysogenic pathway
   b. the lytic pathway   d. both b and c

4. The genetic material of HIV is _____ .

5. Viral genomes can be altered by _____ .
   a. mutation   b. gene exchanges   c. both a and b

6. True or false? Prokaryotic conjugation is a type of asexual reproduction.

Data Analysis Activities

Maternal Transmission of HIV Since the AIDS pandemic began, there have been more than 8,000 cases of mother-to-child HIV transmission in the United States. In 1993, American physicians began giving antiretroviral drugs to HIV-positive women during pregnancy and treating both mother and infant in the months after birth. Only about 10 percent of mothers were treated in 1993, but by 1999 more than 80 percent got antiviral drugs.

Figure 19.14 shows the number of AIDS diagnoses among children in the United States. Use the information in this graph to answer the following questions.

1. How did the number of children diagnosed with AIDS change during the late 1980s?
2. What year did new AIDS diagnoses in children peak, and how many children were diagnosed that year?
3. How did the number of AIDS diagnoses change as the use of antiretrovirals in mothers and infants increased?

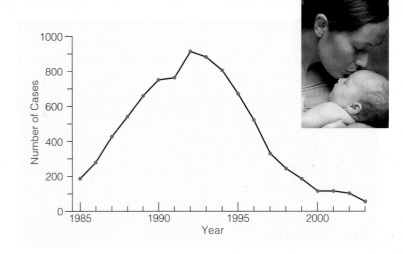

Figure 19.14 Number of new AIDS diagnoses in the United States per year among children exposed to HIV during pregnancy, birth, or by breast-feeding.

7. How many chromosomes does an archaean have?

8. All _____ are oxygen-releasing photoautotrophs.
   a. spirochetes   c. cyanobacteria
   b. archaeans   d. bacteria

9. Nitrogen-fixing bacteria produce _____ .
   a. methane   b. ammonia   c. nitrogen gas

10. Vitamin-producing *E. coli* cells in your gut are _____ .
    a. normal flora   c. bacteria
    b. chemoheterotrophs   d. all of the above

11. Some soil bacteria such as *Bacillus anthracis* survive harsh conditions by forming a(n) _____ .
    a. endospore   b. heterocyst   c. plasmid

12. A Gram-positive coccus is _____ .
    a. spherical   b. rod-shaped   c. spiral-shaped

13. The vector for Lyme disease is a(n) _____ .
    a. spirochete   b. archaean   c. tick

14. Production of _____ by archaeans may contribute to global warming.
    a. methane   b. nitrogen   c. oxygen

15. Match the terms with their most suitable description.
    ___ archaean   a. plant-infecting RNA
    ___ cyanobacteria   b. noncellular infectious particle;
    ___ virus   nucleic acid core, protein coat
    ___ viroid   c. likes it hot
    ___ plasmid   d. site of bacterial chromosome
    ___ extreme   e. sister group to the eukaryotes
        halophile   f. evolved into chloroplasts
    ___ nucleoid   g. small circle of bacterial DNA
    ___ extreme   h. salt lover
        thermophile

Additional questions are available on **CENGAGENOW.**

Critical Thinking

1. Viruses that do not have a lipid envelope tend to remain infectious outside the body longer than enveloped viruses. "Naked" viruses are also less likely to be rendered harmless by soap and water. Can you explain why?

2. Methanogens have been found in the human gut and deep-sea sediments, but not in the human mouth or the surface waters of the ocean. What physiological trait of methanogens could explain this distribution?

3. Review the description of Fred Griffith's experiments with *Streptococcus pneumoniae* in Section 18.3. Using your knowledge of bacterial biology, explain the process by which the harmless bacteria became dangerous.

4. The antibiotic penicillin acts by interfering with the production of new bacterial cell wall. Cells treated with penicillin do not die immediately, but they cannot reproduce. Explain how penicillin halts binary fission. Explain also why the cancer drug taxol, which stops eukaryotic division by interfering with spindle formation, has no effect on bacterial cells.

5. About 1 percent of Europeans are homozygous for an allele that provides protection against infection by HIV. Would you expect more or fewer heterozygous for this allele? Explain your reasoning.

Animations and Interactions on **CENGAGENOW:**
> Bacteriophage replication; Bacterial structure; Binary fission, Conjugation.

‹ Links to Earlier Concepts

This chapter covers the protists, a diverse collection of lineages introduced in Section 1.4. You will learn about how protists reproduce (11.2) and how they are classified (17.14). We reexamine eukaryotic cell structures such as chloroplasts (6.4), pseudopods, and flagella (4.10); reconsider the effects of osmosis on cells (5.6); and return to photosynthetic pigments (6.2). We also delve again into evolution of organelles by endosymbiosis (18.6).

Key Concepts

A Collection of Lineages
Protists include many lineages of eukaryotic organisms, some autotrophs and others heterotrophs. The protists are not a clade; some groups are more closely related to plants, or to fungi and animals, than to other protists.

Single–Celled Lineages
Most protist lineages are entirely single-celled. These groups include flagellated protozoans, shelled cells called foraminiferans and radiolarians, and the alveolates (ciliates, dino-flagellates, and apicomplexans).

20 The Protists

20.1 Harmful Algal Blooms

If you sample water from just about any aquatic habitat, you will find a variety of single-celled protists. A **protist** is a eukaryotic organism that is not a fungus, plant, or animal. Aquatic protists include single-celled and multicellular autotrophs and heterotrophs. Photosynthetic protists play an important ecological role by taking up carbon dioxide and releasing oxygen. They also serve as food for aquatic animals.

However, some of these single-celled producers occasionally become a threat. When conditions are unusually favorable for growth, as occurs when extra nutrients are present, the cells multiply fast. The result is an **algal bloom**, a higher than normal concentration of aquatic microorganisms. The photo at the *left*, taken in 2007, shows an enormous algal bloom in coastal waters near Pensacola, Florida. Algal blooms are commonly known as "red tides" because the protists involved often have a reddish pigment. However, the term is misleading: Not all algal blooms color the water red, and the event is not related to tidal changes.

Algal blooms in the Gulf of Mexico frequently involve the protist *Karenia brevis*, a dinoflagellate (Figure 20.1). This species makes brevetoxin. A **toxin** is a substance that is produced by one organism and is harmful to others. Brevetoxin interferes with animal nerve cells by binding to a protein in their cell membrane. It sickens and even kills marine invertebrates, fish, sea turtles, sea birds, dolphins, and manatees.

Human nerve cells have the same kinds of membrane proteins as those of other vertebrates and are harmed in the same way. Eating shellfish tainted by brevetoxin causes intestinal problems and nervous system symptoms such as headache, vertigo, loss of coordination, and temporary paralysis. People also are exposed to brevetoxin in spray from onshore winds during an algal bloom. When inhaled, brevetoxin irritates nasal membranes and constricts airways, making breathing difficult. Inside the lungs, metabolic breakdown of brevetoxins creates chemicals

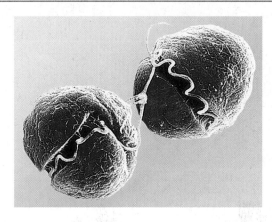

Figure 20.1 Colorized, scanning electron micrograph of the photosynthetic protist *Karenia brevis*. This dinoflagellate benefits us by taking up carbon dioxide and releasing oxygen. However, it also produces a toxin that can sicken people who inhale or ingest it.

that damage DNA. Thus, repeated inhalation of the toxin might increase the risk of lung cancer.

The effects of brevetoxin on humans are interesting, and may be a threat to public health. However, these effects are an evolutionary accident; *K. brevis* does not gain any advantage by sickening humans. Most likely, brevetoxin benefits *K. brevis* by providing protection against potential predators such as heterotrophic protists and tiny animals. Keep this point in mind as you read the chapter: Although we will often mention the ways that protists affect humans, the lineages and traits we describe here had their origins long before humans evolved.

---

algal bloom Population explosion of tiny aquatic producers.
protist Eukaryote that is not a fungus, animal, or plant.
toxin Chemical that is made by one organism and harms another.

Brown Algae and Relatives
Brown algae are an entirely multicellular group of protists. They are members of the same lineage as single-celled photosynthetic cells called diatoms, and filamentous heterotrophs called water molds.

Red Algae, Green Algae
Red algae and green algae are single-celled and multicelled aquatic producers. Red algae have pigments that allow them to live in deep waters. Green algae are the closest relatives of land plants and have the same pigments as them.

Amoebozoans
A lineage of unwalled heterotrophic protists that includes the amoebas and slime molds is the protist group most closely related to the fungi and animals. Members of this lineage live in aquatic habitats or on the forest floor.

> Protists are members of many separate eukaryotic lineages, some only distantly related to one another.
< Links to Asexual Reproduction 11.2, Cladistics 17.14

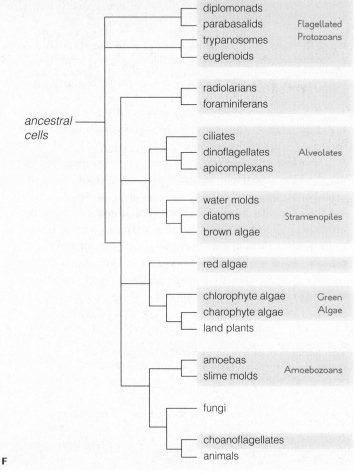

Protists are collection of lineages, rather than a clade, or monophyletic group (Section 17.14). In fact, some protists are more closely related to plants, fungi, or animals than to other protists.

Figure 20.2 shows a few protists and illustrates where the groups we cover in this chapter fit in the eukaryotic family tree. There are many additional protist lineages, but this sampling of major groups will suffice to demonstrate protist diversity, ecological importance, and the ways that protists can affect human health.

Most protist lineages include only unicellular species (Figure 20.2A–C). However, some lineages include colonial cells, and true multicellularity evolved independently in several groups (Figure 20.2D,E). A multicellular organism consists of cells that cannot survive and reproduce on their own. Cells of colonial organisms live together, but retain an ability to survive and reproduce independently.

Protists have diverse life cycles, but most reproduce both sexually and asexually. Depending on the group, asexually reproducing individuals may be haploid or diploid. For example, the parasite that causes malaria has a haploid stage that divides in human blood and liver cells. The only diploid stage in the parasite's life cycle is the zygote, which undergoes meiosis within a few hours of fertilization. By contrast, the silica-shelled cells called diatoms have a diploid-dominant life cycle. Most of the time, diploid cells divide by mitosis and only the gametes are haploid.

Like plants, some multicelled algae have a life cycle with both haploid and diploid multicelled bodies. Protists are often described as "simple," but the multicelled bodies of some algae can be large and complex.

Protists may be heterotrophs, autotrophs, or switch between nutritional modes. Heterotrophic protists serve as decomposers, prey on smaller organisms such as bacteria, or live inside larger organisms. Protistan autotrophs have chloroplasts. In the lineage that includes red algae and green algae, the chloroplasts evolved from bacteria as described in Section 18.6. We call this event *primary* endosymbiosis. All other protists have chloroplasts that evolved through *secondary* endosymbiosis: A protist that had chloroplasts was engulfed by a heterotrophic one and evolved into an organelle.

Figure 20.2 Protist diversity. The single-celled protists include **A** amoebas, **B** euglenoids, and **C** diatoms. Most red algae **D** and all brown algae **E** are multicelled. **F** One proposed eukaryotic family tree with traditional protist groups indicated by *tan* boxes. The protists are not a single lineage.

>> **Figure It Out** Are land plants more closely related to the red algae or the brown algae?

Answer: Red algae

Take-Home Message What are protists?

> Protists are a collection of eukaryotic lineages. Most are single-celled, but there are some multicelled species.

> Some protists are autotrophs; other are heterotrophs. A few can switch between modes.

> Most protists reproduce both sexually and asexually.

❭ Flagellated protozoans swim through lakes, seas, and the body fluids of animals.
❰ Links to Flagella 4.10, Tonicity 5.6

Flagellated protozoans are unwalled cells with one or more flagella. All groups are entirely or largely heterotrophic. A **pellicle**, a layer of elastic proteins beneath the plasma membrane, helps the cells retain their shape. Haploid cells dominate the life cycle of these groups.

Diplomonads and parabasalids have multiple flagella and are among the few protists that can live in places where oxygen is scarce. Instead of mitochondria, they have organelles that produce ATP by an anaerobic pathway. These organelles evolved from mitochondria and are an adaptation to anaerobic habitats. Free-living diplomonads and parabasalids thrive deep in seas and lakes. Others live inside the bodies of animals.

Diplomonads are unusual in that they have two more or less identical nuclei. The diplomonad *Giardia lamblia* (*left*) causes giardiasis, a waterborne human disease. The protist attaches to the intestinal lining and sucks out nutrients. Symptoms of giardiasis include cramps, nausea, and severe diarrhea. Infected people and animals excrete cysts (a hardy resting stage) of *G. lamblia* in their feces. Drinking water contaminated with the cysts spreads the infection.

The parabasalid *Trichomonas vaginalis* (Figure 20.3A) infects human reproductive tracts and causes the disease trichomoniasis. *T. vaginalis* does not make cysts, so it cannot survive very long outside the human body. Fortunately for the parasite, sexual intercourse puts it directly

---

contractile vacuole In freshwater protists, an organelle that collects and expels excess water.
euglenoid Flagellated protozoan with multiple mitochondria; may be heterophic or have chloroplasts decended from algae.
flagellated protozoan Protist belonging to an entirely or mostly heterotrophic lineage with no cell wall and one or more flagella.
pellicle Layer of proteins that gives shape to many unwalled, single-celled protists.
trypanosome Parasitic flagellate with a single mitochondrion and a membrane-encased flagellum.

Take-Home Message What are flagellated protozoans?

❭ Parabasalids and diplomonads are heterotrophs that lack mitochondria. Some are important human pathogens.

❭ Trypanosomes are parasites with a large mitochondrion. Biting insects transmit some that cause human disease.

❭ Euglenoids include heterotrophs and photoautotrophs with chloroplasts that evolved from green algae.

A The parabasilid *Trichomonas vaginalis* causes a sexually transmitted disease.

B The trypanosome *Trypanosoma brucei*, causes African sleeping sickness.

Figure 20.3 Two parasitic flagellates.

into hosts. In the United States, about 6 million people are infected.

Trypanosomes are long tapered cells, with a single large mitochondrion. A flagellum encased in a membrane runs the length of the cell (Figure 20.3B). Action of the flagellum causes a wavelike motion in the membrane around it and moves the cell.

All trypanosomes are parasites and insects serve as vectors for some that cause human disease. Tsetse flies in sub-Saharan Africa carry the trypanosome that causes African sleeping sickness, a nervous system disease that can be fatal. In Central and South America, bloodsucking bugs transmit a trypanosome that causes Chagas disease, which can damage the heart. Desert sandflies are the vector for leishmaniasis, another trypanosome disease. Untreated leishmaniasis can produce disfiguring scars and harm the liver.

Euglenoids are flagellated cells closely related to the trypanosomes. Unlike trypanosomes, euglenoids have many mitochondria. Some also have chloroplasts that evolved from a green alga (Figure 20.4). Photosynthetic euglenoids have an eyespot near the base of a long flagellum that helps the cell detect light. The pellicle consists of translucent strips of protein that spiral around the cell.

The interior of a euglenoid is saltier than its freshwater habitat, so water tends to diffuse into the cell. Like many other freshwater protists, euglenoids have one or more **contractile vacuoles**, organelles that collect excess water, then contract and expel it to the outside through a pore.

Figure 20.4 Animated Body plan of a euglenoid (*Euglena*), a freshwater species with chloroplasts that evolved from a green alga. See also Figure 20.2B.

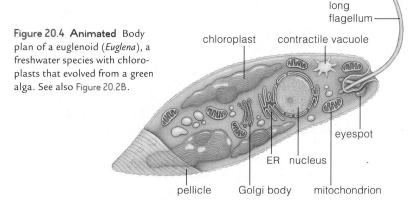

long flagellum
chloroplast
contractile vacuole
eyespot
ER
nucleus
mitochondrion
Golgi body
pellicle

> Heterotrophic single cells with chalky or glassy shells live in great numbers in the world's oceans.

< Link to Microtubules 4.10

Foraminiferans and radiolarians are single-celled marine protists with sieve-like shells. They feed by capturing food with microtubule reinforced cytoplasmic extensions that protrude through the shell's many openings.

Foraminiferans have a calcium carbonate shell. They typically prey on bacteria and smaller protists in ocean sediments. Others are **plankton**; a general term for micro-organisms that drift or swim in an aquatic habitat. Plank-tonic foraminiferans often have small photosynthetic protists inside them (Figure 20.5A). Chalk and limestone deposits (Figure 20.5B) include foraminiferan shells that accumulated on the sea floor over billions of years.

Radiolarians secrete a glassy silica shell (Figure 20.5C). They are a major component of the marine plankton in tropical waters. They capture food with their cytoplasmic extensions and, like planktonic foraminiferans, some have photosynthetic protists living inside them.

---

foraminiferan Heterotrophic single-celled protist with a porous calcium carbonate shell and long cytoplasmic extensions.
plankton Community of tiny drifting or swimming organisms.
radiolarian Heterotrophic single-celled protist with a porous shell of silica and long cytoplasmic extensions.

Take-Home Message **What are foraminiferans and radiolarians?**

> Two related lineages of heterotrophic marine cells have porous secreted shells. The chalky-shelled foraminiferans live in sediments or drift as part of the plankton. The silica-shelled radiolarians are planktonic.

> Dinoflagellates, ciliates, and apicomplexans are single cells that belong to the alveolate lineage.
> Most dinoflagellates and ciliates are aquatic and free-living, but all apicomplexans are parasites.

< Link to Endocytosis 5.8

Dinoflagellates, ciliates, and apicomplexans belong to a lineage known as the **alveolates**. "Alveolus" means sac, and the characteristic trait of alveolates is a layer of sacs beneath the plasma membrane.

Dinoflagellates

The name **dinoflagellate** means "whirling flagellate." These single-celled protists typically have two flagella, one at the cell's tip and the other running in a groove around its middle like a belt. Combined action of the two flagella causes the cell to rotate as it moves forward. Most dinofla-gellates deposit cellulose in the sacs beneath their plasma membrane, and the deposits form thick protective plates. The vast majority of dinoflagellates are marine plank-ton. They are especially abundant in tropical waters. Some prey on bacteria, and others have chloroplasts that evolved from algae. As Section 20.1 explained, runaway growth of some dinoflagellates such as *Karenia brevis* sometimes results in an algal bloom.

Photosynthetic dinoflagellates live inside reef-building coral (a type of invertebrate animal). The protists supply the coral with sugars and oxygen, and receive shelter and carbon dioxide in return. A coral depends on its protist partners. If the coral loses them it will die.

A few marine dinoflagellates are bioluminescent. Like fireflies, they convert ATP energy into light (Figure 20.6). Emitting light may protect a cell by startling a predator

Figure 20.5 Protists with secreted shells. **A** Live foraminiferan with algal cells inside it (*yellow* dots). **B** Chalk cliffs of Dover, England are remains of calcium carbonate–rich shells of marine protists that accumulate on the sea floor. **C** Silica shell of a radiolarian.

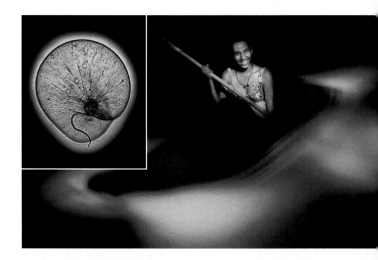

Figure 20.6 Dinoflagellate bioluminescence. A tropical dinoflagel-late (*inset*) emits light when disturbed, as by the motion of an oar. The flash of light may benefit the cell by warding off predators.

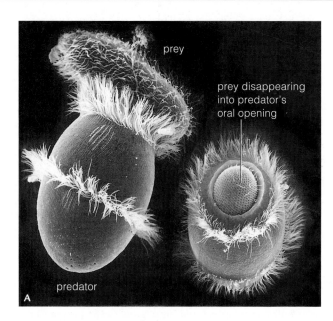

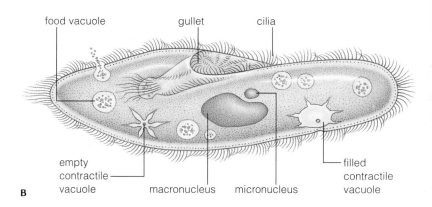

Figure 20.7 **Animated** Freshwater ciliates. **A** *Didinium*, a barrel-shaped ciliate with tufts of cilia, catching and engulfing a *Paramecium* covered with cilia. **B** Body plan of *Paramecium*. Mitochondria are present but not shown.

that was about to eat it. Alternatively the flash of light may function like a car alarm, attracting the attention of other organisms, including predators that pursue the potential dinoflagellate eaters.

Ciliates

Ciliates, or ciliated protozoans, are unwalled, heterotrophic cells that use their many cilia in locomotion and feeding. Most ciliates are aquatic predators that feed on bacteria, algae, and one another (Figure 20.7A).

Figure 20.7B shows the body plan of *Paramecium*, a free-living ciliate common in lakes and ponds. The cell feeds by using its cilia to sweep water laden with bacteria, algae, and other food particles into a gullet. Food particles enter the cell by endocytosis and are digested inside food vacuoles. Digestive waste is then expelled by exocytosis. Like euglenoids, *Paramecium* has contractile vacuoles that collect and squirt out excess water.

There are two types of nuclei, a macronucleus that controls daily activities and one or more smaller micronuclei that function in sexual reproduction. Like most single-celled protists, *Paramecium* reproduces asexually: A cell duplicates its DNA and organelles, then divides in half, producing two identical cells. Sexual reproduction

occurs by a process that involves meiosis and a swap of micronuclei between two cells.

Although most ciliates are free-living and aquatic, some have adapted to life in the animal gut. Gut-dwelling ciliates help cattle, sheep, and related grazing animals break down the cellulose in plant material. Similarly, ciliates that live in the termite gut help these insects break down wood.

Only one ciliate is known to be a human pathogen. It usually lives in the gut of pigs, but can also survive in the human gut, where it causes nausea and diarrhea. Human infections occur when pig feces get into drinking water.

Apicomplexans

Apicomplexans are parasitic protists that spend part of their life inside cells of their hosts. Their name refers to a complex of microtubules at their apical (top) end that allows them to pierce and enter a host cell. They are also sometimes called sporozoans. Apicomplexans infect a variety of animals, from worms and insects to humans. Their life cycle often involves more than one host species. The next section looks in detail at the disease malaria, and the biology of the apicomplexan that causes it.

---

alveolate Member of a protist lineage having small sacs beneath the plasma membrane; dinoflagellate, ciliate, or apicomplexan.
apicomplexan Single-celled protist that lives as a parasite inside animal cells.
ciliate Single-celled, heterotrophic protist with many cilia.
dinoflagellate Single-celled, aquatic protist with cellulose plates and two flagella; may be heterotrophic or photosynthetic.

Take-Home Message What are alveolates?

> Alveolates are single cells with an array of membrane-bound sacs (alveoli) beneath the plasma membrane.

> Dinoflagellates are common in plankton. These flagellated heterotrophs or photoautotrophs have cellulose plates and move with a whirling motion.

> The ciliates are heterotrophs. Cilia cover all or part of the cell surface and function in locomotion and feeding.

> Apicomplexans are intracellular parasites with a special host-piercing device.

> Malaria, caused by an apicomplexan, has the highest death toll of any protist disease.
< Link to Balanced polymorphism 17.7

Malaria is a leading cause of human death, killing more than 1.3 million people every year. *Plasmodium*, a single-celled apicomplexan, causes malaria. Mosquitoes carry the protist from one human host to another.

Figure 20.8 shows the *Plasmodium* life cycle. A bite from a female mosquito transmits the infective stage of *Plasmodium*, a haploid sporozoite, to a human host **1**. The sporozoite travels through blood vessels to liver cells, where it reproduces asexually **2**. Offspring, called merozoites, enter red blood cells and produce more merozoites **3**. Merozoites also can enter red blood cells and develop into immature gametes, or gametocytes **4**. When a mosquito bites an infected person, gametocytes are taken up with blood and mature in the mosquito gut. Gametes fuse and form zygotes **5**. Meiosis produces cells that develop into new sporozoites that migrate to the insect's salivary glands and await transfer to a new host **6**.

Malaria symptoms usually start a week or two after a bite, when the infected liver cells rupture and release merozoites, metabolic wastes, and cellular debris into blood. Shaking, chills, a burning fever, and sweats follow. After the initial episode, symptoms may subside for a few weeks or even months. Infected people often feel healthy. However, ongoing infection damages the liver, spleen, and kidneys, clogs blood vessels, and cuts blood flow to the brain. The result is convulsions, coma, and eventual death.

Plasmodium cannot survive at low temperatures, so malaria is mainly a tropical disease. It remains common in Mexico, South and Central America, as well as Asia and the Pacific Islands, but the greatest toll is in Africa. One African child dies of malaria every 30 seconds.

Malaria has been a potent selective force on humans in Africa. As Section 17.7 explained, the allele that produces sickle-cell anemia was favored in African populations because it also provides protection against malaria.

Natural selection also acts on *Plasmodium*. The protist has recently become resistant to several antimalarial drugs. Over a longer time frame, it has evolved an amazing capacity to alter the behavior of its hosts. *Plasmodium* makes the mosquitoes that carry it more likely to feed several times a night, and thus more likely to bite several people. It also makes infected humans especially appetizing to a hungry mosquito when gametocytes are present. By manipulating its insect and human hosts, the protist maximizes chances that its offspring will reach a new host.

Take-Home Message What is malaria?

> Malaria is a deadly tropical disease caused by an apicomplexan, a parasite that is transmitted by mosquitoes.

> The parasite reproduces asexually in human liver cells and blood cells. Death of these cells causes the symptoms that characterize the disease.

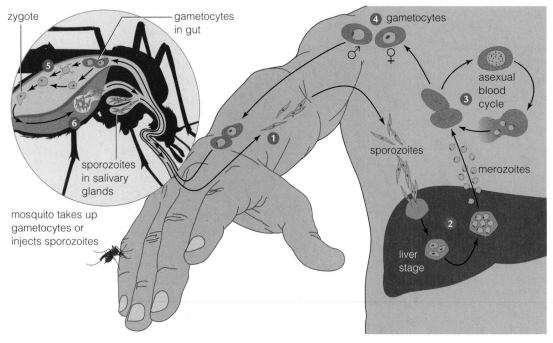

zygote

gametocytes in gut

sporozoites in salivary glands

mosquito takes up gametocytes or injects sporozoites

4 gametocytes

♂ ♀

asexual blood cycle

3

sporozoites

merozoites

1

2

liver stage

Figure 20.8 Animated Life cycle of one of the *Plasmodium* species that cause malaria.

1 Infected mosquito bites a human. Sporozoites enter blood, which carries them to the liver.

2 Sporozoites reproduce asexually in liver cells, mature into merozoites. Merozoites leave the liver and infect red blood cells.

3 Merozoites reproduce asexually in some red blood cells.

4 In other red blood cells, merozoites differentiate into gametocytes.

5 A female mosquito bites and sucks blood from the infected person. Gametocytes in blood enter her gut and mature into gametes, which fuse to form zygotes.

6 Meiosis of zygotes produces cells that develop into sporozoites. The sporozoites migrate to the mosquito's salivary glands.

> The stramenopile lineage includes heterotrophic water molds as well as single-celled diatoms and multicelled brown algae (both photosynthetic).
< Link to Accessory pigments 6.2

Stramenopile means "straw-haired" and refers to a shaggy flagellum that occurs during the life cycle of some members of this group. However, stramenopiles are defined mainly by genetic similarities, rather than visible traits.

Water molds are filamentous heterotrophs that were once mistakenly classified as fungi. Like fungi, they form a mesh of nutrient-absorbing filaments, but the two groups differ in many structural and genetic traits.

Most water molds decompose organic debris in aquatic habitats, but a few are parasites that have significant economic effects. Some grow as fuzzy white patches on fish in fish farms and aquariums (*left*). Others infect land plants, destroying crops and forests. Members of the genus *Phytophthora* are especially notorious. Their name means "plant destroyer." In the mid-1800s, one species destroyed Irish potato crops, causing a famine that killed and displaced millions of people. Today, another species is causing an epidemic of sudden oak death in Oregon, Washington, and California.

The closest relatives of water molds are two photosynthetic groups: diatoms and brown algae. Both have chloroplasts with a brown accessory pigment (fucoxanthin) that

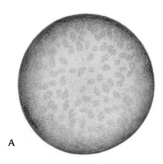

A

Figure 20.9 A Living diatom. Chloroplasts are visible through the glassy silica shell. See also Figure 20.2C. **B** Diatomaceous earth is shells of ancient diatoms. It kills crawling insects by scratching their surface so they lose water, dry out, and die. **B**

brown alga Multicelled marine protist with a brown accessory pigment in its chloroplasts.
diatom Single-celled photosynthetic protist with brown accessory pigments in its choroplasts and a two-part silica shell.
water mold Protist that grows as nutrient-absorbing filaments.

Figure 20.10 Kelp "forest." These brown algae are the largest protists.

tints them olive, golden, or dark brown (Figure 20.9A). The chloroplasts evolved from a red alga.

Diatoms have a two-part silica shell, with upper and lower parts that fit together like a shoe box. Some cells live individually and others form chains. Diatoms can be found in damp soil, lakes, and seas. They are particularly abundant in cool waters. When marine diatoms die, their shells fall to the sea floor. Ancient sea floor that has been uplifted and is now land is the source of silica-rich diatomaceous earth. This substance is an inert powder of glassy bits. It is used in filters, abrasive cleaners, and as an insecticide that does not harm vertebrates (Figure 20.9B).

Brown algae are multicelled inhabitants of temperate or cool seas. In size, they range from microscopic filaments to giant kelps 30 meters (100 feet) tall. Giant kelps form forestlike stands in coastal waters of the Pacific Northwest (Figure 20.10). Like trees in a forest, kelps shelter a wide variety of other organisms.

Brown algae have commercial uses. Alginic acid from their cell walls is used to produce algins that serve as thickeners, emulsifiers, and suspension agents. You will find algins in ice cream, pudding, jelly beans, toothpaste, cosmetics, and many other products.

Take-Home Message **What are stramenopiles?**

> Stramenopiles include diverse lineages that are united on the basis of their genetic similarity, rather than any visible traits.

> Water molds are filamentous decomposers and parasites.

> Diatoms are silica-shelled cells. Brown algae are multicellular. Both groups have a brown accessory pigment in chloroplasts that evolved from a red alga.

> Red algae and green algae belong to the same lineage as the land plants.
‹ Links to Plant cell walls 4.11, Photosynthetic pigments 6.2 and 6.3, Plant cell division 11.4

Red Algae Do It Deeper

Red algae are photosynthetic protists that are typically multicellular and tropical, although there are some single-celled species. Coralline algae (red algae with cell walls that contain calcium carbonate) are a component of some tropical coral reefs. Compared to brown algae or green algae, red algae can live at greater depths. In addition to chlorophyll *a*, they have phycobilins, pigments that absorb green light. This light penetrates deepest into water. Shallow-water red algae tend to have little phyco-bilin and appear green. Deep dwellers are almost black.

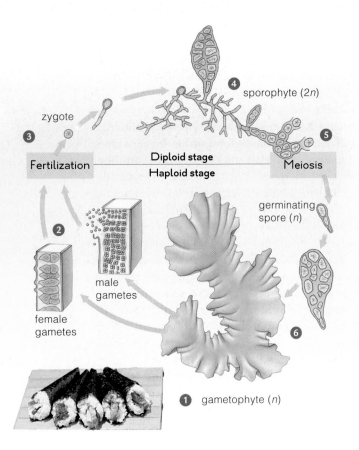

Figure 20.11 Animated Life cycle of a multicellular red alga (*Porphyra*).

❶ The haploid gametophyte is sheetlike.

❷ Gametes form at its edges.

❸ Fertilization produces a diploid zygote.

❹ The zygote develops into a diploid sporophyte.

❺ Haploid spores form by meiosis on the sporophyte body, and are released.

❻ Spores germinate and develop into a new gametophyte.

>> **Figure It Out** In the sheets of algae used to wrap sushi, are the cells haploid or diploid?   Answer: haploid

Red algae have many commercial uses. Agar, a poly-saccharide extracted from cell walls of red algae, is used to keep baked goods and cosmetics moist, to set jellies, and as a vegetarian substitute for gelatin. Carrageenan, another polysaccharide, is added to soy milk and dairy products. Nori, dry sheets of the red alga *Porphyra*, wraps some kinds of sushi. Nori production is big business, with more than 130,000 tons harvested annually.

Like many other multicelled algae, *Porphyra* has an **alternation of generations**, a life cycle that alternates between haploid and diploid multicelled bodies (Figure 20.11). The sheetlike seaweed used as nori is the haploid body, or **gametophyte** ❶. Gametes form along edges of the sheets ❷. Fertilization of gametes produces a diploid zygote ❸. The zygote grows by mitosis into the diploid body, or **sporophyte**. In *Porphyra*, the sporophyte is a microscopic branching filament that grows on the shells of mollusks ❹. Some cells of the sporophyte undergo meiosis and produce haploid spores ❺. Germination of the spore is followed by growth and development of a new gametophyte, thus completing the life cycle ❻.

Green Algae

Green algae are photosynthetic protists with chloroplasts that have chlorophylls *a* and *b*. Most green algae live in fresh water, but some live in the ocean, in the soil, or on surfaces such as tree trunks. Single-celled green algae also partner with fungi to form the composite organisms called lichens.

Chlamydomonas, a flagellated single-celled green alga, lives in standing fresh water (Figure 20.12A). Haploid cells reproduce asexually when conditions favor growth. When nutrients become scarce, gametes form by mitosis. Fusion of two gametes produces a diploid zygote that has a thick protective wall (*right*). When conditions become favorable again, the zygote undergoes meiosis and produces four haploid, flagellated cells.

The colonial green alga *Volvox* also lives in lakes and ponds. A colony con-sists of flagellated cells joined together by thin cytoplasmic strands to form a whirling sphere (Figure 20.12B). New colonies form inside the parental sphere, which eventually ruptures and releases them.

Sheets of the multicelled species *Ulva* cling to rocks along marine coasts (Figure 20.12C). The sheets grow longer than your arm, but are no more than 40 microns thick. *Ulva* is commonly known as sea lettuce and is a popular food in Scotland. It has an alternation of genera-tions with large, sheetlike bodies in both the haploid and diploid generations.

Studies of green algae helped biologists understand the mechanisms of photosynthesis. Section 6.3 explained how

A *Chlamydomonas* is a single celled species that uses its two flagella to swim in fresh water.

B *Volvox* colonies, each with many flagellated cells joined by thin strands of cytoplasm. New colonies are visible as bright green spheres inside each parent colony.

C Multicellular sheets of sea lettuce (*Ulva*). The long thin sheets are common along coasts in zones where there is little wave action.

Theodor Engelmann used filaments of a green alga to determine the most effective wavelengths of light for photosynthesis. Melvin Calvin used *Chlorella*, a single-celled green alga, to clarify the steps in the reactions we now call the Calvin–Benson cycle.

Evolutionary Connections to Land Plants

Red algae, green algae, and land plants all have a cell wall made of cellulose, store sugars as starch, and have chloroplasts that evolved from a cyanobacterial ancestor by primary endosymbiosis. Thus they are thought to be descended from a common ancestor.

The closest relatives of the land plants belong to a subgroup of freshwater green algae known as the charophyte algae. One modern member of this group, *Chara*, is native to freshwater habitats in Florida (Figure 20.12D). Like plants, and unlike most other green algae, *Chara* cells divide their cytoplasm by cell plate formation (Section 11.4) and have plasmodesmata, cytoplasmic connections between neighboring cells (Section 4.11).

"Green algae" does not refer to a clade because the term excludes land plants, which are on the same evolutionary branch as the charophyte algae.

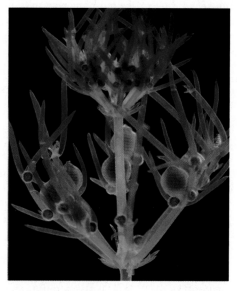

D *Chara*, a charophyte alga known as muskgrass or stinkweed for its strong odor.

Figure 20.12 A sampling of green algal diversity.

alternation of generations Of land plants and some protists, a life cycle in which haploid and diploid multicelled bodies form.
gametophyte Gamete-producing haploid body that forms in the life cycle of land plants and some protists.
green alga Photosynthetic protist that deposits cellulose in its cell wall, stores sugars as starch, and has chloroplasts containing chlorophylls *a* and *b*.
red alga Photosynthetic protist that deposits cellulose in its cell wall, stores sugars as starch, and has chloroplasts containing chlorophyll *a* and red pigments called phycobilins.
sporophyte Spore-forming diploid body that forms in the life cycle of land plants and some protists.

> ## Take-Home Message What are red algae and green algae?
>
> ❯ Red algae and green algae are protists that belong to the same clade as the land plants. All members of this clade deposit cellulose in their cell wall and store excess sugars as starch.
>
> ❯ Red algae are mostly multicelluar and marine. They have red accessory pigments that allow them to live at greater depths than other algae.
>
> ❯ Green algae include single-celled, colonial, and multicelled species. One subgroup, the charophyte algae, includes the closest living relatives of land plants.

> The amoebas and their relatives are shape-shifting heterotrophs. Many are solitary, but some display communal behavior and cell differentiation that hint at complexities to come in animals.
< Link to Pseudopods 4.10

Amoebozoans are one of the monophyletic groups now being carved out of the former kingdom Protista. Few amoebozoans have a cell wall, shell, or pellicle; nearly all undergo dynamic changes in shape. A compact blob of a cell can quickly send out pseudopods, move about, and capture food (Section 4.10).

Solitary Amoebas

The **amoebas** live as single cells. Figure 20.13 shows *Amoeba proteus*. Like most amoebas, it is a predator in freshwater habitats. Other amoebas can live in the gut of humans and other animals. Some gut-dwelling species do no harm or aid their host's digestive process. Others can cause disease. Each year, about 50 million people are affected by amebic dysentery after drinking water contaminated with cysts of a pathogenic amoeba.

Slime Molds

Slime molds are sometimes described as "social amoebas." There are two types, and both are common on the floor of temperate forests.

Plasmodial slime molds spend most of their life cycle as a plasmodium, a slimy, multinucleated mass. The plasmodium forms when a diploid cell undergoes mitosis many times without cytoplasmic division. It streams out along the forest floor feeding on microbes and organic matter (Figure 20.14). The plasmodium can be as big as a dinner plate. When food runs out, a plasmodium develops into spore-bearing fruiting bodies.

Figure 20.14 Plasmodial slime mold (*Physarum*). **A** These protists feed as a large multinucleated mass—a plasmodium—that oozes along the forest floor and over logs, devouring bacteria. **B** When food runs low, the mass forms spore-bearing structures.

Cellular slime molds such as *Dictyostelium discoideum* spend most of their existence as individual amoeba-like cells (Figure 20.15). Each cell eats bacteria and reproduces by mitosis ❶. If food runs out, thousands of cells stream together, forming a multicelled mass ❷. Environmental gradients in light and moisture induce the mass to crawl along as a cohesive multicelled unit often referred to as a "slug" ❸. Cells of the slug are held together by adhesion proteins and a secreted extracellular maxtrix. When the slug reaches a suitable spot, the cells differentiate and develop into a fruiting body: A stalk forms and a group of cells at its tip become spores ❹. Germination of a spore releases an amoeboid cell that starts the life cycle anew ❺.

Dictyostelium and other amoebozoans provide clues to how signaling pathways of multicelled organisms evolved. Coordinated behavior—an ability to respond to stimuli as a unit—is a hallmark of multicellularity. It requires cell-to-cell communication, which may have originated in

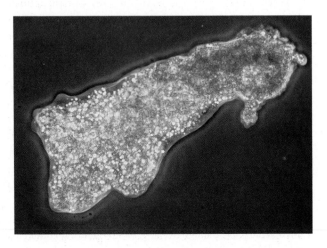

Figure 20.13 An amoeba. The cell has no fixed shape. It feeds or shifts position by extending lobes of cytoplasm (pseudopods).

amoeba Single-celled protist that extends pseudopods to move and to capture prey.
amoebozoan Shape-shifting heterotrophic protist with no pellicle or cell wall; an amoeba or slime mold.
cellular slime mold Amoeba-like protist that feeds as a single predatory cell; joins with others to form a multicellular spore-bearing structure when conditions are unfavorable.
plasmodial slime mold Protist that feeds as a multinucleated mass; forms a spore-bearing structure when enviromental conditions become unfavorable.

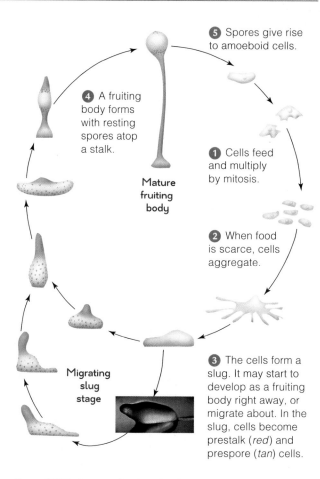

⑤ Spores give rise to amoeboid cells.

④ A fruiting body forms with resting spores atop a stalk.

Mature fruiting body

① Cells feed and multiply by mitosis.

② When food is scarce, cells aggregate.

③ The cells form a slug. It may start to develop as a fruiting body right away, or migrate about. In the slug, cells become prestalk (*red*) and prespore (*tan*) cells.

Migrating slug stage

Figure 20.15 Animated Life cycle of *Dictyostelium discoideum*, a cellular slime mold.

amoeboid ancestors. In *Dictyostelium*, a nucleotide called cyclic AMP is the signal that induces solitary amoeboid cells to stream together. It also triggers changes in gene expression. The changes cause some cells to differentiate into components of a stalk or into spores. Cyclic AMP also functions as the signal among cells in an animal body.

Intriguingly, molecular comparisons suggest that the fungi and animals are a monophyletic group. They also suggest that animals and fungi descended from an ancient amoebozoan-like ancestor.

Take-Home Message **What are amoebozoans?**

❭ Amoebozoans are heterotrophic protists with cells that lack a cell wall or pellicle, and so can constantly change shape.

❭ Amoebas live as single cells, usually in fresh water.

❭ Slime molds live on forest floors. Plasmodial slime molds feed as a big multinucleated mass. Cellular slime molds feed as single cells, but come together as a multicelled mass when conditions are unfavorable. Both types of slime molds form fruiting bodies that release spores.

▪ Harmful Algal Blooms (revisited)

Harmful algal blooms affect every coastal region of the United States. As described in Section 20.1, blooms of the dinoflagellate *Karenia brevis* are common in the Gulf of Mexico. Other toxin-producing dinoflagellates cause problems along the Atlantic coast. Along the Pacific coast, population explosions of diatoms that produce domoic acid have a similar effect. Like breve-toxin, domoic acid binds to nerve cells and interferes with their function. It kills fish, seabirds, and marine mammals. Humans who eat domoic acid–tainted shellfish or crabs typically suffer from headaches, dizziness, and con-fusion. High doses of domoic acid can kill brain cells, causing permanent loss of short-term memory and, in rare cases, coma or death.

Keeping harmful algal toxins out of the human food supply requires constant vigilance. These toxins have no color or odor, and are unaffected by heating or freezing. Government agencies use laboratory tests to detect harmful algal toxins in water samples and shellfish. When the toxins reach a threatening level, a shore is closed to shellfishing and warning signs are posted (*above right*).

Harmful algal blooms have devastating economic effects. A 2005 bloom of toxin-producing dinoflagellates caused an estimated $18 million loss in shellfish sales in Massachusetts alone. Blooms of *Karenia brevis* cost Florida $19–32 million a year.

Even blooms of nontoxic algae can have negative environmental effects. When huge numbers of these cells die, bacterial decomposers go to work breaking down their remains. Metabolic activity of the bacteria can deplete the water of oxygen, causing fish and other aquatic animals to smother.

What causes algal blooms? The nutrient content of the water plays a major role. Just like land plants, algae require certain nutrients. If you fertilize a houseplant or a lawn, a spurt of growth results. Similarly, the addition of nutrients to an aquatic habitat encourages algal cell divisions. We do not deliberately fertilize our waters, but fertilizers that drain off from croplands and lawns, and nutrient-rich wastes from animals at factory farms, get into rivers and are carried to the sea. Add sewage from human communities, and you have the nutrients required for runaway algal growth.

How Would You Vote? Preventing nutrient pollutions from entering coastal water could help reduce the incidence of algal blooms, but taking preventative mea-sures can be costly for farmers, developers, industries, and water treatment plants. Do you favor tightening regulations governing nutrient discharges into coastal waters? See CengageNow for details, then vote online (cengagenow.com).

Summary

Section 20.1 An **algal bloom** is a population explosion of an aquatic **protist**, or another aquatic microorganism. **Toxins** released during some algal blooms can harm wildlife and endanger human health.

Section 20.2 Protists are a collection of mostly single-celled eukaryotes. Many have chloroplasts that evolved from cyanobacteria or from another protist. The dominant stage of the life cycle may be haploid or diploid. The protists are not a natural group, but rather a collection of lineages, some of them only distantly related to one another.

Section 20.3 Flagellated protozoans are single cells with no cell wall. A protein covering, or **pellicle**, helps maintain the cell's shape.

Diplomonads and parabasalids are adapted to oxygen-poor habitats and do not have mitochondria. Members of both groups include species that infect humans.

Trypanosomes are parasites with a single mitochondrion. Insects transmit trypanosomes that cause human diseases. The related **euglenoids** typically live in fresh water. They have a **contractile vacuole** that squirts out excess water. Some have chloroplasts that evolved by secondary endosymbiosis from a green alga.

Section 20.4 Foraminiferans are single cells with a chalky shell. Deposits of their remains are mined for chalk and limestone. **Radiolarians** are single-celled and have a glassy shell. Both groups are marine heterotrophs and may be part of **plankton**. Long cytoplasmic extensions stick out through the porous shell and capture prey.

Sections 20.5, 20.6 Tiny sacs (alveoli) beneath the plasma membrane characterize **alveolates**. All members of this group are single-celled. **Dinoflagellates** are whirling aquatic heterotrophs and autotrophs with cellulose plates. Some are bioluminescent. The **ciliates** are aquatic predators and parasites with many cilia. **Apicomplexans** live as parasites in the cells of animals. Mosquitoes transmit the apicomplexan that causes malaria.

Section 20.7 Water molds are decomposers and parasites that grow as a mesh of absorptive filaments. Some parasitic species are important plant pathogens. Genetic similarities unite the water molds with diatoms and brown algae as stramenopiles.

Diatoms are silica-shelled photosynthetic cells. Deposits of ancient diatom shells are mined as diatomaceous earth. Diatoms contain the pigment fucoxanthin, as do **brown algae**, which include microscopic strands and giant kelps, the largest protists. Brown algae are the source of algins, compounds used as thickeners and emulsifiers.

Section 20.8 Most **red algae** are multicelled and marine. Accessory pigments called phycobilins allow them to capture light even in deep waters. Red algae are commercially important as the source of agar, carrageenan, and as dry sheets (nori) used for wrapping sushi. **Green algae** may be single cells, colonial, or multicelled. They are the closest relatives of land plants.

Like land plants, some of the multicelled algae have an **alternation of generations**. In this type of life cycle, two kinds of multicelled bodies form: a diploid, spore-producing **sporophyte** and a haploid gamete-producing **gametophyte**.

Section 20.9 Amoebozoans are heterotrophic free-living **amoebas** and slime molds. The **plasmodial slime molds** feed as a multinucleated mass. Amoeba-like cells of **cellular slime molds** aggregate when food is scarce and form multicelled fruiting bodies that disperse resting spores. Animal signaling mechanisms may have started in amoebozoan ancestors.

Self-Quiz   Answers in Appendix III

1. All flagellated protozoans _____ .
   - a. lack mitochondria
   - b. are photosynthetic
   - c. live as single cells
   - d. cause disease

2. Deposits of shells from ancient _____ are mined as chalk and limestone.
   - a. dinoflagellates
   - b. diatoms
   - c. radiolarians
   - d. foraminiferans

3. The presence of a contractile vacuole indicates that a single-celled protist _____ .
   - a. is marine
   - b. lives in fresh water
   - c. is photosynthetic
   - d. secretes a toxin

4. Cattle benefit when _____ in their gut help them digest plant material.
   - a. trypanosomes
   - b. diatoms
   - c. ciliates
   - d. foraminiferans

5. An insect bite can transmit a disease-causing _____ to a human host.
   - a. trypanosome
   - b. apicomplexan
   - c. ciliate
   - d. both a and b

6. _____ are the closest protistan relatives of the fungi and animals.
   - a. Stramenopiles
   - b. Radiolarians
   - c. Apicomplexans
   - d. Amoebozoans

7. Accessory pigments of _____ allow them to carry out photosynthesis at greater depths than other algae.
   - a. euglenoids
   - b. green algae
   - c. brown algae
   - d. red algae

8. The protist that causes malaria is _____ .
   - a. multicellular   b. a single cell   c. colonial

Data Analysis Activities

Tracking Changes in Algal Blooms Reports of fish kills along Florida's southwest coast date back to the mid-1800s, suggesting that algal blooms are a natural phenomenon in this region. However, University of Miami researchers suspected that a rise in nutrient delivery from land has contributed to an increase in the abundance of the dinoflagellate *K. brevis*. Since the 1950s, the population of coastal cities in southwestern Florida has soared and the amount of agriculture has increased. Did these changes add nutrients that favor *K. brevis* growth to nearshore waters? To find out, the researchers looked at records for coastal waters that have been monitored for more than 50 years. Figure 20.16 shows the average abundance and distribution of *K. brevis* during two time periods: 1954–1963 and 1994–2002.

1. How did the average concentration of *K. brevis* in waters less than 5 kilometers from shore change between the two time periods?
2. How did the average concentration of *K. brevis* in waters more than 25 kilometers from shore change between the two time periods?
3. Does this data support the hypothesis that human activity increased the abundance of *K. brevis* by adding nutrients to coastal waters?
4. Suppose the two graph lines became farther apart as distance from the shore increased. What would that suggest about the nutrient source?

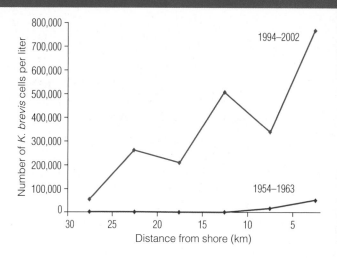

Figure 20.16 Average concentration of *K. brevis* cells detected at various distances from the shore during two time periods: 1954–1963 (*blue* line) and 1994–2002 (*red* line). Samples were collected from offshore waters between Tampa Bay and Sanibel Island.

9. The _____ are important plant pathogens.
   a. dinoflagellates   c. water molds
   b. ciliates   d. slime molds

10. Silica-rich shells of ancient diatoms are the source of diatomaceous earth that can be used _____ .
    a. to thicken foods   c. as a fertilizer
    b. as a gelatin substitute   d. as an insecticide

11. The sporophyte of a multicellular alga _____ .
    a. is haploid   c. produces spores
    b. is a single cell   d. produces gametes

12. Where would you find a cellular slime mold?
    a. on the forest floor   c. in an animal gut
    b. in a tropical sea   d. in a mountain lake

13. The organism that causes the sexually transmitted disease trichomoniasis is a _____ .
    a. flagellated protozoan   c. ciliate
    b. radiolarian   d. apicomplexan

14. All green algae _____ .
    a. have a cell wall   c. are multicellular
    b. are marine   d. all of the above

15. Match each item with its description.
    ____ diplomonad   a. cause of leishmaniasis
    ____ apicomplexan   b. silica-shelled producer
    ____ trypanosome   c. unwalled cell with pseudopods
    ____ diatom   d. anaerobic, no mitochondria
    ____ brown alga   e. closest relative of land plants
    ____ red alga   f. multicelled, with fucoxanthin
    ____ green alga   g. cause of malaria
    ____ amoeba   h. deep dweller with phycobilins

Additional questions are available on **CENGAGENOW**.

Critical Thinking

1. Suppose you vacation in a developing country where sanitation is poor. Having read about parasitic flagellates in water and damp soil, what would you consider safe to drink? What foods might be best to avoid or which food preparation methods might make them safe to eat?

2. The water in abandoned swimming pools often turns green. If you took a drop of this water and examined it under the microscope, you would see many green flagellated cells. What additional information would you need to determine which protist group the cells belong to?

3. The most common "snow alga," *Chlamydomonas nivalis* (*right*), lives on glaciers. It is a green alga, but it has so many carotenoid pigments that it appears red. Think about the intense sunlight striking its icy habitat 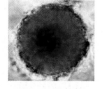 during the summer. Besides their role in photosynthesis, what other function might these light-absorbing carotenoids serve?

Animations and Interactions on **CENGAGENOW**:
> Body plan of a euglenoid; Action of a contractile vacuole; Body plan of a ciliate; Life cycle of an apicomplexan; Life cycle of a red alga; Life cycle of a cellular slime mold.

< Links to Earlier Concepts

Section 20.8 introduced the green algal group that is the closest relative of plants. Section 12.5 introduced you to the general plant life cycle. Here you will see specfic examples. You will learn about the evolution of cell walls with lignin (4.11) and a waxy cuticle perforated by stomata (6.8). You will learn about plant fossils (16.5) and coevolution with pollinators (17.13).

Key Concepts

Adaptive Trends Among Plants

Plants evolved from an aquatic green alga. Over time, new traits evolved that made them increasingly adapted to life in dry climates. The process of adaptation involved changes in plant structure, life cycle, and reproductive processes.

The Bryophytes

Bryophytes are three lineages of low-growing plants. They are the only modern plants in which the gamete-producing body dominates the life cycle, and the spore-producing body is dependent on it. Bryophytes require water for fertilization to occur, and disperse by releasing spores.

21 Plant Evolution

21.1 Speaking for the Trees

The world's great forests influence life in profound ways. They take up carbon dioxide from the air and release oxygen. At the same time, forests act like giant sponges that absorb water from rain, then release it slowly. By wicking up water and holding soil in place, forests prevent erosion, flooding, and sedimentation that can disrupt rivers, lakes, and reservoirs. Remove trees, and exposed soil washes away or loses nutrients.

All countries rely on their forests for fuel and lumber. The photo at the *left* shows a heavily logged forest in the Canadian province of British Columbia. The developing nations of Asia, Africa, and Latin America have fast-growing populations and high demands for food, fuel, and lumber. To meet their own needs and to create products for sale in the global market, these nations turn to their main resource: local tropical forests. These forests are a biological treasure; they have held 50 percent or more of all land-dwelling species for at least 10,000 years. In less than four decades, humans have destroyed more than half of the world's tropical forests by logging.

Deforestation affects evaporation rates, runoff, and regional patterns of rainfall. In tropical forests, most of the water vapor in the air is released from trees. In heavily logged regions, annual rainfall declines, Rain that does fall swiftly drains away from the exposed, nutrient-poor soil. As the region gets hotter and drier, the fertility and moisture content of the soil decrease.

Like many environmental issues, large-scale deforestation can seem like an overwhelming problem. Wangari Maathai, a Kenyan biologist, suggested a simple solution: Plant new trees (Figure 21.1). Maathai founded the Green Belt Movement, which began by giving poor women in Kenya the resources they need to plant and care for trees. Maathai reasoned that a woman who plants a tree seedling, cares for it, and watches it grow, will value and protect that tree. As many women plant trees, they will observe the improved environment and understand that they can each be a force for positive change.

Figure 21.1 Biologist and Nobel Peace Prize winner Wangari Maathai (*right*) holding one of the tree seedlings her Green Belt Movement raised and planted to restore forests in Kenya.

Maathai began her tree-planting campaign in 1977. By 2007, members of her organization had planted 40 million trees throughout Africa. These trees provide shade, reduce erosion, and have begun to restore the diverse forests that had been disappearing. Just as important, Maathai's practical solution to a seemingly overwhelming problem has inspired others to do what they can to stem the tide of environmental degradation. In 2004, the Nobel Committee honored Maathai's efforts with the Nobel Peace Prize. As Maathai recognized, we are dependent on plants for our well-being. This chapter explains how these green partners of ours evolved and diversified.

Seedless Vascular Plants
Vascular plants have internal pipelines that carry materials through the body and provide structural support. A body that makes spores dominates their life cycle. Seedless vascular plants such as ferns require water for fertilization to occur and they disperse by releasing spores.

Gymnosperms
Seed plants make pollen grains that allow fertilization to occur even in dry times. They also make eggs inside an ovule that develops into a seed. Gymnosperms such as pine trees are seed plants with "naked" seeds, meaning their seeds do not form inside an ovary.

Angiosperms
Angiosperms, or flowering plants, are the most recently evolved seed plants. They alone make flowers, and their seeds form inside a floral ovary that develops into a fruit. Angiosperms are the most widely dispersed and diverse group of plants.

> Plants evolved from a green alga and underwent an adaptive radiation on land. Structural and developmental changes adapted plants to increasingly drier habitats.
< Link to Green algae 20.8

The invasion of land got under way about 500 million years ago. Green algae grew at the water's edge, and one lineage, the charophytes, gave rise to the first land plants. The defining trait of land plants is a multicelled embryo that develops within tissues of the parental plant and is nourished by the parent. For this reason, the clade of land plants is called the **embryophytes**.

Structural Adaptations to Life on Land

Plants have unique traits that adapt them to life on land. A multicelled green alga absorbs the water it needs across its body surface. Water also buoys algal parts, helping an alga stand upright. In contrast, land plants face the threat of drying out and must hold themselves upright.

The earliest evolving plant lineages are referred to as bryophytes. Mosses are one example. Some bryophytes, and all other land plants, have a **cuticle**, a secreted waterproof coating that reduces evaporative water loss. Tiny pores called **stomata** (singular stoma) extend across the cuticle and outer leaf surface. Stomata open and close as needed to balance the competing demands for water conservation and gas exchange with air outside the plant.

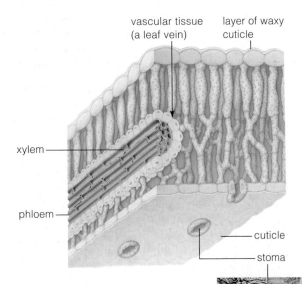

vascular tissue (a leaf vein)

layer of waxy cuticle

xylem

phloem

cuticle

stoma

Figure 21.2 Diagram of a vascular plant leaf in cross-section, showing some traits that contribute to the success of this group. The micrograph on the *right* shows a stoma, a pore at the surface of a squash plant's leaf. Photosynthetic cells on either side of the stoma control its opening and closing to regulate water loss and gas exchange.

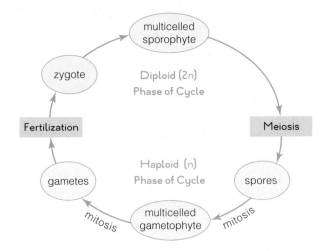

multicelled sporophyte

zygote

Diploid (2n) Phase of Cycle

Fertilization

Meiosis

Haploid (n) Phase of Cycle

gametes

spores

mitosis

multicelled gametophyte

mitosis

Figure 21.3 Generalized plant life cycle.

Bryophytes have structures that hold them in place, but these are not true roots; those evolved later. True roots anchor plants and also contain **vascular tissue**: internal pipelines that transport water and nutrients through a plant body. **Xylem** is the vascular tissue that distributes water and mineral ions. **Phloem** is the vascular tissue that distributes sugars made by photosynthetic cells. Of the 295,000 or so modern plant species, more than 90 percent have xylem and phloem. They are known as the **vascular plants**.

A variety of adaptations contribute to the success of vascular plants. Vascular tissues are reinforced by **lignin**, an organic compound that stiffens cell walls. Vascular tissues with lignin not only distribute materials, they also help plants stand upright and allow them to branch. Vascular tissue extends through stem, roots, and leaves. Evolution of leaves increased the surface area for capturing sunlight and exchanging gases. Figure 21.2 shows a cross-section through a vascular plant leaf.

The Plant Life Cycle

Life cycle changes also adapted vascular plants to life in drier habitats. Like some algae, all land plants have an alternation of generations (Figure 21.3). A diploid body, the sporophyte, makes spores. A plant spore is a walled haploid cell. The spore germinates, then divides by mitosis and develops into a haploid body, or gametophyte. A gametophyte makes gametes (eggs and sperm). When gametes come together at fertilization, they form a zygote that develops into a new sporophyte.

The gametophyte is the largest and longest-lived part of the bryophyte life cycle, but the sporophyte dominates the life cycle of vascular plants. Dry conditions favor an increased emphasis on spore production; spores withstand drying out better than gametes do.

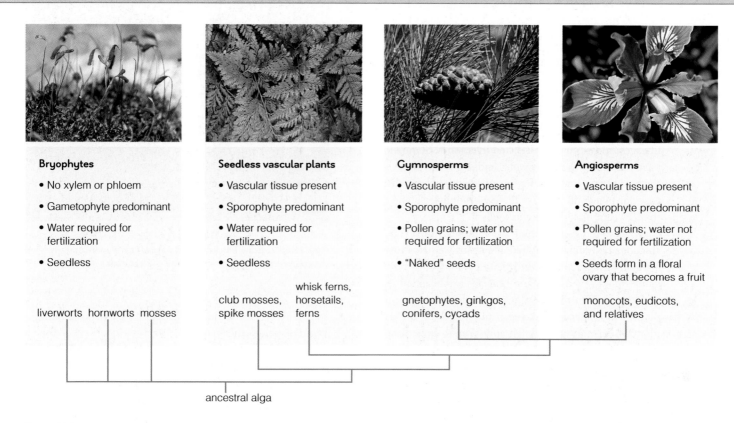

Bryophytes
- No xylem or phloem
- Gametophyte predominant
- Water required for fertilization
- Seedless

liverworts hornworts mosses

Seedless vascular plants
- Vascular tissue present
- Sporophyte predominant
- Water required for fertilization
- Seedless

club mosses, whisk ferns,
spike mosses horsetails,
ferns

Gymnosperms
- Vascular tissue present
- Sporophyte predominant
- Pollen grains; water not required for fertilization
- "Naked" seeds

gnetophytes, ginkgos,
conifers, cycads

Angiosperms
- Vascular tissue present
- Sporophyte predominant
- Pollen grains; water not required for fertilization
- Seeds form in a floral ovary that becomes a fruit

monocots, eudicots,
and relatives

ancestral alga

Figure 21.4 Traits of the four major plant groups and relationships among them. Note that the bryophytes are a collection of lineages, rather than a clade.

Pollen and Seeds

Evolution of new reproductive traits gave the lineage of vascular plants known as the seed plants, a competitive edge in many habitats. Seed plants do not release spores. Instead, their spores give rise to gametophytes inside specialized structures on the sporophyte body.

A **pollen grain** is a walled, immature male gametophyte of a seed plant. It is released into the environment, where it can be transported to another plant by winds or animals even in the driest of times. By contrast, plants that do not make pollen (bryophytes and seedless vascular plants) can only reproduce when a film of water allows their sperm to swim to eggs.

Fertilization of a seed plant also takes place in structures on the sporophyte body. It results in development of a **seed**—an embryo sporophyte with a supply of nutritive tissue inside a waterproof seed coat. Seed plants disperse a new generation by releasing seeds.

There are two lineages of seed plants. The gymnosperms were the first to evolve. A pine tree is a familiar gymnosperm. Angiosperms, the flowering plants, branched off from a gymnosperm lineage and became the most widely distributed and diverse plant group. They alone make flowers and release their seeds inside a fruit. Figure 21.4 summarizes the relationships among the major plant groups and the traits of each group.

---

cuticle Secreted covering at a body surface.
embryophyte Member of the land plant clade.
lignin Material that stiffens cell walls of vascular plants.
phloem Plant vascular tissue that distributes sugars.
pollen grain Male gametophyte of a seed plant.
seed Of seed plants, embryo sporophyte and nutritive tissue inside a waterproof coat; a mature ovule.
stoma Opening across a plant's cuticle and epidermis; can be opened for gas exchange or closed to prevent water loss.
vascular plant Plant with xylem and phloem.
vascular tissue Internal pipelines of xylem and phloem in the body of a vascular plant.
xylem Plant vascular tissue that distributes water and dissolved mineral ions.

Take-Home Message **What adaptive trends shaped plants?**

> Early plant lineages (bryophytes) have a life cycle dominated by a haploid gametophyte. In most bryophytes, a waxy cuticle and stomata minimize evaporative water loss.

> Most plants have a life cycle dominated by a sporophyte with vascular tissue. Like bryophytes, the oldest vascular plant lineages have flagellated sperm and disperse by releasing spores.

> Seed plants, the most recently evolved plant lineage, can reproduce even in dry times because they make pollen. They disperse by releasing seeds, not spores.

> Bryophytes are low-growing plants that require the presence of water droplets to reproduce sexually.
> A haploid gametophyte dominates the bryophyte life cycle.

Bryophyte Characteristics

The term **bryophyte** refers to members of three separate lineages: mosses, hornworts, and liverworts. Some mosses have tubes that transport water and sugar, but none of the bryophytes have the lignin-stiffened pipelines that the vascular plants do. As a result, few bryophytes stand more than 20 centimeters (8 inches) tall.

The haploid gametophyte is the largest, most conspicuous body in a bryophyte life cycle. Gametes form in a chamber (a gametangium) that develops in or on the gametophyte's surface. Some species are bisexual; eggs and sperm form in different gametangia on the same plant. Other species make eggs and sperm on different plants. Sperm are flagellated and swim to eggs, although insects and mites sometimes aid their journey.

The diploid sporophyte is unbranched and remains attached to the gametophyte even when mature. It makes

Figure 21.6 Cutting blocks of peat for use as fuel in Ireland. Peat is the compressed, carbon-rich remains of peat moss.

spores by meiosis inside a chamber called a sporangium. Wind disperses the spores.

Many of the 24,000 or so bryophytes live in constantly moist places. Others can tolerate periodic drought. They shrivel up and become dormant when water gets scarce,

Figure 21.5 **Animated** Life cycle of a common moss (*Polytrichum*).

❶ The leafy green part of a moss is the gametophyte.

❷ It supports a sporophyte (stalk and capsule).

❸ Spores form by meiosis in the capsule, are released, and drift with the winds.

❹ Spores develop into gametophytes that produce eggs or sperm in gametangia at their tips.

❺ Sperm released from tips of sperm-producing gametophytes swim through water to eggs at tips of egg-producing gametophytes.

❻ Fertilization produces a zygote.

❼ The zygote grows and develops into a sporophyte while remaining attached to and nourished by its egg-producing parent.

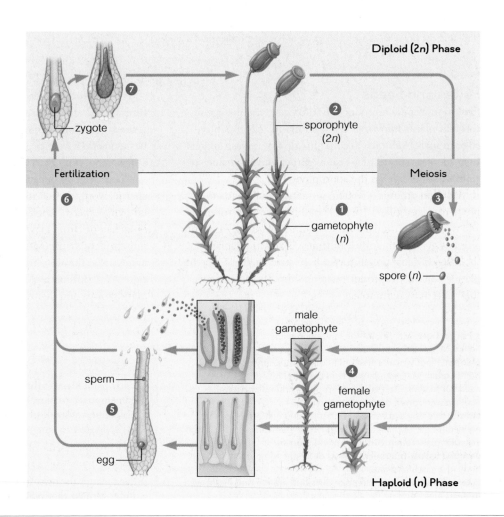

then resume growth when rains return. Drought tolerance and wind-dispersed spores make bryophytes important pioneer species in rocky places. Decomposition of the bryophytes helps form and improve soil, allowing less hardy plants to take root.

Most bryophytes can reproduce asexually by fragmentation. A piece breaks off and develops into a new plant. Unlike sexual reproduction, fragmentation can take place even during dry times.

Mosses

Mosses are the most familiar bryophytes. Figure 21.5 shows the life cycle of one common group (*Polytrichum*). The gametophyte has leaflike green parts that grow from a central stalk ❶. Threadlike **rhizoids** hold the gametophyte in place. Unlike the roots of vascular plants, rhizoids do not distribute water or nutrients. These resources must be absorbed across the gametophyte's leafy surface.

The moss sporophyte spends its life attached to the gametophyte, which it depends on for nourishment. The sporophyte consists of a capsule (a sporangium) on a stalk ❷. Meiosis of cells inside the capsule yields spores ❸. After dispersal by the wind, a spore germinates and grows into a gametophyte. The moss we are using as our example has separate sexes, with each gametophyte producing eggs or sperm ❹. Other mosses are bisexual. In either case, rain triggers the release of flagellated sperm that swim through a film of water to eggs ❺. Fertilization occurs in the egg chamber, producing a zygote ❻ that develops into a new sporophyte ❼.

Mosses include about 15,000 species. Of these, the 350 or so species of peat moss (*Sphagnum*) are the most economically important. Peat mosses are the dominant plants in peat bogs that cover hundreds of millions of acres in high-latitude regions of Europe, Asia, and North America. Many peat bogs have persisted for thousands of years, and layer upon layer of plant remains have become compressed as a carbon-rich material called **peat**. Blocks of peat are cut, dried, and burned as fuel, especially in Ireland (Figure 21.6).

Freshly harvested peat moss is also an important commercial product. The moss is dried and added to planting mixes to help soil retain moisture. Dead strands of peat moss absorb water, then release it slowly, keeping plant roots moist.

Figure 21.7 A liverwort (*Marchantia*).
A Ribbonlike gametophyte with stalked structures that produce eggs for sexual reproduction.
B Asexual reproduction. Clumps of cells form by mitosis in cups on the gametophyte surface. After raindrops knock the clumps out of the cup, they develop into new plants.

Figure 21.8 Hornwort, with a photosynthetic sporophyte that can survive even after death of the gametophyte.

Liverworts and Hornworts

Liverworts commonly grow in moist places, often alongside mosses. In most of the 6,000 or so species, the gametophyte is flattened and attaches to soil by rhizoids. The widespread genus *Marchantia* forms eggs and sperm on separate plants in gametangia that are elevated on stalks above the main body (Figure 21.7A). In addition to fragmentation, members of this genus also reproduce asexually by producing small clumps of cells in cups on the gametophyte surface (Figure 21.7B). Some *Marchantia* species can be pests in commercial greenhouses. Liverwort infestations are difficult to eradicate because the tiny spores can persist even after all plants are killed.

The 150 species of hornworts are named for a pointy, hornlike sporophyte that can be several centimeters tall (Figure 21.8). The base of the sporophyte is embedded in gametophyte tissues and spores form in a capsule at its tip. Unlike the sporophytes of mosses and liverworts, those of hornworts grow continually from their base, and can survive even after the death of the gametophyte.

---

bryophyte Member of an early plant lineage with a gametophyte-dominant life cycle; for example, a moss.
moss Nonvascular plant with a leafy green gametophyte and an attached, dependent sporophyte consisting of a capsule on a stalk.
peat Carbon-rich moss remains; can be dried for use as fuel.
rhizoid Threadlike structure that anchors a bryophyte.

Take-Home Message **What are bryophytes?**

❯ Bryophytes include three lineages of plants: mosses, liverworts, and hornworts. All are low-growing plants with no lignin-reinforced vascular tissues. All have flagellated sperm that require a film of water to swim to eggs, and all disperse by releasing spores.

❯ Among land plants, only the bryophytes have a life cycle in which the gametophyte is dominant. The sporophyte remains attached to the gametophyte even when mature.

> Lignified vascular tissue allowed the evolution of larger sporophyte bodies in the ferns and their relatives.
< Link to Plant cell walls 4.11

Some mosses have internal pipelines that transport fluid within their body. However, only the vascular plants have lignin-strengthened vascular tissue with xylem and phloem. This innovation in support and plumbing allowed the evolution of the larger, more complex sporophytes that dominate the life cycle of all vascular plants.

Vascular plants evolved from a bryophyte, and the oldest lineages, the **seedless vascular plants**, share traits with this group. They have flagellated sperm that swim to eggs, and they disperse a new generation by releasing spores, not seeds. Club mosses, horsetails, and ferns are examples of seedless vascular plants.

Figure 21.10 Horsetail (*Equisetum*). **A** Photosynthetic stems with leaflike branches. **B** Spore-bearing strobilus.

Club Mosses

The 400 or so species of club mosses occur worldwide. Members of the genus *Lycopodium* frequently grow on the floor of temperate forests. The sporophytes resemble miniature pine trees and are known as ground pines (Figure 21.9A). The plant has a horizontal stem, or **rhizome**, that runs along the ground. Roots and upright stems with tiny leaves grow from the rhizome. When a *Lycopodium* plant is several years old, it begins making spores seasonally. The spores form on a strobilus, a soft cone-shaped structure composed of modified leaves.

Lycopodium is gathered from the wild for a variety of uses. Stems are used in wreaths and bouquets. Spores have a waxy coating and are sold as "flash powder" for creating special effects. When spores are sprayed out as a fine mist and ignited, they produce a bright flame (Figure 21.9B). *Lycopodium* may also have medicinal properties. It is used as an herbal medicine, and compounds made by the plant may help fight cancer or memory loss.

Horsetails and Rushes

The 25 or so species of the genus *Equisetum* are commonly known as horsetails or rushes. The sporophyte has a rhizome that gives rise to underground roots and to upright hollow stems with tiny nonphotosynthetic leaves. In some species the stems have few or no branches, while other species have thin, leafless branches that extend out in whorls around the stem (Figure 21.10A). Depending on the species, strobili form at tips of photosynthetic stems or on special chlorophyll-free stems (Figure 21.10B).

Silica deposits inside cell walls support the plant and give stems a sandpapery texture that helps deter grazers. Before scouring powders and pads were widely available, people used some *Equisetum* stems as pot scrubbers; thus the common name "scouring rush."

Ferns—The Most Diverse Seedless Plants

With 12,000 or so species, ferns are the most diverse seedless vascular plants. All but 380 or so species live in the tropics. A typical fern sporophyte has fronds (leaves) and roots that grow from rhizomes (Figure 21.11 ❶). Ferns do not have strobili. Instead, **sori** (singular, sorus), clusters of spore-forming capsules, develop on the lower surface of fronds ❷. When the capsules pop open, wind disperses

Figure 21.9 Club moss. **A** *Lycopodium* sporophyte. Waxy yellow spores are produced on the surface of strobili, conelike structures composed of modified leaves. **B** *Lycopodium* spores in use as a special effect at a concert. The towers of fire are created by a device that blows a puff of spores out and ignites them.

epiphyte Plant that grows on another plant but does not harm it.
rhizome Stem that grows horizontally along or under the ground.
seedless vascular plant Plant such as a fern or horsetail that has vascular tissue and disperses by producing spores.
sorus Cluster of spore-producing capsules on a fern leaf.

Figure 21.11 Animated
Life cycle of a fern.

1 The familiar leafy form is the diploid sporophyte.

2 Meiosis in cells on the underside of fronds produces haploid spores.

3 After their release, the spores germinate and grow into tiny gametophytes that produce eggs and sperm.

4 Sperm swim to eggs and fertilize them, forming a zygote.

5 The sporophyte begins its development attached to the gametophyte, but it continues to grow and live independently after the gametophyte dies.

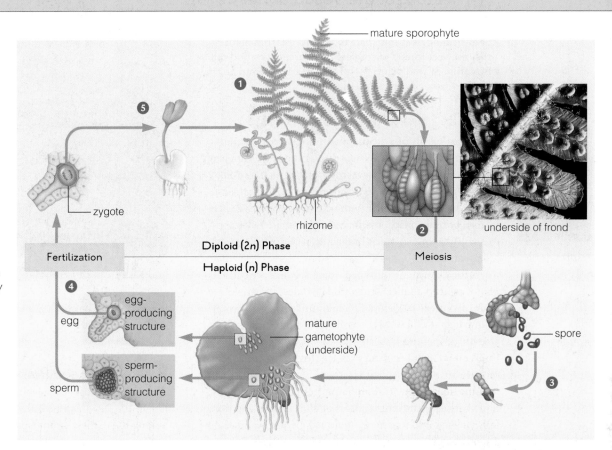

the spores. Each spore grows into a tiny gametophyte a few centimeters wide **3**. Eggs and sperm form in chambers on the gametophyte's underside. When it rains, sperm swim through a film of water to reach and fertilize the eggs **4**. The resulting zygote develops into a new sporophyte, and its parental gametophyte dies **5**.

Fern sporophytes vary greatly in size (Figure 21.12). Some floating ferns have fronds only 1 millimeter long, but tree ferns can be 25 meters (80 feet) high. Many tropical ferns are **epiphytes**, plants that attach to and grow on a trunk or branch of another plant but do not withdraw any nutrients from it.

Ferns are sold as potted plants, and the trunks of some tree ferns are shredded to produce a potting medium used to grow orchids. Fiddleheads, the young, tightly coiled fronds of some ferns, are harvested from the wild as food.

Take-Home Message What are seedless vascular plants?

> Sporophytes with vascular tissue in their roots, stems, and leaves dominate the life cycle of seedless vascular plants.

> Tiny, independent gametophytes produce sperm that swim to eggs through films or droplets of water.

> Ferns are the most diverse seedless vascular plants.

Figure 21.12 A sampling of fern diversity. **A** The floating fern *Azolla pinnata*. The whole plant is not as wide a finger. Chambers in the leaves shelter nitrogen-fixing cyanobacteria. Southeast Asian farmers grow this species in rice fields as a natural alternative to chemical fertilizers. **B** Bird's nest fern (*Asplenium nidus*), one of the epiphytes. **C** Lush forest of tree ferns (*Cyathea*) in Australia's Tarra-Bulga National Park. A tree fern's "trunk" is an enlarged rhizome.

> Seed-bearing trees dominate today's forests, but before they evolved, forests of seedless vascular plants stood tall.
> Evolution of seed plants set off a great adaptive radiation.
⟨ Link to Geologic time scale 16.6

From Tiny Branchers to Coal Forests

Fossil *Cooksonia*

The oldest fossils of vascular plants are spores that date to about 450 million years ago, during the late Ordovician period (Figure 21.13). Early vascular plants such as *Cooksonia* (*left*) stood only a few centimeters high and had a simple branching pattern, with no leaves or roots. Spores formed at branch tips. By the early Devonian, taller species with a more complex branching pattern were common worldwide (Figure 21.14).

As the Devonian continued, some seedless vascular plants evolved a taller structure. The oldest forest we know about existed about 385 million years ago in what is now upstate New York. Fossil stumps and fronds discovered at this site indicate that the main plants of this forest stood about 8 meters (26 feet high) and resembled modern tree ferns in their structure.

Later, during the Carboniferous period (350–299 million years ago), ancient relatives of club mosses and horsetails evolved into massively stemmed giants (Figure 21.15). Some stood 40 meters (more than 130 feet) high. After these forests first formed, climates changed, and the sea level rose and fell many times. When the waters receded, the forests flourished. After the sea moved back in, submerged trees became buried in sediments that protected them from decomposers. Layers of sediments accumulated one on top of the other. Their weight squeezed the water out of the saturated, undecayed remains, and the compaction generated heat. Over time, pressure and heat transformed the compacted organic remains into **coal**.

Coal is one of our premier fossil fuels. It took millions of years of photosynthesis, burial, and compaction to form each seam of coal. Often you will hear about annual production rates for coal or some other fossil fuel. In fact, we do not "produce" these materials. No one is making more coal. Coal is a nonrenewable source of energy, and one that we are well on the way to depleting.

Figure 21.14 Painting of a Devonian swamp, with branching seedless vascular plants.

Rise of the Seed Plants

The first gymnosperms evolved from a seedless ancestor late in the Devonian period. Cycads and ginkgos were among the earliest gymnosperm lineages. Conifers such as pine trees evolved a bit later. Angiosperms, or flowering plants, descended from a gymnosperm ancestor by about 120 million years ago, during the reign of the dinosaurs (Figure 21.16).

Reproductive traits of seed-bearing plants put them at an advantage in dry habitats. Gametophytes of seedless vascular plants develop from spores that were released into the environment. By contrast, gametophytes of seed plants form inside reproductive parts on a sporophyte body (Figure 21.17). Sperm-producing gametophytes (pollen grains) develop from spores that form inside **pollen sacs**. These spores are called **microspores**. Egg-producing gametophytes develop from **megaspores** that form inside a protective chamber, or **ovule**.

A sporophyte releases pollen grains, but holds onto its eggs. Wind or animals can deliver pollen from one seed plant to the ovule of another, a process called **pollination**.

Bryophytes evolve, diversify; seedless vascular plants evolve.	Diversification of seedless vascular plants.	First treelike plants (fern relatives), first seed plants.	Giant horsetails, club mosses, and relatives in swamp forests. Conifers arise late in the period.	Ginkgos, cycads evolve. Most horsetails and club mosses die off by end of the period.	Adaptive radiations of ferns, cycads, conifers; by start of Cretaceous, conifers are dominant trees.	Flowering plants appear in the early Cretaceous, undergo adaptive radiation, and become dominant.
Ordovician	Silurian	Devonian	Carboniferous	Permian	Triassic	Jurassic	Cretaceous	Tertiary
488	443	416	359	299	251	200	146	66

Millions of years ago (mya)

Figure 21.13 A time line for major events in plant evolution.

Figure 21.15 Painting of a Carboniferous "coal forest." An understory of ferns is shaded by tree-sized relatives of modern horsetails and club mosses.

Figure 21.16 Early angiosperms such as magnolias (*foreground*) evolved while dinosaurs walked on Earth.

Because the sperm of seed plants do not need to swim through a film of water to reach eggs, these plants can reproduce even during dry times.

After pollination and fertilization, the ovule develops into a seed. Dispersing seeds puts the seed plants at an advantage over spore-bearing plants. A seed contains a multicelled embryo sporophyte and enough food to nourish it until conditions favor growth. By contrast, seedless plants release single-celled spores without stored food. The spores cannot wait out a dry period.

Structural traits also gave seed plants an advantage over seedless lineages. Many seed plants undergo secondary growth (growth in diameter) and produce wood. Wood is tissue produced by adding new lignin-stiffened xylem. The giant nonvascular plants that lived in Carboniferous forests did undergo secondary growth. However, their trunks were softer and more flexible than those of seed plants. Only seed plants produce true wood.

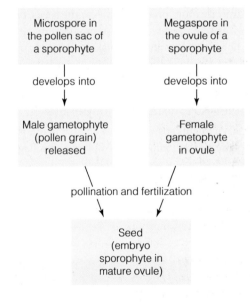

Figure 21.17 How a seed forms.

coal Fossil fuel formed over millions of years by compaction and heating of plant remains.
megaspore Haploid spore formed in ovule of seed plants; develops into an egg-producing gametophyte.
microspore Haploid spore formed in pollen sacs of seed plants; develops into a sperm-producing gametophyte (a pollen grain).
ovule Of seed plants, reproductive structure in which egg-bearing gametophyte develops; after fertilization, matures into a seed.
pollen sac Of seed plants, reproductive structure in which sperm-bearing gametophytes (pollen grains) develop.
pollination Delivery of a pollen grain to the egg-bearing part of a seed plant.

Take-Home Message What are the major events in the evolution of vascular plants?

❭ Seedless vascular plants arose before the seed-bearing lineages. They were widespread during the Carboniferous.

❭ Seed plants produce sperm-producing gametophytes as pollen grains that wind or animals can carry to eggs. As a result, seed plants can reproduce under drier conditions than other plants.

❭ The egg-bearing gametophyte of a seed plant forms in an ovule that becomes a seed after fertilization. Dispersing seeds rather than spores increases the reproductive success of seed plants in dry climates.

› The gymnosperms are the less diverse of the two modern lineages of seed-bearing plants.

Gymnosperms are vascular seed plants that produce seeds on the surface of ovules. The seeds are said to be "naked" because unlike those of angiosperms they are not inside a fruit. (*Gymnos* means naked and *sperma* is taken to mean seed.) However, many gymnosperms enclose their seeds in a fleshy or papery covering.

The Conifers

The 600 or so species of **conifers** are trees and shrubs with woody cones. Conifers typically have needlelike or scalelike leaves with a thick cuticle. They tend to be more resistant to drought and cold than flowering plants and are the main plants in cool Northern Hemisphere forests. Conifers include the tallest trees in the Northern Hemisphere (redwoods), as well as the most abundant (pines). They also include the long-lived bristlecone pines (Figure 21.18A). One of these trees is now 4,600 years old.

We use fir bark as mulch in our gardens, use oils from cedar in cleaning products, and eat the seeds, or "pine nuts," of some pines. Pines also provide lumber for building homes, and some pines make a sticky resin that deters insects from tunneling into them. We use this resin to make turpentine, a paint solvent.

Lesser Known Gymnosperms

Cycads and ginkgos were most diverse in dinosaur times. They are the only modern seed plants with flagellated sperm. Sperm emerge from pollen grains, then swim in fluid produced by the plant's ovule.

The 130 species of modern **cycads** live mainly in the dry tropics and subtropics (Figure 21.18B). Cycads often resemble palms but the two groups are not close relatives. The "sago palms" commonly used in landscaping and as houseplants are actually cycads.

The only living **ginkgo** species is *Ginkgo biloba*, the maidenhair tree (Figure 21.18C,D). It is a deciduous native of China. Deciduous plants drop all their leaves at once seasonally. The ginkgo's attractive fan-shaped leaves and resistance to insects, disease, and air pollution make it a popular tree along urban streets. Dietary supplements made from ginkgo may slow memory loss in people who have Alzheimer's disease.

Gnetophytes include tropical trees, leathery vines, and desert shrubs. Extracts from the stems of *Ephedra* (Figure 21.18E) are sold as an herbal stimulant and a weight loss aid. Such supplements can be dangerous; a few people have died while using them.

A Representative Life Cycle

A ponderosa pine tree is a sporophyte, and its life cycle is typical of conifers (Figure 21.19). Pollen cones and ovule-bearing cones develop on the same tree. Ovules form on cone scales ❶. Inside the ovule, a megaspore forms by meiosis and develops into a female gametophyte ❷. Male cones hold pollen sacs ❸, where microspores form and develop into pollen grains ❹.

The pollen grains are released and drift with the winds. Pollination occurs when one lands on an ovule ❺. The pollen grain germinates: Some cells develop into a pollen tube that grows through the ovule tissue and delivers sperm to the egg ❻. Pollen tube growth is an astonishingly leisurely process. It typically takes about a year for the tube to grow through the ovule tissue to the egg. When fertilization finally occurs, it produces a zygote ❼. Over about six months, the zygote develops into an

Figure 21.18 Gymnosperm diversity. **A** Bristlecone pine on a mountaintop in the Sierra Nevada. **B** An Australian cycad with its fleshy seeds. **C** Ginkgo's fan-shaped leaves and **D** its fleshy seeds. **E** The gnetophyte *Ephedra*. Yellow structures on stems are pollen-bearing cones.

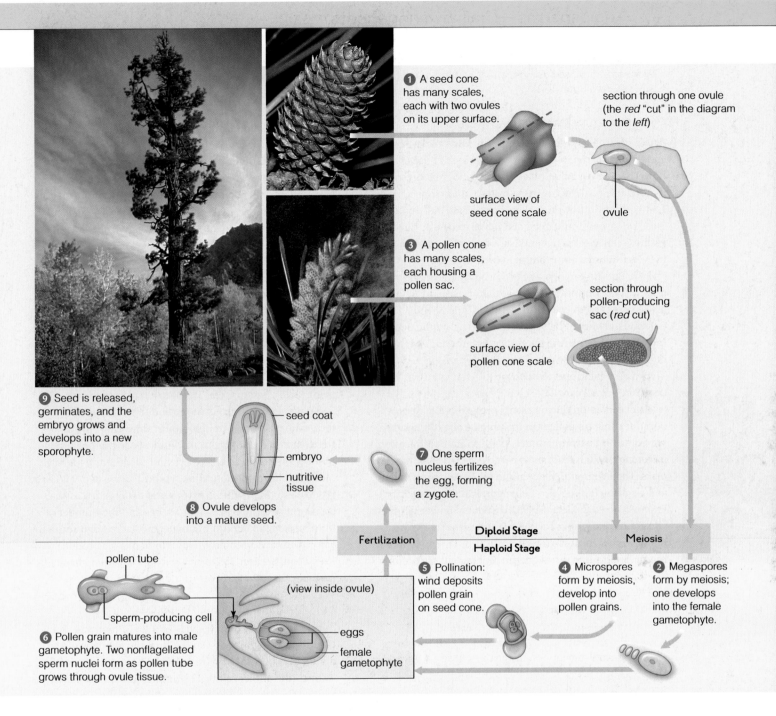

1 A seed cone has many scales, each with two ovules on its upper surface.

surface view of seed cone scale

section through one ovule (the *red* "cut" in the diagram to the *left*)

ovule

3 A pollen cone has many scales, each housing a pollen sac.

surface view of pollen cone scale

section through pollen-producing sac (*red* cut)

9 Seed is released, germinates, and the embryo grows and develops into a new sporophyte.

seed coat

embryo

nutritive tissue

8 Ovule develops into a mature seed.

7 One sperm nucleus fertilizes the egg, forming a zygote.

| Fertilization | Diploid Stage | Meiosis |
| --- | Haploid Stage | --- |

pollen tube

sperm-producing cell

6 Pollen grain matures into male gametophyte. Two nonflagellated sperm nuclei form as pollen tube grows through ovule tissue.

(view inside ovule)

eggs

female gametophyte

5 Pollination: wind deposits pollen grain on seed cone.

4 Microspores form by meiosis, develop into pollen grains.

2 Megaspores form by meiosis; one develops into the female gametophyte.

embryo sporophyte that, along with tissues of the ovule, becomes a seed **8**. The seed is released, germinates, then grows and develops into a new sporophyte **9**.

Figure 21.19 Animated Life cycle of a conifer, the ponderosa pine.

>> Figure It Out Does the pollen tube grow through tissue of a male cone or a female cone?

Answer: A female cone. It grows after a pollen grain alights on a female cone.

conifer Gymnosperm with nonmotile sperm and woody cones; for example, a pine.
cycad Tropical or subtropical gymnosperm with flagellated sperm, palmlike leaves, and fleshy seeds.
ginkgo Deciduous gymnosperm with flagellated sperm, fan-shaped leaves, and fleshy seeds.
gnetophyte Shrubby or vinelike gymnosperm, with nonmotile sperm; for example *Ephedra*.
gymnosperm Seed plant that does not make flowers or fruits.

Take-Home Message What are gymnosperms?

> Gymnosperms are seed-bearing plants. Their eggs and seeds form on the surface of an ovule, not in ovaries as in flowering plants.

> All gymnosperms produce pollen. In the cycads and ginkgos, sperm emerge from pollen grains and swim through fluid released by the ovule. In other groups, the sperm are nonmotile.

> Angiosperms are seed plants that produce flowers. Their seeds form inside an ovary that matures into a fruit.
< Link to Coevolution 17.13

Angiosperm Traits and Diversity

Angiosperms are vascular seed plants that make flowers and fruits. Nearly 90 percent of modern plant species are flowering plants. What accounts for their great diversity? For one thing, many can grow faster than gymnosperms. Think of how a flowering plant such as a dandelion or grass can grow from a seed and make seeds of its own within a few weeks. In contrast, most gymnosperms grow for years before they produce their first seeds.

Flowers also provide a selective advantage. A **flower** is a specialized reproductive shoot that consists of modified leaves arranged in concentric whorls (Figure 21.20). The outermost whorl is the sepals. Inside that are the petals, then the pollen-bearing parts called **stamens**. The innermost part of the flower is the **carpel**, where eggs form. The base of the carpel is the **ovary**, a chamber that contains one or more ovules.

After pollen-producing plants evolved, some insects began feeding on pollen. In the process, they sometimes served as **pollinators**: animals that facilitate fertilization by moving pollen from male parts of a flower to female parts. The bee in the photo at the *left* is gathering pollen and also functioning as a pollinator. Plants and their pollinators coevolve (Section 17.13). In many flowering plants vibrant colors, strong fragrances, and plentiful nectar that attracted insects evolved. Instead of randomly looking for food, some insects began seeking out particular flowers. These insects made more direct pollen deliveries. Not coincidentally, the great adaptive radiation of flowering plants coincided with an adaptive radiation of insects. Some plants such as grasses continued to rely on wind for pollination. Today, these plants have small, pale, unscented flowers without any fragrance or nectar.

After fertilization, an ovule matures into a seed and the ovary around it becomes a **fruit**. (*Angio*– means enclosed chamber, and *sperma*, seed.) Some gymnosperms have seeds with winglike structures or a fleshy coat, but fruits come in a far greater variety. Many fruits attract animals with sweet taste and bright color. Other fruits float in water, ride the winds, or stick to animal fur. All of these dispersal-related traits help distribute seeds to new habitats where they can thrive, thus increase a parent plant's reproductive success.

Major Lineages

The vast majority of flowering plants belong to one of two lineages. The 80,000 or so **monocots** include orchids, palms, lilies, and grasses. The 170,000 or so **eudicots** include most herbaceous (nonwoody) plants such as tomatoes, cabbages, roses, poppies, most flowering shrubs and trees, and cacti. Monocots and eudicots are named for the number of seed leaves, or cotyledons, their embryos have. Monocots have one cotyledon; eudicots have two. The two groups also differ in some structural details such as the arrangement of their vascular tissues and the number of flower petals. In addition, some eudicots undergo secondary growth and become woody. No monocots produce true wood. Section 25.2 discusses the traits that characterize eudicots and monocots in more detail.

A Representative Life Cycle

Figure 21.21 shows the life cycle of a lily, a monocot. A lily plant is a sporophyte ❶. Pollen forms in stamens of the lily flower ❷. A thin filament holds the upper part of the stamen (the anther) aloft. The anther contains pollen sacs where microspores develop into pollen grains. Eggs form

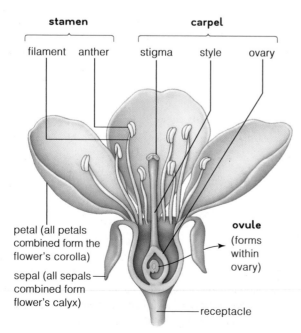

Figure 21.20
Animated
A cherry flower, with pollen-bearing stamens and ovule-bearing carpels. Flowers of some other plants have either carpels or stamens.

stamen
filament   anther

carpel
stigma   style   ovary

petal (all petals combined form the flower's corolla)

sepal (all sepals combined form flower's calyx)

ovule (forms within ovary)

receptacle

---

angiosperms Largest seed plant lineage. Only group that makes flowers and fruits.
carpel Female reproductive part of a flower.
endosperm Nutritive triploid tissue in angiosperm seeds.
eudicots Largest lineage of angiosperms; includes herbaceous plants, woody trees, and cacti.
flower Specialized reproductive shoot of a flowering plant.
fruit Mature flowering plant ovary; encloses a seed or seeds.
monocots Lineage of angiosperms that includes grasses, orchids, and palms.
ovary Of flowering plants, floral chamber that encloses ovule.
pollinator Animal that moves pollen, thus facilitating pollination.
stamen Male reproductive part of a flower.

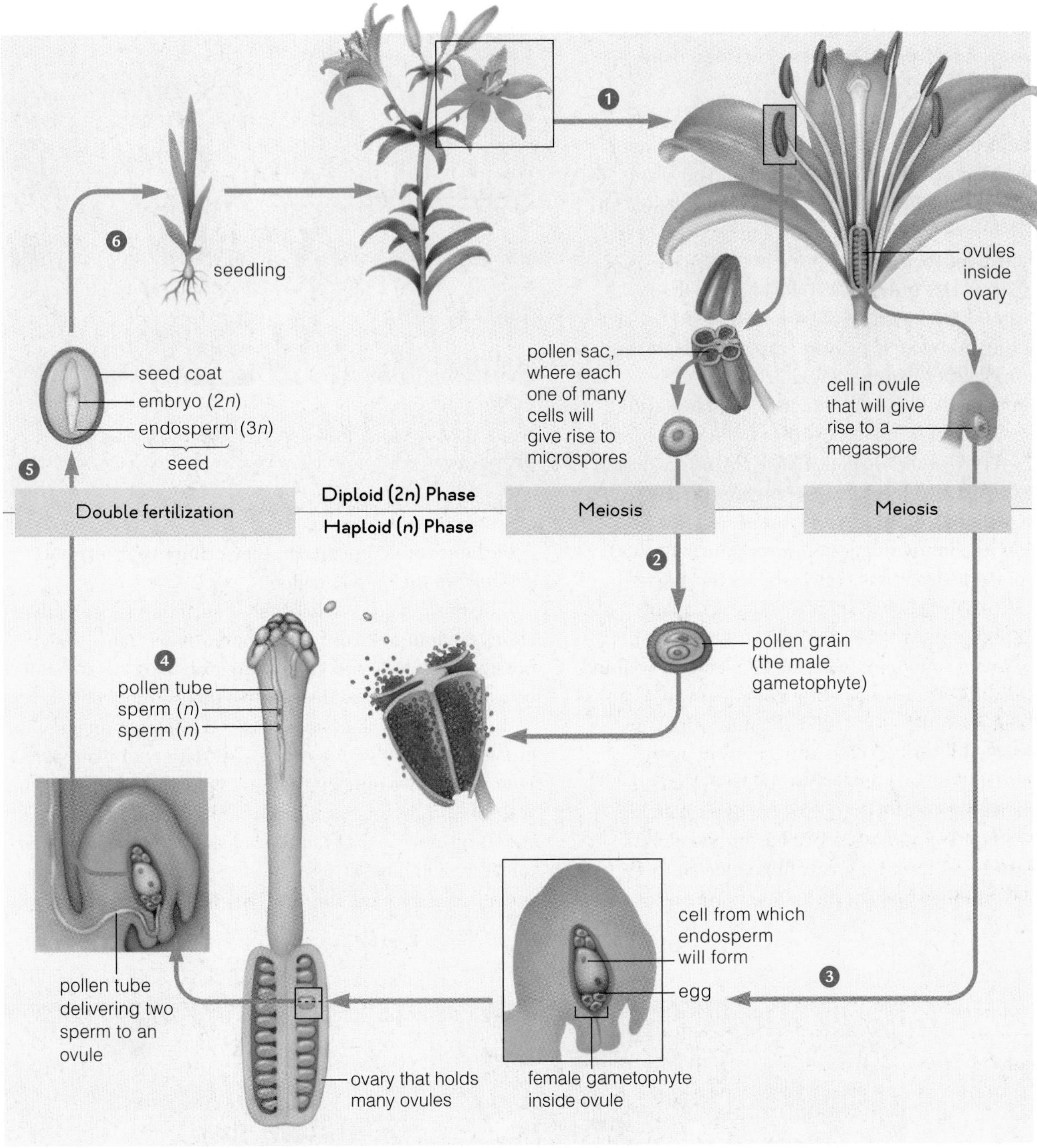

1
6
seedling
ovules inside ovary
seed coat
embryo (2n)
endosperm (3n)
seed
5
pollen sac, where each one of many cells will give rise to microspores
cell in ovule that will give rise to a megaspore
Double fertilization
Diploid (2n) Phase
Haploid (n) Phase
Meiosis
Meiosis
2
4
pollen tube
sperm (n)
sperm (n)
pollen grain (the male gametophyte)
cell from which endosperm will form
egg
3
pollen tube delivering two sperm to an ovule
ovary that holds many ovules
female gametophyte inside ovule

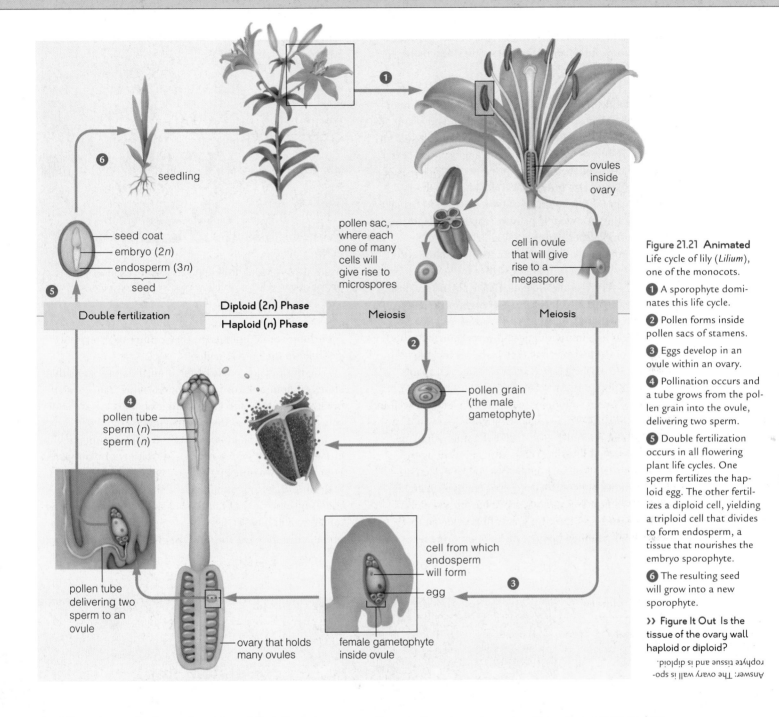

Figure 21.21 Animated Life cycle of lily (*Lilium*), one of the monocots.

❶ A sporophyte dominates this life cycle.

❷ Pollen forms inside pollen sacs of stamens.

❸ Eggs develop in an ovule within an ovary.

❹ Pollination occurs and a tube grows from the pollen grain into the ovule, delivering two sperm.

❺ Double fertilization occurs in all flowering plant life cycles. One sperm fertilizes the haploid egg. The other fertilizes a diploid cell, yielding a triploid cell that divides to form endosperm, a tissue that nourishes the embryo sporophyte.

❻ The resulting seed will grow into a new sporophyte.

>> Figure It Out Is the tissue of the ovary wall haploid or diploid?

Answer: The ovary wall is sporophyte tissue and is diploid.

in a flower's carpels. A carpel consists of the stigma (a slender projection where pollen will land) and the style (the part where the ovary is located). Egg-bearing gametophytes form from megaspores in the ovary ❸.

Pollination occurs when a pollen grain reaches a receptive stigma. After pollination, a pollen tube forms ❹. The tube grows down through the style and into the ovary. It then delivers two sperm to the ovule, where double fertilization occurs ❺. One sperm fertilizes the egg, forming what will become the embryo. The other sperm fertilizes a cell that has two nuclei, forming a triploid cell. This cell will divide and become the **endosperm**, a nutrient-rich tissue unique to angiosperms. Ovary tissue matures into a fruit with seeds that can grow into a new sporophyte ❻.

Take-Home Message **What are angiosperms?**

- Angiosperms are flowering plants. A flower is a special reproductive shoot that makes pollen in stamens and eggs in carpels. The base of the carpel is an ovary, a chamber that encloses the ovules and develops into a fruit after fertilization.
- Angiosperms are the most diverse plants. A short life cycle, coevolution with pollinators, and many fruit dispersal mechanisms contribute to their success.
- Eudicots and monocots are the main lineages of flowering plants.

❯ Angiosperms feed and shelter animals, and they provide us with food, fabric, oils, medicines, drugs, and more.
❮ Link to C3 and C4 plants 6.8

It would be nearly impossible to overestimate the importance of the angiosperms. As the most abundant plants in the majority of habitats, they provide food and shelter for a variety of animals. Animals feed on angiosperm roots, shoots, and fruits. They also sip floral nectar (Figure 21.22).

Angiosperms also provide nearly all human food, either directly or by serving as food for livestock. Cereal grains are the most widely planted crops. All are grasses. Like other monocots, cereals store nutrients in their endosperm mainly as starch. In the United States, corn is the top cereal crop, with hundreds of millions of tons harvested every year. Worldwide, rice feeds the greatest number of people. Rye, oats, barley, sorghum, and wheat are other commonly grown cereal crops (Figure 21.23A).

Variation in photosynthetic pathways contributes to the widespread use of cereal crops. Grasses include both C3 plants (rice, wheat, oats, and barley) and C4 plants (corn and sorghum). As Section 6.8 explained, C3 plants grow best in cool conditions, whereas C4 plants do well in hot, dry climates. Thus, people throughout the world can grow a cereal grain that is well suited to their climate.

Soybeans, lentils, peas, and peanuts are among the legumes, the second most important source of human food. Legumes are eudicots and their endosperm stores nutrients as proteins and oils, as well as carbohydrates. Legumes can be paired with grains to provide all the amino acids a human body needs to build proteins.

Figure 21.22 Flowering plants feed animals. This hummingbird is sipping nectar, which is mainly sucrose, from a columbine flower.

All legumes are C3 plants, so they cannot be grown successfully in the hottest regions.

The third major staple of the human diet is plants that store starch in roots or underground stems. Yams, sweet potatoes, potatoes, and cassava are examples. All are eudicots and C3 plants, so their distribution is limited.

In addition to plants that serve as staples, humans enliven their diet with a remarkable variety of plant parts. For example, we dine on the leaves of lettuce and spinach, the stems of asparagus and celery, the immature floral shoots of broccoli and cauliflower, and the fleshy fruits of tomatoes and blueberries. The stamens of crocus flowers provide the spice saffron, and we grate the bark of a tropi-

Figure 21.23 Angiosperms as crop plants. **A** Mechanized harvesting of wheat, a monocot. **B** Field of cotton, a eudicot. **C** Marijuana (a eudicot) discovered growing illegally in Oregon.

cal tree to make cinnamon. Maple syrup is sap that has been collected from maple tree xylem and boiled until it reaches a syrupy consistency.

Angiosperms are also an important source of fiber. Fabrics made from plant fibers include linen, ramie, hemp, burlap, and cotton (Figure 21.23B). A cotton boll is the fruit of a cotton plant. Cotton fibers are nearly pure cellulose. In addition to fabrics, we use fibers from seed plants to weave rugs. For example, fibers from the leaves of agave plants are used to make sisal rugs.

Oils extracted from the seeds of some eudicots including rapeseed and hemp are used in detergents, skin care products, and as industrial lubricants. Plant oils may also be used as a source of biodiesel fuel.

Angiosperm woods are referred to as "hardwoods," as opposed to gymnosperm "softwoods." Furniture and flooring are often made from oak or other hardwoods. Hardwoods are also preferred as firewood.

Some flowering plants make secondary metabolites that we use as medicines or as mood-altering drugs. A **secondary metabolite** is a compound that has no known metabolic role in the organism that produces it. Many plant secondary metabolites evolved as a defense against grazing animals. Aspirin is derived from a compound discovered in willows. Digitalis from foxglove strengthens a weak heartbeat. The caffeine in coffee beans and nicotine in tobacco leaves are widely used stimulants. Marijuana, a drug illegal in the United States, is one of the country's most valuable cash crops (Figure 21.23C). Worldwide, the cultivation of opium poppies (the source of heroin) and coca (the source of cocaine) have wide-reaching health, economic, and political effects.

Some secondary metabolites of flowering plants are extracted and sold as insecticides that are safe for humans and other vertebrates. For example, chrysanthemums make natural insecticides called pyrethrums. Neem tree oil is another natural insecticide. Compared to synthetic insecticides, plant-derived compounds tend to be less persistent in the environment.

---

secondary metabolite Chemical that has no known role in an organism's normal metabolism; often deters predation.

Take-Home Message What are the ecological and economic roles of angiosperms?

> Angiosperms serve as food and shelter for land animals.

> Humans depend on angiosperms as food plants, and as sources of fiber, wood, oils, medicines, and insecticides.

> Cereal grains are the most widely grown food plants. They are monocots with a starchy endosperm. Different members of the group are adapted to different climates.

■ Speaking for the Trees (revisited)

Plants make organic compounds by absorbing energy from the sun, carbon dioxide from the air, and water and dissolved minerals from soil. By the noncyclic pathway of photosynthesis, they split water molecules and release oxygen. Their oxygen output and carbon uptake sustains the atmosphere. Amazingly, every atom of carbon in a tree that stands a hundred meters high (328 feet) and weighs thousands of tons was taken up from the air in the form of carbon dioxide.

A tree is about 20 percent carbon by weight, so enormous amounts of carbon are stored in the living trees of Earth's forests. Additional carbon remains tied up in the leaf litter and decaying trees on the forest floor. The United Nations Food and Agriculture Organization estimates that forest plants and soils hold about one and a half times as much carbon as the air. Trees take up most carbon when they actively growing. Therefore, an acre of tropical forest where trees grow continually stores more carbon than an acre of forest in a region with a limited growing season.

The great carbon-storing capacity of tropical forests means that threats to these forests have global implications. When a tropical forest is converted to cropland, the rate of carbon dioxide storage per acre goes down. In addition, if the forest is cleared by burning, the carbon that was stored in wood and leaf litter enters the atmosphere (Figure 21.24). Rising levels of carbon dioxide in Earth's atmosphere contribute to global warming, by a mechanism we explain in detail in Section 42.8.

Figure 21.24 Burning forests to make way for agriculture or other uses adds carbon dioxide to the air, reduces carbon uptake, and thus contributes to global warming.

How Would You Vote? One proposed way to discourage tropical deforestation is through international carbon offsets. For example, a company in the United States could agree to offset its greenhouse gas production by funding a tropical reforestation or forest preservation project in South America or Africa. Proponents of such agreements say they minimize a company's cost of reducing greenhouse gases and benefit all involved. Some opponents would rather see companies make pollution-reducing changes at home, rather than pay for programs elsewhere. Do you think international carbon offsets are a good idea? See CengageNow for details, then vote online (cengagenow.com).

Summary

Section 21.1 Deforestation has far-reaching environmental effects, including increased erosion and climate change. Individual efforts to reverse deforestation can collectively have a large positive impact.

Section 21.2 Land plants evolved from green algae. These plants are **embryophytes**; they form a multicelled embryo on the parental body. Key adaptations that allowed plants to move into dry habitats include a waterproof **cuticle** with **stomata**, and internal pipelines of **vascular tissue** (**xylem** and **phloem**) reinforced by **lignin**.

Plant life cycles include two multicelled bodies: the haploid gametophyte and the diploid sporophyte. The gametophyte dominates in early-evolving lineages, but in **vascular plants**, the sporophyte is larger and longer lived. **Seeds** and male gametophytes that can be dispersed without water (**pollen grains**) are important adaptations that contribute to the success of seed plants.

sporophyte —

gametophyte —

Section 21.3 Bryophytes are nonvascular (no xylem or phloem). Their sperm swim through water droplets to eggs. The sporophyte remains attached to the gametophyte. **Rhizoids** attach a gametophyte to the soil or a surface.

Mosses are the most diverse bryophytes. The remains of some mosses form **peat**, which is dried and burned as fuel.

Section 21.4 In **seedless vascular plants**, sporophytes have vascular tissues and are the larger, longer-lived phase of the life cycle. Typically the sporophyte's roots and shoots grow from a horizontal stem, or **rhizome**. Tiny free-living gametophytes make flagellated sperm that require water for fertilization. Ferns, the most diverse group of seedless vascular plants, produce spores in **sori**. Many ferns grow as **epiphytes**. Club mosses and horsetails produce spores in conelike structures.

Section 21.5 Forests of giant seedless vascular plants thrived during the Carboniferous period. Heat and pressure transformed the remains of these forests to **coal**.

The first seed plants evolved during the Carboniferous. Seed plant sporophytes have **pollen sacs**, where **microspores** form and develop into male gametophytes (pollen grains). They also have **ovules**, where **megaspores** form and develop into female gametophytes. **Pollination** unites the egg and sperm of a seed plant.

Section 21.6 Gymnosperms include **conifers**, **cycads**, **ginkgos**, and **gnetophytes**. Many are well adapted to dry climates. In conifers, ovules form on the surfaces of woody cones.

Section 21.7 Angiosperms are the most diverse and widespread land plants. They alone make **flowers** and **fruits**. The **stamens** of a flower produce pollen. The female part of a flower is the **carpel**. An **ovary** at the base of the carpel holds one or more ovules.

Many flowering plants coevolved with **pollinators**. After pollination, the flower's ovary becomes a fruit that contains one or more seeds. A flowering plant seed includes an embryo sporophyte and **endosperm**, a nutritious tissue. The two main types, **eudicots** and **monocots**, differ in seed structure and other traits.

Section 21.8 As the dominant plants in most land habitats, flowering plants are ecologically important, as well as essential to human existence. The nutrient-rich endosperm of angiosperm seeds makes them staples of human diets throughout the world. Angiosperms also supply us with fiber, furniture, oils, medicines, and mood-altering drugs. Many useful plant compounds are secondary metabolites, compounds that probably help defend the plant against predation.

Self-Quiz   Answers in Appendix III

1. The first plants were _____ .
   a. ferns   c. bryophytes
   b. flowering plants   d. conifers

2. Which of the following statements is false?
   a. Ferns produce seeds inside sori.
   b. Bryophytes do not have xylem or phloem.
   c. Gymnosperms and angiosperms produce seeds.
   d. Only angiosperms produce flowers.

3. In bryophytes, eggs are fertilized in a chamber on the _____ and a zygote develops into a _____ .
   a. sporophyte; gametophytes
   b. gametophyte; sporophyte
   c. sorus; cone

4. Horsetails and ferns are _____ plants.
   a. multicelled aquatic   c. seedless vascular
   b. nonvascular seed   d. seed-bearing vascular

5. Coal consists primarily of compressed remains of the _____ that dominated Carboniferous swamp forests.
   a. seedless vascular plants c. flowering plants
   b. conifers   d. a and c

6. The _____ produce flagellated sperm.
   a. mosses   d. monocots
   b. ferns   e. a and b
   c. conifers   f. a through c

7. A seed is a _____ .
   a. female gametophyte   c. mature pollen grain
   b. mature ovule   d. modified microspore

A Global View of Deforestation The U.N. Food and Agriculture Organization (UNFAO) recognizes the importance of forests to human populations and maintains records of forest abundance. Figure 21.25 shows UNFAO data on the amount of forested land throughout the world in 1990, 2000, and 2005.

1. How many hectares of forested land were there in North America in 2005?

2. In which region(s) did the amount of forested land increase between 1990 and 2005?

3. How many hectares of forest did the world lose between 1990 and 2005?

4. In 2002, China embarked on an ambitious campaign to add 76 million hectares of trees over a ten-year period. Do you see any indication that this campaign is successful?

Region	Forested Area (in millions of hectares)		
	1990	**2000**	**2005**
Africa	699	656	635
Asia	574	567	572
Central America	28	24	22
Europe	989	988	1,001
North America	678	678	677
Oceania	233	208	206
South America	891	853	832
World total	4,077	3,988	3,952

Figure 21.25 Changes in forested area by region from 1990 to 2005. One hectare is 2.47 acres. The full report on the world's forests is online at www.fao.org/forestry/en/.

8. True or false? Both spores and sperm of a seedless vascular plant are haploid.

9. Only angiosperms produce _____ .
   a. pollen   c. fruits
   b. seeds   d. all of the above

10. Lignin is not found in stems of _____ .
    a. mosses   b. ferns   c. monocots   d. a and b

11. A waxy cuticle helps land plants _____ .
    a. conserve water   c. reproduce
    b. take up carbon dioxide d. stand upright

12. Cereal crops such as rice and corn are _____ .
    a. monocots   c. seed plants
    b. vascular plants   d. all of the above

13. Some plant secondary metabolites are used as _____ .
    a. fuel   c. textiles
    b. medicines   d. flooring

14. Match the terms appropriately.
    ___ bryophyte   a. seeds, but no fruit
    ___ seedless   b. flowers and fruits
        vascular plant   c. no xylem or phloem
    ___ gymnosperm   d. xylem and phloem,
    ___ angiosperm   but no ovule

15. Match the terms appropriately.
    ___ ovule   a. gamete-producing body
    ___ cuticle   b. spore-producing body
    ___ gametophyte   c. becomes seed
    ___ sporophyte   d. horizontal stem
    ___ fruit   e. mature ovary
    ___ endosperm   f. nutritive tissue in seed
    ___ rhizome   g. where fern spores form
    ___ sorus   h. single haploid cell
    ___ microspore   i. waterproofing layer

Additional questions are available on **CENGAGENOW**.

Critical Thinking

1. Early botanists admired ferns but found their life cycle perplexing. In the 1700s, they learned to propagate ferns by sowing what appeared to be tiny dustlike "seeds" that they collected from the undersides of fronds. Despite many attempts, the botanists could not locate the pollen source, which they assumed must stimulate these "seeds" to develop. Imagine you could write to these botanists. Compose a note that explains the fern life cycle and clears up their confusion.

2. In most plants the largest, longest-lived body is a diploid sporophyte. By one hypothesis, diploid dominance was favored because it allowed a greater level of genetic diversity. Suppose that a recessive mutation arises. It is mildly disadvantageous now, but it will be useful in some future environment. Explain why such a mutation would be more likely to persist in a fern than in a moss.

3. The photo at the *right* is a micrograph of a longitudinal section through the stem of a squash plant. The stem has been dyed with a substance that tints lignin red. Can you identify the red-ringed structures? Would you expect to find similar structures in the stem of a corn plant? Would you find them in the leafy green part of a moss?

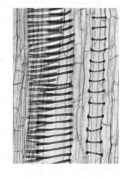

Animations and Interactions on **CENGAGENOW**:
❯ Moss life cycle; Fern life cycle; Conifer life cycle; Lily life cycle.

❮ Links to Earlier Concepts

Before starting, review Figure 20.2 to get a sense of where fungi fit in the eukaryotic family tree. In this chapter you will learn about fungal interactions with cyanobacteria (19.7), and green algae (20.8). You will also draw on your knowledge of the organic molecules chitin (3.3) and lignin (21.2), and your understanding of the processes of fermentation (7.6) and nutrient cycling (1.3).

Key Concepts

Traits and Classification

Fungi are single-celled and multicelled heterotrophs that feed by secreting digestive enzymes onto organic matter, then absorbing the released nutrients. Multicelled species form a mesh of absorptive filaments. In the most diverse groups, cross-walls reinforce the filaments.

The Oldest Lineages

The oldest fungal lineages have filaments with few or no cross-walls. The mostly aquatic chytrid fungi make flagellated spores. Zygote fungi make thick-walled zygotes that release sexual spores. Glomeromycetes are fungi that branch inside plant cell walls.

22 Fungi

22.1 High-Flying Fungi

Fungi are not known for their mobility. You probably do not think of mushrooms and their relatives as world travelers, but some do get around. Fungi produce microscopic spores that drift on air currents. The spores can also lodge in crevices on tiny dust particles. When winds lift these particles aloft, spores go along for the ride. Dust borne fungal spores can disperse great distances, riding winds that swirl high above Earth's surface.

We know, for example, that dust storms in the deserts of North Africa lift fungus-laden particles more than 4.5 kilometers (3 miles) above the desert floor. Winds then transport the dust and spores long distances. Some dust particles travel as far as the Caribbean and the east coast of the Americas.

Most fungi that hitchhike as spores on African dust are harmless. However, fungi that cause plant disease occasionally make the journey. For example, winds of a 1978 cyclone introduced sugar cane rust (a fungal disease) from Cameroon to the Dominican Republic. Similarly, winds probably facilitated the spread of coffee rust fungus from Angola to Brazil in 1980.

Today, an African outbreak of an old fungal foe has agricultural officials around the world on edge. The fungus is wheat stem rust (*upper right*). An infection begins when a spore lands on a wheat plant and fungal filaments invade the plant through its stomata. Within about a week, the fungus grows through the plant's tissues, matures, and produces ten of thousands of rust-colored spores at the stem surface. Each spore can disperse and infect a new plant.

Wheat stem rust was common worldwide until the 1960s, when plant breeders developed new varieties of wheat that resist infection by this fungus. Global distribution of the rust-resistant wheats provided a respite from outbreaks of wheat stem rust for decades. Then, in 1999, a new strain of wheat stem rust (called Ug99) was discovered in Uganda, a country in eastern Africa. Mutations allow Ug99 to infect most wheat varieties that were previously considered resistant to wheat stem rust.

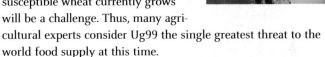

By 2009, windborne spores of Ug99 had dispersed to Kenya, Ethiopia, and Sudan, crossed the Red Sea to Yemen, and from there crossed the Persian Gulf to Iran. India, the world's second largest wheat producer, is expected to be affected next. Most likely, winds will eventually distribute spores of Ug99 throughout the world.

With one of the world's most important food crops under threat, scientists are hard at work screening plants for genes that provide protection against Ug99. They have already bred some resistant wheats. However, creating a resistant wheat suited to each region where Ug99-susceptible wheat currently grows will be a challenge. Thus, many agricultural experts consider Ug99 the single greatest threat to the world food supply at this time.

This chapter discusses the pathogenic fungi that threaten our crops and infect our bodies, but it also introduces many beneficial fungi. Keep in mind that most fungi are decomposers with an important ecological role. Fungi break down organic wastes and remains into inorganic subunits, thereby making essential nutrients available to plants. Fungi also associate with plant roots, helping them to grow better. Other fungi partner with single photosynthetic cells, forming the composite organisms known as lichens. We value fungi as sources of medicine and as food. Some single-celled fungi help us make bread and beer, and countless mushrooms end up atop our pizzas and in our salads and sauces.

Sac Fungi
Sac fungi, the most diverse fungal lineage, include single cells such as the yeasts we use in baking, and multicelled species that have filaments reinforced with cross-walls. When sac fungi reproduce sexually, they produce spores inside a saclike structure.

Club Fungi
Most club fungi are multi-celled. Like the closely related sac fungi, they have filaments reinforced with cross walls. Sexual reproduction involves formation of spores in club-shaped cells. A mushroom is an example of a club fungus fruiting body.

Interactions With Other Organisms
Some fungi partner with photosynthetic cells as lichens. Others live in or on plant roots and benefit the plant by sharing nutrients. A minority of fungi are parasites. Some of these cause diseases in plants or in humans.

> Like animals, fungi are heterotrophs. Unlike animals, fungi have walled cells and digest food outside their body.
< Link to Chitin 3.3

Structure and Function

Fungi are spore-making eukaryotic heterotrophs that have cell walls reinforced with chitin. This polysaccharide also hardens the body covering of animals such as insects or crabs. Like animals, fungi make digestive enzymes, but fungi do not digest food inside their body. Instead, they secrete digestive enzymes onto organic matter and absorb the released breakdown products.

A typical fungus is a free-living **saprobe**, an organism that feeds on organic wastes and remains. Some fungi live inside other organisms. The fungus may benefit its host, have no effect, or be a harmful parasite.

The fungi that live as single cells are informally called yeasts. More often the fungus is multicellular. Molds and mushrooms are familiar examples of multicellular fungi (Figure 22.1). Such fungi grow as a mesh of threadlike filaments collectively called a **mycelium** (plural, mycelia). Each filament is one **hypha** (plural, hyphae), consisting of cells attached end to end (Figure 22.2). The fungus grows by adding cells to the tip of its hyphae.

---

fungus Eukaryotic heterotroph with cell walls of chitin; obtains nutrients by digesting them outside the body and absorbing them.
dikaryotic Having two genetically distinct nuclei.
hypha Component of a fungal mycelium; a filament made up of cells arranged end to end.
mycelium Mass of threadlike filaments (hyphae) that make up the body of a multicelled fungus.
saprobe Organism that feeds on wastes and remains.

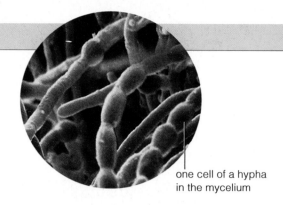

one cell of a hypha in the mycelium

Figure 22.2 Close-up of a mycelium, showing the threadlike hyphae, each consisting of cells arranged end to end.

Depending on the fungal group, there may or may not be cross-walls, or septa (singular, septum), between cells of a hypha. When cross-walls do exist, they are porous, so materials can still flow among hyphal cells. Thus nutrients or water taken up in one part of the mycelium can be shared with cells in other regions of the fungal body.

Life Cycles

In fungi, the diploid stage is the least conspicuous part of the life cycle. Only the zygote is diploid. Depending on the fungal group, a haploid stage or a dikaryotic stage dominates the cycle. **Dikaryotic** means that a cell contains two genetically different nuclei (*n+n*).

Most fungi disperse by producing spores. A fungal spore is a cell or cluster of cells, often with a thick wall that allows it to survive harsh conditions. Spores may form by mitosis (asexual spores) or by meiosis (sexual spores). Each fungal group produces sexual spores in a distinctive way, a trait that aids in classification.

Fungal Diversity

More than 70,000 species of fungi have been named, and there may be a million more. Three relatively small groups—chytrids, glomeromycetes, and zygote fungi—have a haploid mycelium that consists of hyphae with few or no cross-walls.

Most fungi belong to two lineages: sac fungi and club fungi. Hyphae with cross-walls at regular intervals evolved in the common ancestor of these lineages and contribute to the success of both groups. Cross-walls strengthen hyphae, allowing development of larger, more complex spore-producing bodies. In addition, cross-walls divide the cytoplasm, so damage to one part of a hypha does not cause the whole hypha to dry out and die.

A Green mold on grapefruit. A mold is a fungus that grows as a mat of microscopic hyphae that reproduce asexually.

B Scarlet hood mushroom in a Virginia forest. A mushroom forms during sexual reproduction of many club fungi.

Figure 22.1 Examples of multicelled fungi.

Take-Home Message What are fungi?

> Fungi are heterotrophs that secrete digestive enzymes onto organic matter and absorb the released nutrients. They live as single cells or as a multicelled mycelium and disperse by producing spores.

❭ Chytrids, zygote fungi, and glomeromycetes are the oldest and least diverse groups of fungi.

Chytrids

Chytrids include the oldest fungal lineages, and are the only living fungi that make flagellated spores. Most of the 1,000 or so species are aquatic decomposers. A few kinds swim about in the gut of sheep, cattle, and other herbivores and help them digest cellulose. Others are parasites.

A chytrid that parasitizes amphibians has been implicated in the global decline in frog populations. The parasite was discovered in the late 1990s when scientists investigated plummeting frog numbers in Australia and South America. Since then, the parasite has been detected in frogs in the Americas, Europe, Africa, and Asia. Chytrid infections may push some frog species to extinction.

Zygote Fungi

Zygote fungi (zygomycetes) include about 1,100 species that produce a thick-walled zygospore during sexual reproduction. Many grow as molds, a mat of asexually reproducing hyphae. Black bread mold (*Rhizopus stolonifer*) has a typical life cycle (Figure 22.3). When food is plentiful, a haploid mycelium grows and produces spores asexually at the tips of specialized hyphae ❶.

There are two mating strains. Sexual reproduction occurs when hyphae of different strains meet ❷. Special cells form on the hyphae and their cytoplasm fuses, producing a young dikaryotic zygospore ❸. Fusion of nuclei produces a mature diploid zygospore, with a thick, protective wall ❹. Meiosis occurs as the zygospore germinates. A hypha emerges, bearing a sac with haploid spores at its tip ❺. After release, the spores germinate and each can give rise to a haploid mycelium.

Most zygote fungi are saprobes, but some are parasites. One species infects flies. Fungal hyphae grow into the fly's brain and alter its behavior. The fly climbs to a high perch, holds on with its proboscis, and dies. Its posture and height facilitate dispersal of the fungal spores. Another zygote fungus infects people who have a weak immune system. Hyphae grow in blood vessels and cause zygomycosis, an often fatal disease. "Mycosis" is a general term for a disease that is caused by a fungus.

Glomeromycetes

The 150 or so species of **glomeromycetes** were previously placed among the zygote fungi, but are now considered a separate group. All form an association with plant

---

chytrid Fungus that makes flagellated spores.
glomeromycete Fungus with hyphae that grow inside the wall of a plant root cell.
zygote fungus Fungus that forms a zygospore during sexual reproduction.

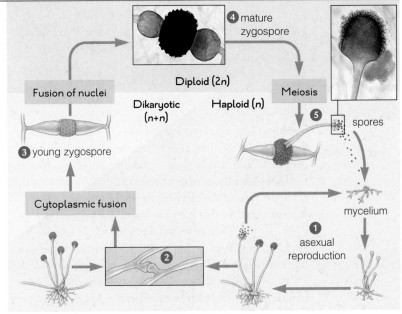

Figure 22.3 Animated Life cycle of a zygote fungus, the black bread mold (*Rhizopus*).

❶ The fungus reproduces asexually as it grows through a slice of bread.

❷ Sexual reproduction begins when hyphae of two different mating strains meet.

❸ Cytoplasmic fusion of hyphal cells yields an immature zygospore that is dikaryotic.

❹ Fusion of the zygospore's haploid nuclei produces the mature diploid zygospore.

❺ The zygospore undergoes meiosis, germinates, and produces an aerial hypha with haploid spores at its tip. Each spore can give rise to a new mycelium.

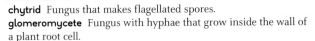

Figure 22.4 Glomeromycete hypha branching inside a root cell wall.

roots. A hypha grows into a root and branches inside the wall of a root cell (Figure 22.4). The relationship is mutually beneficial. The plant shares sugars with its fungal roommate, which provides it with nutrients taken up from the soil.

Take-Home Message **What traits distinguish chytrids, zygote fungi, and glomeromycetes?**

❭ The mostly aquatic chytrids are the only fungi with flagellated spores.

❭ Zygote fungi produce a unique walled zygospore during sexual reproduction.

❭ Glomeromycetes have hyphae that branch inside cells of plant roots.

❯ Sac fungi are the most diverse fungal group. They include single-celled species and multicelled species that have hyphae reinforced by cross-walls at regular intervals.
❮ Link to Fermentation 7.6

Life Cycle

Sac fungi (ascomycetes) are the most diverse group of fungi, with more than 32,000 named species. Sac fungi may be multicelled or single-celled. Hyphae of multicelled sac fungi have cross-walls at regular intervals. The characteristic trait of the group is the ascus (plural, asci), a baglike structure that encloses sexually produced spores.

As in zygote fungi, sexual reproduction of multicelled sac fungi begins when hyphae of different mating strains meet. However, in sac fungi, cytoplasmic fusion is followed by mitosis that produces dikaryotic hyphae. These hyphae intertwine with haploid hyphae to form a fruiting body called an ascocarp. The spore-forming asci develop on the ascocarp (Figure 22.5A).

Multicelled sac fungi also produce spores asexually. In this case, spores form on specialized hyphae that develop from the haploid mycelium.

A Sampling of Diversity

Morels (Figure 22.5B) and truffles (Figure 22.5C) are among the edible ascocarps. A truffle forms underground. When spores mature, the fungus produces an odor like that of a sexually excited male pig. Female pigs detect

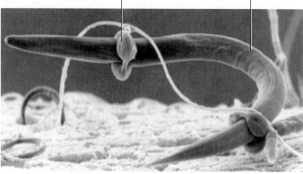

part of one hypha that forms a nooselike ring   roundworm

Figure 22.6 Animated A predatory fungus (*Arthrobotrys*) that captures and feeds on roundworms. Rings on the hyphae constrict and entrap the worms, then hyphae grow into the captive and digest it.

this scent and disperse truffle spores as they root through the soil in search of their seemingly subterranean suitor. Dogs can also be trained to snuffle out truffles. The search can be worthwhile. In 2006, a single 1.5-kilogram (about 3-pound) Italian truffle sold for $160,000.

Many sac fungi grow as molds. Some species of the sac fungus *Penicillium* spoil fruits, such as the grapefruit in Figure 22.1. We use other members of this genus to give blue cheese its blue veins and tangy taste. Yet another *Penicillium* species lives in soils and was the original source of the antibiotic drug penicillin.

The most familiar yeasts are sac fungi. These single cells reproduce asexually by budding, as shown in the photo at the *right*. A packet of baking yeast holds spores of a sac fungus. When bread dough is set out to rise in a warm place, the spores germinate and release cells that use sugar in fermentation reactions. Bubbles of carbon dioxide produced by the fermentation cause dough to expand. Fermentation by other sac fungus yeasts helps us produce beer and wine. Still other sac fungi cause human yeast infections.

Figure 22.6 shows an unusual predatory sac fungus. It has special hyphae with loops that can trap roundworms. After feeding on a worm, the fungus makes asexual spores. Spreading these spores in fields can help control populations of roundworms that infect and harm crops.

sac fungi Most diverse fungal group; sexual reproduction produces ascospores inside a saclike structure (an ascus).

haploid spore in ascus

Figure 22.5 Sac fungus fruiting bodies (ascocarps). **A** The cup of the scarlet cup fungus is an ascocarp. Asci, each containing eight ascospores (sexual spores) form on the cup's inner surface. **B** Morels and **C** truffles are edible ascocarps. Morels develop aboveground. Truffles form underground.

Take-Home Message **What are sac fungi?**

❯ Sac fungi make sexual spores inside sac-shaped asci. They include multicelled species with hyphae divided by internal walls, and yeasts that can reproduce asexually by budding.

› Most club fungi are multicelled, with cross-walled hyphae. Their fruiting bodies include the familiar mushrooms.

‹ Link to Lignin 21.2

A **club fungus** (basidiomycete) is typically multicelled. Its sexual spores form inside club-shaped cells on a fruiting body, or basidiocarp, composed of interwoven dikaryotic hyphae. Fruiting bodies of club fungi can be quite large and have diverse shapes (Figure 22.7).

The button mushrooms common in markets are fruiting bodies of a club fungus. Haploid hyphae of this fungus grow underground. When hyphae of two different mating strains meet, they fuse, forming a dikaryotic mycelium (Figure 22.8 ❶). The mycelium grows through the soil and, when conditions favor sexual reproduction, forms mushrooms ❷. Thin tissue sheets (gills) fringed with club-shaped cells line the underside of the mushroom's cap ❸. Fusion of the nuclei in these dikaryotic cells forms a diploid zygote ❹. The zygote undergoes meiosis, forming four haploid spores ❺. These spores are dispersed by wind, germinate, and start a new cycle ❻.

Club fungi play an essential role as decomposers in forests. They are the only fungi capable of breaking down the lignin in wood. Some forest fungi are long-lived giants. In one Oregon forest, mycelium of a honey mushroom fungus extends through more than 2,000 acres of soil. By one estimate, this fungus is 2,400 years old.

Smuts and rusts are important plant pathogens. Unlike most club fungi, these do not produce a big fruiting body. Wheat stem rust (Section 22.1) is one example.

club fungi Fungi that produce spores in club-shaped cells.

Puffball Shelf fungus

Coral fungus Chanterelles

Figure 22.7 Club fungus fruiting bodies (basidiocarps).

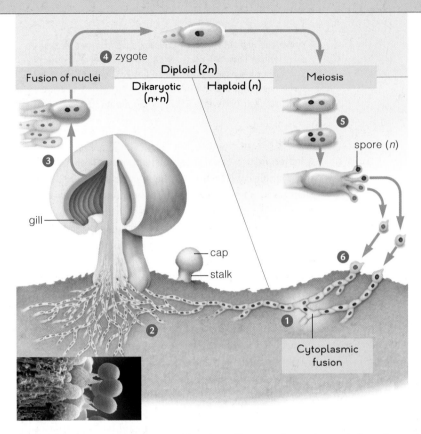

Figure 22.8 Animated Club fungus life cycle. Different mating strains are indicated here by *red* or *blue* nuclei. The photo at *lower left* shows spores at the edge of gills.

❶ Haploid hyphal cells meet and their cytoplasm fuses, forming a dikaryotic (n+n) cell.

❷ Mitotic cell divisions form a mycelium that produces a mushroom.

❸ Spore-making cells form at the edges of the mushroom's gills.

❹ Inside these dikaryotic cells, nuclei fuse, producing diploid (2n) cells.

❺ The diploid cells undergo meiosis, forming haploid (n) spores.

❻ Spores are released and give rise to a new haploid mycelium.

›› **Figure It Out** How many nuclei are in each cell of a mushroom's stalk?

Answer: Two. The cells are dikaryotic.

Fungi cannot run away from predators, but toxins protect some species. Each year thousands of people are sickened after eating poisonous mushrooms that they mistook for an edible species. In some cases, mushroom poisoning is fatal. Edible mushrooms often have poisonous look-alikes, so eating mushrooms found in the wild is risky unless you are expert in identifying species.

Some mushroom toxins alter mood and cause hallucinations. The drug LSD is a compound that was initially isolated from such a mushroom. Psilocybin-containing mushrooms are also eaten for their mind-altering effects. Both LSD and psilocybin-containing mushrooms are illegal in the United States.

Take-Home Message What are club fungi?

› Club fungi produce their sexual spores at the tips of club-shaped cells. These cells form on a short-lived reproductive body, such as a mushroom.

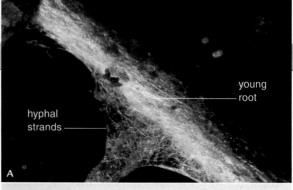

young root

hyphal strands

A

B

> Symbiosis refers to two organisms that interact closely throughout their life cycle. Many fungal species are symbiotic with photosynthetic cells in lichens, while others are symbionts that live in or on plants.
< Links to Cyanobacteria 19.7, Green algae 20.8

Lichens

A **lichen** is a composite organism composed of a fungus and a single-celled photosynthetic partner (Figure 22.9). Most often it consists of a sac fungus and cyanobacteria or green algae. Some club fungi also form lichens.

The fungus makes up most of the lichen's mass. Its hyphae entwine around the photosynthetic cells that provide it with sugars. Cyanobacteria also provide fixed nitrogen. A lichen may be a **mutualism**, a symbiotic interaction that benefits both partners. However, in some cases the fungus may be parasitically exploiting captive photosynthetic cells that would do better on their own.

Lichens disperse by fragmentation or by releasing small packages with cells of both partners. In addition, the fungus can release spores. The fungus that germinates from a spore can form a new lichen only if it alights near an appropriate photosynthetic cell. This is not as unlikely as it may seem; the required algae and bacteria are common as free-living cells in many environments.

Lichens colonize many places too hostile for most organisms, such as newly exposed bedrock. They help break down the rock by releasing acids and by holding water that freezes and thaws. When soil conditions improve, plants move in and take root. Long ago, lichens may have preceded the invasion of plants onto land.

Figure 22.10 **A** Mycorrhiza formed by a fungus and young hemlock tree root. **B** Experimental demonstration of the effects of mycorrhizae on plant growth in phosphorus-poor, sterilized soil. Seedlings at left were grown without a fungus. Seedlings at right, the experimental group, were grown with a partner fungus.

Mycorrhizae: Fungus + Roots

Nearly all plants have a mutually beneficial relationship with a fungus. A mutualism between a plant root and a fungus is called a **mycorrhiza** (plural, mycorrhizae). In some cases, fungal hyphae form a dense net around roots but do not penetrate them (Figure 22.10A). In other cases, the hyphae enter roots and branch inside root cell walls, as shown earlier in Figure 22.4. About 80 percent of plants form such relationships with fungi.

Hyphae of both kinds of mycorrhizae increase the absorptive surface area of their plant partner. The fungus shares minerals it takes up with the plant, and the plant gives up sugars to the fungus. Many plants do not as grow well when deprived of their fungal partner (Figure 22.10B).

---

lichen Composite organisms, consisting of a fungus and a single-celled alga or a cyanobacterium.
mutualism Mutually beneficial relationship between two species.
mycorrhiza Mutually beneficial partnership between a fungus and a plant root.

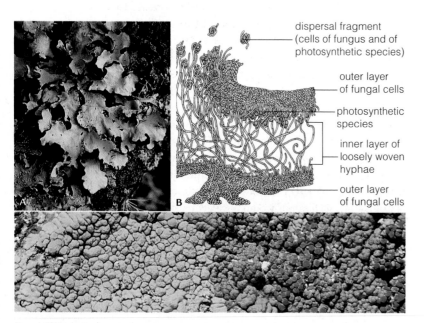

dispersal fragment (cells of fungus and of photosynthetic species)

outer layer of fungal cells

photosynthetic species

inner layer of loosely woven hyphae

outer layer of fungal cells

A

B

C

Figure 22.9 Lichens. **A** Leaflike lichen on a birch tree. **B** Cross-section of a leaflike lichen. **C** Encrusting lichens on granite.

Take-Home Message What types of partnerships do fungi form with other organisms?

> In lichens, a fungus shelters one or more photoautotrophs and shares carbon dioxide and mineral ions with them, while receiving some carbohydrates in return. Often the photosymbionts are nitrogen fixers, such as cyanobacteria.

> In mycorrhizae, a fungus living in or on a plant's young roots increases the plant's uptake of water and dissolved mineral ions and helps protect it from pathogens. The fungus withdraws some nutrients from its partner.

22.7 Fungi as Pathogens

> Most fungi feed on organic wastes and remains, but some are pathogens that attack living plants or animals.

Plant Pathogens

Most fungi decompose dead plant material, but some species are parasites that feed on living plants. Such fungi can be ecologically and economically important plant pathogens. As noted earlier, airborne spores spread wheat stem rust, a type of club fungus. A wheat stem rust infection can reduce crop yield by up to 70 percent. As another example, rose growers must contend with powdery mildews. A whitish powder that appears on leaves is the spores of sac fungi that feed on the leaf tissues.

One fungus, accidentally introduced to the United States from Asia in the early 1900s, caused a dramatic change in the country's eastern forests. The fungus infects the American chestnut, causing chestnut blight. Before the fungus arrived, the American chestnut was an important timber species and its nuts nourished forest animals. The chestnut blight first appeared in New York. As winds spread fungal spores westward, most chestnut trees east of the Mississippi became infected and died. Today, few stands of mature trees remain. Shoots occasionally sprout from old root systems, but they cannot reach mature size.

Fungal pathogens of plants have even affected human history. One notorious sac fungus, *Claviceps purpurea*, parasitizes rye and other cereal grains (Figure 22.11). It produces toxic alkaloids that, when eaten, cause ergotism. Symptoms of ergotism include hallucinations, hysteria, and convulsions. Ergotism epidemics were common in Europe in the Middle Ages. They thwarted Peter the Great, a Russian czar who became obsessed with conquering ports along the Black Sea. Soldiers laying siege to those ports ate rye bread and fed it to their horses. The soldiers went into convulsions and the horses into "blind staggers." Ergotism may also have played a role in witch hunts that occurred in Salem, Massachusetts. Descriptions of the behavior of the supposedly bewitched women resemble the symptoms of ergotism.

Human Pathogens

In humans, fungal infections most frequently involve the skin. Typically the fungus feeds on the outermost dead

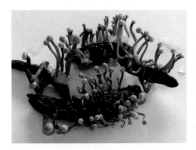

Figure 22.11 Grains of rye with spore-bearing structures of *Claviceps purpurea*. Eating grain infected by this species causes a type of food poisoning called ergotism.

High-Flying Fungi (revisited)

Species of *Fusarium*, a sac fungus, cause plant and human diseases. In 2006, *Fusarium* spores got into contact lens solution and caused eye infections worldwide. Of 122 infections in the United States, a third were so bad that patients required surgery to replace the cornea, the eye's clear layer. David Schmale of Virginia Tech (*above*) has found spores of many *Fusarium* species in air samples he collected using remote-controlled model aircraft.

How Would You Vote? One *Fusarium* strain kills opium poppies. Should spores of this fungus be sprayed in Afghanistan to reduce supplies of opium and heroin? See CengageNOW for details, then vote online (cengagenow.com).

---

skin layers by secreting enzymes that dissolve keratin, the main skin protein. Infected areas become raised, red, and itchy. For example, several species of fungus can take up residence in the skin between toes and on the sole of the foot, causing what is commonly called "athlete's foot" (*right*). Fungi also cause the skin infections misleadingly known as "ringworm." No worm is involved. Instead, a ring-shaped lesion appears and expands as fungal hyphae grow outward from the site of the initial infection.

Low numbers of single-celled fungi normally live in the vagina, but sometimes their population explodes, resulting in fungal vaginitis (a vaginal yeast infection). Symptoms usually include itching or burning sensations and a thick, odorless, whitish vaginal discharge. Intercourse is often painful. Disrupting the normal populations of bacteria in the vagina by douching or using antibiotics increases the risk of fungal vaginitis, as does use of oral contraceptives. Nonprescription medication placed into the vagina usually cures the infection. If this is not effective, a woman should consult a doctor.

Inhaled fungal spores also cause disease. Soil in the midwestern and south central United States holds spores that cause histoplasmosis. Most people that inhale the spores have only a brief episode of coughing. However, in immune-suppressed people, the fungus may spread from the lungs, through the blood, and into other organs, with fatal results. Similarly, spores in soil of the American Southwest can cause coccidioidomycosis, or valley fever.

Take-Home Message What are the effects of fungal pathogens?

> Fungal pathogens of plants reduce crop yield and sometimes kill plants.

> In humans, fungi cause annoying skin diseases and vaginitis. Fungal lung diseases can be deadly in people with weakened immune systems.

Summary

Section 22.1 Fungi produce microscopic spores that winds sometimes disperse long distances. As a result, it is difficult to prevent the spread of a fungal pathogen, such as wheat stem rust.

Section 22.2 Fungi are heterotrophs that secrete digestive enzymes on organic matter and absorb released nutrients. Most are **saprobes** that feed on organic remains; they serve as essential decomposers in most ecosystems. Other fungi live in or on organisms. They may be harmless, beneficial, or parasitic.

Fungi include single-celled yeasts and multicelled species. They disperse by releasing spores. In a multicelled species, spores germinate and give rise to filaments called **hyphae**. The filaments typically grow as an extensive mesh called a **mycelium**. Depending on the group, the cells of a hyphae may be haploid (*n*) or **dikaryotic** (*n+n*). In the oldest lineages, the hyphae have few or no cross-walls (septa). More recently evolved lineages have hyphae with porous cross-walls at regular intervals. In all groups, water and nutrients move freely between cells of a hypha.

Section 22.3 The oldest fungal lineage, the **chytrids**, are a mostly aquatic group and the only fungi with flagellated spores. **Zygote fungi** include many familiar molds that grow on fruits, breads, and other foods. Their hyphae are continuous tubes that have no cross-walls. Sexual reproduction produces a thick-walled zygospore. Zygote fungi also produce asexual spores atop specialized hyphae. **Glomeromycetes** are relatives of the zygote fungi that live in soil and extend their hyphae into plant roots. The hyphae branch inside plant cell walls and deliver nutrients to the cell.

Section 22.4 Sac fungi are the most diverse group. They include single-celled yeasts and multicelled species such as cup fungi and morels. Sac fungi produce sexual spores in asci. In multicelled species, these saclike structures form on an ascocarp, a fruiting body composed of haploid and dikaryotic hyphae that have internal cross-walls at regular intervals. Yeasts that are sac fungi help us make wine, breads, and alcoholic beverages. The antibiotic penicillin was originally isolated from a mold that is a sac fungus. Related molds are used to flavor cheeses.

Section 22.5 Like sac fungi, multicelled **club fungi** have hyphae with cross-walls and can produce complex reproductive structures, such as mushrooms. Club fungi are the only fungi that can break down lignin in wood, and are important as decomposers in forest habitats.

Typically, a dikaryotic mycelium dominates the life cycle of multicelled species. It grows by mitosis and, in some species, extends through a vast volume of soil.

When conditions favor reproduction, a reproductive structure, also made up of dikaryotic hyphae, develops. A mushroom is an example. Nuclei fuse in club-shaped cells at the edges of sheets of tissue called gills. Meiosis of the resulting diploid cell produces haploid spores at the tips of the club-shaped cells.

Section 22.6 Many fungi take part in some type of **mutualism**.

A **lichen** is a composite organism composed of a fungus and photosynthetic cells of a green alga or cyanobacterium. The fungus makes up the bulk of the lichen body and obtains a supply of nutrients from its photosynthetic partner. Lichens disperse by releasing fragments that include cells of both partners. The fungus also produces spores. Lichens are important as pioneers in new habitats. They facilitate the breakdown of rock to form soil.

A **mycorrhiza** is an interaction between a fungus and a plant root. Fungal hyphae surround or penetrate the roots and supplement the root's absorptive surface area. The fungus shares absorbed mineral ions with the plant and gets some photosynthetic sugars in return. Most plants have mycorrhizal fungi and grow less well without them.

Section 22.7 Some fungi are pathogens. Fungal pathogens cause powdery mildew of roses and led to the demise of American chestnuts. A fungus that infects cereal grains makes a toxin that causes ergotism, a type of food poisoning. Fungi also infect the human body, most frequently the skin. Yeast infections of the vagina are also common. Inhaled fungal spores can germinate in lungs and cause diseases that can be fatal to people with a weakened immune system.

Self-Quiz   Answers in Appendix III

1. All fungi _____ .
   a. are multicelled   c. are heterotrophs
   b. form flagellated spores   d. all of the above

2. Saprobic fungi obtain nutrients from _____ .
   a. nonliving organic matter c. living animals
   b. living plants   d. both b and c

3. In _____ , a hypha has few or no cross-walls.
   a. zygomcyetes   c. club fungi
   b. sac fungi   d. all of the above

4. The yeasts whose fermentation reactions produce carbon dioxide that makes bread rise are a type of _____ .
   a. chytrid   c. sac fungus
   b. zygote fungus   d. club fungus

5. In many _____ , an extensive dikaryotic mycelium is the most conspicuous phase of the life cycle.
   a. chytrids   c. sac fungi
   b. zygote fungi   d. club fungi

6. The mycelium of a multicelled fungus is a mesh of filaments, each called a _____ .

Data Analysis Activities

Fighting a Forest Fungus The club fungus *Armillaria ostoyae* infects living trees and acts as a parasite, withdrawing nutrients from them. After the tree dies, the fungus continues to feed on its remains. Fungal hyphae grow out from the roots of infected trees as well as the roots of dead stumps. If these hyphae contact roots of a healthy tree, they can invade and cause a new infection.

Canadian forest pathologists hypothesized that removing fungus-infected stumps after logging could help prevent tree deaths. To test this hypothesis, they carried out an experiment. In half of a forest they removed stumps after logging. In a control area, they left stumps behind. For more than 20 years, they recorded tree deaths and whether *A. ostoyae* caused them. Figure 22.12 shows the results.

1. Which tree species was most often killed by *A. ostoyae* in control forests? Which was least affected by the fungus?

2. For the most affected species, what percentage of deaths did *A. ostoyae* cause in control and in experimental forests?

3. Looking at the overall results, do the data support the hypothesis? Does stump removal reduce tree mortality from *A. ostoyae*?

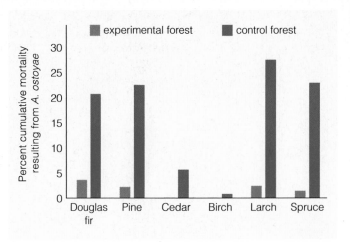

Figure 22.12 Results of a long-term study of how logging practices affect tree deaths caused by the fungus *A. ostoyae*. In the experimental forest, whole trees—including stumps—were removed (*brown* bars). The control half of the forest was logged conventionally, with the stumps left behind (*blue* bars).

---

7. A mushroom is _____ .
   a. the digestive organ of a club fungus
   b. the only part of the fungal body made of hyphae
   c. a reproductive structure that releases sexual spores
   d. the only diploid phase in the club fungus life cycle

8. Spores released from a mushroom's gills are _____ .
   a. diploid   b. haploid   c. dikaryotic

9. _____ are fungi that produce flagellated spores.
   a. Chytrids   c. Zygote fungi
   b. Sac fungi   d. Club fungi

10. Nitrogen-fixing cyanobacteria often interact with a fungus as a _____ .
    a. mycelium   c. mycorrhiza
    b. lichen   d. mycosis

11. _____ are mycorrhizal fungi with hyphae that grow into a root cell and branch inside it.
    a. Glomeromycetes   c. Zygote fungi
    b. Chytrids   d. Club fungi

12. Histoplasmosis is an example of a _____ .
    a. mycelium   c. mycorrhiza
    b. lichen   d. mycosis

13. Chestnut blight _____ .
    a. altered the species composition of eastern forests
    b. was caused by an introduced fungal pathogen
    c. was spread by wind-dispersed spores
    d. all of the above

14. Ergotism _____ .
    a. results from inhalation of fungal spores
    b. produces hallucinations and convulsions
    c. affects only those with weak immune systems
    d. all of the above

15. Match the terms appropriately.
    ___ hypha   a. produces flagellated spores
    ___ chitin   b. component of fungal cell walls
    ___ chytrid   c. partnership between a fungus and
    ___ sac fungus   one or more photoautotrophs
    ___ club fungus d. filament of a mycelium
    ___ lichen   e. fungus–root partnership
    ___ mycorrhiza f. forms sexual spores in an ascus
                      g. many form mushrooms

Additional questions are available on **CENGAGENOW.**

Critical Thinking

1. Fungal skin diseases are persistent. Ointments and creams may not reach the deepest infected skin layers. Oral antifungal drugs are far less common than antibacterials and often have bad side effects. Reflect on the evolutionary relationships among bacteria, fungi, and humans. Explain why it is more difficult to create drugs against fungi than bacteria.

2. Certain toxic mushrooms have bright, distinctive colors that mushroom-eating animals learn to recognize. Once sickened, day-feeding animals tend to avoid them. Other toxic mushrooms are as dull-looking as edible ones, but they have a much stronger odor. Some scientists think the strong odors aid in defense against predators that are active at night. Explain their reasoning.

Animations and Interactions on **CENGAGENOW:**
> Zygote fungus life cycle; Predatory fungus; Club fungus life cycle.

‹ Links to Earlier Concepts

This chapter draws on your knowledge of animal tissues and organs (Section 1.2) and life cycles (12.5), and of homeotic genes (10.3), fossils (16.2, 16.5), analogous structures (16.8), speciation (17.10) and biomarkers (18.5). You will see another use of chitin (3.3) and more effects of osmosis (5.6). You will learn how animals interact with dinoflagellates (20.5), with protists that cause malaria (20.6), and with flowering plants (21.8).

Key Concepts

Introducing the Animals
Animals are multicelled heterotrophs that digest food inside their body. Most animals can move from place to place during some or all of the life cycle. Animals probably evolved from a colonial protist. The vast majority of modern animals are invertebrates.

Structurally Simple Invertebrates
Sponges filter food from the water and have an asymmetrical body with no true tissues. Cnidarians are predators. Their body is radially symmetrical, with two tissue layers and a gelatinous matrix between the two.

23 Animals I: Major Invertebrate Groups

When asked to think of an animal, you probably envision one of the **vertebrates**, an animal that has a backbone. Mammals, birds, reptiles, amphibians, and fishes fall into this category. However, the vast majority of animals (more than 97 percent) have no backbone and are commonly called **invertebrates**.

The many invertebrate species represent a vast and largely unexplored reservoir of genetic diversity. Consider for example the cone snails (genus *Conus*), a group most common in warm, nearshore waters of the South Pacific. Humans have long valued the snails for their elaborately patterned shells (*right*). But cone snails interest biologists for additional reasons. The snails are predators and, like some snakes, they produce venom that helps them subdue their prey.

Studies of how cone snail venoms function have led to the development of new drugs. For example, one fish-eating cone snail, *C. magus*, captures prey by harpooning it, pumping it full of venom, then reeling it in. The snail's venom contains a peptide that interferes with the ability to feel pain. When injected into a fish, this peptide keeps the fish from writhing in pain and possibly damaging the snail's harpoon. An injectable pain-killing drug called ziconotide (sold under the brand name Prialt) is a synthetic version of the *C. magus* peptide. Unlike morphine and similar drugs, ziconotide is nonaddictive. Peptides from the venom of another fish-eating cone snail, *C. geographus* (shown at the *left*), may help control convulsions and seizures.

Studies of cone snail peptides also shed light on the evolution of animal genomes. For example, one gene that encodes an enzyme in the venom production pathway of *C. geographus* also exists in fruit flies and humans. Thus, this gene must have existed in the common ancestor of snails, insects, and vertebrates, which lived about 500 million years ago. After these groups diverged, the gene mutated independently in each lineage, and its product diverged in function. In humans, the gene product helps repair blood vessels. We have yet to figure out what role the gene plays in fruit flies.

This example supports an organizing principle in the study of life. Look back through time, and you discover that all organisms are related. At each branch point in the animal family tree, mutations gave rise to changes in biochemistry, body plans, or behavior. The mutations were the source of unique traits that define each lineage.

This chapter focuses on the major groups of invertebrates. The next chapter continues the story, focusing on vertebrates and their closest invertebrate relatives. Even with two chapters, we can scarcely do justice to so many animal species. We can only consider the scope of diversity and the major evolutionary trends. Keep in mind that although invertebrates are structurally simpler than vertebrates, they are not evolutionarily inferior. The diversity and longevity of the invertebrate lineages attests to how well they are adapted to their environment.

---

invertebrate   Animal that does not have a backbone.
vertebrate   Animal with a backbone.

Bilateral Invertebrates
Most animals have a body with bilateral symmetry and a concentration of nerve cells in the head. All bilateral animals have organ systems. In flatworms, the organs are hemmed in by tissue. In other bilateral animals, the organs reside inside a fluid-filled body cavity.

The Most Successful Animals
In terms of diversity, arthropods are the most successful animals. Crustaceans abound in the seas and insects are the most diverse animal group on land. Insects play essential roles in ecosystems. They also have economic and health effects.

On the Road to Chordates
Echinoderms are on the same branch of the animal family tree as the vertebrates (animals that have a backbone). Echinoderm adults have a spiny skin and a radially symmetrical body. The larvae show bilateral symmetry, indicating descent from a bilateral ancestor.

> No one trait defines all animals. It takes a list of features to see what sets them apart from all other organisms.
< Links to Animal life cycle 12.5, Tissues 1.2

What Is an Animal?

An **animal** is a multicelled heterotroph with unwalled cells. In contrast to fungi, animals typically ingest food (take it in) and digest it inside their body. Nearly all animals are motile (can move from place to place) during all or some part of their life cycle. Cells of an animal body are diploid. During sexual reproduction, meiosis of specialized germ cells produces haploid gametes. The flagellated sperm swim to eggs, and fertilization produces a zygote. Cells differentiate as the animal develops from an embryo (an early developmental stage) to an adult.

Evolution of Animal Body Plans

Figure 23.1 shows an evolutionary tree for the major animal groups covered in this book. All animals are descended from a common multicelled ancestor ❶ and constitute the clade Metazoa.

The earliest animals were probably aggregations of cells, and sponges still show this level of organization. However, most animals have tissues ❷. Each tissue consists of specific cell types arranged in a way that allows the tissue to carry out a particular task. Tissues develop from embryonic germ layers. In the early animal lineages,

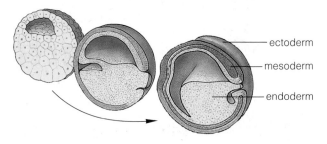

Figure 23.2 How a three-layer animal embryo forms. Most animals have this type of embryo.

embryos had two such layers: an outer **ectoderm** and an inner **endoderm**. In later lineages, cells began to rearrange themselves and form a middle embryonic layer called **mesoderm** (Figure 23.2). The evolution of a three-layer embryo allowed an increase in structural complexity. Most animal groups have organs derived from mesoderm.

The structurally simplest animals such as sponges are asymmetrical. You cannot divide their body into halves that are mirror images. Jellies, sea anemones, and other cnidarians have **radial symmetry**: Body parts are repeated around a central axis, like the spokes of a wheel ❸. Radial animals have no front or back end. They attach to an underwater surface or drift along. Thus, their food can arrive from any direction. Most animals have **bilateral symmetry**. They have many paired structures, so the

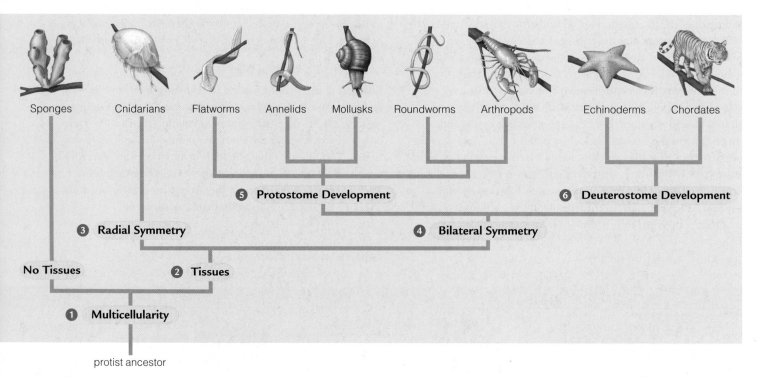

Figure 23.1 One proposed evolutionary tree for major animal groups based on morphological, developmental, and genetic comparisons. Like all such trees, it is a hypothesis and open to revision in light of new data.

>> Figure It Out What invertebrate group is most closely related to chordates?

Answer: Echinoderms

right and left half of the body are mirror images ④. Bilateral animals typically have **cephalization**. Many nerve cells and sensory structures have become concentrated at the anterior (front) end of the body. Bilateral animals move through the world head first, and they use the sensory structures on their head to seek out food and detect potential threats.

The head end of most bilateral animals has an opening for taking in food. Flatworms have a saclike digestive cavity. Food and wastes leave through the same body opening. However, most bilateral animals have a tubular gut, with a mouth at one end and an anus at the other. A tubular gut, known as a complete digestive system, has advantages. Parts of the tube can be specialized for taking in food, digesting food, absorbing nutrients, or compacting the wastes. Unlike a saclike cavity, a complete digestive system can carry out all of these tasks simultaneously.

Two lineages of bilateral animals differ in their embryonic development. In **protostomes**, the first opening that appears on the embryo becomes the mouth ⑤. *Proto–* means first and *stoma* means opening. In **deuterostomes**, the mouth develops from the second embryonic opening, and the first becomes the anus ⑥.

A mass of tissues and organs surrounds the flatworm gut (Figure 23.3A), but a fluid-filled body cavity surrounds the gut of most bilateral animals. A few animals such as roundworms enclose their gut in an unlined cavity called a **pseudocoelom** (Figure 23.3B). More typically, bilateral animals have a **coelom**, a body cavity lined with a tissue derived from mesoderm (Figure 23.3C).

A fluid-filled coelom or pseudocoelom provides three advantages. First, materials can diffuse through the fluid to body cells. Second, muscles can redistribute the fluid to alter the body shape and aid locomotion. Third, organs are not hemmed in by a mass of tissue, so they can grow larger and move more freely.

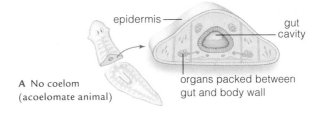

A No coelom (acoelomate animal) / organs packed between gut and body wall / epidermis / gut cavity

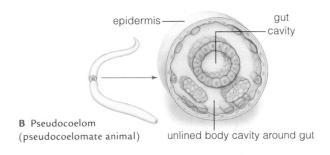

B Pseudocoelom (pseudocoelomate animal) / epidermis / gut cavity / unlined body cavity around gut

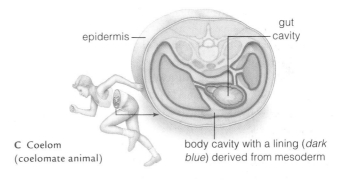

C Coelom (coelomate animal) / epidermis / gut cavity / body cavity with a lining (*dark blue*) derived from mesoderm

Figure 23.3 Animated Main body cavity of bilateral animals. **A** Flatworm, with no body cavity. **B** Roundworm, with an unlined, fluid-filled body cavity (a pseudocoelom). **C** Vertebrate (human) with a body cavity lined with tissue derived from mesoderm (*dark blue*).

---

animal Multicelled heterotroph with unwalled cells. Most ingest food and are motile during at least part of the life cycle.
bilateral symmetry Having paired structures so the right and left halves are mirror images.
cephalization Having a concentration of nerve and sensory cells at the head end.
coelom Body cavity lined with tissue derived from mesoderm.
deuterostomes Lineage of bilateral animals in which the second opening on the embryo surface develops into a mouth.
ectoderm Outermost tissue layer of an animal embryo.
endoderm Innermost tissue layer of an animal embryo.
mesoderm Middle tissue layer of a three-layered animal embryo.
protostomes Lineage of bilateral animals in which the first opening on the embryo surface develops into a mouth.
pseudocoelom Unlined body cavity.
radial symmetry Having parts arranged around a central axis, like spokes around a wheel.
segmentation Having a body composed of similar units that repeat along its length.

Most bilaterally symmetrical animals have some degree of **segmentation**, a division of a body into similar units, repeated one after the other along the main axis. We clearly see body segments in annelids such as earthworms. Segmentation opened the way to evolutionary innovations in body form. When many segments have structures that carry out the same function, some segments can become modified without endangering the animal's survival.

Take-Home Message **What are animals?**

› Animals are multicelled heterotrophs with bodies made of diploid, unwalled cells. Animals develop in stages and most are motile.

› Early animals had no body symmetry. Later, radially symmetrical, and then bilaterally symmetrical bodies evolved. Most animals with bilateral symmetry have a coelom, a fluid-filled body cavity with a lining derived from mesoderm.

> ❯ Fossils and gene comparisons among modern species
> provide insights into how animals arose and diversified.
> ❮ Links to Homeotic gene 10.3, Biomarker 18.5, Speciation 17.10

Colonial Origins

According to the colonial theory of animal origins, the first animals evolved from a colonial protist. At first, all cells in the colony were similar. Each could reproduce and carry out all other essential tasks. Later, mutations produced cells that specialized in some tasks and did not carry out others. Perhaps these cells captured food more efficiently, but did not make gametes. Colonies that had interdependent cells and a division of labor were at a selective advantage, and new specialized cell types evolved. Eventually this process produced the first animal.

What was the ancestral cell like? **Choanoflagellates**, the modern protists most closely related to animals, provide clues. Their name means "collared flagellate." Each choanoflagellate cell has a collar of threadlike structures, called microvilli, surrounding a flagellum (Figure 23.4A). Movement of the flagellum directs food-laden water past microvilli, which filter out food. As you will see, sponges feed the same way. Some choanoflagellates live as single cells, while others form colonies (Figure 23.4B).

The Simplest Living Animal

Studies of the modern representatives of ancient animal lineages can give us an idea of what the earliest animals may have been like. For example, **placozoans** are the oldest group with living representatives. The single species, *Trichoplax adhaerens*, is an asymmetrical marine animal about 2 millimeters in diameter and 2 micrometers thick (Figure 23.4C). It has the simplest body structure of any animal, with only four types of cells. It also has the smallest genome. Ciliated cells on the placozoan's surface allow it to move about. It feeds on bacteria and algae.

Genetic studies of *T. adhaerens* revealed the genetic foundations for traits that evolved in more recent animal lineages. For example, *T. adhaerens* has no nerve cells, but it has genes like those that encode signaling molecules in human nerves. It also has a gene similar to the homeotic genes that regulate development of body parts in more complex animals.

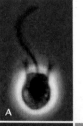

Figure 23.4 **A** One free-living choanoflagellate. A collar of microvilli rings its flagellum. **B** Colony of choanoflagellates. Some researchers view it as a model for the origin of animals. **C** The placozoan, *Trichoplax adhaerens*. Its asymmetrical, two-layered body has ciliated surfaces. The red color of this individual results from a diet of red algae.

Figure 23.5 Fossil of *Spriggina*, an Ediacaran, that lived about 570 million years ago. It was about 3 centimeters (1 inch) long and may have been a soft-bodied ancestor of arthropods, a major group that diversified during the Cambrian.

Fossil Evidence

Biomarkers characteristic of certain sponges have been found in deposits that date back more than 600 million years. Morphological fossils show that by 570 million years ago, a diverse collection of multicelled organisms, including some animals, lived in the seas (Figure 23.5). These groups are collectively called Ediacarans because their fossils were first discovered in Australia's Ediacara Hills.

Animals underwent a dramatic adaptive radiation during the Cambrian (542–488 million years ago). By the end of this period, all major animal lineages were present in the seas. What caused this Cambrian explosion in diversity? Rising oxygen levels and changes in global climate may have played a role. Also, supercontinents were breaking up. Movement of land masses isolated populations, thus increasing opportunities for allopatric speciation. Biological factors also encouraged speciation. Once the first predators arose, mutations that produced protective hard parts were favored. Evolution of homeotic genes may have sped things along. Mutations in these genes would have allowed adaptive changes to body form in response to predation or altered habitat conditions.

---

choanoflagellate Member of the protist group most closely related to animals.
placozoan The simplest modern animal, with an asymmetrical flat body, four types of cells, and a small genome.

Take-Home Message How did animals originate and diversify?

> ❯ The ancestor of all animals was probably a colonial protist. It may have resembled choanoflagellates, the modern protist group most closely related to animals.

> ❯ The oldest known animal lineage, placozoans, includes one marine species. It is tiny, with only four types of cells.

> ❯ Animals may have originated more than 600 million years ago. A great adaptive radiation that occurred during the Cambrian period was probably encouraged by a combination of environmental and biological factors.

> Sponges have no tissues or organs. They filter food from the water that flows through their porous body.
< Link to Phagocytosis 5.8

General Characteristics

Sponges (phylum Porifera) are aquatic animals with no symmetry, tissues, or organs. They resemble a colony of choanoflagellates but with more kinds of cells and a greater division of labor. Most live in tropical seas.

An adult sponge lives attached to a surface and has a body riddled with pores. Flat, nonflagellated cells form the outer surface, flagellated collar cells line the inner one, and a jellylike matrix lies in between (Figure 23.6A).

Most sponges feed on bacteria filtered from water. As in choanoflagellates, movement of flagella drives food-laden water past a collar of microvilli. As water flows through pores in the body wall, collar cells trap bits of food and engulf them by phagocytosis. All digestion is intracellular. Amoeba-like cells in the matrix receive food from the collar cells and distribute it through the body.

In many species, cells in the matrix secrete fibrous proteins or glassy silica spikes. These materials structurally support the body and discourage predators. Some of the protein-rich sponges are commercially important. These sponges are harvested from the sea, dried, cleaned, and bleached. The rubbery protein remains are then sold for use in bathing and cleaning (Figure 23.6B). About $40 million worth of sponges are harvested each year.

Sponge Reproduction

A typical sponge is a **hermaphrodite**; it produces both eggs and sperm. Sperm are released into the water (Figure 23.7A). Eggs are held in the parental body. Fertilization produces a zygote that develops into a ciliated larva (Figure 23.7B). A **larva** (plural, larvae) is a free-living, sexually immature stage in an animal life cycle. Sponge larvae swim briefly, then settle and become adults.

Many sponges reproduce asexually when small buds or fragments break away and grow into new sponges. Some freshwater species survive oxygen-poor water, drying out, or freezing by producing gemmules: tiny clumps of resting cells encased in a hardened coat. Wind can disperse gemmules. Those that land in a hospitable habitat become active again and grow into new sponges.

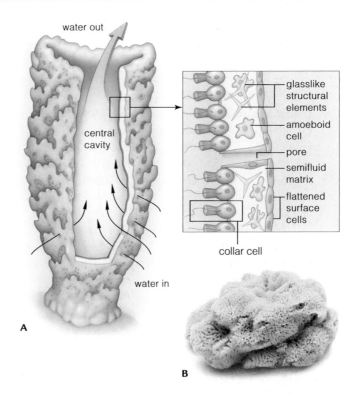

Figure 23.6 **Animated** The sponge body. **A** Body plan of a simple sponge. Flattened cells cover the outer surface and line pores. Flagellated collar cells line inner canals and chambers. Microvilli of these cells act like a sieve that strains food from the water. Cells in the matrix distribute nutrients and secrete structural elements. **B** Dried protein-rich remains of a bath sponge.

Figure 23.7 Sexual reproduction in sponges. **A** Barrel sponge releasing sperm. **B** Colorized scanning electron micrograph of a sponge larva.

Take-Home Message What are sponges?

> Sponges are filter-feeders that have no symmetry, tissues, or organs. Fibers and glassy spikes in the body wall support the body and help deter predators.

> Sponges are hermaphrodites. Their ciliated larvae swim about briefly before settling and developing into adults.

---

hermaphrodite Animal that makes both eggs and sperm.
larva Preadult stage in some animal life cycles.
sponge Aquatic invertebrate that has no tissues or organs and filters food from the water.

> Nearly all cnidarians live in the seas. They have radially symmetrical bodies and a mouth surrounded by tentacles.
< Link to Dinoflagellates 20.5

General Characteristics

Cnidarians (phylum Cnidaria) include radially symmetrical animals such as corals, sea anemones, and jellyfish (also called jellies). Nearly all are marine predators. There are two cnidarian body shapes, the medusa and the polyp (Figure 23.8). In both, a tentacle-ringed mouth opens onto a saclike gastrovascular cavity that functions in digestion and gas exchange.

Medusae such as jellyfish are shaped like a bell or umbrella, with a mouth on the lower surface. Most swim or drift about. Polyps such as sea anemones are tubular and one end usually attaches to a surface.

Both medusae and polyps consist of two tissue layers. The outer layer (epidermis) develops from ectoderm, and the inner layer (gastrodermis) from endoderm. Mesoglea, a jellylike secreted matrix, fills the space between the tissue layers. Medusae tend to have a lot of mesoglea. Polyps usually have somewhat less.

The name Cnidaria is from *cnidos*, the Greek word for nettle, a kind of stinging plant. Cnidarian tentacles have stinging cells (cnidocytes) with unique organelles called **nematocysts**. Nematocysts help capture prey and also

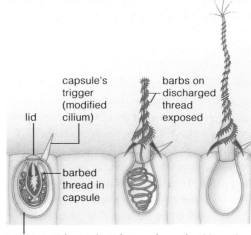

nematocyst (capsule at free surface of epidermal cell)

Figure 23.9 Animated Nematocyst action. In response to a touch of the trigger, the capsule around the nematocyst becomes more permeable. As water diffuses in, pressure builds up, and the thread is forced to turn inside out. The barbed tip pierces prey.

function as a defense. Like a jack-in-the-box, a nematocyst holds a coiled thread beneath a hinged lid (Figure 23.9). When something brushes against the nematocyst's trigger, the lid opens. The thread inside pops out and ejects a barbed thread. Typically, the barbs snag prey. Tentacles then move the prey through the mouth and deposit it in the gastrovascular cavity. Gland cells in the lining of this cavity secrete enzymes that digest the prey.

Occasionally, people trigger nematocyst release by brushing against a coral or jellyfish while swimming. Nematocysts irritate human skin and the stings of one group of jellyfish, known as box jellies, can be deadly.

Coordinated movement of body parts such as tentacles requires interactions among nerves and contractile cells. Cnidarians have a decentralized nervous system called a **nerve net**. This network of nerve cells extends through the body. Tentacles and other body parts move when nerve cells signal contractile cells to shorten. Contraction of these cells redistributes mesoglea, just as a water-filled balloon changes shape when you squeeze it. Mesoglea is a **hydrostatic skeleton**, an enclosed fluid that contractile cells exert force on to bring about movement.

Cnidarian Life Cycles and Diversity

Depending on the group, a cnidarian life cycle may include both polyps and medusae, only polyps, or only medusae. Figure 23.10 shows the life cycle of *Obelia*, a colonial hydroid that produces both polyps and medusae. Sexual reproduction occurs after medusae form on special polyps. Medusae release gametes that fuse, forming a zygote. As in most cnidarians, bilateral ciliated larvae develop. They swim briefly, then settle and develop into

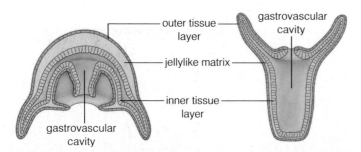

A Diagram showing cross-section of the medusa (*left*) and the polyp (*right*).

B Medusa of the hydroid *Obelia*.

C A sea anemone, a polyp.

Figure 23.8 Animated Cnidarian body plans.

Figure 23.10 Animated Life cycle of *Obelia*, a hydroid with medusa and polyp generations. Figure 23.8B is a photo of the medusa.

1 Medusae are the sexual generation. Each medusa makes either eggs or sperm and releases them into the water. The gametes combine to form a zygote.

2 The zygote develops into a ciliated bilateral larva.

3 The larva swims about for a while, then settles and develops into a polyp.

4 The polyp grows and reproduces asexually, eventually producing a branching colony.

5 Some branches of the colony are specialized for capturing and eating prey.

6 Other branches produce and release medusae that begin the sexual phase of the life cycle again.

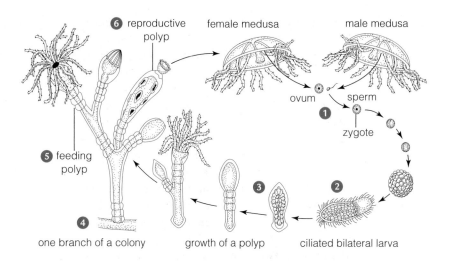

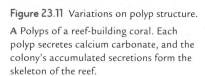

6 reproductive polyp — female medusa — male medusa

ovum — sperm

1

zygote

5 feeding polyp

3   **2**

4

one branch of a colony — growth of a polyp — ciliated bilateral larva

polyps. The colony grows asexually when new polyps branch from older ones.

Sea anemones (Figure 23.8C), corals (Figure 23.11A), and hydras (Figure 23.11B) do not have a medusa stage in their life cycle. During sexual reproduction, gametes form on polyps. Polyps also reproduce asexually by budding.

Coral reefs are colonies of polyps enclosed in a skeleton of secreted calcium carbonate. In a mutually beneficial relationship, photosynthetic dinoflagellates live inside each polyp's tissues. The protists receive shelter and carbon dioxide from the coral, which gets sugars and oxygen in return. If a reef-building coral loses its protist symbionts, an event called "coral bleaching," it may die.

Siphonophores are cnidarian colonies that float or drift in the seas. For example, the Portuguese man-of-war (*Physalia*) in Figure 23.11C appears to be a single animal. In fact, it consists of many specialized polyps. The gas-filled float is one specialized polyp. Polyps that catch prey, digest prey, or produce gametes cluster beneath it.

Decline of the Corals, Rise of the Jellies

Increasing sea temperature and marine pollution may be contributing to two opposing trends among cnidarian populations. Reef-building coral populations are in decline, while population explosions of jellyfish have become increasingly common. The reef declines threaten not only corals, but also the many species that live in, on, or around reefs. Increasing jellyfish numbers present problems too. They sting swimmers, force beach closures, and can clog underwater pipes.

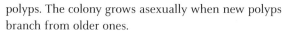
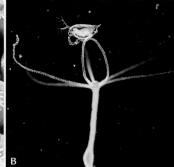

Figure 23.11 Variations on polyp structure.

A Polyps of a reef-building coral. Each polyp secretes calcium carbonate, and the colony's accumulated secretions form the skeleton of the reef.

B A hydra, one of the relatively few freshwater cnidarians, shown here with a captured water flea.

C The Portuguese man-of-war (also known as the blue bubble) consists of four kinds of polyp: the gas-filled float, prey-catching polyps, food-digesting polyps, and gamete-producing polyps. Swimmers who brush against the tentacles of this species end up with painful red welts.

cnidarian Radially symmetrical invertebrate with two tissue layers; uses tentacles with stinging cells to capture food.
hydrostatic skeleton Of soft-bodied invertebrates, a fluid-filled chamber that muscles act on, redistributing the fluid.
nematocyst Stinging organelle unique to cnidarians.
nerve net Mesh of nerve cells that allows movement and other behavior in cnidarians.

Take-Home Message What are cnidarians?

❯ Cnidarians are radially symmetrical predators with two layers of tissue.

❯ The two body forms, medusa and polyp, have tentacles with nematocysts around the mouth, which opens into a gastrovascular cavity.

❯ Movements occur when a nerve net signals contractile cells.

> Most flatworms live in the sea, but some live in fresh water or inside the bodies of other animals.
‹ Links to Organs and organ systems 1.2, Osmosis 5.6

With this section, we begin our survey of protostomes, one of the two lineages of bilaterally symmetrical animals. **Flatworms** (phylum Platyhelminthes) are bilateral worms with an array of organ systems, but (as shown in Figure 23.3A) no body cavity. Like other protostomes, flatworms develop from an embryo with three layers.

Turbellarians, flukes, and tapeworms are the main flatworm lineages. The typical turbellarian lives in tropical marine waters (Figure 23.12). However, a lesser number live in fresh water, and a few live in damp places on land. All flukes and tapeworms are parasites of animals.

Structure of a Free-Living Flatworm

Planarians, a type of turbellarian, commonly live in ponds. They glide along, propelled by the movement of cilia that cover the body surface. The planarian's saclike gut is highly branched (Figure 23.13A). Nutrients and oxygen diffuse from the gut's fine branches to all body cells. There is no anus; food enters and wastes leave through the mouth. Surprisingly, the mouth is not at the planarian's head end. Instead, it is at the tip of a muscular tube that extends from the animal's ventral (lower) surface.

There is clear cephalization. The planarian head has chemical receptors and light-detecting eyespots. These sensory structures send messages to a simple brain consisting of paired nerve cell bodies (ganglia). Nerve cords

Figure 23.12 A marine flatworm. Brightly colored flatworms are common inhabitants of coral reef ecosystems.

extend out from the head and run the length of the body (Figure 23.13B).

A planarian is a hermaphrodite, with both female and male sex organs (Figure 23.13C). However, individuals cannot fertilize their own eggs. They must mate. Planarians also reproduce asexually. The body splits in two near the middle, then each piece regrows the missing parts. Regrowth also occurs if a planarian is cut in two.

The body fluid of a planarian holds more solutes than its freshwater environment, so water tends to move into the body by osmosis. A system of tubes regulates internal water and solute levels. Flame cells with a tuft of "flickering" cilia drive any excess water into these tubes, which open to the body surface at a pore (Figure 23.13D–F).

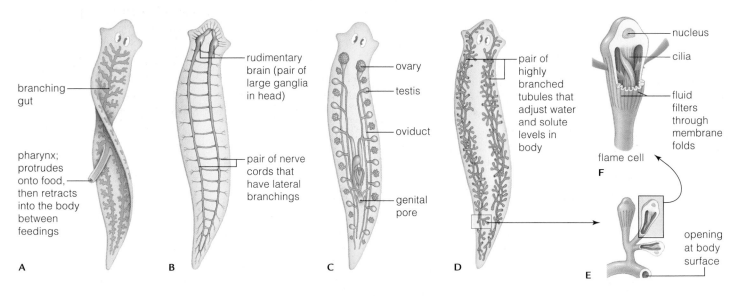

Figure 23.13 Animated Organ systems of a planarian, one of the flatworms. **A** Digestive system. This branching saclike gut is connected to a pharynx, which protrudes onto food, then retracts into a chamber between feedings. **B** Nervous system. **C** Reproductive system. **D–F** Water-regulating system.

Figure 23.14 Animated Life cycle of the beef tapeworm.

1 A person eats undercooked beef containing the resting stage of the tapeworm.

2 In the human intestine, the tapeworm uses its barbed scolex to attach to the intestinal wall. The worm grows by adding new body units called proglottids. Over time, it can become many meters long.

3 Each proglottid produces both eggs and sperm, which combine. Proglottids with fertilized eggs leave the body in feces.

4 Cattle eat grass contaminated with proglottids or early larvae.

5 The larval tapeworm forms a cyst in the cattle muscle.

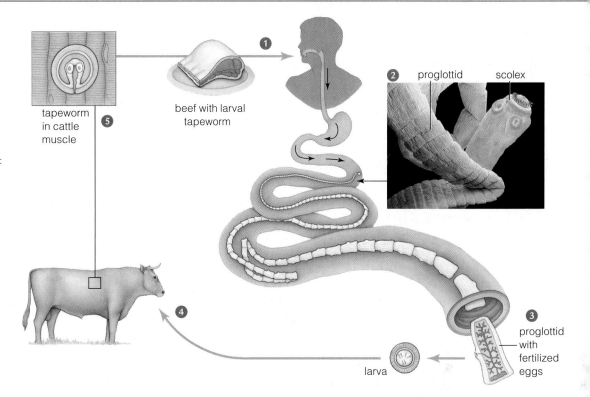

tapeworm in cattle muscle

beef with larval tapeworm

proglottid   scolex

proglottid with fertilized eggs

larva

Flukes and Tapeworms—The Parasites

Flukes and tapeworms are parasitic flatworms. Immature stages often live in one or more intermediate hosts, then reproduction occurs in the main host. For example, aquatic snails are the intermediate host for the blood fluke (*Schistosoma*). But the fluke can only reproduce inside a mammal, such as a human. Humans become infected when they wash or walk in water where infected snails live. The blood fluke crosses their skin and reproduces in their body, causing schistosomiasis. Worldwide, about 200 million people have this disease. It is most common in Asia.

Like flukes, tapeworms are parasites that live and reproduce in the vertebrate gut. The tapeworm head has a scolex, a structure with hooks or suckers that allow the worm to attach to the intestinal wall. Behind the head are body units called proglottids.

Unlike planarians and flukes, the tapeworm has no gut. Nutrients from the food its host ingests reach tapeworm cells by diffusing across the worm's body wall.

Figure 23.14 shows the life cycle of a beef tapeworm. Larvae enter the body when a person eats undercooked beef that contains larvae **1**. In the intestine, the tapeworm grows as new proglottids form in the region behind the scolex **2**. Each proglottid is hermaphroditic, and sperm from one can fertilize eggs in another. Older proglottids (farthest from the scolex) contain fertilized eggs. The oldest proglottids break off and exit the body in feces **3**. Fertilized eggs can survive for months before being eaten by cattle, the intermediate host **4**. Inside cattle, the tapeworm forms a cyst in muscle **5**.

Eating undercooked pork or fish can also cause a human tapeworm infection. Several outbreaks of human infection by fish tapeworms have been traced to raw salmon. Freezing fish for a week before it is served raw can eliminate the risk of tapeworms.

---

flatworm Bilaterally symmetrical invertebrate with organs but no body cavity; for example, a planarian or tapeworm.
planarian Free-living freshwater flatworm.

> **Take-Home Message** **What are flatworms?**
>
> ❯ Flatworms are bilateral, cephalized animals with a saclike gut, simple nervous system, and a system for regulating the content of internal body fluids. The organs develop from three tissue layers that form in the embryo. Flatworms do not have a body cavity.
>
> ❯ Free-living flatworms include many tropical marine species and the freshwater planarians.
>
> ❯ Flukes and tapeworms are parasitic flatworms. Both groups include species that infect humans.

> Annelids are segmented worms that live in aquatic environments or burrow through the soil on land.
< Link to Chitin 3.3

Annelids are bilateral worms with a coelom and a body that is distinctly segmented, inside and out. The majority of species are marine worms called polychaetes. Other groups are oligochaetes (including earthworms) and leeches. In both polychaetes and oligochaetes nearly all segments have chitin-reinforced bristles called chaetae. Hence the names polychaete and oligochaete (*poly–*, many; *oligo–*, few).

Marine Polychaetes

The best-known polychaetes are the sandworms (Figure 23.15A). They are often sold as bait for saltwater fishing. Sandworms use their chitin-strengthened jaws to capture other soft-bodied invertebrates. Each body segment has a pair of paddlelike appendages that help the worm burrow in sediments and pursue prey.

Other polychaetes have modifications of this basic body plan. The fan worms and feather duster worms live inside a tube made of secreted mucus and sand grains. The head end protrudes from the tube and its elaborate tentacles capture food that drifts by (Figure 23.15B).

Leeches

Leeches live most commonly in fresh water, but some are marine or dwell in damp habitats on land. The leech body lacks conspicuous bristles and has a sucker at either end. Most leeches are scavengers and predators of small invertebrates. The infamous minority attach to a vertebrate, pierce its skin, and suck blood (Figure 23.16).

The saliva of bloodsucking leeches contains a protein that prevents blood from clotting while the leech feeds. For this reason, doctors who reattach a severed finger or

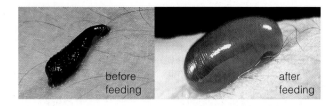

Figure 23.16 The leech *Hirudo medicinalis* feeding on human blood. A leech sticks to the skin with suckers at either end of the body, then draws blood with chitin-hardened jaws.

ear sometimes apply leeches to the reattached part. The presence of the leeches prevents unwanted clots from forming inside blood vessels of the reattached part.

Earthworms—Oligochaetes That Burrow

Oligochaetes include marine and freshwater species, but the land-dwelling earthworms are most familiar. We consider their body in detail as our example of annelid structure (Figure 23.17).

An earthworm body is segmented inside and out. The outer layer is a cuticle of secreted proteins. Visible grooves on its surface correspond to internal partitions. A fluid-filled coelom runs the length of the body. It is divided into coelomic chambers, one per segment.

Gases are exchanged across the body surface, and a **closed circulatory system** helps distribute oxygen. In such a system, blood flows through a continuous system of vessels. All exchanges between blood and the tissues take place across a vessel wall. Hearts in the anterior of the worm provide the pumping power to move the blood.

The earthworm has a complete digestive system that extends through all coelomic chambers. Earthworms are scavengers that eat their way through the soil and digest organic debris. The worms improve soil by loosening its particles and excreting tiny bits of organic matter that

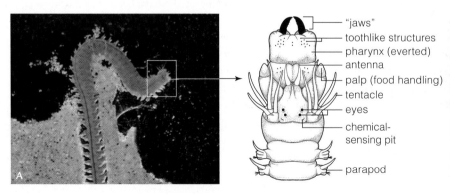

"jaws"
toothlike structures
pharynx (everted)
antenna
palp (food handling)
tentacle
eyes
chemical-sensing pit
parapod

Figure 23.15 Polychaetes. **A** The sandworm uses its many appendages to burrow into sediment on marine mudflats. It is an active predator with hard jaws. **B** The feather duster worm secretes and lives inside a hard tube. Feathery extensions on the head capture food from the water. Cilia move captured food to the mouth.

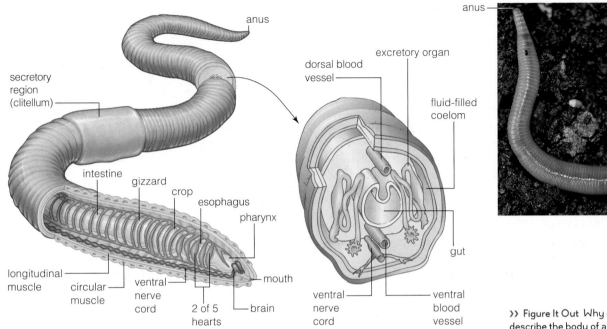

Figure 23.17 **Animated** Earthworm body plan. Each segment contains a coelomic chamber full of organs. A gut, ventral nerve cord, and dorsal and ventral blood vessels run through all coelomic chambers.

›› **Figure It Out** Why do people sometimes describe the body of a coelomate animal such as an earthworm as "a tube inside a tube"?

Answer: One tube, the gut, is inside another tube, the body wall. The fluid-filled coelom lies between the tubes.

decomposers can easily break down. Excreted earthworm "castings" are sold as a fertilizer.

Most segments have paired excretory organs that regulate the solute composition and volume of coelomic fluid. Each organ collects coelomic fluid, adjusts its composition, and expels waste through a pore in the body wall.

Earthworm behavior is coordinated by a simple brain. A pair of nerve cords extend from the brain and run the length of the body. The brain receives information from sensory organs, such as light-sensing organs in the body wall. It also coordinates locomotion.

The earthworm has two sets of muscle. Longitudinal muscles parallel the body's long axis, and circular muscles ring the body. Contraction of these muscles puts pressure on coelomic fluid trapped inside body segments, causing the segments to change shape. When a segment's longitudinal muscles contract, the segment gets shorter and fatter. When circular muscles contract, a segment gets longer and thinner. The worm moves as coordinated waves of contraction run along the body (Figure 23.18).

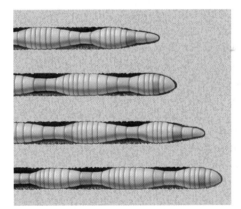

Figure 23.18 How earthworms move through soil. An earthworm's coelomic fluid serves as a hydrostatic skeleton. Circular and longitudinal muscles that work in opposition exert pressure on this fluid. Coordinated activity of these muscles alters segment shape and produces waves of contraction. The worm's front end is pushed forward, then bristles are extended to anchor the front as rear segments are pulled up behind it.

Earthworms are hermaphrodites, but they cannot fertilize themselves. During mating, a secretory organ (the clitellum) produces mucus that glues two worms together while they swap sperm. Later, that organ secretes a silky case that protects the fertilized eggs.

annelid Segmented worm with a coelom, complete digestive system, and closed circulatory system.
closed circulatory system Circulatory system in which blood flows through a continuous network of vessels; all materials are exchanged across the walls of those vessels.

Take-Home Message **What are annelids?**

› Annelids are bilateral, coelomate, segmented worms that have a complete digestive system and a closed circulatory system.

› The group includes aquatic species and the land-dwelling earthworms.

❯ Is there a "typical" mollusk? No. The group has more than 100,000 named species, including tiny snails in treetops, burrowing clams, and giant predators of the open ocean.
❮ Link to Analogous structures 16.8

Mollusks (phylum Mollusca) are bilaterally symmetrical invertebrates with a reduced coelom. Most dwell in seas, but some live in fresh water or on land. All have a mantle, a skirtlike extension of the upper body wall that covers a mantle cavity (Figure 23.19). In shelled mollusks, the shell consists of a calcium-rich, bonelike material secreted by the mantle. Aquatic mollusks typically have one or more respiratory organs called gills inside their fluid-filled mantle cavity. The mollusk digestive system is always complete. In most mollusks, the mouth contains a radula, a tonguelike organ hardened with chitin.

With more than 100,000 living species, mollusks are second only to arthropods in level of diversity. There are three main groups: gastropods, bivalves, and cephalopods (Figure 23.20).

Gastropods

Gastropods are the most diverse mollusk lineage. Their name means "belly foot," and most species glide about on the broad muscular foot that makes up most of the lower body mass (Figure 23.20A). A gastropod shell, when present, is one-piece and often coiled.

Gastropods have a distinct head that usually has eyes and sensory tentacles. In many aquatic species, a part of the mantle forms an inhalant siphon, a tube through which water is drawn into the mantle cavity. Gastropods and bivalves have an **open circulatory system**. In such a system, blood vessels do not form a continuous loop. Instead, blood leaves vessels and seeps around tissues before returning to the heart. Cells exchange substances with the blood while it is outside of vessels.

Gastropods include the only terrestrial mollusks. In land-dwelling snails and slugs (Figure 23.21A,B), a lung

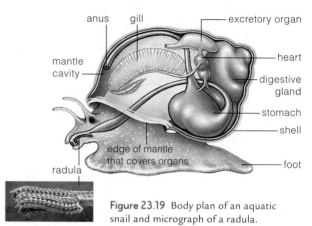

Figure 23.19 Body plan of an aquatic snail and micrograph of a radula.

replaces the gill. Glands on the foot continually secrete mucus that protects the animal as it moves across dry, abrasive surfaces. Most mollusks have separate sexes, but land-dwellers tend to be hermaphrodites. Unlike other mollusks, which produce a swimming larva, the embryos of these groups develop directly into adults.

Lacking a shell, slugs and sea slugs must defend themselves in other ways. Some make and secrete distasteful substances. Certain sea slugs eat cnidarians such as jellyfish and store undischarged nematocysts that serve as a defense. For example, frilly extensions on the back of a Spanish shawl nudibranch function in gas exchange and hold nematocysts (Figure 23.21C).

Bivalves

Bivalves include many mollusks that end up on dinner plates, including mussels, oysters, clams, and scallops (Figure 23.20B). All bivalves have a hinged, two-part shell. Powerful muscles hold the valves together. Contraction of these muscles pulls the two valves shut, enclosing the body and protecting it from predation or drying out. A bivalve has no obvious head. In many, simple eyes arrayed around the edge of the mantle help detect dangers. Most bivalves feed by drawing water into their mantle cavity

A Gastropods (belly footed)　　　B Bivalves (two-part shell)　　　C Cephalopods (jet-propelled)

Figure 23.20 Representatives of three main mollusk groups.

Figure 23.21 Variations on the gastropod body plan. Land snails **A** and slugs **B** are adapted to life in a dry habitat. They have a lung in place of gills. The slime trail left behind after they move over a surface is mucus secreted by their big foot.

C Two Spanish shawl nudibranchs (*Flabellina iodinea*). These sea slugs feed on cnidarians and store undischarged nematocysts inside their bright red respiratory organs.

and trapping food in mucus on the gills. Movement of cilia directs particle-laden mucus to the mouth.

Cephalopods

Cephalopods include squids (Figure 23.20C), nautiluses (Figure 23.22A), octopuses (Figure 23.22B), and cuttlefish. All are predators and most have beaklike, biting mouthparts in addition to a radula. Cephalopods move by jet propulsion. They draw water into the mantle cavity, then force it out through a funnel-shaped siphon. The foot has been modified into arms and/or tentacles that extend from the head. Cephalopod means "head-footed."

Five hundred million years ago, large cephalopods with a long conelike shell were the top predators in the seas. Today, nautiluses have a coiled external shell, but other cephalopods have a highly reduced shell or none at all. Jawed fishes evolved 400 million years ago, and competition with this group may have favored a shift in body form. Cephalopods with the smallest shell could be fastest and most agile. A speedier life-style required other changes as well. Of all mollusks, only cephalopods have a closed circulatory system. Competition with fishes also favored improved eyesight. Like vertebrates, cephalopods have eyes with lenses that focus light. Because mollusks and vertebrates are not closely related, this type of eye is assumed to have evolved independently in these groups.

Cephalopods include the fastest (squids), biggest (giant squid), and smartest (octopuses) invertebrates. Of all invertebrates, octopuses have the largest brain relative to body size, and show the most complex behavior.

Figure 23.22 Cephalopods. **A** Nautilus and **B** an octopus.

---

bivalve Mollusk with a hinged two-part shell.
cephalopod Predatory mollusk with a closed circulatory system; moves by jet propulsion.
gastropod Mollusk in which the lower body is a broad "foot."
mollusk Invertebrate with a reduced coelom and a mantle.
open circulatory system System in which blood leaves vessels and seeps through tissues before returning to the heart.

Take-Home Message What are mollusks?

❯ Mollusks are bilateral, soft-bodied, coelomate animals, the only ones with a mantle over the body mass.

❯ They include snails and slugs that glide about on a huge foot, bivalves with a hinged two-part shell, and the jet-propelled cephalopods.

❯ Roundworms are among the most abundant animals. Sediments in shallow water may hold a million per square meter; a cupful of topsoil teems with them.

Roundworms, or nematodes (phylum Nematoda), are bilateral, unsegmented animals that have a pseudocoelom and a cuticle-covered, cylindrical body (Figure 23.23). The collagen-rich cuticle is pliable, and it is repeatedly molted between growth spurts. **Molting** is the shedding of a body covering (or hair, or feathers).

Most roundworms are free-living species less than 5 millimeters long. The soil roundworm *Caenorhabditis elegans* is a favorite for genetic experiments. It has the same tissue types as complex animals. Yet it is transparent, it has only 959 body cells, and it reproduces fast. Its genome is about 1/30 the size of ours. Such traits make it easy for scientists to monitor each cell's fate during development.

Although most roundworms are free-living decomposers in soil or water, some parasitize plants and animals. Some of the animal parasites reach impressive lengths. One that lives inside sperm whales can be 13 meters long (42.65 feet).

Some parasitic roundworms impair human health, especially in developing countries. Eating undercooked pork or wild game can lead to an infection by *Trichinella spiralis*. The roundworm moves from the intestines, through blood, and into muscles. The result is trichinosis, a painful disease that can be fatal.

Wuchereria bancrofti and certain other roundworms cause lymphatic filariasis. Repeated infections injure lymph vessels, so lymph pools inside the legs and feet (Figure 23.24A). Elephantiasis, the common name for this disease, refers to fluid-filled, "elephant-like" legs. Mosquitoes carry the larval roundworms to new hosts.

Ascaris lumbricoides, a large roundworm, currently infects more than 1 billion people, primarily in Asia and in Latin America (Figure 23.24B). People become infected when they eat its eggs, which survive in soil and get on hands and into food. When enough adults occupy a host, they can clog the digestive tract.

---

molting Periodic shedding of an outer body layer or part.
roundworm Unsegmented worm with a pseudocoel and a cuticle that is molted as the animal grows.

Figure 23.24 Parasitic roundworms. **A** The grossly enlarged leg of the man on the *left* is symptomatic of lymphatic filariasis. The damage to lymphatic vessels is permanent. Fluid will continue to pool in the leg even if the roundworms that caused the damage are killed. **B** Live roundworms (*Ascaris lumbricoides*). These intestinal parasites cause stomach pain, vomiting, and appendicitis.

Hookworms, too, infect more than 1 billion people. Juveniles in the soil cut into human skin and migrate through blood vessels to the lungs. They climb up the windpipe, then enter the digestive tract when their host swallows. Once inside the small intestine, they attach to the intestinal wall and suck blood.

Pinworms commonly infect children. Female worms less than a millimeter long leave the rectum at night and lay eggs near the anus. The migration causes itching, and scratching puts eggs under fingernails. From there, they get into food and onto toys. If swallowed, the eggs spread the infection.

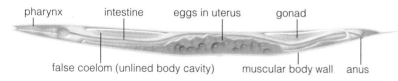

pharynx   intestine   eggs in uterus   gonad

false coelom (unlined body cavity)   muscular body wall   anus

Figure 23.23 Animated Body plan of *Caenorhabditis elegans*, one of the free-living roundworms. Sexes are separate, and this is a female.

Take-Home Message **What are roundworms?**

❯ Roundworms are bilateral worms with a cuticle that they molt as they grow. Organ systems almost fill their false coelom. Some are agricultural pests or human parasites.

> Arthropods are the most diverse invertebrate group. A variety of features contribute to their success.

< Link to Chitin 3.3

Major Groups

Arthropods (phylum Arthropoda) are bilateral animals with a hardened, jointed external skeleton, a complete gut, an open circulatory system, and a reduced coelom. Modern lineages include chelicerates (spiders and close relatives), crustaceans (such as lobsters, shrimps, and crabs), millipedes and centipedes, and insects. Before we discuss arthropod diversity, consider six key adaptations that contributed to their evolutionary success.

Key Arthropod Adaptations

Hardened Exoskeleton Arthropods have a cuticle (a secreted covering) of chitin, proteins, and waxes. It is an **exoskeleton**, an external skeleton. The exoskeleton protected aquatic species against predators. When some groups invaded land, the exoskeleton also helped them conserve water and supported their body.

A hard exoskeleton does not restrict size increases, because arthropods, like roundworms, molt periodically. Hormones control molting. They encourage the formation of a soft, new cuticle under the old one, which is then shed (Figure 23.25A).

Jointed Appendages If a cuticle were uniformly hard and thick like a plaster cast, it would prevent movement. Arthropod cuticles thin at joints, the regions where hardened body parts meet. Contraction of muscles that span joints makes the cuticle bend, so parts move relative to one another. *Arthropod* means jointed leg (Figure 23.25B).

Modified Segments In early arthropods, body segments were more or less alike. In many of their descendants, segments became fused together or modified for specialized tasks. For example, among insects, thin extensions of the wall of some segments evolved into wings (Figure 23.25C).

Respiratory Structures Aquatic arthropods have a gill or gills for gas exchange. Those that live on land have air-conducting tubes that start at surface pores, branch into finer tubes, and deliver oxygen to all internal tissues.

Figure 23.25 Some features that contribute to arthopod success.

A Hardened exoskeleton. This centipede is molting its old exoskeleton (*gray*).

B Jointed appendages, such as crab legs.

C Modified segments. This wing attached to one segment on the thorax of a fly.

D Specialized developmental stages. This larva of a monarch butterfly feeds on plant leaves. The adult monarch sips nectar.

Sensory Specializations Most arthropods have one or more pairs of eyes: organs that sample the visual world. With the exception of chelicerates, they also have paired **antennae** (singular, antenna) that detect touch and odors.

Specialized Stages of Development Especially among insects, the tasks of surviving and reproducing are divided among different stages of development. Individuals of many species undergo **metamorphosis**: Tissues get remodeled as embryonic stages make the transition to the adult form. A typical immature stage specializes in eating and growing rapidly, whereas the adult specializes in reproduction and dispersal. For instance, caterpillars chew leaves and metamorphose into butterflies, which fly about, sip nectar, mate, then deposit eggs (Figure 23.25D). Differences among stages of development are adaptations to specific conditions in the environment, including seasonal shifts in food, water, and shelter. Having such different bodies also prevents adults and juveniles from competing for the same resources.

---

antenna Of some arthropods, sensory structure on the head that detects touch and odors.
arthropod Invertebrate with jointed legs and a hard exoskeleton that is periodically molted.
exoskeleton Hard external parts that muscles attach to and move.
metamorphosis Dramatic remodeling of body form during the transition from larva to adult.

Take-Home Message **What traits contribute to the evolutionary success of the arthropods?**

> As a group, arthropods are abundant and widespread. A hardened, jointed exoskeleton, modified body segments, and specialized appendages, respiratory structures, and sensory structures underlie their evolutionary success.

> In many species success also arises from a division of labor among different stages of the life cycle, such as larvae, juveniles, and adults.

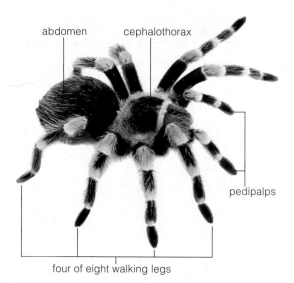

abdomen   cephalothorax

pedipalps

four of eight walking legs

A

B

C

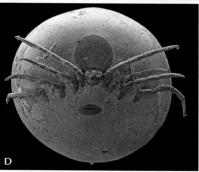

D

E

Figure 23.26 Examples of chelicerates. **A** Tarantula, a spider. **B** Horseshoe crab (*Limulus*). The long spine helps it steer as it swims. **C** Scorpion. Its last segment has a stinger and the pedipalps are claws that can seize prey. **D** A blood-engorged tick. **E** Dust mite, less than 0.5 millimeter in length.

❭ Chelicerates include the marine horseshoe crabs, but most are arachnids that live on land.
❰ Link to Tick-borne bacterial diseases 19.7

Chelicerates have a body with a cephalothorax (fused head and thorax) and abdomen (Figure 23.26A). Paired walking legs attach to the cephalothorax. The head has eyes, but no antennae. Paired feeding appendages called chelicerae near the mouth give the group its name. Another pair of appendages, called pedipalps, lie between the chelicerae and the first legs. The appearance and function of the pedipalps varies among chelicerate groups.

Horseshoe crabs are an ancient chelicerate lineage and the only one that is marine. They eat clams and worms. A horseshoe-shaped shield covers the cephalothorax (Figure 23.26B). The last body segment, the telson, has evolved into a long spine that acts as a rudder when the horseshoe crab swims. It also helps the animal right itself if a wave knocks it over. The spine does not produce venom. Each spring, horseshoe crab eggs laid along Atlantic and Gulf coasts serve as food for migratory shorebirds.

Arachnids include spiders, scorpions, ticks, and mites. All have four pairs of walking legs and live on land. Scorpions and spiders produce venom that helps them subdue their prey. Scorpions dispense venom through a stinger on their telson (Figure 23.26C). Their pedipalps have evolved into large claws. Spiders deliver venom with a bite. Their fanglike chelicerae have poison glands. Of 38,000 spider species, about 30 produce venom that can harm humans.

Most spiders indirectly help us by eating insect pests. Some are sit-and-wait predators. They use paired spinners on their abdomen to eject silk and weave prey-catching webs. Others such as tarantulas and jumping spiders are active hunters that sneak up and grab unwary prey.

All ticks are parasites that suck blood from vertebrates (Figure 23.26D). Some that bite humans can transmit the bacteria that cause Lyme disease or other diseases.

Mites are a highly diverse group that include parasites, predators, and scavengers (Figure 23.26E). Most are less than a millimeter long.

---

arachnids Land-dwelling arthropods with four pairs of walking legs; spiders, scorpions, mites, and ticks.
chelicerates Arthropod subgroup with specialized feeding structures (chelicerae) and no antennae.

Take-Home Message **What are chelicerates?**

❭ Chelicerates include horseshoe crabs and arachnids. The horseshoe crabs are an ancient marine lineage. Arachnids are land-dwellers with eight walking legs and no antennae.

❯ Most marine arthropods are crustaceans. Their amazing diversity and abundance has given them the nickname "the insects of the seas."

Crustaceans are a group of mostly marine arthropods with two pairs of antennae. Some live in fresh water. A few such as wood lice live on land.

Small crustaceans reach great numbers in the seas and are an important food source for larger animals. Krill (euphausiids) have a shrimplike body a few centimeters long and swim in upper ocean waters (Figure 23.27A). They are so plentiful and nutritious that a blue whale weighing more than 100 tons can subsist almost entirely on krill it filters from seawater. Most copepods are also marine, but others live in fresh water (Figure 23.27B).

Larval barnacles swim, but adults are enclosed in a calcified shell and live attached to piers, rocks, and even whales. They filter food from the water with feathery legs (Figure 23.27C). As adults, they cannot move about, so you would think that mating might be tricky. But barnacles tend to settle in groups, and most are hermaphrodites. An individual extends a penis, often several times its body length, out to neighbors.

Lobsters, crayfish, crabs, and shrimps all belong to the same crustacean subgroup (the decapods). All are bottom feeders with five pairs of walking legs (Figure 23.28). In some lobsters, crayfish, and crabs, the first pair of legs has become modified as a pair of claws.

Like all arthropods, crabs molt as they grow (Figure 23.29). Some spider crabs do quite a bit of growing. With legs that can reach more than a meter in length, these crabs are the largest living arthropods.

Figure 23.27 Examples of crustaceans. **A** Antarctic krill no more than 6 centimeters long. **B** Free-living female copepod from the Great Lakes, about 1 millimeter long. Note the eggs. **C** Goose barnacles capturing food with their jointed legs.

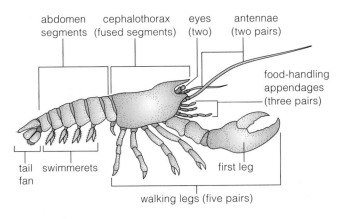

Figure 23.28 Body plan of a clawed lobster (*Homarus americanus*).

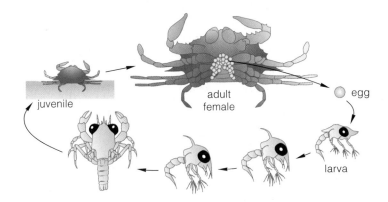

Figure 23.29 Animated Crab life cycle. Larval and juvenile stages molt repeatedly and grow in size before they are mature adults.

Take-Home Message What are crustaceans?

❯ Crustaceans are a group of mostly marine arthropods with antennae. They include the largest arthropods. Many species serve as food for other animals, including humans.

crustaceans Mostly marine arthropod group with two pairs of antennae; for example shrimp, crabs, lobsters, and barnacles.

❯ Arthropods are the most diverse animal phylum, and insects are the most diverse arthropods.

Characteristic Features

Insects have a three-part body plan, with a head, thorax, and abdomen (Figure 23.30). The head has one pair of antennae and two compound eyes. Such eyes consist of many individual units, each with a lens. Near the mouth are jawlike mandibles and other feeding appendages.

Three pairs of legs attach to the insect thorax. In some groups, the thorax also has one or two pairs of wings. Insects are the only winged invertebrates.

A few insects spend some time in the water, but the group is overwhelmingly terrestrial. A respiratory system consisting of tracheal tubes carries air from openings at the body surface to tissues deep inside the body. An insect abdomen contains digestive organs and sex organs.

Insect developmental patterns vary. Depending on the group, a fertilized egg either hatches into a small version of the adult, or a juvenile that will later undergo metamorphosis. During metamorphosis, tissues of a juvenile are reorganized. Incomplete metamorphosis means that the alterations in body form take place a bit at a time. Juveniles, called nymphs, change a little with each molt (Figure 23.31A). Complete metamorphosis is more dramatic. The juvenile, called a larva, grows and molts with no change in body plan. Then, the juvenile becomes a pupa. The pupa undergoes tissue remodeling that produces an adult (Figure 23.31B).

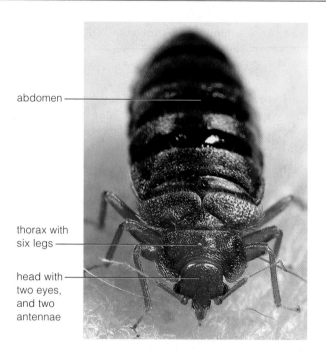

Figure 23.30 A bedbug illustrates the basic insect body plan: a head, thorax, and abdomen. The bug is 7 millimeters long.

abdomen

thorax with six legs

head with two eyes, and two antennae

Diversity and Abundance

With more than a million species, insects are the most diverse arthropod group. They are also breathtakingly abundant. By some estimates, the ants alone make up about 10 percent of the world's terrestrial animal biomass (the total weight of all living land animals).

Figure 23.32 shows a few of the more than 800,000 insect species. Silverfish are the most ancient lineage of modern insects (Figure 23.32A). They do not undergo metamorphosis. Their young are simply miniature versions of adults. Silverfish are scavengers that require a humid environment.

Insects that undergo incomplete metamorphosis include bugs, earwigs, lice, cicadas, damselflies, termites, and grasshoppers. Earwigs are wingless scavengers that have a flattened body (Figure 23.32B). Males have large pincers on their abdomen. Lice are wingless parasites that suck blood from warm-blooded animals (Figure 23.32C). Fleas are also wingless bloodsucking insects. Cicadas (Figure 23.32D) and the related leafhoppers and aphids suck juices from plants. Damselflies (Figure 23.32E) and the related dragonflies are agile aerial predators of other insects. Termites are social insects. They live in big family groups. Bacterial, archaean, and protistan symbionts in their gut allow them to digest wood (Figure 23.32F). They are unwelcome when they devour buildings or decks, but can be important decomposers. Grasshoppers cannot eat wood, but they can chew through tough, nonwoody plant parts (Figure 23.32G).

A Incomplete metamorphosis

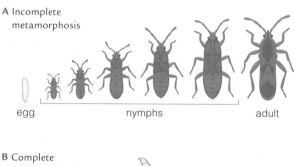

egg   nymphs   adult

B Complete metamorphosis

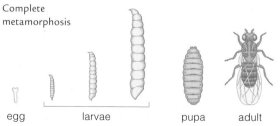

egg   larvae   pupa   adult

Figure 23.31 Insect development. **A** Incomplete metamorphosis of a bug. Small changes occur with each molt. **B** Complete metamorphosis of a fruit fly. A larva develops into a pupa that undergoes a dramatic remodeling into the adult form.

The four most successful insect lineages all have wings and undergo complete metamorphosis. There are approximately 150,000 species of flies (Figure 23.32H), and at least as many beetles (Figure 23.32I,J). The wasp in Figure 23.32K is one of about 130,000 hymenopterans. This group also includes the bees and ants. The moths and butterflies (Figure 23.32L) constitute a group of about 120,000 species. As a comparison, consider that there are about 4,500 species of mammals.

insects Most diverse arthropod group; members have six legs, two antennae and, in some groups, wings.

Take-Home Message What are insects?

❯ Insects are arthropods that have six legs, two antennae, and compound eyes. They are diverse and abundant on land.

❯ The most widespread and diverse insect groups have wings and undergo complete metamorphosis.

Figure 23.32 A sampling of insect diversity.
A Silverfish, an ancient lineage that does not undergo metamorphosis.

Insects that undergo incomplete metamorphosis:
B European earwig, a common household pest.
C Duck louse. It eats bits of feathers and skin.
D Cicada. Male cicadas are among the loudest of all insects.
E Damselfly, one of the insects that has aquatic larvae.
F Termites. These sterile soldiers can shoot a gluelike substance from their head to protect their colony.
G Grasshopper.

Members of the four most diverse insect orders. All are winged and undergo complete metamorphosis:
H A fly, dusted with pollen.
I Ladybird beetles with a distinctive red and black spotted wing cover.
J Staghorn beetle from New Guinea.
K Bald-faced hornet, a type of wasp. This is a fertile female, or queen. She lives in a papery nest with her many offspring.
L Swallowtail butterfly, a lovely lepidopteran, shown acting as pollinator.

❯❯ **Figure It Out** Are insect wings modified legs?

Answer: No. Insect wings are extensions from the body wall. Winged insects have the same number of legs (6) as wingless ones.

> It would be hard to overestimate the importance of insects, for either good or ill.
< Links to Pollinators 21.8, Malaria 20.6

Ecological Services

As you learned in Section 21.8, the flowering plants coevolved with insect pollinators. The vast majority of these plants are pollinated by members of one of the four most diverse insect groups. Other groups contain few or no pollinators. By one hypothesis, the close interactions between pollinator lineages and flowering plants contributed to an increased rate of speciation in both.

Today, declines in populations of insect pollinators concern many biologists. Development of natural areas, use of pesticides, and the spread of newly introduced diseases are reducing the populations of insects that pollinate native plants and agricultural crops. We discuss this problem in more detail in Chapter 27.

Insects serve as food for a variety of wildlife. Most songbirds nourish their nestlings on an insect diet. Those that migrate often travel long distances in order to nest and raise their young in areas where insect abundance is seasonally high. Some insects such as dragonflies and mayflies have aquatic larvae that serve as food for trout and other freshwater fish. Most amphibians and reptiles feed mainly on insects. Even humans eat insects. In many cultures, they are considered a tasty source of protein.

Insects dispose of wastes and remains. Flies and beetles quickly discover an animal corpse or a pile of feces (Figure 23.33). They lay their eggs in or on this organic material, and the larvae that hatch devour it. By their actions, the insects keep wastes and remains from piling up, and help distribute nutrients through the ecosystem.

Figure 23.33 Cleanup crew. A dung beetle gathers a ball of dung (feces). The beetle will lay its eggs in the dung and the larval beetles will feed on it as they grow.

Competitors for Crops

Insects are our main competitors for food and other plant products. They devour about a quarter to a third of all crops grown in the United States. Sadly, in an age of global trade and travel, we have more than just home-grown pests to worry about. Consider the Mediterranean fruit fly (Figure 23.34). The Med fly, as it is known, lays eggs in citrus and other fruits, as well as many vegetables. Damage done to plants and fruits by larvae of the Med fly can cut crop yield in half. Med flies are not native to the United States and there is an ongoing inspection program for imported produce. Even so, some Med flies still slip in. So far, all introduced Med fly populations have been successfully eradicated, but these control efforts have cost hundreds of millions of dollars. This amount is probably a bargain. If the Med fly were to become established, crop losses would likely run into the billions.

Vectors for Disease

Some insects spread human pathogens. As you learned earlier, mosquitoes transmit malaria, which kills more than a million people each year (Figure 23.35). Mosquitoes are also vectors for viruses and for roundworms that cause disease. Biting flies transmit African sleeping sickness; biting bugs spread Chagas disease. Fleas that bite rats and then bite humans can transmit bubonic plague. Body lice transmit typhus. So far as we know, bedbugs (Figure 23.30) do not cause disease. However, a heavy bedbug infestation can cause weakness as a result of blood loss, especially in children.

Figure 23.34 Mediterranean fruit fly (Med fly), a major threat to fruit crops.

Figure 23.35 The deadliest animal. When it comes to bites, no animal is more dangerous than the mosquito. Mosquito-borne diseases kill more than 1 million people each year.

Take-Home Message **What role do insects play in ecosystems?**

> Insects pollinate plants, serve as food for wildlife, and dispose of wastes and remains.

> Insects harm us by eating our crops and spreading diseases.

> With this section we begin our survey of deuterostomes. We will continue this survey in the next chapter.

Echinoderm means "spiny skinned." The **echinoderms** (phylum Echinodermata) have interlocking spines and plates of calcium carbonate embedded in their skin. Adults have a radial body plan, with five parts (or multiples of five) around a central axis. However, gene comparisons and the bilateral body plan of the larvae indicate echinoderms evolved from a bilateral ancestor.

Sea stars (also called starfish) are the most familiar echinoderms, and we will use them to illustrate the body plan (Figure 23.36A,B). Sea stars do not have a brain, but do have a decentralized nervous system. Eyespots at the tips of the arms detect light and movement.

A typical sea star is an active predator that moves about on tiny, fluid-filled tube feet. Tube feet are part of a **water–vascular system** unique to echinoderms. Fluid-filled canals extend into each arm, and side canals deliver coelomic fluid into muscular ampullae that function like the bulb on a medicine dropper. Contraction of an ampulla forces fluid into a tube foot, extending the foot.

Sea stars usually prey on bivalve mollusks. They can slide their stomach out through their mouth and into the bivalve's shell. The stomach secretes acid and enzymes that kill the mollusk and begin to digest it. Partially digested food is taken into the stomach and digestion is completed with the aid of digestive glands in the arms.

Gas exchange occurs by diffusion across the tube feet and tiny skin projections at the body surface. There are no specialized excretory organs.

A sea star is either male or female. Gonads (reproductive organs) in the arms make eggs or sperm that are released into the water. Fertilization produces a zygote that develops into a ciliated, bilateral larva. The larva swims about and develops into the adult form.

Sea urchins and sea cucumbers are less familiar echinoderms. In sea urchins, calcium carbonate plates form a stiff, rounded cover from which spines protrude (Figure 23.36C). Sea urchin roe (eggs) are used in some sushi. Overharvesting for markets in Asia threatens species that produce the most highly prized roe. Sea cucumbers are cucumber-shaped, with microscopic plates embedded in a soft body (Figure 23.36D). Some filter food from the seawater. Most feed like earthworms, by burrowing through sediments and ingesting them. If attacked, a sea cucumber can expel organs through the anus to distract the predator. If this defense works and the sea cucumber escapes, its missing parts grow back.

echinoderms Invertebrates with a water–vascular system and hardened plates and spines embedded in the skin or body.
water–vascular system Of echinoderms, a system of fluid-filled tubes and tube feet that function in locomotion.

Old Genes, New Drugs (revisited)

Marine invertebrates are important components of ecosystems, a source of food, and a treasure trove of molecules with potential for use in industrial applications or as medicines. Various species of cone snails, sponges, corals, crabs, and sea cucumbers make compounds that show promise as drugs. However, even as we begin to explore this potential, marine biodiversity is on the decline as a result of habitat destruction and overharvesting.

How Would You Vote? Bottom trawling is a fishing technique that helps keep seafood prices low, but can destroy invertebrate habitats. Should it be banned? See CengageNow for details, then vote online (cengagenow.com).

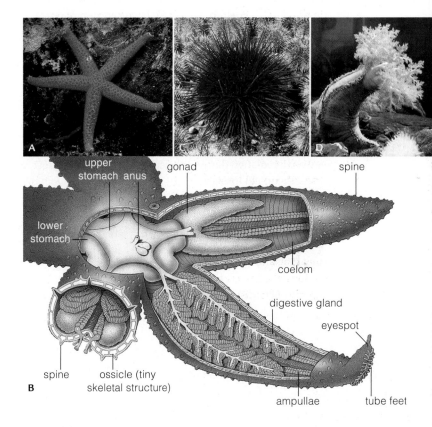

Figure 23.36 Animated Echinoderms. **A,B** Sea star. **C** Sea urchin. **D** Sea cucumber. Other members of this group include brittle stars and sand dollars.

>> **Figure It Out** Does the sea star have a complete digestive system?

Answer: Yes. It has a mouth and an anus.

Take-Home Message **What are echinoderms?**

> Echinoderms are spiny-skinned invertebrates that are radial as adults, but probably had bilateral ancestors. They have a decentralized nervous system and a unique water–vascular system that functions in locomotion.

Summary

Section 23.1 Invertebrates (with no backbone) are the most diverse animal group, and **vertebrates** (with a backbone) evolved from an invertebrate ancestor. As lineages evolved, genes that were present in ancestral lineages took on new functions. Thus, the same gene that functions in the synthesis of venom in a cone snail can play a role in the clotting of human blood.

Section 23.2 Animals are multicelled heterotrophs that digest food inside their body. Most have an embryo with three layers: **ectoderm**, **endoderm**, and **mesoderm**. Early animals had no body symmetry. Cnidarians have **radial symmetry**, but most animals have **bilateral symmetry** and **cephalization**. Bilateral animals typically have a body cavity, either a **pseudocoelom** or (more often) a **coelom**. Some also have **segmentation**, with repeating body units. There are two lineages of bilateral animals, **protostomes** and **deuterostomes**. Table 23.1 summarizes the traits of major animal lineages.

Section 23.3 The first animals evolved from a colonial protist. **Choanoflagellates** are the closest protistan relative of animals. Among living animals, **placozoans** are the oldest and structurally simplest lineage. An adaptive radiation during the Cambrian gave rise to most modern animal lineages.

Section 23.4 Sponges have a porous body with no tissues or organs. They filter food from water, and each is a **hermaphrodite** that produces both eggs and sperm. The ciliated **larva** is the only motile stage.

Section 23.5 Cnidarians such as jellyfishes, corals, and sea anemones are carnivores. They have two tissue layers with a jellylike layer in between. Only cnidarians make **nematocysts**, which they use to capture prey. A gastrovascular cavity functions in both respiration and digestion. Cnidarians have a **hydrostatic skeleton**. A **nerve net** gives commands to contractile cells that redistribute fluid and change the body's shape.

Section 23.6 Flatworms, the simplest animals with organ systems, include marine species and the freshwater **planarians**, as well as parasitic tapeworms and flukes. Some tapeworms and flukes infect humans.

Section 23.7 Annelids are segmented worms. The **closed circulatory system** and digestive, solute-regulating, and nervous systems extend through coelomic chambers. Annelids move when muscle action exerts force on coelomic fluid, altering the shape of all segments in a coordinated manner. Oligochaetes include aquatic species and the familiar earthworms. Polychaetes are predatory marine worms. Leeches are scavengers, predators, or bloodsucking parasites.

Section 23.8 Mollusks have a reduced coelom and a sheetlike mantle that drapes back over itself. They include **gastropods** (such as snails), **bivalves** (such as scallops), and **cephalopods** (such as squids and octopuses). Except in the cephalopods, which are adapted to a speedy, predatory life-style, blood flows through an **open circulatory system**.

Section 23.9 The **roundworms** (nematodes) have an unsegmented, cylindrical body covered by a cuticle that is **molted** as the animal grows. Most are decomposers in soil, but some are parasites.

Sections 23.10–23.14 There are more than 1 million **arthropod** species. Their diversity is attributable to traits such as a hardened **exoskeleton**, jointed appendages, specialized segments, sensory structures such as **antennae**, and **metamorphosis**. The **chelicerates** include horseshoe crabs and **arachnids** (spiders, scorpions, ticks, and mites). They are predators, parasites, or scavengers. The mostly marine **crustaceans** include crabs, lobsters, barnacles, krill, and copepods. **Insects** include the only winged invertebrates. Some serve as decomposers and pollinators; others harm crops and transmit diseases.

Section 23.15 Echinoderms such as sea stars belong to the deuterostome lineage. Spines and other hard parts embedded in the skin support the body. There is no central nervous system. A **water-vascular system** with tube feet functions in locomotion. Adults are radial, but larvae are bilateral, implying a bilateral ancestry.

Table 23.1 Comparative Summary of Animal Body Plans

Group	Body Symmetry	Germ Layers	Digestive System	Main Cavity	Segmented
Placozoans	None	None	None	None	No
Sponges	None	None	None	None	No
Cnidarians	Radial	2	Gut cavity	None	No
Protostomes					
Flatworms	Bilateral	3	Incomplete	None*	No
Annelids	Bilateral	3	Complete	Coelom	Yes
Mollusks	Bilateral	3	Complete	Coelom	No
Roundworms	Bilateral	3	Complete	Pseudocoelom	No
Arthropods	Bilateral	3	Complete	Coelom	Yes
Deuterostomes					
Echinoderms	Radial**	3	Complete	Coelom	No
Chordates	Bilateral	3	Complete	Coelom	Yes

* An ancestor may have had a coelom.
** Radial with some bilateral features; probably had a bilateral ancestor.

Data Analysis Activities

Sustainable Use of Horseshoe Crabs Eggs of Atlantic horseshoe crabs are essential food for many migratory shorebirds. People catch horseshoe crabs for use as bait. In addition, the blood of these crabs is used to test injectable drugs for potentially deadly bacterial toxins. To keep horseshoe crab populations stable, blood is extracted from captured animals, which are then returned to the wild. Concerns about the survival of animals after bleeding led researchers to do an experiment. They compared survival of animals captured and maintained in a tank with that of animals captured, bled, and kept in a similar tank. Figure 23.37 shows the results.

1. In which trial did the most control crabs die? In which did the most bled crabs die?
2. Looking at the overall results, how did the mortality of the two groups differ?
3. Based on these results, would you conclude that bleeding harms horseshoe crabs more than capture alone does?

	Control Animals		Bled Animals	
Trial	Number of crabs	Number that died	Number of crabs	Number that died
1	10	0	10	0
2	10	0	10	3
3	30	0	30	0
4	30	0	30	0
5	30	1	30	6
6	30	0	30	0
7	30	0	30	2
8	30	0	30	5
Total	200	1	200	16

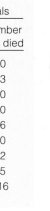

Figure 23.37 Mortality of young male horseshoe crabs kept in tanks during the two weeks after their capture. Blood was taken from half the animals on the day of their capture. Control animals were handled, but not bled. This procedure was repeated eight times with different sets of horseshoe crabs.

Self-Quiz   Answers in Appendix III

1. True or false? Animal cells do not have walls.

2. A coelom is a _____ .
   a. type of bristle   c. sensory organ
   b. resting stage   d. lined body cavity

3. Cnidarians alone have _____ .
   a. nematocysts   c. a hydrostatic skeleton
   b. a mantle   d. a radula

4. Flukes are most closely related to _____ .
   a. tapeworms   c. arthropods
   b. roundworms   d. echinoderms

5. Which group has six legs and two antennae?
   a. crustaceans   c. spiders
   b. insects   d. horseshoe crabs

6. The _____ are mollusks with a hinged shell.
   a. bivalves   c. gastropods
   b. barnacles   d. cephalopods

7. A spider's chelicerae _____ .
   a. detect light   c. produce silk
   b. inject venom   d. eliminate excess water

8. Which of these groups includes the most species?
   a. protostomes   c. arthropods
   b. roundworms   d. mollusks

9. The _____ include the only winged invertebrates.
   a. cnidarians   c. arthropods
   b. echinoderms   d. placozoans

10. The _____ have a coelom and are radial as adults.
    a. cnidarians   c. annelids
    b. echinoderms   d. placozoans

11. Annelids and cephalopods have a(n) _____ circulatory system.

12. Match the organisms with their descriptions.
    ___ choanoflagellates   a. complete gut, pseudocoelom
    ___ placozoan   b. protists closest to animals
    ___ sponges   c. simplest organ systems
    ___ cnidarians   d. body with lots of pores
    ___ flatworms   e. jointed exoskeleton
    ___ roundworms   f. mantle over body mass
    ___ annelids   g. segmented worms
    ___ arthropods   h. tube feet, spiny skin
    ___ mollusks   i. nematocyst producers
    ___ echinoderms   j. simplest known animal

Additional questions are available on **CENGAGENOW**.

Critical Thinking

1. Most animals that are hermaphrodites cannot fertilize their own eggs, but tapeworms can. Explain the advantages and disadvantages of self-fertilization.

2. Both birds and squids have eyes with a lens. Are these structures analogous or homologous?

3. Why are pesticides designed to kill insects more likely to harm lobsters and crabs than fish?

Animations and Interactions on **CENGAGENOW**:
❯ Types of animal body cavity; Sponge body plan; Cnidarian body plan and life cycle; Nematocyst action; Planarian organs; Tapeworm life cycle; Annelid body plan; Roundworm body plan; Crab life cycle; Sea star body plan; Sea star tube feet.

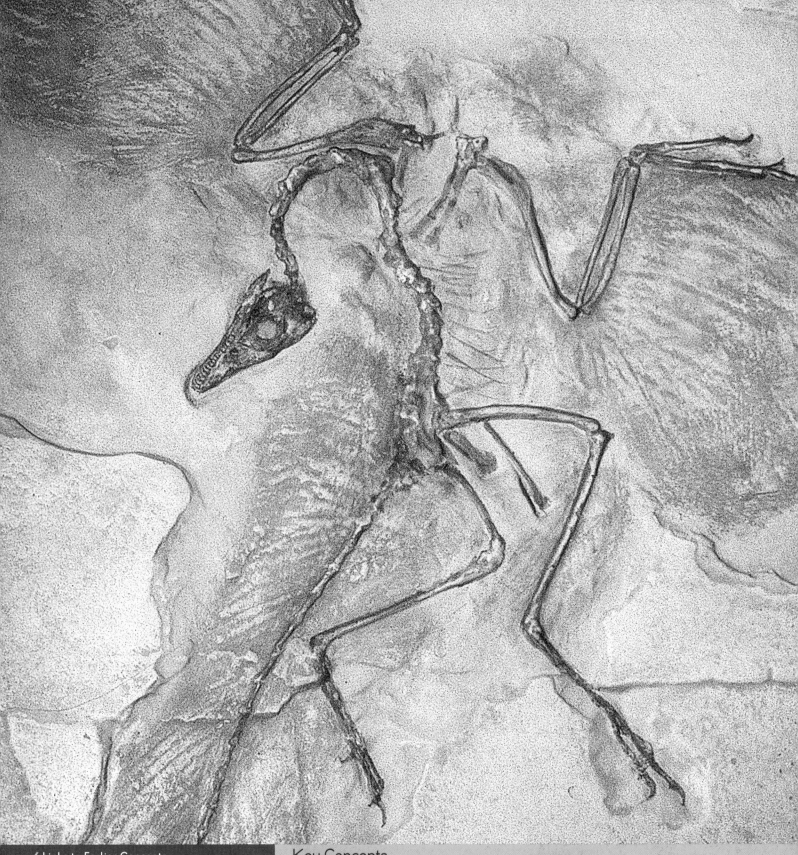

‹ Links to Earlier Concepts

As we continue our survey of animals you may wish to review the discussion of vertebrates and the animal evolutionary tree (Sections 23.1, 23.2). The geologic time scale (16.6) will help put the events discussed here in perspective. You will see more examples of fossils (16.5) and of homologous structures (16.8). You will draw on your understanding of speciation, adaptive radiation, and extinctions (17.10–17.13) and of cladistics (17.14).

Key Concepts

Characteristics of Chordates

Four traits characterize chordate embryos: a supporting rod (notochord), a dorsal nerve cord, a pharynx with gill slits in the wall, and a tail that extends past the anus. Certain invertebrates and all vertebrates belong to this group.

The Fishes

The first vertebrates were jawless fishes. Most modern fishes have jaws. Sharks and their relatives are jawed fishes with a cartilage skeleton. Ray-finned fishes and lobe-finned fishes are jawed bony fishes. Ray-finned fishes are the most diverse vertebrate group.

24 Animals II: The Chordates

24.1 Windows on the Past

In Darwin's time, the presumed absence of transitional fossils was an obstacle to acceptance of his new theory of evolution. Skeptics wondered, if new species evolve from existing ones, then where are the fossils that document these transitions? Where are the fossils that bridge major groups? In fact, one of these "missing links" was unearthed by workers at a limestone quarry in Germany just one year after Darwin's *On the Origin of Species* was published.

That fossil, about the size of a large crow, looked like a small meat-eating dinosaur. It had a long, bony tail, three clawed fingers on each forelimb, and a heavy jaw with short, spiky teeth. It also had feathers. The new fossil species was named *Archaeopteryx* (ancient winged one). Eventually, eight *Archaeopteryx* fossils were unearthed. One is shown at the *left*. Radiometric dating indicates that *Archaeopteryx* lived 150 million years ago.

Archaeopteryx is the most widely known transitional fossil in the bird lineage, but there are others. In 1994, a farmer in China discovered a fossil dinosaur with small forelimbs and a long tail. Unlike most dinosaurs, this one (depicted in the reconstruction at the *upper right*) had downy feathers like those that cover modern chicks. Researchers named the farmer's fuzzy find *Sinosauropteryx prima*, which means first Chinese feathered dragon. Given its shape and lack of long feathers, *S. prima* was certainly flightless. The feathers probably helped the animal stay warm.

Both *S. prima* and *Archaeopteryx* had a long tail and a mouth full of sharp teeth. In contrast, modern birds have a small tailbone and their toothless jaws are covered with hard layers of protein that form a beak. Another fossil from China, *Confuciusornis sanctus* (sacred Confucius bird), is the earliest known bird with a beak. As shown at the *lower right*, it had a bird's typically short tail with long tail feathers. Even so, its dinosaur ancestry remains apparent. Unlike the wings of modern birds, the wings of *C. sanctus* had digits with claws at their tips.

No human witnessed the major transitions that led to modern animal diversity. However, fossils are physical evidence of

Sinosauropteryx prima

Confuciusornis sanctus

changes, and radiometric dating assigns the fossils to places in time. The structure, biochemistry, and genetic makeup of living organisms provide information about branchings. Biological evolution is not a "theory" in the common sense of the word—an idea off the top of someone's head. It is a *scientific* theory. Its predictive power has been tested many times in the natural world, and the theory has withstood these tests (Section 1.9).

Evolutionists often argue among themselves. They argue over how to interpret data and which of the known mechanisms can best explain life's history. At the same time, they eagerly look to new evidence to support or disprove hypotheses. As you will see, fossils and other evidence form the foundation for this chapter's account of vertebrate evolution, including our own origins.

The Transition From Water to Land
Tetrapods, animals that walk on four legs, evolved from a lineage of lobe-finned fishes. Amphibians were the first tetrapod lineage. They can live on land, but their eggs must develop in water and their skin is not waterproof.

The Amniotes
Amniotes have waterproof skin, and their eggs contain membranes that enclose the embryo in fluid. These and other traits allowed the amniotes to expand into dry habitats. Reptiles (including birds) and mammals are the two modern amniote lineages.

Early Primates to Humans
Primates have grasping hands with nails instead of claws. Within the group, there is a trend toward increased brain size and manual dexterity, a flatter face, and upright posture. Fossils provide information about lineages related to our species, which most likely evolved in Africa.

> Chordates are distinguished by their embryonic traits.
> The group includes two lineages of marine invertebrates, as well as the vertebrates.
< Link to Animal classification 23.1

Chordate Characteristics

The previous chapter ended with the echinoderms, a deuterostome lineage. The other major deuterostome lineage, the **chordates**, is defined by four embryonic traits:

1. A **notochord**, a rod of stiff but flexible connective tissue, extends the length of the body and supports it.
2. A dorsal, hollow nerve cord parallels the notochord.
3. Gill slits (narrow openings) extend across the wall of the pharynx (the throat region).
4. A muscular tail extends beyond the anus.

Depending on the group, some, none, or all of these embryonic traits persist in the adult.

Most chordate species are vertebrates. However, the group also includes some marine invertebrates.

Invertebrate Chordates

In **lancelets**, all characteristic distinguishing chordate features are present in the adult (Figure 24.1). A lancelet is about 3 to 7 centimeters long. Its dorsal nerve cord extends into the head, where a group of nerve cells serves as a simple brain. An eyespot at the end of the nerve cord detects light, but there are no paired sensory organs like those of fishes. The lancelet wiggles backward into sediments until it is buried up to its mouth. Movement of cilia lining its pharynx moves water into the pharynx and

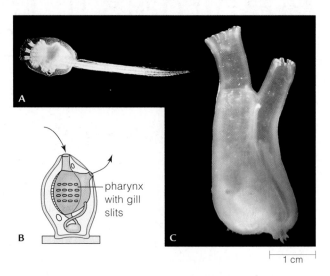

Figure 24.2 Animated Tunicates. **A** Larval tunicate. It is free-swimming and has all characteristic chordate features. **B** Diagram and **C** photo of the adult tunicate. It traps food in gill slits of its pharynx.

pharynx with gill slits

out through gill slits. Gill slits filter food particles out of the water and also function in gas exchange.

In **tunicates**, larvae have typical chordate traits, but adults retain only the pharynx with gill slits (Figure 24.2). Larvae swim about briefly, then undergo metamorphosis. The tail breaks down and other parts become rearranged into the adult body form. A secreted carbohydrate-rich covering, or "tunic," encloses the adult body and gives the group its common name. The adult feeds by drawing water in through an oral opening, past the gill slits, then expelling it through a second opening. Cilia on the gill slits capture food particles and move them to the gut.

Overview of Chordate Evolution

Until recently, lancelets were considered the closest invertebrate relatives of vertebrates. An adult lancelet looks more like fish than an adult tunicate does, but superficial similarities can be deceiving. Studies of developmental processes and gene sequences have revealed that tunicates are closer to the vertebrates, as illustrated in Figure 24.3.

A braincase, or cranium, evolved in the common ancestor of all fishes, amphibians, reptiles, birds, and mammals. We refer to the members of this clade as **craniates** ❶. The braincase may be entirely cartilage, or, like your skull, it may be a combination of cartilage and bone.

With the exception of one group of fish (hagfishes) all craniates are **vertebrates**. As Section 23.1 explained, the defining trait of this group is a backbone ❷. Biologists refer to the backbone as the vertebral column. It consists of stacked vertebrae. Vertebrae and other skeletal elements are components of the vertebrate **endoskeleton**. This internal skeleton consists of cartilage and, in most groups, bone.

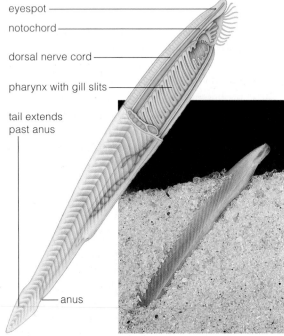

eyespot

notochord

dorsal nerve cord

pharynx with gill slits

tail extends past anus

anus

Figure 24.1 Animated Photo and body plan of a lancelet. Adults retain the set of four traits that define chordates.

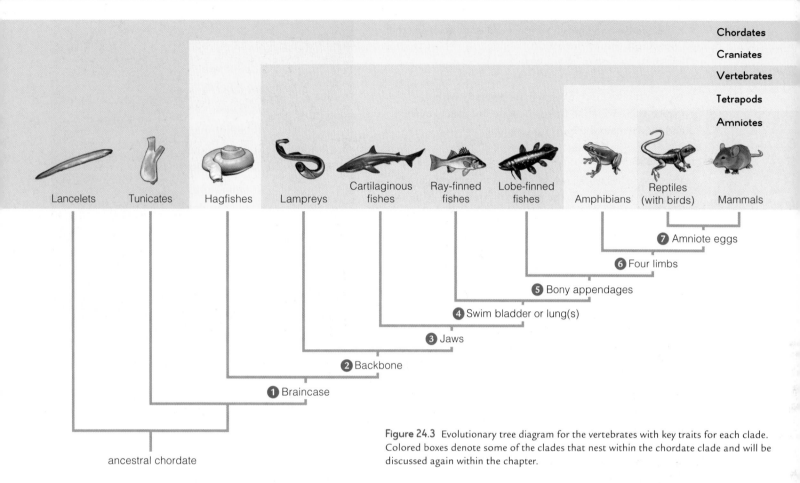

Figure 24.3 Evolutionary tree diagram for the vertebrates with key traits for each clade. Colored boxes denote some of the clades that nest within the chordate clade and will be discussed again within the chapter.

Early fishes sucked up or scraped up food. Then, the evolution of hinged skeletal elements called jaws opened the way to new feeding strategies ❸. Jaws evolved by expansion of the bony parts that structurally supported the gill slits of early jawless fishes. The vast majority of modern fishes have jaws, as do all other vertebrates.

Evolutionary modifications allowed animals to move from water onto the land. In one group of fishes, two small outpouchings on the side of the gut wall evolved into lungs: moist, internal sacs that enhance gas exchange with the air ❹. A subgroup of these fishes also evolved fins with bony supports inside them ❺. These fins would later evolve into the limbs of the first four-legged walkers, or **tetrapods** ❻.

Early tetrapods remained tied to the water. Later, evolution of eggs in which the embryo developed within a series of waterproof membranes allowed the animals known as **amniotes** to disperse widely on land ❼.

amniote Vertebrate in which the embryo develops surrounded by fluid enclosed by membranes inside the egg.
chordate Animal with an embryo that has a notochord, dorsal nerve cord, pharyngeal gill slits, and a tail that extends beyond the anus. For example, a lancelet or a vertebrate.
craniate Chordate with a braincase.
endoskeleton Internal skeleton made up of hardened components such as bones.
lancelet Invertebrate chordate that has a fishlike shape and retains the defining chordate traits into adulthood.
notochord Stiff rod of connective tissue that runs the length of the body in chordate larvae or embryos.
tetrapod Vertebrate with four legs, or a descendant thereof.
tunicate Invertebrate chordate that loses its defining chordate traits during the transition to adulthood.
vertebrate Animal with a backbone.

Take-Home Message **What are chordates and what traits defined the major subgroups?**

❯ All chordate embryos have a notochord, a dorsal tubular nerve cord, a pharynx with gill slits in its wall, and a tail that extends past the anus. These traits are legacies from a shared ancestor, an early invertebrate chordate.

❯ A braincase evolved early in chordate evolution and defines the craniates. Most craniates also have a backbone and are vertebrates. Some of the vertebrates evolved four limbs. These tetrapods colonized the land. Amniotes, a tetrapod subgroup with specialized eggs, are now the main land animals.

❯ The number and diversity of fishes exceed those of all other vertebrate groups combined.
❮ Link to Adaptive radiation 17.13

We begin our survey of vertebrate diversity with the fishes. These were the first vertebrate lineages to evolve and they remain the most fully aquatic.

Jawless Fishes

The first fishes were jawless, and two lineages of jawless fishes have survived to the present. Both have a cylindrical body with no fins or scales (Figure 24.4). Like a lancelet, they move with a wiggling motion.

Hagfishes have a cranium, but no backbone. Thus they are craniates but not vertebrates. A flexible notochord supports the body. Sensory tentacles near the mouth help a hagfish find food such as soft invertebrates and dead or dying fish. Hagfishes secrete slimy mucus when threatened, a useful defense for a soft-bodied animal. The slime deters most predators, but it has not kept humans from harvesting hagfish. Most products sold as "eelskin" are actually hagfish skin.

Lampreys have a backbone made of cartilage, and so are vertebrates. Some lampreys have a parasite-like feeding style. They attach to another fish using an oral disk with horny teeth made of the protein keratin. Once attached, the lamprey uses a tooth-covered tongue to scrape up bits of the host's flesh.

Figure 24.6 Two cartilaginous fishes. **A** Galápagos sharks are fast-swimming predators. **B** Manta rays glide along slowly. Two fleshy projections on the head funnel small organisms into the mouth. Note the gill slits visible at the body surface of both fishes.

A Hagfish. It feeds on worms and scavenges on the sea floor.

B Parasitic lamprey. It attaches to another fish with its oral disk and scrapes up flesh.

Figure 24.4 Two modern jawless fishes.

Fishes With Jaws

Jawed fishes likely evolved from jawless ancestors when some of the gill supports became modified. Gill supports are skeletal elements that sit beside the gill slits and hold them open (Figure 24.5). Jawed fishes typically have paired fins that help them maneuver and move quickly through the water. Most also have a covering of scales.

Cartilaginous Fishes The mostly marine **cartilaginous fishes** (Chondrichthyes) such as sharks and rays are jawed fishes with a skeleton of cartilage. Their gills are visible at the body surface. Scales do not grow with the fish; instead new new scales grow between old ones. The scales consist of dentin and enamel, the same hard materials that make up teeth. Shark teeth grow in rows and are continually shed and replaced. Sharks include streamlined

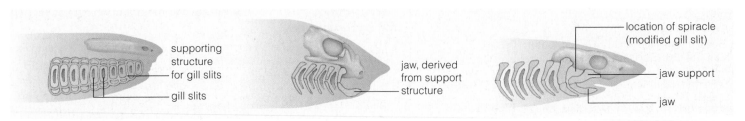

supporting structure for gill slits

gill slits

jaw, derived from support structure

location of spiracle (modified gill slit)

jaw support

jaw

Figure 24.5 Animated Proposed steps in the evolution of jaws.

predators that chase down prey (Figure 24.6A), placid plankton feeders that strain food from the water, and bottom feeders. Rays are cartilaginous fish with a body that is flattened from top to bottom. Manta rays filter plankton from water and some reach 6 meters (20 feet) wide (Figure 24.6B). Stingrays are bottom feeders. Their barbed tail has a venom gland.

Bony Fishes The **bony fishes** (Osteichthyes) have an adult skeleton consisting largely of bone. Typically, a bony covering also protects the gills. Scales are flattened, and they increase in size as a fish grows. The unique trait of bony fishes is a gas-filled organ derived from an outpouching of the gut. In some bony fishes this organ serves as a lung, and functions in gas exchange. In others, it acts as a swim bladder, an adjustable flotation device. By regulating the volume of gas inside its swim bladder, a bony fish can stay suspended in water at different depths.

Figure 24.7A shows the internal anatomy of a bony fish. Like all vertebrates, it has a closed circulatory system with one heart and a urinary system with paired kidneys that filter blood, adjust its volume and solute composition, and eliminate wastes. The digestive system is complete. In most bony fishes, urinary waste, digestive waste, and gametes exit the body through three separate openings. By contrast, cartilaginous fishes, amphibians, and reptiles have one multipurpose opening called a cloaca.

There are two lineages of bony fishes. **Ray-finned fishes** have flexible fin supports derived from skin. With more than 21,000 freshwater and marine species, they are the most diverse vertebrate group. They include most fish that end up on dinner plates, such as salmon, sardines, bass, swordfish, trout, tuna, halibut, carp, and cod. Some have a highly modified body plan (Figure 24.7B–D).

Lobe-finned fishes include coelacanths and lungfishes (Figure 24.8). Their pelvic and pectoral fins are fleshy body extensions that have supporting bones inside them. As the name suggests, lungfishes have gills and one or two air-filled lungs. A lung consists of an air-filled sac or sacs with an associated network of tiny blood vessels. A lungfish fills its lungs by surfacing and gulping air, then oxygen diffuses from the lungs into the blood.

As the next section explains, bony fins and simple lungs proved selectively advantageous when descendants of some lobe-finned fishes ventured onto land.

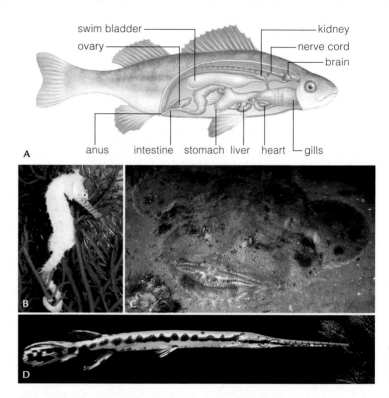

A

swim bladder / ovary / kidney / nerve cord / brain / anus / intestine / stomach / liver / heart / gills

Figure 24.7 Animated Ray-finned bony fishes. **A** Body plan of a perch. **B** Sea horse. **C** Monkfish, a stealthy predator. **D** Long-nose gar, a speedy predator.

pelvic fin   pectoral fin

Figure 24.8 Lungfish, a lobe-finned fish. Its pelvic and pectoral fins are supported by bones inside the fins.

bony fish Fish with a lung or swim bladder and a skeleton consisting largely of bone.
cartilaginous fish Fish with a skeleton of cartilage.
hagfish Jawless fish with a skull case but no backbone.
lamprey Jawless fish with a backbone of cartilage.
lobe-finned fish Fish with fleshy fins that contain bones.
ray-finned fish Fish with fins supported by thin rays derived from skin; member of most diverse lineage of fishes.

Take-Home Message What are the different groups of fishes?

❯ Hagfishes and lampreys are jawless fishes with no fins or scales. Hagfishes do not have a backbone, and so are not considered vertebrates.

❯ Cartilaginous fishes include the sharks and rays. They have a backbone and other skeletal elements made of cartilage.

❯ There are two lineages of bony fishes. The ray-finned fishes are the most diverse group. The lobe-finned fishes are the closest relatives of tetrapods.

> Amphibians often spend much of their life out of water, but return to their aquatic roots to reproduce.
< Links to Homologous structures 16.8, Chytrid fungi 22.3, Flukes 23.6

The Move Onto Land

Amphibians are land-dwelling predators that require water to breed. The recent discovery of fossil footprints indicates that a four-legged body plan had evolved by 395 million years ago, during the Devonian. Fossilized bones reveal how fishes, adapted to swimming, evolved into four-legged walkers, or tetrapods (Figure 24.9). Bones of a lobe-finned fish's pectoral fins and pelvic fins are homologous with bones of an amphibian's front and hind limbs.

The transition to land was not simply a matter of skeletal changes. Division of the heart into three chambers allowed blood flow in two circuits, one to the body and one to the increasingly important lungs. Changes to the inner ear improved detection of airborne sounds. Eyes became protected from drying out by eyelids.

What was the selective advantage to living on land? An ability to spend some time out of water would have been favored in seasonally dry places. On land, individuals escaped aquatic predators and had a new source of food, insects, which also evolved during the Devonian.

Modern Amphibians

Salamanders and newts resemble the early tetrapods in body form. They have a long tail, and their forelimbs are about the same size as their back limbs (Figure 24.10A). When a salamander walks, the movements of its limbs and bending of its trunk resemble movements made by

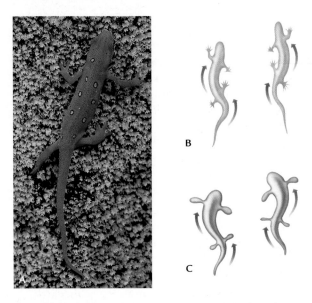

Figure 24.10 Animated Salamander gait. **A** Forelimbs and hindlimbs of a salamander are similar sizes and project from the body at right angles. **B** The salamander walks by bending its trunk back and forth, as it moves its limbs in a coordinated fashion. These movements resemble motions made by a swimming fish **C**.

a fish swimming through water (Figure 24.10B,C). Early tetrapods probably used this motion to walk in the water before one lineage ventured onto land.

Larval salamanders look like small versions of adults, except for the presence of gills. In a few species, called axolotls, gills are retained into adulthood. More typically gills disappear as the animal matures and lungs develop.

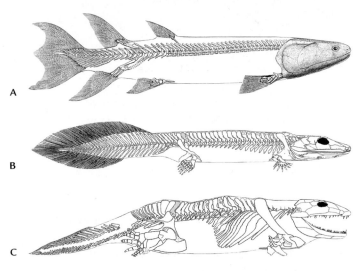

Figure 24.9 Transition to tetrapods. Skeleton of a Devonian lobe-finned fish **A**, and two early amphibians, *Acanthostega* **B**, and *Ichthyostega* **C**. The painting **D** shows what *Acanthostega* (foreground), and *Ichthyostega* (background) may have looked like.

Figure 24.11 Frogs.
A Long, musular hind limbs allow the adult frog to leap. **B** The larval frog, commonly known as a tadpole, is a swimmer with a long tail and no legs.

Figure 24.12 American toad. The flattened disk visible behind the eye is the eardrum.

Figure 24.13 Frog deformity. Infection by a fluke caused abnormal limb development.

Frogs and toads belong to the most diverse amphibian lineage, with more than 5,000 species. Long, muscular hindlimbs allow the tailless adults to swim, hop, and make spectacular leaps (Figure 24.11A). The much smaller forelimbs help absorb the impact of landings. Toads tend to have somewhat shorter hind legs than frogs and to be better adapted to dry conditions (Figure 24.12).

Larval frogs and toads, commonly called tadpoles or pollywogs, have gills and a tail but no limbs (Figure 24.11B). All frogs and toads have lungs as adults.

Declining Amphibian Diversity

Amphibian populations throughout the world are declining or disappearing. Researchers correlate many declines with shrinking or deteriorating habitats. People commonly fill in low-lying ground where rains once collected and formed pools of standing water. Nearly all amphibians need to deposit their eggs and sperm in water, and their larvae must develop in water.

Other factors that contribute to amphibian declines include introduction of new species in amphibian habitats, long-term shifts in climate, increases in ultraviolet radiation, and the spread of pathogens and parasites. Section 22.3 noted the effects of chytrid fungus infections on amphibians. Figure 24.13 provides an example of the deforming effects of a parasitic fluke (a subgroup of flatworms). Chemical pollution of aquatic habitats also harms amphibians. Their thin skin, unprotected by scales, allows waste carbon dioxide to diffuse out of their body. Unfortunately, it also allows chemical pollutants to enter.

---

amphibian Tetrapod with a three-chambered heart and scaleless skin; it typically develops in water, then lives on land as a carnivore with lungs.

Take-Home Message How did amphibians evolve and what traits define them?

> The tetrapod lineage branched off from the lobe-finned fishes. Amphibians are the oldest tetrapod lineage.

> Lungs and a three-chambered heart adapt amphibians to life on land, but nearly all must have access to water to complete their life cycle.

> Salamanders, frogs, and toads are the best-known modern amphibians.

❯ Amniotes are vertebrates that have adapted to a life lived entirely on land.
❮ Links to Asteroid impact 16.1, Cladistics 17.14

About 300 million years ago, during the Carboniferous, amniotes branched off from an amphibian ancestor. A variety of traits that evolved in this lineage adapted its members to life in dry places. **Amniote eggs** have membranes inside them that keep an embryo moist even away from water (Figure 24.14). In addition, amniote skin is rich in keratin, a protein that makes it waterproof. A pair of well-developed kidneys help conserve water, and fertilization usually takes place inside the female's body.

Shortly after the amniotes arose, the lineage branched in two. One branch gave rise to the mammals. The other gave rise to the members of the clade that biologists call **reptiles** (Figure 24.15). This clade includes turtles, lizards, snakes, crocodilians, and birds.

amniote egg Egg with internal membranes that allow the amniote embryo to develop away from water.
dinosaur Reptile lineage abundant in the Jurassic to Cretaceous; now extinct with the exception of birds.
reptile Amniote subgroup that includes lizards, snakes, turtles, crocodilians, and birds.

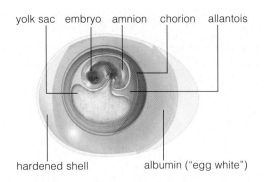

yolk sac   embryo   amnion   chorion   allantois

hardened shell   albumin ("egg white")

Figure 24.14 Animated Amniote egg. A bird's egg has a hard shell that encloses the embryo and the characteristic amniote membranes (yolk sac, amnion, chorion, and allantois).

Dinosaurs are extinct members of the reptile clade. Biologists define them by skeletal features, such as the shape of their pelvis and hips. As Section 24.1 explained, evidence suggests that the first birds evolved from a group of feathered dinosaurs. Figure 24.16 shows a Jurassic scene, with the early bird *Archaeopteryx* and an early mammal. With the exception of the birds, all members of the dinosaur lineage disappeared by the end of the Cretaceous. An asteroid impact is considered the most likely cause of their demise (Section 16.1).

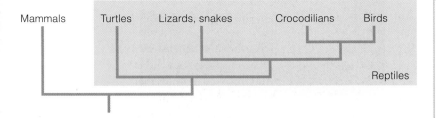

Mammals   Turtles   Lizards, snakes   Crocodilians   Birds

Reptiles

Figure 24.15 One proposed evolutionary tree for the living amniote groups. The *blue* box denotes members of the reptile clade. Note that it includes the birds.

Take-Home Message What are the amniotes?

❯ Amniotes are animals that produce eggs in which the young can develop away from water. They are also adapted to life on land by waterproof skin, highly efficient kidneys, and internal fertilization.

❯ An early divergence separated the ancestors of mammals from the ancestors of reptiles, a group in which biologists include turtles, lizards, snakes, crocodilians, and birds.

Figure 24.16 Painting of a Jurassic scene. In the *middle foreground*, the early bird *Archaeopteryx* flies by. *Behind* the birds, a meat-eating dinosaur sizes up a larger plant-eating dinosaur. At the *far right*, an early mammal surveys the scene from its perch on a tree.

❯❯ **Figure It Out** Which of the animals mentioned above do biologists consider amniotes? Which do they group as reptiles?

Answer: All are amniotes. The dinosaurs and birds are reptiles.

Figure 24.17 Examples of reptiles. **A** Hognose snakes emerging from leathery amniote eggs. **B** Galápagos tortoise. **C** Spectacled caiman, a crocodilian. All crocodilians spend much of their time in water.

❯ All nonbird reptiles—turtles, lizards, snakes, crocodiles, and alligators—have a scale-covered body and acquire heat from their environment.

❮ Link to Metabolic heat 7.5

Modern reptiles include turtles, lizards, snakes, crocodilians, and birds. With the exception of birds, which we consider in the next section, all of these groups are **ectotherms**, meaning "heated from outside." People commonly describe such animals as "cold-blooded," but ectotherms adjust their internal temperature by their behavior. They bask on a warm rock to heat up, or burrow into the soil to cool off. Fish and amphibians are also ectotherms. In contrast, **endotherms** such as birds and mammals produce their own heat through metabolic processes.

Major Groups

Lizards and Snakes Lizards and snakes constitute the

most diverse group of modern reptiles. Overlapping scales cover the body. The smallest lizard can fit on a dime (*left*). The largest, the Komodo dragon, grows up to 3 meters (10 feet) long. Most lizards are predators, but iguanas are herbivores.

The first snakes evolved during the Cretaceous, from short-legged, long-bodied lizards. Some modern snakes have bony remnants of hindlimbs, but most lack limb bones. All are carnivores. Many have flexible jaws that help them swallow prey whole. All snakes have teeth, but not all have fangs. Rattlesnakes and other fanged types bite and subdue prey with venom made in modified salivary glands. On average, only about 2 of the 7,000 snake bites reported annually in the United States are fatal.

Most lizards and snakes lay eggs (Figure 24.17A), but females of some species brood eggs in their body and give birth to well-developed young. The live-bearers do not provide any nourishment to the offspring inside their body, as some mammals do.

Turtles Turtles and tortoises have a bony shell attached to their skeleton (Figure 24.17B). When threatened, many withdraw into the shell for protection. A fossil recently discovered in China revealed an intermediate stage in turtle evolution. The fossil turtle lived in the sea about 200 million years ago and had a shell only on its ventral (belly) side. Turtles do not have teeth. Instead a horny "beak" made of keratin covers their jaws. Some feed on plants and others are predators.

Crocodilians Crocodiles, alligators, and caimans are the closest living relatives of birds. Like birds, they have a four-chambered heart. (Lizards, snakes, and turtles have an amphibian-like three-chambered heart.) Like most birds, crocodilians display complex parental behavior. For example, they make and guard a nest, then feed and care for the hatchlings. All crocodilians spend much of their time in water. They have powerful jaws, a long snout, and sharp teeth (Figure 24.17C). They clench prey, drag it under water, tear it apart as they spin around, and gulp the chunks. The feared saltwater crocodile and Nile crocodile weigh up to 1,000 kilograms (2,200 pounds).

---

ectotherm Animal that controls its internal temperature by altering its behavior; for example a lizard.
endotherm Animal that controls its internal temperature by adjusting its metabolism; for example a bird or mammal.

Take-Home Message **What are the traits of nonbird reptiles?**

❯ All nonbird reptiles are ectotherms, meaning they adjust their temperature by altering their behavior.

❯ Turtles have a bony shell and a horny beak. Lizards and snakes have overlapping scales on an elongated body. Crocodiles and alligators are the closest relatives of birds, and like birds they have a four-chambered heart.

> Birds are the descendants of feathered dinosaurs. They have a body highly modified inside and out for flight.
> Links to Aerobic respiration 7.2, Honeycreepers 17.11

Birds are the only living animals with feathers. These lightweight structures derived from scales have roles in insulation, flight, and courtship displays. They also shed water and thus help keep the skin beneath dry.

The bird body has many adaptations related to flight (Figure 24.18). Like humans, birds stand upright on their hindlimbs. The bird forelimbs have evolved into wings. Each wing is covered with long feathers that extend outward and increase its surface area. The feathers give the wing a shape that helps lift the bird as air passes over it. Air cavities inside bones keep body weight low, and make it easier for a bird to become and remain airborne. Flight muscles connect an enlarged breastbone, or sternum, to bones of the upper limb. When the muscles contract, they produce a powerful downstroke.

Flight requires a lot of energy, which is provided by aerobic respiration (Section 7.2). A unique system of air sacs keeps air flowing continually through lungs, ensuring an adequate oxygen supply. A four-chambered heart pumps blood quickly from the lungs to wing muscles and back. A bird also has a rapid heartbeat, and its heart is large relative to its body size.

Archaeopteryx had teeth and a long tail, but modern birds have neither. Layers of the protein keratin cover their jawbones, forming a beak, or bill. Differences in the size and shape of the bill adapt birds to different diets. For example, some honeycreepers have beaks adapted to feed on nectar while other have beaks suited to eating seeds or insects (Section 17.11).

bird An animal with feathers.

Figure 24.19 Mating house sparrows. Lacking a penis, the male must bend his body so his cloaca covers his mate's.

Most nonbird reptiles have a penis, an organ that the male uses to insert sperm into his partner. Male birds do not have a penis. To mate, a male must balance atop a female's back and press his cloaca to hers, a maneuver sometimes referred to as a cloacal kiss (Figure 24.19).

Birds vary in size, proportions, coloration, and capacity for flight. A bee hummingbird weighs 1.6 grams (0.06 ounce). The ostrich, a flightless sprinter, weighs 150 kilograms (330 pounds). More than half of all bird species belong to the subgroup of perching birds. Birds seen at backyard feeders usually belong to this group. Examples include sparrows, jays, starlings, swallows, finches, robins, warblers, orioles, and cardinals.

Take-Home Message **What are birds?**

> Birds are amniotes with feathers. Lightweight bones and highly efficient respiratory and circulatory systems also adapt a bird to flight.

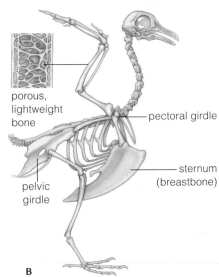

Figure 24.18 Adapted to flight. **A** Among modern vertebrates, only birds and bats fly by flapping a pair of wings. **B** Birds have lightweight bones and a large sternum.

porous, lightweight bone

pectoral girdle

pelvic girdle

sternum (breastbone)

> Mammals scurried about while dinosaurs dominated the land, then diversified after they were gone.

Mammalian Traits

Mammals are hairy or furry animals in which females nourish their offspring with milk secreted from mammary glands. The group name is derived from the Latin *mamma*, meaning breast. Like birds, mammals are endotherms. Their coat of fur or head of hair helps them retain internally produced heat.

Only mammals have four differently shaped kinds of teeth (Figure 24.20). In other vertebrates, an individual's teeth may vary in size, but they are all the same shape. Having teeth of multiple shapes allows mammals to process a wider variety of foods than most other vertebrates.

Three Mammalian Lineages

Monotremes, the oldest mammalian lineage, lay eggs that have a leathery shell similar to that of reptiles. Only three monotreme species survive, the duck-billed platypus (Figure 24.21A) and two kinds of spiny anteater. Monotreme offspring hatch in a relatively undeveloped state, while still tiny, hairless, and blind. They cling to the mother or are held in a skin fold on her belly. Milk oozes from openings on the mother's skin. There are no nipples.

Marsupials are pouched mammals. Young marsupials develop briefly in their mother's body, nourished by

---

mammal Animal with hair or fur; females secrete milk from mammary glands.
marsupial Mammal in which young are born at an early stage and complete development in a pouch on the mother's surface.
monotreme Egg-laying mammal.
placental mammal Mammal in which a mother and her embryo exchange materials by means of an organ called the placenta.

---

egg yolk and by nutrients that diffuse from the mother's tissues. They are born at an early developmental stage, and crawl along their mother's body to a permanent pouch on her ventral surface. A nipple inside the pouch supplies milk that nourishes them as they continue to develop.

Most of the 240 modern species of marsupials live in Australia and on nearby islands. Kangaroos and koalas are the best known. The opossum (Figure 24.21B) is the only marsupial native to North America.

In **placental mammals**, maternal and embryonic tissues combine as a placenta. This organ allows for a highly efficient transfer of nutrients and gases between mother and embryo. Placental embryos grow faster than those of other mammals, and the offspring are born more fully formed. After birth, the young suckle milk secreted by nipples on their mother's surface (Figure 24.21C).

Placental mammals tend to outcompete other mammalian lineages and are now dominant on all continents except Australia. They are the only mammals that live in the seas. Of the approximately 4,000 placental mammals, nearly half are rodents such as rats and mice.

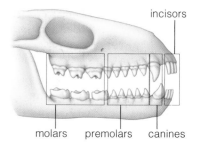

Figure 24.20 Generalized mammalian skull, showing the four differently shaped types of teeth. Compare the teeth of the caiman in Figure 24.17.

Take-Home Message What are mammals?

> Mammals are endothermic amniotes that have hair or fur. The females feed their young milk secreted by mammary glands.

> Egg-laying monotremes, pouched marsupials, and placental mammals are subgroups. Placental mammals are now the dominant lineage in most regions.

A Platypuses are monotremes.

B Opossums are marsupials.

C Humans are placental mammals.

Figure 24.21 Mammalian mothers and their young. All female mammals produce milk to feed offspring.

> Primates include humans and their closest mammalian relatives. As a group, they put their grasping hands to many uses.
〈 Link to Geologic time scale 16.6

Primates, the mammalian subgroup to which humans belong, have five-digit hands and feet capable of grasping (Figure 24.22). The digits are tipped by flattened nails, rather than the sharp claws of related mammals.

Primates first appear in the fossil record about 55 million years ago in Europe, North America, and Asia. Prosimians ("before monkeys") are the most ancient of the surviving lineages. Lemurs and tarsiers are examples. See Table 24.1 and Figure 24.23. An early branching separated the anthropoid lineage to which monkeys, apes, and humans belong, from the prosimians. Further branchings produced hominoids—apes and humans. Our closest living relatives are chimpanzees and bonobos (previously called pygmy chimpanzees). Humans and a variety of extinct humanlike species are referred to as **hominids**.

Figure 24.22 A chimpanzee shows off its grasping five-digit hands and feet. Primates have nails on their digits, rather than claws.

Table 24.1 Primate Classification

Prosimians	Lemurs, tarsiers
Anthropoids	New World monkeys (e.g., spider monkeys, howler monkeys)
	Old World monkeys (e.g., baboons, rhesus monkey)
	Hominoids:
	Hylobatids (gibbons, siamangs)
	Pongids (orangutans, gorillas, chimpanzees, bonobos)
	Hominids (humans, extinct humanlike species)

Key Trends in Primate Evolution

Five trends define the lineage that led to primates, hominids, and then to humans. First, the structure of the face changed (Figure 24.24). Primates are close relatives of tree shrews, and like this group early primates had a long, pointy snout and eyes at the side of the head. Later groups evolved a flattened face with both eyes facing forward. Forward-facing eyes provide better depth perception and probably helped the animals as they made leaps from branch to branch. As the face flattened, the amount of brain area devoted to odor detection declined.

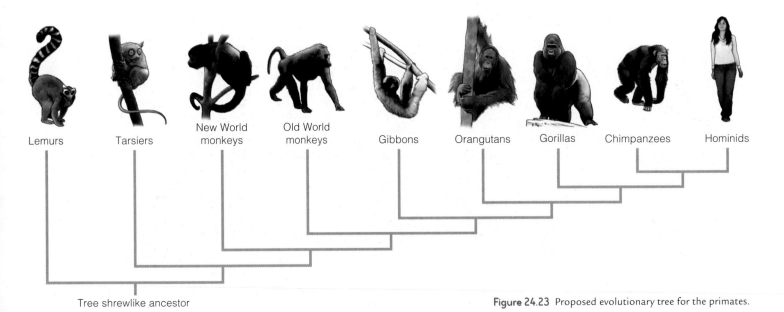

Lemurs   Tarsiers   New World monkeys   Old World monkeys   Gibbons   Orangutans   Gorillas   Chimpanzees   Hominids

Tree shrewlike ancestor

Figure 24.23 Proposed evolutionary tree for the primates.

Tree shrew, a close relative of primates, has a pointy face and its eyes are on either side of its head.

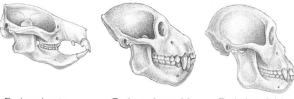

Early primate
(*Plesiadapis*)

Early anthropoid
(*Aegyptopithecus*)

Early hominid
(*Proconsul*)

Figure 24.24 Evolution of the primate skull. One trend is toward a more flattened face, with eyes at the front, rather than the sides of the skull. In hominids, teeth became smaller and more uniform.

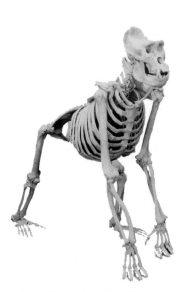

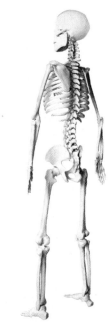

A Gorillas are knuckle-walkers. They walk on all fours, with their hands in a partially closed fist. The backbone is straight and the arms are proportionately longer than those of a human.

B Humans are bipedal. They walk upright. Their backbone has a distinct S shape and their skull has a hole at the bottom where it attaches to the spinal cord. In gorillas, this hole is toward the back of the skull.

Figure 24.25 Some skeletal differences associated with differences in posture and locomotion between a human and a gorilla.

Second, teeth became less specialized. Modifications to the jaws and teeth accompanied a shift from eating insects to eating fruits and leaves, then to a mixed diet. Rectangular jaws and lengthy canines evolved in monkeys and apes. A bow-shaped jaw and smaller, more uniform teeth evolved in early hominids.

Third, skeletal changes allowed upright standing and walking, known as **bipedalism**. The backbone developed an S-shaped curve that kept the body aligned over the feet (Figure 24.25). The skull became modified so it perched atop the backbone, rather than connecting to it at the rear.

Fourth, freed from a primary role in locomotion, hands became adapted for novel tasks. Changes to the arrangement of bone and muscle in the hands provided an ability to grasp objects in a fist, or to pick them up between fingertips and the thumb. These abilities allowed some primates to manipulate objects in unique ways, and to use them as tools:

power grip

precision grip

Fifth, the braincase and the brain increased in size and complexity. As brain size increased, so did the length of pregnancy and extent of maternal care. Compared to early primates, later ones had fewer offspring and invested more in each one.

Early primates were solitary. Later, some started to live in small groups. Social interactions and cultural traits began to affect reproductive success. **Culture** is the sum of all learned behavioral patterns transmitted among members of a group and between generations.

---

bipedalism Standing and walking on two legs.
culture Learned behavior patterns transmitted among members of a group and between generations.
hominid Human or extinct humanlike species.
primate Mammal having grasping hands with nails; includes prosimians, monkeys, apes, and hominids such as humans.

Take-Home Message **What traits and trends characterize the primate lineage?**

> Primates have grasping hands with nails on their five digits.

> Early primates were solitary and shrewlike. Some later lineages evolved a flatter face with more uniformly sized teeth, an upright stance, hands with more dexterity, a larger brain, and a tendency to live in groups.

> Fossils from central, eastern, and southern Africa show that the Miocene through the Pliocene was a "bushy" time of hominid evolution. In other words, many forms were rapidly evolving. We still do not know how they are related.
< Link to Adaptive radiation 17.13

Early Hominids

Genetic comparisons indicate that hominids diverged from apelike ancestors about 6 to 8 million years ago. Fossils that may be hominids are about 6 million years old. *Sahelanthropus tchadensis* had a hominid-like flat face, prominent brow, and small canines, but its brain was relatively small, about the size of a chimpanzee's (Figure 24.26A). *Ardipithecus ramidus* was a hominid that lived 4.4 million years ago. Analysis of a recently discovered fossil female, nicknamed Ardi, suggests that this species may have walked upright when on the ground, but more often used all four limbs to climb in trees.

An indisputably bipedal hominid, *Australopithecus afarensis*, was walking in Africa by about 3.9 million years ago. Remarkably complete skeletons reveal that it habitu-ally walked upright (Figure 24.27A). About 3.7 million years ago, two *A. afarensis* individuals walked across a layer of newly deposited volcanic ash. A light rain fell, and it transformed the powdery ash into a fast-drying cement, which preserved their footprints (Figure 24.27B,C).

A. afarensis was one of the **australopiths** ("southern apes"). This informal group includes *Australopithecus* and *Paranthropus* species. *Australopithecus* species were petite; they had a small face and teeth (Figure 24.26B). One or more species are suspected to be ancestral to modern humans. In contrast, *Paranthropus* species had a stockier build, a wider face, and larger molars. Jaw muscles attached to a pronounced bony crest at the top of their skull (Figure 24.26C). The big molars and strong jaw muscles indicate that fibrous, difficult-to-chew plant parts accounted for a large part of the diet. *Paranthropus* died out about 1.2 million years ago.

Early Humans

What do the fossilized fragments of early hominids tell us about human origins? The record is still too sketchy

Sahelanthropus tchadensis	Australopithecus africanus	Paranthropus boisei	Homo habilis	Homo erectus
6 million years ago	3.2–2.3 million years ago	2.3–1.2 million years ago	1.9–1.6 million years ago	1.9 million to 53,000 years ago

Figure 24.26 A sampling of fossilized hominid skulls from Africa, all to the same scale.

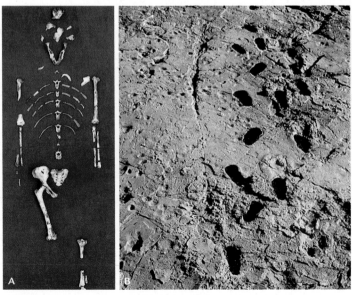

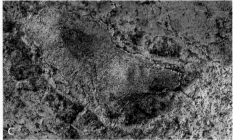

Figure 24.27 A Fossilized bones of Lucy, a female australopith (*Australopithecus afarensis*). This bipedal hominid lived in Africa 3.2 million years ago.

B,C At Laetoli in Tanzania, Mary Leakey discovered footprints made in soft, damp, volcanic ash 3.7 million years ago. The arch, big toe, and heel marks of these footprints are signs of bipedal hominids.

for us to interpret how all the diverse forms were related, let alone which might have been our ancestors. Besides, exactly which traits should we use to define **humans**—members of the genus *Homo*?

Well, what about brains? Our brain is the basis of unsurpassed analytical skills, verbal skills, complex social behavior, and technological innovations. Early hominids did use simple tools, but so do chimpanzees and bonobos. So how did an early hominid make the evolutionary leap to becoming human?

Comparing the brains of modern primates can give us clues. Genes for some brain proteins underwent repeated duplication as the primate lineage evolved. Further studies of how these proteins function may provide additional insight into how our uniquely human traits arose.

Until then, we are left to speculate on the evidence of physical traits among diverse fossils. They include a skeleton that permitted bipedalism, a smaller face, larger cranium, and smaller teeth with more enamel. These traits emerged during the late Miocene and can be observed in *Homo habilis*. The name of this early human means "handy man" (Figure 24.28).

Most of the early known forms of *Homo* are from the East African Rift Valley. Fossil teeth indicate that these early humans ate hard-shelled nuts, dry seeds, soft fruits and leaves, and insects. *H. habilis* may have enriched its diet by scavenging carcasses left behind by carnivores such as saber-tooth cats, but it did not have teeth adapted to a diet rich in meat.

Our close relatives, the chimpanzees and bonobos, use sticks and other natural objects as tools (Section 39.7). Early hominids most likely did the same. They smashed nuts open with rocks and used sticks to dig into termite nests and capture insects. By 2.5 million years ago, some hominids were modifying rocks in ways that made them better tools. Pieces of volcanic rock chipped to a sharp edge were found with animal bones that look like they were scraped by such tools.

The layers of Tanzania's Olduvai Gorge document refinements in toolmaking abilities (Figure 24.29). The layers that date to about 1.8 million years ago hold crudely chipped pebbles. More recent layers contain more complex tools, such as knifelike cleavers.

Olduvai Gorge also held hominid fossils. At the time of their discovery, these fossils were classified as *Homo erectus*. The name means "upright man." Today, some researchers reserve that name for fossils in Asia, and they assign the African fossils to *H. ergaster*. In our discus-

Figure 24.28 Painting of a band of *Homo habilis* in an East African woodland. Two australopiths are shown in the distance at the *left*.

Figure 24.29 A sample of stone tools from Olduvai Gorge in Africa. From *left* to *right*, crude chopper, more refined chopper, hand ax, and cleaver.

sions, we will adopt a traditional approach using "*H. erectus*" in reference to African populations and to descendant populations who, over generations, migrated into Europe and Asia.

H. erectus adults averaged about 5 feet (1.5 meters) tall, and had a larger brain than *H. habilis*. Improved hunting skills may have helped *H. erectus* get the food needed to maintain a large body and brain. Also, *H. erectus* built fires. Cooking probably broadened their diet by softening previously inedible hard foods.

---

australopiths Collection of now-extinct hominid lineages, some of which may be ancestral to humans.
humans Members of the genus *Homo*.

Take-Home Message **What do we know about the australopiths and the early species of the genus *Homo*?**

> Australopiths and certain hominids that preceded them walked upright. Some *Australopithecus* species were close to or on the lineage that led to humans.

> Like some hominids that preceded them, *Homo habilis*, the earliest known human species, walked upright. *Homo erectus* had a larger brain and dispersed out of Africa.

> Judging from the fossil record, the earliest members of the human lineage emerged about 2.5 million years ago, in the great East African Rift Valley.

< Link to Surface-to-volume ratio 4.2

Branchings of the Human Lineage

By 1.7 million years ago *Homo erectus* populations had become established in places as far away from Africa as the island of Java and eastern Europe. At the same time, African populations continued to thrive. Over thousands of generations, genetic divergences led to adaptations to local conditions. Some populations became so different from parental *H. erectus* groups that we call them new species: *H. neanderthalensis* (commonly called Neandertals), *H. floresiensis*, and *H. sapiens*, or modern humans (Figure 24.30).

A fossil from Ethiopia shows that *Homo sapiens* had evolved by 195,000 years ago. Compared to *H. erectus*, *H. sapiens* had smaller teeth, facial bones, and jawbones. Many fossils reveal another new feature: a prominent chin. Compared to earlier hominids, *H. sapiens* had a higher, rounder skull, a larger brain, and the capacity for spoken language.

From 200,000 to 30,000 years ago, Neandertals lived in Africa, the Middle East, Europe, and Asia. They had a stocky body that may have been an adaptation to colder climates. A stocky body has a lower surface-to-volume ratio than a thin one and loses heat to the environment less quickly.

Neandertals had a big brain. Did they believe in an afterlife? Did they have a spoken language? We do not know. They vanished when *H. sapiens* entered the same regions. Did the new arrivals drive them to extinction through direct warfare or indirectly, by outcompeting Neandertals for the same resources? We do not know.

Homo neanderthalensis

Homo sapiens

Homo floresiensis

Figure 24.30
Recent *Homo* species. Whether *H. floresiensis* belongs in this genus is still debated.

Even if a few matings between the species did take place, sequencing data indicates that Neandertal genes did not contribute to the gene pool of modern humans. Even so, because of our common ancestry, modern humans share about 99.5 percent of their genes with Neandertals.

In 2003, scientists discovered human fossils about 18,000 years old on the Indonesian island of Flores. Like *H. erectus*, they had a heavy brow and a relatively small brain for their body size. However, adults would have stood only a meter tall. Scientists who found the fossils assigned them to a new species, *H. floresiensis*. Not everyone is convinced. Some think the fossils belong to *H. sapiens* individuals who had a disease or disorder.

Where Did Modern Humans Originate?

Neandertals evolved from *H. erectus* populations in Europe and western Asia. *H. floresiensis* may have evolved from *H. erectus* in Indonesia. Where did *H. sapiens* originate? Two major models agree that *H. sapiens* evolved from *H. erectus* but differ in the details of where and how fast these events took place. Both attempt to explain the distribution of *H. erectus* and *H. sapiens* fossils, as well as genetic differences among modern humans who live in different regions.

Multiregional Model According to the **multiregional model**, populations of *H. erectus* living in Africa and other regions evolved into populations of *H. sapiens* gradually, over more than a million years. Gene flow among the populations maintained the species through the transition to fully modern humans (Figure 24.31A).

If this model is correct, much of the genetic variation present among modern Africans, Asians, and Europeans began to build up soon after their ancestors branched from an ancestral *H. erectus* population. The multiregional model is based primarily on interpretation of fossils. For example, faces of *H. erectus* fossils from China are said to look more like modern Asians than *H. erectus* that lived in Africa. The idea is that much variation seen among modern *H. sapiens* evolved long ago, in *H. erectus*.

Replacement Model By the more widely accepted **replacement model**, *H. sapiens* arose from a single *H. erectus* population in sub-Saharan Africa within the past 200,000 years. Later, bands of *H. sapiens* entered regions already occupied by *H. erectus* populations, and drove them all to extinction (Figure 24.31B). If this model is correct, then the regional variations observed among modern *H. sapiens* populations arose relatively recently. This model emphasizes the enormous degree of genetic similarity among all living humans.

Fossils support the replacement model. *H. sapiens* fossils date back to 195,000 years ago in East Africa and

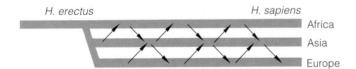

A Multiregional model. *H. sapiens* slowly evolves from *H. erectus* in many regions.

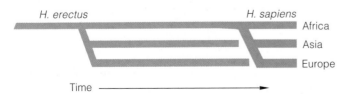

Time

B Replacement model. *H. sapiens* rapidly evolves from one *H. erectus* population in Africa, then disperses and replaces *H. erectus* populations in all regions.

Figure 24.31 Two models for the origin of *H. sapiens*. Arrows represent ongoing gene flow among populations.

100,000 years ago in the Middle East. In Australia, the oldest date to 60,000 years ago and, in Europe, to 40,000 years ago. Global comparisons of markers in mitochondrial DNA, and in the X and Y chromosomes, place the modern Africans closest to the root of the family tree. They also reveal that the most recent common ancestor of all modern humans was alive in Africa 60,000 years ago.

Leaving Home

Fossils and genetic evidence allow scientists to trace human dispersal routes (Figure 24.32). About 120,000 years ago, Africa's interior was becoming cooler and drier. As patterns and amounts of rainfall changed, so did the distribution of herds of grazing animals and the humans who hunted them. A few hunters may have journeyed north from East Africa and into Israel, where fossils 100,000 years old were found in a cave. They have no living descendants. Eruption of Mount Toba in Indonesia may have killed them, along with other ancient travelers.

Later waves of travelers had better luck, as some individuals left established groups and ventured into new territory. Successive generations continued along the coasts of Africa, then Australia and Eurasia. In the Northern Hemisphere, much of Earth's water became locked in vast ice sheets, which lowered the sea level by hundreds of meters. Previously submerged land was drained off between some regions. About 15,000 years ago, one small band of humans crossed such a land bridge from Siberia into North America.

With each step of their journey, humans faced and overcame extraordinary hardships. During this time, they devised cultural means to survive in inhospitable environments. Unrivaled capacities for modifying the habitat

multiregional model Model that postulates *H. sapiens* populations in different regions evolved from *H. erectus* in those regions.
replacement model Model for origin of *H. sapiens*; humans evolved in Africa, then migrated to different regions and replaced the other hominids that lived there.

Windows on the Past (revisited)

Interpreting vertebrate fossils can be a challenge. Scientists often disagree over whether a fossil represents a new species, a variant of a known species, or a collection of bones from more than one species. Such disputes are typically sorted out over time, as many scientists examine the fossils and publish their findings.

For example, consider what happened after one group of researchers uncovered hominid bones in a cave in Indonesia and declared their find a new species, *Homo floresiensis*. Some other scientists suggested that the small stature of the fossils could indicate that they were remains of *H. erectus* or *H. sapiens* with a genetic or nutritional disorder. A scientist who examined the fossil wrist disagreed, concluding that the bones are unlike those of *H. erectus* or *H. sapiens*, and so do represent a distinct species. Still other scientists noted that the fossil skull resembles that of some australopiths. They suggest that the fossils could be a new species that descended from australopiths who migrated out of Africa. Further study of existing fossils, a search for more fossils, and possibly DNA studies will help test the competing hypotheses.

How Would You Vote? Collecting fossils is big business, but it encourages poaching on public lands and decreases access to fossils. Should the government regulate fossil sales? See CengageNow for details, then vote online (cengagenow.com).

and for language served them well. Cultural evolution is ongoing. Hunters and gatherers persist in a few parts of the world, but others moved on from "stone-age" technology to the age of "high tech." This continued coexistence of such diverse groups is a tribute to the deep behavioral plasticity of the human species.

Take-Home Message What are modern humans?

> Fossils and genetic evidence indicate that modern humans, *H. sapiens*, most likely evolved from a *H. erectus* population in Africa.

> As modern humans dispersed out of Africa, geographic and climate factors affected their journey.

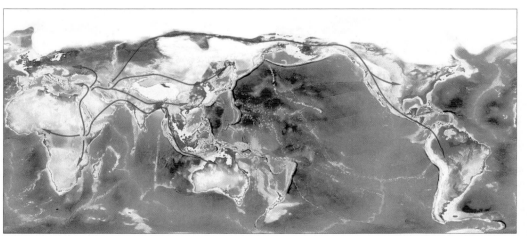

Figure 24.32 Some dispersal routes for small bands of *Homo sapiens*. This map shows ice sheets and deserts that prevailed about 60,000 years ago. Fossils provide evidence for the dates when modern humans appeared in these regions:

Africa	by 195,000 years ago
Israel	100,000
Australia	60,000
China	50,000
Europe	40,000
North America	14,000

The oldest evidence of humans in North America is fossil feces from a cave in Oregon.

Summary

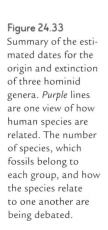

Section 24.1 Fossils are an important source of information about evolutionary transitions between major groups. For example, fossils reveal that birds evolved from feathered dinosaurs, and that some early birds had a bony tail and teeth.

Section 24.2 Four features define **chordate** embryos: a **notochord**, a dorsal hollow nerve cord, a pharynx with gill slits, and a tail extending past the anus. These features may or may not persist in adults. Invertebrate chordates include **tunicates** and **lancelets**. The **craniates** are chordates with a braincase of cartilage or bone. Most craniates are **vertebrates**. Their backbone is part of an **endoskeleton**. The **tetrapods** are vertebrates with four limbs. **Amniotes** are a subset of tetrapods that produce eggs that allow embryos to develop away from water.

Section 24.3 **Hagfishes** are jawless fishes with a cranium but no backbone. **Lampreys** are jawless vertebrates. Some are parasites of other fish. **Cartilaginous fishes** such as sharks have jaws with rows of teeth. **Ray-finned** fish are **bony fishes** and the most diverse vertebrate group. **Lobe-finned** fishes, also bony fishes, have fins reinforced with bones. Aspects of the fish body plan adapt the fish to life in water. For example, a swim bladder allows a ray-finned fish to adjust its buoyancy.

Section 24.4 **Amphibians**, the first lineage of tetrapods, branched off from the lobe-finned fishes. Modern amphibians such as frogs, toads, and salamanders are carnivores. Most spend some time on land, but return to the water to reproduce.

Section 24.5 Amniotes, the first vertebrates able to complete their life cycle on dry land, have water-conserving skin and kidneys, and **amniote eggs**. Fertilization usually takes place inside the female's body. There are two lineages. **Reptiles** (including birds and the extinct **dinosaurs**) belong to one lineage, and mammals to the other.

Section 24.6 The most diverse reptile lineage includes lizards and snakes. All are carnivores with flattened scales. Turtles have a bony shell and a horny beak rather than teeth. Crocodilians are carnivores and the closest relatives of birds. Snakes, lizards, turtles, and crocodilians are **ectotherms**: They adjust their temperature by their behavior. By contrast, birds and mammals are **endotherms**, which produce heat by metabolic activity.

Section 24.7 **Birds** are reptiles with feathers. The body plan of most species has been highly modified for flight, with lightweight bones and a highly efficient respiratory system. Male birds do not have a penis.

Section 24.8 **Mammals** have hair or fur, and the females nourish young with milk secreted from mammary glands. They include egg-laying **monotremes**, pouched **marsupials**, and **placental mammals**, which are the most diverse and widespread group.

Section 24.9 **Primates** have hands and feet capable of grasping objects. Prosimians are the oldest lineages. **Hominids** are the most recent. Early primates were shrewlike. Better depth perception, **bipedalism**, increased dexterity, bigger brains, social behavior, extended parental care, and **culture** evolved in some hominid lineages.

Sections 24.10, 24.11 **Australopiths** are early hominids known from Africa. The **human** lineage (*Homo*) arose by 2 million years ago (Figure 24.33), with *H. habilis* as an early toolmaking species. Neandertals and modern humans are relatives, but have distinct gene pools. By the **multiregional model**, *H. erectus* populations in far-flung regions evolved into *H. sapiens*. The **replacement model** has modern humans evolving from *H. erectus* in Africa, then dispersing into regions already occupied by *H. erectus* and driving them to extinction.

Figure 24.33 Summary of the estimated dates for the origin and extinction of three hominid genera. *Purple* lines are one view of how human species are related. The number of species, which fossils belong to each group, and how the species relate to one another are being debated.

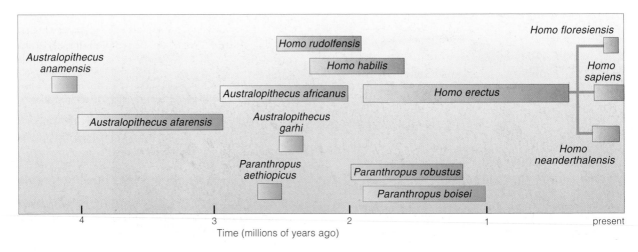

Data Analysis Activities

Primate Life History Trends One trend in primate evolution involved changes in life history traits, such as the length of infancy and the time it takes to reach adulthood. Figure 24.34 compares five primate lineages, from most ancient to most recent. It graphs life spans starting with years spent as "infants," when ongoing maternal care is required. It shows time spent as subadults, when individuals are no longer dependent on their mother for care but have not yet begun to breed. It shows the length of the reproductive years, and the length of time lived after reproductive years have passed.

1. What is the average life span for a lemur? A gibbon?
2. Which group reaches adulthood most quickly?
3. Which group has the longest expanse of reproductive years?
4. Which groups survive past their reproductive years?

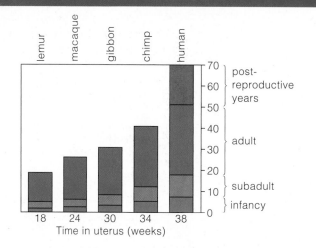

Figure 24.34 Trend toward longer life spans and greater dependency of offspring on adults for five primate lineages.

Self-Quiz Answers in Appendix III

1. All chordates have (a) _____ .
   a. backbone b. jaws c. notochord d. both b and c

2. The lancelet pharynx functions in _____ .
   a. respiration   c. reproduction
   b. filter-feeding   d. both a and b

3. Vertebrate jaws evolved from _____ .
   a. gill supports b. ribs c. scales d. teeth

4. Lampreys and sharks both have _____ .
   a. jaws   d. a swim bladder
   b. a bony skeleton   e. a four-chambered heart
   c. a cranium   f. lungs

5. A divergence from _____ gave rise to tetrapods.
   a. ray-finned fishes   c. cartilaginous fishes
   b. lizards   d. lobe-finned fishes

6. Reptiles, including birds, belong to one major lineage of amniotes, and _____ belong to another.
   a. sharks   c. mammals
   b. frogs and toads   d. salamanders

7. Reptiles are adapted to life on land by _____ .
   a. tough skin   d. amniote eggs
   b. internal fertilization e. both a and c
   c. good kidneys   f. all of the above

8. The closest modern relatives of birds are _____ .
   a. crocodilians b. prosimians c. turtles d. lizards

9. Only birds have _____ .
   a. a cloaca   c. feathers
   b. a four-chambered heart d. amniote eggs

10. An australopith is a(n) _____ .
    a. craniate   d. amniote
    b. vertebrate   e. placental mammal
    c. hominoid   f. all of the above

11. *Homo erectus* _____ .
    a. was the earliest member of the genus *Homo*
    b. was one of the australopiths
    c. evolved in Africa and dispersed to many regions
    d. disappeared as the result of an asteroid impact

12. True or false? Primates are endotherms.

13. Match the organisms with the appropriate description.
    ___lancelets   a. pouched mammals
    ___fishes   b. invertebrate chordates
    ___amphibians   c. feathered amniotes
    ___primates   d. egg-laying mammals
    ___birds   e. humans and close relatives
    ___monotremes   f. have grasping hands with nails
    ___marsupials   g. first land tetrapods
    ___placental   h. most diverse mammal
       mammals   lineage
    ___hominids   i. most diverse vertebrates

Additional questions are available on **CENGAGENOW**.

Critical Thinking

1. Monotremes and marsupials evolved before the supercontinent Pangea broke up. Placental mammals evolved after the breakup. How does this explain why Australia has no native placental mammals?

2. Does Alaska have more native birds or reptiles?

3. Describe some mechanisms that could have prevented hybrid births even if humans and Neandertals did occasionally meet and have intercourse.

Animations and Interactions on **CENGAGENOW**:
> Lancelet body plan; Salamander gait; Structure of an amniote egg.

❮ Links to Earlier Concepts

This chapter builds on what you learned in Sections 21.2 and 21.7, which introduced plant structure and growth, and correlated them with present and past functions. You will revisit carbohydrates (3.3), lignin (4.11), stomata and other surface specializations of plant cells (6.8), and differentiation in eukaryotic tissues (10.2).

Key Concepts

Overview of Plant Tissues
Seed-bearing vascular plants have a shoot system with stems, leaves, and reproductive parts. Most also have a root system. These systems consist of ground, vascular, and dermal tissues. Plants lengthen or thicken at meristems.

Primary Shoots
The two main flowering plant groups (monocots and eudicots) differ in the organization of tissues in stems and leaves. Leaf specializations for sunlight interception, water conservation, and gas exchange support photosynthesis.

25 Plant Tissues

25.1 Sequestering Carbon in Forests

Carbon in the atmosphere occurs mainly as carbon dioxide gas (CO_2). As you learned in Chapter 6, humans release a lot of carbon dioxide by burning fossil fuels and other plant-derived materials. As a result of these activities, the amount of CO_2 in the atmosphere is increasing exponentially, with unintended and potentially catastrophic effects on Earth's climate.

Efforts are now under way to reduce global carbon dioxide emissions, and to reduce the amount of carbon dioxide already in the atmosphere. Part of these climate change mitigation efforts include carbon offsets, which are financial instruments designed to reduce global emissions of carbon dioxide and other carbon-based greenhouse gases. Companies and individuals buy carbon offsets to "offset" activities that release these green-house gases. The funds are then used to support projects aimed at reducing current emissions of the gases, or activities that remove them from the atmosphere.

Any process by which carbon dioxide is removed from the atmosphere is called carbon sequestration. Carbon can be sequestered by capturing carbon dioxide directly from activities that release it, before it even enters the atmosphere. Carbon dioxide already in the atmosphere can also be removed from it. Plants do this naturally. Plants are specialized to absorb carbon dioxide from the air and lock it in their tissues via photosynthesis. Remember from Section 6.1 that photosynthesis harnesses the energy of sunlight to build sugars from carbon dioxide and water. In an actively growing plant, photosynthetically produced sugars are continually remodeled into other carbon-containing compounds that end up in plant parts such as tree trunks, branches, foliage, and roots.

In wood, for example, carbon is a major part of cellulose (Section 3.3) and lignin (Section 4.11). These sturdy materials reinforce the specialized cells and structures that allow a plant to grow tall, and taller plants win the competition for sunlight (Figure 25.1).

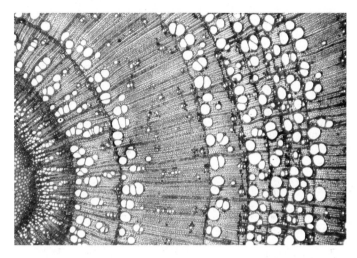

Figure 25.1 Wood. Wood's porous structure (*above*) allows it to transport water and nutrients throughout the plant. Despite its porosity, wood is strong enough to support a plant body that can be hundreds of feet tall (*opposite*). Carbon that gets incorporated into sturdy plant tissues such as wood can stay out of the atmosphere for centuries.

Some carbon offsets support forestry activities such as increasing plant density in forests, and replanting deforested areas. Such activities typically cost less than other methods of removing carbon from the atmosphere, in part because they take advantage of the natural ability of plants to sequester carbon. Forests worldwide absorb about 3 billion tons of CO_2 from the atmosphere every year, about one-third of the amount released by human activities. Earth's forests currently contain more than twice as much carbon as its atmosphere.

This unit examines plant function, growth, development, and reproduction, with special emphasis on angiosperms (flowering plants). In this chapter, you will consider how plants are built, including the adaptations that allow them to sequester carbon.

Primary Roots Monocots and eudicots also differ in the organization of tissues in their roots. Roots anchor a plant and provide a large surface area for absorbing water and minerals from soil.

Secondary Growth In many plants, secondary growth thickens branches and roots during successive growing seasons. Extensive secondary growth of eudicots and conifers produces wood. Tree rings can be used to study past environmental conditions.

Modified Stems Certain types of stem special-izations are adaptations for storing water or nutrients, and for reproduction.

> Most plants consist of roots, stems, and shoots.
> All plant parts consist mainly of ground tissue with vascular tissues threading through it. Dermal tissues cover exposed surfaces of the plant.
> Monocots and eudicots differ in tissue organization.
⟨ Links to Differentiation 10.2, Plant evolution 21.2, Angiosperms 21.7

The Basic Body Plan

With more than 260,000 species (and counting), flowering plants dominate the plant kingdom. Its major groups are the magnoliids, eudicots (true dicots), and monocots (Section 21.7). We focus mainly on eudicots and monocots, many of which have a body plan similar to the one shown in Figure 25.2.

A plant's aboveground shoots consist of stems together with leaves. Stems provide a structural framework for the plant's growth. Pipelines inside stems conduct water, nutrients, and photosynthetically produced sugars between leaves and roots. Roots are structures that absorb water and dissolved minerals as they grow down and outward through the soil. They often serve to anchor the plant as well. All plant cells store carbohydrates for their own use, but some plants have specialized root cells that store carbohydrates for the plant as a whole.

Overview of Plant Tissue Systems

All plant parts consist of the same three types of tissues (Figure 25.3). The **ground tissue system**, which makes up the bulk of the plant, has many functions such as photosynthesis and food storage. Pipelines of the **vascular tissue system** thread through ground tissue. They distribute water and nutrients to all parts of the plant body.

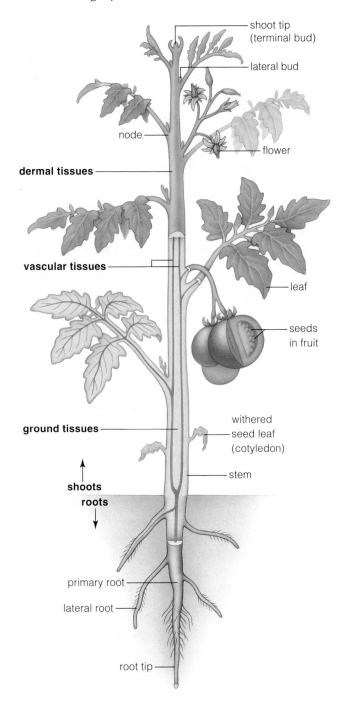

shoot tip (terminal bud)

lateral bud

node

dermal tissues

flower

vascular tissues

leaf

seeds in fruit

ground tissues

withered seed leaf (cotyledon)

stem

shoots
roots

primary root

lateral root

root tip

Figure 25.2 Animated Body plan of a tomato plant. Vascular tissues (*purple*) conduct water and solutes. They thread through ground tissues that make up most of the plant body. Epidermis, a type of dermal tissue, covers root and shoot surfaces.

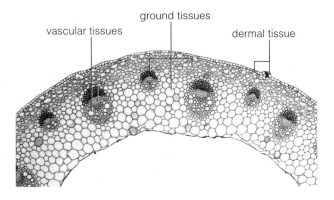

vascular tissues

ground tissues

dermal tissue

Figure 25.3 Location of three plant tissue types in the cross-section of a stem from a buttercup plant.

---

cotyledon Seed leaf; part of a flowering plant embryo.
dermal tissue system Tissue system that covers and protects the plant body.
ground tissue system Tissue system that makes up the bulk of the plant body; includes most photosynthetic cells.
meristem Zone of undifferentiated plant cells; cells of meristem tissue can divide rapidly.
primary growth Plant growth from apical meristems in root and shoot tips.
secondary growth Thickening of older stems and roots.
vascular tissue system Tissue system that distributes water and nutrients through a plant body.

The **dermal tissue system** covers and protects the plant's exposed surfaces. These three plant tissue systems are made up of cells organized as simple and complex tissues. Simple tissues consist primarily of one type of cell; examples include parenchyma, collenchyma, and sclerenchyma. Complex tissues have two or more cell types. Xylem, phloem, and epidermis are examples. You will learn more about all of these tissues in the next section.

Eudicots Versus Monocots

The same tissues form in all flowering plants, but they do so in different patterns. Monocots and eudicots are named after their **cotyledons**, which are leaflike structures that contain food for a plant embryo. These "seed leaves" wither after the seed germinates and the developing plant begins to make its own food by photosynthesis. Cotyledons consist of the same types of tissues in all plants that have them, but the seeds of eudicots have two cotyledons and those of monocots have only one.

Figure 25.4 shows some other differences between monocots and eudicots. Most shrubs and trees, such as rose bushes and maple trees, are eudicots. Lilies, orchids, and grasses are examples of typical monocots.

Introducing Meristems

Plant tissues arise at **meristems**, each a region of undifferentiated cells that can divide rapidly. Cells just under a meristem differentiate and give rise to all plant tissues. Development in plants is similar to development in animals in that plant cells, like animal cells, differentiate according to where they are located in the body. In plants, differentiating cells form tissues that become shoots, stems, and roots.

New plant parts lengthen by activity at meristems in shoot tips and root tips. The lengthening of young shoots and roots is called **primary growth**. Some plants also undergo **secondary growth**, which means that their stems and roots thicken over time. In woody eudicots and in gymnosperms such as pine trees, secondary growth occurs when the cells of a thin cylindrical layer of meristem divide.

Take-Home Message **What is the basic structure of a flowering plant?**

> Plants typically have aboveground shoots and stems, and belowground roots, all of which consist of ground, vascular, and dermal tissue systems.

> Ground tissues make up most of a plant. Vascular tissues that thread through ground tissue distribute water and solutes. Dermal tissues cover and protect plant surfaces.

> Eudicots and monocots have the same types of tissues, but differ somewhat in their pattern of tissue organization.

A Eudicots

In seeds, two cotyledons (seed leaves of embryo)

Flower parts in fours or fives (or multiples of four or five)

Leaf veins usually forming a netlike array

Pollen grains with three pores or furrows

Vascular bundles organized in a ring in ground tissue of stem

B Monocots

In seeds, one cotyledon (seed leaf of embryo)

Flower parts in threes (or multiples of three)

Leaf veins usually running parallel with one another

Pollen grains with one pore or furrow

Vascular bundles throughout ground tissue of stem

Figure 25.4 Animated Comparing some of the characteristics of **A** eudicots and **B** monocots.

> Different types of complex plant tissues form from varying arrangements of simple tissues.
< Links to Polysaccharides 3.3, Plant cell surface specializations 4.11, Stomata 6.8, Lignin in plant evolution 21.2

Simple Tissues

Parenchyma is a simple tissue that consists mainly of parenchyma cells (*left*). These cells have various shapes depending on their function, but all are typically thin-walled, flexible, and can continue to divide because they are alive in mature tissue. Parenchyma makes up most of the soft parts inside roots, stems, leaves, and flowers. This tissue has roles in storage, secretion, wound repair, photosynthesis, and other specialized tasks.

radial: tangential: transverse:

Figure 25.5 Terms that identify how tissue specimens are cut from a plant. Longitudinal cuts along a stem or root radius give radial sections. Cuts at right angles to the radius give tangential sections. Cuts perpendicular to the long axis of a stem or root give transverse sections—that is, cross-sections.

>> **Figure It Out** Was the section of a sunflower stem shown in Figure 25.6 cut along a radial, tangential, or transverse plane?

Answer: Transverse

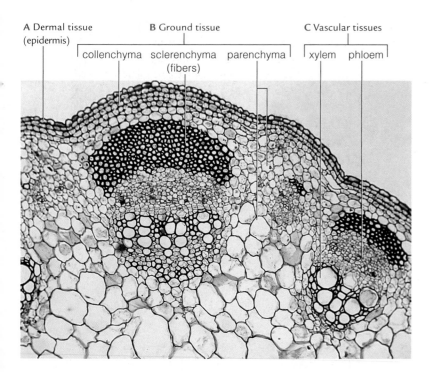

A Dermal tissue (epidermis)   B Ground tissue   C Vascular tissues
collenchyma   sclerenchyma (fibers)   parenchyma   xylem   phloem

Figure 25.6 Locations of complex tissues in the stem of a sunflower (*Helianthus*).

Collenchyma is a simple tissue that consists mainly of collenchyma cells, which are elongated and alive at maturity. A polysaccharide called pectin imparts flexibility to these cells' primary wall, which is thickened unevenly where three or more of the cells abut (*above*). Collenchyma is stretchable, and it supports rapidly growing plant parts such as young stems and leaf stalks.

Variably shaped cells of **sclerenchyma** are dead at maturity, but the lignin-containing cell walls that remain help this tissue resist compression. Fibers (*left*) and sclereids are typical sclerenchyma cells. Fibers are long, tapered cells that structurally support vascular tissues in some stems and leaves. They flex and twist, but resist stretching. We use fibers of some plants in cloth, rope, paper, and other commercial products. The stubbier and often branched sclereids strengthen hard seed coats, such as peach pits, and make pear flesh gritty.

Complex Tissues

A plant's complex tissues (dermal, ground, and vascular tissues) consist of simple tissues in varying arrangements and proportions (Table 25.1). Plant sections are typically cut along standard planes for micrographs (Figure 25.5).

Dermal Tissues The first dermal tissue to form on a plant is **epidermis** (Figure 25.6A), which is usually a single layer of cells on the plant's outer surface (*left*). Epidermal cells secrete substances such as cutin, a polymer of fatty acids, on their outward-facing cell walls. The waxy deposits form a cuticle that helps the plant conserve water and repel pathogens. The epidermis of leaves and young stems includes specialized cells and, often, hairs and other epidermal cell outgrowths. Pairs of specialized epidermal cells form stomata (small gaps

collenchyma Simple plant tissue composed of living cells with unevenly thickened walls; provides flexible support.
companion cell In phloem, parenchyma cell that loads sugars into sieve tubes.
epidermis Outermost tissue layer of a plant.
mesophyll Photosynthetic parenchyma.
parenchyma Simple plant tissue made up of living cells; main component of ground tissue.
phloem Complex vascular tissue of plants; distributes sugars through its sieve tubes.
sclerenchyma Simple plant tissue that is dead at maturity; its lignin-reinforced cell walls structurally support plant parts.
sieve tube Conducting tube of phloem.
tracheid Tapered cell that forms water-conducting tubes in xylem.
vessel member Cell that forms water-conducting tubes in xylem.
xylem Complex vascular tissue of plants; its tracheids and vessel members distribute water and mineral ions.

Table 25.1 Overview of Flowering Plant Tissues

Tissue Type	Main Components	Main Functions
Simple Tissues		
Parenchyma	Parenchyma cells	Photosynthesis, storage, secretion, tissue repair, other tasks
Collenchyma	Collenchyma cells	Pliable structural support
Sclerenchyma	Fibers or sclereids	Structural support
Complex Tissues		
Vascular		
Xylem	Tracheids and vessel members; parenchyma cells; sclerenchyma cells	Water-conducting tubes; structural support
Phloem	Sieve-tube members, parenchyma cells; sclerenchyma cells	Sugar-conducting tubes and their supporting cells
Dermal		
Epidermis	Epidermal cells, their secretions and outgrowths	Secretion of cuticle; protection; control of gas exchange and water loss
Periderm	Cork cambium; cork cells; parenchyma	Protective cover on older stems, roots

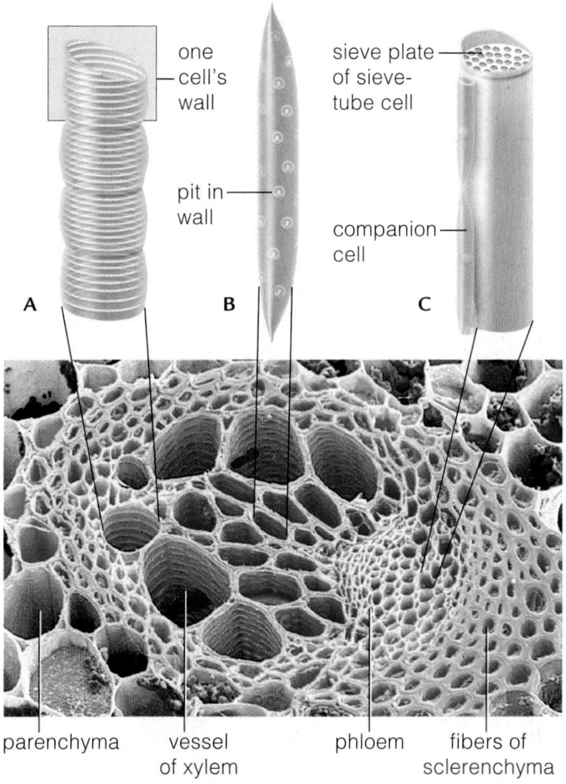

Figure 25.7 Vascular tissues. In xylem, **A** part of a column of vessel members, and **B** a tracheid. **C** One of the living cells that interconnect as sieve tubes in phloem. Companion cells closely associate with sieve tubes. Fibers of sclerenchyma and parenchyma cells are also visible in the micrograph.

›› **Figure It Out** What are the green structures inside the cells?

Answer: Chloroplasts

across the epidermis) when the cells swell with water (Section 6.8). Plants control the diffusion of water vapor, oxygen, and carbon dioxide gases across their epidermis by opening and closing stomata. A dermal tissue called periderm replaces epidermis in woody stems and roots.

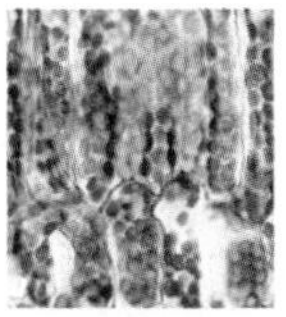

Ground Tissues Ground tissue, which is defined as everything other than dermal and vascular tissue, accounts for the bulk of a plant. It consists mostly of parenchyma, but can also include other simple tissues depending on where it is (Figure 25.6B). **Mesophyll**, the only photosynthetic ground tissue, consists of chloroplast-containing parenchyma cells (*inset*).

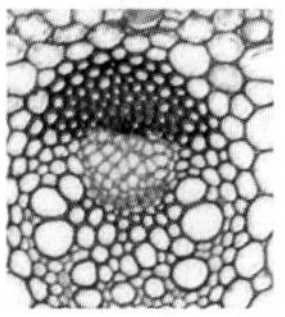

Vascular Tissues Xylem and phloem are vascular tissues (Figure 25.6C). Both are composed of elongated conducting tubes often surrounded by sclerenchyma fibers and parenchyma (*left*). **Xylem**, which conducts water and mineral ions, consists of two types of cells, **tracheids** and **vessel members**, that are dead in

mature tissue (Figure 25.7A,B). The walls of these cells are stiffened and waterproofed with lignin. These walls interconnect to form tubes, and they lend structural support to the plant. Water can move laterally between the tubes as well as upward through them.

Phloem conducts sugars and other organic solutes. Its main cells, sieve-tube members, are alive in mature tissue. They connect end to end at sieve plates, forming **sieve tubes** that transport sugars from photosynthetic cells to all parts of the plant (Figure 25.7C). Phloem's **companion cells** are parenchyma cells that load sugars into the tubes.

Take-Home Message What are the main types of plant tissues?

> Dermal tissues cover and protect plant surfaces. Epidermis, which includes epidermal cells, their secretions, and outgrowths, covers young plant surfaces.

> Ground tissues make up most of a plant. Cells of parenchyma have diverse roles, including photosynthesis. Collenchyma and sclerenchyma support and strengthen plant parts.

> Vascular tissues distribute water and solutes through ground tissue. In xylem, water and ions flow through tubes of dead tracheid and vessel member cells. In phloem, sieve tubes that consist of living cells distribute sugars.

> The organization of ground, vascular, and dermal tissues inside stems differs between monocots and eudicots.

> Meristem lays down tissue behind itself during primary growth in a lengthening stem or root.

Internal Structure of Stems

Inside the stems of most flowering plants, the cells of primary xylem and phloem are bundled together as long, multistranded cords called **vascular bundles**. The main function of these bundles is to conduct water, ions, and nutrients between different parts of the plant. Fibers and lignin-reinforced walls of tracheids in the bundles also play an important role in supporting upright stems.

Vascular bundles extend through the ground tissue system of all stems, leaves, and roots, but the arrangement of the bundles in these plant parts differs between eudicots and monocots. The vascular bundles of most eudicot stems form a cylinder that runs parallel with the long axis of the shoot (Figure 25.8A). In eudicots, the vascular cylinder divides the ground tissue into cortex and pith. Cortex is all tissue between the vascular cylinder and epidermis; pith is all tissue inside the cylinder. Pith is typi-

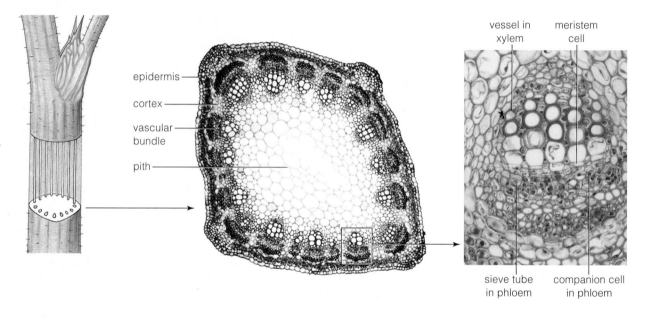

A Cross-section of an alfalfa stem. The cylindrical arrangement of vascular bundles is a characteristic of eudicot stems.

epidermis

cortex

vascular bundle

pith

vessel in xylem

meristem cell

sieve tube in phloem

companion cell in phloem

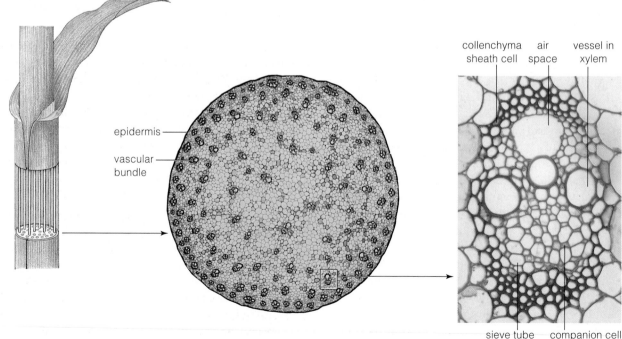

B Cross-section of a corn stem. The vascular bundles in monocot stems are not arranged in a cylinder.

epidermis

vascular bundle

collenchyma sheath cell

air space

vessel in xylem

sieve tube in phloem

companion cell in phloem

Figure 25.8 Animated Primary structure of **A** a eudicot stem and **B** a monocot stem.

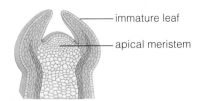

A Shoot tip, tangential cut.

immature leaf

apical meristem

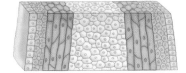

B Same tissue region later, after the shoot tip has lengthened above it.

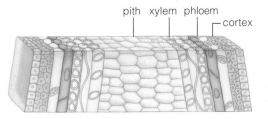

pith xylem phloem ┌ cortex

C Later still, different lineages of cells are differentiating and forming complex tissues.

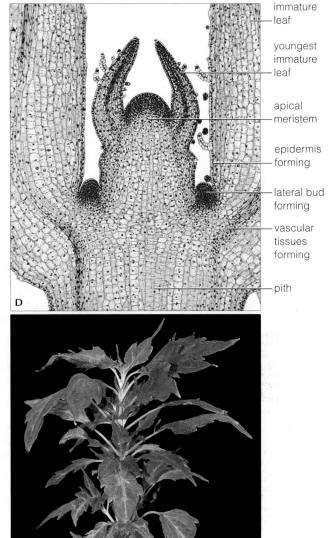

immature leaf

youngest immature leaf

apical meristem

epidermis forming

lateral bud forming

vascular tissues forming

pith

D

Figure 25.9
Growth in the stem of *Coleus*, a eudicot.

A–C Successive stages of this stem's primary growth.

D The micrograph shows a longitudinal cut through the stem's center. The tiers of leaves in the photograph below it formed in this linear pattern of development.

>> **Figure It Out** What is the transparent layer of cells on the outer surface of the stem illustrated in C?

Answer: Epidermis

cally spongy parenchyma specialized for storage. Monocot stems have a different pattern. Their vascular bundles do not form a cylinder; rather, they are distributed throughout the ground tissue (Figure 25.8B).

Primary Growth of a Stem

Most of a plant's primary growth occurs by cell divisions of apical meristem in terminal buds (Figure 25.9A). Just beneath the surface of a terminal bud, meristem cells divide continually during the growing season. Some of those cells differentiate and give rise to dermal tissues, primary vascular tissues, and ground tissues that form behind the meristem (Figure 25.9B,C). The differentiating cells enlarge and lengthen under the bud, so undifferentiated meristem cells remain at the tip of the lengthening root or shoot. A bud may be naked or encased in modified

leaves called bud scales. Small regions of tissue that bulge out near the sides of a bud's apical meristem will develop into new leaves. As the stem lengthens, the leaves form and mature in orderly tiers, one after the next. A region of stem where one or more leaves form is called a node; the region between two successive nodes is an internode.

A lateral (or axillary) bud is a dormant shoot that forms in a leaf axil, the place where a leaf attaches to a stem (Figure 25.9D). Depending on hormonal signals from other regions of the plant, divisions of meristem inside a lateral bud can give rise to either a branch, a leaf, or a flower.

---

vascular bundle Multistranded, sheathed cord of primary xylem and phloem in a stem or leaf.

Take-Home Message **How do plant tissues form inside stems?**

❯ A shoot lengthens as cells of apical meristem in a terminal bud divide and differentiate. New leaves or branches develop from lateral buds.

❯ Vascular bundles are multistranded cords of vascular tissue. The arrangement of vascular bundles in stems differs between eudicots and monocots.

› All leaves are metabolic factories where photosynthetic cells churn out sugars, but they vary in size, shape, surface specializations, and internal structure.

‹ Links to Plasmodesmata 4.11, Plant adaptations for photosynthesis 6.8, Water conservation adaptations 21.2

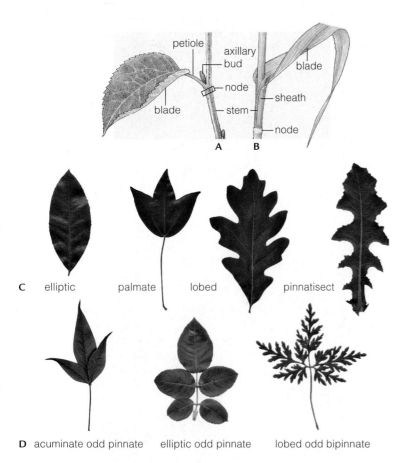

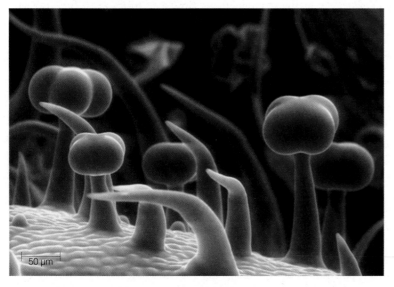

Figure 25.10 Common leaf forms of **A** eudicots and **B** monocots, and a few examples of **C** simple leaves and **D** compound leaves.

Similarities and Differences

A leaf of duckweed is 1 millimeter (0.04 inch) across; leaves of one palm (*Raphia regalis*) can be 25 meters (82 feet) long. Leaves are shaped like cups, needles, blades, spikes, tubes, and feathers. They differ in color, odor, and edibility; many form toxins.

Leaf shapes and orientations are adaptations that help a plant intercept sunlight and exchange gases. Most leaves are thin, with a high surface-to-volume ratio; many reorient themselves during the day so that they stay perpendicular to the sun's rays. Typically, adjacent leaves project from a stem in a pattern that allows sunlight to reach them all. However, the leaves of plants native to arid regions may stay parallel to the sun's rays, reducing heat absorption and thus conserving water. Thick or needlelike leaves of some plants also conserve water.

A typical leaf has a flat blade and, in eudicots, a petiole (or stalk) attached to the stem (Figure 25.10A). The leaves of grasses and most other monocots are flat blades, the base of which forms a sheath around the stem (Figure 25.10B). Simple leaves are undivided, although many are lobed; compound leaves have blades divided as leaflets (Figure 25.10C,D).

Fine Structure

A leaf's internal structure is adapted to intercept the sun's light and to enhance gas exchange. Many leaves also have surface specializations.

Epidermis Epidermis covers every leaf surface exposed to the air. This surface tissue may be smooth, sticky, or slimy, with hairs, scales, spikes, hooks, and other specializations (Figure 25.11). A translucent, waxy secreted cuticle slows water loss from the sheetlike array of epidermal cells (Figure 25.12 ❶).

Typically, most of a leaf's stomata occur on its lower surface ❷. The guard cells on either side of each stoma are the only photosynthetic cells in leaf epidermis. As you will see in Section 26.5, shape changes of the guard cells close the stomata to prevent water loss, or open the stomata to allow gases to cross the epidermis. Carbon

Figure 25.11 Example of leaf cell surface specialization: hairs on a tomato leaf. The lobed heads are glandular structures that occur on the leaves of many plants; they secrete aromatic chemicals that deter plant-eating insects. Those on marijuana plants secrete the psychoactive chemical tetrahydrocannabinol (THC).

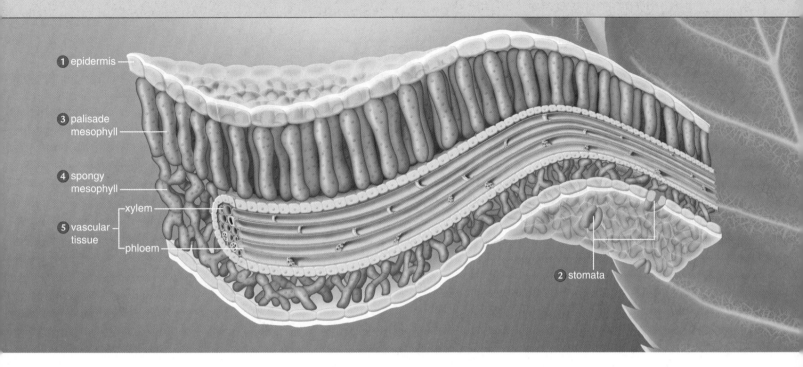

- **①** epidermis
- **③** palisade mesophyll
- **④** spongy mesophyll
- **⑤** vascular tissue
  - xylem
  - phloem
- **②** stomata

Figure 25.12 Animated Anatomy of a eudicot leaf.

① The upper leaf surface is epidermis with a secreted layer of cuticle.

② The lower leaf surface is also covered with cuticle. Gas exchanges between air inside and outside of the leaf occur at stomata in the lower leaf epidermis.

③ The bulk of the leaf is mesophyll, a type of photosynthetic parenchyma. In many leaves, mesophyll occurs in two distinct forms: elongated palisade mesophyll attached to the upper epidermis, with **④** spongy mesophyll below it.

⑤ Vascular bundles of xylem (*blue*) and phloem (*pink*) form the leaf's veins.

dioxide needed for photosynthesis enters the leaf through stomata, then diffuses through air in the spaces between mesophyll cells. Oxygen released by photosynthesis diffuses in the opposite direction.

Mesophyll The bulk of a leaf consists of mesophyll, which is photosynthetic parenchyma with air spaces between cells. Plasmodesmata connect the cytoplasm of adjacent cells. Substances can flow rapidly across the walls of adjoining cells through these cell junctions (Section 4.11). In leaves oriented perpendicular to the sun, mesophyll is arranged in two layers. Palisade mesophyll is attached to the upper epidermis **③**. The elongated parenchyma cells of this tissue have more chloroplasts than cells of the spongy mesophyll layer below **④**. Blades of grass and other monocot leaves that grow vertically can intercept light from all directions. The mesophyll in such leaves is not divided into two layers.

Veins—The Leaf's Vascular Bundles Leaf **veins** are vascular bundles typically strengthened with fibers **⑤**. Inside the bundles, continuous strands of xylem rapidly transport water and dissolved ions to mesophyll. Continuous strands of phloem rapidly transport the products of photosynthesis (sugars) away from mesophyll. In most eudicots, large veins branch into a network of minor veins embedded in mesophyll. In most monocots, all veins are similar in length and run parallel with the leaf's long axis (Figure 25.13).

vein A vascular bundle in the stem or leaf of a plant.

A A grape leaf, with the netlike array of veins typical of eudicots. A stiffened midrib runs from the petiole to the leaf tip. Ever smaller veins branch from it.

B An *Agapanthus* leaf, with the strong parallel orientation of veins typical of monocots.

Figure 25.13 Typical vein patterns in eudicot and monocot leaves. Like umbrella ribs, stiffened veins help maintain leaf shape.

Take-Home Message **How does a leaf's structure contribute to its function?**

❯ Leaves are structurally adapted to intercept sunlight and distribute water and nutrients. Components include mesophyll (photosynthetic cells), veins (bundles of vascular tissue), and cuticle-secreting epidermis.

> Roots mainly function to provide plants with a large sur-
face area for absorbing water and dissolved mineral ions.

A Taproot system of the
California poppy, a eudicot.

B Fibrous root system of
a grass plant, a monocot.

Figure 25.14 Different types of root systems.

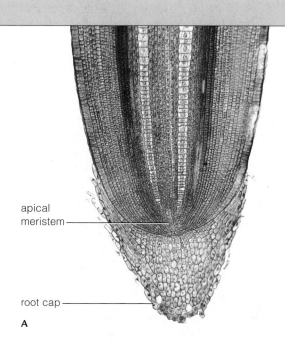

apical
meristem

root cap

A

Root Systems

Unless tree roots start to buckle a sidewalk or clog a
sewer line, roots tend not to occupy our thoughts. Yet
the underground portion of a plant, the root system, is
just as extensive and essential as the shoot system. Roots
grow through the soil and take up water and mineral
nutrients for the plant. Primary growth of roots results in
one of two kinds of root systems. The **taproot system** of
eudicots consists of a primary root and its lateral branchings. Carrots, oak trees, and poppies are among the plants
with a taproot system (Figure 25.14A). By comparison, the
primary root of most monocots is replaced by adventitious roots, which grow outward from the stem. Lateral
roots that are similar in diameter and length branch off of
adventitious roots. Together, adventitious and lateral roots
of such plants form a **fibrous root system** (Figure 25.14B).

The roots of most plants extend less than 5 meters
(16.4 feet) from the plant, but some grow much larger
than that. For example, the roots of mesquite (*Prosopis*), a
drought-tolerant shrub, can be more than 50 meters (164
feet) deep. The shallow roots of some types of cactus radi-
ate 15 meters (49 feet) away from the plant. The surface
area of root systems are impressive. Laid out as a sheet,
the root system of a single young rye plant would cover
about 600 square meters, or 6,548 square feet!

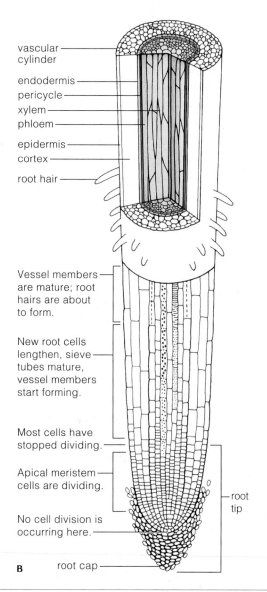

vascular
cylinder

endodermis
pericycle
xylem
phloem

epidermis
cortex

root hair

Vessel members
are mature; root
hairs are about
to form.

New root cells
lengthen, sieve
tubes mature,
vessel members
start forming.

Most cells have
stopped dividing.

Apical meristem
cells are dividing.

No cell division is
occurring here.

root
tip

root cap

B

Figure 25.15 Animated Tissue formation and organization in a typical root. **A** Root tip of corn (*Zea mays*), a monocot. **B** The oldest cells
are farthest from the apical meristem, which a root cap protects.

>> **Figure It Out** Where does most cell division occur in a primary
root?   Answer: At the apical meristem

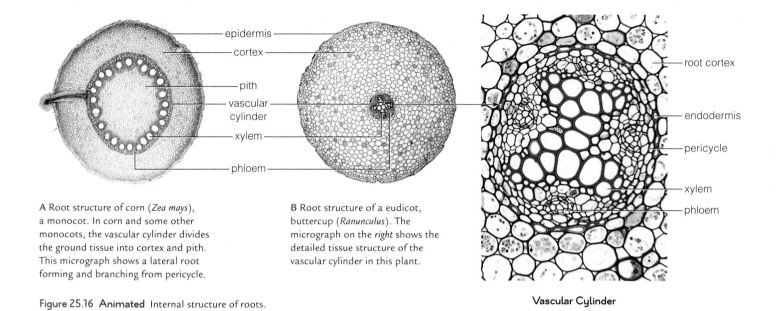

A Root structure of corn (*Zea mays*), a monocot. In corn and some other monocots, the vascular cylinder divides the ground tissue into cortex and pith. This micrograph shows a lateral root forming and branching from pericycle.

B Root structure of a eudicot, buttercup (*Ranunculus*). The micrograph on the *right* shows the detailed tissue structure of the vascular cylinder in this plant.

Vascular Cylinder

Figure 25.16 Animated Internal structure of roots.

Internal Structure of Roots

A root's structural organization begins in a seed. As the seed germinates, a primary root pokes through the seed coat. In nearly all eudicot seedlings, that young root thickens over time.

Look at the root tip in Figure 25.15. Some descendants of root apical meristem give rise to a root cap, a dome-shaped mass of cells that protects the soft, young root as it grows through soil. Other descendants lengthen, widen, or flatten when they differentiate as part of the dermal, ground, and vascular tissue systems.

Ongoing divisions push cells away from the active root apical meristem. Some of their descendants form epidermis. The root epidermis is the plant's absorptive interface with soil. Many of its specialized cells send out fine extensions called **root hairs**, which collectively increase the surface area available for taking up soil water along with dissolved oxygen and mineral ions. Chapter 26 looks at the role of root hairs in plant nutrition.

Descendants of meristem cells also form the root's **vascular cylinder**, a central column of conductive tissue. In a typical monocot, the vascular cylinder divides the ground tissue into two zones, cortex and pith (Figure 25.16A). By contrast, the root vascular cylinder of typical eudicots consists mainly of primary xylem and phloem (Figure 25.16B).

Pericycle, a layer of parenchyma one or two cells thick, encloses the vascular cylinder of both monocots and eudicots. Pericycle cells are differentiated, but they can divide repeatedly in a direction perpendicular to the axis of the root. Pericycle cells can erupt through the cortex and epidermis as the start of new lateral roots (a new lateral root branching from pericycle is shown in the micrograph in Figure 25.16A).

As you will see in Chapter 26, water entering a root moves from cell to cell until it reaches the **endodermis**, a layer of cells that encloses the pericycle. Wherever endodermal cells abut, their walls are waterproofed. Water must pass through the cytoplasm of endodermal cells to reach the vascular cylinder. Transport proteins in the plasma membrane control the uptake of water and dissolved substances.

endodermis In plant roots, a layer of cells just outside the pericycle; separates vascular cylinder from cortex.
fibrous root system Root system composed of an extensive mass of similar-sized roots; typical of monocots.
root hairs Hairlike, absorptive extensions of a young cell of root epidermis.
taproot system In eudicots, a primary root and all of its lateral branchings.
vascular cylinder Sheathed, cylindrical array of primary xylem and phloem in a root.

> **Take-Home Message** **What is the basic structure and function of a plant's root?**
>
> ❯ Taproot systems consist of a primary root and lateral branchings. Fibrous root systems consist of adventitious and lateral roots that replace the primary root.
>
> ❯ Inside each root is a vascular cylinder that contains long strands of primary xylem and phloem.
>
> ❯ Roots provide a plant with a tremendous surface area for absorbing water and dissolved minerals from soil.

❯ Secondary growth occurs at two types of lateral meristem, vascular cambium and cork cambium.

Roots and shoots of many plants thicken and become woody over time. Such thickening is called secondary growth. Cell divisions in lateral meristems give rise to

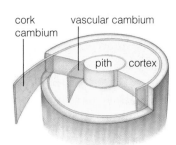

A Secondary growth (thickening of older stems and roots) occurs at two lateral meristems, vascular cambium and cork cambium. Vascular cambium gives rise to secondary tissues; cork cambium, to periderm.

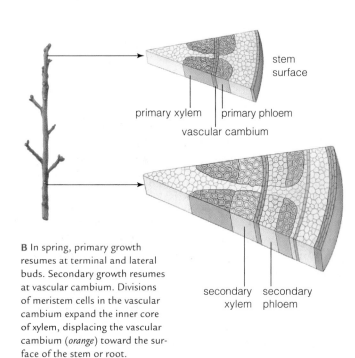

B In spring, primary growth resumes at terminal and lateral buds. Secondary growth resumes at vascular cambium. Divisions of meristem cells in the vascular cambium expand the inner core of xylem, displacing the vascular cambium (*orange*) toward the surface of the stem or root.

secondary growth. **Lateral meristems** are cylindrical layers of meristem that run lengthwise through stems and roots (Figure 25.17A). Woody plants have two types of lateral meristems: vascular cambium and cork cambium (*cambium* is the Latin word for change). Both arise from pericycle. **Vascular cambium** is the lateral meristem that produces secondary vascular tissue, a few cells thick, inside older stems and roots. Divisions of vascular cambium cells produce secondary xylem on the cylinder's inner surface, and secondary phloem on its outer surface (Figure 25.17B). As the core of xylem thickens, it also displaces the vascular cambium toward the surface of the stem or root. The displaced cells of the vascular cambium divide in a widening circle, so the tissue's cylindrical form is maintained (Figure 25.17C).

Vascular cambium consists of two types of cells. Long, narrow cells give rise to the secondary tissues that extend lengthwise through a stem or root: tracheids, fibers, and parenchyma in secondary xylem; and sieve tubes, companion cells, and fibers in secondary phloem. Small, rounded cells that divide perpendicularly to the axis of the stem give rise to "rays" of parenchyma, radially oriented like spokes of a bicycle wheel. Secondary xylem and phloem in the rays conduct water and solutes radially through the stems and roots of older plants.

A core of secondary xylem, or **wood**, contributes up to 90 percent of the weight of some plants. Thin-walled, living parenchyma cells and sieve tubes of secondary phloem lie in a narrow zone outside of the vascular cambium. Bands of thick-walled reinforcing fibers are often interspersed through this secondary phloem. The only living sieve tubes are within a centimeter or so of the vascular cambium; the rest are dead, but they help protect the living cells behind them.

As seasons pass, the expanding inner core of xylem continues to put pressure on the stem or root surface. In time, the pressure ruptures the cortex and the outer sec-

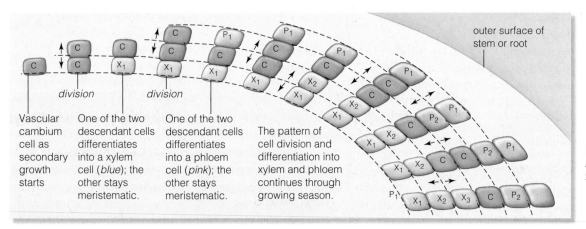

| Vascular cambium cell as secondary growth starts | One of the two descendant cells differentiates into a xylem cell (*blue*); the other stays meristematic. | One of the two descendant cells differentiates into a phloem cell (*pink*); the other stays meristematic. | The pattern of cell division and differentiation into xylem and phloem continues through growing season. |

Figure 25.17
Animated
Secondary growth.

C Overall pattern of growth at vascular cambium.

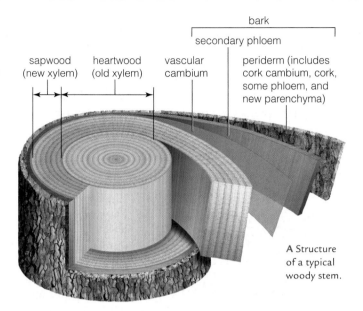

A Structure of a typical woody stem.

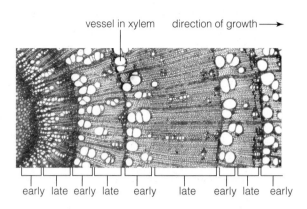

B Early and late wood in an ash tree. Early wood forms during wet springs. Late wood indicates that a tree did not waste energy making large-diameter xylem cells for water uptake during a dry summer or drought.

Figure 25.18 Structure of a woody stem. The rings visible in the heartwood and sapwood are regions of early and late wood. In most temperate zone trees, one ring forms each year.

ondary phloem. Then, another lateral meristem, the **cork cambium**, forms and gives rise to periderm. Periderm is a dermal tissue that consists of parenchyma and cork, as well as the cork cambium that produces them. What we call bark is secondary phloem and periderm. **Bark** consists of all of the living and dead tissues outside of the vascular cambium (Figure 25.18A). The **cork** component of bark has densely packed rows of dead cells with walls thickened by a waxy substance called suberin. Cork protects, insulates, and waterproofs a stem or root surface. Cork also forms over wounded tissues. When leaves drop from the plant, cork forms at the places where petioles had attached to stems.

Wood's appearance and function change as a stem or root ages. Metabolic wastes, such as resins, tannins, gums, and oils, clog and fill the oldest xylem so much that it no longer is able to transport water and solutes. These substances darken and strengthen the wood, which is called **heartwood**. **Sapwood** is moist, still-functional xylem between heartwood and vascular cambium. In trees of temperate zones, dissolved sugars travel from roots to buds through sapwood's secondary xylem in spring. The sugar-rich fluid is sap. Each spring, New Englanders collect maple tree sap to make maple syrup.

Vascular cambium is inactive during cool winters or long dry spells. When the weather warms or moisture returns, the vascular cambium gives rise to early wood, with large-diameter, thin-walled cells. Late wood, with small-diameter, thick-walled xylem cells, forms in dry summers. In regions where growth rates vary seasonally, a cross-section through an older trunk reveals alternating bands of early and late wood (Figure 25.18B). Each band is called a growth ring, or "tree ring." Trees native to regions where growth slows or stops during the winter tend to add one growth ring per year. In tropical regions where rainfall is continuous, trees grow at the same rate year-round and do not have growth rings.

Oak, hickory, and other eudicot trees that evolved in temperate and tropical zones are hardwoods, with vessels, tracheids, and fibers in xylem. Pines and other conifers are called softwoods because they are less dense than the hardwoods. Their xylem has tracheids and parenchyma rays but no vessels or fibers.

bark Secondary phloem and periderm of woody plants.
cork Component of bark; waterproofs, insulates, and protects the surfaces of woody stems and roots.
cork cambium In plants, a lateral meristem that gives rise to periderm.
heartwood Dense, dark accumulation of nonfunctional xylem at the core of older tree stems and roots.
lateral meristem Vascular cambium or cork cambium; sheetlike cylinder of meristem that gives rise to plant secondary growth.
sapwood Functional secondary xylem between the vascular cambium and heartwood in an older stem or root.
vascular cambium Ring of meristematic tissue that produces secondary xylem and phloem.
wood Accumulated secondary xylem.

Take-Home Message **What is secondary growth in plants?**

❯ Secondary growth thickens the stems and roots of older plants.

❯ Wood is mainly accumulated secondary xylem.

❯ Secondary growth occurs at two types of lateral meristem: vascular cambium and cork cambium. Secondary vascular tissues form at a cylinder of vascular cambium. A cylinder of cork cambium gives rise to periderm, which is part of a protective covering of bark.

> Many plants have modified stem structures that function in storage and reproduction.

The structure of a typical stem is shown in Figure 25.2, but there are many variations on that structure in different types of plants. Specialized stems allow some plants to store nutrients, to reproduce asexually, or both.

Stolons Stolons are stems that branch from the main stem of the plant. Stolons may look like roots, but they have nodes, and roots have no nodes. Adventitious roots and leafy shoots that sprout from the nodes develop into new plants. Stolons are commonly called runners because in many plants they "run" along the surface of the soil. The strawberry plants (*Fragaria*) in the photo *above* are reproducing asexually by sending out runners.

Rhizomes Rhizomes are fleshy stems that typically grow under the soil and parallel to its surface. A rhizome is the main stem of the plant, and it also serves as the plant's primary storage tissue. Branches that sprout from nodes grow aboveground for photosynthesis and flowering. Ginger, irises, many ferns, and some grasses have rhizomes. The main stems of turmeric plants (*Curcuma longa*), shown in the photo *above*, are undergound rhizomes.

Bulbs A bulb is a short section of underground stem encased by overlapping layers of thickened, modified leaves called scales. The photo (*left*) shows clearly visible scales surrounding the stem at the center of an onion, which is the bulb of an *Allium cepa* plant. The scales contain starch and other substances that a plant holds in reserve when conditions in the environment are unfavorable for growth. When favorable conditions return, the plant uses these stored substances to sustain rapid growth. The scales develop from a basal plate, as do roots. The dry, paperlike outer scale of many bulbs serves as a protective covering.

Corms A corm is a thickened underground stem that stores nutrients. Like a bulb, a corm has a basal plate from which roots grow. Unlike a bulb, a corm is solid rather than layered, and it has nodes from which new plants develop. Taro, also known as arrowroot, is the corm of *Colocasia esculenta* plants (*above*).

Tubers Tubers are thickened portions of underground stolons; they are the plant's primary storage tissue. Tubers are like corms in that they have nodes from which new shoots and roots sprout, but they do not have a basal plate. The photo (*left*) shows how potatoes, which are tubers, grow on stolons of *Solanum tuberosum* plant. Potato "eyes" are the nodes of the tubers.

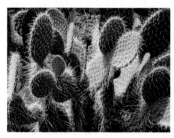

Cladodes Cactuses and other succulents have cladodes, which are flattened, photosynthetic stems that store water. New plants form at the nodes. The cladodes of some plants appear quite leaflike, but most are unmistakably fleshy. The photo (*above*) shows spiky cladodes of prickly pear (*Opuntia*).

Take-Home Message Are all stems alike?

> Many plants have modified stems that function in storage and reproduction. Stolons, rhizomes, bulbs, corms, tubers, and cladodes are examples.

25.9 Tree Rings and Old Secrets

❭ The relative thicknesses of a tree's rings hold clues to environmental conditions during its lifetime.

year: 1 2 3

Tree rings can be used to estimate average annual rainfall; to date archaeological ruins; to gather evidence of wildfires, floods, landslides, and glacier movements; and to study the ecology and effects of parasitic insect populations. How? Some tree species, such as redwoods and bristlecone pines, add wood over centuries, one ring per year (*above*). Count an old tree's rings, and you have an idea of its age. If you know the year in which the tree was cut, you can find out when a particular ring formed by counting the rings backward from the outer edge. Thickness and other features of the rings offer clues about environmental conditions during those years (Figure 25.19).

For an example of how we use the information in tree rings, consider the history of early English settlers of the United States. In 1587, 150 settlers arrived at Roanoke Island, which is off the coast of North Carolina. Supply ships that returned in 1589 found the island abandoned, and searches up and down the coast failed to turn up the missing colonists. About twenty years later, a second set of English settlers arrived, this time at Jamestown, Virginia. Although this colony survived, the initial years were difficult. In the summer of 1610 alone, more than 40 percent of the colonists died, many of them from starvation.

Differences in the thicknesses of tree rings from nearby bald cypress trees (*Taxodium distichum*) that had been growing at the time of the Roanoke and Jamestown colonies revealed that both sets of settlers had terrible timing (Figure 25.19D). The Roanoke settlers arrived just in time for the worst drought in 800 years, and nearly a decade of severe drought struck Jamestown. We know that the corn crop of the Jamestown colony failed, so similar drought-related crop failures probably occurred at Roanoke. The Jamestown settlers also would have had difficulty finding fresh water, because their colony had been established at the head of an estuary. When the river levels dropped during the drought, the settlers' drinking water would have mixed with ocean water and become salty. Piecing together these bits of evidence gives us an idea of what life must have been like for the early settlers.

Take-Home Message Can tree rings reveal more than just the age of a tree?

❭ Tree rings reflect conditions in the environment during the time they formed.

Sequestering Carbon in Forests (revisited)

A tree can live for centuries. After it dies, its tissues decompose, and the carbon in them is re-released to the atmosphere. However, plant matter decomposes more slowly than other organic materials, because the lignin, cellulose, and other molecules that waterproof and reinforce plants are relatively stable. Most of the carbon in a forest is locked in plant parts and their remains. In some forests, as much as 80 percent of the total carbon is in the soil, as part of plant matter in various stages of decay. Thus, a forest stops accumulating carbon as its trees mature.

How Would You Vote? Are carbon offsets a good idea, or do they just give companies and individuals an excuse to continue emitting greenhouse gases? See CengageNow for details, then vote online (cengagenow.com).

direction of growth ⟶

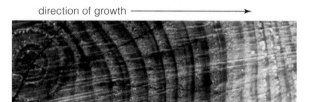

A Pine is a softwood. It grows fast, so it tends to have wider rings than slower-growing species. Note the difference between the appearance of heartwood and sapwood.

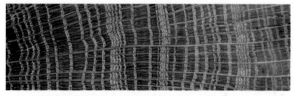

B The rings of this oak tree show dramatic differences in yearly growth patterns over its lifetime.

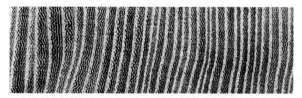

C An elm made this series between 1911 and 1950.

1587–1589   1606–1612

D A section of a bald cypress tree that was living near English colonists when they first settled in North America. Narrower annual rings mark years of severe drought.

Figure 25.19 Tree rings. In most species, each ring corresponds to one year, so the number of rings indicate the age of the tree. Relative thickness of the rings can be used to estimate data such as average annual rainfall long before such records were kept.

Summary

Section 25.1 By the process of photosynthesis, plants naturally remove carbon from the atmosphere and sequester it in their tissues. Carbon that is locked in molecules of wood and other durable plant tissues can stay out of the atmosphere for centuries.

Section 25.2 Most flowering plants have aboveground shoots, including stems, leaves, and flowers. Most kinds also have belowground roots. Shoots and roots consist of **ground**, **vascular**, and **dermal tissue systems**. Ground tissues store materials, function in photosynthesis, and structurally support the plant. Tubes in vascular tissues conduct substances to all living cells. Dermal tissues protect plant surfaces.

Simple plant tissues have only one type of cell. Complex plant tissues have two or more. Monocots and eudicots have the same tissues organized in different ways. For example, monocots and eudicots differ in how xylem and phloem are distributed through ground tissue, in the number of petals in flowers, and in the number of **cotyledons**.

Plant tissues originate at **meristems**, which are regions of undifferentiated cells that retain their ability to divide. **Primary growth** (or lengthening) arises from apical meristems. **Secondary growth** (or thickening) arises from lateral meristems.

Section 25.3 The simple plant tissues are **parenchyma**, **collenchyma**, and **sclerenchyma**. The living, thin-walled cells in parenchyma have diverse roles in ground tissue. **Mesophyll** is photosynthetic parenchyma. Living cells in collenchyma have sturdy, flexible walls that support fast-growing plant parts. Cells in sclerenchyma die at maturity, but their lignin-reinforced walls remain and support the plant.

Vascular tissues (**xylem** and **phloem**) and dermal tissues (**epidermis**) are examples of complex plant tissues. **Vessel members** and **tracheids** of xylem are dead at maturity; their perforated, interconnected walls conduct water and dissolved minerals. Phloem's sieve-tube members remain alive at maturity. These cells interconnect to form **sieve tubes** that conduct sugars. **Companion cells** load sugars into the sieve tubes. Epidermis covers and protects the outer surfaces of young plant parts.

Section 25.4 New shoots form at terminal buds and lateral buds on stems. **Vascular bundles** of xylem and phloem extend through stems. The bundles conduct water, ions, and nutrients between different parts of the plant, and also function in support.

In most herbaceous and young woody eudicot stems, a ring of vascular bundles divides the ground tissue into cortex and pith. Monocot stems often have vascular bundles distributed throughout ground tissue.

Section 25.5 Leaves are photosynthesis factories that contain mesophyll and vascular bundles (**veins**) between their upper and lower epidermis. Air spaces around mesophyll cells allow gas exchange. Water vapor and gases cross the cuticle-covered epidermis at stomata.

Section 25.6 Roots absorb water and mineral ions for the entire plant. Inside each is a **vascular cylinder** enclosed by **endodermis**. **Root hairs** increase the surface area of roots. Most eudicots have a **taproot system** of a primary root with lateral branchings; many monocots have a **fibrous root system** that consists of adventitious and lateral roots.

Section 25.7 Woody plants thicken (add secondary growth) by cell divisions in **lateral meristems**. Cell divisions in **vascular cambium** add secondary xylem (**wood**) and secondary phloem. Cell divisions in **cork cambium** add periderm. **Bark** is periderm and secondary phloem of a woody stem. Wood is classified by its location and function, as in **heartwood** or **sapwood**.

Section 25.8 Stem specializations such as rhizomes, corms, tubers, bulbs, cladodes, and stolons are adaptations that function in storage or reproduction in many types of plants.

Section 25.9 In many trees, one ring forms each season. Tree rings hold information about environmental conditions that prevailed while the rings were forming. For example, the relative thicknesses of the rings reflect the relative availability of water.

Self-Quiz   Answers in Appendix III

1. Roots and shoots lengthen through activity at _____ .
   a. apical meristems   c. vascular cambium
   b. lateral meristems   d. cork cambium

2. In many plant species, older roots and stems thicken by activity at _____ .
   a. apical meristems   c. vascular cambium
   b. cork cambium   d. both b and c

3. _____ conducts water and minerals through a plant, and _____ conducts sugars.
   a. Phloem; xylem   c. Xylem; phloem
   b. Cambium; phloem   d. Xylem; cambium

4. Mesophyll consists of _____ .
   a. waxes and cutin   c. photosynthetic cells
   b. lignified cell walls   d. cork but not bark

5. Which of the following cell types are alive in mature tissue? Choose all that apply.
   a. collenchyma cells   d. tracheids
   b. sieve tubes   e. companion cells
   c. vessel members   f. sclerenchyma cells

Tree Rings and Droughts Douglas fir trees (*Pseudotsuga menziesii*) are exceptionally long-lived, and particularly responsive to rainfall levels. Researcher Henri Grissino-Mayer sampled Douglas firs in El Malpais National Monument, in west central New Mexico. Pockets of vegetation in this site have been surrounded by lava fields for about 3,000 years, so they have escaped wildfires, grazing animals, agricultural activity, and logging. Grissino-Mayer compiled tree ring data from old, living trees, and dead trees and logs to generate a 2,129-year annual precipitation record (Figure 25.20).

1. The Mayan civilization began to suffer a massive loss of population around 770 A.D. Do these tree ring data reflect a drought condition at this time? If so, was that condition more or less severe than the "dust bowl" drought?

2. One of the worst population catastrophes ever recorded occurred in Mesoamerica between 1519 and 1600 A.D., when around 22 million people native to the region died. Which period between 137 B.C. and 1992 had the most severe drought? How long did that drought last?

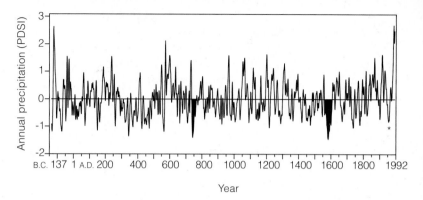

Figure 25.20 A 2,129-year annual precipitation record compiled from tree rings in El Malpais National Monument, New Mexico. Data was averaged over 10-year intervals; graph correlates with other indicators of rainfall collected in all parts of North America.

PDSI: Palmer Drought Severity Index: 0, normal rainfall; increasing numbers mean increasing excess of rainfall; decreasing numbers mean increasing severity of drought.

* A severe drought contributed to a series of catastrophic dust storms that turned the midwestern United States into a "dust bowl" between 1933 and 1939.

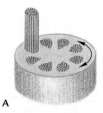 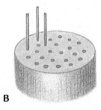

6. Which of the two distribution patterns for vascular tissues *above* is common among eudicots? Which is common among monocots? Are these stem or root sections?

7. In phloem, organic compounds flow through _____ .
   a. collenchyma cells   c. vessels
   b. sieve tubes   d. tracheids

8. Xylem and phloem are _____ tissues.
   a. ground   b. vascular   c. dermal   d. both b and c

9. Tree rings occur when _____ .
   a. there are droughts during the time the rings form
   b. environmental conditions influence xylem cell size
   c. heartwood alternates with sapwood

10. Bark is mainly _____ .
    a. periderm and cork   c. periderm and phloem
    b. cork and wood   d. cork cambium and phloem

11. Match the plant parts with the best description.
    ___ apical meristem   a. secondary growth
    ___ lateral meristem   b. source of primary growth
    ___ xylem   c. distribution of sugars
    ___ phloem   d. source of secondary growth
    ___ vascular cylinder   e. distribution of water
    ___ wood   f. central column in roots

Additional questions are available on CENGAGENOW.

Critical Thinking

1. Is the plant with the yellow flower *below* a eudicot or a monocot? What about the plant with the purple flower?

2. Oscar and Lucinda meet in a tropical rain forest and fall in love, and he carves their initials into the bark of a tiny tree. They never do get together, though. Ten years later, still heartbroken, Oscar searches for the tree. Given what you know about primary and secondary growth, will he find the carved initials higher relative to ground level? If he goes berserk and chops down the tree, what kinds of growth rings will he see?

3. Are the stems shown at *right* stolons, rhizomes, bulbs, corms, or tubers? (*Hint:* Notice where the shoots are growing from.)

Animations and Interactions on CENGAGENOW:
❯ Plant body plan; Monocot and eudicot tissues; Stem structure; Leaf structure; Root structure and function; Secondary growth; Wood structure.

‹ Links to Earlier Concepts

The movement of fluids through plants depends on hydrogen bonding (Sections 2.4 and 2.5), membrane transport (5.7), and osmosis (5.6). You may wish to review nutrients (1.3), ions (2.3), carbohydrates (3.3), photosynthesis (6.4, 6.7), and aerobic respiration (7.2). You will revisit vascular tissues (25.3), leaves (25.5), roots (25.6), nitrogen fixation (19.7), mycorrhiza (22.6), and adaptations of land plants (21.2), including plasmodesmata and cuticle (4.11), and stomata (6.8).

Key Concepts

Plant Nutrients and Soil
Many plant structures are adaptations to limited amounts of water and essential minerals. The amount of water and minerals available for plants depends on the composition of soil, which is vulnerable to leaching and erosion.

Water Uptake and Movement Through Plants
Certain specializations allow vascular plants to selectively take up water and minerals through their roots. Xylem transports absorbed water and solutes from roots to other parts of the plant.

26 Plant Nutrition and Transport

From World War I until the 1970s, the United States Army tested and disposed of weapons at J-Field, Aberdeen Proving Ground in Maryland (Figure 26.1). Obsolete chemical weapons and explosives were burned in open pits, together with plastics, solvents, and other wastes. Lead, arsenic, mercury, and other toxic metals heavily contaminated the soil and groundwater. So did highly toxic organic compounds, including trichloroethylene (TCE). TCE damages the nervous system, lungs, and liver, and exposure to large amounts can be fatal. Today, the toxic groundwater is seeping toward nearby marshes and the Chesapeake Bay.

To protect the bay and clean up the soil, the Army and the Environmental Protection Agency turned to phytoremediation: the use of plants to take up and concentrate or degrade environmental contaminants. They planted hybrid poplar trees (*Populus trichocarpa* × *deltoides*) that cleanse groundwater by taking up TCE and other organic pollutants from it.

Like other vascular plants, poplar trees take up soil water through their roots. Along with the water come nutrients and chemical contaminants, including TCE. Although TCE is toxic to animals, it does not harm the trees. The poplars break down some of the toxin, but they release most of it into the atmosphere. Airborne TCE is the lesser of two evils: It breaks down much more quickly in air than it does in groundwater.

In other types of phytoremediation, groundwater contaminants accumulate in tissues of the plants, which are then harvested for safer disposal elsewhere.

Phytoremediation is cheaper and easier on the environment than mechanical methods of cleaning up toxic waste, which include excavating contaminated soil and pumping out contaminated groundwater. In general, the best plants for phytoremediation take up many contaminants, grow fast, and grow big. Not very many species with those traits tolerate toxic substances, but genetically engineered ones may increase our number of choices. Alpine pennycress (*Thlaspi caerulescens*) absorbs zinc, cadmium, and other potentially toxic minerals dissolved in soil

Figure 26.1 Toxic waste cleanup. *Above*, J-Field during its use as a testing site for weapons. *Opposite*, J-Field today. Poplar trees are helping to remove toxic waste left behind in the soil.

water. Unlike typical cells, the cells of pennycress plants store zinc and cadmium inside a central vacuole. Isolated inside these organelles, the toxic elements are kept safely away from the rest of the cells' activities. Pennycress is a small, creeping plant, so its usefulness for phytoremediation is limited. Researchers are working to transfer a gene that confers its toxin-storing ability to larger, faster-growing plants.

Phytoremediation takes advantage of the ability of plants to take up water and solutes (including toxic waste) from the soil. Plants have this capacity not for our benefit, but rather to meet their own metabolic needs. Water and solutes needed by all living cells must move from soil, into roots, and then to other parts of a plant. Sugars that fuel metabolism must also move from where they are produced to where they are used. The structures that allow this movement, and the processes that drive it, are the subject of this chapter.

Water Loss Versus Gas Exchange
A cuticle and stomata help plants conserve water. Closed stomata stop water loss but also stop gas exchange. Some plant adaptations are trade-offs between water conservation and gas exchange.

Distribution of Organic Molecules Through Plants
Phloem distributes organic products of photosynthesis from leaves to living cells throughout the plant. Organic compounds are actively loaded into conducting tubes at sources, then unloaded in sinks.

> Plants require elemental nutrients in soil, water, and air.
> Different types of soil affect the growth of plants.
< Links to Nutrients 1.3, Ions 2.3

O Horizon
Fallen leaves and other organic material littering the surface of mineral soil

A Horizon
Topsoil, with decomposed organic material; variably deep [only a few centimeters in deserts, elsewhere extending as far as 30 centimeters (1 foot) below the soil surface]

B Horizon
Compared with A horizon, larger soil particles, not much organic material, more minerals; extends 30 to 60 centi-meters (1 to 2 feet) below soil surface

C Horizon
No organic material, but partially weathered fragments and grains of rock from which soil forms; extends to underlying bedrock

Bedrock

Figure 26.2 From a habitat in Africa, an example of soil horizons.

Plant Nutrients

Plant growth requires the sixteen elements listed in Table 26.1. Nine elements are macronutrients, which means they are required in amounts above 0.5 percent of the plant's dry weight (its weight after all the water has been removed from the plant body). Seven other elements are micronutrients, which make up traces of the plant body. Carbon, oxygen, and hydrogen are abundant in air and water. Plants obtain the other nutrients when their roots take up minerals dissolved in soil water.

Properties of Soil

Soil consists of mineral particles mixed with variable amounts of decomposing organic material, or **humus**. The mineral particles form by the weathering of rocks. Humus forms from dead organisms and organic litter: fallen leaves, feces, and so on. Water and air occupy spaces between the mineral particles and the organic bits.

Soils differ in the proportions of mineral particles. The particles, which are primarily sand, silt, and clay, differ in size and chemical properties. The biggest sand grains are big enough to see with the naked eye, about one millime-ter in diameter. Silt particles are hundreds or thousands of times smaller than sand grains, so they are too small to see. Clay particles are even smaller. A clay particle consists of thin, stacked layers of negatively charged crys-tals with sheets of water alternating between the layers. Because clay is negatively charged, it attracts positively

Table 26.1 Plant Nutrients and Symptoms of Deficiency

Macronutrient	Functions	Symptoms of Deficiency	Micronutrient	Functions	Symptoms of Deficiency
Carbon Hydrogen Oxygen	Raw materials for photosynthesis	None; all are abundantly available in water and carbon dioxide	Chlorine	Role in root and shoot growth and photolysis	Wilting; chlorosis; some leaves die
Nitrogen	Component of proteins, coenzymes, chlorophyll, nucleic acids	Stunted growth; young leaves pale green; older leaves yellow and die (these are symptoms of a condition called chlorosis)	Iron	Roles in chlorophyll synthesis and in electron transport	Chlorosis; yellow and green striping in leaves of grass species
Potassium	Activation of enzymes; contributes to water–solute balances that influence osmosis*	Reduced growth; curled, mot-tled, or spotted older leaves; burned leaf edges; weakened plant	Boron	Roles in germination, flowering, fruiting, cell division, nitrogen metabolism	Terminal buds, lateral branches die; leaves thicken, curl, become brittle
Calcium	Component in control of many cell functions; cementing cell walls	Terminal buds wither, die; deformed leaves; poor root growth	Manganese	Chlorophyll synthesis; coenzyme action	Dark veins, but leaves whiten and fall off
Magnesium	Chlorophyll component; activation of enzymes	Chlorosis; droopy leaves	Zinc	Role in forming auxin, chloroplasts, starch; enzyme component	Chlorosis; mottled or bronzed leaves; root abnormalities
Phosphorus	Component of nucleic acids, ATP, phospholipids	Purplish veins; stunted growth; fewer seeds, fruits	Copper	Component of several enzymes	Chlorosis; dead spots in leaves; stunted growth
Sulfur	Component of most proteins, two vitamins	Light-green or yellowed leaves; reduced growth	Molybdenum	Part of enzyme used in nitrogen metabolism	Pale green, rolled or cupped leaves

* All mineral elements contribute to water–solute balances.

Figure 26.3 Runaway erosion in Providence Canyon, Georgia, a result of poor farming practices combined with soft soil. Settlers that arrived in the area around 1800 plowed the land straight up and down the hills. The furrows made excellent conduits for rainwater, which proceeded to carve out deep crevices that made even better rainwater conduits. The area became useless for farming by 1850. It now consists of about 445 hectares (1,100 acres) of deep canyons that continue to expand at the rate of about 2 meters (6 feet) per year.

charged mineral ions dissolved in the soil water. Thus, clay retains dissolved nutrients that would otherwise trickle past roots too quickly to be absorbed.

Even though they do not bind mineral ions as well as clay, sand and silt are also necessary for growing plants. Water and air occupy spaces between the particles and organic bits in soil. Without enough sand and silt to intervene between tiny particles of clay, soil packs so tightly that it excludes air, and without air, root cells cannot secure enough oxygen for aerobic respiration.

Soils with the best oxygen and water penetration are **loams**, which have roughly equal proportions of sand, silt, and clay. Most plants grow best in loams. Humus also affects plant growth because it releases nutrients, and its negatively charged organic acids can trap the positively charged mineral ions in soil water. Humus swells and shrinks as it absorbs and releases water, and these changes in size aerate soil by opening spaces for air to penetrate.

Most plants grow well in soils that contain between 10 and 20 percent humus. Soil with less than 10 percent humus may not have enough nutrients to support plant growth. Soil with more than 90 percent humus, such as that in swamps and bogs, stays so saturated with water that air (and the oxygen in it) is excluded. Very few kinds of plants can grow in these soils.

How Soils Develop Soils develop over thousands of years. They are in different stages of development in different regions. Most form in layers, or horizons, that are distinct in color and other properties (Figure 26.2). The layers help us characterize soil in a given place, and compare it with soils in other places. For instance, the A horizon is **topsoil**. This layer typically contains the greatest amount of organic matter, so the roots of most plants grow most densely in it. Topsoil is deeper in some places than in others. For example, tropical forests tend to have little topsoil; grasslands have a lot.

---

humus Decaying organic matter in soil.
leaching Process by which water moving through soil removes nutrients from it.
loam Soil with roughly equal amounts of sand, silt, and clay.
soil Mixture of various mineral particles and humus.
soil erosion Loss of soil under the force of wind and water.
topsoil Uppermost soil layer; contains the most nutrients for plant growth.

Minerals, salts, and other molecules dissolve in water as it filters through soil. **Leaching** is the process by which water removes soil nutrients and carries them away. Leaching is fastest in sandy soils, which do not bind nutrients as well as clay soils. During heavy rains, more leaching occurs in forests than in grasslands. Why? Grass plants absorb water more quickly than trees.

Soil erosion is a loss of soil under the force of wind and water. Strong winds, fast-moving water, sparse vegetation, and poor farming practices cause the greatest losses (Figure 26.3). For example, each year, about 25 billion metric tons of topsoil erode from croplands in the midwestern United States. The topsoil enters the Mississippi River, which then dumps it into the Gulf of Mexico. Nutrient losses because of this erosion affect not only plants that grow in the region, but also the other organisms that depend on the plants for survival.

Take-Home Message Where do plants get nutrients they need?

> Plants require nine macronutrients and seven micronutrients, all elements that are available from water, air, and soil.

> Soil consists mainly of mineral particles: sand, silt, and clay. Clay attracts and reversibly binds dissolved mineral ions.

> Soil contains humus, a reservoir of organic material rich in organic acids.

> Most plants grow best in loams (soils with equal proportions of sand, silt, and clay) that contain 10 to 20 percent humus.

> Leaching and erosion remove nutrients from soil.

❯ Root specializations such as hairs, mycorrhizae, and nodules help plants absorb water and nutrients.

❮ Links to Plasmodesmata 4.11, Osmosis 5.6, Membrane crossing mechanisms 5.7, Nitrogen fixation 19.7, Fungal symbionts 22.6, Root structure 25.6

Root Specializations

In actively growing plants, new roots grow into different patches of soil. The roots are not "exploring" the soil; their growth is simply greater in areas where the water and nutrient concentrations best match the requirements of the particular plant. Certain specializations of roots help plants take up water and minerals. Mycorrhizae and root hairs absorb water and ions from the soil, and bacteria in root nodules absorb nitrogen from the air.

Root Hairs As a plant adds primary growth, its root system may develop billions of tiny root hairs (Figure 26.4A). Collectively, these thin extensions of root epidermal cells enormously increase the surface area available for absorbing water and nutrients. Root hairs are fragile structures no more than a few millimeters long. They do not develop into new roots, and last only a few days.

Mycorrhizae As Section 22.6 explains, a **mycorrhiza** (plural, mycorrhizae) is a mutually beneficial interaction between a young root and a fungus. Hyphae, which are filaments of the fungus, form a velvety cloak around roots, or penetrate the root cells. Collectively, the hyphae have a large surface area, so they absorb mineral ions from a larger volume of soil than roots alone (Figure

26.4B). The fungus absorbs some sugars and nitrogen-rich compounds from root cells. The root cells get some scarce minerals that the fungus is better able to absorb.

Root Nodules Some types of soil bacteria form mutually beneficial relationships with clover, peas, and other legumes. Like all other plants, legumes require nitrogen for growth. Nitrogen gas (N≡N, or N_2) is abundant in air, but plants do not have an enzyme that can break it apart. The bacteria do. The bacterial enzyme uses ATP to convert nitrogen gas to ammonia (NH_3), a metabolic conversion that is called **nitrogen fixation**. Other soil bacteria convert ammonia to nitrate (NO_3^-), a form of nitrogen that plants readily absorb. You will read more about nitrogen fixation in Section 42.9.

Nitrogen-fixing bacteria infect roots and thus become symbionts in localized swellings called **root nodules** (Figure 26.4C). The bacteria, which are anaerobic, share their fixed nitrogen with the plant (Figure 26.4D). In return, the plant provides the bacteria with an oxygen-free environment, and shares its photosynthetically produced sugars with them.

Control Over Uptake

After mineral-rich soil water diffuses into a root, it can reach the root's vascular cylinder in one of two ways: by moving through cytoplasm, or by diffusing through cell walls. Water can enter cytoplasm by diffusing across the plasma membrane of a cell in the root's epidermis or cortex. Dissolved minerals cannot, because membranes are not permeable to ions (Section 5.7). Mineral ions can

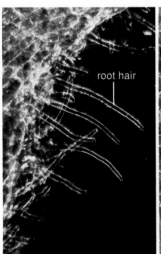
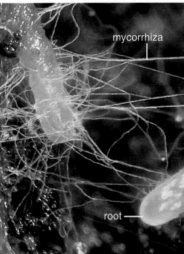

A The hairs on this root of a white clover plant are about 0.2 mm long.

B Mycorrhizae (*white* hairs) extending from the tip of these roots (*tan*) greatly enhance their surface area.

C Anaerobic bacteria in root nodules on this soybean plant fix nitrogen from the air. The nitrogen is shared with the plant.

D Soybean plants growing in nitrogen-poor soil show how root nodules affect growth. Only the *darker green* plants in the *right-hand* rows were infected with nitrogen-fixing bacteria.

Figure 26.4 Examples of root specializations that help plants take up nutrients.

Figure 26.5 Animated In most flowering plants, transport proteins in the plasma membranes of root cells control the plant's uptake of water and dissolved mineral ions from soil.

only enter cytoplasm through active transporters in the plasma membranes. Once in cytoplasm, water and ions diffuse cell to cell through plasmodesmata (Section 4.11), until they cross into the vascular cylinder. After the water enters the vascular cylinder, pipelines of xylem distribute it to the rest of the plant.

Plant cell walls are rigid, but they are permeable to water and ions. They are also shared between adjacent cells. Thus, soil water can reach the vascular cylinder by diffusing directly through cell walls. The walls of the tightly-packed parenchyma cells in the root cortex form a continuous pathway for water and mineral ions to diffuse from epidermis to vascular cylinder.

However, even though soil water can reach the vascular cylinder by diffusing through cell walls, it cannot enter the cylinder the same way. Why not? A vascular cylinder is separated from the root cortex by endodermis. Remember from Section 25.6 that this tissue consists of a single layer of parenchyma cells around the pericycle (Figure 26.5A). Endodermal cells secrete a waxy substance into their walls wherever they abut. The substance forms a **Casparian strip**, a waterproof band between the plasma membranes of endodermal cells (Figure 26.5B). The Casparian strip prevents water from diffusing through endodermal cell walls. So, soil water diffusing through cell walls in root cortex can only enter the vascular cylinder by passing through the cytoplasm of an endodermal cell.

Soil water has to cross the plasma membrane of at least one root cell before it enters a vascular cylinder. Thus, transport proteins in the cells' plasma membranes control the types and amounts of mineral ions that move from soil water into the body of the plant (Figure 26.5C).

Casparian strip Waxy, waterproof band that seals abutting cell walls of root endodermal cells.
mycorrhiza Fungus–plant root partnership.
nitrogen fixation Conversion of nitrogen gas to ammonia.
root nodules Swellings of some plant roots that contain nitrogen-fixing bacteria.

Take-Home Message How do roots take up water and nutrients?

❯ Root hairs, mycorrhizae, and root nodules greatly enhance a root's ability to take up water and nutrients.

❯ Transport proteins in root cell plasma membranes control the uptake of water and ions into the vascular cylinder.

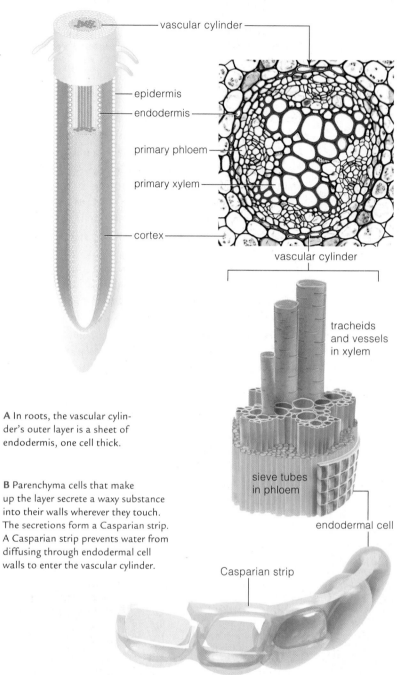

vascular cylinder

epidermis
endodermis

primary phloem

primary xylem

cortex

vascular cylinder

tracheids and vessels in xylem

sieve tubes in phloem

endodermal cell

Casparian strip

A In roots, the vascular cylinder's outer layer is a sheet of endodermis, one cell thick.

B Parenchyma cells that make up the layer secrete a waxy substance into their walls wherever they touch. The secretions form a Casparian strip. A Casparian strip prevents water from diffusing through endodermal cell walls to enter the vascular cylinder.

C Soil water can only enter the vascular cylinder by moving through the cytoplasm of endodermal cells. Water and ions enter the cells via plasmodesmata or via transport proteins in the cells' plasma membranes.

Water and ions must cross at least one cell's plasma membrane before entering a vascular cylinder. Thus, plasma membrane transport proteins control the movement of these substances into the rest of the plant.

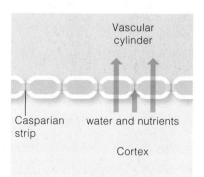

Vascular cylinder

Casparian strip

water and nutrients

Cortex

❯ Evaporation from leaves and stems drives the upward movement of water through pipelines of xylem inside of a vascular plant.

❯ Water's cohesion allows it to be pulled from roots into all other parts of the plant.

❮ Links to Hydrogen bonding 2.4, Properties of water 2.5, Photosynthesis 6.4, Xylem 25.3, Stems 25.4, Roots 25.6

The Cohesion–Tension Theory

Xylem in a root's vascular cylinder connects with xylem in vascular bundles that weave through stems and leaves. Thus, water that enters a root travels to the rest of the plant body inside continuous pipelines of xylem. How does it move all the way from roots to leaves that may be more than 100 meters (330 feet) above the soil?

The tracheids and vessel members of xylem are dead at maturity; only their lignin-impregnated walls remain (Figure 26.6). Being dead, these cells cannot be expending any energy to pump water upward against gravity. The movement of water in vascular plants does not occur by active pumping. Rather, it is driven by two features of water that you learned about in Section 2.5: evaporation and cohesion. Figure 26.7 illustrates the **cohesion–tension theory**, first proposed by botanist Henry Dixon in 1939. By this theory, water in xylem is pulled upward by air's drying power, which creates a continuous negative pressure called tension. The tension extends all the way from leaves to roots that may be hundreds of feet below.

Most of the water a plant takes up is lost by evaporation, typically from stomata on the plant's leaves and stems ❶. The evaporation of water from aboveground plant parts, particularly at stomata, is called **transpiration**.

Transpiration's effect on water inside a plant is a bit like what happens when you suck a drink through a straw. Transpiration puts negative pressure (pulls) on continuous columns of water that fill the narrow conductive tubes of xylem. A column of water resists breaking into droplets as it moves through a narrow conduit such as a straw or a xylem tube. Why? Water molecules are connected by hydrogen bonds (Section 2.5), so a pull on one pulls on all of them. Thus, the negative pressure created by transpiration (tension) pulls on the entire column of water that fills a xylem tube ❷. Because of the water's cohesion, the tension extends from leaves that may be hundreds of feet in the air, down through stems, into young roots where water is being absorbed from soil ❸.

The movement of water through vascular plants is driven mainly by transpiration, but evaporation is only one of many other processes in plants that involve the loss of water molecules. Metabolic pathways that use water also contribute to the negative pressure that results in water movement. For example, water molecules delivered by xylem are used as electron donors in the light reactions of photosynthesis. The reactions split the water molecules into oxygen and hydrogen ions (Section 6.4).

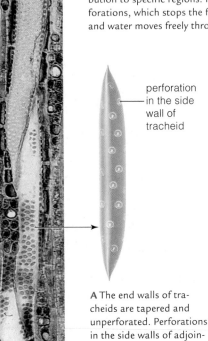

Figure 26.6 Tracheids and vessel members from xylem. Interconnected, perforated walls of dead cells form these water-conducting tubes. The pectin-coated perforations may help control water distribution to specific regions. Hydrated pectins swell and plug the perforations, which stops the flow. Pectins shrink during dry periods, and water moves freely through the open perforations.

cohesion–tension theory Explanation of how transpiration creates a tension that pulls a cohesive column of water through xylem, from roots to shoots.
transpiration Evaporation of water from plant parts.

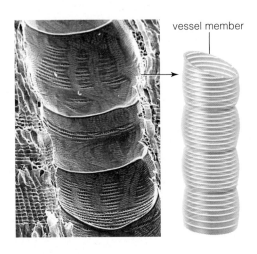

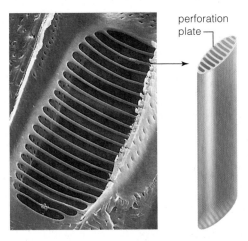

A The end walls of tracheids are tapered and unperforated. Perforations in the side walls of adjoining tracheids match up.

B The thick, finely perforated end walls of dead vessel member cells connect to make long conducting tubes of xylem. The micrograph shows three adjoining vessel members.

C Vessel members vary in shape. A perforation plate imparts strength to a junction where vessel members meet, while also allowing water to flow freely through the xylem tube.

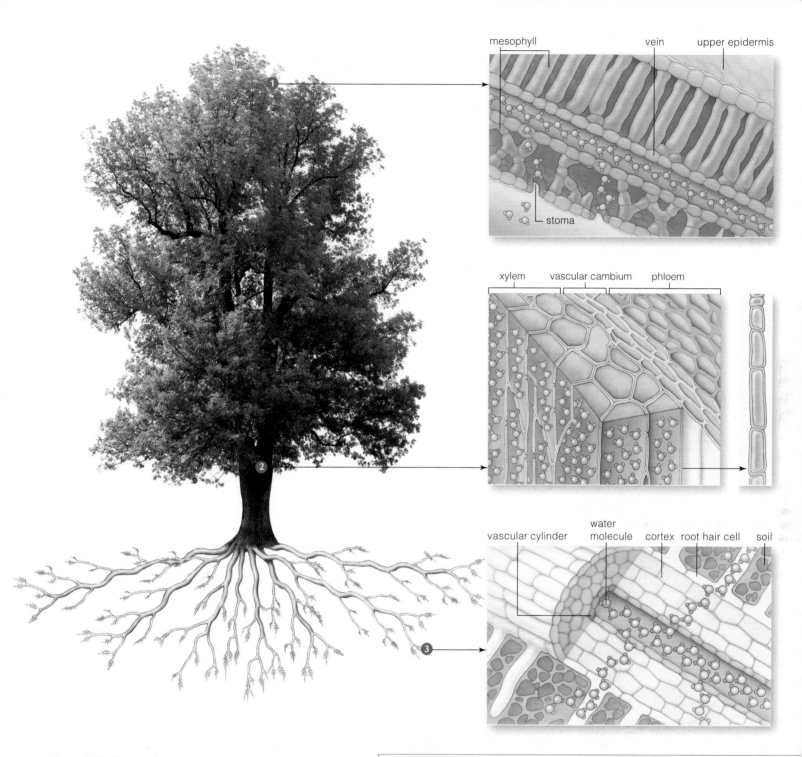

mesophyll   vein   upper epidermis

stoma

xylem   vascular cambium   phloem

vascular cylinder   water molecule   cortex   root hair cell   soil

Figure 26.7 Animated Cohesion–tension theory of water transport in vascular plants.

❶ Water evaporates from aboveground plant parts (a process called transpiration).

❷ The evaporation exerts tension (pulls) on the narrow columns of water that fill xylem tubes. The tension extends from leaves to roots because liquid water has cohesion. Hydrogen bonds among water molecules together impart cohesion to liquid water.

❸ As long as evaporation continues, the tension it creates drives the uptake of water molecules from soil.

Take-Home Message **What makes water move inside plants?**

❭ Transpiration is the evaporation of water from leaves, stems, and other plant parts.

❭ By a cohesion–tension theory, transpiration puts water in xylem into a continuous state of tension from leaves to roots.

❭ Tension pulls columns of water in xylem upward through the plant. The collective strength of many hydrogen bonds (cohesion) keeps the water from breaking into droplets as it rises.

❯ Water is an essential resource for all land plants. Thus, water-conserving structures and processes are key to the survival of these plants.

❮ Links to Plant cuticle 4.11, Osmosis 5.6, Gases in photosynthesis 6.4, Stomata 6.8, Gases in aerobic respiration 7.2, Land plant adaptations 21.2, Leaf structure 25.5

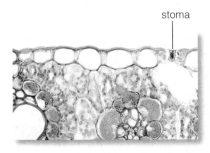

stoma

A Cuticle (*gold*) and stoma on a leaf. The stoma is an opening between two specialized epidermal cells called guard cells.

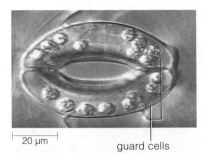

20 μm

guard cells

B This stoma is open. When the guard cells swell with water, they bend so that a gap opens between them.

The gap allows the plant to exchange gases with air. The exchange is necessary to keep metabolic reactions running.

C This stoma is closed. The guard cells, which are not plump with water, are collapsed against each other so there is no gap between them.

A closed stoma limits water loss, but it also limits gas exchange, so photosynthesis and respiration reactions slow.

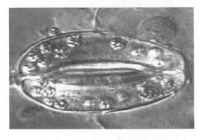

solutes   water

D How do stomata open and close? When a stoma is open, the guard cells are maintaining a relatively high concentration of solutes by pumping solutes into their cytoplasm. Water diffusing into the hypertonic cytoplasm keeps the cells plump.

ABA signal   solutes   water

E When water is scarce, a hormone (ABA) activates a pathway that lowers the concentrations of solutes in guard cell cytoplasm. Water follows its gradient and diffuses out of the cells, and the stoma closes.

In land plants, at least 90 percent of the water transported from roots to leaves evaporates right out. Only about 2 percent is used in metabolism, but that amount must be maintained or photosynthesis, growth, membrane functions, and other processes will shut down.

If a plant is running low on water, it cannot move around to seek out more, as most animals can. A cuticle and stomata (Sections 4.11 and 21.2) help the plant conserve the water it already holds in its tissues. Both of these structures restrict the amount of water vapor that diffuses out of the plant's surfaces.

However, the cuticle and stomata also restrict gas exchanges between the plant and the air. Why is that important? The concentrations of carbon dioxide and oxygen gases in air spaces inside the plant affect the rate of critical metabolic pathways (such as photosynthesis and aerobic respiration) in the plant's cells. If a plant became impermeable to gases, it could not take in carbon dioxide to run photosynthesis. Neither could it sustain aerobic respiration for very long, because it could not take in enough oxygen to run the reactions. Thus, a plant's water-conserving structures and mechanisms must balance its needs for water with its requirements to exchange gases with the air.

Cuticle

Even mildly water-stressed plants would wilt and die without a cuticle. This water-impermeable layer coats the walls of all plant cells exposed to air (Figure 26.8A). It consists of epidermal cell secretions: a mixture of waxes, pectin, and cellulose fibers embedded in cutin, an insoluble lipid polymer. The cuticle is translucent, so it does not prevent light from reaching photosynthetic tissues.

Stomata

A pair of specialized epidermal cells defines each stoma. When these two **guard cells** swell with water, they bend slightly so a gap forms between them (Figure 26.8B). The gap is the stoma. When the cells lose water, they collapse against each other, so the gap closes (Figure 26.8C).

Figure 26.8 Water-conserving structures in plants.

A Cuticle and stoma in a cross-section of basswood (*Tilia*) leaf.

B–E Stomata in action. Whether a stoma is open or closed depends on how much water is plumping up these guard cells. The amount of water in guard cell cytoplasm is influenced by hormonal signals.

The round structures inside the cells are chloroplasts. Guard cells are the only type of plant epidermal cell with these organelles.

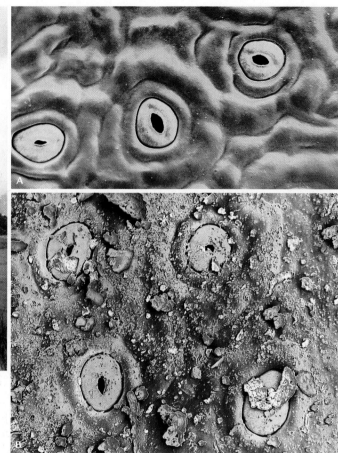

Figure 26.9 Stomata and smog. **A** Normal stomata. **B** Stomata at the leaf surface of a holly plant growing in a smoggy, industrialized region. Airborne pollutants not only block sunlight from photosynthetic cells, they also clog stomata, and can damage them so much that they close permanently.

Environmental cues such as water availability, the level of carbon dioxide inside the leaf, and light intensity affect whether stomata open or close. These cues trigger osmotic pressure changes in the cytoplasm of guard cells. For example, when the sun comes up, the light causes guard cells to begin pumping solutes (in this case, potassium ions) into their cytoplasm. The resulting buildup of potassium ions causes water to enter the cells by osmosis. The guard cells plump up, so the gap between them opens (Figure 26.8D). Carbon dioxide from the air diffuses into the plant's tissues, and photosynthesis begins.

As another example, root cells release the hormone abscisic acid (ABA) when soil water becomes scarce. ABA travels through the plant's vascular system to leaves and stems, where it binds to receptors on guard cells. The binding causes solutes to exit these cells. Water follows by osmosis, the guard cells lose plumpness and collapse against each other, and the stomata close (Figure 26.8E).

The stomata of most plants close at night. Water is conserved, and carbon dioxide builds up inside leaves as cells make ATP by aerobic respiration. The stomata of CAM plants, including most cactuses, open at night, when the plant takes in and fixes carbon from carbon dioxide. Stomata close during the day, and the plant uses the carbon that it fixed during the night to sustain photosynthesis (Section 6.8).

Stomata also close in response to some of the chemicals in polluted air. The closure protects the plant from chemical damage, but it also prevents the uptake of carbon dioxide for photosynthesis, and so inhibits growth. Think about it on a smoggy day (Figure 26.9).

Take-Home Message How do land plants conserve water?

> A waxy cuticle covers all epidermal surfaces of the plant exposed to air. It restricts water loss from plant surfaces.

> Plants conserve water by closing their many stomata. Closed stomata also prevent gas exchanges necessary for photosynthesis and aerobic respiration.

> A stoma stays opens when the guard cells that define it are plump with water. It closes when the cells lose water and collapse against each other.

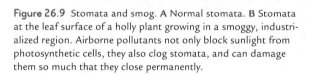

guard cell One of a pair of cells that define a stoma across the epidermis of a leaf or stem.

❭ Xylem distributes water and minerals through plants, and phloem distributes the organic products of photosynthesis.
❰ Links to Carbohydrates 3.3, Osmosis and turgor 5.6, Active transport 5.7, Light-independent reactions of photosynthesis 6.7, Stomata 6.8, Plant vascular tissues 25.3

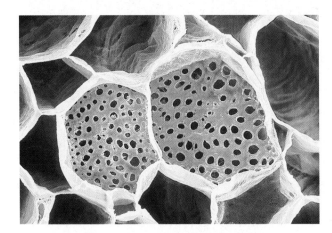

Figure 26.10 Scanning electron micrograph of the sieve plates on the ends of two side-by-side sieve-tube members.

Phloem is a vascular tissue with organized arrays of conducting tubes, fibers, and strands of parenchyma cells. Unlike conducting tubes of xylem, sieve tubes in phloem consist of living cells. Sieve-tube cells are positioned side by side and end to end, and their abutting end walls (sieve plates) are porous (Figure 26.10). Dissolved organic compounds flow through the tubes (Figure 26.11).

Some of the organic products of photosynthesis (sugars) are used by the cells that make them. The rest are actively transported into sieve tubes by companion cells pressed against the tubes. After they are loaded into sieve tubes, the sugars travel to all other parts of the plant, where they are broken down for energy, remodeled into other compounds, or stored for later use.

Plants store their carbohydrates mainly as starch, but starch molecules are too big and too insoluble to transport across plasma membranes. Cells break down starch molecules to sucrose and other small molecules that are easily transported through the plant. Experiments with plant-sucking insects demonstrated that sucrose is the main carbohydrate transported in phloem. Aphids feeding on the juices in the conducting tubes of phloem were anesthetized with high levels of carbon dioxide. Then their bodies were detached from their mouthparts, which remained attached to the plant. Researchers collected and analyzed fluid exuded from the aphids' mouthparts. For most of the

pressure flow theory Explanation of how flow of fluid through phloem is driven by differences in pressure and sugar concentration between a source and a sink.
translocation The movement of organic compounds through phloem.

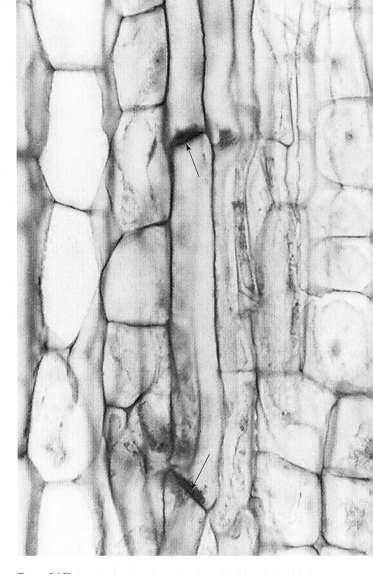

Figure 26.11 Longitudinal section of a stem of milkweed (*Asclepias*), showing sieves tubes in phloem. Arrows point to perforated sieve plates on the ends of sieve-tube members.

plants studied, sucrose was the most abundant carbohydrate in the fluid.

Pressure Flow Theory

Translocation is the formal name for the movement of organic molecules through phloem. The molecules flow from a source (any region of the plant where companion cells are loading organic compounds into the sieve tubes) to a sink (any region where molecules are being used or stored). Photosynthetic tissues in leaves are usually a plant's main source region; roots are sink regions, as are growing fruits.

Why do organic compounds in phloem flow from source to sink? A pressure gradient drives the movement of fluid in phloem (Section 5.6). According to the

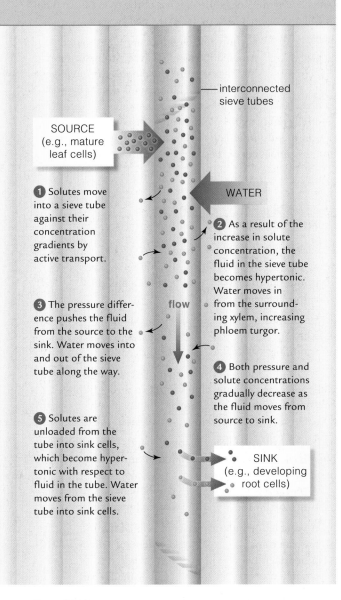

interconnected sieve tubes

SOURCE (e.g., mature leaf cells)

WATER

1 Solutes move into a sieve tube against their concentration gradients by active transport.

2 As a result of the increase in solute concentration, the fluid in the sieve tube becomes hypertonic. Water moves in from the surrounding xylem, increasing phloem turgor.

3 The pressure difference pushes the fluid from the source to the sink. Water moves into and out of the sieve tube along the way.

flow

4 Both pressure and solute concentrations gradually decrease as the fluid moves from source to sink.

5 Solutes are unloaded from the tube into sink cells, which become hypertonic with respect to fluid in the tube. Water moves from the sieve tube into sink cells.

SINK (e.g., developing root cells)

Figure 26.12 Translocation of organic compounds in phloem. In this diagram, water molecules are represented by *blue* balls; solutes, by *red* balls.

pressure flow theory, internal pressure builds up at the source end of the sieve tube. The pressure pushes the solute-rich fluid toward a sink (Figure 26.12). By energy-requiring reactions, companion cells in leaf veins load sugars into sieve-tube members **1**. The sugars increase the solute concentration of the fluid inside the tubes, so water also moves into the tubes, by osmosis **2**. The water increases the fluid volume, which in turn increases the internal pressure (turgor) inside the sieve tubes **3**. Pressure in phloem can be five times higher than the pressure in an automobile tire (Figure 26.13). That high pressure pushes the sugar-laden fluid toward sink regions, where the internal pressure of the phloem is lower **4**. The pressure in phloem decreases at sink regions because sugars are unloaded there; water follows, again by osmosis **5**.

Mean Green Cleaning Machines (revisited)

With elemental pollutants such as lead or mercury, the best phytoremediation strategies use plants that take up toxins and store them in aboveground tissues. The toxin-laden plant parts can then be harvested for safe disposal. Researchers have genetically modified some plants to enhance their absorptive and storage capacity. In the photo *above*, Dr. Kuang-Yu Chen is analyzing zinc and cadmium levels in plants that take up these elements. In the case of organic toxins such as TCE, the best phytoremediation strategies use plants with biochemical pathways that break down the compounds to less toxic molecules. Researchers are beefing up these pathways in many plants. Some are transferring genes from bacteria or animals into plants; others are enhancing expression of genes that encode molecular participants in the plants' own detoxification pathways.

How Would You Vote? Plants can be genetically engineered to take up toxins more effectively. Do you support the use of such plants to help clean up toxic waste sites? See CengageNow for details, then vote online (cengagenow.com).

Figure 26.13 Honeydew exuding from an aphid after the insect's mouthparts penetrated a sieve tube. High pressure in phloem forced this droplet of sugary fluid out through the terminal opening of the aphid's gut.

Take-Home Message **How do organic molecules move through vascular plants?**

> Plants store carbohydrates as starch, and distribute them as sucrose and other small, water-soluble molecules.

> Concentration and pressure gradients move organic compounds inside sieve tubes of phloem.

> The gradients are set up by companion cells moving organic molecules into sieve tubes at sources, and the unloading of the molecules at sinks.

Summary

Section 26.1 The ability of plants to take up substances from soil and water is the basis for phytoremediation, the removal of toxic substances from a contaminated area with the help of plants.

Section 26.2 Plant growth requires steady sources of elemental nutrients obtainable from carbon dioxide, water, and **soil**. The availability of water and mineral ions in soil depends on its proportions of sand, silt, and clay; and its **humus** content. **Loams** have roughly equal proportions of sand, silt, and clay. **Leaching** and **soil erosion** deplete nutrients from soil, particularly **topsoils**.

Section 26.3 Root hairs greatly increase a root's surface area for absorption of soil water. Fungi are symbionts with young roots in **mycorrhizae**, which enhance a plant's ability to absorb mineral ions from soil. **Nitrogen fixation** by bacteria in **root nodules** gives a plant extra nitrogen. In both cases, the microorganisms receive some of the plant's sugars.

Roots control the movement of water and dissolved mineral ions into the vascular cylinder. Endodermal cells that form a layer around the cylinder deposit a waterproof band, a **Casparian strip**, in their abutting walls. The strip keeps water from diffusing around endodermal cells. Water and nutrients enter a root vascular cylinder only by moving through the plasma membrane of at least one root cell. Thus, the uptake of water and ions is controlled by active transport proteins in root cell plasma membranes.

Section 26.4 Water and dissolved mineral ions flow through xylem. The interconnected, perforated walls of tracheids and vessel members form the tubes. **Transpiration** is the evaporation of water from plant parts into air. Evaporation mainly occurs at stomata.

The **cohesion–tension theory** explains how water moves through plants: transpiration pulls water upward by creating a continuous negative pressure (tension) inside xylem from leaves to roots. Hydrogen bonds among water molecules keep the columns of fluid continuous inside the narrow vessels.

Section 26.5 A cuticle and stomata balance a plant's loss of water with its needs for gas exchange. A stoma, which is a gap across the cuticle-covered epidermis of leaves and other plant parts, is defined by a pair of **guard cells**. Closed stomata limit the loss of water, but also prevent the gas exchange required for photosynthesis and aerobic respiration.

Environmental signals cause stomata to open or close. The signals trigger guard cells to pump ions into or out of their cytoplasm; water follows the ions (by osmosis). Water moving into guard cells plumps them, which opens the gap between them. Water diffusing out of the cells causes them to collapse against each other, so the gap closes.

Section 26.6 Organic compounds are distributed through a plant by **translocation**. By the **pressure flow theory**, the movement of fluid through phloem is driven by pressure and solute gradients. The gradients are set up by companion cells actively transporting the products of photosynthesis into sieve tubes at source regions, and the unloading of these sugars at sink regions.

Self-Quiz   Answers in Appendix III

1. Carbon, hydrogen, and oxygen are _____ for plants.
   a. macronutrients   d. required elements
   b. micronutrients   e. both a and d
   c. trace elements

2. Decomposing matter in soil is called _____ .
   a. loam   c. topsoil
   b. humus   d. nutrients

3. A _____ strip between abutting endodermal cell walls forces water and solutes to move through these cells rather than around them.
   a. cutin   b. Casparian   c. cohesion   d. cellulose

4. The nutrition of some plants depends on a mutually beneficial association between a root and a fungus. The association is known as a _____ .
   a. root nodule   c. root hair
   b. mycorrhiza   d. root hypha

5. A vascular cylinder consists of cells of the _____ .
   a. exodermis   d. xylem and phloem
   b. endodermis   e. b and d
   c. root cortex   f. all of the above

6. Water evaporation from plant parts is called _____ .
   a. translocation   c. transpiration
   b. expiration   d. tension

7. Water transport from roots to leaves occurs mainly because of _____ .
   a. pressure flow
   b. differences in source and sink solute concentrations
   c. the pumping force of xylem vessels
   d. transpiration, tension, and cohesion of water
   e. a and b

8. A waxy cuticle is secreted by _____ .
   a. ground tissue   c. a stoma
   b. epidermal cells   d. root hairs

9. When guard cells swell, _____ .
   a. transpiration ceases   c. a stoma opens
   b. sugars enter phloem   d. root cells die

10. Stomata open in response to light when _____ .
    a. guard cells pump ions into their cytoplasm
    b. guard cells pump ions out of their cytoplasm
    c. water evaporates out of guard cells

Transgenic Plants for Phytoremediation Plants used for phytoremediation take up organic pollutants from the soil or air, then transport the chemicals to plant tissues, where they are stored or broken down. Researchers are now designing transgenic plants that have enhanced ability to take up or break down toxins. In 2007, Sharon Doty and her colleagues published the results of their efforts to design plants useful for phytoremediation of soil and air containing organic solvents. The researchers used *Agrobacterium tumefaciens* (Section 15.7) to deliver a mammalian gene into poplar plants. The gene encodes cytochrome P450, an enzyme involved in the metabolism of a range of organic molecules, including solvents such as TCE. The results of a test on these transgenic plants are shown in Figure 26.14.

1. How many transgenic plants did the researchers test?
2. In which group did the researchers see the slowest rate of TCE uptake? The fastest?
3. On day 6, what was the difference between the TCE content of air around transgenic plants and that around vector control plants?
4. Assuming no other experiments were done, what two explanations are there for the results of this experiment? What other control might the researchers have used?

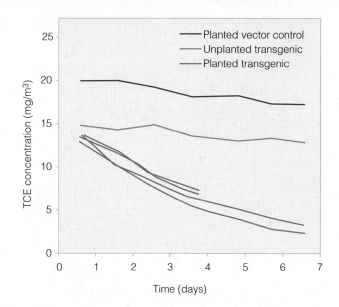

Figure 26.14 TCE uptake by transgenic poplar trees. Planted trees were incubated in sealed containers with an initial 15 milligrams of TCE (trichloroethylene) per cubic meter of air. Samples of the air in the containers were taken daily and measured for TCE content. Controls included a tree transgenic for a Ti plasmid with no cytochrome P450 in it (vector control), and a bare-root transgenic tree (one that was not planted in soil).

11. Tracheids are part of _____ .
   a. cortex   c. phloem
   b. mesophyll   d. xylem

12. Sieve tubes are part of _____ .
   a. cortex   c. phloem
   b. mesophyll   d. xylem

13. When soil is dry, guard cells respond to _____ by collapsing against one another, so stomata close.
   a. air temperature c. abscisic acid
   b. low humidity   d. oxygen

14. Transport of photosynthetically produced sugars from leaves to roots occurs by _____ .
   a. pressure flow
   b. differences in source and sink solute concentrations
   c. the pumping force of xylem vessels
   d. transpiration, tension, and cohesion of water
   e. a and b

15. Match the concepts of plant nutrition and transport.
   ____ stomata   a. evaporation from plant parts
   ____ nutrient   b. harvests soil water and nutrients
   ____ sink   c. balance water loss with gas
   ____ root system   exchange
   ____ hydrogen   d. cohesion in water transport
        bonds   e. sugars unloaded from sieve tubes
   ____ transpiration   f. organic compounds distributed
   ____ translocation   through the plant body
                         g. required element

Additional questions are available on **CENGAGENOW**.

Critical Thinking

1. Successful home gardeners, like farmers, make sure that their plants get enough nitrogen from either fertilizer or nitrogen-fixing bacteria. Which biological molecules incorporate nitrogen? Nitrogen deficiency stunts plant growth; leaves yellow and then die. How would nitrogen deficiency cause these symptoms?

2. You just returned home from a three-day vacation. Your severely wilted plants tell you they were not watered before you left. Being aware of the cohesion–tension theory of water transport, explain what happened to them.

3. When you dig up a plant to move it from one spot to another, the plant is more likely to survive if some of the soil around the roots is transferred along with the plant. Formulate a hypothesis that explains this observation.

4. If a plant's stomata are made to stay open at all times, or closed at all times, it will die. Why?

5. Allen is studying the rate of water uptake by tomato plant roots. He notices that several environmental factors, including wind and relative humidity, affect the rate. Explain how they might do so.

Animations and Interactions on **CENGAGENOW**:
❯ Nutrient uptake in vascular plants; Water transport in vascular plants; Distribution of organic compounds through vascular plants.

❮ Links to Earlier Concepts

A review of plant tissues and structures (Sections 4.9, 4.11, 25.2, 25.4, 25.8, 26.3, 26.5), functions (6.5, 6.7), life cycles (12.5, 21.2), and evolution (21.5, 21.7) will be helpful in understanding some of the reproductive adaptations of flowering plants. In this chapter, you will revisit carbohydrates (3.3), pigments (6.2), cloning (8.7), gene expression and control (9.2, 10.2–10.4), asexual versus sexual reproduction (12.2), meiosis (12.3), genetics (14.6), and evolutionary processes (17.12, 17.13).

Key Concepts

Structure and Function of Flowers
Flowers are shoots that are specialized for reproduction. Modified leaves form their parts. Gamete-producing cells develop in their reproductive structures; other parts such as petals are adapted to attract and reward pollinators.

Plant Sexual Reproduction
In flowering plants, pollination is followed by double fertilization. After fertilization, ovules mature into seeds. As seeds develop, tissues of the ovary and other parts of the flower mature into fruits, which function to disperse seeds.

27 Plant Reproduction and Development

27.1 Plight of the Honeybee

In the fall of 2006, commercial beekeepers worldwide began to notice something was amiss in their honeybee hives. Adult worker bees seemed to be missing, despite abundant pollen, honey, and larvae in their hives. A third of the colonies did not survive the following winter. By spring, the phenomenon had a name: colony collapse disorder. Farmers and biologists began to worry about what would happen if the honeybee populations continued to decline. Honey production would suffer, but many commercial crops would fail too.

Nearly all of our food crops are flowering plants, or angiosperms (Section 21.7). These plants make pollen grains. Honeybees are **pollinators**, which means they carry pollen from one plant to another, pollinating flowers as they do. The flowers of many plant species will not develop into fruits unless they receive pollen from other flowers. Even angiosperm species with flowers that can self-pollinate tend to make bigger fruits and more of them when they are cross-pollinated (Figure 27.1).

Many types of insects pollinate plants, but honeybees are especially efficient pollinators of a variety of plant species. They are also the only ones that tolerate living in man-made hives that can be loaded onto trucks and carted wherever crops require pollination. Loss of their portable pollination service is a huge threat to our agricultural economy.

We still do not know what causes colony collapse disorder. A variety of parasites and diseases that infect honeybees may be part of the problem. Parasitic mites and Israeli acute paralysis virus have been detected in many affected hives. Pesticides may also be taking a toll. In the past few years, pesticides called neonicotinoids have become the most widely used insecticides in the United States. Neonicotinoids are systemic, which means they are taken up by all plant tissues, including the nectar and pollen that honeybees collect. Neonicotinoids are also highly toxic to honeybees.

Colony collapse disorder is currently in the spotlight because it affects our food supply. However, other pollinator populations are also dwindling. Habitat loss is probably the main factor, but pesticides that harm honeybees also harm other pollinators.

Figure 27.1 Importance of insect pollinators. *Opposite*, honeybees are efficient pollinators of a variety of flowers, including those of berry plants. *Above*, raspberry flowers can pollinate themselves, but the fruit that forms from a self-fertilized flower is of lower quality than that of a cross-pollinated flower. The two raspberries on the *left* formed from self-pollinated flowers. The one on the *right* formed from an insect-pollinated flower.

Flowering plants rose to dominance in part because they coevolved with animal pollinators. Most flowers are specialized to attract and be pollinated by a specific type of pollinator or even a specific species. Those adaptations put the plants at risk of extinction if coevolved pollinator populations decline. Wild animal species that depend on the plants for fruits and seeds will also be affected. Recognizing the prevalence and importance of these interactions is our first step toward finding workable ways to protect them.

---

pollinator An organism that moves pollen from one plant to another.

Asexual Reproduction of Plants
Many species of plants reproduce asexually by vegetative reproduction. Humans take advantage of this natural tendency by propagating plants asexually for agriculture and research.

Growth and Development
Plant development includes seed germination and other events of the life cycle, such as root and shoot development, flowering, fruit formation, and dormancy. These events have a genetic basis, and are influenced by the environment.

Responses to Environmental Cues
Plants respond to environmental cues, including gravity, sunlight, and seasonal shifts in night length and temperatures, by altering patterns of growth. Cyclic patterns of growth are responses to seasons and other recurring environmental patterns.

❯ Specialized reproductive shoots called flowers consist of whorls of modified leaves.

❮ Links to ABC model of flowering 10.4, Gametes 12.2, Plant life cycles 12.5 and 21.2, Coevolution 17.13, Coevolution of flowers and pollinators 21.7, Lateral buds 25.4

The life cycle of flowering plants is dominated by the sporophyte, a diploid spore-producing plant body that grows by mitotic cell divisions of a fertilized egg (Sections 12.5 and 21.2). **Flowers** are the specialized reproductive shoots of angiosperm sporophytes. Spores that form by meiosis inside flowers develop into haploid gametophytes. Gametophytes produce gametes (Section 12.2).

Anatomy of a Flower

A flower forms when a lateral bud along a stem develops into a short, modified branch called a receptacle. The petals and other parts of a typical flower are modified leaves that form in four spirals or rings (whorls) at the end of the floral shoot. The outermost whorl develops into a calyx, which is a ring of leaflike sepals (Figure 27.2A). The sepals of most flowers are photosynthetic and inconspicuous; they serve to protect the flower's reproductive parts. Just inside the calyx, petals form in a whorl called the corolla (from the Latin *corona*, or crown). Petals are usually the largest and most brightly colored parts of a flower. They function mainly to attract pollinators.

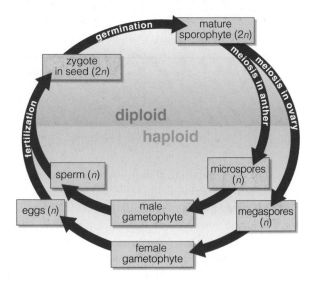

Figure 27.3 Animated Life cycle of a typical flowering plant.

A whorl of stamens forms inside the ring of petals. **Stamens** are the structures that produce the plant's male gametophytes. In most flowers, stamens consist of a thin filament with an anther at the tip. Inside a typical anther are four elongated pouches called pollen sacs. Meiosis of diploid cells in each sac produces haploid, walled spores. The spores differentiate into pollen grains, which are immature male gametophytes.

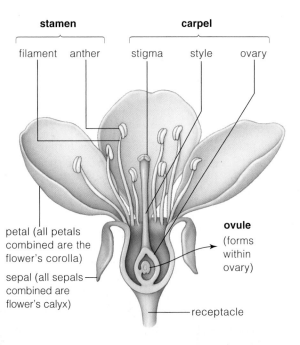

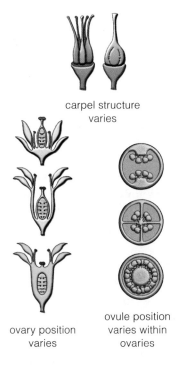

A Like many flowers, the blossom of a cherry plant (*Prunus*) has several stamens and one carpel. The structures that produce male gametophytes are stamens, which consist of pollen-bearing anthers atop slender filaments. The structure that produces female gametophytes is the carpel, which consists of stigma, style, and ovary.

B Flower structure varies among different plant species.

Figure 27.2 Animated Anatomy of a typical flower.

The innermost whorl of modified leaves are folded and fused into **carpels**, structures that produce the plant's female gametophytes. Carpels are sometimes called pistils. Many flowers have one carpel; others have several carpels, or several groups of carpels, that may be fused (Figure 27.2B). The upper region of a carpel is a sticky or hairy stigma that is specialized to trap pollen grains. Typically, the stigma sits on top of a slender stalk called a style. The lower, swollen region of a carpel is the **ovary**, which contains one or more ovules. An **ovule** is a tiny bulge of tissue inside the ovary. A cell in the ovule undergoes meiosis and develops into the haploid female gametophyte.

At fertilization, a diploid zygote forms when male and female gametes meet inside an ovary. The ovule then matures into a seed. The life cycle of the plant is completed when the seed germinates, and a new sporophyte forms and matures (Figure 27.3). We return to fertilization and seed development in later sections.

Pollinators

Sexual reproduction in flowering plants involves the transfer of pollen, typically from one plant to another. Plants cannot move about to find a mate, so they depend on factors in the environment that can move pollen around for them. Many plants are pollinated by wind, which is entirely nonspecific in where it dumps pollen. Wind-pollinated plants often release pollen grains by the billions, a type of insurance in numbers that some of their pollen will reach a receptive stigma.

Other plants enlist the help of pollinators. An animal that is attracted to a particular flower often picks up pollen on a visit, then inadvertently transfers it to the flower of a different plant on a later visit. The more specific the attraction, the more efficient the transfer of pollen among plants of the same species. Given the selective advantage for flower traits that attract specific pollinators, it is not surprising that about 90 percent of flowering plants have coevolved animal pollinators.

A flower's shape, pattern, color, and fragrance are adaptations that attract specific animal pollinators. For example, pollinators such as bats and moths have an excellent sense of smell, and can follow airborne chemicals to a

Figure 27.4 Intimate connections between pollinator and flower.

A A zebra orchid mimics the scent of a female wasp. Male wasps follow the scent to the flower, then try to copulate with and lift the dark red mass of tissue on the lip. The wasp's movements trigger the lip to tilt upward, which brushes the wasp's back against the flower's stigma and pollen.

B Female burnet moths perch on purple flowers—preferably those of field scabious—when they are ready to mate. The combination of colors and patterns attracts males.

flower that is emitting them. Not all flowers smell sweet; odors like dung or rotting flesh beckon beetles and flies.

An animal's reward for a visit to a flower may be nectar (a sweet fluid exuded by flowers), oils, nutritious pollen, or even sex (Figure 27.4). Nectar is the only food for most adult butterflies, and it is the food of choice for hummingbirds. Honeybees collect nectar and convert it to honey, which helps feed the bees through the winter. Pollen is an even richer food, with more protein, vitamins, and minerals than nectar.

Some flowers have specializations that prevent pollination by everything other than a specific pollinator species. For example, nectar (and pollen) at the bottom of a long floral tube may be accessible only to an insect with a matching feeding device. A flower that captivates the attention of an animal has a pollinator that spends its time seeking out (and pollinating) only those flowers. Pollinators also benefit when they receive an exclusive supply of the reward offered by the plant.

carpel Floral reproductive structure that produces female gametophytes; a sticky or hairlike stigma together with an ovary and a style.

flower Specialized reproductive shoot of a flowering plant.

ovary In flowering plants, the enlarged base of a carpel, inside which one or more ovules form and eggs are fertilized.

ovule In a seed-bearing plant, a structure in which a haploid, egg-producing female gametophyte forms; after fertilization, matures into a seed.

stamen Floral reproductive structure that produces male gametophytes; in most plants it consists of a pollen-producing anther on the tip of a filament.

Take-Home Message **What are flowers?**

> Flowers are short reproductive branches of sporophytes. The different parts of a flower (sepals, petals, stamens, and carpels) are modified leaves.

> The parts of flowers that produce male gametophytes are stamens, which typically have a filament with an anther at the tip. Pollen forms inside anthers.

> The parts of flowers that produce female gametophytes are carpels, which typically consist of stigma, style, and ovary. The haploid, egg-producing female gametophytes form in an ovule inside the ovary.

> Most flowering plants coevolved with animal pollinators. The shape, pattern, color, and fragrance of a flower attract the plant's coevolved pollinator.

> Pollinators are often rewarded for visiting a flower, for example, by obtaining nutritious pollen or sweet nectar.

> In flowering plants, fertilization has two outcomes: It results in a zygote, and it is the start of endosperm.
‹ Links to Life cycle of flowering plants 21.2, Evolution of seed-bearing plants 21.5

Figure 27.5 zooms in on the reproduction part of a flowering plant life cycle.

The production of female gametes begins when a mass of tissue—the ovule—starts growing on the inner wall of an ovary ➊. One cell in the middle of the mass undergoes meiosis and cytoplasmic division, forming four haploid **megaspores** ➋. Three of the four megaspores typically disintegrate. The remaining megaspore undergoes three rounds of mitosis without cytoplasmic division, the outcome being a single cell with eight haploid nuclei ➌. The cytoplasm of this cell divides unevenly, forming a seven-celled embryo sac that constitutes the female gametophyte ➍. The gametophyte is enclosed and protected by layers of cells, or integuments, that developed from the outer layers of the ovule. One of the cells in the gametophyte, the endosperm mother cell, has two nuclei ($n + n$). Another cell is the egg.

The production of male gametes begins when masses of diploid, spore-producing cells form by mitosis inside anthers. Walls typically develop around the cell masses to form four pollen sacs ➎. Each cell inside the pollen sacs undergoes meiosis and cytoplasmic division to form four haploid **microspores** ➏. Mitosis and differentiation of a microspore produces a pollen grain.

A pollen grain consists of two cells, one inside the cytoplasm of the other, enclosed by a durable coat ➐. The coat protects the cells inside on their journey to meet an egg. After the pollen grains form, they enter **dormancy**, a period of suspended metabolism, before being released from the anther when the pollen sacs split open ➑.

Pollination refers to the arrival of a pollen grain on a receptive stigma. Interactions between the two structures stimulate the pollen grain to resume metabolic activity. One of the two cells in the pollen grain then develops into a tubular outgrowth called a pollen tube. The other cell undergoes mitosis and cytoplasmic division, producing two sperm cells (the male gametes) within the pollen tube. A pollen tube together with its contents of male gametes constitutes the mature male gametophyte ➒.

The pollen tube grows from its tip down through the style and ovary toward the ovule, carrying with it the two sperm cells. Chemical signals secreted by the female gametophyte guide the tube's growth to the embryo sac within the ovule. Many pollen tubes may grow down into a carpel, but usually only one penetrates an embryo sac. The sperm cells are then released into the sac ➓. Flowering plants undergo **double fertilization**: One of the sperm cells from the pollen tube fuses with (fertilizes) the egg and forms a diploid zygote. The other fuses with the endosperm

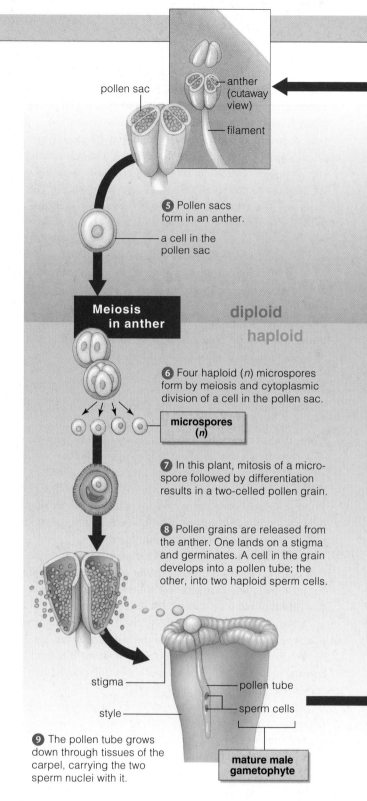

➎ Pollen sacs form in an anther.

a cell in the pollen sac

Meiosis in anther

diploid
haploid

➏ Four haploid (*n*) microspores form by meiosis and cytoplasmic division of a cell in the pollen sac.

microspores (*n*)

➐ In this plant, mitosis of a microspore followed by differentiation results in a two-celled pollen grain.

➑ Pollen grains are released from the anther. One lands on a stigma and germinates. A cell in the grain develops into a pollen tube; the other, into two haploid sperm cells.

pollen sac

anther (cutaway view)

filament

stigma

style

pollen tube

sperm cells

mature male gametophyte

➒ The pollen tube grows down through tissues of the carpel, carrying the two sperm nuclei with it.

Figure 27.5 Animated Life cycle of cherry (*Prunus*), a eudicot.

>> **Figure It Out** What structure gives rise to a pollen grain by mitosis?

Answer: A microspore

mother cell, forming a triploid (3*n*) cell. This cell gives rise to triploid **endosperm**, a nutritious tissue that forms only in seeds of flowering plants. When a seed sprouts, endosperm sustains rapid growth of the sporophyte seedling until true leaves form and photosynthesis begins.

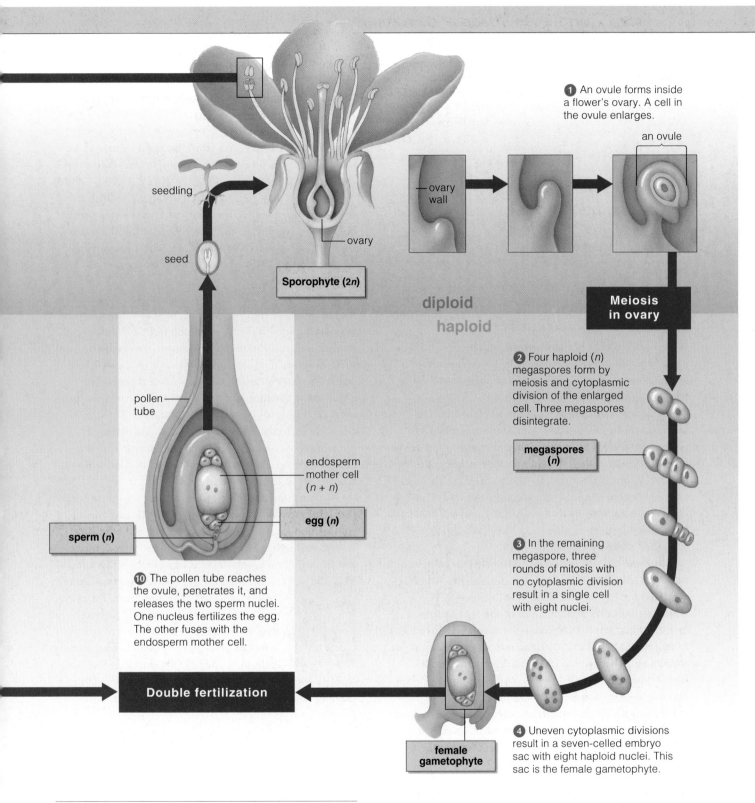

1 An ovule forms inside a flower's ovary. A cell in the ovule enlarges.

an ovule

ovary wall

seedling

seed

ovary

Sporophyte (2n)

diploid

haploid

Meiosis in ovary

2 Four haploid (*n*) megaspores form by meiosis and cytoplasmic division of the enlarged cell. Three megaspores disintegrate.

megaspores (n)

pollen tube

endosperm mother cell (*n + n*)

sperm (n)

egg (n)

3 In the remaining megaspore, three rounds of mitosis with no cytoplasmic division result in a single cell with eight nuclei.

10 The pollen tube reaches the ovule, penetrates it, and releases the two sperm nuclei. One nucleus fertilizes the egg. The other fuses with the endosperm mother cell.

Double fertilization

female gametophyte

4 Uneven cytoplasmic divisions result in a seven-celled embryo sac with eight haploid nuclei. This sac is the female gametophyte.

dormancy Period of temporarily suspended metabolism.
double fertilization Mode of fertilization in flowering plants in which one sperm cell fuses with the egg, and a second sperm cell fuses with the endosperm mother cell.
endosperm Nutritive tissue in the seeds of flowering plants.
megaspore Haploid spore that forms in ovule of seed plants; gives rise to an egg-producing gametophyte.
microspore Walled haploid spore of seed plants; gives rise to a sperm-producing gametophyte.
pollination The arrival of pollen on a receptive stigma.

Take-Home Message **How does sexual reproduction work in flowering plants?**

› A pollen grain that germinates on a receptive stigma develops into a pollen tube and two male gametes. As the tube grows into the carpel and enters an ovule, it carries the gametes along with it.

› Double fertilization occurs when one of the male gametes fuses with the egg, the other with the endosperm mother cell.

> After fertilization, mitotic cell divisions transform a zygote into an embryo sporophyte encased in a seed.
> As embryos develop inside the ovules of flowering plants, tissues around them form fruits.
> Water, wind, and animals disperse seeds in fruits.

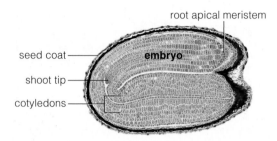

Figure 27.6 **Animated** Seed of shepherd's purse (*Capsella*), a eudicot.

The Embryo Sporophyte Forms

In flowering plants, double fertilization produces a zygote and a triploid ($3n$) cell. By mitotic cell divisions, the zygote develops into an embryo sporophyte, and the triploid cell develops into endosperm.

As the embryo develops, the parent plant transfers nutrients to the ovule. These nutrients accumulate in endosperm mainly as starch with some lipids, proteins, or other molecules. When the embryo approaches maturity, the ovule's layers of integuments separate from the ovary wall and become layers of a protective seed coat. The embryo sporophyte, its reserves of food, and the seed coat have now become a mature ovule, a self-contained package called a **seed** (Figure 27.6).

Nutrients in seeds also nourish humans and other animals. We cultivate cereals—grasses such as rice, wheat, rye, oats, and barley—for their nutritious seeds. The embryo (the germ) contains most of the seed's protein and vitamins, and the seed coat (the bran) contains most of the minerals and fiber. Milling removes bran and germ, leaving only the starch-packed endosperm. Maize, or corn, is the most widely grown cereal crop. Popcorn pops because the moist endosperm steams when heated; pressure builds inside the seed until it bursts. The cotyledons of bean and pea seeds are valued for their starch and protein; those of coffee and cacao, for their stimulants.

Fruits

Seedlings are nourished by nutrients in endosperm and cotyledons, but not by nutrients in fruits. A **fruit** is a seed-containing mature ovary, often with fleshy tissues that develop from the ovary wall (Figure 27.7). In some plants, fruit tissues develop from parts of the flower other than the ovary wall (such as petals, sepals, stamens, or receptacles). Apples, oranges, and grapes are familiar fruits, but so are many "vegetables" such as beans, peas, tomatoes, grains, eggplant, and squash.

Fruits are categorized by how they originate, their tissues, and appearance. A simple fruit, such as a pea pod, acorn, or shepherd's purse, is derived from a single ovary. Strawberries and other aggregate fruits form from separate ovaries of one flower; they mature as a cluster of fruits. Multiple fruits form from the fused ovaries of separate flowers. The pineapple is a multiple fruit that forms from fused ovary tissues of many flowers.

Fruits also may be categorized in terms of which tissues they incorporate. True fruits such as cherries consist only of the ovary wall and its contents. Other floral parts, such as the receptacle, expand along with the ovary in accessory fruits. Most of the flesh of an apple, an accessory fruit, is an enlarged receptacle.

To categorize a fruit based on appearance, we describe it as dry or juicy (fleshy). Acorns and grains (such as corn and wheat) are dry fruits, as are the fruits of sunflowers, maples, and strawberries. Strawberries are not berries and their fruits are not juicy. A strawberry's red flesh is an accessory to the dry fruits on its surface (Figure 27.8).

Cherries, almonds, and olives are fleshy fruits, as are individual fruits of boysenberries and other *Rubus* species. Grapes, tomatoes, and citrus fruits are berries, as are pumpkins, watermelons, and cucumbers. Apples and pears are pomes, in which fleshy tissues derived from the receptacle enclose a core derived from the ovary.

The function of a fruit is to protect and disperse seeds. Dispersal increases reproductive success by minimizing competition for resources among parent and offspring. Fruits are dispersed by mobile organisms such as birds, and by nonliving agents such as wind or moving water.

 Adaptations to a specific dispersal vector may be reflected in a fruit's form. For example, fruits dispersed by water have water-repellent outer layers. The fruits of sedges native to American marshlands have seeds encased in a bladderlike envelope that floats (*above*). Fruits of the coconut palm have thick, tough husks that can float for thousands of miles in seawater.

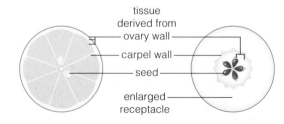

Figure 27.7 Parts of a fruit that develop from parts of a flower. *Left*, the tissues of an orange (*Citrus*) develop from the ovary wall. *Right*, the flesh of an apple (*Malus*) is an enlarged receptacle.

>> **Figure It Out** How many carpels were there in the flower that gave rise to this orange? Answer: Eight

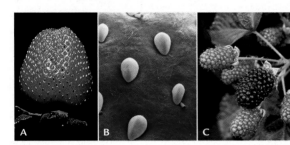

Figure 27.8 Aggregate fruits. **A** A strawberry (*Fragaria*) is not a berry. The flower's carpels turn inside out as the fruits form. The red, juicy flesh is an expanded receptacle; the hard "seeds" on the surface are individual dry fruits **B**.

C Boysenberries and other *Rubus* species are not berries, either. Each is an aggregate of many fleshy, one-seeded fruits.

Fruits dispersed by the wind tend to be lightweight with breeze-catching specializations. For example, part of a maple fruit is a dry, winglike outgrowth of the ovary wall (*left*). The fruit breaks in two when it drops from the tree. As the halves drop to the ground, wind catches the wings and spins the attached seeds away. Tufted fruits such as those of thistle, cattail, dandelion, and milkweed may be blown as far as 10 kilometers (6 miles) from the parent plant.

The fruits of cocklebur (*left*), bur clover, and many other plants have hooks or spines that stick to the feathers, feet, fur, or clothing of more mobile species. The dry, podlike fruit of plants such as California poppy propel their seeds through the air when they pop open explosively.

Colorful, fleshy, fragrant fruits attract animals that disperse seeds. The animal may eat the fruit and discard the seeds, or eat the seeds along with the fruit. Abrasion of the seed coat by digestive enzymes in an animal's gut can facilitate germination after the seeds depart in feces.

---

fruit Mature ovary, often with accessory parts, of a flowering plant.
seed Embryo sporophyte of a seed plant packaged with nutritive tissue inside a protective coat.

Take-Home Message How do seeds and fruits develop?

> After fertilization, the zygote develops into an embryo, the endosperm becomes enriched with nutrients, and the ovule's integuments develop into a seed coat.

> A seed is a mature ovule. It contains an embryo sporophyte.

> A fruit is a mature ovary, with or without accessory tissues that develop from other parts of a flower.

> A fruit's function is seed dispersal. Fruit specializations are adaptations to particular dispersal vectors.

> Asexual reproduction permits plants to rapidly produce genetically identical offspring.
< Links to Cloning 8.7, Asexual versus sexual reproduction 12.2, Meiosis 12.3, Aneuploidy 14.6, Speciation by polyploidy in plants 17.12, Modified stems 25.8

Most flowering plants can reproduce asexually by **vegetative reproduction**, a process in which new roots and shoots grow from extensions or pieces of a parent plant. Each new plant is a clone of its parent. The oldest known plant is a clone: the one and only population of King's holly (*Lomatia tasmanica*), which consists of several hundred stems growing along 1.2 kilometers (0.7 miles) of a river gully in Tasmania. Radiometric dating of the plant's fossilized leaf litter show that the clone is at least 43,600 years old—predating the last ice age! The ancient *Lomatia* is triploid. With three sets of chromosomes, it is sterile, so it can only reproduce asexually. Why? During meiosis, an odd number of chromosome sets cannot be divided equally between the two spindle poles. If meiosis does not fail entirely, the unequal segregation of chromosomes results in aneuploid offspring, which rarely survive.

King's holly probably arose by a chance genetic event. Many other types of plants reproduce asexually when new roots and shoots sprout from an existing plant (*inset*). For thousands of years, humans have been taking advantage of this natural capacity of plants to reproduce asexually. Almost all houseplants, woody ornamentals, and orchard trees are clones that have been grown from stem fragments (cuttings) of a parent plant. Other plants are grafted. Grafting means inducing a cutting to fuse with the tissues of another plant. Often, the stem of a desired plant is spliced onto the roots of a hardier one. For example, French vintners routinely graft their prized grapevines onto the roots of insect-resistant American vines.

An entire plant may be cloned from a single cell with **tissue culture propagation**, in which a body cell is coaxed to divide and form an embryo. The technique is currently being used to improve food crops and to propagate rare ornamental plants such as orchids.

---

tissue culture propagation Laboratory method in which body cells are induced to divide and form an embryo.
vegetative reproduction Growth of new roots and shoots from extensions or fragments of a parent plant; form of asexual reproduction in plants.

Take-Home Message How do plants reproduce asexually?

> Many plants propagate asexually when new shoots grow from a parent plant or pieces of it. Offspring of such vegetative reproduction are clones.

> Humans propagate plants asexually for agricultural or research purposes by grafting, tissue culture, or other methods.

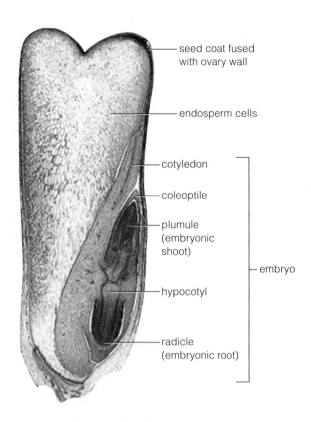

Figure 27.9 Anatomy of a corn seed (*Zea Mays*).

During germination, cell divisions resume mainly at apical meristems of the plumule (the embryonic shoot) and radicle (the embryonic root). A plumule consists of an apical meristem and two tiny leaves. In grasses such as corn, the growth of this delicate structure through soil is protected by a sheathlike coleoptile (Figure 27.10).

❭ Development in plants has a genetic basis, and it is also influenced by the environment.
❮ Links to Carbohydrates 3.3, Gene control 10.2, Meristems 25.2

What happens to a seed after it has been dispersed from the parent plant? An embryonic plant complete with shoot and root apical meristems formed as part of the seed (Figure 27.9). A seed typically dries out as it matures. Drying out causes the embryo's cells to stop dividing, so the embryo enters dormancy. An embryo may remain dormant in its protective seed coat for many years before it resumes metabolic activity. The resumption of an embryo sporophyte's development after a period of dormancy is a process called **germination**. Eudicot embryos transfer nutrients in endosperm to their two cotyledons before germination; monocot embryos tap endosperm only after the seed germinates.

The process of germination begins with water seeping into a seed. The water activates hydrolysis enzymes that break down stored starches into sugars (Section 3.3). As the seed's tissues swell with water, the seed coat ruptures, and oxygen diffuses into the seed's internal tissues. Meristem cells in the embryo use the sugars and the oxygen for aerobic respiration as they begin to divide rapidly, and the embryonic plant begins to grow. Germination ends when the first part of the embryo—the embryonic root, or radicle—breaks out of the seed coat.

Seed dormancy is a climate-specific adaptation that allows germination to occur when conditions in the environ-

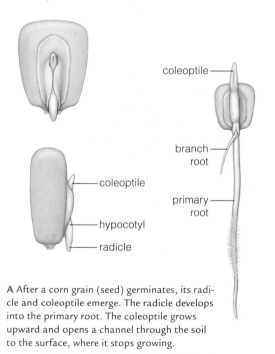

A After a corn grain (seed) germinates, its radicle and coleoptile emerge. The radicle develops into the primary root. The coleoptile grows upward and opens a channel through the soil to the surface, where it stops growing.

B The plumule develops into the seedling's primary shoot, which pushes through the coleoptile and begins photosynthesis. In corn plants, adventitious roots that develop from the stem afford additional support for the rapidly growing plant.

Figure 27.10 Animated Early growth of a typical monocot, corn (*Zea mays*).

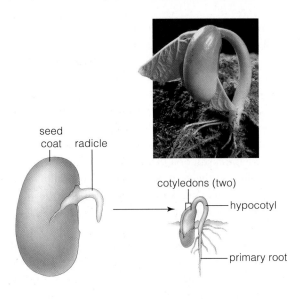

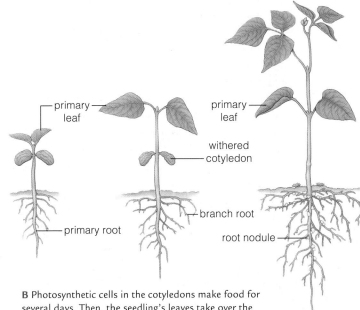

A After a bean seed germinates, its radicle emerges and bends in the shape of a hook. Sunlight causes the hypocotyl to straighten, which pulls the cotyledons up through the soil.

B Photosynthetic cells in the cotyledons make food for several days. Then, the seedling's leaves take over the task and the cotyledons wither and fall off.

Figure 27.11 Animated Early growth of a typical eudicot, common bean (*Phaseolus vulgaris*).

ment are most likely to support the growth of a seedling. For example, seeds of many annual plants native to cold winter regions are dispersed in autumn. If the seeds germinated immediately, the tender seedlings would not survive the coming winter. Instead, the seeds stay dormant until spring, when milder temperatures and longer daylength are more suitable for growth of the seedlings. By contrast, the weather in regions near the equator does not vary by season. Seeds of most plants native to such regions do not enter dormancy; they germinate as soon as they mature.

How does a dormant embryo sporophyte "know" when to germinate? The triggers, other than the presence of water, differ by species, and all have a genetic basis. For example, some seed coats are so dense that they must be abraded or broken (by being chewed, for example) before water can even enter the seed. The seeds of many cool-climate plants require exposure to freezing temperatures; those of some species of lettuce must be exposed to bright light. Germination of the seeds of California poppy and other wildflower species native to regions that experience periodic wildfires is inhibited by light and enhanced by smoke. Seeds of other species native to such regions will not germinate unless they have been previously burned.

Germination requirements are evolutionary adaptations to life in a particular environment. All maximize a seedling's chance of survival, for example by ensuring seeds only germinate at times when competition from established plants is at a minimum.

germination The resumption of growth after dormancy.

Plant Development

Germination is the first step in plant development. As a sporophyte grows and matures, its tissues and organs develop in characteristic patterns. Leaves form in predictable shapes and sizes, stems and roots lengthen and thicken, flowering occurs at a certain time of year, and so on. These patterns are outcomes of gene expression, and they also have an environmental component. As you will see in the next section, gene expression in plants is regulated by hormones and other signaling molecules, just as it is in animals.

The pattern of early growth that occurs after germination varies by species. Figures 27.10 and 27.11 show two examples. Development includes growth, which is an increase in cell number and size. In plants, growth occurs primarily in the direction of cell division—and cell division occurs primarily at meristems. Behind meristems, cells differentiate and form specialized tissues. However, unlike animal cell differentiation, plant cell differentiation is often reversible, as when new shoots or roots sprout from a mature stem.

Take-Home Message What is plant development?

> In plants, development includes growth and differentiation. It results in the formation of plant tissues and organs in predictable patterns.

> Germination and other steps in plant development are an outcome of gene expression and environmental influences.

❯ Plant development depends on cell-to-cell communication, which is mediated by plant hormones.

❮ Links to Transcription factors 10.2, Cell communication in animal development 10.3, Function of stomata 26.5

Plant Hormones

Plant development depends on extensive coordination among individual cells, just as it does in animals. A plant is an organism, not just a collection of cells, and as such it develops as a unit. Cells in different parts of a plant coordinate their activities by communicating with one another. Such communication means, for example, that roots and shoots can be triggered to grow at the same time.

Plant cells use **hormones** to communicate with one another. Plant hormones stimulate or inhibit development, including growth. Environmental cues such as availability of water, length of night, temperature, and gravity influence plants by triggering the production and dispersal of hormones. When a plant hormone binds to a target cell, it may modify gene expression, change solute concentrations, affect enzyme activity, or activate another molecule in cytoplasm. Five types of plant hormones (gibberellins, auxins, abscisic acid, cytokinins, and ethylene) interact in plant development (Table 27.1).

Gibberellins Growth and other processes of development in all flowering plants, gymnosperms, mosses, ferns, and some fungi are regulated in part by **gibberellins**. These hormones induce cell division and elongation in stem tissue; thus, they cause stems to lengthen between the nodes. This effect can be demonstrated by application of gibberellin to the leaves of young plants (Figure 27.12). The short stems of Mendel's dwarf pea plants (Section 13.3) are the result of a mutation that reduces the rate of gibberellin synthesis in these plants. Gibberellins are also involved in breaking dormancy of seeds, seed germination, and, in some plants, flowering.

Auxins **Auxins** are plant hormones that influence cell division and elongation. Their effect depends on the target tissue. Auxins produced in apical meristems result in elongation of shoots (Figure 27.13). They also induce cell division and differentiation in vascular cambium, fruit development in ovaries, and lateral root formation in roots. Auxins also have inhibitory effects. For example, auxin produced in a shoot tip prevents the growth of lateral buds along a lengthening stem, an effect called apical dominance. Gardeners routinely pinch off shoot tips to make a plant bushier. Pinching the tips ends the supply of auxin in a main stem, so lateral buds give rise to branches. Auxins also inhibit abscission, which is the dropping of leaves, flowers, and fruits from the plant.

Figure 27.12 Effect of gibberellins. The three tall cabbage plants were treated with gibberellins. The two short plants in front of the ladder were not treated.

Table 27.1 Major Plant Hormones and Some of Their Effects

Hormone	Primary Source	Effect	Site of Effect
Gibberellins	Stem tip, young leaves	Stimulates cell division, elongation	Stem internode
	Embryo	Stimulates germination	Seed
	Embryo (grass)	Stimulates starch hydrolysis	Endosperm
Auxins	Stem tip, young leaves	Stimulates cell elongation	Growing tissues
		Initiates formation of lateral roots	Roots
		Inhibits growth (apical dominance)	Lateral buds
		Stimulates differentiation of xylem	Cambium
		Inhibits abscission	Leaves, fruits
	Developing embryos	Stimulates fruit development	Ovary
Abscisic acid	Leaves	Closes stomata	Guard cells
		Stimulates formation of dormant buds	Stem tip
	Ovule	Inhibits germination	Seed coat
Cytokinins	Root tip	Stimulates cell division	Stem tip, lateral buds
		Inhibits senescence (aging)	Leaves
Ethylene	Damaged or aged tissue	Inhibits cell elongation	Stem

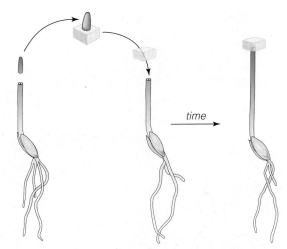

A A coleoptile stops growing after its auxin-producing tip has been removed. A block of agar that absorbs auxin from a cut tip can stimulate a de-tipped coleoptile to resume growth.

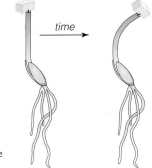

B If an auxin-containing agar block is placed to one side of a cut tip, the coleoptile will continue to grow, but bend as it does.

Figure 27.13 Animated
A coleoptile lengthens in response to auxin produced in its tip.

Abscisic Acid Abscisic acid (ABA) is a hormone that was misnamed; it inhibits growth, and has little to do with abscission. ABA is part of a stress response that causes stomata to close, as you learned in Section 26.5. It also diverts photosynthetic products from leaves to seeds, an effect that overrides growth-stimulating effects of other hormones as the growing season ends. ABA is responsible for inhibiting seed germination in some species, such as apple. Apple seeds do not germinate before most of the ABA they contain has been broken down, for example by a long period of cold, wet conditions.

---

abscisic acid Plant hormone that stimulates stomata to close in response to water stress; induces dormancy in buds and seeds.
auxin Plant hormone that stimulates cell division and elongation.
cytokinin Plant hormone that promotes cell division. Releases lateral buds from apical dominance.
ethylene Gaseous plant hormone that inhibits cell division in stems and roots; also promotes fruit ripening.
gibberellin Plant hormone; induces stem elongation, helps seeds break dormancy. Role in flowering in some species.
hormone Signaling molecule that is released into the body by one type of cell and alters the activity of other cells.

Table 27.2 Some Commercial Uses of Plant Hormones
Gibberellins Increase fruit size; delay citrus fruit ripening
Synthetic auxins Promote root formation in cuttings; induce seedless fruit production before pollination; keep mature fruit on trees until harvest time; widely used as herbicides
ABA Induces nursery stock to enter dormancy before shipment to minimize damage during handling
Cytokinins Tissue culture propagation; prolong shelf life of cut flowers
Ethylene Allows shipping of green, still-hard fruit (minimizes bruises, rotting). Carbon dioxide application stops ripening of fruit in transit to market, then ethylene is applied to ripen distributed fruit quickly.

Cytokinins Plant **cytokinins** form in roots and travel via xylem to shoots, where they induce cell divisions in the apical meristems. These hormones also release lateral buds from apical dominance, and inhibit the normal aging process in leaves. Cytokinins signal to shoots that roots are healthy and active. When roots stop growing, they stop producing cytokinins, so shoot growth slows and leaves begin to deteriorate.

Ethylene The only gaseous hormone, **ethylene**, is produced by damaged cells. It is also produced in autumn in deciduous plants, or near the end of the life cycle as part of a plant's normal process of aging. Ethylene inhibits cell division in stems and roots. It also induces fruit and leaves to mature and drop. Ethylene is widely used to artificially ripen fruit that has been harvested while still green (Table 27.2).

Other Signaling Molecules

As we now know, other signaling molecules have roles in various aspects of plant development. For example, brassinosteroids stimulate cell division and elongation, so stems remain short in their absence. Fatty acid-derived jasmonates interact with other hormones in inhibiting germination and root growth, and in tissue defense. FT protein is part of a signaling pathway in flower formation. Salicylic acid, a molecule similar to aspirin, interacts with nitric oxide in regulating transcription of gene products that help plants resist attacks by pathogens. Systemin is a polypeptide that forms as insects feed on plant tissues; it enhances transcription of genes that encode insect toxins.

Take-Home Message **How is development controlled in plants?**

> Plant hormones are signaling molecules that influence plant development.

> The five main classes of plant hormones are gibberellins, auxins, abscisic acid, cytokinins, and ethylene.

> Interactions among hormones and other kinds of signaling molecules stimulate or inhibit cell division, elongation, differentiation, and other processes.

❯ Plants alter growth in response to environmental stimuli. Hormones are typically part of this effect.
❮ Links to Plastids 4.9, Cytoskeleton 4.10, Pigments 6.2

Plants respond to environmental stimuli by adjusting the growth of roots and shoots. These responses are called tropisms, and they are typically mediated by hormones. For example, a root or shoot "bends" because of differences in auxin concentration. Auxin that accumulates in cells on one side of a shoot causes the cells to elongate more than the cells on the other side (Figure 27.13B). The result is that the shoot bends away from the side with more auxin. Auxin has the opposite effect in roots: It inhibits elongation of root cells. Thus, a growing root will bend toward the side that contains more auxin.

Responses to Gravity When a seed germinates, its primary root always grows downward, and its primary shoot always grows upward. Even if a seedling is turned upside down just after germination, the primary root and shoot will curve so the root grows downward and the shoot grows upward (Figure 27.14). Any growth response to gravity is called **gravitropism**.

Gravity-sensing mechanisms of many organisms are based on statoliths. Plant statoliths are plastids stuffed with dense grains of starch that occur in root cap cells, and also in specialized cells at the periphery of vascular tissues in the stem. Starch grains are heavier than cytoplasm, so statoliths tend to sink to the lowest region of the cell, wherever that is (Figure 27.15). The shift causes auxin to be redistributed to the downward-facing side of roots and shoots.

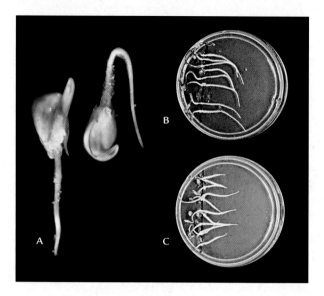

Figure 27.14 Animated Gravitropism. **A** Regardless of how a corn seed is oriented in soil, the seedling's primary root grows down, and its primary shoot grows up.

B These seedlings were rotated 90° counterclockwise after they germinated. The plant adjusts to the change by redistributing auxin, and the direction of growth shifts as a result. **C** In the presence of auxin transport inhibitors, seedlings do not adjust their direction of growth after a 90° counterclockwise rotation. Mutations in genes that encode auxin transport proteins have the same effect.

Responses to Light Light streaming in from one direction causes a stem to curve toward its source. This response, **phototropism**, orients certain parts of the plant in the direction that maximizes the amount of light intercepted by its photosynthetic cells. Phototropism in plants occurs in response to blue light. Nonphotosynthetic pigments called phototropins absorb blue light, and use its

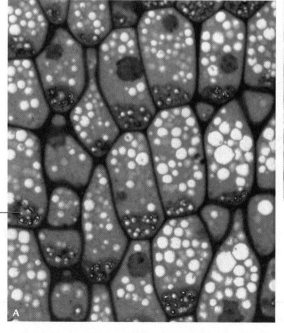

statoliths

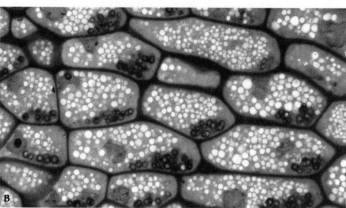

Figure 27.15 Animated Gravity, statoliths, and auxin.

A This micrograph shows heavy, starch-packed statoliths settled on the bottom of gravity-sensing cells in a corn root cap.

B This micrograph was taken ten minutes after the root in **A** was rotated 90°. The statoliths are already settling to the new "bottom" of the cells.

❯❯ **Figure It Out** In which direction was this root rotated?

Answer: Counterclockwise

A Sunlight strikes only one side of a coleoptile.

B More auxin is produced on the shaded side, so cells on that side lengthen faster.

Figure 27.16 Animated Phototropism. Auxin-mediated differences in cell elongation between two sides of a coleoptile induce bending toward light. The photo on the *right* shows a phototropic response of flowering shamrock (*Oxalis*).

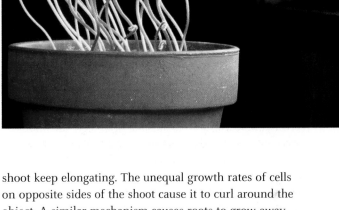

energy to initiate a cascade of intracellular signals. The ultimate effect of this cascade is that more auxin is made on the shaded side of a shoot or coleoptile. As a result, cells on the shaded side elongate faster than cells on the illuminated side. Differences in growth rates between cells on opposite sides of a shoot or coleoptile cause the entire structure to bend toward the light (Figure 27.16).

In a response called **solar tracking**, a leaf or flower changes position in response to the changing angle of the sun throughout the day. The mechanism for solar tracking is not well understood, but it may be similar to phototropism because it has been observed to coincide with differential growth of stems, and it also occurs in response to blue light.

Responses to Contact A plant's contact with an object may result in a change in the direction of its growth, a response called **thigmotropism** (*thigma* means touch in Greek). The mechanism that gives rise to this response is not well understood, but it involves calcium ions and the products of at least five genes that are called *TOUCH*.

We see thigmotropism when a vine's tendril touches an object (*inset*). The cells near the area of contact stop elongating, and the cells on the opposite side of the

shoot keep elongating. The unequal growth rates of cells on opposite sides of the shoot cause it to curl around the object. A similar mechanism causes roots to grow away from contact, so they "feel" their way around rocks and other impassable objects in the soil. Mechanical stress, such as by wind exposure, inhibits stem lengthening in a response related to thigmotropism (Figure 27.17).

Figure 27.17 Effect of mechanical stress on tomato plants. **A** This plant, the control, was never shaken. **B** Each day for twenty-eight days, this plant was mechanically shaken for thirty seconds. **C** This one had two shakings per day.

---

gravitropism Plant growth in a direction influenced by gravity.
phototropism Change in the direction of cell movement or growth in response to a light source.
solar tracking Plant parts change position in response to the sun's changing angle through the day.
thigmotropism Directional growth of a plant in response to contact with a solid object.

Take-Home Message How do plants respond to their environment?

❯ Via hormones, plants adjust the direction and rate of growth in response to gravity, light, contact, mechanical stress, and other environmental stimuli.

> Seasonal shifts in night length, temperature, and light trigger seasonal shifts in plant development.
< Links to Photosynthesis 6.5 and 6.7, Gene expression 9.2 and 10.2, Master genes in flowering 10.4

Biological Clocks

Most organisms have a **biological clock**, which is an internal mechanism that governs the timing of rhythmic cycles of activity. For example, a bean plant holds its leaves horizontally during the day but folds them close to its stem at night. A plant exposed to constant light or constant darkness for a few days will continue to move its leaves in and out of the "sleep" position at the time of sunrise and sunset (Figure 27.18). The response might help reduce heat loss at night, when air cools, and so maintain the plant's internal temperature.

Rhythmic leaf movements of a bean plant are one example of a **circadian rhythm**, which is a cycle of biological activity that starts anew every twenty-four hours or so. Circadian means "about a day." Similar mechanisms cause flowers of some plants to open only at certain times of day. For example, the flowers of many bat-pollinated plants unfurl, secrete nectar, and release fragrance only at night. Periodically closing flowers protects the delicate reproductive parts when the likelihood of pollination is typically lowest.

Setting the Clock

Like a mechanical clock, a biological one can be reset. Sunlight resets biological clocks in plants by activating and inactivating photoreceptors called **phytochromes**. These blue-green pigments absorb red light at 660 nanometers, and far-red light at 730 nanometers. The relative amounts of these wavelengths in sunlight that reaches a given environment vary during the day and with the season. Red light causes phytochromes to change from an inactive form to an active form. Far-red light causes them to change back to their inactive form (Figure 27.19).

Active phytochromes bring about transcription of many genes, including some that encode components of rubisco, photosystem II, ATP synthase, and other proteins used in photosynthesis; phototropin for phototropic

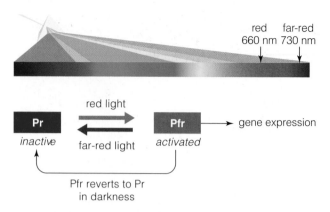

Figure 27.19 Animated Phytochromes. Red light changes the structure of a phytochrome from inactive (Pr) to active (Pfr) form; far-red light changes it back to the inactive form. Activated phytochromes control important processes such as germination and flowering.

responses; and molecules involved in flowering, gravitropism, and germination.

When to Flower?

Photoperiodism refers to an organism's response to changes in the length of night relative to the length of day. Except at the equator, night length varies with the season. Nights are longer in winter than in summer, and the difference increases with latitude (Figure 27.20A).

You have probably noticed that different species of plants flower at different times of the year. In these plants, flowering is photoperiodic. Long-day plants such as irises flower only when the hours of darkness fall below a critical value. Chrysanthemums and other short-day plants flower only when the hours of darkness are greater than some critical value. Sunflowers and other day-neutral plants flower when they mature, regardless of night length.

Figure 27.20B,C shows two experiments that demonstrated how phytochromes play a role in photoperiodism. In the first experiment, a long-day and a short-day plant were exposed to long "nights," interrupted by a brief pulse of red light (which activates phytochrome). Both plants responded in their typical way to a season of short nights. In the second experiment, the pulse of red light (which

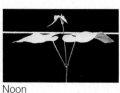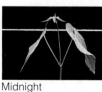

1 A.M.   6 A.M.   Noon   3 P.M.   10 P.M.   Midnight

Figure 27.18 Animated Rhythmic leaf movements by a young bean plant (*Phaseolus*). Physiologist Frank Salisbury kept this plant in darkness for twenty-four hours. Despite the lack of light cues, the leaves kept on folding and unfolding at sunrise (6 A.M.) and sunset (6 P.M.).

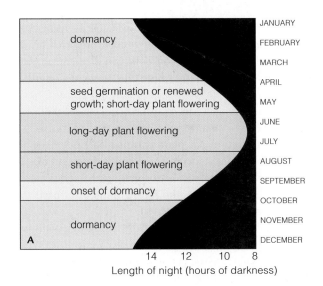

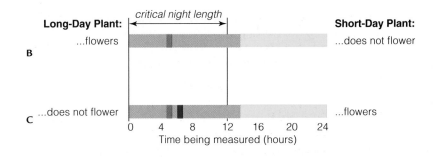

Long-Day Plant: ...flowers | **Short-Day Plant:** ...does not flower

critical night length

B

C ...does not flower | ...flowers

Time being measured (hours)

B A flash of red light interrupting a long night activates phytochrome. It causes plants to respond as if the night were short, and long-day plants flower.

C A pulse of far-red light, which inactivates phytochrome, cancels the effect of the red flash, and short-day plants flower. *Blue* bars indicate night length; *yellow* bars, day length.

Figure 27.20 Animated Photoperiodism and flowering. **A** Plant growth and development correlated with seasonal climate changes in northern temperate zones. **B,C** Two experiments showing how different plant species flower in response to different night lengths. Each horizontal bar represents 24 hours.

❯❯ **Figure It Out** What is the critical length of night that triggers a short–day plant's flowering response in the experiments shown in **B**?

Answer: 12 hours

activates phytochrome) was followed by a pulse of far-red light (which deactivates phytochrome). Both plants responded in their typical way to a season of long nights.

Leaves detect night length and produce signals that travel through the plant. In one experiment, a single leaf was left on a cocklebur, a short-day plant. The leaf was shielded from light for 8-1/2 hours every day, which is the threshold amount of darkness required for flowering in short-day plants. The plant flowered. Later, the leaf was grafted onto another cocklebur plant that had not been exposed to long hours of darkness. After grafting, the recipient plant flowered, too.

How does a leaf signal another part of the plant to flower? In response to night length and other cues, companion cells in leaf phloem transcribe more or less of a gene called *Flowering locus T*, or *FT* for short. The cells export the gene's product, FT protein, into sieve tubes, where it migrates from leaves to shoot tips. The protein interacts with a transcription factor to begin transcription of floral identity genes (Section 10.4) in cells that are differentiating behind the meristem.

The length of night is not the only cue for flowering. Some plants flower only after exposure to cold winter temperatures (Figure 27.21), a process called **vernalization** (from Latin *vernalis*, "to make springlike"). In these plants, expression of the *FT* gene is silenced by a repressor. The repressor stops being produced only after the plant is exposed to a period of cold temperatures.

Figure 27.21 Local effect of cold on dormant buds of lilac (*Syringa*). For this experiment, a single branch was positioned to protrude from a greenhouse through a cold winter. The rest of the plant was kept inside the warm greenhouse. Only buds exposed to the low outside temperatures resumed growth and flowered in springtime.

biological clock Internal time-measuring mechanism by which individuals adjust their activities seasonally, daily, or both in response to environmental cues.
circadian rhythm A biological activity that is repeated about every 24 hours.
photoperiodism Biological response to seasonal changes in the relative lengths of day and night.
phytochrome A light-sensitive pigment that helps set plant circadian rhythms based on length of night.
vernalization Stimulation of flowering in spring by low temperature in winter.

Take-Home Message Do plants have a biological clock?

❯ Some plants respond to recurring cues from the environment with recurring cycles of activity such as rhythmic leaf movements.

❯ The main environmental cue for flowering is the length of night relative to the length of day, which varies by the season in most places. Low winter temperatures stimulate the flowering of many plant species in spring.

❯ Dropping of plant parts and dormancy are triggered by seasonal changes in environmental conditions.
❮ Links to Plant cell walls and extracellular matrix 4.11, Rubisco 6.7, Root symbionts 26.3

Unlike animals, plants cannot run away from attacks by predators. However, even firmly rooted individuals protect themselves. Many types of plants have thorns or nasty-tasting chemicals that directly deter herbivores (plant-eating animals). Some even get help from wasps. Damage to a leaf, such as occurs when an insect chews on it, triggers a stress response in the plant. The wounding results in the cleavage of certain polypeptides (such as systemin) in mesophyll cells. Thus activated, the polypeptides stimulate synthesis of jasmonates, which in turn enhance transcription of a variety of genes.

Some of the resulting gene products break down molecules used in normal activities, such as rubisco, so growth temporarily slows. Other gene products produce chemicals that the plant releases into the air. The chemicals are detected by wasps that parasitize insect herbivores (Figure 27.22). The signaling is quite specific: A leaf releases a different set of chemicals depending on which herbivore is chewing on it. Certain wasp species recognize these chemicals as a signal leading to their preferred prey. The insects follow airborne concentration gradients of the chemicals back to the plant, where they attack the offending herbivores.

Jasmonates are also part of systemic responses that help plants fend off pathogens. The presence of a virus, bacteria, or fungus in one plant part increases pathogen resistance in the entire plant. For example, infection of a plant's roots by beneficial nitrogen-fixing bacteria or a mycorrhizal fungus causes the aboveground plant parts to become more resistant to infection.

A whole-body, long-term defense response in a plant is called **systemic acquired resistance**. Such responses typically begin when a tissue is under attack by a pathogen, for example, after a fungal infection kills leaf cells. The affected tissue releases as yet unknown molecular signals that travel to other parts of the plant, where they trigger cells to produce compounds that include hydrogen peroxide, salicylic acid, and jasmonates. Plant cells respond to these compounds by modifying their cell walls, and by producing antifungal and antiviral proteins. Both activities result in increased resistance to pathogenic attack.

Senescence

The molecules that a plant produces when it is under attack may cause the plant to drop its leaves. The process by which leaves or other plant parts are shed is called **abscission**. Abscission may be induced by any stress: infection, injury, water or nutrient deficiency, or high temperature. It also occurs in the normal life cycle of flowering plants, as part of senescence. **Senescence** is the phase of the plant life cycle between full maturity and death. Some plants respond to conditions that vary seasonally with recurring cycles of growth and inactivity. Plants that periodically drop their leaves are typically native to regions that are too dry or too cold for optimal growth during part of the year.

Abscission in the normal plant life cycle is mediated by hormones, a different mechanism than in stress responses. Let's use a deciduous fruit tree as an example. As leaves and fruits grow in early summer, their cells produce auxin. The auxin moves into the stems, where it helps maintain growth. By midsummer, the nights are longer. The tree begins to divert nutrients away from its leaves, stems, and roots, and into flowers, fruits, and

Figure 27.22 Jasmonates in plant defenses.

A Consuelo De Moraes studies chemical signaling in plants.

B A caterpillar chewing on a tobacco leaf (*Nicotiana*) triggers a stress response from the leaf's cells. The cells release certain chemicals into the air.

C,D A parasitoid wasp follows the chemicals back to the stressed leaves, then attacks the caterpillar and deposits an egg inside it. When the egg hatches, it will release a caterpillar-munching larva. De Moraes discovered that such interactions are highly specific: Leaf cells release different chemicals in response to different caterpillar species. Each chemical attracts only the wasps that parasitize the particular caterpillar that triggered the chemical's release.

---

abscission Process by which plant parts are shed in response to seasonal change, drought, injury, or nutrient deficiency.
senescence Phase in a life cycle from maturity until death.
systemic acquired resistance In plants, a long-term, systemic resistance to pathogens.

Figure 27.23 Abscission in the horse chestnut tree (*Aesculus hippocastanum*). *Left*, leaves change color in autumn before dropping. *Right*, a leaf scar is all that remains of an abscission zone that formed before a leaf detached from the stem. The tree is named after these horseshoe-shaped scars.

seeds. As the growing season comes to a close, nutrients are routed to twigs, stems, and roots, and auxin production declines in leaves and fruits.

The auxin-deprived structures release ethylene that diffuses into nearby twigs, petioles, and fruit stalks. The ethylene is a signal for cells in these abscission zones to produce enzymes that digest their own walls. The cells bulge as their walls soften, and separate from one another as the extracellular matrix that cements them together dissolves. Tissue in the abscission zone weakens, and the structure above it drops (Figure 27.23).

For many species, growth stops in autumn as a plant enters dormancy. Long nights, cold temperature, and dry, nitrogen-poor soil are strong cues for dormancy in many plants. Dormancy-breaking cues usually operate between fall and spring.

Dormant plants do not resume growth until certain conditions in the environment occur. A few species require exposure to many hours of cold temperature. More typical cues include the return of milder temperatures and plentiful water and nutrients. With the return of favorable conditions, life cycles begin to turn once more as buds resume growth.

Take-Home Message Why does a flowering plant drop its leaves?

> Plants produce protective compounds in response to attack by herbivores or pathogens. Abscission may occur during such stress responses.

> In some plants, abscission and dormancy are triggered by environmental cues such as seasonal changes.

Plight of the Honeybee (revisited)

New research into colony collapse disorder has revealed that bees in affected hives have unusually large amounts of ribosomal RNA fragments in their guts. The fragments mean that the bees' ability to make proteins has been compromised, and that could account for their reduced survival rate. Bees that cannot make proteins cannot defend themselves against bacterial or fungal infection, and they are also vulnerable to starvation. The source of the rRNA bits in the bees' guts has not yet been determined, but the researchers are suspicious of picorna-like viruses, which hijack their hosts' protein synthesis machinery. These viruses disrupt normal translation of host cell mRNA, so an infected cell's ribosomes translate only viral RNA. Ribosomes unable to perform their normal function would likely be degraded. Picorna-like viruses, including Israeli acute paralysis virus, are carried by mites that parasitize bees.

How Would You Vote? Systemic pesticides are easy to apply and effective for long periods. They also get into plant nectar and pollen eaten by honeybees. To protect bees and other pollinators, should the use of these pesticides on flowering plants be restricted? See CengageNow for details, then vote online (cengagenow.com).

Summary

Section 27.1 Colony collapse disorder is killing honeybees. Declines in populations of bees and other **pollinators** negatively affect plant populations as well as other animal species that depend on the plants.

Section 27.2 Flowers consist of modified leaves (sepals, petals, **stamens**, and **carpels**) at the ends of specialized branches of angiosperm sporophytes.

An ovule develops from a mass of **ovary** wall tissue inside carpels. Spores produced by meiosis in **ovules** develop into female gametophytes; those produced in anthers develop into immature male gametophytes (pollen grains). A flower's shape, pattern, color, and fragrance may reflect an evolutionary relationship with an animal pollination vector. Pollinators typically receive a reward for visiting a flower.

Section 27.3 Meiosis of diploid cells inside pollen sacs of anthers produces haploid **microspores**. Each microspore develops into a pollen grain that is released from pollen sacs after a period of **dormancy**.

Mitosis and cytoplasmic division of a cell in an ovule produces four **megaspores**, one of which gives rise to the female gametophyte. One of the seven cells of the gametophyte is the egg; another is the endosperm mother cell.

A pollen grain arrives on a receptive stigma at **pollination**. The pollen grain germinates and forms a pollen tube that contains two sperm cells. The tube grows through carpel tissues to the egg. In **double fertilization**, one of the sperm cells in the pollen tube fertilizes the egg, forming a zygote; the other fuses with the endosperm mother cell and gives rise to **endosperm**.

Section 27.4 As a zygote develops into an embryo, endosperm collects nutrients from the parent plant, and the ovule's protective layers develop into a seed coat. A **seed** is a mature ovule: an embryo sporophyte and endosperm enclosed within a seed coat. Nutrients in endosperm or cotyledons make seeds a nutritious food source.

As an embryo sporophyte develops, the ovary wall and sometimes other tissues mature into a **fruit** that encloses the seeds. Fruit specializations are adaptations to dispersal by specific vectors. A fruit is categorized by tissue of origin, composition, and whether it is dry or fleshy.

Section 27.5 Many species of flowering plants are able to reproduce asexually by **vegetative reproduction**. Asexual reproduction produces clones of the parent. Many agriculturally valuable plants are produced by cuttings, grafting, and **tissue culture propagation**.

Section 27.6 Gene expression and cues from the environment coordinate plant development, which is the formation and growth of tissues and parts. **Germination** is one process of development in plants.

Section 27.7 **Hormones** secreted by one cell alter the activity of a different cell. **Gibberellins** lengthen stems, break dormancy in seeds and buds, and stimulate flowering. **Auxins** lengthen coleoptiles, shoots, and roots by promoting cell enlargement. **Cytokinins** stimulate cell division, release lateral buds from apical dominance, and inhibit senescence. **Ethylene** promotes senescence and abscission. It also inhibits growth of roots and stems. **Abscisic acid** promotes bud and seed dormancy, and it limits water loss by causing stomata to close.

Section 27.8 Plants adjust the direction and rate of growth in response to environmental cues. In **gravitropism**, roots grow down and stems grow up in response to gravity. In **phototropism** and **solar tracking**, plant parts bend in response to light. Some plants respond to contact (**thigmotropism**) or mechanical stress.

Section 27.9 Internal timing mechanisms such as **biological clocks** (including **circadian rhythms**) are set by daily and seasonal variations in environmental conditions. **Photoperiodisms** are responses to changes in length of night relative to length of day; they involve phytochromes. **Vernalization** is flowering stimulated by cold weather.

Section 27.10 Systemic acquired resistance refers to long-term plant defenses against pathogens. **Abscission** may occur during a stress response. It is also triggered by seasonal changes during **senescence**.

Self-Quiz   Answers in Appendix III

1. A flower's _____ has one or more ovaries in which eggs develop, fertilization occurs, and seeds mature.
    a. pollen sac   c. receptacle
    b. carpel   d. sepal

2. Name one reward that a pollinator may receive in return for a visit to a flower of a coevolved plant.

3. Meiosis of cells in pollen sacs forms haploid _____ .
    a. megaspores   c. stamens
    b. microspores   d. sporophytes

4. Meiosis in an ovule produces _____ megaspores.
    a. two   b. four   c. six   d. eight

5. Seeds are mature _____ ; fruits are mature _____ .
    a. ovaries; ovules   c. ovules; ovaries
    b. ovules; stamens   d. stamens; ovaries

6. The seed coat forms from the _____ .
    a. integuments   c. endosperm
    b. coleoptile   d. sepals

7. A new plant forms from a stem that broke off of the parent plant. This is an example of _____ .
    a. parthenogenesis   c. vegetative reproduction
    b. exocytosis   d. nodal growth

8. Cotyledons develop as part of _____ .
    a. carpels   c. embryo sporophytes
    b. accessory fruits   d. petioles

9. Plant hormones _____ .
    a. may have multiple effects
    b. are active in developing plant embryos
    c. are active in adult plants
    d. all of the above

10. _____ is the stimulus for phototropism.
    a. Red light   c. Green light
    b. Far-red light   d. Blue light

11. Sunlight resets biological clocks in plants by activating and inactivating _____ .
    a. phototropins   c. photoperiodisms
    b. phytochromes   d. far-red light

12. Solar tracking is similar to _____ .
    a. phototropism   c. photoperiodism
    b. gravitropism   d. thigmotropism

13. In some plants, flowering is a _____ response.
    a. phototropic   c. photoperiodic
    b. gravitropic   d. thigmotropic

Data Analysis Activities

Searching for Pollinators *Massonia depressa* is a low-growing succulent plant native to the desert of South Africa. The dull-colored flowers of this monocot develop at ground level, have tiny petals, emit a yeasty aroma, and produce a thick, jellylike nectar. These features led researchers to suspect that desert rodents such as gerbils pollinate this plant (Figure 27.24).

To test their hypothesis, the researchers trapped rodents in areas where *M. depressa* grows and checked them for pollen. They also put some plants in wire cages that excluded mammals, but not insects, to see whether fruits and seeds would form in the absence of rodents. The results are shown in Figure 27.25.

1. How many of the 13 captured rodents showed some evidence of pollen from *M. depressa*?
2. Would this evidence alone be sufficient to conclude that rodents are the main pollinators of this plant?
3. How did the average number of seeds produced by caged plants compare with that of control plants?
4. Do these data support the hypothesis that rodents are required for pollination of *M. depressa*? Why or why not?

Figure 27.25 *Right*, results of experiments testing rodent pollination of *M. depressa*. **A** Evidence of visits to *M. depressa* by rodents. **B** Fruit and seed production of *M. depressa* with and without visits by mammals. Mammals were excluded from plants by wire cages with openings large enough for insects to pass through. 23 plants were tested in each group.

Figure 27.24 The dull, petal-less, ground-level flowers of *Massonia depressa* are accessible to rodents, who push their heads through the stamens to reach the nectar at the bottom of floral cups. Note the pollen on the gerbil's snout.

A

Type of rodent	Number caught	# with pollen on snout	# with pollen in feces
Namaqua rock rat	4	3	2
Cape spiny mouse	3	2	2
Hairy-footed gerbil	4	2	4
Cape short-eared gerbil	1	0	1
African pygmy mouse	1	0	0

B

	Mammals allowed access to plants	Mammals excluded from plants
Percent of plants that set fruit	30.4	4.3
Average number of fruits per plant	1.39	0.47
Average number of seeds per plant	20.0	1.95

14. Match the observation with the hormone.
    ___ ethylene
    ___ cytokinin
    ___ auxin
    ___ gibberellin
    ___ abscisic acid

    a. Your cabbage plants bolt (they form elongated flowering stalks).
    b. The philodendron in your room is leaning toward the window.
    c. The last of your apples is getting really mushy.
    d. The seeds of your roommate's marijuana plant do not germinate no matter what he does to them.
    e. Lateral buds on your *Ficus* plant are sprouting branch shoots.

15. Match the terms with the most suitable description.
    ___ ovule
    ___ receptacle
    ___ double fertilization
    ___ anther
    ___ carpel
    ___ mature female gametophyte
    ___ mature male gametophyte

    a. pollen tube together with its contents
    b. embryo sac of seven cells, one with two nuclei
    c. starts out as cell mass in ovary; may become a seed
    d. stigma, style, and ovary
    e. pollen sacs inside
    f. base of floral shoot
    g. formation of zygote and first cell of endosperm

Additional questions are available on **CENGAGENOW**.

Critical Thinking

1. The oat coleoptiles on the *right* have been modified: either cut or placed in a light-blocking tube. Which ones will still bend toward a directional light source?

2. Belgian scientists discovered that certain mutations in common wall cress (*Arabidopsis thaliana*) cause excess auxin production. Predict the mutations' impact on the plant's phenotype.

Animations and Interactions on **CENGAGENOW**:
❯ Flower structure; Flowering plant life cycle; Life cycle of a eudicot; Development of a eudicot embryo; Monocot versus eudicot growth and development; Coleoptile transplant experiments; Gravity and statoliths; Phototropism; Circadian rhythm in plants; Phytochrome.

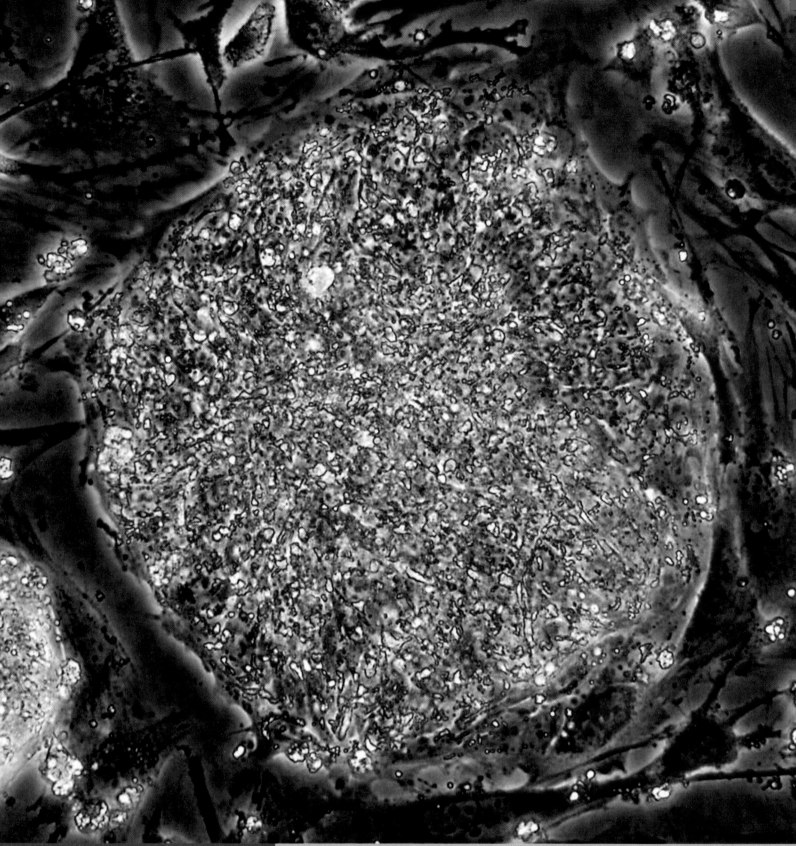

‹ Links to Earlier Concepts

The chapters in this unit focus on the structure and function of animal bodies. This chapter applies what you learned about levels of organization (Section 1.2) to animal bodies and expands on the nature of multicelled body plans (23.2). You may wish to review cell junctions (4.11), the surface-to-volume ratio (4.2), diffusion and transport proteins (5.6, 5.7), aerobic respiration (7.2), and energy conversion pathways (7.7).

Key Concepts

Organization of Animal Bodies
In most animal bodies, cells are organized as tissues, organs, and organ systems. The structure of animal bodies has been shaped by natural selection, but because evolution modifies existing structures, body plans are often less than optimal.

Animal Tissues
Epithelial tissues cover external surfaces and line cavities and tubes. Connective tissues bind, support, strengthen, protect, and insulate other tissues. Contraction of muscle tissue moves body parts. Nervous tissue provides local and long-distance lines of communication.

28 Animal Tissues and Organ Systems

28.1 Stem Cells

Manipulating stem cells may one day make it possible to grow new body parts to replace lost or diseased ones. Stem cells are self-renewing cells (Figure 28.1). A stem cell can divide and produce more stem cells ❶, or it can differentiate into one of the specialized cells that characterize specific body parts ❷. In short, all cells in your body "stem" from **stem cells**.

Embryonic stem cells form soon after fertilization, when cell divisions produce a cluster of cells. Before the cluster has implanted in the wall of the mother's womb, each cell can develop into any of the cell types found in the adult body.

By contrast, stem cells present in adults are specialists. The descendant cells they produce can differentiate into a limited variety of cells. For example, stem cells in adult bone marrow can become blood cells, but not muscle cells or nerve cells.

Cell types that your body replaces on an ongoing basis, such as skin cells and blood cells, arise continually from adult stem cells. However, adults have few stem cells capable of making muscle or nerve cells. Thus, unlike skin and blood, nerves and muscles are not replaced if they get damaged or die. This is why an injury to the nerves of the spinal cord can cause permanent paralysis. New nerves do not grow to replace the damaged ones.

In theory, embryonic stem cells could be used to provide new nerve cells. The first clinical trials to test the feasibility of such treatments are now under way. One trial involves injecting stem cells into patients who recently injured their spinal cord. Another involves using stem cells to save the vision of people with a common degenerative condition that leads to blindness. Embryonic stem cell treatments might also help people with other nerve and muscle disorders such as heart disease, muscular dystrophy, or Parkinson's disease.

Despite the promise of embryonic stem cell treatments, some people oppose them because obtaining these cells requires destruction of pre-implantation human embryos. Opponents of embryonic stem cell research would prefer that scientists focus their attention on adult stem cells.

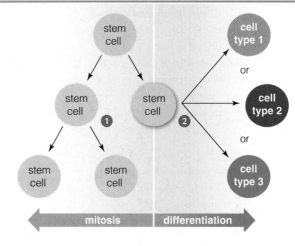

Figure 28.1 Stem cells. Each stem cell can divide to form new stem cells or differentiate to form specialized cell types.

Opposite page, a colony of human embryonic stem cells growing in the laboratory at the University of Pittsburgh.

Recently scientists have discovered that they can coax some adult cells to turn back the clock. By injecting genes or proteins into the cells, they can make them resemble embryonic stem cells. Are these modified adult cells really equivalent to embryonic stem cells? Could using them do away with the need to harvest embryonic stem cells? Perhaps, but many questions remain. We do not yet know if the treated cells can be safely transplanted into a human body, or whether they will behave the same way as embryonic cells in that context.

For now, researchers are keeping their options open. Some are developing and testing embryonic stem cell treatments, while others are looking for ways to make adult cells into a safe and effective embryonic cell substitute.

stem cell Cell that can divide to produce more stem cells or differentiate into specialized cell types.

Animal Organ Systems
Vertebrate body cavities hold many of the body's organs. Organs work cooperatively to carry out the tasks essential to survival and reproduction. All vertebrates have the same set of organ systems.

The Skin
Human skin is an example of an organ system. It has epithelial layers, connective tissue, adipose tissue, glands, blood vessels, and sensory receptors. It protects a body from injury and infection, detects external stimuli, and helps conserve water and control temperature.

Integrating Activities
Negative feedback helps the body with homeostasis. Intracellular communication makes it possible for different body parts to cooperate in tasks. Such communication often involves chemical messages that bind to membrane receptors and change a cell's behavior.

> Most animal bodies have multiple levels of organization, with cells organized as tissues, organs, and organ systems.
> Physical constraints and evolutionary history influence the structure and function of body parts.
< Links to Life's levels of organization 1.2, Cell junctions 4.11, Diffusion 5.6, Adapting to life on land 24.4

Levels of Organization

In all animals, development produces a body with cells of several to many types (Figure 28.2A). In most animals, cells of different types, and often an extracellular matrix, form tissues (Figure 28.2B). Cell junctions of the types described in Section 4.11 typically connect the cells of a tissue. They hold cells in place and allow them to cooperate in carrying out a specific task or tasks.

Four types of tissue occur in all vertebrate bodies:

1. Epithelial tissue covers body surfaces and lines the internal cavities such as the gut.
2. Connective tissue holds body parts together and provides structural support.
3. Muscle tissue moves the body or its parts.
4. Nervous tissue detects stimuli and relays information.

Different types of cells characterize different tissues. For example, contractile cells are found in muscle tissue, but not in nervous tissue or epithelial tissue.

Typically, animal tissues are organized into organs. An organ is a structural unit of two or more tissues organized in a specific way and capable of carrying out specific tasks. For example, a human heart is an organ, and it includes all four tissue types (Figure 28.2C). The wall of the heart is made up mostly of cardiac muscle tissue. A sheath of connective tissue encloses the heart, and the heart's internal chambers are lined with epithelial tissue. Nervous tissue delivers signals to and from the heart.

In organ systems, two or more organs and other components interact physically, chemically, or both in a common task. For example, the force generated by a beating heart moves blood through a system of blood vessels that extends throughout the body (Figure 28.2D). Multiple organ systems sustain the organism (Figure 28.2E).

The Internal Environment

By weight, an animal body consists mainly of water with dissolved salts, proteins, and other solutes. The bulk of this body fluid resides inside cells. The remainder is **extracellular fluid (ECF)**, the fluid that serves as the body's internal environment. ECF bathes cells and provides them with the substances they require to stay alive. It also functions as a dumping ground for cellular waste. In vertebrates, extracellular fluid consists mainly of **interstitial fluid** (the fluid in spaces between cells) and **plasma**, the fluid portion of the blood.

Cells can only survive if solute concentrations and temperature of the internal environment remain within a narrow range. Maintaining conditions of the internal environment within this range is an important aspect of the process we call homeostasis.

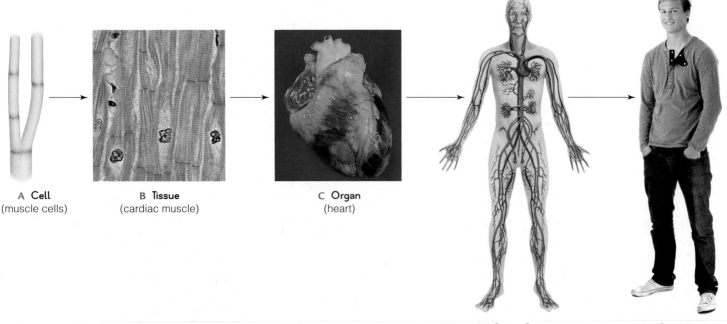

A Cell
(muscle cells)

B Tissue
(cardiac muscle)

C Organ
(heart)

D Organ System
(circulatory system)

E Organism
(human)

Figure 28.2 Levels of organization in a vertebrate (human) body.

Evolution of Animal Structure

An animal's structural traits (its anatomy) evolve in concert with its functional traits (its physiology). Both types of traits are genetically determined and vary among individuals. In each generation, those traits that best help individuals survive and reproduce in their environment are preferentially passed on. Over many generations, these traits become optimized in ways that reflect their function in a specific environment.

Physical laws constrain the evolution of body structure. For example, dissolved substances travel through extracellular fluid by diffusion. Diffusion alone could not sustain a large or thick body because gases, nutrients, and wastes would not move quickly enough through the body to keep up with cellular metabolism. Thus, mechanisms that speed the distribution of materials evolved along with increases in body size. In vertebrates, a circulatory system serves this purpose. The system includes a network of extensively branched blood vessels that extends through the body (Figure 28.3A). Every living cell is close enough to a blood vessel to exchange substances with it by diffusion.

As another example, animals evolved in water and faced new physical challenges when they moved onto land (Section 24.4). Gases can only enter or leave an animal's body by diffusing across a moist surface. In aquatic organisms, body surfaces are always moist, but on land, evaporation causes surfaces to dry out. Evolution of lungs allowed land animals to exchange gases with air across a moist surface inside their body (Figure 28.3B).

Lungs are not modified fish gills. Rather, lungs evolved from outpouchings of the gut in fishes ancestral to land vertebrates. As this example illustrates, evolution typically does not produce entirely new tissues or organs. Instead, it alters the structure and function of existing ones.

As anyone who has remodeled a home knows, modifying an existing structure, rather than designing and building a new one, requires compromises. Similarly, we detect evidence of evolutionary compromises in many animal structural traits. For example, as a legacy of the lungs' ancestral connection to the gut, the human throat connects to both the digestive tract and respiratory tract. As a result, food sometimes goes where air should and a person chokes. It would be safer if food and air entered the body through separate passageways. However, because evolution modifies existing structures, it often does not produce the most optimal body plan.

---

extracellular fluid Of a multicelled organism, body fluid outside of cells; serves as the body's internal environment.
interstitial fluid Of a multicelled organism, body fluid in spaces between cells.
plasma Fluid portion of blood.

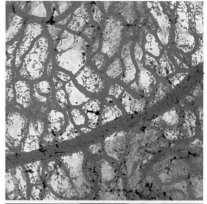

A The branching vessels of a vertebrate circulatory system distribute essential substances to cells throughout a large body.

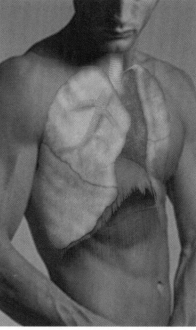

B Lungs with a moist internal surface provide a site where gases can dissolve and move into and out of the body.

Figure 28.3 Anatomical solutions to physiological challenges in the human body.

Take-Home Message How are animal bodies organized?

> In most animals, cells are organized as tissues. Each tissue consists of cells of a specific type that cooperate in carrying out a particular task. Tissues are organized into organs, which in turn are components of organ systems.

> The animal body consists largely of fluid. The bulk of this fluid is inside cells. The fluid outside cells (extracellular fluid) is the body's internal environment. Maintaining the solute concentration and temperature of this fluid is an important facet of homeostasis.

> Many anatomical traits evolved as solutions to physical challenges. However, these solutions are often imperfect because evolution modifies existing structures, rather than building a body plan from the ground up.

> Most of what you see when you look in a mirror—skin, hair, and nails—is epithelial tissue or structures derived from it. Epithelium also lines internal tubes and cavities, such as your blood vessels and gut.
> ❮ Links to Cilia 4.10, Cell junctions 4.11

General Characteristics

Epithelium (plural, epithelia), or epithelial tissue, is a sheetlike tissue consisting of cells with little extracellular material between them. One free surface is exposed to the environment or to some body fluid (Figure 28.4). At the opposite surface, epithelial cell secretions form a noncellular **basement membrane** that glues the epithelium to an underlying tissue. Blood vessels do not run through epithelium, so nutrients reach cells by diffusing from vessels in an adjacent tissue.

Tight junctions (Section 4.11) occur only in epithelial tissue. These junctions join the plasma membranes of adjacent cells so securely that fluids cannot seep between the cells. An epithelium with tight junctions can keep

free surface of a simple epithelium —

basement membrane (material secreted by epithelial cells) —

underlying connective tissue

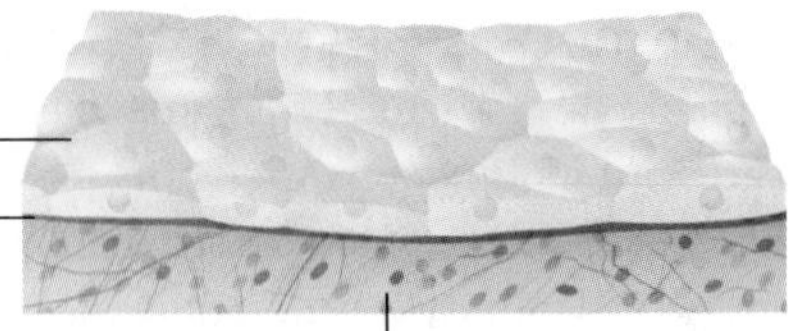

Figure 28.4 Animated Generalized structure of an epithelium. This tissue has a free surface that faces either the outside environment or some internal body fluid. A noncellular basement layer secreted by the epithelial cells attaches the epithelium to another tissue layer, most often connective tissue, which we discuss in the next section.

fluid contained within a particular body compartment. For example, tight junctions join the epithelial cells that line the gut. This epithelium serves as a barrier that prevents acid secreted into the stomach from leaking out and damaging the underlying tissue.

Epithelial tissues that are subject to mechanical stress such as those of the skin have many adhering junctions. These junctions function a bit like buttons that hold a shirt closed. They connect the plasma membranes of cells at distinct points but do not form a tight seal.

Types of Epithelium

Epithelial tissues vary in their number of cell layers and the shape of their cells. A simple epithelium is one cell thick, whereas stratified epithelium has multiple cell layers. Cells in squamous epithelium are flattened or scalelike. (*Squama* is the Latin word for scale.) Cells of cuboidal epithelium are short cylinders that look like cubes when viewed in cross-section. Cells in columnar epithelium are taller than they are wide. Figure 28.5 shows the three types of simple epithelium.

The number of layers affects the function of an epithelium. Simple squamous epithelium functions in the exchange of materials. It is the thinnest type of epithelium, and gases and nutrients diffuse across it easily. This type of epithelium lines blood vessels and the inner surface of the lungs. In contrast, thicker stratified squamous epithelium has a protective function. This type of tissue makes up the outer layer of human skin.

Cells of cuboidal and columnar epithelium act in absorption and secretion. In some tissues, such as the lining of the kidneys and the small intestine, fingerlike projections called **microvilli** extend from the free surface of epithelial cells. These projections increase the surface area across which substances can be absorbed.

In other tissues, such as those of the upper airways and oviducts, the free epithelial surface is ciliated. The action of cilia in the airways moves mucus with inhaled particles away from the lungs. The action of cilia in the oviducts propels an egg toward the uterus (the womb).

Simple squamous epithelium
- Lines blood vessels, the heart, and air sacs of lungs
- Allows substances to cross by diffusion

Simple cuboidal epithelium
- Lines kidney tubules, ducts of some glands, reproductive tract
- Functions in absorption and secretion, movement of materials

Simple columnar epithelium
- Lines some airways, parts of the gut
- Functions in absorption and secretion, protection

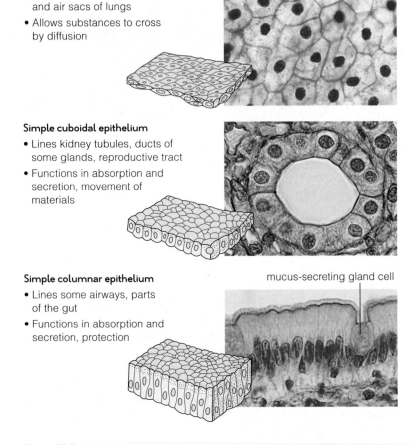

mucus-secreting gland cell

Figure 28.5 Micrographs and drawings of three types of simple epithelia, with examples of their functions and locations.

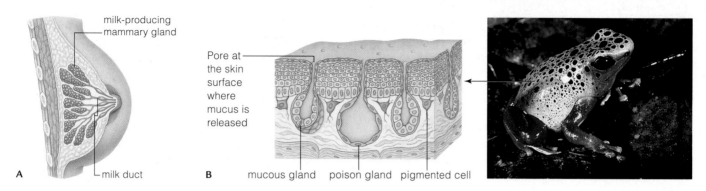

A

B   mucous gland   poison gland   pigmented cell

Figure 28.6 Glandular epithelium. **A** Breast of a lactating woman showing the milk-producing mammary glands. These glands and the milk ducts that deliver milk to the body surface are epithelial tissue.

B Glandular epithelium of a tropical frog (*Dendrobates azureus*) that secretes a paralyzing poison. The pigment-rich skin of all poisonous frogs has vivid colors and patterns that evolved as a warning signal. In essence, it says to predators, "Don't even think about it."

Only epithelial tissue contains gland cells. These cells secrete some substance that functions outside the cell. In most animals, some secretory cells cluster as multicellular glands that release substances onto the skin, or into a body cavity or fluid.

Exocrine glands have ducts or tubes that deliver their secretions onto an internal or external surface. Exocrine secretions include mucus, saliva, tears, digestive enzymes, earwax, and breast milk (Figure 28.6A). In some animals, such as poison dart frogs, exocrine ducts secrete venom that protects the animal from predators (Figure 28.6B).

Endocrine glands have no ducts. They release signaling molecules called hormones into some body fluid. Most commonly, hormones enter the bloodstream, which distributes them throughout the body.

Claws, nails, hooves, fur, hair, beaks, and feathers are all derived from epithelium. These structures form when specialized epithelial cells produce large amounts of the protein keratin. The visible part of a hoof, hair, or feather consists of the remains of such cells (Figure 28.7).

Figure 28.7 This bird's feathers and the horny covering of its beak are keratin-rich structures produced by an epithelium. Hair, fur, claws, and hooves are also derived from epithelium.

Carcinomas—Epithelial Cell Cancers

Adult animals make few new muscle cells or nerve cells, but they constantly renew their epithelial cells. For example, each day you lose skin cells and grow new ones to replace them. An adult sheds about 0.7 kilogram (1.5 pounds) of skin each year. Similarly, the lining of your intestine is replaced every four to six days. All those cell divisions provide lots of opportunities for DNA replication errors that can lead to cancer. As a result, epithelium is the animal tissue most likely to become cancerous.

An epithelial cell cancer is called a carcinoma. About 95 percent of skin cancers are carcinomas. Breast cancers are usually carcinomas of epithelial cells that line the milk ducts, or of the breast's glandular epithelium. Most lung cancers arise in the lung's epithelial lining.

basement membrane Secreted layer that attaches an epithelium to an underlying tissue.
endocrine gland Ductless gland that secretes hormones into a body fluid.
epithelial tissue Sheetlike animal tissue that covers outer body surfaces and lines internal tubes and cavities.
exocrine gland Gland that secretes milk, sweat, saliva, or some other substance through a duct.
microvilli Thin projections from the plasma membrane; increase the surface area of some epithelial cells.

Take-Home Message **What are the functions of epithelial tissue?**

❯ Epithelia are sheetlike tissues that line the body's surface and its cavities, ducts, and tubes. They function in protection, absorption, and secretion. Some epithelia have cilia or microvilli at their surface.

❯ Glands are secretory organs derived from epithelium. Exocrine glands secrete material through a duct onto a body surface or into a body cavity. Endocrine glands secrete hormones into the blood.

❯ Specialized epithelial cells that produce large amounts of the protein keratin are the source of hair, nails, hooves, and feathers.

❯ Epithelial tissues undergo continual turnover and are the most frequent site for cancers.

❭ Connective tissues have "connecting" roles in the body. They structurally or functionally support, bind, separate, and in one case insulate other tissues. They are the body's most abundant and widely distributed tissues.

❰ Links to Fat synthesis 7.7, Extracellular matrix 4.11

Connective tissues consist of cells scattered within an extracellular matrix of their own secretions. Connective tissues are described by the types of cells that they include and the composition of their extracellular matrix. There are two kinds of soft connective tissues—loose and dense. Cartilage, bone tissue, adipose tissue, and blood are specialized connective tissues.

Soft Connective Tissues

Loose and dense connective tissues have the same components but in different proportions and arrangements. In both tissues, the most abundant cells are fibroblasts, cells that secrete complex carbohydrates and fibers of the structural proteins collagen and elastin.

The most common type of connective tissue in the vertebrate body is **loose connective tissue**. It holds organs and epithelia in place, and its fibroblasts and fibers are dispersed widely through the matrix (Figure 28.8A).

In **dense, irregular connective tissue**, the matrix is packed full of fibroblasts and collagen fibers that are oriented every which way, as in Figure 28.8B. Dense, irregular connective tissue makes up deep skin layers. It supports intestinal muscles and also forms capsules around organs that do not stretch, such as kidneys.

Dense, regular connective tissue has fibroblasts in orderly rows between parallel, tightly packed bundles of fibers (Figure 28.8C). This organization helps keep the tissue from being torn apart when placed under mechanical stress. Tendons and ligaments are mainly dense, regular connective tissue. Tendons connect skeletal muscle to bones. Ligaments attach one bone to another and are stretchier than tendons. Elastic fibers in the ligament matrix facilitate movements around joints.

Specialized Connective Tissues

Cartilage has a matrix of collagen fibers and rubbery, compression-resistant glycoproteins. Cells secrete the matrix, which imprisons them (Figure 28.8D). Sharks have a cartilage skeleton. In human embryos, cartilage forms a model for the developing skeleton, then bone replaces most of it. After birth, cartilage still supports the outer ears, nose, and throat. It cushions joints and is a shock absorber between vertebrae. Blood vessels do not extend through cartilage, as they do in other connective tissues. As a result, nutrients and oxygen must diffuse from blood vessels in nearby tissues. Cartilage cells do not divide often in adults. Therefore, injured cartilage does not repair itself, as other connective tissues do.

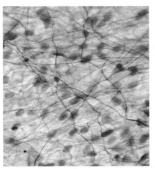
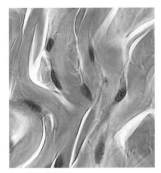
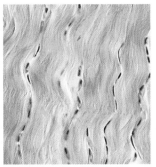
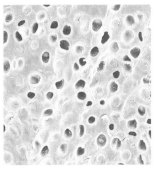

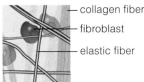
— collagen fiber
— fibroblast
— elastic fiber

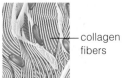
— collagen fibers

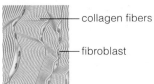
— collagen fibers
— fibroblast

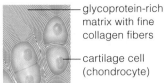
— glycoprotein-rich matrix with fine collagen fibers
— cartilage cell (chondrocyte)

A Loose connective tissue Fibroblasts and other cells scattered in a squishy matrix with relatively few fibers

Common Locations Beneath skin and most epithelia

Functions Elasticity, diffusion

B Dense, irregular connective tissue Fibroblasts in a semisolid matrix with many loosely interwoven collagen fibers

Common Locations In skin and in capsules around some organs

Function Structural support

C Dense, regular connective tissue Fibroblasts in rows between tight parallel bundles of many collagen fibers

Common Locations Tendons, ligaments

Functions Strength, elasticity

D Cartilage Chondrocytes and collagen fibers in a rubbery matrix

Common Locations Nose, ends of long bones, airways, skeleton of cartilaginous fish, vertebrate embryo

Functions Support, protection, low-friction surface for joint movements

Figure 28.8 Animated Connective tissues.

Adipose tissue is the body's main energy reservoir. Most cells can convert excess sugars and lipids into droplets of fat (Section 7.7). However, only the cells of adipose tissue bulge with so much stored fat that the nucleus gets pushed to one side and flattened (Figure 28.8E). Adipose cells have little matrix between them. Small blood vessels that run through the tissue carry fats to and from cells. In addition to its energy-storage role, adipose tissue cushions and protects body parts, and a layer of adipose tissue under the skin functions as insulation that helps keep the body's internal temperature within an optimal range.

Bone tissue is a connective tissue with living cells imprisoned in their own calcium-hardened secretions

(Figure 28.8F). This is the main tissue of bones, the organs that interact with muscles to move a body. Bones also support and protect soft internal organs. Blood cells form inside the spongy interior of some bones.

Blood is considered a connective tissue because its cellular components (red blood cells, white blood cells, and platelets) are descended from stem cells in bone (Figure 28.8G). Red blood cells filled with hemoglobin transport oxygen. White blood cells defend the body against pathogens. Platelets function in clot formation. The cellular components of blood drift along in the plasma, a fluid extracellular matrix consisting mostly of water, with dissolved proteins, nutrients, gases, and other substances.

adipose tissue Connective tissue specializing in fat storage.
blood Fluid connective tissue consisting of plasma and cells that form inside bones.
bone tissue Connective tissue with cells surrounded by a mineral-hardened matrix of their own secretions.
cartilage Connective tissue with cells surrounded by a rubbery matrix of their own secretions.
connective tissue Animal tissue with an extensive extracellular matrix; provides structural and functional support.
dense, irregular connective tissue Connective tissue with asymmetrically arranged fibers and scattered fibroblasts.
dense, regular connective tissue Connective tissue with fibroblasts arrayed between parallel arrangements of fibers.
loose connective tissue Connective tissue with relatively few fibroblasts and fibers scattered in its matrix.

Take-Home Message What are connective tissues?

❯ Connective tissues support, protect, organize, and insulate other tissues. They consist of cells within an extracellular matrix. Except for blood, each contains fibroblasts.

❯ The matrix of soft connective tissues contains characteristic proportions and arrangements of fibroblasts and fibers.

❯ Cartilage, bone, adipose tissue, and blood are specialized connective tissues. Cartilage and bone are both structural materials. Adipose tissue is a reservoir of stored energy. Blood, a fluid connective tissue, transports materials.

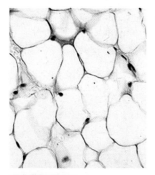

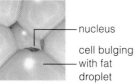
nucleus
cell bulging with fat droplet

E Adipose tissue Large, tightly packed fat cells with little extracellular matrix

Common Locations Under skin, around the heart and the kidneys

Functions Energy storage, insulation, padding

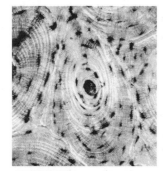

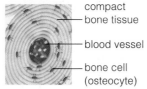
compact bone tissue
blood vessel
bone cell (osteocyte)

F Bone tissue Collagen fibers, osteocytes in chambers inside an extensive, calcium-hardened extracellular matrix

Location All bony vertebrate skeletons

Functions Movement, support, protection

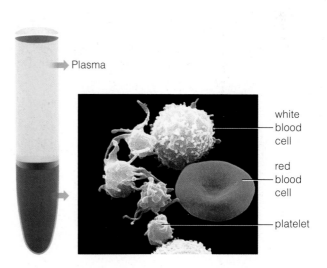
Plasma
white blood cell
red blood cell
platelet

G Blood Protein-rich fluid plasma with cellular components

Location Inside blood vessels

Functions Distributes essential gases, nutrients; removes waste

> Vertebrates have three types of muscle tissue: skeletal muscle, cardiac muscle, and smooth muscle. Each type has unique properties that reflect its functions.
< Link to Gap junctions 4.11

In muscle tissues, cells contract (shorten) in response to stimulation, then they relax and passively lengthen. Coordinated contractions of layers or rings of muscles move the body or move material through the body. Muscle tissue occurs in most animals, but we focus here on the arrangements of muscle found in vertebrates.

Skeletal Muscle

Skeletal muscle tissue, the functional partner of bone (or cartilage), helps move and maintain the positions of the body and its parts. Skeletal muscle tissue has parallel arrays of long, cylindrical muscle fibers (Figure 28.9A). The fibers are not single cells. They form during embryonic development when groups of cells fuse together. Each fiber contains multiple nuclei between long strands with row after row of contractile units. These rows give skeletal muscle a striated, or striped, appearance.

Skeletal muscle tissue makes up 40 percent or so of the weight of an average human. Reflexes activate it, but we can also make it contract when we want to. Thus skeletal muscles are commonly called "voluntary" muscles.

Cardiac Muscle

Cardiac muscle tissue is found only in the heart wall (Figure 28.9B). Like skeletal muscle, it appears striated. Unlike skeletal muscle tissue, it has branch-shaped cells. Cardiac muscle cells abut at their ends, where adhering junctions prevent them from being ripped apart during forceful contractions. Signals to contract pass swiftly from cell to cell at gap junctions along their length. The signals make all cells in cardiac muscle tissue contract as a unit.

Cardiac muscle and smooth muscle are said to be "involuntary" muscle, because most people cannot make these types of muscle contract at will.

Smooth Muscle

We find layers of **smooth muscle tissue** in the wall of some blood vessels and soft internal organs, such as the stomach, uterus, and bladder. This tissue's unbranched cells contain a nucleus at their center and are tapered at both ends (Figure 28.9C). Contractile units are not arranged in an orderly repeating fashion, so smooth muscle tissue does not appear striated. Smooth muscle contracts more slowly than skeletal muscle, but its contractions can be sustained longer. Smooth muscle contractions propel material through the gut, adjust the diameter of some blood vessels, and close sphincters. A sphincter is a ring of muscle in a tubular organ.

---

cardiac muscle tissue Muscle of the heart wall.
skeletal muscle tissue Muscle that interacts with bone to move body parts; under voluntary control.
smooth muscle tissue Muscle that lines blood vessels and forms the wall of hollow organs.

Take-Home Message **What is muscle tissue?**

> Muscle tissue consists of cells that contract when stimulated.

> Skeletal muscle attaches to and pulls on bones. Cardiac muscle is found only in the heart wall. Blood vessels and many soft internal organs have smooth muscle in their walls.

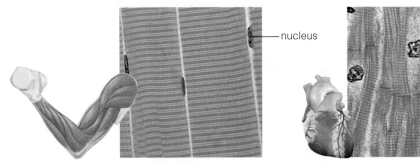

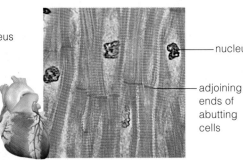

nucleus

nucleus

adjoining ends of abutting cells

A Skeletal muscle

• Long, multinucleated, cylindrical cells with conspicuous striping (striations)

• Interacts with bone to bring about movement, maintain posture

• Reflex activated, but also under voluntary control

B Cardiac muscle

• Striated cells attached end to end, each with a single nucleus

• Found only in the heart wall

• Contraction is not under voluntary control

C Smooth muscle

• Cells with a single nucleus, tapered ends, and no striations

• Found in the walls of some blood vessels, the digestive tract, the reproductive tract, the bladder, and other hollow organs

• Contraction is not under voluntary control

Figure 28.9 Animated Muscle tissues.

> Nervous tissue detects changes in the internal or external environment, integrates information, and controls the activity of muscle and glands.

Nervous tissue consists of specialized signaling cells called **neurons**, and the cells that support them. A neuron has a cell body with a nucleus and other organelles (Figure 28.10). Projecting from the cell body are long cytoplasmic extensions that allow the cell to receive and send electrochemical signals.

When a neuron receives sufficient stimulation, an electrical signal travels along its plasma membrane to the ends of certain cytoplasmic extensions. Here, the electrical signal causes release of chemical signaling molecules. These molecules diffuse across a small gap to an adjacent neuron, muscle fiber, or gland cell, and alter that cell's behavior.

Your nervous system has more than 100 billion neurons. There are three types. Sensory neurons are excited by specific stimuli, such as light or pressure. Interneurons receive and integrate sensory information. They store information and coordinate responses to stimuli. In vertebrates, interneurons occur mainly in the brain and spinal cord. Motor neurons relay commands from the brain and spinal cord to glands and muscle cells (Figure 28.11).

Neuroglial cells, also called neuroglia, keep neurons positioned where they should be, and provide metabolic support. They constitute a significant portion of nervous tissue. More than half of your brain volume is neuroglia. Neuroglial cells also wrap around the signal-sending cytoplasmic extensions of most motor neurons. They act as insulation and speed the rate at which signals travel.

---

nervous tissue Animal tissue composed of neurons and supporting cells; detects stimuli and controls responses to them.
neuron One of the cells that make up communication lines of nervous systems; transmits electrical signals along its plasma membrane and communicates with other cells through chemical messages.

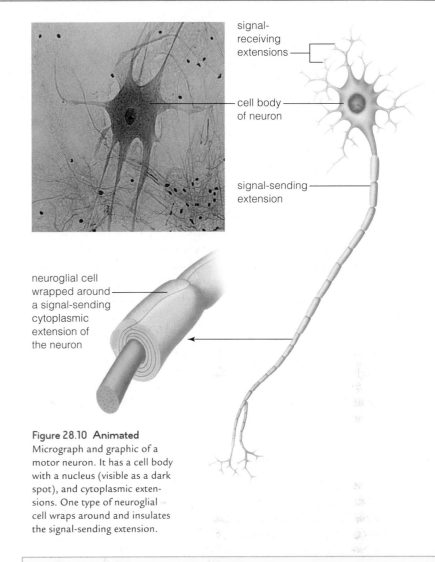

signal-receiving extensions

cell body of neuron

signal-sending extension

neuroglial cell wrapped around a signal-sending cytoplasmic extension of the neuron

Figure 28.10 Animated
Micrograph and graphic of a motor neuron. It has a cell body with a nucleus (visible as a dark spot), and cytoplasmic extensions. One type of neuroglial cell wraps around and insulates the signal-sending extension.

Take-Home Message **What is nervous tissue?**

> Nervous tissue consists of neurons and the cells that support them. Different kinds of neurons detect specific stimuli, integrate information, and issue or relay commands to other tissues.

> The supporting cells in nervous tissue are collectively referred to as neuroglial cells or neuroglia. They make up much of the volume of nervous tissue.

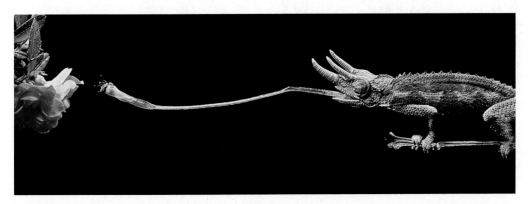

Figure 28.11 An example of the coordinated interaction between skeletal muscle tissue and nervous tissue.

Interneurons in the brain of this lizard, a chameleon, calculate the distance and the direction of a tasty fly.

In response to this stimulus, signals from the interneurons flow along certain motor neurons and reach muscle fibers inside the lizard's long, coiled-up tongue. The tongue uncoils swiftly and precisely to reach the very spot where the fly is perched.

> Organs typically include all four types of tissues and are components of an organ system.
< Link to Animal body plans 23.2

Organs in Body Cavities

Many human organs are located inside body cavities. Like other vertebrates, humans are bilateral and have a lined body cavity known as a coelom (Section 23.2). A sheet of smooth muscle, the diaphragm, divides the human coelom into an upper thoracic cavity and a lower cavity that has abdominal and pelvic regions (Figure 28.12). The heart and lungs are in the thoracic cavity. Digestive organs such as the stomach, intestines, and liver lie in the abdominal cavity. The bladder and reproductive organs are in the pelvic cavity. A cranial cavity in the head holds the brain, and a spinal cavity in the back holds the spinal cord. These last two cavities are not derived from the coelom.

Vertebrate Organ Systems

Figure 28.13 introduces the eleven organ systems of the human body. Other vertebrates have the same systems, which carry out the same functions. Collectively, organ systems of the vertebrate body have a division of labor, or a compartmentalization of functions.

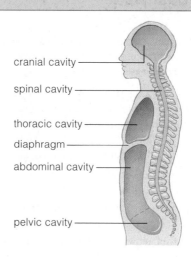

cranial cavity

spinal cavity

thoracic cavity

diaphragm

abdominal cavity

pelvic cavity

Figure 28.12 Main body cavities that hold human organs.

>> **Figure It Out** Which organs lie in body cavities that are not part of the coelom?   Answer: The spinal cord and brain

Organ systems work cooperatively to carry out tasks. For example, several organ systems interact to provide cells with essential raw materials and remove wastes (Figure 28.14). Food and water enter the body by way of the digestive system, which includes digestive organs such as the stomach and intestine, as well as organs that aid

Figure 28.13 Animated *Below*, human organ systems and their functions.

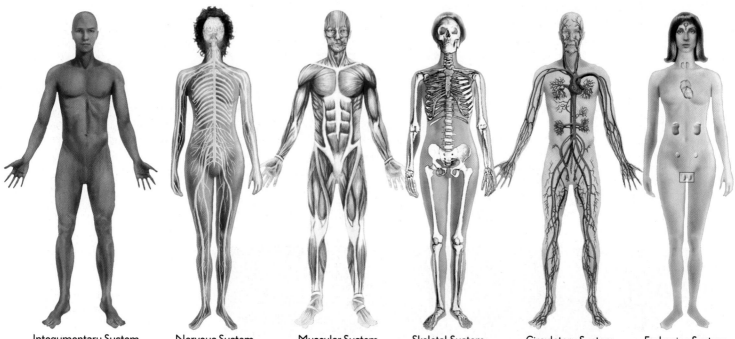

Integumentary System

Protects body from injury, dehydration, and some pathogens; controls its temperature; excretes certain wastes; receives some external stimuli.

Nervous System

Detects external and internal stimuli; controls and coordinates the responses to stimuli; integrates all organ system activities.

Muscular System

Moves body and its internal parts; maintains posture; generates heat by increases in metabolic activity.

Skeletal System

Supports and protects body parts; provides muscle attachment sites; produces red blood cells; stores calcium, phosphorus.

Circulatory System

Rapidly transports many materials to and from interstitial fluid and cells; helps stabilize internal pH and temperature.

Endocrine System

Hormonally controls body functioning; with nervous system integrates short- and long-term activities. (Male testes added.)

digestion such as the pancreas and gallbladder. The digestive system also eliminates undigested waste.

Oxygen enters the body by way of the respiratory system, which includes lungs and airways that lead to them. The heart and blood vessels of the circulatory system deliver nutrients and oxygen to cells, and remove waste carbon dioxide and solutes from them. The circulatory system delivers carbon dioxide to the respiratory system, which eliminates it. The circulatory system also moves excess water, salts, and soluble wastes to the urinary system. Organs of the urinary system include kidneys that filter wastes from the blood and produce urine, and a bladder that stores the urine until it can be eliminated from the body.

Figure 28.14 does not include the nervous, endocrine, muscular, and skeletal systems, but these too help vertebrates obtain essential substances and eliminate wastes. For example, the nervous system detects changes in internal levels of water, solutes, and nutrients. Signals from the nervous and endocrine systems to the kidneys encourage conservation or elimination of water. Signals from the nervous system also result in movement, as when your skeletal muscles interact with bones to move you toward a source of food or water.

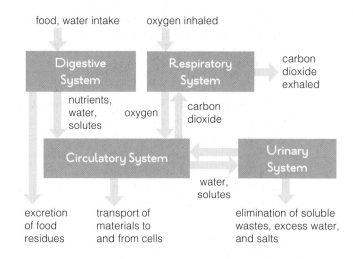

Figure 28.14 Some of the ways that organ systems interact to keep the body supplied with essential substances and eliminate unwanted wastes. Other organ systems that are not shown also take part in these tasks.

Take-Home Message What are organs and organ systems?

> Organs consist of multiple tissues and are themselves components of organ systems. Cooperative action of organ systems sustains the body.

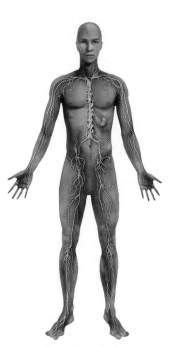

Lymphatic System

Collects and returns some tissue fluid to the bloodstream; defends the body against infection and tissue damage.

Respiratory System

Rapidly delivers oxygen to the tissue fluid that bathes all living cells; removes carbon dioxide wastes of cells; helps regulate pH.

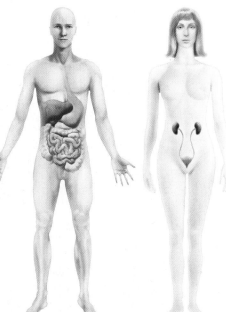

Digestive System

Ingests food and water; mechanically, chemically breaks down food and absorbs small molecules into internal environment; eliminates food residues.

Urinary System

Maintains the volume and composition of internal environment; excretes excess fluid and bloodborne wastes.

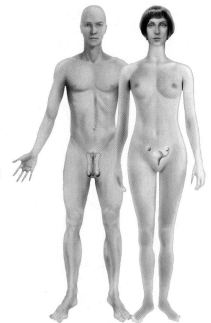

Reproductive System

Female: Produces eggs; after fertilization, affords a protected, nutritive environment for the development of new individuals. Male: Produces and transfers sperm to the female. Hormones of both systems also influence other organ systems.

> In vertebrates, the integumentary system consists of skin, structures derived from skin, and an underlying layer of connective and adipose tissue.
❮ Link to Adaptations to life on land 24.4

Of all vertebrate organs, the outer body covering called skin has the largest surface area. Skin consists of two layers, a thin upper epidermis and the dermis beneath it (Figure 28.15). The dermis connects to the hypodermis, a layer of connective and adipose tissue.

Vertebrate skin has many functions. It contains sensory receptors that keep the brain informed of external conditions. Skin serves as a barrier to keep out pathogens and it helps control internal temperature. In land vertebrates, skin also helps conserve water. In humans, reactions that take place in the skin produce vitamin D.

Structure of the Skin

Epidermis is a stratified squamous epithelium with an abundance of adhering junctions and no extracellular matrix. Human epidermis consists mainly of keratinocytes, cells that make the waterproof protein keratin.

Mitotic cell divisions in deep epidermal layers continually produce new keratinocytes that displace older cells upward toward the skin's surface. As cells move upward, they become flattened, lose their nucleus, and die. Dead cells at the skin surface form an abrasion-resistant layer that helps prevent water loss.

Melanocytes, another type of epidermal cell, make pigments called melanins and donate them to keratinocytes.

Variations in skin color arise from differences in the distribution and activity of melanocytes, and in the type of melanin they produce. One melanin is brown to black. Another is red to yellow. The effect of melanocytes can be seen with vitiligo, a skin disorder in which destruction of these cells results in light patches of skin (Figure 28.16).

Dermis consists primarily of dense connective tissue with stretch-resistant elastin fibers and supportive collagen fibers. Blood vessels, lymph vessels, and sensory receptors weave through the dermis. Dermis is much thicker than epidermis, and more resistant to tearing. Leather is animal dermis that has been treated with chemicals to preserve it.

Sweat glands consist of epidermal cells that migrated into the dermis during early development. The sweat they secrete is mostly water. Evaporation of sweat can help cool the body surface when the temperature is high.

Epithelial tissue embedded in the dermis also forms hair follicles. The base of a hair follicle holds living hair cells. They divide every 24 to 72 hours, making them among the fastest-dividing cells in the body. As the cells divide, they push cells above them up, lengthening the hair. The part of a hair that extends beyond the skin surface is keratin-rich remains of dead cells. A smooth muscle attaches to each hair. Hair stands upright when this muscle reflexively contracts in response to cold or fright.

Secretions from an oil gland next to each hair follicle keep the hair and the skin surrounding hair soft and silky. Oil glands also consist of epithelial tissue that migrated into the dermis during development.

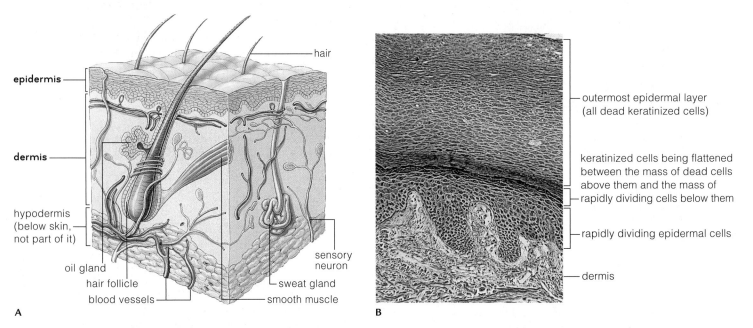

A

epidermis

dermis

hypodermis (below skin, not part of it)

oil gland
hair follicle
blood vessels
smooth muscle
sweat gland
sensory neuron
hair

B

outermost epidermal layer (all dead keratinized cells)

keratinized cells being flattened between the mass of dead cells above them and the mass of rapidly dividing cells below them

rapidly dividing epidermal cells

dermis

Figure 28.15 Animated A Skin structure. Its components, such as glands, differ from one body region to the next. **B** Section through human skin.

Sun and the Skin

The melanin produced by skin functions as a sunscreen, absorbing ultraviolet (UV) radiation that could otherwise damage underlying skin layers. When a patch of skin is exposed to sunlight, melanocytes in that region make more of the brownish-black melanin, resulting in a "tan."

A bit of UV exposure is a good thing; it stimulates production of a molecule that the body later converts to vitamin D. We need vitamin D to absorb calcium ions from food. However, UV exposure also causes the breakdown of folate, one of the B vitamins. Among other problems, a deficiency in folate during development damages the nervous system.

Variations in skin color among human populations probably evolved as adaptations to differences in sunlight exposure. Humans arose in equatorial Africa, where the sun's rays are intense and daylength does not decline dramatically in the winter. In this environment, melanin-rich skin protected folate and still made enough vitamin D. Later, some humans moved to more northerly regions, where sunlight is less intense, winter days are short, and more time is spent indoors or bundled in clothing. Under these circumstances, skin with fewer melanocytes is advantageous. Such skin makes it easier to get enough sunlight to encourage adequate vitamin D production even during cold, dark winters.

People with light skin often get a sunburn before they tan. Dark skin protects people better than light skin. But in anyone, prolonged or repeated UV exposure damages collagen and causes elastin fibers to clump. Chronically tanned skin gets less resilient and starts to look like shoe leather. UV harms DNA, and the damage increases the risk of skin cancer. Melanoma, the most dangerous skin cancer, arises when melanocytes divide uncontrollably.

Effects of Age

As people age, their epidermal cells divide less frequently. Glandular secretions that kept the skin soft and supple dwindle. Thickness and elasticity of the dermis decline as collagen and elastin fibers become sparse. Permanent wrinkles appear in places where facial expressions once produced temporary creases.

As noted above, excessive tanning accelerates skin aging. Smoking has a similar effect. It lessens the flow of blood to the skin, depriving it of oxygen and nutrients. Damaging effects of sun are localized to sun-exposed regions, but smoking harms skin throughout the body.

Figure 28.16 Vitiligo. Lee Thomas, an African American television reporter, has vitiligo. Death of melanocytes has turned his hands white and produced white blotches on his face and arms.

Farming Skin

Skin is the only organ that is already grown artificially for widespread medical uses. Cultured skin substitutes are made using infant foreskins that were removed during routine circumcisions. The foreskin (a tissue that covers the tip of the penis) provides a rich source of keratinocytes and fibroblasts. These cells are grown in culture with other biological materials, and the resulting products are used to close wounds, help burns heal, and cover sores on patients with skin disorders.

Unlike natural skin, cultured skin substitutes do not include melanocytes, sweat glands, oil glands, and other differentiated structures. Use of adult epidermal stem cells may one day allow production of cultured skin as complex as real skin.

Researchers also have ambitious hopes for making stem cells from adult skin. Manipulating skin cells so they dedifferentiate could provide starting material to replace other types of tissues, without the controversy raised by use of embryonic stem cells.

---

dermis Deep layer of skin that consists of connective tissue with nerves and blood vessels running through it.
epidermis Outermost tissue layer; in animals, the epithelial layer of skin.

Take-Home Message **What is skin and what are its functions?**

❯ Skin is a component of the integumentary system. It consists of an epidermis of epithelial tissue, and a dermis of dense connective tissue with blood vessels and nerves running through it.

❯ Skin has sensory receptors that inform the brain about the environment. It also serves as a barrier against pathogens, aids in production of vitamin D, and functions in temperature regulation.

❭ A negative feedback system involving multiple organ systems allows the body to maintain its internal temperature.
❭ Communication between cells is essential to homeostasis.
❬ Link to Metabolic heat 7.5

Detecting and Responding to Change

Homeostasis, again, is the process of keeping conditions in the body within limits. In vertebrates, homeostasis involves interactions among sensory receptors, the brain, and muscles and glands. A **sensory receptor** is a cell or cell component that detects a specific stimulus. Sensory receptors involved in homeostasis function like internal watchmen; they monitor the body for changes. Information from sensory receptors throughout the body flows to an integrator. In vertebrates, the brain usually serves as the integrator. It evaluates the incoming information, then signals effectors—muscles and glands—to take any necessary actions to keep the body functioning properly.

Negative Feedback Control of Body Temperature

Homeostasis often involves **negative feedback**, a process in which a change causes a response that reverses the change. A familiar nonbiological example of negative feedback is the way a furnace with a thermostat works. A person sets the thermostat to a desired temperature. When the temperature falls below this preset point, the furnace turns on and emits heat. When the temperature

rises to the desired level, the thermostat turns off the furnace. Similarly, a negative feedback mechanism keeps your internal temperature near 37°C (98.6°F).

Think about what happens when you exercise on a hot day. Muscle activity generates heat, and the body's internal temperature rises (Figure 28.17). Sensory receptors in the skin detect the increase and signal the brain. The brain sends signals that bring about the body's response. Blood flow shifts, so more blood from the body's hot interior flows to the skin. The shift maximizes the amount of heat given off to the surrounding air. At the same time, sweat glands in the skin increase their output. Evaporation of sweat helps cool the body surface. You breathe faster and deeper, speeding the transfer of heat from the blood in your lungs to the air. Hormonal changes make you feel more sluggish. As your activity level slows and your rate of heat loss increases, your temperature falls.

Receptors in the skin also notify the brain when the external environment becomes chilly. The brain then sends out signals that cause diversion of blood flow away from the skin, lessening movement of heat to the body surface, where it would be lost to the surrounding air. At the same time, reflex contractions of smooth muscle in the skin pull on hair so that it stands up, causing "goose bumps." Erect hair insulates the body better than hair that lies flat. With prolonged cold, the brain commands skeletal muscles to contract ten to twenty times a second. This shivering response dramatically increases heat production by muscles.

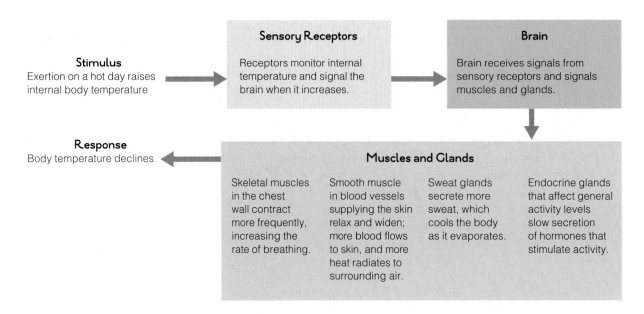

Figure 28.17 Animated Negative feedback mechanism that reduces body temperature when it rises.

Intercellular Communication

For a body to function as an integrated whole, cells must communicate with their neighbors and often with cells some distance away. Gap junctions allow substances to pass quickly between adjoining cells. Communication among more distant cells involves special molecular signals. Some of these signals diffuse through the interstitial fluid to nearby cells. Others such as hormones travel long distances through blood vessels.

Intracellular signaling typically involves three steps. First, a signaling molecule reversibly binds to a receptor. The receptor is often membrane proteins. Second, the signal is transduced, meaning it is converted to a form that acts inside the signal-receiving cell. Third, the cell responds to the signal.

For example, **apoptosis**, a process of cellular suicide, starts when molecular signals bind to receptors at the cell surface (Figure 28.18). Binding sets in motion a chain of reactions that result in the activation of self-destructive enzymes. Some of these enzymes chop up structural proteins such as cytoskeleton proteins and histones that organize DNA. Other enzymes snip apart nucleic acids. As a result of these activities, the cell dies.

During development, apoptosis helps sculpt body parts. For example, cells committed suicide as your hands were developing. An embryonic hand starts out as a paddlelike structure, then apoptosis of cells divides the paddle into individual fingers (Figure 28.19).

---

apoptosis Mechanism of cell suicide.
negative feedback A change causes a response that reverses the change; important mechanism of homeostasis.
sensory receptor Cell or cell component that detects a specific stimulus.

Stem Cells (revisited)

In vitro fertilization, the process of uniting egg and sperm outside the body, is a common practice in fertility clinics. After fertilization, a few cell divisions produce a cell cluster smaller than a grain of sand. The cluster is implanted in a woman's uterus or frozen for later use. An estimated 500,000 such pre-implantation embryos are now frozen. Many will never be implanted in their mother. They are a potential source of stem cells, or a potential child—if another woman is willing to carry the embryo to term.

How Would You Vote? Should scientists be allowed to start new embryonic stem cell lines from early human cell clusters produced in fertility clinics but not used? See CengageNow for details, then vote online (cengagenow.com).

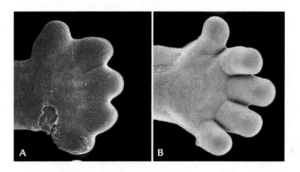

Figure 28.19 Animated How human fingers form. **A** Forty-eight days after fertilization, tissue webs connect embryonic digits. **B** Three days later, after apoptosis by cells making up the tissue webs, the digits are separated.

Take-Home Message **How does the body maintain homeostasis and function as an integrated whole?**

❯ Negative feedback prevents conditions within the body from deviating too far from optimal conditions.

❯ Cells within the body communicate with one another by way of gap junctions and production of short-range and long-range signaling molecules.

Signal to die docks at receptor.

Signal leads to activation of protein-destroying enzymes.

Figure 28.18 Artist's depiction of apoptosis, with enzymes shown as knife-like blades.

Summary

Section 28.1 A **stem cell** can divide to form more stem cells or differentiate as a specialized cell. Embryonic stem cells can produce any cell type in the body. After birth, stem cells are less versatile; they produce fewer cell types. Researchers are testing ways to use stem cells to make tissues that normally do not regenerate.

Section 28.2 Most animals have four types of tissues organized as organs and organ systems. **Extracellular fluid** serves as the body's internal environment. In humans, it consists mainly of **interstitial fluid** and **plasma**. Animal structure is influenced by physical constraints and evolutionary history.

Section 28.3 **Epithelium** has one free surface. The other surface secretes a **basement membrane**. Some epithelial cells have cilia or **microvilli** at their free surface. Glands are epithelial tissue. **Endocrine glands** are ductless and secrete hormones into blood. **Exocrine glands** secrete products such as milk or digestive enzymes through ducts.

Section 28.4 **Connective tissues** "connect" tissues to one another, both functionally and structurally. Different types bind, organize, support, strengthen, protect, and insulate other tissues. All contain cells scattered within an extracellular matrix of their own secretions.

Loose connective tissue; **dense, regular connective tissue**; and **dense, irregular connective tissue** have the same components but differ in the proportions. They are classified as soft connective tissues. **Cartilage, bone tissue, adipose tissue,** and **blood** are classified as specialized connective tissues.

Section 28.5 Muscle tissues contract (shorten) when stimulated. They help move the body and its component parts. The three types are **skeletal muscle**, **cardiac muscle**, and **smooth muscle tissue**. Skeletal muscle and cardiac muscle tissues appear striated. Only skeletal muscle is under voluntary control.

Section 28.6 **Neurons** in **nervous tissue** make up communication lines through the body. Different kinds detect, integrate, and assess information about internal and external conditions, and deliver commands to muscles and glands that carry out responses. Nervous tissue also contains neuroglial cells, or neuroglia. These cells protect and support the neurons.

Section 28.7 An organ system consists of two or more organs that interact chemically, physically, or both in tasks that help keep individual cells as well as the whole body functioning. All vertebrates have the same set of organ systems. Many internal organs reside inside a body cavity derived from the coelom. Organ systems interact to provide cells with the materials that they need and rid the body of wastes.

Section 28.8 The integumentary system functions in protection, temperature control, detection of shifts in external conditions, vitamin D production, and defense. It consists of skin (outer **epidermis** and deeper **dermis**), structures derived from skin (such as hair), and the underlying connective and adipose tissues.

Section 28.9 Homeostasis requires **sensory receptors** that detect changes, an integrating center (the brain), and effectors (muscles and glands) that bring about responses. **Negative feedback** often plays a role in homeostasis: A change causes the body to respond in a way that reverses the change. Intercellular signals allow a body to behave in an integrated manner, as when signals cause **apoptosis**, or progammed cell death.

Self-Quiz   Answers in Appendix III

1. _____ tissues are sheetlike with one free surface.
   a. Epithelial   c. Nervous
   b. Connective   d. Muscle

2. _____ form a waterproof seal between cells.
   a. Tight junctions   c. Gap junctions
   b. Adhering junctions   d. all of the above

3. Glands are derived from _____ tissue.
   a. epithelial   c. muscle
   b. connective   d. nervous

4. Most _____ have many collagen and elastin fibers.
   a. epithelial tissues   c. muscle tissues
   b. connective tissues   d. nervous tissues

5. _____ is mostly plasma.
   a. Adipose tissue   c. Cartilage
   b. Blood   d. Bone

6. Your body converts excess carbohydrates and proteins to fats. _____ specializes in storing the fats.
   a. Epithelial tissue   c. Adipose tissue
   b. Dense connective tissue   d. both b and c

7. Cells of _____ can shorten (contract).
   a. epithelial tissue   c. muscle tissue
   b. connective tissue   d. nervous tissue

8. _____ muscle tissue has a striated appearance and is under voluntary control.
   a. Skeletal   c. Cardiac
   b. Smooth   d. a and c

9. _____ detects and integrates information about changes and controls responses to those changes.
   a. Epithelial tissue   c. Muscle tissue
   b. Connective tissue   d. Nervous tissue

Data Analysis Activities

Growing Skin to Heal Wounds Diabetes is a disorder in which the blood sugar level is not properly controlled. Among other complications, this disorder reduces blood flow to the lower legs and feet. As a result, about 3 million diabetes patients have ulcers—open wounds that do not heal—on their feet. Each year, about 80,000 diabetics require amputations.

Fibroblasts and keratinocytes can be grown to produce a cultured skin substitute that is placed over wounds to help them heal. Figure 28.20 shows the results of a clinical experiment that tested the effect of the cultured skin product versus standard treatment for diabetic foot wounds. Patients were randomly assigned either to the experimental treatment group or to the control group. Their progress was monitored for 12 weeks.

1. What percentage of wounds had healed at 8 weeks when treated the standard way? When treated with cultured skin?
2. What percentage of wounds had healed at 12 weeks when treated the standard way? When treated with cultured skin?
3. How early was the healing difference between the control and treatment groups obvious?

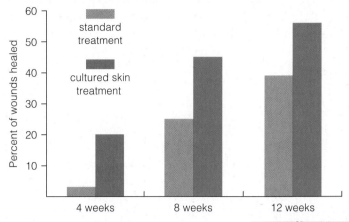

Figure 28.20 Results of a study comparing standard treatment for diabetic foot ulcers to use of a cultured cell product. Bars show the percentage of foot ulcers that had completely healed. The photo at the *right* shows a sheet of the cultured cell product.

10. Skin darkens when exposed to sunlight because _____ produce more pigment.
    a. melanocytes
    b. keratinocytes
    c. neuroglial cells
    d. neurons

11. The heart and lungs are in the _____ cavity.
    a. thoracic
    b. pelvic
    c. cranial
    d. abdominal

12. With negative feedback, a change induces a response that _____ that change.
    a. increases
    b. reverses

13. Hair and nails are keratin-rich structures formed by _____ tissue.
    a. muscle
    b. nervous
    c. epithelial
    d. connective

14. Match the terms with the most suitable description.
    ___ exocrine gland   a. strong, pliable; like rubber
    ___ endocrine gland   b. secretion through duct
    ___ cartilage   c. deep skin layer
    ___ dermis   d. contracts, not striated
    ___ smooth muscle   e. tendons and ligaments
    ___ bone   f. epithelial cell cancer
    ___ carcinoma   g. lining of lungs
    ___ blood   h. cells in a hardened matrix
    ___ adipose tissue   i. fluid connective tissue
    ___ dense, regular   j. ductless secretion
        connective tisue   k. stores fat
    ___ simple, squamous
        epithelium

Additional questions are available on **CENGAGENOW**.

Critical Thinking

1. Many people oppose the use of animals for testing the safety of cosmetics. They argue that alternative test methods are available, such as use of lab-grown tissues. Given what you learned in this chapter, speculate on the advantages and disadvantages of tests that use specific lab-grown tissues as opposed to living animals.

2. Radiation and chemotherapy drugs preferentially kill cells that divide frequently, most notably cancer cells. These cancer treatments also cause hair to fall out. Why?

3. Each level of biological organization has emergent properties that arise from the interaction of its component parts. For example, cells have a capacity for inheritance that the molecules making up the cell do not. Can you think of an emergent property of a tissue? Of an organ that contains that tissue?

4. The micrograph to the *right* show cells from the lining of an airway leading to the lungs. The *gold* cells are ciliated and the darker *brown* ones secrete mucus. What type of tissue is this? How can you tell?

Animations and Interactions on **CENGAGENOW**:
❯ Function of vertebrate organ systems; Structure of human skin; Apoptosis during development of a hand

In this chapter, you will draw on your knowledge of potential energy (Section 5.2), diffusion and concentration gradients (5.6), passive and active transport (5.7), and exocytosis (5.8). You will revisit the body plans of a few invertebrates (23.2, 23.5, 23.6) and vertebrates (24.2). You will increase your knowledge of nervous tissue (28.6) and be reminded of alcohol's negative effects (5.1), and the promise of stem cell research (28.1).

Key Concepts

Nervous Systems
Excitable cells called neurons form communication lines in animal nervous systems. Neurons of radially symmetrical animals connect as a nerve net. Neurons of bilaterally symmetrical animals are concentrated in the head, and one or more nerve cords run the length of the body.

How Neurons Work
Messages flow along a neuron's plasma membrane, from input to output zones. The messages are brief, self-propagating reversals in the distribution of electric charge across the membrane. At an output zone, chemical signals are sent to other neurons, muscles, or glands.

29 Neural Control

29.1   In Pursuit of Ecstasy

Ecstasy is an illegal drug that can make you feel socially accepted, less anxious, and more aware of your surroundings and of sensory stimuli. It also can leave you dying in a hospital, bleeding from all body openings as your temperature skyrockets. It can send your family and friends spiraling into horror and disbelief as they watch you stop breathing. Lorna Spinks ended life that way at age 19. Her anguished parents released the photographs in Figure 29.1 because they wanted others to know what their daughter did not: Ecstasy can kill.

Ecstasy is a psychoactive drug, meaning it alters brain function. The active ingredient, MDMA (3,4-methylenedioxymethamphetamine), is a type of amphetamine. As one effect, MDMA makes some neurons in the brain release an excess of a signaling molecule called serotonin. The serotonin binds to receptors at the plasma membrane of other brain cells and puts these cells into a state of overstimulation.

The abundance of serotonin promotes feelings of energy, empathy, and euphoria. Recreational users take the drug to enjoy these sensations. Unfortunately, MDMA also has less desirable effects. It increases the heart rate and respiratory rate, dilates the eyes, causes muscles to cramp, and restricts urine formation. Blood pressure soars, and the body's internal temperature sometimes rises out of control. Spinks became dizzy, flushed, and incoherent after taking two Ecstasy tablets. She died because her body temperature became too high, causing her organ systems to shut down.

Ecstasy overdoses seldom end in death. More common short-term effects include panic attacks and fleeting psychosis. In addition, researchers have begun to discover some unpleasant long-term effects. Many animal studies have demonstrated that multiple doses of MDMA alter the structure of neurons that secrete serotonin and decrease their number. Ecstasy users have impaired verbal memory, and the more often they use the drug, the worse the impairment becomes.

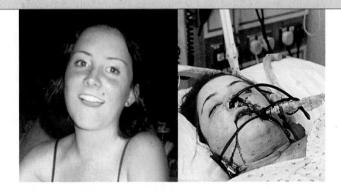

Figure 29.1 Lorna Spinks in life (*left*), and just after her death from side effects of Ecstasy (*right*). If you suspect someone is having a bad reaction to Ecstasy or any other drug, get medical help fast and be honest about the cause of the problem. Immediate, informed medical action may save a life.

It is difficult to do a controlled study of an illegal drug that may harm the brain. However, researchers in Amsterdam got around this difficulty by allowing their subjects to sort themselves into experimental and control groups on the basis of their behavior. The researchers began their study by scanning the brains of about 200 people who had never used Ecstasy, but said they might try the drug. About two years later, the researchers scanned the brains of their subjects once again. By this time, some people had begun using Ecstasy; they served as the experimental group. Others had still not tried the drug; they served as a control group. The brain scans of the control group showed no change. However, the scans of new Ecstasy users showed that the pattern of blood flow in the brain had been significantly altered. Whether the observed blood flow changes negatively affect brain function remains under investigation.

The more you know about the nervous system, the easier it is to understand the effects of psychoactive drugs, both legal and illegal. What you learn here can help you make educated choices about how you treat your brain.

Disrupted Signaling
Some common neurological disorders cause symptoms by interfering with the flow of signals through the nervous system. Psychoactive drugs also affect nervous system activity by raising or lowering the amount of signaling chemicals in the brain.

Vertebrate Nervous System
The central nervous system of vertebrates consists of the brain and spinal cord. The peripheral nervous system includes many pairs of nerves that connect the brain and spinal cord to the rest of the body.

A Closer Look at the Human Brain
The cerebral cortex is the part of the brain that evolved most recently. In humans, it governs conscious behavior and interacts with the limbic system in forming and retrieving memories.

> Signals move quickly through most animal bodies along an information highway consisting of neurons.

‹ Links to Animal body plans 23.2, Cnidarians 23.5, Flatworms 23.6, Chordate traits 24.2, Nervous tissue 28.6

Of all multicelled organisms, animals respond fastest to external stimuli. Specialized cells called neurons are the key to these rapid responses. As Section 28.6 explained, a **neuron** is a cell that relays electrical signals along its plasma membrane and communicates with other cells by way of chemical messages. **Neuroglial cells** provide structural and functional support to the neurons.

The Cnidarian Nerve Net

Cnidarians such as hydras and jellyfishes have a **nerve net**, a mesh of interconnected neurons (Figure 29.2A). Information flows in any direction among cells of a nerve net, and there is no centralized, controlling organ that functions like a brain. By causing cells in the body wall to contract, the nerve net can alter the size of the animal's mouth, change the body's shape, or shift the position of the tentacles. A nerve net allows cnidarians to respond to food or threats that can arrive from any direction.

Bilateral, Cephalized Invertebrates

Most animals are bilateral, with paired organs arranged on either side of the main body axis (Section 23.2). Evolu-tion of bilateral body plans was accompanied by cepha-lization: Neurons that detect and process information became concentrated at the body's anterior end, or "head."

Animals with a bilateral, cephalized body plan have three types of neurons. **Sensory neurons** are stimulated by specific stimuli such as light, touch, or heat. They signal interneurons or motor neurons. **Interneurons** integrate signals from sensory neurons and other inter-neurons, and send signals of their own to motor neurons. **Motor neurons** control muscles and glands.

Planarians are bilateral animals with a simple nervous system. A pair of ganglia in the head serves as an integrat-ing center (Figure 29.2B). Each **ganglion** (plural, ganglia), consists of a cluster of neuron cell bodies. (The cell body is the part of the neuron that holds the nucleus and other organelles.) The ganglia receive signals from eyespots and chemical-detecting cells on the planarian's head. They also connect to a pair of nerve cords that run along the animal's ventral (lower) surface. Each **nerve cord** consists of nerve fibers (cytoplasmic extensions) of many neurons, and runs the length of the body. Nerves branch from the nerve cord and cross the body. A **nerve** consists of nerve fibers inside a connective tissue sheath.

Annelids and arthropods have paired nerve cords that run along the ventral (lower) surface and connect to a simple brain (Figure 29.2C–E). In each segment, a pair of ganglia controls that segment's muscles.

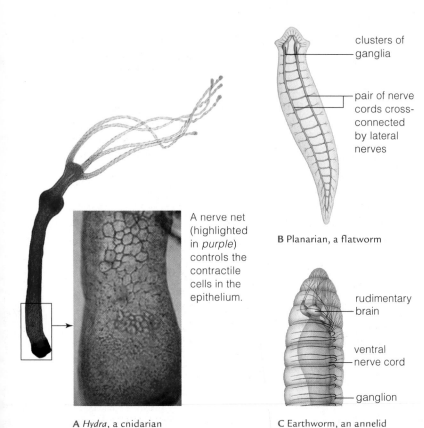

clusters of ganglia

pair of nerve cords cross-connected by lateral nerves

A nerve net (highlighted in *purple*) controls the contractile cells in the epithelium.

B Planarian, a flatworm

rudimentary brain

ventral nerve cord

ganglion

A *Hydra*, a cnidarian

C Earthworm, an annelid

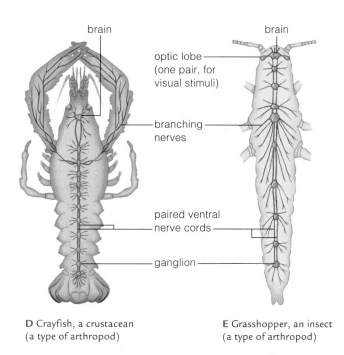

brain

brain

optic lobe (one pair, for visual stimuli)

branching nerves

paired ventral nerve cords

ganglion

D Crayfish, a crustacean (a type of arthropod)

E Grasshopper, an insect (a type of arthropod)

Figure 29.2 Gallery of invertebrate nervous systems. Hydras and other radial animals have a nerve net. The other representatives shown have a bilateral, cephalized nervous system.

The Vertebrate Nervous System

A dorsal nerve cord is one of the defining features of chordates (Section 24.2). In vertebrates, the dorsal nerve cord evolved into a brain and spinal cord, which together constitute the animal's **central nervous system** (Figure 29.3). Most interneurons reside in the central nervous system. Nerves that extend from the central nervous system through the rest of the body constitute the **peripheral nervous system**. Nerves of the peripheral system fall into two functional categories. Autonomic nerves monitor and regulate the internal state of the body. They control smooth muscle, cardiac muscle, and glands. Somatic nerves monitor the body's position and the external environment, and they control skeletal muscle.

The peripheral nervous system consists of 12 cranial nerves that connect to the brain and 31 spinal nerves that connect to the spinal cord (Figure 29.4). Most cranial nerves, and all spinal nerves, include fibers of both sensory and motor neurons. For instance, a sciatic nerve runs from the sacral region of the spinal cord, through the buttock, and down the leg. When someone touches your thigh, the sciatic nerve carries signals from receptors in the skin to the spinal cord. When you move your leg, the sciatic nerve carries commands for the movement from the spinal cord to skeletal muscles in the leg.

---

central nervous system Brain and spinal cord.
ganglion Cluster of nerve cell bodies.
interneuron Neuron that receives signals from and sends signals to other neurons.
motor neuron Neuron that receives signals from another neuron and sends signals to a muscle or gland.
nerve Neuron fibers bundled inside a sheath of connective tissue.
nerve cord Bundle of nerve fibers running the length of a body.
nerve net Of cnidarians, a mesh of interacting neurons with no central control organ.
neuroglial cell Cell that supports neurons.
neuron A cell that transmits electrical signals along its plasma membrane and sends chemical messages to other cells.
peripheral nervous system Nerves that extend through the body and carry signals to and from the central nervous system.
sensory neuron Neuron that responds to a specific internal or external stimulus and signals another neuron.

Take-Home Message What are the features of animal nervous systems?

❯ Cnidarians have a nerve net, a meshlike array of neurons with no central control point.

❯ Bilateral animals typically are cephalized, with ganglia or a brain in the head. Bilateral invertebrates usually have a pair of ventral nerve cords. In contrast, the chordates have a dorsal nerve cord.

❯ The vertebrate nervous system includes a well-developed brain, a spinal cord, and peripheral nerves.

Central Nervous System

Brain   Spinal Cord

Peripheral Nervous System
(cranial and spinal nerves)

Autonomic Nerves	**Somatic Nerves**
Nerves that carry signals to and from smooth muscle, cardiac muscle, and glands	Nerves that carry signals to and from skeletal muscle, tendons, and the skin

Sympathetic Division   Parasympathetic Division
Two sets of nerves that often signal the same effectors and have opposing effects

Figure 29.3 Functional divisions of vertebrate nervous systems. The sympathetic and parasympathetic divisions of the autonomic system are discussed in Section 29.8.

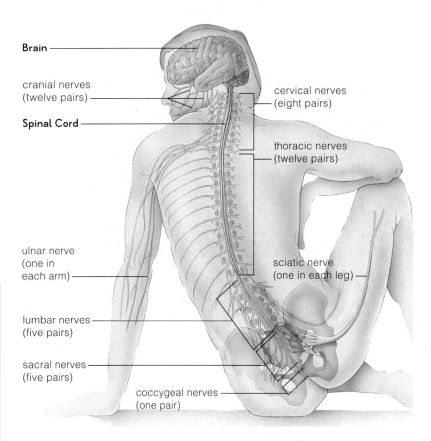

Brain

cranial nerves
(twelve pairs)

Spinal Cord

cervical nerves
(eight pairs)

thoracic nerves
(twelve pairs)

ulnar nerve
(one in
each arm)

sciatic nerve
(one in each leg)

lumbar nerves
(five pairs)

sacral nerves
(five pairs)

coccygeal nerves
(one pair)

Figure 29.4 Some of the major nerves of the human nervous system.

> The structure of neurons reflects their function as building blocks of nervous systems.
< Link to Nervous tissue 28.6

Neurons are signaling cells of a nervous system. Figure 29.5 shows the structure of a motor neuron, a cell that controls a muscle or a gland. The neuron's nucleus and other organelles are in its cell body. Cytoplasmic extensions allow the neuron to receive and send messages. **Dendrites** are cytoplasmic branches that receive the chemical signals sent by other neurons. By contrast, the single **axon** is specialized for sending messages. It has a trigger zone close to the cell body. Electrical signals that originate in the trigger zone travel along the conducting zone of an axon to its endings, or axon terminals. Axon terminals are an output zone; they release signaling molecules that influence the activity of other cells.

Information typically flows from sensory neurons, to interneurons, to motor neurons (Figure 29.6). The three types of neurons differ somewhat in their structure. A sensory neuron typically has no dendrites. One end of its axon has receptor endings that detect a specific stimulus (Figure 29.6A). Axon terminals at the other end send chemical signals, and the cell body lies in between. An interneuron has many signal-receiving dendrites and one axon (Figure 29.6B). Some vertebrate interneurons have many thousands of dendrites. A motor neuron has multiple dendrites and one axon that signals a muscle or gland cell (Figure 29.6C).

axon Neuron cytoplasmic extension that transmits electrical signals along its length and secretes chemical signals at its endings.
dendrite Of a motor neuron or interneuron, a cytoplasmic extension that receives chemical signals sent by other neurons and converts them to electrical signals.

Take-Home Message How does neuron structure relate to function?

> All neurons have a cell body with organelles and an axon that sends chemical signals to other cells.

> Interneurons and motor neurons have signal-receiving dendrites. A sensory neuron has no dendrites. One end of its axon has receptor endings that detect a specific stimulus.

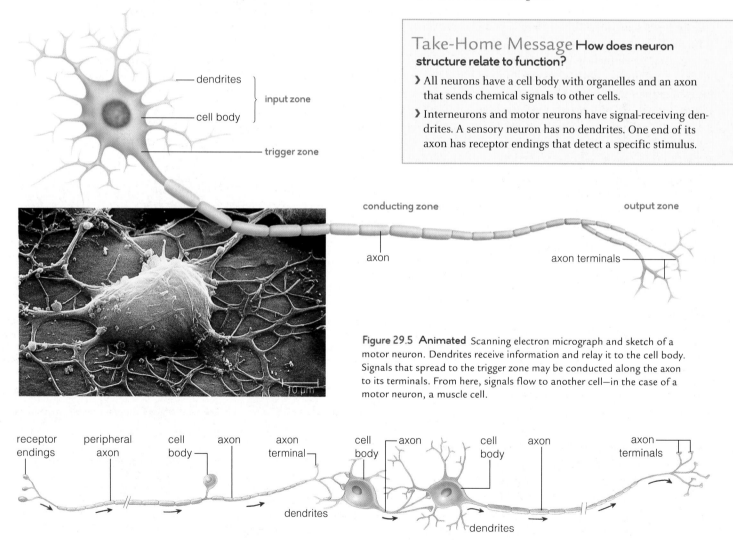

Figure 29.5 Animated Scanning electron micrograph and sketch of a motor neuron. Dendrites receive information and relay it to the cell body. Signals that spread to the trigger zone may be conducted along the axon to its terminals. From here, signals flow to another cell—in the case of a motor neuron, a muscle cell.

A Sensory neuron with no dendrites; receptor endings of the peripheral axon respond to a specific stimulus.

B Interneuron with short dendrites and a short axon.

C Motor neuron with short dendrites, and a longer axon with terminals that signal a muscle or gland cell.

Figure 29.6 Structure of the three types of neurons. Arrows indicate direction of information flow.

❯ A neuron's ability to send and receive messages stems from properties of its membrane.

❮ Links to Potential energy 5.2, Gradients 5.6, Membrane transport proteins 5.7

Resting Potential

All cells have an electric gradient across their plasma membrane. The cytoplasmic fluid near this membrane has more negatively charged ions and proteins than the interstitial fluid outside the cell does. As in a battery, the separation of charge constitutes potential energy. The voltage difference across a cell membrane is a **membrane potential**. Researchers measure membrane potential across a neuron's axon by inserting one electrode into the axon and another into the fluid just outside of it:

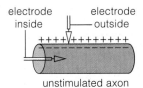

The membrane potential across the axon of a neuron that is not being excited is the **resting potential**. This potential is usually about –70 millivolts. (A millivolt [mV] is one-thousandth of a volt.) The negative sign indicates that the inside of the neuron is more negative than the outside.

Resting potential arises from the distribution of charged proteins and small ions. The cytoplasm of a neuron has more negatively charged proteins than the interstitial fluid. Being large and charged, these proteins cannot diffuse across the lipid bilayer of the cell membrane. Positively charged potassium ions (K^+) and positively charged sodium ions (Na^+) also influence resting potential. These ions move in and out of the neuron with the assistance of transport proteins.

Sodium–potassium pumps move two potassium ions into the cell for every three sodium ions they move out (Figure 29.7A). Because these pumps move more positively charged ions out of the cell than into it, they increase the charge gradient across the membrane. The pumps also contribute to concentration gradients for sodium and potassium ions across the membrane.

Sodium ions cannot cross a resting neuron's membrane, but potassium ions can. Some potassium ions leave the cell through passive transport proteins in the membrane (Figure 29.7B). Leaking of potassium (K^+) outward increases the charge difference across the membrane.

action potential Brief reversal of the charge difference across a neuron membrane.
membrane potential Potential energy of charges separated by a cell membrane.
resting potential Membrane potential of a neuron at rest.

In summary, the cytoplasm of a resting neuron has more negatively charged proteins than interstitial fluid. It also has fewer sodium ions (Na^+) and more potassium ions (K^+). We illustate the difference below, with the green ball representing negatively charged proteins:

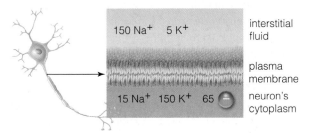

interstitial fluid

plasma membrane

neuron's cytoplasm

Action Potential

Neurons and muscle cells are said to be "excitable" because they can undergo an **action potential**—an abrupt reversal in the electric gradient across the plasma membrane. The charge reversal occurs when sodium and potassium flow down their concentration gradients through voltage-gated ion channels. Neurons have voltage-gated channels in the membrane of their trigger zone and conducting zone (Figure 29.7C). The channels are closed in a resting neuron, but open during an action potential. We discuss action potentials in detail in the next section.

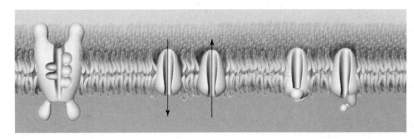

A Sodium–potassium pumps actively transport 3 sodium ions (Na^+) out of a neuron for every 2 potassium ions (K^+) they pump in.

B Passive transporters allow K^+ ions to leak across the plasma membrane, down their concentration gradient.

C Voltage-sensitive channels for K^+ or Na^+ ions are shut when a neuron is at rest (*left*). They open during an action potential (*right*).

Figure 29.7 Animated Protein channels and pumps that span a neuron membrane.

Take-Home Message How do ion gradients across the neuron membrane contribute to neuron function?

❯ The cytoplasm of a resting neuron is more negatively charged than extracellular fluid. The impermeability of the membrane to negatively charged proteins and the activity of transport proteins contribute to this resting potential.

❯ A resting neuron also has concentration gradients for sodium and potassium across its membrane, with more sodium outside and more potassium inside.

❯ When stimulated, a neuron undergoes an action potential. Voltage-gated channels open and the membrane potential briefly reverses as a result of ion flow.

> Action potentials occur when positively charged ions cross a neuron membrane through voltage-gated ion channels.
> ❮ Link to Diffusion 5.6

Approaching Threshold

Tap your wrist, and the pressure stimulates the receptor endings of sensory neurons in your skin. The pressure deforms the plasma membrane at the input zone of these neurons and ions slip across. The ion flow causes a local, graded potential—a slight shift in the voltage difference across the neuron's membrane. "Graded" means the size of this voltage shift varies. With a stronger, longer-lasting stimulus, more ions cross the membrane, so the voltage shift is larger.

When a stimulus is intense or long-lasting, graded potentials spread into a trigger zone (Figure 29.8❶). The membrane in the trigger zone has many voltage-gated sodium (Na⁺) channels. When a stimulus shifts the voltage difference across the membrane to a certain level, called the **threshold potential**, these channels open and an action potential gets under way.

As the gates open, sodium streams in from interstitial fluid into the axon ❷. The influx makes the cytoplasm more positively charged and causes more gates to open and more ions to enter. This increasing inward flow of more and more ions is an example of **positive feedback**, whereby an activity intensifies because of its own occurrence. Once threshold is reached and an action potential has started, the stimulus strength no longer affects the outcome. The rush of sodium into the neu-ron, not diffusion of ions from the input zone, drives the feedback cycle:

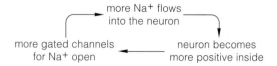

An All-or-Nothing Spike

The rush of sodium ions into the axon causes the inside of the axon to become positive relative to the outside. Thus a graph of membrane potential during an action potential shows a voltage reversal with a distinct peak (Figure 29.9). Once threshold level is reached, and an action potential begins, membrane potential always rises to the same peak level. For this reason, an action potential is described as an all-or-nothing event.

The reversal of charge difference across the membrane during an action potential lasts only milliseconds. The potential declines because, above a certain voltage, gates on sodium channels begin to swing shut. At about the same time, the gates on voltage-gated potassium (K⁺) channels open ❸. The resulting outflow of potassium makes the cytoplasm near the plasma membrane once again more negative than the interstitial fluid. In fact, so much potassium flows out that the membrane potential briefly dips below resting potential.

Immediately after an action potential, diffusion of ions within the axon restores the resting potential. Potassium ions diffuse to the potassium-depleted area just beneath the membrane from deeper within the axon. Over the longer term, sodium–potassium pumps maintain the ion gradients (and the resting membrane potential) by pumping sodium out of the cell and potassium into it ❹.

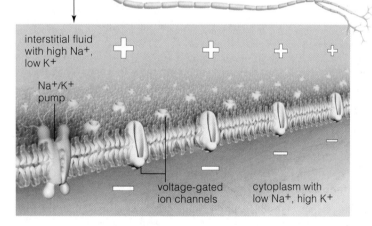

❶ Close-up of the trigger zone of a neuron. One sodium–potassium pump and some of the voltage-gated ion channels are shown. At this point, the membrane is at rest and the voltage-gated channels are closed. The cytoplasm's charge is negative relative to interstitial fluid.

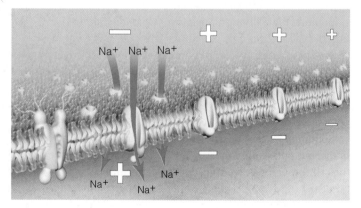

❷ Arrival of a sufficiently large signal in the trigger zone raises the membrane potential to threshold level. Gated sodium channels open and sodium (Na⁺) flows down its concentration gradient into the cytoplasm. Sodium inflow reverses the voltage across the membrane.

Figure 29.8 Animated Propagation of an action potential along part of a motor neuron's axon.

A Resting membrane potential is –70 mV.

B Stimulation causes an influx of positive ions and a rise in the membrane potential.

C Once potential exceeds threshold (–60 mV), the voltage-gated sodium (Na$^+$) gates begin to open, and Na$^+$ rushes in. This causes more gates to open, and so on. Voltage shoots up rapidly.

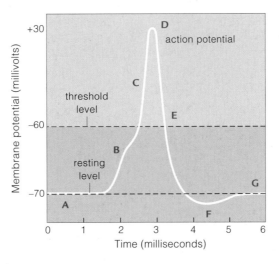

D Every action potential peaks at +33 mV; no more, no less. At this point, Na$^+$ gates have closed and potassium (K$^+$) gates have opened.

E Flow of K$^+$ out of the neuron causes the potential to fall.

F So much K$^+$ exits that potential declines below resting potential.

G Action of the Na$^+$/K$^+$ pump restores resting ion concentrations.

Figure 29.9 Animated Changes in membrane potential during an action potential.

>> Figure It Out How long does the increase in potential last? Answer: About 2 milliseconds

Propagation Along the Axon

An action potential is self-propagating; it moves along an axon without declining in strength. Some of the sodium that enters one region of an axon during an action potential diffuses into an adjoining region, driving that region to threshold and opening gated sodium channels. As these gates open in one region after the next, the action potential spreads without weakening.

Action potentials travel in one direction: toward axon terminals. Gated sodium channels cannot reopen right after they close. As a result, the diffusion of sodium ions only opens gated sodium channels in regions farther along the axon.

---

positive feedback A response intensifies the conditions that caused its occurrence.
threshold potential Neuron membrane potential at which gated sodium channels open, causing an action potential to occur.

Take-Home Message What happens during an action potential?

> An action potential begins in the neuron's trigger zone. When membrane potential is reached, gated sodium channels open and the charge difference across the membrane reverses. Self-propagating voltage reversals that do not weaken with distance occur at consecutive patches of membrane as the action potential travels toward axon terminals.

> At each patch of membrane, an action potential ends when potassium ions flow out of the neuron, and the voltage difference across the membrane is restored. Action potentials move toward axon terminals, because gated sodium channels cannot reopen immediately after an action potential.

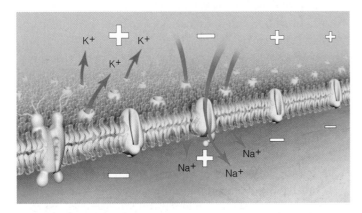

❸ The charge reversal makes gated Na$^+$ channels shut and gated K$^+$ channels open. The K$^+$ outflow restores the voltage difference across the membrane. The action potential is propagated along the axon as positive charges spreading from one region push the next region to threshold.

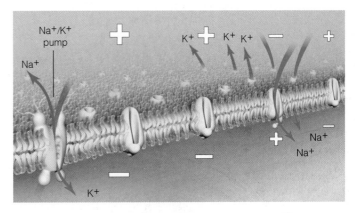

❹ After an action potential, gated Na$^+$ channels are briefly inactivated, so the action potential moves one way only, toward axon terminals. Na$^+$ and K$^+$ gradients disrupted by action potentials are restored by diffusion of ions that were put into place by activity of sodium–potassium pumps.

❯ At axon terminals, action potentials trigger exocytosis of signaling molecules that affect an adjacent cell.

❮ Links to Receptor protein 4.4, Exocytosis 5.8

Sending Signals at Synapses

Action potentials cannot pass directly from a neuron to another cell. Instead, chemicals relay signals between neurons or between a neuron and a muscle or gland. The

1 Action potentials flow along the axon of a motor neuron to neuromuscular junctions, where an axon terminal forms a synapse with a muscle fiber.

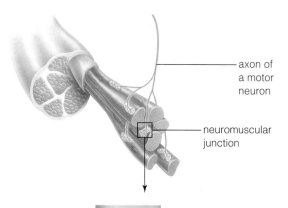

axon of a motor neuron

neuromuscular junction

2 The axon terminal stores chemical signaling molecules (*green*) called neurotransmitter inside synaptic vesicles.

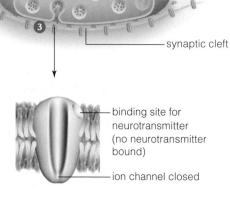

axon terminal of motor neuron

plama membrane of muscle fiber

synaptic vesicle

2

3 Arrival of an action potential causes exocytosis of synaptic vesicles, and neurotransmitter enters the synaptic cleft.

3

synaptic cleft

4 The plasma membrane of the muscle fiber has receptors for neurotransmitter.

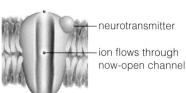

binding site for neurotransmitter (no neurotransmitter bound)

ion channel closed

5 Binding of neurotransmitter opens a channel through the receptor. The opening allows ions to flow into the postsynaptic cell.

neurotransmitter

ion flows through now-open channel

Figure 29.10 Animated Communication at a synapse. This example depicts what occurs at a neuromuscular junction, a synapse between an axon terminal of a motor neuron and the muscle fiber that the neuron controls.

region where an axon terminal sends chemical signals to a neuron, a muscle fiber, or a gland cell is a **synapse**. The signal-sending neuron at a synapse is referred to as the presynaptic cell. A fluid-filled synaptic cleft about 20 nanometers wide separates it from the input zone of a postsynaptic cell, the cell that receives the signal. For comparison, a hair is 100,000 nanometers wide.

Figure 29.10 illustrates how signals are transmitted at a **neuromuscular junction**, a synapse between a motor neuron and a skeletal muscle fiber. Action potentials arrive at a neuromuscular junction by traveling along the axon of a motor neuron to axon terminals **1**. Axon terminals have vesicles filled with **neurotransmitter**, a signaling molecule that relays messages between presynaptic and postsynaptic cells **2**. Arrival of an action potential at an axon terminal causes exocytosis; the vesicles move to the plasma membrane and fuse with it, releasing neurotransmitter into the synaptic cleft **3**.

The plasma membrane of a postsynaptic cell has receptor proteins that can reversibly bind a molecule of neurotransmitter **4**. An axon terminal at a neuromuscular junction releases the neurotransmitter **acetylcholine** (ACh). Binding of ACh to a receptor in the muscle fiber's plasma membrane causes the receptor to change shape. The receptor is also a channel protein that spans the membrane, and the shape change allows sodium ions to travel through the receptor, from the interstitial fluid into the muscle cell **5**.

Like a neuron, a muscle fiber can undergo an action potential. The rise in sodium caused by the binding of ACh drives the muscle fiber's membrane toward threshold potential. Once threshold is reached, action potentials stimulate muscle contraction by a process described in detail in Section 32.6.

Cleaning the Cleft

After neurotransmitter molecules do their work, they must be removed from synaptic clefts to make way for new signals. Membrane pumps transport some neurotransmitter back into presynaptic cells or into nearby neuroglial cells. In addition, postsynaptic cells have enzymes that break down neurotransmitter. For example, at a neuromuscular junction, the muscle cell membrane contains acetylcholinesterase, an enzyme that breaks down ACh.

Drugs or toxins that interfere with reabsorption or breakdown of a neurotransmitter can cause it to accumulate in the synaptic cleft and disrupt the signaling pathway. For example, nerve gases such as sarin exert their deadly effects by binding to acetylcholinesterase and thus inhibiting ACh breakdown. As a result, ACh accumulates, causing skeletal muscle paralysis, confusion, headaches, and, when the dosage is high enough, death.

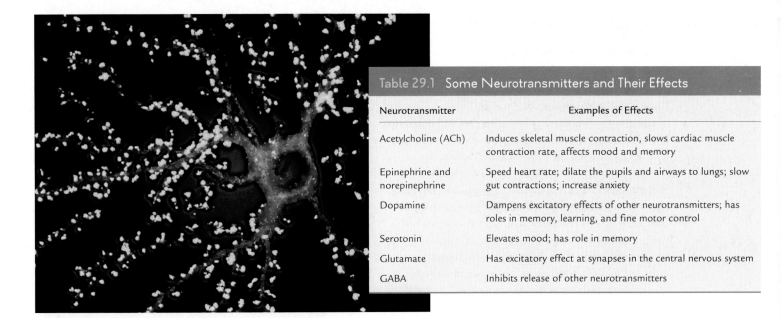

Figure 29.11 Synaptic density. An interneuron stained with a yellow fluorescent dye that indicates the many locations where other neurons synapse on this cell.

Table 29.1 Some Neurotransmitters and Their Effects	
Neurotransmitter	Examples of Effects
Acetylcholine (ACh)	Induces skeletal muscle contraction, slows cardiac muscle contraction rate, affects mood and memory
Epinephrine and norepinephrine	Speed heart rate; dilate the pupils and airways to lungs; slow gut contractions; increase anxiety
Dopamine	Dampens excitatory effects of other neurotransmitters; has roles in memory, learning, and fine motor control
Serotonin	Elevates mood; has role in memory
Glutamate	Has excitatory effect at synapses in the central nervous system
GABA	Inhibits release of other neurotransmitters

Synaptic Integration

A typical neuron or effector cell receives messages from many neurons. An interneuron in the brain can be on the receiving end of hundreds to thousands of synapses (Figure 29.11). At each synapse, the incoming signal may be excitatory and push the membrane potential closer to threshold. Or the signal may be inhibitory and nudge the potential away from threshold.

How does a postsynaptic cell respond to all of this information, some of it conflicting? Through **synaptic integration**, a neuron sums all inhibitory and excitatory signals arriving at its input zone. Incoming synaptic signals can amplify, dampen, or cancel one another's effects.

Competing signals cause the membrane potential at the postsynaptic cell's input zone to rise and fall. When excitatory signals outweigh inhibitory ones, ions diffuse from the input zone into the trigger zone and drive the postsynaptic cell to threshold. Gated sodium channels swing open, and an action potential occurs. Neurons also add up signals that arrive in quick succession from a single presynaptic cell.

acetylcholine (ACh) Neurotransmitter released at neuromuscular junctions, and at synapses in the heart and brain.
neuromuscular junction Synapse between a neuron and a muscle.
neurotransmitter Chemical signal released by axon terminals.
synapse Region where a neuron's axon terminals transmit chemical signals to another cell.
synaptic integration The summation of excitatory and inhibitory signals by a postsynaptic cell.

Neurotransmitter and Receptor Diversity

Different kinds of neurons release different neurotransmitters. Table 29.1 provides some examples of neurotransmitters and their effects. Norepinephrine and epinephrine (commonly known as adrenaline) prepare the body to respond to stress or excitement. Dopamine influences reward-based learning and acts in fine motor control. Serotonin influences mood and memory. Glutamate is the main excitatory signal in the central nervous system. GABA (gamma aminobutyric acid) has a general inhibitory effect; it slows release of other neurotransmitters.

Different kinds of postsynaptic cells have receptors that respond differently to the same neurotransmitter. For example, binding of ACh at a neuromuscular junction stimulates contraction of skeletal muscle, but binding of ACh to the receptors in cardiac muscle discourages contraction. ACh also binds to receptors on interneurons in the brain, where it affects mood and memory.

Take-Home Message How does a neuron signal another cell?

❯ Action potentials travel to a neuron's output zone. There, these signals typically are transduced to a chemical form—neurotransmitter—that can be sent onward to another cell.

❯ Neurotransmitters are signaling molecules secreted into a synaptic cleft from a neuron's output zone. They may have excitatory or inhibitory effects on a postsynaptic cell.

❯ Synaptic integration is the summation of all excitatory and inhibitory signals arriving at a postsynaptic cell's input zone at the same time.

> Neurological disorders and use of psychoactive drugs can interfere with the flow of information at synapses.

‹ Link to Alcohol's effects 5.1

Neurological Disorders

Disorders of the nervous system often involve disruption of signaling at synapses. A few examples will illustrate how our understanding of neurotransmission can help us develop treatments for these disorders.

Parkinson's Disease Damage to dopamine-secreting neurons in the part of the brain that governs motor control results in Parkinson's disease. Former heavyweight boxer Muhammad Ali and actor Michael J. Fox are among those affected (Figure 29.12). Tremors are often the earliest symptom. Later, sense of balance may be impaired, and any voluntary movement, including speech, can be difficult. Because symptoms arise from a dopamine shortage, patients are often treated with a drug (levodopa) that the body converts to this neurotransmitter.

Multiple factors contribute to development of Parkinson's disease. A head injury that causes a loss of consciousness increases the risk, as does exposure to pesticides in drinking water. There are also some rare inherited forms of the disease.

Attention Deficit Hyperactivity Disorder A lower than normal dopamine level also seems to play a role in attention deficit hyperactivity disorder (ADHD). Affected people have trouble concentrating, are unusually impulsive, and tend to fidget when required to remain seated. Drugs used to treat ADHD are stimulants that increase dopamine availability in the brain. For example, Ritalin (methylphenidate) acts by preventing the reuptake of dopamine after it has been released at a synapse.

Alzheimer's Disease Alzheimer's disease is the leading cause of dementia (loss of ability to think). It typically begins with forgetfulness. As the disease progresses, a person becomes increasingly confused, cannot communicate, and eventually is incapable of living independently. Affected people have a lower than normal level of ACh in the brain. Drugs that increase the level of this neurotransmitter by inhibiting acetylcholinesterase can slow the mental decline in some people.

Mood Disorders Interactions among several neurotransmitters, including serotonin, dopamine, and norepinephrine, affect mood. Low levels of one or more of these neurotransmitters can result in depression. Feelings of sadness persist and the person is unable to experience pleasure. Some depressed people become irritable and agitated, whereas others lack energy. The most widely

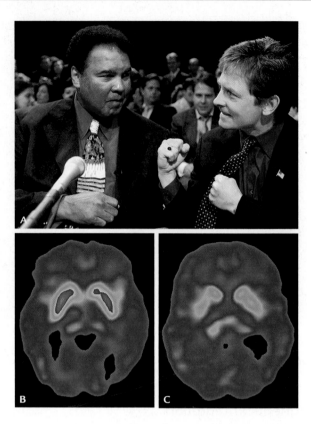

Figure 29.12 Battling Parkinson's disease. **A** This neurological disorder affects former heavyweight champion Muhammad Ali, actor Michael J. Fox, and about half a million other people in the United States. **B** A normal PET scan and **C** one from an affected person. *Red* and *yellow* indicate high metabolic activity in dopamine-secreting neurons. Section 2.2 explains PET scans.

prescribed antidepressants, including Prozac (fluoxetine) and Paxil (paroxetine), increase the level of serotonin by preventing its reuptake.

Depression has a genetic component, and many families predisposed to depression are also prone to anxiety disorders. People with anxiety disorders can be paralyzed with worry or stricken with panic under what most of us would view as ordinary circumstances. Anxiety disorders often respond to treatment with antidepressants that prevent serotonin reuptake. Tranquilizers such as Xanax (alprazolam) and Valium (diazepam) alleviate anxiety by enhancing the inhibitory effect of GABA.

Effects of Psychoactive Drugs

People take psychoactive drugs, both legal and illegal, to alleviate pain, relieve stress, or feel pleasure. Users may develop tolerance (reduced responsiveness), which means that a desired effect requires larger or more frequent doses of the drug. In addition, many drugs are habit-forming.

Habituation and tolerance can lead to drug addiction, in which a drug takes on a vital biochemical role. Table 29.2 lists the main warning signs of addiction. One or more signs may be cause for concern.

All major addictive drugs stimulate the release of dopamine. Under natural circumstances, dopamine release provides pleasurable feedback when an animal engages in behavior that enhances survival or reproduction. This response is adaptive; it helps animals learn to repeat the behaviors that benefit them. When drugs cause dopamine release, they tap into this ancient learning pathway. Drug users inadvertently teach themselves that the drug is essential to their well-being.

Stimulants Stimulants make users feel alert but also anxious, and they can interfere with fine motor control. Nicotine is a stimulant that blocks brain receptors for ACh. The caffeine in coffee, tea, and many soft drinks is also a stimulant. It blocks receptors for adenosine, which acts as a signaling molecule to suppress brain cell activity.

Cocaine, a powerful stimulant, is inhaled or smoked. Users feel elated and aroused, then become depressed and exhausted. Cocaine prevents reuptake of dopamine, serotonin, and norepinephrine from synaptic clefts. When norepinephrine is not cleared away, blood pressure soars. An overdose may cause a stroke or heart attack that can end in death. Cocaine is also highly addictive. Heavy cocaine use alters the brain so that only cocaine can bring about pleasurable sensations.

Amphetamines reduce appetite and energize users by increasing the secretion of serotonin, norepinephrine, and dopamine in the brain. Various types of amphetamine are ingested, smoked, or injected. Section 29.1 focused on the synthetic amphetamine known as Ecstasy. Crystal meth is another widely abused amphetamine. As with cocaine, users require more and more to get high or just to feel okay. Long-term use shrinks the brain areas involved in memory and emotions.

Analgesics **Endorphins** are natural painkillers produced by the central nervous system. They also promote feelings of pleasure. Narcotic analgesics, including morphine, codeine, heroin, fentanyl, and oxycodone, mimic the effects of endorphins. These drugs not only alleviate pain, but also cause a rush of euphoria. They are highly addictive. Ketamine and PCP (phencyclidine) belong to a different class of analgesics. They give users an out-of-body experience and numb the extremities by slowing the clearing of synapses. Use of either drug can lead to seizures, kidney failure, and fatal heat stroke. PCP can induce a violent, agitated psychosis that sometimes lasts more than a week.

---

endorphin Painkiller produced by the central nervous system.

Table 29.2 Warning Signs of Drug Addiction

1. Tolerance; takes increasing amounts of the drug to get the same effect.
2. Habituation; takes continued drug use over time to maintain the self-perception of functioning normally.
3. Inability to stop or curtail drug use, even if desire to do so persists.
4. Concealment; not wanting others to know of the drug use.
5. Extreme or dangerous actions to get and use a drug, as by stealing, by asking more than one doctor for prescriptions, or by jeopardizing employment by using drugs at work.
6. Deterioration of professional and personal relationships.
7. Anger and defensiveness if someone suggests there may be a problem.
8. Drug use preferred over previous customary activities.

Depressants Depressants such as alcohol (ethyl alcohol) and barbiturates slow motor responses by inhibiting ACh output. Alcohol stimulates the release of endorphins and GABA, so users typically experience a brief euphoria followed by depression. Combining alcohol with barbiturates can be deadly. As Section 5.1 explains, alcohol abuse damages the brain, liver, and other organs. Alcoholics deprived of the drug frequently undergo tremors, seizures, nausea, and hallucinations.

Hallucinogens Hallucinogens distort sensory perception and bring on a dreamlike state. LSD (lysergic acid diethylamide) resembles serotonin and binds to receptors for it. Tolerance develops, but LSD is not addictive. However, users can get hurt, and even die, because they do not perceive and respond to hazards such as oncoming cars. Flashbacks, or brief distortions of perceptions, may occur years after the last intake of LSD. Two related drugs, mescaline and psilocybin, have weaker effects.

Marijuana consists of parts of *Cannabis* plants. Smoking a lot of marijuana can cause hallucinations. More often, users become relaxed and sleepy as well as uncoordinated and inattentive. The active ingredient, THC (delta-9-tetrahydrocannabinol), alters levels of dopamine, serotonin, norepinephrine, and GABA. Chronic use can impair short-term memory and decision-making ability.

Take-Home Message **How do neurological disorders and drugs affect the flow of information in the nervous system?**

> Neurological disorders lower the amount of a neurotransmitter or the balance among neurotransmitters.

> Psychoactive drugs act by stimulating release, inhibiting breakdown, or mimicking the action of natural neurotransmitters. Their use can alter the body's ability to produce neurotransmitter and can cause addiction.

❯ Peripheral nerves that run through your body carry signals to and from the spinal cord and brain.

Axons Bundled as Nerves

Information flows rapidly through the vertebrate body along nerves of the peripheral nervous system. Each nerve consists of axons of sensory neurons, motor neurons, or both bundled inside a sheath of connective tissue. Figure 29.13A shows the structure of a vertebrate nerve.

Neuroglial cells called Schwann cells wrap like jelly rolls around axons of most peripheral nerves, one after another. They form a sheath of **myelin**: an insulator that makes action potentials flow faster. Ions cannot cross the neural membrane where it is sheathed. As a result, ion disturbances associated with an action potential spread quickly through the axon cytoplasm until they reach a node, a small, exposed gap between two Schwann cells (Figure 29.13B–D). At each node, the membrane is loaded with gated sodium channels. When these gates open, an action potential occurs. By jumping from node to node in long axons, a signal can move as fast as 120 meters per second. In unmyelinated axons, the maximum speed is about 10 meters per second.

Somatic and Autonomic Divisions

The vertebrate peripheral nervous system has two divisions. Nerves of the **somatic nervous system** relay commands to skeletal muscles. This is the only part of the nervous system that is normally under voluntary control. The somatic nervous system also relays information from sensory receptors in the skin and joints to the central nervous system. You can feel a pinch or wiggle your toes because of signals that travel along somatic nerves.

Nerves of the **autonomic nervous system** relay signals to smooth and cardiac muscle, and to glands. They also relay signals about internal conditions to the central nervous system. Signals traveling along autonomic nerves keep you breathing, adjust your heart rate, and inform your brain of your blood pressure.

A two-neuron pathway delivers autonomic signals to organs. The first neuron has its cell body in the brain or spinal cord. Axons of this neuron synapse on a second neuron at a ganglion. Axons of the second neuron synapse with the organ that they control.

Both sympathetic and parasympathetic nerves signal most organs. The two types of nerves work antagonistically, meaning that the signals from one division oppose signals from the other (Figure 29.14). For instance, both types of nerves signal smooth muscle cells in the gut wall. At the same time that sympathetic neurons are signaling these muscles to contract less, parasympathetic neurons are telling the same cells to contract more. The response of the smooth muscle cell to the competing signals is determined through synaptic integration.

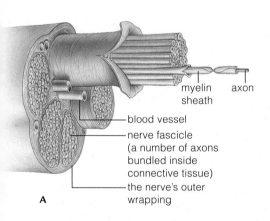

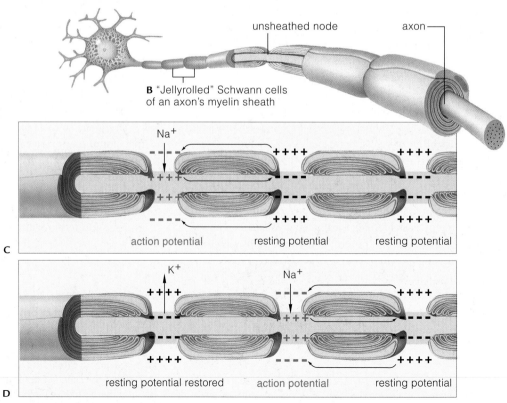

B "Jellyrolled" Schwann cells of an axon's myelin sheath

Figure 29.13 Animated **A** Structure of a vertebrate nerve. **B** Most nerves are wrapped in myelin, an insulating material produced by Schwann cells. **C** Action potentials occur only at nodes, where there are gated ion channels and no myelin. After an action potential occurs at a node, positive ions diffuse quickly through the cytoplasm to the next node because the myelin prevents them from leaking out across the membrane.

D The arrival of positive ions at the next node pushes the region to threshold, and an action potential occurs.

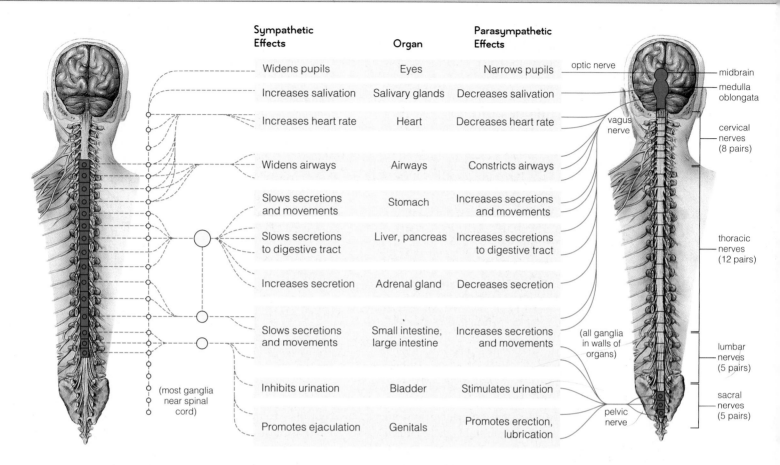

Sympathetic Effects	Organ	Parasympathetic Effects
Widens pupils	Eyes	Narrows pupils
Increases salivation	Salivary glands	Decreases salivation
Increases heart rate	Heart	Decreases heart rate
Widens airways	Airways	Constricts airways
Slows secretions and movements	Stomach	Increases secretions and movements
Slows secretions to digestive tract	Liver, pancreas	Increases secretions to digestive tract
Increases secretion	Adrenal gland	Decreases secretion
Slows secretions and movements	Small intestine, large intestine	Increases secretions and movements
Inhibits urination	Bladder	Stimulates urination
Promotes ejaculation	Genitals	Promotes erection, lubrication

(most ganglia near spinal cord)

optic nerve · vagus nerve · (all ganglia in walls of organs) · pelvic nerve

midbrain · medulla oblongata · cervical nerves (8 pairs) · thoracic nerves (12 pairs) · lumbar nerves (5 pairs) · sacral nerves (5 pairs)

Figure 29.14 Animated Autonomic nerves and their effects. Autonomic signals travel to organs by a two-neuron path. The first neuron has its cell body in the brain or spinal region (indicated in *red* in the two body sketches). This neuron synapses on a second neuron at a ganglion. Sympathetic ganglia are close to the spinal cord. Parasympathetic ganglia are in or near the organs they affect. Axons of the second neuron in the pathway synapse with the organ.

›› **Figure It Out** What effect does simulation of the vagus nerve (a parasympathetic nerve) have on the heart?   Answer: It decreases heart rate.

Sympathetic neurons are most active in times of excitement or danger. Their axon terminals release norepinephrine. **Parasympathetic neurons** are most active in times of relaxation. Release of ACh from their axon terminals promotes housekeeping tasks such as digestion and urine formation.

When something startles or scares you, the output of parasympathetic nerves decreases and the output of sympathetic nerves increases. As a result, your heart rate and blood pressure rise, you sweat more and breathe faster. Your adrenal glands secrete epinephrine (also known as adrenaline). You become intensely aroused, so you are ready to fight or make a fast getaway. This reaction is commonly described as the "fight-or-flight" response.

autonomic nervous system Set of nerves that relay signals to and from internal organs and to glands.
myelin Insulating material that wraps most axons and increases the speed of signal transmission.
parasympathetic neurons Neurons of the autonomic system that encourage housekeeping tasks.
somatic nervous system Set of nerves that control skeletal muscle and relay signals from joints and skin.
sympathetic neurons Neurons of the autonomic system that prepare the body for danger or excitement.

Take-Home Message **What is the function of the peripheral nervous system?**

❯ The peripheral nervous system consists of nerves that extend through the body and connect with the central nervous system.

❯ Neurons of the somatic portion of this system control skeletal muscle, and convey information about the external environment to the central nervous system.

❯ Neurons of the autonomic portion of this system control smooth muscle, cardiac muscle, and glands; and they convey information about the internal environment to the central nervous system.

❯ Most organs receive both sympathetic and parasympathetic stimulation from the autonomic system. Sympathetic stimulation prepares a body for "fight-or-flight." Parasympathetic stimulation encourages housekeeping tasks.

❯ The spinal cord serves as a highway for information traveling to and from the brain, and as the integrating center for reflexes that do not involve the brain.

❮ Link to Stem cells 28.1

Structure of the Spinal Cord

Your **spinal cord** is about as thick as your thumb. It runs through the vertebral column and connects peripheral nerves with the brain (Figure 29.15). The brain and spinal cord together constitute the central nervous system (CNS). Three membranes, called **meninges**, cover and protect the central nervous system. **Cerebrospinal fluid** fills the space between the meninges, the central canal of the spinal cord, and cavities within the brain. This clear fluid cushions blows, thus protecting these organs.

The outer part of the spinal cord is **white matter**. It consists of bundles of myelin-sheathed axons. In the CNS, such bundles are called tracts, rather than nerves. The tracts carry information from one part of the central nervous system to another. **Gray matter** makes up the bulk of the CNS. It consists of cell bodies, dendrites, and many neuroglial cells. In cross-section, the spinal cord's gray matter has a butterfly-like shape.

The peripheral nervous system's spinal nerves connect to the spinal cord at dorsal and ventral branches called "roots." A spinal nerve has both sensory and motor components. Sensory information travels to the spinal cord through a dorsal root. Cell bodies of sensory neurons are found in dorsal root ganglia. Motor signals travel away

from the spinal cord through a ventral root. Cell bodies of motor neurons are in the spinal cord's gray matter.

Reflex Pathways

Reflexes are the simplest and most ancient paths of information flow. A reflex is an automatic response to a stimulus, a movement or other action that does not require thought. Basic reflexes do not require any learning. With such reflexes, sensory signals flow to the spinal cord or the brain stem, which then calls for a response by way of motor neurons.

The stretch reflex is a spinal reflex that causes a muscle to contract after some force stretches it (Figure 29.16). For example, suppose you hold a bowl as someone drops fruit into it ❶. The addition of a piece of fruit increases the weight of the bowl, causing the biceps muscle in your upper arm to passively lengthen. Lengthening of the biceps excites receptor endings of sensory neurons in muscle spindles that wrap around a muscle. As a result of this excitation, action potentials flow along the axon of the sensory neuron toward the spinal cord ❷.

Inside the spinal cord, axons of the sensory neuron synapse on dendrites of one of the motor neurons that controls the stretched muscle ❸. Signals from the sensory neurons cause action potentials to travel along the axon of the motor neuron ❹. The motor neurons release ACh at the neuromuscular junction ❺. In response to this signal, the biceps contracts and steadies the arm against the additional weight ❻.

The knee-jerk response is another example of the stretch reflex. A tap just below the knee stretches the thigh muscle. The stretch is detected by muscle spindles in this muscle. The muscle spindles send signals to the spinal cord, where motor neurons become excited. As a result, signals flow from the spinal cord back to the leg, and the leg jerks in response.

Spinal Cord Injury and Multiple Sclerosis

An injury that interrupts tracts of the spinal cord can cause a loss of sensation and paralysis. Symptoms depend on what portion of the cord is damaged. Nerves carrying signals to and from the upper body lie higher in the cord than nerves that govern the lower body. An injury to the lumbar region of the cord often paralyzes the legs. An injury to higher cord regions can paralyze all limbs, as well as muscles used in breathing.

Unlike axons of peripheral nerves, those of the spinal cord do not grow back and restore function. Thus, spinal injuries cause permanent disability. Worldwide, more than 2.5 million people are disabled by a spinal injury.

As Section 28.1 explained, researchers are testing the use of embryonic stem cells in the treatment of spinal cord injuries. The first human trial of this treatment has

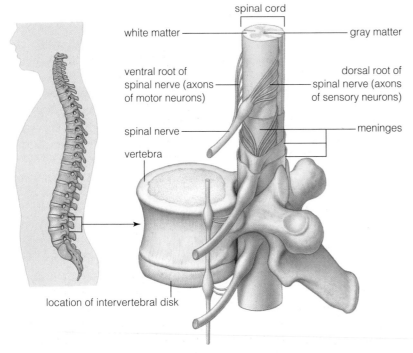

spinal cord

white matter — gray matter

ventral root of spinal nerve (axons of motor neurons) — dorsal root of spinal nerve (axons of sensory neurons)

spinal nerve — meninges

vertebra

location of intervertebral disk

Figure 29.15 Animated Location and organization of the spinal cord.

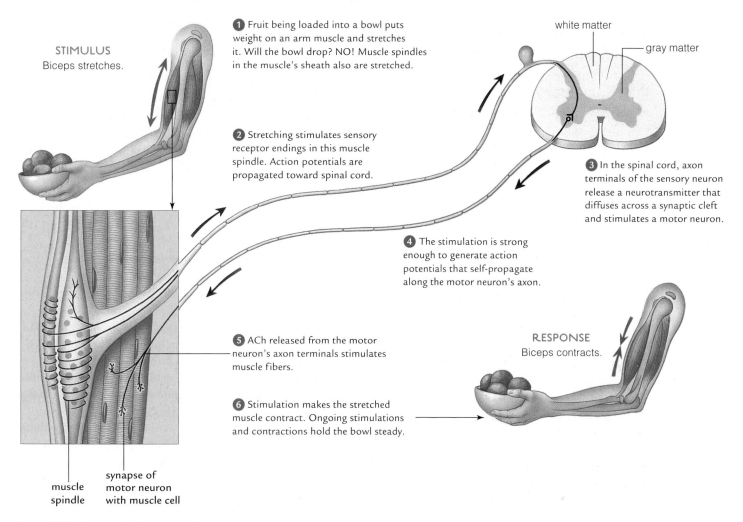

STIMULUS
Biceps stretches.

1 Fruit being loaded into a bowl puts weight on an arm muscle and stretches it. Will the bowl drop? NO! Muscle spindles in the muscle's sheath also are stretched.

2 Stretching stimulates sensory receptor endings in this muscle spindle. Action potentials are propagated toward spinal cord.

white matter

gray matter

3 In the spinal cord, axon terminals of the sensory neuron release a neurotransmitter that diffuses across a synaptic cleft and stimulates a motor neuron.

4 The stimulation is strong enough to generate action potentials that self-propagate along the motor neuron's axon.

5 ACh released from the motor neuron's axon terminals stimulates muscle fibers.

RESPONSE
Biceps contracts.

6 Stimulation makes the stretched muscle contract. Ongoing stimulations and contractions hold the bowl steady.

muscle spindle

synapse of motor neuron with muscle cell

Figure 29.16 **Animated** The stretch reflex, an example of a spinal reflex.

just begun. It is not aimed at restoring function to people with long-standing paralysis, but rather at helping those who have just become injured. Even so, if this trial is successful, it may open the door to a similar treatment for people who have been paralyzed for a long time.

The autoimmune disorder multiple sclerosis (MS) also impairs spinal cord function. White blood cells of affected people attack and destroy oligodendrocytes—neuroglial

cells that are the central nervous system's equivalent of Schwann cells. Oligodendrocytes produce the insulating myelin that wraps around axons in the spinal cord and brain. As MS progresses, this myelin becomes replaced by scar tissue, so affected axons transmit signals more and more slowly. Symptoms of MS frequently include dizziness, numbness of hands and feet, muscle weakness, fatigue, and visual problems.

cerebrospinal fluid Fluid that surrounds the brain and spinal cord and fills ventricles within the brain.
gray matter Tissue in brain and spinal cord that consists of cell bodies, dendrites, and neuroglial cells.
meninges Membranes that enclose the brain and spinal cord.
reflex Automatic response that occurs without conscious thought or learning.
spinal cord Portion of central nervous system that connects peripheral nerves with the brain.
white matter Tissue of brain and spinal cord consisting of myelinated axons.

Take-Home Message What are the functions of the spinal cord?

> Tracts of the spinal cord relay information between peripheral nerves and the brain. The axons involved in these pathways make up the bulk of the cord's white matter. Cell bodies, dendrites, and neuroglia make up gray matter.

> The spinal cord also has a role in some simple reflexes. Signals from sensory neurons enter the cord through the dorsal root of spinal nerves. Commands for responses go out along the ventral root of these nerves.

⟩ The brain is the main integrating organ in the vertebrate nervous system.
⟨ Link to Fishes 24.3

Brain Development and Evolution

In all vertebrates, the embryonic neural tube develops into a spinal cord and brain. During development, the brain becomes organized as three functional regions: the forebrain, midbrain, and hindbrain (Figure 29.17).

The hindbrain is continuous with the spinal cord and is largely responsible for reflexes and coordination. Fishes and amphibians have the most pronounced midbrain (Figure 29.18). Their forebrain sorts out sensory input and initiates motor responses. The midbrain is reduced in birds and mammals; their expanded forebrain took over what were previously midbrain functions.

Ventricles and the Blood–Brain Barrier

The space inside the embryonic neural tube persists in adult vertebrates as a system of cavities and canals filled with cerebrospinal fluid. This clear fluid forms when water and small molecules are filtered out of the blood

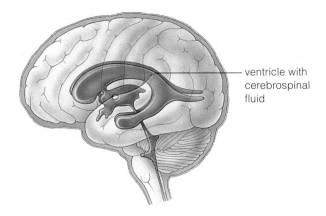

Figure 29.19 Cerebrospinal fluid. This clear fluid, shown here in *blue*, forms inside ventricles (chambers) within the brain.

ventricle with cerebrospinal fluid

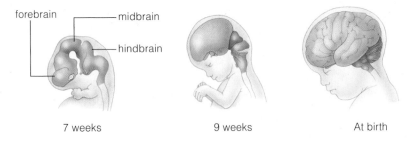

7 weeks   9 weeks   At birth

Figure 29.17 Development of the human brain. At 7 weeks, there is a hollow neural tube with regions that will develop into the forebrain, midbrain, and hindbrain.

into brain cavities called ventricles (Figure 29.19). The fluid then seeps out and bathes the brain and spinal cord. It returns to the bloodstream by entering veins.

A **blood–brain barrier** controls the composition and concentration of the cerebrospinal fluid. No other portion of extracellular fluid has solute concentrations maintained within such narrow limits. The blood–brain barrier is formed by the walls of blood capillaries that service the brain. In most parts of the brain, tight junctions form a seal between adjoining cells of the capillary wall. As a result, water-soluble substances must pass through the cells to reach the brain. Transport proteins in the plasma membrane of these cells allow essential nutrients to cross. Oxygen and carbon dioxide diffuse across the barrier, but most waste products such as urea cannot breach it.

The blood–brain barrier is not perfect; some toxins such as nicotine, alcohol, caffeine, and mercury can cross it. Also, inflammation or a blow to the head can damage it, allowing unwanted substances to slip across.

The Human Brain

The average human brain weighs 1,240 grams, or 3 pounds. It contains about 100 billion interneurons, and neuroglia make up more than half of its volume.

The portion of the hindbrain just above the spinal cord is the **medulla oblongata** (Figure 29.20). It influences the strength of heartbeats and the rhythm of breathing. It also controls reflexes such as swallowing, coughing, vomiting, and sneezing. Above the medulla oblongata lies the **pons**, which also affects breathing. Pons means "bridge," a reference to the tracts that extend through the pons to the midbrain.

The **cerebellum** lies at the back of the brain and is about the size of a plum. It is densely packed with neurons, having more than all other brain regions combined. The cerebellum controls posture and coordinates

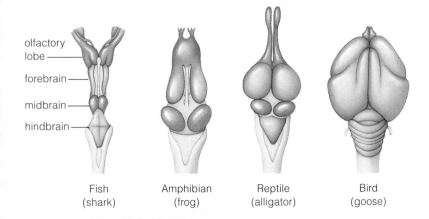

olfactory lobe
forebrain
midbrain
hindbrain

Fish (shark)   Amphibian (frog)   Reptile (alligator)   Bird (goose)

Figure 29.18 Verebrate brains, dorsal views. The sketches are not to the same scale.

Figure 29.20 The human brain.

A View of a whole brain from above. The meninges have been removed.

B The right half of a brain that has been sectioned along the central fissure.

C Major components of the brain and their functions.

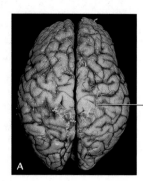

A

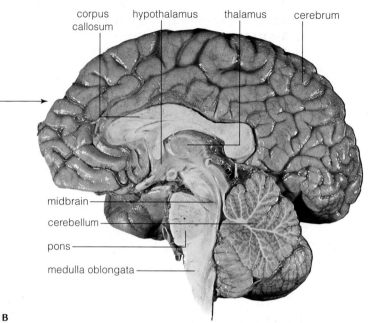

corpus callosum   hypothalamus   thalamus   cerebrum

midbrain
cerebellum
pons
medulla oblongata

B

voluntary movements. Among other effects, excessive alcohol consumption disrupts coordination by impairing the function of neurons in the cerebellum. Police officers evaluate the extent of this impairment by asking a person suspected of being drunk to walk a straight line.

In humans, the midbrain is the smallest of the three brain regions. It plays an important role in reward-based learning. The pons, medulla, and midbrain are collectively referred to as the brain stem.

The forebrain contains the **cerebrum**, the largest part of the human brain. A fissure divides the cerebrum into right and left hemispheres. A thick band of tissue, the corpus callosum, connects the hemispheres. Each hemisphere has an outer layer of gray matter called the cerebral cortex. Our large cerebral cortex is the part of the brain responsible for our unique capacities such as language and abstract thought.

Most sensory signals destined for the cerebrum pass through the adjacent **thalamus**, which sorts them and sends them to the proper region of the cerebral cortex.

The **hypothalamus** ("under the thalamus") is the center for homeostatic control of the internal environment. It receives signals about the state of the body and regulates thirst, appetite, sex drive, and body temperature. It also interacts with the adjacent pituitary gland as a central control center for the endocrine system.

Forebrain

Cerebrum	Localizes, processes sensory inputs; initiates, controls skeletal muscle activity; governs memory, emotions, abstract thought
Thalamus	Relays sensory signals to and from cerebral cortex; has a role in memory
Hypothalamus	With pituitary gland, functions in homeostatic control. Adjusts volume, composition, temperature of internal environment; governs behaviors that ensure homeostasis (e.g., thirst, hunger)

Midbrain   Relays sensory input to the forebrain

Hindbrain

Pons	Bridges cerebrum and cerebellum, also connects spinal cord with forebrain. With the medulla oblongata, controls rate and depth of respiration
Cerebellum	Coordinates motor activity for moving limbs and maintaining posture, and for spatial orientation
Medulla oblongata	Relays signals between spinal cord and pons; functions in reflexes that affect heart rate, blood vessel diameter, and respiratory rate. Also involved in vomiting, coughing, other reflexive functions

C

blood–brain barrier Protective barrier that prevents unwanted substances from entering cerebrospinal fluid.
cerebellum Hindbrain region responsible for coordinating voluntary movements.
cerebrum Forebrain region that controls higher functions.
hypothalamus Forebrain region that controls processes related to homeostasis; control center for endocrine functions.
medulla oblongata Hindbrain region that controls breathing rhythm and reflexes such as coughing and vomiting.
pons Hindbrain region between medulla oblongata and midbrain; helps control breathing.
thalamus Forebrain region that relays signal to the cerebrum.

Take-Home Message How does the vertebrate brain develop, and what are its functional regions?

❯ The vertebrate brain develops from a hollow neural tube, the interior of which persists in adults as a system of cavities and canals filled with cerebrospinal fluid. The fluid cushions nervous tissue from sudden, jarring movements.

❯ Tissue of the embryonic neural tube develops into the hindbrain, forebrain, and midbrain. The hindbrain controls reflexes and coordination. The unique capacities of humans arise in regions of their enlarged forebrain.

> Our capacity for voluntary action, language, and conscious thought arise from activity of the cerebral cortex.
> The cortex interacts with other brain regions in shaping emotional responses and making memories.

Functions of the Cerebral Cortex

The **cerebral cortex** is the outer layer of the cerebrum, a 2-millimeter-thick, highly folded outer layer of gray matter. Prominent folds in the cortex are used as landmarks to define the cerebrum's frontal, parietal, temporal, and occipital lobes (Figure 29.21).

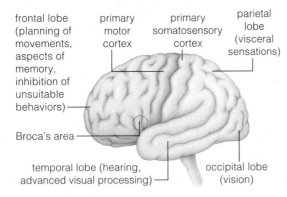

Figure 29.21 Primary receiving and integrating centers of the human cerebral cortex. Association areas coordinate and process sensory input from diverse receptors.

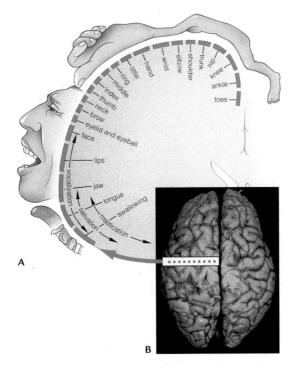

Figure 29.22 A Slice of the primary motor cortex through the region identified in **B**. Sizes of body parts draped over the slice are distorted to indicate which ones get the most precise control.

Much of each frontal lobe consists of association areas. These areas are devoted to integrating information and governing conscious actions. Planning ahead and interacting with other people requires the function of this lobe.

During the 1950s, more than 20,000 people had their frontal lobes deliberately damaged by having a frontal lobotomy. This surgical procedure was intended to treat mental illness, personality disorders, and even severe headaches. One physician carried out the procedure using an ice pick. He inserted this sharp metal spike through the bone at the back of the eye and wiggled it to destroy tissue of the frontal lobe. Frontal lobotomies sometimes made patients calmer, but the procedure also blunted emotions and impaired their ability to plan, concentrate, and behave appropriately in social situations.

A **primary motor cortex**, a region that controls skeletal muscles, lies near the rear of each frontal lobe (Figure 29.22). Each hemisphere controls and receives signals from the opposite side of the body. For example, signals to move your right arm originate in the motor cortex of your left cerebral hemisphere.

The two hemispheres differ somewhat in their function. For example, in the 90 percent of people that are right-handed, the left hemisphere is more active in controlling movement and in language. Broca's area, which is the part of the frontal lobe that helps us translate thoughts into speech, is usually in the left hemisphere. It controls the tongue, throat, and lip muscles and gives humans our capacity to speak complex sentences. Damage to Broca's area often prevents normal speech, although an affected person can still understand language.

The abilities of each hemisphere are flexible. When a stroke or injury damages tissue on one side of the brain, the other hemisphere often can take on new tasks. People can even function with a single hemisphere.

Sensory areas of the cerebral cortex allow us to perceive sensations. The primary somatosensory cortex of the parietal lobe is the receiving area for sensory input from the skin and joints. When someone taps you on your left shoulder, signals that arrive in the primary somatosensory cortex of your right parietal lobe alert you to the tap. Another sensory area in the parietal lobe receives signals about taste. In the occipital lobe, the primary visual cortex integrates incoming signals from both eyes. The perception of sounds and odors arises in the primary sensory areas of the temporal lobe.

Connections With the Limbic System

The **limbic system** encircles the upper brain stem. It governs emotions, assists in memory, and correlates organ activities with self-gratifying behavior such as eating and sex. That is why the limbic system is called our emotional-visceral brain. It can put a heart on fire with

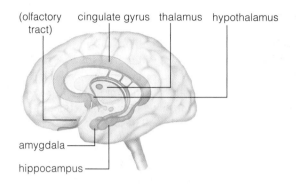

Figure 29.23 Limbic system components.

(olfactory tract) cingulate gyrus thalamus hypothalamus

amygdala

hippocampus

In Pursuit of Ecstasy (revisited)

Now that you know a bit about how the brain functions, take a moment to reconsider the effects of MDMA, the active ingredient in Ecstasy. MDMA harms and possibly kills brain interneurons that produce the neurotransmitter serotonin. Neurons do not divide, so damaged ones are not replaced. MDMA also damages the blood–brain barrier. In one study of rats, the blood–brain barrier remained impaired for as long as 10 weeks after MDMA use. Damage to this protective mechanism allows harmful molecules to slip into the cerebrospinal fluid.

How Would You Vote? Some argue that addiction is a disorder and that it is better to send drug addicts to be treated in rehabilitation programs than to jail. Do you agree? See CengageNow for details, then vote online (cengagenow.com).

passion and a stomach on fire with indigestion. These and other "gut reactions" can be overridden by signals from the prefrontal cortex at the very front of the brain.

The hypothalamus, hippocampus, amygdala, and adjacent structures make up the limbic system (Figure 29.23). The hypothalamus is the major control center for homeostatic responses; it also correlates emotions with visceral activities. The hippocampus helps store and retrieve memories of earlier threats. The almond-shaped amygdala helps us interpret social cues, and it contributes to the sense of self. The amygdala is highly active during episodes of fear and anxiety, and it is typically overactive in people with panic disorders.

Evolutionarily, the limbic system is related to the olfactory lobes. Olfactory input causes signals to flow to the hippocampus, amygdala, and hypothalamus as well as to the olfactory cortex. That is one reason why you feel warm and fuzzy when you recall a scent you associate with someone special. Signals about taste also travel to the limbic system and can call up emotional responses.

Making Memories

The cerebral cortex receives information continually, but only a fraction of it becomes memories. Memory forms in stages. Short-term memory lasts seconds to hours. This type of memory holds a few bits of information, a set of numbers, words of a sentence, and so forth. In long-term memory, larger chunks of information get stored more or less permanently (Figure 29.24).

Different types of memories are stored and brought to mind by different mechanisms. Repetition of motor tasks creates highly persistent skill memories. Once you learn to ride a bicycle, drive a car, dribble a basketball, or play

the piano, you are unlikely to forget how. Skill memories involve the cerebellum, which controls motor activity.

Declarative memory stores facts and impressions. It helps you remember how a lemon smells and that a quarter is worth more than a dime. Input to declarative memory begins with signals from the sensory cortex to the amygdala. The amygdala screens the input, and sends some signals along to the hippocampus. Declarative memory also involves the temporal lobe. Damage to this lobe can cause amnesia, a loss of declarative memory.

Emotions influence memory retention. For instance, epinephrine released during times of stress helps place short-term memories into long-term storage.

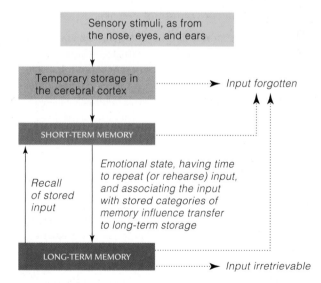

Figure 29.24 Stages in memory processing.

cerebral cortex Outer gray matter layer of the cerebrum; region responsible for most complex behavior.
limbic system Group of brain structures that govern emotion.
primary motor cortex Region of frontal lobe that controls voluntary movement.

Take-Home Message **What is the cerebral cortex?**

> The cerebral cortex, the outer layer of gray matter, has areas that receive and integrate sensory information. It also controls conscious thought and actions.

> The cerebral cortex interacts with the limbic system, a set of brain structures that collectively affect emotions and contribute to memory.

Summary

Section 29.1 Psychoactive drugs such as Ecstasy provide enjoyable sensations by disrupting normal function of brain cells. Sometimes these disruptions cause unwanted effects or even death.

Section 29.2 All **neurons** are excitable cells. **Sensory neurons** detect stimuli and send signals to motor neurons or interneurons. **Interneurons** integrate information and send signals to one another or to motor neurons. **Motor neurons** carry signals to effectors (muscles and glands). **Neuroglial cells** provide functional and structural support to neurons.

Cnidarians have a **nerve net** with no central control point. Simple bilateral animals have a nervous system with a cluster of **ganglia** at the head end, ventral **nerve cords** running the length of the body, and **nerves** extending out from the nerve cords. In vertebrates, the **central nervous system** is a brain and spinal cord. Paired nerves of the **peripheral nervous system** connect the brain and spinal cord to the rest of the body.

Sections 29.3, 29.4 Dendrites of a neuron receive signals; endings of **axons** transmit signals. The voltage difference across a neuron plasma membrane is a **membrane potential**. The **resting potential** arises from the action of transport proteins in the neuron membrane, and the higher concentration of negatively charged proteins inside the cell. An **action potential** is a brief reversal of the charge difference across the membrane.

Section 29.5 An action potential occurs only if a disturbance causes membrane potential to rise to **threshold potential**. The action potential begins when gated sodium channels open. **Positive feedback** causes even more sodium gates to open, so sodium rushes into the axon. The influx makes the axon's cytoplasm more positive than extracellular fluid. As a result, gated potassium channels open and potassium rushes out. Activity of sodium–potassium pumps maintains the ion gradients that are required for the resting membrane potential.

Section 29.6 Neurons signal other neurons, muscle fibers, or gland cells at **synapses**. When an action potential arrives at the presynaptic cell's axon terminals, it triggers release of **neurotransmitter**. These molecules diffuse across the synaptic cleft and bind to receptors on the postsynaptic cell. For example, at a **neuromuscular junction**, a motor neuron releases **acetylcholine** that binds to receptors on a muscle fiber.

Neurotransmitter can have an inhibitory or excitatory effect on a postsynaptic cell. The postsynaptic cell's response is determined by **synaptic integration** of messages arriving at the same time.

Section 29.7 Symptoms of many neurological disorders arise from lowered levels of neurotransmitter. These disorders are treated with drugs that raise the level of the appropriate neurotransmitter. Psychoactive drugs mimic neurotransmitters or disrupt their release or uptake. For example, morphine mimics the action of **endorphins**, a type of natural painkiller.

Section 29.8 Nerves are bundles of axons that carry signals through the body. **Myelin** sheaths that enclose most axons increase the rate of signal conduction. The peripheral nervous system is functionally divided into the **somatic nervous system**, which controls skeletal muscles, and the **autonomic nervous system**, which controls internal organs and glands. **Sympathetic neurons** of the autonomic system increase their output in times of stress or danger. During less stressful times, signals from **parasympathetic neurons** dominate.

Section 29.9 Like the brain, the **spinal cord** consists of **white matter** (with myelinated axons) and **gray matter** (with cell bodies, dendrites, and neuroglia). The spinal cord and brain are enclosed by membranous **meninges** and cushioned by **cerebrospinal fluid**. Spinal reflexes involve peripheral nerves and the spinal cord. A **reflex** is an automatic response to stimulation; it does not require conscious thought.

Sections 29.10, 29.11 A vertebrate embryo's neural tube develops into the spinal cord and brain. A **blood–brain barrier** prevents harmful substances from reaching the brain. The **pons** and **medulla oblongata** control reflexes involved in breathing and other vital tasks. The **cerebellum** coordinates motor activity. The **thalamus** and **hypothalamus** function in homeostasis.

The **cerebral cortex**, the outermost portion of the **cerebrum** governs complex functions. Its **primary motor cortex** controls voluntary movement. The cerebral cortex interacts with the **limbic system** in emotions and memory.

Self-Quiz   Answers in Appendix III

1. _____ relay messages from the brain and spinal cord to muscles and glands.
   a. Motor neurons   c. Interneurons
   b. Sensory neurons   d. Neuroglia

2. When a neuron is at rest, _____ .
   a. it is at threshold potential
   b. gated sodium channels are open
   c. it holds less sodium than the interstitial fluid
   d. both a and c

Data Analysis Activities

Prenatal Effects of Ecstasy Animal studies are often used to assess effects of prenatal exposure to illicit drugs. For example, Jack Lipton used rats to study the behavioral effect of prenatal exposure to MDMA, the active ingredient in Ecstasy. He injected female rats with either MDMA or saline solution when they were 14 to 20 days pregnant and their offspring's brains were forming. When those offspring were 21 days old, Lipton tested their response to a new environment. He placed each young rat in a new cage and used a photobeam system to record how much each rat moved around before settling down. Figure 29.25 shows his results.

1. Which rats moved most (caused the most photobeam breaks) during the first 5 minutes in a new cage, those prenatally exposed to MDMA or the controls?
2. How many photobeam breaks did the MDMA-exposed rats make during their second 5 minutes in the new cage?
3. Which rats moved most during the last 5 minutes of the study?
4. Does this study support the hypothesis that exposure to MDMA affects a developing rat's brain?

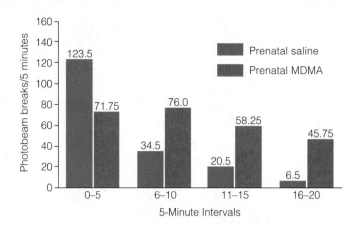

Figure 29.25 Effect of prenatal exposure to MDMA on activity levels of 21-day-old rats placed in a new cage. Movements were detected when the rat interrupted a photobeam. Rats were monitored at 5-minute intervals for a total of 20 minutes. *Blue* bars are average numbers of photobeam breaks for rats whose mothers received saline; *red* bars are rats whose mothers received MDMA.

3. Action potentials occur when _____ .
   a. potassium gates close
   b. a stimulus pushes membrane potential to threshold
   c. sodium–potassium pumps kick into action
   d. neurotransmitter is reabsorbed

4. Neurotransmitters are released by _____ .
   a. axon terminals   c. dendrites
   b. the cell body   d. the myelin sheath

5. Myelin around peripheral nerves is made by _____ .
   a. Schwann cells   c. sensory neurons
   b. motor neurons   d. interneurons

6. Skeletal muscles are controlled by _____ .
   a. sympathetic signals   c. somatic nerves
   b. parasympathetic signals d. both a and b

7. When you sit quietly on the couch and read, output from the _____ system prevails.
   a. sympathetic   c. Both prevail.
   b. parasympathetic   d. Neither prevails.

8. Skeletal muscles contract in response to _____ .
   a. acetylcholine (ACh)   c. serotonin
   b. dopamine   d. endorphins

9. The cerebrum is part of the _____ .
   a. forebrain   c. hindbrain
   b. midbrain   d. brain stem

10. Morphine mimics natural painkillers called _____ .

11. Parkinson's disease involves low levels of _____ .
    a. acetylcholine (ACh) c. serotonin
    b. dopamine   d. endorphins

12. Myelinated axons make up the _____ matter of the brain and spinal cord.

13. Match the terms with their descriptions.
    ___ muscle spindle   a. coordinates motor activity
    ___ neurotransmitter   b. connects the hemispheres
    ___ limbic system   c. protects brain and spinal
    ___ corpus callosum   cord from some toxins
    ___ cerebral cortex   d. type of signaling molecule
    ___ cerebellum   e. support team for neurons
    ___ neuroglia   f. wrap brain and spinal cord
    ___ ganglion   g. roles in emotion, memory
    ___ blood–brain   h. most complex integration
        barrier   i. cluster of neuron cell bodies
    ___ meninges   j. stretch-sensitive receptor

Additional questions are available on **CENGAGENOW**.

Critical Thinking

1. In human newborns, especially premature ones, the blood–brain barrier is not yet fully developed. Why is this one reason to pay careful attention to the diet of infants?

2. Injections of botulinum toxin (Botox) into facial muscles can prevent movements that cause facial skin to wrinkle. Botox acts by preventing the release of a neurotransmitter. Which one?

Animations and Interactions on **CENGAGENOW**:
❯ Neuron's structure; Membrane properties; Action potential; Synapse function; Nerve structure and function; Sympathetic and parasympathetic effects; Structure of the spinal cord; Stretch reflex; Human cerebral cortex.

‹ Links to Earlier Concepts

This chapter draws on your understanding of sensory neurons (Section 29.2), action potentials (29.4), and reflexes (29.9). The discussion of vision touches on pigments and properties of light (6.2). There are examples of X-linked inheritance (14.4) and morphological convergence (16.8). You will also reconsider the sensory changes involved in the vertebrate move onto land (24.4).

Key Concepts

Sensory Pathways
Sensory systems consist of sensory receptors, nerves that carry signals, and brain regions that receive and process sensory input. Each type of sensory receptor reacts to a specific stimulus. Information about stimuli is encoded in the number and frequency of action potentials.

Somatic and Visceral Sensation
Somatic sensations include touch, pressure, pain, temperature, and muscle sense. They start at mechanoreceptors in skin, muscles, and near joints. Visceral sensations arise from stimulation of receptors in walls of soft internal organs.

30 Sensory Perception

Consider the sensory world of a whale, swimming 200 meters (650 feet) beneath the ocean surface. Almost no sunlight penetrates this deep, so the whale sees little as its moves through water. Many fishes detect motion with a lateral line system that responds to differences in water pressure. Fishes also use dissolved chemicals as navigational cues. However, a whale has no lateral line, and it has a very poor sense of smell. How does it know where it is going?

Whales rely heavily on their sense of hearing. Water is an ideal medium for transmitting sound waves; sound moves five times faster in water than in air. Whale ears are adapted to detecting underwater sounds. There are no external ear flaps. Instead, the whale jaw vibrates in response to sound waves traveling through the water. Vibrations are transmitted from the jaws, through a layer of fat, to the internal portion of the ears.

Whales use sound to communicate, locate food, and find their way around underwater. Killer whales and some other species of toothed whales are capable of echolocation. The whale emits high-pitched sounds and then listens as the echoes bounce off objects, including prey. Its ears are especially sensitive to sounds of high frequencies. Baleen whales, including the humpback whale, shown at the *left*, communicate using very low-pitched sounds that can travel across an entire ocean basin. Their ears are adapted to detect those very low sounds.

As the ocean becomes noisier, the superb acoustical adaptations of whales put them at risk. In 2001, some whales beached themselves near an area where the United States Navy was conducting training exercises using a new sonar system (Figure 30.1). This sonar system uses echoes of low-frequency sounds to detect submarines. The sounds are outside the range of human hearing, but whales perceive the sonar signals as intense blasts of sound and react in fear. As autopsies later revealed, the beached whales had blood in their ears and in acoustic fat. Apparently the sounds made them race to the surface. The rapid change in pressure damaged their tissues.

Naval sonar is just one source of underwater sound pollution. Massive tankers that carry oil and other goods between continents emit low-frequency sounds that frighten whales or drown out acoustical cues. Exploitation of undersea resources such as minerals adds to the distracting din.

In this chapter, we turn to sensory systems. With these organ systems, animals receive signals from inside and outside the body, decode them, and become aware of touches, sounds, sights, odors, and other sensations. As you will see, animals differ in the type and number of sensory receptors that sample the environment, and differ in their perception of it. Finding ways to meet human needs without needlessly disrupting the sensory worlds of other species is an ongoing challenge.

Figure 30.1 A few children drawn to one of the whales that beached itself during military testing of a new sonar system. Of sixteen stranded whales, six died on the beach. Volunteers pushed the remainder out to sea. Their fate is unknown.

Vision
Vision requires eyes with a dense array of photoreceptors and a brain that integrates signals from the receptors. The vertebrate eye works like a film camera; a single adjustable opening lets in light. A sensory pathway starts at the eye's retina and ends in the visual cortex.

Chemical Senses
Binding of specific chemicals activates chemoreceptors in the lining of the nose and mouth. Many animals also have organs that detect pheromones: chemicals that one member of a species uses to communicate with another member of the same species.

Balance and Hearing
The ear functions in the senses of balance and hearing. In both cases, movements stimulate mechanoreceptors. In balance, body movements are the source of the stimulation. In hearing, pressure from sound waves causes movement that stimulates mechanoreceptors.

> An animal's sensory receptors determine what features of the environment it can detect and respond to.

❮ Links to Sensory neurons 29.2, Action potential 29.4

Excitation of Sensory Neurons

The sensory portion of a vertebrate nervous system consists of sensory neurons that detect stimuli, nerves that carry information about the stimulus to the brain, and brain regions that process the information. A **stimulus** (plural, stimuli) is a form of energy that excites receptor endings of a sensory neuron. When this excitation is sufficiently powerful, an action potential occurs. The action potential travels along a peripheral nerve to the central nervous system.

Different types of sensory neurons respond to different stimulation. Figure 30.2 shows sensory receptors in human skin. **Mechanoreceptors** are excited by mechanical energy such as pressure, a change in position, or acceleration. **Pain receptors** respond to tissue damage. **Thermoreceptors** detect a temperature change. The skin has receptors that respond to heat and to cold. Other types of sensory receptors are found in other parts of the body. **Chemoreceptors** in the lining of the nose and mouth detect chemical energy of specific substances dissolved in the fluid that bathes them. The eyes detect light when **photoreceptors** become excited.

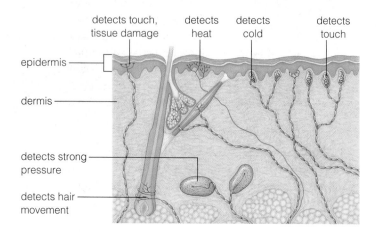

Figure 30.2 Sensory receptors in the skin.

labels:
detects touch, tissue damage
detects heat
detects cold
detects touch
epidermis
dermis
detects strong pressure
detects hair movement

chemoreceptor Sensory receptor that responds to a chemical.
mechanoreceptor Sensory receptor that responds to pressure, position, or acceleration.
pain receptor Sensory receptor that responds to tissue damage.
perception The meaning a brain derives from a sensation.
photoreceptor Sensory receptor that responds to light.
sensation Detection of a stimulus.
sensory adaptation Slowing or cessation of a sensory receptor's response to an ongoing stimulus.
stimulus Form of energy that a sensory receptor detects.
thermoreceptor Temperature-sensitive sensory receptor.

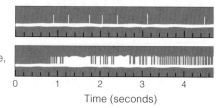

Light touch, low firing rate

Increased pressure, higher firing rate

Time (seconds)

Figure 30.3 Animated Recordings of action potentials from a mechanoreceptor in skin. The stronger the pressure, the more action potentials (*white* bars) occur per second.

Sources of Information About a Stimulus

Three variables allow your brain to determine the location and intensity of a stimulus. First, the brain takes into consideration the nerve that delivers the action potentials. For example, action potentials arriving via the optic nerve are interpreted as arising from visual stimuli. That is why you "see stars" if you press on your eye in a dark room.

Second, the brain assesses the frequency of action potentials. A higher frequency denotes a stronger stimulus. Press lightly on skin and a mechanoreceptor responds with a few action potentials per second. Press harder and the receptor increases its firing rate (Figure 30.3).

Finally, the number of sensory receptors firing gives the brain information about stimulus intensity. A gentle tap on the arm activates fewer receptors than a slap.

Stimulus duration also affects response. In **sensory adaptation**, sensory neurons cease firing in spite of continued stimulation. Walk into a house where an apple pie is in the oven and you notice the sweet scent of baking apples immediately. Then, within a few minutes, the scent seems to lessen. The odor does not actually change in intensity, but chemoreceptors in your nose adapt to it.

Sensation and Perception

Sensation is the detection of sensory signals. **Perception** arises when the brain assigns meaning to those signals. Consider what happens when you watch a plane fly away. As the distance between you and the plane increases, the image of the plane on your eye becomes smaller and smaller. You perceive this change in sensation as indicative of increasing distance, rather than a shrinking plane.

Take-Home Message What is the basis for sensation and perception?

> Sensation arises when sensory receptors detect specific stimuli. Nerves carry signals to brain regions that assess a stimulus according to which nerve pathway is signaling it, the frequency of action potentials, and the number of axons that have been excited by the stimulus.

> Perceptions arise when the brain interprets sensations.

> Sensory receptors involved in somatic and visceral sensations are distributed throughout the body.

< Link to Stretch reflex 29.9

Sensory neurons responsible for **somatic sensations** are located in skin, muscle, tendons, and joints. Somatic sensations are easily localized to a specific part of the body. In contrast, **visceral sensations** arise from neurons in the walls of soft internal organs and are often difficult to pinpoint. It is easy to determine exactly where someone is touching you, but difficult to localize a stomachache, a feeling of bloating, or nausea.

The Somatosensory Cortex

Signals from the sensory neurons involved in somatic sensation travel along axons to the spinal cord, then along tracts in the spinal cord to the brain. The signals end up in the somatosensory cortex, a part of the cerebral cortex. Like the motor cortex (Section 29.11), the somatosensory cortex has neurons arrayed like a map of the body (Figure 30.4). Body parts shown as disproportionately large in the "body" mapped onto this brain correspond to body regions with the most sensory receptors, such as the fingertips, face, and lips.

Sensory receptors depicted in Figure 30.2 report to the somatosensory cortex about touch, pain, and temperature. The fourth somatosensory sense is muscle sense, which relates to the positioning of body parts. Muscle spindles (Section 29.9) are mechanoreceptors that contribute to this sense. The more a muscle stretches, the more frequently these receptors fire. Mechanoreceptors near joints and tendons also contribute to muscle sense.

Pain

Pain is the perception of a tissue injury. Somatic pain arises as a response to signals from pain receptors in skin, skeletal muscles, joints, and tendons. Visceral pain is associated with organs inside body cavities. It occurs as a response to a smooth muscle spasm, inadequate blood flow to an organ, overstretching of a hollow organ such as the stomach, and other abnormal conditions.

Injured or distressed body cells release local signaling molecules such as histamine and prostaglandins. These molecules stimulate neighboring pain receptors, causing action potentials to travel along axons of sensory neurons to the spinal cord. Here, the axons synapse with the

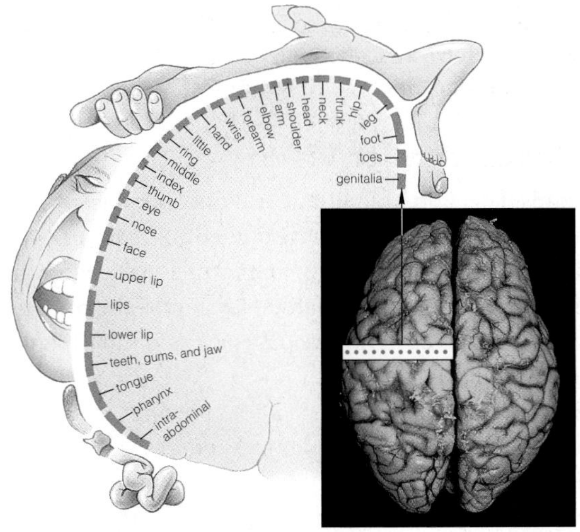

Figure 30.4 A map of where different body regions are represented in the human primary somatosensory cortex. This narrow strip of the cerebral cortex runs from the top of the head to just above the ear. Compare the representation of body parts in the motor cortex (Section 29.11).

spinal interneurons that relay signals about pain to the brain. The signals proceed to the somatosensory cortex.

Numerous substances affect signal transmission at the synapse between a pain-detecting sensory neuron and a spinal interneuron. For example, natural painkillers called endorphins impair flow of signals along the pain pathway. By contrast, substance P enhances pain perception, It is a neuromodulator, a signaling substance that alters neuron behavior. Substance P acts on spinal interneurons, making them more likely to send signals about pain to the sensory cortex. High levels of substance P likely contribute to fibromyalgia, a disorder characterized by chronic pain in muscles and joints throughout the body.

Pain relieving drugs, or analgesics, interfere with steps in the pain pathway. For example, aspirin reduces pain by slowing production of prostaglandins. As another example, synthetic opiates such as morphine mimic the activity of endorphins.

---

pain Perception of tissue injury.
somatic sensations Sensations such as touch and pain that arise when sensory neurons in skin, muscle, or joints are activated.
visceral sensations Sensations that arise when sensory neurons associated with organs inside body cavities are activated.

Take-Home Message **What are somatic and visceral sensations?**

> Somatic sensations are signals from sensory receptors in skin, skeletal muscle, and joints. They travel along sensory neuron axons to the spinal cord, then to the somatosensory cortex.

> Visceral sensations begin with the stimulation of sensory neurons in the walls of organs inside the body. These signals are relayed to the spinal cord, and then the brain.

> Pain is the sensation associated with tissue damage. Endorphins and synthetic pain relievers interfere with signals that inform the brain about pain.

> Most animals are sensitive to light, but only those with a camera eye form images as you do.
< Link to Morphological convergence 16.8

Requirements for Vision

Vision is detection of light in a way that provides a mental image of objects in the environment. It requires eyes and a brain with the capacity to interpret visual stimuli. Image perception arises when the brain integrates signals regarding shapes, brightness, positions, and movement of visual stimuli.

Eyes are sensory organs that hold photoreceptors. Pigment molecules inside the photoreceptors absorb light energy. That energy is converted to the excitation energy of action potentials that are sent to the brain.

Some invertebrates, such as earthworms, do not have eyes, but they do have photoreceptors dispersed under the epidermis or clustered in parts of it. They use light as a cue to orient the body, detect shadows, or adjust their biological clock, but they do not have a true sense of vision.

Formation of an image requires an eye with a **lens**, a transparent body that bends light rays so they converge on photoreceptors. Insects have **compound eyes** with many lens-containing units (Figure 30.5A). The insect brain constructs images based on the light intensities detected by the different units. Compound eyes do not provide the clearest vision, but they are highly sensitive to movement.

Cephalopod mollusks such as squids and octopuses have the most complex eyes of any invertebrate (Figure 30.5B). Their **camera eyes** have an adjustable opening that allows light to enter a dark chamber. Each eye has a single lens that focuses incoming light onto a **retina**, a

tissue densely packed with photoreceptors. The retina of a camera eye is analogous to the light-sensitive film used in a traditional film camera. Compared to compound eyes, camera eyes provide a more defined and detailed image.

Vertebrates also have camera eyes. Because vertebrates are not closely related to cephalopod mollusks, camera eyes are presumed to have evolved independently in the two lineages. This is an example of morphological convergence (Section 16.8).

The Human Eye

A human eyeball sits in a protective, cuplike, bony cavity called the orbit. Skeletal muscles that run from the rear of the eye to bones of the orbit move the eyeball.

Eyelids, eyelashes, and tears protect the eye tissues. Periodic blinking spreads a film of tears over the eyeball's exposed surface. A protective mucous membrane, the conjunctiva, lines the inner surface of the eyelids and folds back to cover most of the eye's outer surface. Conjunctivitis, commonly called pinkeye, is an inflammation of this membrane. It is caused by a viral or bacterial infection.

The eyeball is spherical with a three-layered structure (Figure 30.6). A **cornea** made of transparent crystallin protein covers the front of the eyeball. A dense, white, fibrous sclera covers the rest of the eye's outer surface.

The eye's middle layer includes a blood vessel–rich choroid darkened by the brownish pigment melanin. This dark layer prevents light reflection within the eyeball. Attached to the choroid and suspended behind the cornea is a muscular, doughnut-shaped **iris**. Whether your eyes are blue, brown, or green depends on the amount of melanin in your iris. Light enters the eye's interior through the **pupil**, an opening at the center of the iris. Smooth muscles of the iris adjust the pupil's diameter. In bright light the pupil shrinks, so less light gets in. In low light the pupil widens, so more light enters the eye. Sympathetic stimulation also widens the pupil, allowing a better look at the source of danger or excitement.

A ciliary body of muscle, fibers, and secretory cells attaches to the choroid and holds the lens in place behind the pupil. The stretchable, transparent lens is shaped like a bulging disk about 1 centimeter (1/2 inch) across.

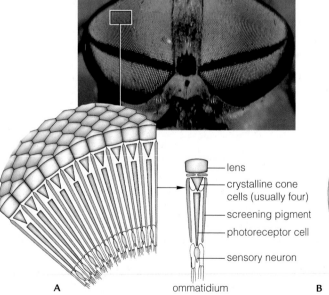

lens
crystalline cone cells (usually four)
screening pigment
photoreceptor cell
sensory neuron

A ommatidium

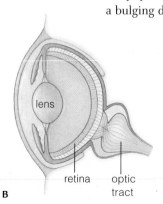

lens

retina   optic tract

B

Figure 30.5 Invertebrate eyes. There are far more photoreceptors than can be shown in the simple diagrams. **A** Compound eye of a deerfly. A lens in each of many units directs light onto a crystalline cone, which focuses it onto one photoreceptor cell. **B** Camera eye of squid (a cephalopod).

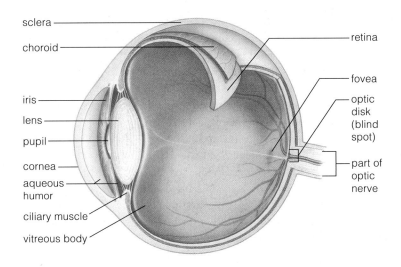

Figure 30.6 **Animated** Components and structure of the human eye.

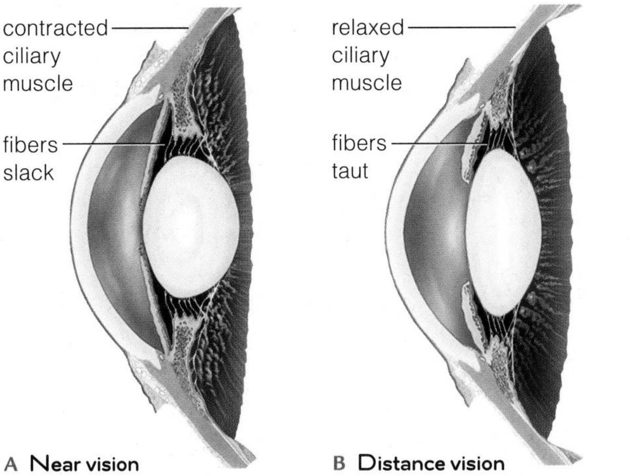

contracted ciliary muscle
fibers slack

relaxed ciliary muscle
fibers taut

A **Near vision** Contraction of ciliary muscle allows fibers to slacken and the lens becomes fatter and rounder.

B **Distance vision** Relaxation of ciliary muscle pulls fibers taut stretching the lens so it becomes thinner and flatter.

Figure 30.7 **Animated** Visual accommodation in the human eye.

❯❯ **Figure It Out** The thicker a lens, the more it bends light. Does the lens bend light more in distance vision or close vision?

Answer: Light rays are bent more with close vision.

Focusing Mechanisms

The eye has two internal chambers. The ciliary body produces aqueous humor, the fluid in the anterior chamber. A jellylike vitreous body fills the chamber behind the lens. The innermost layer of the eye, the retina, lines the rear of this chamber. It contains the photoreceptors.

When you see an object, you are perceiving light rays reflected from that object. Light rays reflected from near and distant objects hit the eye at different angles. By the process of **visual accommodation**, the ciliary muscle adjusts the shape of the lens so that all rays become focused on the retina.

The curvature of the lens determines the extent to which light rays will bend. A flat lens will focus light from a distant object onto the retina. However, the lens must be rounder to focus light from nearby objects. When you read, the ciliary muscle contracts, and fibers that connect this muscle to the lens slacken. The decreased tension on the lens allows it to round up enough to focus light from the page onto your retina (Figure 30.7A). Gaze

into the distance and the ciliary muscle around the lens relaxes, allowing the lens to flatten (Figure 30.7B). Continual viewing of a close object keeps the ciliary muscle contracted. To reduce muscle fatigue, take breaks and focus on more distant objects.

Because of the way the cornea and lens bend light rays, an image formed on the retina is an upside-down mirror image of the real world (Figure 30.8). However, the brain interprets the image so you perceive the world correctly.

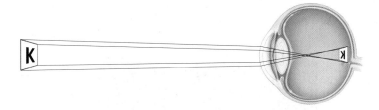

Figure 30.8 The pattern of light rays falling on the retina. The image on the retina is upside down and reversed left to right.

camera eye Eye with an adjustable opening and a single lens that focuses light on a retina.
compound eye Eye with many units, each having its own lens.
cornea Clear, protective covering at front of vertebrate eye.
iris Circular muscle that adjusts the shape of the pupil to regulate how much light enters the eye.
lens Disk-shaped structure that bends light rays so they fall on an eye's photoreceptors.
pupil Adjustable opening that allows light into a camera eye.
retina Layer of eye that contains photoreceptors.
visual accommodation Process of making adjustments to lens shape so light from an object falls on the retina.

Take-Home Message What structures contribute to vision?

❯ Vision requires eyes with a dense array of photoreceptors and image formation in the brain.

❯ A lens allows formation of a detailed image. A compound eye consists of many units, each with its own lens. A camera eye with a single lens evolved independently in cephalopods and vertebrates.

❯ Your eye adjusts the shape of the lens depending on whether you are focusing on a near object or a distant one.

> Processing of visual information begins in the retina and continues along the pathway to the brain.
< Link to Pigments and properties of light 6.2

As explained in the previous section, the cornea and lens bend light rays so the rays fall on the retina, where they can excite photoreceptors.

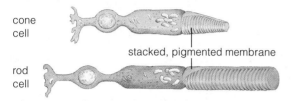

A The two types of photoreceptors in the retina

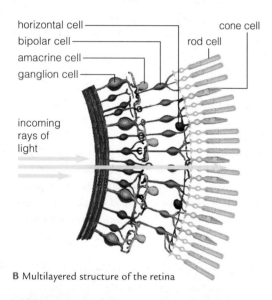

B Multilayered structure of the retina

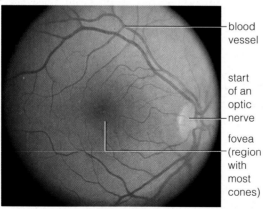

C Magnified view of the retina as seen through the pupil

Figure 30.9 Animated Structure of the retina. The two types of photoreceptors, rods and cones, lie at the very rear of the retina, beneath layers of signal-processing neurons.

The retina has two types of photoreceptors. Each has stacks of membranous disks that contain pigment (Figure 30.9A). Visual pigments (called opsins) are derived from vitamin A, which is why a deficiency in this vitamin can impair vision. **Rod cells**, the most abundant photoreceptors, detect dim light and respond to changes in light intensity across the visual field. In the human eye, rods tend to be concentrated at the edges of the retina. All rods have the same pigment (rhodopsin), which is most excited by exposure to green-blue light.

Cone cells provide acute daytime vision and allow us to detect colors. There are three types of cone cell, each with a slightly different form of the cone pigment photopsin. One cone pigment absorbs mainly red light, another absorbs mainly blue, and a third absorbs green. Normal human color vision requires all three kinds of cones. The **fovea**, a pit in the central region of the retina, has the greatest density of cones. With normal vision, most light rays are focused on the fovea.

Signal integration and processing begin in the retina, which has a multilayered structure (Figure 30.9B). When a pigment in a rod or cone photoreceptor absorbs light, signals flow from the photoreceptor to neurons in the layer above. These neurons process signals from the photoreceptors and send signals to ganglion cells. Bundled axons of ganglion cells constitute the optic nerve (Figure 30.9C). The region of the retina through which the optic nerve exits lacks photoreceptors. It cannot respond to light and thus is a "blind spot." We all have a blind spot in each eye, but usually do not notice it because the information missed by one eye is provided to the brain by the other.

Signals from the right visual field of each eye travel along an optic nerve to the brain's left hemisphere. Signals from the left visual field travel to the right hemisphere. Each optic nerve ends in a brain region (the lateral geniculate nucleus) that processes signals. From here, the signals are conveyed to the visual cortex where the final integration process produces visual sensations.

---

cone cell Photoreceptor that provides sharp vision and allows detection of color.
fovea Retinal region where cone cells are most concentrated.
rod cell Photoreceptor that is active in dim light; provides coarse perception of image and detects motion.

Take-Home Message **How do we detect and process visual information?**

> When stimulated by light, rods and cones send signals to neurons in the layer of retina above them. These neurons process signals, then send messages to the brain. Visual sensations arise from the activity of the visual cortex.

> Vision is impaired when light is not focused properly, when photoreceptors do not respond as they should, or when some aspect of visual processing breaks down.
< Links to X-linked traits 14.4, Embryonic stem cells 28.1

Color Blindness Sometimes one or more types of cones fail to develop or function improperly. The outcome is one form or another of color blindness. In red–green color blindness, a person has trouble distinguishing reds from greens. Red–green color blindness is an X-linked recessive trait (Section 14.4). As is the case for other X-linked traits, it shows up more often in males. In the United States, 7 percent of males and 0.4 percent of females are affected.

Lack of Focus About 150 million Americans have disorders in which light rays do not converge as they should. Astigmatism results from an unevenly curved cornea, which cannot properly focus incoming light on the lens. Nearsightedness occurs when the distance from the front to the back of the eye is longer than normal or when ciliary muscles react too strongly. With either disorder, images of distant objects get focused in front of the retina instead of on it (Figure 30.10A).

In farsightedness, the distance from front to back of the eye is unusually short or ciliary muscles are too weak. Either way, light rays from nearby objects get focused behind the retina (Figure 30.10B). Also, the lens loses its flexibility as a person ages. That is why most people who are over age forty have relatively impaired close vision.

Glasses, contact lenses, or surgery can correct most focusing problems. About 1.5 million Americans undergo laser surgery (LASIK) annually. Typically, LASIK can eliminate the need for glasses during most activities, although some older adults still need reading glasses. Chronic eye irritation is a common complication.

Age–Related Disorders As people age, changes in the structure of proteins in their lens can result in a cataract, a clouding of the lens. Excessive exposure to ultraviolet radiation, smoking, use of steroids, and some diseases such as diabetes promote cataract formation. Typically, both eyes are affected. At first, a cataract scatters light and blurs vision (Figure 30.11A). Eventually, the lens may become fully opaque, causing blindness. Cataract surgery, a common procedure in developed countries, restores normal vision by replacing a clouded lens with a clear plastic implant. Worldwide, an estimated 16 million people remain blinded by age-related cataracts.

Age-related macular degeneration (AMD) is the leading cause of age-related blindness in the United States. The macula, the part of the retina around and including the fovea, is essential to clear vision. Destruction of photoreceptors in the macula clouds the center of the visual field

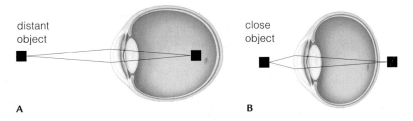

Figure 30.10 Focusing problems. **A** In nearsightedness, light rays from distant objects converge in front of the retina. **B** In farsightedness, light rays from close objects have not yet converged when they arrive at the retina.

A With cataracts   **B** With macular degeneration

Figure 30.11 Photos simulating vision with two common visual disorders.

more than the periphery (Figure 30.11B). Some mutations increase the risk of AMD, as do smoking, obesity, and high blood pressure. A vegetable-rich diet may help protect against it. Damage caused by AMD usually cannot be reversed, but drug injections and laser therapy can slow its progression. A treatment involving embryonic stem cells is also being tested.

Glaucoma results when too much aqueous humor builds up inside the eyeball. The increased fluid pressure damages blood vessels and ganglion cells. It can also interfere with peripheral vision and visual processing. Although we often associate chronic glaucoma with old age, the conditions that give rise to the disorder start to develop long before symptoms arise. Screening for glaucoma allows doctors to detect the increased fluid pressure before the damage becomes severe. They can then manage the disorder with medication, surgery, or both.

Take-Home Message **What causes common visual disorders?**

> Defective cone cells of one or more types result in color blindness.

> A misshapen eyeball can cause nearsightedness or farsightedness. A lens that has become inflexible with age also causes farsightedness.

> Other age-related vision disorders include cataracts (clouding of the lens), macular degeneration (loss of photoreceptors), and glaucoma (excessive aqueous fluid).

❯ Chemoreceptors in the linings of the nose and mouth provide our senses of smell and taste.

❮ Link to Epithelium 28.3

Sense of Smell

Olfaction, the sense of smell, starts with chemoreceptors that bind substances dissolved in the fluid around them. The binding triggers action potentials that olfactory nerves transmit to the brain.

Olfactory receptors detect water-soluble or volatile (easily vaporized) chemicals. A human nose has about 5 million receptors (Figure 30.12). Some dogs have billions.

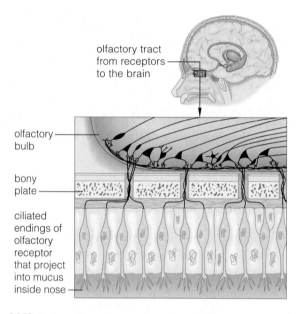

Figure 30.12 Pathway from the sensory endings of olfactory receptors in the human nose to the cerebral cortex and limbic system. Axons of these sensory receptors pass through holes in a bony plate between the lining of the nasal cavities and the brain.

Receptor axons carry signals to one of two olfactory bulbs in the brain.

Many animals use olfactory cues to navigate, find food, and communicate, as with pheromones. **Pheromones** are signaling molecules secreted by one individual that change the social behavior of other individuals of its species. As one example, olfactory receptors on the antennae of a male silk moth help him find a pheromone-secreting female that may be more than a kilometer upwind. We return to the topic of pheromones and their effects on animal behavior in Chapter 39.

In the nasal cavity of reptiles and most mammals, a cluster of sensory neurons forms a **vomeronasal organ** that responds to pheromones. Humans have a reduced version of this organ, but studies suggest that we do make and respond to some pheromones.

Sense of Taste

Taste receptors are chemoreceptors that detect chemicals dissolved in fluid, but they have a different structure and location than olfactory receptors. Taste receptors help animals locate food and avoid poisons. An octopus "tastes" potential food with receptors in suckers on its tentacles; a fly does this with receptors in its antennae and feet. In humans, many taste buds are embedded in the upper surface of the tongue (Figure 30.13). These sensory organs are located in specialized epithelial structures, or papillae, that look like raised bumps or red dots on the tongue's upper surface.

You perceive many tastes, but all are a combination of five main sensations: *sweet* (elicited by glucose and the other simple sugars), *sour* (acids), *salty* (sodium chloride or other salts), *bitter* (plant toxins, including alkaloids), and *umami* (elicited by amino acids such as glutamate, which has a savory taste found in cheese and aged meat). You have probably heard of MSG, or monosodium glutamate. This common food additive enhances flavor by stimulating the taste receptors that contribute to the sensation of umami.

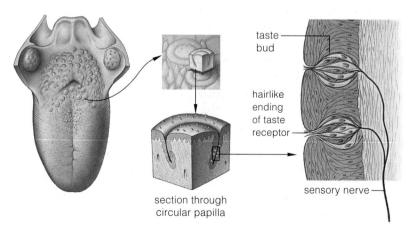

Figure 30.13 Taste receptors in the human tongue. Taste buds are clusters of receptor cells and supporting cells inside special epithelial papillae. One type, a circular papilla, is shown in the section here. The tongue has about 5,000 taste buds, each enclosing as many as 150 taste receptor cells.

---

olfactory receptors Chemoreceptors involved in sense of smell.
pheromones Signaling molecules that affect another member of the same species.
taste receptors Chemoreceptors involved in taste.
vomeronasal organ Pheromone-detecting organ of vertebrates.

Take-Home Message **How do the senses of smell and taste arise?**

❯ The senses of smell and taste start at chemoreceptors. Both involve sensory pathways that lead to processing regions in the cerebral cortex.

> Organs deep inside your ear are essential to maintaining posture and sense of balance.

Organs of equilibrium are parts of sensory systems that monitor the body's positions and motions. Each vertebrate ear has these organs inside a fluid-filled sensory structure called the **vestibular apparatus**. The organs are located in the vestibular apparatus's three semicircular canals and in two sacs, the saccule and utricle (Figure 30.14A).

Organs of the vestibular apparatus have **hair cells**, a type of mechanoreceptor with modified cilia at one end. Fluid pressure inside the canals and sacs makes the cilia bend. That mechanical energy deforms the hair cell plasma membrane just enough to let ions slip across and stimulate an action potential. A vestibular nerve carries the sensory input to the brain. As you will see, another type of hair cell functions in hearing.

The three semicircular canals are oriented at right angles to one another, so rotation of the head in any combination of directions—front/back, up/down, or left/right—moves the fluid inside them. An organ of equilibrium rests on the bulging base of each canal. The cilia of its hair cells are embedded in a jellylike mass (Figure 30.14B). When fluid moves in the canal, it pushes against the mass and generates enough pressure to initiate action potentials in hair cells.

The brain receives signals from semicircular canals on both sides of the head. By comparing the frequency of action potentials and the number of sensory neurons responding on nerves from both sides, the brain can sense *dynamic* equilibrium: any angular movement and rotation of the head. Among other things, you can use this sense to keep your eyes locked on an object even as your head is swiveling or nodding about.

Organs in the saccule and utricle function in *static* equilibrium. They help the brain monitor the head's position and how fast the head is moving in a straight line. They also help keep the head upright and maintain posture. A jellylike mass weighted with calcite crystals lies just above hair cells in these organs. Tilt your head, or start or stop moving, and gravity causes this mass to shift. As the mass shifts, the hair cells bend and alter their rate of action potentials.

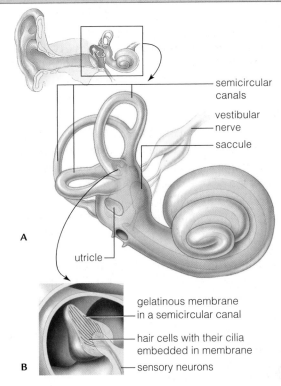

A

semicircular canals

vestibular nerve

saccule

utricle

gelatinous membrane in a semicircular canal

hair cells with their cilia embedded in membrane

sensory neurons

B

Figure 30.14 Animated

A Vestibular apparatus inside a human ear. Organs of equilibrium in its fluid-filled sacs and canals contribute to a sense of balance.

B Components of one of the organs inside a semicircular canal.

The brain also takes into account information from the eyes, and from receptors in skin, muscles, and joints. Integration of the information allows perception of the body's position and motion in space.

A stroke, an inner ear infection, or loose particles in the semicircular canals cause vertigo, a sensation that the world is moving or spinning around. Vertigo also arises from conflicting sensory inputs, as when you stand at a height and look down. The vestibular apparatus reports that you are motionless, but your eyes report that your body is floating in space.

Mismatched signals also cause motion sickness. On a curvy road, passengers in a car experience changes in acceleration and direction that scream "motion" to their vestibular apparatus. At the same time, signals from their eyes about objects in the car tell their brain that the body is at rest. Driving minimizes motion sickness because the driver focuses on sights outside the car, so visual signals are consistent with vestibular signals.

hair cell Mechanoreceptor that is activated when movement of overlying membrane causes its hairlike cilia to bend.
organs of equilibrium Sensory organs that respond to body position and motion.
vestibular apparatus System of fluid-filled sacs and canals in the inner ear; contains organs of equilibrium.

Take-Home Message **How does the sense of balance arise?**

> Organs of equilibrium help keep the body balanced in relation to gravity, velocity, acceleration, and other forces that influence its position and movement.

> Many arthropods and most vertebrates can hear sounds. In land vertebrates, ear flaps capture sounds traveling through air and internal parts of the ear sort them out.
< Link to Vertebrate move onto land 24.4

Properties of Sound

Hearing is the detection of sound, a form of mechanical energy. Sounds arise when a vibrating object causes pressure variations in air, water, or some other medium. When you clap your hands or shout, you create pressure waves that move through the air. The amplitude (height) of those waves determines how loud a sound is (Figure 30.15A). We measure loudness in decibels. The human ear can detect a 1-decibel difference between sounds. Normal conversation is about 60 decibels, a food blender operating at high speed produces a sound of about 90 decibels, and a chain saw produces a 100-decibel racket.

A sound's frequency, the number of wave cycles per second, determines its pitch. The more waves per second, the higher the pitch (Figure 30.15B).

Vertebrate Hearing

Water readily transfers vibrations to body tissues, so fishes do not need elaborate ears to detect sound waves. However, vertebrates that moved onto land faced a sensory challenge; transfer of sound waves from air to body tissues is inefficient. Mammalian ears evolved features that maximize the efficiency of this transfer (Figure 30.16).

Unlike amphibians and reptiles, most mammals have an **outer ear** that funnels sound inward ❶. A skin-covered flap of cartilage (the pinna) projects from the side of the head. The pinna collects sound waves and directs them into the auditory canal, an air-filled passage that connects to the middle ear.

The **middle ear** amplifies and transmits air waves to the inner ear. An eardrum evolved in early reptiles as a shallow depression on each side of the head that vibrates in response to pressure waves. In mammals, sound is amplified by the movement of three tiny bones referred to as the hammer, anvil, and stirrup ❷. The bone transmits the force of sound waves from the eardrum onto the surface of the oval window, an elastic membrane that is the boundary between the middle and inner ear.

The **inner ear** contains the vestibular apparatus that functions in the sense of balance. It also contains the **cochlea**, which in humans is a pea-sized, fluid-filled structure that resembles a coiled snail shell (the Greek *koklias* means snail). The transduction of sound waves into action potentials takes place in the cochlea.

Membranes divide the interior of the cochlea into three fluid-filled ducts ❸. Pressure from movement of the stirrup against the oval window produces pressure waves in the fluid in these ducts. As these waves travel through the cochlear fluid, they cause the membranous walls of the ducts to vibrate.

The organ of Corti, the organ responsible for hearing, sits on the base of the membrane in the middle duct ❹. Inside the organ of Corti are arrays of hair cells with specialized cilia that extend into an overlying membrane. When pressure waves cause the membrane to move, cilia of hair cells bend and the cells undergo action potentials that travel along an auditory nerve to the brain.

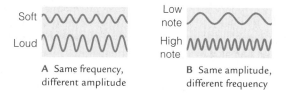

Soft

Loud

A Same frequency, different amplitude

Low note

High note

B Same amplitude, different frequency

Figure 30.15 **Animated** Wavelike properties of sound.

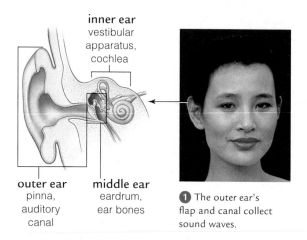

inner ear
vestibular apparatus, cochlea

outer ear
pinna, auditory canal

middle ear
eardrum, ear bones

❶ The outer ear's flap and canal collect sound waves.

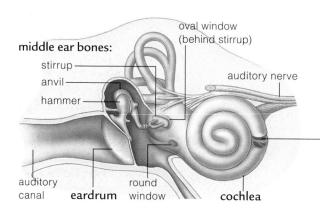

oval window (behind stirrup)

middle ear bones:
stirrup
anvil
hammer

auditory nerve

auditory canal

eardrum

round window

cochlea

❷ The eardrum and middle ear bones amplify sound.

Figure 30.16 **Animated** Anatomy of the ear and how we hear.

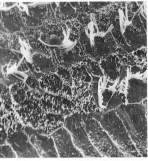

Figure 30.17 Sound-induced damage to hair cells. *Left*, from a guinea pig ear, two rows of normal hair cells. *Right*, hair cells of the same organ after twenty-four hours of exposure to noise levels comparable to extremely loud music.

A Whale of a Dilemma (revisited)

Animal sensory systems evolved over countless generations in a world without human activity. Now, our technology has dramatically altered the sensory landscape for us and the animals. The world has become noisier and more brightly lit. Our communication systems fill the air with radio waves. In humans, a high level of environmental noise also impairs concentration and interferes with sleep patterns. It raises anxiety and increases the risk of high blood pressure and other cardiovascular problems. Sound pollution can also harm animals by disorienting them, interfering with their ability to find prey, or disrupting courtship.

How Would You Vote? Excessive noise can harm marine organisms. Should we regulate the maximum allowable noise that ships and underwater mining operations can produce? See CengageNow for details, then vote online (cengagenow.com).

Hearing Loss

Exposure to loud sounds such as a few hours at a rock concert can cause temporary hearing loss. The loss occurs because the unusually large pressure waves caused by the loud sound damage the cilia of hair cells (Figure 30.17). In most cases, the cilia are replaced and hearing recovers in a day or two. However, repeated exposure to a specific loud sound can kill hair cells that respond to that sound. These cells are not replaced, so permanent hearing loss for sounds within a particular frequency range results. Some antibiotics such as streptomycin can harm hair cells, and the number of hair cells declines with age.

Hearing loss can also result from damage to the auditory nerve, loss of fluid in the inner ear, damage to the small bones of the middle ear, or even an excess amount of earwax.

cochlea Coiled, fluid-filled structure in the inner ear that holds the sound-detecting organ of Corti.
inner ear Fluid-filled vestibular apparatus and cochlea.
middle ear Eardrum and the tiny bones that transfer sound to the inner ear.
outer ear External ear and the air-filled auditory canal.

Take-Home Message How do vertebrates detect sounds?

❯ Ears that collect and amplify sound waves evolved in some vertebrates that live on land.

❯ Mammals have an outer ear that collects sound waves and directs them to the middle ear. In the middle ear, vibrations of the eardrum are amplified via movement of small bones. These bones set up pressure waves in fluid inside the inner ear's cochlea.

❯ The organ of Corti inside the cochlea has hair cells that transduce the pressure waves into action potentials that travel along the auditory nerve to the brain.

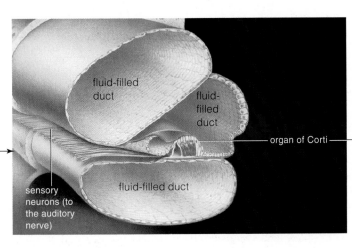

❸ One coil of the cochlea in cross-section. The organ of Corti detects pressure waves in fluid-filled ducts inside the cochlea.

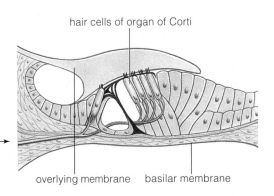

❹ Pressure waves cause the basilar membrane beneath the organ of Corti to move upward. The movement pushes hair cells against an overlying membrane. The resulting action potentials travel along the auditory nerve to the brain.

Summary

Section 30.1 Animal sensory systems evolved as adaptations to the natural environment. Human activities can alter an animal's sensory environment, causing detrimental effects.

Section 30.2 Sensory receptors respond to the energy of a specific **stimulus**, such as pressure, heat, or light, with action potentials. Types of sensory receptors include, **chemoreceptors, thermoreceptors, pain receptors, mechanoreceptors**, and **photoreceptors**.

The brain evaluates action potentials from sensory receptors based on which of the nerves delivers them, their frequency, and the number of axons firing. Continued stimulation of a receptor can lead to **sensory adaptation**. **Sensation** is detection of a stimulus, whereas **perception** involves assigning meaning to a sensation.

Section 30.3 Somatic sensations arise from receptors in the skin and in skeletal muscles. Signals reach the sensory areas of the cerebral cortex, where interneurons are organized like maps of individual parts of the body surface. **Visceral sensations** originate from receptors in the walls of soft organs and are less easily pinpointed. **Pain** is the perception of tissue damage. In vertebrates, a variety of neuromodulators enhance or lessen signals about pain.

Sections 30.4–30.6 An eye is a sensory organ that contains a dense array of photoreceptors. Insects have a **compound eye**, with many individual units. Each unit has a **lens**, a structure that bends light rays so they fall on the photoreceptors. Like squids and octopuses, humans have **camera eyes**, with an adjustable opening that lets in light, and a single lens that focuses the light on a photoreceptor-rich **retina**. A transparent **cornea** covers the front of the eye and helps bend light rays. The light enters the eye through the **pupil** at the center of the **iris**. **Visual accommodation** is the altering of lens shape to focus on objects at different distances.

Rod cells and **cone cells** are the eyes' photoreceptors. Cone cells provide color vision and detailed images. They are most abundant in the **fovea**. Common vision disorders result from defective or degenerating photoreceptors, misshapen eyes, a clouded lens, or excess aqueous humor.

Section 30.7 The senses of taste and smell involve chemoreceptors. In humans, **taste receptors** are concentrated in taste buds of the tongue and mouth. **Olfactory receptors** line the nasal passages. **Pheromones** are chemical signals that act as social cues among many animals that have the means to detect them. Most vertebrates have a **vomeronasal organ** that responds to pheromones.

Section 30.8 The **vestibular apparatus**, a system of fluid-filled sacs and canals in the inner ear, houses **organs of equilibrium**. These organs detect gravity, acceleration, and other forces that related to the body's position and motion. They include **hair cells**, mechanoreceptors with specialized pressure sensitive cilia.

Section 30.9 Sound is a form of mechanical energy—pressure waves that vary in amplitude and frequency. Humans have a pair of ears with three regions. The **outer ear** collects sound waves. The **middle ear** amplifies sound waves and transmits them to the **inner ear**, which includes the vestibular apparatus and the **cochlea**: a coiled, fluid-filled structure with three ducts.

Pressure waves traveling through the fluid inside the cochlea bend hair cells embedded in one of the cochlear membranes. Sounds get sorted out according to their frequency. Mechanical energy of pressure waves is transduced to action potentials that are relayed along auditory nerves to the brain.

Self-Quiz   Answers in Appendix III

1. The pain of heartburn is an example of a _____ .
   a. somatic sensation   c. sensory adaptation
   b. visceral sensation   d. spinal reflex

2. _____ is defined as a decrease in the response to an ongoing stimulus.
   a. Perception   c. Sensory adaptation
   b. Visual accommodation   d. Somatic sensation

3. Which is a somatic sensation?
   a. taste   c. touch   e. a through c
   b. smell   d. hearing   f. all of the above

4. Chemoreceptors play a role in the sense of _____ .
   a. taste   c. touch   e. both a and b
   b. smell   d. hearing   f. all of the above

5. In the _____ , neurons are arranged like maps that correspond to different parts of the body surface.
   a. cerebral cortex   c. basilar membrane
   b. retina   d. all of the above

6. Mechanoreceptors in the _____ send signals to the brain about the body's position relative to gravity.
   a. eye   b. ear   c. tongue   d. nose

7. The middle ear functions in _____ .
   a. detecting shifts in body position
   b. amplifying and transmitting sound waves
   c. sorting sound waves out by frequency

8. Substance P _____ .
   a. increases pain-related signals
   b. is a natural painkiller
   c. is the active ingredient in aspirin

Data Analysis Activities

Occupational Noise and Hearing Loss Frequent exposure to noise of a particular pitch can cause loss of hair cells in the part of the cochlea's coil that responds to that pitch. Many workers are at risk for such frequency-specific hearing loss because they work with or around noisy machinery. Taking precautions such as using ear plugs to reduce sound exposure is important. Noise-induced hearing loss can be prevented, but once it occurs it is irreversible. Dead hair cells are not replaced.

Figure 30.18 shows the threshold decibel levels at which sounds of different frequencies can be detected by an average 25-year-old carpenter, a 50-year-old carpenter, and a 50-year-old who has not been exposed to on-the-job noise. Sound frequencies are given in hertz (cycles per second). The more cycles per second, the higher the pitch.

Figure 30.18 Effects of age and occupational noise exposure. The graph shows the threshold hearing capacities (in decibels) for sounds of different frequencies (given in hertz, or cycles per second) in a 25-year-old carpenter (*blue*), a 50-year-old carpenter (*red*), and a 50-year-old who did not have any on-the-job noise exposure (*brown*).

1. Which sound frequency was most easily detected by all three people?
2. How loud did a 1,000-hertz sound have to be for the 50-year-old carpenter to detect it?
3. Which of the three people had the best hearing in the range of 4,000 to 6,000 hertz? Which had the worst?
4. Based on these data, would you conclude that the hearing decline in the 50-year-old carpenter was caused by age or by job-related noise exposure?

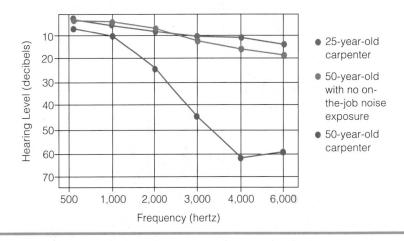

- 25-year-old carpenter
- 50-year-old with no on-the-job noise exposure
- 50-year-old carpenter

9. The organ of Corti contains receptors that signal in response to _____ .
   a. heat   b. sound   c. light   d. pheromones

10. Color vision begins with stimulation of _____ .
    a. hair cells   b. rod cells   c. cone cells   d. neuroglia

11. Visual accommodation involves adjustment to the shape of the _____ .
    a. conjunctiva   b. retina   c. orbit   d. lens

12. When you view a close object your lens gets _____ .
    a. flatter   b. rounder   c. darker   d. cloudier

13. Defective _____ cause color blindness.
    a. hair cells   b. rod cells   c. cone cells   d. neuroglia

14. _____ stimulation causes the pupil to widen.
    a. Sympathetic   b. Parasympathetic

15. Match each structure with its description.
    ____ rod cell   a. protects eyeball
    ____ cochlea   b. functions in balance
    ____ lens   c. detects pheromones
    ____ sclera   d. detects dim light
    ____ cone cell   e. contains chemoreceptor
    ____ taste bud   f. focuses rays of light
    ____ vestibular   g. sorts out sound waves
         apparatus   h. detects color
    ____ pinna   i. collects sound waves
    ____ vomeronasal organ

Additional questions are available on **CENGAGENOW.**

Critical Thinking

1. The strength of Earth's magnetic field and its angle relative to the surface vary with latitude. Many animals sense these differences and use them as cues for assessing their location and direction of movement. Behavioral experiments have shown that sea turtles, salamanders, and spiny lobsters use information from Earth's magnetic field during their migrations. Whales and some burrowing rodents also seem to have a magnetic sense. Evidence about humans is contradictory. Devise an experiment to test whether humans can detect a magnetic field.

2. In humans, photoreceptors are most concentrated at the very back of the eyeball. In birds of prey, including owls and hawks, the greatest density of photoreceptors is in a region closer to the eyeball's roof. When these birds are on the ground, they cannot see objects even slightly above them unless they turn their head almost upside down as shown at the *right*. What is the adaptive advantage of this type of organization of the retina? Would you expect a hummingbird that feeds on flower nectar to have photoreceptors distributed in the same way?

Animations and Interactions on **CENGAGENOW:**
❯ Action of pressure receptors in skin; Anatomy of the human eye; Visual accommodation; Image formation on the retina; Structure of the retina; Organs of equilibrium; Wavelike properties of sound; Human hearing.

‹ Links to Earlier Concepts

This chapter expands on the introduction to cell signaling in Section 28.9. It discusses steroids (3.4), proteins (3.5), cell membranes (5.6), and how the body extracts energy from food (7.7). We touch on promoters (9.3) and sex determination (8.2) and draw on your knowledge of the brain (29.10), synapses (29.6), sympathetic nerves (29.8), and the optic nerve (30.5). There is much on negative feedback (28.9), and we revisit insect molting (23.10).

Key Concepts

The Vertebrate Endocrine System

Hormones and other signaling molecules regulate the pathways that control metabolism, growth, development, and reproduction. Nearly all vertebrates have an endocrine system composed of the same hormone-producing structures.

Signaling Mechanisms

A hormone travels through the blood and acts on any cell that has receptors for it. Receptor activation leads to transduction of the signal and a response in the targeted cell. Hormones are derived from either cholesterol or amino acids.

31 Endocrine Control

31.1 | Hormones in the Balance

We live in a world awash in synthetic chemicals. We drink from plastic bottles, wear clothing made of synthetic fabrics, slather ourselves with synthetic skin products, and dose our food with synthetic pesticides. Enormous numbers of man-made compounds are used to make our computers and other electronic gadgets. What do we know about the safety of these substances?

We have learned by sad experience that some synthetic chemicals harm the environment and threaten human health. For example, DDT (a pesticide) and PCBs (used to make electronic products, caulking, and solvents) are endocrine disrupters. **Endocrine disrupters** are molecules that interfere with action of hormones, the signaling molecules secreted by endocrine glands. DDT was banned in 1972 and PCBs in 1979. However, because both chemicals were in wide use for years and are highly stable, they still persist in the environment.

Some chemicals still in use may also be endocrine disrupters. For example, atrazine has been a popular herbicide for more than 50 years. Americans use about 36,000 metric tons (almost 80 million pounds) of this synthetic chemical each year, mostly to prevent growth of weeds in cornfields. Atrazine-contaminated runoff from fields seeps into soil and waterways. Atrazine can persist in the evironment for more than a year. It accumulates in ponds and lakes, and taints supplies of drinking water.

Biologist Tyrone Hayes (Figure 31.1) discovered that atrazine affects frog development, causing frogs that are genetically male to develop both male and female sex organs. Since Hayes sounded the alarm, scientific scrutiny of atrazine has increased. A team headed by Holly Ingraham found that atrazine also has feminizing effects on zebrafish. Exposure of fish embryos to atrazine at a level comparable to that in runoff from atrazine-treated fields, dramatically increased the percentage that become

---

endocrine disrupter Synthetic chemical that adversely affects hormone production or function.

female. Hayes and Ingraham also investigated the mechanism by which atrazine affects sexual development. Both found that atrazine affects the expression of a gene that encodes the enzyme aromatase. Aromatase converts a male sex hormone (testosterone) into a female sex hormone (estrogen).

Figure 31.1 Benefits and costs of herbicide applications. Tyrone Hayes (*opposite*) discovered that atrazine interferes with amphibian hormonal signals. *Above*, farmers value atrazine for its ability to keep cornfields nearly weed-free with no need for laborious tilling that causes soil erosion.

Atrazine's effect on sexual development could help push some fish and frogs to extinction. If this seems a remote concern, consider this: The results of Hayes's and Ingraham's studies of aquatic vertebrates suggest that atrazine could also disrupt human hormones. Keep this point in mind when you think about endocrine disrupters and their effects: All vertebrates have similar hormone-secreting glands and endocrine systems. What you learn in this chapter will help you evaluate the costs and benefits of synthetic chemicals that affect hormone action.

A Master Integrating Center
In vertebrates, the hypothalamus and pituitary gland deep inside the brain are connected structurally and functionally. Together, they coordinate activities of many other glands.

Other Hormone Sources
Endocrine glands throughout the body respond to signals from the hypothalamus and the pituitary. Others secrete hormones in response to internal changes such as a shift in blood glucose level. Poor diet, immune problems, and genetic factors can cause hormone disorders.

Invertebrate Hormones
Hormones control molting and other events in invertebrate life cycles. Vertebrate hormones and receptors for them first evolved in ancestral lineages of invertebrates.

> Vertebrate endocrine glands and cells release hormones into the blood. The blood distributes the hormones, which often affect cells far from their source.

< Links to Gap junction 4.11, Ectoderm 23.2, Glanular epithelium 28.3, Synapse 29.6, Hypothalamus 29.10, Prostaglandins 30.3

Mechanisms of Intercellular Signaling

In all animals, cells constantly signal one another. Gap junctions allow chemical signals to move directly from the cytoplasm of one cell to that of an adjacent cell. Other cell–cell communication involves signaling molecules that are secreted into interstitial fluid (the fluid between cells). These molecules exert effects only when they bind to a receptor on or inside another cell. A cell with receptors that bind and respond to a specific signaling molecule is a "target" of that molecule.

Some secreted signaling molecules diffuse a short distance through interstitial fluid and bind to nearby cells. For example, neurons secrete neurotransmitters into the synaptic cleft that separates them from their target—a postsynaptic cell. Only neurons release neurotransmitters, but many cells secrete **local signaling molecules** that affect their neighbors. Prostaglandins are one type of local signal. When released by injured cells, they activate pain receptors and increase local blood flow.

Animal hormones are longer-range communication molecules. After being secreted into interstitial fluid, they enter the blood and circulate throughout the body. Compared to neurotransmitters or local signaling molecules, hormones last longer, travel farther, and exert their effects on a greater number of cells.

Discovery of Hormones

Hormones were first discovered in the early 1900s. Physiologists W. Bayliss and E. Starling were trying to determine what triggers the secretion of pancreatic juices when food travels through a dog's gut. As they knew, acids mix with food in the stomach. Arrival of the acidic mixture inside the small intestine causes the pancreas to secrete bicarbonate, a buffer that reduces the acidity. Was the nervous system stimulating this pancreatic response, or was some other signaling mechanism at work?

To find out how the small intestine communicated with the pancreas, Bayliss and Starling did an experiment. They surgically altered a laboratory animal, cutting the

nerves that carry signals to and from its small intestine. Even with these nerves cut, the small intestine responded to the presence of acid by secreting bicarbonate. This result indicated that the signal calling for bicarbonate secretion did not travel along nerves.

Starling and Bayliss hypothesized that the small intestine produces a signal that travels in the blood. To test this idea, the researchers exposed small intestinal cells to acid, then made an extract of those cells. Injecting this extract into a vein in an animal's neck, caused the animal's pancreas to secrete bicarbonate. The researchers concluded that exposure to acid causes the small intestine to release a substance into the blood. The blood-borne substance then causes the pancreas to secrete bicarbonate.

That substance is now called secretin. Identifying its mode of action supported a hypothesis that dated back centuries: The blood carries internal secretions that influence the activities of the body's organs.

Starling coined the term "hormone" for glandular secretions such as secretin (the Greek *hormon* means to set in motion). Other researchers identified additional hormones and their sources. Endocrine glands and other structures that secrete hormones are collectively referred to as an animal's **endocrine system**. Figure 31.2 shows the main glands of the human endocrine system.

Neuroendocrine Interactions

The endocrine system and nervous system are so closely linked that scientists sometimes refer to them collectively as the neuroendocrine system. Both neurons and endocrine cells develop from an embryo's ectodermal layer. Both respond to signals from the hypothalamus, a command center in the forebrain, and both affect the activity of the same organs. Hormones influence brain development, both before and after birth. Hormones can also affect nervous processes such as sleep/wake cycles, emotion, mood, and memory. Conversely, the nervous system affects hormone secretion. For example, in a stressful situation, nervous signals call for increased secretion of some hormones and decreased secretion of others.

---

animal hormone Intercellular signaling molecule secreted by an endocrine gland or cell.
endocrine system Hormone-producing glands and secretory cells of a vertebrate body.
local signaling molecule Chemical signal, such as a prostaglandin, that is secreted by one cell and affects neighboring cells in an animal body.

Take-Home Message **How do cells of an animal body communicate with one another?**

> In all animals, signaling molecules integrate cell activities. Each type of signal acts on all cells that have receptors for it. The targeted cells may alter their activities in response.

> Most vertebrates have the same types of hormones produced by similar structures. Collectively, hormone-secreting glands and cells make up an endocrine system.

> Integrated interactions between the nervous system and nearly all endocrine glands coordinate many different functions for the body as a whole.

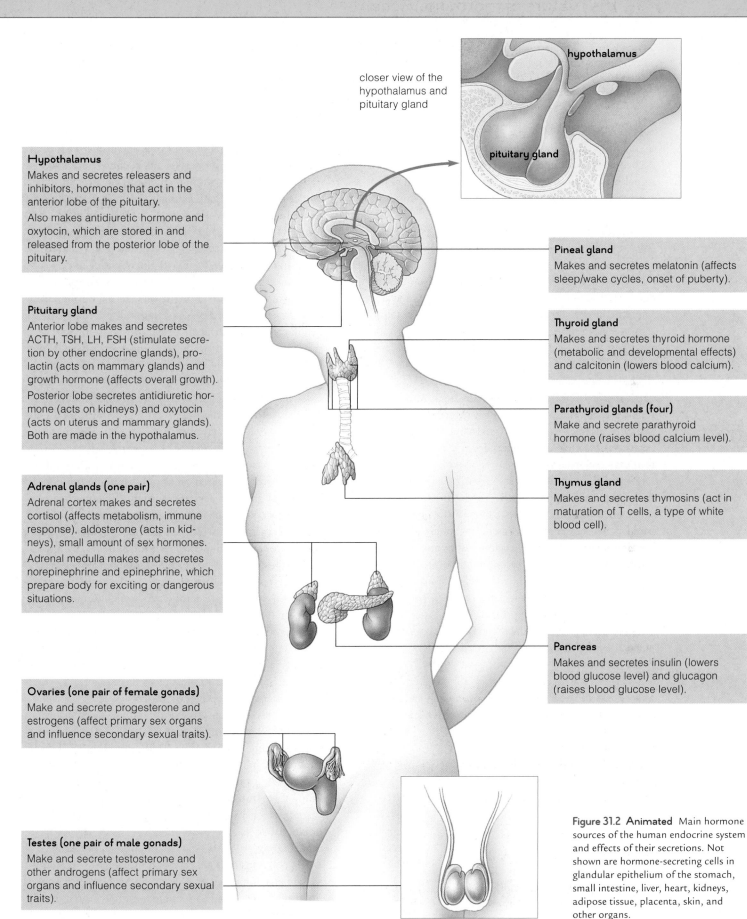

Hypothalamus

Makes and secretes releasers and inhibitors, hormones that act in the anterior lobe of the pituitary.

Also makes antidiuretic hormone and oxytocin, which are stored in and released from the posterior lobe of the pituitary.

Pituitary gland

Anterior lobe makes and secretes ACTH, TSH, LH, FSH (stimulate secretion by other endocrine glands), prolactin (acts on mammary glands) and growth hormone (affects overall growth).

Posterior lobe secretes antidiuretic hormone (acts on kidneys) and oxytocin (acts on uterus and mammary glands). Both are made in the hypothalamus.

Adrenal glands (one pair)

Adrenal cortex makes and secretes cortisol (affects metabolism, immune response), aldosterone (acts in kidneys), small amount of sex hormones.

Adrenal medulla makes and secretes norepinephrine and epinephrine, which prepare body for exciting or dangerous situations.

Ovaries (one pair of female gonads)

Make and secrete progesterone and estrogens (affect primary sex organs and influence secondary sexual traits).

Testes (one pair of male gonads)

Make and secrete testosterone and other androgens (affect primary sex organs and influence secondary sexual traits).

closer view of the hypothalamus and pituitary gland

hypothalamus

pituitary gland

Pineal gland

Makes and secretes melatonin (affects sleep/wake cycles, onset of puberty).

Thyroid gland

Makes and secretes thyroid hormone (metabolic and developmental effects) and calcitonin (lowers blood calcium).

Parathyroid glands (four)

Make and secrete parathyroid hormone (raises blood calcium level).

Thymus gland

Makes and secretes thymosins (act in maturation of T cells, a type of white blood cell).

Pancreas

Makes and secretes insulin (lowers blood glucose level) and glucagon (raises blood glucose level).

Figure 31.2 Animated Main hormone sources of the human endocrine system and effects of their secretions. Not shown are hormone-secreting cells in glandular epithelium of the stomach, small intestine, liver, heart, kidneys, adipose tissue, placenta, skin, and other organs.

> For a hormone to have an effect, it must bind to receptors on or inside a target cell.

< Links to Steroids 3.4, Proteins 3.5, Cell membranes 5.6, Promoters 9.3, Sex determination 8.2, Cell signaling 28.9

Signal Reception, Transduction, Response

Signal processing is a three-step process (Section 28.9). A chemical signal binds to a receptor on a target cell, the signal is transduced (converted to a form that acts in the receiving cell), and the target cell makes a response:

Animal hormones are chemical signals derived from either cholesterol or amino acids. Cholesterol is the starting material for **steroid hormones** such as the sex hormones testosterone and estrogen. Amine hormones are modified amino acids. Peptide hormones are short chains of amino acids; protein hormones are longer chains. Table 31.1 lists a few examples of each.

Hormones initiate responses in different ways, but in all cases, binding to a receptor is reversible and the response declines over time. The decline in response occurs because the body breaks down and eliminates hormone molecules.

Intracellular Receptors Being lipids, the steroid hormones easily diffuse across the bilayer of a plasma membrane. They form a hormone–receptor complex by binding to a receptor in the cytoplasm or nucleus. Most often, this complex binds to a promoter near a hormonally regulated gene. RNA polymerase, recall, binds to promoters before it transcribes genes (Section 9.3). Transcription and translation result in a protein product, such as an enzyme, that carries out the target cell's response to the signal. Figure 31.3A is a generalized illustration of steroid hormone action.

Receptors at the Plasma Membrane Peptide and protein hormones bind to receptor proteins that span a target cell's plasma membrane. Often, binding sets off a cascade of reactions, as when blood glucose level falls below a set point and cells in the pancreas respond by secreting glucagon. This peptide hormone binds to receptors in the plasma membrane of target cells (Figure 31.3B). The binding activates an enzyme that catalyzes the conversion of ATP to cyclic AMP (short for cyclic adenosine monophosphate). The cyclic AMP serves as a **second messenger**: a molecule that is formed in response to an external signal and causes more cellular changes. In this case, cyclic AMP turns on enzymes, which activate more enzymes, and so on. The last enzyme to be activated speeds the breakdown of glycogen into glucose. As glucose enters interstitial fluid, and then blood, the blood glucose level rises.

Some cells have receptors for steroid hormones at the plasma membrane. Binding of a steroid hormone to such a membrane receptor does not cause a change in gene expression. It triggers a faster response by way of a second messenger or by altering a property of the membrane. As an example, when the steroid hormone aldosterone binds to membrane receptors on its target cells inside kidneys, these target cells quickly become more permeable to sodium ions.

Receptor Function and Diversity

Hormone receptors are proteins. Mutations can result in receptors with a lowered capacity for binding hormone, or none at all. In such cases, the hormone will have a lesser or no effect even though the hormone that targets the mutated receptor is present in normal amounts. For example, typical male genitals will not form in an XY embryo without testosterone, one of the steroid hormones (Section 8.2). XY individuals who have androgen insensitivity syndrome secrete testosterone, but a mutation alters their receptors for it. Without functional receptors, it is as if testosterone is not present. As a result, the embryo forms testes, but they do not descend into the scrotum, and the genitals appear female. Such individuals are often raised as females.

Variations in receptor structure also affect responses to hormones. Different tissues often have receptor proteins that respond in different ways to binding of the same hormone. For example, ADH (antidiuretic hormone) from the posterior lobe of the pituitary acts on kidney cells and helps maintain solute concentrations in the internal environment. ADH is sometimes referred to as vasopressin, because it also binds to receptors in the wall of blood vessels and causes these vessels to narrow. In many mammals, ADH helps maintain blood pressure. ADH also binds to brain cells and influences sexual and social behavior, as we will discuss in Section 39.2. This diversity of responses to a single hormone is an outcome of variations in ADH receptors. In each kind of cell, a dif-

Table 31.1 Categories and Examples of Hormones	
Steroid hormones	Testosterone and other androgens, estrogens, progesterone, aldosterone, cortisol
Amines	Melatonin, epinephrine, thyroid hormone
Peptides	Glucagon, oxytocin, antidiuretic hormone, calcitonin, parathyroid hormone
Proteins	Growth hormone, insulin, prolactin, follicle-stimulating hormone, luteinizing hormone

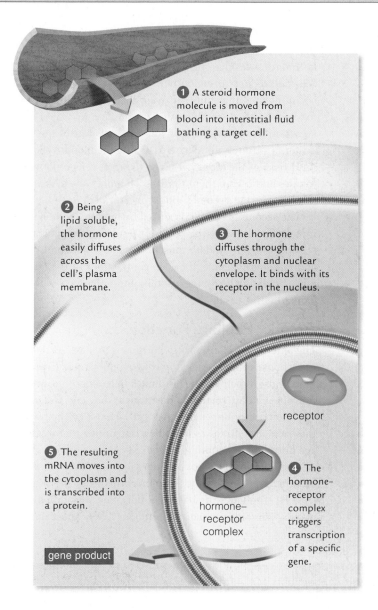

1 A steroid hormone molecule is moved from blood into interstitial fluid bathing a target cell.

2 Being lipid soluble, the hormone easily diffuses across the cell's plasma membrane.

3 The hormone diffuses through the cytoplasm and nuclear envelope. It binds with its receptor in the nucleus.

receptor

5 The resulting mRNA moves into the cytoplasm and is transcribed into a protein.

hormone–receptor complex

4 The hormone–receptor complex triggers transcription of a specific gene.

gene product

A Example of steroid hormone action inside a target cell.

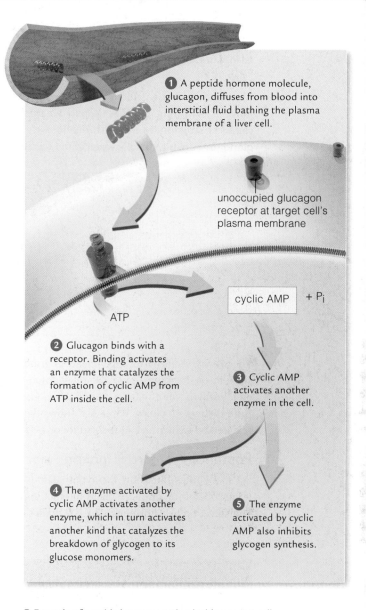

1 A peptide hormone molecule, glucagon, diffuses from blood into interstitial fluid bathing the plasma membrane of a liver cell.

unoccupied glucagon receptor at target cell's plasma membrane

cyclic AMP + P_i

ATP

2 Glucagon binds with a receptor. Binding activates an enzyme that catalyzes the formation of cyclic AMP from ATP inside the cell.

3 Cyclic AMP activates another enzyme in the cell.

4 The enzyme activated by cyclic AMP activates another enzyme, which in turn activates another kind that catalyzes the breakdown of glycogen to its glucose monomers.

5 The enzyme activated by cyclic AMP also inhibits glycogen synthesis.

B Example of peptide hormone action inside a target cell.

Figure 31.3 Animated Mechanisms of hormone action.

›› Figure It Out Which example shows formation of a second messenger and what substance serves as the second messenger? Answer: Cyclic AMP serves as a second messenger in the peptide hormone example.

ferent kind of receptor summons up a different cellular response. You will come across many examples of hormone action in this unit. They will give you a sense of the variations in cellular responses.

second messenger Molecule that forms inside a cell when a hormone binds at the cell surface; sets in motion reactions that alter activity inside the cell.
steroid hormone Hormone such as testosterone that is derived from cholesterol.

Take-Home Message **How do hormones exert their effects on target cells?**

› Hormones exert their effects by binding to receptors, either inside a cell or at the plasma membrane.

› Most steroid and thyroid hormones bind with receptors inside cells and alter gene expression.

› Peptide and protein hormones bind to membrane receptors. Often, a second messenger in the cytoplasm relays the signal into the cell's interior.

› Variations in receptor structure influence the effect a hormone has on a cell.

> The hypothalamus and pituitary gland deep inside the brain control the activities of many other endocrine glands.
< Links to Exocrine gland 28.3, Human brain 29.10

The **hypothalamus** is the main center for control of the internal environment. It lies deep inside the forebrain and connects, structurally and functionally, with the **pituitary gland**. In humans, this pea-sized gland has two lobes:

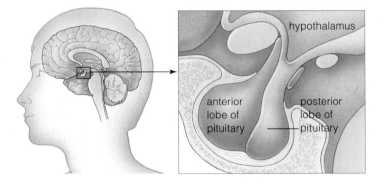

The pituitary's posterior lobe releases hormones synthesized in the hypothalamus. The anterior lobe makes its own hormones but releases them in response to signals from the hypothalamus. Table 31.2 shows the hormones released by each lobe and their functions.

Posterior Pituitary Function

Figure 31.4 illustrates how the hypothalamus and the posterior lobe of the pituitary gland interact in production and secretion of a hormone. The cell bodies of specialized neurons in the hypothalamus produce the hormones that are released by the posterior pituitary. One of these hormones—antidiuretic hormone (ADH)—targets kidney cells and reduces urine output. ADH made by cell bodies in the hypothalamus gets transported through axons to axon terminals in the posterior pituitary. Arrival of an action potential at these axon terminals causes the posterior pituitary to release ADH into the blood. When ADH reaches the kidney, it binds to target cells in kidney tubules and causes them to reabsorb more water. As a result, the urine becomes more concentrated.

The second hormone produced by hypothalamic neurons and released by the posterior pituitary is oxytocin. Oxytocin targets smooth muscle of the uterus (womb) and mammary glands. It causes contractions of the uterus during childbirth and moves milk into milk ducts when a woman nurses a child.

Anterior Pituitary Function

The anterior pituitary makes hormones that are secreted in response to hormones from the hypothalamus (Figure 31.5). Most hypothalamic hormones that act on the anterior pituitary are **releasing hormones**; they encourage secretion of hormones by target cells. However, the hypothalamus also produces **inhibiting hormones** that discourage hormone secretion by target cells.

The anterior pituitary produces four hormones that act on other endocrine glands. Adrenocorticotropic hormone (ACTH) stimulates the release of hormones by adrenal glands. Thyroid-stimulating hormone (TSH) causes the

Table 31.2 Hormones Secreted by the Pituitary Gland				
Pituitary Lobe	**Hormone**	**Designation**	**Main Targets**	**Primary Actions**
Posterior Nervous tissue (extension of hypothalamus)	Antidiuretic hormone (or vasopressin)	ADH	Kidneys	Induces water conservation as required during control of extracellular fluid volume and solute concentrations
	Oxytocin	OT	Mammary glands	Induces milk movement into secretory ducts
			Uterus	Induces uterine contractions during childbirth
Anterior Glandular tissue	Adrenocorticotropic hormone	ACTH	Adrenal cortex	Stimulates release of cortisol, an adrenal steroid hormone
	Thyroid-stimulating hormone	TSH	Thyroid gland	Stimulates release of thyroid hormones
	Follicle-stimulating hormone	FSH	Ovaries, testes	In females, stimulates estrogen secretion, egg maturation; in males, helps stimulate sperm formation
	Luteinizing hormone	LH	Ovaries, testes	In females, stimulates progesterone secretion, ovulation, corpus luteum formation; in males, stimulates testosterone secretion, sperm release
	Prolactin	PRL	Mammary glands	Stimulates and sustains milk production
	Growth hormone	GH	Most cells	Promotes growth in young; induces protein synthesis, cell division; roles in glucose, protein metabolism in adults

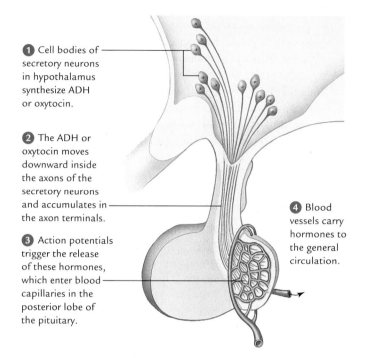

① Cell bodies of secretory neurons in hypothalamus synthesize ADH or oxytocin.

② The ADH or oxytocin moves downward inside the axons of the secretory neurons and accumulates in the axon terminals.

③ Action potentials trigger the release of these hormones, which enter blood capillaries in the posterior lobe of the pituitary.

④ Blood vessels carry hormones to the general circulation.

Figure 31.4 Animated Posterior pituitary function. The posterior pituitary makes hormones and secretes them in response to hormones from the hypothalamus.

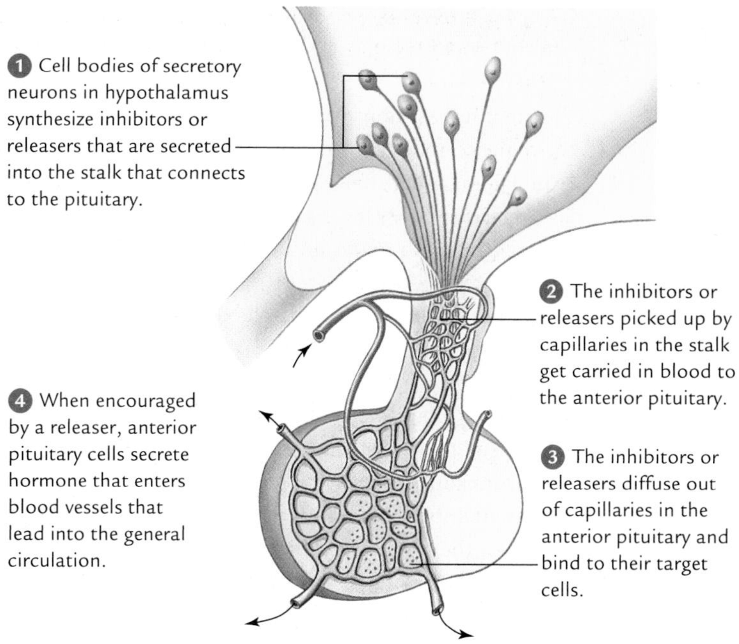

① Cell bodies of secretory neurons in hypothalamus synthesize inhibitors or releasers that are secreted into the stalk that connects to the pituitary.

② The inhibitors or releasers picked up by capillaries in the stalk get carried in blood to the anterior pituitary.

③ The inhibitors or releasers diffuse out of capillaries in the anterior pituitary and bind to their target cells.

④ When encouraged by a releaser, anterior pituitary cells secrete hormone that enters blood vessels that lead into the general circulation.

Figure 31.5 Animated Anterior pituitary function. The anterior pituitary secretes hormones synthesized in the hypothalamus.

thyroid gland to secrete thyroid hormone. Follicle-stimulating hormone (FSH) and luteinizing hormone (LH) affect sex hormone secretion and production of gametes by gonads—a male's testes or a female's ovaries.

The anterior pituitary also produces two other hormones. Prolactin encourages milk production by mammary glands. Growth hormone (GH) has widespread effects throughout the body. It encourages the growth of bone and soft tissues in the young, and it influences metabolism in adults.

Oversecretion of growth hormone in children leads to pituitary gigantism. An affected adult has a normal body form, but is unusually tall, as in Figure 31.6. Growth hormone oversecretion during adulthood causes acromegaly. Continued deposition of new bone and cartilage enlarges and eventually deforms the hands, feet, and face. The skin thickens, and the lips and tongue increase in size. Internal organs are also affected; the heart may become enlarged. Acromegaly is usually caused by a pituitary tumor. If untreated, it can cause serious health problems.

Low GH secretion in childhood causes pituitary dwarfism. Affected adults have the form of an average person but are smaller. Pituitary dwarfism can be inherited, or it can result from a pituitary tumor or injury. Injections of human growth hormone produced through genetic engineering can increase an affected child's growth rate. However, this treatment is controversial. Many people object to the idea that short stature is a defect to be "cured."

Figure 31.6 Pituitary gigantism. Bao Xishun, one of the world's tallest men, stands 2.36 meters (7 feet 9 inches) tall.

---

hypothalamus Forebrain region that controls processes related to homeostasis and has endocrine functions.
inhibiting hormone Hormone that is secreted by one endocrine gland and discourages secretion by another.
pituitary gland Pea-sized endocrine gland in the forebrain that interacts closely with the adjacent hypothalamus.
releasing hormone Hormone that is secreted by one endocrine gland and stimulates secretion by another.

Take-Home Message **How do the hypothalamus and pituitary gland interact?**

❯ Some secretory neurons of the hypothalamus make hormones (ADH, OT) that move through axons into the posterior pituitary, which releases them.

❯ Other hypothalamic neurons produce releasers and inhibitors that are carried by the blood into the anterior pituitary. These hormones regulate the secretion of anterior pituitary hormones (ACTH, TSH, LH, FSH, PRL, and GH).

> A cell in a vertebrate body is a target for a diverse array of hormones secreted by endocrine glands and secretory cells.

The next few sections of this chapter describe effects of the main vertebrate hormones that are released by endocrine glands other than the pituitary. Table 31.3 provides an overview of this information.

In addition to major endocrine glands, vertebrates have hormone-secreting cells in some organs. You learned in Section 31.2 that cells of the small intestine make secretin, a hormone that stimulates the pancreas to secrete bicarbonate. Parts of the gut secrete hormones that affect appetite and digestion. In addition, adipose (fat) tissue makes leptin, a hormone that acts in the brain and suppresses appetite. We discuss hormones related to appetite in Chapter 36.

When the oxygen level in blood falls, the kidneys secrete erythropoietin, a hormone that stimulates maturation and production of red blood cells. The heart makes a hormone, atrial natriuretic peptide, that stimulates the kidneys to excrete water and salt.

As you learn about the effects of specific hormones, keep in mind that cells in most tissues have receptors for more than one hormone. The response called up by one hormone may oppose or reinforce that of another. For example, every skeletal muscle fiber has receptors for glucagon, insulin, cortisol, epinephrine, estrogen, testosterone, growth hormone, somatostatin, and thyroid hormone, as well as others. Thus, blood levels of all of these hormones affect the activity of muscle fibers.

Take-Home Message **What are the sources of vertebrate hormones and how do hormones interact?**

> Hormones are secreted from the pituitary gland, hypothalamus, endocrine glands, and endocrine cells. The gut, kidneys, and heart are among the organs that are not considered glands, but do include hormone-secreting cells.

> Most cells have receptors for multiple hormones, and the effect of one hormone can be enhanced or opposed by that of another hormone.

Table 31.3 Sources and Actions of Vertebrate Hormones Covered in Sections 31.6–31.10

Source	Examples of Secretion(s)	Main Targets	Primary Actions
Thyroid gland	Thyroid hormone	Most cells	Regulates metabolism; has roles in growth, development
	Calcitonin	Bone	Lowers calcium level in blood
Parathyroids	Parathyroid hormone	Bone, kidney	Elevates calcium level in blood
Adrenal cortex	Glucocorticoids (including cortisol)	Most cells	Promote breakdown of glycogen, fats, and proteins as energy sources; thus help raise blood level of glucose
	Mineralocorticoids (including aldosterone)	Kidney	Promote sodium reabsorption (sodium conservation); help control the body's salt–water balance
Adrenal medulla	Epinephrine (adrenaline)	Liver, muscle, adipose tissue	Raises blood level of sugar, fatty acids; increases heart rate and force of muscle contraction
	Norepinephrine	Smooth muscle of blood vessels	Promotes constriction or dilation of certain blood vessels; thus affects distribution of blood volume to different body regions
Pancreatic islets	Insulin	Liver, muscle, adipose tissue	Promotes cell uptake of glucose; thus lowers glucose level in blood
	Glucagon	Liver	Promotes glycogen breakdown; raises glucose level in blood
	Somatostatin	Insulin-secreting cells	Inhibits digestion of nutrients, hence their absorption from gut
Gonads			
Testes (in males)	Androgens (including testosterone)	General	Required in sperm formation, development of genitals, maintenance of sexual traits, growth, and development
Ovaries (in females)	Estrogens	General	Required for egg maturation and release; preparation of uterine lining for pregnancy and its maintenance in pregnancy; genital development; maintenance of sexual traits; growth, development
	Progesterone	Uterus, breasts	Prepares, maintains uterine lining for pregnancy; stimulates development of breast tissues
Pineal gland	Melatonin	Gonads (indirectly)	Influences daily biorhythms, seasonal sexual activity
Thymus gland	Thymopoietin, thymosin	T lymphocytes	Encourages maturation of T lymphocytes (T cells)

> The thyroid regulates metabolic rate, and the adjacent parathyroids regulate calcium levels.

< Link to Negative feedback 28.9

Feedback Control of Thyroid Function

The human **thyroid gland** lies at the base of the neck, attached to the trachea, or windpipe. The thyroid secretes two iodine-containing molecules (triiodothyronine and thyroxine) that are referred to collectively as thyroid hormone. Thyroid hormone increases the metabolic activity of tissues throughout the body. The thyroid gland also secretes calcitonin, a hormone that causes deposition of calcium in the bones of growing children. Normal human adults produce little calcitonin.

The anterior pituitary gland and hypothalamus regulate thyroid hormone secretion by a negative feedback loop (Figure 31.7). A low level of thyroid hormone causes the hypothalamus to secrete thyroid-releasing hormone (TRH) ❶. This releasing hormone causes the anterior pituitary to secrete thyroid-stimulating hormone (TSH) ❷. TSH in turn stimulates the secretion of thyroid hormone ❸. When the blood level of thyroid hormone rises, secretion of TRH and TSH declines ❹.

A diet deficient in iodine can cause thyroid hormone deficiency, or hypothyroidism. Hypothyroidism can also arise when the body's immune system mistakenly attacks the thyroid. In either case, ongoing stimulation of the thyroid can lead to thyroid enlargement, or goiter (Figure 31.8A). Because thyroid hormone increases the body's metabolic rate, a deficiency typically causes fatigue, increased sensitivity to cold temperature, and weight gain. Dietary hypothyroidism in infancy or early childhood often causes cretinism, a syndrome of stunted growth and impaired mental capacity.

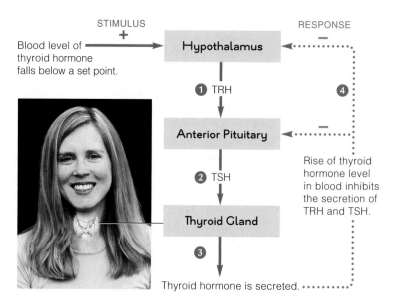

Figure 31.7 Negative feedback loop to the hypothalamus and the pituitary's anterior lobe that governs thyroid hormone secretion.

>> **Figure It Out** What effect does a high thyroid hormone level have on the hypothalamus?

Answer: It inhibits secretion of TRH.

By contrast, an excess of thyroid hormone causes nervousness and irritability, chronic fever, and weight loss. Often, altered metabolism induces tissues behind the eyeball to swell, causing eyes to bulge.

Parathyroid Glands and Calcium Levels

Parathyroid glands regulate the blood's calcium level. There are four of these glands, each about the size of a grain of rice, on the thyroid's rear surface. When calcium level in the blood declines, the glands release parathyroid hormone (PTH). PTH increases the breakdown of bone, thus putting calcium into the blood. It also encourages calcium reabsorption by kidneys and activation of vitamin D, which helps the intestine take up calcium from food.

Children who do not get enough vitamin D absorb too little calcium to build healthy new bone. Their low blood calcium level causes oversecretion of PTH, which encourages breakdown of existing bone. The resulting disorder is called rickets. Typical symptoms include bowed legs and pelvic deformities (Figure 31.8B).

In adults, a decrease in parathyroid hormone, as sometimes occurs as a result of a parathyroid tumor, is one cause of osteoporosis. In this disorder, bone deposition falters, causing bones to become weak and easily broken.

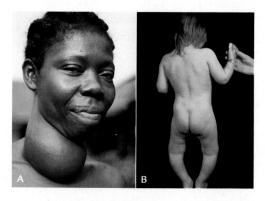

Figure 31.8 Diet-hormone interactions. **A** Goiter caused by a dietary iodine deficiency. **B** Rickets caused by a lack of vitamin D. Parathyroid hormone softed the child's bones, causing bowed legs.

parathyroid glands Four small endocrine glands whose hormone product increases the level of calcium in blood.
thyroid gland Endocrine gland at the base of the neck; produces thyroid hormone, which increases metabolism.

Take-Home Message What are the functions of the thyroid and parathyroid glands?

> The thyroid gland has roles in regulation of metabolism and in development.

> Parathyroid glands regulate blood calcium level.

> An adrenal gland has two functional zones: Its outer cortex secretes steroid hormones. Its inner medulla releases molecules that function as neurotransmitters.

‹ Links to Body's energy sources 7.7, Sympathetic nerves 29.8

There are two **adrenal glands**, one above each kidney. (In Latin *ad*– means near, and *renal* refers to the kidney.) Each adrenal gland is about the size of a big grape. Its outer layer is the **adrenal cortex** and its inner portion is the **adrenal medulla**. The two parts of the gland are controlled by different mechanisms, and they secrete different substances.

The Adrenal Cortex

The adrenal cortex releases steroid hormones. One of these hormones, aldosterone, controls sodium and water reabsorption in the kidneys. Chapter 37 explains this process in detail. The adrenal cortex also produces and secretes small amounts of sex hormones, which we discuss in Section 31.10. Here we will focus on **cortisol**, a hormone that affects metabolism and immune responses.

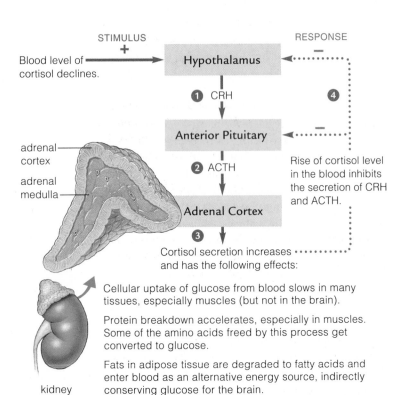

Cellular uptake of glucose from blood slows in many tissues, especially muscles (but not in the brain).

Protein breakdown accelerates, especially in muscles. Some of the amino acids freed by this process get converted to glucose.

Fats in adipose tissue are degraded to fatty acids and enter blood as an alternative energy source, indirectly conserving glucose for the brain.

Figure 31.9 Animated Structure of the human adrenal gland. An adrenal gland rests on top of each kidney. The diagram shows a negative feedback loop that governs cortisol secretion.

›› Figure It Out What effect would a decrease in ACTH have on the rate of fat breakdown in adipose tissue?

Answer: A decrease in ACTH would cause a decrease in cortisol secretion and less fat breakdown.

Figure 31.9 illustrates the negative feedback loop that governs the cortisol level in blood. A decrease in cortisol triggers secretion of CRH (corticotropin-releasing hormone) by the hypothalamus ❶. CRH then stimulates the secretion of ACTH, an anterior pituitary hormone ❷. ACTH causes release of cortisol ❸ from the adrenal cortex. The level of cortisol increases, causing the hypothalamus and anterior pituitary secrete less CRH and ACTH, and cortisol secretion slows ❹.

Cortisol helps maintain the blood glucose available to the brain by inducing liver cells to break down their store of glycogen, and suppressing uptake of glucose by most cells. Cortisol also induces adipose cells to degrade fats, and skeletal muscles to degrade proteins. The breakdown products of these reactions—fatty acids and amino acids—function as alternative energy sources (Section 7.7).

With injury, illness, or anxiety, the nervous system overrides the feedback loop, and cortisol in blood can soar. In the short term, this response helps get enough glucose to the brain when food supplies are likely to be low. The heightened cortisol also suppresses inflammatory responses, thus lessening inflammation-related pain.

The Adrenal Medulla

The adrenal medulla contains specialized neurons of the sympathetic division (Section 29.8). Like other sympathetic neurons, those in the adrenal medulla release norepinephrine and epinephrine. However, in this case, the norepinephrine and epinephrine enter the blood and function as hormones, rather than acting as neurotransmitters at a synapse. Epinephrine and norepinephrine released into the blood have the same effect on a target organ as direct stimulation by a sympathetic nerve.

Remember that sympathetic stimulation plays a role in the fight–flight response. Epinephrine and norepinephrine dilate the pupils, increase breathing rate, and make the heart beat faster. They prepare the body to deal with an exciting or dangerous situation.

Stress, Elevated Cortisol, and Health

When an animal is frightened or under physical stress, commands from the nervous system trigger increased secretion of cortisol, epinephrine, and norepinephrine. As these secretions find their targets, they help the body deal with the immediate threat by diverting resources from longer-term tasks. This stress response is highly adaptive for short periods of time, as when an animal is fleeing from a predator.

However, elevation of cortisol level can become detrimental if stress does not end. Physiological responses to chronic stress interfere with growth, the immune system, sexual function, and cardiovascular function. Chronically high cortisol levels harm cells in the hippocampus, a brain

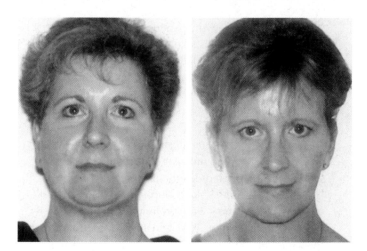

Figure 31.10 Cushing syndrome. *Left*, Woman with elevated cortisol levels as a result of a adrenal gland tumor. She has the characteristic puffy moon face. *Right*, The same woman after removal of the tumor lowered her cortisol to normal levels.

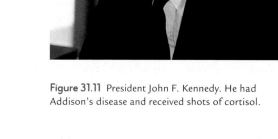

Figure 31.11 President John F. Kennedy. He had Addison's disease and received shots of cortisol.

region central to memory and learning (Section 29.11). We know from studies of primates that animals subjected to chronic social stress have higher than normal cortisol levels and poorer than average health.

Elevated cortisol may help explain the observed link between low social status and poor health. People who are low in a socioeconomic hierarchy do tend to have more health problems—obesity, hypertension, and diabetes—than those who are better off. These differences persist even after researchers factor out the obvious causes, such as variations in diet and access to health care.

We see the impact of long-term elevated cortisol levels in humans affected by Cushing's syndrome, or hypercortisolism. This metabolic disorder can be triggered by an adrenal gland tumor, oversecretion of ACTH by the anterior pituitary, or ongoing use of the drug cortisone. Doctors often prescribe cortisone to relieve chronic inflammation. The body converts it to cortisol, which dampens immune responses.

The symptoms of hypercortisolism include a puffy, rounded "moon face" (Figure 31.10) and increased fat deposition around the torso. Blood pressure and blood glucose become unusually high. White blood cell counts are low, so affected people are more prone to infections. Thin skin, decreased bone density, and muscle loss are

common. Wounds may be slow to heal. Hair on the scalp thins. Women's menstrual cycles are erratic or nonexistent. Men may be impotent. Often, the hippocampus shrinks. Patients with the highest cortisol level also have the greatest reduction in the volume of their hippocampus, and the most impaired memory.

Adrenal Insufficiency

Tuberculosis and other infectious diseases can damage the adrenal glands, and slow or halt cortisol secretion. The result is Addison's disease, or hypocortisolism. In developed countries, this hormonal disorder more often arises after autoimmune attacks on the adrenal glands. President John F. Kennedy (Figure 31.11) had the autoimmune form of the disorder. Symptoms often include fatigue, weakness, depression, weight loss, and darkening of the skin. If cortisol levels get too low, blood sugar and blood pressure can fall to life-threatening levels. Addison's disease is treated with a synthetic form of cortisone.

---

adrenal cortex Outer portion of adrenal gland; secretes aldosterone and cortisol.
adrenal gland Endocrine gland that is located atop the kidney.
adrenal medulla Inner portion of adrenal gland; secretes epinephrine and norepinephrine.
cortisol Adrenal cortex hormone that influences metabolism and immunity; secretions rise with stress.

Take-Home Message **What are the functions of the hormones secreted by the adrenal glands?**

> The adrenal cortex secretes aldosterone, cortisol, and small amounts of sex hormones. Aldosterone affects urine concentration and cortisol affects metabolism and the stress response.

> The adrenal medulla releases epinephrine and norepinephrine, which prepare the body for excitement or danger.

> Cortisol secretion is governed by a feedback loop to the hypothalamus and pituitary, but in times of stress that loop is broken and the level of cortisol in the blood rises.

> Long-term elevation of cortisol harms health. Insufficient cortisol can be fatal.

❭ Two pancreatic hormones with opposing effects work together to regulate the level of sugar in the blood.
❬ Link to Glycogen 3.3

The **pancreas** lies in the abdominal cavity, behind the stomach (Figure 31.12) and has both exocrine and endocrine functions. Its exocrine cells secrete digestive enzymes into the small intestine. Its endocrine cells are grouped in clusters called pancreatic islets. Each islet contains three types of hormone-secreting cells.

Beta cells, the most abundant cells in the pancreatic islets, secrete insulin—the only hormone that causes target cells to take up and store glucose. After a meal, the blood level of glucose rises, which stimulates beta cells to release insulin. The main targets are liver, fat, and skeletal muscle cells. Insulin especially stimulates muscle and fat cells to take up glucose. In all target cells, it activates enzymes that function in protein and fat synthesis, and it inhibits the enzymes that catalyze protein and fat breakdown. As a result of its actions, insulin lowers the level of glucose in the blood (Figure 31.12 ❶–❺).

pancreas Organ that secretes digestive enzymes into the small intestine and hormones into the blood.

Alpha cells secrete the peptide hormone glucagon. Between meals, all cells take up glucose from blood. When the glucose level falls below a set point, alpha cells secrete glucagon. Glucagon binds to cells in the liver and causes the activation of enzymes that break glycogen into glucose subunits. By its action, glucagon raises the level of glucose in blood (Figure 31.12 ❻–❿).

Delta cells secrete somatostatin. This hormone helps control digestion and nutrient absorption. Also, it can inhibit the secretion of insulin and glucagon.

Some shifts in the blood glucose level are typical of all animals that show a discontinuous eating pattern. By working in opposition, glucagon and insulin from the pancreas maintain that level within a range that keeps cells functioning properly.

Take-Home Message **How do pancreatic hormones maintain the level of glucose in the blood?**

❭ Insulin helps cells take up and store more glucose; it lowers the blood level of glucose.

❭ Glucagon triggers the breakdown of glycogen; it raises the blood level of glucose.

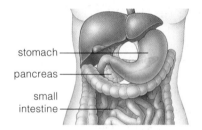

stomach
pancreas
small intestine

Figure 31.12 Animated *Above,* the location of the pancreas. *Right,* how cells that secrete insulin and glucagon work antagonistically to adjust the level of glucose in the blood.

❶ After a meal, glucose enters blood faster than cells can take it up, so blood glucose increases.

❷ The increase stops pancreatic cells from secreting glucagon, and ❸ stimulates other cells to secrete insulin.

❹ In response to insulin, adipose and muscle cells take up and store glucose; cells in the liver and muscle make more glycogen.

❺ As a result, the blood level of glucose declines to its normal level.

❻ Between meals, blood glucose declines as cells take it up and use it for metabolism.

❼ The decrease encourages glucagon secretion, and ❽ slows insulin secretion.

❾ In the liver, glucagon causes cells to break glycogen down into glucose, which enters the blood.

❿ As a result, blood glucose increases to the normal level.

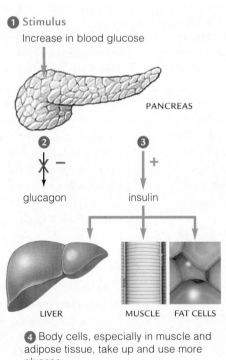

❶ Stimulus
Increase in blood glucose

PANCREAS

❷ — glucagon   ❸ + insulin

LIVER   MUSCLE   FAT CELLS

❹ Body cells, especially in muscle and adipose tissue, take up and use more glucose.

Cells in skeletal muscle and liver store glucose in the form of glycogen.

❺ Response
Decrease in blood glucose

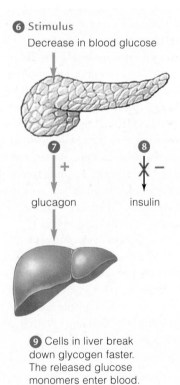

❻ Stimulus
Decrease in blood glucose

❼ + glucagon   ❽ — insulin

❾ Cells in liver break down glycogen faster. The released glucose monomers enter blood.

❿ Response
Increase in blood glucose

> Glucose is the main energy source for brain cells and the only one for red blood cells. Having too much or too little glucose in blood causes problems throughout the body.
‹ Link to Body's alternative energy sources 7.7

Diabetes mellitus is a common metabolic disorder. Its name can be loosely translated as "passing honey-sweet water." Diabetics produce sweet urine because their liver, fat, and muscle cells do not take up and store glucose as they should. The resulting high blood sugar, or hyperglycemia, disrupts normal metabolism. When cells do not take up glucose, they have to break down proteins and fats for energy (Section 7.7). Breakdown of these substances yields harmful waste products. At the same time, high blood sugar causes some cells to overdose on glucose and produce other harmful substances. Accumulation of harmful molecules causes the complications associated with diabetes (Table 31.4).

Type 1 Diabetes There are two main types of diabetes mellitus. Type 1 develops after the body has mounted an autoimmune response against its insulin-secreting beta cells. Certain white blood cells wrongly identify the cells as foreign (nonself) and destroy them. Environmental factors add to a genetic predisposition to the disorder. Symptoms usually appear during childhood and adolescence, which is why this metabolic disorder is known as juvenile-onset diabetes. All affected individuals require injections of insulin, and must monitor their blood sugar level carefully. New devices called insulin pumps dispense a continuous supply of insulin (Figure 31.13).

Type 1 diabetes accounts for only 5 to 10 percent of all reported cases, but it is the most dangerous in the short term. In the absence of a steady supply of glucose, the body of affected people uses fats and proteins as energy

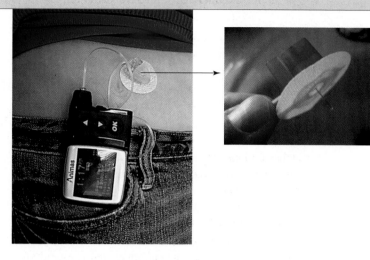

Figure 31.13 An insulin pump. The device is programmed to deliver insulin through a hollow tube that projects through the skin and into the body. The pump helps smooth out fluctuations in blood sugar, thus lowering risk of complications that arise from excessively low or high blood sugar.

sources. Two outcomes are weight loss and ketone accumulation in the blood and urine. Ketones are normal acidic products of fat breakdown, but too many can alter the acidity and solute levels of body fluids. This condition, which is called ketosis, can interfere with normal brain function, and extreme cases may lead to coma or death.

Type 2 Diabetes With type 2 diabetes, the most common form of the disorder, insulin levels are normal or even high. However, target cells do not respond to the hormone as they should, and blood sugar levels remain elevated. Symptoms typically start to develop in middle age, when insulin production declines. Genetics is a factor, but obesity increases the risk.

Diet, exercise, and oral medications can control most cases of type 2 diabetes. Even so, if glucose levels are not lowered, pancreatic beta cells receive continual stimulation. Eventually they will falter, and so will insulin production. When that happens, a type 2 diabetic may require insulin injections.

Worldwide, rates of type 2 diabetes are soaring. By one estimate, more than 150 million people are now affected. Western diets and sedentary life-styles are contributing factors. The prevention of diabetes and its complications is acknowledged to be among the most pressing public heath priorities around the world.

Table 31.4	Some Complications of Diabetes
Eyes	Changes in lens shape and vision; damage to blood vessels in retina; blindness
Skin	Increased susceptibility to bacterial and fungal infections; patches of discoloration; thickening of skin on the back of hands
Digestive system	Gum disease; delayed stomach emptying that causes heartburn, nausea, vomiting
Kidneys	Increased risk of kidney disease and failure
Circulatory system	Increased risk of heart attack, stroke, high blood pressure, and atherosclerosis
Hands and feet	Impaired sensations of pain; formation of calluses, foot ulcers; possible amputation of a foot or leg because of necrotic tissue that formed owing to poor circulation

Take-Home Message **What is diabetes?**

> Diabetes is a metabolic disorder in which the body does not make insulin or the body does not respond to it. As a result, cells do not take up sugar as they should, causing complications throughout the body.

> Outputs from the gonads, pineal gland, and thymus all change as an individual enters puberty.

< Link to Optic nerve 30.5

The Gonads

Gonads, or primary reproductive organs, produce gametes (eggs or sperm) as well as sex hormones. The gonads of male vertebrates are testes (singular, testis) and the main hormone they secrete is testosterone, the male sex hormone. Female gonads are the ovaries. They secrete mainly estrogens and progesterone, the female sex hormones. Figure 31.14 shows the location of human gonads.

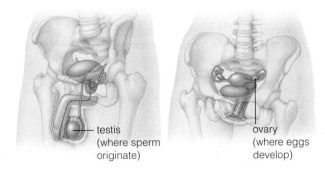

testis (where sperm originate)

ovary (where eggs develop)

Figure 31.14 Location of human gonads. These organs produce gametes and secrete sex hormones.

Puberty is a post-embryonic stage of development when the reproductive organs and structures mature. At puberty, a female mammal's ovaries increase their estrogen production, which causes breasts and other female secondary sexual traits to develop. Estrogens and progesterone control egg formation and ready the uterus for pregnancy. In males, a rise in testosterone output triggers the onset of sperm formation and the development of secondary sexual traits.

The hypothalamus and anterior pituitary control the secretion of sex hormones (Figure 31.15). In both males and females, the hypothalamus produces GnRH (gonadotropin-releasing hormone). This releaser causes the anterior pituitary to secrete follicle-stimulating hormone (FSH) and luteinizing hormone (LH). FSH and LH cause the gonads to secrete sex hormones.

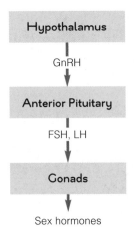

Hypothalamus

GnRH

Anterior Pituitary

FSH, LH

Gonads

Sex hormones

Figure 31.15 Control of sex hormone secretion.

The Pineal Gland

The **pineal gland** lies deep within the vertebrate brain. This small, pinecone-shaped gland secretes melatonin. Melatonin secretion declines when the retina detects light and sends action potentials along the optic nerve to the brain (Section 30.5).

Melatonin targets neurons that lower body temperature and make us drowsy in dim light. The blood level of melatonin peaks in the middle of the night. Exposure to bright light sets a biological clock that controls sleeping versus arousal. Travelers who cross many time zones are advised to spend time in the sun after reaching a destination to minimize jet lag. In winter, a high melatonin level causes seasonal affective disorder in some people.

Affected people become tired and depressed. They may crave foods rich in carbohydrates. Bright artificial light in the morning decreases pineal gland activity and can improve mood.

Melatonin may affect human gonads. A decline in the production of this hormone starts at puberty and may help trigger it. Some pineal gland disorders are known to accelerate or delay puberty.

Melatonin also has a protective effect against some cancers. It directly inhibits division of cancer cells and also suppresses production of sex hormones, which can encourage the growth of some cancers. Routinely working night shifts or having poor sleep habits can disrupt melatonin secretion and raise the risk of cancer.

The Thymus

The **thymus** lies beneath the breastbone. The hormones it secretes (thymosins) help infection-fighting white blood cells called T lymphocytes, or T cells, to mature. The thymus grows until a person reaches puberty, when it is about the size of an orange. Then, the surge of sex hormones causes it to shrink, and its secretions decline. In most people, this decline is not a problem because they have plenty of T cells made earlier in life. However, the disease AIDS kills T cells. Thus, researchers are now looking for ways to restore the immune function of AIDS patients by reactivating the thymus.

gonads Primary reproductive organs (ovaries or testes) that produce gametes and sex hormones.
pineal gland Endocrine gland deep inside the brain that secretes melatonin when the retina is not stimulated by light.
thymus Endocrine gland beneath the breastbone; secretes hormones that encourage maturation of T lymphocytes (T cells).

Take-Home Message What are the endocrine functions of the gonads, pineal gland, and thymus?

> A female's ovaries or a male's testes are gonads that make sex hormones as well as gametes.

> The pineal gland inside the brain produces melatonin, which influences sleep/wake cycles and the onset of puberty. Melatonin also protects against some cancers.

> The thymus is in the chest and it secretes thymosins that are necessary for the maturation of white blood cells called T cells.

31.11 Invertebrate Hormones

> This chapter focused on vertebrate hormones, but hormones and their receptors also coordinate the activity of invertebrate bodies.

< Links to Introns 9.3, Arthropod molting 23.10

Evolution of Receptor Diversity

How did vertebrates evolve so many diverse hormones and hormone receptors? Molecular evidence points to gene duplications and subsequent divergences by way of mutations. Genetic analysis has revealed the invertebrate ancestry of some vertebrate hormone receptors. For example, sea anemones have receptors at the plasma membrane that are structurally similar to vertebrate receptors for TSH, LH, FSH, and other signaling molecules. In addition to this structural similarity, the genes that encode these receptors have similar nucleotide sequences in both vertebrates and invertebrates, and they have the same number and type of introns in similar regions. These many similarities are taken as evidence that the gene for this group of receptor proteins arose millions of years ago in the common ancestor of sea anemones and vertebrates.

Control of Molting

In arthropods, hormones regulate the periodic molting of the cuticle that allows the animal to grow. A soft new cuticle forms beneath the old one, which is then shed (Section 23.10). Before the new cuticle hardens, the body mass increases by the rapid uptake of air or water and by continuous mitotic cell divisions. Details vary among groups. In all cases, however, molting is largely under the control of ecdysone, a steroid hormone unique to invertebrates.

In arthropods, a molting gland produces and stores ecdysone and releases it for distribution through the body at molting time. Hormone-secreting neurons in the brain seem to control its release. They respond to a combination of internal signals and environmental cues, including light and temperature.

Figure 31.16 shows how molting is controlled in crabs and other crustaceans. Right before and during an episode of molting, coordinated interactions among ecdysone and other hormones bring about structural and physiological changes. The interactions make the old cuticle separate from the epidermis and muscles. They also induce changes that dissolve inner layers of the cuticle and recycle the remnants, and promote rapid cell divisions, secretions, and pigment formation that help make a new cuticle.

The steps differ a bit in insects, which do not have a molt-inhibiting hormone. Rather, stimulation of the insect brain sets in motion a cascade of signals that trigger the production of molt-inducing ecdysone. Chemicals that mimic or interfere with ecdysone and other hormones can be used as insecticides. Nematodes also require ecdysone to molt and can be killed with ecdysone inhibitors.

Hormones in the Balance (revisited)

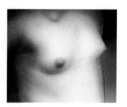

The photo at the *right* shows breast development in a girl a little less than two years old. Exposure to high levels of synthetic chemicals called phthalates (pronounced THAL-aytes) may be one cause of such premature breast enlargement. Phthalates are endocrine disrupters that mimic the female sex hormone estrogen and suppress the testosterone secretion required for normal male development. In recognition of these effects, the United States Congress recently banned the use of phthalates in toys for children under age twelve. But phthalates are still used in other soft plastic items and they provide the scent in many personal care products such as shampoos and lotions. The American Academy of Pediatrics recommends that parents use only unscented products on infants and young children.

How Would You Vote? A variety of synthetic chemicals are suspected of being endocrine disrupters. Should such chemicals remain in use while researchers investigate their effects? See CengageNow for details, then vote online (cengagenow.com).

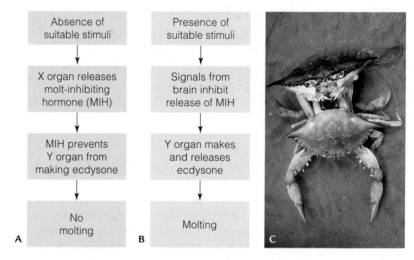

Figure 31.16 Hormonal control of molting in crustaceans such as crabs. Two hormone-secreting organs play a role. The X organ is in the eye stalk. The Y organ is at the base of the crab's antennae.

A In the absence of environmental cues for molting, secretions from the X organ prevent molting. **B** When stimulated by proper environmental cues, the brain sends nervous signals that inhibit X organ activity. With the X organ suppressed, the Y organ releases the ecdysone that stimulates molting.

C A newly molted blue crab with its old shell. The new shell remains soft for about 12 hours, making it a "soft-shelled crab." During this time, the crab is highly vulnerable to predators, including human seafood lovers.

Take-Home Message Do invertebrates make hormones?

> We can trace the evolutionary roots of some vertebrate hormone receptors to invertebrates.

> Invertebrates also produce hormones that have no vertebrate counterparts. Hormones that control molting in arthropods are an example.

Summary

Section 31.1 Endocrine disrupters are synthetic substances that interfere with action of the endocrine system. Atrazine, a widely used herbicide that often seeps into water, disrupts amphibian endocrine function.

Section 31.2 Animal cells communicate with adjacent body cells by way of gap junctions, neurotransmitters, and **local signaling molecules**. Animal hormones travel through the blood and can carry signals between cells in distant parts of the body. All hormone-secreting glands and cells in a body constitute the animal's **endocrine system**.

Section 31.3 The **steroid hormones** are lipid soluble and derived from cholesterol. They can enter cells and bind to receptors inside them. Peptide and protein hormones are derived from amino acids. They bind to receptors in the cell membrane. Often binding triggers the formation of a **second messenger**, a molecule that elicits changes inside the cell. For a cell to respond to a hormone, it must have functional receptors for that hormone.

Section 31.4 The **hypothalamus**, deep in the forebrain, is structurally and functionally linked with the **pituitary gland**.

Axons of neurons in the hypothalamus extend into the posterior pituitary, where they release oxytocin and antidiuretic hormone. Oxytocin targets smooth muscle in mammary glands and the uterus. Antidiuretic hormone concentrates the urine by acting in the kidney.

Releasing hormones and **inhibiting hormones** secreted by the hypothalamus control the secretion of hormones made by the anterior lobe of the pituitary. Four anterior pituitary hormones target other glands (the adrenal cortex, the thyroid gland, and gonads). Another anterior pituitary hormone encourages milk production. Growth hormone (GH) secreted by the anterior pituitary acts throughout the body. Pituitary gigantism and dwarfism result from mutations that affect GH secretion or receptors for this hormone.

Section 31.5 In addition to major endocrine glands, the human body has hormone-secreting cells in many organs. A cell has receptors for many hormones and its behavior depends on the effect of all hormones acting on it.

Section 31.6 A negative feedback loop to the anterior pituitary and hypothalamus governs the **thyroid gland**. Iodine is required for synthesis of thyroid hormone, which increases metabolic rate and plays an important role in development. Four **parathyroid** glands located at the rear of the thyroid gland are the main regulators of calcium levels in the blood. Their secretions act on bone and the kidneys.

Section 31.7 There is an **adrenal gland** atop each kidney. The **adrenal cortex** secretes two of the steroid hormones: aldosterone, which acts on the kidneys, and cortisol. **Cortisol** secretion affects metabolism and it is governed by a negative feedback loop to the anterior pituitary gland and hypothalamus. In times of stress, the central nervous system overrides the feedback controls so that cortisol levels rise. Over the long term, excess cortisol has negative impacts on health.

Norepinephrine and epinephrine released by neurons of the **adrenal medulla** influence organs as sympathetic stimulation does; they cause a fight–flight response.

Sections 31.8, 31.9 Insulin and glucagon secreted by islet cells of the **pancreas** are the main regulators of blood glucose level. Insulin stimulates glucose uptake by muscle and liver cells and thus lowers the blood glucose. Glucagon stimulates the release of glucose, which increases blood levels. Diabetes mellitus is a disorder in which the body does not make or does not respond to insulin.

Section 31.10 Gonads (ovaries and testes) secrete estrogens, progesterone, and testosterone. These sex hormones are steroids with roles in reproduction and in development of secondary sexual traits.

Melatonin secretion by the vertebrate **pineal gland** affects the daily sleep/wake cycle and the onset of puberty. Exposure to light suppresses melatonin production.

The **thymus** beneath the breastbone secretes hormones that encourage maturation of infection-fighting white blood cells called T cells.

Section 31.11 Vertebrate hormone receptor proteins often resemble similar receptor proteins in invertebrates and probably evolved from them. Invertebrates also have hormones with no vertebrate counterpart. For example, the steroid hormone ecdysone regulates molting in arthropods such as crabs and insects.

Self-Quiz   Answers in Appendix III

1. _____ are signaling molecules that travel through the blood to target cells.
   a. Hormones   c. Local signaling molecules
   b. Neurotransmitters d. all of the above

2. Antidiuretic hormone is produced in the hypothalamus but distributed from the _____ .
   a. anterior pituitary   c. kidney
   b. posterior pituitary d. pineal gland

3. Overproduction of _____ causes gigantism.
   a. growth hormone   c. insulin
   b. cortisol   d. melatonin

Sperm Counts Down on the Farm Contamination of water by agricultural chemicals affects the reproductive function of some animals. Are there effects on humans? Epidemiologist Shanna Swan and her colleagues studied sperm collected from men in four cities in the United States (Figure 31.17). The men were partners of women who had become pregnant and were visiting a prenatal clinic, so all were fertile. Of the four cities, Columbia, Missouri, is located in the county with the most farmlands. New York City in New York represents an area with no agriculture.

1. Where did researchers record the highest and lowest sperm counts?
2. In which cities did samples show the highest and lowest sperm motility (ability to move)?
3. Aging, smoking, and sexually transmitted diseases adversely affect sperm. Could differences in any of these variables explain the regional differences in sperm count?
4. Do these data support the hypothesis that living near farmlands can adversely affect male reproductive function?

	Location of Clinic			
	Columbia, Missouri	Los Angeles, California	Minneapolis, Minnesota	New York, New York
Average age	30.7	29.8	32.2	36.1
Percent nonsmokers	79.5	70.5	85.8	81.6
Percent with history of STD	11.4	12.9	13.6	15.8
Sperm count (million/ml)	58.7	80.8	98.6	102.9
Percent motile sperm	48.2	54.5	52.1	56.4

Figure 31.17 Data from a study of sperm collected from men who were partners of pregnant women that visited prenatal health clinics in one of four cities. STD stands for sexually transmitted disease.

4. Steroid hormones are synthesized from _____ .
   a. amines
   b. peptides
   c. proteins
   d. cholesterol

5. _____ lowers blood sugar levels; _____ raises it.

6. The pituitary detects a rising hormone concentration in blood and inhibits the gland secreting the hormone. This is a _____ feedback loop.

7. The _____ has endocrine and exocrine functions.
   a. hypothalamus
   b. pancreas
   c. pineal gland
   d. parathyroid gland

8. Secretion of _____ stimulates breakdown of bone.
   a. glucagon
   b. melatonin
   c. thyroid hormone
   d. parathyroid hormone

9. Exposure to light _____ melatonin levels.
   a. raises
   b. lowers
   c. does not affect

10. During stressful times, the adrenal glands increase their secretion of _____ .
    a. iodine
    b. antidiuretic hormone
    c. cortisol
    d. secretin

11. The male sex hormone testosterone is secreted in response to a hormone secreted by the _____ .
    a. testes
    b. ovaries
    c. pituitary gland
    d. pancreas

12. The _____ shrinks in size after puberty.
    a. pancreas
    b. pituitary
    c. pineal gland
    d. thymus

13. True or false? The steroid hormone ecdysone is made by insects, crabs, and most vertebrates.

14. People who have type 1 diabetes cannot make _____ and must inject it.

15. Match the term with its most suitable description.
    ___ adrenal medulla
    ___ thyroid gland
    ___ parathyroid glands
    ___ pancreatic islets
    ___ pineal gland
    ___ prostaglandin

    a. affected by day length
    b. potent local effects
    c. raise(s) blood calcium level
    d. source of epinephrine
    e. secrete(s) insulin, glucagon
    f. hormones require iodine

Additional questions are available on **CENGAGENOW**.

Critical Thinking

1. A tumor in the pituitary gland can cause a woman to produce milk even when she is not pregnant. In which lobe of the pituitary would such a tumor be located, and what hormone would it affect?

2. Women who are completely blind tend to undergo puberty at an earlier age than sighted women. They also are less likely to have breast cancer. By one hypothesis, these differences arise as a result of the blind women's melatonin level. Based on this information, would you expect the average melatonin level in blind women to be higher or lower than that of sighted women?

Animations and Interactions on **CENGAGENOW**:
❯ Human endocrine glands; Mechanisms of hormone action; Pituitary–hypothalamus interactions; Negative feedback control of cortisol secretion; Hormonal control of blood glucose level.

‹ Links to Earlier Concepts

This chapter elaborates on some animal traits and evolutionary trends introduced in Chapters 23 and 24. You will build on your knowledge of connective tissue (28.4), muscle tissue (28.5), neuromuscular junctions (29.6), and hormonal effects on bone (31.6). There are examples of active transport (5.7) and of X-linked inheritance (14.4). The discussion of muscle contraction involves a detailed look at the role of actin and myosin (4.10).

Key Concepts

Animal Skeletons
An animal's muscles exert force against a skeleton. In some invertebrates, muscles change the shape of a fluid-filled body cavity. Other invertebrates have a hard external skeleton. Vertebrates have a hard internal skeleton made of cartilage and (in most groups) bone.

Bone Structure and Function
Bones are mineral-rich organs that a vertebrate's muscles pull on to move the body. Bones also protect and support soft organs, and store minerals. Blood cells form in some bones. Ligaments connect bones to one another at most joints.

32 Structural Support and Movement

32.1 Muscles and Myostatin

Like neurons, skeletal muscle cells generally do not divide after birth. A muscle becomes bulkier not by adding cells, but by enlarging existing cells. Inside each muscle fiber, the protein filaments involved in muscle contraction are continually built and broken down. Exercise tilts this process in favor of synthesis, so muscle cells get bigger and the muscle gets stronger. Your body is like a machine that improves with use. The more you use your muscles, the more powerful they become.

Certain hormones also encourage increases in muscle mass. For example, one effect of the sex hormone testosterone is to encourage muscle cells to build more proteins. Men make much more testosterone than women, which is why men tend to be more muscular. Human growth hormone also stimulates synthesis of muscle proteins.

Synthetic versions of natural muscle-building, or "anabolic," hormones can bulk up muscles and thus enhance athletic ability. However, most sport organizations consider the use of these drugs to be cheating and penalize athletes who use them.

Some people have a natural genetic advantage when it comes to putting on muscle. Liam Hoekstra, the Michigan boy shown at the *left* is one of them. By his first birthday, Liam could do chin ups. By age 3, he could easily lift 5-pound (2.3-kilogram) dumbbells. The boy is homozygous for a mutation in the gene for myostatin, a regulatory protein that normally acts as a brake on production of muscle proteins. Without myostatin to suppress protein production, an individual has bulky muscles, little body fat, and remarkable strength.

More evidence of myostatin's effect on athletic ability comes from a study of whippets, a type of dog bred for racing. Some whippets are homozygous for a mutation that prevents them from making myostatin. Such dogs, called bully whippets, are heavily muscled (Figure 32.1). Because bully whippets do not match the breeders' ideal of how a whippet should look, they are not usually raced or bred. However, dogs that are heterozygous

for the mutant myostatin allele do race. Compared to normal whippets, they make less myostatin, are more muscular, and are more likely to win races.

Drug companies hope to develop drugs that inhibit myostatin production or activity for use in treatment of muscle-wasting disorders such as muscular dystrophy. Lowering myostatin levels discourages fat deposition, so myostatin inhibitors might also help treat obesity. Will some athletes use myostatin-inhibiting drugs to push their bodies beyond natural limits? No doubt they will. Some are already buying nutritional supplements that purport to bind to myostatin and reduce its activity. In clinical tests, these supplements had no effect on strength, but hope for an extra edge keeps moving them off the shelf.

Figure 32.1 *Opposite page*, A genetic deficiency in the protein myostatin makes this young Michigan boy unusually muscular and strong.

Above, A heavily muscled bully whippet (*left*) also lacks myostatin. A whippet with normal mystatin genes (*right*) has a lightly muscled body.

The Muscle–Bone Partnership
Skeletal muscles interact with bones and with one another. Muscles can only pull on a bone; they cannot push. Many muscles work in pairs, with the action of one reversing the action of the other. Tendons attach skeletal muscles to bones.

How Skeletal Muscle Contracts
A muscle fiber contains many myofibrils, each divided crosswise into sarcomeres. Sarcomeres contain parallel filaments of the proteins actin and myosin. Muscle contracts when ATP-driven interactions between these proteins shortens sarcomeres.

Factors Affecting Contraction
Muscle fibers in a muscle are organized into motor units that contract in response to signals from a motor neuron. Diseases or disorders can interfere with muscle function. Exercise improves muscle strength and endurance.

> An animal skeleton can be internal or external. In either case, muscles exert force against it to move body parts.
‹ Links to Arthropod traits 23.10; Evolution of jaws 24.3, limbs 24.4, and bipedalism 24.9; Connective tissue 28.4

Types of Skeletons

Muscles bring about movement of body parts by interacting with a skeleton. Many soft-bodied invertebrates have a **hydrostatic skeleton**, an internal, fluid-filled chamber or chambers that muscles exert force against. For example, earthworms have a coelom divided into many fluid-filled chambers, one per segment. Section 23.7 explained how muscles alter the shape of these segments by squeezing the fluid-filled chambers. By analogy, think about how squeezing a water-filled balloon can change its shape. Coordinated changes in the shape of body segments move the worm. Similarly, a sea anemone can change its shape by redistributing water trapped inside its gastrovascular cavity (Figure 32.2).

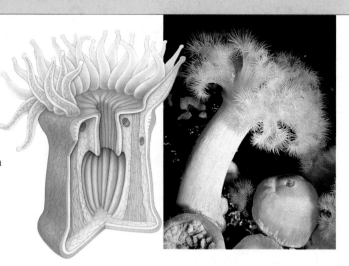

Figure 32.2 Animated Sea anemone, with a hydrostatic skeleton. Muscles run around the body and up and down its length. The body changes shape when these muscles exert force on water trapped in the central gastrovascular cavity.

An **exoskeleton** is a shell, cuticle, or other hard external body part that receives the force of a muscle contraction. A clam's muscles act on its shell to pull the shell shut. Similarly, muscles attach to and pull on an arthropod's hinged exoskeleton. A fly takes flight when the action of muscles attached to its thorax cause its wings to flap up and down (Figure 32.3). Recall that an arthropod exoskeleton is nonliving secreted material. As an animal grows, it periodically molts its exoskeleton and forms a new, larger one.

An **endoskeleton** is an internal framework of hard elements. Echinoderms such as sea stars have an endoskeleton made of hardened calcium-rich plates. Vertebrates also have an endoskeleton.

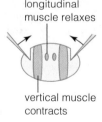

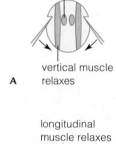

longitudinal muscle contracts

vertical muscle relaxes

A

longitudinal muscle relaxes

vertical muscle contracts

B

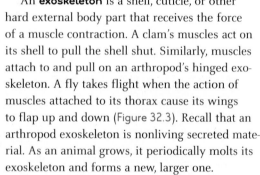

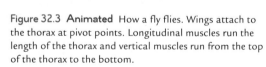

Figure 32.3 Animated How a fly flies. Wings attach to the thorax at pivot points. Longitudinal muscles run the length of the thorax and vertical muscles run from the top of the thorax to the bottom.

A Wings pivot down as relaxation of vertical muscle and contraction of longitudinal muscle pulls in sides of thorax.

B Wings pivot up as vertical muscles contract and longitudinal muscles relax.

Features of the Vertebrate Endoskeleton

The skeleton of sharks and other cartilaginous fishes consists of cartilage, a rubbery connective tissue. Other vertebrate skeletons include some cartilage, but consist mostly of bone tissue (Section 28.4).

The term "vertebrate" refers to the **vertebral column**, or backbone, a feature common to all members of this group (Figures 32.4 and 32.5). The backbone supports the body, serves as an attachment point for muscles, and protects the spinal cord that runs through a canal inside it. Bony segments called **vertebrae** (singular, vertebra) make up the backbone. **Intervertebral disks** of cartilage between vertebrae act as shock absorbers and flex points.

The vertebral column and bones of the head and rib cage constitute the **axial skeleton**. The **appendicular skeleton** consists of the pectoral (shoulder) girdle, the pelvic (hip) girdle, and limbs (or bony fins).

As you know, vertebrate skeletons have evolved over time. For example, jaws are derived from gill supports of ancient jawless fishes (Section 24.3), and bones in the limbs of land vertebrates are homologous to bones in fins of lobe-finned fishes (Section 24.4).

axial skeleton

appendicular skeleton

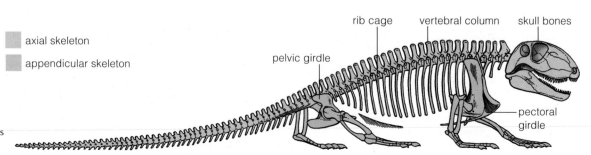

rib cage   vertebral column   skull bones

pelvic girdle

pectoral girdle

Figure 32.4 Skeletal elements typical of early reptiles.

The Human Skeleton

The human skeleton has typical vertebrate features, as well as modifications related to upright posture (Figure 32.5). The human skull's flattened cranial bones form a braincase that surrounds and protects the brain. The brain and spinal cord connect through an opening at the base of the skull. Facial bones include cheekbones and other bones around the eyes, the bone that forms the bridge of the nose, and the jaw bones.

Both men and women have twelve pairs of ribs. Ribs and the breastbone, or sternum, form a protective cage that encloses the heart and lungs.

The vertebral column extends from the base of the skull to the pelvic girdle. The shape of the human vertebral column is another adaptation to upright posture. Viewed from the side, our backbone has an S shape, which keeps the head and torso centered over the feet.

The lowest portions of the backbone are the sacrum and the coccyx. The sacrum consists of five vertebrae that have become fused as a large triangular structure. The coccyx is four fused vertebrae derived from the embryonic tail. At five weeks, the human embryo has a tail of 12 vertebrae. As development proceeds, most of the tail is resorbed, leaving the shorter, smaller coccyx.

The pectoral girdle consists of the scapula (shoulder blade) and clavicle (collarbone). The thin clavicle transfers force from the arms to the axial skeleton. When a person falls on an outstretched arm, excessive force transferred to the clavicle frequently causes it to break.

The upper arm has one bone, the humerus. The forearm has two bones, the radius and ulna. Carpals are bones of the wrist, metacarpals are bones of the palm, and phalanges (singular, phalanx) are finger bones.

The pelvic girdle consists of two sets of fused bones, one set on each side of the body. It protects organs inside the pelvic cavity and supports the weight of the upper body when a person stands upright.

The largest bone of the body is the femur (thighbone). It attaches to the bones of the lower leg (the tibia and fibula) at the knee, which is protected by the patella (kneecap). Tarsals are ankle bones, and metatarsals are bones of the sole of the foot. Like the bones of the fingers, those of the toes are called phalanges.

---

appendicular skeleton Of vertebrates, limb or fin bones and the bones that attach them to the trunk.
axial skeleton Of vertebrates, bones of the head, trunk, and tail.
endoskeleton Internal skeleton.
exoskeleton Of some invertebrates, hard external parts that muscles attach to and move.
hydrostatic skeleton Of soft-bodied invertebrates, a fluid-filled chamber that muscles exert force against, redistributing the fluid.
intervertebral disk Cartilage disk between two vertebrae.
vertebrae Bones of the backbone, or vertebral column.
vertebral column Backbone.

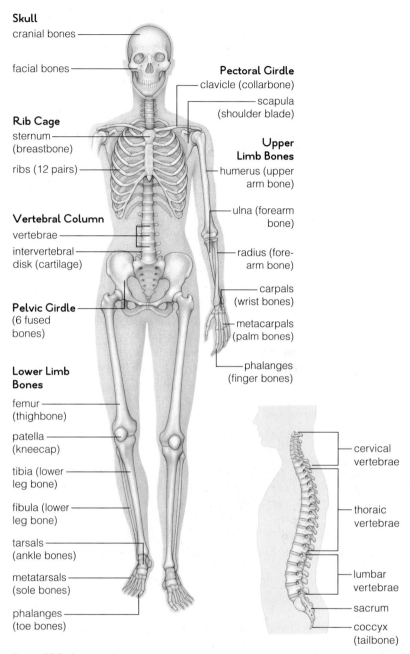

Figure 32.5 Animated *Left*, Major bone (*tan*) and cartilage (*light blue*) elements of the human skeleton. *Right*, side view of the vertebral column showing the curve and the different regions.

Take-Home Message What types of skeletons do animals have?

❯ Soft-bodied invertebrates such as sea anemones and earthworms have a hydrostatic skeleton, an enclosed fluid that contracting muscles redistribute.

❯ Some mollusks and all arthropods have a hard, secreted external skeleton, or exoskeleton.

❯ Echinoderms and vertebrates have an internal skeleton, or endoskeleton.

❯ The vertebrate skeleton consists of cartilage and, in most groups, bone. Evolutionary modifications of the ancestral vertebrate skeleton allow humans to walk upright.

> Bones consist of living cells in a secreted extracellular matrix that is continually remodeled.

❮ Links to Connective tissue 28.4, Parathyroid hormone 31.6

Bone Structure and Function

The 206 bones of an adult human's skeleton range in size from middle ear bones as small as a grain of rice to the massive thighbone, or femur, which weighs about a kilogram (2 pounds). The femur and other bones of the arms and legs are long bones. The ribs, the sternum, and most bones of the skull are flat bones. Still other bones, such as the carpals in the wrists, are short and roughly squarish in shape.

Each bone is wrapped in a dense connective tissue sheath that has nerves and blood vessels running through it. Bone tissue consists of bone cells in an extracellular matrix (Section 4.11). The matrix is mainly collagen (a protein) with calcium and phosphorus salts.

A long bone such as a femur includes two types of bone tissue, compact bone and spongy bone (Figure 32.6). **Compact bone** forms the femur's outer layer. It consists of

many concentric rings of mineralized bone tissue, with living bone cells in spaces between the rings. Nerves and blood vessels run through a canal in the center of each ring. The knobby ends of long bones are filled with **spongy bone**. Spongy bone is strong yet lightweight; its matrix has many cavities.

Bone marrow is the soft tissue that fills the cavities inside a bone. The **red marrow** that fills spaces in spongy bone is the major site of blood cell formation. **Yellow marrow** fills the central cavity of an adult femur and most other mature long bones. It consists mainly of fat. In cases of extreme blood loss, yellow marrow can convert to red marrow. Table 32.1 summarizes the functions of bone.

Bone Formation and Turnover

A cartilage skeleton forms in all vertebrate embryos. In sharks and other cartilaginous fishes, the cartilage skeleton persists into adulthood. In other vertebrates, embryonic cartilage serves as a model for a bony skeleton. Before birth, mineralization of this model transforms most of it to bone.

Until a person is about twenty-four years old, bone cells secrete more matrix than they break down, so bone mass increases. Later in life, bone-producing cells become less active and bone mass gradually declines. However, bone remodeling occurs throughout life. The body must constantly repair the microscopic fractures that result from normal movements. In addition, bone is built and broken down in response to hormonal signals. Bones store most of the body's calcium. Parathyroid hormone, the main regulator of blood calcium, increases calcium concentration in blood by encouraging calcium uptake from the gut and calcium release from bone. Other hormones also affect bone turnover. The sex hormones estrogen and testosterone encourage bone deposition. Cortisol, the stress hormone, slows it.

Osteoporosis is a disorder in which bone loss outpaces bone formation. As a result, the bones become weaker and more likely to break (Figure 32.7). Osteoporosis is

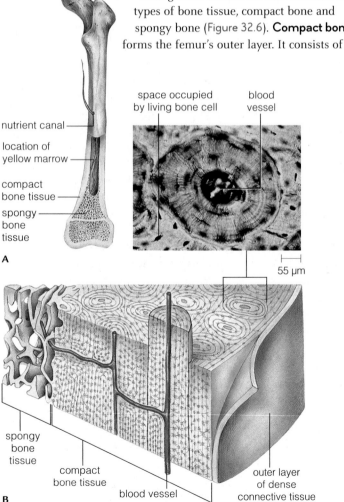

space occupied by living bone cell

blood vessel

nutrient canal

location of yellow marrow

compact bone tissue

spongy bone tissue

A

⊢ ⊣ 55 μm

spongy bone tissue

compact bone tissue

blood vessel

outer layer of dense connective tissue

B

Figure 32.6 Animated **A** Structure of a human femur, or thighbone, and **B** a section through its spongy and compact bone tissues.

Table 32.1 Functions of Bone
1. **Movement.** Bones interact with skeletal muscle to change or maintain positions of the body and its parts.
2. **Support.** Bones support and anchor muscles.
3. **Protection.** Many bones form hardened chambers or canals that enclose and protect soft internal organs.
4. **Mineral storage.** Bones are a reservoir for calcium and phosphorus ions. Deposits and withdrawals of these ions help maintain their concentrations in body fluids.
5. **Blood cell formation.** Only certain bones contain the tissue where blood cells form.

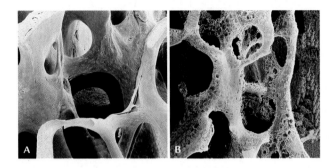

Figure 32.7 **A** Normal spongy bone. **B** Effect of osteoporosis.

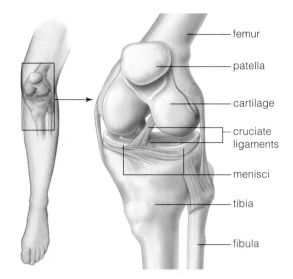

Figure 32.8 Anatomy of the knee, a hinge-type synovial joint. Ligaments hold bones in place. Wedges of cartilage called menisci (singular, meniscus) provide additional stability.

most common in postmenopausal woman because they produce little of the sex hormones that encourage bone deposition. However, about 20 percent of osteoporosis cases occur in men.

To reduce your risk of osteoporosis, ensure that your diet provides adequate levels of calcium and of vitamin D, which facilitates calcium absorption from the gut. Avoid smoking and excessive alcohol intake; they slow bone deposition. Get regular exercise to encourage bone renewal, and avoid excessive intake of cola soft drinks. Several studies have shown that women who drink more than two such soft drinks a day have a slightly lower than normal bone density.

Where Bones Meet—Skeletal Joints

A **joint** is an area where bones come together. Connective tissue holds bones securely in place at fibrous joints such as those between cranial bones. Pads or disks of cartilage connect bones at cartilaginous joints, forming a flexible connection that allows a bit of movement. Cartilaginous joints connect vertebrae to one another and connect some ribs to the sternum. Joints of the hip, shoulder, wrist, elbow, and knee are synovial joints, the most common type of joint. In a synovial joint, the ends of bones are covered with cartilage and enclosed in a fluid-filled capsule. Cords of dense connective tissue called **ligaments** hold the bones of a synovial joint in place (Figure 32.8).

Different kinds of synovial joints allow different movements. Ball-and-socket joints at the shoulders and hips allow rotational motion. At other joints, including some in the wrists and ankles, bones glide over one another. Joints

at elbows and knees function like a hinged door, allowing the bones to move back and forth in one plane only.

Sprains, the most common joint injury, occur when ligaments become overstretched or torn. Athletes often tear cruciate ligaments in the knee joint and require surgery. "Cruciate" means cross, and these ligaments cross one another in the center of the joint and stabilize the knee. When they are torn completely, bones may shift so the knee gives out when a person tries to stand.

A dislocation occurs when the bones of a joint move out of place. It is usually highly painful and requires immediate treatment. Bones must be placed in proper position and immobilized for a time to allow healing.

Arthritis is chronic inflammation of a joint. The most common type of arthritis, osteoarthritis, usually appears in old age, after cartilage wears down at a specific joint or joints. For example, women who often wear high-heeled shoes increase their risk of osteoarthritis of the knees in their later years. Rheumatoid arthritis is a disorder in which the immune system mistakenly attacks all of the body's synovial joints. Rheumatoid arthritis can occur at any age, and women are two to three times more likely than men to be affected.

---

compact bone Dense bone that makes up the shaft of long bones.
joint Region where bones meet.
ligament Strap of dense connective tissue that holds bones together at a joint.
red marrow Bone marrow that makes blood cells.
spongy bone Lightweight bone with many internal spaces; contains red marrow.
yellow marrow Bone marrow that is mostly fat; fills cavity in most long bones.

Take-Home Message What are the structural and functional features of bones?

❯ Bones are collagen-rich, mineralized organs that function in movement, support, protection, storage of mineral ions, and blood cell formation. They are constantly remodeled.

❯ Bones meet at joints. At synovial joints, such as the knee, ligaments of dense connective tissue hold bones in place.

› Skeletal muscles move body parts by pulling on bones.
‹ Links to Muscle tissue 28.5, Connective tissue 28.4, Stretch reflex 29.9

Skeletal Muscle Function

Skeletal muscles allow us to dance, to smile, and to speak. They are sometimes referred to as voluntary muscles because we control their action at will. However, skeletal muscles also take part in reflex actions such as the stretch reflex described in Section 29.9.

A sheath of dense connective tissue encloses each skeletal muscle and extends beyond it to form a cordlike or straplike **tendon**. Most often, tendons attach each end of the muscle to a bone. Muscles and bones act like a lever system, in which a rigid rod attaches to and moves about a fixed point. Muscles connect to bones (rigid rods) near a joint (a fixed point). When a muscle contracts, it transmits force to the bone it attaches to and moves it.

Figure 32.9 shows the muscles of the upper arm: the biceps and the triceps. Two tendons attach the upper part of the biceps to the scapula (shoulder blade). At the opposite end of the muscle, a tendon attaches the biceps to the radius in the forearm. When the biceps contracts (shortens), the forearm is pulled toward the shoulder. You can feel the contraction happen if you extend your arm outward, place your other hand over the biceps, then slowly bend the elbow. Feel the biceps contract? Although the biceps shortens only a bit, it causes a large motion of the bone to which it is connected.

Muscles can only pull; they cannot push. Often two muscles work in opposition, with the action of one resisting or reversing action of another. For example, the triceps in the upper arm opposes the biceps. When you pull your arm toward your shoulder, the triceps muscle relaxes as the biceps contracts. Contraction of the triceps coupled with relaxation of the biceps reverses this movement, extending the arm.

The human body has close to 700 skeletal muscles, some near the surface, others deep inside the body wall. Figure 32.10 shows a sample of some of the largest muscles and describes their functions. Collectively, skeletal muscles account for about 40 percent of the body weight of a young man of average fitness.

Most skeletal muscles move bones, but some have other functions. Skeletal muscles that pull on facial skin cause changes in expression. Others attach to and move the eyeball, or open and close eyelids. The tongue is skeletal muscle, and sphincters of skeletal muscle provide voluntary control of defecation and urination. Skeletal muscles function in respiration and help keep blood circulating through the body. Skeletal muscle activity also generates heat that helps keep the body warm.

Bear in mind, only *skeletal* muscle interacts with bone. Smooth muscle is mainly a component of soft internal organs, such as the stomach. Cardiac muscle forms only in the heart wall. Later chapters consider the structure and function of smooth muscle and cardiac muscle.

Tender or Torn Tendons

Tendon injuries are common and notoriously slow to heal. Short-term overuse of a tendon, such as an overly vigorous run, can cause tendonitis, in which minor damage leads to pain and inflammation. Tendonitis is treated with rest and anti-inflammatory drugs. Chronic overuse of a tendon can change its structure and cause tendonosis. In this case, anti-inflammatory drugs do not help resolve the problem and may even slow healing. Tendonosis is treated with rest and physical therapy. When a tendon tears, and the injury impairs normal function, the only treatment is surgery.

---

tendon Strap of dense connective tissue that connects a skeletal muscle to bone.

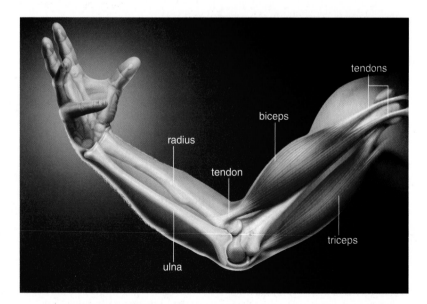

Figure 32.9 **Animated** Opposing muscles of the upper arm. When the biceps contracts and the triceps relaxes, the forearm is pulled toward the upper arm. When the triceps contracts and the biceps relaxes, the arm straightens at the elbow.

›› **Figure It Out** What bone does the biceps pull on?   Answer: The radius

Take-Home Message How do muscles and tendons interact with bones?

› Cordlike or straplike tendons of dense connective tissue attach skeletal muscles to bones.

› Skeletal muscles transmit contractile force to bones. Small muscle movements can bring about large movements of bones. Many muscle groups have opposing actions.

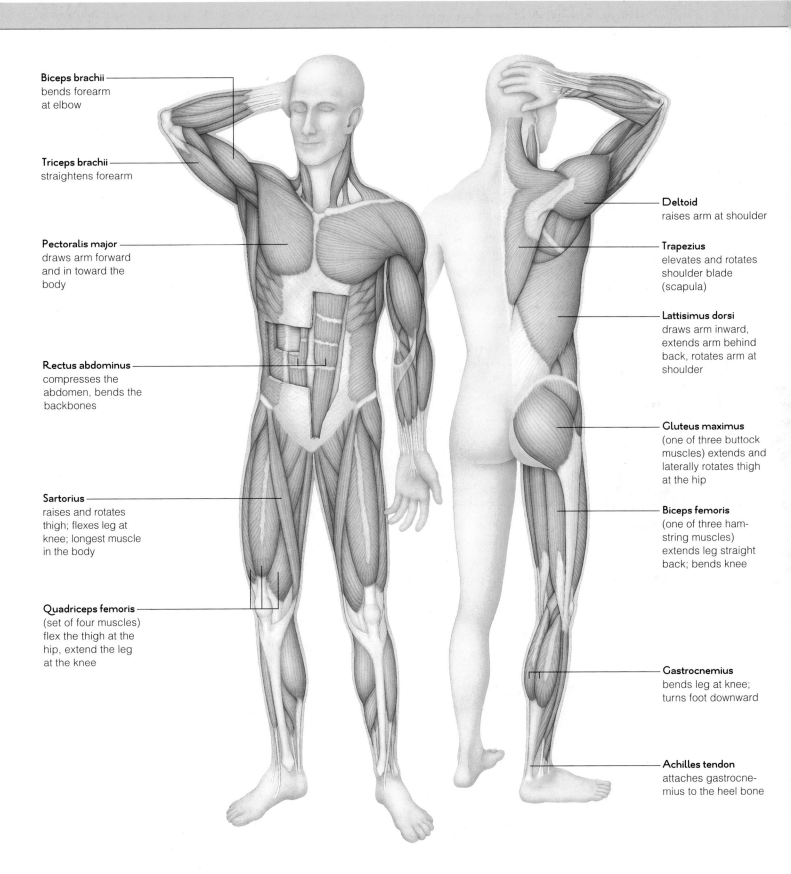

Biceps brachii
bends forearm
at elbow

Triceps brachii
straightens forearm

Pectoralis major
draws arm forward
and in toward the
body

Rectus abdominus
compresses the
abdomen, bends the
backbones

Sartorius
raises and rotates
thigh; flexes leg at
knee; longest muscle
in the body

Quadriceps femoris
(set of four muscles)
flex the thigh at the
hip, extend the leg
at the knee

Deltoid
raises arm at shoulder

Trapezius
elevates and rotates
shoulder blade
(scapula)

Lattisimus dorsi
draws arm inward,
extends arm behind
back, rotates arm at
shoulder

Gluteus maximus
(one of three buttock
muscles) extends and
laterally rotates thigh
at the hip

Biceps femoris
(one of three ham-
string muscles)
extends leg straight
back; bends knee

Gastrocnemius
bends leg at knee;
turns foot downward

Achilles tendon
attaches gastrocne-
mius to the heel bone

Figure 32.10 Animated Major muscles of the human musculoskeletal system. These are the skeletal muscles that gym enthusiasts are familiar with; many more are not shown.

Tendons are shown in *light blue*. For example, the Achilles tendon, the largest tendon in the body,
attaches muscles in the calf to the heel bone.

❯ Bones of a human in motion move when skeletal muscles attached to them shorten. A muscle shortens when muscle fibers, and the contractile units inside the fibers, shorten.
❮ Link to Actin 4.10

outer sheath of one skeletal muscle

one bundle of many muscle fibers in parallel inside the sheath

A

B one myofibril, made up of sarcomeres arranged end to end

sarcomere — sarcomere

Z line   Z line   Z line

C one sarcomere, with parallel actin and myosin filaments

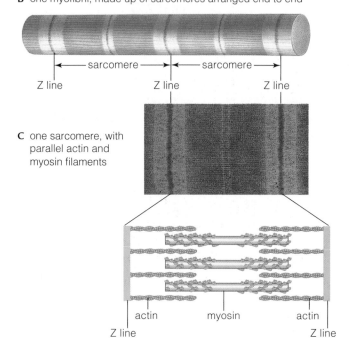

actin   myosin   actin

Z line   Z line

Figure 32.11 **Animated** Zooming down through skeletal muscle from a biceps to the actin and myosin filaments of a single contractile unit.

Structure of Skeletal Muscle

A skeletal muscle's function arises from its internal organization. **Skeletal muscle fibers** run parallel with the muscle's long axis (Figure 32.11A). The multinucleated fibers form during early development, when embryonic muscle cells fuse. Many **myofibrils**—bundles of protein filaments—run the length of the fiber. Each myofibril has light-to-dark crossbands that show up when a muscle is stained for microscopy. The bands give the muscle fiber its striated, or striped, appearance. The bands also define the units of muscle contraction, or **sarcomeres** (Figure 32.11B). The ends of a sarcomere are anchored to its neighbors at a mesh of cytoskeletal elements called Z lines.

The sarcomere has parallel arrays of thin and thick filaments (Figure 32.11C). Thin filaments attached to Z lines extend inward, toward the sarcomere's center. Each thin filament consists of two chains of **actin**, a globular protein. Thicker filaments are centered in the sarcomere. A thick filament consists of **myosin**, a protein that has a clublike head. Each myosin head is positioned just a few nanometers away from a thin filament.

Muscle fibers, myofibrils, thin filaments, and thick filaments all have the same orientation; they all run parallel with a muscle's long axis. What function does this repetitive orientation serve? It focuses the force of contraction; all sarcomeres in all fibers of a muscle work together to pull a bone in the same direction.

The Sliding–Filament Model

The **sliding–filament model** explains how interactions between thick and thin filaments bring about muscle contraction. Neither actin nor myosin filaments change length and the myosin filaments do not change position. Instead, myosin heads bind to actin filaments and slide them toward the center of a sarcomere. As the actin filaments are pulled inward, the ends of the sarcomere are drawn closer together, and the sarcomere shortens.

Figure 32.12 provides a step-by-step look at sarcomere contraction, starting with the sarcomere in a resting muscle ❶. Part of the myosin head can bind ATP and break it into ADP and phosphate. This reaction readies myosin for action ❷. By analogy, binding of ATP to a myosin head

actin Protein that is the main component of thin filaments of muscle fibers.
myofibrils Threadlike, cross-banded skeletal muscle components that consist of sarcomeres arranged end to end.
myosin Protein in thick filaments of muscle fibers.
sarcomere Contractile unit of skeletal and cardiac muscle.
skeletal muscle fiber Multinucleated skeletal muscle cell.
sliding–filament model Explanation of how interactions among actin and myosin filaments shorten a sarcomere and bring about muscle contraction.

Figure 32.12 Animated The sliding-filament model for skeletal muscle contraction.

1 A sarcomere in a muscle at rest. Actin and myosin filaments lie next to one another, but do not interact.

2 Myosin heads of thick filaments were activated by a phosphate-group transfer from ATP. ADP and phosphate remain attached to the myosin.

3 Release of calcium from intracellular storage allows myosin to bind to actin. Cross-bridges form when myosin heads bind to sites on adjacent actin filaments.

4 The myosin head releases bound ADP and phosphate as it tilts toward the sarcomere center and slides the attached actin filaments along with it.

5 New ATP binds to the myosin heads, causing them to release their grip on actin and return to their original orientation, ready to act again.

6 Many myosin heads repeatedly bind to and pull on adjacent actin filaments. Their collective action cause the sarcomere to shorten (contract).

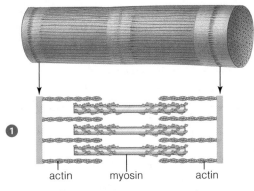

1

actin   myosin   actin

Sarcomere between contractions

energizes the myosin like pulling back the rubber band of a slingshot prepares it for action.

Muscle contraction occurs when nervous signals cause a rise in calcium level. The rise in calcium allows the myosin heads to form cross-bridges with actin filaments **3**. The ADP and phosphate bound to myosin earlier are released, and each myosin head tilts like a slingshot snapping back to its unstretched position. As a myosin head tilts, it pulls an actin filament and the attached Z line toward the sarcomere center **4**.

Binding of a new ATP and its breakdown to ADP and phosphate detaches the myosin head from actin, and the head returns to its original position **5**. If calcium is still present, the head attaches to another binding site on the actin, tilts in another stroke, and repeats the process. The contraction of a sarcomere occurs when hundreds of myosin heads perform a series of repeated strokes all along the length of the actin filaments **6**.

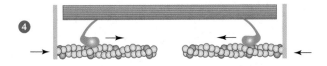

2

myosin head with bound ADP and P

one of many myosin-binding sites on actin

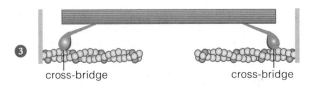

3

cross-bridge   cross-bridge

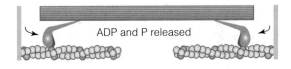

4

ADP and P released

Take-Home Message How does a muscle's structure affect its function?

> Sarcomeres are the basic units of contraction in skeletal muscle. Sarcomeres are lined up end to end in myofibrils that run parallel with muscle fibers. These fibers, in turn, run parallel with the whole muscle.

> The parallel orientation of skeletal muscle components focuses a muscle's contractile force in a particular direction.

> By energy-driven interactions between myosin and actin filaments, the many sarcomeres of a muscle cell shorten and bring about the muscle's contraction.

> During muscle contraction, the length of actin and myosin filaments does not change, and the myosin filaments do not change position. Sarcomeres shorten because myosin filaments pull neighboring actin filaments inward toward the center of the sarcomere.

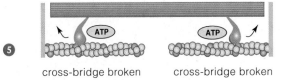

5

ATP   ATP

cross-bridge broken   cross-bridge broken

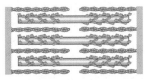

6   Same sarcomere, contracted

> A motor neuron signals a muscle and calls for muscle contraction.
> Muscle contraction requires energy in the form of ATP.
< Links to Active transport 5.7, Energy-releasing pathways Chapter 7, Neuromuscular junction 29.6

Nervous Control of Contraction

Motor neurons in the brain and spinal cord control skeletal muscle contraction (Figure 32.13 ❶). The axon of a motor neuron synapses on a muscle at a neuromuscular junction ❷. When an action potential reaches the axon terminals at the neuromuscular junction, it induces the release of the neurotransmitter acetylcholine (ACh). Like neurons, muscle fibers are excitable. When ACh binds to receptors in a muscle fiber's plasma membrane, action potentials travel along the membrane, then down membrane extensions called T tubules ❸.

A Brief stimulation causes a twitch.

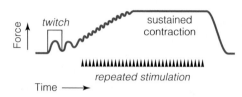

B Repeated stimulation within a short interval causes a sustained contraction with greater force.

Figure 32.14 Animated Recordings of force (muscle tension) generated in response to stimulation of a motor unit.

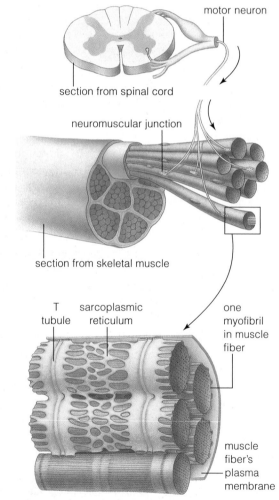

❶ A signal travels along the axon of a motor neuron, from the spinal cord to a skeletal muscle.

❷ The signal is transferred from the motor neuron to the muscle at neuromuscular junctions. Here, ACh released by the neuron's axon terminals diffuses into the muscle fiber and causes action potentials.

❸ Action potentials propagate along a muscle fiber's plasma membrane down to T tubules, then to the sarcoplasmic reticulum, which releases calcium ions. The ions promote interactions of myosin and actin that result in contraction.

Figure 32.13 Animated Pathway by which the nervous system controls skeletal muscle contraction. A muscle fiber's plasma membrane encloses many individual myofibrils. Tubelike extensions of the membrane connect with part of the sarcoplasmic reticulum, which wraps around the myofibrils.

T tubules deliver action potentials to the **sarcoplasmic reticulum**, a specialized endoplasmic reticulum that wraps around the myofibrils. The sarcoplasmic reticulum stores and releases calcium ions.

Arrival of action potentials causes the sarcoplasmic reticulum to release calcium ions. Sites on actin where myosin heads can bind are blocked in resting muscle, but an influx of calcium ions clears these binding sites. Rising calcium concentration allows actin and myosin to interact, and muscle contraction gets under way. When contraction ends, calcium pumps actively transport calcium ions back into the sarcoplasmic reticulum, preparing the muscle to respond to the next signal.

Motor Units and Muscle Tension

A motor neuron has many axon endings that synapse on different fibers in a muscle. A motor neuron and all of the muscle fibers it synapses with constitute one **motor unit**. Stimulate a motor neuron, and all the muscle fibers that it synapses on contract. The nervous system cannot make only some fibers in a motor unit contract.

The mechanical force generated by a contracting muscle—the **muscle tension**—depends on the number of muscle fibers contracting. Some tasks require more muscle tension than others, so the number of muscle fibers controlled by a single motor neuron varies. In motor units that bring about small, fine movements such as those that control eye muscles, one motor neuron synapses with only 5 or so muscle fibers. By contrast, the biceps of the arm has about 700 muscle fibers per motor unit. Having many fibers contract at once increases the force a motor unit can generate.

For any given motor unit, the strength of contraction varies with the type of stimulation. A single brief stimulus causes a short contraction, a **muscle twitch** (Figure

32.14A). Repeatedly stimulating a motor unit during a short interval causes a sustained contraction that generates muscle tension three or four times greater than that of a single twitch (Figure 32.14B).

Energy for Contraction

The availability of ATP affects whether and how long a muscle can contract. ATP is the first energy source a muscle uses, but muscle has a limited amount of ATP. It has a larger store of creatine phosphate, which can transfer a phosphate to ADP and form ATP (Figure 32.15 ❶). Such phosphate transfers can fuel muscle contraction until other pathways increase ATP output.

Some athletes take creatine supplements to increase the amount of creatine phosphate available to muscle. Research suggests that creatine supplements can enhance performance of tasks that require a quick burst of energy. However, they have no effect on endurance, and the side effects of using such supplements are not fully known.

Aerobic respiration yields most of the ATP used by a muscle during prolonged, moderate activity ❷. Glucose derived from stored glycogen fuels five to ten minutes of activity, then glucose and fatty acids that the blood delivers to muscle fibers are burned. Fatty acids are the main fuel for activities that last more than half an hour.

Lactate fermentation is the third source of energy ❸. Some pyruvate is converted to lactate by the fermentation pathway even in resting muscle, but lactate fermentation steps up during exercise. This pathway produces less ATP than aerobic respiration, but unlike aerobic respiration, it operates even when the oxygen level in a muscle is low.

During strenuous exercise, lactate accumulation raises the acidity in muscles, exciting adjacent pain receptors and causing an uncomfortable burning sensation. After the exercise ends, accumulated lactate enters mitochondria, where it is quickly converted to pyruvate and used in aerobic respiration. Thus, lactate buildup does not cause the muscle fatigue and soreness that persists for a day or two after strenuous exercise. Researchers continue to investigate the causes of this postexercise effect.

Types of Muscle Fibers

As you learned in Section 7.6, we divide muscle fibers into two categories based on how they produce ATP. Red fibers have an abundance of mitochondria and produce ATP mainly by aerobic respiration. They are colored bright red by myoglobin, a protein that, like hemoglobin, reversibly binds oxygen. During periods of muscle activity, myoglobin releases the stored oxygen, allowing aerobic respiration to occur even if blood oxygen is low. By contrast, white fibers have no myoglobin, few mitochondria, and make ATP mainly by lactate fermentation. Red fibers fatigue less easily than white fibers, so red fibers tend to predominate in muscles that carry out sustained activities.

Muscle fibers can also be subdivided into fast fibers or slow fibers based on the ATPase activity of their myosin. Myosin of fast fibers splits ATP more efficiently than that of slow fibers. Thus, fast fibers contract more quickly when stimulated.

All white fibers are fast fibers; they react fast and fatigue easily. The muscles that move your eye are mostly white fibers. Red fibers can be either fast or slow. Fast red fibers predominate in the triceps muscle, which must often react quickly. Muscles that play a role in maintaining posture such as some muscles of the back have many slow red fibers.

The mix of fiber types in each muscle varies between individuals and has a genetic basis. Successful sprinters tend to have a higher than average percentage of fast, white fibers in their leg muscles. Marathoners tend to have more slow, red fibers than average.

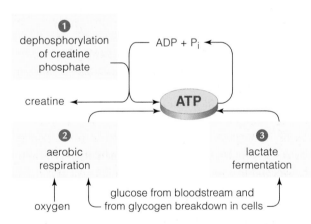

Figure 32.15 **Animated** Three metabolic pathways that muscles use to obtain the ATP that fuels their contraction.

---

motor unit One motor neuron and the muscle fibers it controls.
muscle tension Force exerted by a contracting muscle.
muscle twitch Brief muscle contraction.
sarcoplasmic reticulum Specialized endoplasmic reticulum in muscle cells; stores and releases calcium ions.

Take-Home Message **What factors are required for muscle contraction to occur?**

❭ In muscle fibers, signals from motor neurons initiate action potentials that cause the release of calcium ions from storage. Contraction cannot proceed without this calcium release.

❭ A muscle's response to stimulation varies in speed, strength, and duration. Repeated stimulation and the type of muscle fiber affect muscle response.

> Muscle function is enhanced by exercise and impaired by genetic disorders, infectious disease, and some toxins.
< Links to X-linked inheritance 14.4, Bacterial endospores 19.7

Effects of Exercise

In humans, all muscle fibers form before birth. Exercise cannot stimulate the addition of new ones. It does, however, have other benefits. Aerobic exercise, which is low in intensity and long in duration, makes muscles more resistant to fatigue (Figure 32.16). It increases their blood supply and the number of mitochondria. Impact exercise such as running or walking also promotes healthy bones by encouraging bone deposition.

Strength training (intense, short-duration exercise such as weight lifting) stimulates formation of more actin and myosin, as well more enzymes of glycolysis. Strong, bulging muscles develop, but these muscles do not have much endurance. They fatigue rapidly. Strength training involves two types of muscle contractions. Isotonically contracting muscles shorten and move some load, as when you lift an object (Figure 32.17A). Isometrically contracting muscles tense but do not shorten, as when you try to lift an object but fail because its weight exceeds the muscle's capacity (Figure 32.17B).

As people age, their muscles generally begin to shrink. The number of muscle fibers declines and the remaining fibers increase their diameter more slowly in response to exercise. Muscle injuries take longer to heal. Even so, exercise can be helpful at any age. Strength training can slow the loss of muscle tissue. Aerobic exercise can improve circulation. In addition, aerobic exercise by the middle-aged and elderly can help lift depression. Aerobic exercise helps improve memory and the capacity to plan

Figure 32.16 Aerobic exercise increases the number of mitochondria in muscles, thus increasing endurance.

and organize complex tasks. No matter what your age, exercise is good for more than just muscles. It is also good for the brain.

Muscular Dystrophy

Muscular dystrophies are a class of genetic disorders in which skeletal muscles progressively weaken. A mutation of a gene on the X chromosome causes Duchenne muscular dystrophy. The affected gene encodes dystrophin, a protein found in the plasma membrane of muscle fibers. The mutated dystrophin allows foreign material to enter a muscle fiber, causing the fiber to break down (Figure 32.18). Muscular dystrophy arises in about 1 in 3,500 males. Like other X-linked disorders, it rarely causes symptoms in females, who nearly always have a normal version of the gene on their other X chromosome. People affected by Duchenne muscular dystrophy usually begin to show signs of weakness by the time they are three years old, and require a wheelchair in their teens. Most die in their twenties of respiratory failure when the skeletal muscles involved in breathing stop functioning.

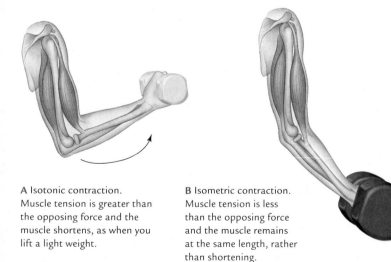

A Isotonic contraction. Muscle tension is greater than the opposing force and the muscle shortens, as when you lift a light weight.

B Isometric contraction. Muscle tension is less than the opposing force and the muscle remains at the same length, rather than shortening.

Figure 32.17 Two types of muscle contraction.

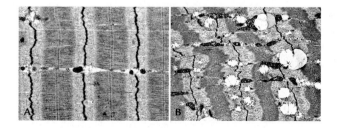

Figure 32.18 Electron micrographs of **A** normal skeletal muscle and **B** muscle of a person affected by muscular dystrophy.

Motor Neuron Disorders

Damage to motor neurons sometimes affects their ability to signal muscles to contract. For example, polio is a viral disease that impairs motor neuron function. It most often affects children and can be fatal. People who survive a poliovirus infection may become paralyzed or develop a weakened voluntary muscle response. Polio vaccines have been available since the 1950s, and no new cases have arisen in the United States since 1979. However, sporadic outbreaks continue in developing countries. In addition, survivors of polio are at risk for post-polio syndrome, a disorder characterized by muscle fatigue and progressive muscle weakness.

Amyotrophic lateral sclerosis (ALS) also affects motor neurons. It is sometimes called Lou Gehrig's disease, after a famous baseball player whose career was cut short by the disease in the late 1930s. ALS usually causes death by respiratory failure within three to five years of diagnosis, but some people survive much longer. For example, the astrophysicist Stephen Hawking was diagnosed with ALS in 1963. Though now confined to a wheelchair and unable to speak, he continues to write and to lecture with the assistance of a voice synthesizer. The causes of ALS are not fully understood. There is a genetic component, but most cases arise in previously unaffected families.

Botulism and Tetanus

Some bacteria make toxins that disrupt the flow of signals from nerves to muscles. Resting spores (endospores) of *Clostridium botulinum* sometimes survive in improperly canned food. When the spores germinate, bacteria grow and make an odorless toxin called botulinum. If a person eats the toxin, it affects motor neurons, preventing them from releasing acetylcholine (ACh). Muscles cannot contract without this neurotransmitter, so the result is temporary paralysis. Botulinum poisoning can be fatal if skeletal muscles involved in breathing become paralyzed.

Spores of a related bacterium, *Clostridium tetani*, last for years in soil. If the spores get into a wound and germinate, the bacteria that grow make a toxin that affects the central nervous system. In the spinal cord, the toxin blocks release of neurotransmitters that inhibit motor neurons. As a result, nothing dampens signals to contract, and symptoms of the disease known as tetanus appear. Muscles stiffen and cannot be released from contraction. The fists and jaw stay clenched; hence, the common name for the disease, lockjaw. The backbone may become locked in an abnormally arching curve (Figure 32.19). Death occurs when respiratory and cardiac muscles become locked in contraction. Vaccines have eradicated tetanus in the United States, but worldwide, the annual death toll is over 200,000. Most are newborns infected during an unsanitary delivery.

Muscles and Myostatin (revisited)

Researchers have long been searching for ways to slow the muscle loss that results from muscular dystrophy, ALS, and even normal aging. Drugs that inhibit myostatin production or prevent myostatin activity may help them reach this goal. One way to learn what sort of effects such drugs might have is to study mice in which the myostatin gene was knocked out. As one example, the larger mouse in the photo at the *left* is a knockout mutant. Its myostatin gene was disabled by genetic engineering, so it is bigger and more muscular than the unaltered mouse beside it. The bad news is that such mice have unusually small, stiff, easily torn tendons. Thus, it is likely that increased tendon injuries will be a side effect of myostatin-inhibiting drugs and conditions.

How Would You Vote? Dietary supplements that claim to block myostatin are already for sale. Should dietary supplements be required to demonstrate effectiveness and safety? See CengageNow for details, then vote online (cengagenow.com).

Figure 32.19 An 1809 painting showing a casualty of a battle wound as he lay dying of tetanus in a military hospital. A bacterial toxin locked his muscles in contraction.

Take-Home Message What effects do exercise, diseases, and disorders have on muscles?

> Exercise cannot add new muscle fibers, but it can increase the number of protein filaments and mitochondria in existing ones.

> Muscle function can be adversely affected by genetic disorders, motor neuron disorders, infectious disease, and certain toxins that interfere with the flow of signals to muscle.

Summary

Section 32.1 The body is like a machine that becomes stronger with use. Exercise encourages muscle fibers to enlarge by making more proteins. Androgens such as testosterone also encourage synthesis of muscle proteins. Myostatin, a regulatory protein, slows protein synthesis in muscles. Individuals with a mutation in the myostatin gene are unusually muscular and strong.

Section 32.2 Animals move when their muscles apply force to skeletal elements. Soft-bodied invertebrates have a **hydrostatic skeleton**, a confined fluid that muscle contractions redistribute. Arthropods have an **exoskeleton** consisting of noncellular hard parts at the body surface. An **endoskeleton** consists of hardened parts inside the body. Echinoderms and vertebrates have an endoskeleton.

The vertebrate skull, vertebral column, and ribs constitute the **axial skeleton**. The **vertebral column** consists of **vertebrae** with **intervertebral disks** between them. Bony fins or limbs and the bones that attach them to the backbone constitute the **appendicular skeleton**. Evolution of an upright posture in human ancestors involved skeletal modifications, such as curving of the backbone into an S shape that keeps the head aligned over the feet.

Section 32.3 Bones consist of living cells in a secreted matrix rich in collagen, calcium, and phosphorus. In addition to having a role in movement, bones store minerals and protect organs. The shaft of a long bone such as a femur consists of **compact bone** that contains **yellow marrow**. Lightweight **spongy bone** contains **red marrow** that makes blood cells. In a human embryo, bones develop from a cartilage model. Even in adults, bones are continually remodeled. A **joint** is an area of close contact between bones. At most joints, one or more **ligaments** of dense connective tissue hold bones in place.

Section 32.4 Bones move when skeletal muscles pull on them. Skeletal muscles are surrounded by a sheath of connective tissue that extends beyond the muscle as a **tendon**. Tendons attach a muscle to a bone or, in some cases, to skin. A muscle can only exert force in one direction; it can pull but not push. Thus, some skeletal muscles work as pairs that have opposing actions. The biceps and triceps of the arm are an example.

Section 32.5 The internal organization of a skeletal muscle promotes a strong, directional contraction. A **skeletal muscle fiber** contains many **myofibrils**. Each consists of **sarcomeres**, basic units of muscle contraction, lined up end to end. A sarcomere has parallel arrays of **actin** and **myosin** filaments. The **sliding-filament model** describes how ATP-driven sliding of actin filaments past myosin filaments shortens the sarcomere. Shortening of all sarcomeres in all myofibrils of all muscle fibers brings about muscle contraction.

Section 32.6 Motor neurons control skeletal muscle. Each motor neuron and the skeletal muscle fibers it synapses on constitute a **motor unit**. Release of ACh at a neuromuscular junction causes action potentials in muscle fibers. The action potential is propagated along the plasma membrane of the muscle cell, and along T tubules to the **sarcoplasmic reticulum**. Calcium ions released by this organelle allow actin and myosin heads to interact so muscle contraction occurs.

Brief stimulation of a motor unit causes a **muscle twitch**, whereas repeated stimulation causes a sustained contraction that generates more force, or **muscle tension**.

Muscle fibers produce the ATP needed for contraction by way of three pathways: dephosphorylation of creatine phosphate, aerobic respiration, and lactate fermentation. Red fibers have many mitochondria and oxygen-storing myoglobin. They make ATP mainly by aerobic respiration. White fibers have no myoglobin and they make ATP mainly by fermentation. Muscles have a mix of red and white fibers.

Section 32.7 Exercise can increase muscle strength and endurance. Muscular dystrophy is a genetic disorder that causes muscles to break down. Muscle function can also be impaired by malfunction of motor neurons, or by toxins that disrupt nervous control of muscle.

Self-Quiz Answers in Appendix III

1. A hydrostatic skeleton consists of _____ .
   a. a fluid in an enclosed space
   b. hardened plates at the surface of a body
   c. internal hard parts
   d. none of the above

2. Bones are _____ .
   a. mineral reservoirs
   b. skeletal muscle's partners
   c. sites where blood cells form (some bones only)
   d. all of the above

3. Bones move when _____ muscles contract.
   a. cardiac
   b. skeletal
   c. smooth
   d. all of the above

4. A ligament connects _____ .
   a. bones at a joint
   b. a muscle to a bone
   c. a muscle to a tendon
   d. a tendon to bone

5. Bone breakdown is stimulated by _____ .
   a. parathyroid hormone
   b. estrogen
   c. vitamin D
   d. cortisol

6. Action of the _____ muscle is opposed by action of the triceps muscle.

Data Analysis Activities

Building Stronger Bones Tiffany (*right*) was born with multiple fractures in her limbs. By age six, she had undergone surgery to correct more than 200 bone fractures. Her fragile bones are symptoms of osteogenesis imperfecta (OI), a genetic disorder caused by a mutation in a gene for collagen. As bones develop, collagen forms a scaffold for deposition of mineralized bone tissue. The scaffold forms improperly in children with OI. Figure 32.20 shows the results of an experimental test of a new drug for OI. Bones of treated children, all less than two years old, were compared to bones of a control group of similarly affected, same-aged children who did not receive the drug.

Treated child	Vertebral area in cm^2 (Initial)	(Final)	Fractures per year
1	14.7	16.7	1
2	15.5	16.9	1
3	6.7	16.5	6
4	7.3	11.8	0
5	13.6	14.6	6
6	9.3	15.6	1
7	15.3	15.9	0
8	9.9	13.0	4
9	10.5	13.4	4
Mean	11.4	14.9	2.6

Control child	Vertebral area in cm^2 (Initial)	(Final)	Fractures per year
1	18.2	13.7	4
2	16.5	12.9	7
3	16.4	11.3	8
4	13.5	7.7	5
5	16.2	16.1	8
6	18.9	17.0	6
Mean	16.6	13.1	6.3

Figure 32.20 Results of a clinical trial of a drug treatment for osteogenesis imperfecta (OI), also known as brittle bone disease. Nine children with OI received the drug. Six others were untreated controls. Surface area of specific vertebrae was measured before and after treatment. An increase in vertebral area during the 12-month period of the study indicates bone growth. Researchers also recorded the number of fractures occurring during the 12-month trial.

1. How many treated children had bone growth (an increase in vertebral area)? How many untreated children?
2. How did the rate of fractures in the two groups compare?
3. Do the results support the hypothesis that this drug increases bone growth and reduce fractures in children with OI?

---

7. Skeletal muscle can only _____ bones.
   a. pull on   b. push against   c. exert force on

8. In sarcomeres, phosphate-group transfers from ATP activate _____ .
   a. actin   b. myosin   c. both   d. neither

9. A sarcomere shortens when _____ .
   a. thick filaments shorten
   b. thin filaments shorten
   c. both thick and thin filaments shorten
   d. none of the above

10. ATP for muscle contraction can be formed by _____ .
    a. aerobic respiration
    b. lactate fermentation
    c. creatine phosphate breakdown
    d. all of the above

11. Red muscle fibers get their color from _____ .
    a. ATP   b. myosin   c. myoglobin   d. collagen

12. A motor unit is _____ .
    a. a muscle and the bone it moves
    b. two muscles that work in opposition
    c. the amount a muscle shortens during contraction
    d. a motor neuron and the muscle fibers it controls

13. True or false? Aerobic exercise increases the number of muscle fibers in a muscle.

14. Muscular dystrophy _____ .
    a. is a genetic disorder
    b. is a type of food poisoning
    c. can be prevented with a vaccine
    d. is caused by a virus
    e. both c and d

15. Match the terms with their definitions.
    ___ tendon
    ___ muscle twitch
    ___ myoglobin
    ___ joint
    ___ myosin
    ___ red marrow
    ___ metacarpals
    ___ ligament
    ___ sarcoplasmic reticulum

    a. stores and releases calcium
    b. all in the hands
    c. blood cell production
    d. holds bones in place at joint
    e. area of contact between bones
    f. motor unit response
    g. reversibly binds oxygen
    h. connects muscle to bone
    i. binds ATP and converts it to ADP and phosphate

Additional questions are available on **CENGAGENOW**.

Critical Thinking

1. A friend is training for a marathon. Knowing you are taking biology, she asks if creatine supplements will improve her performance. What is your response?

2. Zachary's older brother had Duchenne muscular dystrophy and died at age 16. Zachary is 26 years old, healthy, and planning to start a family. He worries that his sons might be at risk for muscular dystrophy. His wife's family has no history of this disorder. Review Section 14.4 and decide whether Zachary's concern is well-founded.

Animations and Interactions on **CENGAGENOW**:
❯ Sea anemone skeleton; Fly flight; Human skeleton; Bone tissue; Human muscles; Muscle structure and contraction; Nervous control of contraction; Energy for contraction.

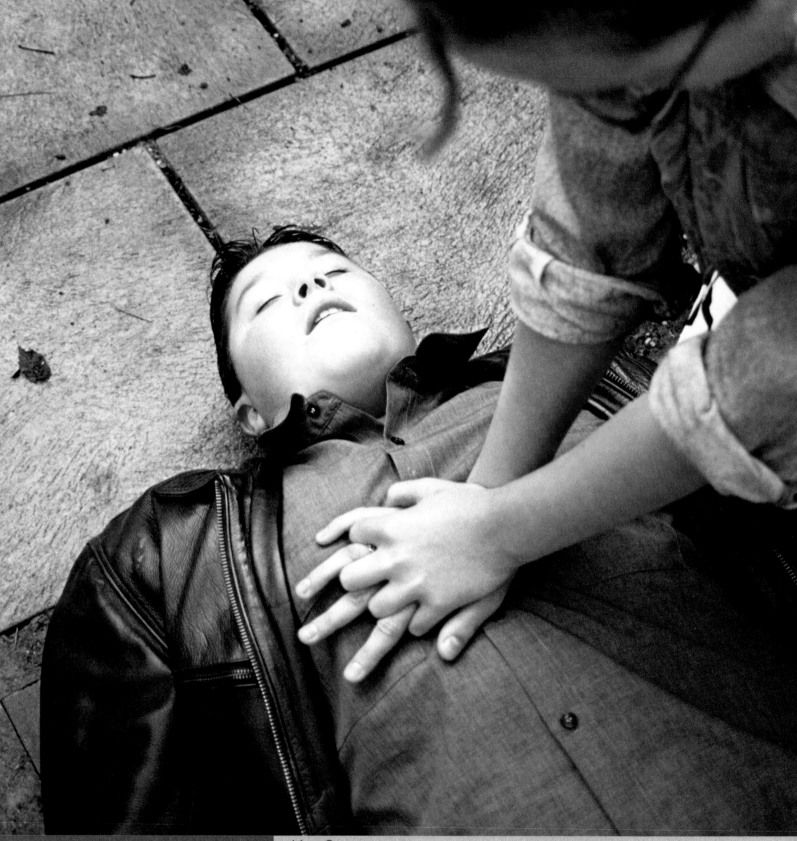

‹ Links to Earlier Concepts

This chapter expands on earlier introductions to circulatory systems (Sections 23.7, 23.8, 24.3, 24.4, 24.7), cardiac muscle (28.5), and muscle contraction (32.5). You will draw on your knowledge of hemoglobin (3.5), diffusion and osmosis (5.6), endocytosis (5.8), autonomic nerves (29.8), and cell junctions (4.11). You will be reminded of the health effects of sickle-cell anemia (9.6, 17.7), malaria, and a high blood cholesterol level (3.5).

Key Concepts

Animal Transport Systems
Most animals have a circulatory system that transports materials through their body. In an open system, blood leaves vessels and mixes with interstitial fluids. In a closed system, blood remains in vessels and exchanges substances with cells by diffusion across the vessel walls.

The Heart
A human heart is a muscular pump that has four chambers (two atria and two ventricles). It pumps blood through two separate circuits: one to the lungs and back, and the other extending through the body. The cardiac pacemaker stimulates heart muscle to contract.

33 Circulation

33.1 And Then My Heart Stood Still

The heart is the body's most durable muscle. It starts to beat during the first month of human development, and keeps on going for a lifetime. An electrical signal generated by a natural pacemaker in the heart wall sets each heartbeat in motion. In sudden cardiac arrest, this pacemaker malfunctions, electrical signaling is disrupted, the heart stops beating, and blood flow halts. In the United States, sudden cardiac arrest strikes more than 300,000 people per year. An inborn heart defect causes most cardiac arrests in people under age 35. In older people, heart disease usually causes the heart to stop functioning.

The chance of surviving sudden cardiac arrest rises by 50 percent when cardiopulmonary resuscitation (CPR) is started within four to six minutes of the arrest. With this technique, a person alternates mouth-to-mouth respiration with chest compressions that keep a victim's blood moving.

CPR cannot restart the heart. That requires a defibrillator, a device with paddles that deliver an electric shock to the chest and reset the natural pacemaker. You have probably seen this procedure depicted in hospital dramas.

Matt Nader (Figure 33.1) owes his life to CPR and defibrillation. He went into sudden cardiac arrest while playing in a high school football game. Nader's parents, who were watching the game, rushed from their seats and began CPR on their son. At the same time, someone ran to get the school's automated external defibrillator (AED). This device, about the size of a laptop computer, provides simple voice commands that direct the user to attach electrodes to a person in distress. The AED then checks for a heartbeat and, if required, shocks the heart.

The AED restarted Nader's heart, and he went on to testify before the Texas Legislature about his experience. Thanks in part to his efforts, Texas passed a law requiring all high schools to have AEDs at athletic events and practices.

Most cardiac arrests do not occur in a hospital, so the presence of a bystander willing to carry out CPR and use an AED often means the difference between life and death. Yet studies show only about 15 percent of sudden cardiac arrest victims get CPR before professional medical personnel arrive. The problem is that most people do not know how to administer CPR or use an AED. A half-day course given by the American Red Cross or another community health organization can teach you both skills. Taking the time to learn these skills is something we can all do for one another.

Figure 33.1 Surviving sudden cardiac arrest. *Opposite page*, CPR keeps blood moving and oxygenated when the heart malfunctions. *Above*, Matt Nader learned he had a heart defect when his heart stopped during a high school football game. CPR and use of a defibrillator saved Nader's life. He now has a defibrillator implanted in his chest to restart his heart when it stops.

 Blood and Blood Vessels Vertebrate blood is a fluid connective tissue with red blood cells, white blood cells, and platelets suspended in plasma. Blood flows through vessels that vary in structure and function. Exchanges between blood and interstitial fluid take place across walls of the smallest vessels.

 Cardiovascular Disorders Circulatory function declines when the heart's rhythm is disrupted or blood vessels become clogged by atherosclerosis. Heart disease arises when vessels that supply blood to heart muscle are narrowed. A healthy life-style can lessen the risk of cardiovascular disorders.

 Links With the Lymphatic System Fluid that diffuses out of capillaries enters the lymph vascular system, which returns it to the blood. As fluid flows through lymphatic vessels, lymphoid organs monitor it for infectious agents and other threats to health.

❭ Most invertebrates and all vertebrates have a circulatory system that speeds the distribution of materials through the body.

❬ Links to Diffusion 5.6, Morphological convergence 16.8

All animals must keep their cells supplied with nutrients and oxygen, and all must get rid of cellular wastes. Some invertebrates, including cnidarians and flatworms, rely on diffusion alone to accomplish these tasks. In such animals, nutrients and gases reach cells by diffusing across a body surface and then diffusing through the interstitial fluid (the fluid between cells). Diffusion only works over short distances to move materials quickly, so animals that rely on diffusion to distribute materials have a body plan in which all cells lie close to a body surface.

Open and Closed Circulatory Systems

The evolution of circulatory systems made possible more complex animal body plans. A **circulatory system** is an organ system that speeds the distribution of materials within an animal body. It includes one or more **hearts** (muscular pumps) that propel **blood** (the circulatory fluid) through vessels that extend through the body.

Different types of circulatory systems evolved in different animal lineages. Arthropods and most mollusks have an **open circulatory system**, in which a heart or hearts pump blood into large vessels that empty into spaces around body tissues (Figure 33.2A). The blood of an open circulatory system mixes with the interstitial fluid and makes direct exchanges with cells before it gets drawn back into the heart.

By contrast, annelids, cephalopod mollusks, and all vertebrates have a **closed circulatory system** in which a heart or hearts pump blood through a continuous series of vessels (Figure 33.2B). A closed circulatory system distributes substances faster than an open one. It is "closed" because blood does not flow out of blood vessels to bathe the tissues. Instead, most transfers between blood and the cells of other tissues take place by diffusion across the smallest-diameter blood vessels, the **capillaries**.

Evolution of Vertebrate Circulation

All vertebrates have a closed circulatory system, with a single heart. However, the structure of the heart and the circuits through which blood flows vary among vertebrate groups. In most fishes, the heart has two main chambers, and blood flows in one circuit (Figure 33.3A). One chamber of the heart, an atrium (plural, atria), receives blood. From there, the blood enters a ventricle, a chamber that pumps blood out of the heart. Pressure exerted by ventricular contractions drives blood through a series of vessels, into capillaries inside each gill, through capillaries in body tissues and organs, and back to the heart. The pressure imparted to the blood by the ventricle's contraction is dissipated as blood travels through capillaries, so the blood is not under much pressure when it leaves the gill capillaries, and even less as it travels back to the heart.

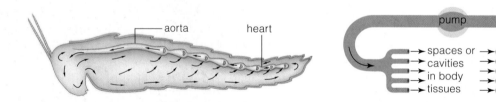

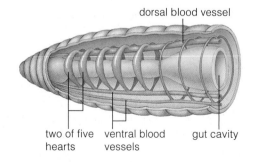

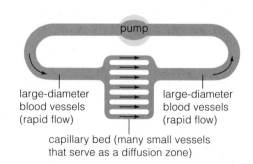

A Open circulatory system. A grasshopper's heart pumps blood through a large vessel and out into tissue spaces. Blood mingles with interstitial fluid, exchanges materials, and reenters the heart through openings in the heart wall.

Figure 33.2 Animated
Comparison of open and closed circulatory systems.

B Closed circulatory system. An earthworm's hearts pumps blood through vessels that extend through the body. Exchanges between blood and the tissues take place across the wall of the smallest vessels.

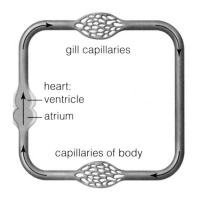

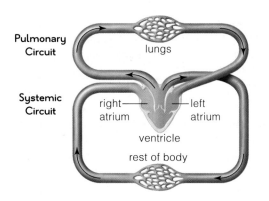

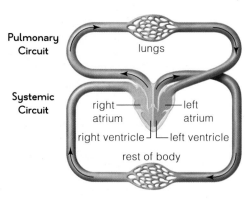

A The fish heart has one atrium and one ventricle. The force of the ventricle's contraction propels blood through the single circuit.

B In amphibians and most reptiles, the heart has three chambers: two atria and one ventricle. Blood flows in two partially separated circuits. Oxygenated blood and oxygen-poor blood mix a bit in the ventricle.

C In crocodilians, birds, and mammals, the heart has four chambers: two atria and two ventricles. Oxygenated blood and oxygen-poor blood do not mix.

Figure 33.3 **Animated** Variation in vertebrate circulatory systems.

Adapting to life on land involved coordinated modifications of respiratory and circulatory systems. Amphibians and most reptiles have a three-chambered heart, with two atria emptying into one ventricle (Figure 33.3B). The three-chambered heart speeds the rate of flow by moving blood through two partially separated circuits. The force of one contraction propels blood through the **pulmonary circuit**, to the lungs and then back to the heart. (The Latin word *pulmo* means lung.) A second contraction sends the oxygenated blood through the **systemic circuit**. This circuit extends through capillaries in body tissues and returns to the heart.

The ventricle became fully separated into two chambers in birds and mammals. Their four-chambered heart has two atria and two ventricles (Figure 33.3C). With two fully separate circuits, only oxygen-rich blood flows to tissues. As an additional advantage, blood pressure can be regulated independently in each circuit. Strong con-

traction of the heart's left ventricle moves blood quickly through the long systemic circuit. At the same time, the right ventricle can contract more gently, protecting the delicate lung capillaries that would be blown apart by higher pressure.

The four-chambered heart of mammals and birds is an example of morphological convergence. The two groups do not share an ancestor with a four-chambered heart; this trait evolved independently in the two groups. The enhanced blood flow associated with a four-chambered heart supports the high metabolism of these endothermic (heated from within) animals. As Section 24.6 explained, endotherms have higher energy needs than comparably sized ectotherms because they lose more energy as heat. Rapid blood flow in an endotherm's body delivers the large amount of oxygen required to keep heat-generating aerobic reactions going nonstop.

blood Circulatory fluid; in vertebrates it is a fluid connective tissue consisting of plasma and cells that form inside bones.
capillaries Small blood vessels where exchanges with the interstitial fluid take place.
circulatory system Organ system consisting of a heart or hearts and blood-filled vessels that distribute substances through a body.
closed circulatory system Circulatory system in which blood flows through a continuous system of vessels, and substances are exchanged across the walls of the smallest vessels.
heart Muscular organ that pumps blood through a body.
open circulatory system Circulatory system in which blood leaves vessels and flows among body tissues.
pulmonary circuit Circuit through which blood flows from the heart to the lungs and back.
systemic circuit Circuit through which blood flows from the heart to the body tissues and back.

Take-Home Message **How do animals distribute essential substances to cells throughout their body?**

❯ Most animals have a circulatory system that speeds the distribution of substances through the body.

❯ Some invertebrates have an open circulatory system, in which blood flows out of vessels and seeps around tissues.

❯ Other invertebrates and all vertebrates have a closed circulatory system, in which blood always remains enclosed within the heart or blood vessels.

❯ Fish have a one-circuit circulatory system. All other vertebrates have a short pulmonary circuit that carries blood to and from the lungs, and a longer systemic circuit that moves blood to and from the body's other tissues.

❯ A four-chambered heart evolved independently in birds and mammals. It allows strong contraction of one ventricle to speed blood through the systemic circuit, while a weaker contraction of the other ventricle protects lung tissue.

❯ The term "cardiovascular" comes from the Greek *kardia* (for heart) and Latin *vasculum* (vessel). In the human cardiovascular system, the heart pumps blood in two circuits: one to the lungs, the other to all body tissues.

❮ Links to Glycogen storage 3.3, Alcohol metabolism 5.1

Like other mammals, humans have a four-chambered heart that pumps blood through two circuits. Each circuit includes a network of blood vessels that carry blood from the heart to a capillary bed and back to the heart. Figure 33.4 shows the location and function of some major blood vessels.

In each circuit, the heart pumps blood out of a ventricle and into branching arteries. **Arteries** are wide-diameter blood vessels that carry blood away from the heart and to organs. Within an organ, arteries branch into smaller vessels called **arterioles**. Arterioles in turn branch into capillaries, the smallest vessels. As noted earlier, exchanges between the blood and interstitial fluid take place as blood flows through the capillaries. Several capillaries join up to form a **venule**, a vessel that carries blood to a vein. **Veins** are large-diameter vessels that return blood to the heart. Blood from veins empties into one of the two atria.

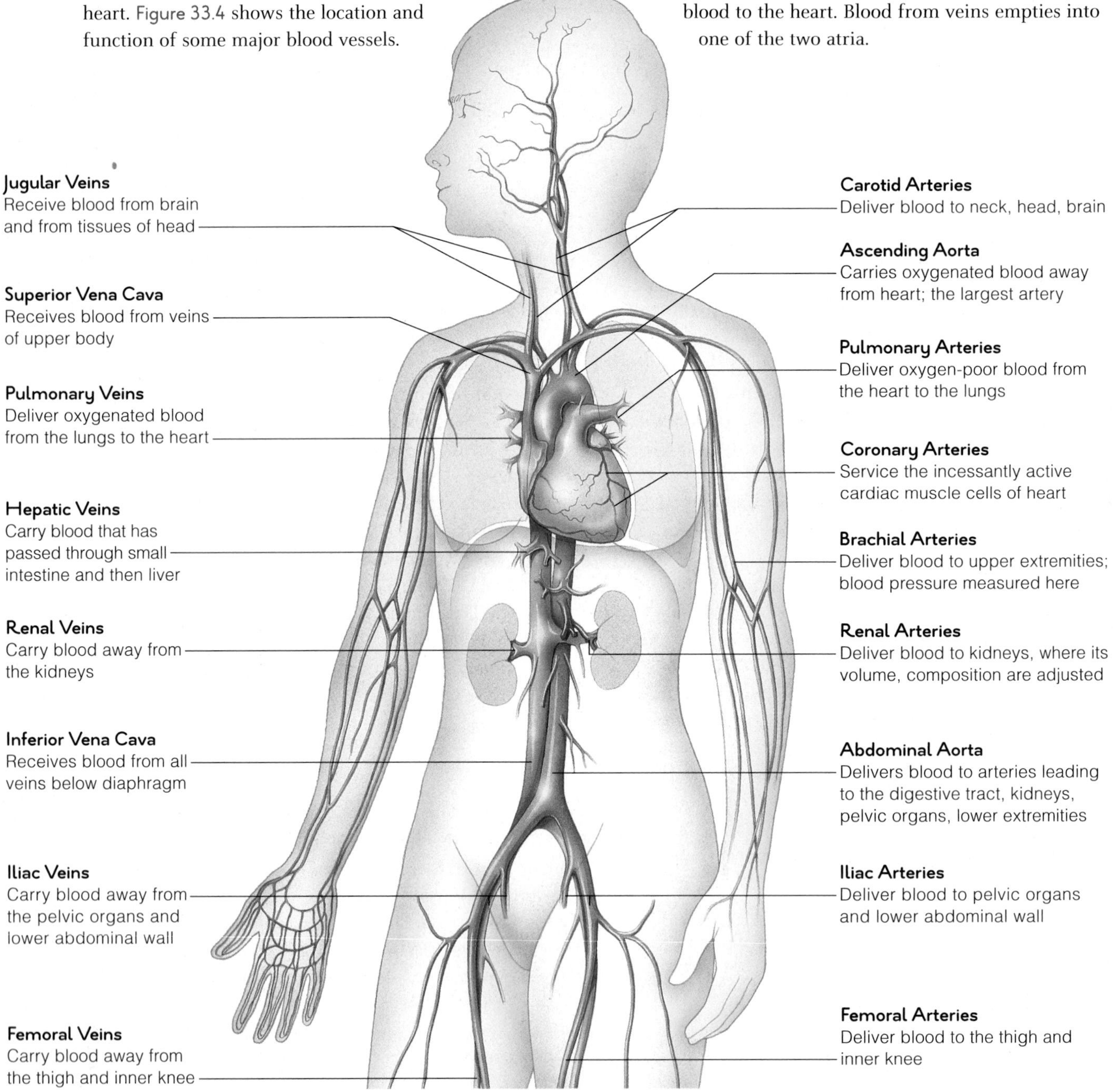

Jugular Veins
Receive blood from brain and from tissues of head

Superior Vena Cava
Receives blood from veins of upper body

Pulmonary Veins
Deliver oxygenated blood from the lungs to the heart

Hepatic Veins
Carry blood that has passed through small intestine and then liver

Renal Veins
Carry blood away from the kidneys

Inferior Vena Cava
Receives blood from all veins below diaphragm

Iliac Veins
Carry blood away from the pelvic organs and lower abdominal wall

Femoral Veins
Carry blood away from the thigh and inner knee

Carotid Arteries
Deliver blood to neck, head, brain

Ascending Aorta
Carries oxygenated blood away from heart; the largest artery

Pulmonary Arteries
Deliver oxygen-poor blood from the heart to the lungs

Coronary Arteries
Service the incessantly active cardiac muscle cells of heart

Brachial Arteries
Deliver blood to upper extremities; blood pressure measured here

Renal Arteries
Deliver blood to kidneys, where its volume, composition are adjusted

Abdominal Aorta
Delivers blood to arteries leading to the digestive tract, kidneys, pelvic organs, lower extremities

Iliac Arteries
Deliver blood to pelvic organs and lower abdominal wall

Femoral Arteries
Deliver blood to the thigh and inner knee

Figure 33.4 Animated Major blood vessels of the human cardiovascular system. Vessels carrying oxygenated blood are color-coded *red* and those carrying oxygen-poor blood are color-coded *blue*.

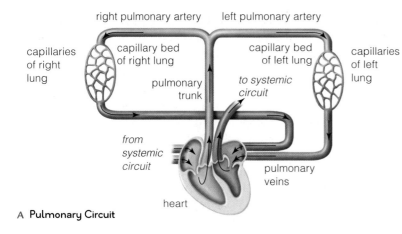

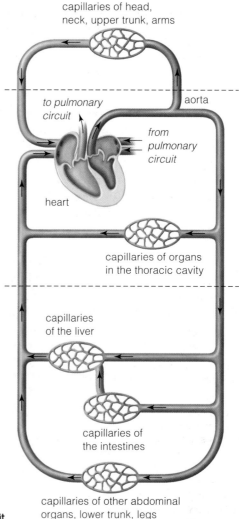

A Pulmonary Circuit

We now take a detailed look at each of the two circuits. The shorter pulmonary circuit carries blood to and from the lungs (Figure 33.5A). Oxygen-poor blood is pumped out of the heart's right ventricle into pulmonary arteries. One pulmonary artery delivers blood to each lung. As blood flows through pulmonary capillaries, it picks up oxygen and gives up carbon dioxide. Oxygen-rich blood then returns to the heart by way of the pulmonary veins, which empty into the left atrium.

Oxygenated blood pumped out of the heart travels through the longer systemic circuit (Figure 33.5B). The heart's left ventricle pumps blood into the body's largest artery, the **aorta**. Arteries that branch from the aorta carry blood to various body parts. For example, the renal artery delivers blood to the kidneys, and the coronary arteries supply heart cells. Each artery branches into arterioles and then capillaries. Blood gives up oxygen and picks up carbon dioxide as it flows through the capillaries. The oxygen-poor blood that leaves the capillaries flows through venules and veins to the heart's right atrium.

Most blood moving through the systemic circuit flows through only one capillary bed. However, after blood passes through the capillaries in the small intestine, it flows through a vein (the hepatic portal vein) to a capillary bed in the liver. This two-capillary journey allows the blood to pick up glucose and other substances absorbed from the gut, and deliver them to the liver. The liver stores some of the absorbed glucose as glycogen. It also breaks down some absorbed toxins, including alcohol.

B Systemic Circuit

Figure 33.5 Animated Circuits of the human cardiovascular system.

» Figure It Out Do all veins carry oxygen–poor blood?

Answer: No. Pulmonary veins carry oxygen-rich blood from lung capillaries to the heart.

---

aorta Large artery that receives blood pumped out of the heart's left ventricle.
arteriole Vessel that carries blood from an artery to a capillary.
artery Large-diameter blood vessel that carries blood away from the heart.
vein Large-diameter vessel that returns blood to the heart.
venule Small-diameter vessel that carries blood from capillaries to a vein.

Take-Home Message What are the two circuits of the human cardiovascular system?

> The pulmonary circuit carries oxygen-poor blood from the heart through the pulmonary arteries to arterioles and then capillaries in the lungs. Pulmonary veins return oxygenated blood to the heart.

> The systemic circuit carries oxygenated blood from the heart out the aorta, through branching arteries and to capillaries throughout the body. It returns oxygen-poor blood to the heart by way of venules and veins.

> Most blood traveling through the systemic circuit passes through one capillary bed, but blood that flows through capillaries in the intestines also flows through capillaries in the liver.

> ❯ The heart contracts in response to signals from a natural pacemaker and keeps blood flowing through the body.
> ❮ Links to Gap junctions 4.11, Epithelium 28.3, Cardiac muscle 28.5

Structure of the Heart

The heart lies in the thoracic cavity, beneath the breastbone and between the lungs (Figure 33.6A). It is protected and anchored by pericardium, a sac of connective tissue. Fluid between the sac's two layers provides lubrication for the heart's continual motions. A layer of fat offers additional protection (Figure 33.6B). The heart's wall consists mostly of cardiac muscle cells, and its chambers and blood vessels are lined with endothelium, a type of epithelium.

Each side of the human heart has two chambers: An **atrium** receives blood from veins, and a **ventricle** pumps blood into arteries (Figure 33.6C). Pressure-sensitive valves function like one-way doors to control the flow of blood through the heart. High fluid pressure forces a valve open. When fluid pressure declines, the valve shuts and prevents blood from moving backward.

Flow To, Through, and From the Heart

Two big veins deliver oxygen-poor blood from the body to the right atrium. The **superior vena cava** delivers blood from the upper regions of the body. The **inferior vena cava** delivers blood from lower regions. Blood from the right atrium flows through the right atrioventricular (AV) valve into the right ventricle. The right ventricle pumps it through the pulmonary valve and into the pulmonary trunk, a vessel that branches into two pulmonary arteries. Each **pulmonary artery** carries blood to a lung.

After passing through the lung, the now-oxygenated blood returns to the left atrium via **pulmonary veins**. The blood then flows through the left atrioventricular (AV) valve into the left ventricle. The left ventricle pumps the blood through the aortic valve into the aorta, and from there it flows to tissues of the body.

The Cardiac Cycle

The events that occur from the onset of one heartbeat to another are collectively called the **cardiac cycle** (Figure 33.7). During this cycle, the heart's chambers alternate through **diastole** (relaxation) and **systole** (contraction). First, the relaxed atria expand with blood ❶. Fluid pressure forces AV valves to open and blood to flow into the relaxed ventricles, which expand as the atria contract ❷. Once filled, the ventricles contract. Contraction raises the fluid pressure inside the ventricles and forces the aortic and pulmonary valves to open. Blood flows through these

A The heart is located between the lungs in the thoracic cavity.

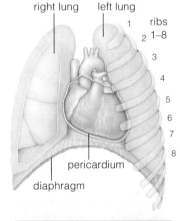

right lung left lung
ribs 1–8
1 2 3 4 5 6 7 8
pericardium
diaphragm

B Heart's external surface. Some fat at the heart's surface is normal.

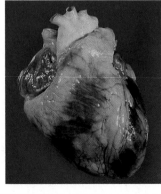

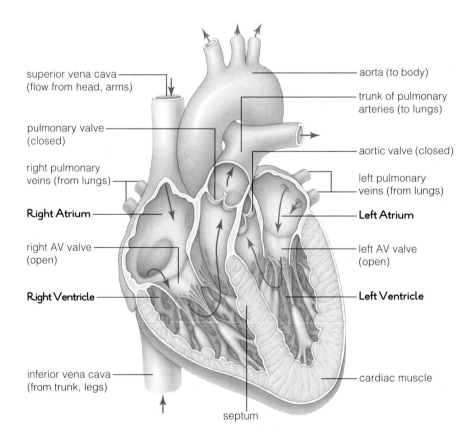

superior vena cava (flow from head, arms)
aorta (to body)
trunk of pulmonary arteries (to lungs)
pulmonary valve (closed)
aortic valve (closed)
right pulmonary veins (from lungs)
left pulmonary veins (from lungs)
Right Atrium
Left Atrium
right AV valve (open)
left AV valve (open)
Right Ventricle
Left Ventricle
inferior vena cava (from trunk, legs)
cardiac muscle
septum

C Cutaway view, showing the heart's internal organization. Arrows indicate the path taken by oxygenated (*red*) and oxygen-poor (*blue*) blood.

Figure 33.6 Animated The human heart.

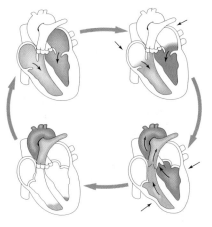

1 Relaxed atria fill. Fluid pressure opens AV valves and blood flows into the relaxed ventricles.

2 Contracting atrial squeeze more blood into the still-relaxed ventricles.

4 As blood flows into the arteries, pressure in the ventricles declines and the aortic and pulmonary valves close.

3 Ventricles start contracting and rising pressure pushes AV valves shut. A further rise in pressure opens aortic and pulmonary valves.

Figure 33.7 Animated Cardiac cycle.

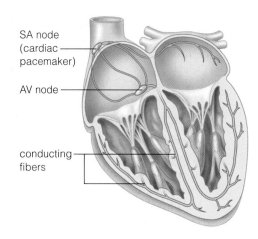

SA node (cardiac pacemaker)

AV node

conducting fibers

Figure 33.8 Animated Cardiac conduction system.

valves and out of the ventricles **3**. Now emptied, the ventricles relax while the atria fill again **4**.

Contraction of ventricles drives circulation; atrial contraction only helps fill ventricles. The structure of the cardiac chambers reflects their different functions. Atria need only generate enough force to squeeze blood into the ventricles, so they have relatively thin walls. Ventricle walls are more thickly muscled because their contraction has to generate enough pressure to propel blood through an entire cardiovascular circuit. The left ventricle, which pumps blood throughout the long systemic circuit, has thicker walls than the right ventricle, which pumps blood only to the lungs and back.

During the cardiac cycle a "lub-dup" sound can be heard through the chest wall. Each "lub" is the heart's AV valves closing. Each "dup" is the heart's aortic and pulmonary valves closing. If a valve does not close properly, blood is forced backward through the defective valve, making a whooshing sound known as a heart murmur. Most valve defects that cause heart murmurs do not threaten health. Those that do require a surgical repair.

---

atrioventricular (AV) node Clump of cells that is the electrical bridge between the atria and ventricles.
atrium Heart chamber that receives blood from veins.
cardiac cycle Sequence of contraction and relaxation of heart chambers that occurs with each heartbeat.
diastole Relaxation phase of the cardiac cycle.
inferior vena cava Vein that delivers blood from the lower body to the heart.
pulmonary artery Vessel carrying blood from the heart to a lung.
pulmonary vein Vessel carrying blood from a lung to the heart.
sinoatrial (SA) node Cardiac pacemaker; group of heart cells that spontaneously emits rhythmic signals that cause contraction.
superior vena cava Vein that delivers blood from the upper body to the heart.
systole Contractile phase of the cardiac cycle.
ventricle Heart chamber that pumps blood into arteries.

Setting the Pace for Contraction

Like skeletal muscle, cardiac muscle has orderly arrays of sarcomeres that contract by a sliding-filament mechanism. Unlike skeletal muscle cells, cardiac muscle cells have gap junctions that connect the cytoplasm of adjacent cells. The connection allows action potentials to spread swiftly between cardiac muscle cells.

Signals for contraction originate at the **sinoatrial (SA) node**, a clump of specialized cells in the wall of the right atrium (Figure 33.8). The SA node is known as the cardiac pacemaker because it generates an action potential about seventy times a minute. The defibrillators you read about in Section 33.1 work by resetting an SA node that has somehow malfunctioned.

A signal from the SA node spreads across the atria, causing them to contract. Simultaneously, specialized noncontractile muscle fibers conduct the signal to the **atrioventricular (AV) node**. This clump of cells is the only electric bridge to the ventricles. The time it takes for a signal to cross this bridge allows blood from atria to fill the ventricles before the ventricles contract.

From the AV node, the signal travels along conducting fibers in the septum between the heart's left and right halves. The fibers extend to the heart's lowest point and up the ventricle walls. In response to the signals, ventricles contract from the bottom up, with a twisting motion.

Take-Home Message How does the structure of the human heart relate to its function?

> The four-chambered heart is a muscular pump partitioned into two halves, each with an atrium and a ventricle. Forceful contraction of the ventricles provides the driving force for blood circulation.

> The SA node is the cardiac pacemaker. Its spontaneous, rhythmic signals make cardiac muscle fibers of the heart wall contract in a coordinated fashion.

> Plasma, the protein-rich fluid portion of blood, distributes essential nutrients and solutes to cells. Blood cells tumbling along in plasma carry oxygen and defend the body.

< Links to Hemoglobin 3.5, Blood as connective tissue 28.4, Red marrow 32.3, Hemophilia 14.4

Functions of Blood

Vertebrate blood is a fluid connective tissue with many functions. It carries essential oxygen and nutrients to cells, and carries their metabolic wastes to the organs that dispose of them. It facilitates internal communications by distributing hormones and serves as a highway for cells and proteins that protect and repair tissues. In birds and mammals, blood helps maintain a stable internal temperature by distributing heat generated by muscle activity to the skin, where it can be lost to the surroundings.

Human Blood Volume and Composition

Body size and the concentrations of solutes determine blood volume. Average-sized adults hold about 5 liters of blood (a bit more than 10 pints), and blood accounts for about 6 to 8 percent of their body weight. Blood is—as the saying goes—thicker than water. Dissolved substances and suspended cells contribute to its greater viscosity. Figure 33.9 describes its components.

Plasma The fluid portion of the blood, known as the **plasma**, constitutes about 50 to 60 percent of the blood volume. Plasma is mostly water with hundreds of different plasma proteins dissolved in it. Some plasma proteins transport lipids and fat-soluble vitamins; others have a role in blood clotting or immunity. Dissolved sugars, amino acids, vitamins, and some gases travel through the bloodstream in plasma.

Red Blood Cells The cellular portion of blood consists of blood cells and platelets. All arise from stem cells in the red marrow of bones.

Erythrocytes, or **red blood cells**, transport oxygen from lungs to aerobically respiring cells and facilitate movement of carbon dioxide to the lungs. In all mammals, red blood cells lose their nucleus and other organelles as they develop. Mature red blood cells are flexible disks with a depression at their center. Their flexibility allows them to slip easily through narrow blood vessels, and their flattened shape facilitates gas exchange.

Hemoglobin fills the interior of the mature red blood cell. You learned about this protein in Section 3.5. Most oxygen that enters the blood travels to the tissues while bound to the heme group of hemoglobin. In addition to hemoglobin, a mature red blood cell has enough stored

Components	Amounts	Main Functions
Plasma Portion (50–60% of total blood volume)		
1. Water	91–92% of total plasma volume	Solvent
2. Plasma proteins (albumins, globulins, fibrinogen, etc.)	7–8%	Defense, clotting, lipid transport, extracellular fluid volume controls
3. Ions, sugars, lipids, amino acids, hormones, vitamins, dissolved gases, etc.	1–2%	Nutrition, defense, respiration, extracellular fluid volume controls, cell communication, etc.
Cellular Portion (40–50% of total blood volume; numbers per microliter)		
1. Red blood cells	4,600,000–5,400,000	Oxygen, carbon dioxide transport to and from lungs
2. White blood cells:		
Neutrophils	3,000–6,750	Fast-acting phagocytosis
Lymphocytes	1,000–2,700	Immune responses
Monocytes (macrophages)	150–720	Phagocytosis
Eosinophils	100–380	Killing parasitic worms
Basophils	25–90	Secretions promote inflammation
3. Platelets	250,000–300,000	Roles in blood clotting

red blood cell   white blood cell   platelet

Figure 33.9 Typical components of human blood. The sketch of a test tube shows what happens when you prevent a blood sample from clotting. The sample separates into straw-colored plasma that floats above a reddish cellular portion. The scanning electron micrograph shows some cellular components of blood.

sugars, RNAs, and other molecules to live about 120 days. In a healthy person, ongoing replacements keep red blood cell numbers at a fairly stable level. A **cell count** is a measure of the quantity of cells of one type in 1 microliter (1/1,000,000 liter) of blood. During their reproductive years, women have a lower red blood cell count than men, because women lose blood during menstruation.

Anemia is a disorder in which the red blood cell count declines or red blood cells are defective. As a result, oxygen delivery and metabolism falter. Anemia can arise as a result of a dietary iron deficiency (iron is needed to make hemoglobin), destruction of red blood cells by pathogens (as occurs in malaria), excessive blood loss (as from unusually heavy menstrual bleeding), and genetic disorders. Sickle-cell anemia arises from a mutation that causes hemoglobin to form large clumps at low oxygen levels. The clumps distort red blood cells so they get stuck in small blood vessels.

White Blood Cells Leukocytes, or **white blood cells**, carry out ongoing housekeeping tasks and function in defense. The cells differ in their size, nuclear shape, and staining traits, as well as function.

We discuss the role of white blood cells in detail in the next chapter, but here is a brief preview. Neutrophils, the most abundant white cells, are phagocytes that engulf bacteria and debris. Eosinophils attack larger parasites, such as worms. Basophils secrete chemicals that have a role in inflammation. Monocytes are white cells that circulate in the blood for a few days, then move into the tissues, where they develop into phagocytic cells known as macrophages. Macrophages interact with lymphocytes to bring about immune responses. There are two types of lymphocytes, B cells and T cells. B cells mature in bone, whereas T cells mature in the thymus. Both protect the body against specific threats.

Leukemias are cancers that originate in stem cells of the bone marrow. They cause overproduction of abnormal white blood cells that do not function properly. Lymphomas are cancers that originate from B or T lymphocytes in lymph glands. Division of the cancerous lymphocytes produces tumors in lymph nodes and other parts of the lymphatic system.

Stimulus

A blood vessel is damaged.

Phase 1 response

The vessel constricts.

Phase 2 response

Platelets stick together, plugging the site.

Phase 3 response

Clot formation:

1. Enzyme cascade results in activation of the enzyme thrombin.
2. Thrombin converts fibrinogen, a plasma protein, to fibrin threads.
3. Fibrin forms a net that entangles cells and platelets, forming a clot.

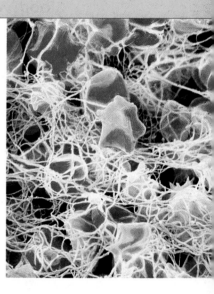

Figure 33.10 Hemostasis. The micrograph shows the result of the final clotting phase—blood cells and platelets in a fibrin net.

Platelets and Hemostasis A **platelet** is a membrane-wrapped fragment of cytoplasm that arises when a large cell (a megakaryocyte) breaks up. Once formed, a platelet will remain functional for up to nine days. Hundreds of thousands of platelets circulate in the blood, ready to take part in **hemostasis**. This process stops blood loss from an injured vessel and provides a framework to begin repairs.

When a vessel is injured, it constricts (narrows), reducing blood loss. Platelets adhere to the injured site and release substances that attract more platelets. Plasma proteins convert blood to a gel and form a clot. During clot formation, fibrinogen, a soluble plasma protein, is converted to insoluble threads of fibrin. **Fibrin** forms a mesh that traps cells and platelets (Figure 33.10).

Clot formation involves a cascade of enzyme reactions. Fibrinogen is converted to fibrin by an enzyme, thrombin, which circulates in blood as the inactive precursor prothrombin. Prothrombin is activated by an enzyme that is activated by another enzyme, and so on. If a mutation affects any one of the enzymes that acts in the cascade of clotting reactions, the blood may not clot properly. Such mutations cause the genetic disorder hemophilia. A vitamin K deficiency can also impair clotting, because this vitamin plays a role in the cascade of enzyme reactions.

---

cell count Number of cells of one type per microliter of blood.
fibrin Threadlike protein formed during blood clotting from the soluble plasma protein fibrinogen.
hemostasis Process by which blood clots in response to injury.
plasma Fluid portion of blood.
platelet Cell fragment that helps blood clot.
red blood cell Hemoglobin-filled blood cell that carries oxygen.
white blood cell Blood cell with a role in housekeeping and defense.

Take-Home Message **What are the components of blood and their functions?**

> Blood consists mainly of plasma, a protein-rich fluid that carries wastes, nutrients, and some gases.

> Blood cells and platelets form in bone marrow and are transported in plasma. Red blood cells contain hemoglobin that carries oxygen from lungs to tissues. White cells help defend the body from pathogens. Platelets are cell fragments that, like some plasma proteins, have a role in clotting.

> As blood flows through a circuit, it passes through a series of vessels that differ in their structure and function.
< Link to Autonomic nerves 29.8

Rapid Transport in Arteries

Blood pumped out of ventricles enters arteries. These large-diameter vessels have a muscular wall reinforced with elastic tissue (Figure 33.11A). Artery structure helps keep blood flowing, even when the ventricles are not contracting. With each ventricular contraction, the pressure exerted by blood forced into an artery causes the artery to bulge a bit. Then, as the ventricle relaxes, the artery wall springs back like a rubber band that has been stretched. As the artery wall recoils, it pushes blood inside the artery a bit farther away from the heart.

The bulging of an artery with each ventricular contraction is referred to as the **pulse**. You can feel a person's pulse if you place your finger on a pulse point, a place where an artery runs near the body surface. For example, to feel the pulse in your radial artery, put your fingers on your inner wrist near the base of your thumb.

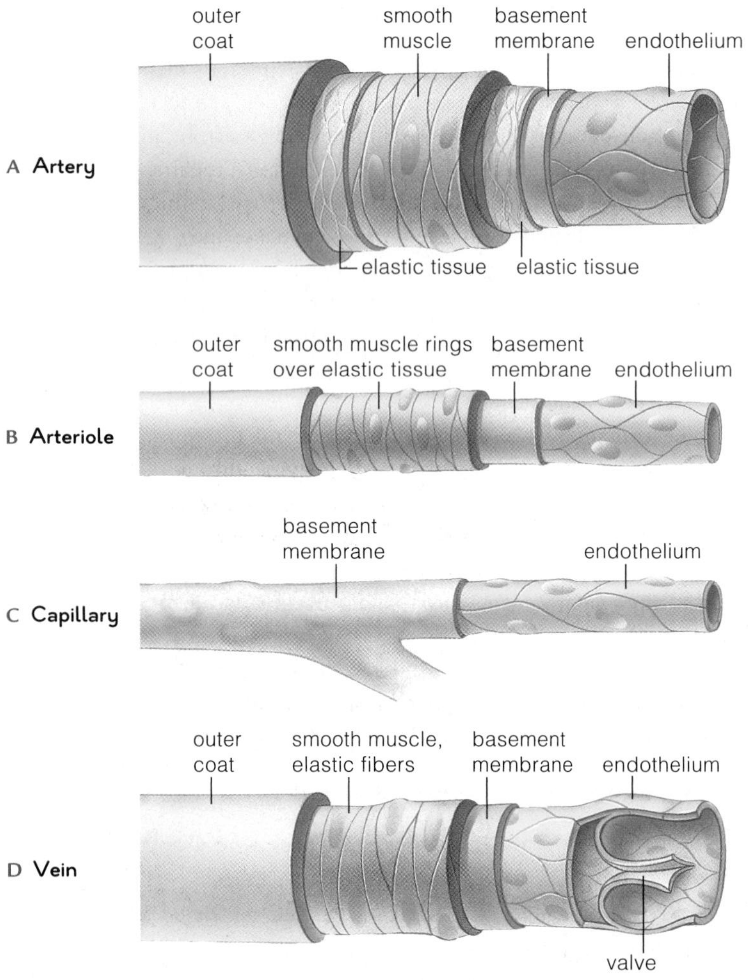

Adjusting Flow at Arterioles

All blood from the right half of your heart flows to your lungs. In the systemic circuit, however, the body adjusts the distribution of blood by altering the diameter of arterioles. Smooth muscle that rings each arteriole (Figure 33.11B) responds to commands from the central nervous system. For example, sympathetic stimulation causes **vasodilation** (widening) of arterioles in the extremities and **vasoconstriction** (narrowing) of arterioles of the gut.

Arterioles also respond to metabolic activity in nearby tissue. When you run, skeletal muscle in your legs uses up oxygen, and releases carbon dioxide. Arterioles delivering blood to leg muscles widen in response to these changes.

Exchanges at Capillaries

A capillary is a cylinder of endothelial cells, one cell thick, wrapped in basement membrane (Figure 33.11C). Its thin wall and narrow diameter, barely wider than a red blood cell, facilitate exchanges between the blood and the interstitial fluid. Oxygen-carrying red blood cells are forced right up against the capillary wall. We discuss capillary exchange in detail in Section 33.8.

Return to the Heart—Venules and Veins

Blood from several capillaries flows into each venule. These thin-walled vessels join together to form veins, the large-diameter, low-resistance transport tubes that carry blood to the heart. Many veins, especially in the legs, have flaplike valves that help prevent backflow (Figure 33.11D). These valves automatically shut when blood in the vein starts to reverse direction.

---

pulse Brief stretching of artery walls that occurs when ventricles contract.
vasoconstriction Narrowing of a blood vessel when smooth muscle that rings it contracts.
vasodilation Widening of a blood vessel when smooth muscle that rings it relaxes.

A Artery — outer coat, smooth muscle, basement membrane, endothelium, elastic tissue, elastic tissue

B Arteriole — outer coat, smooth muscle rings over elastic tissue, basement membrane, endothelium

C Capillary — basement membrane, endothelium

D Vein — outer coat, smooth muscle, elastic fibers, basement membrane, endothelium, valve

Figure 33.11 Structural comparison of human blood vessels. Drawings are not to scale. Venules (not shown) have a structure like capillaries.

Take-Home Message How do blood vessels differ in their structure and function?

> Arteries are thick-walled, large-diameter vessels. Stretching and recoil of arteries helps keep blood moving.

> Smooth muscle in the wall of arterioles allows adjustments to blood flow in the systemic circuit.

> Capillaries are narrow tubes of epithelial cells. They are the site of exchanges with interstitial fluid.

> Veins have valves that prevent backflow of blood.

› Ventricular contractions are the source of blood pressure, which declines throughout a cardiovascular circuit.
‹ Link to Medulla oblongata 29.10

Blood pressure is pressure exerted by blood against the wall of the vessel that encloses it. The right ventricle contracts less forcefully than the left ventricle, so blood entering the pulmonary circuit is under less pressure than blood entering the systemic circuit. In both circuits, blood pressure is highest in arteries, and declines as blood flows through the circuit, being lowest in veins (Figure 33.12).

Blood pressure is usually measured in the brachial artery of the upper arm (Figure 33.13). Two pressures are recorded. **Systolic pressure**, the highest pressure of a cardiac cycle, occurs as contracting ventricles force blood into the arteries. **Diastolic pressure**, the lowest blood pressure of a cardiac cycle, occurs when ventricles are relaxed. Blood pressure is measured in millimeters of mercury (mm Hg), a standard unit for measuring pressure. It is written as systolic value/diastolic value. Normal blood pressure is about 120/80 mm Hg, or "120 over 80."

Blood pressure depends on the total blood volume, how much blood the ventricles pump out (cardiac output), and the degree of arteriole dilation. Receptors in the aorta and in carotid arteries of the neck signal the medulla (a part of the hindbrain) when blood pressure rises or falls. In a reflexive response, the medulla calls for appropriate changes in cardiac output and arteriole diameter. Vasodilation of arterioles lowers blood pressure; vasoconstriction raises it. The reflex response adjusts blood pressure over the short term. Over the longer term, the kidneys adjust blood pressure by regulating the amount of fluid lost as urine and thus determining the total blood volume.

Inability to regulate blood pressure can result in hypertension, in which resting blood pressure remains above 140/90. Often the cause of hypertension is unknown. Heredity is a factor, and African Americans have an elevated risk. Diet also plays a role; in some people high salt intake causes water retention that raises blood pressure. High blood pressure makes the heart and kidneys work harder, increasing risk of heart disease or kidney failure.

---

blood pressure Pressure exerted by blood against a vessel wall.
diastolic pressure Blood pressure when ventricles are relaxed.
systolic pressure Blood pressure when ventricles are contracting.

Take-Home Message How is blood pressure recorded and regulated?

› Blood pressure is the fluid pressure exerted against a vessel wall. It is recorded as systolic/diastolic pressure.

› Adjustments to arteriole diameter, cardiac output, and blood volume regulate blood pressure.

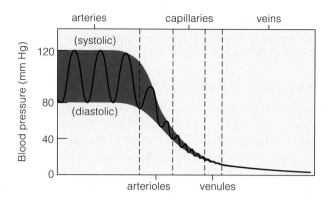

Figure 33.12 Plot of fluid pressure changes as a volume of blood flows through the systemic circuit. Systolic pressure occurs when ventricles contract, diastolic when ventricles are relaxed.

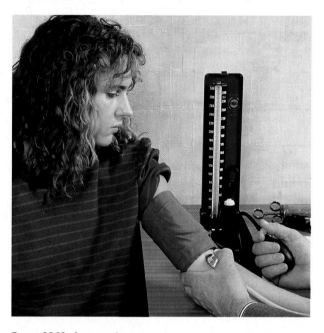

Figure 33.13 Animated Measuring blood pressure. A hollow inflatable cuff attached to a pressure gauge is wrapped around the upper arm. A stethoscope is placed over the brachial artery, just below the cuff.

The cuff is inflated with air to a pressure above the highest pressure of the cardiac cycle, when ventricles contract. Above this pressure, you will not hear sounds through the stethoscope, because no blood is flowing through the vessel.

Air in the cuff is slowly released until the stethoscope picks up soft tapping sounds. Blood flowing into the artery under the pressure of the contracting ventricles—the systolic pressure—causes the sounds. When these sounds start, a gauge typically reads about 120 mm Hg. That amount of pressure will force mercury (Hg) to move up 120 millimeters in a glass column of a standardized diameter.

More air is released from the cuff. Eventually the sounds stop. Blood is now flowing continuously, even when the ventricles are the most relaxed. The pressure when the sounds stop is the lowest during a cardiac cycle, the diastolic pressure, which is usually about 80 mm Hg.

Right, compact monitors are now available that automatically record the systolic/diastolic blood pressure.

> As blood flows through a capillary, it slows down and exchanges substances with interstitial fluid.
< Links to Diffusion and osmosis 5.6, Exocytosis 5.8

Slowdown at Capillaries

As blood flows through a circuit, it moves fastest through arteries, slower in arterioles, and slowest in capillaries. The velocity then picks up a bit as the blood returns to the heart. The slowdown in capillaries occurs because the body has tens of billions of capillaries, and their collective cross-sectional area is far greater than that of the arterioles that deliver blood to them, or the veins that carry blood away. By analogy, think about what happens if a narrow river (representing few larger vessels) delivers water to a wide lake (representing the many capillaries):

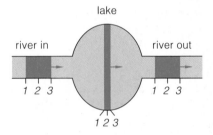

The flow rate is constant, with an identical volume moving from points 1 to 3 in each interval, but flow velocity decreases in the lake. Why? When the volume spreads out through a larger cross-sectional area it flows forward a shorter distance during the specified interval.

Slow flow through narrow vessels enhances the rate of exchanges between the blood and interstitial fluid. The more time blood spends in a capillary, the more time there is for exchanges to take place.

How Substances Cross Capillary Walls

To move between the blood and interstitial fluid, a substance must cross a capillary wall. Oxygen, carbon dioxide, and small lipid-soluble molecules diffuse across endothelial cells of a capillary. Some larger molecules enter endothelial cells by endocytosis, diffuse through the cell, then escape by exocytosis into the interstitial fluid.

Substances also enter the interstitial fluid when a bit of fluid is forced out of capillaries through spaces between cells of the capillary wall. Blood pressure is highest at the arterial end of a capillary bed, and it is here that pressure forces fluid out between cells (Figure 33.14A). The fluid that exits has high levels of oxygen, ions, and nutrients. As blood continues toward the venous end of the capillary, blood pressure falls. Now osmotic pressure is the predominant force. It causes water to move from the interstitial fluid into the hypertonic, protein-rich plasma (Figure 33.14B).

Normally, there is a small net outward flow of fluid from capillaries. The lymphatic system (described in Section 33.11) returns the escaped fluid to the blood. If high blood pressure forces excess fluid out of capillaries, or something prevents fluid return, interstitial fluid pools in tissues. The tissue swelling that results is called edema.

Take-Home Message **How does the blood exchange materials with interstitial fluid?**

> Small molecules cross cells of a capillary by diffusion and larger ones move across by exocytosis.

> Fluid rich in oxygen and nutrients also leaks out between cells of the capillary wall.

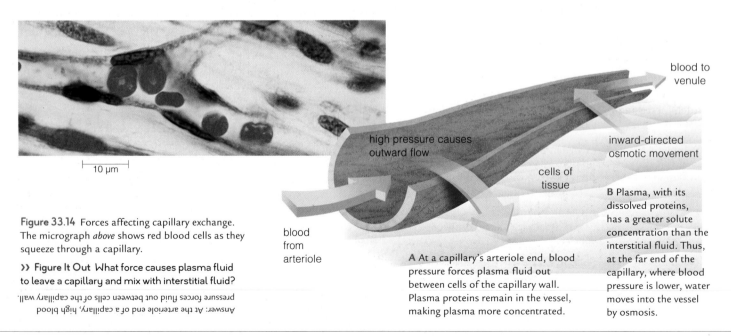

Figure 33.14 Forces affecting capillary exchange. The micrograph *above* shows red blood cells as they squeeze through a capillary.

>> **Figure It Out** What force causes plasma fluid to leave a capillary and mix with interstitial fluid?

Answer: At the arteriole end of a capillary, high blood pressure forces fluid out between cells of the capillary wall.

10 μm

blood to venule

high pressure causes outward flow

inward-directed osmotic movement

cells of tissue

blood from arteriole

A At a capillary's arteriole end, blood pressure forces plasma fluid out between cells of the capillary wall. Plasma proteins remain in the vessel, making plasma more concentrated.

B Plasma, with its dissolved proteins, has a greater solute concentration than the interstitial fluid. Thus, at the far end of the capillary, where blood pressure is lower, water moves into the vessel by osmosis.

lake

river in

river out

1 2 3   1 2 3

1 2 3

> Veins are the body's largest blood reservoir.
> Skeletal muscle activity helps move blood at low pressure through veins and back to the heart.
< Link to Smooth muscle 28.5

Moving Blood to the Heart

Veins carry blood through the final stretch of a circuit and return it to the heart. By the time blood reaches veins, most of the pressure imparted by ventricular contractions has dissipated. Of all blood vessels, veins have the lowest blood pressure. The vein wall can bulge quite a bit under pressure, much more so than an arterial wall. Thus, veins act as reservoirs for great volumes of blood. When you rest, they hold about 60 percent of the total blood volume.

Several mechanisms help blood at low pressure move through veins and back toward the heart. First, veins have flaplike valves that help prevent backflow. These valves automatically shut when blood in the vein starts to reverse direction. For example, the valves in the large veins of the leg prevent blood from moving downward in response to gravity when you stand (Figure 33.15).

Also, smooth muscle inside a vein's wall contracts in response to signals from the nervous system. The contraction causes the vein to stiffen so it cannot hold as much blood, and the pressure inside it rises, forcing blood toward the heart.

Skeletal muscles used in limb movements also help move blood through veins. When these muscles contract, they bulge and press on veins, squeezing blood toward the heart (Figure 33.16).

Exercise-induced deep breathing also raises pressure inside veins. As the lungs and the thoracic cavity expand during inhalation, adjacent organs are forced against veins. As with muscle contraction, pressure from contact forces blood in a vein forward through a valve.

When Venous Flow Slows

Sometimes one or more valves in a vein become damaged, causing blood to accumulate in that vein. Damaged valves in the legs cause varicose veins, in which bulging veins become visible at the skin surface. Damage to valves in the veins of the rectum or anus cause hemorrhoids. High blood pressure raises the risk for valve damage, but there is also a genetic component.

When blood pools in veins because of valve damage or prolonged inactivity, a clot may form in the vein. A clot that forms in a blood vessel and stays there is called a thrombus. A clot that breaks loose and travels through vessels to a new location is an embolus. Both types of clot pose a heath risk because they can slow or halt blood flow. For example, an embolus that blocks blood flow in the brain can cause a stroke, in which brain cells die. An embolus in the lung can also be life-threatening.

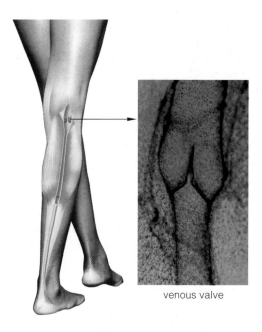
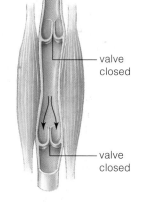

venous valve

Figure 33.15 Valves in veins prevent the backflow of blood.

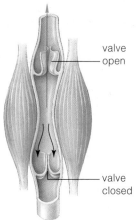

blood flow to heart

valve open

valve closed

valve closed

valve closed

When skeletal muscles contract, they bulge and press on neighboring veins. This puts pressure on the blood in the vein, forcing it forward through the pressure-sensitive valves.

When skeletal muscles relax, the pressure in neighboring veins declines and pressure-sensitive valves shut, preventing blood from moving backward.

Figure 33.16 How skeletal muscle activity encourages blood flow through veins.

Take-Home Message What are the functions of veins?

> Veins are the body's main blood reservoir. The amount of blood residing in the veins is adjusted depending on activity level.

> Blood pressure in veins is low. One-way valves, the activity of skeletal muscle, and respiratory muscle action all help move the blood toward the heart.

> Blood flow keeps cells alive, so disorders that disrupt it have severe impacts on health. Fortunately the risk of many cardiovascular disorders can be lowered by choosing a healthy life-style.

‹ Link to HDL and LDL 3.5

Rhythms and Arrhythmias

Electrocardiograms, or ECGs, record the electrical activity of a beating heart (Figure 33.17A). They can also reveal arrhythmias, which are abnormal heart rhythms (Figure 33.17B–D). Malfunction of the SA node causes arrhythmias.

Bradycardia is a below-average resting cardiac rate. Implanting an artificial pacemaker can speed the heart rate if it falls to the point where slow flow impairs health. However, not all bradycardia is a problem. Athletes often have a low resting heart rate. In response to ongoing exercise, the nervous system has adjusted the firing rate of their cardiac pacemaker downward.

Tachycardia is a faster than normal heart rate. Many people experience palpitations, or occasional episodes of tachycardia. Palpitations can be brought on by stress, drugs such as caffeine, an overactive thyroid, or an underlying heart problem.

Atrial fibrillation is an arrhythmia in which the atria do not contract normally, but instead quiver. This slows blood flow and increases the risk of clot formation. People with atrial fibrillation are often given anticlotting medication to lower their risk of stroke. A stroke is an interruption of blood flow that kills brain cells. Most strokes arise when a clot blocks a blood vessel in the brain.

Ventricular fibrillation is an even more dangerous arrhythmia. Ventricles quiver, and pumping falters or stops, causing loss of consciousness and—if a normal rhythm is not restored—death. A defibrillator often can restore the heart's normal rhythm by resetting the SA node. A person who has had ventricular fibrillation may be treated by implantation of a defibrillator that can restore normal rhythm should another episode occur.

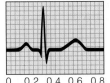

one normal
heartbeat

0 0.2 0.4 0.6 0.8
A time (seconds)

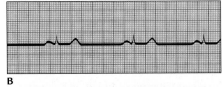

bradycardia (here, 46 beats per minute)

B

tachycardia (here, 136 beats per minute)

C

ventricular fibrillation

D

Figure 33.17 Normal and abnormal ECGs. An ECG uses electrodes placed on the chest to monitor electrical activity of the heart during the cardiac cycle.

Atherosclerosis and Heart Disease

In atherosclerosis, buildup of lipids in the arterial wall narrows the lumen, or space inside the vessel. As you may know, cholesterol plays a role in this "hardening of the arteries." The human body requires cholesterol to make cell membranes, myelin sheaths, bile salts, and steroid hormones (Section 3.4). The liver makes enough cholesterol to meet these needs, but more is absorbed from food in the gut. Genetics affects how different people's bodies deal with an excess of dietary cholesterol.

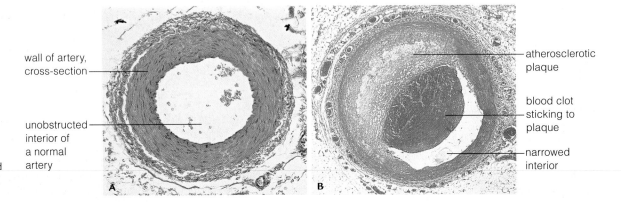

wall of artery, cross-section

unobstructed interior of a normal artery

atherosclerotic plaque

blood clot sticking to plaque

narrowed interior

Figure 33.18 Sections from **A** a normal artery and **B** an artery with a lumen narrowed by an atherosclerotic plaque. A clot clogged this one.

Most of the cholesterol dissolved in blood is bound to protein carriers. The complexes are known as low-density lipoproteins, or LDLs, and most cells can take them up. A lesser amount is bound up in high-density lipoproteins, or HDLs. Cells in the liver metabolize HDLs, using them in the formation of bile, which the liver secretes into the gut. Eventually, the bile leaves the body in the feces.

When the LDL level in blood rises, so does the risk of atherosclerosis. The first sign of trouble is a buildup of lipids in an artery's endothelial lining. Fibrous connective tissue proliferates in the affected area. Eventually, a mass, called an atherosclerotic plaque, bulges into the vessel's interior, narrowing its diameter and slowing blood flow (Figure 33.18). A hardened plaque can abrade an artery wall, thereby triggering clot formation.

With heart disease, atherosclerosis affects vessels that supply blood to heart muscle, the vessels that are shown in Figure 33.19. A heart attack occurs when a coronary artery is completely blocked, most commonly by a clot. If the blockage is not removed fast, cardiac muscle cells die. Clot-dissolving drugs can restore blood flow if they are given within an hour of the onset of an attack, so a suspected heart attack should receive prompt attention.

In coronary bypass surgery, doctors open a person's chest and use a blood vessel from elsewhere in the body (usually a leg vein) to divert blood around the clogged coronary artery (Figure 33.20A). In laser angioplasty, laser beams vaporize plaques. In balloon angioplasty, doctors inflate a small balloon in a blocked artery to flatten the plaques. A wire mesh tube called a stent is then inserted to keep the vessel open (Figure 33.20B).

Risk Factors

Cardiovascular disorders are the leading cause of death in the United States, where such disorders kill about one million people a year. Tobacco smoking tops the list of risk factors. Other factors include a family history of such disorders, hypertension, a high cholesterol level, diabetes mellitus, and obesity. Physical inactivity increases the risk. Regular exercise helps lower the risk of cardiovascular disorders even when the exercise is not particularly strenuous. Gender and age also play a role. Until about age fifty, males are at greater risk. In both sexes, the risk of cardiovascular disorders increases with age.

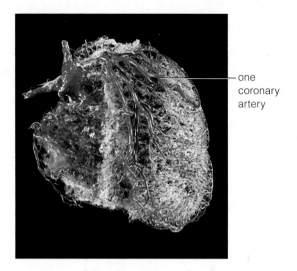

one coronary artery

Figure 33.19 Blood vessels that service the heart. To make this three-dimensional cast, resins were injected into vessels, then the cardiac tissues were dissolved.

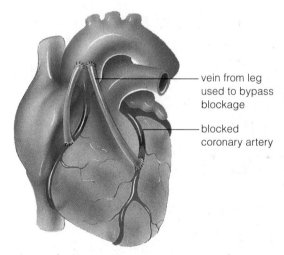

vein from leg used to bypass blockage

blocked coronary artery

A Coronary bypass surgery. Veins from another part of the body are used to divert blood past the blockages. This illustration shows a "double bypass," in which veins are placed to divert blood around two blocked coronary arteries.

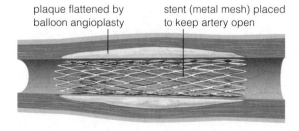

plaque flattened by balloon angioplasty

stent (metal mesh) placed to keep artery open

B Ballon angioplasty and placement of a stent. A balloonlike device is inflated in an artery to open it and flatten the plaque, then a tube of metal (the stent) is left in place to keep the artery open.

Figure 33.20 Two ways of treating blocked coronary arteries, the main cause of heart attacks.

Take-Home Message **What types of disorders affect the cardiovascular system?**

❯ Problems with the cardiac pacemaker cause arrhythmias.

❯ Atherosclerosis narrows blood vessels, increasing the risk of heart attack and stroke.

> Excess fluid that leaves capillaries returns to blood by way of the lymphatic system. This system also plays a major role in immunity, a topic we turn to in the next chapter.
< Link to Thymus gland 31.10

Lymph Vascular System

A portion of the lymphatic system, the **lymph vascular system**, consists of vessels that collect water and solutes from interstitial fluid, then deliver them to the circulatory system. The lymph vascular system includes lymph capillaries and vessels (Figure 33.21A). Fluid that moves through these vessels is the **lymph**.

The lymph vascular system serves three functions. First, its vessels are drainage channels for plasma fluid that leaked out of capillaries and must be returned to the circulatory system. Second, it delivers fats absorbed by the small intestine to the blood. Third, it transports cellular debris, pathogens, and foreign cells to the lymph nodes that serve as the system's disposal sites.

Tonsils
Defense against bacteria and other foreign agents

Right lymphatic duct
Drains right upper portion of the body

Thymus gland
Site where certain white blood cells acquire means to chemically recognize specific foreign invaders

Thoracic duct
Drains most of the body

Spleen
Major site of antibody production; disposal site for old red blood cells and foreign debris; site of red blood cell formation in the embryo

Some of the lymph vessels
Return excess interstitial fluid and reclaimable solutes to the blood

Some of the lymph nodes
Filter bacteria and many other agents of disease from lymph

Bone marrow
Marrow in some bones is production site for infection-fighting blood cells (as well as red blood cells and platelets)

A

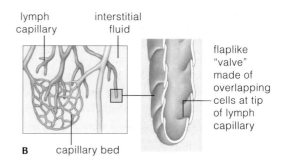

lymph capillary   interstitial fluid

flaplike "valve" made of overlapping cells at tip of lymph capillary

B   capillary bed

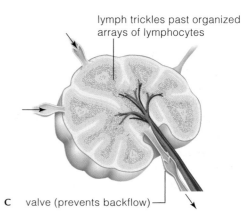

lymph trickles past organized arrays of lymphocytes

C   valve (prevents backflow)

Figure 33.21 Animated A Components of the human lymphatic system and their functions. Not shown are patches of lymphoid tissue in the small intestine and in the appendix. **B** Diagram of lymph capillaries at the start of a drainage network, the lymph vascular system. **C** Cutaway view of a lymph node. Its inner compartments are packed with organized arrays of infection-fighting white blood cells.

Lymph capillaries lie near blood capillaries (Figure 33.21B). Fluid that leaks out of blood capillaries and does not reenter them moves into lymph capillaries through clefts between cells of the lymph capillary wall.

Lymph capillaries merge into larger diameter lymph vessels. Two mechanisms move lymph through these vessels. First, slow wavelike contractions of smooth muscle in the walls of large lymph vessels propel lymph forward. Second, as with veins, the bulging of adjacent skeletal muscles helps move fluid along. Like veins, lymph vessels have one way valves that prevent backflow.

The largest lymph vessels converge on collecting ducts that empty into veins in the lower neck. Each day these ducts deliver about 3 liters of fluid to the circulation.

Lymphoid Organs and Tissues

The other portion of the lymphatic system has roles in the body's defense responses to injury and attack. We call its components lymphoid organs and tissues. These components include the lymph nodes, spleen, and thymus, as well as the tonsils, and some patches of tissue in the wall of the small intestine and appendix.

Lymph nodes are strategically located at intervals along lymph vessels (Figure 33.21C). Before entering blood, lymph trickles through at least one node and gets filtered. Masses of lymphocytes that formed in the bone marrow take up stations inside the nodes. When something identified as nonself reaches a node, lymphocytes divide rapidly and form armies that destroy that threat.

The **spleen** is the largest lymphoid organ, about the size of a fist in an average adult. In embryos only, it functions as a site of red blood cell formation. After birth, the spleen filters pathogens and worn-out red cells and platelets from the blood vessels that branch through it. Phagocytic white blood cells that reside in the spleen engulf and digest altered body cells and alert the immune system to threats. The spleen also holds antibody-producing cells. People can survive without a spleen; it is often removed after being damaged in highway accidents. However, lack of a spleen makes a person more vulnerable to infections.

The thymus is central to immunity. We discussed its hormone secreting capacity in Section 31.10. T lymphocytes, a type of white blood cell, must travel through the thymus to differentiate and become capable of recognizing and responding to particular pathogens.

---

lymph Fluid in the lymph vascular system.
lymph node Small mass of lymphatic tissue through which lymph filters; contains many lymphocytes (B and T cells).
lymph vascular system System of vessels that takes up interstitial fluid and carries it (as lymph) to the blood.
spleen Large lymphoid organ that filters blood.

▉ And Then My Heart Stood Still (revisited)

Saving lives just keeps getting easier. Traditional CPR alternates blowing into a person's mouth to inflate their lungs and compressing the chest. The requirement for mouth-to-mouth contact makes many people reluctant to use this method on strangers. A new method called CCR (cardiocerebral resuscitation) relies on chest compressions alone. This method may be as good as or even better than traditional CPR to treat most people with sudden cardiac arrest or a heart attack.

Increasing availability of automated external defibrillators (AEDs) also helps save lives (Figure 33.22). Having AEDs in airports, shopping malls, schools, gyms, and other sites ensures that a person who suffers a cardiac arrest does not have to wait for an ambulance to receive life-saving defibrillation.

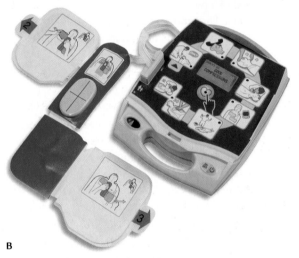

Figure 33.22 Automatic external defibrillators. **A** Signs in public places indicate where AEDs are available. **B** An AED is designed to be used by nonmedical personel. Voice commands and pictures instruct the user in how and where to place the electric paddles. The device then assesses the need for defibrillation and, if necessary, delivers an electric shock.

How Would You Vote? Knowing how to perform CPR and use an AED can save lives. Should learning these skills be a standard part of high school curricula? See CengageNow for details, then vote online (cengagenow.com).

Take-Home Message **What are the functions of the lymphatic system and how does it interact with the circulatory system?**

❯ The lymphatic system supports the circulatory system and helps defend the body against pathogens.

❯ The lymph vascular portion of the system consists of many tubes that collect and deliver excess water and solutes from interstitial fluid to blood. It also delivers absorbed fats to the blood, and delivers disease agents to lymph nodes.

❯ The system's lymphoid organs, including lymph nodes, have specific roles in body defenses.

Summary

Section 33.1 When the heart stops pumping, blood flow halts and cells begin to die from lack of oxygen. CPR can keep some oxygenated blood moving to cells, but it cannot restart a heart. Reestablishing the normal rhythm requires a shock from a defibrillator.

Section 33.2 A **circulatory system** moves substances through a body by way of a fluid transport medium called **blood**. Some invertebrates have an **open circulatory system**; their blood leaves vessels and seeps around tissues. In vertebrates, a **closed circulatory system** confines blood inside a **heart** and blood vessels. All exchanges between blood and interstitial fluid take place across the walls of **capillaries**.

In fish, blood flows through a single circuit. Evolution of a two-circuit system accompanied the evolution of lungs. The **pulmonary circuit** moves blood to the lungs and back. The longer **systemic circuit** moves blood to other body tissues and back.

Section 33.3 Humans have a closed, two-circuit circulatory system. Each circuit consists of **arteries** that move blood from the heart to **arterioles**. The largest artery in the body is the **aorta**. Arterioles supply capillaries. Capillaries drain into **venules** that feed into veins. **Veins** carry blood back to the heart. Most blood flows through only one capillary system, but blood in intestinal capillaries will later flow through liver capillaries. The liver stores nutrients and neutralizes some bloodborne toxins.

Section 33.4 A human heart is a double pump partitioned into two halves, each with an **atrium** above a **ventricle**. The **superior vena cava** and **inferior vena cava** are large veins that fill the right atrium. Blood from the right atrium flows into the right ventricle, which pumps blood into **pulmonary arteries** that carry it to the lungs. Blood returns to the heart's left atrium through **pulmonary veins**. The left atrium empties into the left ventricle, which pumps blood into the aorta and through the systemic circuit.

During one **cardiac cycle**, all heart chambers undergo rhythmic relaxation (**diastole**) and contraction (**systole**). Contraction of the ventricles alone provides the force that powers movement of blood through blood vessels. Contraction of atria only fills the ventricles.

The **sinoatrial (SA) node** in the right atrium wall is the cardiac pacemaker. Action potentials spread across the atria by way of gap junctions. They spread to ventricles through the **atrioventricular (AV) node**, which serves as an electrical bridge. The delay between atrial contraction in response to SA signaling and ventricular contraction in response to signals arriving via the AV node allows the ventricles to fill fully before they contract.

Section 33.5 Blood consists of **plasma**, blood cells, and platelets. Plasma is mostly water. **Red blood cells** contain hemoglobin that functions in oxygen transport. **White blood cells** have roles in day-to-day tissue maintenance and repair and in defenses against pathogens. The number of cells in a given volume is the **cell count**. **Platelets** and **fibrin** act in **hemostasis** (clotting). Platelets and all blood cells arise from stem cells in bone.

Sections 33.6–33.9 **Blood pressure** results from the force exerted by ventricular contraction. It declines as blood proceeds through a circuit and is usually recorded as **systolic pressure** over **diastolic pressure**. Thick-walled arteries smooth out pressure variations. A **pulse** is a brief expansion of an artery caused by ventricular contraction. Arterioles are the main site for regulation of flow through the systemic circuit. **Vasodilation** of arterioles supplies more blood to a region. **Vasoconstriction** decreases the blood supply.

Capillaries, the site of exchanges with interstitial fluid, are one cell layer thick. Blood flow slows in capillaries because of their collectively large cross-sectional area. Substances leave a capillary by diffusion, by exocytosis, or in fluid that seeps out between cells. Fluid that seeps out of a capillary at the arterial end is balanced by osmotic uptake of water nearer the vein end.

Veins return blood to the heart. Blood pressure in veins is low, but one-way valves, action of skeletal muscles, and pressure exerted during respiration keep blood moving.

Section 33.10 Abnormal heart rhythms can slow or halt blood flow. Flow is also impaired when atherosclerosis narrows the interior of a blood vessel. A healthy lifestyle can help prevent atherosclerosis.

Section 33.11 The lymphatic system interacts with the circulatory system. The **lymph vascular system** takes up excess water from interstitial fluid, as well as fats absorbed from the gut, and delivers them to blood. It also delivers bloodborne pathogens to lymph nodes. **Lymph nodes** filter the **lymph**, and white blood cells in the nodes attack any pathogens. The **spleen** filters the blood and removes old red blood cells. The thymus gland is a hormone-secreting organ inside which T lymphocytes (a kind of white blood cell) mature.

Self-Quiz   Answers in Appendix III

1. All vertebrates have _____ .
   a. an open circulatory system
   b. a two-chambered heart
   c. lungs
   d. none of the above

The Stroke Belt Epidemiologists refer to a swath of states in the Southeast as the "stroke belt" because of the unusually high incidence of stroke deaths there. By one hypothesis, the high rate of deaths from stroke in this region results largely from a lack of access to immediate medical care. Compared to other parts of the country, more stroke-belt residents live in rural settings with few medical services. Figure 33.23 compares the rate of stroke deaths in stroke-belt states (Alabama, Arkansas, Georgia, Mississippi, North Carolina, South Carolina, and Tennessee) with that of New York State. It also breaks down the death risk in each region by ethnic group and by sex.

1. How does the rate of stroke deaths among blacks living in the stroke-belt compare with whites in the same region?
2. How does the rate of stroke deaths among blacks living in New York compare with whites in the same region?
3. Which group has the higher rate of stroke deaths, blacks living in New York, or whites living in the stroke belt?
4. Do these data support the hypothesis that poor access to care causes the high rate of death by stroke in the stroke belt?

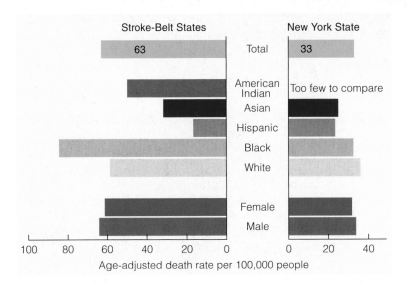

Figure 33.23 Comparison of the age-adjusted rate of deaths by stroke in southeastern "stroke-belt" states and in New York State. *Source: National Vital Statistics System—Mortality (NVSS-M), NCHS, CDC.*

---

2. In _____ blood flows through two completely separate circuits.
   a. birds   b. mammals   c. fish   d. both a and b

3. The _____ circuit carries blood to and from lungs.

4. The fluid portion of blood is _____ .

5. Platelets function in _____ .
   a. oxygen transport   c. thermal regulation
   b. hemostasis   d. both a and b

6. Most oxygen in blood is transported _____ .
   a. in red blood cells   c. bound to hemoglobin
   b. in white blood cells   d. both a and c

7. Blood flows directly from the left atrium to _____ .
   a. the aorta   c. the right atrium
   b. the left ventricle   d. the pulmonary arteries

8. Contraction of _____ drives the flow of blood through the aorta and pulmonary arteries.
   a. atria   b. ventricles

9. Blood pressure is highest in the _____ and lowest in the _____ .
   a. arteries; veins   c. veins; arteries
   b. arterioles; venules   d. capillaries; arterioles

10. At rest, the largest volume of blood is in _____ .
    a. arteries   b. veins   c. capillaries   d. arterioles

11. Which human heart chamber has the thickest wall?

12. A pulmonary _____ carries oxygen-poor blood.

13. Lymph nodes filter _____ .
    a. blood   c. plasma
    b. lymph   d. all of the above

14. Which is more dangerous?
    a. atrial fibrillation   b. ventricular fibrillation

15. Match the components with their functions.
    ___ capillary   a. filters out pathogens
    ___ lymph node   b. cardiac pacemaker
    ___ atrium   c. vessels with valves
    ___ ventricle   d. largest artery
    ___ SA node   e. thin-walled heart chamber
    ___ veins   f. exchange site
    ___ aorta   g. contractions drive
                            blood circulation

Additional questions are available on **CENGAGENOW**.

Critical Thinking

1. Some studies suggest that long-distance flights raise the risk of thrombus formation. Physicians suggest that air travelers drink plenty of fluids and periodically walk around the cabin. Explain why these precautions lower the risk of clot formation.

2. Cardiac muscle has more mitochondria than skeletal muscle or smooth muscle. Why might cardiac muscle have higher energy requirements than other muscle?

Animations and Interactions on **CENGAGENOW**:
❯ Animal circulatory systems; Human cardiovascular system; Human heart; Cardiac cycle; Cardiac conduction; Measuring blood pressure; Lymphatic system.

‹ Links to Earlier Concepts

This chapter revisits bacteria and viruses (Sections 4.5, 19.2) and their hosts (19.1). Concepts of cell structure and function (4.4, 4.8, 4.11, 5.8, 7.6), osmosis (5.6), cDNA (15.2), cell signaling (28.9, 31.2), and genetics (13.5) are important to understanding how the body fights invaders of the internal environment (28.2). Epithelia (28.3) and skin (28.8), and the circulatory and lymphatic systems (33.5, 33.8, 33.11), come up again in the context of diseases (15.10, 29.9, 33.10).

Key Concepts

Immune Defenses
The vertebrate body has three lines of immune defenses: surface barriers, and innate and adaptive immunity. White blood cells and signaling molecules function in all immune responses.

Surface Barriers
External surfaces of the body come into constant contact with microbial pathogens. Physical, mechanical, and chemical barriers prevent most microbes from entering body tissues.

34 Immunity

34.1 Frankie's Last Wish

In October of 2000, Frankie McCullough had known for a few months that something was not quite right. She had not had an annual checkup in many years; after all, she was only 31 and had been healthy her whole life. It never occurred to her to doubt her own invincibility until the moment she saw the doctor's face change as he examined her cervix.

The cervix is the lowest part of the uterus, or womb. Cervical cells can become cancerous, but the process is usually slow. The cells pass through several precancerous stages that are detectable by routine Pap tests. Precancerous and even early-stage cancerous cells can be removed from the cervix before they spread to other parts of the body. However, plenty of women like Frankie do not take advantage of regular exams. Those who end up at the gynecologist's office with pain or bleeding may be experiencing symptoms of advanced cervical cancer, the treatment of which offers only about a 9 percent chance of survival. About 3,600 women die of cervical cancer each year in the United States. Many more than that die in places where routine gynecological testing is not common.

What causes cancer? At least in the case of cervical cancer, we know the answer to that question: Healthy cervical cells are transformed into cancerous ones by infection with human papillomavirus (HPV). HPV is a DNA virus that infects skin and mucous membranes. There are about 100 different types of HPV; a few cause warts on the hands or feet, or in the mouth. About 30 others that infect the genital area sometimes cause genital warts, but usually there are no symptoms of infection. Genital HPV is spread very easily by sexual contact. At least 80 percent of women have been infected with HPV by the age of 50.

A genital HPV infection usually goes away on its own, but not always. A persistent infection with one of about 10 strains is the main risk factor for cervical cancer (Figure 34.1). Types 16 and 18 are particularly dangerous: One of the two is found in more than 70 percent of all cervical cancers.

In 2006, the FDA approved Gardasil, a vaccine against four types of genital HPV, including types 16 and 18. The vaccine prevents cervical cancer caused by these HPV strains. It is most effective in girls who have not yet become sexually active, because they are least likely to have become infected with HPV.

Figure 34.1 HPV and cervical cancer.

Opposite, a Pap test reveals HPV-infected cervical cells among normal ones. Infected cells have enlarged, often multiple nuclei surrounded by a clear area. These changes sometimes lead to cervical cancer, which is treatable if detected early enough.

Left, Frankie McCullough (waving) died of cervical cancer in 2001.

The HPV vaccine came too late for Frankie McCullough. Despite radiation treatments and chemotherapy, her cervical cancer spread quickly. She died in 2001, leaving a wish for other people: awareness. "If there is one thing I could tell a young woman to convince them to have a yearly exam, it would be not to assume that your youth will protect you. Cancer does not discriminate; it will attack at random, and early detection is the answer." She was right. Almost all women newly diagnosed with invasive cervical cancer have not had a Pap test in five years. Many have never had one.

Pap tests, HPV vaccines, and all other medical tests and treatments are direct benefits of our increasing understanding of the interplay of the human body with its pathogens.

Innate Immunity
Innate immune responses involve a set of general, immediate defenses. Phagocytic white blood cells, plasma proteins, inflammation, and fever quickly rid the body of most invaders.

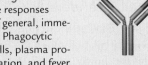

Adaptive Immunity
In an adaptive immune response, white blood cells interact to destroy specific pathogens or altered cells. Antibodies and other antigen receptors are central to these responses.

Immunity in Our Lives
Vaccines are an important part of any health program. Allergies, immune deficiencies, and autoimmune disorders are the result of faulty immune mechanisms or infections.

> In vertebrates, the innate and adaptive immune systems work together to combat infection and injury.
< Links to Phagocytosis 5.8, Coevolution of pathogens and hosts 19.1, Intercellular signaling 31.2, Blood cells 33.5

Evolution of the Body's Defenses

Humans continually cross paths with a tremendous array of viruses, bacteria, fungi, parasitic worms, and other pathogens, but you need not lose sleep over this. Humans coevolved with these pathogens, so you have defenses that protect your body from them.

The evolution of **immunity**, an organism's capacity to resist and combat infection, began well before multicelled eukaryotes evolved from free-living cells. Mutations in the genes for membrane proteins introduced new molecular patterns that were unique in cells of a given type. As multicellularity evolved, so did mechanisms of identifying the patterns as self, or belonging to one's own body.

By 1 billion years ago, nonself recognition had also evolved. Cells of all modern multicelled eukaryotes bear a set of receptors that collectively can recognize around 1,000 different nonself cues, which are called pathogen-associated molecular patterns (PAMPs). As their name suggests, PAMPs occur mainly on or in pathogens. They include some components of bacterial cell walls, flagellum and pilus proteins, double-stranded RNA unique to some viruses, and so on. When a cell's receptors bind to a PAMP, they trigger a set of immediate, general defense responses. In mammals, for example, binding triggers activation of complement. **Complement** is a set of proteins that circulate in inactive form throughout the body. Activated complement can destroy invading cells or mark them for uptake by phagocytic cells.

Pattern receptors and the responses they initiate are part of **innate immunity**, a set of fast, general defenses against infection. All multicelled organisms start out life with these defenses, which normally do not change within the individual's lifetime.

Vertebrates have another set of defenses carried out by interacting cells, tissues, and proteins. This **adaptive**

Figure 34.2 One physical barrier to infection: mucus and the mechanical action of cilia keep pathogens from getting a foothold in the airways to the lungs. Bacteria and other particles get stuck in mucus secreted by goblet cells (*gold*). Cilia (*pink*) on other cells sweep the mucus toward the throat for disposal.

immunity tailors immune defenses to a vast array of specific pathogens that an individual may encounter during its lifetime. Adaptive immunity is triggered by **antigen**, which is a PAMP or any other molecule or particle recognized by the body as nonself. Table 34.1 compares adaptive and innate immunity.

Three Lines of Defense

The mechanisms of adaptive immunity evolved within the context of innate immunity. The two systems were once thought to operate independently of each other, but we now know they function together. Thus, we describe both systems together in terms of three lines of defense. The first line includes the physical, chemical, and mechanical barriers that keep pathogens on the outside of the body (Figure 34.2). Innate immunity, the second line of defense, begins after tissue is damaged, or after antigen is detected inside the body. Its general response mechanisms quickly rid the body of many invaders. Activation of innate immunity triggers the third line of defense, adaptive immunity. Leukocytes, or white blood cells, divide to form huge populations that target a specific antigen and destroy anything bearing it. Some of the white blood cells persist after infection ends. If the same antigen returns, these memory cells mount a secondary response.

The Defenders

White blood cells carry out all immune responses. Many kinds circulate through the body in blood and lymph; a few are shown in Figure 34.3. Others populate the lymph nodes, spleen, and other tissues. Some white blood cells

Table 34.1	Innate and Adaptive Immunity Compared	
	Innate Immunity	**Adaptive Immunity**
Response time	Immediate	About a week
How antigen is detected	Fixed set of receptors for pathogen-associated molecular patterns (PAMPs)	Antigen receptors produced by gene recombinations
Specificity	About 1,000 PAMPs	Billions of antigens
Persistence	None	Long-term

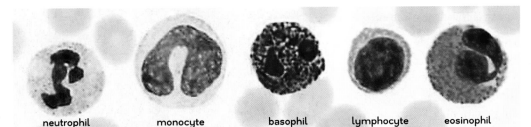

Figure 34.3 Lineup of leukocytes (white blood cells). Staining shows structural details such as cytoplasmic granules that contain enzymes, toxins, and signaling molecules.

neutrophil   monocyte   basophil   lymphocyte   eosinophil

are phagocytic, which means they move about and engulf other cells. All are secretory. The secretions include **cytokines**, which are polypeptides and proteins used by cells of the immune system to communicate with one another. Intercellular communication allows white blood cells to coordinate their activities during immune responses. Vertebrate cytokines include interleukins, interferons, and tumor necrosis factors.

Different types of white blood cells are specialized for specific tasks, such as phagocytosis. **Neutrophils** are the most abundant of the circulating phagocytic cells. Phagocytic **macrophages** that patrol tissue fluids are mature monocytes, which patrol the blood. **Dendritic cells** are phagocytes that alert the adaptive immune system to the presence of antigen in solid tissues.

Some white blood cells have granules, which are secretory vesicles that contain cytokines, destructive enzymes, toxins, and local signaling molecules. The cell releases the contents of its granules (degranulates) in response to a trigger such as binding to antigen. **Eosinophils** target parasites too big for phagocytosis. **Basophils** and **mast cells**

release substances contained by their granules in response to injury or antigen. Unlike most other leukocytes, mast cells do not wander; they stay anchored in tissues (Figure 34.4). Mast cells are often closely associated with nerves, and degranulate in response to somatostatin and other polypeptides active in the endocrine and nervous systems.

Lymphocytes are a special category of white blood cells that are central to adaptive immunity. **B cells** (B lymphocytes) and **T cells** (T lymphocytes) have the collective capacity to recognize billions of specific antigens. There are several kinds of T cells. One type, the **cytotoxic T cell**, can kill infected or cancerous body cells. Lymphocytes called **NK cells** (natural killer cells) kill cancerous body cells that are undetectable by cytotoxic T cells.

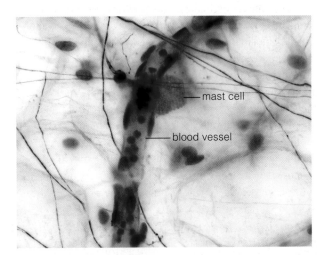

— mast cell

— blood vessel

Figure 34.4 Mast cells. Anchored in tissues near blood vessels, nerves, and mucous membranes that border external surfaces, mast cells degranulate in response to injury or antigen, and also to signals from the nervous system. The micrograph shows a mast cell in loose connective tissue.

adaptive immunity In vertebrates, set of immune defenses that can be tailored to specific pathogens encountered by an organism during its lifetime.
antigen A molecule or particle that the immune system recognizes as nonself. Triggers an immune response.
B cell B lymphocyte. Lymphocyte that can make antibodies.
basophil Circulating white blood cell; role in inflammation.
complement A set of proteins that circulate in inactive form in blood as part of innate immunity.
cytokines Signaling molecules secreted by vertebrate immune cells.
cytotoxic T cell Lymphocyte that kills infected or cancerous cells.
dendritic cell Phagocytic white blood cell that patrols tissue fluids; main type of antigen-presenting cell.
eosinophil White blood cell that targets multicelled parasites.
immunity The body's ability to resist and fight infections.
innate immunity Set of inborn, fixed general defenses against infection.
macrophage Phagocytic white blood cell that patrols tissue fluids.
mast cell White blood cell that is anchored in many tissues; factor in inflammation.
NK cell Natural killer cell. Lymphocyte that can kill cancer cells undetectable by cytotoxic T cells.
neutrophil Circulating phagocytic white blood cell.
T cell T lymphocyte. Lymphocyte central to adaptive immunity; some kinds target sick body cells.

Take-Home Message What is immunity?

❯ The innate immune system is a set of general defenses against a fixed number of antigens. It does not change during an individual's lifetime.

❯ Vertebrate adaptive immunity is a system of defenses that can specifically target billions of different antigens.

❯ White blood cells are central to both systems; signaling molecules such as cytokines integrate their activities.

❯ A pathogen can cause infection only if it enters the internal environment by penetrating skin or other protective barriers at the body's surfaces.

❮ Links to Bacteria and biofilms 4.5, Tight junctions 4.11, Fermentation 7.6, Internal environment 28.2, Epithelium 28.3, Hair follicles and skin 28.8, Atherosclerosis 33.10

Your skin is in constant contact with the external environment, so it picks up many microorganisms. It normally teems with about 200 different kinds of yeast, protozoa, and bacteria. If you showered today, there are probably thousands of microorganisms on every square inch of your external surfaces. If you did not, there may be billions. They tend to flourish in warm, moist areas, such as between the toes. Huge populations inhabit cavities and tubes that open out on the body's surface, including the eyes, nose, mouth, and anal and genital openings.

Microorganisms that typically live on human surfaces, including the interior tubes and cavities of the digestive and respiratory tracts, are called **normal flora** (Figure 34.5). Our surfaces provide them with a stable environment and nutrients. In return, their populations deter more dangerous species from colonizing (and penetrating) body surfaces; help us digest food; and make nutrients that we depend on, including a vitamin (B_{12}) made only by bacteria.

Normal flora are helpful only on body surfaces; they can cause or worsen many conditions when they invade tissues. Consider a major constituent of normal flora, *Propionibacterium acnes* (Figure 34.5B), a bacterium that feeds on sebum. Sebum is a greasy mixture of fats, waxes, and glycerides that lubricates hair and skin. Glands in the skin secrete sebum into hair follicles. During puberty, higher levels of steroid hormones trigger an increase in sebum production. Excess sebum combines with dead, shed skin cells and blocks the openings of hair follicles. *P. acnes* can survive on the surface of the skin, but far

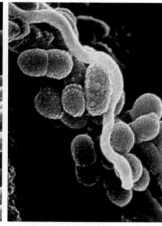

Figure 34.6 Plaque. *Left*, micrograph of toothbrush bristles scrubbing plaque on a tooth surface. *Right*, one of the bacterial species in plaque, *Streptococcus mutans*, is a major contributor to tooth decay and periodontitis.

prefers anaerobic habitats such as the interior of blocked hair follicles. There, they multiply to tremendous numbers. Secretions of the flourishing *P. acnes* populations leak into internal tissues, which causes inflammation in the tissue around the follicles. The resulting pustules are called acne.

A few of the 400 or so species of normal flora in the mouth cause dental **plaque**, a thick biofilm of various bacteria and occasional archaeans, their extracellular products, and saliva glycoproteins. Plaque sticks tenaciously to teeth (Figure 34.6). Some of the bacteria in plaque are fermenters. The lactic acid they produce dissolves minerals that make up the tooth, resulting in holes called cavities.

In young, healthy people, tight junctions normally seal gum epithelium to teeth. The tight seal prevents oral microorganisms from entering gum tissue. As we age, the connective tissue beneath the epithelium thins, so the seal between gums and teeth weakens. Deep pockets form, and

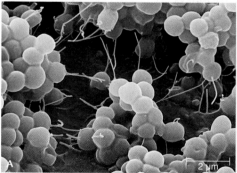
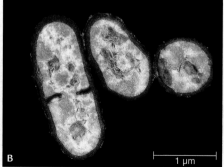
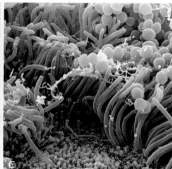

Figure 34.5 Examples of normal flora. **A** *Staphylococcus epidermidis*, a common colonizer of human skin. **B** *Propionibacterium acnes*, the bacterial cause of acne.

C *Staphylococcus aureus* cells (*yellow*) stuck in mucus secreted by human nasal epithelial cells. *S. aureus* is a common inhabitant of human skin and linings of the mouth, nose, throat, and intestines. It is also the leading cause of bacterial disease in humans.

Table 34.2	Examples of Surface Barriers
Physical	Intact skin and epithelia that line tubes and cavities such as the gut and eye sockets; established populations of normal flora
Mechanical	Mucus; broomlike action of cilia; flushing action of tears, saliva, urination, diarrhea
Chemical	Secretions (sebum, other waxy coatings); low pH of urine, gastric juices, urinary and vaginal tracts; lysozyme

a nasty collection of anaerobic bacteria and archaeans tends to accumulate in them. The microorganisms secrete destructive enzymes and acids that cause inflammation of the surrounding gum tissues, a condition called periodontitis. *Porphyromonas gingivalis* is one of those anaerobic species. Along with every other species of oral bacteria associated with periodontitis, *P. gingivalis* is also found in atherosclerotic plaque. Periodontal wounds are an open door to the circulatory system and its arteries. Atherosclerosis is now known to be a disease of inflammation. What role oral microorganisms play in atherosclerosis is not yet clear, but one thing is certain—they contribute to the inflammation that fuels coronary artery disease.

Other serious illnesses associated with normal flora include pneumonia; ulcers; colitis; whooping cough; meningitis; abscesses of the lung and brain; and colon, stomach, and intestinal cancers. The bacterial agent of tetanus, *Clostridium tetani*, passes through our intestines so often that it is considered a normal inhabitant. The bacteria responsible for diphtheria, *Corynebacterium diphtheriae*, was normal skin flora before widespread use of the vaccine eradicated the disease. *Staphylococcus aureus*, a resident of human skin and linings of the mouth, nose, throat, and intestines (Figure 34.5C), is also a leading cause of human bacterial disease. Antibiotic-resistant strains of *S. aureus* are now widespread. A particularly dangerous kind, MRSA (methicillin-resistant *S. aureus*), is resistant to a wide range of antibiotics. MRSA is now a permanent resident of most hospitals around the world.

Barriers to Infection

In contrast to body surfaces, blood and tissue fluids of healthy people are typically sterile (free of microorganisms). Surface barriers (Table 34.2) usually prevent normal flora from entering the body's internal environment.

lysozyme Antibacterial enzyme that occurs in body secretions such as mucus.
normal flora Microorganisms that typically live on human surfaces, including the interior tubes and cavities of the digestive and respiratory tracts.
plaque On teeth, a thick biofilm composed of bacteria, their extracellular products, and saliva proteins.

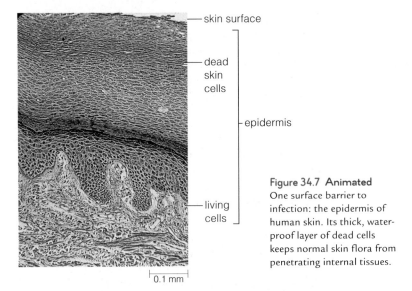

Figure 34.7 Animated One surface barrier to infection: the epidermis of human skin. Its thick, waterproof layer of dead cells keeps normal skin flora from penetrating internal tissues.

0.1 mm

Epidermis, the tough outer layer of vertebrate skin, is one example of a surface barrier (Figure 34.7). Microorganisms flourish on skin's waterproof, oily surface, but they rarely penetrate the thick epidermis.

The thinner epithelial tissues that line the body's interior tubes and cavities also have surface barriers. Sticky mucus secreted by cells of these linings can trap microorganisms. The mucus contains **lysozyme**, an enzyme that kills bacteria. In the sinuses and respiratory tract, the coordinated beating of cilia sweeps trapped microorganisms away before they have a chance to breach the delicate walls of these structures.

Microorganisms that normally inhabit the mouth resist lysozyme in saliva. Those that are swallowed are typically killed by gastric fluid, a potent brew of protein-digesting enzymes and acid in the stomach. Those that survive to reach the small intestine are usually killed by bile salts. The hardy ones that make it to the large intestine must compete with about 500 resident species that are specialized to live there and have already established large populations. Any that displace normal flora are typically flushed out by diarrhea.

Lactic acid produced by *Lactobacillus* helps keep the vaginal pH outside the range of tolerance of most fungi and other bacteria. Urination's flushing action usually stops pathogens from colonizing the urinary tract.

Take-Home Message What prevents ever-present microorganisms from entering the body's internal environment?

> Surface barriers keep microorganisms that contact or inhabit vertebrate surfaces from invading the internal environment.

> Skin's tough epidermis is one barrier. Mucus, lysozyme, and often the sweeping action of cilia protect the softer linings of internal tubes and cavities.

> Resident populations of normal flora deter more dangerous microorganisms from colonizing the body's internal and external surfaces.

> Animals are born with innate immunity, which includes phagocytosis, complement, inflammation, and fever.
> Innate immune mechanisms are fast, general defenses against invading microorganisms.
< Links to Osmosis 5.6, Lysis 19.2, Effectors 28.9, Local signaling molecules 31.2, Blood 33.5, Capillary function 33.8

What happens if a pathogen slips by surface defenses and enters the body's internal environment? All animals are normally born with a set of fast-acting, off-the-shelf immune defenses that can keep an invading pathogen from establishing a population in the body's internal environment. These innate immune defenses include phagocyte and complement action, inflammation, and fever. All are general defense mechanisms that normally do not change over an individual's lifetime.

Phagocytes and Complement

Macrophages engulf and digest essentially everything except undamaged body cells (Figure 34.8A). They patrol interstitial fluid, and are often the first white blood cells to encounter an invading pathogen. When receptors on a macrophage bind antigen, the cell begins to secrete cytokines. The signaling molecules attract more macrophages, neutrophils, and dendritic cells to the site of invasion.

Antigen also triggers the activation of complement. In vertebrates, about 30 different types of complement protein circulate in inactive form throughout the body in blood and interstitial fluid. Some become activated when they encounter antigen, or an antibody bound to antigen (we will return to antibodies in the next section). Activated complement proteins activate other complement proteins, which activate other complement proteins, and so on. These cascading reactions quickly produce huge concentrations of activated complement at the site of invasion. The proteins attach directly to antigen, and typically form a coating on invading cells.

Activated complement is like a dinner invitation to phagocytic leukocytes. These cells follow complement gradients back to the affected tissue. Phagocytes have complement receptors, so a pathogen coated with complement is recognized and engulfed faster than an uncoated pathogen. Other activated complement proteins assemble into complexes that puncture cell walls or plasma membranes of foreign cells (Figure 34.8B,C).

Activated complement proteins also work in adaptive immunity, by guiding the maturation of immune cells and mediating some interactions among them.

Inflammation

Activated complement and cytokines trigger **inflammation**, which is a local response to tissue damage or infection (Figure 34.9 ❶). Inflammation begins when pattern receptors on basophils, mast cells, or neutrophils bind to antigen, or when mast cells directly bind to activated complement. In response to the binding, the leukocytes release the contents of their granules into the affected tissue ❷.

Among the substances released by mast cells are prostaglandins and histamines. These local signaling molecules have two effects. First, they cause nearby arterioles to widen. As a result, blood flow to the area increases. The increased flow speeds the arrival of more phagocytes, which are attracted to the cytokines. Second, the signaling molecules widen the spaces between cells in capillary walls, so they make capillaries in an affected tissue leakier to plasma proteins. Phagocytes can squeeze through leaky capillary walls, out of the blood vessel and into interstitial fluid ❸. By the time this occurs, any invading cells have become coated with activated complement ❹, which makes them easy targets for the phagocytes ❺.

The symptoms of inflammation include redness and warmth that are outward indications of the area's increased blood flow. Swelling and pain occur because

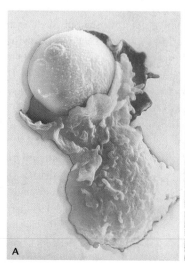

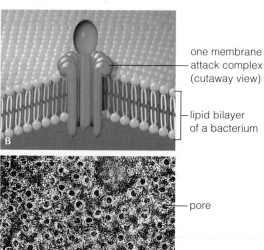

one membrane attack complex (cutaway view)

lipid bilayer of a bacterium

pore

Figure 34.8 Two mechanisms of adaptive immunity, phagocytosis and complement activation.

A Macrophage caught in the act of phagocytosis.

B Activated complement assembles into membrane attack complexes that insert themselves into the lipid bilayer of foreign cells. The resulting pores **C** bring about lysis of the cell.

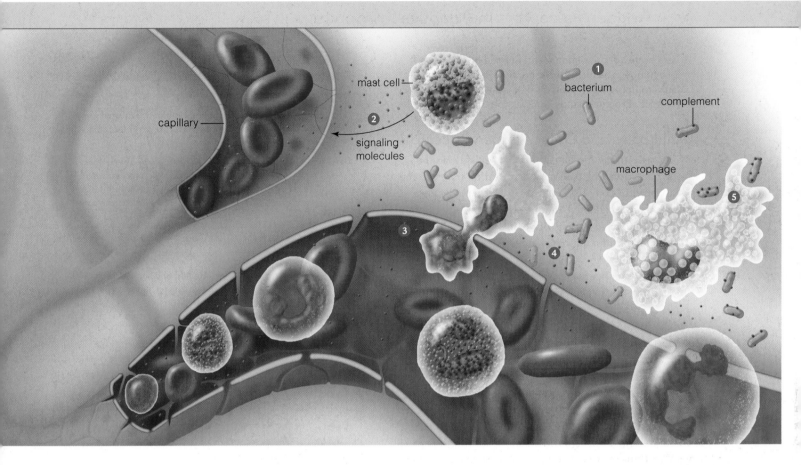

Figure 34.9 Animated Inflammation, an innate immune response to a bacterial infection.

1 Bacteria invade a tissue.

2 Pattern receptors on mast cells in the tissue recognize and bind to bacterial antigen. The mast cells release signaling molecules (*blue* dots) that cause arterioles to widen. The resulting increase in blood flow reddens and warms the tissue.

3 The signaling molecules also increase capillary permeability, which allows phagocytes to squeeze through the vessel walls into the tissue. Plasma proteins leak out of the capillaries, and the tissue swells with fluid.

4 Bacterial antigens activate complement (*purple* dots). Activated complement binds to the bacteria.

5 Phagocytes in the tissue recognize and engulf the complement-coated bacteria.

plasma proteins that escape from the leaky capillaries make interstitial fluid hypertonic with respect to blood. More water diffuses into the tissue and makes it swell. The swelling puts pressure on nerves and causes pain.

Fever

Fever is a temporary rise in body temperature above the normal 37°C (98.6°F) that often occurs in response to infection. Some cytokines stimulate brain cells to make and release prostaglandins, which act on the hypothalamus to raise the body's internal temperature set point. As long as the temperature of the body is below the new set point, the hypothalamus sends out signals that cause blood vessels in the skin to constrict, which reduces heat loss from the skin. The signals also trigger reflexive movements called shivering, or "chills," that increase the metabolic heat output of muscles. Both responses raise the body's internal temperature.

Fever enhances immune defenses by increasing the rate of enzyme activity, thus speeding up metabolism, tissue repair, and the formation and activity of phagocytes. Some pathogens multiply more slowly at the higher tem-

perature, so white blood cells can get a head start in the proliferation race against them.

A fever is a sign that the body is fighting something, so it should never be ignored. However, a fever of 40.6°C (105°F) or less does not necessarily require treatment in an otherwise healthy adult. Body temperature usually will not rise above that value, but if it does, immediate hospitalization is recommended because a fever of 42°C (107.6°F) can result in brain damage or death.

fever An internally induced rise in core body temperature above the normal set point as a response to infection.
inflammation A local response to tissue damage or infection; characterized by redness, warmth, swelling, and pain.

Take-Home Message What is innate immunity?

❯ Innate immunity is the body's built-in set of general immune defenses.

❯ Complement, phagocytes, inflammation, and fever quickly eliminate most invaders from the body.

❯ Vertebrate adaptive immunity is defined by self/nonself recognition, specificity, diversity, and memory.
❮ Links to Recognition proteins 4.4, Lysosomes 4.8, Endocytosis 5.8, Exocrine glands 28.3, Lymphatic system 33.11

Antibodies and Other Antigen Receptors

If innate immune mechanisms do not quickly rid the body of an invading pathogen, an infection may become established in body tissues. By that time, long-lasting mechanisms of adaptive immunity have already begun to target the invaders specifically.

Innate immune responses are triggered by leukocytes that detect antigen via their antigen receptors. What, exactly, is an antigen receptor? Your T cells bear one type, antigen receptors called **T cell receptors**, or TCRs. Part of a TCR recognizes antigen as nonself. Another part recognizes proteins in the plasma membrane of body cells as self. In humans, these recognition proteins are called **MHC markers**, after the genes that encode them.

Antibodies are another type of antigen receptor. **Antibodies** are Y-shaped proteins made only by B cells. Each antibody can bind to the antigen that prompted its synthesis. Many antibodies circulate in blood and enter interstitial fluid during inflammation, but they do not kill pathogens directly. Instead, they activate complement, facilitate phagocytosis, prevent pathogens from attaching to body cells, and neutralize toxins.

An antibody molecule consists of four polypeptides: two identical "light" chains and two identical "heavy" chains (Figure 34.10A). Each chain has a variable and a constant region. When the chains fold up together as an intact antibody, the variable regions form two antigen-binding sites that have a specific distribution of bumps, grooves, and charge. These binding sites are the antigen receptor part of an antibody: They bind only to antigen with a complementary distribution of bumps, grooves, and charge (Figure 34.10B). In addition to antigen-binding

sites, each antibody also has a constant region that determines its structural identity, or class. There are five antibody classes—IgG, IgA, IgE, IgM, and IgD (Ig stands for immunoglobulin, another name for antibody). The different classes serve different functions (Table 34.3).

Most of the antibodies circulating in the bloodstream and tissue fluids are IgG, which binds pathogens, neutralizes toxins, and activates complement. IgG is the only

Table 34.3 Structural Classes of Antibodies

Secreted antibodies

IgG		Main antibody in blood; activates complement, neutralizes toxins; protects fetus and is secreted in early milk.
IgA		Abundant in exocrine gland secretions (e.g., tears, saliva, milk, mucus), where it occurs in dimeric form (*shown*). Interferes with binding of pathogens to body cells.

Membrane-bound antibodies

IgE		Anchored to surface of basophils, mast cells, eosinophils, and some dendritic cells. IgE binding to antigen induces anchoring cell to release histamines and cytokines. Factor in allergies and asthma.
IgD		B cell receptor.
IgM		B cell receptor, as a monomer. Also is secreted as pentamer (group of five, *shown*).

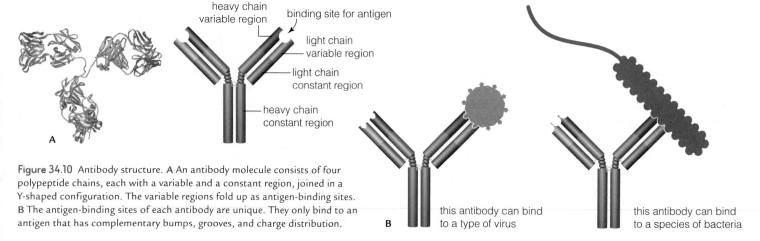

Figure 34.10 Antibody structure. **A** An antibody molecule consists of four polypeptide chains, each with a variable and a constant region, joined in a Y-shaped configuration. The variable regions fold up as antigen-binding sites. **B** The antigen-binding sites of each antibody are unique. They only bind to an antigen that has complementary bumps, grooves, and charge distribution.

heavy chain variable region
binding site for antigen
light chain variable region
light chain constant region
heavy chain constant region

B this antibody can bind to a type of virus

this antibody can bind to a species of bacteria

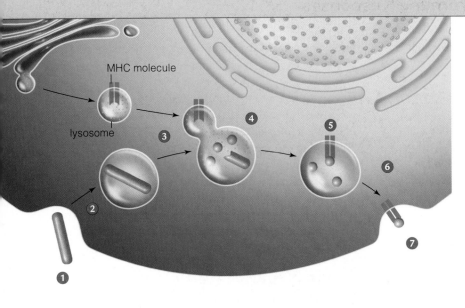

Figure 34.11 Antigen processing. What happens when a B cell, macrophage, or dendritic cell engulfs an antigenic particle—in this case, a bacterium.

1. A phagocytic cell engulfs a bacterium.

2. An endocytic vesicle forms around the bacterium.

3. The vesicle fuses with a lysosome, which contains enzymes and MHC molecules.

4. Lysosomal enzymes digest the bacterium to molecular bits.

5. Inside the vesicle, bits of bacterium bind to MHC molecules.

6. The vesicle fuses with the cell membrane by exocytosis. When it does, the bacterial antigen–MHC complex becomes part of the plasma membrane.

7. The antigen–MHC complex displayed on the surface of the phagocyte is a signal that provokes an adaptive immune response targeting the antigen.

≫ **Figure It Out** From what organelles do the lysosomes bud?

Answer: From Golgi bodies

antibody that can cross the placenta to protect a fetus before its own immune system is active. IgA is the main antibody in mucus and other exocrine gland secretions (Section 28.3). IgA is secreted as a dimer (two antibodies bound together), which makes the molecule stable enough to patrol harsh environments such as the interior of the digestive tract. There, IgA encounters pathogens before they contact body cells. Bound to antigen, IgA interacts with mast cells, basophils, macrophages, and NK cells to initiate inflammation. IgE made and secreted by B cells gets incorporated into the plasma membrane of mast cells, basophils, and some types of dendritic cells. Binding of antigen to membrane-bound IgE triggers the anchoring cell to release the contents of its granules. **B cell receptors** are IgM or IgD antibodies bound to B cell membranes. Secreted IgM pentamers (polymers of five) efficiently bind antigen and activate complement.

Antigen Receptor Diversity

Most humans can make billions of unique antigen receptors. This diversity arises because the genes that encode the receptors do not occur in a continuous stretch on one chromosome; instead, they occur in several segments on different chromosomes, and there are several different versions of each segment. The segments are spliced together during B and T cell differentiation, but which version of each segment gets spliced into the antigen receptor gene of a particular cell is random. As a B or T cell differentiates, it ends up with one out of about 2.5 billion different combinations of gene segments.

antibody Y-shaped antigen receptor protein made only by B cells.
B cell receptor Membrane-bound antibody on a B cell.
MHC markers Self-recognition protein on the surface of body cells. Triggers adaptive immune response when bound to antigen fragments.
T cell receptor (TCR) Antigen receptor on the surface of a T cell.

Before a new B cell leaves bone marrow, it already is making its unique antigen receptors. The constant region of each receptor is embedded in the lipid bilayer of the cell's plasma membrane, and the two arms project above the membrane. In time the B cell bristles with more than 100,000 antigen receptors. T cells also form in bone marrow, but they mature only after they take a tour in the thymus gland (Section 33.11). There, they encounter hormones that stimulate them to make T cell receptors.

Antigen Processing

Recognizing a specific antigen is the first step of the adaptive immune response. A new B or T cell is "naive," which means that no antigen has bound to its receptors yet. Once the cell binds to an antigen, it begins to divide by mitosis, and tremendous populations of clones form.

T cell receptors do not recognize antigen unless it is presented by an antigen-presenting cell. Macrophages, B cells, and dendritic cells do the presenting (Figure 34.11). First, one of these cells engulfs something bearing antigen 1. A vesicle that contains the antigen-bearing particle forms in the cell's cytoplasm 2 and fuses with a lysosome 3. Lysosomal enzymes then digest the particle into molecular bits 4. The lysosomes also contain MHC markers that bind to some of the antigen bits 5. The resulting antigen–MHC complexes become displayed at the cell's surface when the vesicles fuse with (and become part of) the plasma membrane 6. The display of MHC markers paired with antigen fragments serves as a call to arms 7.

Take-Home Message **What are antigen receptors?**

❯ The adaptive immune system has the potential to recognize billions of different antigens via receptors on B cells and T cells.

❯ Antibodies are secreted or membrane-bound antigen receptors. They are made only by B cells. Different classes of antibodies serve different functions.

❯ Vertebrate adaptive immunity is defined by self/nonself recognition, specificity, diversity, and memory.
❮ Link to Lymphatic system 33.11

Unlike innate immunity, the adaptive immune system changes: It "adapts" to different antigens an individual encounters during its lifetime. Lymphocytes and phagocytes interact to effect the four defining characteristics of adaptive immunity, which are self/nonself recognition, specificity, diversity, and memory.

Self/Nonself Recognition is based on the ability of T cell receptors to recognize self in the form of MHC markers. TCRs and other antigen receptors recognize nonself, in the form of antigen.

Specificity means the adaptive immune response can be tailored to combat specific antigens.

Diversity refers to the antigen receptors on a body's collection of B and T cells. There are potentially billions of different antigen receptors, so an individual has the potential to counter billions of different threats.

Memory refers to the capacity of the adaptive immune system to "remember" an antigen. It takes about a week for for B and T cells to respond in force the first time they encounter an antigen. If the same antigen shows up again, they make a faster, stronger response.

First Step: Antigen Alert

Recognizing a specific antigen is the first step of the adaptive immune response. Typically, a naive T cell recognizes and binds to an antigen–MHC complex displayed on the surface of a white blood cell. Once bound, the T cell starts secreting cytokines, which signal all other B or T cells with the same antigen receptor to divide again and again. Huge populations of B and T cells form after a few days; all of the cells recognize the same antigen. Most are **effector cells**, differentiated lymphocytes that act at once. Some are **memory cells**, long-lived B and T cells reserved for future encounters with the antigen. Memory cells can persist for decades after the initial infection ends. If the same antigen enters the body at a later time, these memory cells will initiate a secondary response (Figure 34.12). In a secondary immune response, larger populations of effector cell clones form much more quickly than they did in the primary response.

Two Arms of Adaptive Immunity

Like a boxer's one-two punch, adaptive immunity has two separate arms: the antibody-mediated and the cell-mediated immune responses (Figure 34.13). These two responses work together to eliminate diverse threats.

Why two arms? Not all threats present themselves in the same way. For example, bacteria, fungi, or toxins can circulate in blood or interstitial fluid. These threats are intercepted quickly by B cells and other phagocytes that interact in an **antibody-mediated immune response**. In this response, B cells produce antibodies targeting a specific invader.

antibody-mediated immune response Immune response in which antibodies are produced in response to an antigen.
cell-mediated immune response Immune response involving cytotoxic T cells and NK cells that destroy infected or cancerous body cells.
effector cell Antigen-sensitized B cell or T cell that forms in an immune response and acts immediately.
memory cell Long-lived, antigen-sensitized B or T cell that can act in a secondary immune response.

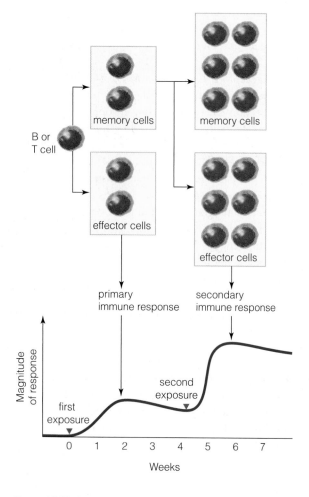

Figure 34.12 Animated Primary and secondary immune responses.

A first exposure to an antigen causes a primary immune response in which effector cells fight the infection.

Memory cells also form in a primary response but are set aside, sometimes for decades. If the antigen returns at a later time, the memory cells initiate a faster, stronger secondary response.

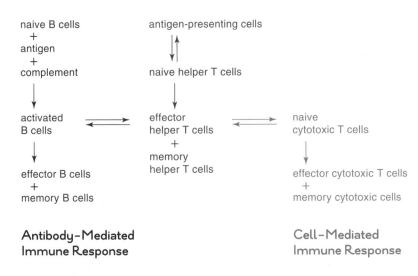

naive B cells
+
antigen
+
complement

↓

activated
B cells

↓

effector B cells
+
memory B cells

**Antibody–Mediated
Immune Response**

antigen-presenting cells

⇅

naive helper T cells

↓

effector
helper T cells
+
memory
helper T cells

naive
cytotoxic T cells

↓

effector cytotoxic T cells
+
memory cytotoxic cells

**Cell–Mediated
Immune Response**

Figure 34.13 Overview of key interactions between antibody-mediated and cell-mediated responses—the two arms of adaptive immunity.

An antibody-mediated immune response is not the most effective way of countering some types of threats. For example, viruses, bacteria, fungi, and protists that reproduce inside body cells are vulnerable to an antibody-mediated response only when they slip out of one cell to infect others. Such intracellular pathogens are targeted primarily by the **cell–mediated immune response**. In this response, cytotoxic T cells and NK cells detect and destroy infected body cells, or those that have been altered by cancer.

Intercepting and Clearing Out Antigen

After engulfing an antigen-bearing particle, a dendritic cell or macrophage migrates to a lymph node, where it presents antigen to T cells. Every day, about 25 billion T cells filter through each node. As you will see shortly, T cells that recognize and bind to antigen presented by a phagocyte initiate an adaptive response.

Antigen-bearing particles in interstitial fluid flow through lymph vessels to a lymph node, where they meet up with resident B cells, dendritic cells, and macrophages (Figure 34.14). These phagocytes engulf, process, and present antigen to T cells passing through the node. During an infection, the lymph nodes swell because T cells accumulate inside of them. When you are ill, you may notice your swollen lymph nodes as tender lumps under the jaw or elsewhere in your body.

The tide of battle turns when the effector cells and their secretions destroy most antigen-bearing agents. With less antigen present, fewer immune fighters are recruited. Complement proteins assist in the cleanup by binding

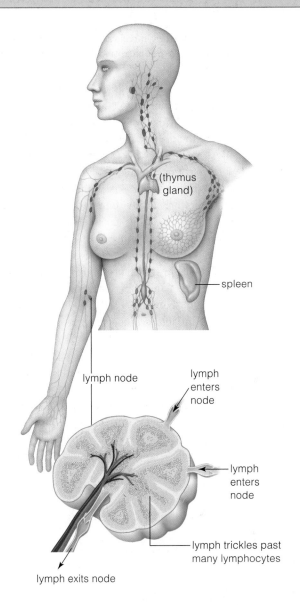

(thymus gland)

spleen

lymph node

lymph enters node

lymph enters node

lymph trickles past many lymphocytes

lymph exits node

Figure 34.14 Animated Battlegrounds of the adaptive immune system. Macrophages, dendritic cells, and B cells present antigen to T cells filtering through lymph nodes. The spleen filters antigenic particles from the blood.

antibody–antigen complexes, forming large clumps that can be quickly cleared from the blood by the liver and spleen. Immune responses subside after the antigen-bearing particles have been cleared from the body.

Take-Home Message **What is an adaptive immune response?**

❯ Phagocytes and lymphocytes interact to bring about vertebrate adaptive immunity, which has four defining characteristics: self/nonself recognition, specificity, diversity, and memory.

❯ The two arms of adaptive immunity work together. Antibody-mediated responses target antigen in blood or interstitial fluid; cell-mediated responses target infected or cancerous body cells.

> In an antibody-mediated immune response, effector B cells form and produce antibodies targeting a specific antigen.

If we liken B cells to assassins, then each one has a genetic assignment to liquidate one particular target: an antigen-bearing extracellular pathogen or toxin. Antibodies are their molecular bullets, as the following example illustrates (Figure 34.15). Suppose that you accidentally nick your finger. Being opportunists, some *Staphylococcus aureus* cells on your skin immediately enter the cut, invading your internal environment. Complement in interstitial fluid quickly attaches to carbohydrates in the bacterial cell walls, and cascading complement activation reactions begin. Within an hour, complement-coated bacteria tumbling along in lymph vessels reach a lymph node in your elbow. There, they filter past an army of naive B cells.

One of the naive B cells residing in that lymph node makes antigen receptors that recognize a polysaccharide in *S. aureus* cell walls ❶. Via those receptors, the B cell binds to the polysaccharide on one of the bacteria. The complement coating stimulates the B cell to engulf the bacterium. The B cell is now activated.

Meanwhile, more *S. aureus* cells have been secreting metabolic products into interstitial fluid around your cut. The secretions are attracting phagocytes. One of the phagocytes, a dendritic cell, engulfs several bacteria, then migrates to the lymph node in your elbow. By the time it gets there, it has digested the bacteria and is displaying their fragments as antigens bound to MHC markers on its surface ❷.

Every hour, about 500 different naive T cells travel through the lymph node, inspecting resident dendritic cells displaying antigen. Within a couple of hours, one of your T cells has recognized and bound to the *S. aureus* antigen on the dendritic cell ❸. This T cell is called a helper T cell because it helps other lymphocytes produce antibodies and kill pathogens. The helper T cell and the dendritic cell interact for about 24 hours. When the two cells disengage, the helper T cell returns to the circulatory system and begins to divide. A gigantic population of identical helper T cells forms. These clones differentiate into effector and memory cells, each of which has receptors that recognize the same *S. aureus* antigen.

By the theory of clonal selection, the T cell was "selected" because its receptors bind to the *S. aureus* antigen. T cells with receptors that do not bind the antigen do not divide to form huge clonal populations (Figure 34.16).

Let's go back to that B cell in the lymph node. By now, it has digested the bacterium, and it is displaying bits of *S. aureus* bound to MHC molecules on its plasma membrane. The new helper T cells recognize the antigen–MHC complexes displayed by the B cell. One of these helper

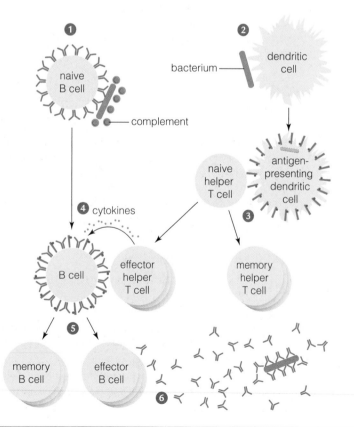

Figure 34.15 Animated Example of an antibody-mediated immune response.

❶ The B cell receptors on a naive B cell bind to an antigen on the surface of a bacterium. The bacterium's complement coating triggers the B cell to engulf it. Fragments of the bacterium bound to MHC markers become displayed at the surface of the B cell.

❷ A dendritic cell engulfs the same kind of bacterium that the B cell encountered. Fragments of the bacterium bound to MHC markers become displayed at the surface of the dendritic cell.

❸ The antigen–MHC complexes on the dendritic cell are recognized by TCRs on a naive helper T cell. The two cells interact, and then the T cell begins to divide. Its descendants differentiate into effector helper T cells and memory helper T cells.

❹ TCRs on one of the effector helper T cells recognize and bind to the antigen–MHC complexes on the B cell. Binding makes the T cell secrete cytokines.

❺ The cytokines induce the B cell to divide. Its descendants differentiate into effector B cells and memory B cells.

❻ The effector B cells begin making and secreting huge numbers of antibodies, all of which recognize the same antigen as the original B cell receptor. The new antibodies circulate throughout the body and bind to any remaining bacteria.

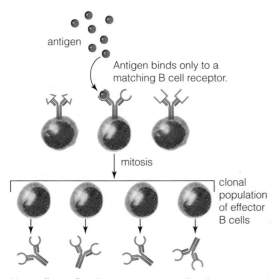

antigen

Antigen binds only to a matching B cell receptor.

mitosis

clonal population of effector B cells

Many effector B cells secrete many antibodies.

Figure 34.16 Animated Clonal selection. Only lymphocytes with receptors that bind to antigen divide and differentiate. This example shows clonal selection of B cells.

T cells binds to the B cell. Like long-lost friends, the two cells stay together for a while and communicate ④.

One of the messages that is communicated consists of cytokines secreted by the helper T cell. The cytokines stimulate the B cell to begin mitosis after the two cells disengage. The B cell divides again and again to form a huge population of genetically identical cells, all with receptors that can bind to the same *S. aureus* antigen ⑤.

The B cell clones differentiate into effector and memory cells. The effector B cells start working immediately. Instead of making membrane-bound B cell receptors, they switch classes and start making secreted antibodies ⑥. The new antibodies recognize the same *S. aureus* antigen as the original B cell receptor. Antibodies now circulate throughout the body and attach themselves to any remaining bacterial cells. An antibody coating prevents bacteria from attaching to body cells and brings them to the attention of phagocytic cells for quick disposal. Antibodies also glue the foreign cells together into clumps, a process called **agglutination**. The clumps are quickly removed from the circulatory system by the spleen.

---

agglutination The clumping together of foreign cells bound by antibodies; the clumps attract phagocytic cells.

Take-Home Message **What happens during an antibody–mediated immune response?**

❯ Antigen-presenting cells, T cells, and B cells carry out an antibody-mediated immune response.

❯ During an antibody-mediated immune response, populations of B cells form; these cells make and secrete antibodies that recognize and bind antigen.

34.8 Blood Typing

❯ Antigens on red blood cells are the basis of blood typing.
❮ Links to Membrane proteins 4.4, ABO genetics 13.5

As you learned in Section 13.5, a carbohydrate on red blood cell membranes occurs in two forms. This carbohydrate is called H antigen. People with one form of the H antigen have type A blood; people with the other form have type B blood. People with both forms have type AB blood; those with neither are type O.

Early in life, each individual starts making antibodies that recognize molecules foreign to the body, including any nonself form of the H antigen:

ABO Type	H Antigen Form on Red Cells	Antibodies Present
A	A	anti-B
B	B	anti-A
AB	both A and B	none
O	neither A nor B	anti-A, anti-B

These antibodies are especially important if a blood transfusion becomes necessary. A transfusion of incompatible red blood cells can cause a potentially fatal transfusion reaction in which the recipient's antibodies recognize and bind to antigens on the transfused cells. The binding activates complement, which punctures the cell membranes. Massive amount of hemoglobin released from the destroyed blood cells can very quickly cause the kidneys to fail.

Identifying red blood cell surface antigens helps prevent pairing of incompatible transfusion donors and recipients, and also alerts physicians to blood incompatibility problems that may arise during pregnancy. A typical blood typing test involves mixing drops of a patient's blood with antibodies to the different forms of red blood cell antigens. Agglutination occurs when the cells bear antigens recognized by the antibodies (Figure 34.17).

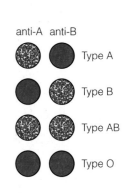

anti-A anti-B

Type A

Type B

Type AB

Type O

anti-Rh D

Rh⁻

Rh+

Figure 34.17 Animated Blood typing test. Such tests determine blood type by mixing samples of a patient's blood with antibodies to molecules on red blood cells.

Right, an agglutination test shows the presence or absence of H antigens and a membrane protein called Rh D.

❯❯ **Figure It Out** A person with what blood type can receive a transfusion of blood from anyone else?   Answer: Type AB, Rh+

Take-Home Message **Why do people get their blood type tested?**

❯ The presence or absence of different antigens on a person's red blood cells determines their blood type.

❯ Receiving a transfusion of blood of an incompatible type is dangerous because a person makes antibodies to molecules not present on their own blood cells.

❯ In a cell-mediated immune response, cytotoxic T cells and NK cells are stimulated to kill infected or altered body cells.
❮ Link to Apoptosis 28.9

If B cells are like assassins, then cytotoxic T cells are specialists in cell-to-cell combat. Antibody-mediated immune responses target pathogens that circulate in blood and interstitial fluid, but they are not as effective against pathogens inside cells. As part of a cell-mediated immune response, cytotoxic T cells kill ailing body cells that may be missed by an antibody-mediated response.

Ailing body cells typically display certain antigens. For example, cancer cells display altered body proteins, and body cells infected with intracellular pathogens display polypeptides of the infecting agent. Both types of cell are detected and killed by cytotoxic T cells. Cytotoxic T cells also recognize foreign body cells (these T cells are responsible for rejection of transplanted organs).

A typical cell-mediated response begins in interstitial fluid during inflammation when a dendritic cell recognizes, engulfs, and digests a sick body cell or the remains of one (Figure 34.18). The dendritic cell begins to display antigen that was part of the sick cell, and migrates to the spleen or a lymph node. There, the dendritic cell presents its antigen–MHC complexes to huge populations of naive helper T cells and naive cytotoxic T cells ❶. Some of the naive cells have T cell receptors that recognize the complexes on the dendritic cell. Helper T cells ❷ and cytotoxic T cells ❸ that bind the antigen–MHC complexes displayed by the dendritic cell become activated.

The activated helper T cells divide, and their descendants differentiate into effector and memory helper T cells. The effector cells immediately begin to secrete cytokines ❹. Activated cytotoxic T cells recognize the cytokines as a signal to divide and differentiate, and tremendous populations of effector and memory cytotoxic T cells form. All of them recognize and bind the same antigen—the one displayed by that first ailing cell.

The effector cytotoxic T cells start working immediately. They circulate throughout blood and interstitial fluid, and bind to any other body cell displaying the original antigen together with MHC markers ❺. After it is bound to an ailing cell, a cytotoxic T cell releases perforin and proteases (Figure 34.19). These toxins poke holes in the sick cell and cause it to die by apoptosis.

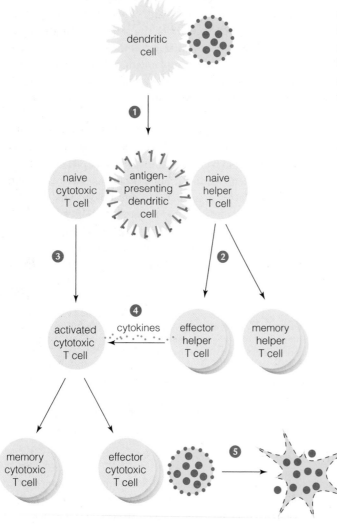

Figure 34.18 Animated A cell-mediated immune response.

❶ A dendritic cell engulfs a virus-infected cell. Digested fragments of the virus bind to MHC markers, and the complexes become displayed at the dendritic cell's surface. The dendritic cell, now an antigen-presenting cell, migrates to a lymph node.

❷ Receptors on a naive helper T cell bind to antigen–MHC complexes on the dendritic cell. The interaction activates the helper T cell, which then begins to divide.

A large population of descendant cells forms. Each cell has T cell receptors that recognize the same antigen. The cells differentiate into effector and memory cells.

❸ Receptors on a naive cytotoxic T cell bind to the antigen–MHC complexes on the surface of the dendritic cell. The interaction activates the cytotoxic T cell.

❹ The activated cytotoxic T cell recognizes cytokines secreted by the effector helper T cells as signals to divide. A large population of descendant cells forms. Each cell bears T cell receptors that recognize the same antigen. The cells differentiate into effector and memory cytotoxic T cells.

❺ The new effector cytotoxic T cells circulate through the body. They recognize and kill any body cell that displays the viral antigen–MHC complexes on its surface.

❯❯ **Figure It Out** What do the large red spots represent?

Answer: Viruses

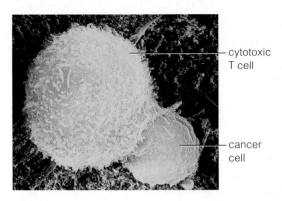

Figure 34.19 Cytotoxic T cell caught in the act of killing a cancer cell.

As occurs in an antibody-mediated response, memory cells form in a primary cell-mediated response. These long-lasting cells do not act immediately. If the antigen returns at a later time, the memory cells will mount a faster, stronger secondary response.

In order to kill a body cell, cytotoxic T cells must recognize the MHC molecules on the surface of the cell (Figure 34.20). However, some infections or cancer can alter a body cell so that it is missing part or all of its MHC markers. NK cells are crucial for fighting such cells. Unlike cytotoxic T cells, NK cells can kill body cells that lack

MHC markers. Cytokines secreted by helper T cells also stimulate NK cell division. The resulting populations of effector NK cells attack body cells tagged by antibodies for destruction. They also recognize certain proteins displayed by body cells that are under stress. Stressed body cells with normal MHC markers are not killed; only those with altered or missing MHC markers are destroyed.

Figure 34.20 T cell receptor function. A TCR (green) on a T cell binds to an MHC marker (tan) on an antigen-presenting cell. An antigen (red) is bound to the MHC marker.

Take-Home Message What happens during a cell-mediated immune response?

> Antigen-presenting cells, T cells, and NK cells interact in a cell-mediated immune response targeting infected body cells or those that have been altered by cancer.

> An allergy is an immune response to something that is ordinarily harmless to most people.

In millions of people, exposure to harmless substances stimulates an immune response. Any substance that is ordinarily harmless yet provokes such responses is an **allergen**. Sensitivity to an allergen is called an **allergy**. Drugs, foods, pollen, dust mites, fungal spores, poison ivy, and venom from bees, wasps, and other insects are among the most common allergens.

Some people are genetically predisposed to having allergies. Infections, emotional stress, and changes in air temperature can trigger reactions. A first exposure to an allergen stimulates B cells to make and secrete IgE, which becomes anchored to mast cells and basophils. With later exposures, antigen binds to the IgE. Binding triggers the anchoring cell to release the contents of its granules, and inflammation is initiated. If the antigen is detected by mast cells in the lining of the respiratory tract, a copious amount of mucus is secreted and the airways constrict; sneezing, stuffed-up sinuses, and a drippy nose result. Contact with an allergen that penetrates the skin's outer layers causes the skin to redden, swell, and become itchy.

Antihistamines relieve allergy symptoms by acting on histamine receptors to dampen the effects of released histamines. Other drugs can inhibit mast cell degranulation, thus preventing the release of histamines.

In people who are hypersensitive to an allergen, a subsequent exposure can cause a severe, whole-body allergic reaction called anaphylactic shock. Huge amounts of cytokines and histamines released in all parts of the body provoke an immediate, systemic reaction. Too much fluid leaks from blood into tissues, so blood pressure drops too

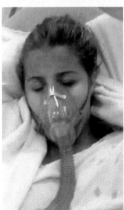

much (a reaction called shock) and tissues swell dramatically. The swelling tissue constricts airways and may block them. Anaphylactic shock is rare but life-threatening and requires immediate treatment (the girl pictured on the *left* is being treated for anaphylactic shock). It may occur at any time, upon exposure to even a tiny amount of allergen. Risks include any prior allergic reaction.

allergen A normally harmless substance that provokes an immune response in some people.
allergy Sensitivity to an allergen.

Take-Home Message What is an allergy?

> Allergens are normally harmless substances that induce an immune response. Sensitivity to an allergen is called an allergy.

❯ Vaccines are designed to elicit immunity to a disease.

Immunization refers to processes designed to induce immunity. In active immunization, a preparation that contains antigen—a **vaccine**—is administered orally or injected. The first immunization elicits a primary immune response, just as an infection would. A second immunization, or booster, elicits a secondary immune response for enhanced immunity.

In passive immunization, a person receives antibodies purified from the blood of another individual. The treatment offers immediate benefit for someone who has been exposed to a potentially lethal agent, such as tetanus, rabies, Ebola virus, or a venom or toxin. Because the antibodies were not made by the recipient's lymphocytes, effector and memory cells do not form, so benefits last only as long as the injected antibodies do.

The first vaccine was the result of desperate attempts to survive smallpox epidemics that swept repeatedly through cities all over the world. Smallpox is a severe disease that kills up to one-third of the people it infects (Figure 34.21). Before 1880, no one knew what caused infectious diseases or how to protect anyone from getting them, but there were clues. In the case of smallpox, survivors seldom contracted the disease a second time. They were immune (protected from infection).

The idea of acquiring immunity to smallpox was so appealing that people had been risking their lives on it for

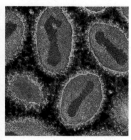

Figure 34.21
Young survivor and the cause of her disease, smallpox viruses. Worldwide use of the vaccine eradicated naturally occurring cases of smallpox; vaccinations for it ended in 1972.

over two thousand years. Many people poked into their skin bits of smallpox scabs or threads soaked in pus from smallpox sores. Some survived the crude practices and became immune to smallpox, but many others did not.

By the late 1700s, it was widely known that dairymaids did not get smallpox if they had already recovered from cowpox (a mild disease that affects cattle as well as humans). In 1796, Edward Jenner, an English physician, injected liquid from a cowpox sore into the arm of a healthy boy. Six weeks later, Jenner injected the boy with liquid from a smallpox sore. Luckily, the boy did not get smallpox. Jenner's experiment showed directly that the agent of cowpox elicits immunity to smallpox. Jenner named his procedure "vaccination," after the Latin word for cowpox (*vaccinia*). The use of Jenner's vaccine spread quickly through Europe, then to the rest of the world. The last known case of naturally occurring smallpox was in 1977, in Somalia. The vaccine had eradicated the disease.

We now know that the cowpox virus is an effective vaccine for smallpox because the antibodies it elicits also recognize smallpox virus antigens. Our knowledge of how the immune system works has allowed us to develop many other vaccines that save millions of lives yearly. These vaccines are an important part of worldwide public health programs (Table 34.4).

immunization Any procedure designed to promote immunity to a specific disease.
vaccine A preparation introduced into the body in order to elicit immunity to a specific antigen.

Table 34.4 Recommended Immunization Schedule for Children	
Vaccine	**Age of Vaccination**
Hepatitis B	Birth to 2 months
Hepatitis B boosters	1–4 months and 6–18 months
Rotavirus	2, 4, and 6 months
DTP: diphtheria, tetanus, and pertussis (whooping cough)	2, 4, and 6 months
DTP boosters	15–18 months, 4–6 years, and 11–12 years
HiB (*Haemophilus influenzae*)	2, 4, and 6 months
HiB booster	12–15 months
Pneumococcal	2, 4, and 6 months
Pneumococcal booster	12–15 months
Inactivated poliovirus	2 and 4 months
Inactivated poliovirus boosters	6–18 months and 4–6 years
Influenza	Yearly, 6 months to 18 years
MMR (measles, mumps, rubella)	12–15 months
MMR booster	4–6 years
Varicella (chicken pox)	12–15 months
Varicella booster	4–6 years
Hepatitis A series	1–2 years
HPV series	11–12 years
Meningococcal	11–12 years

Centers for Disease Control and Prevention (CDC), 2010

Take-Home Message **How do vaccines work?**

❯ Immunization is the administration of an antigen-bearing vaccine designed to elicit immunity to a specific disease.

> The immune system of some people does not function properly. The outcome is often severe or lethal.

< Links to SCID 15.10, Multiple sclerosis 29.9

People usually do not make antibodies to molecules that occur on their own, healthy body cells, in part because thymus cells have a built-in quality control mechanism that weeds out "bad" T cell receptors. Thymus cells snip small polypeptides from a variety of body proteins and attach them to MHC markers. Maturing T cells that bind too strongly to one of these peptide–MHC complexes have TCRs that recognize a self protein; those that do not bind at least weakly to the complexes do not recognize MHC markers. Both types of cells die.

Despite such built-in quality controls and the redundancies of immune system functions, immunity does not always work as well as it should. Its sheer complexity is part of the problem, because there are simply more opportunities for failure to occur in systems that have many components. Even small failures in immune system function can have catastrophic consequences on health. Autoimmune disorders occur when an immune response is misdirected against a person's own healthy body cells. In immunodeficiency, the immune response is insufficient to protect a person from disease.

Autoimmune Disorders

Sometimes lymphocytes and antibody molecules fail to discriminate between self and nonself. When that happens, they mount an **autoimmune response**: an immune response that targets one's own tissues. Autoimmunity can be beneficial, for example when a cell-mediated response targets cancer cells, but in most cases it is not (Table 34.5).

Antibodies to self proteins may bind to hormone receptors, as in the case of Graves' disease. In this disease, self antibodies that bind stimulatory receptors on the thyroid gland cause it to produce excess thyroid hormone, which quickens the body's overall metabolic rate. Antibodies are not part of the feedback loops that normally regulate thyroid hormone production. So, antibody binding continues unchecked, the thyroid continues to release too much hormone, and the metabolic rate spins out of control. Symptoms of Graves' disease include uncontrollable weight loss; rapid, irregular heartbeat; sleeplessness; pronounced mood swings; and bulging eyes.

A neurological disorder, multiple sclerosis, occurs when self-reactive T cells attack the myelin sheaths of axons in the central nervous system (Section 29.9). Symptoms range from weakness and loss of balance to paraly-

autoimmune response Immune response that inappropriately targets one's own tissues.

Table 34.5 Examples of Autoantibodies Associated With Autoimmune Disorders

Disorder	Autoantibody Target	Affected Tissue
Crohn's disease	Neutrophil granule proteins	Gastrointestinal tract
Dermatomyositis	tRNA synthesis enzyme	Muscles, skin
Diabetes mellitus type 1	Islet proteins or insulin	Pancreas
Goodpasture's syndrome	Type IV collagen	Kidney, lung
Graves' disease	TSH receptor	Thyroid
Guillain-Barré syndrome	Lipids of ganglia	Peripheral nervous system
Hashimoto's disease	TH synthesis proteins	Thyroid
Idiopathic thrombo-cytopenic purpura	Platelet glycoproteins	Blood (platelets)
Lupus erythematosus	DNA, nuclear proteins	Connective tissue
Multiple sclerosis	Myelin proteins	Central nervous system
Myasthenia gravis	Acetylcholine receptors	Neuromuscular junctions
Pemphigus vulgaris	Cadherin	Skin
Pernicious anemia	Parietal cell glycoprotein	Stomach epithelium
Polymyositis	tRNA synthesis enzyme	Muscles
Primary biliary cirrhosis	Nuclear pore proteins, mitochondria	Liver
Rheumatoid arthritis	Constant region of IgG	Joints
Scleroderma	Topoisomerase	Arteriole endothelium
Ulcerative colitis	Enzyme in neutrophil granules	Large intestine
Wegener's granulomatosis	Enzyme in neutrophil granules	Blood vessels

sis and blindness. Specific MHC gene alleles increase susceptibility, but a bacterial or viral infection may trigger the disorder.

Immunodeficiency

Impaired immune function is dangerous and sometimes lethal. Immune deficiencies render individuals vulnerable to infections by opportunistic agents that are typically harmless to those in good health. Primary immune deficiencies, which are present at birth, are the outcome of mutations. Severe combined immunodeficiencies (SCIDs) are examples. Secondary immune deficiency is the loss of immune function after exposure to an outside agent, such as a virus. AIDS (acquired immunodeficiency syndrome, described in the next section) is the most common secondary immune deficiency.

Take-Home Message What happens when the immune system does not function as it should?

> Misdirected or compromised immunity, which sometimes occurs as a result of mutation or environmental factors, can have severe or lethal outcomes.

❯ AIDS is an outcome of interactions between the HIV virus and the human immune system.

❮ Links to cDNA 15.2, Viruses and HIV replication 19.2

Acquired immune deficiency syndrome, or **AIDS**, is a constellation of disorders that occur as a result of infection with HIV, the human immunodeficiency virus (Figure 34.22A). This virus cripples the immune system, so it makes the body very susceptible to infections and rare forms of cancer. Worldwide, approximately 39.5 million individuals currently have AIDS (Figure 34.22B).

There is no way to rid the body of HIV, no cure for those already infected. At first, an infected person appears to be in good health, perhaps fighting "a bout of flu." But symptoms eventually emerge that foreshadow AIDS: fever, many enlarged lymph nodes, chronic fatigue and weight loss, and drenching night sweats. Then, infections caused by normally harmless microorganisms strike. Yeast infections of the mouth, esophagus, and vagina often occur, as well as a form of pneumonia caused by the fungus *Pneumocystis jirovecii*. Colored lesions erupt; the lesions are evidence of Kaposi's sarcoma, a type of cancer that is common among AIDS patients (Figure 34.22C).

HIV Revisited HIV is a retrovirus with a lipid envelope. Remember, this type of envelope consist of a small piece of plasma membrane acquired by a virus as it buds from a cell (Section 19.2). Proteins jut from the envelope, span it, and line its inner surface. Just beneath the envelope, more viral proteins enclose two RNA strands and reverse transcriptase enzymes. When a retrovirus particle infects a cell, the reverse transcriptase copies the viral RNA into DNA, which becomes integrated into the host cell's DNA.

A Titanic Struggle HIV mainly infects macrophages, dendritic cells, and helper T cells. When virus particles enter the body, dendritic cells engulf them. The dendritic cells then migrate to lymph nodes, where they present processed HIV antigen to naive T cells. An army of HIV-neutralizing IgG antibodies and HIV-specific cytotoxic T cells forms.

We have just described a typical adaptive immune response. It rids the body of most—but not all—of the virus. In this first response, HIV infects a few helper T cells in a few lymph nodes. For years or even decades, the IgG antibodies keep the level of HIV in the blood low, and the cytotoxic T cells kill HIV-infected cells.

Patients are contagious during this stage, although they might show no symptoms of AIDS. HIV persists in a few of their helper T cells, in a few lymph nodes. Eventually, the level of virus-neutralizing IgG in their blood plummets, and the production of T cells slows. Why IgG decreases is still a major topic of research, but its effect is certain: The adaptive immune system becomes less and less effective at fighting the virus. The number of virus particles rises; up to 1 billion viruses are built each day. Up to 2 billion helper T cells become infected. Half of the virus particles are destroyed and half of the helper T cells are replaced every two days. Lymph nodes begin to swell with infected T cells.

Eventually, the battle tilts as the body makes fewer replacement helper T cells and the body's capacity for adaptive immunity is destroyed. Other types of viruses may make more particles per day, but the immune system eventually demolishes them. HIV demolishes the immune system. Secondary infections and tumors kill the patient.

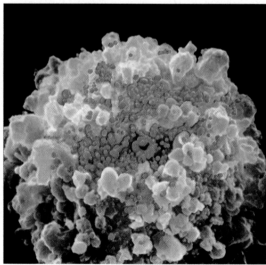

A A human T cell (*blue*), infected with HIV (*red*).

Figure 34.22 Three views of AIDS.

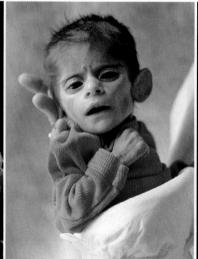

B This Romanian baby contracted AIDS from his mother's breast milk.

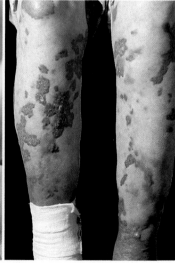

C Lesions of Kaposi's sarcoma, a cancer that is a common symptom of HIV infection in older AIDS patients.

Transmission HIV is not transmitted by casual contact. Most HIV infections are the result of having unprotected sex with an infected partner. The virus occurs in semen and vaginal secretions, and can enter a partner through epithelial linings of the penis, vagina, rectum, and the mouth. The risk of transmission increases by the type of sexual act; for example, anal sex carries 50 times the risk of oral sex. Infected mothers can transmit HIV to a child during pregnancy, labor, delivery, or breast-feeding. HIV also travels in tiny amounts of infected blood in the syringes shared by intravenous drug abusers, or by patients in hospitals of poor countries.

Testing Most AIDS tests check blood, saliva, or urine for antibodies that bind to HIV antigens. These antibodies are detectable in 99 percent of infected people within three months of exposure to the virus. One test can detect viral RNA at about eleven days after exposure. Currently, the only reliable tests are performed in clinical laboratories; home test kits may result in false negatives, which may cause an infected person to unknowingly transmit the virus.

Drugs and Vaccines Drugs cannot cure AIDS, but they can slow its progress. Of the twenty or so FDA-approved AIDS drugs, most target processes unique to retroviral replication. For example, RNA nucleotide analogs such as AZT inhibit reverse transcriptase. They interrupt HIV replication when they substitute for normal nucleotides in the viral RNA-to-DNA synthesis process (Sections 15.2 and 19.2). Other drugs such as protease inhibitors affect different parts of the viral replication cycle.

A three-drug "cocktail" of one protease inhibitor plus two reverse transcriptase inhibitors is currently the most successful AIDS therapy, and has changed the course of the disease from a short-term death sentence to a long-term, often manageable illness.

Researchers are using several strategies to develop an HIV vaccine. At this writing, organizations around the world are testing 25 different HIV vaccines. Most of them consist of isolated HIV proteins or polypeptides, and many deliver the antigens in viral vectors. Live, weakened HIV virus is an effective vaccine in chimpanzees, but the risk of HIV infection from the vaccines themselves far outweighs their potential benefits in humans. Other types of HIV vaccines are notoriously ineffective. IgG antibody exerts selective pressure on the virus, which has a very high mutation rate because it replicates so fast. The human immune system just cannot produce antibodies fast enough to keep up with the mutations.

AIDS Acquired immune deficiency syndrome. A collection of diseases that develops after a virus (HIV) weakens the immune system.

Frankie's Last Wish (revisited)

The Gardasil vaccine consists of viral capsid proteins that self-assemble into virus-like particles (VLPs). The proteins are produced by a genetically engineered yeast, *Saccharomyces cerevisiae*. This yeast carries genes for one surface protein from each of four strains of HPV, so the VLPs carry no viral DNA. Thus, the VLPs are not infectious, but the antigenic proteins they consist of elicit an immune response at least as strong as infection with HPV virus (*left*).

How Would You Vote? Clinical trials of some vaccines take place in countries that have fewer regulations governing human testing than the United States does. Should all clinical trials be held to the same ethical standards no matter where they take place? See CengageNow for details, then vote online (cengagenow.com).

Table 34.6 Global HIV and AIDS Cases

Region	AIDS Cases	New HIV Cases
Sub-Saharan Africa	22,400,000	1,900,000
South/Southeast Asia	3,800,000	280,000
Latin America	2,000,000	170,000
Central Asia/East Europe	1,500,000	110,000
North America	1,400,000	55,000
East Asia	850,000	75,000
Western/Central Europe	850,000	30,000
Middle East/North Africa	310,000	35,000
Caribbean Islands	240,000	20,000
Australia/New Zealand	59,000	3,900
Approx. worldwide total	33,400,000	2,700,000

Source: Joint United Nations Programme HIV/AIDS, 2009 report

At present, our best option for halting the spread of HIV is prevention, by teaching people how to avoid being infected. The best protection against AIDS is to avoid unsafe behaviors. In most circumstances, HIV infection is the consequence of a choice: either to have unprotected sex, or to use a shared needle for intravenous drugs. Educational programs around the world are having an effect on the spread of the virus: In many (but not all) countries, the incidence of new cases of HIV each year is beginning to slow. Overall, however, our global battle against AIDS is not being won (Table 34.6).

Take-Home Message What is AIDS?

❯ AIDS occurs as a result of infection by HIV, a virus that infects lymphocytes and so cripples the human immune system.

Summary

Section 34.1 Screenings, treatments, and vaccines for diseases such as cervical cancer are a direct outcome of our increasing understanding of the way the human body interacts with pathogens.

Section 34.2 Three lines of immune defense protect vertebrates against infection. A pathogen that breaches surface barriers triggers **innate immunity**, a set of general defenses that usually prevents pathogens from becoming established in the body. **Adaptive immunity**, which can specifically target billions of different **antigens**, follows. **Complement** and signaling molecules such as **cytokines** help coordinate the activities of white blood cells—**dendritic cells**, **macrophages**, **neutrophils**, **basophils**, **mast cells**, **eosinophils**, **B** and **T cells** (including **cytotoxic T cells**) and **NK cells**—in **immunity**.

Section 34.3 Vertebrates can fend off pathogens such as those that cause dental **plaque** at body surfaces with physical, mechanical, and chemical barriers (including **lysozyme**). Most **normal flora** do not cause disease unless they penetrate inner tissues.

Section 34.4 An innate immune response includes fast, general responses that can eliminate invaders before they can become established in the body. Complement attracts phagocytes, and punctures invading cells. **Inflammation** begins when mast cells in tissue release histamine, which increases blood flow and also makes capillaries leaky to phagocytes and plasma proteins. **Fever** fights infection by increasing the metabolic rate.

Section 34.5 Antigen receptors (**T cell receptors**, and **B cell receptors** and other **antibodies**) recognize specific antigens. These receptors allow the adaptive immune system to recognize billions of different antigens. TCRs that recognize **MHC markers** are the basis of self/nonself discrimination.

Section 34.6 B cells and T cells carry out adaptive immunity. **Antibody-mediated** and **cell-mediated immune responses** work together to rid the body of a specific pathogen. Phagocytic lymphocytes present antigen in these responses. **Effector cells** form and target the antigen-bearing particles in a primary response. **Memory cells** that also form are reserved for a later encounter with an antigen, in which case they trigger a faster, stronger secondary response.

naive B cell

Sections 34.7, 34.8 B cells, assisted by T cells and signaling molecules, carry out antibody-mediated immune responses, during which antibodies are produced by B cells. Antibody binding causes **agglutination** of foreign cells, and tags antigen-bearing particles for phagocytosis. Blood typing is an agglutination test.

Section 34.9 Cytotoxic T cells and NK cells are central to cell-mediated responses that target and kill body cells altered by infection or cancer. Effector and memory cells form, as occurs in antibody-mediated responses.

Section 34.10 Allergens are normally harmless substances that induce immune responses; hypersensitivity to an allergen is called **allergy**.

Section 34.11 Immunization with **vaccines** designed to elicit immunity to specific diseases saves millions of lives each year as part of worldwide health programs.

Sections 34.12, 34.13 The consequences of malfunctioning immunity can be deadly. Immune deficiency is a reduced capacity to mount an immune response. In an **autoimmune response**, a body's own cells are inappropriately recognized as foreign and attacked. **AIDS** is caused by HIV, a virus that infects helper T cells.

Self-Quiz   Answers in Appendix III

1. Which of the following is not among the first line of defenses against infection?
   a. resident bacterial populations   d. skin
   b. tears, saliva, gastric fluid   e. complement
   c. urine flow   f. lysozyme

2. Which of the following is not considered to be part of an innate immune response?
   a. phagocytic cells   e. inflammation
   b. fever   f. complement activation
   c. histamines   g. antigen-presenting cells
   d. cytokines   h. none of the above

3. Which of the following is not considered to be part of an adaptive immune response?
   a. phagocytic cells   e. antigen receptors
   b. antigen-presenting cells   f. complement activation
   c. histamines   g. antibodies
   d. cytokines   h. none of the above

4. Activated complement proteins _____ .
   a. form pore complexes   c. attract phagocytes
   b. promote inflammation   d. all of the above

5. _____ trigger immune responses.
   a. Cytokines   d. Antigens
   b. Lysozymes   e. Histamines
   c. Immunoglobulins   f. all of the above

6. Name a defining characteristic of innate immunity.

7. Name a defining characteristic of adaptive immunity.

8. Antibodies are _____ .
   a. antigen receptors   c. proteins
   b. made only by B cells d. all of the above

Data Analysis Activities

Cervical Cancer Incidence in HPV–Positive Women In 2003, Michelle Khan and her coworkers published their findings on a 10-year study in which they followed cervical cancer incidence and HPV status in 20,514 women. All women who participated in the study were free of cervical cancer when the test began. Pap tests were taken at regular intervals, and the researchers used a DNA probe hybridization test to detect the presence of specific types of HPV in the women's cervical cells.

The results are shown in Figure 34.23 as a graph of the incidence rate of cervical cancer by HPV type. Women who are HPV positive are often infected by more than one type, so the data were sorted into groups based on the women's HPV status ranked by type: either positive for HPV16; or negative for HPV16 and positive for HPV18; or negative for HPV16 and 18 and positive for any other cancer-causing HPV; or negative for all cancer-causing HPV.

1. At 110 months into the study, what percentage of women who were not infected with any type of cancer-causing HPV had cervical cancer? What percentage of women who were infected with HPV16 also had cervical cancer?

2. In which group would women infected with both HPV16 and HPV18 fall?

3. Is it possible to estimate from this graph the overall risk of cervical cancer that is associated with infection of cancer-causing HPV of any type?

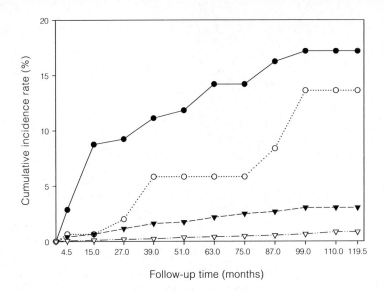

Figure 34.23 Cumulative incidence rate of cervical cancer correlated with HPV status in 20,514 women aged 16 years and older.

The data were grouped as follows: HPV16 positive (closed circles); HPV16 negative and HPV18 positive (open circles); and all other cancer-causing HPV types combined (closed triangles). Open triangles: no cancer-causing HPV type was detected.

---

9. Antibody-mediated responses work against _____ .
   a. intracellular pathogens d. both a and b
   b. extracellular pathogens e. both b and c
   c. cancerous cells   f. a, b, and c

10. Cell-mediated responses work against _____ .
   a. intracellular pathogens d. both a and b
   b. extracellular pathogens e. both a and c
   c. cancerous cells   f. a, b, and c

11. _____ are targets of cytotoxic T cells.
   a. Extracellular virus particles in blood
   b. Virus-infected body cells or tumor cells
   c. Parasitic flukes in the liver
   d. Bacterial cells in pus
   e. Pollen grains in nasal mucus

12. Allergies occur when the body responds to _____ .
   a. pathogens   c. normally harmless substances
   b. toxins   d. all of the above

13. Match the immune cell with the best description.
   ___ mast cell   a. antigen-presenter
   ___ eosinophil   b. targets parasitic worms
   ___ helper T cell   c. activates cytotoxic T cells
   ___ NK cell   d. kills body cells with no
   ___ dendritic cell   MHC markers
                          e. factor in allergic reactions

Additional questions are available on **CENGAGENOW**.

Critical Thinking

1. Monoclonal antibodies are produced by immunizing a mouse with a particular antigen, then removing its spleen. Individual B cells producing mouse antibodies specific for the antigen are isolated from the mouse's spleen and fused with cancerous B cells from a myeloma cell line. The resulting hybrid myeloma cells ("hybridoma" cells) are cloned: Individual cells are grown in tissue culture as separate cell lines. Each line produces and secretes antibodies that recognize the antigen to which the mouse was immunized. These monoclonal antibodies can be purified and used for research or other purposes.

Monoclonal antibodies are sometimes used in passive immunization. They are effective, but only in the immediate term. Antibodies produced by one's own immune system can last up to about six months in the bloodstream, but monoclonals delivered in passive immunization often last for less than a week. Why the difference?

Animations and Interactions on **CENGAGENOW**:
❯ Skin structure; Inflammation; Secondary immune responses; The lymphatic system; An antibody-mediated immune response; Clonal selection; Blood typing; A cell-mediated immune response.

Key Concepts

Respiratory Systems
Respiration is a physiological process that moves oxygen from the outside environment to all metabolically active tissues, and moves carbon dioxide from those tissues to the environment. Respiratory organs such as gills, tracheal tubes, and lungs enhance the rate of gas exchange.

Human Respiratory System
The human respiratory system consists of airways through which gases move to and from paired lungs. In addition to its function in gas exchange, the human respiratory system has roles in speech, temperature regulation, and the sense of smell.

35 Respiration

35.1 Up in Smoke

Each day, 3,000 or so teenagers join the ranks of habitual smokers in the United States. Most are not even fifteen years old. When they first light up, they cough and choke on the irritants in the smoke. Most become dizzy and nauseated, and develop headaches. Sound like fun? Hardly. Why, then, do they ignore signals about threats to the body and work so hard to be a smoker? Mainly to fit in. To many adolescents, a misguided perception of social benefits overwhelms the seemingly remote threats to health.

Despite teenage perceptions, changes that can make the threat a reality start right away. Ciliated cells keep many pathogens and pollutants that enter airways from reaching the lungs. These cells can be immobilized for hours by the smoke from a single cigarette (Figure 35.1). Smoke also kills white blood cells that patrol and defend respiratory tissues. Pathogens multiply in the undefended airways. The result is more colds, more asthma attacks, and more bronchitis.

The highly addictive stimulant nicotine constricts blood vessels, which increases blood pressure. The heart has to work harder to pump blood through the narrowed tubes. Nicotine also triggers a rise in "bad" cholesterol (LDL) and a decline in the "good" kind (HDL) in blood. It makes blood stickier, encouraging clots that can block blood vessels.

Tobacco smoke has more than forty known carcinogens and 80 percent of lung cancers occur in smokers. Women who smoke are more susceptible to cancers than men. On average, women develop cancers earlier, and with lower exposure to tobacco. Fewer than 15 percent of women diagnosed with lung cancer survive five years. Smoking also increases breast cancer risk; females who start to smoke as teenagers are about 70 percent more likely to get breast cancer than those women who never smoked.

Families, coworkers, and friends get unfiltered doses of the carcinogens in tobacco smoke. Each year in the United States, lung cancers arising from secondhand smoke kill about 3,000 people. Children exposed to secondhand smoke also are more likely to develop chronic middle ear infections, asthma, and other respiratory problems later in life.

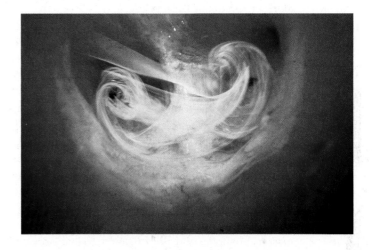

Figure 35.1 Cigarette smoke in the upper airway. Chemicals in smoke paralyze cilia that help clear away mucus and pathogens.

This chapter describes a few respiratory systems. All exchange gases with the outside environment. They also contribute to homeostasis—maintaining the body's internal operating conditions within ranges that cells can tolerate. If you or someone you know smokes, you might use the chapter as a guide to smoking's impact on health. For a more graphic preview, find out what goes on every day with smokers in hospital emergency rooms or intensive care units. There's no glamor there. It is not cool, and it is not pretty.

Respiratory Cycle
Inhalation is always an active process. It occurs when a part of the brain stem signals muscles to contract and increase the size of the thoracic cavity. Exhalation is usually passive. Muscles relax, the chest and lungs shrink, and air flows out of lungs.

Gas Exchanges
Oxygen moves from air in the lungs into pulmonary capillaries, where it binds with hemoglobin. Hemoglobin releases oxygen near active tissues. Carbon dioxide is converted to bicarbonate in blood. In lungs, bicarbonate is converted into carbon dioxide and water that can be exhaled.

Respiratory Problems
Interrupted breathing (apnea), infectious diseases (such as tuberculosis), and inflammatory conditions (such as asthma and bronchitis) interfere with normal respiratory function.

> Respiration is the physiological process that supplies the oxygen needed for aerobic respiration.

‹ Links to Diffusion 5.6, Aerobic respiration 7.2

Gas Exchanges

In Chapter 7 you learned about aerobic respiration, an energy-releasing pathway that requires oxygen (O_2) and produces carbon dioxide (CO_2) as summarized below:

$$C_6H_{12}O_6 \ + \ O_2 \longrightarrow CO_2 \ + \ H_2O$$

glucose   oxygen   carbon dioxide   water

This chapter focuses on **respiration**, the physiological processes that collectively supply body cells with oxygen from the environment, and deliver waste carbon dioxide to the environment.

Gases enter and leave the animal body by diffusing across a thin, moist **respiratory surface** (Figure 35.2A). A typical respiratory surface is one or two cell layers thick. The respiratory surface must be thin because gases diffuse quickly only over very short distances. It must be moist because gases cannot cross the respiratory surface unless they dissolve in fluid and diffuse across.

A second exchange of gases occurs internally, at the plasma membrane of body cells (Figure 35.2B). Oxygen diffuses from the interstitial fluid into a cell and carbon dioxide moves in the opposite direction. In invertebrates without a circulatory system, oxygen that crosses the respiratory surface reaches body cells by diffusion. In most invertebrates and all vertebrates, a circulatory system enhances the movement of gases between body cells and the respiratory surface.

Factors That Affect Gas Exchange

The larger the area of the respiratory surface, the more molecules can cross it in any given time. The area of the respiratory surface is often surprisingly large relative to the animal's body size. Branching and folding allow a large respiratory surface to fit in a small volume.

The concentrations of gases on either side of the respiratory surface also affect the rate of gas exchange. The steeper the concentration gradient across this surface, the faster diffusion proceeds. As a result, many animals have

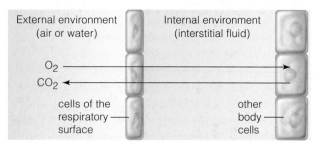

External environment (air or water) Internal environment (interstitial fluid)

O_2
CO_2

cells of the respiratory surface   other body cells

A Cells of the respiratory surface exchange gases with both the external and internal environment.

B Other body cells exchange gases with the internal environment.

Figure 35.2 Two sites of gas exchange during respiration. In some animals, gases simply diffuse between the two sites. However, most animals have a circulatory system and their blood transports gases rapidly between the two exchange sites.

mechanisms that keep oxygen-rich air or water flowing over their respiratory surface. For example, your inhalations and exhalations move air rich in carbon dioxide away from the respiratory surface in your lungs and replace it with oxygen-rich air.

Many animals also have blood with respiratory proteins that steepen the oxygen concentration gradient at the respiratory surface. A **respiratory protein** contains one or more metal ions that bind oxygen when the oxygen concentration is high, and release it when the oxygen concentration is low. Hemoglobin serves as a respiratory pigment in humans and many other animals. When an oxygen atom binds to hemoglobin, that atom no longer contributes to the oxygen concentration of the blood. Thus the presence of hemoglobin lowers the effective oxygen concentration of the blood and encourages diffusion of oxygen from the air into the blood.

---

respiration Physiological process by which an animal body supplies cells with oxygen and disposes of their waste carbon dioxide.
respiratory protein A protein that reversibly binds oxygen when the oxygen concentration is high and releases it when oxygen concentration is low. Hemoglobin is an example.
respiratory surface Moist surface across which gases are exchanged between animal cells and the external environment.

Take-Home Message What is respiration and what factors affect gas exchange?

> Respiration comprises the physiological processes that supply cells with oxygen they need for aerobic respiration, and remove the waste carbon dioxide that this pathway produces.

> A respiratory surface is a thin, moist membrane across which gases diffuse into and out of the internal environment. The larger the respiratory surface area, the faster the rate of gas exchange.

> The steepness of the concentration gradient across the respiratory membrane also affects gas exchange, with a steeper gradient increasing the rate of exchange.

> Steep gradients are maintained by mechanisms that move water or air to and from the respiratory surface and by respiratory proteins that reversibly bind oxygen.

> Most invertebrates are aquatic and exchange gases with their watery surroundings, but some have adapted to life on land and exchange gases with the air.
‹ Link to Arthropod exoskeleton 23.10

Many invertebrates do not have any special respiratory organs. For example, cnidarians and flatworms exchange gases with the environment across the body wall and the lining of a gastrovascular cavity that functions in both respiration and digestion (Figure 35.3A). These animals do not have a circulatory system, so gases must diffuse to and from cells through interstitial fluid.

The evolution of a circulatory system and blood with respiratory pigments increased the efficiency of oxygen distribution through the body. In earthworms, a closed circulatory system carries hemoglobin-containing blood to and from the damp body wall that serves as the respiratory surface. Hemoglobin also transports oxygen in the blood of mollusks and some arthropods. Invertebrates do not have red blood cells; their hemoglobin is in the fluid portion of their blood.

Gills, folded or filamentous respiratory organs that are well supplied with blood vessels, evolved in many aquatic invertebrates with a circulatory system. As blood moves through gills, it picks up oxygen from the water and gives up carbon dioxide. The feathery gills of crustaceans lie beneath the protective exoskeleton. Most aquatic mollusks have gills in their mantle cavity, but some sea slugs have external gills (Figure 35.3B).

The most successful air-breathing land invertebrates are insects and arachnids, such as spiders. A hard exoskeleton helps these animals conserve water, but it also blocks gas exchange across the body surface. Insects and some spiders get around this limitation with a **tracheal system** that delivers air directly to tissues. Chitin-reinforced tracheal tubes start at spiracles, small openings across the exoskeleton, and branch repeatedly (Figure 35.3C). There is usually a pair of spiracles per segment: one on each side of the body. Spiracles can be opened or closed to regulate the amount of oxygen that enters the body. Substances that clog spiracles are used as insecticides. For example, horticultural oils sprayed on fruit trees suffocate scale bugs, aphids, and mites by clogging spiracles.

At the tips of the finest tracheal branches is a bit of fluid in which gases dissolve. The tips of insect tracheal tubes are adjacent to body cells, and oxygen diffuses from air in the tube, through fluid at the tip, and into cells. Carbon dioxide diffuses in the opposite direction, from the cells and into the tracheal tubes.

---

gills Folds or body extensions that increase the surface area for respiration.
tracheal system Of insects and some other land arthropods, tubes that convey gases between the body surface and internal tissues.

A Flatworms exchange gases across the external body wall and the lining of the gastrovascular cavity.

B Sea slugs exchange gases across a gill on the dorsal surface.

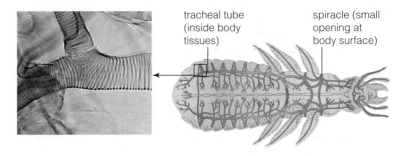

C Insects have a system of chitin-reinforced tracheal tubes that deliver air from an opening at the body surface (a spiracle) to interstitial fluid around tissues deep inside the body.

Figure 35.3 Examples of invertebrate respiratory mechanisms.

Some insects force air into and out of their tracheal tubes. When a grasshopper's abdominal muscles contract, organs press on these pliable tubes and force air out. When the muscles relax, pressure on tracheal tubes decreases, the tubes widen, and air rushes in.

Take-Home Message **How do invertebrates exchange gases with their environment?**

> Cnidarians and flatworms exchange gases across the body wall and the lining of the gastrovascular cavity. Gases move within their body by diffusion.

> Aquatic mollusks and arthropods have gills. Blood that runs through gills transports gases to and from other body cells.

> Insects and some spiders have a system of air-filled tracheal tubes that open at the body surface and end near body cells.

> Depending on the species, vertebrates exchange gases across gills, skin, or the surface of paired internal lungs.
< Links to Fishes 24.3, Evolution of lungs 24.4

Respiration in Fishes

All fishes have gill slits that open across the pharynx (the throat region). In jawless fishes and cartilaginous fishes, the gill slits are visible from the outside, but bony fishes have a gill cover that hides them.

In all fishes, water flows into the mouth, enters the pharynx, then moves out of the body through the gill slits (Figure 35.4A). A bony fish sucks water inward by opening its mouth, closing the cover over each gill, and contracting muscles that enlarge the oral cavity. Water is forced out over the gills when the fish closes its mouth, opens its gill covers, and contracts muscles that reduce the size of the oral cavity.

If you could remove the gill covers of a bony fish, you would see that the gills themselves consist of bony gill

arches, each with many gill filaments attached (Figure 35.4B). Inside each gill filament are many capillary beds where gases in water are exchanged with gases in blood.

Water flowing over gills and blood flowing through gill capillaries move in opposite directions (Figure 35.4C). The result is a **countercurrent exchange**, in which two fluids exchange substances while flowing in opposite directions. For the entire length of the capillary, the water next to the capillary holds more oxygen than the blood flowing inside it (Figure 35.4D). As a result, oxygen continually diffuses from the water into the blood. The blood becomes increasingly oxygenated as it passes through the capillary.

Evolution of Paired Lungs

Most tetrapods have paired lungs. A **lung** is a saclike respiratory organ that lies inside a body cavity. Airways connect the lung to the outside air. The first lungs evolved from outpouchings of the gut wall in some bony fishes. Lungs may have helped these fishes survive short trips between ponds. Gills do not function on land. Without water, their thin filaments collapse under their own weight, dry out, and stick together. Lungs became increasingly important as aquatic tetrapods began moving onto land (Section 24.4).

Amphibian larvae typically have external gills. Most often, as the animal develops, these gills disappear and are replaced by paired lungs. Amphibians also exchange some gases across their thin-skinned body surface. In all amphibians, most carbon dioxide that forms during aerobic respiration leaves the body across the skin.

All frogs have paired lungs. They do not use chest muscles to draw air into their lungs, as you do. Instead, frogs suck in air through their nostrils by lowering the floor of the mouth (Figure 35.5A). Then they close their nostrils and lift the floor of the mouth and throat. Compression of the oral cavity forces air into the lungs (Figure 35.5B).

gill filaments
one gill arch
water is sucked into mouth
Water exits through gill slits

A Bony fish with its gill cover removed. Water flows in through the mouth, over the gills, then out through gill slits. Each gill has bony gill arches with many thin gill filaments attached.

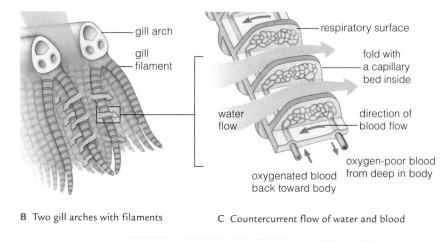

gill arch
gill filament
respiratory surface
fold with a capillary bed inside
water flow
direction of blood flow
oxygenated blood back toward body
oxygen-poor blood from deep in body

B Two gill arches with filaments

C Countercurrent flow of water and blood

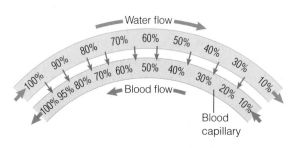

D Oxygen flow from water into a capillary. Percentages indicate the degree of oxygenation of water (*blue*) and blood (*red*). All along the capillary, water flows down its concentration gradient from water into blood.

Figure 35.4 Animated Structure and function of the gills of a bony fish.

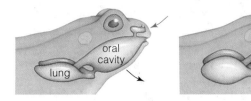

A The frog lowers the floor of its mouth, pulling air into the oral cavity through its nostrils.

B Closing the nostrils and elevating the floor of the mouth pushes air into lungs.

Figure 35.5 Animated How a frog fills its lungs. *Black* arrows show body wall movements. *Blue* arrows show air movement. Frogs push air into their lungs, rather than sucking it in as you do.

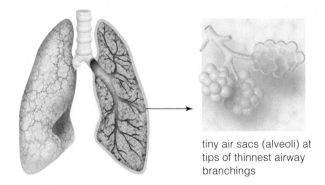

tiny air sacs (alveoli) at tips of thinnest airway branchings

Figure 35.6 Mammalian lungs. Expansion of the chest cavity draws air into branching tubes that end at tiny air sacs called alveoli, where gas exchange occurs.

Reptiles (including birds) and mammals are amniotes with a waterproof skin. Their only respiratory surface is the lining of two well-developed lungs. These lungs inflate when muscles increase the size of the thoracic cavity. As the cavity expands, pressure in the lungs drops and air is sucked inward.

In mammals (as well as amphibians and most reptiles), inhaled air flows through increasingly smaller airways until it reaches tiny sacs called **alveoli** (singular, alveolus). Gas exchange occurs across the wall of alveoli (Figure 35.6). During exhalation, air retraces its steps, flowing back out the same way it came in. The lungs do not deflate completely, so a bit of stale air remains behind even after exhalation.

In birds, there are no alveoli, no "dead ends," and no stale air inside the lung. Birds have small, inelastic lungs that do not expand and contract when the bird breathes. Instead, large air sacs attached to the lungs inflate and deflate (Figure 35.7A). It takes two breaths to move air through this system. Oxygen-rich air flows through tiny tubes in the lung during both inhalations and exhalations. The lining of these tubes serves as the respiratory surface (Figure 35.7B). Continual movement of air over this surface increases the efficiency of gas exchange.

We turn next to the human respiratory system. Its operating principles apply to most vertebrates.

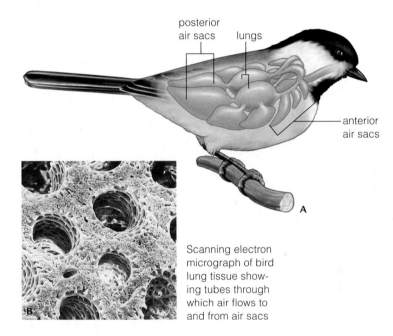

posterior air sacs
lungs
anterior air sacs

A

Scanning electron micrograph of bird lung tissue showing tubes through which air flows to and from air sacs

B

Figure 35.7 Animated Respiratory system of a bird. **A** Large air sacs attach to two small, inelastic lungs. Air flows in through many air tubes inside the lung, and into posterior air sacs. **B** The lining of the tiniest of the air tubes, sometimes called air capillaries, is the respiratory surface.

It takes two respiratory cycles to move air through the lungs and air sacs of a bird's respiratory system:

Inhalation 1— Muscles expand chest cavity, drawing air in through nostrils. Most of the air flowing in through the trachea goes to lungs and some goes to posterior air sacs.

Exhalation 1— Anterior air sacs empty. Air from the posterior air sacs moves into lungs.

Inhalation 2 — Air in lungs moves to anterior air sacs and is replaced by newly inhaled air.

Exhalation 2 — Air in anterior air sacs moves out of the body and air from posterior sacs flows into the lungs.

alveolus Tiny air sac in the mammalian lung; site of gas exchange.
countercurrent exchange Exchange of substances between two fluids moving in opposite directions.
lung A saclike respiratory organ that lies inside a body cavity.

Take-Home Message **How does the structure of vertebrate respiratory organs affect their function?**

❯ Fishes exchange gases with water flowing over their gills. Countercurrent flow of water and blood aids gas exchange.

❯ Amphibians exchange gases across their skin and push air from their mouth into their lungs. Amniotes suck air into their lungs by expanding the size of their thoracic cavity.

❯ Air flows in and out of mammalian lungs, but it flows continually through bird lungs.

❯ Airways, lungs, and some skeletal muscles of the chest play a role in human respiration.

❮ Links to Sense of smell 30.7, Surface defenses 34.3

The System's Many Functions

Figure 35.8 shows parts of the human respiratory system and lists their roles. The respiratory system functions in gas exchange, but it has a wealth of additional tasks. We can speak, sing, or shout as air moves past our vocal cords. We have a sense of smell because airborne molecules stimulate olfactory receptors in the nose. Cells lining nasal passages and other airways of the system help defend the body; they intercept and neutralize airborne pathogens. The respiratory system contributes to the body's acid–base balance by exhaling waste carbon dioxide that can make blood acidic. Controls over breathing also help maintain body temperature, because water evaporating from airways has a cooling effect.

From Airways to Alveoli

The Respiratory Passageways Take a deep breath. Now look at Figure 35.8A to see where the air went. If you are healthy and sitting quietly, air usually enters through your nose, rather than your mouth. As air moves through your nostrils, tiny hairs filter out any large particles. Mucus secreted by cells of the nasal lining captures fine particles and airborne chemicals. Ciliated cells in the nasal lining also help remove inhaled contaminants.

Oral Cavity (Mouth)
Supplemental airway when breathing is labored

Pleural Membrane
Double-layer membrane that separates lungs from other organs; the narrow, fluid-filled space between its two layers has roles in breathing

Intercostal Muscles
At rib cage, skeletal muscles with roles in breathing. There are two sets of intercostal muscles (external and internal)

Diaphragm
Muscle sheet between the chest cavity and abdominal cavity with roles in breathing

A

Nasal Cavity
Chamber in which air is moistened, warmed, and filtered, and in which sounds resonate

Pharynx (Throat)
Airway connecting nasal cavity and mouth with larynx; enhances sounds; also connects with esophagus

Epiglottis
Closes off larynx during swallowing

Larynx (Voice Box)
Airway where sound is produced; closed off during swallowing

Trachea (Windpipe)
Airway connecting larynx with two bronchi that lead into the lungs

Lung (One of a Pair)
Lobed, elastic organ of breathing; site of gas exchange between internal environment and outside air

Bronchial Tree
Increasingly branched airways starting with two bronchi and ending at air sacs (alveoli) of lung tissue

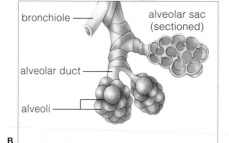

bronchiole — alveolar sac (sectioned)

alveolar duct —

alveoli —

B

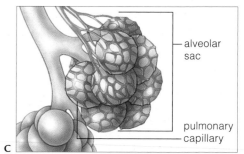

— alveolar sac

pulmonary capillary

C

Figure 35.8 Animated A Components of the human respiratory system and their functions. The diaphragm and intercostal muscles function in breathing.

B,C Location of alveoli relative to bronchioles and lung (pulmonary) capillaries.

❯❯ **Figure It Out** Which has a larger diameter, the trachea or a bronchiole?　Answer: The trachea

Air from the nostrils enters the nasal cavity, where it gets warmed and moistened. It flows next to the **pharynx**, or throat. It continues to the **larynx**, a short airway commonly known as the voice box because of the pair of vocal cords that span it (Figure 35.9). Each vocal cord is skeletal muscle with a cover of mucus-secreting epithelium. Contraction of the vocal cords changes the size of the **glottis**, the gap between them.

When the glottis is wide open, air flows through it silently. When muscle contraction narrows the glottis, outgoing air flowing through the tighter gap makes the vocal cords vibrate, giving rise to sounds. The tension on the cords and changes in the position of the larynx alter the sound's pitch. To get a feel for how that works, place one finger on your "Adam's apple," the laryngeal cartilage that sticks out at the front of your neck. Hum a low note, then a high one. You will feel the vibration of vocal cords and the way that the laryngeal muscles shift the position of the larynx.

A flap of tissue called the **epiglottis** can fold over and cover the larynx. When you are breathing, the epiglottis points up, and air moves through the larynx into the **trachea**, or windpipe. When you swallow, the epiglottis points down and covers the larynx entrance, so food and fluids enter the esophagus. The esophagus connects the pharynx to the stomach.

The cartilage-reinforced trachea branches into two **bronchi** (singular, bronchus). Each bronchus delivers air to a lung. Ciliated and mucus-secreting cells in the epithelial lining of the bronchi help fend off respiratory tract infections. Bacteria and airborne particles get caught in the secreted mucus, then cilia sweep the mucus toward the throat for expulsion.

The Paired Lungs Two cone-shaped lungs reside in the thoracic cavity, one on each side of the heart. The rib cage encloses and protects the lungs. A two-layer thick pleural membrane covers each lung's outer surface and lines the inner thoracic cavity wall.

Inside each lung, air flows through finer and finer branchings of a "bronchial tree." All of these branches are **bronchioles**. The finest bronchioles lead to the respiratory alveoli, the little air sacs where gases are exchanged

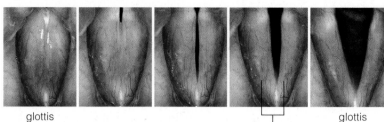

glottis   glottis
closed   open

Figure 35.9 Human vocal cords, inside the larynx. Contraction of skeletal muscle in the cords changes the width of the glottis, the gap between them. The glottis closes tightly when you swallow. It is open during quiet breathing, and narrows when you speak, so that air flow causes the cords to vibrate.

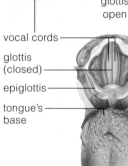

vocal cords
glottis (closed)
epiglottis
tongue's base

(Figure 35.8B,C). The wall of each alveolus is one cell thick. Collectively, alveoli provide an extensive surface for gas exchange. If all 6 million alveoli in your lungs could be stretched out in a single layer, they would cover half a tennis court!

Air in alveoli exchanges gases with blood flowing through pulmonary capillaries (Latin *pulmo*, lung). At this point, a different organ system gets involved. The circulatory system transports oxygen to body tissues and carries carbon dioxide away from them.

Muscles and Respiration A broad sheet of smooth muscle beneath the lungs, the **diaphragm**, separates the thoracic cavity and the abdominal cavity. It is the only smooth muscle that can be controlled voluntarily. You can make your diaphragm contract by deliberately inhaling. The diaphragm and the intercostal muscles—the skeletal muscles between the ribs—act together to change the volume of the thoracic cavity during breathing. There are two sets of intercostal muscles. One set is external to the rib cage and functions in inhalation. The other set is inside the rib cage and acts during forced exhalation.

bronchiole In the lung, a small airway that leads from a bronchus to the alveoli.
bronchus Airway connecting the trachea to a lung.
diaphragm Muscle between the throacic and abdominal cavities; contracts during inhalation.
epiglottis Tissue flap that folds down to prevent food from entering the airways when you swallow.
glottis Opening formed when the vocal cords relax.
larynx Short airway containing the vocal cords (voice box).
pharynx Throat; opens to airways and digestive tract.
trachea Airway to the lungs; windpipe.

Take-Home Message **How does the structure of the human respiratory system reflect its function?**

> The human respiratory system functions in gas exchange. It also has roles in sense of smell, voice production, body defenses, acid–base balance, and temperature regulation.

> Air enters through the nose or mouth. It flows through the pharynx (throat) and larynx (voice box) to a trachea that branches into two bronchi, one leading to each lung.

> Inside each lung, additional branching airways deliver air to the alveoli, where gases are exchanged with pulmonary capillaries. Actions of the diaphragm and muscles between the ribs alter the size of the chest cavity during breathing.

> Rhythmic signals from the brain bring about the muscle contractions that cause air to flow into the lungs.
< Links to pH 2.6, Brain stem 29.10, Chemoreceptors 30.2

The Respiratory Cycle

A **respiratory cycle** is one breath in (inhalation) and one breath out (exhalation). Inhalation is always active, and muscle contractions drive it. Changes in the volume of the lungs and thoracic cavity during a respiratory cycle alter pressure gradients between air inside and outside the respiratory tract (Figure 35.10).

When you start to inhale, the diaphragm contracts, flattening and moving downward (Figure 35.10A). Intercostal muscles on the outside of the rib cage contract, lifting the rib cage up and expanding it outward. As the thoracic cavity expands, so do the lungs. When pressure in the alveoli falls below atmospheric pressure, air flows down the pressure gradient, into the airways.

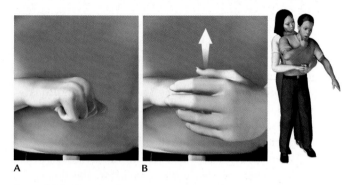

Figure 35.11 Animated How to perform the Heimlich maneuver on an adult who is choking.

1. Determine that the person is actually choking. A person who has an object lodged in their trachea cannot cough or speak.

2. Stand behind the person and place one fist below his or her rib cage, just above the navel, with your thumb facing inward as in **A**.

3. Cover the fist with your other hand as shown in **B** and thrust inward and upward. Repeat until the object is expelled.

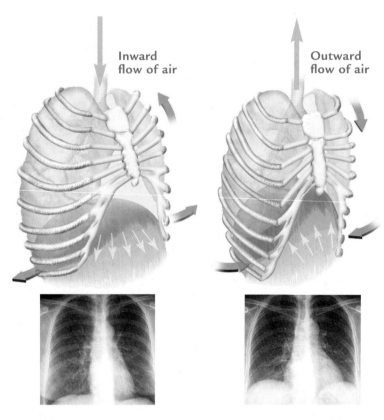

Inward flow of air

Outward flow of air

A Inhalation. Diaphragm contracts, moves down. External intercostal muscles contract, lift rib cage upward and outward. Lung volume expands.

B Exhalation. Diaphragm, external intercostal muscles return to resting positions. Rib cage moves down. Lungs recoil passively.

Figure 35.10 Animated Changes in the size of the thoracic cavity during a single respiratory cycle. The x-ray images reveal how inhalation and expiration change the lung volume.

>> **Figure It Out** What effect does contraction of the diaphragm have on the volume of the thoracic cavity?

Answer: It increases the volume.

Exhalation is usually passive. When the muscles that caused inhalation relax, the lungs passively recoil and lung volume decreases. This compresses alveolar sacs, causing the air pressure inside them to increase above atmospheric pressure. As a result of this increase, air moves out of the lungs (Figure 35.10B).

Exhalation becomes active when you exercise vigorously or consciously attempt to expel more air. During active exhalation, muscles of the abdominal wall contract. Pressure in the abdominal cavity increases, exerting an upward-directed force on the diaphragm. At the same time, contraction of intercostal muscles inside the rib cage pulls the thoracic wall inward and downward. As a result, the volume of the thoracic cavity decreases more than usual and additional air is forced out of the lungs.

In the **Heimlich maneuver**, a rescuer manually raises the intra-abdominal pressure of a choking person to dislodge an object stuck in the trachea (Figure 35.11). By making upward thrusts into the choker's abdomen, a rescuer raises the intra-abdominal pressure, forcing the choker's diaphragm upward. The pressure of air forced out of the lungs by this maneuver can dislodge the object, allowing the victim to resume normal breathing.

Respiratory Volumes

Total lung volume, the maximum volume of air that the lungs can hold, averages 5.7 liters in healthy adult males and 4.2 liters in females. Most of the time, the lungs are about half full. Tidal volume, the volume that moves into and out of lungs during a respiratory cycle, averages about half a liter (Figure 35.12). **Vital capacity**, the maximum volume that moves in and out with forced inhalation and exhalation, is one measure of lung health.

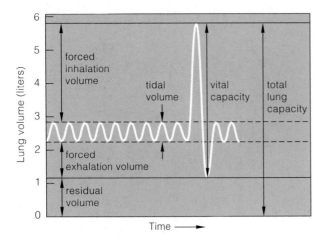

Figure 35.12 **Animated** Respiratory volumes. In quiet breathing, the tidal volume of air entering and leaving the lungs is only 0.5 liter. Lungs never deflate completely. Even with a forced exhalation, a residual volume of air remains in them.

Lungs never fully deflate. As you exhale, the smallest airways collapse, temporarily trapping some air. As a result, air in alveoli is always is a mix of freshly inhaled air and air left behind during the previous exhalation.

Control of Breathing

You do not have to think about breathing. Neurons in the medulla oblongata of the brain stem act as the pacemaker for inhalation, initiating an action potential 10–20 times per minute. Nerves deliver signals calling for contraction to the diaphragm and intercostal muscles and you inhale. Between signals, the muscles relax and you exhale.

Breathing patterns change with activity level. When you are more active, muscle cells produce more CO_2. This CO_2 enters blood, where it combines with water and forms carbonic acid. The acid dissociates, increasing the hydrogen ion (H^+) concentration in the blood and in cerebrospinal fluid. Chemoreceptors in the walls of carotid arteries and the aorta detect the increased acidity and signal the brain, which responds by altering the breathing pattern (Figure 35.13). You breathe faster and deeper, so more carbon dioxide is expelled. As a result, blood pH returns to normal.

Chemoreceptors in the artery walls also signal the medulla oblongata when the O_2 concentration in the blood falls to a life-threatening level. However, this con-

Heimlich maneuver Procedure designed to rescue a choking person; a rescuer presses on a person's abdomen to force air out of the lungs and dislodge an object in the trachea.
respiratory cycle One inhalation and one exhalation.
vital capacity Amount of air moved in and out of lungs with forced inhalation and exhalation.

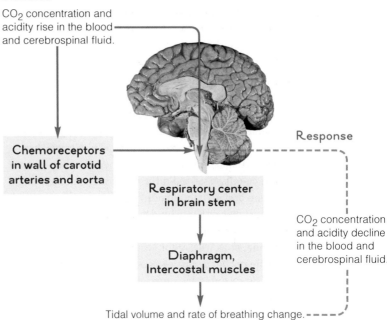

Stimulus

CO_2 concentration and acidity rise in the blood and cerebrospinal fluid.

Chemoreceptors in wall of carotid arteries and aorta

Respose

Respiratory center in brain stem

CO_2 concentration and acidity decline in the blood and cerebrospinal fluid.

Diaphragm, Intercostal muscles

Tidal volume and rate of breathing change.

Figure 35.13 Respiratory response to increased activity levels. An increase in activity raises the CO_2 level in interstitial tissue. It also makes the blood and cerebrospinal fluid more acidic. Chemoreceptors in blood vessels and the medulla sense the changes and signal the brain's respiratory center, also in the brain stem.

In response, the respiratory center sends signals along nerves to the diaphragm and intercostal muscles. These signals result in an increase in the rate and depth of breathing. Excess CO_2 is expelled, which causes the level of this gas and acidity to decline. Chemoreceptors sense the decline and signal the respiratory center, which returns to its resting signaling pattern.

trol mechanism usually comes into play only in people with severe lung diseases and at very high altitudes, where there is little oxygen in the air. (We discuss high-altitude breathing more in the next section.)

Reflexes such as swallowing or coughing can briefly halt breathing. Commands from sympathetic nerves make you breathe faster if you are frightened (Section 29.8). Breathing patterns can also be deliberately altered, as when you hold your breath, or break normal breathing rhythm to talk or sing.

Take-Home Message **What happens when we breathe?**

> Inhalation is always an active process. Contraction of the diaphragm and intercostal muscles increases the volume of the thoracic cavity. As a result, air pressure in alveoli declines below atmospheric pressure and air moves inward.

> Exhalation is usually passive. As muscles relax, the thoracic cavity shrinks back down, air pressure in alveoli rises above atmospheric pressure, and air moves out of the lungs.

> Only some of the air in the lungs is replaced with each breath. The lungs are never fully emptied of air.

> The medulla oblongata in the brain stem controls rate and depth of breathing.

> Gases diffuse between air and fluid at alveoli, and are transported to and from alveoli in the blood.

< Links to Epithelium 28.3, Red blood cells 33.5

The Respiratory Membrane

The oxygen drawn into the lungs by inhalation diffuses from an alveolus into a pulmonary capillary at the lung's **respiratory membrane**. This thin membrane consists of alveolar epithelium, capillary endothelium, and their fused basement membranes (Figure 35.14). Secretions keep the alveolar side of the respiratory membrane moist, so gases can dissolve and diffuse across the membrane.

Oxygen and carbon dioxide diffuse passively across the respiratory membrane. The net direction of movement for these gases depends on concentration gradients across the membrane, or as we say for gases, partial pressure gradient. The **partial pressure** of a gas is its contribution to the pressure exerted by a mix of gases. It is measured in millimeters of mercury (mm Hg). Just as a solute tends to diffuse in response to its concentration gradient, so does a gas. If the partial pressure of a gas differs between two regions, the gas will diffuse from the region of higher partial pressure to the region of lower partial pressure.

Oxygen Transport and Storage

Inhaled air that reaches alveoli has a higher partial pressure of O_2 than does blood in pulmonary capillaries. As a result, O_2 tends to diffuse from the air into the blood of these capillaries. After O_2 enters the blood, most diffuses into red blood cells, where it binds to hemoglobin. An iron-containing heme group associates with each of the four globin subunits in hemoglobin (Figure 35.15). When O_2 is bound to one or more of hemoglobin's heme groups, we refer to the molecule as **oxyhemoglobin**.

Heme binds oxygen only weakly. It releases oxygen in places where the partial pressure of O_2 is lower than that in the alveoli. This occurs in metabolically active tissues, as a comparison of the boxes color-coded *pink* in Figure

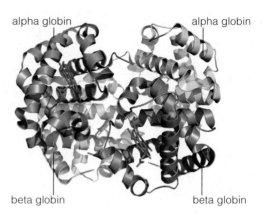

Figure 35.15 Hemoglobin, the oxygen-transporting protein of red blood cells. It consists of four globin chains, each associated with an iron-containing heme group (shown in *red*).

35.16 shows. Metabolically active tissues also have other traits that encourage release of oxygen from heme: high temperature, low pH, and high CO_2 partial pressure.

Carbon Dioxide Transport

Carbon dioxide diffuses into blood capillaries from any tissue with a higher CO_2 partial pressure than blood. As a comparison of the boxes color-coded *blue* in Figure 35.16 shows, metabolically active tissues are such regions.

Carbon dioxide is transported to the lungs in three forms. About 10 percent remains dissolved in plasma. Another 30 percent reversibly binds with hemoglobin and forms carbaminohemoglobin ($HbCO_2$). However, most CO_2 that diffuses into the plasma—60 percent—is transported as bicarbonate (HCO_3^-).

How does bicarbonate form? Carbon dioxide combines with water, forming carbonic acid (H_2CO_3). Carbonic acid then splits into bicarbonate and H^+:

$$CO_2 + H_2O \rightleftharpoons \underset{\text{carbonic acid}}{H_2CO_3} \rightleftharpoons \underset{\text{bicarbonate}}{HCO_3^-} + H^+$$

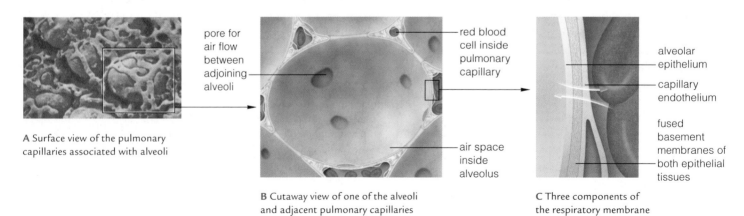

A Surface view of the pulmonary capillaries associated with alveoli

pore for air flow between adjoining alveoli

red blood cell inside pulmonary capillary

air space inside alveolus

alveolar epithelium

capillary endothelium

fused basement membranes of both epithelial tissues

B Cutaway view of one of the alveoli and adjacent pulmonary capillaries

C Three components of the respiratory membrane

Figure 35.14 Zooming in on the respiratory membrane in human lungs.

Carbonic anhydrase, an enzyme inside red blood cells, speeds this reaction. HCO_3^- tends to diffuse out of red blood cells into the plasma. Most of the H^+ binds to hemoglobin. The reverse reactions occur in the alveoli, where the CO_2 partial pressure is lower than that in lung capillaries. Water and CO_2 form inside the alveoli and leave the body in exhalations.

The Carbon Monoxide Threat

Carbon monoxide (CO) is a colorless, odorless gas that is present in the smoke from cigarettes and fossil fuel combustion. Carbon monoxide is dangerous because hemoglobin has a higher affinity for CO than for O_2. When CO builds up in the air, it fills and blocks O_2 binding sites on hemoglobin, preventing transport of O_2 and causing carbon monoxide poisoning. Nausea, headache, confusion, dizziness, and weakness set in as tissues are starved of oxygen. In the United States, accidental CO poisoning kills about 500 people each year. To minimize your risk, be sure fuel-burning appliances are properly vented to the outside, and install a carbon monoxide detector.

Effects of Altitude

Atmospheric pressure decreases with altitude. At about 5,500 meters, or about 18,000 feet, air pressure is half that at sea level. Oxygen makes up the same percentage of the total pressure (21 percent), so there is about half as much oxygen as there is at sea level. Most people live at lower altitudes where there is plenty of oxygen. When they suddenly ascend to high altitudes, their cells get less oxygen than usual and altitude sickness results. Symptoms include shortness of breath, headache, and nausea.

A healthy person who is unaccustomed to life at a high altitude can become physiologically adjusted to such an environment over time. Through acclimatization, the body makes long-term adjustments in cardiac output, and in the rate and magnitude of breathing. In addition, the kidney secretes more erythropoietin, a hormone that stimulates red blood cell formation. Increased numbers of red blood cells improve the oxygen-delivery capacity of blood. However, an altitude-induced increase in red blood cell count can put a strain on the heart. Having more blood cells thickens blood, so the heart has to work harder.

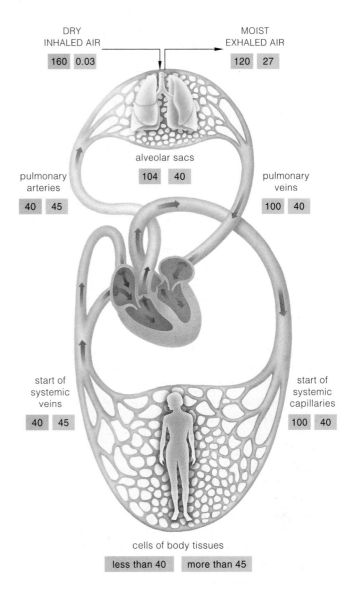

Figure 35.16 Animated Partial pressures (in mm Hg) for oxygen (*pink* boxes) and carbon dioxide (*blue* boxes) in the atmosphere, blood, and tissues.

❯❯ **Figure It Out** Where is the biggest drop in partial pressure of O_2?

Answer: Between the start of the systemic capillaries and systemic veins.

---

carbonic anhydrase Enzyme in red blood cells that speeds the breakdown of carbonic acid into bicarbonate and H^+.
oxyhemoglobin Hemoglobin with oxygen bound to it.
partial pressure Pressure exerted by one gas in a mixture of gases.
respiratory membrane Membrane consisting of alveolar epithelium, capillary endothelium, and their fused basement membranes; site of gas exchange in the lungs.

Take-Home Message How do gases enter blood and how are they transported?

❯ Oxygen enters blood by diffusing across the respiratory membrane and into pulmonary capillaries. It binds to hemoglobin in red blood cells and is carried to tissues, where a low partial pressure of oxygen and other factors encourage its release.

❯ Carbon dioxide diffuses into capillaries in active tissues. Most combines with water to form carbonic acid, which splits into bicarbonate and H^+. Bicarbonate dissolves in plasma. In alveoli, carbon dioxide and water form and are exhaled.

> Interrupted breathing, infectious organisms, and chronic inflammation can impair respiratory function.
< Links to Antibiotic resistance 17.5, Neurotransmitters 29.6, Inflammation 34.4

Interrupted Breathing A tumor or damage to the brain stem's medulla oblongata can affect respiratory controls and cause apnea. In this disorder, breathing repeatedly stops and restarts spontaneously, especially during sleep. More often, sleep apnea occurs when the tongue, tonsils, or other soft tissue obstructs the upper airways. Breathing may stop for up to several seconds many times each night causing daytime fatigue. Risk for heart attacks and strokes rises with sleep apnea, because when breathing stops, blood pressure soars. Obstructive sleep apnea can be reduced by changes in sleeping position or by wearing a mask that delivers pressurized air. Severe cases require surgical removal of tissue that blocks airways.

Sudden infant death syndrome (SIDS) occurs when an infant does not awaken from an episode of apnea. A defect in the medulla oblongata is associated with SIDS. Autopsies reveal that infants who died of SIDS tend to have fewer receptors for the neurotransmitter serotonin than infants who died of other causes. Having fewer of these receptors may impair the medulla's response to potentially deadly respiratory stress. There are also environmental risk factors. Maternal smoking during pregnancy heightens the risk, and infants who sleep on their back are at lower risk than stomach sleepers.

Tuberculosis and Pneumonia Worldwide, about one in three people is currently infected by bacteria that can cause tuberculosis (*Mycobacterium tuberculosis*). Most infected people are carriers who have no symptoms, but about ten percent of those infected develop "active TB." They cough up bloody mucus, have chest pain, and find breathing difficult. If untreated, an active case of TB can be fatal. Antibiotics can cure most infections, but they must be taken diligently for at least six months. In addition, multi-drug-resistant strains of *M. tuberculosis* are increasing in frequency.

Pneumonia is a general term for lung inflammation caused by an infection. Bacteria, viruses, and fungi can infect the lungs and cause pneumonia. Typical symptoms include a cough, an aching chest, shortness of breath, and fever. An x-ray reveals infected tissues filled with fluid and white blood cells instead of air (Figure 35.17). Treatment and outcome depend on the type of pathogen.

Bronchitis, Asthma, and Emphysema Your bronchi are lined by a ciliated, mucus-producing epithelium that helps protect you from respiratory infections. An inflammation of this epithelium is called bronchitis. Inflamed epithelial

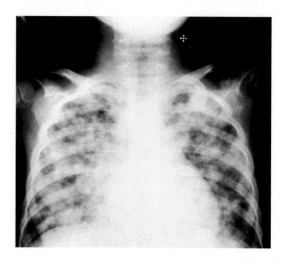

Figure 35.17 X-ray showing pneumonia. Fluid and blood cells fill the lungs. Compare x-rays of clear, healthy lungs in Figure 35.10.

cells secrete extra mucus that triggers the coughing reflex. Bacteria can colonize the mucus, leading to more inflammation, more mucus, and more coughing. Bronchitis often arises after an upper respiratory infection. Repeatedly inhaling irritants such as cigarette smoke can cause chronic bronchitis.

With asthma, an inhaled allergen or irritant triggers inflammation and constriction of the airways, conditions that make breathing difficult. A tendency to have asthma is inherited, but avoiding potential irritants such as cigarette smoke and air pollutants can reduce the frequency of asthma attacks. An acute asthma attack is treated with inhaled drugs that cause dilation of smooth muscle around airways.

With emphysema, tissue-destroying bacterial enzymes digest the thin, elastic alveolar wall. As these walls disappear, the area of the respiratory surface declines. Over time, the lungs become distended and inelastic, leaving the person constantly feeling short of breath. Some people inherit a genetic predisposition to emphysema. They do not have a functioning gene for an enzyme that inhibits bacterial attacks on alveoli. However, tobacco smoking is by far the main risk factor for emphysema.

Take-Home Message What causes common respiratory problems?

> Apnea, or interrupted breathing, is caused by tissue obstructing airways or a defective respiratory pacemaker.

> Tuberculosis is a widespread bacterial disease that can be fatal, although most people have no symptoms. Pneumonia can be caused by many different pathogens.

> In asthma and bronchitis, airways become inflamed and constricted. In emphysema, alveolar sacs become distended and inelastic.

Up in Smoke (revisited)

Globally, cigarette smoking kills 4 million people each year. By 2030, the number may rise to 10 million, with about 70 percent of the deaths occurring in developing countries. In the United States, the direct medical costs of treating smoke-induced disorders drain $22 billion a year from the economy. G. H. Brundtland, a medical doctor and the former director of the World Health Organization, points out that tobacco is the only legal consumer product that kills half of its regular users. If you are a cigarette smoker, you may wish to reflect on the information in Figure 35.18. Consider also that nonsmokers die of cancers and disease brought on by breathing secondhand smoke. Children who breathe cigarette smoke at home have a heightened risk for illnesses such as asthma, bronchitis, and ear infections.

Smoking marijuana (*Cannabis*) also poses some respiratory risks. Marijuana contains fewer toxic particles, or "tar," than tobacco, but marijuana is usually smoked without a filter. Also, people smoking marijuana tend to inhale more deeply than tobacco smokers, to hold hot smoke in their lungs for longer periods, and to smoke their cigarettes down to stubs, where tar accumulates. As a result, long-term marijuana smokers have an increased risk of respiratory problems, and they tend to show lung damage earlier than cigarette smokers. Some recent epidemiological studies have found a heightened risk of lung cancer in long-term marijuana smokers.

How Would You Vote? Should the United States encourage efforts to reduce tobacco use around the world, even if it means less profits for American tobacco companies? See CengageNow for details, then vote online (cengagenow.com).

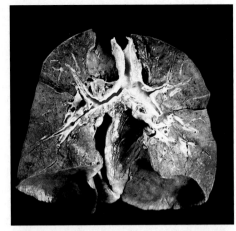

lungs of a nonsmoker

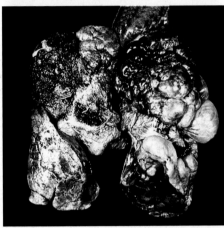

lungs of a smoker

Effects of Smoking

Shortened life expectancy Nonsmokers live about 8.3 years longer than those who smoke two packs a day from their midtwenties on.

Chronic bronchitis, emphysema Smokers have 4–25 times higher risk of dying from these diseases than do nonsmokers.

Cancer of lungs Cigarette smoking is the major cause.

Cancer of mouth 3–10 times greater risk among smokers.

Cancer of larynx 2.9–17.7 times more frequent among smokers.

Cancer of esophagus 2–9 times greater risk of dying from this.

Cancer of pancreas 2–5 times greater risk of dying from this.

Cancer of bladder 7–10 times greater risk for smokers.

Cardiovascular disease Cigarette smoking is a major contributing factor in heart attacks, strokes, and atherosclerosis.

Impact on offspring Women who smoke during pregnancy have more stillbirths, and the weight of liveborns is lower than the average (which makes babies more vulnerable to disease and death).

Impaired immunity More allergic responses, destruction of white blood cells (macrophages) in respiratory tract.

Slow bone healing Surgically cut or broken bones may take 30 percent longer to heal in smokers, perhaps because smoking depletes the body of vitamin C and reduces the amount of oxygen delivered to tissues. Reduced vitamin C and reduced oxygen interfere with formation of collagen fibers in bone (and many other tissues).

Benefits of Quitting

Cumulative risk reduction; after 10–15 years, the life expectancy of ex-smokers approaches that of nonsmokers.

Greater chance of improving lung function and slowing down rate of deterioration.

After 10–15 years, risk approaches that of nonsmokers.

After 10–15 years, risk is reduced to that of nonsmokers.

After 10 years, risk is reduced to that of nonsmokers.

Risk proportional to amount smoked; quitting should reduce it.

Risk proportional to amount smoked; quitting should reduce it.

Risk decreases gradually over 7 years to that of nonsmokers.

Risk for heart attack declines rapidly, for stroke declines more gradually, and for atherosclerosis it levels off.

When smoking stops before fourth month of pregnancy, risk of stillbirth and lower birth weight eliminated.

Avoidable by not smoking.

Avoidable by not smoking.

Figure 35.18 Risks incurred by smokers and benefits of quitting. The photos show normal lung tissue and lungs of a smoker who had emphysema.

Summary

Section 35.1 Smoking damages cells that protect airways from pathogens. It also raises the risk of heart disease and lung and breast cancer. Inhalation of secondhand smoke also has adverse effects.

Section 35.2 Aerobic respiration uses O_2 and yields CO_2 as a product. **Respiration** is a physiological process by which O_2 enters the internal environment and CO_2 leaves it. Both gases diffuse across a **respiratory surface**. **Respiratory proteins** facilitate gas exchange by keeping concentration gradients between the blood and cells steep.

Section 35.3 Small invertebrates that live in aquatic or damp habitats exchange gases mainly across the body surface. Many aquatic invertebrates have **gills** with blood running through them. A **tracheal system** delivers air to cells deep inside the body of insects.

Section 35.4 Water and blood flow in opposing directions at fish gills, allowing a highly efficient **countercurrent exchange** of gases. Most land vertebrates have paired **lungs**, although amphibians also exchange gases across the skin. Frogs pull air into their mouth, then push it into their lungs. Other tetrapods pull air into their lungs. Mammalian gas exchange occurs in tiny sacs called **alveoli**. Birds have a more efficient system; gas exchange occurs as air flows through tiny tubes in their lungs.

Section 35.5 In humans, air flows through two nasal cavities and a mouth into the **pharynx**, then the **larynx**, then the **trachea** (windpipe). The larynx contains the vocal cords, movements of which alter the size of the opening (the **glottis**) between them. When you swallow, the position of the **epiglottis** at the entrance to the larynx shifts, keeping food out of the trachea. The trachea branches into two **bronchi** that enter the lungs. These two airways and finely branching **bronchioles** form the bronchial tree. At the ends of the finest branches of this tree are the thin-walled alveoli. The **diaphragm** at the base of the thoracic cavity and the muscles between ribs are involved in breathing.

Section 35.6 A **respiratory cycle** is one inhalation and one exhalation. Inhalation is active. As muscle contractions expand the chest cavity, pressure in lungs decreases below atmospheric pressure, and air flows into the lungs. These events are reversed during exhalation, which normally is passive. The most air that can move in and out in one cycle is the **vital capacity**. Cells in the brain stem adjust the rate and magnitude of breathing. The **Heimlich maneuver** increases pressure in lungs to diplace an object that is blocking the trachea.

Sections 35.7, 35.8 In human lungs, the alveolar wall, the wall of a pulmonary capillary, and their fused basement membranes form a thin **respiratory membrane** between air inside an alveolus and the internal environment. O_2 following its **partial pressure** gradient diffuses across the respiratory membrane, into the plasma of the blood, and finally into red blood cells.

Where O_2 partial pressure is high, hemoglobin in red blood cells binds O_2 and forms **oxyhemoglobin**. Heme groups release O_2 where its partial pressure is low.

CO_2 follows its partial pressure gradient and diffuses from cells to interstitial fluid, to blood. Most CO_2 reacts with water in red blood cells, forming bicarbonate. The enzyme **carbonic anhydrase** speeds this reaction. The reaction is reversed in the lungs. There, CO_2 diffuses out of blood into air inside alveoli. It is expelled, along with water vapor, in exhalations.

Carbon monoxide poisoning occurs when this gas binds to hemoglobin and prevents oxygen transport.

The amount of oxygen in air decreases with altitude. People from a low altitude can acclimatize to high altitude through altered breathing patterns, increased red blood cell production, and other changes.

Respiratory disorders include apnea and sudden infant death syndrome (SIDS). Tuberculosis, pneumonia, bronchitis, and emphysema are respiratory diseases. Smoking worsens or causes many respiratory problems.

Self-Quiz   Answers in Appendix III

1. Respiratory proteins such as hemoglobin _____ .
   a. contain metal ions
   b. occur only in vertebrates
   c. increase the efficiency of oxygen transport
   d. both a and c

2. In _____ gas exchange occurs at the body surface and gas is distributed by diffusion alone.
   a. earthworms   c. frogs
   b. flatworms   d. insects

3. Countercurrent flow of water and blood increases the efficiency of gas exchange in _____ .
   a. fishes   c. birds
   b. amphibians   d. all of the above

4. In human lungs, gas exchange occurs at the _____ .
   a. two bronchi   c. alveolar sacs
   b. pleural sacs   d. both b and c

5. When you breathe quietly, inhalation is _____ and exhalation is _____ .
   a. passive; passive   c. passive; active
   b. active; active   d. active; passive

6. During inhalation _____ .
   a. the thoracic cavity expands
   b. the diaphragm relaxes
   c. atmospheric pressure declines

Data Analysis Activities

Risks of Radon Radon is a colorless, odorless gas emitted by many rocks and soils. It is formed by the radioactive decay of uranium and is itself radioactive. There is some radon in the air almost everywhere, but routinely inhaling a lot of it raises the risk of lung cancer. Radon also seems to increase cancer risk far more in smokers than in nonsmokers. Figure 35.19 is an estimate of how radon in homes affects risk of lung cancer mortality. Note that these data show only the risk of death for radon-induced cancers. Smokers are also at risk from lung cancers that are caused by tobacco alone.

1. If 1,000 smokers were exposed to a radon level of 1.3 pCi/L over a lifetime (the average indoor radon level) how many would die of a radon-induced lung cancer?
2. How high would the radon level have to be to cause approximately the same number of cancers among 1,000 nonsmokers?
3. The risk of dying in a car crash is about 7 out of 1,000. Is a smoker in a home with an average radon level (1.3 pCi/L) more likely to die in a car crash or of radon-induced cancer?

Radon Level (pCi/L)	Risk of Cancer Death From Lifetime Radon Exposure	
	Never Smoked	Current Smokers
20	36 out of 1,000	260 out of 1,000
10	18 out of 1,000	150 out of 1,000
8	15 out of 1,000	120 out of 1,000
4	7 out of 1,000	62 out of 1,000
2	4 out of 1,000	32 out of 1,000
1.3	2 out of 1,000	20 out of 1,000
0.4	>1 out of 1,000	6 out of 1,000

Figure 35.19 Estimated risk of lung cancer death as a result of lifetime radon exposure. Radon levels are measured in picocuries per liter (pCi/L). The Environmental Protection Agency considers a radon level above 4 pCi/L unsafe. To learn about testing for radon and what to do if the radon level is high, visit the EPA's Radon Information Site at www.epa.gov/radon.

---

7. What type of metal associates with hemoglobin?

8. _____ binds to hemoglobin more strongly than oxygen does.
   a. Carbon dioxide   c. Oxyhemoglobin
   b. Carbon monoxide   d. Carbonic anhydrase

9. Carbonic anhydrase in red blood cells catalyzes formation of bicarbonate from water and _____ .
   a. oxygen   c. oxyhemoglobin
   b. hemoglobin   d. carbon dioxide

10. The hormone erythropoietin causes _____ .
    a. increased heart rate   c. red blood cell formation
    b. deeper breathing   d. all of the above

11. _____ in arteries sense changes in the acidity of the blood.
    a. Mechanoreceptors   c. Photoreceptors
    b. Neurotransmitters   d. Chemoreceptors

12. True or false? Human lungs hold some air even after forced exhalation.

13. The diaphragm is a _____ muscle.
    a. smooth   b. skeletal   c. cardiac

14. What type of organism causes tuberculosis?

15. Match the words with their descriptions.
    ___trachea   a. muscle of respiration
    ___pharynx   b. gap between vocal cords
    ___alveolus   c. between bronchi and alveoli
    ___hemoglobin   d. windpipe
    ___bronchus   e. respiratory protein
    ___bronchiole   f. site of gas exchange
    ___glottis   g. airway leading to lung
    ___diaphragm   h. throat

Additional questions are available on CENGAGENOW.

Critical Thinking

1. The red blood cell enzyme carbonic anhydrase contains the metal zinc. Humans obtain zinc from their diet, especially from red meat and some seafoods. A zinc deficiency does not reduce the number of red blood cells, but it does impair respiratory function by reducing carbon dioxide output. Explain why a zinc deficiency has this effect.

2. Look again at Figure 35.16. Notice that the oxygen and carbon dioxide content of blood in pulmonary veins is the same as at the start of the systemic capillaries. Notice also that systemic veins and pulmonary arteries have equal partial pressures. Explain the reason for the similarities.

3. Respiration supplies cells with the oxygen they need for aerobic respiration. Explain the role of oxygen in this energy-releasing metabolic pathway. Where is it used and what is its function?

4. A developing fetus gets oxygen from its mother's blood. Fetal capillaries run through pools of maternal blood in an organ called the placenta. As fetal blood runs through the capillaries, it exchanges substances with the maternal blood around the capillary. The hemoglobin made by a fetus is different than that made after birth. Fetal hemoglobin binds oxygen more strongly at low oxygen levels than normal hemoglobin. How would fetal hemoglobin's somewhat higher affinity for oxygen benefit the fetus?

Animations and Interactions on CENGAGENOW:
❯ Fish respiratory system; How a frog breathes; Bird respiratory system; Human respiratory system; Respiratory cycle; Heimlich maneuver; Partial pressures of oxygen and carbon dioxide.

‹ Links to Earlier Concepts

This chapter expands the discussion of digestive systems in Chapters 23 and 24. You will consider dietary aspects of organic compounds (Sections 3.3–3.5) and the use and storage of glucose (7.7). You will draw on your knowledge of diffusion, transport mechanisms, and osmosis (5.6–5.8). You will revisit pH and cofactors (2.6, 5.4). Your knowledge of epithelium (28.3) and adipose tissue (28.4) is also relevant here.

Key Concepts

Animal Digestive Systems
Some animal digestive systems are saclike, but most are a tube with two openings. The structure of an animal's digestive system often reveals adaptations that allow the animal to eat a specific diet. For example, antelopes have adaptations that allow them to eat grasses.

Human Digestive System
In humans, food taken into the mouth is mechanically broken down by teeth. Saliva starts the process of chemical digestion. The stomach stores and digests food. Digestion is completed and absorption begins in the small intestine. The large intestine concentrates wastes.

36 Digestion and Human Nutrition

36.1 The Battle Against Bulge

Food supplies your body with raw materials and fuel. When the food you eat contains more energy than you need at the time, you store the excess as bond energy in organic compounds. The body's largest energy store is fat in adipose tissue.

For most of our species' history, an ability to store energy as fat was selectively advantageous. Putting on fat when food is abundant increases the likelihood of survival if food later becomes scarce. However, most people in the United States now have more than enough food all of the time. As a result, about two-thirds of adults are overweight or obese. Excess weight is a matter of concern because, compared to people at a healthy weight, obese people have a higher risk of type 2 diabetes, cardiovascular disease, some cancers, and other disorders.

Why do some people stay slim with no apparent effort, whereas others diet constantly but remain heavy? Genetics certainly plays a role. In terms of weight class (thin, average, overweight, or obese), people raised by adoptive parents resemble their biological parents more than their adoptive ones.

Geneticists have also pinpointed genes that affect weight. The *ob* gene, the first to be discovered, encodes leptin, a hormone made by adipose cells. Leptin acts in the brain and suppresses appetite. Mice that do not have a functional *ob* gene overeat and become fat (Figure 36.1). Similarly, people who cannot make leptin are severely obese. When leptin-deficient mice or humans are injected with leptin, they eat less and slim down.

Human leptin deficiency is extremely rare, but variations in another obesity-related gene, *fto*, are common. About 16 percent of people of European ancestry are homozygous for an *fto* allele that predisposes them to obesity. Compared to people with two low-risk alleles, those homozygous for the high-risk allele are almost twice as likely to be obese. The function of the protein encoded by the *fto* gene is unknown. We do, however, know that the gene is expressed most strongly in the brain and that it affects food intake. People with the high-risk allele tend to eat more food before they feel full.

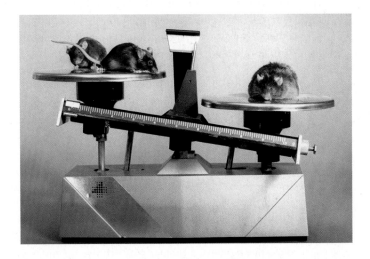

Figure 36.1 Causes of obesity. *Above*, two normal mice (*left*) and a mouse with a mutant *ob* gene (*right*). Genetic differences that affect food intake help explain why some individuals are more likely to become obese.

Opposite page, portion sizes of fast-food meals have increased dramatically since 1980, as have the number of these meals consumed per week.

Genetics can explain why one person is more likely than another to be overweight, but it cannot explain a national trend toward weight gain. Since 1980, the proportion of obese adults in the United States has doubled, and the proportion of obese children has tripled. An increasing dependence on fast food contributes to that trend. Portion sizes of fast-food meals tend to be far above those recommended for a healthy diet, and have risen consistently in the past 20 years.

Discussion of food intake leads us into the world of nutrition. The word encompasses all the processes by which an animal ingests and digests food, then absorbs the released nutrients as energy sources and building blocks for cells. When all works well, all of the body's nutrient needs are met and weight remains at a healthy level.

A Closer Look at Digestive Organs
The stomach is an expandable sac that secretes acidic gastric fluid. The small intestine has a highly folded interior and a huge surface area. It receives enzymes from the pancreas and bile from the gallbladder. The bile is made in the liver.

Human Nutrition
Nutrients absorbed from the gut are raw materials for synthesis of complex carbohydrates, lipids, proteins, and nucleic acids. A healthy diet provides all nutrients, vitamins, and minerals necessary to support metabolism. It is low in salt, simple sugars, and saturated fats.

Maintaining a Healthy Body Weight
Maintaining body weight requires balancing calories taken in with calories burned in metabolism and physical activity. A body weight far above or below normal increases the risk of health problems.

> Animals are heterotrophs that typically digest food inside their body, but outside of their cells.

< Links to Animal body plans 23.2, Bill shape and natural selection 17.6

An animal's digestive system is a body cavity or a tube that mechanically and chemically breaks food down to small particles, then to molecules small enough to get absorbed into the internal environment. The digestive system also expels unabsorbed residues. By interacting with other organ systems, it contributes to homeostasis for the body as a whole.

Incomplete and Complete Systems

Remember from Section 23.2 that some invertebrates have an **incomplete digestive system**. Food enters and wastes leave their saclike gut through a single opening at the body surface. The saclike, branching gut cavity of flatworms opens at the tip of a pharynx, a muscular tube (Figure 36.2A). Food that enters the sac is digested, its nutrients are absorbed, then wastes are expelled. This two-way traffic does not allow for regional specialization.

Most invertebrates and all vertebrates have a **complete digestive system**: a tubular gut with two openings. Food enters at one end; wastes leave through the other. Specialized regions along the tube's length process food, absorb nutrients, and concentrate wastes.

A frog has a complete digestive system (Figure 36.2B). The tubular part includes a mouth, pharynx, esophagus stomach, small intestine, and large intestine. Digestive wastes exit through a cloaca (as do urinary wastes and gametes). The liver, gallbladder, and pancreas assist digestion by secreting substances into the small intestine.

Regardless of the complexity, a complete digestive systems carries out five overall tasks:

1. *Mechanical processing and motility.* Movements that break up, mix, and directionally propel food material.
2. *Secretion.* Release of substances, especially digestive enzymes, into the lumen (the space inside the tube).
3. *Digestion.* Breakdown of food into particles, then to nutrient molecules small enough to be absorbed.
4. *Absorption.* Uptake of digested nutrients and water across the gut wall, into extracellular fluid.
5. *Elimination.* Expulsion of undigested or unabsorbed solid residues.

Diet–Related Structural Adaptations

Features of an animal digestive systems are shaped by natural selection and adapt the animal to a particular diet.

Beaks and Bites Birds do not have teeth; their jawbones are covered with a layer of the protein keratin to form a beak. The size and shape of a bird's beak determines what kinds of food it can process. Section 17.6 discussed the effect of beak size in African finches. As another example, the pigeon shown in Figure 36.2C has a relatively small beak that it uses to pick up seeds from the ground.

Mammals have different types of teeth, depending on their diet. Carnivores tend to have long, sharp canine teeth for killing prey, whereas herbivores typically lack canine teeth and have prominent incisors. The shape of individual teeth also reflects diet. Humans and antelopes both have molars that they use to grind food, but the relative size of the molar crown differs dramatically (Figure 36.3A). The large crown of the antelope molar is an adap-

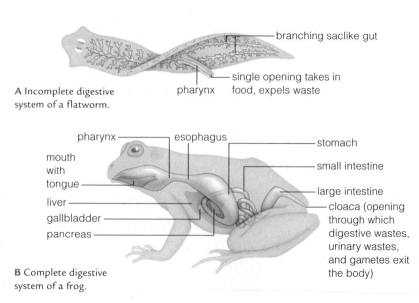

A Incomplete digestive system of a flatworm.

— branching saclike gut

pharynx — single opening takes in food, expels waste

B Complete digestive system of a frog.

pharynx — esophagus
mouth with tongue
liver
gallbladder
pancreas
— stomach
— small intestine
— large intestine
— cloaca (opening through which digestive wastes, urinary wastes, and gametes exit the body)

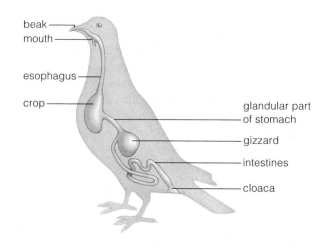

beak
mouth
esophagus
crop
glandular part of stomach
gizzard
intestines
cloaca

C Complete digestive system of a bird. The crop is an expandable organ for storing food. The muscular gizzard grinds up food.

Figure 36.2 Animated Examples of animal digestive systems.

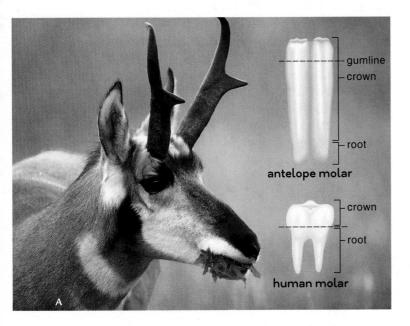

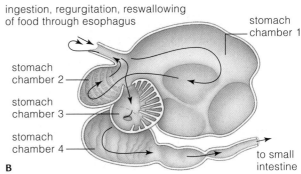

Figure 36.3 Animated Dietary adaptations of an antelope. **A** Human and antelope lower molars. **B** An antelope's multiple stomach chambers. In the first two, food is mixed with fluid and exposed to microbial symbionts that engage in fermentation. Some of the microbes degrade cellulose; others synthesize organic compounds, fatty acids, and vitamins. Partly digested food is regurgitated into the mouth, chewed, then swallowed. It enters the third chamber and is digested again before entering the last stomach chamber.

tation to a diet of plant material that is often mixed with abrasive soil particles. An enlarged crown keeps the soil from wearing an antelope's molars down to nubs.

Gut Specialization Like other seed eaters, a pigeon has a large crop, a saclike food-storing region above the stomach. The bird quickly fills its crop with seeds, then digests them later, in safer places. Birds grind up food inside a gizzard: a stomach chamber lined with hard protein particles. Compared to hawks and other meat-eating birds, seed eaters have larger gizzards relative to their body size.

Cattle, goats, sheep, and antelopes are **ruminants**, hoofed grazers with multiple stomach chambers (Figure 36.3B). This system allows ruminants to maximize the nutrients they extract from plant foods rich in cellulose. Microbes in the first two stomach chambers carry out fermentation reactions that break down cellulose in plant cell walls. Solids accumulate in the second chamber, forming "cud" that is regurgitated—moved back into the mouth for a second round of chewing. Nutrient-rich fluid moves from the second chamber to the third and fourth chambers, and finally to the intestine.

Meat takes less time to digest than plant material, so carnivores typically have a shorter gut than grazers. Their stomach can often stretch enormously. A highly expandable stomach allows them to get food inside their body fast, so competitors cannot access it (Figure 36.4).

Figure 36.4 Python engulfing its prey. Pythons may eat only once or twice a year and some species can swallow a prey animal more than twice their weight.

Take-Home Message What is a digestive system and how does its structure reflect its function?

❯ Digestive systems mechanically and chemically degrade food into small molecules that can be absorbed, along with water, into the internal environment. These systems also expel the undigested residues from the body.

❯ Incomplete digestive systems are a saclike cavity with one opening. Complete digestive systems are a tube with two openings and regional specializations in between.

❯ Structural variations in bills, teeth, and regions of the gut are adaptations that allow an animal to exploit a particular type or types of foods.

complete digestive system Tubelike digestive system; food enters through one opening and wastes leave through another.
incomplete digestive system Saclike digestive system; food enters and leaves through the same opening.
ruminant Hoofed mammal with a multiple-chamber stomach that adapts it to a cellulose-rich diet.

> Humans have a complete digestive system with specialized regions along its length. Glands and secretory organs secrete enzymes and other substances into the tubular portion of this system.

< Links to Epithelium 28.3, Taste 30.7, Trachea 35.5

Like other vertebrates, humans have a complete digestive system; a tubular gut with two openings (Figure 36.5A).

If it were stretched out in a straight line, the gut would extend 6.5 to 9 meters (21 to 30 feet). Different regions specialize in digesting food, absorbing nutrients, and concentrating and storing unabsorbed waste. Salivary glands, the pancreas, and the liver are accessory organs that secrete substances into the tube (Figure 36.5B).

Food enters the body through the mouth, or oral cavity. The tongue, an organ that consists of membrane-covered

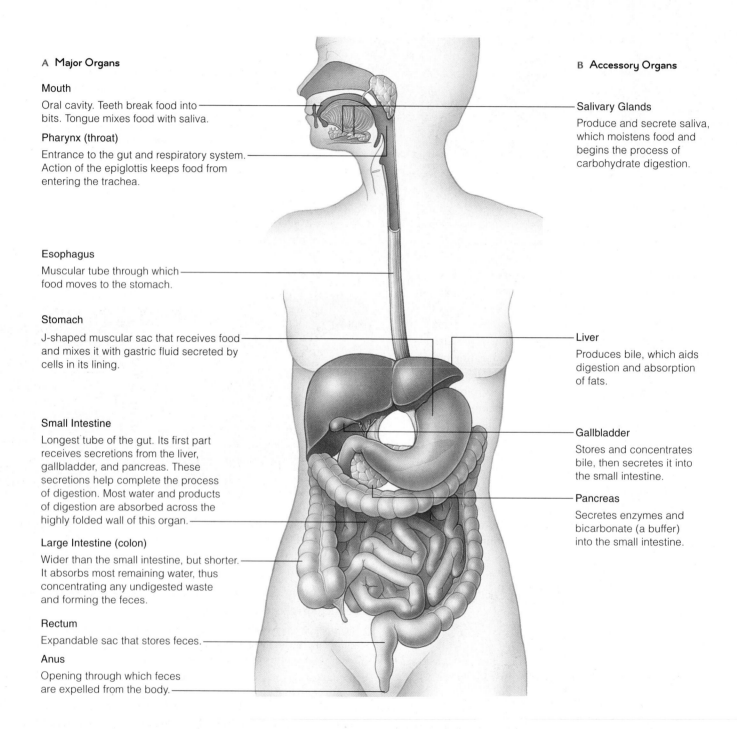

A Major Organs

Mouth

Oral cavity. Teeth break food into bits. Tongue mixes food with saliva.

Pharynx (throat)

Entrance to the gut and respiratory system. Action of the epiglottis keeps food from entering the trachea.

Esophagus

Muscular tube through which food moves to the stomach.

Stomach

J-shaped muscular sac that receives food and mixes it with gastric fluid secreted by cells in its lining.

Small Intestine

Longest tube of the gut. Its first part receives secretions from the liver, gallbladder, and pancreas. These secretions help complete the process of digestion. Most water and products of digestion are absorbed across the highly folded wall of this organ.

Large Intestine (colon)

Wider than the small intestine, but shorter. It absorbs most remaining water, thus concentrating any undigested waste and forming the feces.

Rectum

Expandable sac that stores feces.

Anus

Opening through which feces are expelled from the body.

B Accessory Organs

Salivary Glands

Produce and secrete saliva, which moistens food and begins the process of carbohydrate digestion.

Liver

Produces bile, which aids digestion and absorption of fats.

Gallbladder

Stores and concentrates bile, then secretes it into the small intestine.

Pancreas

Secretes enzymes and bicarbonate (a buffer) into the small intestine.

Figure 36.5 Animated Major organs and accessory organs of the human digestive system.

skeletal muscles, attaches to the floor of the mouth. The tongue positions food for swallowing. It helps us in speech, and the many chemoreceptors at its surface contribute to our sense of taste (Section 30.7).

Swallowing forces food into the pharynx, or throat. The presence of food at the back of the throat triggers a swallowing reflex. When you swallow, the epiglottis flops down and the vocal cords constrict, so the route between the pharynx and larynx is blocked. The reflex keeps food from getting stuck in an airway and choking you.

Swallowed food enters the **esophagus**, the muscular tube between the pharynx and stomach. Wavelike smooth muscle contractions, called **peristalsis**, move food through the esophagus to the stomach, and through the rest of the digestive tract. The **stomach** is a stretchable sac that stores food, secretes acid and digestive enzymes, and mixes them all together.

The stomach empties into the **small intestine**, the part of the gut where most carbohydrates, lipids, and proteins are digested and where most of the released nutrients and water are absorbed. Secretions from the liver and pancreas assist the small intestine in these tasks.

The **colon (large intestine)** absorbs water and ions, thus compacting the wastes. Wastes are briefly stored in a stretchable tube, the **rectum**, before being expelled from the gut's terminal opening, or **anus**.

---

anus Opening through which digestive waste is expelled from a complete digestive system.
colon or **large intestine** Organ that receives digestive waste from the small intestine and concentrates it as feces.
esophagus Muscular tube between the throat and stomach.
peristalsis Wavelike smooth muscle contractions that propel food through the digestive tract.
rectum Region where feces are stored prior to excretion.
small intestine Longest portion of the digestive tract, and the site of most digestion and absorption.
stomach Muscular organ that mixes food with gastric fluid that it secretes.

Take-Home Message **What are the components of the human digestive system?**

> Humans have a complete digestive system. Swallowing forces food and water from the mouth into the pharynx. Food continues through an esophagus to the stomach.

> Food processing starts in the mouth. Most digestion and absorption occurs in the small intestine. The colon absorbs most of the remaining water and ions, which causes the wastes to compact. The rectum briefly stores the wastes before they are expelled through the anus.

> The liver and pancreas have an accessory role in digestion. They produce substances that are secreted into the small intestine.

> Mechanical digestion, the smashing of food into smaller pieces, begins in the mouth. So does chemical digestion, the enzymatic breakdown of food into molecular subunits.
❮ Links to Buffers 2.6, Exocrine glands 28.3

Mechanical digestion begins when teeth rip and crush food. Human adults have thirty-two teeth of four types (Figure 36.6A). Each tooth consists mostly of bonelike dentin (Figure 36.6B). Dentin-secreting cells in a central pulp cavity are serviced by nerves and blood vessels that extend through the tooth's root. Enamel, the hardest material in the body, covers the tooth's exposed crown.

Chemical digestion begins when food mixes with saliva secreted by **salivary glands**, exocrine glands that open into the mouth. Saliva contains enzymes, bicarbonate, and mucins. One enzyme (salivary amylase) begins the breakdown of starch. Another (salivary lipase) starts the chemical digestion of fats. Bicarbonate, a buffer, keeps the pH in the mouth from becoming too acidic. Mucins are proteins that combine with water and form mucus. Mucus makes the chewed-up bits of food stick together in easy-to-swallow clumps.

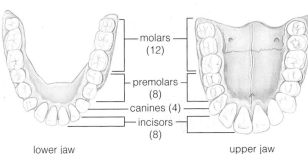

A Adult teeth. Incisors shear off bits of food. Canines tear meats. Premolars and molars have broad, bumpy crowns that are platforms for grinding and crushing food.

molars (12)
premolars (8)
canines (4)
incisors (8)
lower jaw
upper jaw

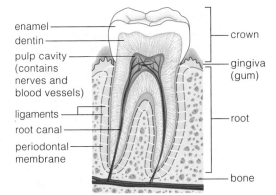

B Cross-section of one human tooth. The crown is the portion extending above the gum; the root is embedded in the jaw.

enamel
dentin
pulp cavity (contains nerves and blood vessels)
ligaments
root canal
periodontal membrane

crown
gingiva (gum)
root
bone

Figure 36.6 Structure and function of human teeth.

---

salivary gland Exocrine gland that secretes saliva into the mouth.

Take-Home Message **What happens to food in the mouth?**

> Teeth mechanically break food into smaller particles. Enzymes in saliva begin the chemical digestion of carbohydrates and fats.

> The stomach stores food and continues the process of digestion that began in the mouth.

❮ Links to pH 2.6, Peptide bond 3.5, Enzymes 5.4, Smooth muscle 28.5, Autonomic nervous system 29.8

Structure and Function of the Stomach

The stomach is a muscular, stretchable sac with a sphincter at either end (Figure 36.7). A **sphincter** is a ring of muscle that controls the passage of material through a tubular organ or a body opening.

The stomach has three functions. First, it stores food and controls the rate of passage to the small intestine. Second, it mechanically breaks down food. Third, it secretes substances that aid in chemical digestion.

When the stomach is empty, its inner surface is highly folded. As it fills with food, these folds smooth out, increasing the stomach's capacity. In an average adult, the stomach can expand enough to hold about 1 liter of fluid (a little bit more than a quart).

A glandular epithelium, or mucosa, lines the stomach's inner wall. Cells of this lining secrete about 2 liters of **gastric fluid** each day. Gastric fluid includes mucus, hydrochloric acid, and pepsinogen, an inactive form of the protein-digesting enzyme pepsin.

Like the heart, the stomach has an internal pacemaker. Spontaneous action potentials generated in the upper portion of the stomach cause the smooth muscle in the stomach wall to contract rhythmically about three times a minute. The contractions mix gastric fluid with food to form a semiliquid mass called **chyme**. They also propel

---

chyme Mix of food and gastric fluid.
gastric fluid Fluid secreted by the stomach lining; contains digestive enzymes, acid, and mucus.
sphincter Ring of muscle that controls passage through a tubular organ or body opening.

a bit of chyme out through the pyloric sphincter and into the first segment of the small intestine.

Chemical digestion of proteins begins in the stomach. The acidity of chyme denatures proteins (makes them unfold) and exposes their peptide bonds. High acidity also converts pepsinogen into pepsin. Pepsin breaks peptide bonds, snipping proteins up into smaller polypeptides.

The stomach steps up or slows down its acid secretion depending on when and what you eat. Arrival of food in the stomach, especially protein, triggers endocrine cells in the stomach lining to secrete the hormone gastrin into the blood. Gastrin acts on acid-secreting cells of the stomach lining, causing them to increase their output. When the stomach is empty, gastrin secretion and acid secretion decline. This prevents excess acidity from damaging the stomach wall.

Stomach Disorders

Gastroesophageal Reflux In some people, the gastroesophageal sphincter at the entrance to the stomach does not close properly or opens when it should not. The result is gastroesophageal reflux. Acidic chyme splashes back into the esophagus, causing a burning pain commonly called heartburn or acid indigestion. Occasional acid reflux can be treated with over-the-counter antacids, but a chronic problem should be discussed with a doctor. Repeated exposure to acid can damage the tissue of the esophagus and raise the risk of esophageal cancer.

Stomach Ulcers When something disrupts the stomach's protective mucus layer, gastric fluid and enzymes can erode the stomach lining, causing an ulcer. Most ulcers occur after an acid-loving species of bacteria (*Helicobacter pylori*) makes its way through gastric mucus and infects cells of the stomach lining. *H. pylori* releases chemicals that increase gastrin secretion, thus making the stomach more acid. The extra acidity stresses cells of the stomach lining so the infection spreads more easily. Antibiotics halt the infection and allow healing to occur.

Continual use of nonsteroidal anti-inflammatory drugs such as ibuprofen or aspirin can also cause a stomach ulcer. These drugs interfere with chemical signals that maintain the health of the stomach lining.

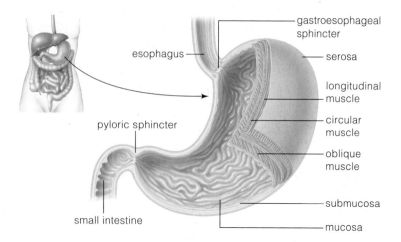

Figure 36.7 Location and structure of the stomach. The folds shown on the inner surface become smoothed out when the stomach fills with food.

> **Take-Home Message** **What are the functions of the stomach?**
>
> ❯ The stomach receives food from the esophagus and stretches to store it.
>
> ❯ Stomach contractions break up food and mix it with gastric fluid. They also move the resulting mixture (the chyme) into the small intestine.
>
> ❯ Chemical digestion of proteins begins in the stomach.

Structure of the Small Intestine

> The small intestine has a highly folded lining with many projections that make its surface area enormous.

< Link to Lymphatic system 33.11

Chyme forced out of the stomach through the pyloric sphincter enters the duodenum, the initial portion of the small intestine. The small intestine is "small" only in terms of its diameter—about 2.5 cm (1 inch). It is the longest segment of the gut. Uncoiled, the small intestine would extend for about 5 to 7 meters (16 to 23 feet). In addition, the small intestine has an immense surface area. Most digestion and absorption takes place at the surface of the small intestine.

Figure 36.8 Structure and function of the small intestine.

>> **Figure It Out** Are microvilli multicelled or smaller than a cell?

Answer: Microvilli are smaller than a cell. Villi are multicellular.

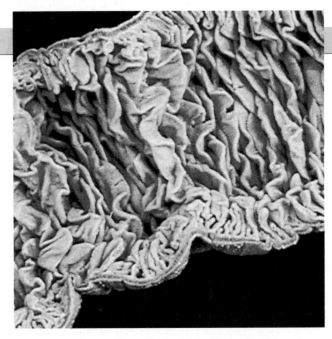

A Longitudinal cross section through the small intestine showing its highly folded lining.

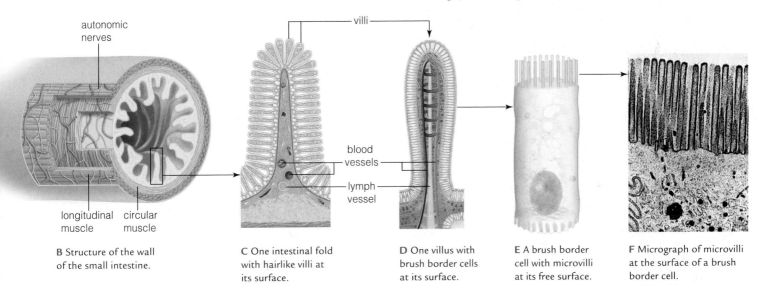

B Structure of the wall of the small intestine.

C One intestinal fold with hairlike villi at its surface.

D One villus with brush border cells at its surface.

E A brush border cell with microvilli at its free surface.

F Micrograph of microvilli at the surface of a brush border cell.

The small intestine has a highly folded lining (Figure 36.8A,B). Unlike the folds of the empty stomach, those of the small intestine are permanent.

The surface of each fold has many **villi** (singular, villus). A villus is a hairlike multicelled projection about 1 millimeter long (Figure 36.8C,D). The millions of villi that project from the intestinal lining give the lining a furry or velvety appearance. Blood vessels and lymph vessels run through the interior of each villus.

Epithelial cells at the surface of a villus have even tinier projections called **microvilli** (singular, microvillus).

brush border cell In the lining of the small intestine, an epithelial cell with microvilli at its surface.
microvilli Thin projections that increase the surface area of some epithelial cells.
villi Multicelled projections at the surface of each fold in the small intestine.

The 1,700 or so microvilli at the surface of a cell make its outer edge look like a brush. Thus, these cells are sometimes called **brush border cells** (Figure 36.8E,F).

Collectively, the many folds and projections of the small intestinal lining increase its surface area by hundreds of times. As a result, the surface area of the small intestine is comparable to that of a tennis court.

Take-Home Message How does the structure of the small intestine affect its function?

> The surface of the small intestine is highly folded and each fold has many projections (villi). Brush border cells at the surface of a villus have tiny projections (microvilli) at their surface.

> The many folds and projections greatly increase the surface area for the two functions of the small intestine—digestion and absorption.

❯ Chemical and mechanical digestion are completed in the small intestine, and most nutrients are absorbed here.
❮ Links to Carbohydrates 3.3, Lipids 3.4, Proteins 3.5, Enzymes 5.4, Osmosis 5.6, Transport proteins 5.7

The process of chemical digestion that began in the mouth and continued in the stomach is completed in the small intestine (Figure 36.9). The small intestine receives chyme from the stomach, enzymes and bicarbonate from the pancreas, and bile from the gallbladder. Pancreatic enzymes work in concert with enzymes at the surface of brush border cells to complete the breakdown of large organic compounds into absorbable subunits. The bicarbonate provided by the pancreas buffers the chyme, raising the pH enough for digestive enzymes to function.

Carbohydrate Digestion and Absorption

In the small intestine, carbohydrates are broken down into monosaccharides, or simple sugars ❶. This process began in the mouth, where salivary amylase broke polysaccharides into disaccharides (two-unit sugars). A pancreatic amylase carries out the same reaction in the small intestine. Disaccharides are substrates for enzymes embedded in the plasma membrane of the brush border cells. The enzymes split disaccharides into monosaccharides. For example, sucrase breaks sucrose into glucose and fructose subunits. Lactase splits lactose into glucose and galactose. Monosaccharides are actively transported into a brush border cell, then out into the interstitial fluid inside a villus ❷. From here they enter the blood.

Protein Digestion and Absorption

Protein digestion began in the stomach, where pepsin broke proteins into polypeptides. It is completed in the small intestine ❸. The pancreas secretes proteases such as trypsin and chymotrypsin that break polypeptides into peptide fragments. Enzymes at the surface of the brush border cell break these fragments into amino acids.

Like monosaccharides, amino acids are actively transported into brush border cells, then out into the interstitial fluid. From here they enter the blood ❹.

Fat Digestion and Absorption

Chemical digestion of fats begins with the action of salivary lipase, but most fat digestion occurs in the small intestine. Here bile increases the effectiveness of lipases

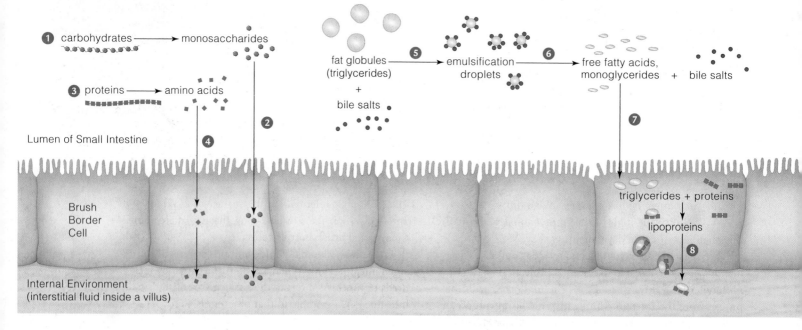

❶ Enzymes break polysaccharides down to simple sugars, or monosaccharides.

❷ Monosaccharides are actively transported into brush border cells, then out into interstitial fluid.

❸ Proteins are broken into polypeptides, then amino acids.

❹ Amino acids are actively transported into brush border cells, then out into interstitial fluid.

❺ Movements of the intestinal wall break up fat globules into small droplets. Bile salts coat the droplets, so that globules cannot form again.

❻ Pancreatic enzymes digest the droplets to fatty acids and monoglycerides.

❼ Monoglycerides and fatty acids diffuse across the plasma membrane's lipid bilayer, into brush border cells.

❽ In a brush border cell, the products of fat digestion form triglycerides, which associate with proteins. The resulting lipoproteins are then expelled by exocytosis into the interstitial fluid inside the villus.

Figure 36.9 Animated Summary of digestion and absorption in the small intestine.

secreted into the small intestine by the pancreas. **Bile** contains salts, pigments, cholesterol, and lipids. It is made in the liver, then stored and concentrated in the **gallbladder**. A fatty meal causes the gallbladder to contract, forcing bile out through a short duct into the small intestine.

Bile enhances fat digestion by aiding **emulsification**, the dispersion of droplets of fat in a fluid. Water-insoluble triglycerides from food tend to clump together as fat globules. Oscillating movements of the small intestine break big globs of fat into smaller droplets, then bile salts coat the droplets so they remain separated ❺. Compared to big globules, the many small droplets present a much greater surface area to the lipases that break triglycerides into fatty acids and monoglycerides ❻.

Being lipid soluble, fatty acids and monoglycerides produced by fat digestion can enter a villus by diffusing across the lipid bilayer of brush border cells ❼. Inside these cells, triglycerides form and become coated with proteins. The resulting lipoproteins are moved by exocytosis into interstitial fluid inside a villus ❽. From the interstitial fluid, triglycerides enter lymph vessels that eventually drain into the general circulation.

Fluid Absorption

Each day, eating and drinking puts 1 to 2 liters of fluid into the lumen of your small intestine. Secretions from your stomach, accessory glands, and the intestinal lining add another 6 to 7 liters. About 80 percent of the water that enters the small intestine moves across the intestinal lining and into the internal environment by osmosis. Transport of salts, sugars, and amino acids across brush border cells creates an osmotic gradient. Water follows that gradient from chyme into the interstitial fluid.

Disorders That Affect Digestion in the Small Intestine

Lactose Intolerance To be able to digest lactose, a person must have functional lactase at the surface of their brush border cells. In many people, including most Asians and African Americans, expression of the gene for lactase declines in adulthood. As a result, lactose is not broken down in the small intestine. When lactose enters the large intestine, resident bacteria break it down in reactions that produce hydrogen gas as a by-product. The result is flatulence, a feeling of bloating, and cramps.

Figure 36.10 A cross-sectioned gallstone. Its light color indicates that is composed mainly of cholesterol.

Gallstones The main components of bile are cholesterol and bilirubin. Bilirubin is a yellow-orange pigment that forms when the liver breaks down hemoglobin. Sometimes cholesterol or bilirubin accumulates as hard pellets called gallstones (Figure 36.10).

Many gallstones cause no symptoms but some may cause pain after a meal, especially one that is high in fat. Gallstones can be dangerous when they block or become lodged in a duct. In this case, the gallbladder or gallstones are usually removed surgically. After removal of the gallbladder, all bile from the liver drains directly into the small intestine.

Pancreatitis Gallstones can irritate the pancreas, leading to an inflammation called pancreatitis. Use of some prescription drugs or alcohol abuse can also cause pancreatitis. As a result of the inflammation, pancreatic enzymes begin to digest the pancreas itself, causing further inflammation and narrowing the duct leading to the small intestine. An affected person often has upper abdominal pain. Weight loss is common, because a lack of pancreatic enzymes in the small intestine prevents normal digestion. People who have chronic pancreatitis may need to take pills that provide enzymes to replace those normally provided by a healthy pancreas.

---

bile Mix of salts, pigments, and cholesterol produced in the liver, then stored and concentrated in the gallbladder; emulsifies fats when secreted into the small intestine.
emulsification Suspension of fat droplets in a fluid.
gallbladder Organ that stores and concentrates bile.

Take-Home Message **What are the roles of the small intestine?**

❯ Chemical digestion is completed in the small intestine. Enzymes from the pancreas and enzymes embedded in the membrane of brush border cells break large molecules into smaller absorbable subunits.

❯ Small subunits (monosaccharides, amino acids, fatty acids, and monoglycerides) enter the internal environment when they are absorbed into the intersitial fluid in a villus. Most fluid that enters the gut is also absorbed across the wall of the small intestine.

> The large intestine completes the process of absorption, then concentrates, stores, and eliminates wastes.
< Links to Osmosis 5.6, Cancer 11.6, Normal flora 19.7, Autonomic nerves 29.8, Autoimmune disorders 34.12

Structure and Function

Not everything that enters the small intestine can be or should be absorbed. Contractions propel indigestible material, dead bacteria and mucosal cells, inorganic substances, and some water from the small intestine into the colon, or large intestine. The large intestine is wider than the small intestine, but much shorter—only about 1.5 meters (5 feet) long.

As the wastes travel through the colon, they become compacted as **feces**. The colon concentrates wastes by actively pumping sodium ions across its wall, into the internal environment. Water follows by osmosis.

The first part of the colon is a cup-shaped pouch called the cecum. The short, tubular **appendix** projects from the cecum. Beyond the cecum, the colon ascends the wall of the abdominal cavity, extends across the cavity, descends, and connects to the rectum (Figure 36.11).

Contraction of the smooth muscle in the colon wall mixes the colon's contents and propels this material along. Compared with other gut regions, wastes move more slowly through the colon, which also has a moderate pH. These conditions favor growth of bacteria such as *Escherichia coli*. This species is part of our normal gut flora. It makes vitamin B_{12} that we absorb across the colon lining.

After a meal, signals from autonomic nerves cause much of the colon to contract forcefully and propel feces to the rectum. The rectum stretches, which activates a defecation reflex to expel feces. The nervous system can override the reflex by calling for contraction of a sphincter at the anus.

Health and the Colon

Healthy adults typically defecate about once a day, on average. Emotional stress, a diet low in fiber, minimal exercise, dehydration, and some medications can lead to constipation. This means defecation occurs fewer than three times a week, is difficult, and yields small, hardened, dry feces. Occasional constipation usually goes away on its own. A chronic problem should be discussed with a doctor. Infection by a viral, bacterial, or protozoan pathogen can cause an episode of diarrhea, the frequent passing of watery feces. Autoimmune disorders that affect the gut, such as Crohn's disease, can cause chronic diarrhea.

Appendicitis—an inflammation of the appendix—requires prompt treatment. It often occurs after a bit of feces lodges in the appendix and infection sets in. Removing an inflamed appendix prevents it from bursting and releasing bacteria into the abdominal cavity. Such a rupture could cause a life-threatening infection.

Some people are genetically predisposed to develop colon polyps, small growths on the colon wall (Figure 36.11B). Most polyps are benign, but some can become cancerous. If detected in time, colon cancer is highly curable. Blood in feces and dramatic changes in bowel habits may be symptoms of colon cancer and should be reported to a doctor. Also, anyone over the age of 50 should have a colonoscopy, a procedure in which clinicians use a camera to examine the colon for polyps or cancer.

appendix Wormlike projection from the first part of the large intestine.
feces Unabsorbed food material and cellular waste that is expelled from the digestive tract.

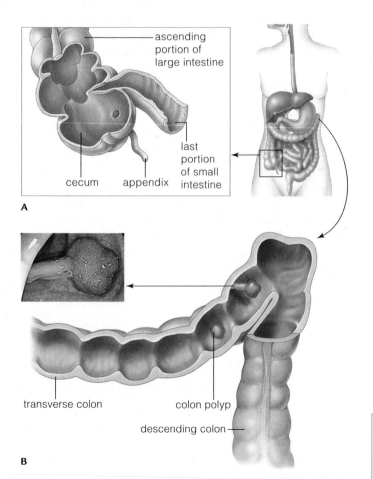

A

B

Figure 36.11 A Cecum and appendix of the large intestine (colon). **B** Sketch and photo of polyps in the transverse colon.

Labels: ascending portion of large intestine; last portion of small intestine; cecum; appendix; transverse colon; colon polyp; descending colon

Take-Home Message What are the functions of the large intestine?

> By absorbing water and mineral ions, the colon compacts undigested residues and other wastes as feces, which are stored briefly in the rectum before expulsion.

❯ Small organic compounds—sugars, amino acids, and triglycerides—are distributed to cells and burned as fuel, stored, or used in synthesis of larger organic compounds.
❮ Links to Alcohol and the liver 5.1, Energy sources 7.7, Human circulation 33.3

Section 7.7 sketched out some mechanisms of control over organic metabolism, the disposition of glucose and other organic compounds in the body as a whole. It introduced the main pathways by which carbohydrates, fats, and proteins are broken down to forms used as intermediates in aerobic respiration. Figure 36.12A rounds out this picture by illustrating the major routes by which the organic compounds obtained from food are shuffled and reshuffled in the body.

Living cells constantly recycle some carbohydrates, lipids, and proteins by breaking them apart. They use the breakdown products as energy sources and raw materials to build other compounds. The nervous and endocrine systems regulate this massive molecular turnover.

When you eat, your body adds to its pool of raw materials. Excess carbohydrates and other organic compounds absorbed from the gut are transformed mostly into fats, which become stored in adipose tissue. Some glucose is converted to glycogen stores in liver and muscle. As long as there is plenty of glucose in the blood, cells use it as their main energy source. They turn to glycogen or fats only when the amount of glucose circulating in the blood is inadequate to meet their needs.

In between meals, the brain takes up two-thirds of the glucose available in the blood, so other body cells tap fat and glycogen stores for energy. Adipose cells degrade fats to glycerol and fatty acids, which enter the bloodstream. Liver cells break down glycogen and release glucose, which also enters blood. Body cells take up the fatty acids as well as the glucose and use it for ATP production.

Figure 36.12B highlights the liver's central role in metabolism and homeostasis. All blood that passes through intestinal capillaries enters capillary beds in the liver before returning to the heart (Section 33.3). The liver takes up and stores essential nutrients. It stores glucose as glycogen and is the body's main reservoir for vitamins A and B_{12}. The liver also detoxifies many potentially dangerous substances. As an example, ammonia (NH_3) is a potentially toxic product of amino acid breakdown. The liver removes much of the ammonia from the blood and converts it to urea, a less toxic compound that is later excreted in urine. The liver also detoxifies alcohol and other ingested drugs.

Take-Home Message What happens to absorbed compounds?

❯ Absorbed sugars are the human body's most accessible energy source. Between meals, the brain draws on glucose in blood; other cells tap fat and glycogen stores. Adipose cells convert and store excess carbohydrates as fats.

❯ Blood with substances absorbed from the gut flows to the liver before returning to the heart. The liver stores glycogen and detoxifies harmful substances.

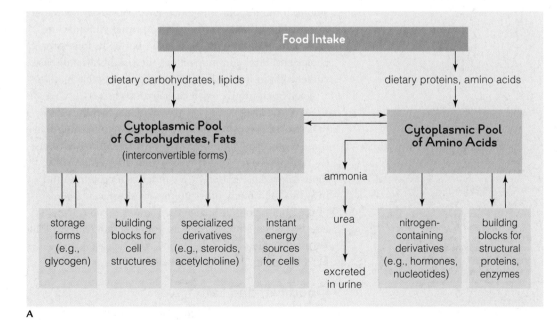

Liver Functions

Forms bile (assists fat digestion), rids body of excess cholesterol and blood's respiratory pigments

Controls amino acid levels in the blood; converts potentially toxic ammonia to urea

Controls glucose level in blood; major reservoir for glycogen

Removes hormones that served their functions from blood

Removes ingested toxins, such as alcohol, from blood

Breaks down worn-out and dead red blood cells, and stores iron

Stores some vitamins

Figure 36.12 A Summary of major pathways of organic metabolism. Cells continually synthesize and tear down carbohydrates, fats, and proteins. Most urea forms in the liver, an organ that is at the crossroads of organic metabolism. **B** Functions of the liver.

> You are what you eat—diet profoundly affects your body's structure and function. So what should you eat?

< Links to Carbohydrates 3.3; Lipids 3.1, 3.4; Proteins 3.5

USDA Dietary Recommendations

The Department of Agriculture and other United States government agencies research diets that may help prevent diabetes, cancers, and other health problems. They periodically update their nutritional guidelines. In 2005, they replaced their traditional one-size-fits-all food pyramid with a new Internet-based program that generates recommendations specific for a person's age, sex, height, weight, and activity level (Figure 36.13). You can generate your own healthy eating plan by visiting the USDA web site: www.mypyramid.gov.

USDA Nutritional Guidelines

Food Group	Amount Recommended
Vegetables	2.5 cups/day
Dark green vegetables	3 cups/week
Orange vegetables	2 cups/week
Legumes	3 cups/week
Starchy vegetables	3 cups/week
Other vegetables	6.5 cups/week
Fruits	2 cups/day
Milk Products	3 cups/day
Grains	6 ounces/day
Whole grains	3 ounces/day
Other grains	3 ounces/day
Fish, poultry, lean meat	5.5 ounces/day
Oils	24 grams/day

Figure 36.13 Example of nutritional guidelines from the United States Department of Agriculture (USDA). These recommendations are for females between ages ten and thirty who get less than 30 minutes of vigorous exercise daily. Portions add up to a 2,000-kilocalorie daily intake.

Table 36.1 Main Types of Dietary Lipids

Polyunsaturated Fatty Acids: Liquid at room temperature; essential for health.
  Omega-3 fatty acids
    Alpha-linolenic acid and its derivatives
    Sources: Nut oils, vegetable oils, oily fish
  Omega-6 fatty acids
    Linoleic acid and its derivatives
    Sources: Nut oils, vegetable oils, meat

Monounsaturated Fatty Acids: Liquid at room temperature. Main dietary source is olive oil. Beneficial in moderation.

Saturated Fatty Acids: Solid at room temperature. Main sources are meat and dairy products, palm and coconut oils. Excessive intake may raise risk of heart disease.

Trans Fatty Acids (Hydrogenated Fats): Solid at room temperature. Manufactured from vegetable oils and used in many processed foods. Excessive intake raises the risk of heart disease.

The new guidelines recommend lowering intake of refined grains, saturated fats, *trans* fatty acids, added sugar or caloric sweeteners, and salt. They also recommend eating more vegetables and fruits with a high potassium and fiber content, fat-free or low-fat milk products, and whole grains. About 55 percent of daily caloric intake should come from carbohydrates.

Energy–Rich Carbohydrates

Fresh fruits, whole grains, and vegetables—especially legumes such as peas and beans—provide abundant complex carbohydrates (Section 3.3). The body breaks the starch in these foods into glucose, your primary source of energy. These foods also provide essential vitamins and fiber. Eating foods high in soluble fiber helps lower one's cholesterol level and may reduce the risk of heart disease. A diet high in insoluble fiber helps prevent constipation.

Foods rich in processed carbohydrates such as white flour, refined sugar, and corn syrup are sometimes said to be full of "empty calories." This is a way of saying that these foods provide little in the way of vitamins or fiber.

You may have noticed breads and other grain-based foods labeled as "gluten-free." Gluten is a protein found in wheat and many other grains. An estimated 1 percent of the population has a genetic disorder called celiac disease, in which gluten causes an autoimmune reaction that harms the small intestine's villi. Celiac disease is treated by eliminating gluten from the diet.

Good Fat, Bad Fat

Your body uses lipids to build cell membranes, as energy stores, and as a reservoir for fat-soluble vitamins.

Linoleic acid and alpha-linolenic acid are **essential fatty acids**, meaning the human body needs them but cannot make them, so they are required in the diet. Both are polyunsaturated fats; their long carbon tails include two or more double bonds (Table 36.1). Unsaturated fats are liquid at room temperature (Section 3.4).

We divide the polyunsaturated fatty acids into two categories: omega-3 fatty acids and omega-6 fatty acids. Omega-3 fatty acids, the main fat in oily fish such as sardines, seem to have special health benefits. Studies suggest that a diet high in omega-3 fatty acids can reduce the risk of cardiovascular disease, lessen the inflammation associated with rheumatoid arthritis, and help diabetics control their blood glucose.

Oleic acid, the main fat in olive oil, may also have health benefits. It is monounsaturated, which means its carbon tails have only one double bond. A diet in which olive oil is substituted for saturated fats helps prevent heart disease.

Dairy products and meats are rich in saturated fats and cholesterol. Overindulging in these foods increases one's risk for heart disease, stroke, and some cancers.

Trans fatty acids, or *trans* fats, are manufactured from vegetable oils. However, they have a molecular structure that makes them even worse for the heart than saturated fats (Section 3.1). All food labels are now required to show the amounts of *trans* fats, saturated fats, and cholesterol per serving (Figure 36.14).

Body-Building Proteins

Amino acids are building blocks of proteins (Section 3.5). Your cells can make some amino acids but you must get eight **essential amino acids** from food. The eight essential amino acids are methionine (or cysteine, its metabolic equivalent), isoleucine, leucine, lysine, phenylalanine, threonine, tryptophan, and valine.

Most proteins in meat are "complete," meaning their amino acid ratios match a human's nutritional needs. By contrast, most plant proteins are "incomplete," meaning they lack one or more amino acids essential for the human diet. The American Dietetic Association states that, with careful planning, a vegetarian diet can provide all essential nutrients for people in any stage of life. To obtain all the required amino acids from plant sources alone, one must combine foods so that the amino acids missing from one component are present in some others. As an example, rice and beans together provide all necessary amino acids, but rice alone or beans alone do not. You do not have to eat the two complementary foods at the same meal, but both should be consumed within a 24-hour period.

essential amino acid Amino acid that the body cannot make and must obtain from food.
essential fatty acid Fatty acid that the body cannot make and must obtain from the diet.

Nutrition Facts
Serving Size 1 cup (228g)
Servings Per Container 2

Amount Per Serving	
Calories 250	Calories from Fat 110

	% Daily Value*
Total Fat 12g	18%
Saturated Fat 3g	15%
Trans Fat 1.5g	
Cholesterol 30mg	10%
Sodium 470mg	20%
Total Carbohydrate 31g	10%
Dietary Fiber 0g	0%
Sugars 5g	
Protein 5g	

Vitamin A	4%
Vitamin C	2%
Calcium	20%
Iron	4%

* Percent Daily Values are based on a 2,000 calorie diet. Your Daily Values may be higher or lower depending on your calorie needs:

	Calories:	2,000	2,500
Total Fat	Less than	65g	80g
Sat Fat	Less than	20g	25g
Cholesterol	Less than	300mg	300mg
Sodium	Less than	2,400mg	2,400mg
Total Carbohydrate		300g	375g
Dietary Fiber		25g	30g

Check serving size. A package often holds more than one serving, but nutritional information is given per serving.

Avoid foods in which a large proportion of calories comes from fat.

Choose foods that provide a low percent of the recommended maximum of saturated fat, *trans* fat, cholesterol, and sodium. 20% or more is high.

Choose foods that are high in dietary fiber and low in sugar.

If you eat meat, you probably get more than enough protein.

Choose foods that provide a high percentage of your daily vitamin and mineral requirements.

This part of the label shows recommended intake of nutrients for two levels of calorie intake. Keeping fat and salt intake below recommended levels and dietary fiber above recommended levels decreases the risk of some chronic health problems.

Figure 36.14 How to read a food label. Information on a food label can be used to ensure that you get the nutrients you need without exceeding recommended limits on less healthy substances such as salt and *trans* fats.

>> Figure It Out This hypothetical label is for a ready-to-eat macaroni and cheese product. What proportion of the fat in a serving of this product comes from the least healthy forms of fat (saturated fat and *trans* fat)?

Answer: Of the total fat content in a serving (12 grams), 3 g are saturated fat and 3 g are *trans* fat. Thus 6 g, or half the fat, is from unhealthy sources.

Take-Home Message **What are the main types of nutrients that humans require and what is the healthiest way to obtain them?**

> A healthy diet provides energy and all necessary building blocks for assembling essential body components.

> Nutritional guidelines are periodically revised in light of new research. Current guidelines call for most calories to come from complex carbohydrates, rather than simple sugars. They also favor fat and protein sources that are low in saturated and *trans* fats.

> A person can obtain all the required nutrients from a vegetarian diet, but doing so requires combining plant foods so amino acids lacking in one are present in the other.

> To function properly, the body requires small amounts of vitamins and minerals in addition to major nutrients.
< Links to Coenzymes 5.4, Thyroid function 31.6

Vitamins are organic substances that are essential in very small amounts; no other substance can carry out their metabolic functions. At a minimum, human cells require the thirteen vitamins listed in Table 36.2. Each has specific roles. For instance, the B vitamin niacin is modified to make NAD, a coenzyme (Section 5.4).

Minerals are inorganic substances that are essential for growth and survival because no other substance can serve their metabolic functions (Table 36.3). As an example, all of your cells use iron as a component of electron transfer chains. Red blood cells require iron to make oxygen-transporting hemoglobin. Iodine is essential for

Table 36.2 Major Vitamins: Sources, Functions, and Effects of Deficiencies or Excesses*

Vitamin	Common Sources	Main Functions	Effects of Chronic Deficiency	Effects of Extreme Excess
Fat-Soluble Vitamins				
A	Its precursor comes from beta-carotene in yellow fruits, yellow or green leafy vegetables; also in fortified milk, egg yolk, fish, liver	Used in synthesis of visual pigments, bone, teeth; maintains epithelia	Dry, scaly skin; lowered resistance to infections; night blindness; permanent blindness	Malformed fetuses; hair loss; changes in skin; liver and bone damage; bone pain
D	Inactive form made in skin, activated in liver, kidneys; in fatty fish, egg yolk, fortified milk products	Promotes bone growth and mineralization; enhances calcium absorption	Bone deformities (rickets) in children; bone softening in adults	Retarded growth; kidney damage; calcium deposits in soft tissues
E	Whole grains, dark green vegetables, vegetable oils	Counters effects of free radicals; helps maintain cell membranes; blocks breakdown of vitamins A and C in gut	Lysis of red blood cells; nerve damage	Muscle weakness, fatigue, headaches, nausea
K	Enterobacteria form most of it; also in green leafy vegetables, cabbage	Blood clotting; ATP formation via electron transport	Abnormal blood clotting; severe bleeding (hemorrhaging)	Anemia; liver damage and jaundice
Water-Soluble Vitamins				
B_1 (thiamin)	Whole grains, green leafy vegetables, legumes, lean meats, eggs	Connective tissue formation; folate utilization; coenzyme action	Water retention in tissues; tingling sensations; heart changes; poor coordination	None reported from food; possible shock reaction from repeated injections
B_2 (riboflavin)	Whole grains, poultry, fish, egg white, milk	Coenzyme action	Skin lesions	None reported
B_3 (niacin)	Green leafy vegetables, potatoes, peanuts, poultry, fish, pork, beef	Coenzyme action	Contributes to pellagra (damage to skin, gut, nervous system, etc.)	Skin flushing; possible liver damage
B_6	Spinach, tomatoes, potatoes, meats	Coenzyme in amino acid metabolism	Skin, muscle, and nerve damage; anemia	Impaired coordination; numbness in feet
Pantothenic acid	In many foods (meats, yeast, egg yolk especially)	Coenzyme in glucose metabolism, fatty acid and steroid synthesis	Fatigue, tingling in hands, headaches, nausea	None reported; may cause diarrhea occasionally
Folate (folic acid)	Dark green vegetables, whole grains, yeast, lean meats; enterobacteria produce some folate	Coenzyme in nucleic acid and amino acid metabolism	A type of anemia; inflamed tongue; diarrhea; impaired growth; mental disorders	Masks vitamin B_{12} deficiency
B_{12}	Poultry, fish, red meat, dairy foods (not butter)	Coenzyme in nucleic acid metabolism	A type of anemia; impaired nerve function	None reported
Biotin	Legumes, egg yolk; colon bacteria produce some	Coenzyme in fat, glycogen formation and in amino acid metabolism	Scaly skin (dermatitis); sore tongue; depression; anemia	None reported
C (ascorbic acid)	Fruits and vegetables, especially citrus, berries, cantaloupe, cabbage, broccoli, green pepper	Collagen synthesis; possibly inhibits effects of free radicals; structural role in bone, cartilage, and teeth; used in carbohydrate metabolism	Scurvy; poor wound healing; impaired immunity	Diarrhea, other digestive upsets; may alter results of some diagnostic tests

* Guidelines for appropriate daily intakes are being worked out by the Food and Drug Administration.

development of a healthy nervous system and to make thyroid hormone (Section 31.6).

Most people can get all the vitamins and minerals they need from a well-balanced diet. Some studies have even found an increased death rate among people who take supplemental antioxidants (beta-carotene, vitamin A, and vitamin E). If you take an over-the-counter multivitamin, be sure the quantities it provides are not excessive.

In addition to vitamins and minerals, a healthy diet should include a variety of phytochemicals, also known as phytonutrients. These organic molecules are found in plant foods and while not essential, they may reduce the risk of certain disorders. For example, eating leafy green vegetables ensures adequate intake of the plant pigments lutein and zeaxanthin. A diet low in these phytochemicals raises the risk of macular degeneration–related blindness.

mineral Inorganic substance that is required in small amounts for normal metabolism.
vitamin Organic substance required in small amounts for normal metabolism.

Take-Home Message What roles do vitamins, minerals, and phytonutrients play?

❯ Normal metabolism requires the organic substances called vitamins and inorganic substances called minerals. In addition, some molecules made by plants are not essential but may reduce the risk of certain disorders.

❯ A balanced diet provides the required amounts of vitamins and minerals for most people. Deficiencies or excesses of either can cause health problems.

Table 36.3 Major Minerals: Sources, Functions, and Effects of Deficiencies or Excesses*

Mineral	Common Sources	Main Functions	Effects of Chronic Deficiency	Effects of Extreme Excess
Calcium	Dairy products, dark green vegetables, dried legumes	Bone, tooth formation; blood clotting; neural and muscle action	Stunted growth; possibly diminished bone mass (osteoporosis)	Impaired absorption of other minerals; kidney stones in susceptible people
Chloride	Table salt (usually too much in diet)	HCl formation in stomach; contributes to body's acid-base balance; neural action	Muscle cramps; impaired growth; poor appetite	Contributes to high blood pressure in certain people
Copper	Nuts, legumes, seafood, drinking water	Used in synthesis of melanin, hemoglobin, and some transport chain components	Anemia, changes in bone and blood vessels	Nausea, liver damage
Fluorine	Fluoridated water, tea, seafood	Bone, tooth maintenance	Tooth decay	Digestive upsets; mottled teeth and deformed skeleton in chronic cases
Iodine	Marine fish, shellfish, iodized salt, dairy products	Thyroid hormone formation	Enlarged thyroid (goiter), with metabolic disorders	Toxic goiter
Iron	Whole grains, green leafy vegetables, legumes, nuts, eggs, lean meat, molasses, dried fruit, shellfish	Formation of hemoglobin and cytochrome (transport chain component)	Iron-deficiency anemia, impaired immune function	Liver damage, shock, heart failure
Magnesium	Whole grains, legumes, nuts, dairy products	Coenzyme role in ATP–ADP cycle; roles in muscle, nerve function	Weak, sore muscles; impaired neural function	Impaired neural function
Phosphorus	Whole grains, poultry, red meat	Component of bone, teeth, nucleic acids, ATP, phospholipids	Muscular weakness; loss of minerals from bone	Impaired absorption of minerals into bone
Potassium	Diet alone provides ample amounts	Muscle and neural function; roles in protein synthesis and body's acid-base balance	Muscular weakness	Muscular weakness, paralysis, heart failure
Sodium	Table salt; diet provides ample to excessive amounts	Key role in body's salt-water balance; roles in muscle and neural function	Muscle cramps	High blood pressure in susceptible people
Sulfur	Proteins in diet	Component of body proteins	None reported	None likely
Zinc	Whole grains, legumes, nuts, meats, seafood	Component of digestive enzymes; roles in normal growth, wound healing, sperm formation, and taste and smell	Impaired growth, scaly skin, impaired immune function	Nausea, vomiting, diarrhea; impaired immune function and anemia

* Guidelines for appropriate daily intakes are being worked out by the Food and Drug Administration.

> Maintaining a healthy weight requires balancing energy inputs with energy expenditures.
‹ Links to Adipose tissue 28.4, Diabetes 31.9

What Is a Healthy Weight?

Figure 36.15 shows one of the widely accepted weight guidelines for women and men. The body mass index (BMI) is another guideline. It is a measurement designed to help assess increased health risk associated with weight gains. You can calculate your body mass index with this formula:

$$\text{BMI} = \frac{\text{weight (pounds)} \times 703}{\text{height (inches)} \times \text{height (inches)}}$$

Generally, individuals with a BMI of 25 to 29.9 are said to be overweight. A score of 30 or more indicates obesity: an overabundance of fat in adipose tissue that may lead to severe health problems. How body fat gets distributed also helps predict the risks. Fat deposits just above the belt, as in a "beer belly," are associated with an increased likelihood of heart problems.

It is difficult to lose weight by dieting alone. When you eat less, your body slows its metabolic rate to conserve energy. So how do you function normally over the long term while maintaining an acceptable weight? You must balance your caloric intake and energy output. For most people, this means eating only recommended portions of low-calorie, nutritious foods and exercising regularly.

Energy stored in food is expressed as kilocalories, or Calories (with a capital C). One kilocalorie equals 1,000 calories, which are units of heat energy.

Here is a way to calculate how many kilocalories you should take in daily to maintain a preferred weight. First, multiply the weight (in pounds) by 10 if you are not active physically, by 15 if you are moderately active, and by 20 if you are highly active. Next, subtract one of the following amounts from the multiplication result:

Age:	25–34	Subtract:	0
	35–44		100
	45–54		200
	55–64		300
	Over 65		400

For example, if you are 25 years old, are highly active, and weigh 120 pounds, you will require 120 × 20 = 2,400 kilocalories daily to maintain weight. If you want to gain weight you will require more; to lose, you will require less. The amount is only a rough estimate. Other factors, such as height, must be considered. A person 5 feet, 2 inches tall and active does not require as much energy as an active 6-footer whose body weight is the same.

Why Is Obesity Unhealthy?

Being obese has a negative effect on health. Among other things, it increases the risk of type 2 diabetes, high blood pressure, heart disease, breast and colon cancer, arthritis, and gallstones.

Why does excess weight have ill effects? As Section 7.7 explained, triglycerides in fat cells are the body's main form of energy storage. Fat cells of people who are at a healthy weight hold a moderate amount of triglycerides and function normally. In obese people, an excess of these

Weight Guidelines for Women		
Starting with an ideal weight of 100 pounds for a woman who is 5 feet tall, add five additional pounds for each additional inch of height. Examples:		
Height (feet)	Weight (pounds)	
5′ 2″	110	
5′ 3″	115	
5′ 4″	120	
5′ 5″	125	
5′ 6″	130	
5′ 7″	135	
5′ 8″	140	
5′ 9″	145	
5′ 10″	150	
5′ 11″	155	
6′	160	

Weight Guidelines for Men		
Starting with an ideal weight of 106 pounds for a man who is 5 feet tall, add six additional pounds for each additional inch of height. Examples:		
Height (feet)	Weight (pounds)	
5′ 2″	118	
5′ 3″	124	
5′ 4″	130	
5′ 5″	136	
5′ 6″	142	
5′ 7″	148	
5′ 8″	154	
5′ 9″	160	
5′ 10″	166	
5′ 11″	172	
6′	178	

Figure 36.15 How to estimate "ideal" weights for adults. Values shown are consistent with a long-term Harvard study into the link between weight and risk of cardiovascular disorders. The "ideal" varies. It is influenced by specific factors such as having a small, medium, or large skeletal frame; bones are heavy.

molecules distends fat cells and impairs their function. Like cells damaged in other ways, the overstuffed fat cells respond by sending out signals that summon up an inflammatory response (Section 34.4). The resulting chronic inflammation harms organs throughout the body and increases risk of cancer.

Overstuffed fat cells also increase secretion of signals that interfere with the action of insulin. Remember that this hormone encourages cells to take up sugar from the blood (Section 31.8). When insulin becomes ineffective, the result is type 2 diabetes (Section 31.9).

Armed with an understanding of how excess weight impairs health, researchers hope to dampen or offset harmful signals secreted by fat cells. One day, it may be possible to keep fat cells from causing inflammation or interfering with insulin function. For now, the only way to prevent these effects is by losing weight.

Eating Disorders

Eating too little can be as dangerous as eating too much, or more so. With anorexia nervosa, a person who has access to food routinely eats too little to maintain a weight within 15 percent of normal. Although the name means "nervous loss of appetite," most affected people are obsessed with food and continually hungry. They typically see themselves as fat, even when they are dangerously thin. Young women are disproportionately affected, and societal pressures to be thin certainly play a role. However, how a person responds to these pressures is determined in part by genetics. Anorexia tends to run in families, and scientists have pinpointed some genes that increase the risk of this disorder.

Anorexia damages organ systems throughout the body. The reproductive system shuts down; affected women stop menstruating. Starved of essential calcium, the body breaks down bone. Inadequate iron intake causes anemia. Heart muscle weakens and heart rhythms can become disrupted. Deaths from anorexia most often occur as a result of sudden cardiac arrest.

Bulimia nervosa is another eating disorder. Like anorexia nervosa it is most common among young women. Bulimics tend to be near normal weight, but perceive themselves as heavy. They "binge and purge" eating far too much, then trying to get rid of the food by inducing vomiting.

Induced vomiting has a variety of harmful effects. It bathes the esophagus and teeth in acidic gastric fluid. The acid harms the enamel of the teeth, causing them to become pitted and brittle. Repeated exposure to acid increases the risk of cancer of the esophagus. Frequent vomiting can also cause life-threatening changes in the body's ion balance. Loss of gastric fluid depletes the body of hydrogen ions, making body fluids more basic than

◼ The Battle Against Bulge (revisited)

Leptin decreases appetite. Another hormone, called ghrelin increases it. Ghrelin is secreted by cells in the stomach lining, as well as cells in the brain. Ghrelin secretion increases when the stomach is empty, and declines after a big meal.

In one study of ghrelin's effects, a group of obese volunteers stayed on a low-fat, low-calorie diet for six months. They lost weight, but the concentration of ghrelin in their bloodstream climbed dramatically— they were hungrier than ever!

Gastric bypass surgery may be more effective than standard weight loss methods in part because, unlike dieting, it lowers ghrelin levels. The surgery is designed for extremely obese people and it effectively reduces the size of the stomach and small intestine. After surgery, decreased stomach size and ghrelin secretion makes people feel full faster and less hungry between meals. In addition, shortening the length of small intestine decreases absorption so fewer calories are absorbed from the food that is eaten. The results can be dramatic. The photos at the *upper right* show a young woman before and after gastric bypass.

Gastric bypass surgery also has some major drawbacks. There are the risks of anesthesia and the possibility of complications. Also, after surgery, the person is at an increased risk for vitamin and mineral deficiencies.

How Would You Vote? Obesity may soon replace smoking as the number one cause of preventable deaths in the United States. Some states now require restaurant chains to include calorie counts on their menu. Do you think this policy will encourage healthier choices? See CengageNow for details, then vote online (cengagenow.com).

normal. The shift in pH can cause apnea (a halt in breathing), disrupt normal heart rhythm, and cause convulsions.

Both anorexia and bulimia are treated with a combination of therapy and medical attention to the damage done by the disorder. Many patients respond favorably to treatment and are able to maintain a normal weight.

Take-Home Message **How does weight affect health?**

> A person who balances caloric intake with energy expenditures will maintain current weight.

> Obesity raises the risk of heart disease, type 2 diabetes, some cancers, and other disorders. These problems may arise because overstuffed adipose cells summon up inflammatory responses in organs throughout the body.

> Anorexia is an eating disorder that results in a lower than normal body weight. It harms organs throughout the body, especially the bones and the heart. Anorexia-induced changes in heart rhythms can be fatal.

Summary

Section 36.1 Humans store excess food energy mainly in the bonds of fat in adipose cells. An ability to store fat, adaptive when food scarcity is common, can become maladaptive when food is always plentiful. Obesity is caused by a combination of genetic and environmental factors.

Section 36.2 A digestive system breaks food down into molecules that are small enough to be absorbed into the internal environment. It also stores and eliminates unabsorbed materials, and promotes homeostasis by its interactions with other organ systems. Some invertebrates have an **incomplete digestive system**: a saclike gut with a single opening. Most animals and all vertebrates have a **complete digestive system**: a tube with two openings and specialized areas between them. Variations in the structure of vertebrate digestive systems are adaptations to particular diets. For example, the multiple stomachs of **ruminants** allow them to digest grasses.

Section 36.3 Food taken into the human mouth is swallowed and moves into the pharynx, which opens onto the **esophagus**. **Peristalsis** of the esophagus conveys food to the **stomach**. A **sphincter** at the junction between the two organs regulates flow between them. From the stomach, food enters the **small intestine**, then the **large intestine**. Wastes are stored in the **rectum** until they are eliminated through the **anus**.

Sections 36.4, 36.5 Digestion starts in the mouth, where food is broken into bits and mixed with saliva from **salivary glands**. Protein digestion begins in the stomach, a muscular sac with a glandular lining that secretes **gastric fluid**. This fluid contains acid, enzymes, and mucus. It mixes with food and forms **chyme**.

Sections 36.6, 36.7 The small intestine is the longest portion of the gut and has the largest surface area. Its highly folded lining has many **villi** at its surface. Each multicelled villus has a covering of **brush border cells**. These cells have **microvilli** that increase their surface area for digestion and absorption.

Chemical digestion is completed in the small intestine through the action of enzymes from the pancreas, bile from the gallbladder, and enzymes embedded in the plasma membrane of brush border cells.

Carbohydrates are broken down to monosaccharides, actively transported across brush border cells, and enter the blood. Similarly, proteins are broken down to amino acids that are actively transported and enter the blood.

Bile made in the liver and stored and concentrated in the **gallbladder** aids in the **emulsification** of fats. Monoglycerides and fatty acids diffuse into the brush border cells. Here, they recombine as triglycerides, get a protein coat, and are moved by exocytosis into interstitial fluid. They then enter lymph vessels that deliver them to blood.

The small intestine is also the site of most water absorption. Water moves out of the gut by osmosis.

Section 36.8 More water and ions are absorbed in the large intestine, or colon, which compacts undigested solid wastes as **feces**. Feces are stored in the rectum, a stretchable region just before the anus. The **appendix** is a short extension from the first part of the large intestine.

Section 36.9 The small organic compounds absorbed from the gut are stored, used in biosynthesis or as energy sources, or excreted by other organ systems. Blood that flows through the small intestine travels next to the liver, which eliminates ingested toxins and stores excess glucose as glycogen.

Sections 36.10, 36.11 Eating a healthy diet can help reduce risk of chronic diseases. Current guidelines emphasize minimizing intake of refined carbohydrates and saturated fat. Food must provide energy and raw materials, including **essential amino acids** and **essential fatty acids**. It must also include two additional types of compounds needed for metabolism: **vitamins**, which are organic, and **minerals**, which are inorganic. Vegetarian diets can meet all these needs, only if foods are carefully combined.

Section 36.12 To maintain body weight, energy intake must balance with energy output. Obesity raises the risk of health problems because stressed adipose cells cause inflammation. Anorexia and bulimia are eating disorders that also threaten health.

Self-Quiz   Answers in Appendix III

1. A digestive system functions in _____ .
   a. secreting enzymes   c. eliminating wastes
   b. absorbing compounds   d. all of the above

2. Protein digestion begins in the _____ .
   a. mouth   c. small intestine
   b. stomach   d. large intestine

3. Most nutrients are absorbed in the _____ .
   a. mouth   c. small intestine
   b. stomach   d. large intestine

4. Bile has a role in _____ digestion and absorption.
   a. carbohydrate   c. protein
   b. fat   d. amino acid

5. Monosaccharides and amino acids absorbed from the gut enter _____ .
   a. blood vessels   c. fat droplets
   b. lymph vessels   d. both b and c

Human Dietary Adaptation The human *AMY-1* gene encodes salivary amylase, an enzyme that breaks down starch. The number of copies of this gene varies, and people who have more copies generally make more of the enzyme. In addition, the average number of *AMY-1* copies differs among cultural groups.

George Perry and his colleagues hypothesized that duplications of the *AMY-1* gene would confer a selective advantage in cultures in which starch is a large part of the diet. To test this hypothesis, the scientists compared the number of copies of the *AMY-1* gene among members of seven cultural groups that differed in their traditional diets. Figure 36.16 shows their results.

1. Starchy tubers are a mainstay of Hadza hunter–gatherers in Africa, whereas fishing sustains Siberia's Yakut. Almost 60 percent of Yakut had fewer than 5 copies of the *AMY-1* gene. What percent of the Hadza had fewer than 5 copies?
2. None of the Mbuti (rain-forest hunter–gatherers) had more than 10 copies of *AMY-1*. Did any European Americans?
3. Do these data support the hypothesis that a starchy diet favors duplications of the *AMY-1* gene?

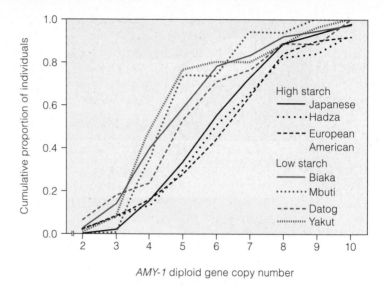

Figure 36.16 Number of copies of the *AMY-1* gene among members of cultures with traditional high-starch or low-starch diets. The Hadza, Biaka, Mbuti, and Datog are tribes in Africa. The Yakut live in Siberia.

6. Bacteria in the _____ make essential vitamins.
   a. stomach c. large intestine
   b. small intestine d. esophagus

7. The pH is lowest in the _____ .
   a. stomach c. large intestine
   b. small intestine d. esophagus

8. Most water that enters the gut is absorbed across the lining of the _____ .
   a. stomach c. large intestine
   b. small intestine d. esophagus

9. _____ are inorganic substances with metabolic roles that no other substance can fulfill.
   a. Fats c. Vitamins
   b. Minerals d. Simple sugars

10. Blood that flows through capillaries in the small intestine flows next through vessels in the _____ .
    a. stomach c. pancreas
    b. heart d. liver

11. Tiny filaments called _____ increase the surface area of a brush border cell.

12. _____ is(are) a good source of omega-3 fatty acids that can reduce the risk of heart disease.
    a. Oatmeal c. Fish
    b. Legumes d. Corn syrup

13. What is the most common cause of anorexia-related death?
    a. cancer c. pneumonia
    b. cardiac arrest d. stroke

14. Which of the following vitamins is fat soluble?
    a. vitamin A c. vitamin C
    b. vitamin B_6 d. all of the above

15. Match each organ with its digestive function.
    ___gallbladder a. makes bile
    ___colon b. compacts undigested residues
    ___liver c. secretes most digestive enzymes
    ___small intestine d. absorbs most nutrients
    ___stomach e. secretes gastric fluid
    ___pancreas f. stores, secretes bile

Additional questions are available on **CENGAGENOW**.

Critical Thinking

1. A python can survive by eating a large meal once or twice a year (Figure 36.4). When it does eat, microvilli in its small intestine lengthen fourfold and its stomach pH drops from 7 to 1. Explain the benefits of these changes.

2. Starch and sugar have the same number of calories per gram. However, not all vegetables are equally calorie dense. For example, a serving of boiled sweet potato provides about 1.2 calories per gram, while a serving of kale yields only 0.3 calories per gram. What do you think accounts for the difference in the calories your body obtains from these two foods?

3. List the foods you ate today. Which item was highest in fat? In protein? In insoluble fiber?

Animations and Interactions on **CENGAGENOW**:
❯ Animal digestive systems; Human digestive system; Digestion and absorption; Calculating BMI.

Key Concepts

Maintaining the Extracellular Fluid

Animals produce metabolic wastes, and gain and lose water and solutes. Yet the composition and volume of extracellular fluid must stay within a tolerable range. Most animals have an organ system that regulates solutes and eliminate wastes.

Human Urinary System

The human urinary system consists of two kidneys, two ureters, a bladder, and a urethra. Inside a kidney, millions of nephrons filter water and solutes from the blood. Most of this filtrate is returned to the blood. Water and solutes that are not returned to the blood become urine.

37 The Internal Environment

37.1 Truth in a Test Tube

Light or dark? Clear or cloudy? A lot or a little? Asking about and examining urine is an ancient art (Figure 37.1). The Sushruta Samhita, an Indian medical text that dates back more than 2,000 years, reports that some people form an excess of sweet-tasting urine that attracts insects. Their disorder, now called diabetes mellitus, loosely translates as "passing honey-sweet water." Doctors still diagnose it by testing the sugar level in urine, although they have replaced the taste test with chemical analysis.

Today, physicians routinely check the pH and solute concentrations of urine to monitor their patients' health. Acidic urine suggests metabolic problems. Alkaline urine can indicate an infection. Damaged kidneys will produce urine high in proteins. An abundance of some salts can result from dehydration or trouble with the hormones that control kidney function. Special urine tests detect chemicals produced by cancers of the kidney, bladder, and prostate gland.

Do-it-yourself urine tests have become popular. If a woman is hoping to become pregnant, she can use one test to keep track of the amount of luteinizing hormone, or LH, in her urine. About midway through a menstrual cycle, LH triggers ovulation, the release of an egg from an ovary. Another urine test can reveal whether she has become pregnant. Still other tests allow older women to check for declining hormone levels in urine, a sign that they are entering menopause.

Urine tests can also reveal the use of various drugs. Olympic athletes have been stripped of their medals when mandatory urine tests reveal they use prohibited drugs. Major League Baseball players agreed to urine tests after repeated allegations that some star players took prohibited steroids. The National Collegiate Athletic Association (NCAA) tests urine samples from about 3,300 student athletes per year for any performance-enhancing substances as well as for "street drugs."

If you take a drug, breakdown products of that drug end up in the urine. For example, if you smoke marijuana, your kidneys

Figure 37.1 Analyzing urine. *Opposite page*, 17th century doctor examines a urine sample. *Left*, preparing urine samples for modern analysis.

filter a breakdown product of the active ingredient (THC) out of the blood and into the urine. It can take as long as ten days for all molecules of the breakdown product to become fully metabolized and removed from the body. Until that happens, urine tests can detect the metabolite's presence.

Urine's usefulness as an indicator of health, hormonal status, and drug use arises from the kidneys' function. Each day, they filter all of the blood in an adult human body more than forty times. When all goes well, the kidneys eliminate excess water as well as unwanted toxins, hormones, and drugs.

So far in this unit, you have considered several organ systems that work to keep cells supplied with oxygen, nutrients, water, and other substances. We turn now to the mechanisms that maintain the composition, volume, and temperature of the internal environment.

What Kidneys Do
Urine begins forming when protein-free plasma filters across capillary walls and into kidney tubules. Reabsorption returns most water, solutes, and nutrients to the blood. Unabsorbed filtrate and secreted substances become the urine. Hormones adjust urine concentration.

Adjusting the Core Temperature
Heat losses to the environment and heat gains from the environment and from metabolic activity determine an animal's body temperature. Ectotherms adjust their temperature behaviorally. Endotherms can adjust production of metabolic heat.

Mammalian Temperature Regulation
Mammals react to heat stress by moving blood to the skin and increasing evaporation by sweating or panting. They react to cold stress by moving blood to their core, fluffing up fur or hair, and increasing metabolic heat production.

❭ All animals constantly acquire and lose water and solutes, yet they must keep the volume and composition of their internal environment—the extracellular fluid—stable.

❮ Links to Osmosis 5.6, Aerobic respiration 7.2, Coelom 23.2, Fate of absorbed compounds 36.9

Gains and Losses of Water and Solutes

By weight, an animal consists mostly of water, with dissolved salts and other solutes. Fluid outside cells—the extracellular fluid (ECF)—serves as the body's internal environment. In vertebrates, interstitial fluid and plasma account for most of the extracellular fluid. Interstitial fluid fills the spaces between cells, and plasma is the fluid portion of the blood (Figure 37.2).

Keeping the solute composition and volume of the extracellular fluid within the range that cells can tolerate is a major aspect of homeostasis. Water and solute gains need to be balanced by water and solute losses. An animal loses water and solutes in feces and urine, in exhalations, and in secretions. It gains water by eating and drinking. In aquatic animals, water also moves into or out of the body by osmosis across the body surface (Section 5.6).

Metabolic reactions also put water and solutes into the ECF. The most abundant metabolic wastes are carbon dioxide and ammonia. Aerobic respiration produces carbon dioxide and water (Section 7.2). Breakdown of amino acids and nucleic acids produces **ammonia** (Section 36.9). Carbon dioxide diffuses out across the body surface or leaves with the help of respiratory organs. In most animals, excretory organs rid the body of ammonia and other unwanted solutes, as well as excess water.

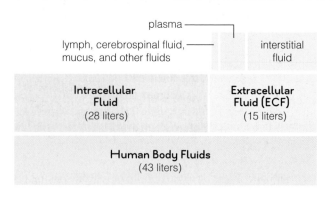

Figure 37.2 Fluid distribution in the human body.

Water–Solute Balance in Invertebrates

Most marine invertebrates have body fluids with the same concentration of solutes as seawater. As a result, osmosis produces no net movement of water into or out of the body.

Planarian flatworms face a problem common to all freshwater animals. Their body fluids have a higher solute concentration than the surrounding water. As a result, water enters the body by osmosis. Excess water and metabolic wastes are eliminated by a pair of branching, tubular excretory organs that run the length of the body.

In animals with a circulatory system, this system interacts with organs that excrete unwanted solutes. For example, an earthworm is a segmented annelid with a fluid-filled body cavity (a coelom) and a closed circulatory system. Most body segments have a pair of tubular excretory organs called nephridia that collect coelomic fluid

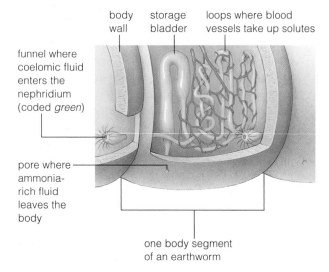

Figure 37.3 Earthworm excretory system. Coelomic fluid enters a nephridium (*green*). As fluid travels through the nephridium, essential solutes leave this tube and enter adjacent blood vessels (*red*). Ammonia-rich waste exits the body through a pore.

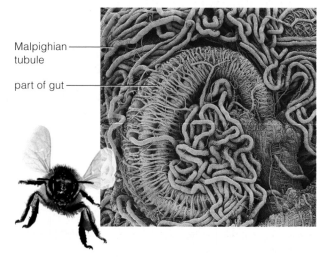

Figure 37.4 Insect excretory function. A honeybee's Malpighian tubules (*gold*) are outpouchings of the gut (*pink*). The tubules are bathed in blood of the open circulatory system. Uric acid and other waste solutes move from blood into the tubules, which deliver the wastes to the gut for elimination through the anus.

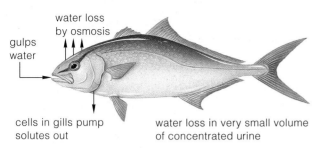

A Marine bony fish with body fluids less salty than the surrounding water; the fish is hypotonic relative to its environment.

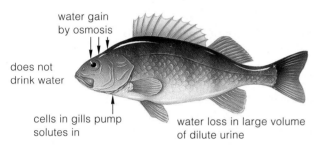

B Freshwater bony fish with body fluids saltier than the surrounding water; the fish is hypertonic relative to its environment.

Figure 37.5 Fluid–solute balance in bony fishes.

Figure 37.6 Two mammals with highly efficient kidneys. Bottlenose dolphins (*left*) and desert kangaroo rats (*right*) live in very different habitats, but they face a common challenge, a lack of fresh water. Both species have large kidneys relative to their body size and produce a highly concentrated urine.

from the adjacent segment (Figure 37.3). As fluid flows through a nephridium, essential solutes and some water leave and enter adjacent blood vessels. Wastes remain in the tube. The ammonia-rich fluid that forms through this process is expelled from the body through a pore.

Land-dwelling arthropods such as insects do not excrete ammonia. Instead, enzymes in their blood convert ammonia to **uric acid**. Uric acid and waste solutes are actively transported into Malpighian tubules, excretory organs that connect to and empty into the gut (Figure 37.4). Ammonia can only be excreted when dissolved in water, but uric acid can be excreted as crystals mixed with just a tiny bit of water to produce a thick paste.

Water–Solute Balance in Vertebrates

Bony fishes have body fluids less salty than seawater, but saltier than fresh water. Thus, they face an osmotic challenge in both environments.

A marine bony fish loses water by osmosis across its body surfaces. To replace this lost water, the fish gulps seawater, then pumps salt out through its gills (Figure 37.5A). Like other vertebrates, bony fishes have a pair of **kidneys**, organs that filter the blood and produce urine. **Urine** consists of water and soluble wastes. Marine bony fishes produce a small amount of urine.

In contrast, a freshwater bony fish produces a large volume of dilute urine because water continually enters its body by osmosis. Solutes lost in the urine are offset by solutes absorbed from the gut, and by sodium ions pumped in across the gills (Figure 37.5B).

Waterproof skin and highly efficient kidneys adapt amniotes to life on land. Birds and other reptiles convert ammonia to uric acid, whereas mammals convert it to **urea**. It takes twenty to thirty times more water to excrete 1 gram of urea than to excrete 1 gram of uric acid. Thus, a typical mammal requires more water than a bird or reptile of similar size. Variations in kidney structure adapt mammals to different habitats. Mammals with limited or no access to fresh water tend to have large kidneys for their size and produce a concentrated urine (Figure 37.6).

---

ammonia Nitrogen-containing compound that is a waste product of amino acid and nucleic acid breakdown.
kidney Organ of the vertebrate urinary system that filters blood, adjusts its composition, and forms urine.
urea Main nitrogen-containing compound in urine of mammals.
uric acid Main nitrogen-containing compound in the urine of insects, as well as birds and other reptiles.
urine Mix of water and soluble wastes formed and excreted by the vertebrate urinary system.

Take-Home Message How do animals maintain the volume and composition of their body fluid?

> In all animals, daily gains in water and solutes must balance daily losses. All must rid the body of waste carbon dioxide and ammonia. Many animals convert ammonia to urea or uric acid before excreting it.

> Most animals have excretory organs that interact with a circulatory system to remove wastes from the blood and excrete them.

> Invertebrate excretory organs include the ammonia-excreting nephridia of earthworms and the uric acid–excreting Malpighian tubules of insects.

> All vertebrates have two kidneys. The volume of urine and the nitrogen-containing wastes excreted (ammonia, urea, or uric acid) varies among groups.

❯ Kidneys filter water, mineral ions, organic wastes, and other substances from the blood. They adjust the volume and composition of this filtrate, and return most of it to the blood. The fluid not returned becomes urine.

❮ Links to Epithelium 28.3, Microvilli 36.6

Components of the System

A human urinary system has two kidneys, two ureters, one urinary bladder, and one urethra (Figure 37.7A). The kidneys filter blood and form urine. The other organs collect and store urine, and convey it to the body surface.

Kidneys are bean-shaped organs about the size of an adult fist. They lie just beneath the peritoneum that lines the abdominal cavity, to the left and right of the backbone (Figure 37.7B). The outermost kidney layer, the renal capsule, consists of fibrous connective tissue (Figure 37.7C). The Latin *renal* means "relating to the kidneys." The bulk of tissue inside the renal capsule is divided into two zones: the outer renal cortex and the inner renal medulla. A renal artery transports blood to each kidney and a renal vein carries blood away from it.

Inside each kidney, urine collects in a central cavity called the renal pelvis. A tubular **ureter** conveys the fluid from each kidney to the **urinary bladder**, a hollow, muscular organ that stores urine. When the bladder is full, a reflex action occurs. Stretch receptors signal motor neurons in the spinal cord. These neurons cause smooth muscle in the bladder wall to contract. At the same time, sphincters encircling the **urethra**, the tube that delivers urine to the body surface, relax. As a result urine flows out of the body. After age two or three, the brain can override the spinal reflex and prevent urine from flowing through the urethra at inconvenient moments.

A male's urethra runs the length of the penis and it conveys urine and semen at different times. A sphincter cuts off urine flow during erections. In females, the urethra opens onto the body surface near the vagina. The female urethra is a short tube, so infectious organisms can easily reach the urinary bladder. That is one reason why women get bladder infections more often than men do.

Introducing the Nephrons

In the next section, we discuss the three processes that rid the body of excess water and solutes in the form of urine. Understanding these processes will be simpler if you first acquaint yourself with the structures involved.

Overview of Nephron Structure A kidney has more than 1 million **nephrons**—microscopically small tubes of cuboidal epithelium associated with capillaries. Kidney tubules are just one cell thick, so substances diffuse easily across them. Each nephron begins in the cortex, where

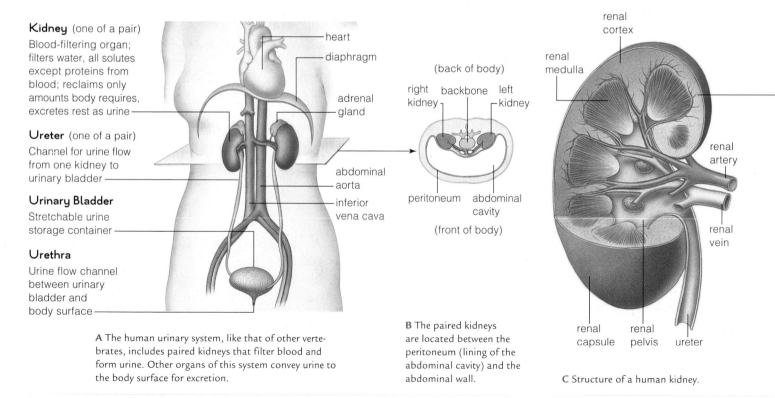

Kidney (one of a pair)
Blood-filtering organ; filters water, all solutes except proteins from blood; reclaims only amounts body requires, excretes rest as urine

Ureter (one of a pair)
Channel for urine flow from one kidney to urinary bladder

Urinary Bladder
Stretchable urine storage container

Urethra
Urine flow channel between urinary bladder and body surface

heart
diaphragm
adrenal gland
abdominal aorta
inferior vena cava

A The human urinary system, like that of other vertebrates, includes paired kidneys that filter blood and form urine. Other organs of this system convey urine to the body surface for excretion.

(back of body)
right kidney   backbone   left kidney
peritoneum   abdominal cavity
(front of body)

B The paired kidneys are located between the peritoneum (lining of the abdominal cavity) and the abdominal wall.

renal cortex
renal medulla
renal artery
renal vein
renal capsule   renal pelvis   ureter

C Structure of a human kidney.

Figure 37.7 Animated Components of the human urinary system and their functions.

its wall balloons out and folds back to form a cup-shaped **Bowman's capsule** (Figure 37.8A,B). Beyond the capsule, the nephron twists a bit and straightens out as a **proximal tubule** (the part nearest the beginning of the nephron). After extending down into the renal medulla, the nephron makes a hairpin turn called the **loop of Henle**. The tubule reenters the cortex and twists again, as the **distal tubule** (the farthest from the start of the nephron). The distal tubules of up to eight nephrons drain into a **collecting tubule**. Many collecting tubules extend through the kidney medulla and open into the renal pelvis.

Like the cells lining the small intestine, cells of kidney tubules have microvilli. These tiny extensions increase the surface area for absorption of substances.

Bowman's capsule Portion of the nephron that encloses the glomerulus and receives filtrate from it.
collecting tubule Kidney tubule that receives filtrate from several nephrons and delivers it to the renal pelvis.
distal tubule Portion of kidney tubule that delivers filtrate to a collecting tubule.
glomerulus Ball of capillaries enclosed by Bowman's capsule.
loop of Henle U-shaped portion of a kidney tubule that extends deep into the renal medulla.
nephron Kidney tubule and glomerular capillaries; filters blood and forms urine.
peritubular capillaries Capillaries that surround and exchange substances with a kidney tubule.
proximal tubule Portion of kidney tubule that receives filtrate from Bowman's capsule.
ureter Tube that carries urine from a kidney to the bladder.
urethra Tube through which urine from the bladder flows out of the body.
urinary bladder Hollow, muscular organ that stores urine.

Blood Vessels Associated With Nephrons Inside each kidney, a renal artery branches into smaller, afferent arterioles. Each arteriole in turn branches into a **glomerulus** (plural, glomeruli), a cluster of capillaries in Bowman's capsule (Figure 37.8C). Glomerular capillaries have gaps between the cells in their walls. These gaps make these capillaries about a hundred times more permeable than a typical capillary. As blood flows through the glomerulus, blood pressure forces some fluid out through the gaps in the capillary wall and into Bowman's capsule. *Glomerulus* is the Greek word for filter.

The unfiltered portion of the blood flows out of the glomerulus and into an efferent arteriole. This arteriole quickly branches into the **peritubular capillaries**, which thread lacily around the nephron (*peri–*, around). These capillaries are the site for exchanges between the fluid flowing through kidney tubules and the blood. From the peritubular capillaries, blood continues into venules that carry it to the renal vein.

> ## Take-Home Message What are the components of the human urinary system and how do they function?
>
> › The human urinary system has two kidneys, two ureters, a urinary bladder, and a urethra. Kidneys filter the blood and form urine. Urine flows out of the kidney through ureters, and into a hollow, muscular bladder. When the bladder contracts, urine flows out of the body through the urethra.
>
> › The functional unit of the kidneys is the nephron, a microscopic tubule that interacts with two systems of capillaries to filter blood and form urine.

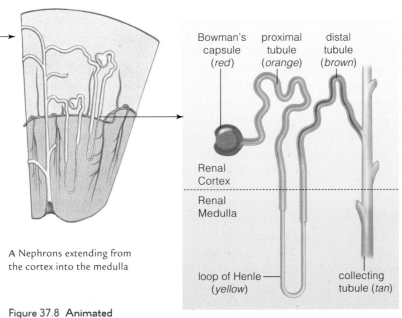

Figure 37.8 Animated
Orientation and structure of a nephron, the functional unit of the kidney.

A Nephrons extending from the cortex into the medulla

B Bowman's capsule and tubular regions of one nephron, cutaway view

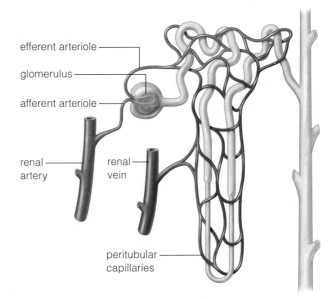

C Blood vessels associated with the nephron. The glomerulus is a ball of capillaries that have unusually leaky walls.

> Urine consists of water and solutes that were filtered from blood and not returned to it, along with solutes secreted from the blood into the nephron's tubular regions.
< Links to Blood pressure 33.7, ADH 31.4, Aldosterone 31.7

This section describes the three processes by which the kidneys form urine. Urine formation begins when blood pressure drives water and small solutes out of the blood and into a nephron. Variations in permeability along the nephron's tubular parts determine whether components of the filtrate return to blood or leave the body in urine.

Glomerular Filtration

The blood pressure generated by a beating heart drives **glomerular filtration**, the first step in urine formation (Figure 37.9 and Figure 37.10 ❶). About 20 percent of the fluid that flows into a glomerulus is forced out through gaps in the capillary walls into Bowman's capsule. Collectively, the glomerular capillary walls and the inner wall of Bowman's capsule function like a filter for the blood. Plasma proteins, blood cells, and platelets cannot pass through the filter. They remain in the blood and leave the glomerulus via the efferent arteriole. The protein-free plasma that enters Bowman's capsule becomes the filtrate that enters the proximal tubule.

Tubular Reabsorption

Only a small fraction of the filtrate actually ends up in urine. **Tubular reabsorption** returns most water and solutes to the blood. Reabsorption begins in the proximal tubule ❷, where transport proteins move sodium ions (Na^+), chloride ions (Cl^-), potassium ions (K^+), and glucose and other nutrients across the tubule wall and into peritubular capillaries. Water follows these solutes by osmosis, so it moves in the same direction.

Most water and nutrients are reabsorbed from the proximal tubule, but reabsorption occurs along the entire length of the kidney tubule. Tubular reabsorption returns about 99 percent of the water filtered into Bowman's capsule to the blood. It returns all the glucose and amino acids, as well as most sodium ions, chloride ions, and bicarbonate ions.

Tubular Secretion

Tubular secretion is the movement of substances from the blood in peritubular capillaries into the filtrate ❸. Membrane proteins in the walls of peritubular capillaries actively transport substances into the interstitial fluid, from which they cross the epithelium of the kidney tubule and enter the filtrate. Secreted substances include hydrogen ions (H^+), potassium ions (K^+) and breakdown products of foreign organic molecules such as drugs, food additives, and pesticides.

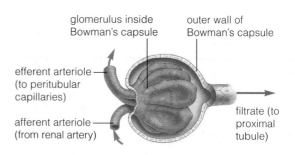

Figure 37.9 Glomerular filtration. Urine formation begins when blood pressure forces protein-free plasma out of the glomerular capillaries and into Bowman's capsule.

Concentrating the Urine

Sip soda all day and your urine will be dilute; sleep eight hours and it will be concentrated. However, even the most dilute urine has far more solutes than the plasma or the typical interstitial fluid.

For water to move out of a nephron by osmosis, the interstitial fluid surrounding the nephron must be saltier than the filtrate inside the nephron. The concentration of solutes in the interstitial fluid is constant throughout the cortex, but increases with the distance into the medulla.

The loop of Henle carries filtrate down into the medulla, then back into the cortex. The ascending and descending arms of the loop differ in their permeabilities to water and sodium. The descending arm of the loop is permeable to water but not to sodium. The loop's ascending arm is impermeable to water and actively pumps salt into the interstitial fluid.

A high solute concentration in interstitial fluid draws water out of filtrate as it flows through the descending loop of Henle. Then, ions are actively transported out of the filtrate as it flows through the ascending loop. These ions that leave the ascending loop contribute to the high solute concentration in the interstitial fluid.

Filtrate entering the distal tubule is less concentrated than normal body fluid. The distal tubule delivers this filtrate to the collecting tubule. Like the descending loop of Henle, this tubule extends down into the medulla. In the deepest part of the medulla, urea is pumped out of the collecting tubule, contributing to the high solute concentration of the interstitial fluid. As urine descends through the collecting tubule, the increasing solute concentration of the interstitial fluid around the tubule draws water outward by osmosis.

The body can adjust how much water is reabsorbed at distal tubules and collecting tubules. When it needs to conserve water, distal tubules and collecting tubules become more permeable to water, so less leaves in urine. When the body needs to rid itself of excess water, the distal tubule and collecting tubules become less permeable to water and the urine remains dilute.

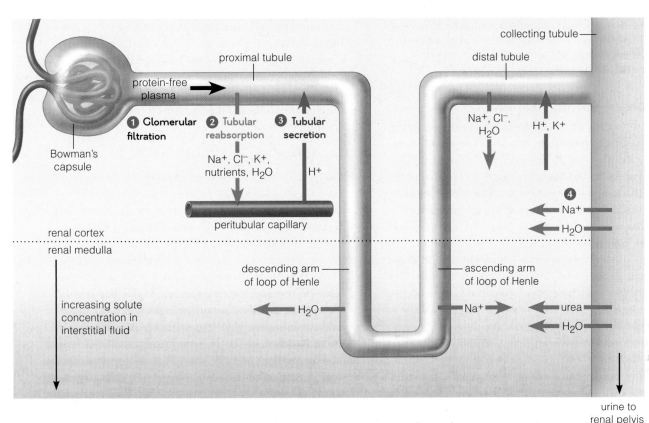

Figure 37.10 **Animated** How urine forms. Only a small segment of peritubular capillary is shown.

>> **Figure It Out** What process moves H$^+$ from peritubular capillaries into the distal tubule? Answer: Tubular secretion

Hormonal Effects on Urine Formation

When the body has too little water, the Na$^+$ concentration in blood rises and blood pressure declines. These changes trigger secretion of hormones that concentrate the urine.

A rise in Na$^+$ concentration causes the hypothalamus to tell the pituitary gland to secrete **antidiuretic hormone** (ADH). ADH makes the distal tubules and collecting tubules more permeable to water. As a result, more water is reabsorbed, and the urine become more concentrated.

When blood pressure declines, cells in the walls of arterioles that carry blood to nephrons release the enzyme renin. Renin sets off a chain of reactions that eventually results in secretion of **aldosterone** by the adrenal glands. This hormone makes distal tubules and collecting tubules

more permeable to Na$^+$ **4**. As a result, more Na$^+$ is reabsorbed, more water follows by osmosis, and the urine becomes more concentrated.

Atrial natriuretic peptide (ANP) is a hormone that makes urine more dilute by inhibiting aldosterone secretion. Muscle cells in the heart's atria release ANP when a high blood volume causes the atrial walls to stretch.

Parathyroid hormone (PTH) regulates blood calcium level in part by adjusting the amount of calcium excreted in urine. When calcium level declines, PTH acts on the kidney to increase reabsorption of this essential ion.

---

aldosterone Adrenal hormone that makes kidney tubules more permeable to sodium; encourages sodium reabsorption, which leads to more water reabsorption and more concentrated urine.
antidiuretic hormone Pituitary hormone that encourages water reabsorption, thus concentrating the urine.
glomerular filtration Protein-free plasma forced out of glomerular capillaries by blood pressure enters Bowman's capsule.
tubular reabsorption Substances move from the filtrate inside a kidney tubule into the peritubular capillaries.
tubular secretion Substances move out of peritubular capillaries and into the filtrate in kidney tubules.

Take-Home Message How is urine formed and concentrated?

> During glomerular filtration, pressure generated by the beating heart drives water and solutes out of glomerular capillaries and into kidney tubules.

> In tubular reabsorption, water, some ions, glucose, and other solutes move out of the filtrate and return to the blood in peritubular capillaries.

> In tubular secretion, transporters move urea, H$^+$, and K$^+$ from peritubular capillaries into the nephron for excretion.

> Flow of filtrate through the loop of Henle sets up a concentration gradient in the interstitial fluid, with this fluid being saltiest deep in the medulla.

> The final urine concentration depends on how much water flows out of the distal tubule and the collecting tubule. Hormones affect urine concentration by their effects on the permeability of these tubules.

> One healthy kidney is enough to filter the blood and regulate fluid content for the body. Unfortunately, failure of both kidneys is not uncommon.
< Link to Fate of absorbed compounds 36.9

Causes of Kidney Failure

The vast majority of kidney problems arise as complications of diabetes mellitus or high blood pressure. These disorders damage small blood vessels, including capillaries that interact with nephrons. Some people are genetically predisposed to infections or conditions that damage kidneys. Kidneys also fail after filtering lead, arsenic, pesticides, or other toxins from the blood. Repeated high doses of aspirin or other drugs can also damage kidneys.

High-protein diets force the kidneys to work overtime eliminating excess urea. Such diets also increase the risk for kidney stones. These hardened deposits form when uric acid, calcium, and other wastes settle out of urine and collect in the renal pelvis. Most kidney stones are washed away in urine, but sometimes one lodges in a ureter or the urethra and causes severe pain. A stone that blocks urine flow raises risk of infections and kidney damage.

Kidney function is measured in terms of the rate of filtration through glomerular capillaries. Kidney failure occurs when the filtration rate falls by half. Failure of both kidneys can be fatal because wastes build up in the blood and interstitial fluid. The pH rises and changes in the concentrations of other ions, most notably Na^+ and K^+, interfere with metabolism.

Treating Kidney Failure

Kidney dialysis can restore proper solute balances in a person who has kidney failure. "Dialysis" refers to exchanges of solutes across a semipermeable membrane between two solutions. With hemodialysis, a dialysis machine is connected to a patient's blood vessel (Figure 37.11A). The machine pumps a patient's blood through semipermeable tubes submerged in a warm solution of salts, glucose, and other substances. As the blood flows through the tubes, wastes dissolved in the blood diffuse out and solute concentrations return to normal levels. Cleansed, solute-balanced blood is returned to the patient's body. Typically a person has hemodialysis three times a week at an outpatient dialysis center. By contrast, peritoneal dialysis can be done at home. Each night, dialysis solution is pumped into a patient's abdominal cavity (Figure 37.11B). Wastes diffuse across the peritoneal lining into the fluid, which is drained out the next morning. Thus this body lining serves as the dialysis membrane.

Kidney dialysis can keep a person alive through an episode of temporary kidney failure. When kidney damage is permanent, dialysis must be continued for the rest of a person's life, or until a donor kidney becomes available for transplant surgery.

Each year in the United States, about 12,000 people are recipients of kidney transplants. More than 40,000 others remain on a waiting list because there is a shortage of donated kidneys. The National Kidney Foundation estimates that every day, 17 people die of kidney failure while waiting for a transplant. Most kidneys for transplant come from deceased donors, but the number of living donors is increasing. One kidney is adequate to maintain good health, so the risks to a living donor are mainly related to the surgery—unless a donor's remaining kidney fails.

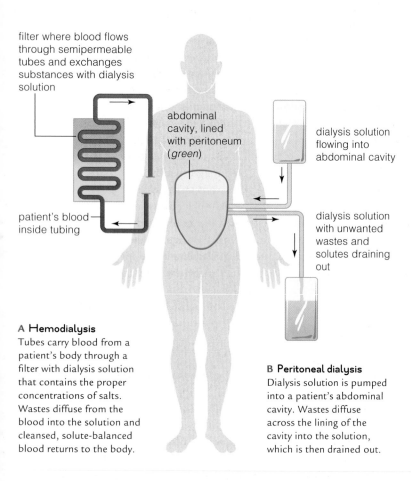

filter where blood flows through semipermeable tubes and exchanges substances with dialysis solution

patient's blood inside tubing

abdominal cavity, lined with peritoneum (*green*)

dialysis solution flowing into abdominal cavity

dialysis solution with unwanted wastes and solutes draining out

A Hemodialysis
Tubes carry blood from a patient's body through a filter with dialysis solution that contains the proper concentrations of salts. Wastes diffuse from the blood into the solution and cleansed, solute-balanced blood returns to the body.

B Peritoneal dialysis
Dialysis solution is pumped into a patient's abdominal cavity. Wastes diffuse across the lining of the cavity into the solution, which is then drained out.

Figure 37.11 Animated Two types of kidney dialysis.

> Take-Home Message **What causes kidney failure, and how does it affect health?**

> Most kidney failure occurs as a complication of diabetes or high blood pressure. Kidney stones, infections, and drug reactions are less common causes.

> Untreated kidney failure is fatal. Dialysis can keep a person with kidney failure alive, but it must be continued until a person dies or receives a kidney transplant.

❯ We turn now to another major aspect of homeostasis. How does the body maintain the core of its internal environment within a tolerable temperature range?

❮ Links to Properties of water 2.5, Amniote traits 24.5

Changes to Core Temperature

The core temperature of an animal body rises when heat from the surroundings or metabolism builds up. A warm body will give up heat to cooler surroundings. The core temperature stabilizes when the rate of heat loss is equal to the rate of heat gain and production. The heat content of any complex animal depends on a balancing act between gains and losses:

$$\text{change in body heat} = \text{heat produced} + \text{heat gained} - \text{heat lost}$$

Heat is gained or lost at body surfaces by processes of radiation, conduction, convection, and evaporation.

Thermal radiation is emission of heat into space around a warm object. Energy radiating from the sun heats animals. Also, metabolic activity produces heat, which radiates from the body. A typical human at rest produces about as much heat as a 100-watt lightbulb.

In conduction, heat is transferred between objects in direct contact with each other. An animal loses heat when it rests on objects cooler than it is. If it contacts objects that are warmer, the animal will gain heat.

In convection, moving air or water transfers heat. Heat moves down a thermal gradient between the body and the cooler air or water next to it. The heated air or water constantly move away from a body, so the thermal gradient remains steep.

In evaporation, an input of heat energy converts a substance from a liquid to a gas. Evaporation of water from body surfaces cools the body (Section 2.5). Evaporative heat loss increases with dry air and a breeze; high humidity and still air slow evaporative heat loss.

Endotherm? Ectotherm? Heterotherm?

Fishes, amphibians, and nonbird reptiles are **ectotherms**, which means "heated from outside." They adjust behaviorally to rising or falling outside temperatures. Most have low metabolic rates and little insulation. A rattlesnake (Figure 37.12A) is an example. When the snake gets cold, it basks in the sun. When hot, it moves into shade.

A Rattlesnake, an ectotherm. **B** Pine grosbeak, an endotherm.

Figure 37.12 Two types of thermoregulation.

Most birds and mammals are **endotherms**, which means "heated from within." When the external temperature falls, an endotherm can increase its metabolic heat production to keep core body temperature stable. The ability to regulate metabolic heat production allows endotherms to remain active at a wider range of external temperatures than ectotherms. Fur, fat, or feathers serve as insulation that minimizes heat transfers (Figure 37.12B).

Ectotherms are at a selective advantage in warm climates because they do not have to spend as much energy as endotherms on maintaining core temperature. They can devote more energy to reproduction and other tasks. In cool or cold regions, endotherms have the selective edge because they can remain active longer. The Arctic has 130 species of mammals and 280 species of birds, but only a few species of reptiles.

Some birds and mammals are **heterotherms**. They maintain a constant core temperature some of the time but allow their temperature to shift at other times. For example, hummingbirds have very high metabolic rates when foraging for nectar during the day. At night, metabolic activity decreases so much that the bird's body may become almost as cool as its surroundings.

ectotherm Animal that controls its internal temperature by altering its behavior; for example, a fish or a lizard.
endotherm Animal that controls its internal temperature by adjusting its metabolism; for example, a bird or a mammal.
heterotherm Animal that maintains its temperature by production of metabolic heat sometimes, and allows its temperature to fluctuate with the environment at other times.

Take-Home Message **How do animals regulate their core body temperature?**

❯ Animals can gain heat from, or lose heat to, the environment. They can also generate heat by metabolic reactions.

❯ Fishes, amphibians, and most reptiles are ectotherms that adjust their body temperature by behavior that facilitates heat exchanges with the environment.

❯ Birds and mammals are endotherms that maintain body temperature by varying their production of metabolic heat.

❯ Most mammals cannot tolerate extreme changes in body temperature. Feedback mechanisms ensure that the core temperature remains within safe limits.
❮ Links to Negative feedback 28.9, Thermoreceptors 30.2

The hypothalamus in the brain interacts with organs throughout the body to maintain the body's core temperature via a negative feedback response (Section 28.9). The hypothalamus receives input from thermoreceptors in the skin and from others deep in the body. When the temperature deviates from a set point, the hypothalamus calls for responses that return temperature to the set point. When the temperature returns to set point, the hypothalamus senses the change and stops calling for a response.

Responses to Heat Stress

High external temperature and metabolic heat production by skeletal muscle can raise core temperature. When a mammal becomes too hot, the hypothalamus issues commands for peripheral vasodilation: The diameter of the blood vessels in skin increases. More blood flows to the skin and delivers more metabolic heat that can be given up to the surroundings (Table 37.1).

Another response to heat stress, evaporative heat loss, occurs at moist respiratory surfaces and across skin. Animals that sweat lose some water this way. For instance, humans and some other mammals have sweat glands that release water and solutes through pores at the skin's surface (Figure 37.13). An average-sized adult human has 2.5 million or more sweat glands. For every liter of sweat produced, about 600 kilocalories of heat energy leave the body by way of evaporative heat loss.

Sweat dripping from skin dissipates little heat. Sweat only cools you if it evaporates from your skin. On humid days, the evaporation rate slows, so sweating is less effective at cooling the body.

Not all mammals sweat. Many drool, lick their fur, or pant to speed cooling. "Panting" refers to shallow, rapid

Figure 37.13 Animated Human sweating response. All primates can sweat, but humans have the greatest number of sweat glands.

breathing. It increases evaporative water loss from the respiratory tract, nasal cavity, mouth, and tongue.

Sometimes increased peripheral blood flow and evaporative heat loss are not enough to counter heat stress, and hyperthermia occurs. A human body temperature above 41.5°C (105°F) is dangerous.

What about a fever? Recall, from Section 34.4, that fever is not itself an illness. It is one of the responses to infection. When activated by a threat, macrophages release signaling molecules that stimulate part of the brain to secrete prostaglandins. These local signaling molecules cause the hypothalamus to allow the core temperature to rise a bit above the normal set point. The rise makes the body less hospitable for pathogens and calls up more immune responses. Generally, the hypothalamus does not let the core temperature rise above 41.5°C (105°F). When a fever exceeds that point or lasts more than a few days, the condition causing it may be life threatening, so medical intervention is essential.

Responses to Cold Stress

Selectively distributing blood flow, fluffing up hair or fur, and shivering help mammals respond to the cold. Some thermoreceptors in skin signal the hypothalamus when things get chilly. The hypothalamus then causes smooth muscle in arterioles that deliver blood to the skin to contract. When arterioles in skin constrict, less metabolic heat reaches the body surface. When your fingers or toes become chilled, all but 1 percent of the blood that would usually flow to the skin is diverted to other regions of the body. At the same time, muscle contractions make hair (or fur) "stand up." This response creates a layer of still air

Table 37.1 Heat and Cold Stress Compared		
Stimulus	Main Responses	Outcome
Heat stress	Widespread vasodilation in skin; behavioral adjustments; in some species, sweating, panting	Dissipation of heat from body
	Decreased muscle action	Heat production decreases
Cold stress	Widespread vasoconstriction in skin; behavioral adjustments (e.g., minimizing surface parts exposed)	Conservation of body heat
	Increased muscle action; shivering; nonshivering heat production	Heat production increases

Figure 37.14 Death by hypothermia. In 1912, the *Titanic* collided with an immense iceberg on her maiden voyage. It took about 2–1/2 hours for the *Titanic* to sink, and rescue ships arrived before it went under. Even so, 1,517 people died. Many died in lifeboats or while afloat in life vests. Hypothermia killed them.

Table 37.2	Impact of Increases in Cold Stress
Core Temperature	Physiological Responses
36–34°C (about 95°F)	Shivering response; faster breathing. Peripheral vasoconstriction moves blood away from body surface. Dizziness, nausea.
33–32°C (about 91°F)	Shivering response ends. Metabolic heat output declines.
31–30°C (about 86°F)	Capacity for voluntary motion is lost. Eye and tendon reflexes inhibited. Consciousness lost. Cardiac muscle action becomes irregular.
26–24°C (about 77°F)	Ventricular fibrillation sets in (Section 33.10). Death follows.

next to skin and helps reduce heat lost by convection and radiation. Minimizing exposed body surfaces can also prevent heat loss, as when polar bear cubs curl up and cuddle against their mother.

With prolonged cold exposure, the hypothalamus commands skeletal muscles to contract ten to twenty times each second. Although this **shivering response** increases metabolic heat production, it has a high energy cost.

brown adipose tissue Adipose tissue that responds to cold by releasing energy as heat, rather than using it to make ATP.
nonshivering heat production Heat-generating mechanism of brown adipose tissue; energy is released as heat, rather than stored in ATP.
shivering response In response to cold, rhythmic muscle contractions generate metabolic heat.

■ Truth in a Test Tube (revisited)

Solutes and nutrients that the body requires are reabsorbed from the filtrate in kidney tubules. Water-soluble drugs and toxins are generally not reabsorbed, so they end up in the urine. How fast the kidneys will remove a substance from blood depends in part on the efficiency of the kidneys, which can vary with age and health. A drug is eliminated from the body of a healthy 35-year-old about twice as fast it is eliminated from the body of a healthy 85-year-old.

How Would You Vote? Should employers be allowed to require potential employees to pass a urine test as a condition of employment? See CengageNow for details, then vote online (cengagenow.com).

Brown adipose tissue helps many mammals warm themselves in a cold environment. This tissue has mitochondria that release energy as heat, rather than storing it in ATP. The result is **nonshivering heat production**. In human infants, brown adipose tissue accounts for about 5 percent of body weight. As a person ages, their amount of brown adipose tissue declines, but some remains around the neck and upper chest. Exposure to cold increases the secretion of thyroid hormone, which binds to brown adipose tissue and encourages nonshivering heat production. This response is impaired in people with hypothyroidism. Because they make too little thyroid hormone, they cannot turn up their nonshivering heat production. As a result, they tend to feel cold more often than other people. They also tend to be heavy because nonshivering heat production burns a lot of calories.

Hypothermia occurs when normal mechanisms fail to keep core temperature from declining and the drop in temperature disrupts normal function. In humans, a decline in body temperature to 95°F (35°C) impairs brain activity. "Stumbles, mumbles, and fumbles" are symptoms of early hypothermia. Severe hypothermia causes loss of consciousness, disrupts heart rhythm, and can be fatal (Figure 37.14 and Table 37.2).

Take-Home Message **How do mammals regulate their core temperature?**

> Temperature shifts are detected by thermoreceptors that send signals to an integrating center in the hypothalamus.

> This center serves as the body's thermostat and calls for adjustments that maintain core temperature.

> Mammals counter heat stress by widespread peripheral vasodilation in skin and evaporative water loss.

> Mammals counter cold stress by vasoconstriction in skin, behavioral adjustments, increased muscle activity, shivering, and nonshivering heat production.

Summary

Section 37.1 Composition of urine provides information about health. Drugs enter the blood, are broken down, and are filtered out by the kidneys. The breakdown products end up in the urine.

Section 37.2 Plasma and interstitial fluid are the main components of the extracellular fluid, or ECF. Maintaining ECF volume and composition is an essential aspect of homeostasis. Organisms must balance solute and fluid gains with solute and fluid losses. They also must eliminate metabolic wastes such as the **ammonia** produced by the breakdown of proteins and nucleic acids.

Animals in different habitats face different challenges. Those living in water lose or gain water by osmosis. On land, the main challenge is avoiding dehydration. Insects and reptiles (including birds) conserve water by converting ammonia to **uric acid**, crystals of which can be eliminated with very little water. Mammals excrete **urea** dissolved in a lot of water. All vertebrates have a pair of **kidneys** that filter the blood and produce **urine**.

Section 37.3 Paired **ureters** carry urine from the kidneys to a muscular **urinary bladder** that stores urine until it is expelled through the **urethra**. Urination is a reflex, but it can be overridden by voluntary control.

A kidney has more than a million **nephrons**, small tubules that interact with capillaries to filter blood and form urine. **Bowman's capsule** in the renal cortex is the start of the kidney tubule. It receives fluid filtered out of the leaky capillaries of the **glomerulus**. The filtered fluid continues through a **proximal tubule**, a **loop of Henle** that descends into and ascends from the renal medulla, and a **distal tubule** that drains into a **collecting tubule**. All collecting tubules drain into the renal pelvis. **Peritubular capillaries** lie in close proximity to the kidney tubules and exchange substances with them.

Section 37.4 Urine formation begins when **glomerular filtration** puts protein-free plasma into a kidney tubule. Most water and solutes are returned to the blood by **tubular reabsorption**. Substances that are not reabsorbed, and substances added to the filtrate by **tubular secretion**, end up in the urine. Hormones regulate the urine's concentration and composition. **Antidiuretic hormone** makes tubules more permeable to water, so more is reabsorbed and urine is more concentrated. **Aldosterone** increases sodium reabsorption. Water follows sodium into the blood, so aldosterone indirectly concentrates the urine.

Section 37.5 Kidneys can be harmed by chronic disease, genetic factors, infections, or drugs. When kidneys fail, frequent dialysis or a kidney transplant is required to sustain life.

Section 37.6 All animals produce metabolic heat. To maintain the core temperature, heat gains by metabolism and by absorption from the environment must balance heat losses to the environment.

For **ectotherms** such as reptiles, the core temperature depends more on heat exchanges with the environment than on any metabolic heat. Such animals control core temperatures mainly by modifications in behavior.

For **endotherms** (most birds and mammals), a high metabolic rate is the primary source of heat. Endotherms regulate their core temperature mainly by controlling the production and loss of metabolic heat.

Heterotherms control core temperature tightly part of the time and allow it to fluctuate with the environmental temperature at other times.

Section 37.7 In mammals, the hypothalamus is the main control center for body temperature. It receives information from thermoreceptors and calls for responses by smooth muscle in arterioles, sweat glands, and other effectors. The core temperature is maintained by behavioral, metabolic, and physiological responses.

A rise in mammalian body temperature induces dilation of blood vessels in the skin. Sweating and panting cool the body by increasing evaporative heat loss.

A decrease in mammalian body temperature causes constriction of blood vessels in skin, **shivering**, raising up hair, feathers, or fur, and an increase in **nonshivering heat production** by **brown adipose tissue**.

Self-Quiz   Answers in Appendix III

1. A freshwater fish gains most of its water by _____ .
   a. drinking   c. osmosis
   b. eating food   d. transport across the gills

2. Breakdown of _____ produces ammonia.
   a. carbohydrates   c. proteins
   b. nucleic acids   d. b and c

3. Insects and birds excrete _____ .
   a. ammonia   c. uric acid
   b. urea   d. nucleic acid

4. Bowman's capsule, the start of the tubular part of a nephron, is located in the _____ .
   a. renal cortex   c. renal pelvis
   b. renal medulla   d. renal artery

5. Fluid filtered into Bowman's capsule flows directly into the _____ .
   a. renal artery   c. distal tubule
   b. proximal tubule   d. loop of Henle

6. Blood pressure forces water and small solutes out of blood and into nephrons during _____ .
   a. glomerular filtration   c. tubular secretion
   b. tubular reabsorption   d. both a and c

Data Analysis Activities

Pesticides and Organic Food A food that carries the USDA's organic label must be produced without pesticides such as malathion and chlorpyrifos that conventional farmers typically use on fruits, vegetables, and grains. Chensheng Lu of Emory University used urine testing to find out if eating organic food significantly affects the level of pesticide residues in a child's body (Figure 37.15). He collected urine of twenty-three children and tested it for metabolites (breakdown products) of pesticides. During the first five days, children ate their standard, nonorganic diet. For the next five days, they ate organic versions of the same types of foods and drinks. Then, for the final five days, the children returned to their nonorganic diet.

1. During which phase of the experiment did the children's urine contain the lowest level of the malathion metabolite?

2. During which phase of the experiment was the maximum level of the chlorpyrifos metabolite detected?

3. Did switching to an organic diet lower the amount of pesticide residues excreted by the children?

Study Phase	No. of Samples	Malathion Metabolite		Chlorpyrifos Metabolite	
		Mean (µg/liter)	Maximum (µg/liter)	Mean (µg/liter)	Maximum (µg/liter)
1. Nonorganic	87	2.9	96.5	7.2	31.1
2. Organic	116	0.3	7.4	1.7	17.1
3. Nonorganic	156	4.4	263.1	5.8	25.3

Figure 37.15 *Above*, levels of metabolites (breakdown products) of malathion and chlorpyrifos in the urine of children taking part in a study of effects of an organic diet. The difference in the mean level of metabolites in the organic and inorganic phases of the study was statistically significant. *Right*, USDA organic food label.

4. Even in the nonorganic phases of this experiment, the highest pesticide metabolite levels detected were far below those known to be harmful. Given these data, would you spend more to purchase organic foods?

---

7. Kidneys return most of the water and small solutes back to blood by way of _____ .
   - a. glomerular filtration
   - b. tubular reabsorption
   - c. tubular secretion
   - d. both a and b

8. Which of the following is secreted into kidney tubules?
   - a. H^+
   - b. glucose
   - c. water
   - d. protein

9. Antidiuretic hormone makes distal tubules and collecting tubules more permeable to _____ .

10. _____ can keep people with kidney failure alive, but it cannot cure them.

11. Match each structure with a function.
    - ____ ureter
    - ____ Bowman's capsule
    - ____ urethra
    - ____ collecting tubule
    - ____ pituitary gland
    - a. start of nephron
    - b. delivers urine to body surface
    - c. carries urine from kidney to bladder
    - d. secretes ADH
    - e. target of aldosterone

12. Which of the following is an endotherm?
    - a. a shark
    - b. a frog
    - c. a monkey
    - d. a snake

13. Match each term with the most suitable description.
    - ____ endotherm
    - ____ ectotherm
    - ____ convection
    - ____ conduction
    - ____ thermal radiation
    - a. environment dictates core temperature
    - b. metabolism dictates core temperature
    - c. heat transfer between objects that are in direct contact
    - d. water, air current transfers heat
    - e. emission of radiant energy

Additional questions are available on **CENGAGENOW**.

Critical Thinking

1. The desert kangaroo rat shown in Figure 37.6 excretes a very small volume of urine relative to its size. Compared with a human nephron, kangaroo rat nephrons have a loop of Henle that is proportionally much longer. Explain how this helps the kangaroo rat conserve water.

2. Drinking too much water can be dangerous. When marathoners or other endurance athletes sweat heavily and drink lots of water, their sodium level drops. The resulting "water intoxication" can be fatal. Why is maintaining sodium balance so important?

3. In cold habitats, ectotherms are few and endotherms often show morphological adaptations to cold. Compared to closely related species that live in warmer areas, cold dwellers tend to have smaller appendages. Also, animals adapted to cool climates tend to be larger than closely related species that dwell in warmer climates. For example, the largest bear species is the polar bear and the largest penguin is Antarctica's emperor penguin. Think about heat transfers between animals and their habitat, then explain why smaller appendages and larger overall body size are advantageous in very cold places.

4. Drinking alcohol inhibits ADH secretion. What effect will drinking a beer have on the permeability of kidney tubules to sodium? To water?

Animations and Interactions on **CENGAGENOW**:
› Human urinary system; Orientation and structure of a nephron; How urine forms.

‹ Links to Earlier Concepts

This chapter continues the discussion of sexual reproduction begun in Section 12.1. We discuss gametes and fertilization (12.5). You will revisit cell differentiation and genes that influence it (10.2, 28.1). We return to primary tissue layers (23.2), amniote eggs (24.5), and the mammalian placenta (24.8). There is more about hormones (31.4, 31.10) and local signaling molecules (31.2). We also reconsider pathogens such as HIV (19.1).

Key Concepts

Modes of Reproduction
Some animals reproduce asexually, but most produce genetically varied offspring by sexual reproduction. Internal fertilization evolved in many land animals, whereas most aquatic animals release gametes into water. Egg yolk typically provides nutrients to developing young.

Human Reproductive Organs
Sperm form in testes and eggs form in ovaries. A series of ducts deliver sperm to the body surface. A female's ovaries produce eggs on a monthly cycle. Her reproductive tract is adapted to receive sperm and to sustain developing offspring.

38 Reproduction and Development

A couple that cannot conceive naturally often turns to **in vitro fertilization** (IVF), an infertility treatment that combines an egg and sperm outside the body. After fertilization, the zygote undergoes mitotic divisions, eventually forming a ball of cells (a blastocyst) that can be placed in a woman's womb to develop.

Louise Brown, the first child conceived by IVF, was born in 1976. At the time, much of the public and many scientists were appalled by the idea of what newspapers referred to as "test tube babies." Scientists worried that this new, unnatural procedure would produce children with psychological and genetic defects. Ethicists warned about the societal implications of manipulating human embryos.

Despite the initial reservations, IVF has become widely accepted and practiced. Worldwide, the technique has produced more than 3 million children. The earliest test tube babies have reached adulthood and begun having children of their own. Louise Brown is among them. In 2007, at age 28, she conceived naturally and gave birth to a healthy son.

One negative impact of IVF has been an increase in multiple births. Embryos conceived by IVF frequently fail to take hold and develop, so doctors often place several embryos in a woman's womb in the hope that one will survive. If all of these embryos do develop, the woman faces the risks associated with a multiple pregnancy. Compared to a woman carrying a single child, she is more likely to undergo miscarriage, stillbirth, or premature birth, and the children she carries are at an increased risk of low birth weight and birth defects.

Professional organizations have urged fertility doctors to limit the number or embryos they place in a woman, but this advice sometimes goes unheeded. In 2009, Nadya Suleman gave birth to octuplets after her doctor put six embryos conceived by IVF inside her and two of them split, forming identical twins (Figure 38.1). Suleman gave birth nine weeks prematurely and her newborns were all underweight. She already had six children conceived by IVF, so she now is the mother of fourteen.

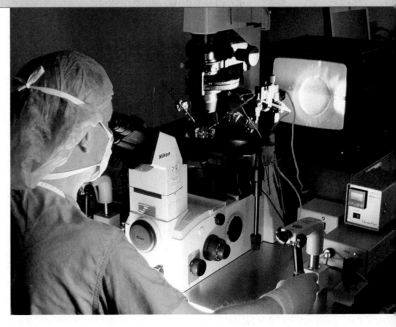

Figure 38.1 *In vitro* fertilization. *Opposite page*, the Suleman octuplets were conceived by IVF. *Above*, a fertility specialist uses a micromanipulator to insert a human sperm into an egg. The video screen shows the view through the microscope.

In this chapter, we delve into the specifics of human reproduction, including the technology that allows us to manipulate human fertility. We consider one of life's most amazing dramas, animal reproduction and development. How does an individual form gametes that convey his or her genetic legacy to the next generation? How does a single fertilized egg develop into a complex body with many specialized kinds of cells? Answers to these questions will emerge as we consider the developmental processes common to all animals.

***in vitro* fertilization** Assisted reproduction technology in which eggs and sperm are united outside the body.

Human Sexual Behavior
Sexual intercourse brings together sperm and eggs, allowing fertilization. A variety of methods allow people who engage in intercourse to prevent fertilization. Intercourse also raises the risk of infection by sexually transmitted bacteria, viruses, or protozoans.

Principles of Development
The same processes regulate development of all animals. Division of the single-celled zygote distributes different materials to different cells. These cells go on to express different master genes that regulate formation of body parts in particular places.

Human Development
Human cleavage forms a blastocyst that implants in the uterus. A placenta connects the developing embryo with its mother. Organs form in the embryonic period, then grow and become functional in the fetal stage. During childbirth, uterine contractions expel the newborn.

❯ Unlike other organ systems, a reproductive system is not necessary for an individual's survival. It is, however, key to passing on genes and ensuring survival of a lineage.
❮ Link to Sexual reproduction 12.1

Asexual Reproduction

With **asexual reproduction**, a single individual produces offspring. All offspring are genetic replicas (clones) of the parent and identical to one another. Asexual reproduction can be advantageous in a stable environment where the gene combination that makes a parent successful is likely to do the same for its offspring.

Invertebrates reproduce by a variety of asexual mechanisms. In some, a new individual grows on the body of its parent, a process called budding. For example, new hydras bud from existing ones (Figure 38.2A). With fragmentation, a piece of the parent breaks off and develops into a new animal. Many corals reproduce by fragmentation. Some flatworms reproduce by transverse fission: The worm divides in two, leaving one piece headless and one tailless. Each piece then grows the missing body parts.

In some animals, offspring can develop from unfertilized eggs, a process called parthenogenesis. For example, a female aphid (a plant-sucking insect) can give birth to several smaller clones of herself every day. Some fishes, amphibians, lizards, and birds can also produce offspring from unfertilized eggs. No mammal is known to reproduce asexually by natural means.

Sexual Reproduction

With **sexual reproduction**, two parents produce haploid gametes (eggs and sperm) that combine at fertilization. Each resulting offspring has a unique combination of paternal and maternal genes.

Sexual reproducers incur higher genetic and energetic costs than asexual reproducers. On average, only half of a sexually reproducing parent's genes end up in each offspring. Producing gametes has costs, and many animals spend time and energy finding and courting a mate. What benefits offset these costs? Most animals live where resources and threats change over time. In such environments, producing offspring that differ from both parents and from one another can be advantageous. By reproducing sexually, a parent increases the likelihood that some of its offspring will have a combination of alleles that suits them to the newly changed environment.

Many animals reproduce asexually under favorable conditions, but switch to sexual reproduction when conditions begin to change. For example, aphids produce sexual offspring when the offspring are likely to disperse to new environments. This occurs with crowding and in autumn.

Gamete Formation and Fertilization

Some animals that reproduce sexually produce both eggs and sperm; they are **hermaphrodites**. Tapeworms and some roundworms are simultaneous hermaphrodites. They produce eggs and sperm at the same time, and can

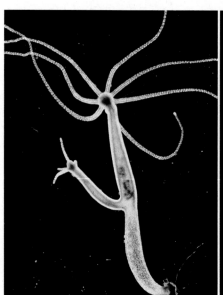

A Asexual reproduction in a hydra. A new individual (*left*) is budding from its parent.

B Sexual reproduction in the barred hamlet, a simultaneous hermaphrodite. Each fish lays eggs and also fertilizes its partner's eggs.

C Sexual reproduction in elephants. The male is inserting his penis into the female. Eggs will be fertilized and the offspring will develop inside the mother's body, nourished by nutrients delivered by her bloodstream.

Figure 38.2 Examples of animal reproduction.

fertilize themselves. Earthworms, land snails, and slugs are simultaneous hermaphrodites too, but they require a partner. So do hamlets, a type of marine fish (Figure 38.2B). During a bout of mating, hamlet partners take turns donating and receiving sperm. Other fishes are sequential hermaphrodites. They switch from one sex to another during the course of a lifetime. More typically, vertebrates have separate sexes that remain fixed for life; each individual is either male or female.

Most aquatic invertebrates, fishes, and amphibians release gametes into the water, where they combine during **external fertilization**. With internal fertilization, sperm fertilize an egg inside the female's body (Figure 38.2C). **Internal fertilization** evolved in most land animals, including insects and the amniotes.

After internal fertilization, a female may lay eggs in the environment or retain them inside her body for some portion of their development. Birds eggs develop outside the mother's body, as do eggs of most insects (Figure 38.3A). In many sharks, snakes, and lizards, embryos develop while enclosed by an egg sac in the mother's body. The eggs hatch inside the mother shortly before she gives birth to well-developed young (Figure 38.3B).

Nourishing the Developing Young

A developing animal requires nutrients. Most animals are nourished by **yolk**, a thick fluid rich in proteins and lipids deposited in the egg during its formation. You have no doubt seen the yolk of a chicken egg. The amount of yolk in a bird egg varies with the time that the egg takes to hatch. Kiwi birds have the longest incubation period of any bird—11 weeks. Their eggs are about two-thirds yolk by volume, whereas an average bird egg is about one-third yolk.

Placental mammals, including humans, have almost yolkless eggs. Instead, the **placenta**, an organ that forms during pregnancy, allows exchange of substances between a mother's blood and that of her developing offspring (Figure 38.3C). Nutrients in the maternal blood diffuse into an offspring's blood and support its development.

A Bug depositing fertilized eggs on a pine needle. Yolk in the eggs provides nutrients that will sustain development of the young.

B Lemon shark giving birth to young that recently hatched in her body. The young were nourished by egg yolk, not by their mother.

C An elk examines her newborn calf. The placenta, the organ that allowed nutrients from the maternal blood to diffuse into the calf's blood, is visible at the *left*. It is expelled at birth.

Figure 38.3 How developing offspring are nourished.

---

asexual reproduction Reproductive mode by which offspring arise from one parent and inherit that parent's genes only.
external fertilization Sperm and egg are released into the external environment and meet there.
hermaphrodite Animal that produces both eggs and sperm, either simultaneously or at different times in its life.
internal fertilization A female retains eggs in her body and sperm fertilize them there.
placenta Of placental mammals, organ that forms during pregnancy and allows diffusion of substances between the maternal and embryonic bloodstreams.
sexual reproduction Reproductive mode by which offspring arise from two parents and inherit genes from both.
yolk Nutritious material in many animal eggs.

Take-Home Message How do animals reproduce?

❯ Some animals produce genetic copies of themselves through asexual reproduction. Most reproduce sexually, producing offspring with unique combinations of parental traits. Some can switch between asexual and sexual reproduction.

❯ A sexually reproducing animal may produce eggs and sperm at the same time, produce both types of gametes at different times, or always produce only one type of gamete.

❯ Animals that live in water often have external fertilization, but land-dwelling groups typically fertilize eggs inside the female's body.

❯ Yolk deposited during egg formation sustains development of most animals. In placental mammals, nutrients diffuse from maternal blood into the blood of the developing young.

> A male's reproductive system produces testosterone and sperm, and delivers sperm to a female reproductive tract.
< Links to Human sex determination 10.4, Gamete formation 12.5, Sex hormones 31.10

Male Reproductive Anatomy

Gametes form in primary reproductive organs, or **gonads**. A human male's gonads are his paired **testes** (singular, testis). They produce sperm and secrete **testosterone**, the main sex hormone in males.

Early in embryonic development, the testes descend from their initial position in the abdominal cavity into the scrotum, a pouch of skin and muscle suspended just below the pelvic girdle (Figure 38.4). Contraction and relaxation of smooth muscle in the scrotum adjusts the position of testes to protect them and keep sperm-making cells from overheating. These cells function best when they are just a bit below typical body temperature.

At **puberty**, the developmental stage when sexual organs mature, the testes enlarge and sperm production begins. Increased testosterone secretion leads to development of secondary sexual traits. Vocal cords thicken, deepening the voice. Coarse hair begins to grow on the face and chest, and in armpits and the pubic region.

A series of ducts convey sperm from testes to the body surface. Cilia push newly formed sperm from a testis into a coiled duct called the **epididymis** (plural, epididymides). Secretions from the epididymis wall help sperm mature. The final portion of each epididymis stores mature sperm and is continuous with a **vas deferens** (plural, vasa deferentia). The vasa deferentia are the longest ducts of the male reproductive tract. Each carries sperm to a short ejaculatory duct that connects to the urethra. The urethra extends through the penis to the body surface.

The **penis** is the male organ of intercourse and also functions in urination. It has a rounded head (the glans) at the end of a narrower shaft. Nerve endings in the glans make it highly sensitive to touch.

The foreskin, a retractable tube of skin, covers the glans when a man is not sexually excited. In the United States, many males undergo circumcision as infants. This elective surgical procedure removes the foreskin. Circumcision reduces the risk of contracting sexually transmitted diseases, but the procedure is painful and rare complications require follow-up surgery or impair sexual function.

Three elongated cylinders of spongy tissue fill the interior of the penis. When a male is sexually excited, blood flows into the spongy tissue faster than it flows out. Fluid

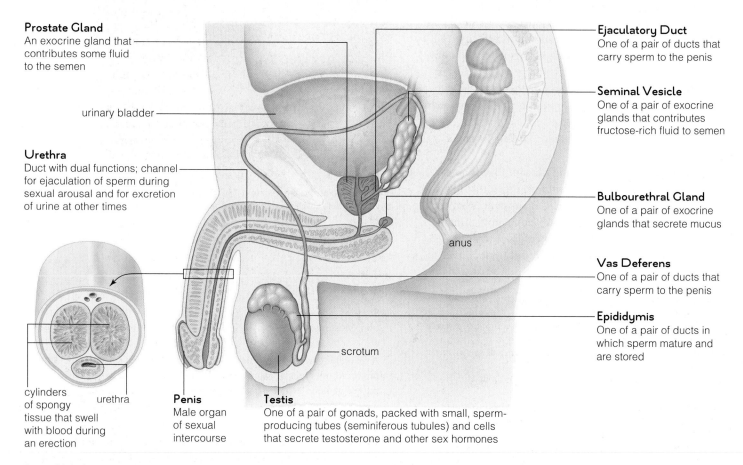

Prostate Gland
An exocrine gland that contributes some fluid to the semen

urinary bladder

Urethra
Duct with dual functions; channel for ejaculation of sperm during sexual arousal and for excretion of urine at other times

Ejaculatory Duct
One of a pair of ducts that carry sperm to the penis

Seminal Vesicle
One of a pair of exocrine glands that contributes fructose-rich fluid to semen

Bulbourethral Gland
One of a pair of exocrine glands that secrete mucus

anus

Vas Deferens
One of a pair of ducts that carry sperm to the penis

Epididymis
One of a pair of ducts in which sperm mature and are stored

scrotum

cylinders of spongy tissue that swell with blood during an erection

urethra

Penis
Male organ of sexual intercourse

Testis
One of a pair of gonads, packed with small, sperm-producing tubes (seminiferous tubules) and cells that secrete testosterone and other sex hormones

Figure 38.4 Animated Components of the human male reproductive system and their functions.

pressure rises and the normally limp penis becomes stiff and erect.

Secretions of exocrine glands that empty into the male reproductive ducts join with sperm to form a thick white fluid called **semen**. Sperm makes up less than 5 percent of the volume of semen. The main component is fructose-rich fluid secreted by seminal vesicles. Sperm use the fructose (a sugar) as their energy source. Secretions from the prostate gland encircling the urethra are the other major contributor to semen volume. Two pea-sized bulbourethral glands secrete a lubricating mucus into the urethra when a male is sexually excited. With sufficient stimulation, a male ejaculates—semen is expelled from the body by contraction of smooth muscles.

During aging, many men develop an enlarged prostate that impairs flow of urine through the urethra. Most prostate enlargement is not dangerous, but it can be a sign of prostate cancer. Doctors diagnose prostate cancer by blood tests and physical examination. By inserting a finger into the rectum, a doctor can feel the adjacent prostate and check it for lumps that could indicate cancer.

Sperm Formation

Although smaller than a golf ball, a testis holds coiled **seminiferous tubules** that would extend more than the length of a football field if stretched out (Figure 38.5A). Testosterone-secreting cells cluster in spaces between the tubules. Diploid male germ cells (spermatogonia) line the tubule's inner wall (Figure 38.5B). These germ cells divide repeatedly by mitosis, and their offspring differentiate to become primary spermatocytes ❶. Meiosis of primary spermatocytes produces spermatids, or immature sperm ❷,❸. Sertoli cells inside the tubules support developing sperm and help regulate sperm production.

A mature sperm is a haploid, flagellated cell (Figure 38.5C). It uses the flagellum, or "tail," to swim toward an egg. Mitochondria in an adjacent midpiece supply the energy required for movement. A sperm's "head" is packed full of DNA and has a cap with enzymes that help the sperm penetrate an egg during fertilization.

---

epididymis One of a pair of ducts where sperm that formed in a testis mature; empties into a vas deferens.
gonads Organs that produce gametes.
penis Male organ of intercourse; also functions in urination.
puberty Period when human reproductive organs mature and begin to function.
semen Sperm mixed with secretions from seminal vesicles and the prostate gland.
seminiferous tubules In testes, tiny tubes where sperm form.
testes In animals, gonads that produce sperm.
testosterone Main hormone produced by testes; causes sperm production and development of male secondary sexual traits.
vas deferens One of a pair of long ducts that carry mature sperm toward the ejaculatory duct.

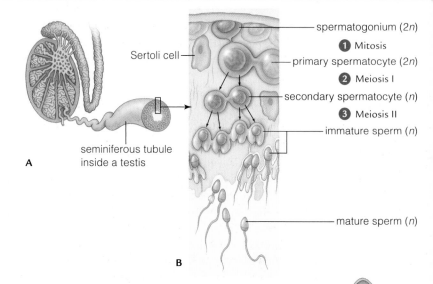

Figure 38.5 Animated Sperm formation.

A Testis cross-section, illustrating seminiferous tubules where sperm form.

B Cross-section of a seminiferous tubule showing the cells within the tubule. Primary germ cells (spermatagonia) give rise to sperm. During development, sperm are supported by Sertoli cells inside the tubule.

C Structure of a mature sperm.

Hormones regulate sperm formation. As Section 31.10 explained, gonadotropin-releasing hormone (GnRH) secreted by the hypothalamus stimulates anterior pituitary cells to secrete luteinizing hormone (LH) and follicle-stimulating hormone (FSH). LH and FSH affect both male and female reproductive function. In males, LH binds to testosterone-secreting cells, causing them to increase their output. Testosterone and FSH bind to Sertoli cells and encourage their sperm-sustaining activity.

A negative feedback loop regulates testosterone secretion and sperm formation. When the concentration of testosterone in the blood falls below a set point, secretion of GnRH by the hypothalamus rises. The resulting rise in testosterone levels then feeds back to the hypothalamus and slows GnRH secretion.

Take-Home Message **What are the structures of the male reproductive system and how do they function?**

❯ A pair of testes, the primary reproductive organs in human males, produce sperm. They also make and secrete the sex hormone testosterone.

❯ A series of ducts convey sperm from the testis to the body surface. Exocrine glands that empty into these ducts provide most of the volume of semen.

❯ The male organ of intercourse, the penis, contains cylinders of spongy tissue that become engorged with blood during sexual excitement.

> In addition to producing gametes and secreting hormones, the female reproductive system encloses and supports developing offspring.

< Links to Gamete formation 12.5, Sex hormones 31.10

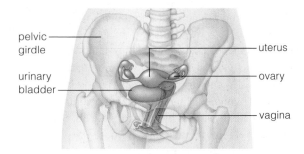

Figure 38.6 Location of the human female reproductive system relative to the pelvic girdle and the urinary bladder.

Female Reproductive Anatomy

A female's gonads, her **ovaries**, lie deep inside her pelvic cavity (Figures 38.6 and 38.7). An ovary is about the size and shape of an almond. It produces and releases immature eggs, called **oocytes**, and secretes estrogens and progesterone, the main sex hormones of females. **Estrogens** maintain the female reproductive tract and cause development of female sexual traits. **Progesterone** prepares the reproductive tract for pregnancy.

Next to each ovary is an **oviduct**, or Fallopian tube. Movement of fingerlike projections around the opening of an oviduct draws an oocyte released by the ovary into the oviduct. Cilia in the oviduct lining then propel the oocyte along the length of the tube. Fertilization usually occurs in an oviduct.

Each oviduct opens into the **uterus**, a hollow, pear-shaped organ above the urinary bladder. A thick layer of smooth muscle, called the myometrium, makes up most of the uterine wall. The uterine lining, or **endometrium**, consists of glandular epithelium, connective tissues, and

blood vessels. When a woman becomes pregnant, the embryo attaches to the endometrium and completes its development inside the uterus.

The lowest portion of the uterus is a narrowed region called the **cervix**. It opens into the vagina. The **vagina** extends from the cervix to the body's surface. It functions as the female organ of intercourse and as the birth canal.

Two pairs of liplike skin folds, referred to as labia, enclose the surface openings of the vagina and urethra. (*Labia* is Latin for lips.) Adipose tissue fills the thick outer folds, which are called the labia majora. The thin inner folds are the labia minora. An erectile organ called the clitoris lies near the anterior junction of the labia minora.

Ovary
One of a pair of gonads that produces oocytes and sex hormones. The hormones work on a monthly basis. They stimulate an oocyte to mature and a corpus luteum (glandular structure) to form, and prime the uterine lining for a potential pregnancy

Oviduct
One of a pair of ciliated channels through which oocytes are propelled from an ovary to the uterus; usual site of fertilization

Uterus
Chamber in which embryo develops; its narrowed portion, the cervix, secretes mucus that helps sperm travel into the uterus and defends the embryo against many bacteria

Myometrium
Thick muscle layers of uterus; stretch greatly during pregnancy

Endometrium
Inner lining of the uterus into which a blastocyst implants itself; gets thicker and has increased blood supply during pregnancy; gives rise to maternal portion of placenta, an organ that metabolically supports embryonic and fetal development

Clitoris
Small organ responsive to sexual stimulation

Labium Minora
One of a pair of innermost, thin skin folds; part of the genitals

Labium Majora
One of a pair of outermost, fat-padded skin folds; part of the genitals

Vagina
Organ of sexual intercourse; also the birth canal

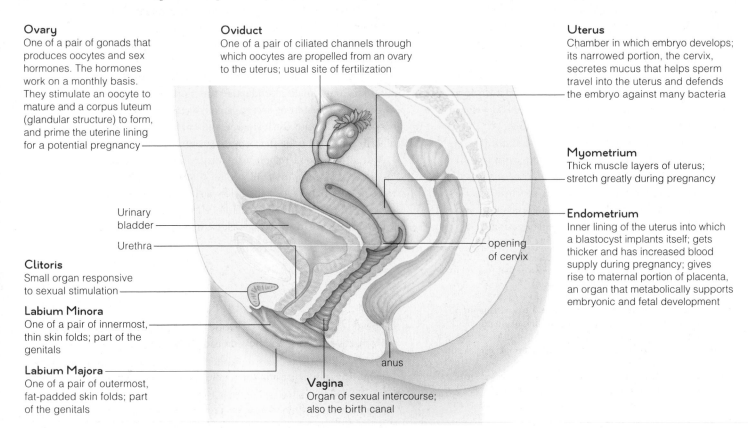

Figure 38.7 **Animated** Components of the human female reproductive system and their functions.

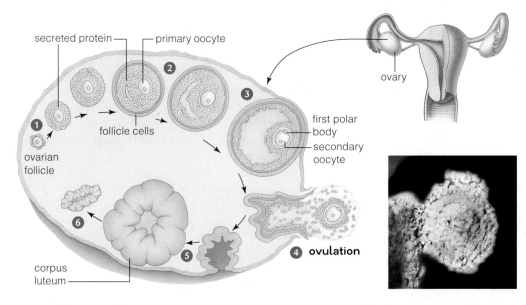

secreted protein — primary oocyte

2

1 ovarian follicle

follicle cells

3

ovary

first polar body
secondary oocyte

6

5

4 **ovulation**

corpus luteum

1 An ovarian follicle begins to mature; the primary oocyte enlarges and secretes proteins, and follicle cells around it divide.

2 A fluid-filled cavity begins to form in the follicle's cell layer.

3 The primary oocyte competes meiosis I and undergoes unequal cytoplasmic division, forming a secondary oocyte and the first polar body.

4 The ovary wall ruptures during ovulation, the release of a secondary oocyte. The oocyte will enter an adjacent oviduct. Its cell cycle is halted in metaphase II and will not continue until fertilization.

5 Follicle cells left behind after ovulation develop into a corpus luteum that secretes hormones.

6 If pregnancy does not occur, the corpus luteum degenerates.

Figure 38.8 Animated Cyclic events in a human ovary, cross-section. The follicle does not "move around" as in this diagram, which simply shows the *sequence* of events. The photo shows ovulation.

The clitoris and penis develop from the same embryonic tissue and both are highly sensitive to tactile stimulation.

Oocyte Maturation and Release

Unlike male germ cells, female germ cells do not divide after birth. A girl is born with about 2 million primary oocytes in her ovaries. A primary oocyte is an immature egg that has entered meiosis but stopped in prophase I. When a female reaches puberty, hormonal changes prompt her primary oocytes to mature, one at a time, in an approximately 28-day ovarian cycle.

Figure 38.8 shows how an oocyte matures over the course of this cycle. A developing oocyte and the cells around it constitute an **ovarian follicle 1**. In the first part of the ovarian cycle, cells around the oocyte divide repeatedly as the oocyte enlarges and secretes a layer of proteins **2**. As the follicle matures, a fluid-filled cavity opens around the oocyte.

More than one follicle may begin to develop, but usually only one becomes fully mature. In that follicle, the primary oocyte completes meiosis I and undergoes unequal cytoplasmic division. This division produces a large secondary oocyte and a tiny polar body **3**. The polar body will eventually disintegrate. The secondary oocyte begins meiosis II, then halts in metaphase II. It will not complete meiosis and become a mature egg until it meets up with a sperm.

About two weeks after the follicle began to mature, its wall ruptures and **ovulation** occurs: The secondary oocyte, polar body, and some surrounding follicle cells are ejected into the adjacent oviduct **4**. The oocyte must meet up with sperm within 24 hours for fertilization to occur.

Meanwhile, back in the ovary, the cells of the ruptured follicle develop into a hormone-secreting structure called the **corpus luteum 5**. (The name means "yellow body" in Latin and refers to its yellowish color.) If pregnancy does not occur, the corpus luteum will break down **6**. After it is gone, a new follicle begins to mature.

cervix Narrow part of uterus that connects to the vagina.
corpus luteum Hormone-secreting structure that forms from follicle cells left behind after ovulation.
endometrium Lining of uterus.
estrogen Hormone secreted by ovaries; causes development of female sexual traits and maintains the reproductive tract.
oocyte Immature egg.
ovarian follicle In animals, immature egg and surrounding cells.
ovary Organ in which eggs form.
oviduct Duct between an ovary and the uterus.
ovulation Release of a secondary oocyte from an ovary.
progesterone Hormone secreted by ovaries; prepares the uterus for pregnancy.
uterus Muscular chamber where offspring develop; womb.
vagina Female organ of intercourse and birth canal.

Take-Home Message What are the structures of the female reproductive system and how do they function?

❯ A pair of ovaries, the primary reproductive organs in human females, produces eggs. They also make and secrete the hormones estrogen and progesterone.

❯ A female does not make new eggs after birth; at puberty her existing oocytes begin to mature, one at a time, on an approximately monthly basis.

❯ An oocyte released from an ovarian follicle by ovulation enters an oviduct. The oviduct conveys the oocyte to the uterus. If fertilization occurs, the resulting embryo will complete development in the uterus.

❯ The vagina serves as the female organ of intercourse and as the birth canal.

> Hormones coordinate cyclic changes in the ovaries and the uterus.

‹ Links to Local signaling molecules 31.2, Hypothalamus and pituitary 31.4, Aldosterone 37.4

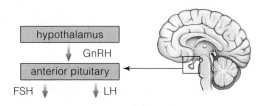

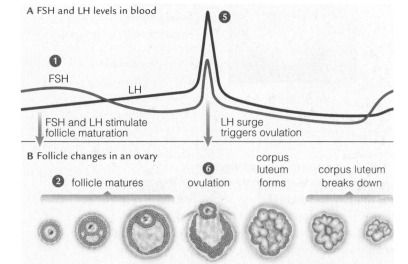

A FSH and LH levels in blood

FSH

LH

FSH and LH stimulate follicle maturation

LH surge triggers ovulation

B Follicle changes in an ovary

follicle matures | ovulation | corpus luteum forms | corpus luteum breaks down

follicle secretes estrogens

corpus luteum secretes estrogens, progesterone

C Estrogen and progesterone levels in blood

Progesterone

Estrogens

low estrogen

estrogens, progesterone, cause uterine lining to thicken

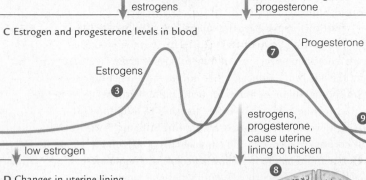

D Changes in uterine lining

menstrual flow

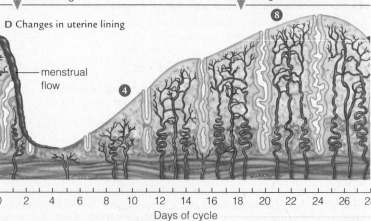

0 2 4 6 8 10 12 14 16 18 20 22 24 26 28
Days of cycle

├── Follicular phase ──┤├── Luteal phase ──┤

The Ovarian and Menstrual Cycles

The previous section described the cyclic changes that occur in ovaries. These events coordinate with cyclic changes in the uterus. We refer to the approximately monthly changes in the uterus as the **menstrual cycle**. The first day of the menstrual cycle is marked by the onset of **menstruation**: the flow of bits of uterine lining and a small amount of blood from the uterus, through the cervix, and out of the vagina.

Table 38.1 describes the phases of the menstrual cycle and Figure 38.9 shows how hormone levels and the thickness of the uterine lining change during those phases. The figure also shows corresponding events in the ovaries.

Like testes, ovaries are under the control of GnRH. As the menstrual cycle begins, increased GnRH secretion from the hypothalamus causes the pituitary to increase its output of FSH and LH ❶. As its name would suggest, follicle-stimulating hormone stimulates an ovarian follicle to begin maturing ❷. The interval of follicle maturation that precedes ovulation is called the follicular phase of the menstrual cycle. As a follicle matures, cells around the oocyte secrete estrogens ❸. Estrogens bind to cells of the endometrium and signal them to begin mitosis. The resulting cell divisions thicken the endometrium ❹.

The pituitary detects the increased level of estrogens in the blood and responds with an outpouring of LH ❺. The surge of LH causes the primary oocyte to complete meiosis I and undergo cytoplasmic division. The LH surge also causes the follicle to swell and burst. Thus, the mid-cycle surge of LH is the trigger for ovulation ❻.

Immediately after ovulation, the estrogen level declines until the corpus luteum forms. During the luteal phase of the cycle, the corpus luteum secretes some estrogen and a lot of progesterone ❼. Estrogens and progesterone cause the uterine lining to thicken and encourage blood vessels to grow through it. The uterus is now prepared for a new pregnancy ❽.

If pregnancy does not occur, the corpus luteum persists for about twelve days. The estrogen and progesterone it

Figure 38.9 Animated Changes in a human ovary and uterus correlated with changing hormone levels. We start with the onset of menstrual flow on day one of a twenty-eight-day menstrual cycle.

A,B The anterior pituitary secretes FSH and LH, which stimulate a follicle to grow and an oocyte to mature in an ovary. A mid-cycle surge of LH triggers ovulation and the formation of a corpus luteum. A decline in FSH after ovulation stops more follicles from maturing.

C,D Early on, estrogen from a maturing follicle calls for repair and rebuilding of the endometrium. After ovulation, the corpus luteum secretes some estrogen and more progesterone that primes the uterus for pregnancy.

>> **Figure It Out** What is the source of the progesterone secreted after ovulation? Answer: The corpus luteum

Table 38.1	Events of a Menstrual Cycle Lasting Twenty-Eight Days	
Phase	Events	Day of Cycle
Follicular phase	Menstruation; endometrium breaks down	1–5
	Follicle matures in ovary; endometrium rebuilds	6–13
Ovulation	Oocyte released from ovary	14
Luteal phase	Corpus luteum forms, secretes progesterone; the endometrium thickens and develops	15–28

secretes keep the hypothalamus from secreting FSH, so no other follicles can start to mature. When the corpus luteum begins to break down, estrogen and progesterone levels plummet ❾. The hypothalamus senses this decline, and stimulates the pituitary to once again increase its secretion of FSH and LH. In the uterus, the decline in estrogens and progesterone causes the thickened lining to break down, and menstruation begins. Blood and endometrial tissue flow out of the vagina for 3 to 6 days.

Menstrual Disorders

Premenstrual Syndrome (PMS) Many women regularly experience discomfort a week or two before they menstruate. Hormones released during the cycle cause milk ducts to widen, making breasts feel tender. Other tissues may swell also, because premenstrual changes influence aldosterone secretion. This hormone stimulates reabsorption of sodium and, indirectly, water (Section 37.4). Cycle-induced changes also cause depression, irritability, anxiety, headaches, and can disrupt sleep.

The regular recurrence of these symptoms is known as premenstrual syndrome (PMS). A balanced diet and regular exercise make PMS less likely and less severe. Use of oral contraceptives minimizes hormone swings and therefore PMS. In some cases, drugs that suppress the secretion of sex hormones entirely are prescribed to alleviate premenstrual discomfort.

Menstrual Pain During menstruation, the secretion of local signaling molecules called prostaglandins stimulates contractions of smooth muscle in the uterine wall. Many women do not feel the muscle contractions, but others experience a dull ache or sharp pains commonly known as menstrual cramps. Women who secrete high levels

of prostaglandins are more likely to feel uncomfortable. Aspirin and nonsteroidal anti-inflammatory drugs such as ibuprofen can prevent or relieve menstruation-related pain because they discourage prostaglandin release.

Endometriosis, the growth of endometrial tissue in the wrong places in the pelvis, is another cause of painful menstruation. The disorder affects about 10 percent of women. The mislocated tissue reacts to hormonal changes in the same way as normal endometrium. However, with this tissue, the repeated bleeding and healing cause pain and scarring. Hormone suppression relieves symptoms, but the only cure is surgical removal of the tissue.

More than one-third of women over age thirty have benign uterine tumors called fibroids. Most uterine fibroids cause no symptoms, but some cause pain during menstruation, unusually long and heavy bleeding, and spotting (minor bleeding) between periods. A woman who needs to change pads or tampons after an hour or two should consult her doctor. Surgical removal of fibroids halts the excessive bleeding and pain.

From Puberty to Menopause

When a woman enters puberty, increased estrogen secretion by her ovaries results in the development of female secondary sexual traits such as breasts and pubic hair. During a woman's reproductive years, her estrogen level fluctuates during her monthly cycle.

A woman enters **menopause** when all the follicles in her ovaries have either been released during a menstrual cycle or disintegrated as a result of normal aging. With no follicles left to mature, the woman no longer produces estrogen or progesterone and menstrual cycles cease.

Hormonal changes around the time of menopause cause many women to have hot flashes. The woman gets abruptly and uncomfortably hot, flushed, and sweaty as blood surges to her skin. When episodes occur at night, they disrupt sleep. Hormone replacement therapy can relieve these symptoms, but may raise the risk of breast cancer and stroke, especially if continued for many years.

---

menopause Permanent cessation of menstrual cycles.
menstrual cycle Approximately 28-day cycle in which the uterus lining thickens and then, if pregnancy does not occur, is shed.
menstruation Flow of shed uterine tissue out of the vagina.

Take-Home Message **What hormone-induced cycles occur in the ovaries and uterus?**

❯ Every 28 days or so, FSH and LH stimulate maturation of an ovarian follicle.

❯ A midcycle surge of LH triggers ovulation—the release of the secondary oocyte into an oviduct.

❯ Prior to ovulation, estrogen secreted by a maturing follicle causes the endometrium to thicken. After ovulation, progesterone secreted by the corpus luteum encourages secretion by endometrial glands.

❯ If pregnancy does not occur, the corpus luteum breaks down, hormone levels drop, the endometrial lining is shed, and the cycle begins again.

❯ Cycles continue until a woman enters menopause.

> When people engage in sexual intercourse, the excitement of the moment may obscure what can happen if a secondary oocyte is in an oviduct.
< Link to Autonomic signals 29.8

Sexual Intercourse

For males, intercourse requires an erection. The penis consists mostly of long cylinders of spongy tissue. When a male is not sexually aroused, his penis is limp, because the blood vessels that transport blood to its spongy tissue are constricted. When he becomes aroused, these vessels dilate and veins that carry blood out of the penis constrict. Inward blood flow exceeds outward flow, and the increasing fluid pressure enlarges and stiffens the penis so it can be inserted into a female's vagina.

The ability to get and sustain an erection peaks during the late teens. As a male ages, he may have episodes of erectile dysfunction. With this disorder, the penis does not stiffen enough for intercourse. Men who have circulatory problems are most often affected. Smoking increases the risk of circulatory problems, so smokers are especially likely to have erectile dysfunction.

Viagra and similar drugs prescribed for erectile dysfunction help blood enter the penis by relaxing the smooth muscle inside of it. The drugs have some side effects. For example, Viagra affects cone cells in the eyes, causing a temporary shift in color perception.

When a woman becomes sexually excited, blood flow to the vaginal wall, labia, and clitoris increases. Glands in the cervix secret mucus and glands on the labia (the equivalent of a male's bulbourethral glands), produce a lubricating fluid. The vagina itself does not have any glandular tissue; it is moistened by mucus from the cervix and plasma fluid that seeps out between the epithelial cells of the vaginal lining.

During intercourse, increased sympathetic stimulation raises the heart rate and breathing rate in both partners. The posterior pituitary steps up its secretion of oxytocin. This hormone acts in the brain, inhibiting signals from the amygdala, the region that controls fear and anxiety.

Continued mechanical stimulation of the penis or clitoris can lead to orgasm. During orgasm, a surge of oxytocin causes rhythmic contractions of muscles in the male reproductive tract and female vagina. Endorphins flood the brain and evoke feelings of pleasure. In males, orgasm is usually accompanied by ejaculation, in which contracting muscles force the semen out of the penis.

Fertilization

An ejaculation can put 300 million sperm into the vagina. Sperm live for about three days after ejaculation, so fertilization can occur even if intercourse takes place a few days before ovulation. About thirty minutes after sperm arrive in the vagina, a few hundred of them reach the oviducts and swim toward the ovaries. Fertilization usually occurs in the upper part of an oviduct (Figure 38.10A).

The secondary oocyte released at ovulation has a wrapping of follicle cells over a layer of secreted proteins that form a jelly coat around it (Figure 38.10B). The sperm makes its way between the follicle cells to the jelly coat. The plasma membrane of the sperm's head has receptors that bind species-specific proteins in the jelly. Binding these proteins triggers the release of protein-digesting enzymes from the cap on the sperm's head. The collective effect of enzyme release from many sperm clears a passage through the jelly coat to the oocyte's plasma membrane. Receptors in this membrane bind a sperm's plasma membrane and the two membranes fuse. The sperm then enters the oocyte. Usually only one sperm enters the secondary oocyte. Its entry triggers changes to the jelly coat that prevent other sperm from binding.

It only takes one sperm to fertilize an egg, but it takes the binding of many to release enough enzyme to clear a way to the egg plasma membrane. This is why a man who has healthy sperm but a low sperm count can be functionally infertile.

Remember that the secondary oocyte had entered meiosis II, then stopped at metaphase II. Binding of the sperm membrane to the oocyte membrane causes completion of meiosis II. The unequal cytoplasmic division that follows produces a single mature egg—an **ovum** (plural, ova)—and a polar body.

Inside the oocyte, a sperm's tail and organelles degenerate, leaving only the sperm nucleus (Figure 38.10C). Chromosomes in the haploid egg and sperm nuclei become the genetic material of the new zygote.

Take-Home Message How does does an egg get fertilized?

> An erection stiffens the penis enough for it to be inserted into the vagina. Ejaculation places sperm in the vagina and they swim toward the oviducts, where fertilization occurs.

> A sperm that meets a secondary oocyte makes its way between follicle cells and binds to proteins in the jelly coat. Enzymes digest a path to the oocyte and the sperm enters.

> Entry of sperm into an oocyte causes the oocyte to complete meiosis II. The haploid sperm nucleus and egg nucleus provide the genetic material of the new zygote.

---

ovum Mature animal egg.

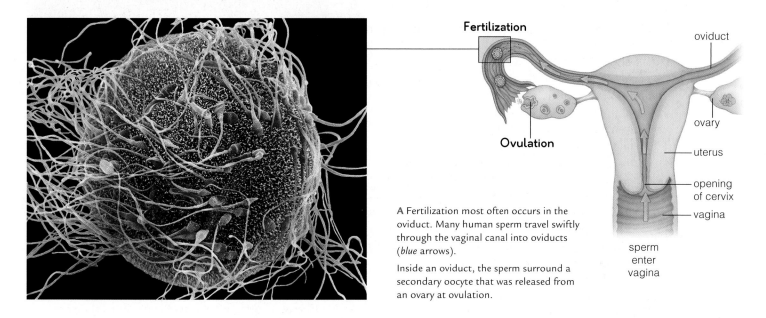

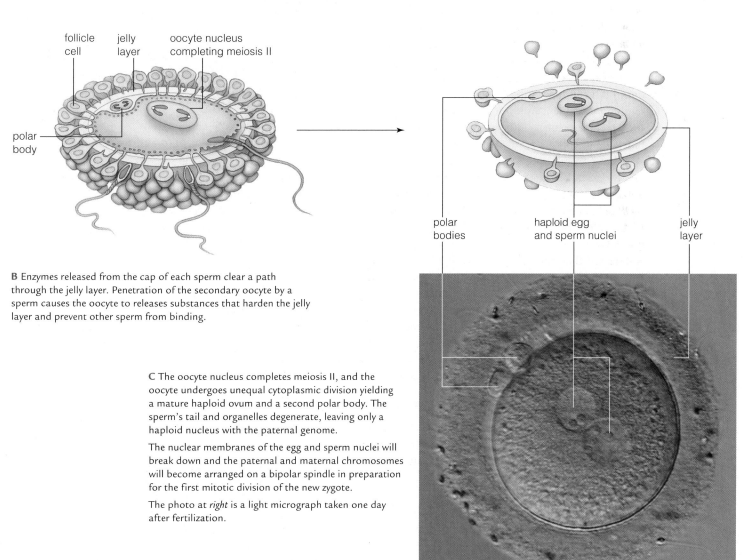

A Fertilization most often occurs in the oviduct. Many human sperm travel swiftly through the vaginal canal into oviducts (*blue* arrows).

Inside an oviduct, the sperm surround a secondary oocyte that was released from an ovary at ovulation.

B Enzymes released from the cap of each sperm clear a path through the jelly layer. Penetration of the secondary oocyte by a sperm causes the oocyte to releases substances that harden the jelly layer and prevent other sperm from binding.

C The oocyte nucleus completes meiosis II, and the oocyte undergoes unequal cytoplasmic division yielding a mature haploid ovum and a second polar body. The sperm's tail and organelles degenerate, leaving only a haploid nucleus with the paternal genome.

The nuclear membranes of the egg and sperm nuclei will break down and the paternal and maternal chromosomes will become arranged on a bipolar spindle in preparation for the first mitotic division of the new zygote.

The photo at *right* is a light micrograph taken one day after fertilization.

Figure 38.10 Animated Fertilization in humans.

> Keeping sperm and egg apart is the goal of contraception.

On average a couple who takes no precautions to prevent pregnancy stands an 85 percent chance of conceiving within a year. A variety of methods can lower the likelihood of pregnancy. The most effective option is abstinence—no intercourse. It is 100 percent effective, but difficult to sustain.

Table 38.2 lists common types of contraception available to people who are sexually active. Rhythm methods are forms of abstinence; a woman avoids intercourse during her fertile period. She calculates when she is ovulating by recording how long her menstrual cycles last, checking her body temperature daily (it usually rises around ovulation), monitoring the thickness of her cervical mucus, or some combination of these indicators.

Withdrawal, or removing the penis from the vagina before ejaculation, requires great willpower. It also may be ineffective because sperm can be in pre-ejaculation fluids. Similarly, rinsing out the vagina (douching) immediately after intercourse is unreliable. Typically, some sperm swim through the cervix within seconds of ejaculation.

Surgical methods are highly effective, but they are meant to make a person permanently sterile. Men may opt for a vasectomy. A doctor makes a small incision into the scrotum, then cuts and ties off each vas deferens. A tubal ligation blocks or cuts a woman's oviducts.

Other fertility control methods use chemical and physical barriers to stop sperm from reaching an egg. Spermicidal foams and jellies poison sperm. They are best used with a physical barrier method such as a diaphragm, a flexible, dome-shaped device that is positioned inside the vagina so it covers the cervix. A cervical cap is similar, but smaller than a diaphragm. Condoms are thin, tight-fitting sheaths worn over the penis. Latex ones have the advantage of also protecting against sexually transmitted diseases (STDs). However, any condom can tear or leak.

An intrauterine device, or IUD, must be inserted into the uterus by a physician. Some IUDs thicken cervical mucus so much that sperm cannot swim through it. Others release copper ions, which keeps an early embryo from implanting in the uterus.

The birth control pill is the most common fertility control method in developed countries. "The Pill" is a mix of synthetic estrogens and progesterone-like hormones that prevents both maturation of oocytes and ovulation. When taken consistently, it is highly effective. Its use lowers risk of ovarian and uterine cancer but raises risk of breast, cervical, and liver cancer. A birth control patch is a small, flat adhesive patch applied to skin. It delivers the same mixture of hormones as an oral contraceptive, and it blocks ovulation the same way. Both birth control pills and the birth control patch raise the risk of stroke and other seri-

Table 38.2 Common Methods of Contraception

Method	Mechanism of Action	Pregnancy Rate*
Abstinence	Avoid intercourse entirely	0% per year
Rhythm method	Avoid intercourse when female is fertile	25% per year
Withdrawal	End intercourse before male ejaculates	27% per year
Vasectomy	Cut or close off male's vasa deferentia	>1% per year
Tubal ligation	Cut or close off female's oviducts	>1% per year
Condom	Enclose penis, block sperm entry to vagina	15% per year
Diaphragm, cervical cap	Cover cervix, block sperm entry to uterus	16% per year
Spermicides	Kill sperm	29% per year
Intrauterine device	Prevent sperm entry to uterus or prevent implantation	>1% per year
Oral contraceptives	Prevent ovulation	>1% per year
Hormone patches, implants, or injections	Prevent ovulation	>1% per year
Emergency contraception pill	Prevent ovulation	15–25% per use**

* Percent of users who get pregnant despite consistent, correct use
** Not meant for regular use

ous cardiovascular disorders, particularly in older women and women who smoke.

Hormone injections or implants can also prevent ovulation. An injection acts for several months; an implant lasts for three years. Both methods are quite effective, but may cause sporadic, heavy bleeding.

Emergency contraception is available if a woman has unprotected sex or a condom breaks. "Morning-after pills" such as Plan B are available without a prescription to women over age 18. They prevent ovulation and work best if taken immediately after intercourse, but they can be effective up to three days later. Morning-after pills are not meant for regular use. They cause nausea, vomiting, abdominal pain, headache, and dizziness.

Take-Home Message How can pregnancy be prevented?

> Couples can prevent pregnancy by avoiding sex entirely or abstaining when the woman is fertile.

> Hormones delivered by pills, patches, or injections can prevent a woman from ovulating.

> Temporary barriers such as condoms or a diaphragm keep sperm and egg apart, as do permanent surgical methods such as a vasectomy or tubal ligation.

> Sexually transmitted pathogens enter a body through the reproductive tract, but some cause bodywide symptoms.
< Links to HIV and AIDS 19.1, 34.13; Bacteria 19.7; Flagellated protozoans 20.3; HPV 34.1

Each year, pathogens that cause sexually transmitted diseases, or STDs, infect about 15 million Americans (Table 38.3). Women are more easily infected than men, and have more complications. For example, pelvic inflammatory disease (PID), a secondary outcome of bacterial STDs, scars the female reproductive tract and can cause infertility, chronic pain, and tubal pregnancies (Figure 38.11A). An infected women can pass an STD on to her newborn (Figure 38.11B). Both men and women can be debilitated by bodywide effects of some infections (Figure 38.11C).

The most common bacterial STD is now chlamydia, caused by *Chlamydia trachomatis*. It often causes no symptoms in women. About half of infected men have discharge from the penis and painful urination. Untreated men risk inflammation of the epididymides and infertility. Like other bacterial STDs, chlamydia can be cured with antibiotics.

Bacteria also cause gonorrhea and syphilis. Less than one week after a male is infected by *Neisseria gonorrhoeae*, yellow pus oozes from his penis. Urination becomes frequent and painful. An infected woman usually has few symptoms at first, but the pathogen can infect oviducts, causing cramps, scarring, and sterility. Syphilis is caused by *Treponema pallidum*, a spiral-shaped bacteria. If the infection is untreated, bacteria that replicate throughout the body cause skin ulcers and damage the liver, bones, and eventually the brain.

Trichomoniasis is caused by the flagellated protozoan *Trichomonas vaginalis*. Infected women have a yellowish discharge and a sore, itchy vagina. Men usually have no symptoms. In both sexes, untreated infection can cause infertility. An antiprotozoal drug can cure the infection.

Infection by human papillomaviruses (HPV) is widespread in the United States. Of about 100 HPV strains, a few cause genital warts. In women, the bumplike growths form on the vagina, cervix, and genitals. In men, they form on the penis and scrotum. Some strains of HPV cause cervical cancer. Sexually active females should have an annual Pap smear, in which cells from the cervix are examined for signs of cancer using a microscope. A recently approved vaccine can prevent HPV infection if given before viral exposure (Section 34.1). Like other viral diseases, HPV cannot be cured by any drug.

About 45 million Americans have genital herpes caused by type 2 herpes simplex virus. An initial infection commonly causes small sores at the site of infection. The sores heal, but the infection can be reactivated, causing tingling or itching, which may or may not be accompanied by visible sores. Antiviral drugs cannot cure the

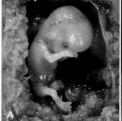
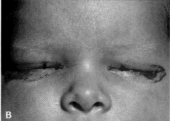
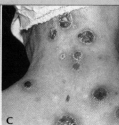

Figure 38.11 Possible consequences of unsafe sex. **A** Increased risk of tubal pregnancy. Scarring from STDs can cause an embryo to implant in an oviduct, instead of the uterus. Untreated tubal pregnancies can rupture an oviduct and cause bleeding, infection, and death. **B** An infant with chlamydia-inflamed eyes. Its mother transmitted the bacterial pathogen to it during birth. **C** Skin sores caused by untreated syphilis.

Table 38.3 Frequency and Causes of STDs		
STD	U.S. Cases	Pathogen
HPV infection	5,500,000	Virus
Trichomoniasis	5,000,000	Protozoan
Chlamydia	3,000,000	Bacteria
Genital herpes	1,000,000	Virus
Gonorrhea	650,000	Bacteria
Syphilis	70,000	Bacteria
AIDS	40,000	Virus

infection, but they promote healing of sores and lessen the likelihood of viral reactivation.

Infection by HIV (or human immunodeficiency virus) can cause AIDS (acquired immune deficiency syndrome). Sections 19.1 and 19.2 described the HIV virus and Section 34.13 explained how its effects on the immune system lead to AIDS. With regard to sexual transmission, oral sex is least likely to transmit an infection. Unprotected anal sex is 5 times more dangerous than unprotected vaginal sex and 50 times more dangerous than oral sex. To reduce the risk of HIV transmission during vaginal or anal sex, doctors recommend use of a latex condom and a lubricant. The lubricant helps prevent small abrasions that could allow the virus to enter the body. If you think you have been exposed to HIV, get tested as soon as possible. Early treatment may prevent development of AIDS.

Take-Home Message What are causes and effects of common STDs and how are they treated?

> Viruses, bacteria, and protozoans cause common STDs.

> Symptoms range from the merely discomforting to the life-threatening. STDs can cause infertility and infected women can pass them to their offspring.

> Bacterial and protozoan pathogens can be killed by drugs. Viral infections cannot be cured with drugs, but antiviral drugs reduce the effects of herpes and HIV infection. An HPV vaccine can prevent infection.

❯ After fertilization occurs, an animal begins developing. The same development steps and processes occur in all vertebrates, a legacy of their common ancestry.
❮ Links to Tight junctions 4.11, Germ layers 23.2

In all sexually reproducing animals, a new individual begins life as a zygote, the diploid cell that forms at fertilization. Development from a zygote to an adult typically proceeds through a series of stages (Figure 38.12). An example of these stages in one vertebrate, the leopard frog, appears in Figure 38.13.

A female frog releases eggs into the water and a male releases sperm onto the eggs. External fertilization produces the zygote. New cells arise when **cleavage** carves up a zygote by repeated mitotic cell divisions ❶. During cleavage, the number of cells increases, but the zygote's

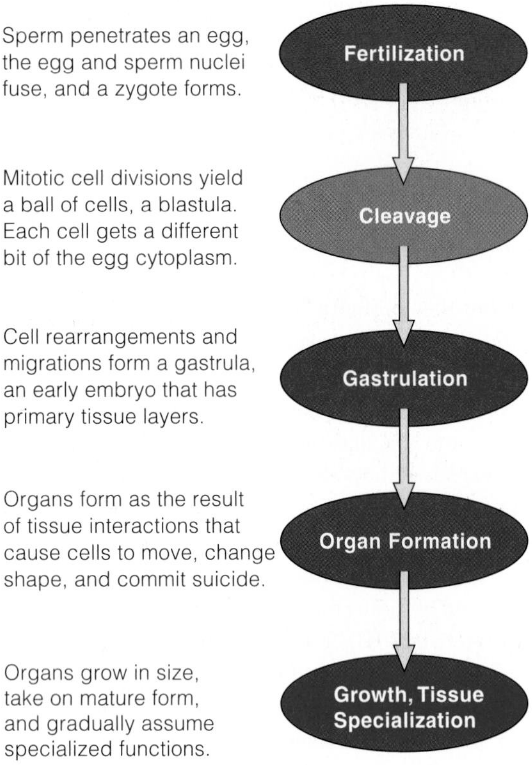

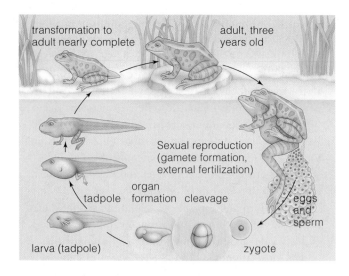

Figure 38.13 Animated *Above*, overview of reproduction and development in the leopard frog. *Opposite*, a closer look at some stages.

Sperm penetrates an egg, the egg and sperm nuclei fuse, and a zygote forms.

→ **Fertilization**

Mitotic cell divisions yield a ball of cells, a blastula. Each cell gets a different bit of the egg cytoplasm.

→ **Cleavage**

Cell rearrangements and migrations form a gastrula, an early embryo that has primary tissue layers.

→ **Gastrulation**

Organs form as the result of tissue interactions that cause cells to move, change shape, and commit suicide.

→ **Organ Formation**

Organs grow in size, take on mature form, and gradually assume specialized functions.

→ **Growth, Tissue Specialization**

Figure 38.12 Stages of development in vertebrates.

original volume remains unchanged. As a result, cells become more numerous but smaller. The cells that form during cleavage are called blastomeres. They typically become arranged as a **blastula**: a ball of cells that enclose a cavity (blastocoel) filled with their secretions ❷. Tight junctions hold the cells of the blastula together.

During **gastrulation**, cells move about and organize themselves as the layers of the **gastrula** ❸. In most animals and all vertebrates, a gastrula consists of three primary tissue layers, or **germ layers**. The three germ layers give rise to the same types of tissues and organs in all vertebrates (Table 38.4). This developmental similarity is evidence of a shared ancestry.

Ectoderm, the outer germ layer, forms first. It gives rise to nervous tissue and to the outer layer of skin or other body covering. **Endoderm**, the inner germ layer, is the start of the respiratory tract and gut linings. A third layer called **mesoderm** forms between the ectoderm and the endoderm. This "middle" layer is the source of all muscles, connective tissues, and the circulatory system.

Organ formation begins after gastrulation. The neural tube and notochord characteristic of all chordate embryos

Table 38.4	Derivatives of Vertebrate Germ Layers
Ectoderm (outer layer)	Outer layer (epidermis) of skin; nervous tissue
Mesoderm (middle layer)	Connective tissue of skin; skeletal, cardiac, smooth muscle; bone; cartilage; blood vessels; urinary system; gut organs; peritoneum (coelom lining); reproductive tract
Endoderm (inner layer)	Lining of gut and respiratory tract, and organs derived from these linings

blastula Hollow ball of cells that forms as a result of cleavage.
cleavage Mitotic division of an animal cell.
ectoderm Outermost tissue layer of an animal embryo.
endoderm Innermost tissue layer of an animal embryo.
gastrula Three-layered developmental stage formed by gastrulation in an animal.
gastrulation Animal developmental process by which cell movements produce a three-layered gastrula.
germ layer One of three primary layers in an early embryo.
mesoderm Middle tissue layer of a three-layered animal embryo.

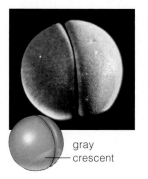
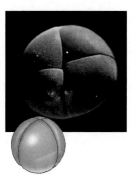

gray crescent

blastocoel

blastula

1 Here we show the first three divisions of cleavage, a process that carves up a zygote's cytoplasm. In this species, cleavage results in a blastula, a ball of cells with a fluid-filled cavity.

2 Cleavage is over when the blastula forms.

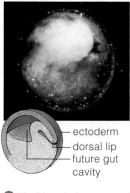
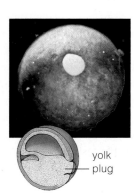

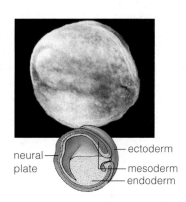

ectoderm
dorsal lip
future gut cavity

yolk plug

neural plate

ectoderm
mesoderm
endoderm

neural tube

notochord

gut cavity

3 The blastula becomes a three-layered gastrula—a process called gastrulation. At the dorsal lip (a fold of ectoderm above the first opening that appears in the blastula) cells migrate inward and start rearranging themselves.

4 Organs begin to form as a primitive gut cavity opens up. A neural tube, then a notochord and other organs, form from the primary tissue layers.

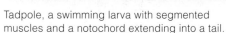

Tadpole, a swimming larva with segmented muscles and a notochord extending into a tail.

Limbs grow and the tail is absorbed during metamorphosis to the adult form.

Sexually mature, four-legged adult leopard frog.

5 The frog's body form changes as it grows and its tissues specialize. The embryo becomes a tadpole, which metamorphoses into an adult.

form early **4**. Many organs incorporate tissues derived from more than one germ layer. For example, the stomach's epithelial lining is derived from endoderm, and the smooth muscle that makes up the stomach wall develops from mesoderm.

In frogs, as in some other animals, a larva (in this case a tadpole) undergoes metamorphosis, a drastic remodeling of tissues into the adult form **5**.

Take-Home Message How does an adult vertebrate develop from a single-celled zygote?

> A zygote undergoes cleavage, which increases the number of cells. Cleavage ends with formation of a blastula.

> Rearrangement of blastula cells forms a three-layered gastrula.

> After gastrulation, organs such as the nerve cord begin forming.

> Continued growth and tissue specialization produce the adult body.

› Positioning of material in the egg cytoplasm sets the stage for development.

‹ Links to Cell cortex 4.10, mRNA transport 10.2, Cytoplasmic division 11.4, Protostomes and deuterostomes 23.2

Components of Eggs and Sperm

A sperm consists of paternal DNA and a bit of equipment that helps it swim to and penetrate an egg. The egg has far more cytoplasm. Egg cytoplasm includes yolk proteins that will nourish a new embryo, mRNA transcripts for proteins that will be translated in early development, tRNAs and ribosomes to translate the mRNA transcripts, and the proteins required to build mitotic spindles.

Cytoplasmic localization is a feature of all oocytes: Some cytoplasmic components are not distributed evenly throughout egg cytoplasm, but rather localized in one particular region or another. For example, in a yolk-rich egg, the vegetal pole has most of the yolk and the animal pole has little. In some amphibian eggs, dark pigment molecules accumulate in the cell cortex, a cytoplasmic region just beneath the plasma membrane. Pigment is the most concentrated close to the animal pole. After a sperm penetrates the egg at fertilization, the cortex rotates. Rotation reveals a gray crescent, a region of the cell cortex that is lightly pigmented (Figure 38.14A).

Early in the 1900s, experiments by Hans Spemann showed that substances essential to development are localized in the gray crescent. In one experiment, he separated the first two blastomeres formed at cleavage. Each had half of the gray crescent and developed normally (Figure 38.14B). In the next experiment, Spemann altered the cleavage plane. One blastomere got all the gray crescent, and it developed normally. The other lacked gray crescent and got stuck in the blastula stage (Figure 38.14C).

Cleavage—The Start of Multicellularity

During cleavage, a furrow appears on the cell surface and defines the plane of the cut. Beneath the plasma membrane, a ring of microfilaments starts contracting, and eventually divides the cell in two (Section 11.4). The plane of division is not random. It dictates what types and proportions of materials a blastomere will get, as well as its size. Cleavage puts different parts of the egg cytoplasm into different blastomeres.

Each species has a characteristic cleavage pattern. Remember the branching of the coelomate lineage into

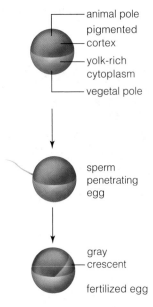

animal pole
pigmented cortex
yolk-rich cytoplasm
vegetal pole

sperm penetrating egg

gray crescent

fertilized egg

A Many amphibian eggs have a dark pigment concentrated in cytoplasm near the animal pole. At fertilization, the cytoplasm shifts, and exposes a gray crescent-shaped region just opposite the sperm's entry point. The first cleavage normally distributes half of the gray crescent to each descendant cell.

Figure 38.14 Animated Experimental evidence of cytoplasmic localization in an amphibian oocyte.

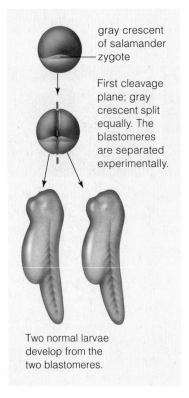

gray crescent of salamander zygote

First cleavage plane; gray crescent split equally. The blastomeres are separated experimentally.

Two normal larvae develop from the two blastomeres.

B In one experiment, the first two cells formed by normal cleavage were physically separated from each other. Each cell developed into a normal larva.

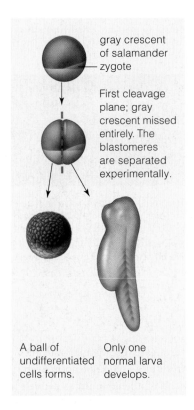

gray crescent of salamander zygote

First cleavage plane; gray crescent missed entirely. The blastomeres are separated experimentally.

A ball of undifferentiated cells forms.

Only one normal larva develops.

C In another experiment, a zygote was manipulated so one descendant cell received all the gray crescent. This cell developed normally. The other gave rise to an undifferentiated ball of cells.

Figure 38.15 Gastrulation in a fruit fly (*Drosophila*). In fruit flies, cleavage is restricted to the outermost region of cytoplasm; the interior is filled with yolk. The series of photographs, all cross-sections, shows sixteen cells (stained gold) migrating inward. The opening the cells move in through will become the fly's mouth. Descendants of the stained cells will form mesoderm. Movements shown in these photos occur during a period of less than 20 minutes.

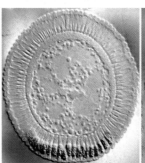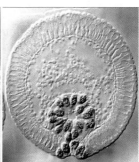

the protostomes and deuterostomes (Section 23.2)? These two groups differ in certain details of cleavage, such as the angle of divisions relative to an egg's polar axis. The amount of yolk also influences the pattern of division. Insects, frogs, fishes, and birds have yolk-rich eggs. In such eggs, a large volume of yolk slows or blocks some of the cuts. The result is fewer divisions of the yolky part of the egg than the less yolky part. By comparison, the cuts slice right through the nearly yolkless eggs of sea stars and mammals.

From Blastula to Gastrula

A hundred to thousands of cells may form at cleavage, depending on the species. Starting with gastrulation, cells migrate about and rearrange themselves. A portion at the embryo's surface moves inward. Figure 38.15 shows one example of this process.

What initiates gastrulation? Hilde Mangold, one of Spemann's students, discovered the answer. She knew that during gastrulation, cells of a salamander blastula move inward through an opening on its surface. Cells in the dorsal (upper) lip of the opening are descended from a zygote's gray crescent. Mangold suspected that signals

from dorsal lip cells caused gastrulation. She predicted that a transplant of dorsal lip material from one embryo to another would cause gastrulation at the recipient site. She did many transplants (Figure 38.16A), and the results supported her prediction. Cells migrated inward at the transplant site, as well as at the usual location (Figure 38.16B). A salamander larva with two joined sets of body parts developed (Figure 38.16C). Apparently, the transplanted cells had signaled their new neighbors to develop in a novel way.

This experiment also explained the results shown in Figure 38.14C. Without any gray crescent cytoplasm, an embryo does not have cells that would normally form the dorsal lip. Without proper signals from these cells, development stops short.

---

cytoplasmic localization Accumulation of different materials in different regions of the egg cytoplasm.

Take-Home Message What are the effects of cytoplasmic localization and cleavage?

❯ Enzymes, mRNAs, yolk, and other materials are localized in specific parts of the cytoplasm of unfertilized eggs. This cytoplasmic localization helps guide early development.

❯ Cleavage divides a fertilized egg into a number of small cells but does not increase its original volume. The cells—blastomeres—inherit different parcels of cytoplasm that make them behave differently, starting at gastrulation.

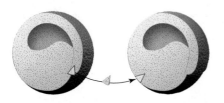  

C The embryo develops into a "double" larva, with two heads, two tails, and two bodies joined at the belly.

A Dorsal lip excised from donor embryo, grafted to novel site in another embryo.

B Graft induces a second site of inward migration.

Figure 38.16 Animated Experimental evidence that signals from dorsal lip cells start amphibian gastrulation. A dorsal lip region of a salamander embryo was transplanted to a different site in another embryo. A second set of body parts started to form.

› Different tissues form as cells turn different genes on or off. Interactions among tissues form organs.

‹ Links to Cell differentiation 10.2, Master genes 10.3, Comparative embryology 16.9, Apoptosis 28.9

Cell Differentiation

All cells in an embryo are descended from the same zygote, so all have the same genes. How then, do the specialized tissue and organs arise? From gastrulation onward, selective gene expression occurs: Different cell lineages express different subsets of genes. Selective gene expression is the key to cell differentiation, the process by which cell lineages become specialized in composition, structure, and function (Section 10.2).

An adult human body has about 200 differentiated cell types. As your eye developed, cells of one lineage turned on genes for crystallin, a transparent protein. These differentiated cells formed the lens of your eye. No other cells in your body make crystallin.

A differentiated cell still retains the entire genome. That is why it is possible to clone an adult animal—to create a genetic copy—from one of its differentiated cells (Section 8.7). As an example, the DNA of Dolly, the first sheep clone, developed from a mammary gland cell of a six-year-old female sheep.

Cell Communication in Development

Intercellular signals can encourage differentiation. For example, certain embryonic cells secrete **morphogens**, molecular signals that are encoded by master genes. A morphogen diffuses out from its source and forms a concentration gradient in the embryo. A morphogen's effects on target cells depends on its concentration. Cells nearest the source of a morphogen will respond differently than distant cells.

Other signals operate at close range, as when cells of a salamander gastrula's dorsal lip cause adjacent cells to migrate inward and become mesoderm. This is an example of **embryonic induction**: Embryonic cells produce signals that alter the behavior of neighboring cells.

Cell Movements and Apoptosis

Long-range and short-range signals regulate the development of tissues and organs. Organs begin to form as cells migrate, entire sheets of tissue fold and bend, and specific cells die on cue.

Cell migrations produce a multilayered embryo during gastrulation. Cells also migrate in the developing brain. Neurons form in the center of the brain, then creep along extensions of glial cells or axons of other neurons until they reach their final position (Figure 38.17A). Once in place, they send out axons of their own.

Controlled assembly and disassembly of microtubules and microfilaments alter cell shape in ways that cause sheets of cells to expand and fold. Such folding produces the vertebrate neural tube (Figure 38.17B). Ectodermal cells at the embryo's midline elongate as microtubules inside them assemble. Then neighboring cells become wedge-shaped as microfilaments at one end constrict, forming a neural groove. Edges of the groove move inward. Eventually flaps of tissue fold over and meet at the midline, forming the neural tube. This tube later develops into the brain and spinal cord.

Cell death helps sculpt body parts. By the process of **apoptosis**, signals from certain cells activate the tools of self-destruction in target cells. Apoptosis causes a tadpole to lose its tail and it separates the digits of the developing human hand (Figure 38.17C).

A Cells migrate. This graphic shows one embryonic neuron (*orange*) at successive times as it migrates along a glial cell (*yellow*). Its adhesion proteins stick to glial cell proteins.

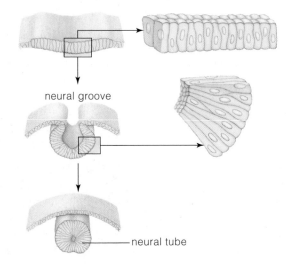

1 Gastrulation produces a sheet of ectodermal cells.

2 A neural groove forms as microtubules constrict or lengthen in different cells, making the cells change shape.

neural groove

3 Edges of the groove meet and detach from the main sheet, forming the neural tube.

neural tube

B Cells change shape. Here, shape changes in ectodermal cells form a neural tube.

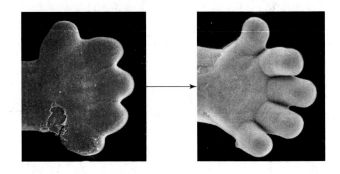

C Cells commit suicide. As a human hand develops, cells in the webs of skin between digits die.

Figure 38.17 Animated How body takes shape.

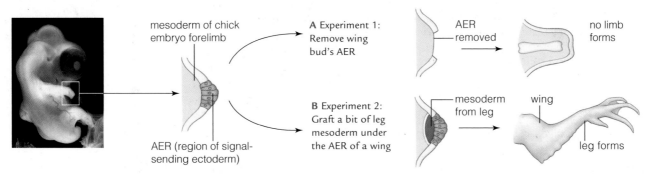

Figure 38.18 Animated Experiments demonstrating interactions between AER (ectoderm) and mesoderm in chick wing development. AER at the tip of a limb bud tells mesoderm under it to form a limb. Whether that limb becomes a wing or a leg depends on positional signals that the mesoderm received earlier.

Pattern Formation

Pattern formation is the process by which certain body parts form in a specific place. As one example, tissue called AER (apical ectodermal ridge) develops at the tips of limb buds in all tetrapods. Signals from the AER induce the mesoderm beneath it to form a limb. Remove the AER from a chick's wing bud, and wing development comes to a halt (Figure 38.18A).

AER stimulates mesoderm to develop, but earlier positional cues have previously determined what that mesoderm will become. Implant a piece of mesoderm from a chick hindlimb under the wing AER, and a leg forms (Figure 38.18B). Early events sent this mesoderm down the developmental road toward leg formation.

Evolution and Development

Through studies of many animals, researchers have come up with a general model for development. The key point is this: Where and when particular genes are expressed determines how an animal body develops.

First, molecules localized in different areas of an unfertilized egg induce localized expression of master genes in the zygote. (Master genes are genes whose products affect expression of many other genes.) Products of master genes become distributed in gradients along the front-to-back and top-to-bottom axis of the developing embryo.

Second, depending on where they fall within these concentration gradients, cells in the embryo activate or suppress other genes. The products of these genes form gradients, and so on.

Third, this positional information affects expression of **homeotic genes**, genes that regulate development of specific body parts, as introduced in Section 10.3. All animals have similar homeotic genes. For example, a mouse's *eyeless* gene guides development of its eyes. Introduce this gene into a fruit fly, and eyes will form in the tissue where it is expressed.

This model helps explain similarities among animal body plans. Body plans are influenced by physical constraints (such as the surface-to-volume ratio). Evolution of large body size requires circulatory and respiratory mechanisms to service body cells that reside far from the body surface. Other constraints are imposed by the existing body framework. For example, the first vertebrates on land had a body plan with four limbs. Evolution of wings in birds and bats occurred through modification of existing forelimbs, not by sprouting new limbs. Although it might be advantageous to have both wings and arms, no vertebrate with both has ever been discovered.

Interactions among genes that regulate development also impose restraints. Once master genes had evolved and began interacting, a major change in any one of them probably would have been lethal. Mutations led to a variety of forms among animal lineages. But they did so by modifying existing developmental pathways, rather than by blazing entirely new genetic trails.

---

apoptosis Mechanism of cell suicide.
embryonic induction Embryonic cells produce signals that alter the behavior of neighboring cells.
homeotic gene Type of master gene; its expression controls formation of specific body parts during development.
morphogen Chemical encoded by a master gene; diffuses out from its source and affects development.
pattern formation Formation of body parts in specific locations.

Take-Home Message **What processes produce specialized cells, tissues, and organs?**

❯ All cells in an embryo have the same genes, but they express different subsets of this genome. Selective gene expression is the basis of cell differentiation. It results in cell lineages with characteristic structures and functions.

❯ Cytoplasmic localization sets the stage for cell signaling. The signals activate sets of master genes, the products of which cause embryonic cells to form tissues and organs in specific places.

❯ Physical, architectural, and evolutionary constraints have limited drastic changes in the basic pattern of development.

❯ Like all animals, humans form a blastula by cleavage. As in other placental mammals, their blastula implants in a uterus and continues its development there.
❮ Links to Amniotes 24.5, Neural tube 29.10

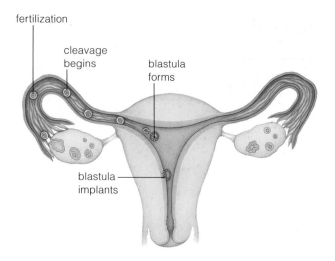

Figure 38.19 Locations of early developmental events.

Cleavage and Implantation

In humans, cleavage usually begins the day after fertilization, while the zygote is still in the oviduct (Figure 38.19). The zygote becomes two cells, which become four, which become eight, and so on (Figure 38.20A). Sometimes a cluster of four or eight cells splits in two and each cluster goes on to develop, so that identical twins form. More typically cells stick together as divisions continue.

By a week after fertilization, a **blastocyst**—the type of blastula formed in mammals—has developed (Figure 38.20B). It consists of an outer layer of cells, a blastocoel filled with their fluid secretions, and an inner cell mass. Of the 200 to 250 cells, only about thirty are part of the inner cell mass.

About a week after fertilization, implantation begins (Figure 38.20C): The blastocyst attaches to the uterine lining, and begins to sink into the endometrium.

An implanted blastula prevents menstruation by the release of human chorionic gonadotropin (HCG). This hormone stimulates the corpus luteum to continue its secretion of progesterone and estrogens. These hormones maintain the uterine lining. HCG can be detected in a

mother's urine as early as the third week of pregnancy. At-home pregnancy tests include a treated "dipstick" that changes color when exposed to urine containing HCG.

Extraembryonic Membranes

During implantation, the inner cell mass develops into two layers. One layer is the embryonic disk that will give rise to the embryo tissues. The other layer will develop into the yolk sac, one of the extraembryonic membranes. A mammal's yolk sac does not contain yolk, as do those

A Days 1–4
Cleavage. Repeated mitotic divisions divide the zygote cytoplasm among an increasing number of cells.

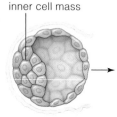

B Days 5–7 Blastocyst forms. It has a layer of flattened outer cells, an inner cell mass, and a fluid-filled blastocoel. To implant, it must first shed the jelly layer that surrounds it.

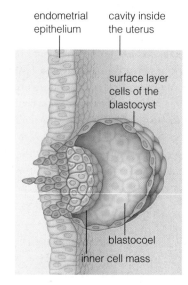

endometrial epithelium   cavity inside the uterus

surface layer cells of the blastocyst

blastocoel

inner cell mass

C Days 8–9 Implantation begins. The blastocyst attaches to the lining of the uterus (the endometrium) and begins to sink into it.

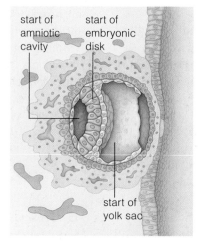

start of amniotic cavity   start of embryonic disk

start of yolk sac

D Days 10–11 There are now two layers to the inner cell mass. The layer near the blastocoel will become the yolk sac. The other, the embryonic disk, will become the embryo proper.

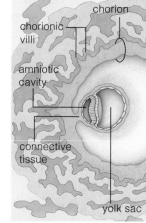

chorionic villi   chorion   chorionic cavity

amniotic cavity

connective tissue

yolk sac

E Day 14 Blood-filled spaces form in the endometrium. Projections from the chorion, called chorionic villi, grow into the spaces. The amniotic cavity is now filled with fluid.

actual size

actual size

Figure 38.20 Animated Cleavage and implantation.

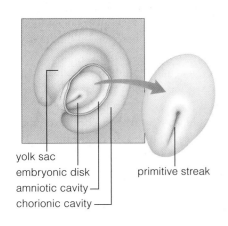

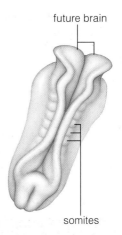
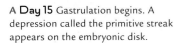
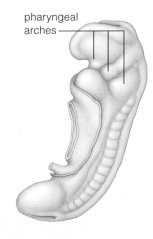

paired neural folds

future brain

pharyngeal arches

yolk sac
embryonic disk
amniotic cavity
chorionic cavity

primitive streak

neural groove (below, notochord is forming)

somites

A Day 15 Gastrulation begins. A depression called the primitive streak appears on the embryonic disk.

B Day 16 Neural groove appears as the neural tube begins forming.

C Days 18–23 Bumps of mesoderm (somites) appear. They will develop into the dermis and most of the axial skeleton.

D Days 24–25 Pharyngeal arches appear. They will contribute to the face, neck, larynx, and pharynx.

Figure 38.21 Gastrulation and the onset of organ formation.

of reptiles and amphibians. Instead, a mammalian yolk sac produces the embryo's blood cells and the individual's germ cells.

In addition to the yolk sac, three other membranes characteristic of amniotes form outside the embryo (Figure 38.20D). A fluid-filled amniotic cavity opens up between the embryonic disk and part of the blastocyst surface. Many cells migrate around the wall of the cavity and form the **amnion**, a membrane that will enclose the embryo. Fluid in the cavity functions as a buoyant cradle in which an embryo can grow, move freely, and be protected from temperature changes and sudden impacts.

Before a blastocyst is fully implanted, spaces open in maternal tissues and become filled with blood that seeps in from ruptured capillaries. In the blastocyst, a new cavity opens up around the amnion and yolk sac. The lining of this cavity becomes the **chorion**, a membrane that is folded into many fingerlike projections that extend into blood-filled maternal tissues (Figure 38.20E). The chorion will become part of the placenta, the organ that functions in exchanges of materials between the bloodstreams of a mother and her child.

After the blastocyst is implanted, an outpouching of the yolk sac produces the fourth extraembryonic membrane, the **allantois**. It becomes part of the umbilical cord that connects the embryo to the placenta.

Gastrulation and Organ Formation

At the start of the third week, gastrulation occurs. Cells migrate inward along a depression, the primitive streak, that forms on the disk's surface (Figure 38.21A).

By the eighteenth day after fertilization, the embryonic disk has two folds that will merge into a neural tube, the precursor of the spinal cord and brain (Figure 38.21B). Beneath the neural tube, mesoderm folds into another tube that develops into a notochord. The human notochord acts as a structural model for the backbone.

By the end of the third week, somites form (Figure 38.21C). These paired segments of mesoderm will develop into bones, skeletal muscles of the head and trunk, and overlying dermis of the skin. Pharyngeal arches that begin to form at this time will later contribute to structures of the head and neck (Figure 38.21D).

---

allantois Extraembryonic membrane that, in mammals, becomes part of the umbilical cord.
amnion Extraembryonic membrane that encloses an amniote embryo and the amniotic fluid.
blastocyst Mammalian blastula.
chorion Outermost extraembryonic membrane of amniotes; major component of the placenta in placental mammals.

Take-Home Message **What occurs during the first month of human development?**

> Cleavage of the human zygote produces a blastocyst that implants in the mother's uterus.

> Projections from the blastocyst's surface invade maternal tissues. An embryonic disk that will give rise to all body tissues forms. Other parts of a blastocyst give rise to external membranes: an amnion, yolk sac, chorion, and allantois.

> The embryonic disk undergoes gastrulation, then organ formation begins. A neural tube and notochord form, and pharyngeal arches develop.

> A human embryo takes on a vertebrate appearance, with its pharyngeal arches and tail, by the fourth week.

> By the beginning of the fetal period, the tail is gone and the developing individual has distinctly human features.

When the fourth week ends, the embryo is 500 times the size of a zygote, but still less than 1 centimeter long (Figure 38.22A). Growth slows as details of organs begin to fill in. Limbs form; paddles are sculpted into fingers and toes. The umbilical cord and the circulatory system develop. Growth of the head now surpasses that of all other regions (Figure 38.22B). Reproductive organs begin

forming. At the end of the eighth week, all organ systems have formed and the individual is described as a human **fetus** (Figure 38.22C).

In the second trimester (months 4 to 6), reflexive movements begin as developing nerves and muscles connect. Legs kick, arms wave about, and fingers grasp. When a fetus is five months old, its heartbeat can be heard clearly through a stethoscope positioned on the mother's abdomen. The mother can sense movements of fetal arms and legs. In the sixth month, eyelids and eyelashes form. Eyes open during the seventh month, the start of the final

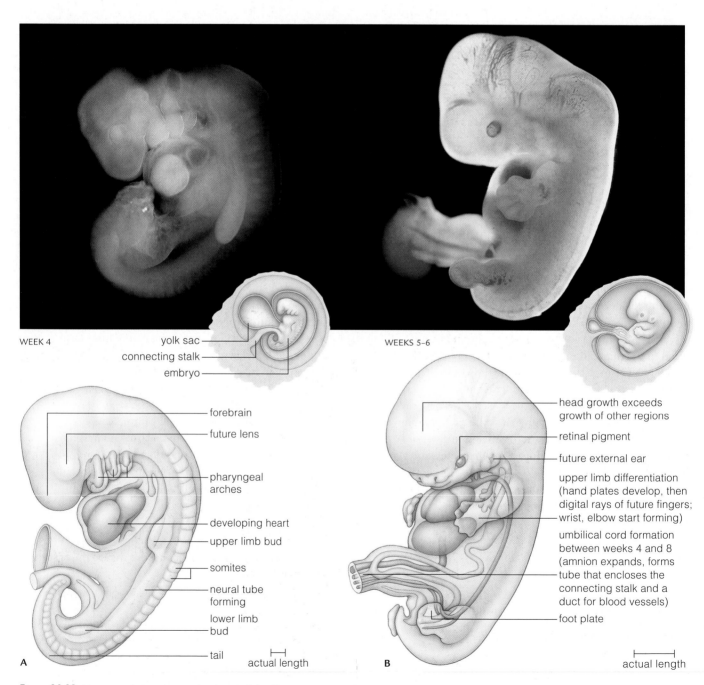

WEEK 4
yolk sac
connecting stalk
embryo

WEEKS 5–6

forebrain
future lens
pharyngeal arches
developing heart
upper limb bud
somites
neural tube forming
lower limb bud
tail

A   actual length

head growth exceeds growth of other regions
retinal pigment
future external ear
upper limb differentiation (hand plates develop, then digital rays of future fingers; wrist, elbow start forming)
umbilical cord formation between weeks 4 and 8 (amnion expands, forms tube that encloses the connecting stalk and a duct for blood vessels)
foot plate

B   actual length

Figure 38.22 Human embryo at successive stages of development.

trimester (months 7–9). By this time, all portions of the brain have formed and have begun to function.

Thirty-eight weeks is considered full term (Figure 38.22D), so births before 37 weeks are defined as premature. A fetus born earlier than 22 weeks rarely survives because its lungs are not yet fully functional.

fetus Human from week 9 of development to birth.

Take-Home Message What occurs during the late embryonic and fetal periods?

> At one month, an embryo is less than a centimeter long, with a tail and tiny limb buds.

> At eight weeks, the tail is gone and the fetus appears human. During the fetal period, organs that formed during the embryonic period begin functioning and growth is rapid.

> Thirty-eight weeks after fertilization, a fetus is ready for birth.

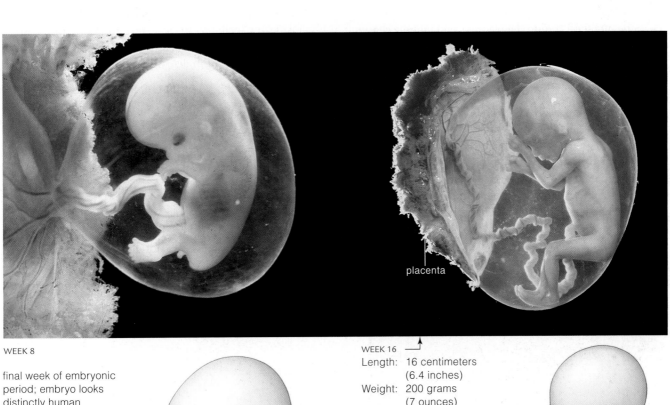

placenta

WEEK 8

final week of embryonic period; embryo looks distinctly human compared to other vertebrate embryos

upper and lower limbs well formed; fingers and then toes have separated

primordial tissues of all internal, external structures now developed

tail has become stubby

C |— actual length —|

WEEK 16 —
Length: 16 centimeters
         (6.4 inches)
Weight: 200 grams
         (7 ounces)

WEEK 29
Length: 27.5 centimeters
         (11 inches)
Weight: 1,300 grams
         (46 ounces)

WEEK 38 (full term) —→
Length: 50 centimeters
         (20 inches)
Weight: 3,400 grams
         (7.5 pounds)

During fetal period, length measurement extends from crown to heel (for embryos, it is the longest measurable dimension, as from crown to rump).

D

> Like other placental mammals, a human mother supplies her developing offspring with nutrients and oxygen by exchanges across the placenta.
< Links to Mammals 24.8, Cretinism 31.6

All exchange of materials between an embryo and its mother takes place by way of the placenta, a pancake-shaped, blood-engorged organ made of uterine lining and extraembryonic membranes (Figure 38.23). The placenta forms early in pregnancy. By the third week, chorionic villi—the tiny fingerlike projections from the chorion—are growing into the pools of maternal blood in the endometrial tissue. Embryonic blood vessels extend through the umbilical cord to the placenta, and into the chorionic villi, where the pooled maternal blood surrounds them. Maternal and embryonic bloodstreams never mix. Instead, substances move between maternal and embryonic blood by diffusing across the walls of the embryonic vessels in the chorionic villi. Oxygen and nutrients diffuse from maternal blood into embryonic blood. Wastes diffuse the other way, and the mother's body disposes of them.

The placenta also has a hormonal role. From the third month on, it produces large amounts of HCG, progesterone, and estrogens. These hormones encourage the ongoing maintenance of the uterine lining.

Maternal diet is important. If a mother is deficient in a vitamin or mineral, her child may not develop properly. For example, a woman who does not get enough folate in early pregnancy puts an embryo at an increased risk of neural tube defects. If she lacks iodine, her child may be born with cretinism (Section 31.6).

An embryo or fetus is exposed to pathogens that infect its mother. If a pregnant woman is infected by the virus that causes rubella (German measles) during the first six weeks of development, there is a 50 percent chance that her embryo will be infected and will have a birth defect. An HIV infected mother can infect her child.

Many toxins can cross the placenta. A mother who drinks alcohol or caffeine-laden drinks, uses illegal drugs, or inhales tobacco smoke exposes her child to chemicals that can interfere with development. Some prescription medicines, including the acne medication retinoic acid (Accutane) and the antidepressant paroxetine (Paxil), also raise the risk of birth defects.

Take-Home Message What is the function of the placenta?

> Vessels of the embryo's circulatory system extend through the umbilical cord to the placenta, where they run through pools of maternal blood.

> Maternal and embryonic blood do not mix; substances diffuse between the maternal and embryonic bloodstreams by crossing vessel walls.

> Some pathogens and chemicals that cause birth defects can cross the plancenta from mother to developing child.

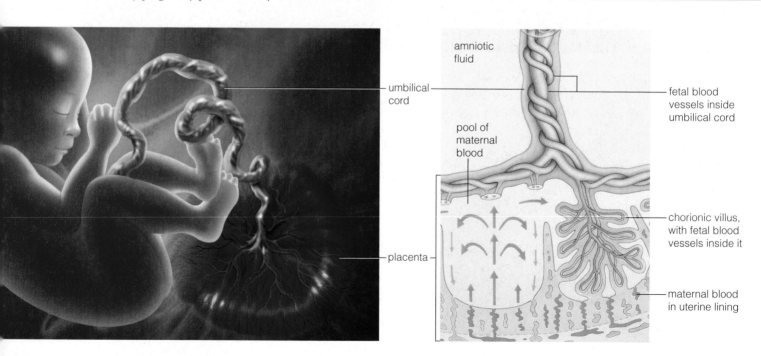

Artist's depiction of the view inside the uterus, showing a fetus connected by an umbilical cord to the pancake-shaped placenta.

Figure 38.23 Animated Life support system of a developing human.

The placenta consists of maternal and fetal tissue. Fetal blood flowing in vessels of chorionic villi exchanges substances by diffusion with maternal blood around the villi. The bloodstreams do not mix.

> Like other placental mammals, human females give birth and nourish their young with nutritious milk secreted from the mother's mammary glands. Shifts in the levels of hormones control both these processes.

‹ Link to Hypothalamus and pituitary 31.4

Giving Birth

A mother's body changes as her fetus nears full term, at about 38 weeks after fertilization. Until the last few weeks, her firm cervix helped prevent the fetus from slipping out of her uterus prematurely. Now cervical connective tissue becomes thinner, softer, and more flexible so it can stretch during childbirth.

The birth process is known as **labor**. Typically, the amnion ruptures right before birth, so amniotic fluid drains out from the vagina. The cervical canal dilates. Strong contractions propel the fetus through the cervix, then out through the vagina (Figure 38.24).

A positive feedback mechanism operates during labor. As the fetus nears full term, it typically shifts position so that its head touches the mother's cervix. Receptors in the cervix sense pressure and signal the hypothalamus, which signals the posterior lobe of the pituitary to secrete oxytocin. In a positive feedback loop, oxytocin binds to

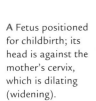

A Fetus positioned for childbirth; its head is against the mother's cervix, which is dilating (widening).

placenta

wall of uterus

umbilical cord

dilating cervix

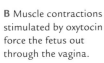

B Muscle contractions stimulated by oxytocin force the fetus out through the vagina.

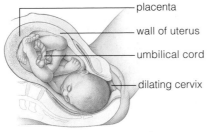

placenta detaching from wall of uterus

umbilical cord

C The placenta detaches from the wall of the uterus and is expelled.

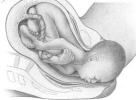

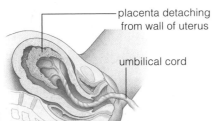

Figure 38.24 Expulsion of fetus and afterbirth during labor. The afterbirth consists of the placenta, tissue fluid, and blood.

Research into IVF opened the way to a variety of assisted reproductive technologies, each with its own ethical considerations. If a woman cannot conceive, she can now obtain donated eggs and use IVF to produce a blastocyst that she can carry to term. A woman who has fertile eggs but cannot, or does not want to, carry them herself can have a blastocyst conceived by IVF implanted in a surrogate mother. Blastocysts created by IVF can be frozen away for later use. Prospective parents can also screen blastocysts before they are implanted to avoid genetic defects, or to choose desirable traits such as a particular sex.

How Would You Vote? Medical organizations have made many recommendations about reproductive technologies, but some fertility doctors ignore these recommendations. Should decisions about use of reproductive technology be left to patients and their doctors, or should we pass laws to limit the use of these technologies? See CengageNow for details, then vote online (cengagenow.com).

smooth muscle of the uterus, causing stronger uterine contractions that put additional pressure on the cervix. The added pressure triggers more oxytocin secretion, causing more cervical stretching, and so on until a fetus is expelled. Synthetic oxytocin is often given to induce labor or to increase strength of contractions.

Strong muscle contractions also detach and expel the placenta from the uterus as the "afterbirth." The umbilical cord that connects the newborn to this mass of expelled tissue is clamped, cut short, and tied. The body's navel marks the former attachment site.

Nourishing the Newborn

Before a pregnancy, a woman's breast tissue is largely adipose tissue. During pregnancy, milk ducts and mammary glands enlarge in preparation for **lactation**, or milk production. Prolactin, a hormone secreted by the mother's anterior pituitary, encourages milk synthesis.

When a newborn suckles, neural signals cause the release of oxytocin. The hormone stimulates muscles around the milk glands to contract and force milk into the ducts. Besides being nutrient-rich, human breast milk has antibodies that protect a newborn from some viruses and bacteria. Nursing mothers should keep in mind that drugs, alcohol, and other toxins also can end up in milk.

labor Expulsion of a placental mammal from its mother's uterus by muscle contractions.
lactation Milk production by a female mammal.

Take-Home Message **What role do hormones play in childbirth and lactation?**

> During birth, hormonally stimulated muscle contractions force a fetus out of its mother's body. Hormones also stimulate milk production and secretion.

Summary

Section 38.1 Reproductive technology such as *in vitro* **fertilization** can help couples have children. However, breakthroughs in reproductive technology raise fears and force us to confront new ethical questions.

Section 38.2 Asexual reproduction produces genetic copies of a parent. **Sexual reproduction** produces varied offspring, and can be advantageous in environments where conditions fluctuate.

Most animals reproduce sexually and have separate sexes, but some are **hermaphrodites** that produce both eggs and sperm. With **external fertilization**, gametes are released into water. Most animals on land have **internal fertilization**; gametes meet in a female's body. Offspring may develop inside or outside the maternal body. **Yolk** helps nourish developing young of most animals. In placental mammals, young are sustained by nutrients delivered across the **placenta**.

Section 38.3 A human male's **gonads** are his **testes**. They produce sperm and the sex hormone **testosterone**. Testosterone influences reproduction and secondary sexual traits that emerge at **puberty**. Sperm form in the **seminiferous tubules** of a testis and mature in an **epididymis** that opens into a **vas deferens**. Secretions from the seminal vesicles and prostate gland join with sperm to form **semen**. Semen is expelled from the body through the **penis**.

Sections 38.4, 38.5 The human female's gonads are **ovaries**. Ovarian **follicles** produce **oocytes** and secrete **estrogens** and **progesterone**. An **oviduct** conveys an oocyte released at **ovulation** to the **uterus**. The **cervix** of the uterus opens into the **vagina**.

From puberty until **menopause**, a woman has an approximately monthly **menstrual cycle**. During each cycle, follicle-stimulating hormone causes a primary oocyte that formed before birth to mature and a follicle to secrete estrogen. The estrogen causes growth of the **endometrium**. A surge of luteinizing hormone triggers ovulation of a secondary oocyte. After ovulation, the **corpus luteum** secretes progesterone that primes the uterus for pregnancy. When the corpus luteum breaks down, **menstruation** occurs.

Sections 38.6–38.8 Intercourse puts sperm into the vagina. Usually only one penetrates the secondary oocyte, causing it to complete meiosis II. The nucleus of the resulting **ovum** and that of the sperm supply the genetic material of the zygote.

Humans prevent pregnancy by abstinence; with surgical, physical, or chemical barriers; and by manipulating female sex hormones. Protozoan, viral, and bacterial pathogens cause sexually transmitted diseases.

Sections 38.9–38.11 All animals go through similar developmental stages. After fertilization takes place, **cleavage** increases the number of cells. Because of **cytoplasmic localization**, different cells receive different components of the maternal cytoplasm. Cleavage ends with production of a **blastula**. In vertebrates, **gastrulation** forms a **gastrula** that has three **germ layers** (**ectoderm**, **mesoderm**, and **endoderm**). Organs begin forming after gastrulation.

Cells become specialized by selective gene expression. Some cells affect gene expression in other cells by secreting **morphogens** or by **embryonic induction**. Organs take shape as cells migrate, tissue layers fold, and cells commit suicide (**apoptosis**). **Homeotic gene** products regulate **pattern formation**: formation of body parts in predictable places on the body.

Sections 38.12–38.14 Human development begins with fertilization, which usually occurs in an oviduct. Cleavage produces a **blastocyst** that implants itself in the uterine wall. After implantation, an **amnion** forms and encloses the embryo in fluid. The **chorion** becomes part of the placenta and the **allantois** becomes part of the umbilical cord. The placenta consists of extraembryonic membranes and endometrium. It allows embryonic blood to exchange substances with the maternal blood by diffusion. Alcohol, drugs, and other harmful substances also cross the placenta, so a mother's health, nutrition, and life-style can affect the growth and development of her future child.

Gastrulation occurs at about 2 weeks; then a neural tube, pharyngeal arches, and a tail form in the early embryo. The tail disappears and organ formation ends before an individual becomes a **fetus** after 8 weeks.

Section 38.15 Hormones induce **labor** at about 38 weeks. Positive feedback controls contractions that expel a fetus and then the afterbirth. Hormones also control maturation of the mammary glands and **lactation**.

Self-Quiz   Answers in Appendix III

1. Most land animals have _____ fertilization.
   a. internal   b. external

2. An animal that makes eggs and sperm is a _____ .
   a. zygote   c. gastrula
   b. hermaphrodite   d. gamete

3. Testosterone is secreted by the _____ .
   a. testes   c. prostate gland
   b. hypothalamus   d. pituitary gland

4. During a menstrual cycle, a midcycle surge of _____ triggers ovulation.
   a. estrogens   b. progesterone   c. LH   d. FSH

Data Analysis Activities

Multiple Births and Birth Defects Fertility treatments raise the risk of multiple pregnancies, which are associated with an increased risk of some birth defects. Figure 38.25 shows the results of Yiwei Tang's study of birth defects reported in Florida from 1996 to 2000. Tang compared the incidence of various defects among single and multiple births. She calculated the relative risk for each type of defect based on type of birth, and corrected for other differences that might increase risk such as maternal age, income, race, and medical care during pregnancy. A relative risk of less than 1 means a defect occurs less often with multiple births than single births. A relative risk greater than 1 means that multiples are more likely to have a defect.

1. What was the most common type of birth defect in the single-birth group?
2. Was that defect more or less common in the multiple-birth group?
3. Tang found that multiples have more than twice the risk of single newborns for one type of defect. Which type?
4. Does a multiple pregnancy increase the relative risk of chromosomal defects in offspring?

	Prevalence of Defect		Relative Risk
	Multiples	Singles	
Total birth defects	358.50	250.54	1.46
Central nervous system defects	40.75	18.89	2.23
Chromosomal defects	15.51	14.20	0.93
Gastrointestinal defects	28.13	23.44	1.27
Genital/urinary defects	72.85	58.16	1.31
Heart defects	189.71	113.89	1.65
Musculoskeletal defects	20.92	25.87	0.92
Fetal alcohol syndrome	4.33	3.63	1.03
Oral defects	19.84	15.48	1.29

Figure 38.25 Prevalence, per 10,000 live births, of various types of birth defects among multiple and single births. Relative risk for each defect is given after researchers adjusted for mother's age, race, previous adverse pregnancy experience, education, Medicaid participation during pregnancy, and the infant's sex and number of siblings.

---

5. The corpus luteum develops from _____ .
   a. a polar body
   b. follicle cells
   c. a secondary oocyte
   d. spermatogonia

6. Sexually transmitted bacteria cause _____ .
   a. trichomoniasis
   b. genital herpes
   c. syphilis
   d. all of the above

7. Match each term with the most suitable description.
   ___ epididymis   a. carries egg to ovary
   ___ vas deferens   b. can be inflated by blood
   ___ vagina   c. where sperm mature
   ___ seminal vesicle   d. target of vasectomy
   ___ penis   e. main contributor to semen
   ___ oviduct   f. sheds lining monthly
   ___ uterus   g. birth canal

8. A homeotic gene regulates _____ .
   a. where body parts form
   b. milk production
   c. secondary sexual traits
   d. sperm formation

9. True or false. All blastomeres have the same cytoplasmic components and express the same genes.

10. What are the three tissues of a vertebrate gastrula?

11. A human blastocyst normally implants in _____ .
   a. an oviduct
   b. a seminiferous tubule
   c. the uterus
   d. the vagina

12. Match each hormone with its effect.
   ___ oxytocin   a. sperm production increases
   ___ testosterone   b. uterus contracts
   ___ LH   c. LH, FSH are released
   ___ GnRH   d. surge causes ovulation
   ___ FSH   e. follicle develops
   ___ HCG   f. milk is produced
   ___ prolactin   g. endometrium is maintained

13. Put these human developmental events in order, from earliest to latest.
   a. blastula forms
   b. heart begins beating
   c. gastrulation
   d. implantation
   e. eyes open
   f. neural tube forms

Additional questions are available on **CENGAGENOW**.

Critical Thinking

1. Based on what you learned in this chapter and the previous one, explain why urine tests can be used to detect a pregnancy.

2. Fraternal twins are nonidentical siblings that form when two eggs mature and are released and fertilized at the same time. Explain why an increased level of FSH raises the likelihood of fraternal twins.

3. The most common ovarian tumors in young women are teratomas. The name comes from the Greek word *teraton*, which means monster. The "monstrous" feature of these tumors is the presence of a variety of tissues, most commonly bones, teeth, fat, and hair. Explain why a tumor that arises from a germ cell can contain a wider assortment of tissues than one derived from a differentiated body cell.

Animations and Interactions on **CENGAGENOW**:
❯ Male reproductive system; How sperm form; Female reproductive system; Ovarian cycle; Menstrual cycle; Fertilization; Frog development; Cytoplasmic localization; Dorsal lip transplants; Neural tube formation; Hand development; Chick limb experiments; Human cleavage to implantation.

‹ Links to Earlier Concepts

This chapter builds on your knowledge of sensory and endocrine systems (Sections 30.2, 31.3). You will revisit pheromones (30.7) and learn about additional effects of the hormone oxytocin (31.4, 38.15). Be sure that you understand the concepts of sexual selection (17.7) and adaptation (16.4). In our discussion of human behavior, you will be reminded once again of the limits of science (1.9).

Key Concepts

Genetic Foundations
Genes that affect the ability to detect stimuli or to respond to nervous or hormonal stimulation influence behavior. Studying behavioral differences within and between species allows scientists to determine the behavior's proximate and ultimate causes.

Instinct and Learning
Instinctive behavior can be performed without any practice. Most behavior has a learned component. Even instinctive behavior is often modified over time. Some learning can only occur during a certain portion of the lifetime.

39 Animal Behavior

39.1 An Aggressive Defense

Many animals sting or bite when threatened. For example, honeybees sting skunks, bears, and other animals that attack their hive. Honeybees live in large family groups, gather nectar from flowers, and store the nectar as honey. A hive's stored honey, along with the protein-rich bodies of developing bees, makes it a tempting target for hungry animals. Stinging is an adaptive response, an evolved defense against the threat posed by potential hive raiders.

Humans tend to label an animal whose defensive response is easily triggered by humans as "aggressive." Africanized honeybees, known in the popular press as "killer bees," are a prime example. These bees are a hybrid between the European honeybees favored by beekeepers and a bee subspecies native to Africa (Figure 39.1). The European bees raised for commercial use have been selectively bred to have a high threat threshold, but Africanized bees retain their normal defensiveness. Both strains of honeybee can sting only once and make the same kind of venom, but the Africanized bees sting with less provocation, respond to threats in greater numbers, and pursue intruders with more persistence.

Africanized honeybees arose in Brazil in the 1950s. Bee breeders there had imported African bees in the hope of breeding an improved pollinator for this region's tropical orchards. Some of the African imports escaped and mated with European honeybees that had already become established there. Descendants of the resulting hybrids expanded northward and reached Texas in 1990. By 2009, Africanized bees had been detected in New Mexico, Nevada, Utah, southern California, Oklahoma, Louisiana, Alabama, and Florida.

Honeybees live in cavities and Africanized bees can take up residence in spaces between walls and in boxes that hold utility meters. Vibrations from machinery such as lawn mowers can trigger a defensive response. In 2008, a Florida man died after being stung about 100 times. He had inadvertently disturbed an Africanized bee colony while demolishing a trailer.

Fatalities from Africanized bee stings are rare (there have been fewer than 20 since the bees arrived in 1990) but even a single sting can be fatal to someone allergic to honeybee venom. In addition, bee stings are highly painful and people who receive a large number of stings often require hospitalization. Thus, the ongoing spread of Africanized bees through the United States is a matter of concern.

What makes Africanized bees so testy? One factor is a greater response to alarm pheromone. A **pheromone** is a chemical signal that transmits information between two individuals of the same species. When a honeybee guarding the entrance to a hive detects a threat, she releases an alarm pheromone. Bees inside the hive detect this chemical signal and rush out to join the guards in driving off the intruder.

Researchers have tested the response of Africanized honeybees and European honeybees to alarm pheromone by positioning a dark cloth near the entrance of hives and releasing an artificial pheromone. Africanized bees flew out of a hive and attacked the cloth faster than European bees and they plunged six to eight times as many stingers into the cloth.

Considering differences among honeybees leads us into the study of animal behavior. Scientists investigate proximate causes of behavior—the genetic and physiological factors that explain how a behavior is carried out. They also investigate ultimate causes—the factors that favored evolution of a behavior. Interactions among animals and between animals and the environment are the focus of **ecology**. Behavioral ecology, our focus here, is one facet of that field. Chapters that follow consider the study of populations, ecosystems, communities, and the biosphere.

Figure 39.1 Africanized honeybees on guard duty. If threatened, they will release an alarm pheromone that stimulates hivemates to join an attack.

ecology Study of organism–environment interactions.
pheromone Intraspecific chemical communication signal.

Animal Communication
Animal communication behavior can only evolve if communication benefits both signal senders and signal receivers. Animals communicate by a variety of mechanisms. Predators sometimes take advantage of the communication behavior of their prey.

Mating and Parenting
Males and females maximize their reproductive success in different ways. Females tend to be choosier about mates than males. Monogamy is rare among animals, but it tends to correlate with parental care by both parents. More often, the female provides parental care.

Costs and Benefits of Social Living
Life in social groups has reproductive benefits and costs. Not all environments favor the evolution of such groups. Self-sacrificing behavior has evolved among a few kinds of animals that live in large family groups.

❯ Much variation in behavior within or among species results from inherited differences. In a few instances, scientists have even pinpointed genes responsible for the variation.
❮ Links to Polygenic inheritance 14.2, Oxytocin 31.4

How Genes Can Influence Behavior

Animal behavior requires a capacity to detect stimuli. A **stimulus** is some type of information about the environment that a sensory receptor detects and responds to. The structure of the nervous system determines the types of stimuli that an animal can detect and the types of responses it can make. Differences in genes that affect the structure and activity of the nervous system cause many differences in behavior.

Genes that affect metabolism or structural traits also influence behavior. For example, some African seedcrackers routinely eat large seeds and others focus on small seeds (Section 17.6). The birds that focus on small seeds do so because their genes specify a beak structure that allows them to easily open small seeds but not larger ones.

Genetic Variation Within a Species

One way to investigate the genetic basis of behavior is to examine behavioral and genetic differences among members of a single species. Studies designed to determine the roots of the difference in defensive behavior between Africanized honeybees and European honeybees fall into this category.

Stevan Arnold's study of snake feeding behavior is another example. Garter snakes that live in coastal forests of the Pacific Northwest prefer to eat the banana slugs that are common on the forest floor (Figure 39.2). Farther inland, there are no banana slugs. Garter snakes

Figure 39.2 Coastal garter snake dining on a banana slug. The snake is genetically predisposed to recognize slugs as prey. By contrast, snakes from inland regions where there are no banana slugs ignore slugs when experimenters offer them.

that live inland prefer to eat fishes and tadpoles. The food preference is inborn. When Arnold offered newborn garter snakes a slug as their first meal, offspring of coastal snakes ate it, but offspring of inland snakes ignored it.

Arnold hypothesized that inland snakes lack the genetically determined ability to associate the scent of slugs with food. He predicted that if coastal garter snakes were crossed with inland snakes, the resulting offspring would make an intermediate response to slugs. Results from his experimental crosses confirmed this prediction. We do not know which gene or genes underlie this difference.

We know more about the genetic basis of differences in foraging behavior among fruit fly larvae. The wingless, wormlike larvae crawl about eating yeast that grows on decaying fruit. In wild fruit fly populations, about 70 percent of the fly larvae are "rovers," which means they tend to move around a lot as they feed. Rovers often leave one patch of food to seek another (Figure 39.3A). The remaining 30 percent of larvae are "sitters"; they tend to move little once they find a patch of yeast (Figure 39.3B). When food is absent, rovers and sitters move the same amount, so both are equally energetic.

The proximate cause of the difference in larval behaviors is a difference in alleles of a gene called *foraging*. Flies with a dominant allele of this gene have the rover phenotype. Sitters are homozygous for the recessive allele. The *foraging* gene encodes an enzyme involved in learning about olfactory cues. Rovers make more of the enzyme than sitters.

The ultimate cause of the behavioral variation in larval foraging behavior is natural selection that arises from competition for food. In experimental populations with

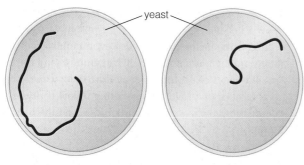

yeast

A Rovers (genotype *FF* or *Ff*) move often as they feed. When a rover's movements on a petri dish filled with yeast are traced for 5 minutes, the trail is relatively long.

B Sitters (genotype *ff*) move little as they feed. When a sitter's movements on a petri dish filled with yeast are traced for 5 minutes, the trail is relatively short.

Figure 39.3 Genetic polymorphism for foraging behavior in fruit fly larvae. When a larva is placed in the center of a yeast-filled plate, its genotype at the *foraging* locus influences whether it moves a little or a lot while it feeds. *Black* lines show a representative larva's path.

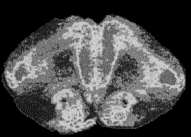

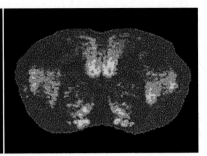

Figure 39.4 Studying the genetic roots of mating and bonding behavior. Voles (*Microtus*) are small rodents in which closely related species vary in their mating and bonding behavior, and in the number and distribution of receptors for the hormone oxytocin.

B PET scan of a monogamous prairie vole's brain with many receptors for the hormone oxytocin (*red*).

C PET scan of a promiscuous prairie vole's brain with few hormone receptors for oxytocin.

limited food, both rovers and sitters are most likely to survive to adulthood when their foraging type is rare. A rover does best when surrounded by sitters, and vice versa. Presumably, when there are lots of larvae of one type they all compete for food in the same way. Under these circumstances, a fly that behaves differently than the majority is at an advantage. As a result, natural selection maintains both alleles of the *foraging* gene.

Genetic Variation Among Species

Comparing behavior of related species can sometimes clarify the genetic basis of a behavior. For example, studies of voles (Figure 39.4) have revealed that inherited differences in the number and distribution of certain hormone receptors influence mating and bonding behavior.

Differences in mating behavior between closely related species of vole make these animals ideal subjects for study of this behavior. Most voles, like most mammals, are promiscuous. However, some voles form lifelong, largely monogamous relationships. For example, in prairie voles, a permanent social bond forms after a night of repeated matings. The hormone oxytocin plays a central role in a female prairie vole's bonding. When females who are part of an established pair are injected with a chemical that interferes with oxytocin action, they dump their long-term partners and mate with other males.

Females of promiscuous vole species may behave as they do because they are less influenced by oxytocin than prairie voles. When researchers compared the brains of promiscuous and monogamous species, they found a striking difference in the number and distribution of oxytocin receptors (Figure 39.4B,C). Monogamous prairie voles have many oxytocin receptors in part of the brain associated with social learning. Promiscuous mountain voles have far fewer of these receptors.

In male voles, variations in the distribution of receptors for another hormone (arginine vasopressin, or AVP) correlate with bonding tendency. Compared to males of promiscuous vole species, males of monogamous species have more AVP receptors. When scientists isolated the prairie vole gene for the AVP receptor and used a virus to transfer copies of this gene into the forebrain of male mice (which are naturally promiscuous), the genetically modified mice gave up their playboy ways. They now preferred a female with whom they had already mated over an unfamiliar female. These results confirmed the role of AVP in fostering monogamy among male rodents.

Human Behavior Genetics

Nearly all human behavioral traits have a polygenic basis and are influenced by the environment. Keep this in mind when you see headlines touting discovery of a gene "for" thrill-seeking behavior, alcoholism, or some other human trait. Generally, such discoveries show a small statistical correlation between a particular allele and an increased tendency toward a particular behavior.

Insights from studies of animal behavior sometimes inspire human studies. For example, investigators are looking at the role oxytocin plays in human social behavior. Impaired oxytocin production or reception may contribute to autism, a disorder in which a person has trouble making social and emotional attachments. Scientists are collecting data about oxytocin receptor genes of autistic children and their unaffected siblings. The aim of the genetic study is to determine whether particular alleles of the oxytocin receptor gene raise the likelihood of autism. Oxytocin is also being tested as a treatment for autism.

---

stimulus Environmental cue that a sensory receptor responds to.

Take-Home Message **How do genes affect behavior?**

❯ Many genes shape the nervous system. Such genes affect traits such as an animal's ability to sense and respond to stimuli.

❯ Genetic differences affect species-specific behavior, and also differences in behavior among individuals of a species.

❯ Most human behaviors are complex, polygenic traits. Studies of genetic differences can shed light on predispositions to particular behaviors.

> Some behaviors are inborn and can be performed without any practice.
> Most behaviors can be modified as a result of experience.

Instinctive Behavior

All animals are born with the capacity for **instinctive behavior**—an innate response to a specific and usually simple stimulus. For example, a newborn coastal garter snake behaves instinctively when it attacks a banana slug in response to the slug's scent.

The life cycle of the cuckoo bird provides several examples of instinct at work. This European bird is a brood parasite, an animal that lays its eggs in another animal's nest. A female cuckoo lays her eggs in the nests of other birds. The newly hatched cuckoo is blind, but contact with an egg laid by its foster parent stimulates an instinctive response. That hatchling maneuvers the egg onto its back, then shoves it out of the nest (Figure 39.5A). This behavior removes any potential competition for the foster parent's attention.

A cuckoo's egg-dumping response is a **fixed action pattern**: a series of instinctive movements, triggered by a specific stimulus, that—once started—continues to completion without the need for further cues. Such fixed behavior has survival advantages when it permits a rapid reaction to an important stimulus. However, a fixed response to a simple stimulus has limitations. For example, the cuckoo's foster parent responds to the simple stimulus of a chick's gaping mouth with the fixed action pattern of parental feeding behavior (Figure 39.5B). It disregards a cuckoo's unusual large size and coloration.

Figure 39.5 Instinctive behavior.
A Young cuckoo shoves its foster parent's eggs out of the nest.
B The foster parent feeds the cuckoo chick in response to one simple cue: a gaping mouth.

Time-Sensitive Learning

Learned behavior is behavior that is altered by experience. Some instinctive behavior can be modified with learning. A garter snake's initial strikes at prey are instinctive, but the snake learns to avoid dangerous or unpalatable prey. Learning may occur throughout an animal's life, or be restricted to a critical period.

Imprinting is a form of learning that occurs during a genetically determined time period early in life. For example, baby geese learn to follow the large object that bends over them in response to their first peep (Figure 39.6). With rare exceptions, this object is their mother. When mature, the geese will seek out a sexual partner that is similar to the imprinted object.

A genetic capacity to learn, combined with actual experiences in the environment, shapes most forms of behavior. For example, a male songbird has an inborn capacity to recognize his species' song when he hears older males singing it. The young male uses these overheard songs as a guide to fill in details of his own song. Males reared with no model or exposed only to songs of other species often sing a simplified version of their species' song.

Many birds can only learn the details of their species-specific song during a limited period early in life. For example, a male white-crowned sparrow will not sing normally if he does not hear a male "tutor" of his own species during his first 50 or so days. Hearing a same-species tutor later in life will not influence his singing.

Most birds must also practice their song to perfect it. In one experiment, researchers temporarily paralyzed throat muscles of zebra finches who were beginning to sing. After being temporarily unable to practice, these birds never mastered their song. In contrast, temporary

Figure 39.6 Nobel laureate Konrad Lorenz with geese that imprinted on him. The inset photograph shows a more typical imprinting pattern.

paralysis of throat muscles in very young birds or adults did not impair later song production. Thus, in this species, there is a critical period for song practice, as well as for song learning.

Conditioned Responses

Nearly all animals are lifelong learners. Most learn to associate certain stimuli with rewards and others with negative consequences.

With classical conditioning, an animal's involuntary response to a stimulus becomes associated with a stimulus that accompanies it. In the most famous example, Ivan Pavlov rang a bell whenever he fed a dog. Eventually, the dog's reflexive response—increased salivation—was elicited by the sound of the bell alone. Taste aversion is a natural variant of classical conditioning. When an animal becomes nauseous after eating a new food, it avoids that food in the future.

With operant conditioning, an animal modifies its voluntary behavior in response to consequences of that behavior. This type of learning was first described in lab animals. A rat that presses a lever and is rewarded with a food pellet becomes more likely to press the lever again. A rat that receives a shock when it enters a particular area will quickly learn to avoid that area. In nature as well, animals learn to repeat behaviors that provide food or mating opportunities and to avoid those that cause pain.

Other Types of Learned Behavior

With **habituation**, an animal learns by experience not to respond to a stimulus that has neither positive nor negative effects. For example, pigeons in cities learn not to flee from the large numbers of people who walk past them.

Many animals learn about the landmarks in their environment and form a sort of mental map. This map may be put to use when the animal needs to return home. For example, a fiddler crab foraging up to 10 meters (30 feet) away from its burrow is able to scurry straight home when it perceives a threat.

Many animals also learn the details of their social landscape; they learn to recognize mates, offspring, or competitors by appearance, calls, odor, or some combination of cues. For example, two male lobsters that meet up for the first time will fight (Figure 39.7). Later, they will recognize one another by scent and behave accordingly, with the loser actively avoiding the winner.

fixed action pattern Series of instinctive movements elicited by a simple stimulus and carried out with little variation once begun.
habituation Learning not to respond to a repeated stimulus.
imprinting Learning that can occur only during a specific interval in an animal's life.
instinctive behavior An innate response to a simple stimulus.
learned behavior Behavior that is modified by experience.

Figure 39.7 Social learning. Two male lobsters battling at their first meeting. Later, the loser will remember the scent of the winner and avoid him. Without another meeting, memory of the defeat lasts up to two weeks.

Figure 39.8 Results of observational learning. A marmoset opens a container using its teeth. After watching one individual successfully perform this maneuver, other marmosets used the same technique. Analysis of videos of their movements showed that the observers closely imitated the behavior they had seen earlier.

With imitation learning, an animal copies the behavior it observes in another individual. Primates often learn new skills in this manner. For example, Ludwig Huber and Bernhard Voelkl allowed marmoset monkeys to watch another marmoset open a container to get the treat inside. Some marmosets used their hands to open the container, others used their teeth. When the observing monkeys were later given a similar container, they used either their hands or their teeth depending on which technique they had observed (Figure 39.8).

Take-Home Message How do instinct and learning shape behavior?

> Instinctive behavior can initially be performed without any prior experience, as when a simple cue triggers a fixed action pattern. Even instinctive behavior may be modified by experience.

> Certain types of learning can only occur at particular times in the life cycle.

> Learning can affect both voluntary and involuntary behaviors.

39.4 Adaptive Behavior

❯ Scientists use experiments to test hypotheses about the adaptive value of a behavior.
❮ Link to Adaptive traits 16.4

The fact that a particular behavior exists does not mean that it is an adaptive trait (Section 16.4). The behavior may have been adaptive under previous conditions, or it could be a consequence of another trait. Thus, behavior researchers often test hypotheses about the adaptive value of a behavior.

For example, starlings and many other birds place sprigs of aromatic plants such as wild carrot in their nest (Figure 39.9A). Larry Clark and Russell Mason hypothesized that nest decorating behavior increases a bird's reproductive success by reducing the number of bloodsucking mites that feed on nestlings. Some plant chemicals deter insect reproduction or maturation. The researchers tested their hypothesis by replacing natural starling nests with artificial ones. Some replacement nests had wild carrot sprigs and others were sprig-free. After the starling chicks had fledged, Clark and Mason counted the mites left behind in nests. As they expected, sprig-free nests had more mites (Figure 39.9B).

In other studies, researchers have found that aromatic plants encourage chick growth and health even when they do not affect parasite numbers. Thus, nest decorating behavior may be adaptive for a variety of reasons.

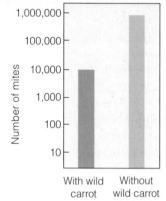

Figure 39.9 Starling nest decorating behavior.
A Starling with nesting material.
B Results of an experiment to test the effect of wild carrot sprigs on the number of mites in nests. Nests with wild carrots had significantly fewer mites than those with no greenery.

39.5 Communication Signals

❯ Communication signals play a role in both cooperative and competitive behavior.
❮ Link to Pheromone 30.7

Communication signals are chemical, acoustical, visual, or tactile cues that transmit information from one member of a species to another. A communication signal evolves and persists only if it benefits both the signal sender and the signal receiver.

Pheromones are chemical cues. Signal pheromones cause a rapid shift in the receiver's behavior. The honeybee alarm pheromone is an example, as are sex attractants that help males and females of many species find each other. Priming pheromones cause longer-term responses, as when a chemical dissolved in the urine of male mice triggers ovulation in females of the same species.

Bird song is an acoustical signal used to attract mates. Territorial animals often use calls or song to advertise their presence to potential competitors. Alarm calls are emitted in response to a threat. Some convey a surprising amount of information. A prairie dog makes one call

Figure 39.10 Visual signals. **A** Male baboon showing his teeth in a threat display. **B** Penguins engaged in a courtship display. **C** A wolf's play bow tells another wolf that behavior that follows is play, not aggression.

Take-Home Message What makes a behavior adaptive?

❯ A behavior is adaptive if it enhances reproductive success.

❯ Biologists can use experiments to test hypotheses about the adaptive value of a behavior.

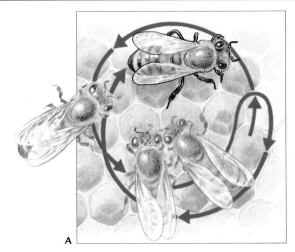

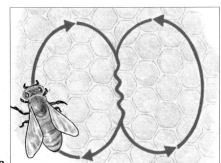

Figure 39.11 Animated Honeybee dances, an example of a tactile display. **A** Bees that have visited a source of food close to their hive return and perform a round dance on the hive's vertically oriented honeycomb. Bees that maintain contact with the dancer fly out and search for food near the hive.

B A bee that visits a feeding source more than 100 meters (110 yards) from her hive performs a *waggle* dance. The orientation of an abdomen-waggling dancer in the straight run of her dance informs other bees about the direction of the food.

C If the food is in line with the sun, the dancer's waggling run proceeds straight up the honeycomb. **D** If food is in the opposite direction from the sun, the dancer's waggle run is straight down. **E** If food is 90 degrees to the right of the direction of the sun, the waggle run is 90 degrees to the right of vertical.

The speed of the dance and the number of waggles in the straight run provide information about distance to the food. When food is 200 meters away, a bee dances much faster, and with more waggles per straight run, compared with a dance inspired by a food source that is 500 meters away.

When bee moves straight up comb, recruits fly straight toward the sun.

When bee moves straight down comb, recruits fly to source directly away from the sun.

When bee moves to right of vertical, recruits fly at 90° angle to right of the sun.

A **B** **C** **D** **E**

when it sees an eagle and a different call when it sights a coyote. Upon hearing the call, other prairie dogs respond appropriately: they either dive into burrows (to escape an eagle's attack) or stand erect (to spot the coyote).

An acoustical signal's characteristics are related to its adaptive value. Courtship-related signals are easily localized. A singing male songbird benefits when a female finds him. By contrast, alarm calls often have properties that make it harder to pinpoint the caller's position.

A male baboon's threat display is a visual signal that communicates readiness to fight a rival (Figure 39.10A). Most bird courtship involves coordinated visual signaling (Figure 39.10B). Selection favors unambiguous signals, so movements that serve as signal often become exaggerated. Body form may evolve in concert with movements, as when bright-colored feathers enhance a courtship display.

Honeybees use tactile signals to communicate the location of food. A honeybee worker who finds food returns to the hive and moves in a defined pattern, jostling a crowd of other bees that surround her. The signals give other bees information about the distance and the direction of the food source (Figure 39.11).

The same signal sometimes functions in more than one context. For example, dogs and wolves solicit play behavior with a play bow (Figure 39.10C). Without the visual cue, a signal receiver may construe behaviors that follow as aggressive or sexual—but not playful.

Predators sometimes tap into signaling systems of their prey. For example, male tungara frogs attract females with complex calls. Frog-eating bats use these calls to zero in on the caller. When bats are near, male frogs make fewer and simpler calls. The subdued signal is a trade-off between locating a partner for mating and the need for immediate survival. As another example, fireflies attract mates by producing flashes of light in a characteristic pattern. Some female fireflies prey on males of other species. When a predatory female sees the flash from a male of the prey species, she flashes back as if she were a female of his own species. When he approaches ready to mate, she captures and eats him.

Take-Home Message **What are animal communication signals?**

❭ A chemical, visual, acoustical, or tactile communication signal transfers information from one individual to another of the same species. Both signaler and receiver benefit from the transfer.

❭ Signals can draw attention of predators as well as intended receivers. Features of signals reflect a balance between the benefit of sending information and the potential cost of signaling.

---

communication signal Chemical, acoustical, visual, or tactile cue that is produced by one member of a species and detected and responded to by other members of the same species.

> Mating and parenting behaviors reflect often opposing selective pressures operating on males and females.
< Link to Sexual selection 17.7

Mating Behavior

Males or females of a species often compete for access to mates, and many are choosy about their partners. Both situations lead to sexual selection. This microevolutionary process favors characteristics that provide a competitive advantage in attracting and retaining mates.

Male animals produce many small sperm, and females produce fewer, larger eggs. Thus a male can produce more offspring than a female. Males usually maximize their reproductive success by mating as many times as possible. Mating with more than one male can benefit a female by increasing the genetic diversity of her offspring. However, given that her number of potential offspring is more limited, she has more at stake in each offspring. Thus females are more likely than males to turn down opportunities to breed with a low-quality individual.

Female hangingflies only breed with males who provide an energy-laden meal. A male hunts and kills a moth or some other insect. Then he releases a sex pheromone that attracts females to him and his "nuptial gift" (Figure 39.12A). Only after a female has been eating for five minutes or so does she start to accept sperm from her partner. Even after mating begins, a female can break off from her suitor, if she finishes her meal. If she does end the mating, she will seek out a new male and his sperm will replace the first male's. Thus, the larger a male's gift, the longer a female will spend eating it and the greater the chance that his sperm will end up fertilizing her eggs.

A female fiddler crab assesses both a male and his real estate. The crabs live along many sandy shores. One of the male's two claws is enlarged; it often accounts for more than half his total body weight (Figure 39.12B). During their breeding season, hundreds of males dig mating burrows in close proximity to one another. Each male stands at the entrance to his burrow, waving his oversized claw. Female crabs stroll along, checking out males. If a female likes what she sees, she inspects her suitor's burrow. Only when a burrow has the right location and dimensions does she mate with its owner and lay eggs in his burrow. One particularly fussy female was observed to visit 106 burrows in an hour. Burrow location and size are important because they affect larval development.

By contrast, male sage grouse converge at a **lek**, a communal display ground that serves only as a dance floor. With tail feathers erect, the males stamp their feet and emit booming calls by puffing and deflating big neck pouches (Figure 39.12C). Each female mates with a single male, but popular males mate many times.

In species in which females cluster around a necessary resource, males may hold a **territory**, a region that they occupy and defend from others. The territory holder mates with all the females in his territory. Lions, elk, elephant seals, and bison (Figure 39.13) have this sort of mating behavior.

When females choose among a group of displaying males or males defend a mating territory, nearly

A Male hangingfly dangling a moth as a nuptial gift for a potential mate.

B Male fiddler crabs (*top*) wave their one enlarged claw to a attract a female (*bottom*).

C Male sage grouse gather on a communal display ground, where they dance, puff out their neck, and make booming calls.

Figure 39.12 How to impress a female.

all females mate but many males do not. Such systems impose strong selective pressure for mating-related traits on males. The result is often **sexual dimorphism**, a difference in morphology between the sexes. The large claw of the male fiddler crab, inflatable neck pouch of a male sage grouse, and the massive body size of a male bison are examples of sexually dimorphic traits.

With the exception of birds, monogamy is rare in animals. An estimated 3 percent of mammals are socially monogamous; they pair bond and cooperate to raise their young. However, genetic analysis nearly always turns up evidence of mixed paternity in offspring of some pairs. Both males and females engage in outside matings.

Parental Care

Parental behavior requires an investment of time and energy that might otherwise be invested in reproducing again. Evolution of parental care requires that the benefit to a parent in terms of increased offspring survival be greater than the cost in lost reproductive opportunity.

In mammals, young are usually born in a relatively helpless state, so parental care is essential. Typically, the female is the sole caregiver (Figure 39.14A). Males can leave immediately after mating and reproduce again while a females is left holding eggs or developing young inside her body. Also, males who care for young may unknowingly end up protecting another male's offspring—an investment with no genetic payoff.

Males seldom serve as sole caregivers. The midwife toad is an interesting exception. A male holds strings of fertilized eggs around his legs until the eggs hatch (Figure 39.14B). A female will mate with multiple males.

Most birds are monogamous, and parents cooperate in caring for young (Figure 39.14C). Sage grouse and other birds in which females alone care for the young tend to hatch when they are more fully developed and require less care.

---

lek Of some birds, a communal mating display area for males.
sexual dimorphism Distinct male and female phenotypes.
territory Region that an animal or animals occupy and defend against competitors.

Take-Home Message How does natural selection affect mating and parental care?

> Each individual's behavior tends to maximize his or her own reproductive success.

> In many species, a few males monopolize mating opportunities either by enticing females to choose them or by defending a territory with many females.

> Monogamy is rare and even socially monogamous species often mate with nonpartners. Male parental care is associated with monogamy.

Figure 39.13 Male bison locked in combat during the breeding season. A few males will mate with many females. Some males will not mate at all.

A Female grizzlies care for their cub for as long as two years. The male takes no part in its upbringing.

B A male midwife toad carries developing eggs.

C A pair of Caspian terns cooperate in the care of their chick.

Figure 39.14 Caring for offspring.

> Many animals benefit by clustering in groups, but social living also has costs.
< Link to Culture 24.9

Defense Against Predators

In some groups, cooperative responses to predators reduce the net risk to all. Multiple individuals can be on the alert for predators, join a counterattack, or engage in more effective defenses.

Birds, monkeys, meerkats, prairie dogs, and many other animals that live in social groups make alarm calls in response to a predator (Figure 39.15A). An alarm call typically provokes such a flurry of activity in the group that the predator ends up looking elsewhere for food.

Sawfly caterpillars that feed in groups on branches benefit by the group's coordinated response to predatory birds. When a bird approaches, the caterpillars all rear up and vomit partly digested eucalyptus leaves (Figure 39.15B). Predatory birds tend to avoid the wiggling mass of fluid-exuding caterpillars, preferring instead to attack individuals. In one experiment, birds offered caterpillars one at a time ate an average of 5.6. When offered a cluster of twenty caterpillars, the birds ate an average of 4.1.

Whenever animals cluster, some individuals shield others from predators (Figure 39.15C). Preference for the center of a group can create a **selfish herd**, in which individuals hide behind one another. Selfish-herd behavior occurs in bluegill sunfishes. A male sunfish builds a nest by scooping out a depression in mud on the bottom of a lake, and females lay eggs in these nests. Snails and fishes prey on the eggs. Competition for the safest sites is greatest near the center of a group, with large males taking the innermost locations. Smaller males cluster around them and bear the brunt of the egg predation. Even so, small males are better off nesting at the edge of the group than on their own.

Improved Feeding Opportunities

Many mammals, including wolves, lions, wild dogs, and chimpanzees, live in social groups and cooperate in hunts (Figure 39.16). Are cooperative hunters more efficient than solitary ones? Often, no. In one study, researchers observed that a solitary lion catches prey about 15 percent of the time. Two lions cooperatively hunting catch prey twice as often as a solitary lion, but having to share the spoils of the hunt means the amount of food per lion is the same. When more lions join a hunt, the success rate per lion falls. Wolves have a similar pattern. Among carnivores that hunt cooperatively, hunting success is not the major advantage of group living. Individuals hunt together, but they also fend off scavengers, care for one another's young, and protect territory together.

Group living also allows transmission of cultural traits, or behaviors learned by imitation. For example, chimpanzees make and use simple tools by stripping leaves from branches. They use thick sticks to make holes in a termite mound, then insert long, flexible "fishing sticks" into the holes (Figure 39.17). The long stick agitates the termites, which attack and cling to it. Chimps withdraw the stick

---

dominance hierarchy Social system in which resources and mating opportunities are unequally distributed within a group.
selfish herd Temporary group that forms when individuals cluster to minimize their individual risk of predation.

A Black-tailed prairie dog giving an alarm call.

B Clump of Australian sawfly caterpillars regurgitating fluid.

C Musk oxen adults forming a ring of horn. Young are protected in the center of the ring.

Figure 39.15 Group defenses.

Figure 39.16 Members of a pack of wolves (*Canis lupus*). Wolves cooperate in hunting, caring for the young, and defending territory. Benefits are not distributed equally. Most of the time, only the highest-ranking individuals, the alpha male and alpha female, breed.

and lick off termites. Different groups of chimpanzees use slightly different tool-shaping and termite-fishing methods. Youngsters learn by imitating the adults.

Dominance Hierarchies

Many animals that live in permanent groups form a **dominance hierarchy**. In this type of social system, dominant animals get a greater share of resources and breeding opportunities than subordinate ones. Typically, dominance is established by physical confrontation. In most wolf packs, one dominant male breeds with one dominant female. The other members of the pack are nonbreeding brothers and sisters, aunts and uncles. All hunt and carry food back to individuals that guard the young in their den.

Why would a subordinate give up resources and often breeding privileges? Challenging a strong individual can be dangerous, as is living on one's own. Subordinates get their chance to reproduce by outliving a dominant peer.

Regarding the Costs

Relatively few animals spend the bulk of their time in social groups. Why? In most habitats, the costs of social living outweigh the benefits. Individuals that live in denser groups compete more for resources (Section 40.5). Penguins and many other seabirds form dense breeding colonies in which competition for space and food is intense (Figure 39.18). Given the opportunity, a pair of breeding herring gulls will cannibalize the eggs and even the chicks of their neighbors. Large social groups also attract more predators, and individuals that live in dense groups are at a higher risk of parasites and contagious diseases that jump from host to host.

Figure 39.17 Chimpanzees (*Pan troglodytes*) using sticks as tools for extracting tasty termites from a nest. This behavior is learned by imitation.

Figure 39.18 A crowded penguin breeding colony.

Take-Home Message **What are the benefits and costs of living in a group?**

> Living in a social group can provide benefits, as through cooperative defenses or shielding against predators.

> Group living has costs such as increased competition and increased vulnerability to infections.

> ❯ Extreme cases of sterility and self-sacrifice have evolved in only two groups of insects and one group of mammals. How are genes of the nonreproducers perpetuated?
> ❮ Link to Insects 23.13

Social Insects

True social insects include honeybees, termites, and ants. All of these eusocial insects stay together for generations in a group that has a division of labor. Groups include permanently sterile workers that care cooperatively for the offspring of just a few breeding individuals. Such workers often are highly specialized in form and function (Figure 39.19).

A queen honeybee is the only fertile female in her hive (Figure 39.20A). She is larger than her worker daughters, partly because of her enlarged ovaries. The queen honeybee secretes a pheromone that makes all other female bees in the hive sterile. The 30,000 to 50,000 sterile female workers feed larvae, clean and maintain the hive, and construct honeycomb from waxy secretions. Workers also gather the nectar and pollen that feeds the colony. They guard the hive and will sacrifice themselves to repel any intruders.

Male stingless drones are produced seasonally and subsist on food gathered by their worker sisters. Each day, drones fly in search of a mate. The occasional lucky one will meet a virgin queen on her one flight away from a colony. He dies after mating. The queen mates with many males, then uses their stored sperm for years.

Like honeybees, termites live in enormous family groups with a queen who specializes in egg production (Figure 39.20B). Unlike the honeybee hive, a termite mound holds sterile males and females. A king supplies the female with sperm. Winged reproductive termites of both sexes develop seasonally.

Social Mole–Rats

Sterility and extreme self-sacrifice are uncommon in vertebrates. The only eusocial mammals are two species of African mole-rat. The best studied is *Heterocephalus glaber*, the naked mole-rat. Clans of this nearly hairless rodent live in burrows in dry parts of East Africa.

A reproducing female dominates the clan and mates with one to three males (Figure 39.20C). Nonbreeding members live to protect and care for the "queen" and "king" (or kings) and their offspring. Sterile diggers excavate tunnels and chambers. When a digger finds an edible root, it hauls a bit back to the main chamber and chirps. Its chirps recruit others, which help carry food back to the chamber. In this way, the queen, her mates, and her young offspring get fed. Other sterile helpers guard the colony. When a predator appears, they chase and attack it at great risk to themselves.

Evolution of Altruism

A sterile worker in a social insect colony or a naked mole-rat clan shows **altruistic behavior**, which is behavior that enhances another individual's reproductive success at the altruist's expense. How did this behavior evolve? According to William Hamilton's **theory of inclusive fitness**, genes associated with altruism are selected if they lead to behavior that promotes the reproductive success of an altruist's closest relatives.

A sexually reproducing, diploid parent caring for offspring is not helping exact genetic copies of itself. Each of its gametes, and each offspring, inherits one-half of its genes. Other individuals of the social group that have the same ancestors also share genes. Siblings (brothers or sisters) are as genetically similar as a parent and offspring. Nephews and nieces share about one-fourth of their uncle's genes.

A Australian honeypot ant worker, a living container for the colony's food reserves.

B Army ant soldier with formidable mandibles.

C Eyeless soldier termite. It bombards intruders with a stream of sticky goo from its nozzle-shaped head.

Figure 39.19 Specialized ways of serving and defending the colony. All of these individuals are sterile.

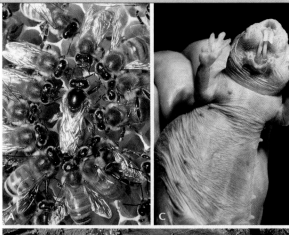

Figure 39.20 Three queens. **A** Queen honeybee with sterile daughters. **B** A termite queen dwarfs her offspring and mate. **C** A naked mole-rat queen.

Sterile workers promote genes for "self-sacrifice" through behavior that can benefit very close relatives. In honeybee, termite, and ant colonies, sterile workers assist fertile relatives with whom they share genes. A guard bee dies after she stings, but her sacrifice preserves copies of her genes in her hivemates.

Inbreeding increases the genetic similarity among relatives and may play a role in mole-rat sociality. A clan is highly inbred as a result of many generations of sibling, mother–son, and father–daughter matings. Researchers are now searching for other factors that select for eusocial behavior in mole-rats. According to one hypothesis, arid habitats and patchy food sources favor mole-rat genes that contribute to cooperation in digging burrows, searching for food, and fending off competitors for resources.

---

altruistic behavior Behavior that benefits others at the expense of the individual.
theory of inclusive fitness Genes associated with altruism can be advantageous if the expense of this behavior to the altruist is outweighed by the reproductive success of relatives.

Take-Home Message How can altruistic behavior evolve?

> Altruistic behavior may arise and persist when altruists help relatives survive and reproduce. The altruist's actions increase the likelihood that copies of its genes for altruism will be passed to the next generation.

An Aggressive Defense (revisited)

When a European queen bee mates with an Africanized drone, her worker offspring are just as aggressive as workers in a pure Africanized colony. In contrast, a cross between an Africanized queen and a European drone yields workers with an intermediate level of aggression. Unfortunately, European queen–Africanized male pairings occur far more frequently than the reciprocal cross. Africanized males outcompete European males for matings.

How Would You Vote? Africanized honeybees continue to increase their range. Should study of their genetics be a high priority? See CengageNow for details, then vote online (cengagenow.com).

Evolution and Human Behavior

> Evolutionary forces shaped and continue to influence human behavior—but humans alone can make moral choices about their actions.
< Link to Philosophy of science 1.9

Many people resist the idea of analyzing the evolutionary basis of human behavior. A common fear is that an objectionable behavior will be defined as "adaptive." To evolutionary biologists, however, "adaptive" does not mean "morally right." It simply means a behavior increases reproductive success. Scientific studies do not address moral issues (Section 1.9).

For example, infanticide is morally repugnant, but it happens in many animal groups and all human cultures. In her book on maternal behavior, primatologist Sarah Blaffer Hrdy cites a study of a village in Papua New Guinea in which parents killed about 40 percent of the newborns. Hrdy argues that when resources and social support are hard to come by, a mother who is unlikely to be able to raise a newborn to adulthood can reduce her costs by killing it. The mother can then allocate her limited resources to her other offspring.

Is such behavior appalling? Yes. Can considering the possible evolutionary advantages of the behavior help us prevent it? Perhaps. An analysis of the conditions under which infanticide occurs tells us this: When mothers lack the resources they need to care for their children, they are more likely to harm them. We as a society can act on such information.

Take-Home Message What do studies of evolution tell us about human behavior?

> Evolutionary analysis of human behavior can tell us why some behaviors occur in many cultures. It does not tell us anything about the morality of a behavior.

Summary

Section 39.1 Scientists study the proximate causes of animal behavior, such as differing responses to **pheromones**. They also investigate a behavior's ultimate cause, its evolutionary roots. The study of environmental effects on behavior are one facet of **ecology**.

Section 39.2 Behavior refers to coordinated responses that an animal makes to **stimuli**. It starts with genes that influence the development and activity of the nervous, endocrine, and muscular systems. Studies of behavioral differences within a species or among closely related species can shed light on the proximate and ultimate causes of a behavior.

Section 39.3 **Instinctive behavior** can occur without a prior experience. A **fixed action pattern** is an instinctive response to a simple stimulus.

Learned behavior arises in response to experience. **Imprinting** is time-sensitive learning. With **habituation**, an animal learns to disregard certain stimuli. Animals also form mental maps, learn the identity of other individuals, develop conditioned responses, and imitate observed behaviors.

Section 39.4 A behavior that has a genetic basis is subject to evolution by natural selection. Experiments can test hypotheses about the current adaptive value of a specific behavior.

Section 39.5 Chemical, acoustical, or tactile **communication signals** are meant to change the behavior of individuals of the same species. Evolution influences properties of signals, as when courtship calls are easily localized, but alarm calls are not. Predators sometimes take advantage of the communication system of their prey.

Section 39.6 Sexual selection favors traits that give an individual a competitive edge in attracting mates. Both males and females benefit by mating with multiple partners, but males tend to be less choosy about mates than females. When large numbers of females cluster in a defensible area, males compete with one another to control the areas. Males may display for females at a **lek** or hold a **territory** with resources that females need. Mating systems in which some males father most offspring favor **sexual dimorphism**. Monogamy is rare; individuals of socially monogamous species often "cheat."

Parental care has reproductive costs in terms of reduced frequency of reproduction. It is adaptive when benefits to a present set of offspring offset the costs. Male parental care is typically associated with monogamy.

Section 39.7 Some animals come together as a **selfish herd** in response to the threat of predation. Others live in social groups and benefit by cooperating in predator detection, defense, and rearing the young. With a **dominance hierarchy,** resource and mating opportunities are distributed unequally among group members. Species that live in large groups incur costs, including increased disease and parasitism, and more intense competition for resources.

Section 39.8 Ants, termites, and some other insects as well as mole-rats are eusocial: They live in colonies with overlapping generations and have a reproductive division of labor. Most colony members do not reproduce; they assist their relatives instead.

The **theory of inclusive fitness** states that **altruistic behavior** can be perpetuated when altruistic individuals help their reproducing relatives. Altruistic individuals help perpetuate the genes that lead to their altruism by promoting reproductive success of close relatives that also carry copies of these genes.

Section 39.9 A human behavior that is adaptive in the evolutionary sense may still be judged by society to be morally wrong. Science does not address morality.

Self-Quiz   Answers in Appendix III

1. Genes affect the behavior of individuals by _____ .
   a. influencing the development of nervous systems
   b. affecting how individuals respond to hormones
   c. determining which stimuli can be detected
   d. all of the above

2. Stevan Arnold offered slug meat to newborn garter snakes from different populations to test his hypothesis that the snakes' response to slugs _____ .
   a. was shaped by indirect selection
   b. is an instinctive behavior
   c. is based on pheromones
   d. is adaptive

3. A behavior is defined as adaptive if it _____ .
   a. varies among individuals of a population
   b. occurs without prior learning
   c. increases an individual's reproductive success
   d. is widespread across a species

4. The honeybee dance transmits information about _____ by way of tactile signals.
   a. predators   c. location of food
   b. mating opportunities   d. amount of honey

5. A _____ is a chemical that conveys information between individuals of the same species.
   a. pheromone   c. hormone
   b. neurotransmitter d. all of the above

Data Analysis Activities

Spread of Africanized Honeybees Honeybees disperse by forming new colonies. An old queen leaves the hive along with a group of workers. These bees find a new nest site, and set up a new hive. Meanwhile, at the old hive, a new queen emerges, mates, and takes over. A new hive can be several kilometers from the old one.

Africanized honeybees form new colonies more often than European ones, a trait that contributes to their spread. Africanized bees also spread by taking over existing hives of European bees. In addition, in areas where European and Africanized hives coexist, European queens are more likely to mate with Africanized males, thus introducing Africanized traits into the colony. Figure 39.21 shows the counties in the United States where Africanized honeybees became established between 1990 and 2009.

1. Where in the United States did Africanized bees first become established?

2. In what states did Africanized bees first appear in 2005?

3. Why is it likely that human transport of bees contributed to the spread of Africanized honeybees to Florida?

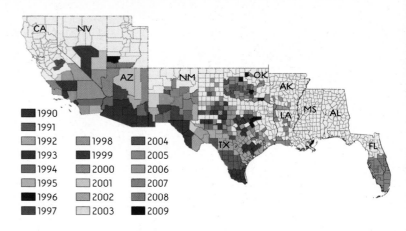

Figure 39.21 The spread of Africanized honeybees in the United States, from 1990 through 2009. The USDA adds a county to this map only when the state officially declares bees in that county Africanized. Bees can be identified as Africanized on the basis of morphological traits or analysis of their DNA.

6. In what group of animals are monogamy and cooperative care of the young by two parents most common?

7. List two possible benefits of living in a group.

8. All honeybee workers are _____ .
   a. male   c. female
   b. sterile   d. b and c

9. Eusocial insects _____ .
   a. live in extended family groups
   b. include termites, honeybees, and ants
   c. show a reproductive division of labor
   d. all of the above

10. Helping other individuals at a reproductive cost to oneself might be adaptive if those helped are _____ .

11. Match the terms with their most suitable description.
   ___ fixed action   a. time-dependent form of
       pattern   learning
   ___ altruism   b. communal display area
   ___ habituation   c. learning to ignore a stimulus
   ___ lek   d. defended area with resources
   ___ territory   e. assisting another individual
   ___ imprinting   at one's own expense
   ___ dominance   f. unequal distribution of
       hierarchy   benefits within group
   ___ classical   g. involuntary response becomes
       conditioning   tied to a stimulus
                          h. series of instinctive movements
                             triggered by a simple stimulus

Additional questions are available on **CENGAGENOW**.

Critical Thinking

1. For billions of years, the only bright objects in the night sky were stars or the moon. Night-flying moths use them to navigate a straight line. Today, the instinct to fly toward bright objects causes moths to exhaust themselves fluttering around streetlights and banging against brightly lit windowpanes. This behavior clearly is not adaptive, so why does it persist?

2. A female chimpanzee engages in sex only during her fertile period, which is advertised by a swelling of her external genitals. Bonobos, the closest relatives of the chimpanzees, are more sexually active than chimpanzees. Female bonobos mate even when they are not fertile. In both species, mating is promiscuous; each female mates with many males. By one hypothesis, female promiscuity may help prevent male infanticide. Chimpanzee males have been observed to kill infants in the wild, but this behavior has never been observed among bonobos.

Explain why it would be disadvantageous for a male to kill the infant of a female with whom he had sex. How might the bonobo female's increased sexual activity lower likelihood of male infanticide still further?

Animations and Interactions on **CENGAGENOW**:
> Honeybee dance language.

< Links to Earlier Concepts

Earlier chapters explored the evolutionary history and genetic nature of populations, including those of humans (Sections 1.4, 16.4, 17.2, 24.11). Now you will consider ecological factors that limit population growth. Again, science cannot address social issues—in this case, how exponential growth affects human lives (1.9). It can only help explain how ecological conditions and events sustain a population's growth, or put a stop to it.

Key Concepts

The Vital Statistics
Ecologists explain population growth in terms of population size, density, distribution, and number of individuals in different age categories. They have methods of estimating population size and density in the field.

Exponential Rates of Growth
A population's size and reproductive base influence its rate of growth. As long as births exceed deaths, a population will grow exponentially. Each generation will be larger than the preceding one.

40 Population Ecology

40.1 A Honking Mess

Canada geese (*Branta canadensis*) were hunted to near extinction in the late 1800s. In the early 1900s, federal laws and international treaties were put in place to protect them and other migratory birds. In recent decades, the number of geese in the United States has soared. For example, Michigan had about 9,000 birds in 1970 and today has more than 300,000. These plant-eating birds often congregate at golf courses and parks (Figure 40.1). They are considered pests because they produce slimy, green feces that soil shoes and stain clothes. Goose feces that gets into water adds nutrients that encourage bacterial and algal growth.

Canada geese also pose a hazard to air traffic. In January of 2009, both engines of a US Airways flight failed shortly after the plane took off from New York's La Guardia airport. The quick-thinking pilot managed to land the plane in the nearby Hudson River, where boats safely unloaded all 155 people aboard. After the crash, investigators determined that the engine failures occurred after Canada geese were sucked into both engines.

Controlling the number of Canada geese poses a challenge, because several different Canada goose populations spend time in the United States. A **population** is a group of organisms of the same species that occupy a particular area. Members of a population breed with one another more than they breed with members of other populations. In the past, nearly all Canada geese seen in the United States were migratory. (A migration is a round-trip between regions, usually in response to expected shifts or gradients in environmental resources.) The geese nested in northern Canada, flew to the United States to spend the winter, then returned to Canada. The common name of the species reflects this tie to Canada.

Most Canada geese still migrate, but some populations have lost this trait. Canada geese breed where they were raised, and nonmigratory birds are generally descendants of geese deliberately introduced to a park or hunting preserve. During the win-

Figure 40.1 Goose troubles. *Opposite page*, a park in Oakland, California, overrun by Canada geese. *Above*, US Airways Flight 1549 floats in the Hudson River after collisions with geese incapacitated both of its engines.

ter, migratory birds often mingle with nonmigratory ones. For example, a bird that breeds in Canada and flies to Virginia for the winter finds itself beside geese that have never left Virginia.

Life is more difficult for migratory geese than for nonmigratory ones. Flying hundreds of miles to and from a northern breeding area takes lots of energy and is dangerous. Compared to a migratory bird, one that stays put can devote more energy to producing young. If the nonmigrant lives in a suburban or urban area, it also benefits from an unnatural abundance of food (grass) and an equally unnatural lack of predators. Not surprisingly, the biggest increases in Canada geese have been among nonmigratory birds that live where humans are plentiful.

In 2006, increasing complaints about Canada geese led the U.S. Fish and Wildlife Service to encourage wildlife managers to look for ways to reduce nonmigratory Canada goose populations, without unduly harming migratory birds. To do so, these biologists need to know which traits characterize each goose population, as well as how populations interact with one another, with other species, and with their physical environment. This sort of information is the focus of the field of population ecology.

---

population Group of organisms of the same species that live in the same area and interbreed.

Limits on Increases in Size Density dependent factors such as competition for resources lead to logistic growth. A population grows exponentially at first, then growth slows as the number approaches the environment's carrying capacity.

Patterns of Survival and Reproduction Life history traits such as age at first reproduction and number of offspring per reproductive event vary and are shaped by natural selection. Adaptive life history traits are those that maximize an individual's lifetime reproductive success.

The Human Population Human populations have sidestepped historical limits to growth by way of global expansion into new habitats, cultural traits, and technological innovations. But no population can expand indefinitely and we are starting to push our limits.

> Ecological factors affect the size, density, distribution, and age structure of a population.
< Links to Sampling error 1.8, Asexual reproduction in plants 27.5, Social groups 39.7

Studying population ecology often involves the use of **demographics**—statistics that describe a population. The demographics of a population often change over time.

Population Size

Population size is the number of individuals in a population. It is often impractical to actually count individuals, so biologists frequently use sampling techniques to estimate population size.

Plot sampling estimates the total number of individuals in an area on the basis of direct counts in a small portion of that area. For example, ecologists might estimate the number of grass plants in a grassland or the number of clams in a mudflat by measuring the number of individuals in several 1 meter by 1 meter square plots. To estimate the total population size, scientists multiply the average number of individuals in the sample plots by the number of these plots that can fit in the area where the population lives. Estimates derived from plot sampling are most accurate when the organisms under study are not very mobile and conditions across the area they occupy are more or less uniform.

Scientists use **mark–recapture sampling** to estimate the population size of mobile animals. Scientists capture animals and mark them with a unique identifier of some sort (Figure 40.2), then release them. After allowing a sufficient time to pass for the marked individuals to meld back into the population, the scientists capture animals again. The proportion of marked animals in the second sample is taken to be representative of the proportion marked in the whole population. For example, if 100 deer are captured and marked, and 50 of them are recaptured, scientists assume that they initially marked half of the population. Thus the total population is estimated at 200.

Information about the traits of individuals in a sample plot or capture group can be used to infer properties of the population as a whole. For example, if half the recaptured geese are of reproductive age, half of the population is assumed to share this trait. This sort of extrapolation is based on the assumption that the individuals in the sample are representative of the general population in terms of age, sex, and other traits under investigation.

Population Density and Distribution

Population density is the number of individuals per unit area or volume. Examples of population density include the number of dandelions per square meter of lawn or the number of euglenas per milliliter of pond water.

Figure 40.2 Florida Key deer marked for a population study.

Population distribution describes the location of individuals relative to one another. Members of a population may be clumped together, be an equal distance apart, or be distributed randomly.

Clumped Distribution Most often members of a population have a clumped distribution; they are closer to one another than would be predicted by chance alone. A patchy distribution of resources encourages clumping. For example, Canada geese tend to congregate in places with suitable food such as grass and a nearby body of water. A cool, damp, north-facing slope may be covered with ferns, whereas an adjacent drier south-facing slope has none.

Limited dispersal ability increases the likelihood of a clumped distribution. As the saying goes, the nut does not fall far from the tree. Asexual reproduction is another source of clusters. It produces colonies of coral (Section 38.2) and vast stands of King's holly clones (Section 27.5).

age structure Of a population, the number of individuals in each of several age categories.
demographics Statistics that describe a population.
mark–recapture sampling Method of estimating population size of mobile animals by marking individuals, releasing them, then checking the proportion of marks among individuals recaptured at a later time.
plot sampling Method of estimating population size of organisms that do not move much by making counts in small plots, and extrapolating from this to the number in the larger area.
population density Number of individuals per unit area.
population distribution Where individuals are clumped, uniformly dispersed, or randomly dispersed in an area.
population size Total number of individuals in a population.
reproductive base Of a population, all individuals who are of reproductive age or younger.

A Clumped distribution of schooling squirrel fish

B Near-uniform distribution of nesting seabirds

C Random distribution of dandelions

Figure 40.3 Population distribution patterns.

Finally, as Section 39.7 explained, many animals benefit by grouping together, as when geese travel in flocks and fish swim in schools (Figure 40.3A).

Near–Uniform Distribution Competition for resources can produce a near uniform distribution, with individuals more evenly spaced than would be expected by chance. Creosote bushes in deserts of the American Southwest typically grow in this pattern. Competition for limited water among the root systems keeps the plants from growing in close proximity. Similarly, seabirds in breeding colonies often show a near uniform distribution. Each bird aggressively repels others that get within reach of its beak as it sits atop its nest (Figure 40.3B).

Random Distribution Members of a population become distributed randomly when environmental resources are uniformly distributed, and proximity to others neither benefits nor harms individuals. For example, when the wind-dispersed seeds of dandelions land on the uniform environment of a suburban lawn, dandelion plants grow in a random pattern (Figure 40.3C). Wolf spider burrows are also randomly distributed relative to one another. When seeking a burrow site, the spiders neither avoid one another nor seek one another out.

Age Structure

The **age structure** of a population refers to the number of individuals in various age categories. Individuals are frequently grouped as pre-reproductive, reproductive, or post-reproductive. Those in the pre-reproductive category

have the capacity to produce offspring when mature. Together with individuals in the reproductive group, they make up the population's **reproductive base**.

Effects of Scale and Timing

The scale of the area sampled and the timing of a study can influence the observed demographics. For example, seabirds are spaced almost uniformly at a nesting site, but the nesting sites are clumped along a shoreline. The birds crowd together during the breeding season, but disperse when breeding is over.

Wildlife managers use demographic information to decide how best to manage populations. For example, to set up a plan for management of nonmigratory Canada geese, wildlife managers began by evaluating the size, density, and distribution of nonmigratory populations. Based on this information, the U.S. Fish and Wildlife Service decided to allow destruction of some eggs and nests, and increased hunting opportunities during times when migratory Canada geese are least likely to be present.

Take-Home Message **What characteristics do we use to describe a population and what factors affect these characteristics?**

❯ Each population has characteristic demographics, such as its size, density, distribution pattern, and age structure.

❯ Characteristics of the population as a whole are often inferred on the basis of a study of the traits of a smaller subsample.

❯ Environmental conditions and interactions among individuals influence a population's demographics, which often change over time.

❯ With exponential growth, the size of the population grows faster and faster over time.

❯ A population will grow exponentially as long as the birth rate is greater than the death rate.

❮ Link to Binary fission 19.6

Gains and Losses in Population Size

Populations continually fluctuate in size. The number of individuals is increased by births and **immigration**, the arrival of new residents that previously belonged to another population. The number is decreased by deaths and **emigration**, the departure of individuals who take up permanent residence elsewhere.

In many animals, young of one or both sexes leave the area where they were born and breed elsewhere. For example, young freshwater turtles typically emigrate from their parental population and become immigrants at another pond some distance away.

By contrast, seabirds typically breed where they were born. However, some individuals may emigrate and end up at breeding sites more than a thousand kilometers away. The tendency of individuals to emigrate to a new breeding site is usually related to resource availability and crowding. As resources decline and crowding increases, the likelihood of emigration rises.

From Zero to Exponential Growth

If we set aside the effects of immigration and emigration, we can define **zero population growth** as an interval during which the number of births is balanced by an equal number of deaths. As a result, population size remains unchanged, with no net increase or decrease in the number of individuals.

We can measure births and deaths in terms of rates per individual, or per capita. *Capita* means head, as in a head count. Subtract a population's per capita death rate (d) from its per capita birth rate (b) and you have the **per capita growth rate**, or r:

$$r = b - d$$

| r (per capita growth rate) | = | b (per capita birth rate) | − | d (per capita death rate) |

Imagine 2,000 mice living in the same cornfield. If 1,000 mice are born each month, then the birth rate is 0.5 births per mouse per month (1,000 births/2,000 mice). If 200 mice die one way or another each month, then the death rate is 200/2,000 or 0.1 deaths per mouse per month. Thus r is 0.5 − 0.1 or 0.4 per mouse per month.

As long as r remains constant and greater than zero, **exponential growth** will occur: Population size will increase by the exact same proportion of its total in every successive time interval. We can calculate population growth (G) for each interval based on the per capita growth rate and the number of individuals (N):

$$G = r \times N$$

| G (population growth per unit time) | = | r (per capita growth rate) | × | N (number of individuals) |

Figure 40.4 Animated Exponential growth in hypothetical population of mice with a per capita rate of growth (r) of 0.4 per mouse per month and a population size of 2,000.

	Starting Population Size	Net Monthly Increase	New Population Size
$G = r \times$	2,000 =	800	2,800
$r \times$	2,800 =	1,120	3,920
$r \times$	3,920 =	1,568	5,488
$r \times$	5,488 =	2,195	7,683
$r \times$	7,683 =	3,073	10,756
$r \times$	10,756 =	4,302	15,058
$r \times$	15,058 =	6,023	21,081
$r \times$	21,081 =	8,432	29,513
$r \times$	29,513 =	11,805	41,318
$r \times$	41,318 =	16,527	57,845
$r \times$	57,845 =	23,138	80,983
$r \times$	80,983 =	32,393	113,376
$r \times$	113,376 =	45,350	158,726
$r \times$	158,726 =	63,490	222,216
$r \times$	222,216 =	88,887	311,103
$r \times$	311,103 =	124,441	435,544
$r \times$	435,544 =	174,218	609,762
$r \times$	609,762 =	243,905	853,667
$r \times$	853,667 =	341,467	1,195,134

A Increases in size over time. Note that the net increase becomes larger with each generation.

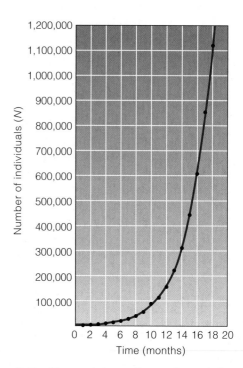

B Graphing numbers over time produces a J-shaped curve.

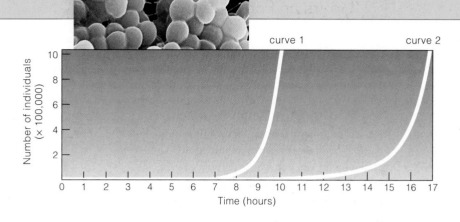

Figure 40.5 Effect of deaths on the rate of increase for two hypothetical populations of bacteria. Plot the population growth for bacterial cells that reproduce every half hour and you get growth curve 1. Next, plot the population growth of bacterial cells that divide every half hour, with 25 percent dying between divisions, and you get growth curve 2. Deaths slow the rate of increase, but as long as the birth rate exceeds the death rate, exponential growth will continue.

After one month, 2,800 mice are in the field (Figure 40.4A). A net increase of 800 fertile mice has made the reproductive base larger. They all reproduce, so the population size expands, for a net increase of 0.4 × 2,800 = 1,120. Population size is now 3,920. At this growth rate, the number of mice would rise from 2,000 to more than 1 million in under two years! Plot the increases against time and you end up with a curve, as shown in Figure 40.4B. Such a J-shaped curve is evidence of exponential growth.

With exponential growth, the number of new individuals added increases each generation, although the per capita growth rate stays the same. Exponential population growth is analogous to the compounding of interest on a bank account. The annual interest *rate* stays fixed, yet every year the *amount* of interest paid increases. Why? The annual interest paid into the account adds to the size of the balance, and the next interest payment will be based on that balance.

In exponentially growing populations, *r* is like the interest rate. Although *r* remains constant, population growth accelerates as the population size increases. When 6,000 individuals reproduce, population growth is three times higher than it was when there were only 2,000 reproducers. With exponential growth, each generation is larger than the prior one.

As another example, think of a single bacterium in a culture flask. After thirty minutes, the cell divides in two. Those two cells divide, and so on every thirty minutes. If no cells die between divisions, then the population size

will double in every interval—from 1 to 2, then 4, 8, 16, 32, and so on. The time it takes for a population to double in size is its doubling time.

After 9–1/2 hours, or nineteen doublings, there are more than 500,000 cells. Ten hours (twenty doublings) later, there are more than a million. Curve 1 in Figure 40.5 is a plot of this increase.

Suppose 25 percent of the descendant cells die every thirty minutes. It now requires about seventeen hours, not ten, for that population to reach 1 million. Deaths slow the rate of increase but do not stop exponential growth (curve 2 in Figure 40.5). Exponential growth will continue as long as birth rates exceed death rates—as long as *r* is greater than zero.

What Is the Biotic Potential?

The growth rate for a population under ideal conditions is its **biotic potential**. This is a theoretical value that would hold if shelter, food, and other essential resources were unlimited and there were no predators or pathogens. Factors that affect biotic potential include the age at which reproduction begins, how long individuals remain reproductive, and the number of offspring that are produced by each reproductive event. Microbes have some of the highest biotic potentials, whereas large-bodied mammals have some of the lowest.

Populations seldom reach their biotic potential because of the effects of limiting factors, a topic we discuss in detail in the next section.

biotic potential Maximum possible population growth rate under optimal conditions.
emigration Movement of individuals out of a population.
exponential growth A population grows by a fixed percentage in successive time intervals; the size of each increase is determined by the current population size.
immigration Movement of individuals into a population.
per capita growth rate For some interval, the added number of individuals divided by the initial population size.
zero population growth Interval in which births equal deaths.

Take-Home Message What determines the size of a population and its growth rate?

> The size of a population depends on its rates of births, deaths, immigration, and emigration.

> Subtract the per capita death rate from the per capita birth rate to get *r*, the per capita growth rate of a population. As long as *r* is constant and greater than zero, a population will grow exponentially. With exponential growth, the number of individuals increases at an ever accelerating rate.

> The biotic potential of a species is its maximum possible population growth rate under optimal conditions.

❯ Many complex interactions take place within and between populations in nature, and it is not always easy to identify all the factors that can restrict population growth.

Environmental Limits on Growth

Most of the time, a population cannot fulfill its biotic potential because of environmental limits. That is why, although a female sea star can produce 2,500,000 eggs a year, oceans do not overflow with sea stars.

Figure 40.6 One example of a limiting factor. **A** Wood ducks build nests only inside hollows of specific dimensions. With the clearing of old-growth forests, the access to natural cavities of the correct size and position is now a limiting factor on wood duck population size. **B** Artificial nesting boxes are being placed in preserves to help ensure the health of wood duck populations.

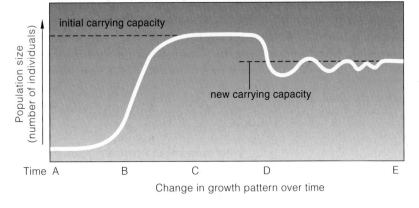

Figure 40.7 Idealized S-shaped curve characteristic of logistic growth. After a rapid growth phase (time B to C), growth slows and the curve flattens as carrying capacity is reached (time C to D).

In the real world, growth curves vary more, as when a change in the environment lowers carrying capacity (time D to E). That happened to the human population of Ireland in the mid-1800s. Late blight, a disease caused by a water mold, destroyed the potato crop that was the mainstay of Irish diets (Section 20.7).

An essential resource that is in short supply acts as a **limiting factor** for population growth. Food, mineral ions, refuge from predators, and safe nesting sites are examples (Figure 40.6). Many factors can potentially limit population growth. Even so, in any environment, one essential factor will kick in first, and it will act as the brake on population growth.

To get a sense of the limits on growth, start again with a bacterial cell in a culture flask, where you can control the variables. First, enrich the culture medium with glucose and other nutrients required for bacterial growth. Next, let many generations of cells reproduce.

Initially, growth will be exponential. Then it slows, and population size remains relatively stable. After a brief stable period, population size plummets until all the bacterial cells are dead. What happened? The larger population required more nutrients. In time, nutrient levels declined, and the cells could no longer divide. Even after cell division stopped, existing cells kept on taking up and using nutrients. Eventually, when the nutrient supply was exhausted, the last cells died out.

Even if you kept freshening the nutrient supply, the population would still eventually collapse. Like other organisms, bacteria generate metabolic wastes. Over time, accumulation of this waste would pollute the habitat and halt growth. Adding nutrients simply substitutes one limiting factor for another. All natural populations run up against limits eventually.

Carrying Capacity and Logistic Growth

Carrying capacity refers to the maximum number of individuals of a population that a given environment can sustain indefinitely. Ultimately, it means that the *sustainable* supply of resources determines population size. We can use the pattern of **logistic growth**, shown in Figure 40.7, to reinforce this point. By this pattern, a small population starts growing slowly in size, then it grows rapidly, then its size levels off as the carrying capacity is reached.

Logistic growth plots out as an S-shaped curve, as shown in Figure 40.7 (time A to C). In equation form:

population growth per unit time	=	maximum per capita population growth rate	×	number of individuals	×	proportion of resources not yet used

Two Categories of Limiting Factors

Factors that affect population growth fall into two categories: density dependent and density independent. **Density–dependent factors** decrease birth rates or increase death rates, and they come into play or worsen with crowding. Competition among members of a population for limited resources leads to density-dependent

Figure 40.8 Overshoot and crash. A reindeer herd introduced to a small island in 1944 increased in size exponentially, then crashed when a low food supply was coupled with an especially cold and snowy winter in 1963–64.

effects, as does infectious disease. Pathogens and parasites spread more easily when hosts are crowded than when host population density is low. The logistic growth pattern results from the effects of density-dependent factors on population size.

Density-independent factors also decrease births or increase deaths, but crowding does not influence the likelihood of their occurrence, or the magnitude of their effects. Fires, snowstorms, earthquakes, and other natural disasters affect crowded and uncrowded populations alike. For example, in December 2004, a powerful tsunami (a giant wave caused by an earthquake) hit Indonesia and killed about 250,000 people. The degree of crowding did not make the tsunami any more or less likely to happen, or to strike any particular island.

Density-dependent and density-independent factors can interact to determine the fate of a population. As an example, consider what happened after a herd of 29 reindeer was introduced to St. Matthew Island off the coast of Alaska in 1944 (Figure 40.8). Biologist David Klein

visited the island in 1957 and found 1,350 well-fed reindeer munching on lichens. In 1963, Klein returned to the island and counted 6,000 reindeer. The population had soared far above the island's carrying capacity. Lichens had become sparser and the average body size of the reindeer had decreased. When Klein returned again in 1966, bleached-out reindeer bones littered the island and only 42 reindeer were alive. Only one was a male; it had abnormal antlers, which made it unlikely to reproduce. There were no fawns. Klein figured out that thousands of reindeer had starved to death during the winter of 1963–1964. The winter had been unusually harsh, in both temperature and amount of snow. The reindeer had already been in poor condition because of the increased competition for food, so most starved when deep snow covered their food source. By the 1980s, there were no reindeer on the island at all.

---

carrying capacity Maximum number of individuals of a species that an environment can sustain.
density-dependent factor Factor that limits population growth and has a greater effect in dense populations than less dense ones.
density-independent factor Factor that limits population growth and arises regardless of population density.
limiting factor A necessary resource, the depletion of which halts population growth.
logistic growth A population grows slowly, then increases rapidly until it reaches carrying capacity and levels off.

Take-Home Message **How do environmental factors affect population growth?**

❯ Carrying capacity is the maximum number of individuals of a population that can be sustained indefinitely by the resources in a given environment.

❯ With logistic growth, population growth is fastest during times of low density, then it slows as the population approaches carrying capacity.

❯ The effects of density-dependent factors such as disease cause a logistic growth pattern. Density-independent factors such as natural disasters also affect population size.

> Age at maturity, the number of offspring produced per reproductive event, and life span affect population growth. Natural selection influences these life history traits.
⟨ Link to Natural selection 16.4

Patterns of Survival and Reproduction

Biologists refer to the reproduction-related events that occur between birth and death as a **life history pattern**. Traits that influence this pattern include the age at which an organism begins to reproduce, how often it reproduces, the number of offspring produced per reproductive event, and the length of the reproductive stage.

We can study life history traits within a population by recording what happens to a specific **cohort**, a group of individuals born at more or less the same time. Table 40.1 shows a life table for a cohort of 1,000 annual plants.

Human life tables are usually not based on a real cohort. Instead, information about current conditions is used to predict life expectancy. Table 40.2 shows the life expectancy for people in the United States based on conditions in the United States during 2006. The table also shows reported births for this year.

Information about age-specific death rates can also be summarized by a **survivorship curve**, a plot that shows how many members of a cohort remain alive over time. Three types of survivorship curves are common.

A type I curve indicates survivorship is high until late in life. Populations of large animals that bear one or, at most, a few offspring at a time and provide extended parental care show this pattern (Figure 40.9A). For example, a female elephant has one calf at a time and cares for it for several years. Type I curves are typical of human populations with access to good health care.

A type II curve indicates that death rates do not vary much with age (Figure 40.9B). In lizards, small mammals, and large birds, old individuals are about as likely to die of disease or predation as young ones.

A type III curve indicates that the death rate for a population peaks early in life. It is typical of species that produce many small offspring and provide little or no parental care. Figure 40.9C shows how the curve plummets for sea urchins. Sea urchin larvae are soft and tiny, so fish, snails, and sea slugs devour most of them before protective hard parts can develop. A type III curve is common for marine invertebrates, insects, fishes, fungi, and for annual plants.

Table 40.1 Life Table for an Annual Plant Cohort*

Age Interval (days)	Survivorship (number surviving at start of interval)	Number Dying During Interval	Death Rate (number dying/ number surviving)	"Birth" Rate During Interval (number of seeds from each plant)
0–63	996	328	0.329	0
63–124	668	373	0.558	0
124–184	295	105	0.356	0
184–215	190	14	0.074	0
215–264	176	4	0.023	0
264–278	172	5	0.029	0
278–292	167	8	0.048	0
292–306	159	5	0.031	0.33
306–320	154	7	0.045	3.13
320–334	147	42	0.286	5.42
334–348	105	83	0.790	9.26
348–362	22	22	1.000	4.31
362–	0	0 / 996	0	0

* *Phlox drummondii*; data from W. J. Leverich and D. A. Levin, 1979.

Table 40.2 Life Table for Humans in the United States (based on 2006 conditions)

Age Interval	Number at Start of Interval	Number Dying During Age Interval	Life Expectancy at Start of Interval	Reported Live Births
0–1	100,000	671	77.7	
1–5	99,329	113	77.2	
5–10	99,216	69	73.3	
10–15	99,147	81	68.4	6,396
15–20	99,065	318	63.4	415,262
20–25	98,747	494	58.6	1,080,437
25–30	98,253	495	53.9	1,181,899
30–35	97,759	546	49.2	950,258
35–40	97,213	718	44.4	498,616
40–45	96,495	1,098	39.7	105,539
45–50	95,397	1,647	35.2	6,480
50–55	93,750	2,398	30.7	494
55–60	91,352	3,295	26.5	
60–65	88,057	4,806	22.4	
65–70	83,251	6,591	18.5	
70–75	76,661	9,329	14.9	
75–80	67,331	13,130	11.6	
80–85	54,201	16,396	98.7	
85–90	37,806	16,906	6.4	
90–95	20,900	12,908	4.6	
95–100	7,992	6,255	3.2	
100+	1,737	1,737	2.3	

cohort Group of individuals born during the same interval.
life history pattern A set of traits related to growth, survival, and reproduction such as life span, age-specific mortality, age at first reproduction, and number of breeding events.
K-selection Individuals who produce offspring capable of outcompeting others for limited resources have a selective advantage; occurs when a population is near carrying capacity.
r-selection Individuals who produce the maximum number offspring as quickly as possible have a selective advantage; occurs when population density is low and resources are abundant.
survivorship curve Graph showing the decline in numbers of a cohort over time.

Figure 40.9 Three generalized survivorship curves. Graphs show the number of members of a cohort still alive as age increases.

>> Figure It Out What type of survivorship curve do phlox plants (Table 40.1) have? *Answer: Type III, with high mortality early in life*

Allocating Reproductive Investment

Natural selection influences the timing of reproduction and how much a parent invests in each offspring. The most adaptive reproductive strategy is that which maximizes a parent's lifetime reproductive success.

Some species such as bamboo and Pacific salmon reproduce once, then die. Others such as oak trees, mice, and humans reproduce repeatedly. A one-shot strategy is evolutionarily favored when an individual is unlikely to have a second chance to reproduce. For Pacific salmon, reproduction requires a life-threatening journey from the sea to a stream. For bamboo, environmental conditions that favor reproduction occur only sporadically.

The most effective reproductive strategy can vary with population density. Reproduction involves trade-offs between offspring quality and quantity. At a low population density, there is little competition for resources and *r*-**selection** predominates: Individuals who turn resources into offspring fastest have a selective advantage. Species that colonize new habitats or are adapted to habitats where environmental changes frequently reduce population size tend to have a small body size and a short generation time.

By contrast, when a stable environment allows population density to near carrying capacity, individuals who can outcompete others for limited resources have the selective edge. Under these circumstances, **K-selection** predominates: Individuals who invest more resources in building their own body and produce fewer, but larger, more competitive offspring, leave the most descendants.

Life history traits of fruit flies and opportunistic weeds such as dandelions reflect mainly *r*-selection, whereas those of large mammals and oak trees reflect mainly *K*-selection. Most organisms are subject to both types of selection at different times and have intermediate traits.

A Elephants have type I survivorship, with low mortality until old age.

B Snowy egrets are a type II population, with a fairly constant death rate.

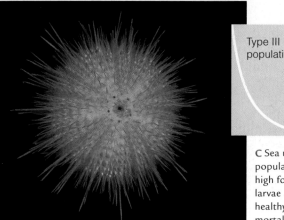

C Sea urchins are type III populations. Mortality is high for tiny soft-bodied larvae and in old age, but healthy spiny adults have low mortality.

Take-Home Message How do researchers study and describe life history patterns?

> Tracking a group of same-aged individuals from birth to death reveals patterns of reproduction and survivorship. This data can be summarized in life tables or survivorship curves.

> Different environmental conditions and population densities can favor different reproductive strategies.

❯ Evolution is an ongoing process. Organisms continually adapt to environmental changes, as when shifts in predation pressure alter life history traits.
❮ Link to Gene flow 17.9

Effect of Predation on Guppies

A long-term study by the evolutionary biologists John Endler and David Reznick illustrates the effect of predation on guppy life history traits. Endler and Reznick studied populations of guppies (*Poecilia reticulata*), small fishes that live in shallow freshwater streams (Figure 40.10) in the mountains of Trinidad.

For their study sites, the scientists focused on streams with many small waterfalls. These waterfalls serve as nat-ural barriers that prevent guppies in one part of a stream from moving easily to another. As a result, each stream holds several populations of guppies that have very little gene flow between them (Section 17.9).

The waterfalls also keep guppy predators from moving from one part of the stream to another. The main guppy predators are killifishes and cichlids. The two differ in size and prey preferences. The killifish is relatively small and preys mostly on immature guppies. It ignores the larger adults. The cichlids are bigger fish. They tend to pursue mature guppies and pass on the small ones. Some parts of the streams hold one type of predator but not the other. Thus, different guppy populations face different predation pressures.

A *Right*, guppy that shared a stream with killifishes (*below*).

B *Right*, guppy that shared a stream with cichlids (*below*).

Figure 40.10 A,B Guppies and two guppy eaters, a killifish A and a cichlid B. C Biologist David Reznick contemplating interactions among guppies and their predators in a freshwater stream in Trinidad.

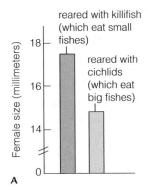
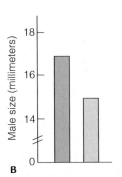
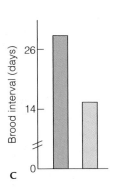
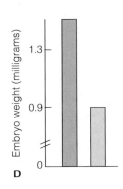

reared with killifish (which eat small fishes)

reared with cichlids (which eat big fishes)

A Female size (millimeters)

B Male size (millimeters)

C Brood interval (days)

D Embryo weight (milligrams)

Figure 40.11 Experimental evidence of natural selection among guppy populations subject to different predation pressures. Compared to the guppies raised with killifish (*green* bars), guppies raised with cichlids (*tan* bars) differed in body size and in the length of time between broods.

Reznick and Endler discovered that guppies in streams with cichlids grow faster and are smaller at maturity than guppies in streams with killifish. Also, guppies hunted by cichlids reproduce earlier, have more offspring at a time, and breed more frequently (Figure 40.11).

Were these differences in life history traits genetic, or did some environmental variation cause them? To find out, the biologists collected guppies from both cichlid- and killifish-dominated streams. They reared the groups in separate aquariums under identical predator-free conditions. Two generations later, the groups continued to show the differences observed in natural populations. The researchers concluded that differences between guppies preyed on by different predators have a genetic basis.

Reznick and Endler hypothesized that the predators act as selective agents that influence guppy life history patterns. They made a prediction: If life history traits evolve in response to predation, then these traits will change when a population is exposed to a new predator that favors different prey traits.

To test their prediction, they found a stream region above a waterfall that had killifish but no guppies or cichlids. They brought in some guppies from a region below the waterfall where there were cichlids but no killifish. At the experimental site, the guppies that had previously lived only with cichlids were now exposed to killifish. The control site was the downstream region below the waterfall, where relatives of the transplanted guppies still coexisted with cichlids.

Reznick and Endler revisited the stream over the course of eleven years and thirty-six generations of guppies. They monitored traits of guppies above and below the waterfall. The recorded data showed that guppies at the upstream experimental site were evolving. Exposure to a previously unfamiliar predator caused changes in their rate of growth, age at first reproduction, and other life history traits. By contrast, guppies at the control site showed no such changes. Reznick and Endler concluded that life history traits in guppies can evolve rapidly in response to the selective pressure exerted by predation.

Effect of Overfishing on Atlantic Cod

The evolution of life history traits in response to predation pressure is not merely of theoretical interest. It has economic importance. Just as guppies evolved in response to predators, a population of Atlantic codfish (*Gadus morhua*) evolved in response to human fishing pressure. Atlantic codfish can grow quite large (Figure 40.12). However, the fishing pressure on the North Atlantic population increased from the mid-1980s to early 1990s. As it did, the age of sexual maturity shifted. The frequency of fast-maturing fish that reproduced while still young and smaller increased. Such individuals were at a selective advantage because both commercial fisherman and sports fishermen preferentially caught and kept the larger fish.

Figure 40.12 Fishermen with a prized catch, a large Atlantic codfish.

In 1992, the declining numbers of cod caused the Canadian government to ban cod fishing in some areas. That ban, and later restrictions, came too late to stop the Atlantic cod population from crashing. The population still has not recovered.

Looking back, it is clear that life history changes were an early sign that the North Atlantic cod population was in trouble. Had biologists recognized the sign, they might have been able to save the fishery and protect the livelihood of more than 35,000 fishers and associated workers. Ongoing monitoring of the life history data for other economically important fishes may help prevent similar disastrous crashes in the future.

Take-Home Message What effect does predation have on life history traits?

> When predators prefer large prey, individual prey who reproduce when still small and young are at a selective advantage. When predators focus on small prey, fast growing-individuals have the selective advantage.

> Human population size surpassed 6.8 billion in 2009. Take a look now at what the number means.

< Links to Human dispersal 24.11, Contraception 38.7

A History of Human Population Growth

For most of its history, the human population grew very slowly. The growth rate began to increase about 10,000 years ago, and during the past two centuries, it soared (Figure 40.13). Three trends promoted the large increases. First, humans were able to migrate into new habitats and expand into new climate zones. Second, humans developed new technologies that increased the carrying capacity of existing habitats. Third, humans sidestepped some limiting factors that restrain growth of other species.

Early humans evolved in the dry woodlands of Africa, then moved into the savannas. Bands of hunter–gatherers moved out of Africa about 2 million years ago. By 40,000 years ago, their descendants were established in much of the world (Section 24.11).

Few species can expand into such a broad range of habitats, but the early humans had large brains that allowed them to develop the necessary skills. They learned how to start fires, build shelters, make clothing, manufacture tools, and cooperate in hunts. With the advent of language, knowledge of such skills did not die with the individual. Compared to most species, humans have a greater capacity to disperse fast over long distances and to become established in a variety of environments.

The invention of agriculture about 11,000 years ago provided a more dependable food supply. A pivotal factor was the domestication of wild grasses, including species ancestral to modern wheat and rice.

Infectious diseases are density-dependent factors that helped dampen human population growth. In the mid-1300s, one-third of Europe's population was lost to a pandemic known as the Black Death. Waterborne diseases such as cholera were widespread. In the past 200 years, improved sanitation and medical advances such as vaccines and antibiotics cut the death toll from disease. Births increasingly outpaced deaths, and exponential growth accelerated.

In the middle of the eighteenth century, people learned to harness the energy of fossil fuels, starting with coal. Fuel-driven, mechanized agriculture provided food to sustain the ever larger population.

In sum, by learning to grow food, controlling disease agents, and tapping into fossil fuels—a concentrated source of energy—the human population has sidestepped many factors that previously limited its rate of increase.

Viewed in a historical context, the rate of population growth in the past two centuries has been astonishing. It took more than 100,000 years for the human population size to reach 1 billion. It took just 123 years to reach 2 billion, 33 more to reach 3 billion, 14 more to reach 4 billion, and then 13 more to get to 5 billion. It took only 12 more years to arrive at 6 billion.

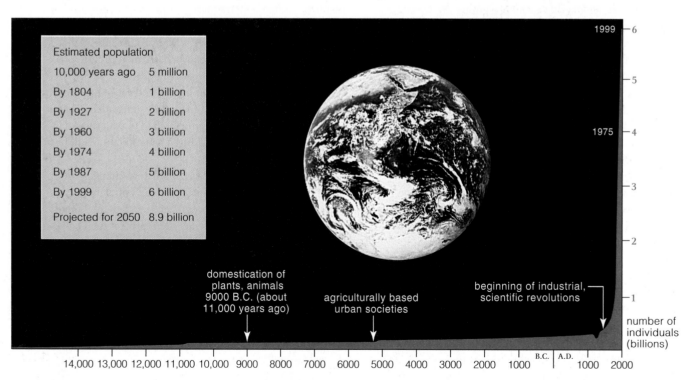

Estimated population

10,000 years ago	5 million
By 1804	1 billion
By 1927	2 billion
By 1960	3 billion
By 1974	4 billion
By 1987	5 billion
By 1999	6 billion
Projected for 2050	8.9 billion

domestication of plants, animals 9000 B.C. (about 11,000 years ago)

agriculturally based urban societies

beginning of industrial, scientific revolutions

number of individuals (billions)

14,000 13,000 12,000 11,000 10,000 9000 8000 7000 6000 5000 4000 3000 2000 1000 | B.C. | A.D. | 1000 2000

Figure 40.13 Growth curve (*red*) for the world human population. To check the current world and U.S. population estimates, visit the U.S. Census web site at www.census.gov/main/www/popclock.html.

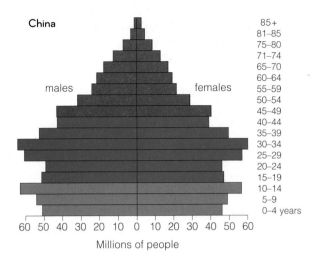
China

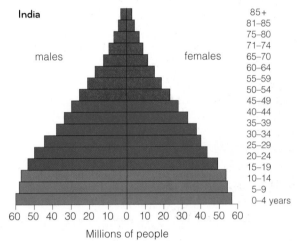
India

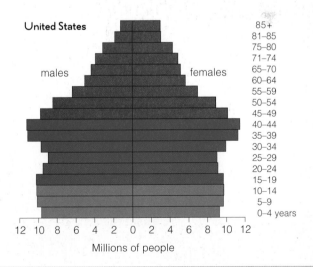
United States

Figure 40.14 Animated Age structure diagrams for the world's three most populous countries. The width of each bar represents the number of individuals in a 5-year age group. *Green* bars represent people in their pre-reproductive years. The *left* side of each chart indicates males; the *right* side, females.

›› Figure It Out Which country has the largest number of men in the 45 to 49 age group?

Answer: China

Fertility Rates and Future Growth

Most governments now acknowledge that population growth, resource depletion, pollution, and quality of life are interconnected. Many offer family planning programs. The United Nations Population Division estimates that globally, more than 60 percent of married women use family planning methods.

The **total fertility rate** of a human population is the average number of children born to a woman during her reproductive years. In 1950, the worldwide total fertility rate averaged 6.5. By 2008, it had declined to 2.6. It remains above the **replacement fertility rate**—the average number of children a woman must bear to replace herself with one daughter of reproductive age. At present, the replacement rate is 2.1 for developed countries and as high as 2.5 in some developing countries. (It is higher in developing countries because more daughters die before reaching the age of reproduction.) A population grows as long as total fertility rate exceeds the replacement rate.

World population is expected to reach 8.9 billion by 2050, and possibly to decline as the century ends. China and India already have more than one billion people apiece. Together they hold 38 percent of the world population. Next in line is the United States, with 307 million.

Age structure diagrams show the distribution of individuals among age groups. Figure 40.14 shows the age structure in the three most populous countries. Notice the size of the age groups that will reproduce during the next fifteen years. The broader the base of an age structure diagram, the greater the proportion of young people, and the greater the expected growth. Government policies that favor couples who have only one child have helped China to narrow its pre-reproductive base.

Even if every couple now alive decides to bear no more than two children, world population growth will not slow for many years, because 1.9 billion people are about to come of reproductive age. More than one-third of the world population is in the broad pre-reproductive base.

---

replacement fertility rate Average number of children women of a population must bear to replace themselves with a daughter of reproductive age.
total fertility rate Average number of children the women of a population bear over the course of a lifetime.

Take-Home Message What is the history of human population growth and what are prospects for future growth?

> Through expansion into new habitats, improved agriculture, and technological innovations, the human population has temporarily skirted environmental resistance to growth.

> Global population growth has slowed somewhat, but large numbers of young people ensure that numbers will continue to rise for the foreseeable future.

> High population growth is correlated with low levels of economic development and low per capita consumption of resources.

Development and Demographics

Demographic factors vary among countries, with the most highly developed countries having the lowest birth rates and infant mortality, and the highest life expectancy (Figure 40.15). The **demographic transition model** describes how changes in population growth often unfold in four stages of economic development (Figure 40.16). Living conditions are harshest during the preindustrial stage,

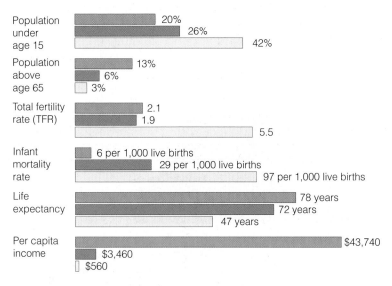

Population under age 15
- 20%
- 26%
- 42%

Population above age 65
- 13%
- 6%
- 3%

Total fertility rate (TFR)
- 2.1
- 1.9
- 5.5

Infant mortality rate
- 6 per 1,000 live births
- 29 per 1,000 live births
- 97 per 1,000 live births

Life expectancy
- 78 years
- 72 years
- 47 years

Per capita income
- $43,740
- $3,460
- $560

Figure 40.15 Key demographic indicators for three countries, mainly in 2006. The United States (*brown* bar) is highly developed, Brazil (*blue* bar) is moderately developed, and Nigeria (*beige* bar) is less developed.

before technological and medical advances become widespread. Birth and death rates are both high, so the growth rate is low ❶.

Next, in the transitional stage, industrialization begins. Food production and health care improve. The death rate drops fast, but the birth rate declines more slowly ❷. As a result, the population growth rate increases rapidly.

During the industrial stage, when industrialization is in full swing, the birth rate declines. People move from the country to cities, where birth control is available and couples tend to want smaller families. The birth rate moves closer to the death rate, and the population grows less rapidly ❸.

In the postindustrial stage, a population's growth rate becomes negative. The birth rate falls below the death rate, and population size slowly decreases ❹.

The United States is in the industrial stage. Developing countries such as Mexico are in the transitional stage, with people continuing to migrate from agricultural regions to cities. Japan and some countries in Europe are in the postindustrial stage.

The demographic transition model is based on analysis of what happened when western Europe and North America industrialized. Whether it can accurately predict changes in modern developing countries remains to be seen. Less developed countries now receive aid from highly developed countries, but must also compete against these countries in a global market.

Runaway population growth is problematic, but negative growth also poses challenges. Some countries in Europe, some members of the former Soviet Union, and Japan currently have negative population growth. Their birth rate has fallen below their death rate. Negative

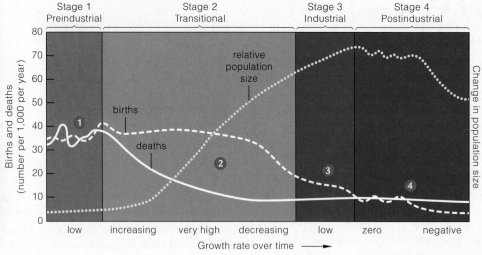

Figure 40.16 Animated Demographic transition model for changes in population growth rates and sizes, correlated with long-term changes in economy.

Table 40.3 Ecological Footprints*	
Country	Hectares per Capita
United States	9.4
Canada	7.1
United Kingdom	5.3
France	4.9
Japan	4.9
Russia	3.7
Mexico	3.4
Brazil	2.4
China	2.1
India	0.9
World Average	2.7
* Global Footprint Network, 2008 data	

growth produces a population with more old people than young. In Japan, people over 65 currently make up about 20 percent of the population. The aging of a population has social implications because older individuals have traditionally been supported by a larger, younger workforce.

Development and Consumption

What is Earth's carrying capacity for humans? There is no simple answer to this question. For one thing, we cannot predict what new technologies may arise or their effects. For another, different types of societies require different amounts of resources to sustain them. On a per capita basis, people in highly developed countries use far more resources than those in less developed countries, and they also generate more waste and pollution.

Ecological footprint analysis is one widely used method of measuring and comparing resource use. An **ecological footprint** is the amount of Earth's surface required to support a particular level of development and consumption in a sustainable fashion. It includes the amount of area required to grow crops, graze animals, produce forest products, catch fish, hold buildings, and take up carbon emissions.

In 2008, the per capita global footprint for the human population was 2.7 hectares, or about 6.5 acres (Table 40.3). The world's two most populous countries, China and India, were below that average, whereas the per capita footprint of the United States was more than three times the average. In other words, the life-style of an average person in the United States requires three times as much of Earth's sustainable resources as the life-style of

demographic transition model Model describing the changes in human birth and death rates that occur as a region becomes industrialized.
ecological footprint Area of Earth's surface required to sustainably support a particular level of development and consumption.

A Honking Mess (revisited)

When a plane crashes or is forced to land unexpectedly, investigators from the Federal Aviation Agency try to determine what went wrong. In 2009, they were asked to find out why both engines of a passenger plane stopped, forcing the plane to land in the Hudson River. The pilot had reported a bird strike, and the plane had bits of feather, bone, and muscle in its wing flaps and its engines (*upper right*).  Samples of this tissue were sent to the Smithsonian Institute, which analyzed the DNA. Unique sequences in the DNA identified the tissue in both engines as Canada goose. One engine had female goose DNA. The other had both male and female DNA, indicating that at least two birds entered it.

Researchers were even able to tell which population of geese these unlucky birds belonged to. As a feather develops, it incorporates hydrogen that the bird took up from its environment. The mix of hydrogen isotopes in the environment varies with latitude, so the isotope mix in a feather provides information about where that feather developed. The isotope mix in the feather bits from the engines indicated that these bird were migratory; their feathers had developed in Canada, not in New York.

How would you vote? One way to decrease the number of nonmigratory Canada geese is by encouraging hunting when migratory birds are unlikely to be present. Would you support relaxing hunting restrictions in regions where Canada geese have become pests? See CengageNow for details, then vote online (cengagenow.com).

an average world citizen. It requires more than ten times the resources of a person in India.

The United States is unlikely to lower its resource consumption to match that of India. In fact, billions of people in India, China, and other less developed nations dream that one day they or their offspring will enjoy the same type of life-style as the average American.

Ecological footprint analysis tells us that, with current technology, Earth may not have enough resources to make those dreams come true. For everyone now alive to live like an average American, four times the sustainable resources available on Earth would be required. Such analysis also suggests that the human population may already be living beyond its ecological means. If we divide the Earth equally, each of us now has 2.1 hectares available to support us sustainably, but our average per capita ecological footprint is 2.7 hectares.

Take-Home Message How does economic development affect population growth and resource consumption?

> Historically, population growth has slowed as countries have industrialized. The same trend may eventually slow growth in currently developing countries.

> The most highly developed countries have relatively low growth rates, but they consume a disproportionately large share of Earth's resources.

> The current global level of resource use is unsustainable, and the pressure on resources is expected to rise.

Summary

Section 40.1 A **population** is a group of individuals of the same species that live in the same area and tend to interbreed. Canada geese in the United States include migratory populations and resident ones.

Section 40.2 Demographics are statistics used to describe a population. We often estimate **population size** by using a sampling method such as **plot sampling** or **mark-recapture sampling**. Other demographics include **population density** and **population distribution**. Most populations have a clumped distribution. **Age structure** describes the proportion of individuals in each age category. The size of the **reproductive base** affects population growth.

Section 40.3 A population's per capita birth rate minus its per capita death rate gives us *r*, the **per capita growth rate**. When birth rate and death rate are equal, there is **zero population growth**.

A population in which *r* is greater than zero undergoes **exponential growth**. The increase in size in any interval is determined by the equation $G = r \times N$, where *G* is population growth and *N* is the number of individuals. With exponential growth, a graph of population size against time produces a J-shaped growth curve. The maximum possible exponential growth rate under optimal conditions is the population's **biotic potential**.

Emigration and **immigration** of individuals can also affect population size.

Section 40.4 The **carrying capacity** is the maximum number of individuals of a given population that can be sustained indefinitely by resources in their environment. **Density-dependent factors** are conditions or events that can lower reproductive success and that get worse with crowding. Disease and competition for food are examples. **Density-independent factors** are conditions or events that can lower reproductive success, but their effect does not vary with crowding. Carrying capacity varies among environments and over time. A population may temporarily overshoot its carrying capacity, then crash.

Density-dependent factors lead to **logistic growth**: a population begins growing exponentially, then growth levels off as **limiting factors** begin to come into play. With logistic growth, population size over time plots out as an S-shaped curve.

Sections 40.5, 40.6 The time to maturity, number of reproductive events, number of offspring per event, and life span are aspects of a **life history pattern**. Such patterns can be studied by following a **cohort**, a group of individuals born at the same time. Three types of **survivorship curves** are common: a high death rate late

in life, a constant death rate at all ages, or a high death rate early in life. Life histories have a genetic basis and are subject to natural selection. At low population density, **r-selection** favors quickly producing as many offspring as possible. At a higher population density, **K-selection** favors investing more time and energy in fewer, higher-quality offspring. Most populations have a mixture of both *r*-selected and *K*-selected traits.

Section 40.7 The human population has surpassed 6.8 billion. Expansion into new habitats and agriculture allowed early increases. Medicine and technology have allowed greater increases. Today, the global **total fertility rate** is declining, but it remains above the **replacement fertility rate**. The very broad pre-reproductive base will cause numbers to increase for at least sixty years.

Section 40.8 The **demographic transition model** predicts economic development will slow population growth. Negative population growth increases the proportion of elderly in a population. World resource consumption will probably continue to rise because a highly developed nation has a much larger **ecological footprint** than a developing one. However, with current technology, Earth does not have enough resources to support the existing population in the style of developed nations.

Self-Quiz Answers in Appendix III

1. Most commonly, individuals of a population show a _____ distribution through their habitat.

2. The rate at which population size grows or declines depends on the rate of _____ .
   - a. births
   - b. deaths
   - c. immigration
   - d. emigration
   - e. a and b
   - f. all of the above

3. Suppose 200 fish are marked and released in a pond. The following week, 200 fish are caught and 100 of them have marks. How many fish are in this pond?

4. A population of worms is growing exponentially in a compost heap. Thirty days ago there were 400 worms and now there are 800. How many worms will there be thirty days from now, assuming conditions remain constant?

5. For a given species, the maximum rate of increase per individual under ideal conditions is its _____ .
   - a. biotic potential
   - b. carrying capacity
   - c. environmental resistance
   - d. density control

6. _____ is a density-independent factor that influences population growth.
   - a. Resource competition
   - b. Infectious disease
   - c. Predation
   - d. Harsh weather

7. A life history pattern for a population is a set of adaptive traits such as _____ .
   - a. longevity
   - b. fertility
   - c. age at reproductive maturity
   - d. all of the above

Data Analysis Activities

Monitoring Iguana Populations In 1989, Martin Wikelski began a long-term study of marine iguana populations in the Galápagos Islands. He marked the iguanas on two islands—Genovesa and Santa Fe—and collected data on how their body size, survival, and reproductive rates varied over time. The iguanas eat algae and have no predators, so deaths typically result from food shortages, disease, or old age. His studies showed that the iguana populations decline during El Niño events, when water surrounding the islands heats up.

In January 2001, an oil tanker ran aground and leaked a small amount of oil into the waters near Santa Fe. Figure 40.17 shows the number of marked iguanas that Wikelski and his team counted just before the spill and about a year later.

1. Which island had more marked iguanas at the time of the first census?

2. How much did the population size on each island change between the first and second census?

3. Wikelski concluded that changes on Santa Fe were the result of the oil spill, rather than a factor common to both islands. How would the census numbers be different from those he observed if an adverse event had affected both islands?

Figure 40.17 Shifting numbers of marked marine iguanas on two Galápagos islands. An oil spill occurred near Santa Fe just after the January 2001 census (*blue* bars). A second census was carried out in December 2001 (*green* bars).

8. The human population is now over 6.8 billion. It reached 6 billion in _____ .
   a. 2007   b. 1999   c. 1802   d. 1350

9. Compared to the less developed countries, the highly developed ones have a higher _____ .
   a. death rate   c. total fertility rate
   b. birth rate   d. resource consumption rate

10. An increase in infant mortality will _____ a population's replacement fertility rate.
   a. raise   b. lower   c. not affect

11. Species that usually colonize empty habitats are more likely to show traits that are favored by _____ .
   a. *r*-selection   b. *K*-selection

12. All members of a cohort are the same _____ .
   a. sex   b. size   c. age   d. weight

13. Match each term with its most suitable description.
   ___carrying   a. maximum rate of increase per
      capacity   individual under ideal conditions
   ___exponential   b. population growth plots out
      growth   as an S-shaped curve
   ___biotic   c. maximum number of individuals
      potential   sustainable by the resources
   ___limiting   in a given environment
      factor   d. population growth plots out
   ___logistic   as a J-shaped curve
      growth   e. essential resource that restricts
                           population growth when scarce

Additional questions are available on **CENGAGENOW**.

Critical Thinking

1. When researchers moved guppies from pools with cichlids that eat large guppies to pools with killifish that eat small ones, life history traits were not the only traits that changed. Over generations, the males became more colorful. Why do you think this change occurred?

2. Each summer, a giant saguaro cactus produces tens of thousands of tiny black seeds. Most die, but the few that land in a sheltered spot sprout the following spring. The saguaro is a slow-growing CAM plant (Section 6.8). After fifteen years, it may be only knee high, and it will not flower for another fifteen years. It may live for 200 years. Saguaros share their habitat with annuals such as poppies, which sprout, form seeds, and die in just a few weeks. Speculate on how these different life histories can both be adaptive in the same desert environment.

3. Age structure diagrams for two hypothetical populations are shown at *right*. Describe the growth rate of each population and discuss the current and future social and economic problems that each is likely to face.

Animations and Interactions on **CENGAGENOW**:
❯ Mark–recapture sampling; Exponential growth; Logistic growth; Age structure; Demographic transition.

‹ Links to Earlier Concepts

In this chapter you will revisit biogeography and take a closer look at global patterns in species richness (Sections 16.2, 16.4). You will be reminded of the effects of pathogens (19.3) and how species interactions lead to natural selection (17.4, 17.13, 21.7). You will see examples of field experiments (1.7) and apply what you know about populations and factors affecting their growth (40.2–40.4).

Key Concepts

Community Characteristics

A community consists of all species in a habitat. A habitat's history, its biological and physical characteristics, and interactions among species in the habitat affect the number of species in the community and their relative abundance.

Forms of Species Interactions

Commensalism, mutualism, competition, predation, and parasitism are interspecific interactions. They influence the population size of participating species, which in turn influences the community's structure.

41 Community Ecology

Red imported fire ants, *Solenopsis invicta*, are native to South America. They arrived in the United States in the 1930s, probably as stowaways on a cargo ship. The ants spread out from the Southeast and in time established colonies as far west as California and as far north as Kansas and Delaware.

Like most ants, *S. invicta* nests in the ground (Figure 41.1). Accidentally step on one of these nests, and you will quickly realize your mistake. Like honeybees, fire ants defend their nest against perceived threats by stinging. Venom injected by the stinger causes a burning sensation, and results in formation of reddened bumps that are slow to heal.

S. invicta nests are an annoyance to people, but they pose a greater threat to wildlife. For example, researchers have demonstrated that the presence of *S. invicta* decreases the number of bobwhites (a type of quail) and vireos (a type of songbird). The ants feed on the birds' eggs and on nestlings. *S. invicta* also disrupts populations of Texas horned lizards. The invasive ants outcompete native ants that are the lizard's main source of food. The lizard cannot eat *S. invicta*. In addition, *S. invicta* can prey on the lizard's eggs.

Invicta means "invincible" in Latin, and *S. invicta* lives up to its name. Pesticides do not slow the spread of this species and may facilitate invasions by wiping out native ant populations. To fight the ants, scientists have turned to biological controls. A biological control agent is a natural enemy of a pest species. It can decrease pest numbers by preying on, infecting, or parasitizing the pest.

Biological control of *S. invicta* involves phorid flies, insects that prey on *S. invicta* in their native habitat. Phorid flies are parasitoids, specialized parasites that kill their host in a rather gruesome way. A female fly pierces the cuticle of an adult ant, then lays an egg in the ant's soft tissues. The egg hatches into a larva, which grows and then eats its way through the tissues to the ant's head. After the larva gets big enough, it secretes an

Figure 41.1 Red imported fire ants and their foes. *Opposite*, nest mounds of fire ants in a Texas pasture. **A** Phorid fly. This fly uses the hooked extension on its abdomen to insert a fertilized egg into an ant's thorax. **B** Parasitized ant that lost its head after a developing fly larva moved into it. The larva will undergo metamorphosis to an adult inside the detached head.

enzyme that makes the ant's head fall off. The fly larva develops into an adult within the shelter of the detached ant head.

Several phorid fly species have been introduced in various southern states. The flies seem to be surviving, reproducing, and increasing their range. They are not expected to kill off all *S. invicta* in affected areas. Rather, the hope is that the flies will reduce the density of invading colonies.

Ecologists are also exploring other options. They are testing effects of imported pathogenic fungi or protists that infect *S. invicta* but not native ants. Another idea is to introduce a parasitic South American ant that invades *S. invicta* colonies and kills the egg-laying queens.

Species interactions such as those between ants and flies or fungi are the focus of community ecology. A **community** is all the species that live in a region. As you will see, species interactions and disturbances can shift community structure (the types of species and their relative abundances) in small and large ways, some predictable, and others unexpected.

---

community All species that live in a particular region.

Long-Term Change in Communities
The array of species in a community changes over time, although the exact outcome of these changes is difficult to predict. When a new community forms, the early-arriving species often alter the habitat in a way that facilitates their own replacement.

Species Effects on Community Stability
Removing a species from a community or adding one to it can have a dramatic effect on other species. Some species are adapted to disturbances and a change in the frequency of disturbances can affect their number.

Global Patterns in Community Structure
Biogeographers identify regional patterns in species distribution. They have shown that tropical regions hold the greatest number of species and that characteristics of islands can be used to predict how many species an area will hold.

> Community structure refers to the number and relative abundances of species in a habitat. It changes over time.
< Link to Coevolution 17.13

The type of place where a species normally lives is its **habitat**, and all species living in a habitat constitute a community. Communities often are nested one inside another. For example, we find a community of microbial organisms inside the gut of a termite. That termite is part of a larger community of organisms living on a fallen log. The log-dwellers are part of a larger forest community.

Even communities that are similar in scale differ in their species diversity. There are two components to species diversity. The first, species richness, refers to the number of species. The second is species evenness, or the relative abundance of each species. For example, a pond that has five fish species in nearly equal numbers has a higher species diversity than a pond with one abundant fish species and four rare ones.

Community structure is dynamic. The array of species and their relative abundances change over time. Communities change over a long time span as they form and then age. They also change suddenly as a result of natural or human-induced disturbances.

Physical factors such as climate and resource availability influence community structure. Habitats that are nutrient-poor and have extreme conditions support fewer species than those that are nutrient-rich and have moderate conditions.

Species interactions also influence the types of species in a community and their relative abundances. In some cases, the effect is indirect. For example, when songbirds

Table 41.1   Direct Two-Species Interactions

Type of Interaction	Effect on Species 1	Effect on Species 2
Commensalism	Helpful	None
Mutualism	Helpful	Helpful
Interspecific competition	Harmful	Harmful
Predation, herbivory parasitism, parasitoidism	Helpful	Harmful

eat caterpillars, the birds indirectly benefit the trees that the caterpillars feed on, while directly reducing the abundance of caterpillars.

We can group direct species interactions by their effects on both participants (Table 41.1). For example, **commensalism** helps one species and has no effect on the other. Commensal ferns live attached to the trunk or branches of a tree (Figure 41.2). Having a perch in the light benefits the fern, and the tree is unaffected. Relationships are considered commensal when one species benefits, and the other neither benefits nor is harmed by the relationship. Should evidence of either effect come to light, the relationship is reclassified.

The other types of species interaction are discussed in detail in sections that follow. Here we will simply note that interactions can be mutually beneficial, mutually harmful, or benefit one species while harming the other.

Species interactions may be fleeting or a long-term relationship. **Symbiosis** means "living together" and biologists use the term to refer to a relationship in which two species have a prolonged close association.

Regardless of whether one species helps or harms another, two species that interact closely for generations can coevolve. As Section 17.13 explained, coevolution is an evolutionary process in which each species acts as a selective agent that shifts the range of variation in the other.

---

commensalism Species interaction that benefits one species and neither helps nor harms the other.
habitat Type of environment in which a species typically lives.
symbiosis One species lives in or on another in a commensal, mutualistic, or parasitic relationship.

Figure 41.2 A tree with a commensal fern. The fern benefits by growing on the tree, which is unaffected by the presence of the fern.

Take-Home Message **What factors affect the types and abundances of species in a community?**

> The types and abundances of species in a community are affected by physical factors such as climate and by species interactions.

> A species can be benefited, harmed, or unaffected by its interaction with another species.

❭ In a mutualistic interaction, two species benefit by taking advantage of one another.
❬ Links to Endosymbiosis 18.6, Lichens and mycorrhizae 22.6, Pollinators 21.7

Mutualism is an interspecific interaction that benefits both species. For example, birds, insects, bats, and other animals pollinate flowering plants. Pollinators eat nectar and pollen. In return, they transfer pollen between plants of the same species. Similarly, many animals that eat fruits disperse the seeds inside the fruits to new sites.

In some mutualisms, neither species can complete its life cycle without the other. Yucca plants and the moths that pollinate them have such an interdependence (Figure 41.3). In other cases, the mutualism is helpful but not a life-or-death requirement. Most plants, for example, have more than one pollinator.

Mutualistic microorganisms help plants obtain the nutrients they require. Nitrogen-fixing bacteria living on roots of legumes such as peas provide the plant with extra nitrogen. Mycorrhizal fungi living in or on plant roots enhance a plant's mineral uptake. Other fungi interact with photosynthetic algae or bacteria in lichens.

There is some conflict between the partners in all mutualisms. For example, in a lichen, the fungus would benefit by getting as much sugar as possible from its photosynthetic partner. The photosynthesizer would benefit by keeping as much sugar as possible for its own use. In each species, natural selection will favor individuals who minimize their cost and maximize their benefit.

For some mutualists, the main benefit is defense. A sea anemone has stinging cells (nematocysts), so most fishes avoid its tentacles. However, an anemone fish can nestle among the tentacles in safety (Figure 41.4). A mucus layer shields the anemone fish from stings, and the tentacles keep it safe from predatory fish. The anemone fish repays its partner by chasing off the few fishes that are able to eat sea anemone tentacles.

Finally, think back on an idea introduced in Section 18.6. According to the theory of endosymbiosis, mutualistic aerobic bacteria that lived in early eukaryotic cells evolved into mitochondria. Cyanobacteria living inside eukaryotic cells evolved into chloroplasts by a similar process. Thus, all eukaryotes are the product of ancient mutualisms.

---

mutualism Species interaction that benefits both species.

Take-Home Message What is a mutualism?

❭ A mutualism is a species interaction in which each species benefits by associating with the other.

❭ In each species, individuals who maximize the benefit they receive while minimizing their costs will be favored.

Figure 41.3 Mutualism on a rocky slope of the high desert in Colorado.

Only one yucca moth species pollinates plants of each *Yucca* species; each moth cannot complete its life cycle with any other plant. The moth matures when yucca flowers blossom. The female has specialized mouthparts that collect and roll sticky pollen into a ball. She flies to another flower and pierces its ovary, where seeds will form and develop, and lays eggs inside. As the moth crawls out, she pushes a ball of pollen onto the flower's pollen-receiving platform.

After pollen grains germinate, they give rise to pollen tubes, which grow through the ovary tissues and deliver sperm to the plant's eggs. Seeds develop after fertilization.

Meanwhile, moth eggs develop into larvae that eat a few seeds, then gnaw their way out of the ovary. Seeds that larvae do not eat give rise to new yucca plants.

Figure 41.4 The sea anemone *Heteractis magnifica* has a mutualistic association with the pink anemone fish (*Amphiprion perideraion*). This tiny but aggressive fish chases away predatory butterfly fishes that would bite off tips of anemone tentacles. The fish cannot survive and reproduce without the protection of an anemone. The anemone does not need a fish to protect it, but it does better with one.

> Resources are limited and individuals of different species often compete for access to them.
> Competition adversely affects both species.
< Links to Darwin's theory 16.4, Directional selection 17.5

As Charles Darwin clearly understood, competition for resources among individuals of the same species can be intense and leads to evolution by natural selection. **Interspecific competition**, competition among members of different species, is not usually as intense. The requirements of two species might be similar, but they are never as close as they are for members of one species.

Each species has a unique set of ecological requirements and roles that we refer to as its **ecological niche**. Both physical and biological factors define the niche. Aspects of an animal's niche include the temperature range it can tolerate, the species it eats, and the places it can breed. A description of a flowering plant's niche would include its soil, water, light, and pollinator requirements. The more similar the niches of two species are, the more intensely the species will compete.

Competition takes two forms. With interference competition, one species actively prevents another from accessing some resource. As an example, one species of scavenger will often chase another away from a carcass (Figure 41.5). As another example, some plants use chemical weapons against potential competition. Aromatic chemicals that ooze from tissues of sagebrush plants, black walnut trees, and eucalyptus trees seep into the soil around these plants. The chemicals prevent other kinds of plants from germinating or growing.

In exploitative competition, species do not interact directly; each reduces the amount of resources available to the other by using that resource. For example, deer and blue jays both eat acorns in oak forests. The more acorns the birds eat, the fewer there are for the deer.

Effects of Competition

Deer and blue jays share a fondness for acorns, but each also has other sources of food. Any two species differ in their resource requirements. Species compete most intensely when the supply of a shared resource is the main limiting factor for both (Section 40.4).

In the 1930s, G. Gause conducted experiments with two species of ciliated protozoans (*Paramecium*) that compete for the same prey: bacteria. He cultured the *Paramecium* species separately and together (Figure 41.6). Within weeks, population growth of one species outpaced the other, which went extinct.

Experiments by Gause and others are the basis for the concept of **competitive exclusion**: Whenever two species require the same limited resource to survive or reproduce, the better competitor will drive the less competitive species to extinction in that habitat.

When resource needs of competitors are not exactly the same, species can coexist, but competition reduces their numbers. For example, Nelson Hairston showed the effects of competition between two species of salamanders that coexist in woodlands. He created experimental plots by removing all members of one species or the other. Another group of plots served as unaltered controls. After five years, the number of salamanders in both types of single-species plots had increased, whereas the number in control plots was unchanged. Hairston concluded that where these salamanders coexist, competitive interactions suppress the population growth of both.

A Golden eagle and a red fox face off over a moose carcass.

Figure 41.5 Interspecific competition among scavengers.

B In a dramatic demonstration of interference competition, the eagle attacks the fox with its talons. After this attack, the fox retreated, leaving the eagle to exploit the carcass.

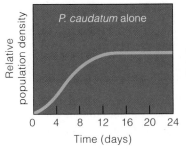

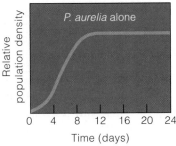

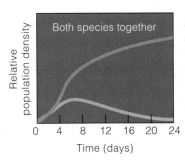

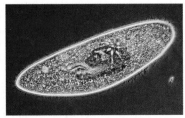

Paramecium

Figure 41.6 Animated Results of competitive exclusion between two *Paramecium* species that compete for the same food. When the two species were grown together, *P. aurelia* drove *P. caudatum* to extinction.

Resource Partitioning

Resource partitioning is an evolutionary process by which species become adapted to use a shared limiting resource in a way that minimizes competition. For example, three species of annual plants are common in abandoned fields. All require water and nutrients, but each is adapted to meet these needs in a slightly different way (Figure 41.7). This variation allows the plants to coexist.

Similarly, eight species of woodpecker coexist in Oregon forests. All feed on insects and nest in hollow trees, but the details of their foraging behavior and nesting preferences vary. Differences in nesting time also help reduce competitive interactions.

Resource partitioning occurs because directional selection occurs when species with similar requirements share a habitat and compete for a limiting resource. In each species, those individuals who differ most from the competing species have the least competition and thus leave the most offspring. Over generations, directional selection leads to **character displacement**: The range of variation for one or more traits is shifted in a direction that lessens the intensity of competition for a limiting resource.

A recently observed directional shift in the beak dimensions of the medium ground finch (*Geospiza fortis*) shows how competition can result in character displacement. The medium ground finch was the only seed-eating bird on one of the Galápagos islands until some large ground finches (*G. magnirostris*) arrived. When a drought caused a food shortage, the medium ground finches with the largest beaks found themselves at a selective disadvantage. These big-beaked, medium-sized birds had to compete with the large new arrivals for the few big seeds. Small-beaked birds had no competition for the smaller seeds they favor. As a result, the large-beaked medium finches were more likely to die of starvation. The differential deaths caused a shift in the average beak size of the medium finch population; beaks became significantly shorter and shallower.

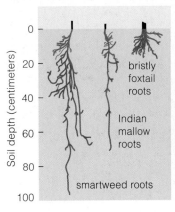

bristly foxtail

Indian mallow   smartweed

Figure 41.7 Resource partitioning among three annual plant species in a plowed but abandoned field.

Roots of each species take up water and mineral ions from a different soil depth. This difference reduces competition among the species and allows them to coexist.

character displacement Outcome of competition between two species; similar traits that lead to competition become dissimilar.
competitive exclusion Process whereby two species compete for a limiting resource, and one drives the other to local extinction.
ecological niche All of a species' requirements and roles in an ecosystem.
interspecific competition Competition between two species.
resource partitioning Species become adapted in different ways to access different portions of a limited resource; allows species with similar needs to coexist.

Take-Home Message What happens when species compete?

❯ In some competitive interactions, one species controls or blocks access to a resource, regardless of whether it is scarce or abundant. In other interactions, one is better than another at exploiting a shared resource.

❯ When two species compete, individuals whose needs are least like those of the competing species are favored.

> Predation and herbivory are short-term interactions in which one species obtains nutrients and energy by feeding on another.
< Links to Ricin 9.1, Coevolution 17.13

Predator and Prey Abundance

With **predation**, one species (the predator) captures, kills, and digests another species (the prey). The abundance of prey species in a community affects how many predators it can support. The number of predators reduces the number of prey, but the extent of this effect depends partly on how the predator responds to changes in prey density.

With some predators, such as web-spinning spiders, the proportion of prey killed is constant, so the number killed in any given interval depends solely on prey density. As the number of flies in an area increases, more and more become caught in a web.

More often, the number of prey killed depends in part on the time it takes predators to capture, eat, and digest prey. As prey density increases, the rate of kills rises steeply at first because there are more prey to catch. Eventually, the rate of increase slows, because a predator is exposed to more prey than it can handle at one time. A wolf that just killed a caribou will not hunt another until it has eaten and digested the first one.

Predator and prey populations sometimes rise and fall in a cyclical fashion. For example, Figure 41.8 shows historical data for the numbers of lynx and their main prey, the snowshoe hare. Both populations rise and fall over an approximately ten-year cycle, with predator abundance lagging behind prey abundance. Field studies indicate that lynx numbers fluctuate mainly in response to hare numbers. However, the size of the hare population is affected by the abundance of the hare's food as well as the number of lynx. Hare populations continued to rise and fall even when predators were experimentally excluded from areas.

Coevolution of Predators and Prey

Predator and prey exert selection pressure on one another. Suppose a mutation gives members of a prey species a more effective defense. Over generations, directional selection will cause this mutation to spread through the prey population. If some members of a predator population have a trait that makes them better at thwarting the improved defense, these predators and their descendants will be at an advantage. Thus, predators exert selection pressure that favors improved prey defense, which in turn exerts selection pressure on predators, and so it goes over many generations.

You have already learned about some defensive adaptations. Many prey species have hard or sharp parts that make them difficult to eat. Think of a snail's shell or a sea urchin's spines. Others contain chemicals that taste bad or sicken predators. Most defensive toxins in animals come from the plants they eat. For example, a monarch butterfly caterpillar takes up chemicals from the milkweed plant that it feeds on. A bird that later eats the butterfly will be sickened by these chemicals.

Well-defended prey often have warning coloration, a conspicuous color pattern that predators learn to avoid. For example, many species of stinging wasps and bees have a pattern of black and yellow stripes (Figure 41.9A). The similar appearance of bees and wasps is an example of **mimicry**, an evolutionary pattern in which one species comes to resemble another. The bees and wasps benefit from their similar appearance. The more often a predator is stung by a black and yellow striped insect, the less likely it is to attack a similar insect.

In another type of mimicry, prey masquerade as a species that has a defense that they lack. For example, some

Figure 41.8 Graph of the number of Canadian lynx (*dashed* line) and snowshoe hares (*solid* line), based on counts of pelts sold by trappers to Hudson's Bay Company during a ninety-year period.

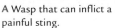
A Wasp that can inflict a painful sting.

B Fly, with no stinger, mimics wasp color pattern.

Figure 41.9 Mimicry. Some edible insect species resemble toxic or unpalatable species that are not closely related.

B Frilly pink body parts of a flower mantis hide it from insect prey attracted to the real flowers.

flies, which cannot sting, resemble stinging bees or wasps (Figure 41.9B). The fly benefits when predators avoid it after a run-in with the better-defended look-alike species.

Another adaptation to predation is a last-chance trick that can startle an attacking predator. Section 1.7 described how eye spots and a hissing sound protect some butterflies. Similarly, a lizard's tail may detach from the body and wiggle a bit as a distraction. Skunks squirt a foul-smelling, irritating repellent.

Many prey have camouflage—a form, patterning, color, and often behavior that allows them to blend into their surroundings and avoid detection (Figure 41.10A).

Predators too benefit from camouflage (Figure 41.10B,C). Other predator adaptations include sharp teeth and claws that can pierce protective hard parts. Speedy prey select for faster predators. For example, the cheetah, the fastest land animal, can run 114 kilometers per hour (70 mph). Its preferred prey, Thomson's gazelles, run 80 kilometers per hour (50 mph).

A Reedlike defensive posture of a bittern. The bird even sways in the wind like a reed.

C Fleshy protrusions give a predatory scorpionfish the appearance of an algae-covered rock. Algae-eating fish that come close for a nibble end up as prey.

Figure 41.10 Camouflage in prey and predators.

Coevolution of Herbivores and Plants

With **herbivory**, an animal eats a plant or plant parts. The number and type of plants in a community can influence the number and type of herbivores present.

Two types of defenses have evolved in response to herbivory. Some plants have adapted to withstand and recover quickly from herbivory. For example, prairie grasses are seldom killed by native grazers such as bison. The grasses have a fast growth rate and store enough resources in roots to grow back lost shoots.

Other plants have traits that deter herbivory. Such traits include spines or thorns, tough leaves that are difficult to chew or digest, and chemicals that taste bad or sicken herbivores. Ricin, the toxin made by castor bean

plants (Section 9.1), makes herbivores ill. Caffeine in coffee beans and nicotine in tobacco leaves are evolved defenses against insects.

The existence of plant defenses favors herbivores capable of overcoming those defenses. For example, eucalyptus leaves contain toxins that make them poisonous to most mammals, but not to koalas. Specialized liver enzymes allow koalas to break down toxins made by a few eucalyptus species.

---

herbivory An animal feeds on plant parts.
mimicry A species evolves traits that make it more similar in appearance to another species.
predation One species captures, kills, and eats another.

Take-Home Message What are the evolutionary effects of predation and herbivory?

❯ Predation benefits predators and harms prey. Predator numbers sometimes fluctuate in response to prey availability.

❯ Predators can coevolve with their prey. Herbivores coevolve with the plants that they eat.

❯ With mimicry, one species benefits by its resemblance to another species.

> Some species benefit by withdrawing nutrients from other species, or by tricking them into providing parental care.
< Links to Lamprey 24.3, Cuckoo 39.3

Parasitism

With **parasitism**, one species (the parasite) benefits by feeding on another (the host), without immediately killing it. Endoparasites such as parasitic roundworms live and feed inside their host (Figure 41.11A). An ectoparasite such as a tick feeds while attached to a host's external surface (Figure 41.11B).

A parasitic life-style has evolved in members of a diverse variety of groups. Bacterial, fungal, protistan, and invertebrate parasites feed on vertebrates. Lampreys (Section 24.3) attach to and feed on other fish. There are even a few parasitic plants that withdraw nutrients from other plants (Figure 41.12).

Parasites usually do not kill their host immediately. In terms of evolutionary fitness, killing a host too fast is bad for the parasite. Ideally, a host will live long enough to give a parasite time to produce some offspring. The

Figure 41.12 Dodder (*Cuscuta*), also known as strangleweed or devil's hair. This parasitic flowering plant has almost no chlorophyll. Leafless stems wrap around a host plant, and modified roots absorb water and nutrients from the host's vascular tissue.

A Endoparasitic roundworms feeding in the intestine of a host pig.

B Ectoparasitic ticks attached to and sucking blood from a finch.

Figure 41.11 Parasites inside and out.

longer the host survives, the more parasite offspring can be produced. Thus parasites with less-than-fatal effects on hosts are at a selective advantage.

Although parasites typically do not kill their hosts, they can still have an important impact on a host population. Many parasites are pathogens; they cause disease in hosts. Even when a parasite does not cause obvious symptoms, its presence can weaken a host so it is more vulnerable to predation or less attractive to potential mates. Some parasites cause their host to become sterile. Others shift the sex ratio among their host's offspring.

Adaptations to a parasitic life-style include traits that allow the parasite to locate hosts and to feed undetected. For example, ticks that feed on mammals or birds will move toward a source of heat and carbon dioxide, which may be a potential host. A chemical in tick saliva acts as a local anesthetic, preventing the host from noticing the feeding tick. Endoparasites often have traits that help them evade a host's immune defenses.

Among hosts, traits that minimize the negative effects of parasites confer a selective advantage. For example, the allele that causes sickle-cell anemia persists at high levels in some human populations because having one copy of the allele increases the odds of surviving malaria. Grooming and preening behavior are adaptations that minimize the impact of ectoparasites. Some animals produce chemicals that interfere with parasite activity. For example, crested auklets (a type of seabird) produce a citrus-scented secretion that repels ticks.

Strangers in the Nest

With **brood parasitism**, one egg-laying species benefits by having another raise its offspring. The European cuckoos described in Section 39.3 are brood parasites, as are North American cowbirds (Figure 41.13). Not having to invest in

Figure 41.13 A cowbird with its foster parent. A female cowbird minimizes the cost of parental care by laying her eggs in the nests of other bird species.

Figure 41.14 Biological control agent: a commercially raised parasitoid wasp about to deposit a fertilized egg in an aphid. This wasp reduces aphid populations. After the egg hatches, a wasp larva devours the aphid from the inside.

parental care allows a female cowbird to produce a large number of eggs, in some cases as many as thirty in a single reproductive season.

The presence of brood parasites decreases the reproductive rate of the host species and favors host individuals that can detect and eject foreign young. Some brood parasites counter this host defense by producing eggs that closely resemble those of the host species. In cuckoos, different subpopulations have different host preferences and egg coloration. Females of each subpopulation lay eggs that closely resemble those of their preferred host.

Brood parasitism also evolved in some bee species. Females of these species lay eggs in the nest of a different bee species. Larvae of the parasitic bee eat food stored by the host for its larvae, as well as the host's larvae.

Parasitoids

Parasitoids are insects that lay their eggs in other insects. Larvae hatch, develop in the host's body, eat its tissue, and eventually kill it. The fire ant–killing phorid flies described in Section 41.1 are an example. As many as 15 percent of all insects may be parasitoids.

Parasitoids reduce the size of a host population in two ways. First, as the parasitoid larvae grow inside their host, they withdraw nutrients and prevent it from reproducing. Second, the presence of these larvae eventually leads to the death of the host.

---

brood parasitism One egg-laying species benefits by having another raise its offspring.
parasitism Relationship in which one species withdraws nutrients from another species, without immediately killing it.
parasitoid An insect that lays eggs in another insect, and whose young devour their host from the inside.

Biological Controls

Parasites and parasitoids are commercially raised and released in target areas as biological control agents. They are promoted as an environmentally friendly alternative to pesticides. Most chemical pesticides kill a wide variety of insects. By contrast, biological control agents usually target a limited number of species. The phorid flies that attack fire ants and the parasitoid wasp in Figure 41.14 are examples of biological control agents.

For a species to be an effective biological control agent, it must be adapted to take advantage of a specific host species and to survive in that species' habitat. The ideal biological control agent is good at finding hosts, has a population growth rate comparable to the host's, and has offspring that are good at dispersing.

Introducing a species into a community as a biological control does entail some risks. The introduced parasites sometimes go after nontargeted species in addition to, or instead of, those species they were expected to control. For example, parasitoids were introduced to Hawaii to control stink bugs that feed on some Hawaiian crops. Instead, the parasitoids decimated the population of koa bugs, Hawaii's largest native bug. Introduced parasitoids have also been implicated in ongoing declines of many native Hawaiian butterfly and moth populations.

Take-Home Message **What are the effects of parasites, brood parasites, and parasitoids?**

❭ Parasites reduce the reproductive rate of host individuals by withdrawing nutrients from them.

❭ Brood parasites reduce the reproductive rate of hosts by tricking them into caring for young that are not their own.

❭ Parasitoids reduce the number of host organisms by preventing reproduction and eventually killing the host.

> The array of species in a community depends on physical factors such as climate, biological factors such as which species arrived first, and the frequency of disturbances.
< Links to Mosses 21.3, Lichens 22.6

Successional Change

Species composition of a community can change over time. Species often alter the habitat in ways that allow other species to come in and replace them. We call this type of change ecological succession.

The process of succession starts with the arrival of **pioneer species**, species that are opportunistic colonizers of new or newly vacated habitats. Pioneer species have high dispersal rates, grow and mature fast, and produce many offspring. Later, other species replace the pioneers. Then the replacements are replaced, and so on.

Primary succession is a process that begins when pioneer species colonize a barren habitat with no soil, such as a new volcanic island or land exposed by the retreat of a glacier (Figure 41.15). The earliest pioneers to colonize a new habitat are often mosses and lichens (Sections 21.3 and 22.6), which are small, have a brief life cycle, and can tolerate intense sunlight, extreme temperature changes, and little or no soil. Some hardy annual flowering plants with wind-dispersed seeds are also typical pioneers.

Pioneers help build and improve the soil. In doing so, they often set the stage for their own replacement. Many pioneer species partner with nitrogen-fixing bacteria, so they can grow in nitrogen-poor habitats. Seeds of later species find shelter inside mats of the pioneers. Organic wastes and remains accumulate and, by adding volume and nutrients to soil, this material helps other species take hold. Later successional species often shade and eventually displace earlier ones.

In **secondary succession**, a disturbed area within a community recovers. If improved soil is still present, secondary succession can occur fast. It commonly occurs in abandoned agricultural fields and burned forests.

Factors That Influence Succession

When the concept of ecological succession was first developed in the late 1800s, it was thought to be a predictable and directional process. Physical factors such as climate, altitude, and soil type were considered to be the main determinants of which species appeared in what order during succession. Also by this view, succession culminates in a "climax community," an array of species that will persist over time and will be reconstituted in the event of a disturbance.

Ecologists now realize that the species composition of a community changes frequently, and in unpredictable ways. Communities do not journey along a well-worn path to some predetermined climax state.

Random events can determine the order in which species arrive in a habitat and thus affect the course of succession. The arrival of a certain species may make it easier or more difficult for others to take hold. As an example, surf grass can grow along a shoreline only if algae have

Figure 41.15 In Alaska's Glacier Bay region, one observed pathway of primary succession. **A** As a glacier retreats, meltwater leaches minerals from the rocks and gravel left behind. **B** Pioneer species include lichens, mosses, and some flowering plants such as mountain avens, which associate with nitrogen-fixing bacteria. Within 20 years, alder, cottonwood, and willow seedlings take hold. Alders also have nitrogen-fixing symbionts. **C** Within 50 years, the alders form dense, mature thickets. **D** After 80 years, western hemlock and spruce crowd out alders. **E** In areas deglaciated for more than a century, tall Sitka spruce dominate.

Figure 41.16 A natural laboratory for succession after the 1980 Mount Saint Helens eruption. **A** The community at the base of this volcano in the Cascades was destroyed. **B** In less than a decade, pioneer species had sprouted. **C** Twelve years later, seedlings of a dominant species, Douglas firs, took hold.

already colonized that area. The algae act as an anchoring site for the grass. In contrast, when sagebrush gets established in a dry habitat, chemicals it secretes into the soil keep most other plants out.

Ecologists had an opportunity to investigate these factors after the 1980 eruption of Mount Saint Helens leveled about 600 square kilometers (235 square miles) of forest in Washington State (Figure 41.16). Ecologists recorded the natural pattern of colonization and carried out experiments in plots inside the blast zone. The results showed that the presence of some pioneers helped other, later-arriving plants become established. Other pioneers kept the same late arrivals out.

Disturbances also influence species composition in communities. According to the **intermediate disturbance hypothesis**, species richness is greatest in communities where disturbances are moderate in their intensity or frequency. In such habitats, there is enough time for new colonists to arrive and become established but not enough for competitive exclusion to cause extinctions:

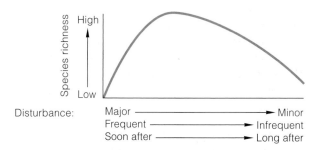

In short, the modern view of succession holds that the species composition of a community is affected by (1) physical factors such as soil and climate, (2) chance events such as the order in which species arrive, and (3) the extent of disturbances in the habitat. Because the second and third factors often vary even between two geographically close regions, it is generally difficult to predict exactly what any given community will look like in the future.

---

intermediate disturbance hypothesis Species richness is greatest in communities where disturbances are moderate in their intensity or frequency.
pioneer species Species that can colonize a new habitat.
primary succession A new community becomes established in an area where there was previously no soil.
secondary succession A new community develops in a site where the soil that supported an old community remains.

Take-Home Message What is ecological succession?

❯ Succession is a process in which one array of species replaces another over time. It can occur in a barren habitat (primary succession), or a region where a community previously existed (secondary succession).

❯ Physical factors affect succession, but so do species interactions and disturbances. As a result, the course that succession will take in any particular community is difficult to predict.

> The loss or addition of even one species may destabilize the number and abundances of species in a community.
> Some species adapted to being disturbed are at a competitive disadvantage if the disturbance does not occur.

The Role of Keystone Species

A **keystone species** is a species that has a disproportionately large effect on a community relative to its abundance. Robert Paine was the first to describe the effect of a keystone species after his experiments on the rocky shores of California's coast. Species in the rocky intertidal zone withstand pounding surf by clinging to rocks. A rock to cling to is a limiting factor. Paine set up control plots with the sea star *Pisaster ochraceus* and its main prey—chitons, limpets, barnacles, and mussels. Then he removed all sea stars from his experimental plots.

Sea stars prey mainly on mussels. With sea stars gone from experimental plots, mussels took over, crowding out seven other species of invertebrates. Paine concluded that sea stars are a keystone species. They normally keep the number of prey species in the intertidal zone high by preventing competitive exclusion by mussels.

The impact of a keystone species can vary between habitats that differ in their species arrays. For example, Jane Lubchenco discovered that periwinkle snails (Figure 41.17A) can increase or decrease the diversity of algal species, depending on the habitat. In tidepools, periwinkles

prefer to eat a certain alga that can outgrow other algal species. By keeping the best algal competitor in check, periwinkles help other, less competitive algal species survive (Figure 41.17B). On rocks of the lower intertidal zone, a tough red alga dominates. Here, periwinkles preferentially graze on competitively weaker algae, giving the dominant competitor an added advantage (Figure 41.17C). Thus, periwinkles promote species diversity in tidepools but reduce it on intertidal rocks.

Keystone species need not be predators. For example, beavers can be a keystone species. These large rodents cut down trees by gnawing through their trunks. The beaver then uses the felled trees to build a dam, thus creating a deep pool where a shallow stream would otherwise exist. By altering the physical conditions in a section of the stream, the beaver affects the types of fish and aquatic invertebrates that can live there.

Adapting to Disturbance

In communities repeatedly subjected to a particular type of disturbance, individuals that withstand or benefit from that disturbance have a selective advantage. For example, some plants in areas subject to periodic fires produce seeds that germinate only after a fire. Seedlings of these plants benefit from the lack of competition for resources. Other plants have an ability to resprout quickly after a fire (Figure 41.18).

Because different species respond differently to fire, the frequency of fire affects competitive interactions. When naturally occurring fires are suppressed, species adapted to fire lose their competitive edge. They can be overgrown by species that devote all of their energy to

A Periwinkle snails (*Littorina littorea*) are a keystone species that affects the number of algal species in different ways in different tidal habitats.

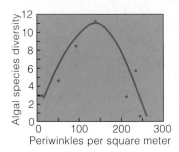

B In tidepools, periwinkles eat the alga that is the most effective competitor and thus promote survival of less competitive algal species that would otherwise be overgrown.

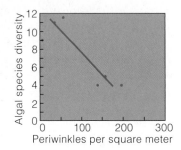

C On rocks exposed only at high tide, periwinkles eat algae that are less effective competitors, thus enhancing the growth of the most competitive species.

Figure 41.17 Effect of competition and predation in an intertidal zone.

Figure 41.18 Adapted to disturbance. Some woody shrubs, such as this toyon, resprout from their roots after a fire. In the absence of occasional fire, toyons are outcompeted and displaced by species that grow faster but are less fire resistant.

A Kudzu native to Asia is overgrowing trees across the southeastern United States.

B Gypsy moths native to Europe and Asia feed on oaks through much of the United States.

C Nutrias native to South America are abundant in freshwater marshes of the Gulf States.

Figure 41.19 Three exotic species that are altering natural communities in the United States. To learn more about invasive species in the United States, visit the National Invasive Species Information Center online at www.invasivespeciesinfo.gov.

growing shoots, rather than storing some in underground parts in readiness for a spurt of regrowth.

Some species are especially sensitive to disturbances to the environment. These **indicator species** are the first to do poorly when conditions change, so they can provide an early warning of environmental degradation. For example, a decline in a trout population can be an early sign of problems in a stream, because trout are highly sensitive to pollutants and cannot tolerate low oxygen levels.

Species Introductions

The arrival of a new species in a community can also cause dramatic changes. When you hear someone speaking enthusiastically about exotic species, you can safely bet the speaker is not an ecologist. An **exotic species** is a resident of an established community that dispersed from its home range and became established elsewhere. Unlike most imports, which never take hold outside the home range, an exotic species becomes a permanent member of its new community.

More than 4,500 exotic species have become established in the United States. An estimated 25 percent of Florida's plant and animal species are exotics. In Hawaii, 45 percent are exotic. Some species were brought in for use as food crops, to brighten gardens, or to provide textiles. Other

species, including red imported fire ants, arrived as stowaways along with cargo from distant regions.

One of the most notorious exotic species is a vine called kudzu (*Pueraria lobata*). Native to Asia, it was introduced to the American Southeast as a food for grazers and to control erosion. It quickly became an invasive weed. Kudzu overgrows trees, telephone poles, houses, and almost everything else in its path (Figure 41.19A).

Gypsy moths (*Lymantria dispar*) are an exotic species native to Europe and Asia. They entered the northeastern United States in the mid-1700s and now range into the Southeast, Midwest, and Canada. Gypsy moth caterpillars (Figure 41.19B) preferentially feed on oaks. Loss of leaves to gypsy moths weakens trees, making them less efficient competitors and more susceptible to parasites.

Large semiaquatic rodents called nutrias (*Myocastor coypus*) were imported from South America for their fur, then released into the wild in the 1940s. In states bordering the Gulf of Mexico, nutrias now thrive in freshwater marshes (Figure 41.19C). Their voracious appetite threatens the native vegetation. In addition, their burrowing contributes to marsh erosion and damages levees, increasing the risk of flooding.

---

exotic species A species that evolved in one community and later became established in a different one.
indicator species Species that is especially sensitive to disturbance and can be monitored to assess the health of a habitat.
keystone species A species that has a disproportionately large effect on community structure.

Take-Home Message What types of changes alter community structure?

> Removing a keystone species can alter the diversity of species in a community.

> Changing the frequency of a disturbance can favor some species over others.

> Introducing an exotic species can threaten native species.

> ❯ The richness and relative abundances of species differ from one region of the world to another.
> ❮ Link to Biogeography 16.2

Biogeography is the study of how species are distributed in the natural world (Section 16.2). **Species richness**, the number of species in an area, correlates with differences in sunlight, temperature, rainfall, and other factors that vary with latitude, elevation, or water depth. The history of a habitat also affects species richness.

Latitudinal Patterns

Perhaps the most striking pattern of species richness corresponds with distance from the equator. For most major plants and animal groups, the number of species is greatest in the tropics and declines from the equator to the poles. Figure 41.20 illustrates one example of this pattern.

Consider just a few factors that help bring about this latitudinal pattern and maintain it. First, tropical latitudes intercept more intense sunlight and receive more rainfall, and their growing season is longer. Thus resource availability tends to be greater and more reliable in the tropics than elsewhere. One result is specialized interrelationships not possible where species are active for shorter periods. Second, tropical communities have been evolving for a long time. Some temperate communities did not start forming until the end of the last ice age. Third, species richness may be self-reinforcing. The number of tree species in tropical forests is much greater than in comparable forests at higher latitudes. Where more plant species compete and coexist, more species of herbivores can coexist, partly because no single herbivore species can overcome all chemical defenses of all plants. In addition, more predators and parasites can evolve in response to more kinds of prey and hosts. The same principles apply to tropical reefs.

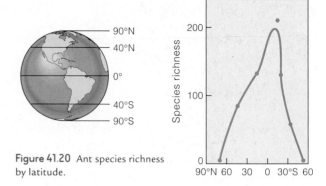

Figure 41.20 Ant species richness by latitude.

> ❯❯ **Figure It Out** Houston, Texas, is at 30 degrees north latitude and Minneapolis, Minnesota, is at 45 degrees north latitude. Where would you expect to find more ant species?
> Answer: Houston

Island Patterns

Islands are natural laboratories for population studies. They have also been laboratories for community studies. For instance, in the mid-1960s volcanic eruptions formed a new island 33 kilometers (21 miles) from the coast of Iceland. The island was named Surtsey (Figure 41.21A). Bacteria and fungi were early colonists. The first vascular plant became established on the island in 1965. Mosses appeared two years later and thrived (Figure 41.21B). The first lichens were found five years after that. The rate of arrivals of new vascular plants picked up after a seagull colony became established in 1986 (Figure 41.21C).

The number of species on Surtsey will not continue increasing forever. How many species will there be when the number levels off? The **equilibrium model of island biogeography** addresses this question. According to this model, the number of species living on any island reflects a balance between immigration rates for new species and extinction rates for established ones. The distance between an island and a mainland source of colonists

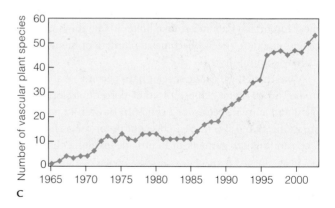

Figure 41.21 Island colonizations. **A** Surtsey, a volcanic island, at the time of its formation. **B** Surtsey in 1983. **C** Graph of the number of vascular plant species found in yearly surveys. Seagulls began nesting on the island in 1986.

affects immigration rates. An island's size affects both immigration rates and extinction rates.

Consider first the **distance effect**: Islands far from a source of colonists receive fewer immigrants than those closer to a source. Most species cannot disperse very far, so they will not turn up far from a mainland.

Species richness is also shaped by the **area effect**: Big islands tend to support more species than small ones. More colonists will happen upon a larger island simply by virtue of its size. Also, big islands are more likely to offer a variety of habitats, such as high and low elevations. These options make it more likely that a new arrival will find a suitable habitat. Finally, big islands can support larger populations of species than small islands. The larger a population, the less likely it is to become locally extinct as the result of some random event.

Figure 41.22 illustrates how interactions between the distance effect and the area effect can influence the equilibrium number of species on islands.

Robert H. MacArthur and Edward O. Wilson first developed the equilibrium model of island biogeography in the late 1960s. Since then the model has been modified and its use has been expanded to help scientists think about habitat islands—natural settings surrounded by a "sea" of habitat that has been disturbed by humans. Many parks and wildlife preserves fit this description. Island-based models can help estimate the size of an area that must be set aside as a protected reserve to ensure survival of a species or a community.

---

area effect Larger islands have more species than small ones.
distance effect Islands close to a mainland have more species than those farther away.
equilibrium model of island biogeography Model that predicts the number of species on an island based on the island's area and distance from the mainland.
species richness Of an area, number of species.

Fighting Foreign Fire Ants (revisited)

Now that you have learned a bit about how communities and ecosystems work, let's take a second look at those red imported fire ants (RIFAs) we discussed at the beginning of this chapter. These ants are an introduced predator that is having a negative effect on competing ants and on prey species. RIFAs did not evolve in North America, so there are few predators, parasites, or pathogens to hold them in check. Freed from these natural restrictions, RIFAs have an easier time outcompeting native ants. As RIFA populations soar, populations of organisms that are their prey decline.

Global climate change is expected to help RIFAs extend their range. For example, RIFAs require mild winters and relatively high rainfall to do well. With winters becoming milder, the ants are expected to move farther north and ascend to higher altitudes. If current climate predictions are correct, the amount of habitat available to red imported fire ants in the United States is expected to increase by about 20 percent before the close of this century.

How Would You Vote? Increased global trade and faster ships are contributing to a rise in the rate of species introductions into North America. Faster ships mean shorter trips, increasing the likelihood that pests will survive a voyage. Wood-eating insects from Asia turn up with alarming frequency in the wood of packing crates and spools for steel wire. Some of these insects, such as the Asian long-horned beetle, now pose a serious threat to North America's forests. Currently, only a small fraction of the crates imported into the United States are inspected to see if they contain exotic species. Should inspections increase? See CengageNow for details, then vote online (cengagenow.com).

Take-Home Message **What are some biogeographic patterns in species richness?**

> Generally, species richness is highest in the tropics and lowest at the poles. Tropical habitats have conditions that more species can tolerate, and tropical communities have often been evolving for longer than temperate ones.

> When a new island forms, species richness rises over time and then levels off. The size of an island and its distance from a colonizing source influence its species richness.

Figure 41.22 Animated Island biodiversity patterns.

Distance effect: Species richness on islands of a given size declines as distance from a source of colonists rises. *Green* circles are values for islands less than 300 kilometers from the colonizing source. *Orange* triangles are values for islands more than 300 kilometers (190 miles) from a source of colonists.

Area effect: Among islands the same distance from a source of colonists, larger islands tend to support more species than smaller ones.

>> **Figure It Out** Which is likely to have more species, a 100-km^2 island more than 300 km from a colonizing source or a 500-km^2 island less than 300 km from a colonist source?

Answer: The 500-km^2 island

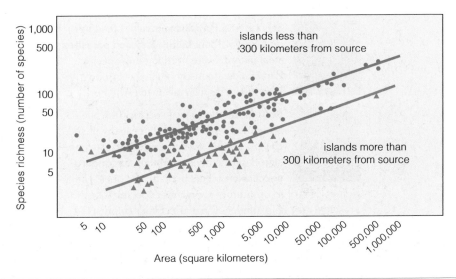

Summary

Section 41.1 All the species that live in an area constitute a **community**. Introduction of a new species to a community can have negative effects on the species that evolved in that community.

Section 41.2 Each species occupies a certain **habitat** characterized by physical and chemical features and by the array of other species living in it. Species interactions affect community structure. **Commensalism** is a common interaction in which one species benefits and the other is unaffected. A **symbiosis** is an interaction in which one species lives in or on another.

Section 41.3 In a **mutualism**, both species benefit from an interaction. Some mutualists cannot complete their life cycle without the interaction. Mutualists who maximize their own benefits while limiting the cost of cooperating are at a selective advantage.

Section 41.4 A species' roles and requirements define its unique **ecological niche**. With **interspecific competition**, two species with similar resource requirements are harmed by one another's presence. When one competitor drives the other to local extinction, we call the process **competitive exclusion**. Competing species with similar requirements become less similar when directional selection causes **character displacement**. This change in traits allows **resource partitioning**.

Section 41.5 Predation benefits a predator at the expense of the prey it captures, kills, and eats. Predators and their prey exert selective pressures on one another. Evolved prey defenses include camouflage and **mimicry**. Prey have traits that allow them to overcome prey defenses. With **herbivory**, an animal eats plant or algae. Plants have traits that discourage herbivory or that allow a quick recovery from it.

Section 41.6 Parasitism involves feeding on a host without killing it. **Brood parasites** steal parental care from another species. **Parasitoids** are insects that lay eggs on a host insect, then larvae devour the host.

Section 41.7 Ecological succession is the sequential replacement of one array of species by another over time. **Primary succession** happens in new habitats. **Secondary succession** occurs in disturbed ones. The first species of a community are **pioneer species**. Their presence may help, hinder, or not affect later colonists.

Modern models of succession emphasize the unpredictability of the outcome as a result of chance events, ongoing changes, and disturbance. The **intermediate disturbance**

hypothesis predicts that a moderate level of disturbance keeps a community diverse.

Section 41.8 Keystone species are especially important in determining the composition of a community. An **indicator species** can provide information about the health of the environment. The removal of a keystone species or introduction of an **exotic species** can alter community structure.

Section 41.9 Species richness, the number of species in a given area, varies with latitude. The **equilibrium model of island biogeography** predicts the number of species that an island will sustain based on the **area effect** and the **distance effect**. Scientists can use this model to predict the number of species that habitat islands such as parks can sustain.

Self-Quiz Answers in Appendix III

1. The type of place where a species typically lives is called its _____ .
   a. niche   c. community
   b. habitat   d. population

2. Which cannot be a symbiosis?
   a. mutualism   c. commensalism
   b. parasitism   d. interspecific competition

3. Lizards that eat flies they catch on the ground and birds that catch and eat flies in the air are engaged in _____ competition.
   a. exploitative   d. interspecific
   b. interference   e. both a and d
   c. intraspecific   f. both b and c

4. _____ can lead to resource partitioning.
   a. Mutualism   c. Commensalism
   b. Parasitism   d. Interspecific competition

5. Match the terms with the most suitable descriptions.
   ___ mutualism   a. one free-living species feeds
   ___ parasitism   on another and usually kills it
   ___ commensalism   b. two species interact and both
   ___ predation   benefit by the interaction
   ___ interspecific   c. two species interact and one
        competition   benefits while the other is
                              neither helped nor harmed
                          d. one species feeds on another
                              but usually does not kill it
                          e. two species access a resource

6. A tick is a(n) _____ .
   a. brood parasite   b. ectoparasite   c. endoparasite

7. By a currently favored hypothesis, species richness of a community is greatest between physical disturbances of _____ intensity or frequency.
   a. low   c. high
   b. intermediate   d. variable

Data Analysis Activities

Biological Control of Fire Ants Ant-decapitating phorid flies are just one of the biological control agents used to battle imported fire ants. Researchers have also enlisted the help of a fungal parasite that infects the ants and slows their production of off-spring. An infected colony dwindles in numbers and eventually dies out.

Are these biological controls useful against imported fire ants? To find out, USDA scientists treated infested areas with either traditional pesticides or pesticides plus biological controls (both flies and the parasite). The scientists left some plots untreated as controls. Figure 41.23 shows the results.

1. How did population size in the control plots change during the first four months of the study?
2. How did population size in the two types of treated plots change during this same interval?
3. If this study had ended after the first year, would you conclude that biological controls had a major effect?
4. How did the two types of treatment (pesticide alone versus pesticide plus biological controls) differ in their longer-term effects? Which is most effective?

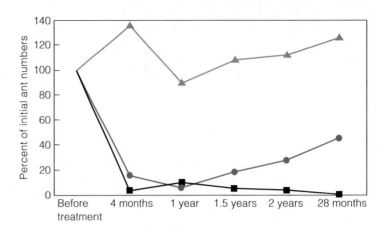

Figure 41.23 Effects of two methods of controlling red imported fire ants. The graph shows the numbers of red imported fire ants over a 28-month period. *Orange* triangles represent untreated control plots. *Green* circles are plots treated with pesticides alone. *Black* squares are plots treated with pesticide and biological control agents (phorid flies and a fungal parasite).

8. _____ species are the first to colonize a new habitat.

9. Growth of a forest in an abandoned corn field is an example of _____ .
   a. primary succession   c. secondary succession
   b. resource partitioning   d. competitive exclusion

10. Species richness is greatest in communities _____ .
    a. near the equator   c. near the poles
    b. in temperate regions   d. that recently formed

11. If you remove a species from a community, the population size of its main _____ is likely to increase.
    a. parasite   b. competitor   c. predator

12. Mammals are least likely to be _____ .
    a. mutualists   c. brood parasites
    b. commensal   d. predators

13. Herbivory benefits _____ .
    a. the herbivore   c. both a and b
    b. the plant   d. neither a nor b

14. Match the terms with the most suitable descriptions.
    ___ area effect   a. greatly affects other species
    ___ pioneer   b. first species established in a
        species   new habitat
    ___ indicator   c. more species on large islands
        species   than small ones at same distance
    ___ keystone   from the source of colonists
        species   d. species that is especially sensitive
    ___ exotic   to changes in the environment
        species   e. allows competitors to coexist
    ___ resource   f. often outcompete, displace native
        partitioning   species of established community

Additional questions are available on **CENGAGENOW**.

Critical Thinking

1. With antibiotic resistance rising, researchers are looking for ways to reduce use of antibiotics. Some cattle once fed antibiotic-laced food now get probiotic feed instead, with cultured bacteria that can establish or bolster populations of helpful bacteria in the animal's gut. The idea is that if a large population of beneficial bacteria is in place, then the harmful bacteria cannot become established or thrive. Which ecological principle is guiding this research?

2. Phasmids are herbivorous insects that mimic leaves or sticks, as shown at the *right*. Most rest during the day, and feed at night. If they do move in daytime, they do so very slowly. If disturbed, a phasmid will fall to the ground and lie motionless. Speculate on selective pressures that could have shaped phasmid morphology and behavior. Suggest an experiment to test a hypothesis about how its appearance or behavior may be adaptive.

Animations and Interactions on **CENGAGENOW**:
> Competitive exclusion; Island biodiversity patterns.

❮ Links to Earlier Concepts

This chapter returns to the concept of a one-way flow of energy in nature (Sections 1.3, 5.2) and looks at nitrogen-fixing microbes, soil erosion, and leaching (19.7, 26.2) in the context of nitrogen cycling. You will be reminded of properties of chemical bonds (2.4) and the pathways of photosynthesis (6.4) and aerobic respiration (7.2). You will see how slow movements of Earth's crust (16.7) influence the cycling of some nutrients.

Key Concepts

Organization of Ecosystems

A one-way flow of energy and the cycling of raw materials among species maintain an ecosystem. Nutrients and energy are transferred in a stepwise fashion through food chains that interconnect as complex food webs.

Biogeochemical Cycles

In a biogeochemical cycle, a nutrient moves relatively slowly among its environmental reservoirs. The reservoirs may include air, water, and rocks. Nutrients move more quickly into, through, and out of food webs.

42 Ecosystems

42.1 Too Much of a Good Thing

All organisms require certain elements to build their bodies and carry out metabolic processes. Phosphorus is one of these essential nutrients. It is part of ATP, phospholipids, nucleic acids, and other biological compounds. Plants meet their phosphorus requirement by taking up dissolved phosphates in soil water. Animals get phosphorus by eating plants or by eating other animals. Thus phosphorus taken up from the environment passes from one organism to another. When a plant or animal dies and decays, the phosphorus in its body returns to the environment. As a result, phosphorus moves continually from the environment, through organisms, and back to the environment.

All organisms in a community together with their environment constitute an ecosystem, and the study of nutrient cycles in ecosystems is one aspect of the science of ecology. Ecology is not the same as environmentalism, which is advocacy for protection of the environment. However, environmentalists often point to the results of ecological studies when drawing attention to a particular environmental concern.

As one example, environmentalists in many parts of the United States are pushing for laws that ban high-phosphorus laundry detergents, dish detergents, and lawn fertilizers. Their desire to reduce phosphate content of these products is based on evidence that phosphorus enrichment of water can disrupt aquatic ecosystems (Figure 42.1).

The addition of nutrients to an aquatic ecosystem is called **eutrophication**. It can occur slowly by natural processes, or fast as a result of human actions. Phosphorus is often a limiting factor for aquatic producers. A sudden addition of phosphorus removes this limitation and allows a population explosion of algae and cyanobacteria. The resulting algal bloom (Section 20.1) clouds water and threatens other aquatic species.

We began using phosphate-rich products when we had no idea of their effects beyond cleaner homes and greener lawns. Today, we confront the simple truth that the daily actions of

Figure 42.1 *Above*, Results of a field experiment in which different types of nutrients were added to two artificially separated portions of a lake. Including phosphorus in the mix (lower region) caused rapid overgrowth of algae.

Opposite page, Graphic from a Washington State Ecology Department campaign designed to remind people that lawn fertilizer often ends up in lakes.

millions of individuals can disrupt nutrient cycles that have been operating since long before humans existed. Our species has a unique capacity to shape the environment to our liking. In doing so, we have become major players in the global flows of energy and nutrients even before we fully comprehend how ecosystems work. Decisions we make today about environmental issues are likely to shape the quality of life and the environment far into the future.

---

eutrophication Nutrient enrichment of an aquatic ecosystem.

The Water Cycle
Most of Earth's water is in its oceans. Only a tiny fraction is fresh water. Evaporation, condensation, precipitation, and flow of rivers and streams move water. Water also plays a role in other nutrient cycles by transporting soluble forms of those nutrients.

The Carbon Cycle
Most of Earth's carbon is tied up in rocks, but organisms take carbon up from water or the air. Carbon dioxide is one of the atmospheric greenhouse gases that help keep Earth's surface warm. Increasing carbon dioxide in the air is the most likely cause of climate change.

Nitrogen and Phosphorus Cycles
Plants take up dissolved forms of nitrogen and phosphorus from soil water. Nitrogen is abundant in air, but only certain bacteria can use the gaseous form. Phosphorus has no major gaseous form; most of it is in rocks.

> All energy captured by the producers in an ecosystem eventually returns to the environment as heat.

> Nutrients taken up by producers are returned to the environment by decomposers, then taken up again.

‹ Links to Acquiring energy and nutrients 1.3, Energy transfers 5.2, Photosynthesis 6.4, Modes of nutrition 19.5

Each ecosystem is an array of organisms and a physical environment, all interacting through a one-way flow of energy and a cycling of nutrients (Figure 42.2). It is an open system, because it requires ongoing inputs of energy to persist. Most ecosystems also gain nutrients from and lose nutrients to other ecosystems.

Primary Producers and Production

An ecosystem runs on the energy captured by **primary producers**. These autotrophs get energy and carbon from a nonliving source to build organic compounds. In most ecosystems, photoautotrophs such as plants, algae, and photosynthetic bacteria are the primary producers. In some dark environments such as deep-sea vent ecosystems, chemoautrophs fill this role.

Primary production is the rate at which producers capture and store energy. Daylength, temperature, and the availability of nutrients, including phosphates, affect producer growth and so influence primary production. As a result, primary production can vary seasonally within an ecosystem and also differs among ecosystems. Per unit area, primary production on land tends to be higher than that in the oceans. However, because oceans cover about 70 percent of Earth's surface, marine producers contribute nearly half of the global net primary production.

The Roles of Consumers

As Section 1.3 explained, **consumers** get energy and carbon by feeding on tissues and remains of producers and one another. We describe consumers by their diets. Herbivores eat plants. Carnivores eat the flesh of animals. Parasites live inside or on a living host and feed on its tissues. Omnivores eat both animals and plants. **Detritivores** such as earthworms eat small bits of decaying organic matter, or detritus. **Decomposers** feed on wastes and remains, breaking them down into inorganic building blocks. Bacteria, archaeans, and fungi serve as decomposers.

---

consumer Organism that obtains energy and carbon by feeding on tissues, wastes, or remains of other organisms.
decomposer Organism that feeds on biological remains and breaks organic material down into its inorganic subunits.
detritivore Consumer that feed on small bits of organic material.
primary producer An organism that obtains energy and nutrients from inorganic sources; an autotroph.
primary production The rate at which an ecosystem's producers capture and store energy.

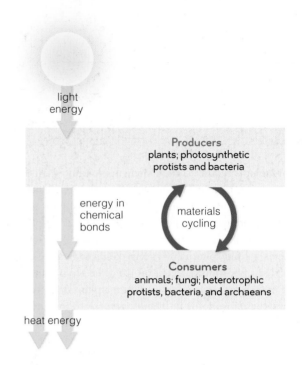

Figure 42.2 Animated One-way flow of energy (*yellow* arrows) and nutrient cycling (*blue* arrows) in the most common type of ecosystem. All light energy that enters the system eventually returns to the environment as heat energy that is not reused. By contrast, nutrients are continually recycled.

Energy Flow and Nutrient Cycling

Energy captured by producers is converted to bond energy in organic molecules. This energy is released by metabolic reactions that give off heat. The flow of energy through living organisms is a one-way process because heat energy is not recycled—producers do not use heat energy to form chemical bonds.

In contrast, nutrients are cycled within an ecosystem. Producers take up hydrogen, oxygen, and carbon from inorganic sources such as the air and water. They also take up dissolved nitrogen, phosphorus, and other necessary minerals. Nutrients move from producers into the consumers who eat them. Decomposition returns nutrients to the environment, from which producers take them up again.

Take-Home Message **What factors characterize ecosystems?**

> An ecosystem is a community of autrophic producers and heterotrophic consumers that interact with one another and the environment by a one-way flow of energy and a cycling of materials.

> Ecosystems vary in their primary production. Marine ecosystems generally have a lower primary production per unit area than land ecosystems.

Food Chains

> Food chains describe how energy and materials are transferred from one organism to another. Inefficient transfers constrain food chain length.

< Links to Energy transfers 5.2, Lignin 4.11, Ectotherms and endotherms 24.6

The hierarchy of feeding relationships within an ecosystem is the ecosystem's trophic structure (*troph* means nourishment). All organisms at the same **trophic level** are the same number of transfers away from the energy input into that system. The primary producers are at the first trophic level. The primary consumers that eat them are at the second trophic level, and so on.

A **food chain** is a sequence of steps by which some energy captured by primary producers is transferred to higher trophic levels. For example, in one tallgrass prairie food chain, energy flows from grasses to grasshoppers, to sparrows, and finally to coyotes (Figure 42.3). At the first trophic level in this food chain, grasses and other plants are the primary producers. At the second trophic level, grasshoppers are primary consumers. At the third trophic level, sparrows that eat grasshoppers are second-level consumers. At the fourth trophic level, coyotes that eat sparrows are third-level consumers.

Energy captured by producers usually passes through no more than four or five trophic levels. Even in ecosystems with many species, the number of participants in each food chain is limited. The length of food chains is constrained by the inefficiency of energy transfers. Only 5 to 30 percent of the energy in tissues of an organism at one trophic level ends up in tissues of an organism at the next trophic level.

Why isn't all the energy an organism takes in available to the organism that eats it? First, not all energy that an organism takes in is used to build body parts. Energy an organism invests in producing offspring or loses as metabolic heat is not available to a consumer. The more heat an organism produces, the less energy goes into building tissues. Thus, food chains that involve ectotherms are typically longer than those that involve endotherms.

Second, energy in some body parts cannot be accessed by most consumers. For example, few herbivores have the ability to break down the lignin that reinforces the bodies of most woody plants. Similarly, many animals have some energy tied up in a difficult-to-digest internal or external skeleton. Hair, feathers, and fur also resist digestion. A coyote that eats a sparrow does not benefit from energy that the bird invested in building bones and feathers.

food chain Description of who eats whom in one path of energy in an ecosystem.
trophic level Position of an organism in a food chain.

Fourth Trophic Level
Third-level consumer

Third Trophic Level
Second-level consumer

Second Trophic Level
Primary consumer

First Trophic Level
Primary producer

Figure 42.3 One food chain in a Kansas tallgrass prairie.

Take-Home Message **What is a food chain?**

> A food chain is a description of a series of trophic interactions among the members of an ecosystem.

> Primary producers are the first trophic level of a food chain. Energy is transferred to primary consumers, and then to higher-level consumers.

> The inefficiency of energy transfers limits the number of steps in food chains. Only a small fraction of the energy that an organism takes in can be accessed by an organism that consumes it.

> Each food web consists of cross-connecting food chains. Its structure reflects environmental constraints and the inefficiency of energy transfers among trophic levels.

An organism that participates in one food chain usually has a role in many others as well. The food chains of an ecosystem cross-connect as a **food web**. Figure 42.4 shows a small sampling of the participants in an arctic food web.

Nearly all food webs include two types of interconnecting food chains. In **grazing food chains**, energy stored in producer tissues flows to herbivores, which often are relatively large animals such as mammals. In a **detrital food chain**, energy in producers flows to detritivores, which tend to be smaller animals such as worms or insects.

In most land ecosystems, the bulk of the energy that gets stored in producer tissues moves through detrital

Higher Trophic Levels

A sampling of carnivores that feed on herbivores and one another

human (Inuk) arctic wolf arctic fox

gyrfalcon snowy owl ermine

Second Trophic Level

Major parts of the buffet of primary consumers (herbivores)

mosquito flea

Parasitic consumers feed at more than one trophic level.

vole arctic hare lemming

First Trophic Level

This is just part of the buffet of primary producers.

grasses, sedges purple saxifrage arctic willow

Detritivores and decomposers (nematodes, annelids, saprobic insects, protists, fungi, bacteria)

Figure 42.4 Animated Some organisms in an Arctic food web. Arrows point from eater to eaten.

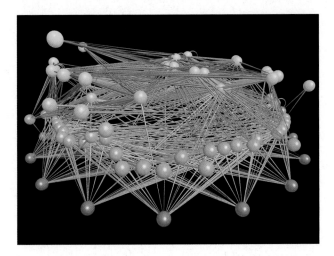

Figure 42.5 Computer model for a land food web in East River Valley, Colorado. Balls signify species. Their colors identify trophic levels, with producers (coded *red*) at the bottom and top predators (*yellow*) at top. The connecting lines thicken, as they go from an eaten species to the eater.

food chains. For example, in an arctic ecosystem, voles, lemmings, and hares eat some living plant parts. However, far more plant material ends up as detritus. Bits of dead plant material sustain detritivores such as nematodes and soil-dwelling insects.

Understanding food webs helps ecologists predict how ecosystems will respond to change. Neo Martinez and his colleagues constructed the food web diagram shown in Figure 42.5. By comparing different food webs, Martinez realized that trophic interactions connect species more closely than people thought. On average, each species in a food web was typically two links away from all other species. Ninety-five percent of species were within three links of one another, even in large communities with many species. As Martinez concluded in his paper on these findings, "Everything is linked to everything else." He cautioned that the extinction of any species in a food web has a potential impact on many other species.

detrital food chain Food chain in which energy is transferred directly from producers to detritivores.
food web Set of cross-connecting food chains.
grazing food chain Food chain in which energy is transferred from producers to grazers (herbivores).

Take-Home Message **What have ecologists learned about food webs?**

❯ Two types of food chains connect in most food webs. Tissues of living producers are the base for grazing food chains. Producer remains are the base for detrital food chains.

❯ Even in complex ecosystems, trophic interactions link each species in a food web with many others.

42.5 Ecological Pyramids

❯ Ecological pyramid diagrams dramatically illustrate the inefficiency of transfers between trophic levels.

A food web diagram is one way of depicting the trophic interactions among species in an ecosystem. Ecological pyramid diagrams are another. A biomass pyramid shows the amount of organic material in the bodies of organisms at each trophic level at a specific time. An energy pyramid shows the amount of energy that flows through each trophic level in a given interval. Figure 42.6 shows ecological pyramids for one freshwater spring ecosystem in Florida.

Most commonly, primary producers make up most of the biomass in a pyramid, and top carnivores make up very little. The Florida ecosystem has lots of aquatic plants but very few gars (a top predator in this ecosystem). Similarly, if you walk through a prairie, you will see more grams of grass than grams of coyote.

An energy pyramid is always broadest at the bottom. This is why people promote a vegetarian diet by touting the ecological benefits of "eating lower on the food chain." They are referring to the energy losses in transfers between plants, livestock, and humans. When a person eats a plant food, he or she gets most of the calories in that food. When the plant food is used to grow livestock, only a small percentage of the food's calories ends up in meat a person can eat. Thus, feeding a population of meat-eaters requires far greater crop production than sustaining a population of vegetarians.

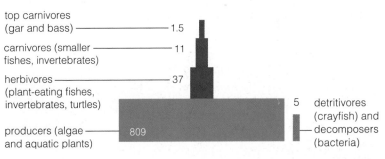

A Biomass pyramid (grams per square meter)

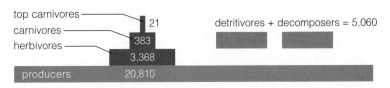

B Energy flow pyramid (kilocalories per square meter per year)

Figure 42.6 Ecological pyramids for Silver Springs, an aquatic ecosystem in Florida.

Take-Home Message **What are ecological pyramids?**

❯ Ecological pyramids depict the distribution of materials and energy among the trophic levels of an ecosystem.

> Elements essential to life move between a community and its environment in a biogeochemical cycle.
< Links to Plate tectonics 16.7, Erosion 26.2

In a **biogeochemical cycle**, an essential element moves from one or more environmental reservoirs, through the biological component of an ecosystem, and then back to the reservoirs (Figure 42.7). Depending on the element, environmental reservoirs may include Earth's rocks and sediments, waters, and atmosphere.

Chemical and geologic processes move elements to, from, and among environmental reservoirs. For example, elements locked in rocks can become part of the atmosphere as a result of volcanic activity. As one of Earth's crustal plates moves under another, rocks on the sea floor can be uplifted so they become part of a land mass. On land, the rocks are exposed to the erosive forces of wind and rain. As the rocks are slowly broken down, elements in them enter rivers, and eventually seas. Compared to the movement of elements among organisms of an ecosystem, the movement of elements among nonbiological reservoirs is far slower. Processes such as erosion and uplifting operate over thousands or millions of years.

biogeochemical cycle A nutrient moves among environmental reservoirs and into and out of food webs.

Take-Home Message **What is a biogeochemical cycle?**

> A biogeochemical cycle is the slow movement of a nutrient among environmental reservoirs and into, through, and out of food webs.

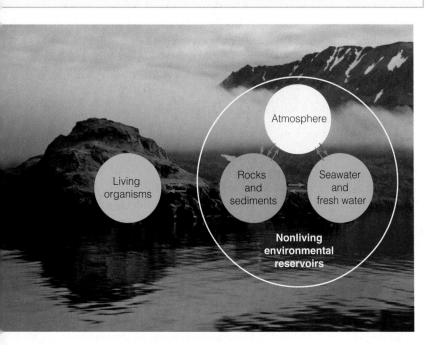

Figure 42.7 Generalized biogeochemical cycle. For all nutrients, the cumulative amount in all environmental reservoirs far exceeds the amount in living organisms.

> Water makes up the bulk of all organisms and serves as a transport medium for many soluble nutrients.
< Links to Properties of water 2.5, Transpiration 26.4

How and Where Water Moves

Most of Earth's water (97 percent) is in its oceans (Table 42.1). The **water cycle** moves water from the ocean to the atmosphere, onto land, and back to the oceans (Figure 42.8). Sunlight energy drives the water cycle by causing evaporation, the conversion of liquid water to water vapor. Water vapor that enters the cool upper layers of the atmosphere condenses into droplets, forming clouds. When droplets get large and heavy enough, they fall as precipitation—as rain, snow, or hail.

Oceans cover about 70 percent of Earth's surface, so most rainfall returns water directly to the oceans. On land, we define a **watershed** as an area in which all precipitation drains into a specific waterway. A watershed may be as small as a valley that feeds a stream, or as large as the 5.88 million square kilometers of the Amazon River Basin. The Mississippi River Basin watershed includes 41 percent of the continental United States.

Most precipitation that enters a watershed seeps into the ground. Some of this water remains between soil particles as **soil water**. Plant roots can tap into this water source. Soils differ in water holding capacity, with clay-rich soils holding the most water and sandy soils the least. Water that drains through soil layers often collects in **aquifers**. These natural underground reservoirs consist of porous rock layers that hold water. **Groundwater** is water in soil and aquifers. Water that falls on impermeable rock or on saturated soil becomes **runoff**: It flows over the ground into streams. The flow of groundwater and surface water slowly returns water to oceans.

The movement of water causes the movement of other nutrients. Carbon, nitrogen, and phosphorus all have soluble forms that can be moved from place to place by flowing water. As water trickles through soil, it carries nutrient particles from topsoil into deeper soil layers. As a stream flows over limestone, water slowly dissolves the rock and carries carbonates back to the seas where the limestone formed. On a less natural note, runoff from heavily fertilized lawns and agricultural fields carries dissolved phosphates and nitrates into streams and lakes.

Limited Fresh Water

The vast majority of Earth's water is too salty to drink or irrigate crops. If all Earth's water filled a bathtub, the amount of fresh water that could be used each year without decreasing the overall supply would fill a teaspoon. In addition, most fresh water is frozen as ice.

Groundwater supplies drinking water to about half of the United States population. Water overdrafts from

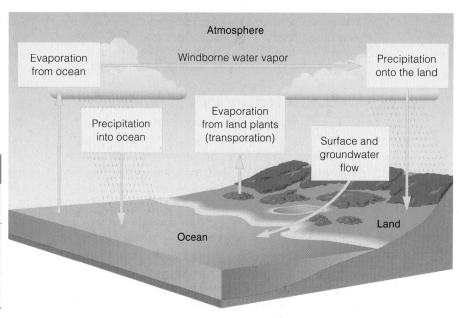

Figure 42.8 **Animated** The water cycle. Water moves from the ocean to the atmosphere, land, and back. The arrows identify processes that move water.

Table 42.1 Environmental Water Reservoirs	
Reservoir	Volume (10^3 cubic kilometers)
Ocean	1,370,000
Polar ice, glaciers	29,000
Groundwater	4,000
Lakes, rivers	230
Atmosphere (water vapor)	14

aquifers are now common; water is drawn from aquifers faster than natural processes replenish it. When too much fresh water is withdrawn from a coastal aquifer, salt water moves in and replaces it. Figure 42.9 shows areas of aquifer depletion and saltwater intrusion in the United States.

Overdrafts have reduced the volume of water in the Ogallala aquifer by about half. This aquifer stretches from South Dakota to Texas and supplies irrigation water for about 20 percent of the nation's crops. For the past thirty years, withdrawals have exceeded replenishment by a factor of ten.

Desalinization, the removal of salt from seawater, can help increase freshwater supplies. However, the process is expensive because it is energy intensive. Most likely, desalinization will never be cost-effective for widespread use in the United States. In addition, the process produces mountains of waste salts that must be disposed of.

A more practical approach to meeting freshwater needs involves more efficient use of this limited resource. In the United States, about 80 percent of the water withdrawn for human use ends up irrigating agricultural fields. Using the most efficient irrigation methods available would go a long way toward ensuring that future generations have an adequate and safe water supply.

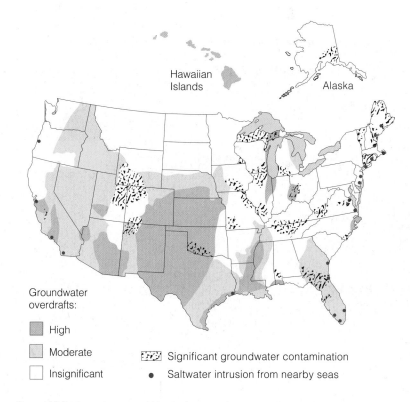

Figure 42.9 Groundwater troubles in the United States.

---

aquifer Porous rock layer that holds some groundwater.
groundwater Soil water and water in aquifers.
runoff Water that flows over soil into streams.
soil water Water between soil particles.
water cycle Movement of water among Earth's oceans, atmosphere, and the freshwater reservoirs on land.
watershed Land area that drains into a particular stream or river.

Take-Home Message **What is the water cycle and how do human activities affect it?**

❯ Water moves slowly from the world ocean—the main reservoir—through the atmosphere, onto land, then back to the ocean.

❯ Fresh water makes up only a tiny portion of the global water supply. Excessive water withdrawals threaten many sources of drinking water.

❯ After water, carbon is the most abundant substance in living things. Most of it is in rocks, but it enters food webs as gaseous carbon dioxide or bicarbonate dissolved in water.

❮ Links to Photosynthesis 6.4, Carbon sequestration 25.1

Carbon Reservoirs and Movements

In the **carbon cycle**, natural processes move carbon among Earth's atmosphere, oceans, soils, and into and out of food webs (Figure 42.10). It is an **atmospheric cycle**, a biogeochemical cycle in which a gaseous form of the element plays a significant role.

On land, plants take up carbon dioxide from the atmosphere and incorporate it into their tissues when they carry out photosynthesis ❶. Plants and most other land organisms release carbon dioxide into the atmosphere by the process of aerobic respiration ❷.

On an annual basis, the greatest flow of carbon between nonbiological reservoirs takes place between the ocean and atmosphere. The ocean holds 38,000–40,000 gigatons of dissolved carbon, primarily in the form of

bicarbonate (HCO_3^-) and carbonate (CO_3^{2-}) ions. The air holds about 750 gigatons of carbon, mainly in the form of carbon dioxide (CO_2).

Bicarbonate ions form when atmospheric carbon dioxide dissolves in water ❸. Aquatic producers take up bicarbonate and convert it to carbon dioxide for use in photosynthesis. As on land, aquatic organisms carry out aerobic respiration and release carbon dioxide ❹.

The single greatest reservoir of carbon is Earth's rocks, which contain about 50 million gigatons of it. Sedimentary rocks such as limestone form over millions of years when sediments containing carbon-rich shells of marine organisms became compacted ❺. The carbon-rich rocks become part of land ecosystems when movements of crustal plates uplift portions of the sea floor. However, carbon locked in rocks does not move in significant amounts into the air or water and it is not available to producers, so it has little effect on ecosystems.

Plant remains in soil hold about 1,600 gigatons of carbon, about twice that in the atmosphere. Fossil fuels such

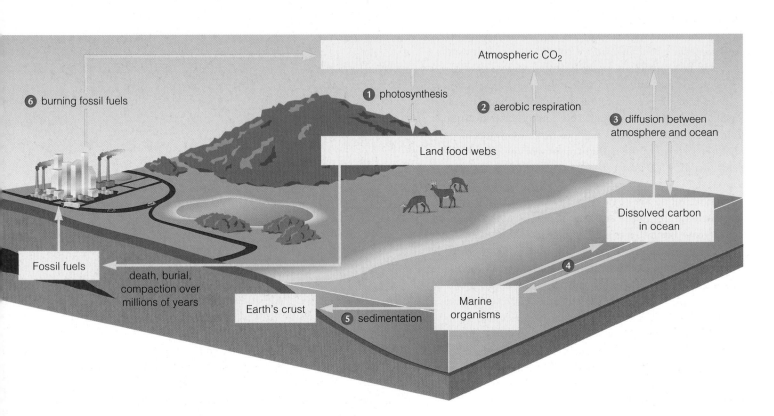

Figure 42.10 Animated The carbon cycle. Most carbon is in Earth's crust, where it is largely unavailable to living organisms.

❶ Carbon enters land food webs when plants take up carbon dioxide from the air for use in photosynthesis.

❷ Carbon returns to the atmosphere as carbon dioxide when plants and other land organisms carry out aerobic respiration.

❸ Carbon diffuses between the atmosphere and the ocean. Bicarbonate forms when carbon dioxide dissolves in seawater.

❹ Marine producers take up bicarbonate for use in photosynthesis, and marine organisms release carbon dioxide from aerobic respiration.

❺ Many marine organisms incorporate carbon into their shells. After they die, these shells become part of the sediments. Over time, the sediments become carbon-rich rocks such as limestone and chalk in Earth's crust.

❻ Burning of fossil fuels derived from the ancient remains of plants puts additional carbon dioxide into the atmosphere.

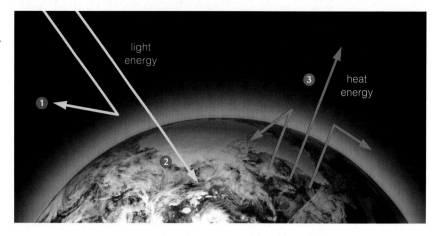

Figure 42.11 **Animated** Greenhouse effect.

❶ Earth's atmosphere reflects some sunlight energy back into space.

❷ More light energy reaches and warms Earth's surface.

❸ Earth's warmed surface emits heat energy. Some of this energy escapes through the atmosphere into space. But some is absorbed and then emitted in all directions by greenhouse gases. The emitted heat warms Earth's surface and lower atmosphere.

>> Figure It Out Do greenhouse gases reflect heat energy toward the Earth?

Answer: No. The gases absorb heat energy, then reemit it in all directions.

as coal and gas are the ancient remains of such organic material. Fossil fuels hold an estimated 5,000 gigatons of carbon. Until recently, this carbon, like the carbon in rocks, had little impact on ecosystems. We now withdraw 4 to 5 gigatons of carbon from fossil fuel reservoirs each year **❻**. Burning this fuel, cutting down forests, and other activities adds more carbon to the air than can dissolve in the oceans. Each year, only about 2 percent of the extra carbon that we put into the atmosphere dissolves in ocean water. The rest of it remains in the atmosphere.

Carbon, the Greenhouse Effect, and Global Warming

Atmospheric carbon dioxide helps keep the Earth warm enough for life. In what is known as the **greenhouse effect**, sunlight heats Earth's surface, then carbon dioxide and other "greenhouse gases" absorb some heat radiating from that surface and reradiate the heat toward Earth (Figure 42.11). Without the greenhouse effect, heat from Earth's surface would escape into space, leaving the planet cold and lifeless.

Scientists can use bubbles of air trapped in ancient glaciers to determine atmospheric conditions when the ice first formed. This and other evidence indicates that the current atmospheric carbon dioxide is the highest in 420,000 years and is climbing.

Given the greenhouse effect, we would predict that increases in the atmospheric concentration of carbon dioxide and other greenhouse gases would raise the tem-

perature of Earth's surface. Evidence supports this prediction. We are in the midst of a period of **global climate change**, with a trend toward rising temperature and shifts in other climate patterns.

Earth's climate has varied greatly over its long history. During ice ages, much of the planet was covered by glaciers. Other periods were warmer than the present, and tropical plants and coral reefs thrived at what are now cool latitudes. Scientists can correlate past large-scale temperature changes with shifts in Earth's orbit, which varies in a regular fashion over 100,000 years, and Earth's tilt, which varies over 40,000 years. Changes in solar output and volcanic eruptions also affect Earth's temperature. However, most scientists do not see evidence that these factors play a major role in the current temperature rise.

In 2007, the Intergovernmental Panel on Climate Change reviewed the results of many scientific studies related to climate change. The panel of hundreds of scientists from all over the world concluded that the warming in recent decades is very likely due to a human-induced increase in atmospheric greenhouse gases. Unlike other possible factors, the rise in greenhouse gases correlates well with the rise in temperature.

We return to the evidence of global climate change and its effects on natural systems in Section 44.8.

---

atmospheric cycle Biogeochemical cycle in which a gaseous form of an element plays a significant role.
carbon cycle Movement of carbon, mainly between the oceans, atmosphere, and living organisms.
global climate change A currently ongoing rise in average temperature that is altering climate patterns around the world.
greenhouse effect Warming of Earth's lower atmosphere and surface as a result of heat trapped by greenhouse gases.

Take-Home Message **How does carbon move between reservoirs, and what are the effects of releasing excess carbon?**

> Most of Earth's carbon is in rocks, but little carbon moves out of this reservoir.

> Oceans and the soil hold more carbon than the air. Carbon flows continually between these three reservoirs and into and out of food webs.

> Burning fossil fuels, cutting forests, and other activities add more carbon to the air than the ocean naturally absorbs.

> Carbon dioxide is a greenhouse gas. Rising levels of these gases correlate with rising global temperature. This indicates to scientists that human induced increases in these gases are the most likely cause of recent global climate change.

> Gaseous nitrogen abounds in the atmosphere, but only bacteria can make it available to other organisms.
< Links to Acids 2.6, Nitrogen fixation 19.7

Nitrogen Reservoirs and Movements

Nitrogen moves in an atmospheric cycle known as the **nitrogen cycle** (Figure 42.12). The main reservoir is the atmosphere, which is about 80 percent nitrogen. Triple covalent bonds hold two atoms of nitrogen gas together (N_2, or $N{\equiv}N$). Plants cannot use gaseous nitrogen because they cannot break these bonds. Only certain bacteria can carry out **nitrogen fixation**. They break the bonds in N_2, and use the nitrogen atoms to form ammonia, which is ionized in water as ammonium (NH_4^+) ❶. Nitrogen-fixing cyanobacteria live in aquatic habitats, soil, and as components of lichens. Another group of nitrogen-fixing bacteria forms nodules on the roots of peas and other legumes.

Plants take up ammonium from soil water ❷ and use it in metabolic reactions. Consumers get nitrogen by eating plants or one another. Bacterial and fungal decomposers that break down wastes and remains return ammonium to the soil ❸.

Nitrification is a two-step process that converts ammonium to nitrates ❹. First, ammonia-oxidizing bacteria and archaeans convert ammonium to nitrite (NO_2^-), then bacteria convert nitrites to nitrates (NO_3^-). Like ammonium, nitrates are taken up and used by producers ❺.

Ecosystems lose nitrogen by **denitrification**. Denitrifying bacteria that use nitrate for energy produce nitrogen gas that escapes into the atmosphere ❻.

Human Effects on the Nitrogen Cycle

Manufactured ammonia fertilizers improve crop yields, but also modify soil chemistry. These fertilizers increase the concentration of hydrogen ions, as well as nitrogen. As a result of the increased acidity, nutrient ions bound to soil particles get replaced by hydrogen ions, and nutrients essential for plant growth trickle away in soil water. Nitrogen fertilizer also causes problems by running off and polluting aquatic habitats.

Burning fossil fuels releases nitrous oxide, which is a greenhouse gas and contributes to acid rain (Section 2.6). Nitrogen deposition in acid rain has the same negative effects on soil as overfertilizing.

---

denitrification Bacteria convert nitrates or nitrites to gaseous forms of nitrogen.
nitrification Bacteria convert ammonium to nitrates.
nitrogen cycle Movement of nitrogen among the atmosphere, soil, and water, and into and out of food webs.
nitrogen fixation Bacteria use nitrogen gas to form ammonia.

Take-Home Message **How does nitrogen cycle in ecosystems?**

> Nitrogen-fixing bacteria convert gaseous nitrogen to ammonium that plants can use. Other bacteria, archaeans, and fungi convert ammonium to nitrites and nitrates. Still other bacteria convert these soluble forms back to nitrogen gas.

> Use of manufactured fertilizers and nitrogen oxides from vehicle emissions add excess nitrogen to ecosystems.

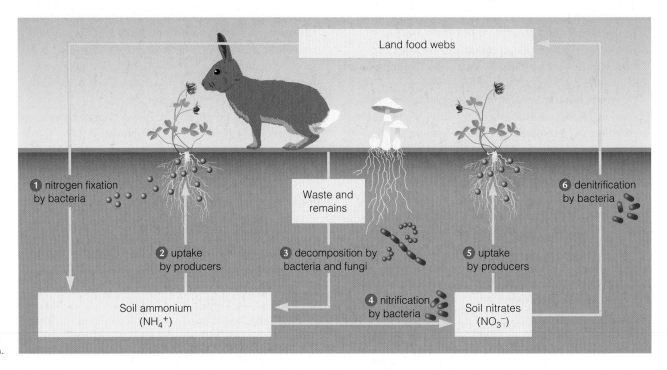

Figure 42.12 Animated Nitrogen cycle in a land ecosystem.

Land food webs

❶ nitrogen fixation by bacteria
❻ denitrification by bacteria
Waste and remains
❷ uptake by producers
❸ decomposition by bacteria and fungi
❺ uptake by producers
Soil ammonium (NH_4^+)
❹ nitrification by bacteria
Soil nitrates (NO_3^-)

42.10 The Phosphorus Cycle

❯ Unlike carbon and nitrogen, phosphorus seldom occurs as a gas. It moves in a sedimentary cycle.
❮ Link to Plate tectonics 16.7

Most of Earth's phosphorus is bonded to oxygen as phosphate (PO_4^{3-}), an ion that occurs in rocks and sediments. In the **phosphorus cycle**, phosphorus passes quickly through food webs as it moves from land to ocean sediments, then slowly back to land (Figure 42.13). There is no commonly occurring gaseous form of phosphorus, so the phosphorus cycle is a **sedimentary cycle**.

Weathering and erosion move phosphates from rocks into soil, lakes, and rivers ❶. Leaching and runoff carry dissolved phosphates to the ocean ❷. Here, most phosphorus comes out of solution and settles as deposits along continental margins ❸. Slow movements of Earth's crust can uplift these deposits onto land ❹, where weathering releases phosphates from rocks once again.

The biological portion of the phosphorus cycle begins when producers take up phosphorus. Land plants take up dissolved phosphate from the soil water ❺. Land animals get phosphates by eating the plants or one another. Phosphorus returns to the soil in wastes and remains ❻.

In the seas, phosphorus enters food webs when producers take up phosphate dissolved in seawater ❼. As on land, wastes and remains replenish the supply ❽.

phosphorus cycle Movement of phosphorus among Earth's rocks and waters, and into and out of food webs.
sedimentary cycle Biochemical cycle in which the atmosphere plays little role and rocks are the major reservoir.

Too Much of a Good Thing (revisited)

Water treatment systems can remove phosphates from detergents out of wastewater before it is discharged, but the additional treatment adds to the cost. Phosphate-rich runoff from lawns usually gets into waterways without going through a treatment plant. Thus, the most effective and economical way to keep aquatic ecosystems healthy is to avoid the use of phosphate-rich products when substitutes are available. Detergents that get their cleaning power from enzymes can replace phosphate-based products and most lawns will thrive even with low-phosphate fertilizer.

How Would You Vote? Is banning high-phosphate products the best way to prevent phosphate-related eutrophication, or are consumer education programs a better option? See CengageNow for details, then vote online (cengagenow.com).

Phosphorus is often a limiting factor for plant growth, so phosphate-rich droppings from seabird or bat colonies are collected and used as fertilizer. Phosphate-rich rock is also mined for this purpose. When fertilizer, detergent, or sewage with lots of phosphate gets into water, it can cause the type of eutrophication described in Section 42.1.

Take-Home Message How does phosphorus cycle in ecosystems?

❯ Rocks are the main phosphorus reservoir. Weathering puts phosphates into water. Producers take up dissolved phosphates.

❯ Phosphate-rich rock or wastes can be used as fertilizer; however, excess phosphate from these or other sources can cause eutrophication if they get into aquatic habitats.

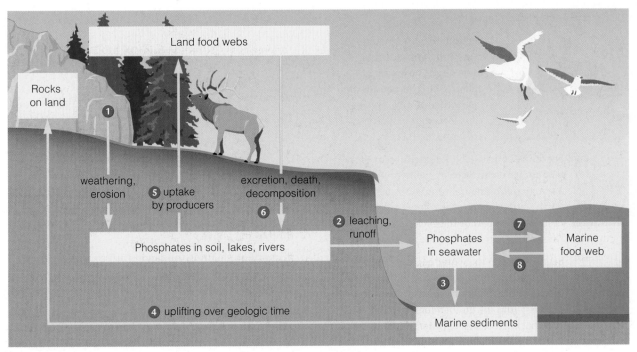

Land food webs

Rocks on land

weathering, erosion

❺ uptake by producers

excretion, death, decomposition ❻

❷ leaching, runoff

Phosphates in soil, lakes, rivers

Phosphates in seawater ❼ Marine food web ❽

❸

❹ uplifting over geologic time

Marine sediments

Figure 42.13 Animated
The phosphorus cycle.

Summary

Section 42.1 Inorganic substances such as phosphorus serve as nutrients for living organisms. Excessive inputs of nutrients as a result of human activities can alter ecosystem dynamics, as when phosphate from detergents or fertilizers causes **eutrophication**.

Section 42.2 There is a one-way flow of energy into and out of an ecosystem, and a cycling of materials among the organisms within it. All ecosystems have inputs and outputs of energy and nutrients.

Primary producers convert energy from an inorganic source (usually light) into chemical bond energy. **Primary production**, the rate at which producers capture and store energy, can vary over time and between locations.

Consumers feed on producers or one another. For example, **detritivores** eat small bits of organic remains and **decomposers** break wastes and remains into their inorganic components.

Section 42.3 A **food chain** shows one path of energy and nutrient flow among organisms. Each organism in a food chain is at a different **trophic level**, with the primary producer being the first level and consumers at higher levels. Efficiency of energy transfers from one trophic level to the next is always low, so most food chains have only four or five links.

Section 42.4 A **food web** consists of interconnecting food chains. Most food webs include both **grazing food chains**, in which herbivores eat producers, and **detrital food chains**, in which producers die and are consumed by detritivores. Because of connections through food webs, a change that affects one species in an ecosystem will have effects on many others.

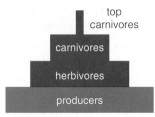

Section 42.5 Energy pyramids and biomass pyramids depict how energy and organic compounds are distributed among organisms. All energy pyramids are largest at their base.

Section 42.6 In the nonbiological part of a **biogeochemical cycle**, an element moves among environmental reservoirs such as Earth's atmosphere, rocks, and waters. In the biological part of the cycle, elements move through food webs, then return to the environment.

Section 42.7 In the **water cycle**, evaporation, condensation, and precipitation move water from its main reservoir—the oceans—into the atmosphere, onto land, then back to oceans. A **watershed** is a region in which all water drains into the same stream or river. Water that falls onto land may become part of the **groundwater**; it may be become **soil water** or be stored in an **aquifer**. Alternatively, it may become **runoff**. The water cycle helps move soluble forms of other nutrients.

Section 42.8 The main reservoir for carbon is rocks, but the **carbon cycle** moves carbon mainly among seawater, the air, soils, and living organisms in an **atmospheric cycle**. Carbon dioxide contributes to the **greenhouse effect**. Greenhouse gases keep Earth's surface warm enough to support life. However, as a result of fossil fuel consumption and other human activities, the levels of these gases are increasing. The increase correlates with and is the most likely cause of **global climate change**.

Section 42.9 The **nitrogen cycle** is an atmospheric cycle. Air is the main reservoir for N_2, a gaseous form of nitrogen that plants cannot use. Plants can take up and use ammonium that bacteria produce by **nitrogen fixation**. Fungi and bacteria that act as decomposers add ammonium derived from remains to the soil. Plants also use nitrates that some bacteria produce from ammonium through **nitrification**. Nitrogen is returned to the air by bacteria that carry out **denitrification** of nitrates. Humans add extra nitrogen to ecosystems by using synthetic fertilizer and by burning fossil fuels, which releases nitrous oxide.

Section 42.10 The **phosphorus cycle** is a **sedimentary cycle** with no significant atmospheric component. Phosphorus from rocks dissolves in water and is taken up by producers. Phosphate-rich rocks and deposits of bird droppings are mined for use as fertilizer.

Self-Quiz   Answers in Appendix III

1. In most ecosystems, the primary producers use energy from _____ to build organic compounds.
   - a. inorganic chemicals
   - b. sunlight
   - c. heat
   - d. lower trophic levels

2. Decomposers are commonly _____ .
   - a. fungi
   - b. bacteria
   - c. plants
   - d. a and b

3. Organisms at the first trophic level _____ .
   - a. capture energy from a nonliving source
   - b. are eaten by organisms at higher trophic levels
   - c. are shown at the bottom of an energy pyramid
   - d. all of the above

4. Primary productivity on land is affected by _____ .
   - a. nutrient availability
   - b. amount of sunlight
   - c. temperature
   - d. all of the above

5. A(n) _____ is an autotroph.
   - a. primary producer
   - b. herbivore
   - c. detritivore
   - d. top carnivore

Rising Atmospheric Carbon To assess the impact of human activity on the carbon dioxide level in Earth's atmosphere, it helps to take a long view. One useful data set comes from deep core samples of Antarctic ice. The oldest ice core that has been fully analyzed dates back a bit more than 400,000 years. Air bubbles trapped in the ice provide information about the gas content in Earth's atmosphere at the time the ice formed. Combining ice core data with more recent direct measurements of atmospheric carbon dioxide—as in Figure 42.14—can help scientists put current changes in the atmospheric carbon dioxide into historical perspective.

1. What was the highest carbon dioxide level between 400,000 B.C. and 0 A.D.?
2. During this period, how many times did carbon dioxide reach a level comparable to that measured in 1980?
3. The industrial revolution occurred around 1800. How much did carbon dioxide levels change in the 800 years prior to this event? In 175 years after it?
4. Did carbon dioxide levels rise more between 1800 and 1975 or between 1980 and 2007?

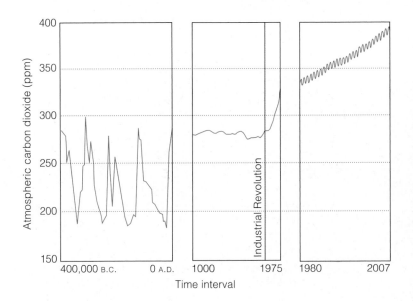

Figure 42.14 Changes in atmospheric carbon dioxide levels (in parts per million). Direct measurements began in 1980. Earlier data are based on ice cores.

6. Most of earth's fresh water is _____ .
   a. in lakes and streams   c. frozen as ice
   b. in aquifers and soil   d. in bodies of organisms

7. Earth's largest carbon reservoir is _____ .
   a. the atmosphere   c. seawater
   b. sediments and rocks   d. living organisms

8. Carbon is released into the atmosphere by _____ .
   a. photosynthesis   c. burning fossil fuels
   b. aerobic respiration   d. b and c

9. Greenhouse gases _____ .
   a. help keep Earth's surface warm enough for life
   b. are released by natural and human activities
   c. are at higher levels than they were 100 years ago
   d. all of the above

10. The _____ cycle is a sedimentary cycle.
    a. phosphorus   c. nitrogen
    b. carbon   d. water

11. Earth's largest phosphorus reservoir is _____ .
    a. the atmosphere   c. sediments and rocks
    b. bird droppings   d. living organisms

12. Plants obtain _____ by taking it up from the air.
    a. nitrogen   c. phosphorus
    b. carbon   d. a and b

13. Nitrogen fixation converts _____ to _____ .
    a. nitrogen gas; ammonia   c. ammonia; nitrates
    b. nitrates; nitrites   d. nitrites; nitrogen oxides

14. Which holds the least carbon?
    a. soils   b. the air   c. seawater   d. rocks

15. Match each term with its most suitable description.
    ___ carbon dioxide   a. contains triple bond
    ___ bicarbonate   b. product of nitrogen fixation
    ___ ammonium   c. marine carbon source
    ___ nitrogen gas   d. greenhouse gas

Additional questions are available on CENGAGENOW.

Critical Thinking

1. Marguerite has a vegetable garden in Maine. Eduardo has one in Florida. List the variables that could cause differences in the primary production of these gardens.

2. Where does your water come from? A well, a reservoir? Beyond that, what area is included within your watershed and what are the current flows? Visit the *Science in Your Watershed* site at water.usgs.gov/wsc and find the answers to these questions.

3. Why do all organisms require both phosphorus and nitrogen? List some of the molecules common to all life that contain these essential elements.

4. Rather than using fertilizer, a farmer may rotate crops, planting legumes one year, then another crop, then legumes again. Explain how crop rotation keeps soil fertile.

Animations and Interactions on CENGAGENOW:
❭ Energy flow and nutrient cycling; Food webs; Water cycle; Carbon cycle; Greenhouse effect; Nitrogen cycle; Phosphorus cycle.

Key Concepts

Air Circulation Patterns
Air circulation starts with latitudinal differences in energy inputs from the sun. Movement of air from the equator toward poles is affected by Earth's rotation and gives rise to major surface winds and latitudinal patterns in rainfall.

Ocean Circulation Patterns
Heating of the tropics also sets ocean waters in motion. As water circulates, it carries and releases heat, and so affects the climate on land. Interactions between oceans, air, and land affect coastal climates.

43 The Biosphere

43.1 Effects of El Niño

Professional surfer Ken Bradshaw has ridden a lot of waves, but one stands out in his memory. In January of 1998, he found himself off the coast of Hawaii riding the biggest wave he had ever seen (Figure 43.1). It towered more than 12 meters (39 feet) high and gave him the ride of a lifetime.

The giant wave was one manifestation of an **El Niño**, a recurring climate event in which equatorial waters of the eastern and central Pacific Ocean warm above their average temperature. The term El Niño means "baby boy" and refers to Jesus; it was first used by Peruvian fishermen to describe local weather changes and a shortage of fishes that occurred in some years around Christmas. Scientists now know that during an El Niño, marine currents interact with the atmosphere in ways that influence weather patterns worldwide.

An El Niño affects marine food webs along eastern Pacific coasts. As unusually warm water flows toward these coasts, it displaces currents that would otherwise bring up nutrients from the deep. Without these nutrients, marine primary producers decline in numbers. The dwindling producer populations and warming water cause a decrease in populations of small, cold-water fishes such as anchovies, as well as the larger consumers that rely on them. During the 1997–1998 El Niño, about half of the sea lions on the Galápagos Islands starved to death. California's population of northern fur seals also suffered a sharp decline.

Rainfall patterns shift during an El Niño. During the winter of 1997–1998, torrential rains caused flooding and landslides along eastern Pacific coasts, while Australia and Indonesia suffered from drought-driven crop failures and raging wildfires. A El Niño typically brings cooler, wetter weather to the American Gulf states, and reduces the likelihood of hurricanes.

An El Niño typically persists for 6 to 18 months. It may be followed by an interval in which the temperature of the eastern Pacific remains near its average, or by a La Niña. During a

Figure 43.1 El Niño. *Opposite*, Ken Bradshaw rides a wave more than 12 meters (39 feet) high, during the most powerful El Niño of the past century. *Above*, interactions among Earth's oceans and atmosphere give rise to an El Niño and to other climate patterns.

La Niña, eastern Pacific waters become cooler than average. As a result, the west coast of the United States gets little rainfall and the likelihood of hurricanes in the Atlantic increases.

The previous three chapters have considered three levels of biological organization—populations, communities, and ecosystems. With this discussion of El Niño/La Niña, we invite you to move to the next level, to consider the factors that influence the properties of the biosphere. The **biosphere** includes all places where we find life on Earth. Many organisms live in the hydrosphere: the ocean, ice caps, and other bodies of water, liquid and frozen. Others live on and in sediments and soils of the lithosphere: Earth's outer, rocky layer. Still others lift off into the lower region of the atmosphere: the gases and airborne particles that envelop Earth.

---

biosphere All regions of Earth where organisms live.
El Niño Periodic warming of equatorial Pacific waters and the associated shifts in global weather patterns.
La Niña Periodic cooling of equatorial Pacific waters and the associated shifts in global weather patterns.

Land Biomes
A biome consists of geographically separated regions that have a similar climate and soils, and so support similar types of vegetation. Biomes include deserts, grasslands, chaparral, various types of forests, and tundra.

Freshwater Ecosystems
Lakes have gradients of light and temperature. They undergo succession, changing over time. In temperate zones, their waters mix in response to seasonal changes in temperature. Rivers vary along their length in their properties, and in the organisms they contain.

Coastal and Marine Ecosystems
Productivity is high in coastal wetlands, on coral reefs, and in the ocean's upper, sunlit water. Life also thrives in the ocean's deeper, darker waters and on the sea floor, especially on undersea mountains and at hydrothermal vents.

> The amount of solar energy that reaches Earth's surface varies from place to place and with the season.
> Latitudinal changes in sunlight give rise to air circulation patterns that determine regional rainfall.

Climate refers to average weather conditions, such as cloud cover, temperature, humidity, and wind speed, over time. Regional climates differ because many factors that influence winds and ocean currents, such as intensity of sunlight, the distribution of land masses and seas, and elevation, vary from place to place.

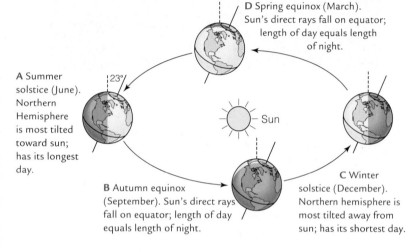

D Spring equinox (March). Sun's direct rays fall on equator; length of day equals length of night.

A Summer solstice (June). Northern Hemisphere is most tilted toward sun; has its longest day.

23°

Sun

B Autumn equinox (September). Sun's direct rays fall on equator; length of day equals length of night.

C Winter solstice (December). Northern hemisphere is most tilted away from sun; has its shortest day.

Figure 43.2 Earth's tilt and yearly rotation around the sun cause seasonal effects. The 23° tilt of Earth's axis causes the Northern Hemisphere to receive more intense sunlight and have longer days in summer than in winter.

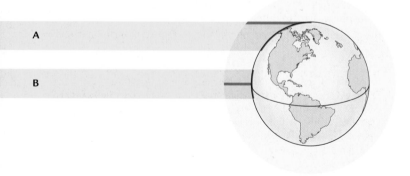

A

B

Figure 43.3 Variation in intensity of solar radiation with latitude. For simplicity, we depict two equal parcels of incoming radiation on an equinox, a day when incoming rays are perpendicular to Earth's axis.

Rays that fall on high latitudes **A** pass through more atmosphere (*blue*) than those that fall near the equator **B**. Compare the length of the *green* lines. Atmosphere is not to scale.

Also, energy in the rays that fall at the high latitude is spread over a greater area than energy that falls on the equator. Compare the length of the *red* lines.

Seasonal Effects

Each year, Earth rotates around the sun in an elliptical path (Figure 43.2). Seasonal changes in daylength and temperature arise because Earth's axis is not perpendicular to the plane of this ellipse, but rather is tilted about 23 degrees. In June, when the Northern Hemisphere is angled toward the sun, it receives more intense sunlight and has longer days than the Southern Hemisphere (Figure 43.2A). In December, the opposite occurs (Figure 43.2C). Twice a year—on spring and autumn equinoxes—Earth's axis is perpendicular to incoming sunlight. On these days, every place on Earth has 12 hours of daylight and 12 hours of darkness (Figure 43.2B,D).

In each hemisphere, the degree of seasonal change in daylength increases with latitude. At 25° north or south of the equator, the longest daylength is a bit less than 14 hours. By contrast, 60° north or south of the equator, the longest daylength is nearly 19 hours.

Air Circulation and Rainfall

On any given day, equatorial regions get more sunlight energy than higher latitudes for two reasons (Figure 43.3). First, fine particles of dust, water vapor, and greenhouse gases absorb some solar radiation or reflect it back into space. Because sunlight traveling to high latitudes passes through more atmosphere to reach Earth's surface than light traveling to the equator, less energy reaches the ground. Second, energy in an incoming parcel of sunlight is spread out over a smaller surface area at the equator than at the higher latitudes. As a result, Earth's surface warms more at the equator than at the poles.

Knowing about two properties of air can help you understand how regional differences in surface warming give rise to global air circulation and rainfall patterns. First, as air warms, it becomes less dense and rises. Hot air balloonists take advantage of this effect when they take off from the ground by heating the air inside their balloon. Second, warm air can hold more water than cooler air. This is why you can "see your breath" in cold weather; water vapor in warm exhaled air condenses into droplets when exposed to the cool external environment.

The global air circulation pattern begins at the equator, where intense sunlight warms the air and causes evaporation from the ocean. The result is an upward movement of warm, moist air (Figure 43.4A). As the air from the equator rises to higher altitudes, it cools and flows north and south, releasing moisture as rain that supports tropical rain forests.

By the time the air has moved to 30° north or south of the equator, it has given up most moisture and cooled off, so it sinks back toward Earth's surface (Figure 43.4B). Many of the world's great deserts, including the Sahara, are about 30° from the equator.

D At the poles, cold air sinks and moves toward lower latitudes.

C Air rises again at 60° north and south, where air flowing poleward meets air coming from the poles.

B As the air flows toward higher latitudes, it cools and loses moisture as rain. At around 30° north and south latitude, the air sinks and flows north and south along Earth's surface.

A Warmed by energy from the sun, air at the equator picks up moisture and rises. It reaches a high altitude, and spreads north and south.

E Major winds near Earth's surface do not blow directly north and south because of the effects of Earth's rotation. Winds deflect to the right of their original direction in the Northern Hemisphere and to the left in the Southern Hemisphere.

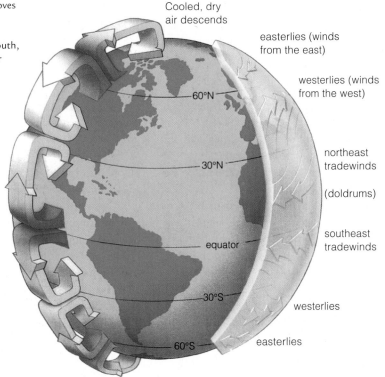

Cooled, dry air descends

easterlies (winds from the east)

westerlies (winds from the west)

60°N

northeast tradewinds

30°N

(doldrums)

southeast tradewinds

equator

30°S

westerlies

60°S   easterlies

Figure 43.4 **Animated** Global air circulation patterns.

As air continues flowing along Earth's surface toward the poles, it again picks up heat and moisture. At a latitude of about 60°, warm, moist air rises again, losing moisture as it does so (Figure 43.4C). The resulting rains support temperate zone forests.

Cold, dry air descends near the poles (Figure 43.4D). Precipitation is sparse, and polar deserts form.

Surface Wind Patterns

Major wind patterns arise as air in the lower atmosphere moves continually from latitudes where air is sinking toward those where air is rising. Earth's rotation affects the apparent trajectory of these winds. Air masses are not attached to Earth's surface, so the Earth spins beneath them, moving fastest at the equator and most slowly at the poles. Thus, as an air mass moves away from the equator, the speed at which the Earth rotates beneath it continually slows. As a result, major winds trace a curved path relative to the Earth's surface (Figure 43.4E). In the

Northern Hemisphere, winds curve toward the right; in the Southern Hemisphere, they curve toward the left. For example, between 30° north and 60° north, surface air traveling toward the North Pole is deflected right, or toward the east. Winds are named for the direction from which they blow, so the prevailing winds in the United States are westerlies.

Winds blow most consistently between regions where air rises. In the regions where the air is rising, winds can be intermittent, as in the doldrums near the equator.

Take-Home Message How does sunlight affect climate?

> Equatorial regions receive more intense sunlight than higher latitudes.

> Sunlight drives the rise of moisture-laden air at the equator. The air cools as it moves north and south, releasing rains that support tropical forests. Deserts form where cool, dry air descends. Sunlight energy also drives moisture-laden air aloft at 60° north and south latitude. This air gives up moisture as it flows toward the equator or the pole.

> Air flow in the lower atmosphere toward latitudes where air is rising and away from latitudes where it is sinking creates major surface winds. These winds trace a curved path relative to Earth's surface because of Earth's rotation.

climate Average weather conditions in a region over a long time period.

> The ocean is a continuous body of water that covers more than 71 percent of Earth's surface. Its water moves in currents that distribute nutrients through marine ecosystems.
< Link to Properties of water 2.5

Ocean Currents

Latitudinal variations in sunlight affect ocean temperature and set major currents in motion. At the equator, where vast volumes of water warm and expand, the sea level is about 8 centimeters (3 inches) higher than at either pole. The volume of water in this "slope" is enough to start sea surface water moving toward the poles. As the water moves, it gives up heat energy to the air above it.

Enormous volumes of water flow as ocean currents. The directional movement of surface currents is shaped by the force of major winds, Earth's rotation, and topography. Surface currents circulate clockwise in the Northern Hemisphere and counterclockwise in the Southern Hemisphere (Figure 43.5).

Swift, deep, and narrow currents of nutrient-poor water flow away from the equator along the east coast of continents. Along the east coast of North America, warm water flows north, as the Gulf Stream. Slower, shallower, broader currents of cold water parallel the west coast of continents and flow toward the equator.

Ocean currents affect climate. For example, Pacific Northwest coasts are cool and foggy in summer because the cold California current chills the air, and water condenses out of the cooled air as droplets. As another example, Boston and Baltimore are warm and muggy in summer because air masses pick up heat and moisture from the warm Gulf Stream, then flow over these cities.

Regional Effects

Mountains, valleys, and other surface features of the land affect climate. Suppose you track a warm air mass after it picks up moisture off California's coast. It moves inland as wind from the west, and piles up against the Sierra Nevada, a high mountain range that parallels the coast. The air cools as it rises in altitude and loses moisture as rain (Figure 43.6). The result is a **rain shadow**, a semiarid or arid region of sparse rainfall on the leeward side of high mountains. *Leeward* is the side facing away from the wind. The Himalayas, Andes, Rockies, and other great mountain ranges cause vast rain shadows.

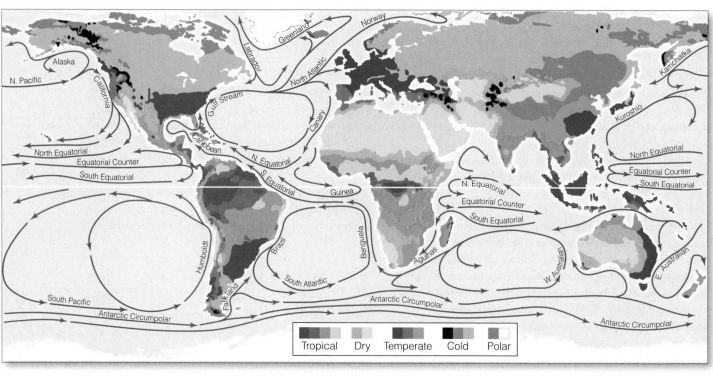

Tropical   Dry   Temperate   Cold   Polar

Figure 43.5 Animated Major climate zones correlated with surface currents and surface drifts of the world ocean. Warm surface currents start moving from the equator toward the poles, but prevailing winds, Earth's rotation, gravity, the shape of ocean basins, and landforms influence the direction of flow. Water temperatures, which differ with latitude and depth, contribute to regional differences in air temperature and rainfall.

↙ **warm surface current**   ↙ **cold surface current**

A Prevailing winds move moisture inland from the Pacific Ocean.

B Clouds pile up and rain forms on side of mountain range facing prevailing winds.

C Rain shadow on side facing away from the prevailing winds makes arid conditions.

4,000/ 75
3,000/ 85   2,000/ 25
1,800/ 125   1,000/ 25
moist habitats
1,000/ 85
15/ 25

Figure 43.6 Animated Rain shadow effect. On the side of mountains facing away from prevailing winds, rainfall is light. *Black* numbers signify annual precipitation, in centimeters, averaged on both sides of the Sierra Nevada, a mountain range. *White* numbers signify elevations, in meters.

Differences in the heat capacity of water and land give rise to coastal breezes. In the daytime, land warms faster than water. As air over land warms and rises, cooler offshore air moves in to replace it (Figure 43.7A). After sundown, land cools more than the water, so the breezes reverse direction (Figure 43.7B).

Differential heating of water and land also causes **monsoons**, which are winds that change direction seasonally. For example, the continental interior of Asia heats up in the summer, so air rises above it. Moist air from over the warm Indian Ocean to the south moves in to replace the rising air, and this north-blowing wind delivers heavy rains. In the winter, the continental interior is cooler than the ocean. As a result, cool, dry wind blowing from the north toward southern coasts causes a seasonal drought.

Proximity to an ocean moderates climate. For example, Seattle, Washington, is slightly farther north than Minneapolis, Minnesota, but Seattle has much milder winters. Air over Seattle draws heat from the adjacent Pacific Ocean, a heat source not available to Minneapolis.

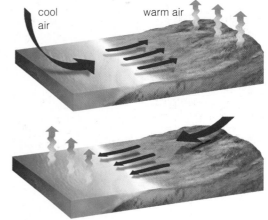

cool air   warm air

A In afternoon; the land is warmer than the sea, so the breeze blows onto shore.

B In the evening, the sea is warmer than the land; the breeze blows out to sea.

Figure 43.7 Coastal breezes.

---

monsoon Wind that reverses direction seasonally.
rain shadow Dry region downwind of a coastal mountain range.

Take-Home Message **How do ocean currents arise and affect regional climates?**

❯ Surface ocean currents are set in motion by latitudinal differences in solar radiation. Currents are affected by winds and by Earth's rotation.

❯ The collective effects of air masses, oceans, and landforms determine regional temperature and moisture levels.

> Similar communities often evolve in widely separated regions as a result of similar environmental factors.
< Links to Carbon-fixing pathways 6.8, Morphological convergence 16.8, Soil 26.2

Differences Between Biomes

Biomes are areas of land characterized by their climate and type of vegetation (Figure 43.8). A biome is discontinuous; most biomes include widely separated areas on different continents. For example, the temperate grassland biome includes areas of North American prairie, South African veld, South American pampa, and Eurasian steppe. In all of these regions, the main vegetation is grasses and other nonwoody flowering plants.

Rainfall and temperature are the main determinants of the type of biome in a given region. Desert biomes get the least annual rainfall, grasslands and shrublands get more, and forests get the most. Deserts occur where temperatures soar the highest and tundra where they drop the lowest.

Soils also influence biome distribution. Soils consist of a mixture of mineral particles and varying amounts of humus. Water and air fill spaces between soil particles. Properties of soils vary depending on the types, proportions, and compaction of particles. Deserts have sandy or gravelly, fast-draining soil with little topsoil. Topsoil tends to be deepest in natural grasslands, where it can be more than one meter thick. For this reason, grasslands are often converted to agricultural uses.

Climate and soils affect primary production, so primary production varies among biomes (Figure 43.9).

Similarities Within a Biome

Unrelated species living in widely separated parts of a biome often have similar body structures that arose by the process of morphological convergence (Section 16.8). For example, cactuses with water-storing stems live in

---

biome Discontinuous region characterized by its climate and dominant vegetation.

- desert
- dry shrubland, dry woodland
- warm grassland (e.g., savanna)
- temperate grassland
- mountain grassland
- tropical broadleaf forest
- temperate deciduous forest
- tropical coniferous forest
- temperate coniferous forest (e.g., rain forest)
- northern coniferous forest (e.g., boreal forest)
- tropical dry forest
- tundra
- mountains, complex zonation
- mangrove swamp
- perpetual ice cover
- marine ecoregions

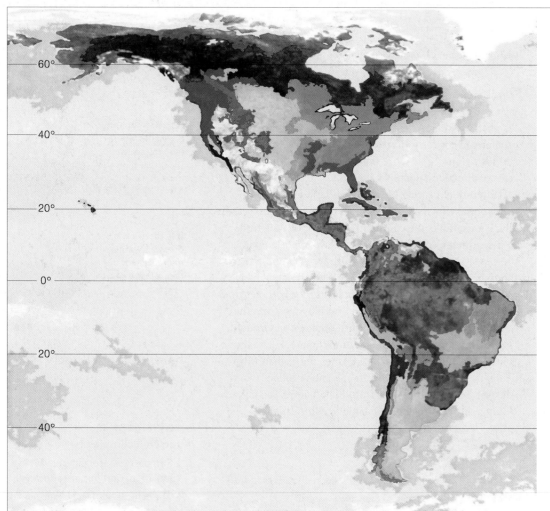

Figure 43.8 Global distribution of major categories of biomes and marine ecoregions.

North American deserts and euphorbs with water-storing stems live in African deserts. Cactuses and euphorbs do not share an ancestor with a water-storing stem. This feature evolved independently in the two groups as a result of similar selection pressures. Similarly, an ability to carry out C4 photosynthesis evolved independently in grasses growing in warm grasslands on different continents. C4 photosynthesis is more efficient than the more common C3 pathway under hot, dry conditions.

Take-Home Message What factors shape marine communities?

❭ Biomes are vast expanses of land dominated by distinct kinds of plants that support characteristic communities. They vary in their primary productivity.

❭ Evolution often produces similar solutions to environmental challenges in different regions of a biome.

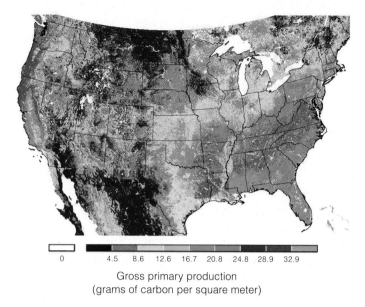

| 0 | | 4.5 | 8.6 | 12.6 | 16.7 | 20.8 | 24.8 | 28.9 | 32.9 |

Gross primary production
(grams of carbon per square meter)

Figure 43.9 Remote satellite monitoring of gross primary productivity across the United States. The differences roughly correspond with variations in soil types and moisture.

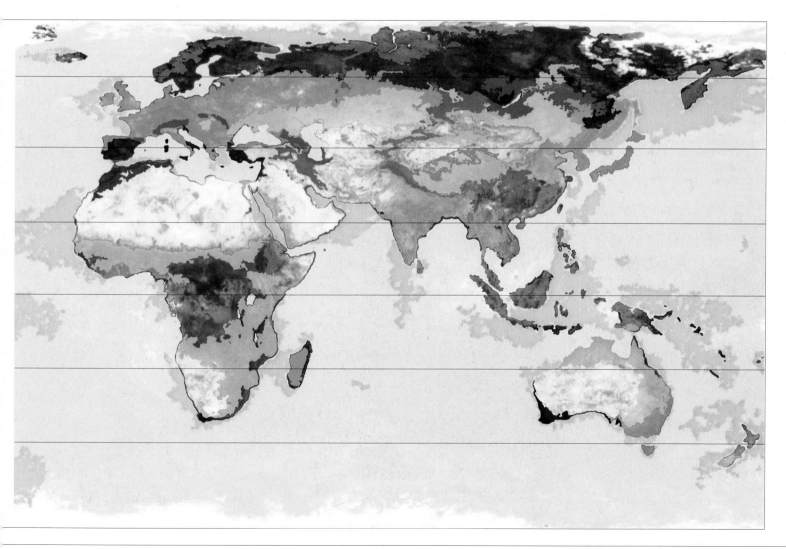

❭ Low rainfall shapes the desert biome.
❬ Links to Carbon-fixing pathways 6.8, Atacama Desert 18.1, Nitrogen fixation 19.7, Kangaroo rat 37.2

Deserts

Desert Locations and Conditions

Deserts receive an average of less than 10 centimeters (4 inches) of rain per year. They cover about one-fifth of Earth's land surface and many are located at about 30° north and south latitude, where global air circulation patterns cause dry air to sink. Rain shadows also reduce rainfall. For example, Chile's Atacama desert is on the leeward side of the Andes. Similarly, the Himalayas prevent rain from falling in China's Gobi desert.

Lack of rainfall keeps the humidity in deserts low. With little water vapor to block rays, intense sunlight reaches the ground and heats it. At night, the lack of insulating water vapor in the air allows the temperature to fall fast. As a result, deserts tend to have larger daily temperature shifts than other biomes.

Desert soils have little topsoil (Figure 43.10), the layer most important for plant growth. The soils often are somewhat salty, because rain that falls usually evaporates before seeping into the ground. Rapid evaporation allows any salt in rainwater to accumulate at the soil surface.

O horizon:
Pebbles, little organic matter

A horizon:
Shallow, poor soil

B horizon:
Evaporation causes salt buildup; leaching removes nutrients

C horizon:
Rock fragments from uplands

Figure 43.10 Desert soil profile.

Adaptations to Desert Conditions

Despite their harsh conditions, most deserts support some plant life. Diversity is highest in regions where moisture is available in more than one season (Figure 43.11).

Many desert plants have adaptations that reduce water loss. For example, some have spines or hairs (Figure 43.12A). In addition to detering herbivory, these structures reduce water loss by trapping some water and keeping the humidity around the stomata high. Where rains fall seasonally, some plants reduce their water loss by leafing out after a rain, then dropping their leaves when dry conditions return (Figure 43.12B).

Some desert plants store water in their tissues during the wet season, for use in drier times. For example, the stem of the barrel cactus in Figure 43.12A has a spongy pulp that holds water. The cactus stem swells after a rain, then shrinks as the plant uses up its store of water.

Woody desert shrubs such as mesquite and creosote have extensive, efficient root systems that take up the little water that is available. Mesquite roots have been found as deep as 60 meters (197 feet) beneath the soil surface.

A Creosote bush is the predominant vegetation in the driest lowlands.

B A greater variety of plants survive in uplands, which are a bit wetter and cooler.

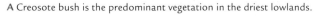

Figure 43.11 Vegetation in Arizona's Sonoran Desert.

A Barrel cactuses are covered by spines that reduce evaporative water loss. The cactus is a CAM plant.

B Ocitillo, a desert shrub, grows leaves on its stems after a rain, then sheds them when conditions become dry again.

Figure 43.12 Perennials adapted to desert conditions.

Figure 43.13 Mojave Desert after the rains. Annual poppies sprout, flower, produce seeds, and die within weeks beneath slow-growing perennial cactuses.

Alternative carbon-fixing pathways also help desert plants conserve water. Cactuses, agaves, and euphorbs are CAM plants. They open their stomata only at night when temperature declines.

Most deserts contain a mix of annuals and perennials (Figure 43.13). The annuals are adapted to desert life by a life cycle that allows them to sprout and reproduce in the short time that the soil is moist.

Animals also have adaptations that allow them to conserve water. For example, the highly efficient kidneys of a desert kangaroo rat minimize its water needs (Section 37.2). Most desert animals are not active at the height of the daytime heat (Figure 43.14).

The Crust Community

In many deserts, the soil is covered by a desert crust, a community that can include cyanobacteria, lichens, mosses, and fungi. The organisms secrete organic molecules that glue them and the surrounding soil particles together. The crust benefits members of the larger desert community in important ways. Its cyanobacteria fix nitrogen and make ammonia available to plants. The crust also holds soil particles in place. When the fragile connections within the desert crust are broken, the soil can blow away. Negative effects of such disturbance are exacerbated when windblown soil buries healthy crust in an undisturbed area, killing additional crust organisms and allowing more soil to take flight.

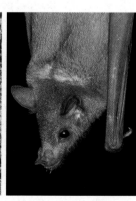

A The Sonoran desert tortoise spends much of its life inactive. In hot summer months, it ventures out of its burrow only in cool mornings to feed. It hibernates during the cold winter, when little food is available.

B Lesser long-nosed bats spend spring and summer in the Sonoran Desert, where they avoid the daytime heat by resting in caves.

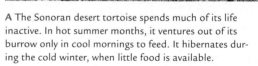

Figure 43.14 Two Sonoran desert animals.

Take-Home Message **What are deserts?**

> A desert gets very little rain and has low humidity. There is plenty of sunlight, but the lack of water prevents most plants from surviving here.

> The plants that dominate in deserts have adaptations that allow them to reduce water lost by transpiration, store water, or access water deep below ground.

> Desert animals often spend the day inactive, sheltering from the heat.

> Desert soils are held in place by a community of organisms that form a desert crust. Disruption of this crust allows wind to strip away soil.

desert Biome with little rain and low humidity; plants that have water-storing and conserving adaptations predominate.

> Perennial grasses adapted to fire and to grazing are the main plants in grasslands.
< Links to Erosion 26.2, Adaptations to herbivory 41.5

Temperate grasslands and tropical savannahs

Figure 43.16 African savanna, a tropical grassland with scattered trees. Africa's savannas are famous for their abundant wildlife such as this immense herd of wildebeests.

Grasslands form in the interior of continents between deserts and temperate forests. Grassland soils are rich, with deep topsoil. Annual rainfall is enough to keep desert from forming, but not enough to support woodlands. Low-growing grasses and other nonwoody plants tolerate strong winds, sparse and infrequent rain, and intervals of drought. Growth tends to be seasonal. Constant trimming by grazers, along with periodic fires, keeps trees and most shrubs from taking hold.

Temperate Grasslands

Temperate grasslands are warm in summer, but cold in winter. Annual rainfall is 25 to 100 centimeters (10–40 inches), with rains throughout the year. Grass roots extend profusely through the thick topsoil and help hold it in place, preventing erosion by the constant winds. North America's grasslands are shortgrass and tallgrass prairies (Figure 43.15).

During the 1930s, much of the shortgrass prairie of the American Great Plains was plowed to grow wheat. The strong winds, a prolonged drought, and unsuitable farm-

ing practices turned much of the region into what the newspapers of that time called the Dust Bowl.

Tallgrass prairie has somewhat richer topsoil and slightly more frequent rainfall than in the shortgrass prairie. Before the arrival of Europeans, it covered about 140 million acres, mostly in Kansas. Nearly all tallgrass prairie has now been converted to cropland. The Tallgrass Prairie National Preserve was created in 1996 to protect the little that remains.

Savannas

Savannas are broad belts of grasslands with scattered shrubs and trees. They lie between tropical forests and deserts. Temperatures are warm year-round, but rainfall is seasonal. Africa has the most extensive savanna; it covers about half the continent. African savanna supports enormous herds of hoofed grazers (Figure 43.16), and the predators such as lions that feed on them.

grassland Biome in the interior of continents where grasses and nonwoody plants adapted to grazing and fire predominate.

Take-Home Message **What are grasslands?**
> Grasslands are biomes dominated by grasses and other nonwoody plants that are adapted to fire and to grazers.

A horizon:
Alkaline, deep, rich in humus

B horizon:
Percolating water enriches layer with calcium carbonates

A Prairie soil profile.

B Tallgrass prairie in Kansas. See Figure 42.3 for a food chain.

C Bison grazing in South Dakota's shortgrass prairie.

Figure 43.15 North America's temperate grasslands.

❭ Regions with cool, rainy winters and hot, dry summers support dry shrublands and woodlands.

❬ Link to Adaptation to fire 41.8

Chaparral

Dry shrublands receive 25 to 60 centimeters (10–24 inches) of rain annually. We see them in South Africa, Greece, Italy, Chile, and California, where they are known as **chaparral**. California has about 6 million acres of chaparral (Figure 43.17A,B).

Rains occur seasonally, and lightning-sparked fires sometimes sweep through shrublands during the dry season. In California, where homes are often built near chaparral, the fires frequently cause property damage (Figure 43.17C). Foliage of many chaparral shrubs has oils that deter herbivores and also make the plant highly flammable. However, the plants are adapted to occasional fires. Some grow back from root crowns after a fire. Seeds of other chaparral species germinate only after they are exposed to heat or smoke, ensuring that the seeds sprout only when young seedlings face little competition.

chaparral Biome of dry shrubland in regions with hot, dry summers and cool, rainy winters.

Figure 43.18 Oak woodland in northern California.

Dry woodlands prevail where the seasonal rainfall is 40 to 100 centimeters (16–40 inches). Examples include the eucalyptus forests of Australia, and California's oak forests (Figure 43.18). Acorns produced by the oaks are an important seasonal source of food for birds and mammals in this community.

Take-Home Message What are dry shrublands and woodlands?

❭ Dry shrublands and woodlands form in areas with a hot, dry season and a cool, rainy one. Plants in these biomes are adapted to seasonal drought and to fire.

❭ Dry shrublands, known as chaparral, get less water than dry woodlands.

A Chaparral is California's most extensive natural community.

Figure 43.17 California's chaparral.

B Evergreen shrubs with leathery leaves predominate. Most are less than 2 meters (6 feet) high.

C Fire on chaparral-covered hills above Malibu. Most fires are now started by people, rather than lightning.

❯ Broadleaf (angiosperm) trees are the main plants in temperate and tropical forests.
❮ Links to Deforestation 21.1, Primary production 42.2

Semi-Evergreen and Deciduous Forests

Semi-evergreen forests occur in the humid tropics of Southeast Asia and India. The forests include broadleaf (angiosperm) trees that retain leaves year-round, and deciduous broadleaf trees. A deciduous plant sheds leaves annually, prior to a season when cold or dry conditions would not favor growth. In semi-evergreen forests, deciduous trees shed their leaves at the start of the dry season.

Where less than 2.5 centimeters (1 inch) of rain falls in the dry season, tropical deciduous forests form. In tropical deciduous forests, most trees shed leaves at the start of the dry season.

Temperate deciduous forests form in the Northern Hemisphere in parts of eastern North America, western and central Europe, and parts of Asia, including Japan. In these regions, 50 to 150 centimeters (about 20–60 inches) of precipitation falls throughout the year. Winters are cool and summers are warm.

Growth of temperate deciduous forests is seasonal. Leaves often turn color before dropping in autumn (Figure 43.19). Winters are cold and trees remain dormant while water is locked in snow and ice. In the spring, when conditions again favor growth, deciduous trees flower and put out new leaves. Also during the spring, leaves that were shed the prior autumn decay to form a rich humus. Rich soil and a somewhat open canopy that lets sunlight through allows understory plants to flourish.

The temperate deciduous forests of North America are the most species-rich examples of this biome. Different tree species characterize different regions of these forests. For example, Appalachian forests include mainly oaks, whereas beeches and maples dominate Ohio's forests. Animals in North American deciduous forests include grazing deer, seed-eating squirrels and chipmunks, as well as omnivores such as raccoons, opossums, and black bears. Native predators such as wolves and mountain lions have been largely eliminated.

Tropical Rain Forests

Tropical rain forests of evergreen broadleaf trees form between latitudes 10° north and south in equatorial Africa, the East Indies, Southeast Asia, South America, and Central America. Rain falls throughout the year for an annual total of 130 to 200 centimeters (50 to 80 inches).

Regular rains, combined with an average temperature of 25°C (77°F) and little variation in daylength, allows photosynthesis to continue year-round. Of all land biomes, tropical forests have the greatest primary production. Per unit area, they remove more carbon from the atmosphere than other forests or grasslands.

Tropical rain forest is the most structurally complex and species-rich biome. The forest has a multilayer structure (Figure 43.20). Its broadleaf trees can stand 30 meters (100 feet) tall. The trees often form a closed canopy that prevents most sunlight from reaching the forest floor.

Temperate deciduous forest

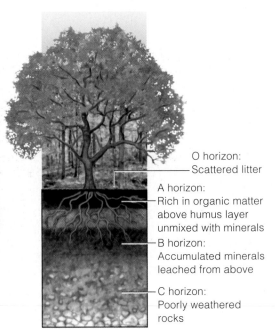

O horizon: Scattered litter

A horizon: Rich in organic matter above humus layer unmixed with minerals

B horizon: Accumulated minerals leached from above

C horizon: Poorly weathered rocks

Figure 43.19 North American temperate deciduous forest. Forest soil profile at *right*.

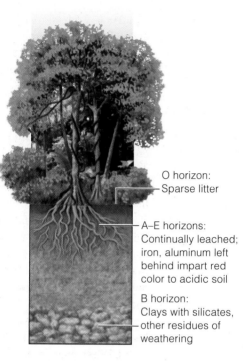

O horizon:
Sparse litter

A–E horizons:
Continually leached;
iron, aluminum left
behind impart red
color to acidic soil

B horizon:
Clays with silicates,
other residues of
weathering

Figure 43.20 Tropical rain forest. Forest soil profile at *right*.

Tropical
rain forest

Vines and epiphytes (plants that grow on another plant, but do not withdraw nutrients from it) thrive in the shade beneath the canopy.

Trees of topical rain forests shed leaves continually, but decomposition and mineral cycling happen so fast in this warm, moist environment that litter does not accumulate. Soils are highly weathered, heavily leached, and are very poor nutrient reservoirs.

Deforestation is an ongoing threat to tropical rain forests. Tropical forests are located in developing countries with fast-growing human populations who look to the forest as a source of lumber, fuel, and potential agricultural land. As human populations expand, more and more trees fall to the ax.

Deforestation in any region leaves fewer trees to remove carbon dioxide from the atmosphere. In rain forests, it also causes the extinction of species found nowhere else in the world. Compared to other land biomes, tropical rain forests have the greatest variety and numbers of insects, as well as the most diverse collection of birds and primates. This great species diversity means many species are affected by the loss of any amount of forest. Among the potential losses are species with chemicals that could save human lives. Two chemotherapy drugs, vincristine and vinblastine, were extracted from the rosy periwinkle, a low-growing plant native to Madagascar's rain forests. Today, these drugs help fight leukemia, lymphoma, breast cancer, and testicular cancer. No doubt other similarly valuable species live in the rain forests and will go extinct before we learn how they can help us.

temperate deciduous forest Northern Hemisphere biome in which the main plants are broadleaf trees that lose their leaves in fall and become dormant during cold winters.
tropical rain forest Highly productive and species-rich biome in which year-round rains and warmth support continuous growth of evergreen broadleaf trees.

Take-Home Message **What are broadleaf forests?**

❯ Temperate broadleaf forests grow in the Northern Hemisphere where cold winters prevent year-round growth. Trees lose their leaves in autumn, then remain dormant during the winter.

❯ Year-round warmth and rains support tropical rain forests, the most productive, structurally complex, and species-rich land biome.

> Conifers withstand harsher conditions than broadleaf
  trees, so they grow farther north and at higher altitudes.
< Link to Conifers 21.6

Conifers (evergreen trees with seed-bearing cones) are the
main plants in coniferous forests. Conifer leaves are typically needle-shaped, with a thick cuticle and stomata that
are sunk below the leaf surface. These adaptations help
conifers conserve water during drought or times when the
ground is frozen. As a group, conifers tolerate poorer soils
and drier habitats than deciduous trees.

The most extensive land biome is the coniferous forest that sweeps across northern Asia, Europe, and North
America (Figure 43.21A). It is known as **boreal forest**, or
taiga, which means "swamp forest" in Russian. The conifers are mainly pine, fir, and spruce. Most rain falls in the
summer, and little evaporates into the cool summer air.
Winters are long, cold, and dry. Moose are dominant grazers in this biome.

Also in the Northern Hemisphere, montane coniferous forests extend southward through the great mountain
ranges (Figure 43.21B). Spruce and fir dominate at the
highest elevations. At lower elevations, the mix becomes
firs and pines.

Conifers also dominate temperate lowlands along the
Pacific coast from Alaska into northern California. These
coniferous forests hold the world's tallest trees, Sitka
spruce to the north and coast redwoods to the south.

We find other conifer-dominated ecosystems in the
eastern United States. About a quarter of New Jersey is
pine barrens, a mixed forest of pitch pines and scrub oaks
that grow in sandy, acidic soil. Pine forest covers about
one-third of the Southeast. Fast-growing loblolly pines
dominate these forests and are a major source of lumber
and wood pulp. The pines can survive periodic fires that
kill most hardwood species. When fires are suppressed,
hardwoods outcompete pines.

boreal forest Extensive high-latitude forest of the Northern
Hemisphere; conifers are the predominant vegetation.

Take-Home Message What are coniferous forests?

> Conifers prevail across the Northern Hemisphere's high-
  latitude forests, at high elevations, and in temperate regions
  with nutrient-poor soils.

Coniferous forests

B Montane coniferous forest near Mount Rainier, Washington.

A Boreal forest (taiga) in Siberia.

Figure 43.21 Coniferous forests.

> Low-growing, cold-tolerant plants have only a brief growing season on the tundra.
< Link to Carbon and global warming 42.8

Arctic Tundra

Arctic tundra forms between the polar ice cap and the belts of boreal forests in the Northern Hemisphere. Most is in northern Russia and Canada. Arctic tundra is Earth's youngest biome, having first appeared about 10,000 years ago when glaciers retreated at the end of the last ice age.

Conditions in this biome are harsh; snow blankets arctic tundra for as long as nine months of the year. Annual precipitation is usually less than 25 centimeters (10 inches), but cold temperature keeps the snow that does fall from melting. During a brief summer, plants grow fast under the nearly continuous sunlight (Figure 43.22). Lichens and shallow-rooted, low-growing plants are the producers for food webs that include voles, arctic hares, caribou, arctic foxes, wolves, and brown bears. Enormous numbers of migratory birds nest here in the summer.

Only the surface layer of tundra soil thaws during summer. Below that lies **permafrost**, a frozen layer 500 meters (1,600 feet) thick in places. Permafrost acts as a barrier that prevents drainage, so the soil above it remains perpetually waterlogged. The cool, anaerobic conditions in this soil slow decay, so organic remains can build up. Organic matter in the permafrost makes the arctic tundra one of Earth's greatest stores of carbon.

As global temperatures rise, the amount of frozen soil that melts each summer is increasing. With warmer temperatures, much of the snow and ice that would otherwise reflect sunlight is disappearing. As a result, newly exposed dark soil absorbs heat from the sun's rays, which encourages more melting.

Alpine Tundra

Alpine tundra occurs at high altitudes throughout the world (Figure 43.23). Even in the summer, some patches of snow persist in shaded areas, but there is no permafrost. The alpine soil is well drained, but thin and nutrient-poor. As a result, primary productivity is low. Grasses, heaths, and small-leafed shrubs grow in patches where soil has accumulated to a greater depth. These low-growing plants can withstand the strong winds that discourage the growth of trees.

Arctic tundra

Figure 43.22 Arctic tundra in the summer. Permafrost underlies the soil.

Figure 43.23 Alpine tundra. Low-growing, hardy plants at a high altitude in Washington's Cascade range.

alpine tundra Biome of low-growing, wind-tolerant plants adapted to high-altitude conditions.
arctic tundra Highest-latitude Northern Hemisphere biome, where low, cold-tolerant plants survive with only a brief growing season.
permafrost Continually frozen soil layer that lies beneath arctic tundra and prevents water from draining.

Take-Home Message What is tundra?

> Arctic tundra prevails at high latitudes, where short, cold summers alternate with long, cold winters.

> Alpine tundra prevails in high, cold mountains across all latitudes.

> ❭ Gradients in light penetration, temperature, and dissolved gases affect the distribution of life in aquatic habitats.
> ❰ Links to Properties of water 2.5, Eutrophication 42.1, Indicator species 41.8

With this section, we turn our attention to Earth's waters. We begin here with freshwater systems, continue to coasts, then dive into the oceans.

Lakes

A lake is a body of standing fresh water. If sufficiently deep, it will have zones that differ in their physical characteristics and species composition (Figure 43.24). Nearest shore is the littoral zone. Here, sunlight penetrates all the way to the lake bottom and aquatic plants are primary producers. A lake's open waters include an upper, well-lit limnetic zone, and a profundal zone where light does not penetrate. Primary producers in the limnetic zone are members of the phytoplankton, a group of photosynthetic microorganisms that includes green algae, diatoms, and cyanobacteria. They serve as food for zooplankton, which are tiny consumers such as copepods. In the profundal zone, there is not enough light for photosynthesis, so consumers here depend on food produced above. Debris that drifts down feeds detritivores and decomposers.

Nutrient Content and Succession Like a habitat on land, a lake undergoes succession; it changes over time. A newly formed lake is oligotrophic: deep, clear, and nutrient-poor, with low primary productivity (Figure 43.25). Over time, the lake becomes eutrophic. Again, eutrophication refers to processes, either natural or artificial, that enrich a body of water with nutrients. Increased nutrients allow producers to grow and productivity rises.

Figure 43.25 An oligotrophic lake. Crater Lake in Oregon is a collapsed volcano that filled with snow melt. It began filling about 7,700 years ago; from a geologic standpoint, it is a young lake.

Seasonal Changes Temperate zone lakes undergo seasonal changes that affect primary productivity. Unlike most substances, water is not most dense in its solid state (ice). As water cools, its density increases until it reaches 4°C (39°F). Below this temperature, any additional cooling decreases water's density—which is why ice floats on water (Section 2.5). In an ice-covered lake, water just under the ice is near freezing and at its lowest density. The densest (4°C) water is at the bottom (Figure 43.26A).

In spring, winds cause vertical currents that lead to a spring overturn, during which oxygen-rich water in the surface layers moves down and nutrient-rich water from the lake's depths moves up (Figure 43.26B). After the spring overturn, longer days and the dispersion of nutrients through the water encourage primary productivity.

In summer, a lake has three layers (Figure 43.26C). The upper layer is warm and oxygen-rich. Below this is a thermocline, a thin layer where temperature falls rapidly. Beneath the thermocline is the coolest water. The upper and lower waters on either side of this boundary do not mix. As a result, decomposers deplete oxygen dissolved near the lake bottom, and nutrients near the lake bottom cannot escape into surface waters. Nutrient shortages limit growth and production declines.

In autumn, the lake's upper waters cool, the thermocline vanishes, and a fall overturn occurs (Figure 43.26D). Oxygen-rich water moves down while nutrient-rich water moves up. The overturn in the fall brings nutrients to the surface and favors a brief burst of primary productivity. However, unlike the spring overturn, it does not lead to sustained production because decreasing light and temperature slows photosynthesis. Primary productivity will not peak again until after the next spring overturn.

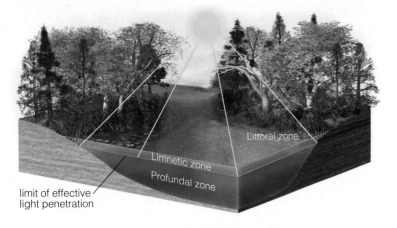

limit of effective
light penetration

Figure 43.24 Animated Lake zonation. A lake's littoral zone extends all around the shore to a depth where rooted aquatic plants stop growing. Its limnetic zone is the open water where light penetrates and photosynthesis occurs. Below the limnetic zone is the cooler, dark water of the profundal zone.

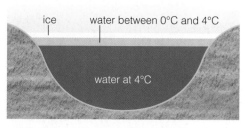

A Winter. Ice covers the thin layer of slightly warmer water just below it. Densest (4°C) water is at bottom. Winds do not affect water under the ice, so there is little circulation.

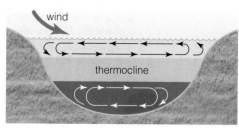

C Summer. Sun warms the upper water, which floats on a thermocline, a layer across which temperature changes abruptly. Waters above and below the thermocline do not mix.

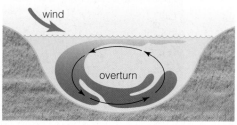

B Spring. Ice thaws. Upper water warms to 4°C and sinks. Winds blow across the lake causing currents that help overturn water, bringing nutrients up from the bottom.

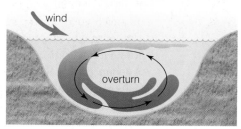

D Fall. Upper water cools and sinks, thus causing the thermocline to disappear. Vertical currents can now mix waters that remained separate during summer.

Figure 43.26 Season changes in a temperate zone lake.

Streams and Rivers

Flowing-water ecosystems start as freshwater springs or seeps. As they flow downslope, they grow and merge. Rainfall, snowmelt, geography, altitude, and shade cast by plants affect flow volume and temperature. The composition of the rocks under flowing water can affect the water's solute concentrations. Because water in different parts of a river moves at different speeds, contains different solutes, and differs in temperature, the species composition of a river varies along its length (Figure 43.27).

The Importance of Dissolved Oxygen

The dissolved oxygen content of water is one of the most important factors affecting aquatic organisms. More oxygen dissolves in cooler, fast-flowing water than in warmer, still water. When water temperature increases or water becomes stagnant, aquatic species that have high oxygen needs suffocate.

In freshwater habitats, aquatic larvae of mayflies and stoneflies are the first invertebrates to disappear when oxygen levels fall. These insect larvae are active predators that demand considerable oxygen, so they serve as indicator species. Gilled snails disappear, too. Declines in populations of invertebrates can have cascading effects on the fishes that feed on them. Fishes can also be more directly affected. Trout and salmon are especially intolerant of low oxygen. Carp (including koi and goldfish) are among the most tolerant; they survive even in tepid algae-rich ponds and in tiny goldfish bowls.

No fishes can survive when the oxygen content of water falls below 4 parts per million. Leeches thrive as most competing invertebrates disappear. In waters with the lowest oxygen concentration, annelids called sludge worms (*Tubifex*) often are the only animals. The worms

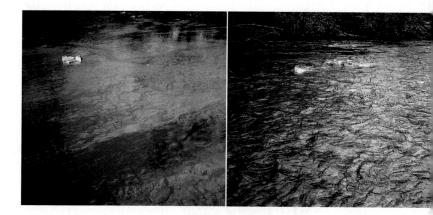

Figure 43.27 Variation in properties along the length of a river. Smoothly flowing water (*left*), holds less oxygen than water that mixes with air as it runs over rocks (*right*).

are colored red by large amounts of hemoglobin. A high affinity for oxygen is an adaptation that allows these worms to exploit low-oxygen habitats where predators and competition for food are scarce.

Take-Home Message **What factors affect life in freshwater habitats?**

> Lakes have gradients in light, dissolved oxygen, and nutrients.

> Primary productivity varies with a lake's age and—in temperate zones—with the season.

> Rivers move nutrients into and out of ecosystems. Characteristics such as temperature and nutrient content typically vary along the length of a river.

> Species differ in their dissolved oxygen needs. Cold, fast-moving water holds more oxygen than still, warm water.

❯ Near the coasts of continents and islands, concentrations of nutrients support some of the world's most productive aquatic ecosystems.
❮ Link to Food chains 42.3

Coastal Wetlands

An **estuary** is an enclosed coastal region where seawater mixes with nutrient-rich fresh water from rivers and streams (Figure 43.28A). Water inflow continually replenishes nutrients, so estuaries are often highly productive. Primary producers include algae and other types of phytoplankton, along with plants that tolerate submergence at high tide. Detrital food chains predominate. Estuaries are marine nurseries; many larval and juvenile invertebrates and fishes live in them.

In tidal flats at tropical latitudes, we find nutrient-rich mangrove wetlands. Mangroves are salt-tolerant woody plants that live in sheltered areas along tropical coasts. The plants have prop roots that extend out from the trunk (Figure 43.28B). Specialized cells at the surface of some roots allow gas exchange.

Rocky and Sandy Shores

Organisms that live along ocean shores are adapted to withstand the force of the waves and repeated tidal changes. Many species are underwater during high tide, but are exposed to the air when the tide is low.

Tide height varies over an approximately monthly cycle as a result of the position of the moon and Earth. Biologists divide a shoreline into three zones (Figure 43.29). The *upper* littoral zone is submerged only at the highest tide of a lunar cycle. It holds the fewest species. The *mid*littoral zone is submerged during the highest regular tide and exposed at the lowest tide. The *lower* littoral zone, exposed only during the lowest tide of the lunar cycle, holds the most diverse collection of species.

Along rocky shores, where waves prevent detritus from piling up, algae that cling to rocks are the producers in grazing food chains. In contrast, waves that continually rearrange loose sediments along sandy shores make it difficult for algae to take hold. Here, detrital food chains start with organic debris from land or offshore.

intertidal zone's upper littoral; submerged only at highest tide of lunar cycle

midlittoral; submerged at each highest regular tide and exposed at lowest tide

lower littoral; exposed only at low tide of lunar cycle

Figure 43.29 Vertical zonation in the intertidal zone.

A South Carolina salt marsh dominated by marsh grass (*Spartina*).

B Florida mangrove wetland.

Figure 43.28 Two types of coastal wetlands.

Take-Home Message What are some coastal ecosystems?

❯ Estuaries are highly productive areas where fresh water and seawater mix.

❯ Mangrove wetlands form along tropical tidal flats.

❯ Grazing food chains predominate on rocky shores, and detrital food chains on sandy shores.

estuary A highly productive ecosystem where nutrient-rich water from a river mixes with seawater.

❯ Coral reefs are highly diverse—and highly threatened—marine ecosystems.
❮ Link to Corals 23.5

Coral reefs are wave-resistant formations that consist primarily of calcium carbonate secreted by generations of coral polyps. Reef-forming corals live mainly in shallow, clear, warm waters between latitudes 25° north and 25° south. About 75 percent of all coral reefs are in the Indian and Pacific oceans. A healthy reef is home to living corals and a huge number of other species (Figure 43.30). Biologists estimate that about a quarter of all marine fish species are associated with coral reefs.

The largest existing reef, Australia's Great Barrier Reef, parallels Queensland for 2,500 kilometers (1,550 miles), and is the largest example of biological architecture. Scientists estimate it began forming about 600,000 years ago. Today it is a string of reefs, some 150 kilometers (95 miles) across. The Great Barrier Reef supports about 500 coral species, 3,000 fish species, 1,000 kinds of mollusks, and 40 kinds of sea snakes.

Photosynthetic dinoflagellates live as symbionts inside the tissues of all reef-building corals (Section 23.5). The dinoflagellates find protection in the coral's tissues and provide the coral polyp with oxygen and sugars that it depends on. When stressed, coral polyps expel the dinoflagellates. Because the dinoflagellates give a coral its color, expelling these protists turns the coral white, an event called **coral bleaching**. When a coral is stressed for more than a short time, the dinoflagellate population in a coral's tissues cannot rebound and the coral dies, leaving only its bleached hard parts behind (Figure 43.31).

The incidence of coral bleaching events has been increasing. Rising sea temperatures and sea level associated with global climate change most likely play a role. People also stress reefs by discharging sewage and other pollutants into coastal waters, by causing erosion that clouds water with sediments, and by destructive fishing practices. Fishing nets break pieces off corals. Fishermen hoping to capture reef fishes for the pet trade use explosives or sodium cyanide to stun the fishes, and destroy corals in the process. Invasive species also threaten reefs. Hawaiian reefs are threatened by exotic algae, including several species imported for cultivation during the 1970s.

Assaults on reefs are taking a huge toll. For example, the Indo-Pacific region, the global center for reef diversity, lost about 3,000 square kilometers (1,160 square miles) of living coral reef each year between 1997 and 2003.

coral bleaching A coral expels its photosynthetic dinoflagellate symbionts in response to stress and becomes colorless.
coral reef Highly diverse marine ecosystem centered around reefs built by living corals that secrete calcium carbonate.

Figure 43.30 Healthy coral reef near Fiji. The coral gets its color from pigments of symbiotic dinoflagellates that live in the its tissues and supply it with sugars.

Figure 43.31 "Bleached" reef near Australia. The coral skeletons shown here belong mainly to staghorn coral (*Acropora*), a genus especially likely to undergo coral bleaching.

Take-Home Message What are coral reefs, how are they threatened, and why is reef loss a matter of concern?

❯ Coral reefs form by the action of living corals that lay down a calcium carbonate skeleton. Photosynthetic dinoflagellates in the coral's tissues are necessary for the coral's survival.

❯ Rising water temperature, pollutants, fishing, and exotic species contribute to loss of reefs.

❯ Declines in coral reefs will affect the enormous number of fishes and invertebrate species that make their home on or near the reefs.

> From its upper, brightly lit waters to the hydrothermal vents on its deep, dark floor, the ocean is filled with life.
< Link to Modes of nutrition 19.5

As in fresh water, gradients of light and temperature affect the distribution of marine life. We divide the ocean into two regions: pelagic and benthic provinces (Figure 43.32).

The Pelagic Province

The ocean's open waters are the **pelagic province**. This province includes the water over continental shelves and the more extensive waters farther offshore. In the ocean's upper, bright waters, phytoplankton such as single-celled algae and bacteria are the primary producers, and grazing food chains predominate.

Depending on the region, some light may penetrate as far as 1,000 meters (more than a half mile) beneath the sea surface. Below that, organisms live in continual darkness, and organic material that drifts down from above serves as the basis of detrital food chains.

The Benthic Province

The **benthic province** is the ocean bottom, its rocks and sediments. Species richness is greatest on continental shelves (the underwater edges of continents). The benthic province also includes largely unexplored species-rich

Figure 43.33 Seamounts.
A Computer model of three seamounts on the sea floor off the coast of Alaska. Patton Seamount, at the rear, stands 3.6 kilometers (about 2 miles) high.

B Flytrap anemone, from Davidson Seamount near the California coast.

regions, including seamounts and the regions around hydrothermal vents.

Seamounts are undersea mountains that stand 1,000 meters or more tall, but are still below the sea surface (Figure 43.33A). They attract large numbers of fishes and are home to many marine invertebrates (Figure 43.33B). Like islands, seamounts often are home to species that evolved there and are found nowhere else.

The abundance of life at seamounts makes them attractive to commercial fishing vessels. Fishes and other organisms are often harvested by trawling, a fishing technique in which a large net is dragged along the bottom, capturing everything in its path. The process is ecologically devastating; trawled areas are stripped bare of life, and the silt stirred up by the giant, weighted nets suffocates filter-feeders in adjacent areas.

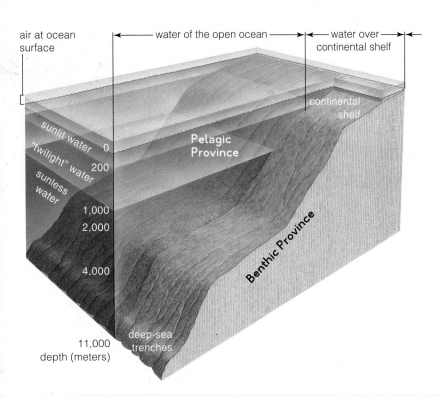

Figure 43.32 Animated Oceanic zones. Dimensions of the zones are not to scale.

benthic province The ocean's sediments and rocks.
hydrothermal vent Place where hot, mineral-rich water streams out from an underwater opening in Earth's crust.
pelagic province The ocean's waters.
seamount An undersea mountain.

Figure 43.34 Members of a hydrothermal vent community. Tube worms (annelids) and crabs are among the consumers. The producers are chemoautotrophic bacteria and archaeans.

At **hydrothermal vents**, superheated water rich in dissolved minerals spews out from an opening on the ocean floor. The water is seawater that seeped into cracks in the ocean floor at the margins of tectonic plates and was heated by geothermal energy. Where this mineral-rich water mixes with the cold deep-sea water, minerals settle out and form extensive deposits. Chemoautotrophic bacteria and archaeans that obtain energy by removing electrons from minerals are the primary producers for food webs that include diverse invertebrates, including large numbers of tube worms (Figure 43.34).

Take-Home Message What factors shape marine communities?

> In the pelagic province's upper waters, photosynthesis supports grazing food chains. In deeper, darker waters of this province, organisms feed mainly on detritus that drifts down from above.

> The benthic province has pockets of high species diversity at undersea mountains (seamounts) and near hydrothermal vents. A hydrothermal vent ecosystem does not run on energy from the sun; the producers are chemoautotrophs rather than photoautotrophs.

Effects of El Niño (revisited)

An El Niño event can have surprising effects. For example, it can increase the number of cases of cholera. During the 1997–1998 El Niño event, 30,000 cases of cholera were reported in Peru alone, compared with only 60 cases from January to August in 1997.

The bacterium, *Vibrio cholerae*, shown at the *right*, causes cholera. Severe diarrhea is the main symptom. During a cholera outbreak, bacteria-contaminated feces often enter the water supply. People who drink or wash food with the tainted water become infected.

Marine biologist Rita Caldwell figured out how a change in the ocean temperature can affect the incidence of cholera in humans. Caldwell discovered that copepods, a type of small crustacean, serve as a reservoir for cholera-causing bacteria between disease outbreaks. During an El Niño, the rise in the temperature of surface waters causes a rise in abundance of the phytoplankton that the copepods feed on, so the number of disease-carrying copepods increases. When Caldwell analyzed records of sea surface temperature and cholera outbreaks in the Bay of Bengal, she found that reports of cholera cases rise four to six weeks after an increase in water temperature. Caldwell continues to research how the environment influences the incidence of cholera. She also investigates simple, inexpensive methods that people can use to avoid the disease (Figure 43.35).

Figure 43.35 Rita Caldwell examining filtered and unfiltered water. She advised women in Bangladesh to use sari cloth as a filter to remove disease-carrying copepods from the water. The copepod hosts are too big to pass through the thin cloths, which can be rinsed in clean water, sun-dried, and used again. This inexpensive, simple method has cut cholera outbreaks by half.

How Would You Vote? Is supporting studies of El Niño and the other long-term climate cycles a good use of government funds? See CengageNow for details, then vote online (cengagenow.com).

Summary

Section 43.1 Interactions among Earth's air and waters, as when oceans warm during an **El Niño** and then cool during a **La Niña**, affect weather throughout the **biosphere**. Such events alter ocean currents and rainfall patterns that influence primary production, thus affecting biological communities.

Sections 43.2, 43.3 Global air circulation patterns affect **climate** and the distribution of communities. The patterns are set into motion when sunlight heats tropical regions more than higher latitudes. Ocean currents distribute heat energy worldwide and influence weather patterns. Interactions between ocean currents, air currents, and landforms determine where regional phenomena such as **rain shadows** or **monsoons** occur.

Section 43.4 Biomes are discontinuous areas characterized by a particular type of vegetation. Differences in climate, elevation, and soil properties affect the distribution of biomes.

Sections 43.5–43.7 Deserts form around latitudes 30° north and south and in rain shadows. **Grasslands** dominated by plants adapted to fire and grazing form in the somewhat moister interior of midlatitude continents. Shrubby, fire-adapted **chaparral** is common in California and other coastal regions with hot, dry summers and cool, wet winters.

Section 43.8 Broadleaf trees are angiosperms. Those in **temperate deciduous forests** shed their leaves all at once just before a cold winter that prevents growth. Broadleaf trees in **tropical rain forests** can grow year-round, and these enormously productive forests are home to a large number of species.

Sections 43.9, 43.10 Conifers withstand cold and drought better than broadleaf trees and dominate Northern Hemisphere high latitude **boreal forests**. Farther north in this hemisphere, **arctic tundra** overlies **permafrost**. **Alpine tundra** occurs at high altitudes.

Section 43.11 Lakes undergo succession, becoming more nutrient-rich over time. In temperate zone lakes, seasonal variation in water circulation patterns affects productivity. Properties of a river vary along its length. Fast-moving, cool water holds the most oxygen.

Sections 43.12, 43.13 Nutrient-rich freshwater mixes with seawater in an **estuary**. In tropical regions, mangrove wetlands form on tidal flats. Along rocky shores, algae form the base for grazing food chains.

On sandy shores, detrital food chains predominate. All tidal zones show a vertical zonation, with most diversity in the region submerged during all but the lowest tides.

The accumulated calcium carbonate skeletons of coral polyps form **coral reefs** in shallow tropical seas, especially in the Indo-Pacific Ocean. Many marine species associate with reefs. When stressed, a coral may eject its photosynthetic symbionts, an event called **coral bleaching**.

Section 43.14 Life exists throughout the ocean. In the **pelagic province**, diversity is highest in sunlit waters, where grazing food chains predominate. Detritus forms the base for food chains in deeper, darker waters. **Seamounts** are regions of high diversity in the **benthic province**. At **hydrothermal vents**, chemoautotrophic bacteria and archaeans are the producers; they get energy from mineral-rich seawater emitted from the vent.

Self-Quiz   Answers in Appendix III

1. The Northern Hemisphere is most tilted toward the sun in _____ .
     a. spring   b. summer   c. autumn   d. winter

2. Which latitude will have the most hours of daylight on the summer solstice?
     a. 0° (the equator)   c. 45° north
     b. 30° north   d. 60° north

3. Warm air _____ and it holds _____ water than cold air.
     a. sinks; less   c. sinks; more
     b. rises; less   d. rises; more

4. A rain shadow is a reduction in rainfall _____ .
     a. on the inland side of a coastal mountain range
     b. during an El Niño event
     c. that results from global warming

5. The Gulf Stream is a current that flows _____ along the eastern coast of the United States.
     a. north to south   b. south to north

6. _____ have a deep layer of humus-rich topsoil.
     a. Deserts   c. Rain forests
     b. Grasslands   d. Seamounts

7. Biome distribution depends on _____ .
     a. climate   c. soils
     b. elevation   d. all of the above

8. Grasslands most often are found _____ .
     a. at 30° north and south   c. in interior of continents
     b. at high altitudes   d. all of the above

9. Permafrost underlies _____ .
     a. arctic tundra   c. boreal forest
     b. alpine tundra   d. all of the above

10. Warm, still water holds _____ oxygen than cold, fast-flowing water.

Changing Sea Temperatures In an effort to predict El Niño or La Niña events in the near future, the National Oceanographic and Atmospheric Administration collects information about sea surface temperature (SST) and atmospheric conditions. Scientists compare monthly temperature averages in the eastern equatorial Pacific Ocean to historical data and calculate the difference (degree of anomaly) to see if El Niño conditions, La Niña conditions, or neutral conditions are developing. El Niño is a rise in the average SST above 0.5°C. A decline of the same amount is La Niña. Figure 43.36 shows data for nearly 39 years.

1. When did the greatest positive temperature deviation occur during this time period?
2. What type of event, if any, occurred during the winter of 1982–1983? What about the winter of 2001–2002?
3. During a La Niña event, less rain than normal falls in the American West and Southwest. In the time interval shown, what was the longest interval without a La Niña event?
4. What type of conditions were in effect in the fall of 2007 when California suffered severe wildfires?

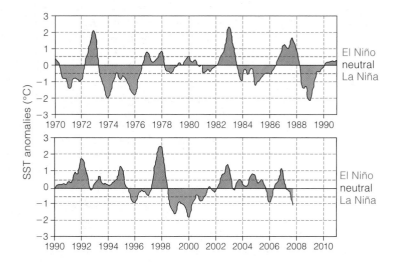

Figure 43.36 Sea surface temperature anomalies (differences from the historical mean) in the eastern equatorial Pacific Ocean. A rise above the dashed *red* line is an El Niño event, a decline below the *blue* line is La Niña.

11. Chemoautotrophic bacteria and archaeans are the primary producers for food webs _____ .
    a. in mangrove wetlands   c. on coral reefs
    b. at seamounts   d. at hydrothermal vents

12. Corals rely on symbiotic _____ for sugars.
    a. fungi   c. dinoflagellates
    b. bacteria   d. green algae

13. What biome borders on boreal forest?
    a. savanna   c. tundra
    b. taiga   d. chaparral

14. Unrelated species living in geographically separated parts of a biome may resemble one another as a result of _____ .
    a. competitive interactions
    b. morphological convergence
    c. morphological divergence
    d. coevolution

15. Match the terms with the most suitable description.
    ___ tundra   a. broadleaf forest near equator
    ___ chaparral   b. partly enclosed by land; where
    ___ desert   fresh water and seawater mix
    ___ savanna   c. African grassland with trees
    ___ estuary   d. low-growing plants at
    ___ boreal forest   high latitudes or elevations
    ___ prairie   e. dry shrubland
    ___ tropical rain   f. at latitudes 30° north and south
        forest   g. mineral-rich, superheated
    ___ hydrothermal   water supports communities
        vents   h. conifers dominate
                         i. North American grassland

Additional questions are available on **CENGAGENOW**.

Critical Thinking

1. On April 26, 1986, a meltdown occurred at the Chernobyl nuclear power plant in Ukraine. Nuclear fuel burned for nearly ten days and released 400 times more radioactive material than the atomic bomb that dropped on Hiroshima. Winds carried radioactive fallout around the globe. By 1998, the rate of thyroid abnormalities in children living downwind from the site was nearly seven times as high as for those upwind; their thyroid gland concentrated the iodine radioisotopes. Chernoboyl is at 51° north latitude. In what direction did the major winds carry the fallout after the accident?

2. Owners of off-road recreational vehicles would like increased access to government-owned deserts. Some argue that it is the perfect place for off-roaders because "There's nothing there." Do you agree?

3. Rita Caldwell, the scientist who discovered why cholera outbreaks often occur during an El Niño, is concerned that global climate change could increase the incidence of this disease. By what mechanism might global warming cause an increase in cholera outbreaks?

Animations and Interactions on **CENGAGENOW**:
❯ Earth's tilt and the seasons; Air circulation patterns; Major ocean currents; Rain shadow effect; Coastal breezes; Biome distribution; Oceanic zones.

‹ Links to Earlier Concepts

This chapter considers the causes of an ongoing mass extinction (Section 16.1) in light of human population growth (40.7). We look again at effects of species introductions (41.8), deforestation (21.1, 43.8), coral bleaching (43.13), and mercury pollution (2.1). You will draw on your knowledge of pH (2.6), the ozone layer (18.5), plant nutrition (26.2), trophic levels (42.3), and water and nutrient cycles (42.7–42.9).

Key Concepts

An Extinction Crisis
Extinction is a natural process, but human activities have increased the frequency of extinction events. Species go extinct when they are adversely affected by habitat destruction, degradation, and fragmentation. The extent of species losses is not fully known.

Harmful Land Uses
Plowing under grasslands and cutting down forests can have long-term and long-range effects. Loss of plants allows soil erosion, raises soil temperature, and affects rainfall patterns in ways that make it difficult to restore plant cover.

44 Human Effects on the Biosphere

44.1 A Long Reach

We began this book with the story of biologists who ventured into a remote forest in New Guinea, and their excitement at the many previously unknown species that they encountered (Section 1.1). At the far end of the globe, a U.S. submarine surfaced in Arctic waters and discovered polar bears hunting on the ice-covered sea. The polar bears were about 445 kilometers (270 miles) from the North Pole and 805 kilometers (500 miles) from the nearest land (Figure 44.1).

Even such seemingly remote regions are no longer beyond the reach of human explorers—and human influence. You already know that the temperature of Earth's atmosphere and seas are rising. In the Arctic, unusually warm temperatures are affecting the seasonal cycle of sea ice melting and formation. In recent years, sea ice has begun to thin and to break up earlier in the spring and to form later in the fall. A decrease in the persistence of sea ice is bad news for polar bears. They can only reach their main prey—seals—by traveling across ice. A longer ice-free period means less time for bears to feed. In addition, an earlier seasonal ice melt raises the risk that polar bears hunting far from land will become stranded and unable to return to solid ground before the ice thaws.

Polar bears face other threats as well. They are top predators and their tissues contain a surprisingly high amount of mercury and organic pesticides. The pollutants entered the water and air far away, in more temperate regions. Winds and ocean currents deliver them to polar realms. Contaminants also travel north in the tissues of migratory animals such as seabirds that spend their winters in temperate regions and nest in the Arctic.

In places less remote than the Arctic, human activities have a more direct effect. As we cover more and more of the world with our dwellings, factories, and farms, less appropriate habitat remains for other species. We also put species at risk by competing with them for resources, overharvesting them, and introducing non-native competitors.

Figure 44.1 Polar bears on ice. *Opposite page*, a polar bear investigates an American submarine that surfaced in ice-covered Arctic waters.

It would be presumptuous to think that we alone have had a profound impact on the world of life. As long ago as the Proterozoic, photosynthetic cells were irrevocably changing the course of evolution by enriching the atmosphere with oxygen. Over life's existence, the evolutionary success of some groups ensured the decline of others. What is new is the increasing pace of change and the capacity of our own species to recognize and affect its role in this increase.

A century ago, Earth's physical and biological resources seemed inexhaustible. Now we know that many practices put into place when humans were largely ignorant of how natural systems operate take a heavy toll on the biosphere. The rate of species extinctions is on the rise and many types of biomes are threatened. These changes, the methods scientists use to document them, and the ways that we can address them, are the focus of this chapter.

Effects of Pollutants
Some pollutants fall to Earth as acid rain that can harm aquatic communities and soils. Other pollutants disrupt metabolism when they accumulate to high levels in animal bodies. Still other pollutants harm the protective ozone layer or contribute to global climate change.

Conserving Biodiversity
Earth's biodiversity is the product of billions of years of evolution, and our well-being depends on sustaining it. Conservation biologists help us prioritize which areas to protect first by assessing which are the most threatened and most biodiverse.

Reducing Negative Impacts
Extraction of fuel and other resources harms the environment. Individual actions that minimize the use of resources and energy can help reduce the threats to the health of the planet and to biodiversity.

> Extinction is a natural process, but human activities are raising the rate at which other species disappear.
‹ Links to Mass extinction 16.1, Codfish population crash 40.6, Human population growth 40.7, Species interactions and introductions 41.8, Nitrogen cycle 42.9

The rising size of the human population and increasing industrialization have far-reaching effects on the biosphere. We begin by discussing effects on individual species, then turn to the wider impacts.

Threatened and Endangered Species

Extinction, like speciation, is a natural process. Species arise and become extinct on an ongoing basis. The rate of extinction picks up dramatically during a mass extinction, when many kinds of organisms in many different habitats become extinct in a relatively short period. We are currently in the midst of such an event. Unlike historical mass extinctions, this one does not stem from some natural catastrophe such as an asteroid impact. Rather, this mass extinction is the outcome of the success of a single species—humans—and their effects on Earth.

An **endangered species** is a species that faces extinction in all or part of its range. A **threatened species** is one that is likely to become endangered in the near future. Keep in mind that not all rare species are threatened or endangered. Some species have always been uncommon. A species is considered endangered when one or more of its populations have declined or are declining.

Causes of Species Declines

When European settlers first arrived in North America, they found between 3 and 5 billion passenger pigeons. In the 1800s, commercial hunting caused a steep decline in the bird's numbers. The last time anyone saw a wild passenger pigeon was 1900—and he shot it. The last captive bird died in 1914.

We are still overharvesting species. The crash of the Atlantic codfish population, described in Section 40.6, is one recent example. Another is the fate of the white abalone, a gastropod mollusk native to kelp forests off the coast of California (Figure 44.2A). Heavy harvesting of this species during the 1970s reduced the population to about 1 percent of its original size. In 2001, it became the first invertebrate to be listed as endangered by the United States Fish and Wildlife Service. Although some white abalone remain in the wild, the population density is too low for effective reproduction. The species' only hope for survival is a program of captive breeding. If this program succeeds, individuals can be reintroduced to the wild.

Overharvest directly reduces population size, but we also affect species by altering their habitat. Each species requires a specific type of habitat, and any degradation,

A White abalone   **B** Panda

Figure 44.2 Some threatened and endangered species. To learn about others, visit the International Union for Conservation of Nature and Natural Resources web site at www.iucnredlist.org/.

fragmentation, or destruction of that habitat reduces population numbers.

An **endemic species**, one that is confined to the limited area in which it evolved, is more likely to go extinct than a species with a more widespread distribution. For example, giant pandas are endemic to China's bamboo forests (Figure 44.2B). As China's human population soared, bamboo was cut for building materials and to make room for farms. As the bamboo forests disappeared, so did the pandas. Their numbers, which may once have been as high as 100,000, have fallen to 1,000 or so animals in the wild.

Similarly, logging of forests in the southeastern United States reduced numbers of ivory-billed woodpeckers endemic to these forests (Figure 44.2C). The birds are listed as endangered, but may be extinct. Other species endangered by logging of their habitat include Costa Rica's harlequin frog (Figure 44.2D) and the gorillas of the Congo (Figure 44.2E).

Habitat loss also affects plants. Plowing of North American prairies threatens prairie fringed orchids (Figure 44.2F). Destruction of the Indonesian rain forest has put all species of *Rafflesia*, the plants with world's largest flowers, on the endangered species list (Figure 44.2G).

Human activities that degrade, rather than destroy, habitats also endanger species. Texas blind salamanders (Figure 44.2H) are among the species endemic to the Edwards aquifer, a series of water-filled, underground limestone formations. Excessive withdrawals of water, along with water pollution, threatens species in the aquifer. Similarly, development of beaches interferes with reproduction of sea turtles such as leatherbacks that lay their eggs on these shores (Figure 44.2I). As a final example, manatees (Figure 44.2J) in Florida's waters are injured by motor boat propellers and get tangled in fishing line.

Deliberate or accidental species introductions also can threaten a species (Section 41.8). Rats that reached islands by stowing away on ships attack and endanger many ground-nesting birds that evolved in the absence of egg-eating ground predators. Exotic species also cause

C Ivory-billed woodpecker

D Harlequin frog

E Gorilla

F Prairie fringed orchid

G *Rafflesia*

H Texas blind salamander

I Leatherback sea turtle

J Manatee

K Koa bug

problems by outcompeting native ones. After European brown trout and eastern brook trout were introduced into California's mountain streams for sport fishing, populations of the native golden trout declined.

The koa bug (Figure 44.2K) is the victim of a biological control program gone awry. This Hawaiian native is threatened by an exotic parasitoid fly introduced to attack another bug that is an agricultural pest. Unfortunately, the fly prefers koa bugs to its intended target.

The decline or loss of one species can endanger others. For example, running buffalo clover and the buffalo that grazed on it were once common in the Midwest. The plants thrived in the open woodlands where soil was enriched by buffalo droppings and periodically disturbed by the animals' hooves. Buffalo helped to disperse the clover's seeds. When buffalo were hunted to near extinction, buffalo clover populations declined.

Like most endangered species, buffalo clover faces a number of simultaneous threats. In addition to the loss of the buffalo, the clover is threatened by conversion of its habitat to housing developments, competition from introduced plants, and attacks by introduced insects.

The Unknown Losses

In November of 2009, the International Union for Conservation of Nature and Natural Resources (IUCN) reported that of the 47,677 species they had assessed, 36 percent were threatened or endangered. We do not know the level of threat for the vast majority of the approximately 1.8 million named species, or for the countless millions of species yet to be discovered and named.

Endangered species listings have historically focused on vertebrates. Scientists have only recently begun to consider the threats to invertebrates and to plants. Our impact on protists and fungi is largely unknown, and the IUCN does not address threats to bacteria or archaeans.

Microbiologist Tom Curtis is among those making a plea for increased research on microbial ecology and microbial diversity. He argues that we have barely begun to comprehend the vast number of microbial species and to understand their importance. Curtis writes, "I make no apologies for putting microorganisms on a pedestal above all other living things. For if the last blue whale choked to death on the last panda, it would be disastrous but not the end of the world. But if we accidentally poisoned the last two species of ammonia-oxidizers, that would be another matter. It could be happening now and we wouldn't even know . . . " Ammonia-oxidizing bacteria play an essential role in the nitrogen cycle by converting ammonia in wastes and remains to nitrites. Without them wastes would pile up and plants would not have the nitrogen they need to grow.

---

endangered species A species that faces extinction in all or a part of its range.
endemic species Species that remains restricted to the area where it evolved.
threatened species Species likely to become endangered in the near future.

Take-Home Message How do humans threaten other species?

> Species often decline when humans destroy or fragment natural habitat by converting it to human use, or degrade it through pollution or withdrawal of an essential resource.

> Humans also directly cause declines by overharvesting species.

> Global travel and trade can introduce exotic species that harm native ones.

> The number of endangered species remains largely unknown.

❯ Around the globe, alteration of habitats as a result of farming, grazing, and logging contributes to species loss.
❮ Links to Deforestation 21.1, Tropical forest loss 43.8

With this section, we begin a survey of some of the ways that human activities threaten species by destroying or degrading habitat.

Desertification

Deserts naturally expand and contract over time as climate conditions vary. However, humans sometimes convert a dry grassland or woodland to desert, a process called **desertification**. Desertification occurs when plowing or grazing removes plants and so exposes topsoil to wind erosion. Plants cannot thrive where topsoil has blown away.

In a positive feedback cycle, drought encourages desertification, which worsens drought. Fewer plants means less transpiration (Section 26.4). Because less water enters the atmosphere, local rainfall decreases.

Drought and poor agricultural practices are currently expanding Africa's Sahara Desert south into the Sahel region. Effects of this expansion are felt as far away as the southern United States and the Caribbean, where dust blown across the Atlantic eventually alights (Figure 44.3). Desertification also threatens China's northwestern region. Dust clouds that often darken skies above Beijing are a symptom of expansion of the Gobi Desert.

Deforestation

The amount of forested land is currently stable or increasing in North America, Europe, and China, but tropical forests continue to disappear at an alarming rate. Deforestation has detrimental effects beyond the immediate destruction of forest organisms. For example, deforestation encourages flooding because water runs off into

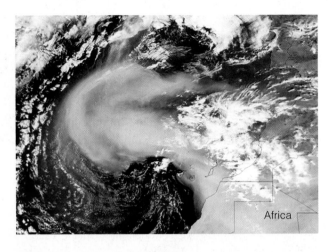

Figure 44.3 One symptom of ongoing desertification. Increasing amounts of dust now blow from Africa's Sahara Desert into and across the Atlantic Ocean.

streams, rather than being taken up by tree roots. Deforestation also raises risk of landslides in hilly areas. Tree roots tend to stabilize the soil. When they are removed, waterlogged soil becomes more likely to slide.

Deforested areas also become nutrient-poor. Figure 44.4 shows results of an experiment in which scientists deforested a region in New Hampshire and monitored the nutrients in runoff. Deforestation caused a spike in loss of essential soil nutrients such as calcium.

Like desertification, deforestation affects local weather. Temperatures in forests are cooler than in adjacent nonforested areas because trees shade the ground and transpiration causes evaporative cooling. When forest is cut down, daytime temperatures rise and reduced transpiration results in less rainfall.

Once a tropical forest is logged, the resulting nutrient losses and drier, hotter conditions can make it impossible for tree seeds to germinate or for seedlings to survive. Thus, deforestation can be difficult to reverse.

Because forests take up and store huge amounts of carbon, ongoing forest losses also contribute to global climate change.

---

desertification Conversion of dry grassland to desert.

A plot of forest was stripped of vegetation as an experiment.

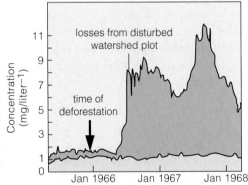

Concentration (mg/liter⁻¹)

losses from disturbed watershed plot

time of deforestation

Jan 1966 Jan 1967 Jan 1968

After experimental deforestation, calcium levels in runoff increased sixfold (*purple*). An undisturbed control plot in the same forest showed no similar increase during this time (*light blue*).

Figure 44.4 Animated Experimental deforestation in the Hubbard Brook watershed.

Take-Home Message **What are the effects of desertification and deforestation?**

❯ With desertification, excess plowing or grazing causes soil to blow away. With fewer plants, rainfall declines.

❯ Deforestation increases flooding and loss of soil nutrients, raises the local temperature, and decreases rainfall. Changes in soil and temperature produced by deforestation make it difficult for new trees to become established.

44.4 Acid Rain

> Acid deposition contributes to forest destruction and degrades aquatic habitats.
< Links to pH 2.6, Plant nutrition 26.2

Pollutants are natural or man-made substances released into soil, air, or water in greater than natural amounts. The presence of a pollutant disrupts the physiological processes of organisms that evolved in its absence, or that are adapted to lower levels of it.

Sulfur dioxides and nitrogen oxides are examples of common air pollutants. Most sulfur dioxide comes from coal-burning power plants. Vehicles and power plants that burn gas and oil emit nitrogen oxides. Synthetic nitrogen-rich fertilizers also contribute nitrogen oxides to the atmosphere.

In dry weather, airborne sulfur and nitrogen oxides coat dust particles. Dry acid deposition occurs when the coated dust falls to the ground. Wet acid deposition, or **acid rain**, occurs when pollutants combine with water and fall as acidic precipitation. The pH of unpolluted rainwater is about 5 (Section 2.6). Acid rain can be ten times more acidic (Figure 44.5).

Acid rain that falls on or drains into waterways, ponds, and lakes affects aquatic organisms. Low pH prevents fish eggs from developing and kills adult fish. In New York's Adirondack Mountains, acid rain has left more than 200 lakes free of fish (Figure 44.5A).

When acid rain falls on forests it burns tree leaves and alters the composition of soils. As acidic water drains through the soil, positively charged hydrogen ions displace positively charged nutrient ions such as calcium, causing nutrient loss. The acidity also causes soil particles to release metals such as aluminum that can harm plants. The combination of poor nutrition and exposure to toxic aluminum weakens trees, making them more susceptible to insects and pathogens, and thus more likely to die (Figure 44.5B). Effects are most pronounced at higher elevations where trees are frequently exposed to clouds of acidic droplets.

---

acid rain Low pH rain that forms when sulfur dioxide and nitrogen oxides mix with water vapor in the atmosphere.
pollutant A substance that is released into the environment by human activities and interferes with the function of organisms that evolved in the absence of the substance or with lower levels.

Take-Home Message What is acid rain?

> Acid rain contains nitric or sulfuric acid. It forms when pollutants mix with water vapor, and it harms plants and aquatic organisms.

A Biologist testing the pH of water in New York's Woods Lake. In 1979, the lake's pH was 4.8. Since then, experimental addition of calcite to soil around the lake has raised the pH to more than 6.

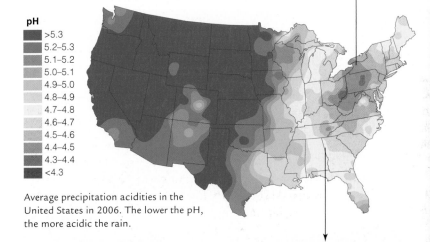

pH
- >5.3
- 5.2–5.3
- 5.1–5.2
- 5.0–5.1
- 4.9–5.0
- 4.8–4.9
- 4.7–4.8
- 4.6–4.7
- 4.5–4.6
- 4.4–4.5
- 4.3–4.4
- <4.3

Average precipitation acidities in the United States in 2006. The lower the pH, the more acidic the rain.

B Dying trees in Great Smoky Mountains National Park, where acid rain harms leaves and causes loss of nutrients from soil.

Figure 44.5 Animated Acid rain.

>> **Figure It Out** Is rain more acidic on the East Coast or the West Coast?

Answer: The East Coast

> ❯ Chemicals released by human activities can accumulate in the bodies of organisms and interfere with metabolism.
> ❮ Links to Mercury 2.1, Phytoremediation 26.1

Accumulation and Magnification

Industrial processes and the burning of fossil fuels put a vast array of chemical pollutants into Earth's air, water, and soils. Some of these pollutants can build up in the tissues of organisms. By the process of **bioaccumulation**, an organism's tissues store a pollutant taken up from the environment, causing the amount in the body to increase over time. The ability of some plants to bioaccumulate toxic substances makes them useful in phytoremediation of polluted soils (Section 26.1).

DDT Residues (In parts per million wet weight of organism)	
Osprey	13.8
Green heron	3.57
Atlantic needlefish	2.07
Summer flounder	1.28
Sheepshead minnow	0.94
Hard clam	0.47
Marsh grass	0.33
Flying insects (mostly flies)	0.30
Mud snail	0.26
Shrimps	0.16
Green alga	0.083
Water	0.00005

Figure 44.6 Biological magnification in an estuary on Long Island, New York, as reported in 1967 by George Woodwell, Charles Wurster, and Peter Isaacson. The *upper* photo shows an osprey, a top predator in this ecosystem.

In animals, hydrophobic chemical pollutants ingested or absorbed across the skin tend to accumulate in fatty tissues. Because the amount of pollutant in an animal's body increases over time, longer-lived species tend to be more affected by fat-soluble pollutants than shorter-lived ones. Within each species, older individuals tend to have a higher pollutant load than younger ones.

Pollutant concentration in tissues also varies with trophic level. By the process of **biological magnification**, the concentration of a chemical increases as the pollutant moves up a food chain. Figure 44.6 provides data documenting the biological magnification of DDT (a pesticide) in a salt marsh ecosystem during the 1960s. Notice that the concentration of DDT in osprey tissues was 276,000 times higher than that in the water. As a result of bioaccumulation and biological magnification, even seemingly low environmental concentrations of pollutants can have major detrimental effects on a species. During the 1960s, ospreys were driven toward extinction by widespread use of DDT, because this chemical interferes with their egg shell formation. A ban on DDT use in the United States has allowed the species to recover.

Bioaccumulation and magnification of other pollutants such as methylmercury continue to pose a threat to wildlife and to human health (Section 2.1).

Point and Nonpoint Sources

Some pollutants come from a few easily identifiable sites, or point sources. A factory that discharges pollutants into the air or water is a point source. Pollutants that come from point sources are the easiest to control: Identify the sources, and you can take action there.

Dealing with pollution from nonpoint sources is more challenging. Such pollution stems from widespread release of a pollutant. For example, oil that drips from vehicles is a nonpoint source of water pollution. Rain washes accumulated oil from pavement into natural waterways or storm drains. In many places, water that flows into storm drains is discharged into a body of water without any treatment. Halting nonpoint source pollution is difficult because it requires large numbers of people to alter their activity.

bioaccumulation An organism accumulates increasing amounts of a chemical pollutant in its tissues over the course of its lifetime.
biological magnification A chemical pollutant becomes increasingly concentrated as it moves up through food chains.

Take-Home Message How do chemical pollutants become concentrated in tissues?

> ❯ Some chemical pollutants can accumulate to high levels over an organism's lifetime. Such chemicals also become more concentrated as they move up food chains.

> Improperly disposing of trash can threaten water supplies and harm wildlife.
❮ Link to Groundwater 42.7

Six billion people use and discard a lot of stuff. Where does all the waste go? Historically, unwanted material was simply buried in the ground or dumped out at sea. Trash was out of sight, and also out of mind. We now know that chemicals in buried trash can contaminate groundwater, as when lead from discarded batteries seeps into the ground. Waste that gets into oceans harms marine life. For example, seabirds often eat floating bits of plastic and feed them to their chicks, with deadly results (Figure 44.7).

In 2006, the United States generated 251 million tons of garbage, which averages out to 2.1 kilograms (4.6 pounds) per person per day. By weight, about a third of that material was recycled, but there is plenty of room for improvement. Two-thirds of plastic soft drink bottles and three-quarters of glass bottles were not recycled. Nonrecycled trash now gets burned in high-temperature incinerators or placed in engineered landfills lined with material that minimizes the risk of groundwater contamination. No solid municipal waste can legally be dumped at sea.

Nevertheless, plastic and other garbage constantly enters our coastal waters. Foam cups and containers from fast-food outlets, plastic shopping bags, plastic water bottles, and other material discarded as litter ends up in storm drains. From there it is carried to streams and rivers that can convey it to the sea. A seawater sample taken near the mouth of the San Gabriel River in southern California had 128 times as much plastic as plankton by weight.

Once in the ocean, trash can persist for a surprisingly long time. Components of a disposable diaper will last for more than 100 years, as will fishing line. A plastic bag will be around for more than 50 years, and a cigarette filter for more than 10.

To reduce the impact of plastic trash, choose more durable objects over disposable ones, and avoid buying plastic when other, less environmentally harmful alternatives exist. If you use plastic, be sure to recycle or dispose of it properly.

Figure 44.7 One effect of discarded plastic in a marine environment.

A Recently deceased Laysan albatross chick, dissected to reveal the contents of its gut.

B Scientists found more than 300 pieces of plastic inside the bird. One of the pieces had punctured its gut wall, resulting in its death. The chick was fed the plastic by its parents, who gathered the material from the ocean surface, mistaking it for food.

Take-Home Message **What are the ecological effects of discarding trash?**

> Trash, especially plastics, often ends up in oceans where it harms marine life.

> You can minimize your environmental impact by avoiding disposable plastic goods and by recycling.

> ❯ Ozone is said to be "good up high, but bad nearby." It forms a protective layer in the upper atmosphere, but is a harmful pollutant in air near the ground.
> ❮ Links to UV as mutagen 9.6, Ozone layer formation 18.5

Depletion of the Ozone Layer

In the upper layers of the atmosphere, between 17 and 27 kilometers (10.5 and 17 miles) above sea level, the ozone (O_3) concentration is so great that scientists refer to this region as the **ozone layer**. The ozone layer benefits living organisms by absorbing most ultraviolet (UV) radiation from incoming sunlight. UV radiation, remember, damages DNA and causes mutations (Section 9.6).

In the mid-1970s, scientists noticed that Earth's ozone layer was thinning. Its thickness had always varied a bit with the season, but now the average level was declining steadily from year to year. By the mid-1980s, the spring ozone thinning over Antarctica was so pronounced that people were referring to the lowest-ozone region as an "ozone hole" (Figure 44.8A).

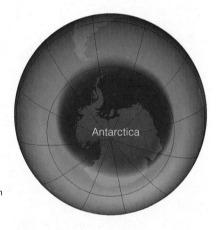

A Ozone levels in the upper atmosphere in September 2007, the Antarctic spring.

Purple indicates the least ozone, with *blue*, *green*, and *yellow* indicating increasingly higher levels.

Check the current status of the ozone hole at NASA's web site (http://ozonewatch .gsfc.nasa.gov/).

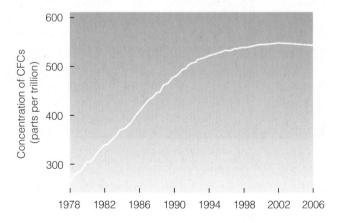

B Concentration of CFCs in the upper atmosphere. These pollutants destroy ozone. A worldwide ban on CFCs has successfully halted the rise in CFC concentration.

Figure 44.8 Animated Ozone and CFCs.

Declining ozone quickly became an international concern. With a thinner ozone layer, people would be exposed to more UV radiation, the main cause of skin cancers. Higher UV levels also harm wildlife, which do not have the option of avoiding sunlight. In addition, exposure to higher than normal UV levels affects plants and other producers, slowing the rate of photosynthesis and the release of oxygen into the atmosphere.

Chlorofluorocarbons, or CFCs, are the main ozone destroyers. These odorless gases were once widely used as propellants in aerosol cans, as coolants, and in solvents and plastic foam. In response to the potential threat posed by ozone thinning, countries worldwide agreed in 1987 to phase out the production of CFCs and other ozone-destroying chemicals. As a result of that agreement (the Montreal Protocol), the concentrations of CFCs in the atmosphere are no longer rising dramatically (Figure 44.8B). However, CFCs break down quite slowly, so scientists expect them to remain at a level that significantly impairs the ozone layer for several decades.

Near–Ground Ozone Pollution

Near the ground, where ozone levels are naturally low, ozone is considered a pollutant. Ozone irritates the eyes and respiratory tracts of humans and wildlife. It also interferes with plant growth.

Ground-level ozone forms when nitrogen oxides and volatile organic compounds released by burning or evaporating fossil fuels are exposed to sunlight. Warm temperatures speed the reaction. Thus, ground-level ozone tends to vary daily (being higher during the daytime) and seasonally (being higher during the summer).

You can help reduce ozone pollution by avoiding actions that put fossil fuels or their combustion products into the air at times that favor ozone production. On hot, sunny, still days, postpone filling your gas tank or using gasoline-powered appliances such as lawn mowers until the evening, when there is less sunlight to power the conversion of pollutants to ozone.

---

ozone layer High atmospheric layer with a high concentration of ozone (O_3) that prevents much ultraviolet radiation from reaching Earth's surface.

Take-Home Message **How do human activities affect ozone levels?**

> ❯ Certain synthetic chemicals destroy ozone in the upper atmosphere's protective ozone layer. This layer serves as a protective shield against UV radiation.

> ❯ Evaporation and burning of fossil fuels increases the amount of ozone in the air near the ground, where ozone is considered a harmful pollutant.

❭ Climate change is the most widespread threat to habitats. Among other effects, it melts ice, causing sea level to rise.

❮ Links to Greenhouse gases 42.8, Coral bleaching 43.13

Ongoing climate change affects ecosystems worldwide. How is the global climate changing? Most notably, average temperatures are increasing (Figure 44.9). Warming is more pronounced at temperate and polar latitudes than at the equator.

A rising temperature raises sea level. Water expands as it is heated, and heating also melts sea ice and glaciers (Figure 44.10). Together, thermal expansion and the addition of meltwater from glaciers cause sea level to rise. In the past century, the sea level has risen about 20 centimeters (8 inches). As a result, some coastal wetlands are disappearing underwater.

The warming climate is having widespread effects on biological systems. Temperature changes are important cues for many temperate zone species. Warmer than normal springs are causing deciduous trees to put leaves out earlier, and spring-blooming flowers to flower earlier. Animal migration times and breeding seasons are also shifting. Species arrays in biological communities are changing as warmer temperatures allow some species

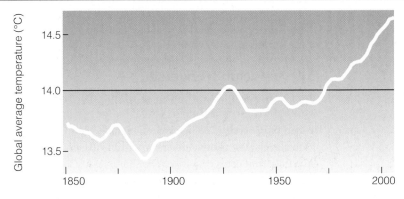
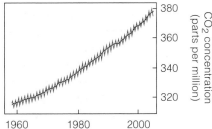

Figure 44.9 Changing global average temperature and atmospheric carbon dioxide. The large graph illustrates the increase in average temperature between 1850 and 2005.

The small graph shows the rise in atmospheric carbon dioxide during the last 45 years of this period. Carbon dioxide is a greenhouse gas.

A 1941 photo of Muir Glacier in Alaska.

B 2004 photo of the same region.

Figure 44.10 Melting glaciers, one sign of a warming world. Water from melting glaciers contributes to rising sea level.

to shift their range to higher latitudes or elevations. Of course, not all species can move or spread quickly, and warmer temperatures are expected to drive some of these species to extinction. For example, warming of tropical waters is already stressing reef-building corals and increasing the frequency of coral bleaching events.

Global warming is just one aspect of global climate change. Because the temperature of the land and seas affects evaporation, winds, and currents, many weather patterns are expected to change as temperature continues to rise. For example, warmer temperatures are correlated with extremes in rainfall patterns: periods of drought interrupted by unusually heavy rains. Also, warmer seas tend to increase the intensity of hurricanes.

As Section 42.8 explained, the most widely accepted explanation for global climate change is the rising levels of greenhouse gases such as carbon dioxide. Reducing greenhouse gas emissions will be a challenge. Fossil fuel combustion is the single biggest source of greenhouse gas emissions. Fossil fuel use is still rising as large nations such as China and India become increasingly industrialized. However, efforts are under way to increase the efficiency of processes that require fossil fuels, shift to alternative energy sources such as solar and wind power, and develop innovative ways to store carbon dioxide.

Take-Home Message How is the global climate changing, and what are some biological effects of the changes?

❭ The rise in global temperature is causing the sea level to rise, and is affecting weather patterns.

❭ These changes are altering the range of some species, threatening others with extinction, and disrupting the structure of biological communities.

> Conservation biologists survey and seek ways to protect the world's existing biodiversity.

The Value of Biodiversity

Every nation has several forms of wealth: material wealth, cultural wealth, and biological wealth. Biological wealth is called **biodiversity**. We measure a region's biodiversity at three levels: the genetic diversity within species, species diversity, and ecosystem diversity. Biodiversity is currently declining at all three levels, in all regions.

Conservation biology addresses these declines. The goals of this relatively new field of biology are (1) to survey the range of biodiversity, and (2) to find ways to maintain and use biodiversity to benefit human populations. The aim is to conserve biodiversity by encouraging people to value it and use it in ways that do not destroy it.

Why should we protect biodiversity? From a selfish standpoint, doing so is an investment in our future. Healthy ecosystems are essential to the survival of our species. Other organisms produce the oxygen we breathe and the food we eat. They remove waste carbon dioxide from the air and decompose and detoxify wastes. Plants take up rain and hold soil in place, preventing erosion and reducing the risk of flooding. Compounds in wild species can serve as medicines. Wild relatives of crop plants are reservoirs of genetic diversity that plant breeders draw on to protect and improve crops.

In addition, there are ethical reasons to preserve biodiversity. As we have emphasized many times, all living species are the result of an ongoing evolutionary process that stretches back billions of years. Each species has a unique combination of traits. The extinction of a species removes its unique collection of traits from the world of life forever.

Setting Priorities

Protecting biological diversity is often a tricky proposition. Even in developed countries, people often oppose environmental protections because they fear such measures will have adverse economic consequences. However, taking care of the environment can make good economic sense. With a bit of planning, people can both preserve and profit from their biological wealth.

Conservation biologists can help us make the difficult choices about which regions should be targeted for protection first. These biologists identify **hot spots**, places that are home to species found nowhere else and are under great threat of destruction. Once identified, hot spots can take priority in worldwide conservation efforts.

On a broader scale, conservation biologists define ecoregions, which are land or aquatic regions characterized by climate, geography, and the species found within them. The most widely used ecoregion system was developed by conservation scientists of the World Wildlife Fund. These scientists defined 867 distinctive land ecoregions. Figure 44.11 shows the locations and conservation status of ecoregions that are considered the top priority for conservation efforts.

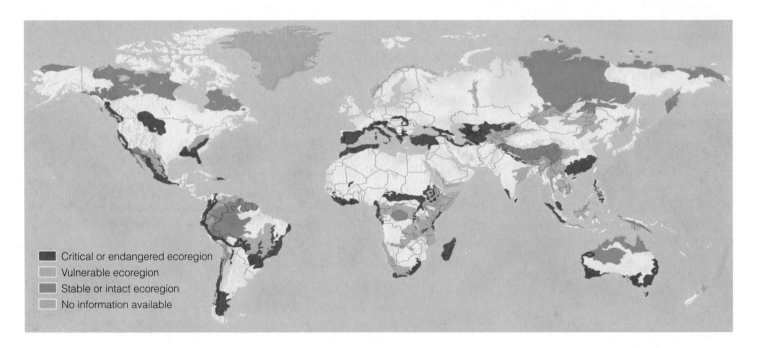

Critical or endangered ecoregion
Vulnerable ecoregion
Stable or intact ecoregion
No information available

Figure 44.11 The location and conservation status of the land ecoregions deemed most important by the World Wildlife Fund.

The Klamath-Siskiyou forest in southwestern Oregon and northwestern California is one of North America's endangered ecoregions. It is home to many rare conifers. Two endangered birds, the northern spotted owl and the marbled murrelet, nest in old-growth parts of the forest, and endangered coho salmon breed in streams that run through the forest. Logging threatens all of these species.

By focusing on hot spots and critical ecoregions rather than on individual endangered species, scientists hope to maintain ecosystem processes that naturally sustain biological diversity.

Preservation and Restoration

Worldwide, many ecologically important regions have been protected in ways that benefit local people. The Monteverde Cloud Forest in Costa Rica is one example. During the 1970s, George Powell was studying birds in this forest, which was rapidly being cleared. Powell decided to buy part of the forest as a nature sanctuary. His efforts inspired individuals and conservation groups to donate funds, and much of the forest is now protected as a private nature reserve (Figure 44.12). The reserve's plants and animals include more than 100 mammal species, 400 bird species, and 120 species of amphibians and reptiles. It is one of the few habitats left for jaguars and ocelots. A tourism industry centered on the reserve provides economic benefits to local people.

Sometimes, an ecosystem is so damaged, or there is so little of it left, that conservation alone is not enough to sustain biodiversity. **Ecological restoration** is work designed to bring about the renewal of a natural ecosystem that has been degraded or destroyed, fully or in part.

Restoration work in Louisiana's coastal wetlands is an example. More than 40 percent of the coastal wetlands in the United States are in Louisiana. The marshes are an ecological and economic treasure, but they are in trouble. Dams and levees built upstream of the marshes keep back sediments that would normally replenish sediments lost to the sea. Channels cut through the marshes for oil exploration and production have encouraged erosion, and the rising sea level threatens to flood the existing plants. Since the 1940s, Louisiana has lost an area of marshland the size of Rhode Island. Restoration efforts now under way aim to reverse some of those losses (Figure 44.13).

Figure 44.12 A conservation success story. Costa Rica's Monteverde Cloud Forest Reserve protects an area of tropical rain forest that is home to many endangered species, including jaguars (*inset photo*).

Figure 44.13 Ecological restoration in Louisiana's Sabine National Wildlife Refuge. In places where marshland has become open water, sediments are barged in and marsh grasses are planted on them. The squares are new sediment with marsh grass.

biodiversity Of a region, the genetic variation within its species, variety of species, and variety of ecosystems.
conservation biology Field of applied biology that surveys biodiversity and seeks ways to maintain and use it.
ecological restoration Actively altering an area in an effort to restore or create a functional ecosystem.
hot spot Threatened region with great biodiversity that is considered a high priority for conservation efforts.

Take-Home Message How do we sustain biodiversity?

> Biodiversity is the genetic diversity of individuals of a species, the variety of species, and the variety of ecosystems.

> Conservation biologists identify threatened regions with high biodiversity and prioritize which should be first to receive protection.

> Through ecological restoration, we re-create or renew a biologically diverse ecosystem that has been destroyed or degraded.

❯ The cumulative effects of individual actions will determine the health of our planet.

Ultimately, the health of our planet depends on our ability to recognize that the principles of energy flow and of resource limitation, which govern the survival of all systems of life, do not change. We must take note of these principles and find a way to live within our limits. The goal is living sustainably, which means meeting the needs of the present generation without reducing the ability of future generations to meet their own needs.

Promoting sustainability begins with recognizing the environmental consequences of one's own life-style. People in industrial nations use enormous quantities of resources, and the extraction, delivery, and use of these resources has negative effects on biodiversity. In the United States, the size of the average family has declined since the 1950s, while the size of the average home has doubled. All of the materials used to build and furnish those larger homes come from the environment. For example, an average new home contains about 500 pounds of copper in its wiring and plumbing.

Where does copper come from? Like most other mineral elements used in manufacturing, most copper is mined from the ground (Figure 44.14). Surface mining strips an area of vegetation and soil, creating an ecological

Figure 44.15 Volunteers restoring the Little Salmon River in Idaho so that salmon can migrate upstream to their breeding grounds.

dead zone. It puts dust into the air, creates mountains of rocky waste, and can contaminate nearby waterways.

Globalization makes it difficult to know the source of the raw materials in products you buy. Resource extraction in developing countries is often carried out under regulations that are less strict or less stringently enforced than those in the United States, so the environmental impact is even greater.

Figure 44.14 Effects of resource extraction. Bingham copper mine near Salt Lake City, Utah. This open pit mine is 4 kilometers (2.5 miles) wide and 1,200 meters (0.75 miles) deep, the largest man-made excavation on Earth.

Nonrenewable mineral resources are also used in electronic devices such as phones, computers, televisions, and MP3 players. Constantly trading up to the newest device may be good for the ego and the economy, but it is bad for the environment. Reducing consumption by fixing existing products is a sustainable resource use, as is recycling. Obtaining nonrenewable materials by recycling reduces the need for extraction of those resources, and it also helps keep materials out of landfills.

Reducing your energy use is another way to promote sustainability. Fossil fuels such as petroleum, natural gas, and coal supply most of the energy used by developed countries. You already know that burning these nonrenewable fuels contributes to global warming and acid rain. In addition, extracting and transporting these fuels have negative impacts. Oil harms many species when it leaks from pipelines or from ships.

Renewable energy sources do not produce greenhouse gases, but they have their own drawbacks. For example, dams in rivers of the Pacific Northwest generate renewable hydroelectric power, but they also prevent endangered salmon from returning to streams above the dam to breed. Similarly, wind turbines can harm birds and bats. Panels used to collect solar energy are made using nonrenewable mineral resources, and manufacturing the panels generates pollutants.

In short, all commercially produced energy has negative environmental impacts, so the best way to minimize your impact is to use less energy. Buy energy-efficient appliances, use fluorescent bulbs instead of incandescent ones, do not overheat or overcool rooms, and turn out unneeded lights. Walking, bicycling, and using public transportation are energy-efficient alternatives to driving. Shopping locally and purchasing locally produced goods also reduces energy use.

If you want to make more of a difference, learn about the threats to ecosystems in your own area. Support efforts to preserve and restore local biodiversity. Many ecological restoration projects are supervised by trained biologists but carried out primarily through the efforts of volunteers (Figure 44.15). Keep in mind that unthinking actions of billions of individuals are the greatest threat to biodiversity. Each of us may have little impact on our own, but our collective behavior, for good or for bad, will determine the future of the planet.

Take-Home Message What can individuals do to reduce their harmful impact on biodiversity?

❭ Resource extraction and usage have side effects that threaten biodiversity.

❭ You can save energy and other resources by reducing energy consumption and recycling and reusing materials.

■ A Long Reach (revisited)

The Arctic is not a separate continent like Antarctica, but rather a region that encompasses the northernmost parts of several continents. Eight countries, including the United States, Canada, and Russia, control parts of the Arctic and have rights to its extensive oil, gas, and mineral deposits. Until recently, ice sheets covered the Arctic Ocean, making it difficult for ships to move to and from the Arctic land mass, but those sheets are breaking up as a result of global climate change (Figure 44.16). At the same time, ice that covered the Arctic land mass is melting. These changes will make it easier for people to remove minerals and fossil fuels from the Arctic. With the world supply of fuel and minerals dwindling, pressure to exploit Arctic resources is rising. However, conservationists warn that extracting these resources will harm Arctic species such as the polar bear that are already threatened by global climate change.

Arctic perennial sea ice in 1979

Arctic perennial sea ice in 2003

Figure 44.16 Declining Arctic perennial ice.

How Would You Vote? The Arctic has extensive deposits of minerals and fossil fuel, but tapping into these resources might pose a risk to species already threatened by global climate change. Should the United States exploit its share of the Arctic resources or advocate for protection of the region? See CengageNow for details, then vote online (cengagenow.com).

Summary

Section 44.1 Human activities affect even remote places such as the Arctic. Polar bears in the Arctic have pollutants in their bodies and are threatened by thinning sea ice, one effect of global climate change.

Section 44.2 We are in the midst of a human-caused mass extinction, with numbers of **threatened species** and **endangered species** rising. **Endemic species** are especially likely to be at risk. Overharvest, species introductions, and habitat destruction, degradation, and fragmentation can push species toward extinction.

We know about only a tiny fraction of the species that are currently under threat.

Section 44.3 Overplowing or overgrazing can cause **desertification**. Both desertification and deforestation affect soil properties and rainfall. Changes caused by deforestation are especially difficult to reverse.

Section 44.4 Burning fossil fuels releases sulfur dioxide and nitrogen oxides. These **pollutants** mix with water vapor in the air, then fall to earth as **acid rain**. The resulting increase in the acidity of soils and waters can sicken or kill organisms.

Section 44.5 With **bioaccumulation**, a pollutant taken up from the environment becomes stored in an organism's tissues. With **biological magnification**, a pollutant increases in concentration as it is passed up a food chain. As a result of bioaccumulation and biological magnification, the tissues of an organism can have a far higher concentration of a pollutant than the environment.

Section 44.6 Human populations discard large amounts of trash that is burned, buried, or dumped in seas. Trash that washes into or is dumped into oceans poses a threat to marine life.

Section 44.7 The **ozone layer** in the upper atmosphere protects against incoming UV radiation. Chemicals called CFCs were banned when they were found to cause thinning of the ozone layer. Near the ground, where the ozone concentration is naturally low, ozone emitted as a result of fossil fuel use is considered a pollutant. It irritates animal respiratory tracts and interferes with photosynthesis by plants.

Section 44.8 Global climate change is causing glaciers to melt, thus raising the sea level. It also is affecting the range of species, allowing some to move into higher elevations or latitudes. Other species such as corals are showing signs of temperature-related stress.

Global climate change is also expected to further alter rainfall patterns and the intensity of hurricanes.

Sections 44.9, 44.10 Genetic diversity, species diversity, and ecosystem diversity are components of **biodiversity**, which is declining in all regions. **Conservation biology** involves surveying biodiversity and developing strategies to protect it and allow its sustainable use. Because resources are limited, biologists often focus on **hot spots**, where many unique species are under threat. They also attempt to ensure that portions of all ecoregions are protected. When an ecosystem has been totally or partially degraded, **ecological restoration** can help restore biodiversity.

Individuals can help sustain biodiversity by limiting their energy use and material consumption. Reuse and recycling help prevent destructive practices required to extract resources.

Self-Quiz   Answers in Appendix III

1. A(n) _____ species has population levels so low it is at great risk of extinction in the near future.
   a. endemic   c. threatened
   b. endangered   d. indicator

2. Species are threatened by habitat _____ .
   a. fragmentation   c. destruction
   b. degradation   d. all of the above

3. Deforestation _____ the amount of minerals that run off from soil.
   a. increases   b. decreases   c. has no effect on

4. Sulfur dioxide released by coal-burning power plants contributes to _____ .
   a. ozone destruction   c. acid rain
   b. sea level rise   d. desertification

5. As a result of _____ , an old animal usually has more pollutants in its body than a young one.
   a. bioaccumulation   b. biological magnification

6. With biological magnification, a _____ will have the highest pollutant load.
   a. producer   c. secondary consumer
   b. primary consumer   d. top-level consumer

7. An increase in the size of the ozone hole would be expected to _____ .
   a. increase skin cancers   c. both a and b
   b. reduce respiratory disorders

8. True or false? All pollutants are synthetic chemicals.

9. Global climate change is causing _____ .
   a. a decrease in sea level   c. acid rain
   b. glacial melting   d. all of the above

10. The Montreal Protocol banned use of _____ , which contribute(s) to ozone depletion.
    a. DDT   c. fossil fuels
    b. CFCs   d. sulfur dioxides

Data Analysis Activities

Arctic PCB Pollution Winds carry chemical contaminants produced and released at temperate latitudes to the Arctic, where the chemicals enter food webs. Top carnivores in arctic food webs, including polar bears and people, end up with high concentrations of these chemicals in their body. Arctic people who eat a lot of local wildlife tend to have unusually high levels of industrial chemicals called polychlorinated biphenyls, or PCBs, in their body. The Arctic Monitoring and Assessment Programme studies the effects of these industrial chemicals on the health and reproduction of Arctic people. Figure 44.17 shows how sex ratio at birth varies with average maternal PCB levels among people native to the Russian Arctic.

1. Which sex was more common in offspring of women with less than 1 microgram per milliliter of PCB in serum?
2. At what PCB concentrations were women more likely to have daughters?
3. In some villages in Greenland, nearly all recent newborns are female. Would you expect PCB levels in those villages to be above or under 4 micrograms per milliliter?

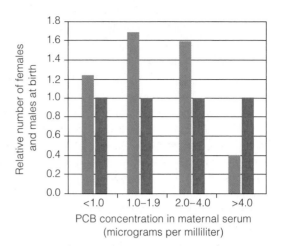

Figure 44.17 Effect of maternal PCB concentration on sex ratio of newborns in human populations native to the Russian Arctic. *Blue* bars indicate the relative number of males born per one female (*pink* bars).

---

11. A highly threatened region that is home to many unique species is a(n) _____ .
    a. ecoregion   c. hot spot
    b. biome   d. community

12. Biodiversity refers to _____ .
    a. genetic diversity   c. ecosystem diversity
    b. species diversity   d. all of the above

13. Restoring a marsh that has been damaged by human activities is an example of _____ .
    a. biological magnification c. ecological restoration
    b. bioaccumulation   d. globalization

14. Individuals help sustain biodiversity by _____ .
    a. reducing resource consumption
    b. reusing materials
    c. recycling materials
    d. all of the above

15. Match the terms with the most suitable description.
    ____ hot spot   a. good up high; bad nearby
    ____ ozone   b. tree loss alters rainfall pattern
    ____ biodiversity   and is difficult to reverse
    ____ acid rain   c. can increase dust storms
    ____ endemic   d. evolved in one region and
         species   remains there
    ____ biological   e. coal-burning is major cause
         magnification   f. results in highest pollution
    ____ global climate   level at top trophic level
         change   g. has lots of threatened species
    ____ deforestation   h. cause of rising sea level
    ____ desertification   i. genetic, species, and ecosystem
                              diversity

Additional questions are available on **CENGAGENOW**.

Critical Thinking

1. In one seaside community in New Jersey, the U.S. Fish and Wildlife Service suggested trapping and removing feral cats (domestic cats that live in the wild). The goal was to protect some endangered wild birds (plovers) that nested on the town's beaches. Many residents were angered by the proposal, arguing that the cats have as much right to be there as the birds. Do you agree? Why or why not?

2. Burning fossil fuel puts excess carbon dioxide into the atmosphere, but deforestation and desertification also affect carbon dioxide concentration. Explain why a global decrease in the amount of vegetation is contributing to the rise in carbon dioxide.

3. The magnitude of acid rain's effects can be influenced by the properties of the rock that the rain runs over. Acid rain is least likely to significantly acidify lakes in regions where the bedrock consists of calcium carbonate–rich limestone or marble. How does the presence of these rocks mitigate the effects of acid rain?

4. Two arctic marine mammals that live in the same waters differ in the level of pollutants in their bodies. Bowhead whales have a lower pollutant load than ringed seals. What are some factors that might explain this difference?

Animations and Interactions on **CENGAGENOW**:
› Acid rain; Ozone and CFCs.

Appendix I. Classification System

This revised classification scheme is a composite of several that microbiologists, botanists, and zoologists use. The major groupings are agreed upon, more or less. However, there is not always agreement on what to name a particular grouping or where it might fit within the overall hierarchy. There are several reasons why full consensus is not possible at this time.

First, the fossil record varies in its completeness and quality. Therefore, the phylogenetic relationship of one group to other groups is sometimes open to interpretation. Today, comparative studies at the molecular level are firming up the picture, but the work is still under way. Also, molecular comparisons do not always provide definitive answers to questions about phylogeny. Comparisons based on one set of genes may conflict with those comparing a different part of the genome. Or comparisons with one member of a group may conflict with comparisons based on other group members.

Second, ever since the time of Linnaeus, systems of classification have been based on the perceived morphological similarities and differences among organisms. Although some original interpretations are now open to question, we are so used to thinking about organisms in certain ways that reclassification often proceeds slowly.

A few examples: Traditionally, birds and reptiles were grouped in separate classes (Reptilia and Aves); yet there are compelling arguments for grouping the lizards and snakes in one group and the crocodilians, dinosaurs, and birds in another. Many biologists still favor a six-kingdom system of classification (archaea, bacteria, protists, plants, fungi, and animals). Others advocate a switch to the more recently proposed three-domain system (archaea, bacteria, and eukarya).

Third, researchers in microbiology, mycology, botany, zoology, and other fields of inquiry inherited a wealth of literature, based on classification systems that have been developed over time in each field of inquiry. Many are reluctant to give up established terminology that offers access to the past.

For example, botanists and microbiologists often use *division*, and zoologists *phylum*, for taxa that are equivalent in hierarchies of classification.

Why bother with classification frameworks if we know they only imperfectly reflect the evolutionary history of life? We do so for the same reasons that a writer might break up a history of civilization into several volumes, each with a number of chapters. Both are efforts to impart structure to an enormous body of knowledge and to facilitate retrieval of information from it. More importantly, to the extent that modern classification schemes accurately reflect evolutionary relationships, they provide the basis for comparative biological studies, which link all fields of biology.

Bear in mind that we include this appendix for your reference purposes only. Besides being open to revision, it is not meant to be complete. Names shown in "quotes" are polyphyletic or paraphyletic groups that are undergoing revision. For example, "reptiles" comprise at least three and possibly more lineages.

The most recently discovered species, as from the mid-ocean province, are not listed. Many existing and extinct species of the more obscure phyla are also not represented. Our strategy is to focus primarily on the organisms mentioned in the text or familiar to most students. We delve more deeply into flowering plants than into bryophytes, and into chordates than annelids.

RELATIONSHIPS AMONG THE THREE DOMAINS

As a general frame of reference, note that almost all bacteria and archaeans are microscopic in size. Their DNA is typically concentrated in a nucleoid (a region of cytoplasm), not in a membrane-bound nucleus. They reproduce by binary fission or budding, transfer genes by conjugation, and none are multicellular.

Bacteria and Archaeans have historically been grouped together as "prokaryotes." However, some microbiologists now advocate avoiding the use of this term. They point out that bacteria and archaeans do not constitute a monophyletic group, and that the two lineages differ in many structural and genetic traits. In deference to this sentiment, we use the term "prokaryote" only in discussion of historical classification systems.

Use of the term "eukaryote" remains noncontroversial. The eukaryotes are a monophyletic group thought to be descended from an archaean ancestor. Eukaryotic cells are typically larger than those of archaeans or bacteria and they have a DNA-enclosing nucleus and other membrane-bound organelles such as endoplasmic reticulum. Mitosis and meiosis occur only in eukaryotes.

Eukaryotes include a diverse array of single-celled organisms, as well as all multicellular species.

DOMAIN OF BACTERIA

KINGDOM BACTERIA The largest, and most diverse group of microorganisms. Includes photosynthetic autotrophs, chemosynthetic autotrophs, and heterotrophs. Includes many important pathogens of vertebrates.

PHYLUM AQIFACAE Most ancient branch of the bacterial tree. Gram-negative, mostly aerobic chemoautotrophs, mainly of volcanic hot springs. *Aquifex.*

PHYLUM DEINOCOCCUS-THERMUS Gram-positive, heat-loving chemoautotrophs. *Deinococcus* is the most radiation resistant organism known. *Thermus* occurs in hot springs and near hydrothermal vents.

PHYLUM CHLOROFLEXI Green nonsulfur bacteria. Gram-negative bacteria of hot springs, freshwater lakes, and marine habitats. Act as nonoxygen-producing photoautotrophs or aerobic chemoheterotrophs. *Chloroflexus.*

PHYLUM ACTINOBACTERIA Gram-positive, mostly aerobic heterotrophs in soil, freshwater and marine habitats, and on mammalian skin. *Propionibacterium, Actinomyces, Streptomyces.*

PHYLUM CYANOBACTERIA Gram-negative, oxygen-releasing photoautotrophs mainly in aquatic habitats. They have chlorophyll *a* and photosystem I. Includes many nitrogen-fixing genera. *Anabaena, Nostoc, Oscillatoria.*

PHYLUM CHLOROBIUM Green sulfur bacteria. Gram-negative nonoxygen-producing photosynthesizers, mainly in freshwater sediments. *Chlorobium.*

PHYLUM FIRMICUTES Gram-positive walled cells and the cell wall-less mycoplasmas. All are heterotrophs. Some survive in soil, hot springs, lakes, or oceans. Others live on or in animals. *Bacillus, Clostridium, Heliobacterium, Lactobacillus, Listeria, Mycobacterium, Mycoplasma, Streptococcus.*

PHYLUM CHLAMYDIAE Gram-negative intracellular parasites of birds and mammals. *Chlamydia.*

PHYLUM SPIROCHETES Free-living, parasitic, and mutualistic gram-negative spring-shaped bacteria. *Borelia, Pillotina, Spirillum, Treponema.*

PHYLUM PROTEOBACTERIA The largest bacterial group. Includes photoautotrophs, chemoautotrophs, and heterotrophs; free-living, parasitic, and colonial groups. All are gram-negative.

Class Alphaproteobacteria. *Agrobacterium, Azospirillum, Nitrobacter, Rickettsia, Rhizobium.*

Class Betaproteobacteria. *Neisseria.*

Class Gammaproteobacteria. *Chromatium, Escherichia, Haemopilius, Pseudomonas, Salmonella, Shigella, Thiomargarita, Vibrio, Yersinia.*

Class Deltaproteobacteria. *Azotobacter, Myxococcus.*

Class Epsilonproteobacteria. *Campylobacter, Helicobacter.*

DOMAIN OF ARCHAEA

KINGDOM ARCHAEA Single cells that are evolutionarily between eukaryotic cells and the bacteria. Most are anaerobes. None are photosynthetic. Originally discovered in extreme habitats, they are now known to be widely dispersed. Compared with bacteria, the archaea have a distinctive cell wall structure and unique membrane lipids, ribosomes, and RNA sequences. Some are symbiotic with animals, but none are known to be animal pathogens.

PHYLUM EURYARCHAEOTA Largest archean group. Includes extreme thermophiles, halophiles, and methanogens. Others are abundant in the upper waters of the ocean and other more moderate habitats. *Methanocaldococcus, Nanoarchaeum.*

PHYLUM CRENARCHAEOTA Includes extreme theromophiles, as well as species that survive in Antarctic waters, and in more moderate habitats. *Sulfolobus, Ignicoccus.*

PHYLUM KORARCHAEOTA Known only from DNA isolated from hydrothermal pools. As of this writing, none have been cultured and no species have been named.

DOMAIN OF EUKARYOTES

KINGDOM "PROTISTA" A collection of single-celled and multicelled lineages, which does not constitute a monophyletic group. Some biologists consider the groups listed below to be kingdoms in their own right.

PARABASALIA Parabasalids. Flagellated, single-celled anaerobic heterotrophs with a cytoskeletal "backbone" that runs the length of the cell. There are no mitochondria, but a hydrogenosome serves a similar function. *Trichomonas, Trichonympha.*

DIPLOMONADIDA Diplomonads. Flagellated, anaerobic single-celled heterotrophs that do not have mitochondria or Golgi bodies and do not form a bipolar spindle at mitosis. May be one of the most ancient lineages. *Giardia.*

EUGLENOZOA Euglenoids and kinetoplastids. Free-living and parasitic flagellates. All with one or more mitochondria. Some photosynthetic euglenoids with chloroplasts, others heterotrophic. *Euglena, Trypanosoma, Leishmania.*

RHIZARIA Formaminiferans and radiolarians. Free-living, heterotrophic amoeboid cells that are enclosed in shells. Most live in ocean waters or sediments. *Pterocorys, Stylosphaera.*

ALVEOLATA Single cells having a unique array of membrane-bound sacs (alveoli) just beneath the plasma membrane.

Ciliata. Ciliated protozoans. Heterotrophic protists with many cilia. *Paramecium, Didinium.*

Dinoflagellates. Diverse heterotrophic and photosynthetic flagellated cells that deposit cellulose in their alveoli. *Gonyaulax, Gymnodinium, Karenia, Noctiluca.*

Apicomplexans. Single-celled parasites of animals. A unique microtubular device is used to attach to and penetrate a host cell. *Plasmodium.*

STRAMENOPHILA Stramenophiles. Single-celled and multicelled forms; flagella with tinsel-like filaments.

Oomycotes. Water molds. Heterotrophs. Decomposers, some parasites. *Saprolegnia, Phytophthora, Plasmopara.*

Chrysophytes. Golden algae, yellow-green algae, diatoms, coccolithophores. Photosynthetic. *Emiliania, Mischococcus.*

Phaeophytes. Brown algae. Photosynthetic; nearly all live in temperate marine waters. All are multicellular. *Macrocystis, Laminaria, Sargassum, Postelsia.*

RHODOPHYTA Red algae. Mostly photosynthetic, some parasitic. Nearly all marine, some in freshwater habitats. Most multicellular. *Porphyra, Antithamion.*

CHLOROPHYTA Green algae. Mostly photosynthetic, some parasitic. Most freshwater, some marine or terrestrial. Single-celled, colonial, and multicellular forms. Some biologists place the chlorophytes and charophytes with the land plants in a kingdom called the Viridiplantae. *Acetabularia, Chlamydomonas, Chlorella, Codium, Udotea, Ulva, Volvox.*

CHAROPHYTA Photosynthetic. Closest living relatives of plants. Include both single-celled and multicelled forms. Desmids, stoneworts. *Micrasterias, Chara, Spirogyra.*

AMOEBOZOA True amoebas and slime molds. Heterotrophs that spend all or part of the life cycle as a single cell that uses pseudopods to capture food. *Amoeba, Entoamoeba* (amoebas), *Dictyostelium* (cellular slime mold), *Physarum* (plasmodial slime mold).

KINGDOM FUNGI

Nearly all multicelled eukaryotic species with chitin-containing cell walls. Heterotrophs, mostly saprobic decomposers, some parasites. Nutrition based upon extracellular digestion of organic matter and absorption of nutrients by individual cells. Multicelled species form absorptive mycelia and reproductive structures that produce asexual spores (and sometimes sexual spores).

PHYLUM CHYTRIDIOMYCOTA   Chytrids. Primarily aquatic; saprobic decomposers or parasites that produce flagellated spores. *Chytridium.*

PHYLUM ZYGOMYCOTA   Zygomycetes. Producers of zygospores (zygotes inside thick wall) by way of sexual reproduction. Bread molds, related forms. *Rhizopus, Philobolus.*

PHYLUM ASCOMYCOTA   Ascomycetes. Sac fungi. Sac-shaped cells form sexual spores (ascospores). Most yeasts and molds, morels, truffles. *Saccharomycetes, Morchella, Neurospora, Claviceps, Candida, Aspergillus, Penicillium.*

PHYLUM BASIDIOMYCOTA   Basidiomycetes. Club fungi. Most diverse group. Produce basidiospores inside club-shaped structures. Mushrooms, shelf fungi, stinkhorns. *Agaricus, Amanita, Craterellus, Gymnophilus, Puccinia, Ustilago.*

"IMPERFECT FUNGI"   Sexual spores absent or undetected. The group has no formal taxonomic status. If better understood, a given species might be grouped with sac fungi or club fungi. *Arthobotrys, Histoplasma, Microsporum, Verticillium.*

"LICHENS"   Mutualistic interactions between fungal species and a cyanobacterium, green alga, or both. *Lobaria, Usnea.*

KINGDOM PLANTAE

Most photosynthetic with chlorophylls *a* and *b*. Some parasitic. Nearly all live on land. Sexual reproduction predominates.

BRYOPHYTES (NONVASCULAR PLANTS)

Small flattened haploid gametophyte dominates the life cycle; sporophyte remains attached to it. Sperm are flagellated; require water to swim to eggs for fertilization.

PHYLUM HEPATOPHYTA   Liverworts. *Marchantia.*

PHYLUM ANTHOCEROPHYTA   Hornworts.

PHYLUM BRYOPHYTA   Mosses. *Polytrichum, Sphagnum.*

SEEDLESS VASCULAR PLANTS

Diploid sporophyte dominates, free-living gametophytes, flagellated sperm require water for fertilization.

PHYLUM LYCOPHYTA   Lycophytes, club mosses. Small single-veined leaves, branching rhizomes. *Lycopodium, Selaginella.*

PHYLUM MONILOPHYTA

Subphylum Psilophyta. Whisk ferns. No obvious roots or leaves on sporophyte, very reduced. *Psilotum.*

Subphylum Sphenophyta. Horsetails. Reduced scalelike leaves. Some stems photosynthetic, others spore-producing. *Calamites* (extinct), *Equisetum.*

Subphylum Pterophyta. Ferns. Large leaves, usually with sori. Largest group of seedless vascular plants (12,000 species), mainly tropical, temperate habitats. *Pteris, Trichomanes, Cyathea* (tree ferns), *Polystichum.*

SEED-BEARING VASCULAR PLANTS

PHYLUM CYCADOPHYTA   Cycads. Group of gymnosperms (vascular, bear "naked" seeds). Tropical, subtropical. Compound leaves, simple cones on male and female plants. Plants usually palm-like. Motile sperm. *Zamia, Cycas.*

PHYLUM GINKGOPHYTA   Ginkgo (maidenhair tree). Type of gymnosperm. Motile sperm. Seeds with fleshy layer. *Ginkgo.*

PHYLUM GNETOPHYTA   Gnetophytes. Only gymnosperms with vessels in xylem and double fertilization (but endosperm does not form). *Ephedra, Welwitchia, Gnetum.*

PHYLUM CONIFEROPHYTA   Conifers. Most common and familiar gymnosperms. Generally cone-bearing species with needle-like or scale-like leaves. Includes pines (*Pinus*), redwoods (*Sequoia*), yews (*Taxus*).

PHYLUM ANTHOPHYTA   Angiosperms (the flowering plants). Largest, most diverse group of vascular seed-bearing plants. Only organisms that produce flowers, fruits. Some families from several representative orders are listed:

BASAL FAMILIES

Family Amborellaceae. *Amborella.*
Family Nymphaeaceae. Water lilies.
Family Illiciaceae. Star anise.

MAGNOLIIDS

Family Magnoliaceae. Magnolias.
Family Lauraceae. Cinnamon, sassafras, avocados.
Family Piperaceae. Black pepper, white pepper.

EUDICOTS

Family Papaveraceae. Poppies.
Family Cactaceae. Cacti.
Family Euphorbiaceae. Spurges, poinsettia.
Family Salicaceae. Willows, poplars.
Family Fabaceae. Peas, beans, lupines, mesquite.
Family Rosaceae. Roses, apples, almonds, strawberries.
Family Moraceae. Figs, mulberries.
Family Cucurbitaceae. Squashes, melons, cucumbers.
Family Fagaceae. Oaks, chestnuts, beeches.
Family Brassicaceae. Mustards, cabbages, radishes.
Family Malvaceae. Mallows, okra, cotton, hibiscus, cocoa.
Family Sapindaceae. Soapberry, litchi, maples.
Family Ericaceae. Heaths, blueberries, azaleas.
Family Rubiaceae. Coffee.
Family Lamiaceae. Mints.
Family Solanaceae. Potatoes, eggplant, petunias.
Family Apiaceae. Parsleys, carrots, poison hemlock.
Family Asteraceae. Composites. Chrysanthemums, sunflowers, lettuces, dandelions.

MONOCOTS

Family Araceae. Anthuriums, calla lily, philodendrons.
Family Liliaceae. Lilies, tulips.
Family Alliaceae. Onions, garlic.
Family Iridaceae. Irises, gladioli, crocuses.
Family Orchidaceae. Orchids.
Family Arecaceae. Date palms, coconut palms.
Family Bromeliaceae. Bromeliads, pineapples.
Family Cyperaceae. Sedges.
Family Poaceae. Grasses, bamboos, corn, wheat, sugarcane.
Family Zingiberaceae. Gingers.

KINGDOM ANIMALIA

Multicelled heterotrophs, nearly all with tissues and organs, and organ systems, that are motile during part of the life cycle. Sexual reproduction occurs in most, but some also reproduce asexually. Embryos develop through a series of stages.

PHYLUM PORIFERA   Sponges. No symmetry, tissues.

PHYLUM PLACOZOA   Marine. Simplest known animal. Two cell layers, no mouth, no organs. *Trichoplax.*

PHYLUM CNIDARIA   Radial symmetry, tissues, nematocysts.
Class Hydrozoa. Hydrozoans. *Hydra, Obelia, Physalia, Prya.*
Class Scyphozoa. Jellyfishes. *Aurelia.*
Class Anthozoa. Sea anemones, corals. *Telesto.*

PHYLUM PLATYHELMINTHES Flatworms. Bilateral, cephalized; simplest animals with organ systems. Saclike gut.

Class Turbellaria. Triclads (planarians), polyclads. *Dugesia.*
Class Trematoda. Flukes. *Clonorchis, Schistosoma.*
Class Cestoda. Tapeworms. *Diphyllobothrium, Taenia.*

PHYLUM ROTIFERA Rotifers. *Asplancha, Philodina.*

PHYLUM MOLLUSCA Mollusks.

Class Polyplacophora. Chitons. *Cryptochiton, Tonicella.*

Class Gastropoda. Snails, sea slugs, land slugs. *Aplysia, Ariolimax, Cypraea, Haliotis, Helix, Liguus, Limax, Littorina.*

Class Bivalvia. Clams, mussels, scallops, cockles, oysters, shipworms. *Ensis, Chlamys, Mytelus, Patinopectin.*

Class Cephalopoda. Squids, octopuses, cuttlefish, nautiluses. *Dosidiscus, Loligo, Nautilus, Octopus, Sepia.*

PHYLUM ANNELIDA Segmented worms.

Class Polychaeta. Mostly marine worms. *Eunice, Neanthes.*

Class Oligochaeta. Mostly freshwater and terrestrial worms, many marine. *Lumbricus* (earthworms), *Tubifex.*

Class Hirudinea. Leeches. *Hirudo, Placobdella.*

PHYLUM NEMATODA Roundworms. *Ascaris, Caenorhabditis elegans, Necator* (hookworms), *Trichinella.*

PHYLUM ARTHROPODA

Subphylum Chelicerata. Chelicerates. Horseshoe crabs, spiders, scorpions, ticks, mites.

Subphylum Crustacea. Shrimps, crayfishes, lobsters, crabs, barnacles, copepods, isopods (sowbugs).

Subphylum Myriapoda. Centipedes, millipedes.

Subphylum Hexapoda. Insects and sprintails.

PHYLUM ECHINODERMATA Echinoderms.

Class Asteroidea. Sea stars. *Asterias.*
Class Ophiuroidea. Brittle stars.
Class Echinoidea. Sea urchins, heart urchins, sand dollars.
Class Holothuroidea. Sea cucumbers.
Class Crinoidea. Feather stars, sea lilies.
Class Concentricycloidea. Sea daisies.

PHYLUM CHORDATA Chordates.

Subphylum Urochordata. Tunicates, related forms.
Subphylum Cephalochordata. Lancelets.

CRANIATES

Class Myxini. Hagfishes.

VERTEBRATES (SUBGROUP OF CRANIATES)

Class Cephalaspidomorphi. Lampreys.

Class Chondrichthyes. Cartilaginous fishes (sharks, rays, skates, chimaeras).

Class "Osteichthyes." Bony fishes. Not monophyletic (sturgeons, paddlefish, herrings, carps, cods, trout, seahorses, tunas, lungfishes, and coelocanths).

TETRAPODS (SUBGROUP OF VERTEBRATES)

Class Amphibia. Amphibians. Require water to reproduce.
Order Caudata. Salamanders and newts.
Order Anura. Frogs, toads.
Order Apoda. Apodans (caecilians).

AMNIOTES (SUBGROUP OF TETRAPODS)

Class "Reptilia." Skin with scales, embryo protected and nutritionally supported by extraembryonic membranes.

Subclass Anapsida. Turtles, tortoises.
Subclass Lepidosaura. *Sphenodon,* lizards, snakes.
Subclass Archosaura. Crocodiles, alligators.

Class Aves. Birds. In some classifications birds are grouped in the archosaurs.

Order Struthioniformes. Ostriches.
Order Sphenisciformes. Penguins.
Order Procellariiformes. Albatrosses, petrels.
Order Ciconiiformes. Herons, bitterns, storks, flamingoes.
Order Anseriformes. Swans, geese, ducks.
Order Falconiformes. Eagles, hawks, vultures, falcons.
Order Galliformes. Ptarmigan, turkeys, domestic fowl.
Order Columbiformes. Pigeons, doves.
Order Strigiformes. Owls.
Order Apodiformes. Swifts, hummingbirds.
Order Passeriformes. Sparrows, jays, finches, crows, robins, starlings, wrens.
Order Piciformes. Woodpeckers, toucans.
Order Psittaciformes. Parrots, cockatoos, macaws.

Class Mammalia. Skin with hair; young nourished by milk-secreting mammary glands of adult.

Subclass Prototheria. Egg-laying mammals (monotremes; duckbilled platypus, spiny anteaters).

Subclass Metatheria. Pouched mammals or marsupials (opossums, kangaroos, wombats, Tasmanian devils).

Subclass Eutheria. Placental mammals.

Order Edentata. Anteaters, tree sloths, armadillos.
Order Insectivora. Tree shrews, moles, hedgehogs.
Order Chiroptera. Bats.
Order Scandentia. Insectivorous tree shrews.
Order Primates.
Suborder Strepsirhini (prosimians). Lemurs, lorises.
Suborder Haplorhini (tarsioids and anthropoids).
Infraorder Tarsiiformes. Tarsiers.
Infraorder Platyrrhini (New World monkeys).
Family Cebidae. Spider monkeys, howler monkeys, capuchin.
Infraorder Catarrhini (Old World monkeys and hominoids).
Superfamily Cercopithecoidea. Baboons, macaques, langurs.
Superfamily Hominoidea. Apes and humans.
Family Hylobatidae. Gibbon.
Family "Pongidae." Chimpanzees, gorillas, orangutans.
Family Hominidae. Existing and extinct human species (*Homo*) and humanlike species, including the australopiths.
Order Lagomorpha. Rabbits, hares, pikas.
Order Rodentia. Most gnawing animals (squirrels, rats, mice, guinea pigs, porcupines, beavers, etc.).
Order Carnivora. Carnivores (wolves, cats, bears, etc.).
Order Pinnipedia. Seals, walruses, sea lions.
Order Proboscidea. Elephants, mammoths (extinct).
Order Sirenia. Sea cows (manatees, dugongs).
Order Perissodactyla. Odd-toed ungulates (horses, tapirs, rhinos).
Order Tubulidentata. African aardvarks.
Order Artiodactyla. Even-toed ungulates (camels, deer, bison, sheep, goats, antelopes, giraffes, etc.).
Order Cetacea. Whales, porpoises.

Appendix II. Annotations to A Journal Article

This journal article reports on the movements of a female wolf during the summer of 2002 in northwestern Canada. It also reports on a scientific process of inquiry, observation and interpretation to learn where, how and why the wolf traveled as she did. In some ways, this article reflects the story of "how to do science" told in section 1.5 of this textbook. These notes are intended to help you read and understand how scientists work and how they report on their work.

(1) ARCTIC

(2) VOL. 57, NO. 2 (JUNE 2004) P. 196–203

(3) Long Foraging Movement of a Denning Tundra Wolf

(4) Paul F. Frame,[1,2] David S. Hik,[1] H. Dean Cluff,[3] and Paul C. Paquet[4]

(5) *(Received 3 September 2003; accepted in revised form 16 January 2004)*

(6) ABSTRACT. Wolves (*Canis lupus*) on the Canadian barrens are intimately linked to migrating herds of barren-ground caribou (*Rangifer tarandus*). We deployed a Global Positioning System (GPS) radio collar on an adult female wolf to record her movements in response to changing caribou densities near her den during summer. This wolf and two other females were observed nursing a group of 11 pups. She traveled a minimum of 341 km during a 14-day excursion. The straight-line distance from the den to the farthest location was 103 km, and the overall minimum rate of travel was 3.1 km/h. The distance between the wolf and the radio-collared caribou decreased from 242 km one week before the excursion to 8 km four days into the excursion. We discuss several possible explanations for the long foraging bout.

(7) *Key words:* wolf, GPS tracking, movements, *Canis lupus*, foraging, caribou, Northwest Territories

(8) RÉSUMÉ. Les loups (*Canis lupus*) dans la toundra canadienne sont étroitement liés aux hardes de caribous des toundras (*Rangifer tarandus*). On a équipé une louve adulte d'un collier émetteur muni d'un système de positionnement mondial (GPS) afin d'enregistrer ses déplacements en réponse au changement de densité du caribou près de sa tanière durant l'été. On a observé cette louve ainsi que deux autres en train d'allaiter un groupe de 11 louveteaux. Elle a parcouru un minimum de 341 km durant une sortie de 14 jours. La distance en ligne droite de la tanière à l'endroit le plus éloigné était de 103 km, et la vitesse minimum durant tout le voyage était de 3,1 km/h. La distance entre la louve et le caribou muni du collier émetteur a diminué de 242 km une semaine avant la sortie à 8 km quatre jours après la sortie. On commente diverses explications possibles pour ce long épisode de recherche de nourriture.

Mots clés: loup, repérage GPS, déplacements, *Canis lupus*, recherche de nourriture, caribou, Territoires du Nord-Ouest

Traduit pour la revue *Arctic* par Nésida Loyer.

(9) Introduction

Wolves (*Canis lupus*) that den on the central barrens of mainland Canada follow the seasonal movements of their main prey, migratory barren-ground caribou (*Rangifer tarandus*) (Kuyt, 1962; Kelsall, 1968; Walton et al., 2001). However, most wolves do not den near caribou calving grounds, but select sites farther south, closer to the tree line (Heard and Williams, 1992). Most caribou migrate beyond primary wolf denning areas by mid-June and do not return until mid-to-late July (Heard et al., 1996; Gunn et al., 2001). Conse-

quently, caribou density near dens is low for part of the summer.

During this period of spatial separation from the main caribou herds, wolves must either search near the homesite for scarce caribou or alternative prey (or both), travel to where prey are abundant, or use a combination of these strategies.

Walton et al. (2001) postulated that the travel of tundra wolves outside their normal summer ranges is a response to low caribou availability rather than a pre-dispersal exploration like that observed in territorial wolves (Fritts and Mech, 1981; Messier, 1985). The authors postulated this because most such travel was directed toward caribou calving grounds. We report details of such a long-distance excursion by a breeding female tundra wolf wearing a GPS radio collar. We discuss the relationship of the excursion to movements of satellite-collared caribou (Gunn et al., 2001), supporting the hypothesis that tundra wolves make directional, rapid, long-distance movements in response to seasonal prey availability.

[1] Department of Biological Sciences, University of Alberta, Edmonton, Alberta T6G 2E9, Canada
[2] Corresponding author: pframe@ualberta.ca
[3] Department of Resources, Wildlife, and Economic Development, North Slave Region, Government of the Northwest Territories, P.O. Box 2668, 3803 Bretzlaff Dr., Yellowknife, Northwest Territories X1A 2P9, Canada; Dean_Cluff@gov.nt.ca
[4] Faculty of Environmental Design, University of Calgary, Calgary, Alberta T2N 1N4, Canada; current address: P.O. Box 150, Meacham, Saskatchewan S0K 2V0, Canada

© The Arctic Institute of North America

196

Margin annotations (right column):

1 Title of the journal, which reports on science taking place in Arctic regions.

2 Volume number, issue number and date of the journal, and page numbers of the article.

3 Title of the article: a concise but specific description of the subject of study—one episode of long-range travel by a wolf hunting for food on the Arctic tundra.

4 Authors of the article: scientists working at the institutions listed in the footnotes below. Note #2 indicates that P. F. Frame is the *corresponding author*—the person to contact with questions or comments. His email address is provided.

5 Date on which a draft of the article was received by the journal editor, followed by date one which a revised draft was accepted for publication. Between these dates, the article was reviewed and critiqued by other scientists, a process called peer review. The authors revised the article to make it clearer, according to those reviews.

6 ABSTRACT: A brief description of the study containing all basic elements of this report. First sentence summarizes the *background* material. Second sentence encapsulates the *methods* used. The rest of the paragraph sums up the *results*. Authors introduce the main *subject* of the study—a female wolf (#388) with pups in a den—and refer to later *discussion* of possible explanations for her behavior.

7 Key words are listed to help researchers using computer databases. Searching the databases using these key words will yield a list of studies related to this one.

8 RÉSUMÉ: The French translation of the abstract and key words. Many researchers in this field are French Canadian. Some journals provide such translations in French or in other languages.

9 INTRODUCTION: Gives the background for this wolf study. This paragraph tells of known or suspected wolf behavior that is important for this study. Note that (a) major species mentioned are always accompanied by scientific names, and (b) statements of fact or *postulations* (claims or assumptions about what is likely to be true) are followed by references to studies that established those facts or supported the postulations.

10 This paragraph focuses directly on the wolf behaviors that were studied here.

11 This paragraph starts with a statement of the *hypothesis* being tested, one that originated in other studies and is supported by this one. The hypothesis is restated more succinctly in the last sentence of this paragraph. This is the *inquiry* part of the scientific process—asking questions and suggesting possible answers.

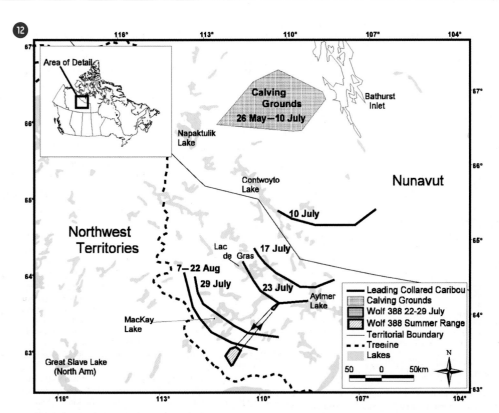

12 This map shows the study area and depicts wolf and caribou locations and movements during one summer. Some of this information is explained below.

13 STUDY AREA: This section sets the stage for the study, locating it precisely with latitude and longitude coordinates and describing the area (illustrated by the map in Figure 1).

14 Here begins the story of how prey (caribou) and predators (wolves) interact on the tundra. Authors describe movements of these nomadic animals throughout the year.

15 We focus on the denning season (summer) and learn how wolves locate their dens and travel according to the movements of caribou herds.

Figure 1. Map showing the movements of satellite radio-collared caribou with respect to female wolf 388's summer range and long foraging movement, in summer 2002.

13 Study Area

Our study took place in the northern boreal forest–low Arctic tundra transition zone (63° 30′ N, 110° 00′ W; Figure 1; Timoney et al., 1992). Permafrost in the area changes from discontinuous to continuous (Harris, 1986). Patches of spruce (*Picea mariana, P. glauca*) occur in the southern portion and give way to open tundra to the northeast. Eskers, kames, and other glacial deposits are scattered throughout the study area. Standing water and exposed bedrock are characteristic of the area.

14 *Details of the Caribou-Wolf System*

The Bathurst caribou herd uses this study area. Most caribou cows have begun migrating by late April, reaching calving grounds by June (Gunn et al., 2001;

Figure 1). Calving peaks by 15 June (Gunn et al., 2001), and calves begin to travel with the herd by one week of age (Kelsall, 1968). The movement patterns of bulls are less known, but bulls frequent areas near calving grounds by mid-June (Heard et al., 1996; Gunn et al., 2001). In summer, Bathurst caribou cows generally travel south from their calving grounds and then, parallel to the tree line, to the northwest. The rut usually takes place at the tree line in October (Gunn et al., 2001). The winter range of the Bathurst herd varies among years, ranging through the taiga and along the tree line from south of Great Bear Lake to southeast of Great Slave Lake. Some caribou spend the winter on the tundra (Gunn et al., 2001; Thorpe et al., 2001).

In winter, wolves that prey on Bathurst caribou do **15** not behave territorially. Instead, they follow the herd throughout its winter range (Walton et al., 2001; Musiani, 2003). However, during denning (May–

Foraging Movement of A Tundra Wolf **197**

16 Other variables are considered—prey other than caribou and their relative abundance in 2002.

17 METHODS: There is no one scientific method. Procedures for each and every study must be explained carefully.

18 Authors explain when and how they tracked caribou and wolves, including tools used and the exact procedures followed.

19 This important subsection explains what data were calculated (average distance ...) and how, including the software used and where it came from. (The calculations are listed in Table 1.) Note that the behavior measured (traveling) is carefully defined.

20 RESULTS: The heart of the report and the *observation* part of the scientific process. This section is organized parallel to the Methods section.

21 This subsection is broken down by periods of observation. Pre–excursion period covers the time between 388's capture and the start of her long–distance travel. The investigators used visual observations as well as telemetry (measurements taken using the global positioning system (GPS)) to gather data. They looked at how 388 cared for her pups, interacted with other adults, and moved about the den area.

Table 1. Daily distances from wolf 388 and the den to the nearest radio-collared caribou during a long excursion in summer 2002.

Date (2002)	Mean distance from caribou to wolf (km)	Daily distance from closest caribou to den
12 July	242	241
13 July	210	209
14 July	200	199
15 July	186	180
16 July	163	162
17 July	151	148
18 July	144	137
19 July[1]	126	124
20 July	103	130
21 July	73	130
22 July	40	110
23 July[2]	9	104
29 July[3]	16	43
30 July	32	43
31 July	28	44
1 August	29	46
2 August[4]	54	52
3 August	53	53
4 August	74	74
5 August	75	75
6 August	74	75
7 August	72	75
8 August	76	75
9 August	79	79

[1] Excursion starts.
[2] Wolf closest to collared caribou.
[3] Previous five days' caribou locations not available.
[4] Excursion ends.

August, parturition late May to mid-June), wolf movements are limited by the need to return food to the den. To maximize access to migrating caribou, many wolves select den sites closer to the tree line than to caribou calving grounds (Heard and Williams, 1992). Because of caribou movement patterns, tundra denning wolves are separated from the main caribou herds by several hundred kilometers at some time during summer (Williams, 1990:19; Figure 1; Table 1).

16 Muskoxen do not occur in the study area (Fournier and Gunn, 1998), and there are few moose there (H.D. Cluff, pers. obs.). Therefore, alternative prey for wolves includes waterfowl, other ground-nesting birds, their eggs, rodents, and hares (Kuyt, 1972; Williams, 1990:16; H.D. Cluff and P.F. Frame, unpubl. data). During 56 hours of den observations, we saw no ground squirrels or hares, only birds. It appears that the abundance of alternative prey was relatively low in 2002.

17 Methods

Wolf Monitoring

18 We captured female wolf 388 near her den on 22 June 2002, using a helicopter net-gun (Walton et al., 2001). She was fitted with a releasable GPS radio collar (Merrill et al., 1998) programmed to acquire locations at 30-

minute intervals. The collar was electronically released (e.g., Mech and Gese, 1992) on 20 August 2002. From 27 June to 3 July 2002, we observed 388's den with a 78 mm spotting scope at a distance of 390 m.

Caribou Monitoring

In spring of 2002, ten female caribou were captured by helicopter net-gun and fitted with satellite radio collars, bringing the total number of collared Bathurst cows to 19. Eight of these spent the summer of 2002 south of Queen Maud Gulf, well east of normal Bathurst caribou range. Therefore, we used 11 caribou for this analysis. The collars provided one location per day during our study, except for five days from 24 to 28 July. Locations of satellite collars were obtained from Service Argos, Inc. (Landover, Maryland).

Data Analysis

Location data were analyzed by ArcView GIS soft- **19** ware (Environmental Systems Research Institute Inc., Redlands, California). We calculated the average distance from the nearest collared caribou to the wolf and the den for each day of the study.

Wolf foraging bouts were calculated from the time 388 exited a buffer zone (500 m radius around the den) until she re-entered it. We considered her to be traveling when two consecutive locations were spatially separated by more than 100 m. Minimum distance traveled was the sum of distances between each location and the next during the excursion.

We compared pre- and post-excursion data using Analysis of Variance (ANOVA; Zar, 1999). We first tested for homogeneity of variances with Levene's test (Brown and Forsythe, 1974). No transformations of these data were required.

Results **20**

Wolf Monitoring

Pre-Excursion Period: Wolf 388 was lactating when **21** captured on 22 June. We observed her and two other females nursing a group of 11 pups between 27 June and 3 July. During our observations, the pack consisted of at least four adults (3 females and 1 male) and 11 pups. On 30 June, three pups were moved to a location 310 m from the other eight and cared for by an uncollared female. The male was not seen at the den after the evening of 30 June.

Before the excursion, telemetry indicated 18 foraging bouts. The mean distance traveled during these bouts was 25.29 km (± 4.5 SE, range 3.1–82.5 km). Mean greatest distance from the den on foraging

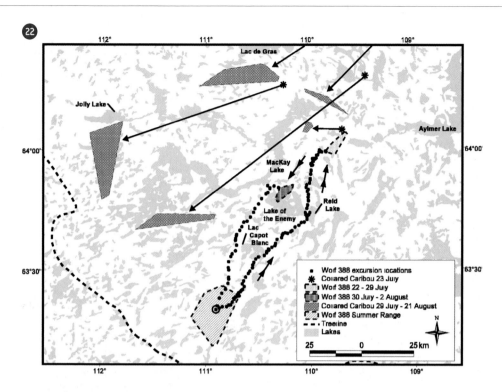

Figure 2. Details of a long foraging movement by female wolf 388 between 19 July and 2 August 2002. Also shown are locations and movements of three satellite radio-collared caribou from 23 July to 21 August 2002. On 23 July, the wolf was 8 km from a collared caribou. The farthest point from the den (103 km distant) was recorded on 27 July. Arrows indicate direction of travel.

22 The key in the lower right-hand corner of the map shows areas (shaded) within which the wolves and caribou moved, and the dotted trail of 388 during her excursion. From the results depicted on this map, the investigators tried to determine when and where 388 might have encountered caribou and how their locations affected her traveling behavior.

23 The wolf's excursion (her long trip away from the den area) is the focus of this study. These paragraphs present detailed measurements of daily movements during her two-week trip—how far she traveled, how far she was from collared caribou, her time spent traveling and resting, and her rate of speed. Authors use the phrase "minimum distance traveled" to acknowledge they couldn't track every step but were measuring samples of her movements. They knew that she went at least as far as they measured. This shows how scientists try to be exact when reporting results. Results of this study are depicted graphically in the map in Figure 2.

bouts was 7.1 km (± 0.9 SE, range 1.7–17.0 km). The average duration of foraging bouts for the period was 20.9 h (± 4.5 SE, range 1–71 h).

The average daily distance between the wolf and the nearest collared caribou decreased from 242 km on 12 July, one week before the excursion period, to 126 km on 19 July, the day the excursion began (Table 1).

Excursion Period: On 19 July at 2203, after spending 14 h at the den, 388 began moving to the northeast and did not return for 336 h (14 d; Figure 2). Whether she traveled alone or with other wolves is unknown. During the excursion, 476 (71%) of 672 possible locations were recorded. The wolf crossed the southeast end of Lac Capot Blanc on a small land bridge, where she paused for 4.5 h after traveling for 19.5 h (37.5 km). Following this rest, she traveled for 9 h (26.3 km) onto a peninsula in Reid Lake, where she spent 2 h before backtracking and stopping for 8 h just off the peninsula. Her next period of travel lasted 16.5 h (32.7 km), terminating in a pause of 9.5 h just 3.8 km from a concentration of locations at the far end of her excursion, where we presume she encountered caribou. The mean duration of these three movement periods was 15.7 h (± 2.5 SE), and that of the pauses, 7.3 h (± 1.5). The wolf required 72.5 h (3.0 d) to travel a minimum of 95 km from her den to this area near caribou (Figure 2). She remained there (35.5 km2) for 151.5 h (6.3 d) and then moved south to Lake of the Enemy, where she stayed (31.9 km^2) for 74 h (3.1 d) before returning to her den. Her greatest distance from the den, 103 km, was recorded 174.5 h (7.3 d) after the excursion

Foraging Movement of A Tundra Wolf **199**

24 Post-excursion measurements of 388's movements were made to compare with those of the pre-excursion period. In order to compare, scientists often use *means*, or averages, of a series of measurements—mean distances, mean duration, etc.

25 In the comparison, authors used statistical calculations (F and df) to determine that the differences between pre- and post-excursion measurements were *statistically insignificant*, or close enough to be considered essentially the same or similar.

26 As with wolf 388, the investigators measured the movements of caribou during the study period. The areas within which the caribou moved are shown in Figure 2 by shaded polygons mentioned in the second paragraph of this subsection.

27 This subsection summarizes how distances separating predators and prey varied during the study period.

28 DISCUSSION: This section is the *interpretation* part of the scientific process.

29 This subsection reviews observations from other studies and suggests that this study fits with patterns of those observations.

30 Authors discuss a prevailing *theory* (CBFT) which might explain why a wolf would travel far to meet her own energy needs while taking food caught closer to the den back to her pups. The results of this study seem to fit that pattern.

began, at 0433 on 27 July. She was 8 km from a collared caribou on 23 July, four days after the excursion began (Table 1).

The return trip began at 0403 on 2 August, 318 h (13.2 d) after leaving the den. She followed a relatively direct path for 18 h back to the den, a distance of 75 km.

The minimum distance traveled during the excursion was 339 km. The estimated overall minimum travel rate was 3.1 km/h, 2.6 km/h away from the den and 4.2 km/h on the return trip.

(24) Post-Excursion Period: We saw three pups when recovering the collar on 20 August, but others may have been hiding in vegetation.

Telemetry recorded 13 foraging bouts in the post-excursion period. The mean distance traveled during these bouts was 18.3 km (+ 2.7 SE, range 1.2–47.7 km), and mean greatest distance from the den was 7.1 km (+ 0.7 SE, range 1.1–11.0 km). The mean duration of these post-excursion foraging bouts was 10.9 h (+ 2.4 SE, range 1–33 h).

When 388 reached her den on 2 August, the distance to the nearest collared caribou was 54 km. On 9 August, one week after she returned, the distance was 79 km (Table 1).

Pre- and Post-Excursion Comparison

(25) We found no differences in the mean distance of foraging bouts before and after the excursion period (F = 1.5, df = 1, 29, *p* = 0.24). Likewise, the mean greatest distance from the den was similar pre- and post-excursion (F = 0.004, df = 1, 29, *p* = 0.95). However, the mean duration of 388's foraging bouts decreased by 10.0 h after her long excursion (F = 3.1, df = 1, 29, *p* = 0.09).

(26) *Caribou Monitoring*

Summer Movements: On 10 July, 5 of 11 collared caribou were dispersed over a distance of 10 km, 140 km south of their calving grounds (Figure 1). On the same day, three caribou were still on the calving grounds, two were between the calving grounds and the leaders, and one was missing. One week later (17 July), the leading radio-collared cows were 100 km farther south (Figure 1). Two were within 5 km of each other in front of the rest, who were more dispersed. All radio-collared cows had left the calving grounds by this time. On 23 July, the leading radio-collared caribou had moved 35 km farther south, and all of them were more widely dispersed. The two cows closest to the leader were 26 km and 33 km away, with 37 km between them. On the next location (29 July), the most southerly caribou were 60 km

farther south. All of the caribou were now in the areas where they remained for the duration of the study (Figure 2).

A Minimum Convex Polygon (Mohr and Stumpf, 1966) around all caribou locations acquired during the study encompassed 85 119 km^2.

Relative to the Wolf Den: **(27)** The distance from the nearest collared caribou to the den decreased from 241 km one week before the excursion to 124 km the day it began. The nearest a collared caribou came to the den was 43 km away, on 29 and 30 July. During the study, four collared caribou were located within 100 km of the den. Each of these four was closest to the wolf on at least one day during the period reported.

(28) Discussion

Prey Abundance

Caribou are the single most important prey of tundra **(29)** wolves (Clark, 1971; Kuyt, 1972; Stephenson and James, 1982; Williams, 1990). Caribou range over vast areas, and for part of the summer, they are scarce or absent in wolf home ranges (Heard et al., 1996). Both the long distance between radio-collared caribou and the den the week before the excursion and the increased time spent foraging by wolf 388 indicate that caribou availability near the den was low. Observations of the pups' being left alone for up to 18 h, presumably while adults were searching for food, provide additional support for low caribou availability locally. Mean foraging bout duration decreased by 10.0 h after the excursion, when collared caribou were closer to the den, suggesting an increase in caribou availability nearby.

Foraging Excursion

One aspect of central place foraging theory (CPFT) **(30)** deals with the optimality of returning different-sized food loads from varying distances to dependents at a central place (i.e., the den) (Orians and Pearson, 1979). Carlson (1985) tested CPFT and found that the predator usually consumed prey captured far from the central place, while feeding prey captured nearby to dependants. Wolf 388 spent 7.2 days in one area near caribou before moving to a location 23 km back towards the den, where she spent an additional 3.1 days, likely hunting caribou. She began her return trip from this closer location, traveling directly to the den. While away, she may have made one or more successful kills and spent time meeting her own energetic needs before returning to the den. Alternatively, it may have taken several attempts to make a kill, which she then fed on before beginning her return trip. We do not know if she returned food to the pups, but such behavior would be supported by CPFT.

[31] Other workers have reported wolves' making long round trips and referred to them as "extraterritorial" or "pre-dispersal" forays (Fritts and Mech, 1981; Messier, 1985; Ballard et al., 1997; Merrill and Mech, 2000). These movements are most often made by young wolves (1–3 years old), in areas where annual territories are maintained and prey are relatively sedentary (Fritts and Mech, 1981; Messier, 1985). The long excursion of 388 differs in that tundra wolves do not maintain annual territories (Walton et al., 2001), and the main prey migrate over vast areas (Gunn et al., 2001).

Another difference between 388's excursion and those reported earlier is that she is a mature, breeding female. No study of territorial wolves has reported reproductive adults making extraterritorial movements in summer (Fritts and Mech, 1981; Messier, 1985; Ballard et al., 1997; Merrill and Mech, 2001). However, Walton et al. (2001) also report that breeding female tundra wolves made excursions.

Direction of Movement

[32] Possible explanations for the relatively direct route 388 took to the caribou include landscape influence and experience. Considering the timing of 388's trip and the locations of caribou, had the wolf moved northwest, she might have missed the caribou entirely, or the encounter might have been delayed.

A reasonable possibility is that the land directed 388's route. The barrens are crisscrossed with trails worn into the tundra over centuries by hundreds of thousands of caribou and other animals (Kelsall, 1968; Thorpe et al., 2001). At river crossings, lakes, or narrow peninsulas, trails converge and funnel towards and away from caribou calving grounds and summer range. Wolves use trails for travel (Paquet et al., 1996; Mech and Boitani, 2003; P. Frame, pers. observation). Thus, the landscape may direct an animal's movements and lead it to where cues, such as the odor of caribou on the wind or scent marks of other wolves, may lead it to caribou.

[33] Another possibility is that 388 knew where to find caribou in summer. Sexually immature tundra wolves sometimes follow caribou to calving grounds (D. Heard, unpubl. data). Possibly, 388 had made such journeys in previous years and killed caribou. If this were the case, then in times of local prey scarcity she might travel to areas where she had hunted successfully before. Continued monitoring of tundra wolves may answer questions about how their food needs are met in times of low caribou abundance near dens.

[34] Caribou often form large groups while moving south to the tree line (Kelsall, 1968). After a large aggregation of caribou moves through an area, its scent can linger for weeks (Thorpe et al., 2001:104). It is conceivable that 388 detected caribou scent on the wind, which was blowing from the northeast on 19–21 July (Environment Canada, 2003), at the same time her excursion began. Many factors, such as odor strength and wind direction and strength, make systematic study of scent detection in wolves difficult under field conditions (Harrington and Asa, 2003). However, humans are able to smell odors such as forest fires or oil refineries more than 100 km away. The olfactory capabilities of dogs, which are similar to wolves, are thought to be 100 to 1 million times that of humans (Harrington and Asa, 2003). Therefore, it is reasonable to think that under the right wind conditions, the scent of many caribou traveling together could be detected by wolves from great distances, thus triggering a long foraging bout.

Rate of Travel

[35] Mech (1994) reported the rate of travel of Arctic wolves on barren ground was 8.7 km/h during regular travel and 10.0 km/h when returning to the den, a difference of 1.3 km/h. These rates are based on direct observation and exclude periods when wolves moved slowly or not at all. Our calculated travel rates are assumed to include periods of slow movement or no movement. However, the pattern we report is similar to that reported by Mech (1994), in that homeward travel was faster than regular travel by 1.6 km/h. The faster rate on return may be explained by the need to return food to the den. Pup survival can increase with the number of adults in a pack available to deliver food to pups (Harrington et al., 1983). Therefore, an increased rate of travel on homeward trips could improve a wolf's reproductive fitness by getting food to pups more quickly.

Fate of 388's Pups

[36] Wolf 388 was caring for pups during den observations. The pups were estimated to be six weeks old, and were seen ranging as far as 800 m from the den. They received some regurgitated food from two of the females, but were unattended for long periods. The excursion started 16 days after our observations, and it is improbable that the pups could have traveled the distance that 388 moved. If the pups died, this would have removed parental responsibility, allowing the long movement.

Our observations and the locations of radio-collared caribou indicate that prey became scarce in

31 Here our authors note other possible explanations for wolves' excursions presented by other investigators, but this study does not seem to support those ideas.

32 Authors discuss possible reasons for why 388 traveled directly to where caribou were located. They take what they learned from earlier studies and apply it to this case, suggesting that the lay of the land played a role. Note that their description paints a clear picture of the landscape.

33 Authors suggest that 388 may have learned in traveling during previous summers where the caribou were. The last two sentences suggest ideas for future studies.

34 Or maybe 388 followed the scent of the caribou. Authors acknowledge difficulties of proving this, but they suggest another area where future studies might be done.

35 Authors suggest that results of this study support previous studies about how fast wolves travel to and from the den. In the last sentence, they speculate on how these observed patterns would fit into the theory of evolution.

36 Authors also speculate on the fate of 388's pups while she was traveling. This leads to . . .

37 Discussion of cooperative rearing of pups and, in turn, to speculation on how this study and what is known about cooperative rearing might fit into the animal's strategies for survival of the species. Again, the authors approach the broader theory of evolution and how it might explain some of their results.

38 And again, they suggest that this study points to several areas where further study will shed some light.

39 In conclusion, the authors suggest that their study supports the hypothesis being tested here. And they touch on the implications of increased human activity on the tundra predicted by their results.

40 ACKNOWLEDGEMENTS: Authors note the support of institutions, companies and individuals. They thank their reviewers ad list permits under which their research was carried on.

41 REFERENCES: List of all studies cited in the report. This may seem tedious, but is a vitally important part of scientific reporting. It is a record of the sources of information on which this study is based. It provides readers with a wealth of resources for further reading on this topic. Much of it will form the foundation of future scientific studies like this one.

the area of the den as summer progressed. Wolf 388 may have abandoned her pups to seek food for herself. However, she returned to the den after the excursion, where she was seen near pups. In fact, she foraged in a similar pattern before and after the excursion, suggesting that she again was providing for pups after her return to the den.

37 A more likely possibility is that one or both of the other lactating females cared for the pups during 388's absence. The three females at this den were not seen with the pups at the same time. However, two weeks earlier, at a different den, we observed three females cooperatively caring for a group of six pups. At that den, the three lactating females were observed providing food for each other and trading places while nursing pups. Such a situation at the den of 388 could have created conditions that allowed one or more of the lactating females to range far from the den for a period, returning to her parental duties afterwards. However, the pups would have been weaned by eight weeks of age (Packard et al., 1992), so nonlactating adults could also have cared for them, as often happens in wolf packs (Packard et al., 1992; Mech et al., 1999).

Cooperative rearing of multiple litters by a pack could create opportunities for long-distance foraging movements by some reproductive wolves during summer periods of local food scarcity. We have recorded multiple lactating females at one or more tundra wolf dens per year since 1997. This reproductive strategy may be an adaptation to temporally and **38** spatially unpredictable food resources. All of these possibilities require further study, but emphasize both the adaptability of wolves living on the barrens and their dependence on caribou.

Long-range wolf movement in response to caribou **39** availability has been suggested by other researchers (Kuyt, 1972; Walton et al., 2001) and traditional ecological knowledge (Thorpe et al., 2001). Our report demonstrates the rapid and extreme response of wolves to caribou distribution and movements in summer. Increased human activity on the tundra (mining, road building, pipelines, ecotourism) may influence caribou movement patterns and change the interactions between wolves and caribou in the region. Continued monitoring of both species will help us to assess whether the association is being affected adversely by anthropogenic change.

40 Acknowledgements

This research was supported by the Department of Resources, Wildlife, and Economic Development, Government of the Northwest Territories; the Department of Biological Sciences at the University of Alberta; the Natural Sciences and Engineering Research Council of Canada; the Department of Indian and Northern Affairs Canada; the Canadian Circumpolar Institute; and DeBeers Canada, Ltd. Lorna Ruechel assisted with den observations. A. Gunn provided caribou location data. We thank Dave Mech for the use of GPS collars. M. Nelson, A. Gunn, and three anonymous reviewers made helpful comments on earlier drafts of the manuscript. This work was done under Wildlife Research Permit – WL002948 issued by the Government of the Northwest Territories, Department of Resources, Wildlife, and Economic Development.

41 References

BALLARD, W.B., AYRES, L.A., KRAUSMAN, P.R., REED, D.J., and FANCY, S.G. 1997. Ecology of wolves in relation to a migratory caribou herd in northwest Alaska. Wildlife Monographs 135. 47 p.

BROWN, M.B., and FORSYTHE, A.B. 1974. Robust tests for the equality of variances. Journal of the American Statistical Association 69:364–367.

CARLSON, A. 1985. Central place foraging in the red-backed shrike (*Lanius collurio* L.): Allocation of prey between forager and sedentary consumer. Animal Behaviour 33:664–666.

CLARK, K.R.F. 1971. Food habits and behavior of the tundra wolf on central Baffin Island. Ph.D. Thesis, University of Toronto, Ontario, Canada.

ENVIRONMENT CANADA. 2003. National climate data information archive. Available online: http://www.climate.weatheroffice.ec.gc.ca/Welcome_e.html

FOURNIER, B., and GUNN, A. 1998. Musk ox numbers and distribution in the NWT, 1997. File Report No. 121. Yellowknife: Department of Resources, Wildlife, and Economic Development, Government of the Northwest Territories. 55 p.

FRITTS, S.H., and MECH, L.D. 1981. Dynamics, movements, and feeding ecology of a newly protected wolf population in northwestern Minnesota. Wildlife Monographs 80. 79 p.

GUNN, A., DRAGON, J., and BOULANGER, J. 2001. Seasonal movements of satellite-collared caribou from the Bathurst herd. Final Report to the West Kitikmeot Slave Study Society, Yellowknife, NWT. 80 p. Available online: http://www.wkss.nt.ca/HTML/08_ProjectsReports/PDF/Seasonal MovementsFinal.pdf

HARRINGTON, F.H., and ASA, C.S. 2003. Wolf communication. In: Mech, L.D., and Boitani, L., eds. Wolves: Behavior, ecology, and conservation. Chicago: University of Chicago Press. 66–103.

HARRINGTON, F.H., MECH, L.D., and FRITTS, S.H. 1983. Pack size and wolf pup survival: Their relationship under varying ecological conditions. Behavioral Ecology and Sociobiology 13:19–26.

HARRIS, S.A. 1986. Permafrost distribution, zonation and stability along the eastern ranges of the cordillera of North America. Arctic 39(1):29–38.

HEARD, D.C., and WILLIAMS, T.M. 1992. Distribution of wolf dens on migratory caribou ranges in the Northwest

Territories, Canada. Canadian Journal of Zoology 70:1504–1510.

HEARD, D.C., WILLIAMS, T.M., and MELTON, D.A. 1996. The relationship between food intake and predation risk in migratory caribou and implication to caribou and wolf population dynamics. Rangifer Special Issue No. 2:37–44.

KELSALL, J.P. 1968. The migratory barren-ground caribou of Canada. Canadian Wildlife Service Monograph Series 3. Ottawa: Queen's Printer. 340 p.

KUYT, E. 1962. Movements of young wolves in the Northwest Territories of Canada. Journal of Mammalogy 43:270–271.

———. 1972. Food habits and ecology of wolves on barren-ground caribou range in the Northwest Territories. Canadian Wildlife Service Report Series 21. Ottawa: Information Canada. 36 p.

MECH, L.D. 1994. Regular and homeward travel speeds of Arctic wolves. Journal of Mammalogy 75:741–742.

MECH, L.D., and BOITANI, L. 2003. Wolf social ecology. In: Mech, L.D., and Boitani, L., eds. Wolves: Behavior, ecology, and conservation. Chicago: University of Chicago Press. 1–34.

MECH, L.D., and GESE, E.M. 1992. Field testing the Wildlink capture collar on wolves. Wildlife Society Bulletin 20:249–256.

MECH, L.D., WOLFE, P., and PACKARD, J.M. 1999. Regurgitative food transfer among wild wolves. Canadian Journal of Zoology 77:1192–1195.

MERRILL, S.B., and MECH, L.D. 2000. Details of extensive movements by Minnesota wolves (*Canis lupus*). American Midland Naturalist 144:428–433.

MERRILL, S.B., ADAMS, L.G., NELSON, M.E., and MECH, L.D. 1998. Testing releasable GPS radiocollars on wolves and white-tailed deer. Wildlife Society Bulletin 26:830–835.

MESSIER, F. 1985. Solitary living and extraterritorial movements of wolves in relation to social status and prey abundance. Canadian Journal of Zoology 63:239–245.

MOHR, C.O., and STUMPF, W.A. 1966. Comparison of methods for calculating areas of animal activity. Journal of Wildlife Management 30:293–304.

MUSIANI, M. 2003. Conservation biology and management of wolves and wolf-human conflicts in western North America. Ph.D. Thesis, University of Calgary, Calgary, Alberta, Canada.

ORIANS, G.H., and PEARSON, N.E. 1979. On the theory of central place foraging. In: Mitchell, R.D., and Stairs, G.F., eds. Analysis of ecological systems. Columbus: Ohio State University Press. 154–177.

PACKARD, J.M., MECH, L.D., and REAM, R.R. 1992. Weaning in an arctic wolf pack: Behavioral mechanisms. Canadian Journal of Zoology 70:1269–1275.

PAQUET, P.C., WIERZCHOWSKI, J., and CALLAGHAN, C. 1996. Summary report on the effects of human activity on gray wolves in the Bow River Valley, Banff National Park, Alberta. In: Green, J., Pacas, C., Bayley, S., and Cornwell, L., eds. A cumulative effects assessment and futures outlook for the Banff Bow Valley. Prepared for the Banff Bow Valley Study. Ottawa: Department of Canadian Heritage.

STEPHENSON, R.O., and JAMES, D. 1982. Wolf movements and food habits in northwest Alaska. In: Harrington, F.H., and Paquet, P.C., eds. Wolves of the world. New Jersey: Noyes Publications. 223–237.

THORPE, N., EYEGETOK, S., HAKONGAK, N., and QITIR-MIUT ELDERS. 2001. The Tuktu and Nogak Project: A caribou chronicle. Final Report to the West Kitikmeot/Slave Study Society, Ikaluktuuttiak, NWT. 160 p.

TIMONEY, K.P., LA ROI, G.H., ZOLTAI, S.C., and ROBINSON, A.L. 1992. The high subarctic forest-tundra of northwestern Canada: Position, width, and vegetation gradients in relation to climate. Arctic 45(1):1–9.

WALTON, L.R., CLUFF, H.D., PAQUET, P.C., and RAMSAY, M.A. 2001. Movement patterns of barren-ground wolves in the central Canadian Arctic. Journal of Mammalogy 82:867–876.

WILLIAMS, T.M. 1990. Summer diet and behavior of wolves denning on barren-ground caribou range in the Northwest Territories, Canada. M.Sc. Thesis, University of Alberta, Edmonton, Alberta, Canada.

ZAR, J.H. 1999. Biostatistical analysis. 4th ed. New Jersey: Prentice Hall. 663 p.

Appendix III. Answers to Self-Quizzes and Genetics Problems

Italicized numbers refer to relevant section numbers

Chapter 1

1. atoms — *1.2*
2. cell — *1.2*
3. Animals — *1.4*
4. energy, nutrients — *1.3*
5. Homeostasis — *1.3*
6. d — *1.3*
7. reproduction — *1.3*
8. d — *1.3*
9. a — *1.2*
   d — *1.3*
   e — *1.4*
10. a — *1.2*
    d — *1.3*
    f — *1.4*
11. b — *1.9*
12. b — *1.6*
13. c — *1.2*
    b — *1.5*
    d — *1.9*
    e — *1.6*
    a — *1.6*
    f — *1.8*

Chapter 2

1. False, H lacks neutrons — *2.2*
2. d — *2.2*
3. b — *2.2*
4. b — *2.3*
5. a — *2.3*
6. a — *2.4*
7. polar covalent — *2.4*
8. c — *2.5*
9. H⁺ and OH⁻ — *2.5*
10. d — *2.6*
11. a — *2.6*
12. c — *2.6*
13. b — *2.5*
14. c — *2.5*
    b — *2.2*
    d — *2.3*
    h — *2.2*
    g — *2.5*
    f — *2.3*
    e — *2.3*
    a — *2.3*

Chapter 3

1. c — *3.2*
2. four — *3.2*
3. a — *3.2*
4. e — *3.3, 3.7*
5. c — *3.4*
6. False — *3.4*
7. b — *3.4*
8. starch, glycogen, cellulose — *3.3*
9. e — *3.4*
10. d — *3.5, 3.7*
11. d — *3.6*
12. d — *3.7*
13. a — *3.7*
14. a, amino acid — *3.5*
    b, carbohydrate — *3.3*
    c, polypeptide — *3.5*
    d, fatty acid — *3.4*

15. c — *3.5*
    e — *3.7*
    f — *3.4*
    d — *3.7*
    a — *3.3*
    b — *3.4*
16. g — *3.5*
    b — *3.5*
    d — *3.7*
    f — *3.7*
    a — *3.4*
    c — *3.4*
    e — *3.3*
    i — *3.5*
    h — *3.3*

Chapter 4

1. cell — *4.2*
2. c — *4.2*
3. False — *4.6*
4. c — *4.4*
5. c — *4.5*
6. c — *4.4*
7. lysosomes — *4.8*
8. c, b, d, a — *4.8*
9. lipids and proteins — *4.8*
10. False. Some secrete a covering — *4.5*
11. a — *4.4*
12. d — *4.11*
13. a — *4.11*
14. c — *4.9*
    f — *4.9*
    a — *4.5*
    e — *4.8*
    d — *4.8*
    b — *4.8*

Chapter 5

1. c — *5.2*
2. a — *5.3*
3. b — *5.2*
4. d — *5.4*
5. c — *5.3, 5.4*
6. temperature, pH, salt concentration are all correct — *5.7*
7. e — *5.5*
8. more, less — *5.6*
9. gases, small nonpolar molecules, and water are all correct — *5.7*
10. b — *5.7*
11. a — *5.6*
12. e — *5.8*
13. c — *5.3*
    d — *5.4*
    e — *5.2*
    b — *5.3*
    a — *5.4*
    f — *5.6*
    g — *5.7*
    i — *5.7*
    h — *5.8*

Chapter 6

1. autotroph: weed; heterotrophs: cat, bird, caterpillar — *6.1*

2. carbon dioxide; sunlight — *6.1*
3. a — *6.2*
4. a — *6.4*
5. d — *6.5*
6. b — *6.5*
7. b — *6.5*
8. c — *6.4, 6.7*
9. b — *6.7*
10. e — *6.7*
11. f — *6.7*
12. f — *6.7*
    h — *6.7*
    g — *6.5*
    d — *6.5, 6.6*
    e — *6.8*
    b — *6.1*
    a — *6.2*
    c — *6.1*

Chapter 7

1. False (plants use aerobic respiration too) — *7.2*
2. d — *7.3*
3. a — *7.2*
4. c — *7.3*
5. b — *7.4*
6. e — *7.3–7.5*
7. b — *7.4*
8. c — *7.5*
9. c — *7.6*
10. b — *7.6*
11. c — *7.5*
12. d — *7.7*
13. f — *7.6*
14. b — *7.3*
    c — *7.6*
    a — *7.4*
    d — *7.5*
15. b — *7.4*
    d — *7.3*
    a — *7.3, 7.6*
    c — *7.5*
    f — *7.5*
    e — *7.4*

Chapter 8

1. b — *8.2*
2. centromere — *8.2*
3. d — *8.2*
4. c — *8.4*
5. d — *8.4*
6. c — *8.4*
7. a — *8.6*
8. d — *8.6*
9. 3'—CCAAAGAAGTTCTCT—5' — *8.6*
10. d — *8.7*
11. c — *8.4*
12. d — *8.7*
13. b — *8.2*
14. therapeutic cloning — *8.7*
15. d — *8.3*
    b — *8.1, 8.7*
    a — *8.4*
    f — *8.2*
    e — *8.6*
    g — *8.6*
    c — *8.2*

Chapter 9

1. c — *9.2*
2. b — *9.3*
3. d — *9.3*
4. a — *9.2*
5. c — *9.2*
6. c — *9.4*
7. a — *9.4*
8. 15 amino acids — *9.4*
9. a — *9.3*
10. a — *9.5*
11. d — *9.5*
12. a — *9.4*
13. c — *9.4*
14. f — *9.6*
15. c — *9.4*
    g — *9.3*
    e — *9.5*
    a — *9.4*
    f — *9.4*
    d — *9.3*
    b — *9.6*

Chapter 10

1. d — *10.2*
2. d — *10.2, 10.3*
3. d — *10.2*
4. transcription factors — *10.2*
5. b — *10.2*
6. h — *10.2*
7. c — *10.3*
8. d — *10.4*
9. d — *10.2*
10. b — *10.4*
11. c — *10.4*
12. c — *10.3*
13. d — *10.3*
14. b — *10.5*
15. f — *10.4*
    a — *10.4*
    b — *10.5*
    e — *10.4*
    c — *10.2*
    d — *10.2*

Chapter 11

1. d — *11.2*
2. b — *11.2*
3. d — *11.2*
4. e — *11.2*
5. b — *11.3*
6. c — *11.3*
7. b — *11.2*
8. a — *11.2*
9. interphase, prophase, metaphase, anaphase, telophase — *11.3*
10. c — *11.2*
11. a — *11.2*
12. The EGF receptor gene (*EGFR*), as well as *BRCA1*, *BRCA2*, and *53BP1* are all named in this chapter — *11.5, 11.6*
13. b — *11.6*

14. c — *11.4*
    f — *11.3*
    a — *11.6*
    g — *11.4*
    b — *11.4*
    e — *11.6*
    d — *11.3*
15. d, b, c, a — *11.3*

Chapter 12

1. b — *12.1*
2. c — *12.2*
3. d — *12.2*
4. b — *12.2*
5. b — *12.4*
6. prophase I — *12.4*
7. c — *12.3, 12.6*
8. a — *12.2*
9. e — *12.4*
10. sister chromatids are still attached — *12.3*
11. b — *12.4, 12.6*
12. b — *12.2*
    c — *12.3*
    a — *12.2*
    e — *12.5*
    d — *12.2*

Chapter 13

1. b — *13.2*
2. a — *13.2*
3. offspring — *13.3*
4. b — *13.3*
5. c — *13.3*
6. a — *13.3*
7. b — *13.3*
8. d — *13.4*
9. c — *13.4*
10. c — *13.5*
11. b — *13.5*
12. continuous variation — *13.6*
13. b — *13.4*
    d — *13.3*
    a — *13.2*
    c — *13.2*

Chapter 14

1. b — *14.2*
2. b — *14.2*
3. a — *14.2*
4. b — *14.3*
5. False — *14.4*
6. c — *14.4*
7. d — *14.4*
8. Y-linked dominant inheritance — *14.4, 14.6*
9. X from mom and dad — *14.6*
10. d — *14.6*
11. b — *14.6*
12. d — *14.6*
13. True — *14.6*
14. c — *14.6*
15. c — *14.6*
    e — *14.5*
    f — *14.6*
    b — *14.5*
    a — *14.2*
    d — *14.6*

Chapter 15

1. c — 15.2
2. a — 15.2
3. b — 15.2
4. b — 15.3
5. c — 15.3
6. DNA sequencing — 15.4
7. b — 15.4
8. d — 15.3, 15.5
9. d — 15.5
10. b — 15.10
    d — 15.2
    e — 15.10
    f — 15.7
11. c — 15.6
12. b — 15.10
13. a — 15.3
    d — 15.2
    c — 15.2
    e — 15.4
    b — 15.2
14. c — 15.5
    g — 15.7
    d — 15.3
    e — 15.10
    b — 15.1
    a — 15.6
    f — 15.6

Chapter 16

1. b — 16.2
2. d — 16.3, 16.4
3. e — 16.5
4. Gondwana — 16.7
5. c — 16.2, 16.8
6. d — 16.5
7. a — 16.2, 16.3
    c — 16.7
    e — 16.7
    f — 16.3
    g — 16.1
    h — 16.7
8. 65.5 — 16.6
9. Archaean Eon — 16.6
10. a — 16.8
11. d — 16.8
12. e — 16.4
    a — 16.2
    d — 16.9
    h — 16.5
    c — 16.8
    b — 16.3
    f — 16.8
    g — 16.4

Chapter 17

1. populations — 17.2
2. b — 17.2
3. a — 17.2
4. c — 17.2, 17.3
5. allopatric speciation — 17.11
6. a, d — 17.6
7. b, c — 17.6
8. b, e — 17.5
9. d — 17.7
10. e — 17.7
11. b — 17.9
12. c — 17.14
13. b — 17.14
14. c — 17.9
    e — 17.4
    g — 17.2
    b — 17.8
    f — 17.13
    a — 17.13
    d — 17.14
    h — 17.14

Chapter 18

1. c — 18.2, 18.5
2. c — 18.4
3. a — 18.4
4. a — 18.4
5. c — 18.5
6. b — 18.6
7. electrodes — 18.3
8. d — 18.5
9. c — 18.5
10. RNA — 18.4
11. a — 18.5
12. d — 18.7
13. c — 18.6
14. DNA — 18.4
15. f — 18.2
    c — 18.4
    d — 18.5
    a — 18.5
    b — 18.7
    e — 18.7

Chapter 19

1. c — 19.2
2. b — 19.4
3. c — 19.2
4. RNA — 19.2
5. c — 19.3
6. False — 19.6
7. one — 19.5
8. c — 19.7
9. b — 19.7
10. d — 19.7
11. a — 19.7
12. a — 19.5
13. c — 19.3, 19.7
14. a — 19.8
15. e — 19.8
    f — 19.7
    b — 19.2
    a — 19.4
    g — 19.6
    h — 19.8
    d — 19.5
    c — 19.8

Chapter 20

1. c — 20.3
2. d — 20.4
3. b — 20.3
4. c — 20.5
5. d — 20.3
6. d — 20.9
7. d — 20.8
8. b — 20.6
9. c — 20.7
10. d — 20.7
11. c — 20.8
12. a — 20.9
13. a — 20.3
14. a — 20.8
15. d — 20.3
    g — 20.6
    a — 20.3
    b — 20.7
    f — 20.7
    h — 20.8
    e — 20.8
    c — 20.9

Chapter 21

1. c — 21.2
2. a — 21.4
3. b — 21.3
4. c — 21.4
5. a — 21.5
6. e — 21.3, 21.4, 21.6
7. b — 21.5
8. True — 21.4
9. c — 21.6, 21.7
10. a — 21.3, 21.4
11. a — 21.2
12. d — 21.8
13. b — 21.8
14. c — 21.3
    a — 21.4
    b — 21.6
    b — 21.7
15. c — 21.5
    i — 21.2
    a — 21.2
    b — 21.2
    e — 21.7
    f — 21.7
    d — 21.4
    g — 21.4
    h — 21.5

Chapter 22

1. c — 22.2
2. a — 22.2
3. a — 22.2
4. c — 22.4
5. d — 22.5
6. hypha — 22.2
7. c — 22.5
8. b — 22.5
9. a — 22.3
10. b — 22.6
11. a — 22.3, 22.6
12. d — 22.3, 22.7
13. d — 22.7
14. b — 22.7
15. d — 22.2
    b — 22.2
    a — 22.3
    f — 22.4
    g — 22.5
    c — 22.6
    e — 22.6

Chapter 23

1. True — 23.2
2. d — 23.2
3. a — 23.5
4. a — 23.6
5. b — 23.13
6. a — 23.8
7. b — 23.11
8. a — 23.2
9. c — 23.10, 23.13
10. b — 23.15
11. closed — 23.7, 23.8
12. b — 23.3
    j — 23.3
    d — 23.4
    i — 23.5
    c — 23.6
    a — 23.2, 23.9
    g — 23.7
    e — 23.10
    f — 23.8
    h — 23.15

Chapter 24

1. c — 24.2
2. d — 24.2
3. a — 24.3
4. c — 24.2, 24.3
5. d — 24.3, 24.4
6. c — 24.5
7. f — 24.5
8. a — 24.5, 24.6
9. c — 24.7
10. f — 24.2, 24.8–24.10
11. c — 24.10, 24.11
12. True — 24.6
13. b — 24.2
    i — 24.3
    g — 24.4
    f — 24.9
    c — 24.7
    d — 24.8
    a — 24.8
    h — 24.8
    e — 24.9

Chapter 25

1. a — 25.2
2. d — 25.2, 25.7
3. c — 25.3
4. c — 25.3
5. a, b, e — 25.3
6. A; B; stem — 25.4
7. b — 25.3
8. b — 25.3
9. b — 25.7
10. c — 25.7
11. b — 25.2
    d — 25.7
    e — 25.3
    c — 25.3
    f — 25.6
    a — 25.7

Chapter 26

1. e — 26.2
2. b — 26.2
3. b — 26.3
4. b — 26.3
5. e — 26.3
6. c — 26.4
7. d — 26.4
8. b — 26.3
9. c — 26.5
10. a — 26.5
11. d — 26.4
12. c — 26.6
13. c — 26.5
14. e — 26.6
15. c — 26.5
    g — 26.2
    e — 26.6
    b — 26.3
    d — 26.5
    a — 26.5
    f — 26.6

Chapter 27

1. b — 27.2
2. nectar, oils, pollen, sex — 27.2
3. b — 27.3
4. b — 27.3
5. c — 27.4
6. a — 27.4
7. c — 27.5
8. c — 27.6
9. d — 27.7
10. d — 27.8
11. b — 27.9
12. a — 27.8
13. c — 27.9
14. c, e, b, a, d — 27.7
15. c — 27.3
    f — 27.2
    g — 27.3
    e — 27.3
    d — 27.2
    b — 27.3
    a — 27.3

Chapter 28

1. a — 28.3
2. a — 28.3
3. a — 28.3
4. b — 28.4
5. b — 28.4
6. c — 28.4
7. c — 28.5
8. a — 28.5
9. d — 28.6
10. a — 28.8
11. a — 28.7
12. b — 28.9
13. c — 28.3
14. b — 28.3
    j — 28.3
    a — 28.4
    c — 28.8
    d — 28.5
    h — 28.4
    f — 28.3
    i — 28.4
    k — 28.4
    e — 28.4
    g — 28.3

Chapter 29

1. a — 29.1
2. b — 29.4
3. b — 29.4
4. a — 29.6
5. a — 29.8
6. c — 29.8
7. b — 29.8
8. a — 29.6
9. a — 29.10
10. endorphins — 29.7
11. b — 29.7
12. white — 29.9
13. f — 29.9
    d — 29.6
    g — 29.11
    h — 29.10
    e — 29.2
    i — 29.2
    c — 29.10
    f — 29.9

Chapter 30

1. b — 30.3
2. c — 30.2
3. c — 30.3
4. e — 30.7
5. a — 30.3
6. b — 30.8
7. b — 30.9
8. a — 30.3
9. b — 30.9
10. c — 30.5
11. d — 30.5
12. b — 30.4
13. c — 30.6
14. a — 30.4
15. d — 30.5
    g — 30.9
    f — 30.4
    a — 30.4
    h — 30.5
    e — 30.7
    b — 30.8
    i — 30.9
    c — 30.7

Chapter 31

1. a — 31.2
2. b — 31.4
3. a — 31.4
4. d — 31.3
5. insulin; glucagon — 31.8
6. negative — 31.6
7. b — 31.8
8. d — 31.6
9. a — 31.10
10. c — 31.7
11. a — 31.4, 31.10
12. d — 31.10
13. f — 31.11
14. insulin — 31.9
15. d — 31.7
    f — 31.6
    c — 31.6
    e — 31.8
    a — 31.10
    b — 31.2

Chapter 32

1. a — 32.2
2. d — 32.3
3. b — 32.4
4. a — 32.3
5. a — 32.3
6. biceps — 32.4
7. a — 32.4
8. b — 32.5
9. d — 32.5
10. d — 32.6
11. c — 32.6
12. d — 32.6
13. False — 32.1, 32.7
14. a — 32.7
15. h — 32.4
    f — 32.6
    g — 32.6
    e — 32.3
    i — 32.5
    c — 32.3
    b — 32.2
    d — 32.3
    a — 32.6

Chapter 33

1. d — 33.2
2. d — 33.2
3. pulmonary — 33.2
4. plasma — 33.5
5. b — 33.5
6. d — 33.5
7. b — 33.4
8. b — 33.4
9. a — 33.7
10. b — 33.9
11. left ventricle — 33.4
12. artery — 33.3
13. b — 33.11
14. b — 33.10
15. f — 33.6
    a — 33.11
    e — 33.4
    g — 33.4
    b — 33.4
    c — 33.6
    d — 33.3

Chapter 34

1. e — 34.2, 34.3
2. g — 34.4, 34.5
3. c — 34.4
4. d — 34.4
5. d — 34.2
6. Fast, fixed, and general are all correct — 34.4
7. Self/nonself discrimination, diversity, specificity, and memory are all correct — 34.6
8. d — 34.5
9. b — 34.7
10. e — 34.9
11. b — 34.9
12. c — 34.10
13. e — 34.10
    b — 34.2
    c — 34.9
    d — 34.9
    a — 34.5

Chapter 35

1. d — 35.2
2. b — 35.3
3. a — 35.4
4. c — 35.4
5. d — 35.6
6. a — 35.6
7. iron — 35.7
8. b — 35.7
9. d — 35.7
10. c — 35.7
11. d — 35.6
12. True — 35.6
13. a — 35.6
14. bacteria — 35.8
15. d — 35.5
    h — 35.5
    f — 35.4
    e — 35.7
    g — 35.5
    c — 35.5
    b — 35.5
    a — 35.6

Chapter 36

1. d — 36.2
2. b — 36.5
3. c — 36.3
4. b — 36.7
5. a — 36.7
6. c — 36.8, 36.11
7. a — 36.5, 36.7
8. b — 36.7
9. b — 36.11
10. d — 36.9
11. microvilli — 36.6
12. c — 36.10
13. b — 36.12
14. a — 36.11
15. f — 36.7
    b — 36.8
    a — 36.7
    d — 36.7
    e — 36.5
    c — 36.7

Chapter 37

1. c — 37.2
2. d — 37.2
3. c — 37.2
4. a — 37.3
5. b — 37.4
6. a — 37.4
7. b — 37.4
8. a — 37.4
9. water — 37.4
10. dialysis — 37.5
11. c — 37.3
    a — 37.3
    b — 37.3
    e — 37.4
    d — 37.4
12. c — 37.6
13. b — 37.6
    a — 37.6
    d — 37.6
    c — 37.6
    e — 37.6

Chapter 38

1. a — 38.2
2. b — 38.2
3. a — 38.3
4. c — 38.5
5. b — 38.4
6. c — 38.8
7. c — 38.3
    d — 38.3
    g — 38.4
    e — 38.3
    b — 38.3
    a — 38.4
    f — 38.5
8. a — 38.11
9. False — 38.10
10. ectoderm, mesoderm, endoderm — 38.9
11. c — 38.12
12. b — 38.15
    a — 38.3
    d — 38.5
    c — 38.5
    e — 38.5
    g — 38.12
    f — 38.15
13. a — 38.12
    d — 38.12
    c — 38.12
    f — 38.12
    b — 38.13
    e — 38.13

Chapter 39

1. d — 39.2
2. b — 39.2
3. c — 39.4
4. c — 39.5
5. a — 39.1
6. birds — 39.6
7. protection from predators; greater ability to protect territory in a group — 39.7
8. d — 39.8
9. d — 39.8
10. close relatives — 39.8
11. h — 39.3
    e — 39.8
    c — 39.3
    b — 39.6
    d — 39.6
    a — 39.3
    f — 39.7
    g — 39.3

Chapter 40

1. clumped — 40.2
2. f — 40.3
3. 400 — 40.2
4. 1600 — 40.3
5. a — 40.3
6. d — 40.4
7. d — 40.5
8. b — 40.7
9. d — 40.8
10. a — 40.7
11. c — 40.5
12. c — 40.5
13. c — 40.4
    d — 40.3
    a — 40.3
    e — 40.4
    b — 40.4

Chapter 41

1. b — 41.2
2. d — 41.2
3. e — 41.4
4. d — 41.4
5. b — 41.3
    d — 41.6
    c — 41.2
    a — 41.5
    e — 41.4
6. b — 41.6
7. b — 41.7
8. Pioneer — 41.7
9. c — 41.7
10. a — 41.9
11. b — 41.4
12. c — 41.6
13. a — 41.5
14. c — 41.9
    b — 41.7
    d — 41.8
    a — 41.8
    f — 41.8
    e — 41.4

CHAPTER 42

1. b — 42.2
2. d — 42.2
3. d — 42.3
4. d — 42.2
5. a — 42.2
6. c — 42.7
7. b — 42.8
8. d — 42.8
9. d — 42.8
10. a — 42.10
11. c — 42.10
12. b — 42.8
13. a — 42.9
14. b — 42.8
15. d — 42.8
    c — 42.8
    b — 42.9
    a — 42.9

Chapter 43

1. b — 43.2
2. d — 43.2
3. d — 43.2
4. a — 43.3
5. b — 43.3
6. b — 43.4
7. d — 43.4
8. c — 43.6
9. a — 43.10
10. less — 43.11
11. d — 43.14
12. c — 43.13
13. c — 43.10
14. b — 43.4
15. d — 43.10
    e — 43.7
    f — 43.5
    c — 43.6
    b — 43.12
    h — 43.9
    i — 43.6
    a — 43.8
    g — 43.14

Chapter 44

1. b — 44.2
2. d — 44.2
3. a — 44.3
4. c — 44.4
5. a — 44.5
6. d — 44.5
7. a — 44.7
8. False — 44.4
9. b — 44.8
10. b — 44.7
11. c — 44.9
12. d — 44.9
13. c — 44.9
14. d — 44.10
15. g — 44.9
    a — 44.7
    i — 44.9
    e — 44.4
    d — 44.2
    f — 44.5
    h — 44.8
    b — 44.3
    c — 44.3

Chapter 13: Genetics Problems

1. yellow

2. a. *AB*
   b. *AB, aB*
   c. Ab, ab
   d. *AB, Ab, aB, ab*

3. a. All offspring will be AaBB.
   b. 1/4 *AABB*   (25% each genotype)
      1/4 *AABb*
      1/4 *AaBB*
      1/4 *AaBb*
   c. 1/4 *AaBb*   (25% each genotype)
      1/4 *Aabb*
      1/4 *aaBb*
      1/4 *aabb*
   d. 1/16 *AABB*   (6.25% of genotype)
      1/8 *AaBB*   (12.5%)
      1/16 *aaBB*   (6.25%)
      1/8 *AABb*   (12.5%)
      1/4 *AaBb*   (25%)
      1/8 *aaBb*   (12.5%)
      1/16 *AAbb*   (6.25%)
      1/8 *Aabb*   (12.5%)
      1/16 *aabb*   (6.25%)

4. monohybrid cross

5. Incomplete dominance; a mating of two *Mm* cats yields 1/2 *Mm*, 1/4 *MM*, and 1/4 *mm*. Because *MM* is lethal, the probability that any one kitten among the survivors will be heterozygous is 2/3.

6. a. Both parents are heterozygous (*Aa*), Children with the albino phenotype are homozygous recessive (*aa*); unaffected children may be homozygous dominant (*AA*) or heterozygous (*Aa*).
   b. All are homozygous recessive (*aa*).
   c. Homozygous recessive (*aa*) father, and heterozygous (*Aa*) mother. The child with the albino phenotype is homozygous recessive (*aa*), the unaffected children, heterozygous (*Aa*).

Chapter 14: Genetics Problems

1. Autosomal dominant. One of the first generation females is unaffected, so the allele cannot be carried on the X chromosome. The autosomal allele cannot be recessive, because if it were, both parents would be homozygous for the allele, so all of their children would be affected.

2. a. With respect to X-linked alleles, a male can produce only one type of gamete. Half of the gametes he produces carry an X chromosome and half carry a Y chromosome. All the gametes that carry the X chromosome will have the X-linked allele.
   b. A female homozygous for an X-linked allele produces only one kind of gamete.
   c. Fifty percent of the gametes of a female who is heterozygous for an X-linked allele carry one of the two alleles at that locus; the other fifty percent carry its partner allele for that locus.

3. a. 25%; b. 25%; c. 50%

4. By translocation, one of the individual's three chromosomes 21 may have become attached to the end of a chromosome 14. The individual would still have 46 chromosomes, but his or her somatic cells would have the translocated chromosome 21 in addition to two normal chromosomes 21.

5. 50%.

Appendix IV. Periodic Table of the Elements

The symbol for each element is an abbreviation of its name. Some symbols for elements are abbreviations for their Latin names. For instance, Pb (lead) is short for *plumbum*; the word "plumbing" is related—ancient Romans made their water pipes with lead.

Elements in each vertical column of the table behave in similar ways. For instance, all of the elements in the far right column of the table are inert gases; they do not interact with other atoms. In nature, such elements occur only as solitary atoms.

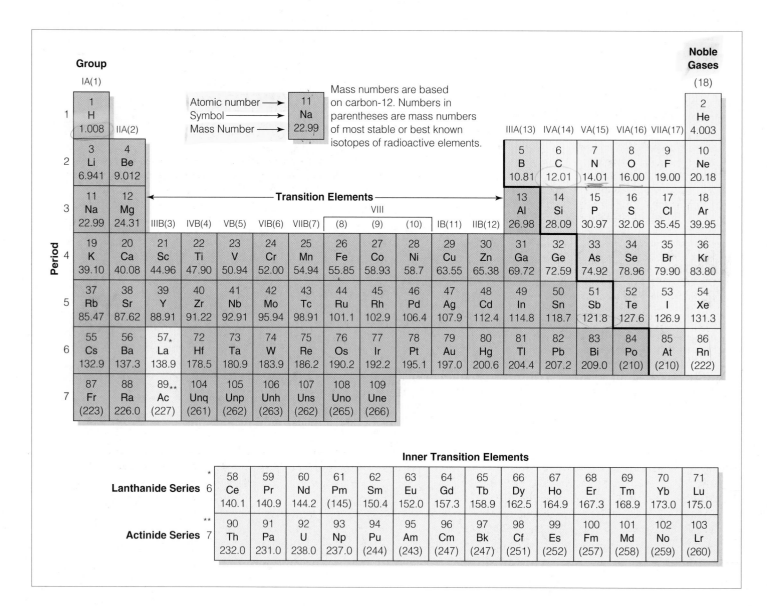

Appendix V. Molecular Models

A molecule's structure can be depicted by different kinds of molecular models. Such models allow us to visualize different characteristics of the same structure.

Structural models show how atoms in a molecule connect to one another:

methane   glucose

In such models, each line indicates one covalent bond: Double bonds are shown as two lines; triple bonds as three lines. Some atoms or bonds may be implied but not shown. For example, carbon ring structures such as those of glucose and other sugars are often represented as polygons. If no atom is shown at the corner of a polygon, a carbon atom is implied. Hydrogen atoms bonded to one of the atoms in the carbon backbone of a molecule may also be omitted:

glucose   glucose

Ball-and-stick models show the relative sizes of the atoms and their positions in three dimensions:

methane   glucose

All types of covalent bonds (single, double, or triple) are shown as one stick. Typically, the elements in such models are coded in standardized colors:

carbon   hydrogen   oxygen   nitrogen

Space-filling models show the outer boundaries of the atoms in three dimensions:

methane   glucose

A model of a large molecule can be quite complex if all the atoms are shown. This space-filling model of hemoglobin is an example:

To reduce visual complexity, other types of models omit individual atoms. Surface models of large molecules can show features such as an active site crevice (Figure 5.10). In this surface model of hemoglobin, you can see two heme groups (red) nestled in pockets of the protein:

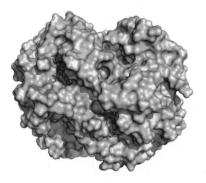

Large molecules such as proteins are often shown as ribbon models. Such models highlight secondary structure such as coils or sheets. In this ribbon model of hemoglobin, you can see the four coiled polypeptide chains, each of which folds around a heme group:

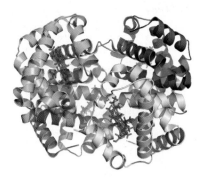

Such structural details are clues to how a molecule functions. Hemoglobin is the main oxygen carrier in vertebrate blood. Oxygen binds at the hemes, so one hemoglobin molecule can hold four molecules of oxygen.

The Amino Acids

Neutral, nonpolar side group

glycine (gly)

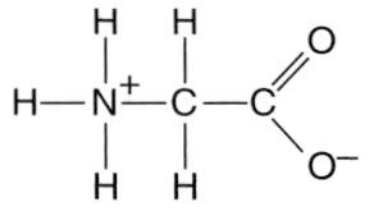

alanine (ala)

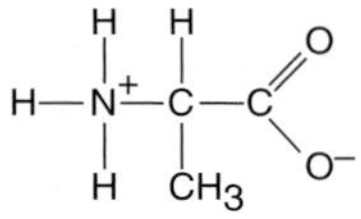

valine (val)

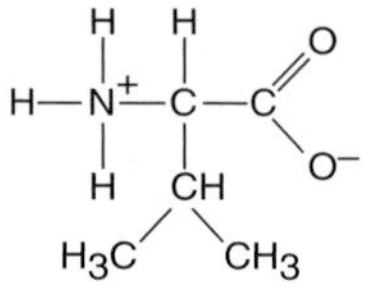

isoleucine (ile)

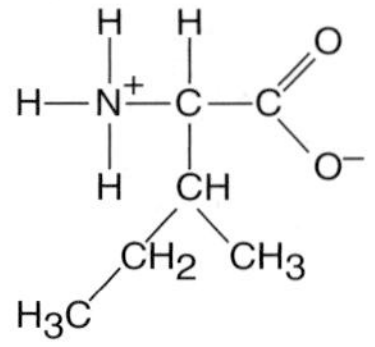

leucine (leu)

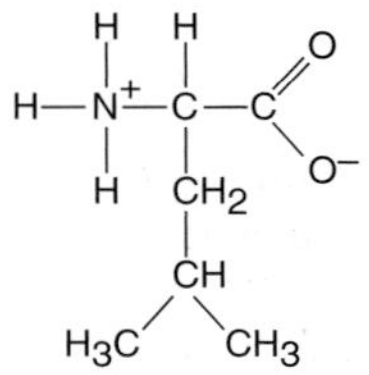

phenylalanine (phe)

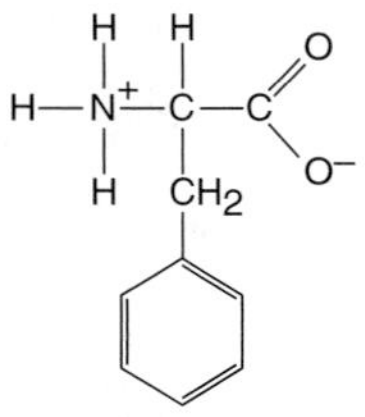

proline (pro)

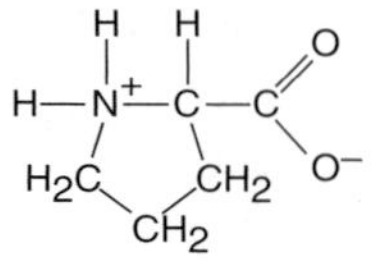

methionine (met)

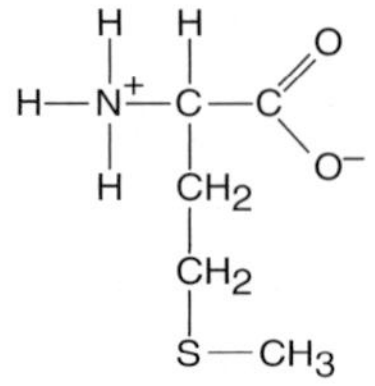

Neutral, polar side group

serine (ser)

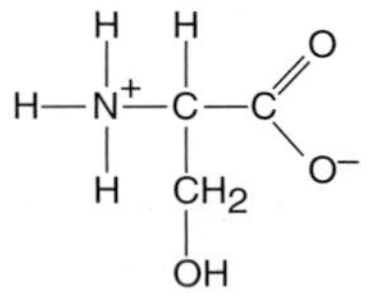

threonine (thr)

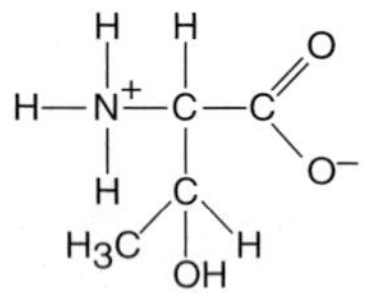

tyrosine (tyr)

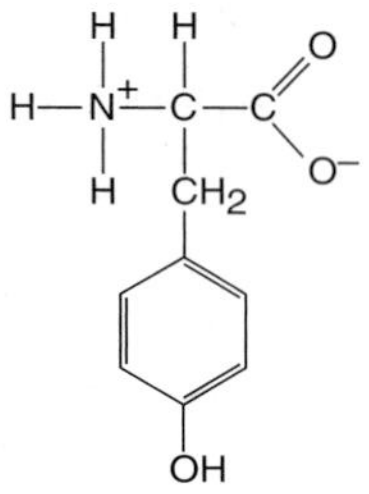

tryptophan (trp)

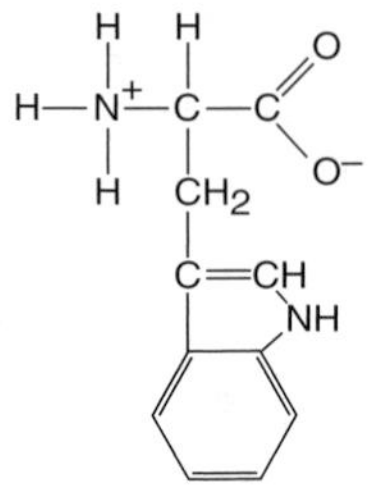

asparagine (asn)

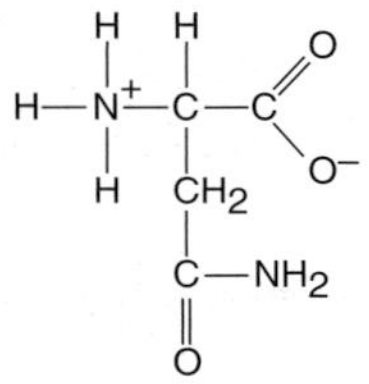

glutamine (gln)

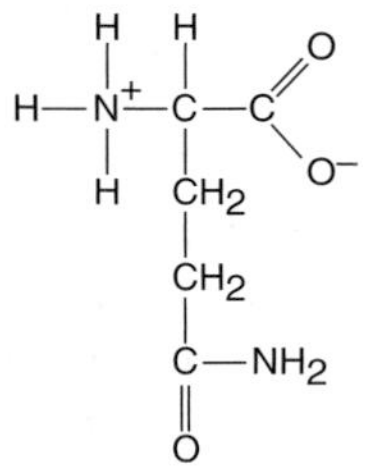

cysteine (cys)

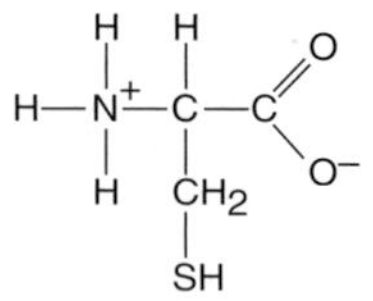

Acidic side group

aspartic acid (asp)

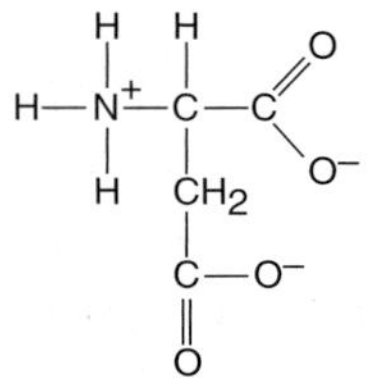

glutamic acid (glu)

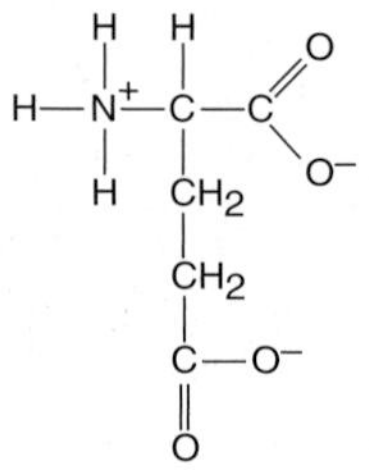

Basic side group

lysine (lys)

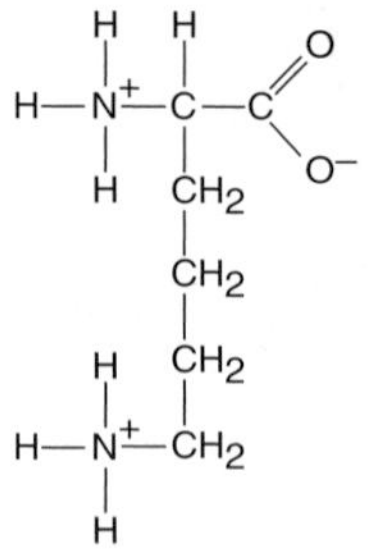

arginine (arg)

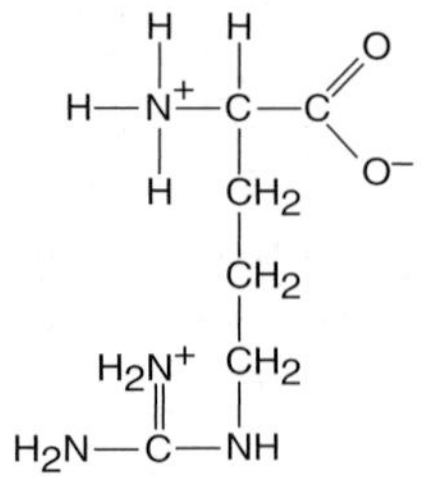

histidine (his)

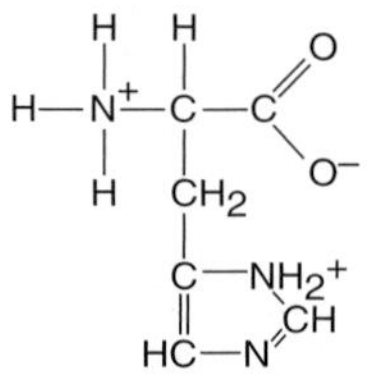

Appendix VI. Closer Look at Some Major Metabolic Pathways

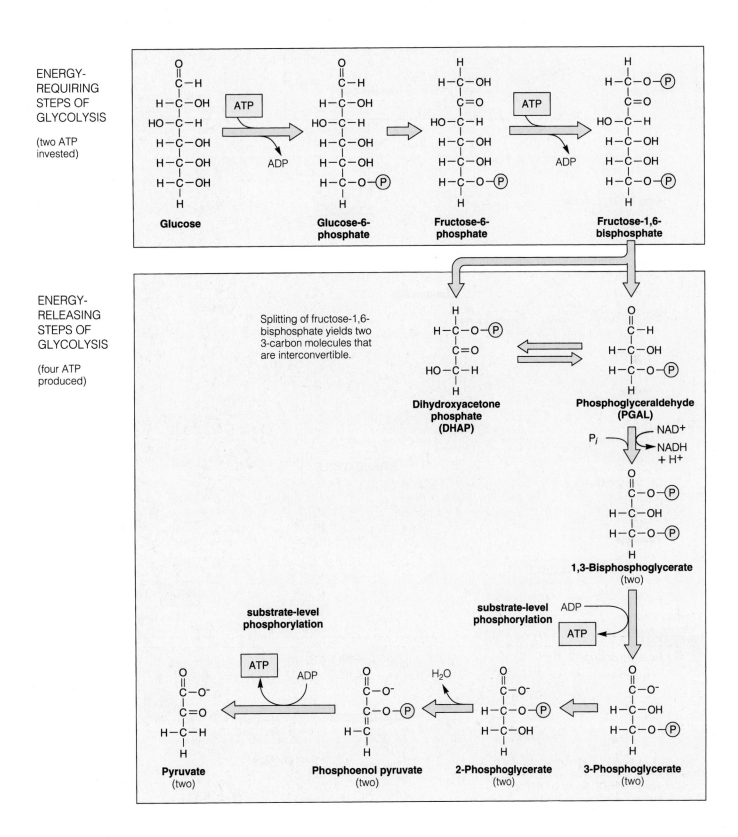

ENERGY-
REQUIRING
STEPS OF
GLYCOLYSIS

(two ATP
invested)

ENERGY-
RELEASING
STEPS OF
GLYCOLYSIS

(four ATP
produced)

Splitting of fructose-1,6-bisphosphate yields two 3-carbon molecules that are interconvertible.

Figure A Glycolysis, ending with two 3-carbon pyruvate molecules for each 6-carbon glucose molecule entering the reactions. The *net* energy yield is two ATP molecules (two invested, four produced).

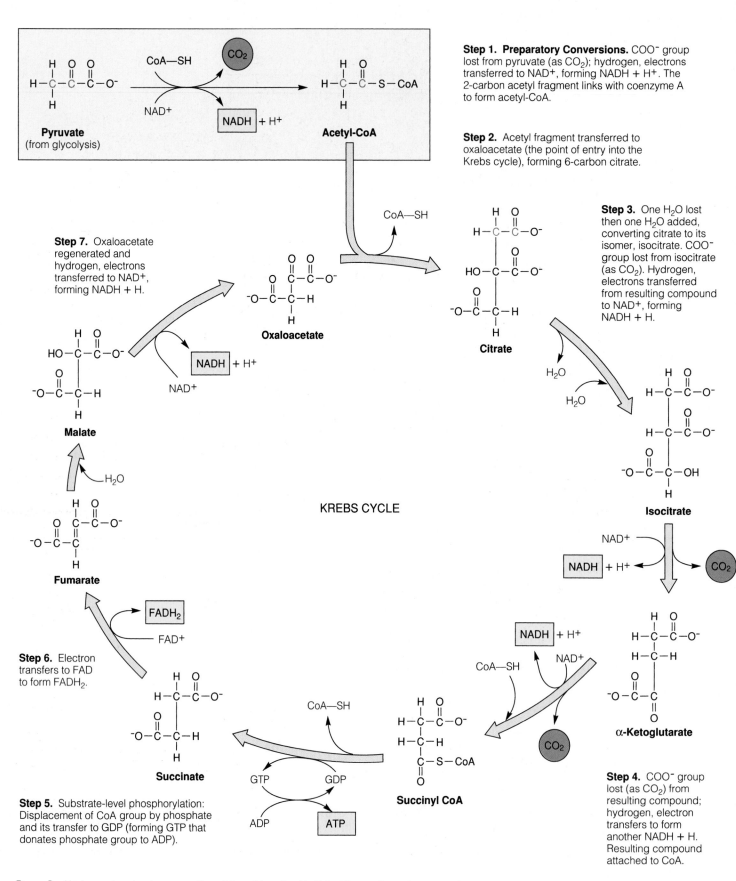

Step 1. Preparatory Conversions. COO^- group lost from pyruvate (as CO_2); hydrogen, electrons transferred to NAD^+, forming $NADH + H^+$. The 2-carbon acetyl fragment links with coenzyme A to form acetyl-CoA.

Step 2. Acetyl fragment transferred to oxaloacetate (the point of entry into the Krebs cycle), forming 6-carbon citrate.

Step 3. One H_2O lost then one H_2O added, converting citrate to its isomer, isocitrate. COO^- group lost from isocitrate (as CO_2). Hydrogen, electrons transferred from resulting compound to NAD^+, forming $NADH + H$.

Step 7. Oxaloacetate regenerated and hydrogen, electrons transferred to NAD^+, forming $NADH + H$.

Step 6. Electron transfers to FAD to form $FADH_2$.

Step 5. Substrate-level phosphorylation: Displacement of CoA group by phosphate and its transfer to GDP (forming GTP that donates phosphate group to ADP).

Step 4. COO^- group lost (as CO_2) from resulting compound; hydrogen, electron transfers to form another $NADH + H$. Resulting compound attached to CoA.

Pyruvate (from glycolysis)
Acetyl-CoA
Oxaloacetate
Citrate
Isocitrate
Malate
α-Ketoglutarate
Fumarate
KREBS CYCLE
Succinate
Succinyl CoA

Figure B Krebs cycle, also known as the citric acid cycle. *Red* identifies carbon atoms entering the cyclic pathway (by way of acetyl-CoA) and leaving (by way of carbon dioxide). These cyclic reactions run twice for each glucose molecule that has been degraded to two pyruvate molecules.

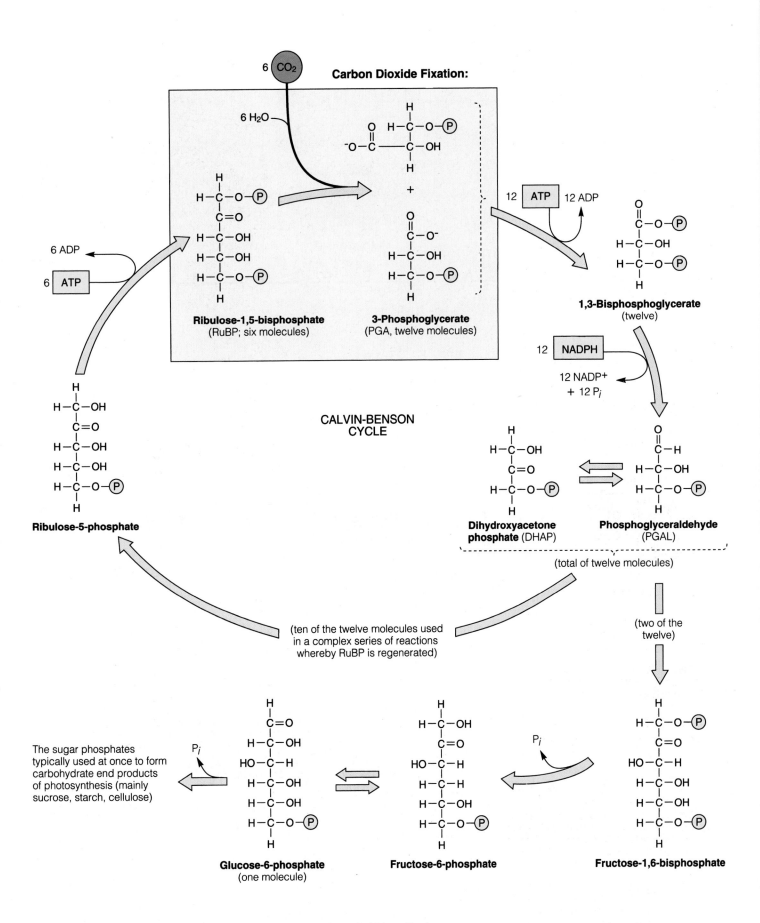

Figure C Calvin–Benson cycle of the light-independent reactions of photosynthesis.

Noncyclic photophosphorylation

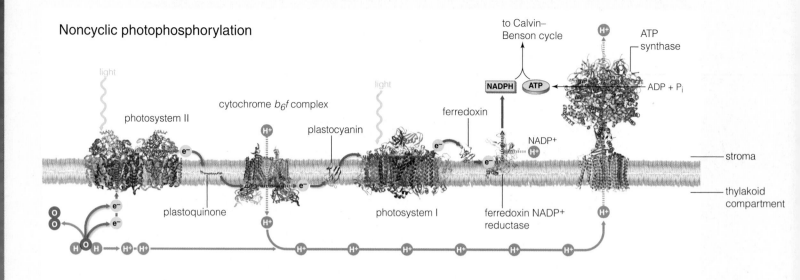

Cyclic photophosphorylation

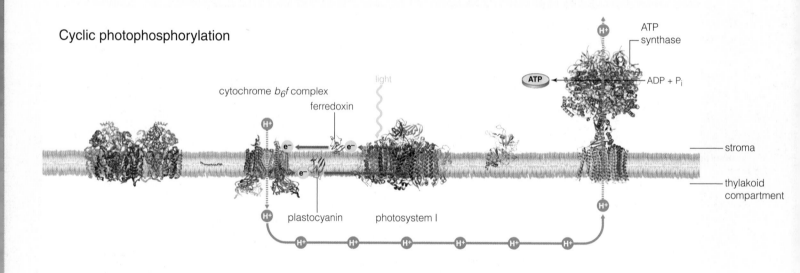

The arrangement of electron transfer chain components in highly folded thylakoid membranes maximizes the efficiency of ATP production. ATP synthases are positioned only on the outer surfaces of the thylakoid stacks, in contact with the stroma and its supply of NADP+ and ADP.

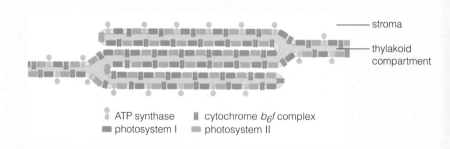

ATP synthase   cytochrome b_6f complex
photosystem I   photosystem II

Figure D Electron transfer in the light-dependent reactions of photosynthesis. Members of the electron transfer chains are densely packed in thylakoid membranes; electrons are transferred directly from one molecule to the next. For clarity, we show the components of the chains widely spaced.

Appendix VII. A Plain English Map of the Human Chromosomes

1
- sweet taste receptors
- Rh blood type
- marijuana receptor
- (anorexia nervosa susceptibility)
- leptin receptor
- TSH β chain
- lamin A (progeria)
- Duffy blood group antigen

2
- LH/choriogonadotropin receptor (micropenis)
- CD8; cytotoxic T cell antigen
- antibody light chain
- lactase
- (cleft palate)
- glucagon

3
- oxytocin receptor
- HIV receptor
- rhodopsin
- (alkaptonuria)
- (sucrose intolerance)
- somatostatin

4
- (achondroplasia)
- (Huntington disease)
- (Ellis-van Creveld syndrome)
- alcohol dehydrogenase (susceptibility to alcoholism)
- red hair color

5
- Cri-du-chat syndrome
- bitter taste receptor
- growth hormone receptor (pituitary dwarfism)
- interleukin-4

6
- (gluten intolerance)
- HLA/MHC
- tumor necrosis factor
- α chains of HCG, FSH, LH, and TSH
- estrogen receptor

7
- cytochrome c
- elastin
- DLX 5/6 homeotic genes
- CFTR (cystic fibrosis)
- leptin (obesity)
- (blue-deficient colorblind)
- TCR β subunit

8
- gonadotropin releasing hormone
- helicase (Werner's syndrome)
- corticotropin releasing hormone

9
- (galactosemia)
- (cerebral palsy)
- (Friedreich ataxia)
- (fructose intolerance)
- ABO blood group

10
- vitamin B-12 receptor
- mannose binding protein
- perforin
- (gluten intolerance)

11
- hemoglobin β chain (sickle cell anemia)
- insulin
- parathyroid hormone
- catalase
- PAX6 (aniridia)
- FSH, β chain
- tyrosinase (albinism)

12
- CD4 helper T cell antigen
- oncogene KRAS2 (lung cancer, bladder cancer, breast cancer)
- keratins
- lysozyme
- (phenylketonuria)
- aldehyde dehydrogenase (alcohol intolerance)

13
- ribosomal RNA
- BRCA 2 (breast cancer)
- (gastroesophageal reflux)

14
- ribosomal RNA
- presinilin (Alzheimer's)
- TSH receptor
- immunoglobulin heavy chains

15
- ribosomal RNA
- fibrillin 1 (Marfan syndrome)
- (Tay-Sachs disease)

16
- hemoglobin α chain
- DNAse I (lupus)

17
- (Canavan disease)
- p53 tumor antigen
- NF1 (neurofibromatosis)
- serotonin transporter
- BRCA 1 (breast, ovarian cancer)
- Growth hormone

18
- B cell apoptosis regulator (B cell lymphoma)
- myelin basic protein

19
- LDL receptor (coronary artery disease)
- insulin receptor
- brown hair color
- green/blue eye color
- (Warfarin resistance)
- HCG, β chain
- LH, β chain

20
- prion protein (Creutzfeld-Jacob disease)
- oxytocin
- GHRH (acromegaly)

21
- ribosomal RNA
- interferon receptors
- (bipolar disorder, early onset)

22
- ribosomal RNA
- immunoglobulin light chains
- myoglobin

X
- dystrophin (muscular dystrophy)
- (anhidrotic ectodermal dysplasia)
- IL2RG (SCID-X1)
- XIST X chromosome inactivation control
- (hemophilia B)
- (hemophilia A)
- (red-deficient colorblind)
- (green-deficient colorblind)

Y
- sex determining region Y (SRY)
- (no sperm)
- male stature

© 2002 Susan Offner/SK45176-02

Haploid set of human chromosomes. The banding patterns characteristic of each type of chromosome appear after staining with a reagent called Giemsa. The locations of some of the 20,065 known genes (as of November, 2005) are indicated. Also shown are locations that, when mutated, cause some of the genetic diseases discussed in the text.

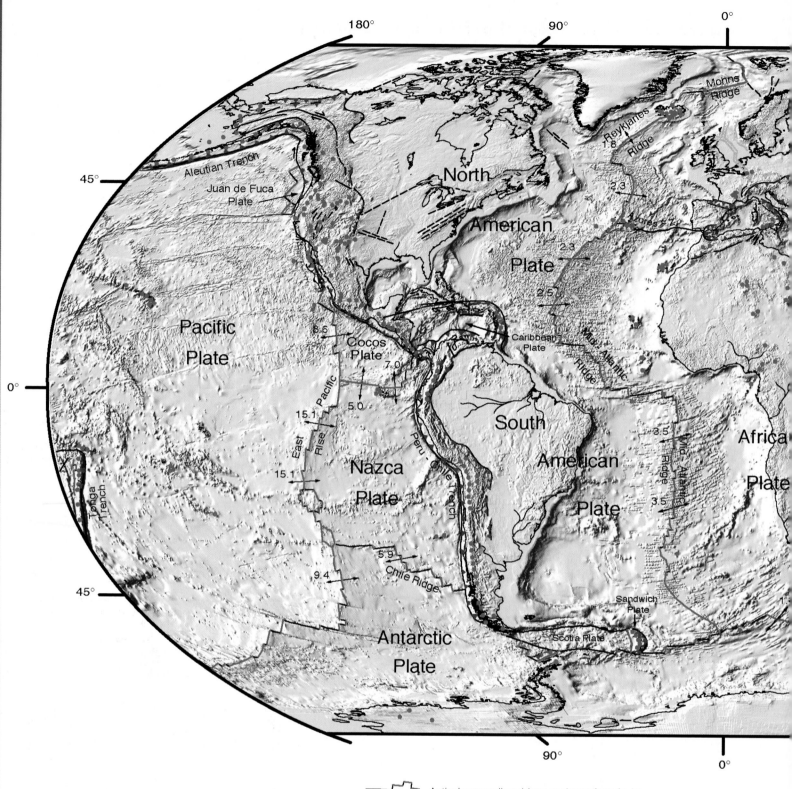

180°   90°   0°

Mohns Ridge

45°

Aleutian Trench

Juan de Fuca Plate

Reykjanes Ridge 1.8

North

American 2.3

Plate Azores F.Z.

2.3

2.5

Pacific 8.6

Plate Cocos Plate 7.0 Caribbean Plate

0° Mid-Atlantic Ridge

5.0

15.1 Pacific

East South

Rise 3.5 Africa

15.1 Nazca American Mid-Atlantic Ridge Plate

Plate Peru - Chile Trench Plate 3.5

Tonga Trench

5.9

9.4 Chile Ridge

45° Sandwich Plate

Scotia Plate

Antarctic

Plate

90°   0°

Appendix VIII.
Restless Earth—Life's Changing Geologic Stage

This NASA map summarizes the tectonic and volcanic activity of Earth during the past 1 million years. The reconstructions at far right indicate positions of Earth's major land masses through time.

Actively-spreading ridges and transform faults

Total spreading rate, cm/year 1.4

Major active fault or fault zone; dashed where nature, location, or activity uncertain

Normal fault or rift; hachures on downthrown side

Reverse fault (overthrust, subduction zones); generalized; barbs on upthrown side

Volcanic centers active within the last one million years; generalized. Minor basaltic centers and seamounts omitted.

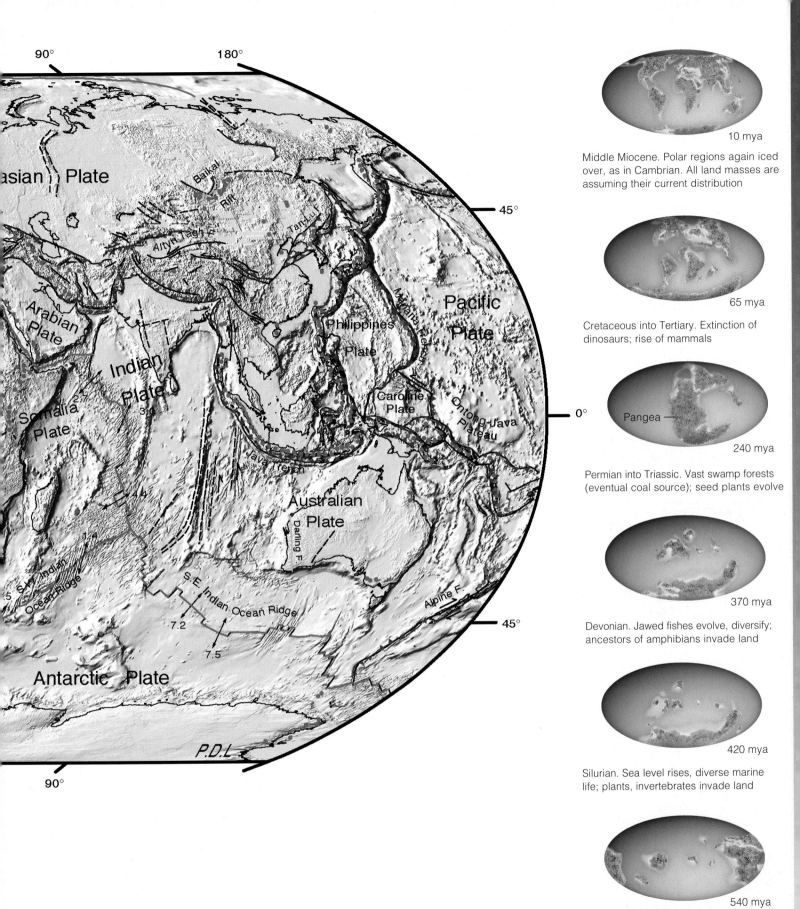

90° **180°**

45°

ar ian) Plate

Baikal
Rift

Tarim-Lu F.

Altyn Tagh F.

Arabian
Plate

Pacific
Plate

Philippines
Plate

Mariana Trench

45°

Somalia
Plate

2.7

Indian
Plate

3.0

Caroline
Plate

Ontong-Java
Plateau

0°

Java Trench

Australian
Plate

.5 S.W. Indian
Ocean Ridge

4.4

Darling F.

Alpine F.

45°

S.E. Indian Ocean Ridge

7.2

Antarctic Plate

7.5

P.D.L

90°

Middle Miocene. Polar regions again iced
over, as in Cambrian. All land masses are
assuming their current distribution

10 mya

Cretaceous into Tertiary. Extinction of
dinosaurs; rise of mammals

65 mya

Pangea

Permian into Triassic. Vast swamp forests
(eventual coal source); seed plants evolve

240 mya

Devonian. Jawed fishes evolve, diversify;
ancestors of amphibians invade land

370 mya

Silurian. Sea level rises, diverse marine
life; plants, invertebrates invade land

420 mya

Cambrian. Fragments of Rodinia, the first
supercontinent. Major adaptive radiations
in equatorial seas; icy polar regions

540 mya

Appendix VIII

Appendix IX. Units of Measure

Length

1 kilometer (km) = 0.62 miles (mi)

1 meter (m) = 39.37 inches (in)

1 centimeter (cm) = 0.39 inches

To convert	multiply by	to obtain
inches	2.25	centimeters
feet	30.48	centimeters
centimeters	0.39	inches
millimeters	0.039	inches

Area

1 square kilometer = 0.386 square miles

1 square meter = 1.196 square yards

1 square centimeter = 0.155 square inches

Volume

1 cubic meter = 35.31 cubic feet

1 liter = 1.06 quarts

1 milliliter = 0.034 fluid ounces = 1/5 teaspoon

To convert	multiply by	to obtain
quarts	0.95	liters
fluid ounces	28.41	milliliters
liters	1.06	quarts
milliliters	0.03	fluid ounces

Weight

1 metric ton (mt) = 2,205 pounds (lb) = 1.1 tons (t)

1 kilogram (kg) = 2.205 pounds (lb)

1 gram (g) = 0.035 ounces (oz)

To convert	multiply by	to obtain
pounds	0.454	kilograms
pounds	454	grams
ounces	28.35	grams
kilograms	2.205	pounds
grams	0.035	ounces

Temperature

Celcius (°C) to Fahrenheit (°F):

$$°F = 1.8 \, (°C) + 32$$

Fahrenheit (°F) to Celsius:

$$°C = \frac{(°F - 32)}{1.8}$$

	°C	°F
Water boils	100	212
Human body temperature	37	98.6
Water freezes	0	32

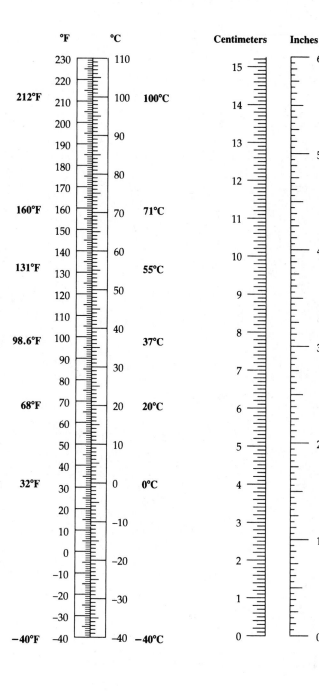

Appendix X. A Comparative View of Mitosis in Plant and Animal Cells

For step-by-step description of the stages of mitosis, refer to Figure 11.5.

Mitosis in a
generalized
animal cell.
For simplicity,
only two
chromosomes
are shown.

  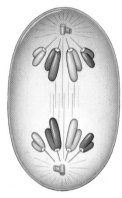 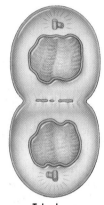

| Prophase | Metaphase | Anaphase | Telophase |

Mitosis in a white-fish cell.

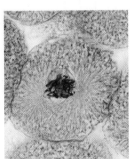 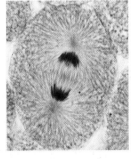 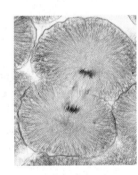

| Prophase | Metaphase | Anaphase | Telophase |

Mitosis in
a lily cell.

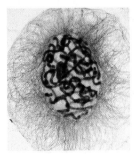   

| Prophase | Metaphase | Anaphase | Telophase |

Glossary

abscisic acid Plant hormone. Stimulates stomata to close in response to water stress and induces dormancy in buds and seeds, among other effects. **439**

abscission Process by which plant parts are shed in response to seasonal change, drought, injury, or nutrient deficiency. **444**

acetylcholine (ACh) Neurotransmitter released by neurons at neuromuscular junctions, and at synapses in the heart and brain. **474**

acid Substance that releases hydrogen ions in water. **32**

acid rain Low pH rain that forms when sulfur dioxide and nitrogen oxides mix with water vapor in the atmosphere. **751**

actin Protein monomer of microfilaments; the main component of thin filaments of muscle fibers. **528**

action potential Brief reversal of the charge difference across a neuron membrane. **471**

activation energy Minimum amount of energy required to start a reaction. **79**

activator Regulatory protein that increases the rate of transcription when it binds to a promoter or enhancer. **152**

active site Pocket in an enzyme where substrates bind and a reaction occurs. **80**

active transport Energy-requiring mechanism in which a transport protein pumps a solute across a cell membrane against its concentration gradient. **87**

adaptation (adaptive trait) A heritable trait that enhances an individual's fitness. **242**

adaptive immunity In vertebrates, set of immune defenses that can be tailored to specific pathogens encountered by an organism during its lifetime. **558**

adaptive radiation A burst of genetic divergences from a lineage gives rise to many new species. **276**

adhering junction Cell junction composed of adhesion proteins; anchors cells to each other and to extracellular matrix. Role in cell signaling. **69**

adhesion protein Membrane protein that is a component of adhering junctions. **56**

adipose tissue Connective tissue that specializes in fat storage. **455**

adrenal cortex Outer portion of adrenal gland; secretes aldosterone and cortisol. **512**

adrenal gland Endocrine gland located atop the kidney; secretes aldosterone, cortisol, epinephrine and norepinephrine. **512**

adrenal medulla Inner portion of adrenal gland; secretes epinephrine and norepinephrine. **512**

aerobic Involving or occurring in the presence of oxygen. **108**

aerobic respiration Oxygen-requiring pathway that breaks down carbohydrates to produce ATP. Occurs in stages: glycolysis; acetyl-CoA formation and the Krebs cycle; electron transfer phosphorylation. **108**

age structure Of a population, the number of individuals in each of several age categories. **675**

agglutination The clumping together of foreign cells bound by antibodies. **569**

AIDS Acquired immune deficiency syndrome. A collection of diseases that develops after a virus (HIV) weakens the immune system. **574**

alcoholic fermentation Anaerobic carbohydrate breakdown pathway that produces ATP and ethanol. Begins with glycolysis; end reactions regenerate NAD$^+$ so glycolysis can continue. **116**

aldosterone Adrenal hormone that makes kidney tubules more permeable to sodium; encourages sodium reabsorption, thus increasing water reabsorption and concentrating the urine. **621**

algal bloom Population explosion of tiny aquatic producers such as dinoflagellates. **311**

allantois Extraembryonic membrane that, in mammals, becomes part of the umbilical cord. **648**

allele frequency Abundance of a particular allele among members of a population. **259**

alleles Forms of a gene that encode slightly different versions of the gene's product. **176**

allergen A normally harmless substance that provokes an immune response in some people. **571**

allergy Sensitivity to an allergen. **571**

allopatric speciation Speciation pattern in which a physical barrier that separates members of a population ends gene flow between them. **272**

allosteric Describes a region of an enzyme other than the active site that can bind regulatory molecules. **82**

alpine tundra Biome of low-growing, wind-tolerant plants adapted to high-altitude conditions. **737**

alternation of generations Of land plants and some algae, a life cycle in which both haploid and diploid multicelled bodies form. **318**

alternative splicing RNA processing event in which some exons are removed or joined in various combinations. **141**

altruistic behavior Behavior that benefits others at the expense of the individual. **668**

alveolate Member of a protist lineage having small sacs beneath the plasma membrane; dinoflagellate, ciliate, or apicomplexan. **314**

alveolus Tiny sac; in the mammalian lung, the site of gas exchange. **583**

amino acid Small organic compound that is a subunit of proteins. Consists of a carboxyl group, an amine group, and a characteristic side group (R), all typically bonded to the same carbon atom. **44**

ammonia Nitrogen-containing waste produced by amino acid and nucleic acid breakdown. **616**

amnion Extraembryonic membrane that encloses an amniote embryo and the amniotic fluid. **648**

amniote Vertebrate whose embryos develop within a fluid enclosed within a membrane (the amnion); a reptile (including birds) or mammal. **379**

amniote egg Egg with internal membranes that allow the amniote embryo to develop away from water. **384**

amoeba Single-celled protist that extends pseudopods to move and to capture prey. **320**

amoebozoan Shape-shifting heterotrophic protist with no pellicle or cell wall; an amoeba or slime mold. **320**

amphibian Tetrapod with a three-chambered heart and scaleless skin; typically develops in water, then lives on land as an air-breathing carnivore. **382**

anaerobic Occurring in the absence of oxygen. **108**

analogous structures Similar body structures that evolved separately in different lineages. **251**

anaphase Stage of mitosis during which sister chromatids separate and move to opposite spindle poles. **166**

aneuploidy A chromosome abnormality in which an individual's cells carry too many or too few copies of a particular chromosome. **212**

angiosperms Largest seed plant lineage. Only group that makes flowers and fruits. **336**

animal Multicelled consumer with unwalled cells; develops through a series of stages and moves about during part or all of the life cycle. **8, 354**

animal hormone Intercellular signaling molecule secreted by an endocrine gland or cell. **504**

annelid Segmented worm with a coelom, complete digestive system, and closed circulatory system. **362**

antenna Of some arthropods, sensory structure on the head that detects touch and odors. **367**

antibody Y-shaped antigen receptor protein made only by B cells. **564**

antibody–mediated immune response Immune response in which antibodies are produced in response to an antigen. **566**

antidiuretic hormone Pituitary hormone that encourages water reabsorption, thus concentrating the urine. **621**

antigen A molecule or particle that the immune system recognizes as nonself. Triggers an immune response. **558**

antioxidant Substance that prevents molecules from reacting with oxygen. **81**

anus Opening through which digestive waste is expelled from a complete digestive system. **599**

aorta Large artery that receives blood pumped out of the heart's left ventricle. **541**

apicomplexan Single-celled protist that lives as a parasite inside animal cells. **315**

apoptosis Mechanism of cell suicide. **463, 646**

appendicular skeleton Of vertebrates, limb or fin bones and bones of the pelvic and pectoral girdles. **522**

appendix Worm-shaped projection from the first part of the large intestine. **604**

aquifer Porous rock layer that holds some groundwater. **714**

arachnid Land-dwelling arthropod with four pairs of walking legs; a spider, scorpion, mite, or tick. **368**

archaean Member of a group of single-celled microorganisms that superficially resemble bacteria, but are genetically and structurally distinct. **8**

arctic tundra Highest-latitude Northern Hemisphere biome, where low, cold-tolerant plants survive with only a brief growing season. **737**

area effect Larger islands have more species than small ones. **705**

arteriole Vessel that carries blood from an artery to a capillary. **540**

artery Large-diameter blood vessel that carries blood away from the heart. **540**

arthropod Invertebrate with jointed legs and a hard exoskeleton that is periodically molted. **367**

artificial selection Selective breeding of animals by humans. **242**

asexual reproduction Reproductive mode by which offspring arise from a single parent only. **165, 630**

astrobiology The scientific study of life's origin and distribution in the universe. **283**

atmospheric cycle Biogeochemical cycle in which a gaseous form of an element plays a significant role. **716**

atom Particle that is a fundamental building block of all matter. **4, 24**

atomic number Number of protons in the atomic nucleus; determines the element. **24**

ATP Adenosine triphosphate. Nucleotide that consists of an adenine base, a five-carbon ribose sugar, and three phosphate groups. Often functions as an energy carrier that couples endergonic with exergonic reactions in cells. **47, 79**

ATP/ADP cycle Process in which cells regenerate ATP. ADP forms when ATP loses a phosphate group, then ATP forms again as ADP gains a phosphate group. **79**

atrioventricular (AV) node Clump of cells that serve as the electrical bridge between the atria and ventricles. **543**

atrium Heart chamber that receives blood from veins. **542**

australopiths Collection of now-extinct hominid lineages; some may be ancestral to humans. **390**

autoimmune response Immune response that inappropriately targets one's own tissues. **573**

autonomic nervous system Set of nerves that relay signals to and from internal organs and to glands. **478**

autosome Any chromosome other than a sex chromosome. **125**

autotroph Organism that makes its own food using carbon from inorganic molecules such as CO_2 and energy from the environment; a producer. **93**

auxin Plant hormone. Stimulates cell division and elongation, among other effects. **438**

axial skeleton Of vertebrates, bones of the head, trunk, and tail. **522**

axon Neuron cytoplasmic extension that transmits electrical signals along its length and secretes chemical signals at its endings. **470**

bacterial chromosome Circle of double-stranded DNA that resides in the bacterial cytoplasm. **302**

bacteriophage Virus that infects bacteria. **127, 298**

bacterium Single-celled organism belonging to the Domain Bacteria; cells are typically walled and do not contain a nucleus. **8**

balanced polymorphism Maintenance of two or more alleles for a trait at high frequency in a population as a result of natural selection against homozygotes. **267**

bark Secondary phloem and periderm of woody plants. **409**

basal body Organelle that develops from a centriole. **67**

base Substance that accepts hydrogen ions in water. **32**

basement membrane Secreted layer that attaches an epithelium to an underlying tissue. **452**

base–pair substitution Type of mutation in which a single base-pair changes. **147**

basophil Circulating granular white blood cell; role in inflammation. **559**

B cell B lymphocyte. Only type of white blood cell that can make antibodies. **559**

B cell receptor Membrane-bound IgM or IgD antibody on a B cell. **565**

bell curve Bell-shaped curve; typically results from graphing frequency versus distribution for a trait that varies continuously. **198**

benthic province The ocean's sediments and rocks. **742**

big bang theory Model describing formation of the universe as a nearly instant distribution of matter through space. **284**

bilateral symmetry Having paired structures so the right and left halves are mirror images. **354**

bile Mix of salts, pigments, and cholesterol produced in the liver, then stored and concentrated in the gallbladder; emulsifies fats when secreted into the small intestine. **603**

binary fission Method of asexual reproduction that divides one bacterial or archaean cell into two identical descendant cells. **303**

bioaccumulation An organism accumulates a chemical pollutant in its tissues over the course of its lifetime. **752**

biodiversity Of a region, the genetic variation within its species, variety of species, and variety of ecosystems. **8, 756**

biofilm Community of microorganisms living within a shared mass of slime. **59**

biogeochemical cycle Movement of a chemical nutrient among environmental reservoirs and into and out of food webs. **714**

biogeography Study of patterns in the geographic distribution of species and communities. **238**

biological clock Internal time-measuring mechanism by which individuals adjust their activities seasonally, daily, or both in response to environmental cues. **442**

biological magnification A chemical pollutant becomes increasingly concentrated as it moves up through food chains. **752**

biology The scientific study of life. **3**

biomarker Molecule produced only by a specific type of cell. **289**

biome Discontinuous region characterized by its climate and dominant vegetation. **728**

biosphere All regions of Earth where organisms live. **5, 723**

biotic potential Maximum possible population growth rate under optimal conditions. **677**

bipedalism Standing and walking on two legs. **389**

bird An animal with feathers. **386**

bivalve Mollusk with a hinged two-part shell. **364**

blastocyst Mammalian blastula. **648**

blastula Hollow ball of cells that forms as a result of cleavage. **642**

blood Circulatory fluid; in vertebrates it is a fluid connective tissue consisting of plasma, red blood cells, white blood cells, and platelets. **455, 538**

blood–brain barrier Protective barrier that prevents unwanted substances from entering cerebrospinal fluid. **482**

blood pressure Pressure exerted by blood against a vessel wall. **547**

bone tissue Connective tissue with cells surrounded by a mineral-hardened matrix of their own secretions. **455**

bony fish Fish with a lung or swim bladder and a skeleton consisting largely of bone. **381**

boreal forest Extensive high-latitude forest of the Northern Hemisphere; conifers are the predominant vegetation; also known as tiaga. **736**

bottleneck Reduction in population size so severe that it reduces genetic diversity. **268**

Bowman's capsule Portion of the nephron that encloses the glomerulus and receives filtrate from it. **619**

bronchiole In the lung, a small airway that leads from a bronchus to the alveoli. **585**

bronchus Airway connecting the trachea to a lung. **585**

brood parasitism One egg-laying species benefits by having another raise its offspring. **698**

brown adipose tissue Adipose tissue that responds to cold by releasing energy as heat, rather than using it to make ATP. **625**

brown alga Multicelled marine protist with a brown accessory pigment in its chloroplasts. **317**

brush border cell In the lining of the small intestine, an epithelial cell with microvilli at its surface. **601**

bryophyte Member of an early-evolving plant lineage with a gametophyte-dominant life cycle; a moss, liverwort, or hornwort. **328**

buffer Set of chemicals that can keep the pH of a solution stable by alternately donating and accepting ions that contribute to pH. **33**

C3 plant Type of plant that uses only the Calvin–Benson cycle to fix carbon. **102**

C4 plant Type of plant that minimizes photorespiration by fixing carbon twice, in two cell types. **102**

Calvin–Benson cycle Light-independent reactions of photosynthesis; cyclic carbon-fixing pathway that forms sugars from CO_2. **101**

camera eye Eye with an adjustable opening and a single lens that focuses light on a retina. **492**

CAM plant Type of C4 plant that conserves water by fixing carbon twice, at different times of day. **102**

cancer Disease that occurs when a neoplasm physically and metabolically disrupts body tissues. **151, 170**

capillary Small, thin-walled blood vessel; exchanges substances with the interstitial fluid. **538**

carbohydrate Molecule that consists primarily of carbon, hydrogen, and oxygen atoms in a 1:2:1 ratio. **40**

carbon cycle Movement of carbon, mainly between the oceans, atmosphere, and living organisms. **716**

carbon fixation Process by which carbon from an inorganic source such as carbon dioxide gets incorporated into an organic molecule. **101**

carbonic anhydrase Enzyme in red blood cells that speeds the breakdown of carbonic acid into bicarbonate and H^+. **589**

cardiac cycle Sequence of contraction and relaxation of heart chambers that occurs with each heartbeat. **542**

cardiac muscle tissue Muscle of the heart wall. **456**

carpel Floral reproductive structure that produces female gametophytes; a sticky or hairlike stigma together with an ovary and a style. **336, 431**

carrying capacity Maximum number of individuals of a species that an environment can sustain. **678**

cartilage Connective tissue with cells surrounded by a rubbery matrix of their own secretions. **454**

cartilaginous fish Jawed fish with a skeleton of cartilage. **380**

Casparian strip Waxy, waterproof band that seals abutting cell walls of root endodermal cells. **419**

catalysis The acceleration of a reaction rate by a molecule that is unchanged by participating in the reaction. **80**

catastrophism Now-abandoned hypothesis that catastrophic geologic forces unlike those of the present day shaped Earth's surface. **240**

cDNA DNA synthesized from an RNA template by the enzyme reverse transcriptase. **221**

cell Smallest unit that has the properties of life; at minimum, consists of plasma membrane, cytoplasm, and DNA. **5, 52**

cell cortex Reinforcing mesh of cytoskeletal elements under a plasma membrane. **66**

cell count Number of cells of one type per microliter of blood. **545**

cell cycle A series of events from the time a cell forms until its cytoplasm divides. **164**

cell junction Structure that connects a cell to another cell or to extracellular matrix. **69**

cell-mediated immune response Immune response involving cytotoxic T cells and NK cells that destroy infected or cancerous body cells. **567**

cell plate After nuclear division in a plant cell, a disk-shaped structure that forms a cross-wall between the two new nuclei. **168**

cell theory Theory that all organisms consist of one or more cells, which are the basic unit of life; all cells come from division of preexisting cells; and all cells pass hereditary material to their offspring. **53**

cell wall Semirigid but permeable structure that surrounds the plasma membrane of some cells. **58**

cellular slime mold Amoeba-like protist that feeds as a single predatory cell; joins with others to form a multicellular spore-bearing structure when conditions are unfavorable. **320**

central nervous system Brain and spinal cord. **469**

central vacuole Fluid-filled vesicle in many plant cells. **63**

centriole Barrel-shaped organelle from which microtubules grow. **67**

centromere Constricted region in a eukaryotic chromosome where sister chromatids are attached. **124**

cephalization A concentration of nerve and sensory cells at the head end. **355**

cephalopod Predatory mollusk with a closed circulatory system; moves by jet propulsion. **365**

cerebellum Hindbrain region responsible for coordinating voluntary movements. **482**

cerebral cortex Outer gray matter layer of the cerebrum. **484**

cerebrospinal fluid Fluid that surrounds the brain and spinal cord and fills ventricles within the brain. **480**

cerebrum Forebrain region that controls higher functions. **483**

cervix Narrow part of uterus that connects to the vagina. **634**

chaparral Biome of dry shrubland in regions with hot, dry summers and cool, rainy winters. **733**

character Quantifiable, heritable characteristic or trait. **278**

character displacement Outcome of competition between two species; similar traits that result in competition become dissimilar. **695**

charge Electrical property. Opposite charges attract and like charges repel. **24**

chelicerates Arthropod subgroup with specialized feeding structures (chelicerae) and no antennae. **368**

chemical bond An attractive force that arises between two atoms when their electrons interact. **27**

chemoreceptor Sensory receptor that responds to a chemical. **490**

chlamydias Tiny round bacteria that are intracellular parasites of eukaryotic cells. **305**

chlorophyll *a* Main photosynthetic pigment in plants. **94**

chloroplast Organelle of photosynthesis in the cells of plants and many protists. **65, 97**

choanoflagellate Member of the protist group most closely related to animals. **356**

chordate Animal with an embryo that has a notochord, dorsal nerve cord, pharyngeal gill slits, and a tail that extends beyond the anus. For example, a lancelet or a vertebrate. **378**

chorion Outermost extraembryonic membrane of amniotes; major component of the placenta in placental mammals. **648**

chromosome A structure that consists of DNA and associated proteins; carries part or all of a cell's genetic information. **124**

chromosome number The sum of all chromosomes in a cell of a given type. **124**

chyme Mix of food and gastric fluid. **600**

chytrid Fungus that makes flagellated spores. **345**

ciliate Single-celled, heterotrophic protist with many cilia. **315**

cilium Short, movable structure that projects from the plasma membrane of some eukaryotic cells. **67**

circadian rhythm A biological activity that is repeated about every 24 hours. **442**

circulatory system Organ system consisting of a heart or hearts and blood-filled vessels that distribute substances through a body. **538**

clade A species or group of species that share a set of characters; ideally, a monophyletic group. **278**

cladistics Method of determining evolutionary relationships by grouping species into clades based on shared characters. **278**

cladogram Evolutionary tree that shows a network of evolutionary relationships among clades. **278**

cleavage Mitotic division of an animal cell. **642**

cleavage furrow In a dividing animal cell, the indentation where cytoplasmic division will occur. **168**

climate Average weather conditions in a region over a long time period. **724**

clone Genetically identical copy of an organism. **123**

cloning vector A DNA molecule that can accept foreign DNA, be transferred to a host cell, and get replicated in it. **221**

closed circulatory system Circulatory system in which blood flows through a continuous system of vessels, and substances are exchanged across the walls of the smallest vessels. **362, 538**

club fungi Fungi that have septate hyphae and produce spores by meiosis in club-shaped cells. **347**

cnidarian Radially symmetrical invertebrate with two tissue layers; uses tentacles with

unique stinging structures (nematocysts) to capture food. **358**

coal Fossil fuel formed over millions of years by compaction and heating of plant remains. **332**

cochlea Coiled, fluid-filled structure in the inner ear that holds the sound-detecting organ of Corti. **498**

codominant Refers to two alleles that are both fully expressed in heterozygous individuals. **196**

codon In mRNA, a nucleotide base triplet that codes for an amino acid or stop signal during translation. **142**

coelom Body cavity lined with tissue derived from mesoderm. **355**

coenzyme An organic cofactor. **81**

coevolution The joint evolution of two closely interacting species; each species is a selective agent for traits of the other. **277**

cofactor A metal ion or an organic molecule that associates with an enzyme and is necessary for its function. **81**

cohesion Tendency of molecules to resist separating from one another. **31**

cohesion–tension theory Explanation of how transpiration creates a tension that pulls a cohesive column of water through xylem, from roots to shoots. **420**

cohort Group of individuals born during the same interval. **680**

collecting tubule Kidney tubule that receives filtrate from several nephrons and delivers it to the renal pelvis. **619**

collenchyma Simple plant tissue composed of living cells with unevenly thickened walls; provides flexible support. **400**

colon or **large intestine** Organ that receives digestive waste from the small intestine and concentrates it as feces. **599**

commensalism Species interaction that benefits one species and neither helps nor harms the other. **692**

communication signal Chemical, acoustical, visual, or tactile cue that is produced by one member of a species and detected and responded to by other members of the same species. **662**

community All populations of all species that live a particular region. **5, 691**

compact bone Dense bone that makes up the shaft of long bones. **524**

companion cell In phloem, parenchyma cell that loads sugars into sieve tubes. **401**

comparative morphology Study of body plans and structures among groups of organisms. **239**

competitive exclusion Process whereby two species compete for a limiting resource, and one drives the other to local extinction. **694**

complement A set of proteins that circulate in inactive form in blood as part of innate immunity. **558**

complete digestive system Tubelike digestive system; food enters through one opening and wastes leave through another. **596**

compound Type of molecule that has atoms of more than one element. **27**

compound eye Eye with many units, each having its own lens. **492**

concentration The number of molecules or ions per unit volume of a solution. **32, 84**

concentration gradient Difference in concentration between adjoining regions of fluid. **84**

condensation Process by which enzymes build large molecules from smaller subunits; water also forms. **39**

cone cell Photoreceptor that provides sharp vision and allows detection of color. **494**

conifer Gymnosperm with nonmotile sperm and woody cones; for example, a pine. **334**

conjugation Mechanism of gene exchange in which one bacterial or archaean cell passes a plasmid to another. **303**

connective tissue Animal tissue with an extensive extracellular matrix; provides structural and functional support. **454**

conservation biology Field of applied biology that surveys biodiversity and seeks ways to maintain and use it. **756**

consumer Organism that gets energy and nutrients by feeding on tissues, wastes, or remains of other organisms; a heterotroph. **6, 710**

continuous variation In a population, a range of small differences in a shared trait. **198**

contractile vacuole In freshwater protists, an organelle that collects and expels excess water. **313**

control group In an experiment, group of individuals who are not exposed to the independent variable that is being tested. **13**

coral bleaching A coral expels its photosynthetic dinoflagellate symbionts in response to stress and becomes colorless. **741**

coral reef Highly diverse marine ecosystem centered around reefs built by living corals that secrete calcium carbonate. **741**

cork Component of bark; waterproofs, insulates, and protects the surfaces of woody stems and roots. **409**

cork cambium In plants, a lateral meristem that gives rise to periderm. **409**

cornea Clear, protective covering at front of the vertebrate eye. **492**

corpus luteum Hormone-secreting structure that forms from the cells of a mature follicle that are left behind after ovulation. **635**

cortisol Adrenal cortex hormone that influences metabolism and immunity; secretions rise with stress. **512**

cotyledon Seed leaf; part of a flowering plant embryo. **399**

countercurrent exchange Exchange of substances between fluids moving in opposite directions. **582**

covalent bond Chemical bond in which two atoms share a pair of electrons. **28**

craniate Chordate with a braincase; a hagfish or a vertebrate. **378**

critical thinking Judging the quality of information before allowing it to guide one's thoughts or actions. **12**

crossing over Process in which homologous chromosomes exchange corresponding segments during prophase I of meiosis. **181**

crustaceans Mostly marine arthropod group with two pairs of antennae; for example, shrimp, crabs, lobsters, and barnacles. **369**

culture Learned behavior patterns transmitted among members of a group and between generations. **389**

cuticle Secreted covering at a body surface. **68, 326**

cyanobacteria Photosynthetic, oxygen-producing bacteria. **304**

cycad Tropical or subtropical gymnosperm with flagellated sperm, palmlike leaves, and fleshy seeds. **334**

cytokine Signaling molecule secreted by vertebrate leukocytes (e.g., interleukins, interferons). **559**

cytokinesis Cytoplasmic division. **168**

cytokinin Plant hormone. Promotes cell division and releases lateral buds from apical dominance, among other functions. **439**

cytoplasm Semifluid substance enclosed by a cell's plasma membrane. **52**

cytoplasmic localization Accumulation of different materials in different regions of the egg cytoplasm. **644**

cytoskeleton Dynamic framework of protein filaments that support, organize, and move eukaryotic cells and their internal structures. **66**

cytotoxic T cell T lymphocyte. White blood cell specialized to kill infected or cancerous cells. **559**

data Experimental results. **13**

decomposer Organism that feeds on biological remains and breaks organic material down into its inorganic subunits. **710**

deductive reasoning Using a general idea to make a conclusion about a specific case. **12**

deletion Mutation in which one or more base pairs are lost from a chromosome. **146, 210**

demographics Statistics that describe the traits of a population. **674**

demographic transition model Model describing the changes in human birth and death rates that occur as a region becomes industrialized. **686**

denature To unravel the shape of a protein or other large biological molecule. **46**

dendrite Of a motor neuron or interneuron, a cytoplasmic extension that receives chemical signals sent by other neurons and converts them to electrical signals. **470**

dendritic cell Phagocytic white blood cell that patrols tissue fluids; main type of antigen-presenting cell. **559**

denitrification Bacteria convert nitrates or nitrites to gaseous forms of nitrogen. **718**

dense, irregular connective tissue Connective tissue with asymmetrically arranged fibers and scattered fibroblasts. **454**

dense, regular connective tissue Connective tissue with fibroblasts arrayed between parallel arrangements of fibers. **454**

density-dependent factor Factor that limits population growth and has a greater effect in dense populations than less dense ones. **678**

density-independent factor Factor that limits population growth and arises regardless of population density. **679**

dependent variable In an experiment, variable that is presumably affected by the independent variable being tested. **13**

dermal tissue system Tissue system that covers and protects the plant body. **399**

dermis Deep layer of skin; connective tissue with nerves and blood vessels running through it. **460**

desert Biome with little rain and low humidity; plants that have water-storing and conserving adaptations predominate. **730**

desertification Conversion of dry grassland to desert. **750**

detrital food chain Food chain in which energy is transferred directly from producers to detritivores. **712**

detritivore Consumer that feeds on small bits of organic material. **710**

deuterostomes Lineage of bilateral animals in which the second opening on the embryo surface develops into a mouth. **355**

development Multistep process by which the first cell of a new individual becomes a multi-celled adult. **7**

diaphragm Muscle between the thoracic and abdominal cavities; contracts during inhalation. **585**

diastole Relaxation phase of the cardiac cycle. **542**

diastolic pressure Blood pressure when ventricles are relaxed. **547**

diatom Single-celled photosynthetic protist with brown accessory pigments in its chloroplasts and a two-part silica shell. **317**

differentiation The process by which cells become specialized. **152**

diffusion Net movement of molecules or ions from a region where they are more concentrated to a region where they are less so. **84**

dihybrid cross Breeding experiment in which individuals identically heterozygous for two genes are crossed. The frequency of traits among the offspring offers information about the dominance relationships between the paired alleles. **194**

dikaryotic Having two genetically distinct nuclei $(n + n)$. **344**

dinoflagellate Single-celled, aquatic protist with cellulose plates and two flagella; may be heterotrophic or photosynthetic. **314**

dinosaur Reptile lineage abundant in the Jurassic to Cretaceous; now extinct with the exception of birds. **384**

diploid Having two of each type of chromosome characteristic of the species $(2n)$. **124**

directional selection Mode of natural selection in which phenotypes at one end of a range of variation are favored. **262**

disruptive selection Mode of natural selection that favors forms of a trait at the extremes of a range of variation; intermediate forms are selected against. **265**

distal tubule Portion of kidney tubule that delivers filtrate to a collecting tubule. **619**

distance effect Islands close to a mainland have more species than those farther away. **705**

DNA Deoxyribonucleic acid. Nucleic acid that carries hereditary information about traits; consists of two nucleotide chains twisted in a double helix. **7, 47**

DNA cloning Set of procedures that uses living cells to make many identical copies of a DNA fragment. **220**

DNA library Collection of cells that host different fragments of foreign DNA, often representing an organism's entire genome. **222**

DNA ligase Enzyme that seals gaps in double-stranded DNA. **131**

DNA polymerase DNA replication enzyme. Uses a DNA template to assemble a complementary strand of DNA. **130**

DNA profiling Identifying an individual by analyzing the unique parts of his or her DNA. **227**

DNA repair mechanism Any of several processes by which enzymes repair damaged DNA. **131**

DNA replication Process by which a cell duplicates its DNA before it divides. **130**

DNA sequence Order of nucleotide bases in a strand of DNA. **130**

DNA sequencing Method of determining the order of nucleotides in DNA. **224**

dominance hierarchy Social system in which resources and mating opportunities are unequally distributed within a group. **666**

dominant Refers to an allele that masks the effect of a recessive allele paired with it. **191**

dormancy Period of temporarily suspended metabolism. **432**

dosage compensation Theory that X chromosome inactivation equalizes gene expression between males and females. **156**

double fertilization Mode of fertilization in flowering plants in which one sperm cell fuses with the egg, and a second sperm cell fuses with the endosperm mother cell. **432**

duplication Repeated section of a chromosome. **210**

echinoderms Invertebrates with hardened plates and spines embedded in the skin or body, and a water–vascular system. **373**

ecological footprint Area of Earth's surface required to sustainably support a particular level of development and consumption. **686**

ecological niche All of a species' requirements and roles in an ecosystem. **694**

ecological restoration Actively altering an area in an effort to restore or create a functional ecosystem. **757**

ecology Study of interactions among organisms and their environment. **657**

ecosystem A community interacting with its environment. **5**

ectoderm Outermost tissue layer of an animal embryo. **354, 642**

ectotherm Animal that controls its internal temperature by altering its behavior; for example, a fish or a lizard. **385, 623**

effector cell Antigen-sensitized B cell or T cell that forms in an immune response and acts immediately. **566**

egg Mature female gamete, or ovum. **182**

electron Negatively charged subatomic particle that occupies orbitals around an atomic nucleus. **24**

electronegativity Measure of the ability of an atom to pull electrons away from other atoms. **27**

electron transfer chain Array of enzymes and other molecules that accept and give up electrons in sequence, thus releasing the energy of the electrons in usable increments. **83**

electron transfer phosphorylation Process in which electron flow through electron transfer chains sets up a hydrogen ion gradient that drives ATP formation. **98**

electrophoresis Technique that separates DNA fragments by size. **224**

element A pure substance that consists only of atoms with the same number of protons. **24**

El Niño Periodic warming of equatorial Pacific waters and the associated shifts in global weather patterns. **723**

embryonic induction Embryonic cells produce signals that alter the behavior of neighboring cells. **646**

embryophyte Member of the land plant clade. **326**

emergent property A characteristic of a system that does not appear in any of the system's component parts. **4**

emerging disease A disease that was previously unknown or has recently begun spreading to a new region. **300**

emigration Movement of individuals out of a population. **676**

emulsification Suspension of fat droplets in a fluid. **603**

endangered species A species that faces extinction in all or a part of its range. **748**

endemic disease Disease that persists at a low level in a region or population. **300**

endemic species Species that remains restricted to the area where it evolved. **748**

endergonic Type of reaction that requires a net input of free energy to proceed. **78**

endocrine disrupter Synthetic chemical that adversely affects hormone production or function. **503**

endocrine gland Ductless gland that secretes hormones into a body fluid. **453**

endocrine system Hormone-producing glands and secretory cells of a vertebrate body. **504**

endocytosis Process by which a cell takes in a small amount of extracellular fluid by the ballooning inward of its plasma membrane. **88**

endoderm Innermost tissue layer of an animal embryo. **354, 642**

endodermis In plant roots, layer of endodermal cells that separates vascular cylinder from cortex. **407**

endomembrane system Series of interacting organelles (endoplasmic reticulum, Golgi bodies, vesicles) between the nucleus and plasma membrane. **62**

endometrium Lining of uterus. **634**

endoplasmic reticulum (ER) Organelle that is a continuous system of sacs and tubes extending from the nuclear envelope. Rough ER is studded with ribosomes; smooth ER is not. **62**

endorphin Painkiller produced by the central nervous system. **477**

endoskeleton Internal skeleton made up of hardened components such as bones. **378, 522**

endosperm Nutritive triploid tissue in angiosperm seeds. **336, 432**

endospore Resistant resting stage of some soil bacteria. **305**

endosymbiosis One species lives and reproduces inside another. **290**

endotherm Animal controls its internal temperature by adjusting its metabolism; for example, a bird or mammal. **385, 623**

energy The capacity to do work. **6, 76**

enhancer Binding site in DNA for proteins that enhance the rate of transcription. **152**

entropy Measure of how much the energy of a system is dispersed. **76**

enzyme Compound (usually a protein) that speeds a reaction without being changed by it. **39**

eosinophil Granular white blood cell; targets multicelled parasites. **559**

epidemic Disease outbreak limited to one region. **300**

epidermis Outermost tissue layer of a plant or an animal. **400, 460**

epididymis One of a pair of ducts where sperm that formed in a testis mature; empties into a vas deferens. **632**

epiglottis Tissue flap that folds down to prevent food from entering airways during swallowing. **585**

epiphyte Plant that grows on another plant but does not harm it. **331**

epistasis Effect in which a trait is influenced by the products of multiple genes. **197**

epithelial tissue Sheetlike animal tissue that covers outer body surfaces and lines internal tubes and cavities. **452**

equilibrium model of island biogeography Model that predicts the number of species on an island based on the island's area and distance from the mainland. **704**

esophagus Muscular tube between the throat and stomach. **599**

essential amino acid Amino acid that the body cannot make and must obtain from food. **607**

essential fatty acid Fatty acid that the body cannot make and must obtain from the diet. **606**

estrogen Hormone secreted by ovaries; causes development of female sexual traits and maintains the reproductive tract. **634**

estuary A highly productive ecosystem where nutrient-rich water from a river mixes with seawater. **740**

ethylene Gaseous plant hormone; inhibits cell division in stems and roots and promotes fruit ripening, among other effects. **439**

eudicots Largest lineage of angiosperms; includes herbaceous plants, woody trees, and cacti. **336**

eugenics Idea of deliberately improving the genetic qualities of the human race. **232**

euglenoid Flagellated protozoan with multiple mitochondria; may be heterophic or have chloroplasts decended from algae. **313**

eukaryote Organism whose cells characteristically have a nucleus. **8**

eutrophication Nutrient enrichment of an aquatic ecosystem. **709**

evaporation Transition of a liquid to a gas. **31**

evolution Change in a line of descent. **240**

evolutionary tree Type of diagram that summarizes evolutionary relationships among a group of species. **278**

exaptation Adaptation of an existing structure for a completely different purpose; a major evolutionary novelty. **276**

exergonic Type of reaction that ends with a net release of free energy. **78**

exocrine gland Gland that secretes milk, sweat, saliva, or some other substance through a duct. **453**

exocytosis Process by which a cell expels a vesicle's contents to extracellular fluid. **88**

exon Nucleotide sequence that is not spliced out of RNA during processing. **141**

exoskeleton Of some invertebrates, hard external parts that muscles attach to and move. **367, 522**

exotic species A species that evolved in one community and later became established in a different one. **703**

experiment A test designed to support or falsify a prediction. **13**

experimental group In an experiment, group of individuals who are exposed to an independent variable. **13**

exponential growth A population grows by a fixed percentage in successive time intervals; the size of each increase is determined by the current population size. **676**

external fertilization Sperm and egg are released into the external environment and meet there. **631**

extinct Refers to a species that has been permanently lost. **276**

extracellular fluid Of a multicelled organism, body fluid outside of cells; serves as the body's internal environment. **450**

extracellular matrix (ECM) Complex mixture of cell secretions that supports cells and tissues. Also has roles in cell signaling. **68**

extreme halophile Organism adapted to life in a highly salty environment. **306**

extreme thermophile Organism adapted to life in a very high-temperature environment. **306**

fat Lipid that consists of a glycerol molecule with one, two, or three fatty acid tails. **42**

fatty acid Organic compound that consists of a chain of carbon atoms with an acidic carboxyl group at one end. Carbon chain of saturated types has single bonds only; that of unsaturated types has one or more double bonds. **42**

feces Unabsorbed food material and cellular waste that is expelled from the digestive tract. **604**

feedback inhibition Mechanism by which a change that results from some activity decreases or stops the activity. **82**

fermentation An anaerobic pathway by which cells harvest energy from carbohydrates to produce ATP. **109**

fertilization Fusion of two gametes to form a zygote. **177**

fetus Human from week 9 of development to birth. **650**

fever An internally induced rise in core body temperature above the normal set point as a response to infection. **563**

fibrin Threadlike protein formed during blood clotting from a soluble plasma protein (fibrinogen). **545**

fibrous root system Root system composed of an extensive mass of similar-sized roots; typical of monocots. **406**

first law of thermodynamics Energy cannot be created or destroyed. **76**

fitness Degree of adaptation to an environment, as measured by an individual's relative genetic contribution to future generations. **242**

fixed Refers to an allele for which all members of a population are homozygous. **268**

fixed action pattern Series of instinctive movements elicited by a simple stimulus and carried out with little variation once begun. **660**

flagellated protozoan Protist belonging to an entirely or mostly heterotrophic lineage with no cell wall and one or more flagella. **313**

flagellum Long, slender cellular structure used for motility. **58**

flatworm Bilaterally symmetrical invertebrate with organs but no body cavity; for example, a planarian or tapeworm. **360**

flower Specialized reproductive shoot of a flowering plant. **336, 430**

fluid mosaic Model of a cell membrane as a two-dimensional fluid of mixed composition. **56**

food chain Description of who eats whom in one path of energy in an ecosystem. **711**

food web Set of cross-connecting food chains. **712**

foraminiferan Heterotrophic single-celled protist with a porous calcium carbonate shell and long cytoplasmic extensions. **314**

fossil Physical evidence of an organism that lived in the ancient past. **239**

founder effect Change in allele frequencies that occurs when a small number of individuals establish a new population. **268**

fovea Retinal region where cone cells are most concentrated. **494**

fruit Mature ovary of a flowering plant, often with accessory parts; encloses a seed or seeds. **336, 434**

functional group A group of atoms bonded to a carbon of an organic compound; imparts a specific chemical property to the molecule. **38**

fungus Eukaryotic heterotroph with cell walls of chitin; obtains nutrients by digesting food outside the body and absorbing them. **8, 344**

gallbladder Organ that stores and concentrates bile. **603**

gamete Mature, haploid reproductive cell; e.g., an egg or a sperm. **177**

gametophyte A haploid, multicelled body that produces gametes in the life cycle of land plants and some protists. **182, 318**

ganglion Cluster of nerve cell bodies. **468**

gap junction Cell junction that forms a channel across the plasma membranes of adjoining animal cells. **69**

gastric fluid Fluid secreted by the stomach lining; contains digestive enzymes, acid, and mucus. **600**

gastropod Mollusk in which the lower body is a broad "foot." **364**

gastrula Three-layered developmental stage formed by gastrulation in an animal. **642**

gastrulation Animal developmental process by which cell movements produce a three-layered gastrula. **642**

gene Part of a DNA base sequence; specifies an RNA or protein product. **138**

gene expression Process by which the information in a gene becomes converted to an RNA or protein product. **139**

gene flow The movement of alleles into and out of a population. **269**

gene pool All of the alleles of all of the genes in a population; a pool of genetic resources. **259**

gene therapy The transfer of a normal or modified gene into an individual with the goal of treating a genetic defect or disorder. **232**

genetic code Complete set of sixty-four mRNA codons. **142**

genetic drift Change in allele frequencies in a population due to chance alone. **268**

genetic engineering Process by which deliberate changes are introduced into an individual's genome. **228**

genetic equilibrium Theoretical state in which a population is not evolving. **259**

genetically modified organism (GMO) Organism whose genome has been modified by genetic engineering. **228**

genome An organism's complete set of genetic material. **222**

genomics The study of genomes. **226**

genotype The particular set of alleles carried by an individual. **191**

genus A group of species that share a unique set of traits; also the first part of a species name. **10**

geologic time scale Chronology of Earth's history. **246**

germ cell Diploid reproductive cell that gives rise to haploid gametes by meiosis. **177**

germination The resumption of growth after dormancy. **436**

germ layer One of three primary layers in an early embryo. **642**

gibberellin Plant hormone. Induces stem elongation and helps seeds break dormancy, among other effects. **438**

gills Folds or body extensions that increase the surface area for respiration. **581**

ginkgo Deciduous gymnosperm with flagellated sperm, fan-shaped leaves, and fleshy seeds. **334**

global climate change A rise in temperature and shifts in other climate patterns. **717**

glomeromycete Fungus with hyphae that grow inside the wall of a plant root cell. **345**

glomerular filtration Protein-free plasma forced out of glomerular capillaries by blood pressure enters Bowman's capsule. **620**

glomerulus Ball of capillaries enclosed by Bowman's capsule. **619**

glottis Opening formed when the vocal cords relax. **585**

glycolysis Set of reactions in which glucose or another sugar is broken down to two pyruvate for a net yield of two ATP. **109**

gnetophyte Shrubby or vinelike gymnosperm, with nonmotile sperm; for example, *Ephedra*. **334**

Golgi body Organelle that modifies polypeptides and lipids; also sorts and packages the finished products into vesicles. **63**

gonads Primary reproductive organs (ovaries or testes); produce gametes and sex hormones. **516, 632**

Gondwana Supercontinent that existed before Pangea, more than 500 million years ago. **249**

Gram staining Process used to prepare bacterial cells for microscopy, and to distinguish groups based on cell wall structure. **304**

grassland Biome in the interior of continents where grasses and nonwoody plants adapted to grazing and fire predominate. **732**

gravitropism Plant growth in a direction influenced by gravity. **440**

gray matter Brain and spinal cord tissue consisting of cell bodies, dendrites, and neuroglial cells. **480**

grazing food chain Food chain in which energy is transferred from producers to grazers (herbivores). **712**

green alga Photosynthetic protist that deposits cellulose in its cell wall, stores sugars as starch, and has chloroplasts containing chlorophylls *a* and *b*. **318**

greenhouse effect Warming of Earth's lower atmosphere and surface as a result of heat trapped by greenhouse gases. **717**

ground tissue system Tissue system that makes up the bulk of the plant body; includes most photosynthetic cells. **398**

groundwater Soil water and water in aquifers. **714**

growth In multicelled species, an increase in the number, size, and volume of cells. **7**

growth factor Molecule that stimulates mitosis. **169**

guard cell One of a pair of cells that define a stoma across the epidermis of a leaf or stem. **422**

gymnosperm Seed plant that does not make flowers or fruits; for example, a conifer. **334**

habitat Type of environment in which a species typically lives. **692**

habituation Learning not to respond to a repeated stimulus. **661**

hagfish Jawless fish with a skull case but no backbone. **380**

hair cell Mechanoreceptor that is activated when movement of overlying membrane causes its hairlike cilia to bend. **497**

half-life Characteristic time it takes for half of a quantity of a radioisotope to decay. **245**

haploid Having one of each type of chromosome characteristic of the species. **177**

heart Muscular organ that pumps blood through a body. **538**

heartwood Dense, dark accumulation of nonfunctional xylem at the core of older stems and roots. **409**

Heimlich maneuver Procedure designed to rescue a choking person; a rescuer presses on a person's abdomen to force air out of the lungs and dislodge an object in the trachea. **586**

hemostasis Process by which blood clots in response to injury. **545**

herbivory An animal feeds on plant parts. **697**

hermaphrodite Animal that produces both eggs and sperm, either simultaneously or at different times in its life. **357, 630**

heterotherm Animal that maintains its temperature by production of metabolic heat sometimes, and allows its temperature to fluctuate with the environment at other times. **623**

heterotroph Organism that obtains energy and carbon from organic compounds assembled by other organisms; a consumer. **93**

heterozygous Having two different alleles of a gene. **191**

histone Type of protein that structurally organizes eukaryotic chromosomes. **124**

homeostasis Set of processes by which an organism keeps its internal conditions within tolerable ranges. **7**

homeotic gene Type of master gene; its expression controls formation of specific body parts during development. **154, 647**

hominid Human or extinct humanlike species. **388**

homologous chromosomes Chromosomes with the same length, shape, and set of genes. **164**

homologous structures Similar body parts that evolved in a common ancestor. **250**

homozygous Having identical alleles of a gene. **191**

horizontal gene transfer Transfer of genetic material among existing individuals. **303**

hormone Signaling molecule that is released into the body by one type of cell and alters the activity of other cells. **438**

hot spot Threatened region with great biodiversity that is considered a high priority for conservation efforts. **756**

humans Members of the genus *Homo*. **391**

humus Decaying organic matter in soil. **416**

hybrid The offspring of a cross between two individuals that breed true for different forms of a trait; a heterozygous individual. **191**

hydrocarbon Compound or region of one that consists only of carbon and hydrogen atoms. **38**

hydrogen bond Attraction that forms between a covalently bonded hydrogen atom and another atom taking part in a separate covalent bond. **29**

hydrolysis Process by which an enzyme breaks a molecule into smaller subunits by attaching a hydroxyl group to one part and a hydrogen atom to the other. **39**

hydrophilic Describes a substance that dissolves easily in water. **30**

hydrophobic Describes a substance that resists dissolving in water. **30**

hydrostatic skeleton Of soft-bodied invertebrates, a fluid-filled chamber that muscles exert force against, redistributing the fluid. **358, 522**

hydrothermal vent Rocky, underwater opening where mineral-rich water heated by geothermal energy streams out. **285, 743**

hypertonic Describes a fluid that has a high overall solute concentration relative to another fluid. **84**

hypha Component of a fungal mycelium; a filament made up of cells arranged end to end. **344**

hypothalamus Forebrain region that controls processes related to homeostasis; control center for endocrine functions. **483, 508**

hypothesis Testable explanation of a natural phenomenon. **12**

hypotonic Describes a fluid that has a low overall solute concentration relative to another fluid. **84**

immigration Movement of individuals into a population. **676**

immunity The body's ability to resist and fight infections. **558**

immunization Any procedure designed to promote immunity to a specific disease. **572**

imprinting Learning that can occur only during a specific interval in an animal's life. **660**

inbreeding Mating among close relatives. **269**

incomplete digestive system Saclike digestive system; food enters and leaves through the same opening. **596**

incomplete dominance Condition in which one allele is not fully dominant over another, so the heterozygous phenotype is between the two homozygous phenotypes. **196**

independent variable Variable that is controlled by an experimenter in order to explore its relationship to a dependent variable. **13**

indicator species Species that is especially sensitive to disturbance and can be monitored to assess the health of a habitat. **703**

induced–fit model Substrate binding to an active site improves the fit between the two. **80**

inductive reasoning Drawing a conclusion based on observation. **12**

inferior vena cava Vein that delivers blood from the lower body to the heart. **542**

inflammation A local response to tissue damage or infection; characterized by redness, warmth, swelling, and pain. **562**

inheritance Transmission of DNA from parents to offspring. **7**

inhibiting hormone Hormone that is secreted by one endocrine gland and discourages secretion by another. **508**

innate immunity Set of inborn, fast-acting defenses against infection. Triggered by a fixed set of molecular patterns found mainly on pathogens. **558**

inner ear Fluid-filled vestibular apparatus and cochlea. **498**

insects Most diverse arthropod group; members have six legs, two antennae, and, in some groups, wings. **370**

insertion Mutation in which one or more base pairs become inserted into DNA. **146**

instinctive behavior An innate response to a simple stimulus. **660**

intermediate disturbance hypothesis Species richness is greatest in communities where disturbances are moderate in their intensity or frequency. **701**

intermediate filament Stable cytoskeletal element structurally supports cells and tissues. **66**

internal fertilization A female retains eggs in her body and sperm fertilize them there. **631**

interneuron Neuron that receives signals from and sends signals to other neurons. **468**

interphase In a eukaryotic cell cycle, the interval between mitotic divisions when a cell grows, roughly doubles the number of its cytoplasmic components, and replicates its DNA. **164**

interspecific competition Competition between two species. **694**

interstitial fluid Of a multicelled organism, body fluid in spaces between cells. **450**

intervertebral disk Cartilage disk between two vertebrae. **522**

intron Nucleotide sequence that intervenes between exons and is excised during RNA processing. **141**

inversion Structural rearrangement of a chromosome in which part of it becomes oriented in the reverse direction. **210**

invertebrate General term for an animal that does not have a backbone. **353**

in vitro fertilization Assisted reproduction technology in which eggs and sperm are united outside the body. **629**

ion Charged atom. **30**

ionic bond Type of chemical bond in which a strong mutual attraction forms between ions of opposite charge. **28**

iris Circular muscle that adjusts the shape of the pupil to regulate how much light enters the eye. **492**

isotonic Describes two fluids with identical solute concentrations. **84**

isotopes Forms of an element that differ in the number of neutrons their atoms carry. **25**

joint Region where bones meet. **525**

karyotype Image of an individual's complement of chromosomes arranged by size, length, shape, and centromere location. **125**

key innovation An evolutionary adaptation that gives its bearer the opportunity to exploit a particular environment more efficiently or in a new way. **276**

keystone species A species that has a disproportionately large effect on community structure. **702**

kidney Organ of the vertebrate urinary system that filters blood, adjusts its composition, and forms urine. **617**

kinetic energy The energy of motion. **76**

knockout An experiment in which a gene is deliberately inactivated in a living organism. **154**

Krebs cycle Cyclic pathway that, along with acetyl–CoA formation, breaks down two pyruvate to carbon dioxide for a net yield of two ATP and many reduced coenzymes. Part of aerobic respiration pathway. **113**

K-selection Individuals who produce offspring capable of outcompeting others for limited resources have a selective advantage; occurs when a population is near carrying capacity. **681**

labor Expulsion of a placental mammal from its mother's uterus by muscle contractions. **653**

lactate fermentation Anaerobic carbohydrate breakdown pathway that produces ATP and lactate. **116**

lactation Milk production by a female mammal. **653**

lamprey Jawless fish with a backbone of cartilage. **380**

lancelet Invertebrate chordate that has a fish-like shape and retains the defining chordate traits into adulthood. **378**

La Niña Periodic cooling of equatorial Pacific waters and the associated shifts in global weather patterns. **723**

large intestine or **colon** Organ that receives digestive waste from the small intestine and concentrates it as feces. **599**

larva Preadult stage in some animal life cycles. **357**

larynx Short airway containing the vocal cords (voice box). **585**

lateral meristem Vascular cambium or cork cambium; sheetlike cylinder of meristem that gives rise to plant secondary growth. **408**

law of independent assortment During meiosis, members of a pair of genes on homologous chromosomes tend to be distributed into gametes independently of other gene pairs. **194**

law of nature Generalization that describes a consistent natural phenomenon for which there is incomplete scientific explanation. **18**

law of segregation The two members of each pair of genes on homologous chromosomes end up in different gametes during meiosis. **193**

leaching Process by which water moving through soil removes nutrients from it. **417**

learned behavior Behavior that is modified by experience. **660**

lek Of some birds, a communal mating display area for males. **664**

lens Disk-shaped structure that bends light rays so they fall on an eye's photoreceptors. **492**

lethal mutation Mutation that drastically alters phenotype; causes death. **258**

lichen Composite organisms, consisting of a fungus and a single-celled alga or a cyanobacterium. **348**

life history pattern A set of traits related to growth, survival, and reproduction such as age-specific mortality, life span, age at first reproduction, and number of breeding events. **680**

ligament Strap of dense connective tissue that holds bones together at a joint. **525**

light-dependent reactions First stage of photosynthesis; a set of reactions that convert light energy to chemical energy of ATP and NADPH. **97**

light-independent reactions Second stage of photosynthesis; a set of reactions that use ATP and NADPH to assemble sugars from water and CO_2. **97**

lignin Material that stiffens cell walls of vascular plants. **68, 326**

limbic system Group of brain structures that govern emotion. **484**

limiting factor A necessary resource, the depletion of which halts population growth. **678**

lineage Line of descent. **240**

linkage group All genes on a chromosome. **195**

lipid Fatty, oily, or waxy organic compound. **42**

lipid bilayer Structural foundation of cell membranes; double layer of lipids arranged tail-to-tail. **56**

loam Soil with roughly equal amounts of sand, silt, and clay. **417**

lobe-finned fish Fish with fleshy fins that contain bones. **381**

local signaling molecule Chemical signal, such as a prostaglandin, that is secreted by one cell and affects neighboring cells in an animal body. **504**

locus Location of a gene on a chromosome. **191**

logistic growth A population grows slowly, then increases rapidly until it reaches carrying capacity and levels off. **678**

loop of Henle U-shaped portion of a kidney tubule that extends deep into the renal medulla. **619**

loose connective tissue Connective tissue with relatively few fibroblasts and fibers scattered in its matrix. **454**

lung A saclike respiratory organ that lies inside a body cavity. **582**

lymph Fluid in the lymph vascular system. **552**

lymph node Small mass of lymphatic tissue through which lymph filters; contains many lymphocytes (B and T cells). **553**

lymph vascular system System of vessels that takes up interstitial fluid and carries it (as lymph) to the blood. **552**

lysogenic pathway Bacteriophage replication path in which viral DNA becomes integrated into the host's chromosome and is passed to the host's descendants. **298**

lysosome Enzyme-filled vesicle that functions in intracellular digestion. **63**

lysozyme Antibacterial enzyme that occurs in body secretions such as mucus. **561**

lytic pathway Bacteriophage replication pathway in which a virus immediately replicates in its host and kills it. **298**

macrophage Phagocytic white blood cell that patrols tissue fluids. **559**

mammal Animal with hair or fur; females secrete milk from mammary glands. **387**

mark–recapture sampling Method of estimating population size of mobile animals by marking individuals, releasing them, then checking the proportion of marks among individuals recaptured at a later time. **674**

marsupial Mammal in which young are born at an early stage and complete development in a pouch on the mother's surface. **387**

mass extinction Simultaneous loss of many lineages from Earth. **237**

mass number Total number of protons and neutrons in the nucleus of an element's atoms. **25**

mast cell Granulatory white blood cell anchored in many tissues; factor in inflammation. **559**

master gene Gene encoding a product that affects the expression of many other genes. **154**

mechanoreceptor Sensory receptor that responds to pressure, position, or acceleration. **490**

medulla oblongata Hindbrain region that controls breathing rhythm and reflexes such as coughing and vomiting. **482**

megaspore Haploid spore formed in ovule of seed plants; develops into an egg-producing gametophyte. **332, 432**

meiosis Nuclear division process that halves the chromosome number. Basis of sexual reproduction. **176**

membrane potential Potential energy of charges separated by a cell membrane. **471**

memory cell Long-lived, antigen-sensitized B or T cell that can act in a secondary immune response. **566**

meninges Membranes that enclose the brain and spinal cord. **480**

menopause Permanent cessation of menstrual cycles. **637**

menstrual cycle Approximately 28-day cycle in which the uterus lining thickens and then, if pregnancy does not occur, is shed. **636**

menstruation Flow of shed uterine tissue out of the vagina. **636**

meristem Zone of undifferentiated plant cells that can divide rapidly. **399**

mesoderm Middle tissue layer of a three-layered animal embryo. **354, 642**

mesophyll Photosynthetic parenchyma. **401**

messenger RNA (mRNA) A type of RNA that carries a protein-building message. **138**

metabolic pathway Series of enzyme-mediated reactions by which cells build, remodel, or break down an organic molecule. **82**

metabolism All the enzyme-mediated chemical reactions by which cells acquire and use energy as they build and break down organic molecules. **39**

metamorphosis Dramatic remodeling of body form during the transition from larva to adult. **367**

metaphase Stage of mitosis at which the cell's chromosomes are aligned midway between poles of the spindle. **166**

metastasis The process in which cancer cells spread from one part of the body to another. **171**

methanogen Organism that produces methane gas as a metabolic by-product. **306**

MHC markers Self-recognition protein on the surface of body cells. Triggers adaptive immune response when bound to antigen fragments. **564**

microevolution Change in allele frequencies in a population or species. **259**

microfilament Reinforcing cytoskeletal element that consists of actin subunits. **66**

microspore Walled, haploid spore formed in pollen sacs of seed plants; develops into a sperm-producing gametophyte (a pollen grain). **332, 432**

microtubule Cytoskeletal element involved in cellular movement; hollow filament of tubulin subunits. **66**

microvilli Thin projections from the plasma membrane of some epithelial cells. **452, 601**

middle ear Eardrum and the tiny bones that transfer sound to the inner ear. **498**

mimicry A species evolves traits that make it similar in appearance to another species. **696**

mineral Inorganic substance that is required in small amounts for normal metabolism. **469, 609**

mitochondrion Double-membraned organelle that produces ATP by aerobic respiration in eukaryotes. **64**

mitosis Nuclear division mechanism that maintains the chromosome number. Basis of body growth and tissue repair in multicelled eukaryotes; also asexual reproduction in some plants, animals, fungi, and protists. **164**

mixture An intermingling of two or more types of molecules. **27**

model Analogous system used for testing hypotheses. **13**

molecule Two or more atoms joined by chemical bonds. **4, 27**

mollusk Invertebrate with a reduced coelom and a mantle; such as a bivalve, gatropod, or cephalopod. **364**

molting Periodic shedding of an outer body layer or part. **366**

monocots Lineage of angiosperms with one seed leaf (cotyledon); includes grasses, orchids, and palms. **336**

monohybrid cross Breeding experiment in which individuals identically heterozygous for one gene are crossed. The frequency of traits among the offspring offers information about the dominance relationship between the alleles. **192**

monomers Molecules that are subunits of polymers. **39**

monophyletic group An ancestor and all of its descendants. **278**

monotreme Egg-laying mammal. **387**

monsoon Wind that reverses direction seasonally. **727**

morphogen Chemical encoded by a master gene; diffuses out from a source, creating a chemical gradient that affects development. **646**

morphological convergence Evolutionary pattern in which similar body parts evolve separately in different lineages. **251**

morphological divergence Evolutionary pattern in which a body part of an ancestor changes in its descendants. **250**

moss Nonvascular plant with a leafy green gametophyte and an attached, dependent sporophyte consisting of a capsule on a stalk. **329**

motor neuron Neuron that receives signals from another neuron and sends signals to a muscle or gland. **468**

motor protein Type of energy-using protein that interacts with cytoskeletal elements to move the cell's parts or the whole cell. **66**

motor unit One motor neuron and the muscle fibers it controls. **530**

multiple allele system Gene for which three or more alleles persist in a population. **196**

multiregional model Model that postulates that *H. sapiens* populations in different regions evolved from *H. erectus* in those regions. **392**

muscle tension Force exerted by a contracting muscle. **530**

muscle twitch Brief muscle contraction. **530**

mutation Permanent change in the sequence of DNA. **131**

mutualism Species interaction that benefits both species. **348, 693**

mycelium Mass of threadlike filaments (hyphae) that make up the body of a multicelled fungus. **344**

mycorrhiza Mutually beneficial partnership between a fungus and a plant root. **348, 418**

myelin Insulating material made by neuroglial cells; wraps most axons and increases the speed of signal transmission. **478**

myofibrils Threadlike, cross-banded skeletal muscle components that consist of sarcomeres arranged end to end. **528**

myosin Motor protein with ATPase activity; makes up thick filaments of muscle fibers. **528**

naturalist Person who observes life from a scientific perspective. **238**

natural selection A process in which environmental pressures result in the differential survival and reproduction of individuals of a population who vary in the details of shared, heritable traits. **243, 261**

negative feedback A change causes a response that reverses the change; important mechanism of homeostasis. **462**

nematocyst Stinging organelle unique to cnidarians. **358**

neoplasm An accumulation of abnormally dividing cells. **169**

nephron Kidney tubule and glomerular capillaries; filters blood and forms urine. **618**

nerve Neuron fibers bundled inside a sheath of connective tissue. **468**

nerve cord Bundle of nerve fibers that runs the length of a body. **468**

nerve net Of cnidarians, a mesh of nerve cells with no central control organ; allows movement and other behavior. **358, 468**

nervous tissue Animal tissue composed of neurons and supporting cells (neuroglia); detects stimuli and controls responses to them. **457**

neuroglial cell Cell that supports neurons. **468**

neuromuscular junction Synapse between a neuron and a muscle. **474**

neuron One of the cells that make up communication lines of nervous systems; transmits electrical signals along its plasma membrane and communicates with other cells through chemical messages. **457, 468**

neurotransmitter Chemical signal released by axon terminals of a neuron. **474**

neutral mutation A mutation that has no effect on survival or reproduction. **258**

neutron Uncharged subatomic particle in the atomic nucleus. **24**

neutrophil Circulating phagocytic white blood cell. **559**

niche See ecological niche.

nitrification Bacteria convert ammonium to nitrates. **718**

nitrogen cycle Movement of nitrogen among the atmosphere, soil, and water, and into and out of food webs. **718**

nitrogen fixation Incorporation of nitrogen gas into ammonia. **304, 418, 718**

NK cell Natural killer cell; a type of lymphocyte. White blood cell that can kill cancer cells undetectable by cytotoxic T cells. **559**

nondisjunction Failure of sister chromatids or homologous chromosomes to separate during nuclear division. **212**

nonshivering heat production Heat-generating mechanism of brown adipose tissue; energy is released as heat, rather than stored in ATP. **625**

normal flora Normally harmless or beneficial microorganisms that typically live in or on a host body. **304, 560**

notochord Stiff rod of connective tissue that runs the length of the body in chordate larvae or embryos. **378**

nuclear envelope A double membrane that constitutes the outer boundary of the nucleus. **61**

nucleic acid Single- or double-stranded chain of nucleotides joined by sugar–phosphate bonds; for example, DNA, RNA. **47**

nucleic acid hybridization Base-pairing between DNA or RNA from different sources. **222**

nucleoid Region of cytoplasm where the DNA is concentrated in a bacterium or archaean. **58, 302**

nucleolus In a cell nucleus, a dense, irregularly shaped region where ribosomal subunits are assembled. **61**

nucleoplasm Viscous fluid enclosed by the nuclear envelope. **61**

nucleosome A length of DNA wound around a spool of histone proteins. **124**

nucleotide Monomer of nucleic acids. Consists of a five-carbon sugar, a nitrogen-containing base, and phosphate groups. **47**

nucleus Of an atom, the core region occupied by protons and neutrons. **24** Of a eukaryotic cell, an organelle with two membranes that holds the cell's DNA. **8, 52**

nutrient Substance that an organism needs for growth and survival, but cannot make for itself. **6**

olfactory receptors Chemoreceptors involved in sense of smell. **496**

oncogene Gene that has the potential to transform a normal cell into a tumor cell. **170**

oocyte Immature egg. **634**

open circulatory system System in which blood leaves vessels and seeps through tissues before returning to the heart. **364, 538**

operator Part of an operon; a DNA binding site for a repressor. **158**

operon Group of genes together with a promoter–operator DNA sequence that controls their transcription. **158**

organ In multicelled organisms, a grouping of tissues engaged in a collective task. **5**

organelle Structure that carries out a specialized metabolic function inside a cell. **52**

organic Type of compound that consists primarily of carbon and hydrogen atoms. **38**

organism Individual that consists of one or more cells. **5**

organs of equilibrium Sensory organs that respond to body position and motion. **497**

organ system In multicelled organisms, set of organs engaged in a collective task that keeps the body functioning properly. **5**

osmosis The diffusion of water across a selectively permeable membrane in response to a differing overall solute concentration. **85**

osmotic pressure Amount of turgor that prevents osmosis into cytoplasm or other hypertonic fluid. **85**

outer ear External ear and the air-filled auditory canal. **498**

ovarian follicle In animals, an immature egg and surrounding cells. **635**

ovary Of flowering plants, the enlarged base of a carpel, inside which one or more ovules form and eggs are fertilized. **336, 431** Of animals, organ in which oocytes form and mature. **634**

oviduct Duct between an ovary and the uterus. **634**

ovulation Release of a secondary oocyte from an ovary. **635**

ovule Of seed plants, reproductive structure in which egg-producing female gametophyte develops; after fertilization, matures into a seed. **332, 431**

ovum Mature animal egg. **638**

oxyhemoglobin Hemoglobin with oxygen bound to it. **588**

ozone layer High atmospheric layer with a great concentration of ozone (O_3); prevents much ultraviolet radiation from reaching Earth's surface. **754**

pain Perception of tissue injury. **491**

pain receptor Sensory receptor that responds to tissue damage. **490**

pancreas Organ that secretes digestive enzymes into the small intestine and hormones (insulin and glucagon) into the blood. **514**

pandemic Outbreak of disease that affects many separate regions and poses a serious threat to human health. **300**

Pangea Supercontinent that formed about 237 million years ago and broke up about 152 million years ago. **248**

parapatric speciation Speciation model in which differing selection pressures lead to divergences within a single population. **275**

parasitism Relationship in which one species withdraws nutrients from another species, without immediately killing it. **698**

parasitoid An insect that lays eggs in another insect, and whose young devour their host from the inside. **698**

parasympathetic neurons Neurons of the autonomic system that encourage housekeeping tasks. **479**

parathyroid glands Four small endocrine glands whose hormone product increases the level of calcium in blood. **511**

parenchyma Simple plant tissue made up of living cells; main component of ground tissue. **400**

partial pressure Pressure exerted by one gas in a mixture of gases. **588**

passive transport Mechanism by which a concentration gradient drives the movement of a solute across a cell membrane through a transport protein. Requires no energy input. **86**

pathogen Disease-causing agent. **300**

pattern formation Process by which a complex body forms from local processes during embryonic development. **155, 647**

PCR See polymerase chain reaction.

peat Carbon-rich moss remains; can be dried for use as fuel. **329**

pedigree Chart showing the pattern of inheritance of a trait through generations in a family. **204**

pelagic province The ocean's waters. **742**

pellicle Layer of proteins that gives shape to many unwalled, single-celled protists. **313**

penis Male organ of intercourse; also functions in urination. **632**

peptide bond A bond between the amine group of one amino acid and the carboxyl group of another. Joins amino acids in proteins. **44**

per capita growth rate For some interval, the added number of individuals divided by the initial population size. **676**

perception The meaning a brain derives from a sensation. **490**

periodic table Tabular arrangement of the known elements by atomic number. **24**

peripheral nervous system Nerves that extend through the body and carry signals to and from the central nervous system. **469**

peristalsis Wavelike smooth muscle contractions that propel food through the digestive tract. **599**

peritubular capillaries Capillaries that surround and exchange substances with a kidney tubule. **619**

permafrost Continually frozen soil layer that lies beneath arctic tundra and prevents water from draining. **737**

peroxisome Enzyme-filled vesicle that breaks down amino acids, fatty acids, and toxic substances. **63**

pH A measure of the number of hydrogen ions in a fluid. **32**

phagocytosis "Cell eating"; an endocytic pathway by which a cell engulfs particles such as microbes or cellular debris. **88**

pharynx Tube connecting mouth and digestive tract; in vertebrates it is the throat. **585**

phenotype An individual's observable traits. **191**

pheromone Intraspecific chemical communication signal. **496, 657**

phloem Plant vascular tissue that distributes sugars through sieve tubes. **326, 401**

phospholipid A lipid with a phosphate group in its hydrophilic head, and two nonpolar fatty acid tails; main constituent of eukaryotic cell membranes. **43**

phosphorus cycle Movement of phosphorus among Earth's rocks and waters, and into and out of food webs. **719**

phosphorylation Addition of a phosphate group to a molecule; occurs by the transfer of a phosphate group from a donor molecule such as ATP. **79**

photoautotroph Photosynthetic autotroph. **108**

photolysis Process by which light energy breaks down a molecule. **98**

photoperiodism Biological response to seasonal changes in the relative lengths of day and night. **442**

photoreceptor Sensory receptor that responds to light. **490**

photorespiration In plants, reaction in which rubisco attaches oxygen instead of carbon dioxide to ribulose bisphosphate. Reduces sugar production by photosynthesis because it diverts carbon atoms from the Calvin–Benson cycle. **102**

photosynthesis Metabolic pathway by which most autotrophs capture light energy and use it to make sugars from CO_2 and water. **6, 93**

photosystem Cluster of pigments and proteins that converts light energy to chemical energy in photosynthesis. **98**

phototropism Change in the direction of cell movement or growth in response to a light source. **440**

phylogeny Evolutionary history of a species or group of species. **278**

phytochrome A light-sensitive pigment that helps set plant circadian rhythms based on length of night. **442**

pigment An organic molecule that can absorb light of certain wavelengths. **94**

pilus Protein filament that projects from the surface of some bacterial cells. **58, 302**

pineal gland Endocrine gland deep inside the brain that secretes melatonin when the retina is not stimulated by light. **516**

pioneer species Species that can colonize a new habitat. **700**

pituitary gland Pea-sized endocrine gland in the forebrain that interacts closely with the adjacent hypothalamus. **508**

placenta Of placental mammals, organ that forms during pregnancy and allows diffusion of substances between the maternal and embryonic bloodstreams. **631**

placental mammal Mammal in which a mother and her embryo exchange materials by means of an organ called the placenta. **387**

placozoan The simplest modern animal known, with an asymmetrical flat body, four types of cells, and a small genome. **356**

planarian Free-living freshwater flatworm. **360**

plankton Community of tiny drifting or swimming organisms. **314**

plant A multicelled, typically photosynthetic autotroph. **8**

plaque On teeth, a thick biofilm composed of bacteria, their extracellular products, and saliva proteins. **560**

plasma Fluid portion of blood. **450, 544**

plasma membrane A cell's outermost membrane. **52**

plasmid Of many bacteria and archaeans, a small ring of nonchromosomal DNA replicated independently of the chromosome. **58, 220, 303**

plasmodesmata Cell junctions that connect the cytoplasm of adjacent plant cells. **69**

plasmodial slime mold Protist that feeds as a multinucleated mass; forms a spore-bearing structure when enviromental conditions become unfavorable. **320**

plastid An organelle that functions in photosynthesis or storage; e.g., chloroplast, amyloplast. **65**

plate tectonics Theory that Earth's outer layer of rock is cracked into plates, the slow movement of which rafts continents to new locations over geologic time. **249**

platelet Cell fragment that helps blood clot. **545**

pleiotropic Refers to a gene whose product influences multiple traits. **197**

plot sampling Method of estimating population size of organisms that do not move much by making counts in small plots, and extrapolating from this to the number in the larger area. **674**

polarity Any separation of charge into distinct positive and negative regions. **29**

pollen grain Male gametophyte of a seed plant. **327**

pollen sac Of seed plants, reproductive structure in which sperm-bearing gametophytes (pollen grains) develop. **332**

pollination Arrival of a pollen grain on the egg-bearing part of a seed plant. **332, 432**

pollinator Animal that moves pollen, thus facilitating pollination. **336, 429**

pollutant A substance that is released into the environment by human activities and interferes with the function of organisms that evolved in the absence of the substance or with lower levels. **751**

polymer Molecule that consists of multiple monomers. **39**

polymerase chain reaction (PCR) Method that rapidly generates many copies of a specific section of DNA. **222**

polypeptide Chain of amino acids linked by peptide bonds. **44**

polyploid Having three or more of each type of chromosome characteristic of the species. **212**

pons Hindbrain region between medulla oblongata and midbrain; helps control breathing. **482**

population A group of organisms of the same species who live in a specific location and breed with one another more often than they breed with members of other populations. **5, 258, 673**

population density Number of individuals per unit area. **674**

population distribution Where individuals are clumped, uniformly dispersed, or randomly dispersed in an area. **674**

population size Total number of individuals in a population. **674**

positive feedback A response intensifies the conditions that caused its occurrence. **472**

potential energy Stored energy. **77**

predation One species captures, kills, and eats another. **344, 696**

prediction Statement, based on a hypothesis, about a condition that should exist if the hypothesis is correct. **12**

pressure flow theory Explanation of how flow of fluid through phloem is driven by differences in pressure and sugar concentration between a source region and a sink region. **425**

primary growth Plant growth from apical meristems in root and shoot tips. **399**

primary motor cortex Region of frontal lobe that controls voluntary movement. **484**

primary producer In an ecosystem, an organism that captures energy from an inorganic source and stores it as biomass; first trophic level. **710**

primary production The rate at which an ecosystem's producers capture and store energy. **710**

primary succession A new community becomes established in an area where there was no soil. **700**

primary wall The first cell wall of young plant cells. **68**

primate Mammal having grasping hands with nails; includes prosimians, monkeys, apes, and hominids such as humans. **388**

primer Short, single strand of DNA designed to hybridize with a particular nucleotide sequence. **223**

prion Infectious protein. **46**

probability The chance that a particular outcome of an event will occur. **17**

probe Short fragment of DNA labeled with a tracer; designed to hybridize with a nucleotide sequence of interest. **222**

producer Organism that makes its own food using energy and simple raw materials from the environment; an autotroph. **6**

product A molecule that remains at the end of a reaction. **78**

progesterone Hormone secreted by ovaries; prepares the uterus for pregnancy. **634**

promoter In DNA, a sequence to which RNA polymerase binds. **140**

prophase Stage of mitosis in which chromosomes condense and become attached to a newly forming spindle. **166**

protein Organic compound that consists of one or more chains of amino acids (polypeptides). **44**

proteobacteria Largest bacterial lineage. **304**

protist General term for a eukaryote that is not a fungus, animal, or plant. **8, 311**

protocell Membranous sac that contains interacting organic molecules; hypothesized to have formed prior to the earliest life forms. **286**

proton Positively charged subatomic particle that occurs in the nucleus of all atoms. **24**

proto-oncogene Gene that can become an oncogene. **170**

protostomes Lineage of bilateral animals in which the first opening on the embryo surface develops into a mouth. **355**

proximal tubule Portion of kidney tubule that receives filtrate from Bowman's capsule. **619**

pseudocoelom Unlined body cavity. **355**

puberty Period when human reproductive organs mature and begin to function. **632**

pulmonary artery Vessel carrying blood from the heart to a lung. **542**

pulmonary circuit Circuit through which blood flows from the heart to the lungs and back. **539**

pulmonary vein Vessel carrying blood from a lung to the heart. **542**

pulse Brief stretching of artery walls that occurs when ventricles contract. **546**

Punnett square Diagram used to predict the genetic and phenotypic outcome of a cross. **192**

pupil Adjustable opening that allows light into a camera eye. **492**

pyruvate Three-carbon organic compound that is the end product of glycolysis. **109**

radial symmetry Having parts arranged around a central axis, like spokes around a wheel. **354**

radioactive decay Process by which atoms of a radioisotope emit energy and/or subatomic particles when their nucleus spontaneously disintegrates. **25**

radioisotope Isotope with an unstable nucleus. **25**

radiolarian Heterotrophic single-celled protist with a porous silica shell and long cytoplasmic extensions. **314**

radiometric dating Method of estimating the age of a rock or fossil by measuring the content and proportions of a radioisotope and its daughter elements. **245**

rain shadow Dry region downwind of a coastal mountain range. **726**

ray-finned fish Fish with fins supported by thin rays derived from skin; member of the most diverse lineage of fishes. **381**

reabsorption See tubular reabsorption.

reactant Molecule that enters a reaction. **78**

reaction Process of chemical change. **78**

receptor protein Plasma membrane protein that binds to a particular substance outside of the cell. **57**

recessive Refers to an allele with an effect that is masked by a dominant allele on the homologous chromosome. **191**

recognition protein Plasma membrane protein that identifies a cell as belonging to self (one's own body); e.g., MHC marker. **56**

recombinant DNA A DNA molecule that contains genetic material from more than one organism. **220**

rectum Region where feces are stored prior to excretion. **599**

red alga Photosynthetic protist that deposits cellulose in its cell wall, stores sugars as starch, and has chloroplasts containing chlorophyll *a* and red pigments (phycobilins). **318**

red blood cell Hemoglobin-filled blood cell that transports oxygen. **544**

red marrow Bone marrow that makes blood cells. **524**

redox reaction Oxidation–reduction reaction in which one molecule accepts electrons (it becomes reduced) from another molecule (which becomes oxidized). Also called electron transfer. **83**

reflex Automatic response that occurs without conscious thought or learning. **480**

releasing hormone Hormone that is secreted by one endocrine gland and stimulates secretion by another. **508**

replacement fertility rate Average number of children women of a population must bear to replace themselves with a daughter of reproductive age. **685**

replacement model Model for origin of *H. sapiens*; humans evolved in Africa, then migrated to different regions and replaced the other hominids that lived there. **392**

repressor Regulatory protein that blocks transcription. **152**

reproduction Processes by which parents produce offspring. **7**

reproductive base Of a population, all individuals who are of reproductive age or younger. **675**

reproductive cloning Technology that produces genetically identical individuals; e.g., SCNT. **132**

reproductive isolation Absence of gene flow between populations; always part of speciation. **270**

reptile Amniote subgroup that includes lizards, snakes, turtles, crocodilians, and birds. **384**

resource partitioning Species become adapted in different ways to access different portions of a limited resource; allows species with similar needs to coexist. **695**

respiration Physiological process by which an animal body supplies cells with oxygen and disposes of their waste carbon dioxide. **580**

respiratory cycle One inhalation and one exhalation. **586**

respiratory membrane Membrane consisting of alveolar epithelium, capillary endothelium, and their fused basement membranes; site of gas exchange in the lungs. **588**

respiratory protein A protein that reversibly binds oxygen when the oxygen concentration is high and releases it when oxygen concentration is low. Hemoglobin is an example. **580**

respiratory surface Moist surface across which gases are exchanged between animal cells and the external environment. **580**

resting potential Membrane potential of a neuron at rest. **471**

restriction enzyme Type of enzyme that cuts specific nucleotide sequences in DNA. **220**

retina Eye layer that contains photoreceptors. **492**

reverse transcriptase A viral enzyme that uses mRNA as a template to make a strand of cDNA. **221**

rhizoid Threadlike structure that anchors a bryophyte. **329**

rhizome Stem that grows horizontally along or under the ground. **330**

ribosomal RNA (rRNA) A type of RNA that is a component of ribosomes. **138**

ribosome Organelle of protein synthesis. **58**

ribozyme RNA that functions as an enzyme. **287**

RNA Ribonucleic acid; e.g., mRNA, rRNA, tRNA. **47**

RNA polymerase Enzyme that carries out transcription. **140**

RNA world Hypothetical early interval when RNA served as the genetic information. **287**

rod cell Photoreceptor that is active in dim light; provides coarse perception of image and detects motion. **494**

root hairs Hairlike, absorptive extensions of a young cell of root epidermis. **407**

root nodules Swellings of some plant roots that contain nitrogen-fixing bacteria. **418**

roundworm Unsegmented worm with a pseudocoel and a cuticle that is molted as the animal grows. **366**

r-selection Individuals who produce the maximum number offspring as quickly as possible have a selective advantage; occurs when population density is low and resources are abundant. **681**

rubisco Ribulose bisphosphate carboxylase. Carbon-fixing enzyme of the Calvin–Benson cycle. **101**

ruminant Hoofed mammal with a multiple-chamber stomach that adapts it to a cellulose-rich diet. **597**

runoff Water that flows over soil into streams. **714**

sac fungi Most diverse fungal group; sexual reproduction produces ascospores inside a saclike structure (an ascus). **346**

salivary gland Exocrine gland that secretes saliva into the mouth. **599**

salt Compound that releases ions other than H^+ and OH^- when it dissolves in water. **30**

sampling error Difference between results derived from testing an entire group of events or individuals, and results derived from testing a subset of the group. **16**

saprobe Organism that feeds on wastes and remains. **344**

sapwood Functional secondary xylem between the vascular cambium and heartwood in an older woody stem or root. **409**

sarcomere Contractile unit of skeletal and cardiac muscle. **528**

sarcoplasmic reticulum Specialized endoplasmic reticulum in muscle cells; stores and releases calcium ions. **530**

science Systematic study of the observable world. **12**

scientific method Systematically making, testing, and evaluating hypotheses. **13**

scientific theory Hypothesis that has not been disproven after many years of rigorous testing. **18**

sclerenchyma Simple plant tissue that is dead at maturity; its lignin-reinforced cell walls structurally support plant parts. **400**

SCNT See somatic cell nuclear transfer.

seamount An undersea mountain. **742**

secondary growth Thickening of older stems and roots; originates at lateral meristems. **399**

secondary metabolite Chemical that has no known role in an organism's normal metabolism; often deters predation. **339**

secondary succession A new community develops in a site where the soil that supported an old community remains. **700**

secondary wall Lignin-reinforced wall that forms inside the primary wall of a plant cell. **68**

second law of thermodynamics Energy tends to disperse spontaneously. **76**

second messenger Molecule that forms inside a cell when a hormone binds at cell surface; sets in motion reactions that alter activity inside the cell. **506**

sedimentary cycle Biochemical cycle in which the atmosphere plays little role and rocks are the major reservoir. **719**

seed Embryo sporophyte of a seed plant packaged with nutritive tissue inside a protective coat. **327, 434**

seedless vascular plant Plant such as a fern or horsetail that has vascular tissue and disperses by producing spores. **330**

segmentation Having a body composed of similar units that repeat along its length. **355**

selfish herd Temporary group that forms when individuals cluster to minimize their individual risk of predation. **666**

semen Sperm mixed with secretions from seminal vesicles and the prostate gland. **633**

seminiferous tubules In testes, tiny tubes where sperm form. **633**

senescence Phase in a life cycle from maturity until death. **444**

sensation Detection of a stimulus. **490**

sensory adaptation Slowing or cessation of a sensory receptor response to an ongoing stimulus. **490**

sensory neuron Neuron that responds to a specific internal or external stimulus and signals another neuron. **468**

sensory receptor Cell or cell component that detects a specific stimulus. **462**

sex chromosome Member of a pair of chromosomes that differs between males and females. **125**

sexual dimorphism Having distinct male and female phenotypes. **664**

sexual reproduction Reproductive mode by which offspring arise from two parents and inherit genes from both. **175, 630**

sexual selection Mode of natural selection in which some individuals outreproduce others of a population because they are better at securing mates. **266**

shell model Model of electron distribution in an atom. **26**

shivering response In response to cold, rhythmic muscle contractions generate metabolic heat. **625**

short tandem repeats In chromosomal DNA, sequences of 4 or 5 bases repeated multiple times in series. **227**

sieve tube Conducting tube of phloem. **401**

sinoatrial (SA) node Cardiac pacemaker; group of heart cells that spontaneously emits rhythmic signals that cause contraction. **543**

sister chromatid One of the two attached DNA molecules of a duplicated eukaryotic chromosome. **124**

sister groups The two lineages that emerge from a node on a cladogram. **278**

skeletal muscle fiber Multinucleated skeletal muscle cell. **528**

skeletal muscle tissue Muscle that pulls on bones to move body parts; under voluntary control. **456**

sliding-filament model Explanation of how interactions among actin and myosin filaments shorten a sarcomere and bring about muscle contraction. **528**

small intestine Longest portion of the digestive tract; site of most digestion and absorption. **599**

smooth muscle tissue Muscle that lines blood vessels and forms the wall of hollow organs. **456**

soil Mixture of various mineral particles and humus. **416**

soil erosion Loss of soil under the force of wind and water. **417**

soil water Water between soil particles. **714**

solar tracking The movement of a plant's parts in response to the sun's changing angle through the day. **441**

solute A dissolved substance. **30**

solvent Liquid that can dissolve other substances. **30**

somatic Relating to the body. **176**

somatic cell nuclear transfer (SCNT) Method of reproductive cloning in which genetic material is transferred from an adult somatic cell into an unfertilized, enucleated egg. **132**

somatic nervous system Set of nerves that control skeletal muscle and relay signals from joints and skin. **478**

somatic sensations Sensations such as touch and pain that arise when sensory neurons in skin, muscle, or joints are activated. **491**

sorus Cluster of spore-producing capsules on a fern leaf. **330**

speciation One of several processes by which new species arise. **270**

species A type of organism. Designated by genus name together with specific epithet. **10**

species richness Of an area, number of species. **704**

specific epithet Second part of a species name. **10**

sperm Mature male gamete. **182**

sphincter Ring of muscle that controls passage through a tubular organ or body opening. **600**

spinal cord Portion of central nervous system that connects peripheral nerves with the brain. **480**

spindle Dynamically assembled and disassembled network of microtubules that moves chromosomes during nuclear division. **166**

spirochetes Bacteria that resemble a stretched-out spring. **305**

spleen Large lymphoid organ that filters blood. **553**

sponge Aquatic invertebrate that has no tissues or organs and filters food from the water. **357**

spongy bone Lightweight bone with many internal spaces; contains red marrow. **524**

sporophyte Diploid, spore-producing body in the life cycle of land plants and some algae. **182, 318**

stabilizing selection Mode of natural selection in which intermediate forms of a trait are favored over extreme forms. **264**

stamen Floral reproductive structure that produces male gametophytes; in most plants it consists of a pollen-producing anther on the tip of a filament. **336, 430**

stasis Evolutionary pattern in which a lineage persists with little or no change over evolutionary time. **276**

statistically significant Refers to a result that is statistically unlikely to have occurred by chance alone. **17**

stem cell Cell that can divide to produce more stem cells or differentiate into specialized cell types. **449**

steroid Type of lipid with four carbon rings and no fatty acid tails. **43**

steroid hormone Hormone such as testosterone that is derived from cholesterol. **506**

stimulus Form of energy that a sensory receptor detects. **490, 658**

stoma Opening across a plant's cuticle and epidermis; can be opened for gas exchange or closed to prevent water loss. **326**

stomach Muscular organ that secretes gastric fluid and mixes food with it. **599**

stomata Closeable gaps defined by guard cells on plant surfaces; when open, they allow water vapor and gases to diffuse across the epidermis. **102**

stroma Semifluid matrix between the thylakoid membrane and the two outer membranes of a chloroplast. **97**

stromatolite Dome-shaped structures composed of layers of bacterial cells and sediments. **288**

substrate A molecule that is specifically acted upon by an enzyme. **80**

substrate-level phosphorylation A reaction that transfers a phosphate group from a substrate directly to ADP, thus forming ATP. **110**

superior vena cava Vein that delivers blood from the upper body to the heart. **542**

surface-to-volume ratio A relationship in which the volume of an object increases with the cube of the diameter, but the surface area increases with the square. **52**

survivorship curve Graph showing the decline in numbers of a cohort over time. **680**

symbiosis One species lives in or on another in a commensal, mutualistic, or parasitic relationship. **692**

sympathetic neurons Neurons of the autonomic system that prepare the body for danger or excitement. **479**

sympatric speciation Pattern in which speciation occurs in the absence of a physical barrier to gene flow. **274**

synapse Region where a neuron's axon terminals transmit chemical signals to another cell. **474**

synaptic integration The summation of excitatory and inhibitory signals by a postsynaptic cell. **475**

systemic acquired resistance In plants, a long-term, systemic resistance to pathogens. **444**

systemic circuit Circuit through which blood flows from the heart to the body tissues and back. **539**

systole Contractile phase of the cardiac cycle. **542**

systolic pressure Blood pressure when ventricles are contracting. **547**

T cell T lymphocyte. Lymphocyte bearing T cell receptors; central to adaptive immunity. **559**

T cell receptor (TCR) Antigen receptor on the surface of a T cell. **564**

taiga See boreal forest.

taproot system In eudicots, a primary root and all of its lateral branchings. **406**

taste receptors Chemoreceptors involved in taste. **496**

taxon Linnaean category; a grouping of organisms. Each consists of a group of the next lower taxon. **10**

taxonomy The science of naming and classifying species. **10**

telophase Stage of mitosis during which chromosomes arrive at the spindle poles and decondense, and new nuclei form. **166**

temperate deciduous forest Northern Hemisphere biome in which the main plants are broadleaf trees that lose their leaves in fall and become dormant during cold winters. **734**

temperature Measure of molecular motion. **31**

tendon Strap of dense connective tissue that connects a skeletal muscle to bone. **526**

territory Region that an animal or animals occupy and defend against competitors. **664**

testcross Method of determining genotype in which an individual of unknown genotype is crossed with one that is known to be homozygous recessive. **192**

testes Male gonads; produce sperm. **632**

testosterone Main hormone produced by testes; causes sperm production and development of male secondary sexual traits. **632**

tetrapod Vertebrate with four legs, or a descendant thereof; an amphibian, reptile (including birds), or mammal. **379**

thalamus Forebrain region that relays signal to the cerebrum. **483**

theory of inclusive fitness Genes associated with altruism can be advantageous if the expense of this behavior to the altruist is outweighed by the reproductive success of relatives. **668**

theory of uniformity Idea that gradual repetitive processes occurring over long time spans shaped Earth's surface. **240**

therapeutic cloning The use of SCNT to produce human embryos for research purposes. **133**

thermoreceptor Temperature-sensitive sensory receptor. **490**

thigmotropism Directional growth of a plant in response to contact with a solid object. **441**

threatened species Species likely to become endangered in the near future. **748**

threshold potential Neuron membrane potential at which gated sodium channels open, causing an action potential to occur. **472**

thylakoid membrane A chloroplast's highly folded inner membrane system; molecules embedded in it carry out the light-dependent reactions of photosynthesis. **97**

thymus Endocrine gland beneath the breastbone; secretes hormones that encourage maturation of T lymphocytes (T cells). **516**

thyroid gland Endocrine gland at the base of the neck; produces thyroid hormone that increases metabolism. **511**

tight junctions Arrays of fibrous proteins that join the plasma membranes of adjacent cells in epithelial tissue; collectively, they prevent fluids from leaking between the cells. **69**

tissue In multicelled organisms, specialized cells organized in a pattern that allows them to perform a collective function. **5**

tissue culture propagation Laboratory method in which body cells are induced to divide and form an embryo. **435**

topsoil Uppermost soil layer; contains the most nutrients for plant growth. **417**

total fertility rate Average number of children the women of a population bear over the course of a lifetime. **685**

toxin Chemical that is made by one organism and harms another. **311**

tracer A molecule that has been labeled with a detectable substance. **25**

trachea Airway to the lungs; windpipe. **585**

tracheal system Of insects and some other land arthropods, tubes that convey gases between the body surface and internal tissues. **581**

tracheid Tapered cell that forms water-conducting tubes in xylem of vascular plants. **401**

transcription Process by which an RNA is assembled from nucleotides using the base sequence of a gene as a template. **138**

transcription factor Regulatory protein that influences transcription; e.g., an activator or repressor. **152**

transfer RNA (tRNA) A type of RNA that delivers amino acids to a ribosome during translation. **138**

transgenic Refers to a genetically modified organism that carries a gene from a different species. **228**

translation Process by which a polypeptide chain is assembled from amino acids in the order specified by an mRNA. **139**

translocation In genetics, chromosomal mutation in which a broken piece gets reattached in the wrong location. **210** In a vascular plant, the movement of organic compounds through phloem. **424**

transpiration The evaporation of water from plant parts. **420**

transport protein Protein that passively or actively assists specific ions or molecules across a membrane. **57**

transposable element Segment of DNA that can spontaneously move to a new location in a chromosome. **147**

triglyceride A fat with three fatty acid tails. **42**

trophic level Position of an organism in a food chain. **711**

tropical rain forest Highly productive, species-rich biome in which year-round rains and warmth support continuous growth of evergreen broadleaf trees. **734**

trypanosome Parasitic flagellate with a single mitochondrion and a membrane-encased flagellum. **313**

tubular reabsorption Substances move from the filtrate inside a kidney tubule into the peritubular capillaries. **620**

tubular secretion Substances move out of peritubular capillaries and into the filtrate in kidney tubules. **620**

tumor A neoplasm that forms a lump. **170**

tunicate Invertebrate chordate that loses its defining chordate traits during the transition to adulthood. **378**

turgor Pressure that a fluid exerts against a wall, membrane, or other structure that contains it. **85**

urea Main nitrogen-containing waste in urine of mammals. **617**

ureter Tube that carries urine from a kidney to the bladder. **618**

urethra Tube through which urine from the bladder flows out of the body. **618**

uric acid Main nitrogen-containing compound in the excretions of insects, as well as birds and other reptiles. **617**

urinary bladder Hollow, muscular organ that stores urine. **618**

urine Mix of water and soluble wastes formed and excreted by the vertebrate urinary system. **617**

uterus Muscular chamber where offspring develop; womb. **634**

vaccine A preparation introduced into the body in order to elicit immunity to a specific antigen. **572**

vacuole A fluid-filled organelle that isolates or disposes of waste, debris, or toxic materials. **63**

vagina Female organ of intercourse and birth canal. **634**

variable In an experiment, a characteristic or event that differs among individuals or over time. **13**

vascular bundle Multistranded, sheathed cord of primary xylem and phloem in the stem or leaf of a plant. **402**

vascular cambium Of vascular plants, a ring of meristematic tissue that produces secondary xylem and phloem. **408**

vascular cylinder Of vascular plant roots, sheathed, cylindrical array of primary xylem and phloem. **407**

vascular plant Plant with xylem and phloem; a seedless vascular plant, gymnosperm, or angiosperm. **326**

vascular tissue Internal pipelines of xylem and phloem in the body of a vascular plant. **326**

vascular tissue system Tissue system that distributes water and nutrients through the body of a vascular plant. **398**

vas deferens One of a pair of long ducts that carry mature sperm toward the ejaculatory duct. **632**

vasoconstriction Narrowing of a blood vessel when smooth muscle that rings it contracts. **546**

vasodilation Widening of a blood vessel when smooth muscle that rings it relaxes. **546**

vector Animal that carries a pathogen from one host to the next. **300**

vegetative reproduction Growth of new roots and shoots from extensions or fragments of a parent plant; form of asexual reproduction in plants. **435**

vein Of vascular plants, a vascular bundle in a stem or leaf. **405** Of circulatory systems, a large-diameter vessel that returns blood to the heart. **540**

ventricle Heart chamber that pumps blood into arteries. **542**

venule Small-diameter vessel that carries blood from capillaries to a vein. **540**

vernalization Stimulation of flowering in spring by low temperature in winter. **443**

vertebrae Bones of the backbone, or vertebral column. **522**

vertebral column Backbone. **522**

vertebrate Animal with a backbone. **353, 378**

vesicle Small, membrane-enclosed, saclike organelle; different kinds store, transport, or degrade their contents. **62**

vessel member Cell that forms water-conducting tubes in xylem of vascular plants. **401**

vestibular apparatus System of fluid-filled sacs and canals in the inner ear that contains the organs of equilibrium. **497**

villi Multicelled projections at the surface of each fold in the small intestine. **601**

viroid Small noncoding RNA that can infect and cause disease in plants. **301**

virus Noncellular, infectious particle of protein and nucleic acid; replicates only in a host cell. **298**

visceral sensations Sensations that arise when sensory neurons associated with organs inside body cavities are activated. **491**

visual accommodation Process of making adjustments to lens shape so light from an object falls on the retina. **493**

vital capacity Amount of air moved in and out of lungs with forced inhalation and exhalation. **586**

vitamin Organic substance required in small amounts for normal metabolism. **608**

vomeronasal organ Pheromone-detecting organ of vertebrates. **496**

water cycle Movement of water among Earth's oceans, atmosphere, and freshwater reservoirs on land. **714**

water mold Heterotrophic protist that grows as nutrient-absorbing filaments. **317**

watershed Land area that drains into a particular stream or river. **714**

water–vascular system Of echinoderms, a system of fluid-filled tubes and tube feet that function in locomotion. **373**

wavelength Distance between the crests of two successive waves of light. **94**

wax Water-repellent mixture of lipids with long fatty acid tails bonded to long-chain alcohols or carbon rings. **43**

white blood cell Of vertebrates, blood cell with a role in housekeeping and defense; leukocyte. **545**

white matter Tissue of brain and spinal cord consisting of myelinated axons. **480**

wood Accumulated secondary xylem. **408**

X chromosome inactivation Shutdown of one of the two X chromosomes in the cells of female mammals. **156**

xenotransplantation Transplantation of an organ from one species into another. **231**

xylem Complex vascular tissue of plants; its tracheids and vessel members distribute water and mineral ions. **326, 401**

yellow marrow Bone marrow that is mostly fat; fills cavity in most long bones. **524**

yolk Nutritious material in many animal eggs. **631**

zero population growth Interval in which births equal deaths. **676**

zygote Cell formed by fusion of two gametes; the first cell of a new individual. **177**

zygote fungus Fungus that forms a zygospore during sexual reproduction. **345**

Art Credits & Acknowledgments

TABLE OF CONTENTS **Page vi** from top, © Raymond Gehman/ Corbis; © Bill Beatty/ Visuals Unlimited. **Page vii** left, © Dylan T. Burnette and Paul Forscher; middle left, Hemoglobin models: PDB ID: 1GZX; Paoli, M., Liddington, R., Tame, J., Wilkinson, A., Dodson, G.; Crystal structure of T state hemoglobin with oxygen bound at all four haems. *J.Mol.Biol.*, v256, pp. 775–792, 1996; middle right, Kristian Peters, http://commons. wikimedia.org/wiki/User:Fabelfroh; right, Lilly M/ http://commons.wikimedia.org. **Page viii** middle bottom, © Maria Samsonova and John Reinitz; bottom, Michael Clayton/ University of Wisconsin, Department of Botany. **Page ix** from left, Francis Leroy, Biocosmos/ Science Photo Library/ Photo Researchers, Inc.; © Michael Stuckey/ Comstock, Inc.; Photo courtesy of The Progeria Research Foundation; Photo courtesy of MU Extension and Agricultural Information. **Page x** from top, USGS; Courtesy of Hopi Hoekstra, University of California, San Diego; © Jack Jeffrey Photography. **Page xi** from left, Courtesy of the University of Washington; SciMAT/ Photo Researchers, Inc.; Dr. Richard Feldmann/ National Cancer Institute; © Kim Taylor/ Bruce Coleman, Inc./ Photoshot; Photo USDA. **Page xii** from top, Eye of Science/ Photo Researchers, Inc.; © Wim van Egmond/ Visuals Unlimited; CDC/ Piotr Naskrecki; © Kenneth Garrett/ National Geographic Image Collection. **Page xiii** from left, © Kevin Schafer/ Tom Stack & Associates; Courtesy of Dr. Thomas L. Rost; © J. C. Revy/ ISM/ Phototake. **Page xiv** from top, © David Goodin; © Richard H. Gross; Biophoto Associates/ Photo Researchers, Inc.; © Michael Shore Photography. **Page xv** from left, From Neuro Via Clinicall Research Program, Minneapolis VA Medical Center; Courtesy of Dr. Bryan Jones, University of Utah School of Medicine; Scott Camazine/ Photo Researchers, Inc.; © David Ryan/ SuperStock. **Page xvi** from top, Professor P. Motta/ Department of Anatomy/ La Sapienza, Rome/ SPL/ Photo Researchers, Inc.; © Lester V. Bergman/ Corbis. **Page xvii** from middle left, CDC/ Dr. Joel D. Meyer; © Gunter Ziesler/ Bruce Coleman, Inc.; © Microslide courtesy Mark Nielsen, University of Utah. **Page xviii** from top, Tom McHugh/ Photo Researchers, Inc.; © Rodger Klein/ Peter Arnold, Inc.; Carolina Biological Supply Company; © Lennart Nilsson/ Bonnierforlagen AB. **Page xix** from left, © Alexander Wild; © Jacques Langevin/ Corbis Sygma; E.R. Degginger/ Photo Researchers, Inc. **Page xx** from top, © D. A. Rintoul; © George H. Huey/ Corbis; © John Easley/ www.johneasley.com. **Page xxi** from left, © George M. Sutton/ Cornell Lab of Ornithology; © Gary Head; Image © Lee Prince, Used under license from Shutterstock.com.

PREFACE **Page xxii** Clockwise from top left, Courtesy East Carolina University; © Steven D. Johnson; Photo courtesy of The Progeria Research Foundation.

CHAPTER 1 **Key Concepts** from left, California Poppy, © 2009, Christine M. Welter; Dr. Marina Davila Ross, University of Portsmouth, Courtesy of Allen W. H. Bé and David A. Caron; © Roger W. Winstead, NC State University; © Adrian Vallin. **1.1** page 2, © Steve Richards; page 3, © Bruce Beehler/ Conservation International. **1.3** (3–4) © Umberto Salvagnin, www .flickr.com/photos/kaibara; (5) California Poppy, © 2009, Christine M. Welter; (6) Lady Bird Johnson Wildflower Center; (7) Michael Szoenyi/ Photo Researchers, Inc.; (8) Photographers Choice RF/ SuperStock; (9) © Sergei Krupnov, www.flickr.com/photos/7969319@ N03; (10) © Mark Koberg Photography; (11) NASA. **1.4** above, © Victoria Pinder, http://www.flickr.com/ photos/vixstarplus. **1.5** Dr. Marina Davila Ross, University of Portsmouth. **1.6** (a) Clockwise from top left,

Dr. Richard Frankel; SPL/ Photo Researchers, Inc.; © Susan Barnes; Tony Brian, David Parker/ SPL/ Photo Researchers, Inc.; www.zahnarzt-stuttgart.com; Sci-MAT/ Photo Researchers, Inc.; (b) left, ArchiMeDes; right, © Dr. Harald Huber, Dr. Michael Hohn, Prof. Dr. K.O.Stetter, University of Regensburg, Germany; (c) Protists, left, © Lewis Trusty/ Animals Animals; right, clockwise from top left, M I Walker/ Photo Researchers, Inc.; Courtesy of Allen W. H. Bé and David A. Caron; © Emiliania Huxleyi photograph, Vita Pariente, scanning electron micrograph taken on a Jeol T330A instrument at Texas A&M University Electron Microscopy center; Oliver Meckes/Photo Researchers, Inc.; © Carolina Biological Supply Company; Plants, left © John Lotter Gurling / Tom Stack & Associates; right © Edward S. Ross; Fungi, left, © Robert C. Simpson/ Nature Stock; right, © Edward S. Ross; Animals, from left, © Martin Zimmerman, Science, 1961, 133:73–79, © AAAS.; Thomas Eisner, Cornell University; © Tom & Pat Leeson, Ardea London Ltd.; © Pixtal/ Super-Stock. **1.7** from left, Joaquim Gaspar; Bogdan; Opiola Jerzy; Ravedave; Luc Viatour. **1.9** From Meyer, A., Repeating Patterns of Mimicry. PLoS Biology Vol. 4, No. 10, e341 doi:10.1371/journal.pbio.0040341. **Page 12** left, © Bruce Beehler/ Conservation International; right, Volker Steger/ SPL/ Photo Researchers, Inc.; **1.10** from left, Cape Verde National Institute of Meteorology and Geophysics and the U.S. Geological Survey; © Roger W. Winstead, NC State University; Photo by Scott Bauer, USDA/ARS. **1.11** © Superstock. **1.12** (a) © Matt Rawlings, www.eurobutterflies.com; (b) © Adrian Vallin; (c) © Antje Schulte. **1.13** © Gary Head. **Page 16** right, © Bruce Beehler/ Conservation International. **1.14** © Adrian Vallin. **Page 18** © Raymond Gehman/ Corbis. **1.15** above, Photo courtesy of Dr. Robert Zingg/ Zoo Zurich; below, Courtesy East Carolina University. **Page 20** Section 1.1, © Bruce Beehler/ Conservation International; Section 1.2, California Poppy, © 2009, Christine M. Welter; Section 1.3, © Victoria Pinder, http://www.flickr.com/photos/vixstarplus; Section 1.4, Courtesy of Allen W. H. Bé and David A. Caron; Section 1.5, From Meyer, A., Repeating Patterns of Mimicry. PLoS Biology Vol. 4, No. 10, e341 doi:10.1371/journal.pbio.0040341; Section 1.6, © Bruce Beehler/ Conservation International; Section 1.7, © Adrian Vallin; Section 1.8, © Gary Head; Section 1.9, © Raymond Gehman/ Corbis. **Page 21** Scientific Paper; Adrian Vallin, Sven Jakobsson, Johan Lind and Christer Wiklund, Proc. R. Soc. B (2005 272, 1203, 1207). Used with permission of The Royal Society and the author.

CHAPTER 2 **Key Concepts** from left, © Michael S. Yamashita/ Corbis; © Bill Beatty/ Visuals Unlimited; © Herbert Schnekenburger; W. K. Fletcher/ Photo Researchers, Inc. **2.1** page 22, © Kim Westerskov, Photographer's Choice/ Getty Images; page 23, © Michael Grecco/ Picture Group. **2.4** Brookhaven National Laboratory. **Page 26** © Michael S. Yamashita/ Corbis. **2.7** (a) © Bill Beatty/ Visuals Unlimited. **2.12** © Herbert Schnekenburger. **2.13** left, © Vicki Rosenberg, www .flickr.com/photos/roseofredrock. **2.14** © JupiterImages Corporation. **2.15** W. K. Fletcher/ Photo Researchers, Inc. **Page 33** © Jared C. Benedict. **Page 34** Section 2.1, © Kim Westerskov, Photographer's Choice/ Getty Images; Section 2.3, © Michael S. Yamashita/ Corbis; Section 2.4, © Bill Beatty/ Visuals Unlimited; Section 2.5, © Herbert Schnekenburger; Section 2.6, W. K. Fletcher/ Photo Researchers, Inc.

CHAPTER 3 **Key Concepts** from left, © JupiterImages Corporation; Tim Davis/ Photo Researchers, Inc.; © JupiterImages Corporation. **3.1** page 36 © ThinkStock/ SuperStock. **3.5** left, © JupiterImages

Corporation. **3.6** (c) © JupiterImages Corporation. **3.11** right, Tim Davis/ Photo Researchers, Inc. **Page 43** Kenneth Lorenzen. **3.14** #5 right, © JupiterImages Corporation. **Page 43** This lipoprotein image was made by Amy Shih and John Stone using VMD and is owned by the Theoretical and Computational Biophysics Group, NIH Resource for Macromolecular Modeling and Bioinformatics, at the Beckman Institute, University of Illinois at Urbana-Champaign. Labels added to the original image by book author. **3.15** (a) © Lily Echeverria/ Miami Herald; (b) Sherif Zaki, MD PhD, Wun-Ju Shieh, MD PhD; MPH/ CDC. **Page 47** © JupiterImages Corporation. **Page 48** Section 3.1, 3.3, 3.5 © JupiterImages Corporation; Section 3.6, Sherif Zaki, MD PhD, Wun-Ju Shieh, MD PhD; MPH/ CDC.

CHAPTER 4 **Key Concepts** from left, Astrid Hanns-Frieder michler/ SPL/ Photo Researchers, Inc.; © JupiterImages Corporation; Courtesy of © Roberto Kolter Lab, Harvard Medical School; Martin W. Goldberg, Durham University, UK. **4.1** page 50 © JupiterImages Corporation; page 51, Stephanie Schuller/ Photo Researchers, Inc. **4.2** from left, Astrid Hanns-Frieder michler/SPL/ Photo Researchers, Inc.; Courtesy of Allen W. H. Bé and David A. Caron ; M I Walker/ Photo Researchers, Inc.; © Dr. Dennis Kunkel/ Visuals Unlimited; © Wim van Egmond/ Visuals Unlimited. **4.5** (a) left, © JupiterImages Corporation; (b) right, Geoff Tompkinson/ Science Photo Library/ Photo Researchers, Inc. **4.6** (a–b, d–e) © Jeremy Pickett-Heaps, School of Botany, University of Melbourne; (c) © Prof. Franco Baldi. **4.7** Virus, CDC; Bacteria, Tony Brian, David Parker/ SPL/ Photo Researchers, Inc.; Eukaryotic cells, Jeremy Pickett-Heaps, School of Botany, University of Melbourne; Louse, Edward S. Ross; Ladybug, Image © Dole, Used under license from Shutterstock.com; Goldfish, Image © Ultrashock, Used under license from Shutterstock. com; Butterfly, From Meyer, A., Repeating Patterns of Mimicry. PLoS Biology Vol. 4, No. 10, e341 doi:10.1371/ journal.pbio.0040341; Guinea pig, Image © Sascha Burkard, Used under license from Shutterstock.com; Dog, Image © Pavel Sazonov, Used under license from Shutterstock.com; Human, Jupiter Images Corporation; Giraffe, © Ingram Publishing, SuperStock; Whale, Jupiter Images Corporation; Tree, Courtesy of © Billie Chandler. **4.10** (a) Rocky Mountain Laboratories, NIAID, NIH; (b) © R. Calentine/ Visuals Unlimited; (c) ArchiMeDes; (d–e) K.O. Stetter & R. Rachel, Univ. Regensburg. **4.12** Courtesy of Roberto Kolter Lab, Harvard Medical School. **4.13** (a) Dr. Gopal Murti/ Photo Researchers, Inc.; (b) M.C. Ledbetter, Brookhaven National Laboratory. **4.14** right, © Kenneth Bart. **4.15** left, © Martin W. Goldberg, Durham University, UK. **4.16** left, © Kenneth Bart; (#1,3) Don W. Fawcett/ Visuals Unlimited; (#4) Micrograph, Gary Grimes. **4.17** left, Keith R. Porter. **4.18** top, Kristian Peters, http://commons.wikimedia. org/wiki/User:Fabelfroh; bottom, Dr. Jeremy Burgess/ SPL/ Photo Researchers, Inc. **4.19** below, © Dylan T. Burnette and Paul Forscher. **4.21** (a) Don W. Fawcett/ Photo Researchers, Inc.; (b) © Mike Abbey/ Visuals Unlimited. **4.22** upper right, Don W. Fawcett/ Photo Researchers, Inc. **4.23** © George S. Ellmore. **4.24** (c) Russell Kightley/ Photo Researchers, Inc. **4.25** © ADVANCELL (Advanced In Vitro Cell Technologies; S.L.) www.advancell.com. **Page 71**, Section 4.1 & 4.2, © Wim van Egmond/ Visuals Unlimited; Section 4.3, © JupiterImages Corporation. **Page 72** Section 4.5, © K.O. Stetter & R. Rachel, Univ. Regensburg; Section 4.6 & 4.7, © Kenneth Bart; Section 4.10, © Dylan T. Burnette and Paul Forscher; Section 4.11, © ADVANCELL (Advanced In Vitro Cell Technologies; S.L.)

www.advancell.com. **4.27** From "Tissue & Cell", Vol. 27, pp.421–427, Courtesy of Bjorn Afzelius, Stockholm University. **Page 73**, Critical Thinking, P.L. Walne and J. H. Arnott, Planta, 77:325–354, 1967.

CHAPTER 5 **Key Concepts** from left, © Martin Barraud/ Stone/ Getty Images; © JupiterImages Corporation; Andrew Lambert Photography/ Science Photo Library/ Photo Researchers, Inc.; Biology Media/ Photo Researchers, Inc. **5.1** page 74, © BananaStock/ SuperStock. **5.2** © Martin Barraud/ Stone/ Getty Images. **5.4** © JupiterImages Corporation. **5.10** Hemoglobin models: PDB ID: 1GZX; Paoli, M., Liddington, R., Tame, J., Wilkinson, A., Dodson, G.; Crystal structure of T state hemoglobin with oxygen bound at all four haems. *J.Mol.Biol.*, v256, pp. 775–792, 1996. **5.12** © Scott McKiernan/ ZUMA Press. **5.13** Perennou Nuridsany / Photo Researchers, Inc. **5.15** Lisa Starr, after David S. Goodsell, RCSB, Protein Data Bank. **5.16** (a) © JupiterImages Corporation. **Page 84** Andrew Lambert Photography/ Science Photo Library/ Photo Researchers, Inc. **5.19** (b–d) Reproduced from *The Journal of Cell Biology*, 1976, vol. 70, pp. 193 by copyright permission of the Rockefeller Univeristy Press. **5.20** PDB files from NYU Scientific Visualization Lab. **5.21** After: David H. MacLennan, William J. Rice, and N. Michael Green, "The Mechanism of $Ca2^{+}$ Transport by Sarco (Endo) plasmic Reticulum $Ca2^{+}$-ATPases." *JBC* Volume 272, Number 46, Issue of November 14, 1997, pp. 28815–28818. **5.24** © R.G.W. Anderson, M.S. Brown and J.L. Goldstein. Cell 10:351 (1977). **5.25** (a, right) Biology Media/ Photo Researchers, Inc. **Page 89** © BananaStock/ SuperStock. **Page 90** Section 5.1, © BananaStock/ SuperStock; Section 5.2, © Martin Barraud/ Stone/ Getty Images; Section 5.5, © JupiterImages Corporation; Section 5.6, Andrew Lambert Photography/ Science Photo Library/ Photo Researchers, Inc.; Section 5.8, Biology Media/ Photo Researchers, Inc.

CHAPTER 6 **Key Concepts** from left, Kristian Peters, http://commons.wikimedia.org/wiki/User:Fabelfroh; © Bill Boch/ FoodPix/ Jupiter Images. **6.1** page 92, Photo by Peggy Greb/ USDA; page 93, © Roger W. Winstead, NC State University. **6.3** above, © Larry West/ FPG/ Getty Images. **6.4** (a) Jason Sonneman. **6.5** above, Kristian Peters, http://commons.wikimedia.org/wiki/User:Fabelfroh. **6.10** (a) Courtesy of John S. Russell, Pioneer High School; (c) © Bill Boch/ FoodPix/ Jupiter Images; (d) Lisa Starr. **6.11** © JupiterImages Corporation. **Page 103** Photo by Peggy Greb/ USDA. **Page 104** Section 6.1, Photo by Peggy Greb/ USDA; Section 6.4, Kristian Peters, http://commons.wikimedia.org/wiki/User:Fabelfroh; Section 6.8 © Bill Boch/ FoodPix/ Jupiter Images. **Page 105** Critical Thinking, © E.R. Degginger.

CHAPTER 7 **Key Concepts** from left, © Dr. Dennis Kunkel/ Visuals Unlimited; © Lois Ellen Frank/ Corbis. **7.1** page 106, Courtesy of © Louise Chalcraft-Frank and FARA; page 107, Professors P. Motta and T Naguro/ SPL/ Photo Researchers, Inc. **7.7** right, SPL/ Photo Researchers, Inc. **7.10** (b–c) © Ben Fink/ Foodpix/ Jupiter Imges; (d) © Dr. Dennis Kunkel/ Visuals Unlimited. **7.11** (b) Courtesy of © William MacDonald, M.D.; (c) © Randy Faris/ Corbis; (d) Lilly M/ http://commons.wikimedia.org. **7.12** left, © Lois Ellen Frank/ Corbis. **Page 119** top, Professors P. Motta and T Naguro/ SPL/ Photo Researchers, Inc.; Section 7.1, Courtesy of © Louise Chalcraft-Frank and FARA. **Page 120** Section 7.6, © Dr. Dennis Kunkel/ Visuals Unlimited; Section 7.7 © Lois Ellen Frank/ Corbis. **7.13** (a) Steve Gschmeissner/ Photo Researchers, Inc.; (b) © Images Paediatr Cardiol.

CHAPTER 8 **Key Concepts** from left, Andrew Syred/ Photo Researchers, Inc., Courtesy of Cyagra, Inc., www.cyagra.com. **8.1** page 122, © James Symington; page 123, Ben Glass, courtesy of © BioArts International. **8.2** inset, Andrew Syred/ Photo Researchers, Inc. **8.3** © University of Washington Department of Pathology. **8.6** (a) below, Eye of Science/ Photo Researchers, Inc. **8.8** C. Barrington Brown, 1968 J. D. Watson; left, A C. Barrington Brown © 1968 J. D. Watson; right, PDB ID: 1BBB; Silva, M. M., Rogers, P. H., Arnone, A.: A third quaternary structure of human hemoglobin A at 1.7-≈ resolution. *J Biol Chem* 267 pp. 17248 (1992). **8.13, 8.14** Courtesy of Cyagra, Inc., www.cyagra.com. **Page 133** Ben Glass, courtesy of © BioArts International. **Page 134** Section 8.1, Ben Glass, courtesy of © BioArts International; Section 8.2, Andrew Syred/ Photo Researchers, Inc.; Section 8.7, Andrew Syred/ Photo Researchers, Inc. **8.15** above, Journal of General Physiology 36(1), September 20, 1952: "Independent Functions of viral Protein and Nucleic Acid in Growth of Bacteriophage;" below, © JupiterImages Corporation.

CHAPTER 9 **Key Concepts** from left, O. L. Miller, © John W. Gofman and Arthur R. Tamplin. From Poisoned Power: The Case Against Nuclear Power Plants Before and After Three Mile Island, Rodale Press, PA, 1979. **9.1** page 136, Vaughan Fleming/ SPL/ Photo Researchers, Inc. **9.5** O. L. Miller. **9.12** inset, © Kiseleva and Donald Fawcett/ Visuals Unlimited. **9.13** (e) Dr. Gopal Murti/ SPL/ Photo Researchers, Inc. **9.14** (a) © John W. Gofman and Arthur R. Tamplin. From Poisoned Power: The Case Against Nuclear Power Plants Before and After Three Mile Island, Rodale Press, PA, 1979. **Page 148** Vaughan Fleming/ SPL/ Photo Researchers, Inc.; Section 9.3, O. L. Miller; Section 9.6, © John W. Gofman and Arthur R. Tamplin. From Poisoned Power: The Case Against Nuclear Power Plants Before and After Three Mile Island, Rodale Press, PA, 1979.

CHAPTER 10 **Key Concepts** from left, From the collection of Jamos Werner and John T. Lis; © Maria Samsonova and John Reinitz; © Jose Luis Riechmann. **10.1** page 150, Courtesy of Robin Shoulla and Young Survival Coalition; page 151, From the archives of www.breastpath.com, courtesy of J.B. Askew, Jr., M.D., P.A. Reprinted with permission, copyright 2004 Breastpath.com. **10.4** From the collection of Jamos Werner and John T. Lis. **10.5** (a–b) David Scharf/ Photo Researchers, Inc.; (c) Eye of Science/ Photo Researchers, Inc.; (d) Courtesy of the Aniridia Foundation International, www.aniridia.net; (e) M. Bloch. **10.6** (a–e) © Maria Samsonova and John Reinitz; (f) © Jim Langeland, Jim Williams, Julie Gates, Kathy Vorwerk, Steve Paddock and Sean Carroll, HHMI, University of Wisconsin-Madison. **10.7** © Dr. William Strauss. **10.8** after Patten, Carlson & others. **10.9** (a) below, Juergen Berger, Max Planck Institute for Developmental Biology, Tuebingen, Germany; (b) © Jose Luis Riechmann. **Page 159** left, © Lowe Worldwide, Inc. as Agent for National Fluid Milk Processor Promotion Board; right, Courtesy of Robin Shoulla and Young Survival Coalition. **Page 160** Section 10.1, From the archives of www.breastpath.com, courtesy of J.B. Askew, Jr., M.D., P.A. Reprinted with permission, copyright 2004 Breastpath.com; Section 10.2, From the collection of Jamos Werner and John T. Lis; Section 10.3, © Maria Samsonova and John Reinitz; Section 10.4, © Dr. William Strauss.

CHAPTER 11 **Key Concepts**, from left © Carolina Biological Supply Company/ Phototake; Michael Clayton/ University of Wisconsin, Department of Botany; From "Expression of the epidermal growth factor receptor (EGFR) and the phosphorylated EGFR in invasive breast carcinomas." http://breast-cancer research.com/content/10/3/R49. **11.1** page 162, Dr. Paul D. Andrews/ University of Dundee; page 163, Courtesy of the family of Henrietta Lacks. **11.4** © Carolina Biological Supply Company/ Phototake. **11.5** left (all), Michael Clayton/ University of Wisconsin, Department of Botany; right (all) Ed Reschke. **11.8** © Phillip B. Carpenter, Department of Biochemistry and Molecular Biology, University of Texas — Houston Medical School. **11.9** From "Expression of the epidermal growth factor receptor (EGFR) and the phosphorylated EGFR in invasive breast carcinomas." http://breast-cancer research.com/content/10/3/R49. **11.11** (a) © Ken Greer/Visuals Unlimited; (b) Biophoto Associates/Science Source/Photo Researchers; (c) James Stevenson/ Photo Researchers, Inc. **Page 171** above, Dr. Paul D. Andrews/ University of Dundee; below, Courtesy of the family of Henrietta Lacks. **Page 172** Section 11.1, Courtesy of the family of Henrietta Lacks; Section 11.2, © Carolina Biological Supply Company/ Phototake; Section 11.3, Michael Clayton/ University of Wisconsin, Department of Botany; Section 11.5, From "Expression of the epidermal growth factor receptor (EGFR) and the phosphorylated EGFR in invasive breast carcinomas." http://breast-cancer research.com/content/10/3/R49; Section 11.6, James Stevenson/ Photo Researchers, Inc. **11.12** Courtesy of © Dr. Thomas Ried, NIH and the American Association for Cancer Research. **Page 173**, Critical Thinking, © micrograph, D. M. Phillips/ Visuals Unlimited.

CHAPTER 12 **Key Concepts** from left, Image courtesy of Carl Zeiss MicroImaging, Thornwood, NY; With thanks to the John Innes Foundation Trustees, computer enhanced by Gary Head; Francis Leroy, Biocosmos/ Science Photo Library/ Photo Researchers, Inc. **12.1** page 174, © JupiterImages Corporation; page 175, Susumu Nishinaga / Photo Researchers, Inc. **12.2** Image courtesy of Carl Zeiss MicroImaging, Thornwood, NY. **12.5** upper, With thanks to the John Innes Foundation Trustees, computer enhanced by Gary Head. **12.8** right, Courtesy of © Billie Chandler. **12.10** Francis Leroy, Biocosmos/ Science Photo Library/ Photo Researchers, Inc. **Page 185** AP/ Wide World Photos. **Page 186** Section 12.1, © JupiterImages Corporation; Section 12.2, Image courtesy of Carl Zeiss MicroImaging, Thornwood, NY; Section 12.3, With thanks to the John Innes Foundation Trustees, computer enhanced by Gary Head; Section 12.5, Francis Leroy, Biocosmos/ Science Photo Library/ Photo Researchers, Inc. **12.12** above, Reprinted from *Current Biology*, Vol 13, (Apr 03), Authors Hunt, Koehler, Susiarjo, Hodges, Ilagan, Voigt, Thomas, Thomas and Hassold, Bisphenol A Exposure Causes Meiotic Aneuploidy in the Female Mouse, pp. 546–553, © 2003 Cell Press. Published by Elsevier Ltd. With permission from Elsevier.

CHAPTER 13 **Key Concepts** from left, © The Moravian Museum, Brno; © Michael Stuckey/ Comstock, Inc. **13.1** page 188, clockwise from top left, Courtesy of © Steve & Ellison Widener and Breathe Hope, http://breathehope.tamu.edu; Courtesy of © The Family of Savannah Brooke Snider; Courtesy of © The Cody Dieruf Benefit Foundation, www.breathinisbelievin.org; Courtesy of © Bobby Brooks and The Family of Jeff Baird; Courtesy of The family of Benjamin Hill, reprinted with permission of © Chappell/Marathonfoto; Courtesy of © the family of Brandon Herriott. **13.2** (a) Jean M. Labat/ Ardea, London. **Page 190**, © The Moravian Museum, Brno. **13.6** right, At Tip, © George Lepp/ Corbis. **13.9** above, © David Scharf/ Peter Arnold, Inc. **13.10** (a) © JupiterImages Corporation. **13.11** dogs, left and right, © Michael Stuckey/ Comstock, Inc.; center, Bosco Broyer, photograph by Gary Head. **13.12** Courtesy of the family of Haris Charalambous and the University of Toledo. **13.13** below, Courtesy of Ray Carson, University of Florida News and Public Affairs. **13.14** left, © JupiterImages Corporation; right, © age photostock/ SuperStock. **Page 198** right, from top, © Frank Cezus/ FPG/ Getty Images; speakslowly, http://commons.wikimedia.org/wiki/File:Heterochromia.jpg; i love images/ Jupiter Images; © Georg Tanner, http://www.flickr.com/photos/gtanner; © Michael Prince/ Corbis; © Medioimages/Photodisc/ Jupiter Images. **13.15** © Dr. Christian Laforsch. **Page 199** © Gary Gaugler/ The Medical File/ Peter Arnold Inc. **Page 200** Section 13.1, Courtesy of © The Cody Dieruf Benefit Foundation, www.breathinisbelievin.org; Section 13.2, © The Moravian Museum, Brno; Section 13.5, © Michael Stuckey/ Comstock, Inc.; Section 13.6, © Dr. Christian Laforsch. **Page 201** © Leslie Faltheisek.

CHAPTER 14 **Key Concepts** from left, Courtesy of Irving Buchbinder, DPM, DABPS, Community Health Services, Hartford CT; Photo courtesy of The Progeria Research Foundation; © Lennart Nilsson/ Bonnierforlagen AB. **14.1** page 202, © Gary Roberts/ worldwidefeatures.com; page 203, Richard A. Sturm, Molecular genetics of human pigmentation diversity, Human Molecular Genetics, 2009 Apr 15;18(R1):R9–17, by permission of Oxford University Press. **14.2** (a) right, Courtesy of Irving Buchbinder, DPM, DABPS, Community Health Services, Hartford CT; (c) © Steve Uzzell. **14.4** (a) © Newcastle Photos and Ivy & Violet Broadhead and family; (b) Photo courtesy of The Progeria Research Foundation. **14.5** (b) © Rick Guidotti/ Positive Exposure; (c) © Conner's Way Foundation, www.connersway.com. **14.7** (a–b) Photo by Gary L. Friedman, www.FriedmanArchives.com. **14.13** Left, L. Willatt, East Anglian Regional Genetics Service / Photo Researchers, Inc; right, Ciarra, photo by © Michelle Harmon. **14.14** (a) http://en.wikipedia.org/wiki/File:Embryo_at_14_weeks_profile.jpg; (b) © Howard Sochurek/ The Medical File/ Peter Arnold, Inc. **14.15** © Lennart Nilsson/ Bonnierforlagen AB. **14.16** © Fran Heyl Associates/ Jacques Cohen, computer-enhanced by © Pix Elation. **Page 215** © Gary Roberts/ worldwidefeatures.com. **Page 216** Section 14.1, © Gary Roberts/ worldwidefeatures.com; Section 14.2, Courtesy of Irving Buchbinder, DPM, DABPS, Community Health Services, Hartford CT; Section 14.3, Photo courtesy of The Progeria Research Foundation; Section 14.7, © Lennart Nilsson/ Bonnierforlagen AB.

CHAPTER 15 **Key Concepts** from left, Professor Stanley Cohen/ SPL/ Photo Researchers, Inc.; Patrick Landmann/ Photo Researchers, Inc.; Photo courtesy of MU Extension and Agricultural Infomation; © Corbis/ SuperStock. **15.1** page 218, © 2009 Jupiter Images Corporation/ Fotosearch; page 219, Courtesy of Affymetrix. **15.3** (a) Professor Stanley Cohen/ SPL/ Photo Researchers, Inc.; (b) with permission of © QIAGEN, Inc. **15.8** left, Volker Steger/ SPL/ Photo Researchers, Inc.; right, Patrick Landmann/ Photo Researchers, Inc. **15.10** left, The Sanger Institute. Wellcome Images; right, Wellcome Trust Sanger Institute. **15.11** Raw STR data courtesy of © Orchid Cellmark, www.orchidcellmark.com. **Page 228** Photo Courtesy of Systems Biodynamics Lab, P.I. Jeff Hasty, UCSD Department of Bioengineering, and Scott Cookson. **15.12** (d) © Lowell Georgis/ Corbis; (e) © Keith V. Wood. **15.13** The Bt and Non-Bt corn photos were taken as part of field trial conducted on the main campus of Tennessee State University at the Institute of Agricultural and Environmental Research. The work was supported by a competitive grant from the CSREES, USDA titled "Southern Agricultural Biotechnology Consortium for Underserved Communities, (2000–2005)." Dr. Fisseha Tegegne and Dr. Ahmad Aziz served as Principal and Co-principal Investigators respectively to conduct the portion of the study in the State of Tennessee. **15.14** (a) © Adi Nes, Dvir Gallery Ltd.; (b) Photo courtesy of MU Extension and Agricultural Infomation; (c) Transgenic goat produced using nuclear transfer at GTC Biotherapeutics. Photo used with permission. **15.15** Courtesy of © Dr. Jean Levit. The Brainbow technique was developed in the laboratories of Jeff W. Lichtman and Joshua R. Sanes at Harvard University. This image has received the Bioscape imaging competition 2007 prize. **15.16** © Jeans for Gene Appeal. **Page 233** left, © Corbis/ SuperStock; right, © 2009 Jupiter Images Corporation/ Fotosearch; Section 15.1, Courtesy of Affymetrix; Section 15.2, Professor Stanley Cohen/ SPL/ Photo Researchers, Inc. **Page 234** Section 15.4, Patrick Landmann/ Photo Researchers, Inc.; Section 15.5, Wellcome Trust Sanger Institute; Section 15.6–15.9, Photo courtesy of MU Extension and Agricultural Infomation; Section 15.10 © Corbis/ SuperStock. **15.17** Courtesy of Dr. S. Thuret, Kings College London.

CHAPTER 16 **Key Concepts** from left, © Wolfgang Kaehler/ Corbis; Courtesy of Stan Celestian/ Glendale Comunity College Earth Science Image Archive; USGS; © Taro Taylor, www.flickr.com/photos/tjt195. **16.1** page 236, © 2009 Mike Park/ www.flickr.com/photos/38869446@N00; page 237, © David A. Kring, NASA/ Univ. Arizona Space Imagery Center. **16.2** (a) © Earl & Nazima Kowall/ Corbis; (b–c) © Wolfgang Kaehler/ Corbis. **16.3** © Edward S. Ross. **16.4** (a) © Dr. John Cunningham/ Visuals Unlimited; (b) © Gary Head. **Page 239** Courtesy of Daniel C. Kelley, Anthony J. Arnold, and William C. Parker, Florida State University Department of Geological Science. **16.5** above, © Gordon Chancellor. **16.6** above, Painting by George Richmond; right, Cambridge University Library. **16.7** (a) © John White; (b) 2004 Arent. **16.8** © Down House and The Royal College of Surgeons of England. **Page 244** top, Courtesy of Stan Celestian/ Glendale Comunity College Earth Science Image Archive. **16.9** (a) W. B. Scott (1894); (b) Doug Boyer in P. D. Gingerich et al. (2001) © American Association for Advancement of Science; (c) © P. D. Gingerich and M. D. Uhen (1996), © University of Michigan. Museum of Paleontology; bottom © P. D. Gingerich, University of Michigan. Museum of Paleontology. **16.10** (b) © Photodisc/ Getty Images. **Page 245** , right, Courtesy of Stan Celestian/ Glendale Comunity College Earth Science Image Archive. **16.11** (c) © Michael Pancier. **16.12** left, USGS. **16.15** (a) © Taro Taylor, www.flickr.com/photos/tjt195; (b) © JupiterImages Corporation; (c) © Linda Bingham. **16.13** (a–e) After A.M. Ziegler, C.R. Scotese, and S.F. Barrett, "Mesozoic and Cenozoic Paleogeographic Maps," and J. Krohn and J. Sundermann (Eds.), *Tidal Frictions and the Earth's Rotation II*, Springer-Verlag, 1983. **16.16** from left, © Lennart Nilsson/ Bonnierforlagen AB; Courtesy of Anna Bigas, IDIBELL-Institut de Recerca Oncologica, Spain; From "Embryonic staging system for the short-tailed fruit bat, Carollia perspicillata, a model organism for the mammalian order Chiroptera, based upon timed pregnancies in captive-bred animals" C.J. Cretekos et al., Developmental Dynamics Volume 233, Issue 3, July 2005, Pages: 721–738. Reprinted with permission of Wiley-Liss, Inc. a subsidiary of John Wiley & Sons, Inc.; Courtesy of Prof. Dr. G. Elisabeth Pollerberg, Institut für Zoologie, Universität Heidelberg, Germany; USGS. **16.17** (a) © Jürgen Berger, Max-Planck-Institut for Developmental Biology, Tübingen; (b) © Visuals Unlimited. **16.18** Courtesy of Ann C. Burke, Wesleyan University. **Page 253** © David A. Kring, NASA/ Univ. Arizona Space Imagery Center. **Page 254** Section 16.1, © 2009 Mike Park/ www.flickr.com/photos/38869446@N00; Section 16.2, © Wolfgang Kaehler/ Corbis; Section 16.3, Painting by George Richmond; Section 16.4, © Down House and The Royal College of Surgeons of England; Section 16.5, Courtesy of Stan Celestian/ Glendale Comunity College Earth Science Image Archive; Section 16.6, © Michael Pancier; Section 16.7, USGS; Section 16.8, © Taro Taylor, www.flickr.com/photos/tjt195; Section 16.9, Courtesy of Ann C. Burke, Wesleyan University. **Page 255** Lawrence Berkeley National Laboratory.

CHAPTER 17 **Key Concepts** from left, © Alan Solem; Courtesy of Hopi Hoekstra, University of California, San Diego; © Jack Jeffrey Photography; © Jeremy Thomas/ Natural Visions; © Jack Jeffrey Photography. **17.1** page 256, © Rollin Verlinde/ Vilda; page 257, © Reuters NewMedia, Inc./ Corbis. **17.2** (a) © Alan Solem; (b) from left, © Roderick Hulsbergen/ http://www.photography.euweb.nl; all others, © JupiterImages Corporation. **17.6** J. A. Bishop, L. M. Cook. **17.7** (a) © James "Bo" Insogna/ Photos.com Plus; (b–d) Courtesy of Hopi Hoekstra, University of California, San Diego. **17.9** above, Peter Chadwick/ Science Photo Library/ Photo Researchers, Inc. **17.11** © Thomas Bates Smith. **17.12** (a) © Ingo Arndt/ Nature Picture Library; (b) © Bruce Beehler; (c) Courtesy of Gerald Wilkinson. **17.13** (a,b) After Ayala and others; (c) © Michael Freeman/ Corbis. **17.14** (a,b) Adapted from S. S. Rich, A. E. Bell, and S. P. Wilson, "Genetic drift in small populations of Tribolium," *Evolution* 33:579–584, Fig. 1, p. 580, 1979. Used by permission of the publisher; below, Photo by Peggy Greb/ USDA. **17.15** © Dr. Victor A. McKusick. **17.16** (a) © David Neal Parks; (b) © W. Carter Johnson. **Page 270** lvin E. Staffan/ Photo Researchers, Inc. **17.18** (a) Courtesy of Dr. James French; (b) Courtesy of © Ron Brinkmann, www.flickr.com/photos/ronbrinkmann; (c) © David Goodin. **17.19** G. Ziesler/ ZEFA. **17.20** right, © Arthur Anker. **17.21** Po'ouli, Bill Sparklin/ Ashley Dayer; All others, © Jack Jeffrey Photography. **17.23** Kevin Bauman, www.african-cichlid.com. **17.24** top, Courtesy of The Virtual Fossil Museum, www.fossilmuseum.net; bottom, © Mark Erdman. **17.26** (a) © David Simcox; (b) © Jeremy Thomas/ Natural Visions. **17.28** (a–b) © Jack Jeffrey Photography; (c) Bill Sparklin/ Ashley Dayer. **Page 279** © Rollin Verlinde/ Vilda. **Page 280** Section 17.1, © Rollin Verlinde/ Vilda; Section 17.2–3, © Alan Solem; Section 17.4–6, Courtesy of Hopi Hoekstra, University of California, San Diego; Section 17.7, © Bruce Beehler; Section 17.8–9, © Dr. Victor A. McKusick; Section 17.10, Alvin E. Staffan/ Photo Researchers, Inc.; Section 17.11, © Jack Jeffrey Photography; Section 17.12, Kevin Bauman, african-cichlid.com; Section 17.13, © Jeremy Thomas/ Natural Visions; Section 17.14, © Jack Jeffrey Photography.

CHAPTER 18 **Key Concepts** from left, Painting by William K. Hartmann; From Hanczyc, Fujikawa, and Szostak, Experimental Models of Primitive Cellular Compartments: Encapsulation, Growth, and Division;www.sciencemag.org Science 24 October 2003; 302;529, Figure 2, page 619. Reprinted with permission of the authors and AAAS; Bruce Runnegar, NASA Astrobiology Institute. **18.1** page 282, © Jeff Hester and Paul Scowen, Arizona State University, and NASA; page 283, Photo by Julio Betancourt/ U.S. Geological Survey. **18.2** Painting by William K. Hartmann. **18.5** Courtesy of the University of Washington. **18.7** © Micheal J. Russell, Scottish Universities Environmental Research Centre. **18.8** (a) © Janet Iwasa; (b) From Hanczyc, Fujikawa, and Szostak, Experimental Models of Primitive Cellular Compartments: Encapsulation, Growth, and Division;www.sciencemag.org Science 24 October 2003; 302;529, Figure 2, page 619. Reprinted with permission of the authors and AAAS; (c) Photo by Tony Hoffman, courtesy of David Deamer. **18.9** (a) © Stanley M. Awramik; (b) Russell Kightley/ Science Photo Library/ Photo Researchers, Inc.; (c) © University of California Museum of Paleontology; (d) Chase Studios/ Photo Researchers, Inc.; (e) Courtesy of © Department of Industry and Resources, Western Australia. **18.10** (a–b) Bruce Runnegar, NASA Astrobiology Institute; (c) © N.J. Butterfield, University of Cambridge. **18.12** (a) Photo provided by M.G. Klotz, Department of Biology, University of Louisville, KY, USA (http://mgkmicro.com); (b) Courtesy of John Fuerst, University of Queensland. originally published in Archives of Microbiology vol 175, p 413–429 (Lindsay MR, Webb RI, Strous M, Jetten MS, Butler MK, Forde RJ, Fuerst JA. Cell compartmentalisation in planctomycetes: novel types of structural organisation for the bacterial cell.Arch Microbiol. 2001 Jun;175(6):413–29). **18.13** (a) CNRI/ Photo Researchers, Inc.; (b) © Robert Trench, Professor Emeritus, University of British Columbia; above, © Courtesy of Isao Inouye, Institute of Biological Sciences, University of Tsukuba. **Page 294**, NASA/ JPL. **Page 294** Section 18.1, Photo by Julio Betancourt/ U.S. Geological Survey; Section 18.3, Courtesy of the University of Washington; Section 18.4, © Janet Iwasa; Section 18.5, Chase Studios/ Photo Researchers, Inc.; Section 18.6, © Courtesy of Isao Inouye, Institute of Biological Sciences, University of Tsukuba.

CHAPTER 19 **Key Concepts** from left, Dr. Richard Feldmann/ National Cancer Institute; SciMAT/ Photo Researchers, Inc.; © Dr. Dennis Kunkel/ Visuals Unlimited; CDC/ Janice Haney Car; © Savannah River Ecology Laboratory. **19.1** page 296, © Lennart Nilsson/ Bonnierforlagen AB; page 297, Micrographs Z. Salahuddin, National Institutes of Health. **19.2**

(b) left, after Stephen L. Wolfe; (c) Dr. Richard Feldmann/ National Cancer Institute; (d) Russell Knightly/ Photo Researchers, Inc. **19.3** top left, Science Photo Library/ Photo Researchers, Inc. **19.5** © WHO, Pierre-Michel Virot, photographer. **Page 300** Sercomi/ Photo Researchers, Inc. **Page 301** left, CDC/ C. S. Goldsmith and A. Balish; right, Electron Microscopy Laboratory, Agricultural Research Service, U. S. Department of Agriculture. **19.7** top, L. Santo. **19.8** © Dr. Dennis Kunkel / Visuals Unlimited. **19.9** (a) © P. W. Johnson and J. MeN. Sieburth, Univ. Rhode Island/ BPS; (b) © Dr. Manfred Schloesser, Max Planck Institute for Marine Microbiology; (c) SciMAT/ Photo Researchers, Inc. **19.10** (a) CDC/ Janice Haney Car; (b) CDC/ Bruno Coignard, M.D./ Jeff Hageman, M.H.S. **19.11** (a) Stem Jems/ Photo Researchers, Inc.; (b) CDC/ James Gathany. **Page 306** © Courtesy Jack Jones, Archives of Microbiology, Vol. 136, 1983, pp. 254–261. Reprinted by permission of Springer-Verlag. **19.13** (a) © Dr. W. Michaelis/ Universitat Hamburg; (b) Courtesy of Benjamin Brunner; (c) © Savannah River Ecology Laboratory. **Page 308** Section 19.1, Micrographs Z. Salahuddin, National Institutes of Health; Section 19.3, WHO, Pierre-Michel Virot, photographer; Section 19.7, © P. W. Johnson and J. MeN. Sieburth, Univ. Rhode Island/ BPS; Section 19.8, © Dr. W. Michaelis/ Universitat Hamburg. **19.14** © Photodisc/ Getty Images.

CHAPTER 20 **Key Concepts**, from left © Dr. David Phillips/ Visuals Unlimited; © Wim van Egmond/ Visuals Unlimited; © Lewis Trusty/ Animals Animals; © Wim van Egmond/ Visuals Unlimited; M I Walker/ Photo Researchers, Inc. **20.1** page 310, © Peter M. Johnson - www.flickr.com/photos/pmjohnso; page 311, © Dr. David Phillips/ Visuals Unlimited. **20.2** (a) Astrid Hanns-Frieder michler/SPL/ Photo Researchers, Inc.; (b) © Dr. David Phillips/ Visuals Unlimited; (c) Science Museum of Minnesota; (d) D. P. Wilson/ Eric & David Hosking; (e) © Jeffrey Levinton, State University of New York, Stony Brook. **Page 313** © Dr. Stan Erlandsen, University of Minnesota. **20.3** (a) © Dr. Dennis Kunkel/ Visuals Unlimited; (b) Oliver Meckes/ Photo Researchers, Inc.. **20.5** (a) Courtesy of Allen W. H. Bé and David A. Caron ; (b) © Ric Ergenbright/ Corbis; (c) © John Clegg/ Ardea, London. **20.6** © Frank Borges Llosa/ www.frankley.com; inset, © Wim van Egmond/ Visuals Unlimited. **20.7** (a) Gary W. Grimes and Steven L'Hernault. **20.8** Based on Fig. 1 from *Genetic linkage and association analyses for trait mapping in Plasmodium falciparum*, by Xinzhuan Su, Karen Hayton & Thomas E. Wellems, *Nature Reviews Genetics* 8, 497–506 (July 2007). **Page 317**, Heather Angel. **20.9** (a) © Wim van Egmond/ Visuals Unlimited; (b) Image courtesy of Woodstream Corporation. **20.10** © Lewis Trusty/ Animals Animals. **20.11** below left, © Photo-Disc/ Getty Images. **Page 318**, Courtesy of Professeur Michel Cavalla. **20.12** (a) Courtesy of © Brian P. Piasecki and Carolyn D. Silflow; (b) © Kim Taylor/ Bruce Coleman, Inc./ Photoshot; (c) © Lawson Wood/ Corbis; (d) © Wim van Egmond/ Visuals Unlimited. **20.13** M I Walker/ Photo Researchers, Inc. **20.14** (a) © Edwards S. Ross; (b) © Courtesy of www.hiddenforest.co.nz. **20.15** below, © Carolina Biological Supply Company. **20.16** Larry Brand. **Page 321**, Woods Hole Oceanographic Institution. **Page 322** Section 20.1, © Peter M. Johnson - www.flickr.com/photos/pmjohnso; Section 20.4, Courtesy of Allen W. H. Bé and David A. Caron ; Section 20.5–6, © Wim van Egmond/ Visuals Unlimited; Section 20.7, Heather Angel; Section 20.8, D. P. Wilson/ Eric & David Hosking; Section 20.9, © Courtesy of www.hiddenforest.co.nz. **Page 323**, Critical Thinking, Courtesy Brian Duval.

CHAPTER 21 **Key Concepts** from left, Courtesy of © Professor T. Mansfield; National Park Services, Paul Stehr-Green; © A. & E. Bomford/ Ardea, London; © Dave Cavagnaro/ Peter Arnold, Inc.; © Sanford/ Agliolo/ Corbis. **21.1** page 324, T. Kerasote/ Photo Researchers, Inc.; page 325, © William Campbell/ TimePix/ Getty Images. **21.2** right, Courtesy of © Professor T. Mansfield. **21.4** Courtesy of © Christine Evers. **21.5** left, © Jane Burton/ Bruce Coleman Ltd. **21.6** © Fred Bavendam/ Peter Arnold, Inc. **21.7** (a) National Park Services, Paul Stehr-Green; (b) Wayne P. Armstrong, Professor of Biology and Botany, Palomar College, San Marcos, California. **21.8** © University of Wisconsin-Madison, Department of Biology, Anthoceros CD. **21.9** (a) © Martin LaBar, www.flickr.com/photos/martinlabar; (b) Courtesy of BOCATEC Sales and Rent GmbH & Co. KG. **21.10** (a) © William Ferguson; (b) Courtesy of © Christine Evers. **21.11** right, © A. & E. Bomford/ Ardea, London. **21.12** (a) © S. Navie; (b) David C. Clegg/ Photo Researchers, Inc.; (c) © Klein Hubert/ Peter Arnold, Inc. **Page 332** © Reprinted with permission from Elsevier. **21.14** © Jeri Hochman and Martin Hochman, Illustration by Zdenek Burian. **21.16** © Karen Carr Studio/ www.karencarr.com. **21.18** (a) © Dave Cavagnaro/ Peter Arnold, Inc.; (b) © M. Fagg, Australian National Botanic Gardens; (c) Michael P. Gadomski/ Photo Researchers, Inc.; (d) © E. Webber/ Visuals Unlimited; (e) © Gerald & Buff Corsi/ Visuals Unlimited. **21.19** far left © Robert Potts, California Academy of Sciences; left upper, © Robert & Linda Mitchell Photography; lower, R. J. Erwin/ Photo Researchers, Inc. **Page 336** Courtesy of © Christine Evers. **21.22** © Sanford/ Agliolo/ Corbis. **21.23** (a–b) Photo USDA; (c) Courtesy of Linn County, Oregon Sheriff's Office. **21.24** R. Bieregaard/Photo Researchers, Inc. **Page 340** Section 21.1, © William Campbell/ TimePix/ Getty Images; Section 21.4, © Martin LaBar, www.flickr.com/photos/martinlabar; Section 21.6, © Robert & Linda Mitchell Photography; Section 21.7, Courtesy of © Christine Evers; Section 21.8, Photo USDA. **21.25** right, © Clinton Webb.

CHAPTER 22 **Key Concepts** from left, Garry T. Cole, University of Texas, Austin/ BPS; © Ed Reschke; © Dr. Dennis Kunkel/ Visuals Unlimited; © Robert C. Simpson/ Nature Stock; Gary Head. **Page 342** © Robert C. Simpson/ Nature Stock. **Page 343** Photo by Yue Jin/ USDA. **22.1** (a) Photo by Scott Bauer/ USDA; (b) © Robert C. Simpson/ Nature Stock. **22.2** Garry T. Cole, University of Texas, Austin/ BPS. **22.3** top, © Ed Reschke. **22.4** © Dr. Mark Brundrett, The University of Western Australia. **22.5** (a) above, © Michael Wood/ mykob.com; below, © North Carolina State University, Department of Plant Pathology; (b) © Bill Beatty/ Visuals Unlimited; (c) © agefotostock/ SuperStock. **22.6** N. Allin and G. L. Barron. **Page 346** right, © Dr. Dennis Kunkel/ Visuals Unlimited. **22.7** top left, Dr. J. O'Brien, USDA Forest Service; all others, © Robert C. Simpson/ Nature Stock. **22.8** below, Eye of Science/ Photo Researchers, Inc.; art, After T. Rost, et al., *Botany*, Wiley 1979. **22.9** (a) Gary Head; (b) After Raven, Evert, and Eichhorn, *Biology of Plants*, 4th ed., Worth Publishers, New York, 1986; (c) © Mark E. Gibson/ Visuals Unlimited. **22.10** (a) © Gary Braasch; (b) © F. B. Reeves. **22.11** © Harry Regin. **Page 349** right, above, Courtesy of D. G. Schmale III Courtesy of D. G. Schmale III; below, Dr. P. Marazzi/ Photo Researchers, Inc. **Page 350** Section 22.1, Photo by Yue Jin/ USDA; Section 22.2, Garry T. Cole, University of Texas, Austin/ BPS; Section 22.3, © Ed Reschke; Section 22.4, © North Carolina State University, Department of Plant Pathology; Section 22.5, Eye of Science/ Photo Researchers, Inc.; Section 22.6, © Mark E. Gibson/ Visuals Unlimited; Section 22.7, Dr. P. Marazzi/ Photo Researchers, Inc. **22.12** After graph from www.pfc.foresty.ca.

CHAPTER 23 **Key Concepts** from left © The Natural History Museum (London); © Wim van Egmond/ Visuals Unlimited; Courtesy of Karen Swain, North Carolina Museum of Natural Sciences; © Fred Bavendam/ Minden Pictures. **Page 352** © K.S. Matz. **Page 353** © Callum Roberts, University of York. **23.4** (a–b) David Patterson, courtesy micro*scope/http://microscope.mbl .edu; (c) © 2003 Ana Signorovitch. **23.5** © The Natural History Museum (London). **23.6** (a) After Eugene Kozloff; (b) Image © ultimathule, Used under license from Shutterstock.com. **23.7** (a) Marty Snyderman/ Planet Earth Pictures; (b) © Dr. Sally Leys/ University of Alberta. **23.8** (b) © Wim van Egmond/ Visuals Unlimited; (c) © Brandon D. Cole/ Corbis. **23.10** After T. Storer, et al., *General Zoology*, Sixth Edition. **23.11** (a) © Jeffrey L. Rotman/ Corbis; (b) © Kim Taylor/ Bruce Coleman, Ltd.; (c) A.N.T./ Photo Researchers, Inc. **23.12** © Cory Gray. **23.14** right, Andrew Syred/ SPL/ Photo Researchers, Inc. **23.15** (a) © J. Solliday/ BPS; center art, adapted from Rasmussen, "Ophelia," Vol. 11, in Eugene Kozloff, *Invertebrates*, 1990; (b) © Jon Kenfield/ Bruce Coleman Ltd. **23.16** © J. A. L. Cooke/ Oxford Scientific Films. **23.17** left art, from Solomon, 8th edition, p. 624, figure 29-4; right, Courtesy of © Christine Evers. **23.19** left, Danielle C. Zacherl with John McNulty. **23.20** (a) © B. Borrell Casals/ Frank Lane Picture Agency/ Corbis; (b) © Frank Park/ ANT Photo Library; (c) © Dave Fleetham/Tom Stack & Associates. **23.21** (a) © Joe McDonald/Corbis; (b) Courtesy of © Christine Evers; (c) © Alex Kirstitch. **23.22** (a) © Alex Kirstitch; (b) J. Grossauer/ ZEFA. **23.24** (a) Courtesy of © Emily Howard Staub and The Carter Center; (b) Sinclair Stammers/ SPL/ Photo Researchers, Inc. **23.25** (a) © Jane Burton/ Bruce Coleman, Ltd.; (b) NOAA; (c) Courtesy of © Christine Evers; (d) Photo by Peggy Greb/ USDA. **23.26** (a) Image © Eric Isselée, Used under license from Shutterstock.com; (b) © Angelo Giampiccolo; (c) © Frans Lemmens/ The Image bank/ Getty Images; (d) Eye of Science/ Photo Researchers, Inc.; (e) Andrew Syred/ Photo Researchers, Inc. **23.27** (a) © David Tipling/ Photographer's Choice/ Getty Images; (b) Herve Chaumeton/ Agence Nature; (c) © Peter Parks/ Imagequestmarine.com. **23.28** After D.H. Milne, *Marine Life and the Sea*, Wadsworth, 1995. **23.30** CDC/ Piotr Naskrecki. **23.32** (a,h) Courtesy of © Christine Evers; (b–c,i) Edward S. Ross (d) Alvin E. Staffan/ Photo Researchers, Inc.; (e) Joseph L. Spencer; (f) John Alcock, Arizona State University; (g) © D. A. Rintoul; (j) © Mark Moffett/ Minden Pictures; (k) CDC/ Harvard University, Dr. Gary Alpert; (l) Courtesy of Karen Swain, North Carolina Museum of Natural Sciences. **23.33** Gregory G. Dimijian, M.D./ Photo Researchers, Inc. **23.34** Photo by Scott Bauer/ USDA. **23.35** Photo by James Gathany, Centers for Disease Control. **Page 373** © Walter Deas/ Seaphot Limited/ Planet Earth Pictures. **23.36** (a) © Herve Chaumeton/ Agence Nature; (c) © Fred Bavendam/ Minden Pictures; (d) © Jan Haaga, Kodiak Lab, AFSC/NMFS. **Page 374** Section 23.1, © Callum Roberts, University of York; Section 23.3, © 2003 Ana Signorovitch; Section 23.4, Marty Snyderman/ Planet Earth Pictures; Section 23.5, © Jeffrey L. Rotman/ Corbis; Section 23.6, © Cory Gray; Section 23.7, © J. Solliday/ BPS; Section 23.8, © B. Borrell Casals/ Frank Lane Picture Agency/ Corbis; Section 23.9, Sinclair Stammers/ SPL/ Photo Researchers, Inc.; Section 23.10–14, CDC/ Harvard University, Dr. Gary Alpert; Section 23.15, © Herve Chaumeton/ Agence Nature. **23.37** right, Jane Burton/ Bruce Coleman, Ltd.

CHAPTER 24 **Key Concepts** from left, Peter Parks/ Oxford Scientific Films/ Animals Animals; © Wernher Krutein/ photovault.com; © Bill M. Campbell, MD; © Z. Leszczynski/ Animals Animals; © Jean-Paul Tibbles, "Book of Life", Ebury Press. **Page 376** © P. Morris/ Ardea London. **Page 377** above, © James Reece, Nature Focus, Australian Museum; below, With permission of the Australian Museum. **24.1** right, Runk & Schoenberger/ Grant Heilman, Inc. **24.2** (a) Peter Parks/ Oxford Scientific Films/ Animals Animals; (b) Redrawn from *Living Invertebrates*, V. & J. Pearse and M. & R. Buchsbaum. The Boxwood Press, 1987. Used by permission; (c) © California Academy of Sciences. **24.4** (a) © Brandon D. Cole/ Corbis; (b) Heather Angel/ Natural Visions. **24.5** Adapted from A.S. Romer and T.S. Parsons, *The Vertebrate Body*, Sixth Edition, Saunders, 1986. **24.6** (a) Jonathan Bird/ Oceanic Research Group, Inc.; (b) © Gido Braase/ Deep Blue Productions. **24.7** (a) from E. Solomon, L. Berg, and D.W. Martin, *Biology*, Seventh Edition, Thomson

Brooks/Cole; (b) Robert & Linda Mitchell Photography; (c) Mark Dixon, NOAA; (d) Patrice Ceisel/ © 1986 John G. Shedd Aquarium. **24.8** © Wernher Krutein/ photovault.com. **24.9** (a–c) © P. E. Ahlberg; (d) © Alfred Kamajian. **24.10** (a) © Bill M. Campbell, MD; (b,c) Adapted from A.S. Romer and T.S. Parsons, *The Vertebrate Body*, Sixth Edition, Saunders, 1986. **24.11** (a) Stephen Dalton/ Photo Researchers, Inc.; (b) © David M. Dennis/ Tom Stack & Associates, Inc. **24.12** © John Serraro/ Visuals Unlimited. **24.13** © Pieter Johnson. **24.16** © Karen Carr Studio/ www.karencarr.com. **24.17** (a) © Z. Leszczynski/ Animals Animals; (b) © Kevin Schafer/Corbis; (c) © Kevin Schafer/ Tom Stack & Associates. **Page 385** © S. Blair Hedges, Pennsylvania State University. **24.18** (a) © Gerard Lacz/ ANTPhoto.com.au. **24.19** http://en.wikipedia.org/wiki/User:Jmgarg1 'Creating awareness of Indian Flora & Fauna' Image Resource of thousands of my images of Birds, Butterflies, Flora etc. (arranged alphabetically & place-wise): http://commons.wikimedia.org/wiki/Category:J.M.Garg. For learning about Indian Flora, visit/ join Google e-group- Indiantreepix:http://groups.google.co.in/group/indiantreepix?hl=en. **24.20** After M. Weiss and A. Mann, *Human Biology and Behavior*, 5th Edition, HarperCollins, 1990. **24.21** (a) Jean Phillipe Varin/ Jacana/ Photo Researchers, Inc.; (b) Jack Dermid; (c) © Sandy Roessler/ FPG/ Getty Images. **24.22** © Kenneth Garrett/ National Geographic Image Collection. **24.24** above, © Rod Williams/ www.bciusa.com. **24.25** (a) © Bone Clones®, www.boneclones.com; (b) © Gary Head. **24.26** (a) © AFP/ Getty Images; (b–e) Pascal Goetgheluck/ Photo Researchers, Inc. **24.27** (a) © Dr. Donald Johanson, Institute of Human Origins; (b) © Louise M. Robbins; (c) © Kenneth Garrett/ National Geographic Image Collection. **24.28** © Jean-Paul Tibbles, "Book of Life", Ebury Press. **24.29** John Reader/ Photo Researchers, Inc. **24.30** top, © Pascal Goetgheluck/ Photo Researchers, Inc.; others, Peter Brown. **24.32** © Christopher Scotese, PALEOMAP Project. **Page 394** Section 24.1, © James Reece, Nature Focus, Australian Museum; Section 24.2, Runk & Schoenberger / Grant Heilman, Inc.; Section 24.3, Robert & Linda Mitchell Photography; Section 24.4, © John Serraro/ Visuals Unlimited; Section 24.5, © Karen Carr Studio/ www.karencarr.com; Section 24.6, © Kevin Schafer/Corbis; Section 24.7, © Gerard Lacz/ ANTPhoto.com.au; Section 24.8, © Sandy Roessler/ FPG/ Getty Images; Section 24.9, © Kenneth Garrett/ National Geographic Image Collection; Section 24.10, Pascal Goetgheluck/ Photo Researchers, Inc.

CHAPTER 25 **Key Concepts** from left, M. I. Walker/ Photo Researchers, Inc.; © Biodisc/ Visuals Unlimited; © David W. Stahle, Department of Geosciences, University of Arkansas; © iStockphoto.com/ mjutabor. **25.1** page 396, Getty Images/Flickr RF; page 397, © Peter Gasson, Royal Botanic Gardens, Kew. **25.3** © Donald L. Rubbelke/ Lakeland Community College. **25.4** (a) from top, © Bruce Iverson; © Ernest Manewal/ Index Stock Imagery; Courtesy of Dr. Thomas L. Rost; © Franz Holthuysen, Making the invisible visible, Electron Microscopist, Phillips Research; (b) from top, Photo by Mike Clayton/ University of Wisconsin Department of Botany; © Darrell Gulin/ Corbis; Gary Head; Courtesy of Janet Wilmhurst, Landcare Research, New Zealand. **25.6** © Ross E. Koning, http://plantphys.info. **Page 400** epidermis, © George S. Ellmore; all others, © Ross E. Koning, http://plantphys.info. **Page 401** Ground Tissues, © ISM/ Phototake. All rights reserved; Vascular Tissues, © Donald L. Rubbelke/ Lakeland Community College. **25.7** below, Andrew Syred/ Photo Researchers, Inc. **25.8** (a) Center, Ray F. Evert; right, James W. Perry; (b) Center, Carolina Biological Supply Company; right, James W. Perry. **25.9** (d) above, M. I. Walker/ Photo Researchers, Inc.; below, © Gary Head. **25.10** (c), (d) left, Benjamin de Bivort; (d) center, Miguel Bugallo; right, Sigman. **25.11** © Kenneth Bart. **25.13** (a) Courtesy of Dr. Thomas L. Rost; (b) © Gary Head. **25.15** (a) © Biodisc/ Visuals Unlimited; (b) after Salisbury and Ross, *Plant Physiology*, Fourth Edition, Wadsworth. **25.16** (a) Michael Clayton/ University of Wisconsin, Department of Botany; (b) left, © Brad Mogen/ Visuals Unlimited; right, © Dr. John D. Cunningham/ Visuals Unlimited. **25.18** (b) © Peter Gasson, Royal Botanic Gardens, Kew. **Page 410** Clockwise from top left, © Anthony Tripodi, www.thecompostbin.com; © Eric Sueyoshi/ Audrey Magazine; Chase Studio/ Photo Researchers, Inc.; © Chris Hellier/ Corbis; © iStockphoto.com/ mjutabor; © Dinodia Photo Library/ Botanica/ Jupiter Images. **25.19** (a) Peter Ryan/ SPL/ Photo Researchers, Inc.; (b) © Jon Pilcher; (c) George Bernard/ SPL/ Photo Researchers, Inc.; (d) © David W. Stahle, Department of Geosciences, University of Arkansas. **Page 412** Section 25.1, Getty Images/Flickr RF; Section 25.3, © Donald L. Rubbelke/ Lakeland Community College; Section 25.4, M. I. Walker/ Photo Researchers, Inc.; Section 25.5, Courtesy of Dr. Thomas L. Rost; Section 25.6, © Biodisc/ Visuals Unlimited; Section 25.7, © Peter Gasson, Royal Botanic Gardens, Kew; Section 25.8, © iStockphoto.com/ mjutabor. **Page 413**, Critical Thinking, #1, © Edward S. Ross; #3, © Ian Young, www.srgc.org.uk.

CHAPTER 26 **Key Concepts** from left, Photo courtesy of Stephanie G. Harvey, Georgia Southwestern State University; Photo courtesy of Iowa State University Plant and Insect Diagnostic Clinic; Jeremy Burgess/ SPL/ Photo Researchers, Inc; © Martin Zimmerman, Science, 1961, 133:73–79, © AAAS. **26.1** page 414, Photo by © Billy Wrobel, courtesy of Argonne National Laboratory; page 415, © OPSEC Control Number #4 077-A-4. **26.2** Photo courtesy of Stephanie G. Harvey, Georgia Southwestern State University. **26.3** © William Ferguson. **26.4** (a) Courtesy of Mark Holland, Salisbury University; (b) Photo courtesy of Iowa State University Plant and Insect Diagnostic Clinic; (c) © Wally Eberhart/ Visuals Unlimited; (d) © NifTAL Project, Univ. of Hawaii, Maui. **26.5** (a) right, © Dr. John D. Cunningham/ Visuals Unlimited. **26.6** (a) © Alison W. Roberts, University of Rhode Island; (b–c) © H. A. Core, W. A. Cote and A. C. Day, Wood Structure and Identification, 2nd Ed., Syracuse University Press, 1979. **26.7** left, Image © Jan Martin Will, Used under license from Shutterstock.com. **26.8** (a) Micrograph by Ken Wagner/ Visuals Unlimited; (b–e) Courtesy of E. Raveh. **26.9** left © Don Hopey/ Pittsburgh Post-Gazette, 2002, all rights reserved. Reprinted with permission; (a–b) Jeremy Burgess/ SPL/ Photo Researchers, Inc. **26.10** © J. C. Revy/ ISM/ Phototake. **26.11** © James D. Mauseth, MCDB. **Page 425** top, Photo by Keith Weller, ARS, Courtesy of USDA. **26.13** © Martin Zimmerman, Science, 1961, 133:73–79, © AAAS. **Page 426** Section 26.1, Photo by © Billy Wrobel, courtesy of Argonne National Laboratory; Section 26.2, Photo courtesy of Stephanie G. Harvey, Georgia Southwestern State University; Section 26.3, Photo courtesy of Iowa State University Plant and Insect Diagnostic Clinic; Section 26.4, © H. A. Core, W. A. Cote and A. C. Day, Wood Structure and Identification, 2nd Ed., Syracuse University Press, 1979; Section 26.5, Jeremy Burgess/ SPL/ Photo Researchers, Inc; Section 26.6 © Martin Zimmerman, Science, 1961, 133:73–79, © AAAS.

CHAPTER 27 **Key Concepts** from left, © Robert Essel NYC/ Corbis; Andrew Syred/ SPL/ Photo Researchers, Inc.; Image © Daniel Gale, Used under license from Shutterstock.com; © Herve Chaumeton/ Agence Nature; © Cathlyn Melloan/ Stone/ Getty Images. **27.1** page 428, © Alan McConnaughey, www.flickr.com/photos/engrpiman; page 429, Courtesy of James H. Cane, USDA-ARS Bee Biology and Systematics Lab, Utah State University, Logan, UT. **27.2** (a) left, © Robert Essel NYC/ Corbis. **27.4** (a) John Alcock, Arizona State University; (b) © David Goodin. **27.6** © Michael Clayton, University of Wisconsin. **Page 434** © T. M. Jones**.** **27.8** (a) © Richard H. Gross; (b) Andrew Syred/ SPL/ Photo Researchers, Inc.; (c) Photo by Stephen Ausmus, USDA, ARS. **Page 435** left, upper, © R. Carr; lower © Robert H. Mohlenbrock © USDA-NRCS PLANTS Database / USDA SCS. 1989. "Midwest wetland flora; field office illustrated guide to plant species." Midwest national Technical Center, Lincoln, NE.; right, Image © Daniel Gale, Used under license from Shutterstock.com. **27.9** © Mike Clayton/ University of Wisconsin Department of Botany. **27.10** (b) left, © Barry L. Runk/ Grant Heilman, Inc.; right, © James D. Mauseth, University of Texas. **27.11** (a) above, © Herve Chaumeton/ Agence Nature. **27.12** © Sylvan H. Wittwer/ Visuals Unlimited. **27.14** (a) © Michael Clayton, University of Wisconsin; (b–c) © Muday, GK and P. Haworth (1994) "Tomato root growth, gravitropism, and lateral development: Correlations with auxin transport." "Plant Physiology and Biochemistry 32, 193–203" with permission from Elsevier Science. **27.15** Micrographs courtesy of Randy Moore from "How Roots Respond to Gravity", M. L. Evans, R. Moore, and K. Hasenstein, Scientific American, December 1986. **27.16** right, © Cathlyn Melloan/ Stone/ Getty Images. **27.17** Cary Mitchell. **Page 441** left, © Gary Head. **27.18** © Frank B. Salisbury. **27.21** © Eric Welzel/ Fox Hill Nursery, Freeport, Maine. **27.22** (a) Courtesy of Dr. Consuelo M. De Moraes; (b–d) © Andrei Sourakov and Consuelo M. De Moraes. **27.23** left, © Roger Wilmshurst, Frank Lane Picture Agency/ Corbis; right, © Adrian Chalkley. **Page 445** top right, © Alan McConnaughey, www.flickr.com/photos/engrpiman; Section 27.1, Courtesy of James H. Cane, USDA-ARS Bee Biology and Systematics Lab, Utah State University, Logan, UT; Section 27.2, © Robert Essel NYC/ Corbis. **Page 446** Section 27.4, Andrew Syred/ SPL/ Photo Researchers, Inc.; Section 27.5, Image © Daniel Gale, Used under license from Shutterstock.com; Section 27.6, © Herve Chaumeton/ Agence Nature; Section 27.7, © Sylvan H. Wittwer/ Visuals Unlimited; Section 27.8, © Cathlyn Melloan/ Stone/ Getty Images; Section 27.9, © Frank B. Salisbury; Section 27.10, © Andrei Sourakov and Consuelo M. De Moraes. **27.34** (a) © Mary Sue Ittner/ Pacific Bulb Society; (b) © Steven D. Johnson.

CHAPTER 28 **Key Concepts** from left, © PhotoDisc/ Getty Images, with art by Lisa Starr; Ed Reschke. **28.1** page 448, Used with permission of University of Wisconsin Board of Regents. **28.2** (b) © Ed Reschke; (c) © C. Yokochi and J. Rohen, Photographic Anatomy of the Human Body, 2nd Ed., Igaku-Shoin, Ltd., 1979; (e) Image © Yuri Arcurs, Used under license from Shutterstock.com. **28.3** (a) Biophoto Associates/ Photo Researchers, Inc.; (b) © PhotoDisc/ Getty Images, with art by Lisa Starr. **28.5** from top, Ray Simmons/ Photo Researchers, Inc.; © Ed Reschke/ Peter Arnold, Inc.; © Don W. Fawcett. **28.6** (b) left, adapted from C.P. Hickman, Jr., L.S. Roberts, and A. Larson, *Integrated Principles of Zoology*, Ninth Edition, Wm. C. Brown, 1995; (b) right, Gregory Dimijian/ Photo Researchers, Inc. **28.7** Rajesh Bedi. **28.8** above, (a) © John Cunningham/ Visuals Unlimited; (b–c) © Ed Reschke; (d) Science Photo Library/ Photo Researchers, Inc.; (e) © University of Cincinnati, Raymond Walters College, Biology; (f) Michael Abbey/ Photo Researchers, Inc.; (g) right, Science Photo Library/ Photo Researchers, Inc. **28.9** (a–b) Ed Reschke; (c) Biophoto Associates/ Photo Researchers, Inc. **28.10** above, © Triarch/ Visuals Unlimited. **28.11** © Kim Taylor/ Bruce Coleman, Ltd. **28.15** right, © John D. Cunningham/ Visuals Unlimited. **28.16** © Michael Shore Photography. **28.19** Courtesy of Dr.Kathleen K. Sulik, Bowles Center for Alcohol Studies, the University of North Carolina at Chapel Hill. **Page 464** Section 28.1, Used with permission of University of Wisconsin Board of Regents; Section 28.2, © C. Yokochi and J. Rohen, Photographic Anatomy of the Human Body, 2nd Ed., Igaku-Shoin, Ltd., 1979; Section 28.3, © Don W. Fawcett; Section 28.4, © John Cunningham/ Visuals Unlimited; Section 28.5, Ed Reschke; Section 28.6, © Triarch/ Visuals Unlimited; Section 28.8, © Michael Shore Photography; Section 28.9, Courtesy of Dr.Kathleen K. Sulik, Bowles Center for Alcohol Studies, the University of

North Carolina at Chapel Hill. **28.20** right, Courtesy of © Organogenesis, Inc., www.organo.com. **Page 465**, Critical Thinking, CNRI/ Photo Researchers, Inc.

CHAPTER 29 **Key Concepts** from left, From Neuro Via Clinicall Research Program, Minneapolis VA Medical Center. **29.1** page 466, © Jamie Baker/ Taxi/ Getty Images; page 477 left, © EMPICS; right, © Manni Mason's Pictures. **29.2** (a) Courtesy Dr. William J. Tietjen, Bellarmine University. **29.5** left, © Manfred Kage/ Peter Arnold, Inc. **29.11** Courtesy of Riken Brain Science Institute. **29.12** (a) AP/ Wide World Photos; (b–c) From Neuro Via Clinicall Research Program, Minneapolis VA Medical Center. **29.20** (a) Colin Chumbley/ Science Source/ Photo Researchers, Inc.; (b) © C. Yokochi and J. Rohen, Photographic Anatomy of the Human Body, 2nd Ed., Igaku-Shoin, Ltd., 1979. **29.22** (a) after Penfield and Rasmussen, *The Cerebral Cortex of Man*, © 1950 Macmillan Library Reference. Renewed 1978 by Theodore Rasmussen; (b) Colin Chumbley/ Science Source/ Photo Researchers, Inc. **Page 486** Section 29.1, © Jamie Baker/ Taxi/ Getty Images; Section 29.3–4, © Manfred Kage/ Peter Arnold, Inc; Section 29.7, From Neuro Via Clinicall Research Program, Minneapolis VA Medical Center; Section 29.10–11, © C. Yokochi and J. Rohen, Photographic Anatomy of the Human Body, 2nd Ed., Igaku-Shoin, Ltd., 1979.

CHAPTER 30 **Key Concepts** from left, © E. R. Degginger. **30.1** page 488, © Phillip Colla, OceanLight.com. All Rights Reserved Worldwide; page 489, © AP/ Wide World Photos. **30.4** left, after Penfield and Rasmussen, *The Cerebral Cortex of Man*, © 1950 Macmillan Library Reference. Renewed 1978 by Theodore Rasmussen; right, Colin Chumbley/ Science Source/ Photo Researchers, Inc. **30.5** (a) above, © E. R. Degginger; art, After M. Gardiner, *The Biology of Vertebrates*, McGraw-Hill, 1972. **30.7** Bo Veisland/ Photo Researchers, Inc. **30.9** (c) Courtesy of Dr. Bryan Jones, University of Utah School of Medicine. **30.11** National Eye Institute, U.S. National Institute of Health. **Page** 497 © AFP Photo/ Timothy A. Clary/ Getty Images. **30.16** (1) right, © Fabian/ Corbis Sygma; (3) Medtronic Xomed. **30.17** © Robert E. Preston, courtesy Joseph E. Hawkins, Kresge Hearing Research Institute, University of Michigan Medical School. **Page 500** Section 30.1, © Phillip Colla, OceanLight.com. All Rights Reserved Worldwide; Section 30.9, © Robert E. Preston, courtesy Joseph E. Hawkins, Kresge Hearing Research Institute, University of Michigan Medical School. **Page 501** Critical Thinking, © Chase Swift.

CHAPTER 31 **Key Concepts** from left, Scott Camazine/ Photo Researchers, Inc.; © Kevin Fleming/ Corbis. **31.1** page 502, © Catherine Ledner; page 503, © David Ryan/ SuperStock. **31.6** China Daily/ Reuters. **31.7** left © Gary Head. **31.8** (a) Scott Camazine/ Photo Researchers, Inc.; (b) Biophoto Associates/ SPL/ Photo Researchers, Inc. **31.10** Permission obtained from Blackwell Publishing © Holt RIG and Hanley NA (2006) Essential Endocrinology & Diabetes, edn 5. **31.11** Photograph by Cecil Stoughton, White House, in the John F. Kennedy Presidential Library and Museum, Boston. **31.13** left, © Elizabeth Musar; right © Manny Hernandez/ Diabetes Hands Foundation, www.tudiabetes.com. **Page 517** Dr. Carlos J. Bourdony. **31.16** (c) © Kevin Fleming/ Corbis. **Page 518** Section 31.1, © Catherine Ledner; Section 31.6, © Gary Head; Section 31.11, © Kevin Fleming/ Corbis.

CHAPTER 32 **Key Concepts** from left, © Linda Pitkin/ Planet Earth Pictures; Professor P. Motta/ Department of Anatomy/ La Sapienza, Rome/ SPL/ Photo Researchers, Inc.; Image courtesy of Department of Pathology, The University of Melbourne. **32.1** page 520, The Muskegon Chronicle; page 521 left, @ Stuart Isett; right, Image © Lakatos Sandor, Used under license from Shutterstock .com. **32.2** right, © Linda Pitkin/ Planet Earth Pictures. **32.3** (a) above. Stephen Dalton/ Photo Researchers, Inc. **32.5** right, Washington University/ www.thalamus .wustl.edu. **32.6** (a) right, © Ed Reschke. **32.7** Professor P. Motta/ Department of Anatomy/ La Sapienza, Rome/ SPL/ Photo Researchers, Inc. **32.8** (a) above, Tony McConnell/ Science Photo Library/Photo Researchers, Inc.; (c) above, © Don Fawcett/ Visuals Unlimited. **32.16** Image © Phase4Photography, Used under license from Shutterstock.com. **32.18** Image courtesy of Department of Pathology, The University of Melbourne. **Page 533** Se-Jin Lee, Johns Hopkins University School of Medicine. **32.19** Painting by Sir Charles Bell, 1809, courtesy of Royal College of Surgeons, Edinburgh. **Page 534** Section 32.1, The Muskegon Chronicle; Section 32.7, Image © Phase4Photography, Used under license from Shutterstock.com. **Page 535** Courtesy of the family of Tiffany Manning.

CHAPTER 33 **Key Concepts** from left, © C. Yokochi and J. Rohen, Photographic Anatomy of the Human Body, 2nd Ed., Igaku-Shoin, Ltd., 1979; © Lennart Nilsson/ Bonnierforlagen AB; Biophoto Associates/ Photo Researchers, Inc. **33.1** page 536, Faye Norman/ SPL/ Photo Researchers, Inc.; page 537, Courtesy of the family of Matt Nadar. **33.2** (a,b) right, after M. Labarbera and S. Vogel, *American Scientist*, 1982, 70:54–60. **33.6** (b) © C. Yokochi and J. Rohen, Photographic Anatomy of the Human Body, 2nd Ed., Igaku-Shoin, Ltd., 1979. **33.9** right, National Cancer Institute/ Photo Researchers, Inc. **33.10** Professor P. Motta/ Department of Anatomy/ University La Sapienca, Rome/ SPL/ Photo Researchers, Inc. **33.11** R. Demarest based on A. Spence, *Basic Human Anatomy*, Benjamin-Cummings, 1982. **33.13** above, Sheila Terry/ Photo Researchers, Inc.; below, Courtesy of Oregon Scientific, Inc. **33.14** left, © Lennart Nilsson/ Bonnierforlagen AB. **33.15** left, Lisa Starr, using © 2001 PhotoDisc, Inc./ Getty Images photograph; right, © Dr. John D. Cunningham/ Visuals Unlimited. **33.18** (a) © Ed Reschke; (b) Biophoto Associates/ Photo Researchers, Inc. **33.19** © Lester V. Bergman/ Corbis. **33.22** (a) Courtesy of © Christine Evers; (b) Courtesy of ZOLL Medical Corporation. **Page 554** Section 33.1, Faye Norman/ SPL/ Photo Researchers, Inc.; Section 33.5, National Cancer Institute/ Photo Researchers, Inc.; Section 33.10, © Lester V. Bergman/ Corbis.

CHAPTER 34 **Key Concepts** from left, © Antonio Zamora, www.scientificpsychic.com; Juergen Berger/ Photo Researchers, Inc.; Biology Media/ Photo Researchers, Inc.; James Hicks, Centers for Disease Control and Prevention. **34.1** page 556, Biomedical Imaging Unit, Southampton General Hospital/SPL/Photo Researchers, Inc; page 557, In memory of Frankie McCullough. **34.2** © Dr. Richard Kessel and Dr. Randy Kardon/ Tissues & Organs/ Visuals Unlimited. **34.3** © Antonio Zamora, www.scientificpsychic.com. **34.4** © Alvin Telser/ Visuals Unlimited, Inc. **34.5** (a) © David Scharf. All rights reserved.; (b) Kwangshin Kim/ Photo Researchers, Inc.; (c) Juergen Berger/ Photo Researchers, Inc. **34.6** www .zahnarzt-stuttgart.com. **34.7** © John D. Cunningham/ Visuals Unlimited. **34.8** (a) Biology Media/ Photo Researchers, Inc.; (b) below, © Robert R. Dourmashkin, courtesy of Clinical Research Centre, Harrow, England. **34.19** Dr. A. Liepins/ SPL/ Photo Researchers, Inc. **Page 571** Hayley Witherell. **34.21** left, James Hicks, Centers for Disease Control and Prevention; right, Eye of Science/ Photo Researchers, Inc. **34.22** (a) NIBSC/ Photo Researchers, Inc.; (b) © Peter Turnley/ Corbis; (c) © Zeva Oelbaum/ Peter Arnold, Inc. **Page 575** CDC. **Page 576** Section 34.1, Biomedical Imaging Unit, Southampton General Hospital/SPL/Photo Researchers, Inc; Section 34.2, © Antonio Zamora, www.scientificpsychic.com; Section 34.3, Juergen Berger/ Photo Researchers, Inc; Section 34.4, Biology Media/ Photo Researchers, Inc.; Section 34.9, Dr. A. Liepins/ SPL/ Photo Researchers, Inc.; Section 34.12–13, © Peter Turnley/ Corbis.

CHAPTER 35 **Key Concepts** from left, CDC/ Dr. Joel D. Meyer. **35.1** page 578, Image © Timothy Large, Used under license from Shutterstock.com; page 579, © Lennart Nilsson/ Bonnierforlagen AB. **35.3** (a) © Peter Parks/ Oxford Scientific Films; (b) John Glowczwski/ University of Texas Medical Branch; (c) left, © Ed Reschke; right, from Biology: *The Dynamic Science*, Russell et al., page 1002, figure 44.5 bottom. **35.7** (a) from *Biology: The Dynamic Science*, Russell et al., page 1003, figure 44.7a; (b) © H. R. Duncker, Justus-Liebig University, Giessen, Germany. **35.9** Photographs, Courtesy of Kay Elemetrics Corporation; art, modified from A. Spence and E. Mason, *Human Anatomy and Physiology*, 4th Ed., 1992, West Publishing Company. **35.10** (a–b) © Charles McRae, MD/ Visuals Unlimited. **35.13** © C. Yokochi and J. Rohen, Photographic Anatomy of the Human Body, 2nd Ed., Igaku-Shoin, Ltd., 1979. **35.14** (a) © R. Kessel/ Visuals Unlimited. **35.17** CDC/ Dr. Joel D. Meyer. **35.18** © O. Auerbach/ Visuals Unlimited. **Page 592** Section 35.1, Image © Timothy Large, Used under license from Shutterstock.com; Section 35.3, © Ed Reschke.

CHAPTER 36 **Key Concepts** from left, © W. Perry Conway/ Corbis; USDA, www.mypyramid.gov; © Gary Head. **36.1** page 594, Simon Law, flickr.com/people/ sfllaw; page 595, Courtesy of Dr. Jeffrey M. Friedman, Rockefeller University. **36.3** (a) © W. Perry Conway/ Corbis; (a,b art) Adapted from A. Romer and T. Parsons, *The Vertebrate Body*, Sixth Edition, Saunders Publishing Company, 1986. **36.4** © Gunter Ziesler/ Bruce Coleman, Inc. **36.7** right, after A. Vander, et al., *Human Physiology: Mechanisms of Body Function*, 5th ed., McGraw-Hill, 1990, used by permission. **36.8** (a) © Microslide courtesy Mark Nielsen, University of Utah; (c–e) art, After Sherwood and others; (f) D. W. Fawcett/ Photo Researchers, Inc. **36.10** © C. James Webb / Phototake. All rights reserved. **36.11** (b) National Cancer Institute. **36.13** USDA, www.mypyramid.gov. **36.14** USDA. **36.15** © Gary Head. **Page 611** Courtesy of Lisa Hyche. **Page 612** Section 36.1, Courtesy of Dr. Jeffrey M. Friedman, Rockefeller University; Section 36.6–7, D. W. Fawcett/ Photo Researchers, Inc.; Section 36.8, National Cancer Institute; Section 36.10–11, USDA, www.mypyramid.gov; Section 36.12, © Gary Head.

CHAPTER 37 **Key Concepts** from left, Susumu Nishinaga / Photo Researchers, Inc.; S. J. Krasemann/ Photo Researchers, Inc.; Evan Cerasoli. **37.1** page 614, © Archivo Iconografico, S.A./ Corbis; page 615, © Ed Kashi/ Corbis. **37.4** left, Stephen Dalton/ Photo Researchers, Inc.; right, Susumu Nishinaga / Photo Researchers, Inc. **37.5** From T Garrison, *Oceanography: An Invitation to Marine Science*, Brooks/Cole, 1993. **37.6** left, © Stuart Westmorland /Stone / Getty Images; right, Tom McHugh/ Photo Researchers, Inc. **37.12** (a) © Bob McKeever/ Tom Stack & Associates; (b) S. J. Krasemann/ Photo Researchers, Inc. **37.13** Evan Cerasoli. **37.14** © Corbis-Bettmann. **Page 625** Lawrence Lawry/ Photo Researchers, Inc. **Page 626** Section 37.1, © Archivo Iconografico, S.A./ Corbis; Section 37.6, S. J. Krasemann/ Photo Researchers, Inc.; Section 37.7, Evan Cerasoli. **37.15** below, USDA.

CHAPTER 38 **Key Concepts** from left, © Rodger Klein/ Peter Arnold, Inc.; Dr. Maria Leptin, Institute of Genetics, University of Koln, Germany; © Lennart Nilsson/ Bonnierforlagen AB. **38.1** page 628, © TTA Media/ Splash News; page 629, © Heidi Specht, West Virginia University. **38.2** (a) Biophoto Associates/ Photo Researchers, Inc.; (b) © Rodger Klein/ Peter Arnold, Inc.; (c) © Staebler/ Jupiter Images. **38.3** (a) R. Scott Cameron, Advanced Forest Protection, Inc., Bugwood.org; (b) © Doug Perrine/ seapics.com; (c) NPS Yellowstone/ Becky Wyman. **38.8** right, © Lennart Nilsson/ Bonnierforlagen AB. **38.10** (a) left, David M. Phillips/ Photo Researchers, Inc.; (c) Courtesy of Elizabeth Sanders, Women's Specialty Center, Jackson, MS. **38.11** (a) Dr. E. Walker/ Photo Researchers, Inc.; (b) Western Ophthalmic Hospital/ Photo Researchers, Inc.; (c) CNRI/ Photo Researchers, Inc. **38.13** (1–4) Carolina Biological Supply Company; (5) Left and center, © David M. Dennis/ Tom Stack & Associates, Inc.; right,

© John Shaw/ Tom Stack & Associates. **38.15** right, © Carolina Biological Supply Company; all others, Dr. Maria Leptin, Institute of Genetics, University of Koln, Germany. **38.16** (c) © Professor Jonathon Slack. **38.17** (c) Courtesy of Dr. Kathleen K. Sulik, Bowles Center for Alcohol Studies, the University of North Carolina at Chapel Hill. **38.18** left, © Peter Parks/ Oxford Scientific Films/ Animals, Animals; right, art by Raychel Ciemma after S. Gilbert, *Developmental Biology*, 4th Ed., Sinauer. **38.22** top, (all) © Lennart Nilsson/ Bonnierforlagen AB. **Page 654** Section 38.1, © Heidi Specht, West Virginia University; Section 38.2, © Staebler/ Jupiter Images; Section 38.6–8, David M. Phillips/ Photo Researchers, Inc.; Section 38.12–14, © Lennart Nilsson/ Bonnierforlagen AB.

CHAPTER 39 **Key Concepts** from left, Reprinted from Trends in Neuroscience, Vol. 21, Issue 2, 1998, L.J.Young, W. Zuoxin, T.R. Insel, "Neuroendocrine bases of monogamy", Pages 71–75, ©1998, with permission from Elsevier Science; © Eric Hosking; © Steve Kaufman/ Corbis; © Alexander Wild. **39.1** © Scott Camazine. **39.2** © Christine Majul/ www.flickr.com/photos/kitkaphotogirl. **39.4** (a) © Robert M. Timm & Barbara L. Clauson, University of Kansas; (b–c) Reprinted from Trends in Neuroscience, Vol. 21, Issue 2, 1998, L.J.Young, W. Zuoxin, T.R. Insel, "Neuroendocrine bases of monogamy", Pages 71–75, ©1998, with permission from Elsevier Science. **39.5** (a) © Eric Hosking; (b) Stephen Dalton/ Photo Researchers, Inc. **39.6** © Nina Leen/ TimePix/ Getty Images; inset, © Robert Semeniuk/ Corbis. **39.7** © Professor Jelle Atema, Boston University. **39.8** © Bernhard Voelkl. **39.9** (a) © Robert Maier/ Animals Animals. **39.10** (a) © Tom and Pat Leeson, leesonphoto.com; (b) © Kevin Schafer/ Corbis; (c) © Monty Sloan, www.wolfphotography.com. **39.12** (a) © John Alcock, Arizona State University; (b) upper, © Pam Gardner, Frank Lane Picture Agency/ Corbis; lower, © Pam Gardner; Frank Lane Picture Agency/ Corbis; (c) © D. Robert Franz/ Corbis. **39.13** © Michael Francis/ The Wildlife Collection. **39.14** (a) John Conrad/ Corbis; (b) © B. Borrell Casals, Frank Lane Picture Agency/ Corbis; (c) © Steve Kaufman/ Corbis. **39.15** (a) © Tom and Pat Leeson, leesonphoto.com; (b) © John Alcock, Arizona State University; (c) © Paul Nicklen/ National Geographic/ Getty Images. **39.16** © Jeff Vanuga/ Corbis. **39.17** © Steve Bloom/ stevebloom.com. **39.18** © A. E. Zuckerman/ Tom Stack & Associates. **39.19** (a) © Australian Picture Library/ Corbis; (b) © Alexander Wild; (c) © Professor Louis De Vos. **39.20** (a) © Kenneth Lorenzen; (b) © Peter Johnson/ Corbis; (c) © Nicola Kountoupes/ Cornell University. **Page 669** © Lynda Richardson/ Corbis. **Page 670** Section 39.1, © Scott Camazine; Section 39.2, Reprinted from Trends in Neuroscience, Vol. 21, Issue 2, 1998, L.J.Young, W. Zuoxin, T.R. Insel, "Neuroendocrine bases of monogamy," Pages 71–75, ©1998, with permission from Elsevier Science; Section 39.3, Stephen Dalton/ Photo Researchers, Inc.; Section 39.4, © Robert Maier/ Animals Animals; Section 39.5, © Tom and Pat Leeson, leesonphoto.com; Section 39.6, © D. Robert Franz/ Corbis; Section 39.7, © A. E. Zuckerman/ Tom Stack & Associates; Section 39.8, © Kenneth Lorenzen.

CHAPTER 40 **Key Concepts** from left, © Cynthia Bateman, Bateman Photography; © G. K. Peck; © Wayne Bennett/ Corbis; © Don Mason/ Corbis. **40.1** page 672, Courtesy of @ Joel Pete; page 673, AP Images/ Steven Day. **40.2** © Cynthia Bateman, Bateman Photography. **40.3** (a) © Amos Nachoum/ Corbis; (b) © Eric and David Hosking/ Corbis; (c) Elizabeth A. Sellers /life.nbii.gov. **40.4** left, Jeff Lepore/ Photo Researchers, Inc. **40.5** above, © David Scharf. All rights reserved. **40.6** (a) © G. K. Peck; (b) © Rick Leche, www.flickr.com/photos/rick_leche. **40.8** right, © Jacques Langevin/ Corbis Sygma. **40.9** (a) © Joe McDonald/ Corbis; (b) © Wayne Bennett/ Corbis; (c) Estuary to Abyss 2004. NOAA Office of Ocean Exploration. **40.10** (a–b) © Hippocampus Bildarchiv; above, © David Reznick/ University of California - Riverside; computer enhanced by Lisa Starr;(c) © Helen Rodd. **40.12** © Bruce Bornstein, www.captbluefin.com. **40.13** NASA; Art by Precision Graphics. **40.16** left, © Adrian Arbib/ Corbis; right, © Don Mason/ Corbis. **Page 687** National Transportation Safety Board. **Page 688** Section 40.1, Courtesy of @ Joel Pete; Section 40.2, © Amos Nachoum/ Corbis; Section 40.5–6, © Wayne Bennett/ Corbis; Section 40.8, © Don Mason/ Corbis. **40.17** © Reinhard Dirscherl/ www.bciusa.com.

CHAPTER 41 **Key Concepts** from left, David C. Clegg/ Photo Researchers, Inc.; © Bob and Miriam Francis/ Tom Stack & Associates; © Duncan Murrell/ Taxi/ Getty Images; Photo by Scott Bauer, USDA/ARS; © Pierre Vauthey/ Corbis Sygma. **41.1** page 690, Photography by B. M. Drees, Texas A&M University. Http://fireant.tamu; page 691 (a) Scott Bauer/ USDA; (b) USDA. **41.2** David C. Clegg/ Photo Researchers, Inc. **41.3** above, Harlo H. Hadow; below, © Bob and Miriam Francis/ Tom Stack & Associates. **41.4** © Thomas W. Doeppner. **41.5** © Pekka Komi. **41.6** right, Michael Abbey/ Photo Researchers, Inc. **41.7** above right, © Joe McDonald/ Corbis; below left, © Hal Horwitz/ Corbis; right © Tony Wharton, Frank Lane Picture Agency/ Corbis. **41.8** Ed Cesar/ Photo Researchers, Inc. **41.9** (a) © Edward S. Ross; (b) © Nigel Jones. **41.10** (a) © JH Pete Carmichael; (b) © Bob Jensen Photography; (c) David Burdick, NOAA. **41.11** (a) © C. James Webb/ Phototake USA; (b) Bill Hilton, Jr., Hilton Pond Center. **41.12** © The Samuel Roberts Noble Foundation, Inc. **41.13** E.R. Degginger/ Photo Researchers, Inc. **41.14** © Peter J. Bryant/ Biological Photo Service. **41.15** (a) © Doug Peebles/ Corbis; (b) © Pat O'Hara/ Corbis; (c–d) © Tom Bean/ Corbis; (e) © Duncan Murrell/ Taxi/ Getty Images. **41.16** (a) R. Barrick/ USGS; (b) USGS; (c) P. Frenzen, USDA Forest Service. **41.17** (a) © Jane Burton/ Bruce Coleman, Ltd. **41.18** © Richard W. Halsey, California Chaparral Institute. **41.19** (a) Angelina Lax/ Photo Researchers, Inc.; (b) Photo by Scott Bauer, USDA/ARS; (c) © Greg Lasley Nature Photography, www.greglasley.net. **41.21** © Pierre Vauthey/ Corbis Sygma. **Page 706** Section 41.1, USDA; Section 41.2, David C. Clegg/ Photo Researchers, Inc.; Section 41.3, © Thomas W. Doeppner; Section 41.4, © Pekka Komi; Section 41.5, Ed Cesar/ Photo Researchers, Inc.; Section 41.6, Bill Hilton, Jr., Hilton Pond Center; Section 41.7, USGS; Section 41.8, Angelina Lax/ Photo Researchers, Inc.; Section 41.9, © Pierre Vauthey/ Corbis Sygma. **Page 707** Critical Thinking, © Anthony Bannister, Gallo Images/ Corbis.

CHAPTER 42 **Key Concepts** from left, © Tom & Pat Leeson, Ardea London Ltd.; Jack Scherting, USC&GS, NOAA. **42.1** page 708, Courtesy of State of Washington Department of Ecology; page 709, Fisheries & Oceans Canada, Experimental Lakes Area . **42.3** bottom, © Van Vives; all others, © D. A. Rintoul. **42.4** from left, top row, © Bryan & Cherry Alexander/ Photo Researchers, Inc.; © Dave Mech; © Tom & Pat Leeson, Ardea London Ltd.; 2nd row, © Tom Wakefield/ Bruce Coleman, Inc.; © Paul J. Fusco/ Photo Researchers, Inc.; © E. R. Degginger/ Photo Researchers, Inc.; 3rd row, © Tom J. Ulrich/ Visuals Unlimited; © Dave Mech; © Tom McHugh/ Photo Researchers, Inc.; mosquito, Photo by James Gathany, Centers for Disease Contro; flea, © Edward S. Ross; 4th row, © Jim Steinborn; © Jim Riley; © Matt Skalitzky; earthworm, © Peter Firus, flagstaffotos.com.au. **42.5** Graphic created by FoodWeb3D program written by Rich Williams courtesy of the Webs on the Web project (www.foodwebs.org). **42.7** Jack Scherting, USC&GS, NOAA. **42.11** NASA. **Page 720** Section 42.1, Courtesy of State of Washington Department of Ecology; Section 42.3, © D. A. Rintoul; Section 42.4, Graphic created by FoodWeb3D program written by Rich Williams courtesy of the Webs on the Web project (www.foodwebs.org); Section 42.6, Jack Scherting, USC&GS, NOAA; Section 42.8, NASA.

CHAPTER 43 **Key Concepts** from left, NASA; Jack Carey; © John Easley, www.johneasley.com. **43.1** page 722, © Hank Fotos Photography; page 723, NASA. **43.5** NASA. **43.6** left, © Sally A. Morgan, Ecoscene/ Corbis; right, © Bob Rowan, Progressive Image/ Corbis. **43.8** NASA. **43.9** NASA's Earth Observatory. **43.11** Courtesy of Jim Deacon, The University of Edinburgh. **43.12** Courtesy of © Christine Evers. **43.13** © George H. Huey/ Corbis. **43.14** (a) Jeff Servos, US Fish & Wildlife Service; (b) Bill Radke, US Fish & Wildlife Service. **43.15** (b) © D. A. Rintoul; (c) © Tom Bean Photography. **43.16** © Jonathan Scott/ Planet Earth Pictures. **43.17** (a) © John C. Cunningham/ Visuals Unlimited; (b) Jack Wilburn/ Animals Animals; (c) AP/ Wide World Photos. **43.18** Anlace, http://en.wikipedia.org/wiki/File:Sonomamtneflank.jpg. **43.19** left, © James Randklev/ Corbis. **43.20** left, © Franz Lanting/ Minden Pictures. **43.21** (a) Image © Serg Zastavkin, Used under license from Shutterstock.com; (b) © Thomas Wiewandt / ChromoSohm Media Inc. / Photo Researchers, Inc. **43.22** © Darrell Gulin/ Corbis. **43.23** © Pat O'Hara/ Corbis. **43.25** Jack Carey. **43.27** © E. F. Benfield, Virginia Tech. **43.28** (a) © Annie Griffiths Belt/ Corbis; (b) © Douglas Peebles/ Corbis. **43.29** Courtesy of J. L. Sumich, Biology of Marine Life, 7th ed., W. C. Brown, 1999. **43.30** © John Easley, www.johneasley.com. **43.31** © Dr. Ray Berkelmans, Australian Institute of Marine Science. **43.33** (a) NOAA; (b) Image courtesy of NOAA and MBARI. **43.34** NOAA/ Photo courtesy of Cindy Van Dover, Duke University Marine Lab; inset, © Peter Batson/ imagequestmarine.com. **43.35** upper, Eye of Science/ Photo Researchers, Inc.; lower, © Raghu Rai/ Magnum Photos. **Page 744** Section 43.1, NASA; Section 43.4, NASA; Section 43.5–7, © Tom Bean Photography; Section 43.8, © James Randklev/ Corbis; Section 43.9–10, Image © Serg Zastavkin, Used under license from Shutterstock.com; Section 43.11, Jack Carey; Section 43.12–13, © John Easley, www.johneasley.com; Section 43.14, NOAA/ Photo courtesy of Cindy Van Dover, Duke University Marine Lab.

CHAPTER 44 **Key Concepts** from left, © George M. Sutton/ Cornell Lab of Ornithology; © Orbimage Imagery. Image provided by GeoEye and processing by NASA Goddard Space Flight Center; © Adolf Schmidecker/ FPG/ Getty Images; Image © Lee Prince, Used under license from Shutterstock.com. **44.1** page 746, U.S. Navy photo by Chief Yeoman Alphanso Braggs; page 747, © Dan Guravich/ Corbis. **44.2** (a) John Butler, NOAA; (b) © Jeffrey Sylvester/ FPG /Getty Images; (c) © George M. Sutton/ Cornell Lab of Ornithology; (d) Courtesy of Ken Nemuras; (e) © Dallas Zoo, Robert Cabello; (f) © Dr. John Hilty; (g) Edward S. Ross; (h) Joe Fries, U.S. Fish & Wildlife Service; (i) Christian Zuber/ Bruce Coleman, Ltd.; (j) Douglas Faulkner/ Photo Researchers, Inc.; (k) © Hawaii Biology Survey. **44.3** © Orbimage Imagery. Image provided by GeoEye and processing by NASA Goddard Space Flight Center. **44.4** left, USDA Forest Service, Northeastern Research Station. **44.5** (a) Ted Spiegel/ Corbis; (b) Frederica Georgia/ Photo Researchers, Inc. **44.6** above, U.S. Department of the Interior, National Park Service; below, © Gary Head. **44.7** Claire Fackler/ NOAA. **44.8** (a) NASA. **44.10** National Snow and Ice Data Center, W. O. Field. **44.12** Hans Renner; inset, © Adolf Schmidecker/ FPG/ Getty Images. **44.13** Diane Borden-Bilot, U.S. Fish and Wildlife Service. **44.14** Image © Lee Prince, Used under license from Shutterstock.com. **44.15** Mountain Visions/ NOAA. **44.16** NASA. **Page 760** Section 44.1, U.S. Navy photo by Chief Yeoman Alphanso Braggs; Section 44.2, © Dr. John Hilty; Section 44.3, © Orbimage Imagery. Image provided by GeoEye and processing by NASA Goddard Space Flight Center; Section 44.4, Frederica Georgia/ Photo Researchers, Inc.; Section 44.5, U.S. Department of the Interior, National Park Service; Section 44.6, Claire Fackler/ NOAA; Section 44.7, NASA; Section 44.8, National Snow and Ice Data Center, W. O. Field; Section 44.9, Diane Borden-Bilot, U.S. Fish and Wildlife Service.